RSMeans

Repair & Remodeling Cost Data

25th Annual Edition

2004

First Printing

Foreword

RSMeans is a product line of Reed Construction Data, a leading provider of construction information, products, and services in North America and globally. Reed Construction Data's project information products include more than 100 regional editions, national construction data, sales leads, and local plan rooms in major business centers. Reed Construction Data's PlansDirect provides surveys, plans and specifications. The First Source suite of products consists of *First Source for Products,* SPEC-DATA™, MANU-SPEC™, CADBlocks, Manufacturer Catalogs and First Source Exchange (www.firstsourceexchange.com) for the selection of nationally available building products. Reed Construction Data also publishes ProFile, a database of more than 20,000 U.S. architectural firms. RSMeans provides construction cost data, training, and consulting services in print, CD-ROM and online. Reed Construction Data, headquartered in Atlanta, is owned by Reed Business Information (www.cahners.com), a leading provider of critical information and marketing solutions to business professionals in the media, manufacturing, electronics, construction and retail industries. Its market-leading properties include more than 135 business-to-business publications, over 125 Webzines and Web portals, as well as online services, custom publishing, directories, research and direct-marketing lists. Reed Business Information is a member of the Reed Elsevier plc group (NYSE: RUK and ENL)—a world-leading publisher and information provider operating in the science and medical, legal, education and business-to-business industry sectors.

Our Mission

Since 1942, RSMeans has been actively engaged in construction cost publishing and consulting throughout North America.

Today, over 50 years after the company began, our primary objective remains the same: to provide you, the construction and facilities professional, with the most current and comprehensive construction cost data possible.

Whether you are a contractor, an owner, an architect, an engineer, a facilities manager, or anyone else who needs a fast and reliable construction cost estimate, you'll find this publication to be a highly useful and necessary tool.

Today, with the constant flow of new construction methods and materials, it's difficult to find the time to look at and evaluate all the different construction cost possibilities. In addition, because labor and material costs keep changing, last year's cost information is not a reliable basis for today's estimate or budget.

That's why so many construction professionals turn to RSMeans. We keep track of the costs for you, along with a wide range of other key information, from city cost indexes . . . to productivity rates . . . to crew composition . . . to contractor's overhead and profit rates.

RSMeans performs these functions by collecting data from all facets of the industry, and organizing it in a format that is instantly accessible to you. From the preliminary budget to the detailed unit price estimate, you'll find the data in this book useful for all phases of construction cost determination.

The Staff, the Organization, and Our Services

When you purchase one of RSMeans' publications, you are in effect hiring the services of a full-time staff of construction and engineering professionals.

Our thoroughly experienced and highly qualified staff works daily at collecting, analyzing, and disseminating comprehensive cost information for your needs. These staff members have years of practical construction experience and engineering training prior to joining the firm. As a result, you can count on them not only for the cost figures, but also for additional background reference information that will help you create a realistic estimate.

The RSMeans organization is always prepared to help you solve construction problems through its five major divisions: Construction and Cost Data Publishing, Electronic Products and Services, Consulting Services, Insurance Services, and Educational Services.

Besides a full array of construction cost estimating books, Means also publishes a number of other reference works for the construction industry. Subjects include construction estimating and project and business management; special topics such as HVAC, roofing, plumbing, and hazardous waste remediation; and a library of facility management references.

In addition, you can access all of our construction cost data through your computer with RSMeans CostWorks 2004 CD-ROM, an electronic tool that offers over 50,000 lines of Means detailed construction cost data, along with assembly and whole building cost data. You can also access Means cost information from our Web site at www.rsmeans.com

What's more, you can increase your knowledge and improve your construction estimating and management performance with a Means Construction Seminar or In-House Training Program. These two-day seminar programs offer unparalleled opportunities for everyone in your organization to get updated on a wide variety of construction-related issues.

RSMeans also is a worldwide provider of construction cost management and analysis services for commercial and government owners and of claims and valuation services for insurers.

In short, RSMeans can provide you with the tools and expertise for constructing accurate and dependable construction estimates and budgets in a variety of ways.

Robert Snow Means Established a Tradition of Quality That Continues Today

Robert Snow Means spent years building his company, making certain he always delivered a quality product.

Today, at RSMeans, we do more than talk about the quality of our data and the usefulness of our books. We stand behind all of our data, from historical cost indexes... to construction materials and techniques... to current costs.

If you have any questions about our products or services, please call us toll-free at 1-800-334-3509. Our customer service representatives will be happy to assist you or visit our Web site at www.rsmeans.com

Table of Contents

Note: Changes to Section contents and line numbers have been made in Unit Cost and Assembly Sections of the 2004 publications to better comply with the industry standard organization systems, MasterFormat95 and UNIFORMAT II. While all the data you need is still here, it may have moved or been renumbered. Using the Index will help you find anything you have trouble locating. For a complete map of changes, visit our web site at **www.rsmeans.com/links** and follow the instructions to obtain the mapping tables.

How the Book Is Built: An Overview

A Powerful Construction Tool

You have in your hands one of the most powerful construction tools available today. A successful project is built on the foundation of an accurate and dependable estimate. This book will enable you to construct just such an estimate.

For the casual user the book is designed to be:

- quickly and easily understood so you can get right to your estimate
- filled with valuable information so you can understand the necessary factors that go into the cost estimate

For the regular user, the book is designed to be:

- a handy desk reference that can be quickly referred to for key costs
- a comprehensive, fully reliable source of current construction costs and productivity rates, so you'll be prepared to estimate any project
- a source book for preliminary project cost, product selections, and alternate materials and methods

To meet all of these requirements we have organized the book into the following clearly defined sections.

Quick Start

This one-page section (see following page) can quickly get you started on your estimate.

How To Use the Book: The Details

This section contains an in-depth explanation of how the book is arranged . . . and how you can use it to determine a reliable construction cost estimate. It includes information about how we develop our cost figures and how to completely prepare your estimate.

Unit Price Section

All cost data has been divided into the 16 divisions according to the MasterFormat system of classification and numbering as developed by the Construction Specifications Institute (CSI) and Construction Specifications Canada (CSC). For a listing of these divisions and an outline of their subdivisions, see the Unit Price Section Table of Contents.

Estimating tips are included at the beginning of each division.

Assemblies Section

The cost data in this section has been organized in an "Assemblies" format. These assemblies are the functional elements of a building and are arranged according to the 7 divisions of the UNIFORMAT II classification system. For a complete explanation of a typical "Assemblies" page, see "How To Use the Assemblies Cost Tables."

Reference Section

This section includes information on Reference Numbers, Change Orders, Crew Listings, Historical Cost Indexes, City Cost Indexes, Location Factors, and a listing of Abbreviations. It is visually identified by a vertical gray bar on the edge of pages.

Reference Numbers: At the beginning of selected major classifications in the Unit Price Section are "reference numbers" shown in bold squares. These numbers refer you to related information in the Reference Section.

In this section, you'll find reference tables, explanations, and estimating information that support how we develop the unit price data. Also included are alternate pricing methods, technical data, and estimating procedures, along with information on design and economy in construction. You'll also find helpful tips on what to expect and what to avoid when estimating and constructing your project.

It is recommended that you refer to the Reference Section if a "reference number" appears within the section you are estimating.

Change Orders: This section includes information on the factors that influence the pricing of change orders.

Crew Listings: This section lists all the crews referenced in the book. For the purposes of this book, a crew is composed of more than one trade classification and/or the addition of power equipment to any trade classification. Power equipment is included in the cost of the crew. Costs are shown both with bare labor rates and with the installing contractor's overhead and profit added. For each, the total crew cost per eight-hour day and the composite cost per labor-hour are listed.

Historical Cost Indexes: These indexes provide you with data to adjust construction costs over time. If you know costs for a project completed in the past, you can use these indexes to calculate a rough estimate of what it would cost to construct the same project today.

City Cost Indexes: Obviously, costs vary depending on the regional economy. You can adjust the "national average" costs in this book to 316 major cities throughout the U.S. and Canada by using the data in this section. How to use information is included.

Location Factors, to quickly adjust the data to over 930 zip code areas, are included.

Abbreviations: A listing of the abbreviations used throughout this book, along with the terms they represent, is included.

Index

A comprehensive listing of all terms and subjects in this book to help you find what you need quickly when you are not sure where it falls in MasterFormat.

The Scope of This Book

This book is designed to be as comprehensive and as easy to use as possible. To that end we have made certain assumptions and limited its scope in three key ways:

1. We have established material prices based on a "national average."
2. We have computed labor costs based on a 30-city "national average" of union wage rates.
3. We have targeted the data for projects of a certain size range.

For a more detailed explanation of how the cost data is developed, see "How To Use the Book: The Details."

Project Size

This book is aimed primarily at residential, commercial and industrial repair/remodeling projects costing $10,000 to $1,000,000.

With reasonable exercise of judgment the figures can be used for any repair/ remodeling work. *However, for civil engineering structures such as bridges, dams, highways, or the like, please refer to* ***Means Heavy Construction Cost Data.***

Quick Start

If you feel you are ready to use this book and don't think you need the detailed instructions that begin on the following page, this Quick Start section is for you.

These steps will allow you to get started estimating in a matter of minutes.

1 First, decide whether you require a Unit Price or Assemblies type estimate. Unit price estimating requires a breakdown of the work to individual items. Assemblies estimates combine individual items or components into building systems.

If you need to estimate each line item separately, follow the instructions for **Unit Prices.**

If you can use an estimate for the entire assembly or system, follow the instructions for **Assemblies.**

2 Find each cost data section you need in the Table of Contents (either for Unit Prices or Assemblies).

Unit Prices: The cost data for Unit Prices has been divided into 16 divisions according to the CSI MasterFormat.

Assemblies: The cost data for Assemblies has been divided into 7 divisions according to the UNIFORMAT II.

3 Turn to the indicated section and locate the line item or assemblies table you need for your estimate. Portions of a sample page layout from both the Unit Price Listings and the Assemblies Cost Tables appear below.

Unit Prices: If there is a reference number listed at the beginning of the section, it refers to additional information you may find useful. See the referenced section for additional information.

- Note the crew code designation. You'll find full descriptions of crews in the Crew Listings including labor-hour and equipment costs.

Assemblies: The Assemblies (*not* shown in full here) are generally separated into three parts: 1) an illustration of the system to be estimated; 2) the components and related costs of a typical system; and 3) the costs for similar systems with dimensional and/or size variations.

4 Determine the total number of units your job will require.

Unit Prices: Note the unit of measure for the material you're using is listed under "Unit."

- Bare Costs: These figures show unit costs for materials and installation. Labor and equipment costs are calculated according to crew costs and average daily output. Bare costs do not contain allowances for overhead, profit or taxes.
- "Labor-hours" allows you to calculate the total labor-hours to complete that task. Just multiply the quantity of work by this figure for an estimate of activity duration.

Assemblies: Note the unit of measure for the assembly or system you're estimating is listed in the System Components section.

5 Then multiply the total units by . . .

Unit Prices: "Total Incl. O&P" which stands for the total cost including the installing contractor's overhead and profit. (See the "How To Use the Unit Price Pages" for a complete explanation.)

Assemblies: The "Total" in the right-hand column, which is the total cost including the installing contractor's overhead and profit. (See the "How To Use the Assemblies Cost Tables" section for a complete explanation.)

Material and equipment cost figures include a 10% markup. For labor markups, see the inside back cover of this book. If the work is to be subcontracted, add the general contractor's markup, approximately 10%.

6 The price you calculate will be an estimate for either an individual item of work, or for a completed assembly or *system.*

7 Compile a list of all items or assemblies included in the total project. Summarize cost information, and add project overhead.

Localize costs by using the City Cost Indexes or Location Factors found in the Reference Section.

For a more complete explanation of the way costs are derived, please see the following sections.

Editors' Note: We urge you to spend time reading and understanding all of the supporting material and to take into consideration the reference material such as Crews Listing and the "reference numbers."

Unit Price Pages

03300 | Cast-In-Place Concrete

	03310 | Structural Concrete	CREW	DAILY OUTPUT	LABOR-HOURS	UNIT	2004 BARE COSTS MAT.	LABOR	EQUIP.	TOTAL	TOTAL INCL O&P	
240	0010 CONCRETE IN PLACE Including forms (4 uses), reinforcing R03310-010										240
	0050 steel and finishing unless otherwise indicated										
	3800 Footings, spread under 1 C.Y.	C-14C	38.07	2.942	C.Y.	105	93	.68	198.68	272	
	3850 Over 5 C.Y.		81.04	1.382		96.50	43.50	.32	140.32	179	
	3900 Footings, strip, 18" x 9", unreinforced		40	2.800		91	88.50	.64	180.14	248	
	3920 18" x 9", reinforced		35	3.200		105	101	.74	206.74	284	
	3925 20" x 10", unreinforced		45	2.489		88	78.50	.57	167.07	229	
	3930 20" x 10", reinforced		40	2.800		99.50	88.50	.64	188.64	257	
	3935 24" x 12", unreinforced		55	2.036		87	64.50	.47	151.97	204	

Assemblies Pages

B10 Superstructure

B1010 Floor Construction

This page illustrates and describes a wood joist floor system including wood joist, oak floor, sub-floor, bridging, sand and finish floor, furring, plasterboard, taped, finished and painted ceiling. Lines within System Components give the unit price and total price per square foot for this system. Prices for alternate wood joist floor systems are on Line Items B1010 263 1700 thru 2300. Both material quantities and labor costs have been adjusted for the system listed.

Factors: To adjust for job conditions other than normal working situations use Lines B1010 263 2700 thru 4000.

Example: You are to install the system during off peak hours, 6 P.M. to 2 A.M. Go to Line B1010 263 3800 and apply this percentage to the appropriate INST. costs.

System Components	QUANTITY	UNIT	COST PER S.F. MAT.	INST.	TOTAL
Wood joists, 2" x 8", 16" O.C.,oak floor (sanded & finished),1/2" sub Floor, 1"x 3" bridging, furring, 5/8" drywall, taped, finished and painted.					
Wood joists, 2" x 8", 16" O.C.	1.000	L.F.	.88	.80	1.68
Subfloor plywood CDX 1/2"	1.000	SF Flr.	.43	.58	1.01
Bridging 1" x 3"	.150	Pr.	.05	.51	.56
Oak flooring, No. 1 common	1.000	S.F.	4.09	2.57	6.66
Sand and finish floor	1.000	S.F.	.76	1.17	1.93
Wood furring, 1" x 3", 16" O.C.	1.000	L.F.	.23	1.25	1.48
Gypsum drywall, 5/8" thick	1.000	S.F.	.24	.49	.73
Tape and finishing	1.000	S.F.	.04	.44	.48
Paint ceiling	1.000	S.F.	.15	.50	.65
TOTAL			6.87	8.31	15.18

B1010 263	Floor-Ceiling, Wood Joists	COST PER S.F. MAT.	INST.	TOTAL

How to Use the Book: The Details

What's Behind the Numbers? The Development of Cost Data

The staff at RSMeans continuously monitors developments in the construction industry in order to ensure reliable, thorough and up-to-date cost information.

While *overall* construction costs may vary relative to general economic conditions, price fluctuations within the industry are dependent upon many factors. Individual price variations may, in fact, be opposite to overall economic trends. Therefore, costs are continually monitored and complete updates are published yearly. Also, new items are frequently added in response to changes in materials and methods.

Costs—$ (U.S.)

All costs represent U.S. national averages and are given in U.S. dollars. The Means City Cost Indexes can be used to adjust costs to a particular location. The City Cost Indexes for Canada can be used to adjust U.S. national averages to local costs in Canadian dollars.

Material Costs

The RSMeans staff contacts manufacturers, dealers, distributors, and contractors all across the U.S. and Canada to determine national average material costs. If you have access to current material costs for your specific location, you may wish to make adjustments to reflect differences from the national average. Included within material costs are fasteners for a normal installation. RSMeans engineers use manufacturers' recommendations, written specifications and/or standard construction practice for size and spacing of fasteners. Adjustments to material costs may be required for your specific application or location. Material costs do not include sales tax.

Labor Costs

Labor costs are based on the average of wage rates from 30 major U.S. cities. Rates are determined from labor union agreements or prevailing wages for construction trades for the current year. Rates along with overhead and profit markups are listed on the inside back cover of this book.

- If wage rates in your area vary from those used in this book, or if rate increases are expected within a given year, labor costs should be adjusted accordingly.

Labor costs reflect productivity based on actual working conditions. These figures include time spent during a normal workday on tasks other than actual installation, such as material receiving and handling, mobilization at site, site movement, breaks, and cleanup.

Productivity data is developed over an extended period so as not to be influenced by abnormal variations and reflects a typical average.

Equipment Costs

Equipment costs include not only rental, but also operating costs for equipment under normal use. The operating costs include parts and labor for routine servicing such as repair and replacement of pumps, filters and worn lines. Normal operating expendables such as fuel, lubricants, tires and electricity (where applicable) are also included. Extraordinary operating expendables with highly variable wear patterns such as diamond bits and blades are excluded. These costs are included under materials. Equipment rental rates are obtained from industry sources throughout North America—contractors, suppliers, dealers, manufacturers, and distributors.

Crew Equipment Cost/Day—The power equipment required for each crew is included in the crew cost. The daily cost for crew equipment is based on dividing the weekly bare rental rate by 5 (number of working days per week), and then adding the hourly operating cost times 8 (hours per day). This "Crew Equipment Cost/Day" is listed in Subdivision 01590.

Mobilization/Demobilization—The cost to move construction equipment from an equipment yard or rental company to the job site and back again is not included in equipment costs. Mobilization (to the site) and demobilization (from the site) costs can be found in Section 02305-250. If a piece of equipment is already at the job site, it is not appropriate to utilize mob/demob costs again in an estimate.

General Conditions

Cost data in this book is presented in two ways: Bare Costs and Total Cost including O&P (Overhead and Profit). General Conditions, when applicable, should also be added to the Total Cost including O&P. The costs for General Conditions are listed in Division 1 of the Unit Price Section and the Reference Section of this book. General Conditions for the *Installing Contractor* may range from 0% to 10% of the Total Cost including O&P. For the *General* or *Prime Contractor*, costs for General Conditions may range from 5% to 15% of the Total Cost including O&P, with a figure of 10% as the most typical allowance.

Overhead and Profit

Total Cost including O&P for the *Installing Contractor* is shown in the last column on both the Unit Price and the Assemblies pages of this book. This figure is the sum of the bare material cost plus 10% for profit, the base labor cost plus total overhead and profit, and the bare equipment cost plus 10% for profit. Details for the calculation of Overhead and Profit on labor are shown on the inside back cover and in the Reference Section of this book. (See the "How to Use the Unit Price Pages" for an example of this calculation.)

Factors Affecting Costs

Costs can vary depending upon a number of variables. Here's how we have handled the main factors affecting costs.

Quality—The prices for materials and the workmanship upon which productivity is based represent sound construction work. They are also in line with U.S. government specifications.

Overtime—We have made no allowance for overtime. If you anticipate premium time or work beyond normal working hours, be sure to make an appropriate adjustment to your labor costs.

Productivity—The productivity, daily output, and labor-hour figures for each line item are based on working an eight-hour day in daylight hours in moderate temperatures. For work that extends beyond normal work hours or is performed under adverse conditions, productivity may decrease. (See the section in "How To Use the Unit Price Pages" for more on productivity.)

Size of Project—The size, scope of work, and type of construction project will have a significant impact on cost. Economies of scale can reduce costs for large projects. Unit costs can often run higher for small projects. Costs in this book are intended for the size and type of project as previously described in "How the Book Is Built: An Overview." Costs for projects of a significantly different size or type should be adjusted accordingly.

Location—Material prices in this book are for metropolitan areas. However, in dense urban areas, traffic and site storage limitations may increase costs. Beyond a 20-mile radius of large cities, extra trucking or transportation charges may also increase the material costs slightly. On the other hand, lower wage rates may be in effect. Be sure to consider both these factors when preparing an estimate, particularly if the job site is located in a central city or remote rural location.

In addition, highly specialized subcontract items may require travel and per diem expenses for mechanics.

Other factors—

- season of year
- contractor management
- weather conditions
- local union restrictions
- building code requirements
- availability of:
  - adequate energy
  - skilled labor
  - building materials
- owner's special requirements/restrictions
- safety requirements
- environmental considerations

Unpredictable Factors—General business conditions influence "in-place" costs of all items. Substitute materials and construction methods may have to be employed. These may affect the installed cost and/or life cycle costs. Such factors may be difficult to evaluate and cannot necessarily be predicted on the basis of the job's location in a particular section of the country. Thus, where these factors apply, you may find significant, but unavoidable cost variations for which you will have to apply a measure of judgment to your estimate.

Rounding of Costs

In general, all unit prices in excess of $5.00 have been rounded to make them easier to use and still maintain adequate precision of the results. The rounding rules we have chosen are in the following table.

Prices from . . .	Rounded to the nearest . . .
$.01 to $5.00	$.01
$5.01 to $20.00	$.05
$20.01 to $100.00	$.50
$100.01 to $300.00	$1.00
$300.01 to $1,000.00	$5.00
$1,000.01 to $10,000.00	$25.00
$10,000.01 to $50,000.00	$100.00
$50,000.01 and above	$500.00

Final Checklist

Estimating can be a straightforward process provided you remember the basics. Here's a checklist of some of the items you should remember to do before completing your estimate.

Did you remember to . . .

- factor in the City Cost Index for your locale
- take into consideration which items have been marked up and by how much
- mark up the entire estimate sufficiently for your purposes
- read the background information on techniques and technical matters that could impact your project time span and cost
- include all components of your project in the final estimate
- make use of Minimum Labor/Equipment Charges for Small Quantities (see the following page for more details)
- double check your figures to be sure of your accuracy
- call RSMeans if you have any questions about your estimate or the data you've found in our publications

Remember, RSMeans stands behind its publications. If you have any questions about your estimate . . . about the costs you've used from our books . . . or even about the technical aspects of the job that may affect your estimate, feel free to call the RSMeans editors at 1-800-334-3509.

Using Minimum Labor/Equipment Charges for Small Quantities

Estimating small construction or repair tasks often creates situations where the quantity of work to be performed is very small. When this occurs, the labor and/or equipment costs to perform the work may be too low to allow for the crew to get to the job, receive instructions, find their materials, get set up, perform the work, clean up and get to the next job. In these situations, the estimator should compare the developed labor and/or equipment costs for performing the work (e.g. quantity x labor and/or equipment costs) against the *"minimum labor/equipment charge"* within that Unit Price section of the book.

If the labor and/or equipment costs developed by the estimator are LOWER THAN the *"minimum labor/equipment charge"* listed at the bottom of specific sections of Unit Price costs, the estimator should adjust the developed costs upward to the *"minimum labor/equipment charge."* The proper use of a *"minimum labor/equipment charge"* results in having enough money in the estimate to cover the contractor's higher cost of performing a very small amount of work during a partial workday.

A *"minimum labor/equipment charge"* should be used only when the task being estimated is the only task the crew will perform at the job site that day. If, however, the crew will be able to perform other tasks at the job site that day, the use of a *"minimum labor/equipment charge"* is not appropriate.

08500 | Windows

	08550 | Wood Windows	CREW	DAILY OUTPUT	LABOR-HOURS	UNIT	2004 BARE COSTS MAT.	LABOR	EQUIP.	TOTAL	TOTAL INCL O&P	
200	0010 **CASEMENT WINDOW** Including frame, screen, and grills										200
	0100 Avg. quality, bldrs. model, 2'-0" x 3'-0" H, dbl. insulated glass	1 Carp	10	.800	Ea.	177	26.50		203.50	239	
	0150 Low E glass		10	.800		234	26.50		260.50	300	
	0200 2'-0" x 4'-6" high, double insulated glass		9	.889		230	29.50		259.50	300	
	0250 Low E glass		9	.889		320	29.50		349.50	400	
	0300 2'-3" x 6'-0" high, double insulated glass		8	1		269	33		302	350	
	0350 Low E glass		8	1		325	33		358	415	
	8100 Metal clad, deluxe, dbl. insul. glass, 2'-0" x 3'-0" high		10	.800		175	26.50		201.50	237	
	8120 2'-0" x 4'-0" high		9	.889		211	29.50		240.50	281	
	8140 2'-0" x 5'-0" high		8	1		240	33		273	320	
	9000 Minimum labor/equipment charge	1 Carp	3	2.667	Job		88		88	146	

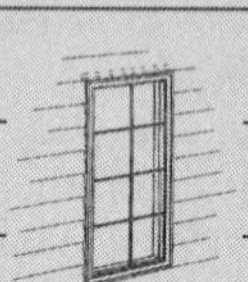

Example:

Establish the bid price to install two casement windows. Assume installation of a 2′ x 4′ metal clad windows with insulating glass [Unit Price line number 08550-200-8120], and that this is the only task this crew will perform at the job site that day.

Solution:

Step One —Develop the Bare Labor Cost for this task:

Bare Labor Cost = 2 windows @ $29.50/each = $59.00

Step Two —Evaluate the *"minimum labor/equipment charge"* for this Unit Price section against the developed Bare Labor Cost for this task:

"minimum labor/equipment charge" = $88.00 (compare with $59.00)

Step Three —Choose to adjust the developed labor cost upward to the *"minimum labor/equipment charge."*

Step Four —Develop the bid price for this task (including O&P):

Add together the marked up Bare Material Cost for this task, and the marked up *"minimum labor/equipment charge"* for this Unit Price section.

2 x ($211.00 + 10%) + ($88.00 + 65.8%)

= 2 x ($211.00 + $21.10) + ($88.00 + $57.90)

= 2 x ($232.10) + $145.90

=$464.20 + $145.90

= $610.10

ANSWER: $610.10 is the correct bid price to use. This sum takes into consideration the Material Cost (with 10% for profit) for these two windows, plus the *"minimum labor/equipment charge"* (with O&P included) for this section of the Unit Price book.

Unit Price Section

Table of Contents

How to Use the Unit Price Pages

The following is a detailed explanation of a sample entry in the Unit Price Section. Next to each bold number below is the item being described with appropriate component of the sample entry following in parenthesis. Some prices are listed as bare costs, others as costs that include overhead and profit of the installing contractor. In most cases, if the work is to be subcontracted, the general contractor will need to add an additional markup (RSMeans suggests using 10%) to the figures in the column "Total Incl. O&P."

1 *Division Number/Title (03300/Cast-In-Place Concrete)*

Use the Unit Price Section Table of Contents to locate specific items. The sections are classified according to the CSI MasterFormat (1995 Edition).

2 *Line Numbers (03310 240 3900)*

Each unit price line item has been assigned a unique 12-digit code based on the CSI MasterFormat classification.

Level One - CSI-MasterFormat Division

Level Two - CSI

03300

03310-240-3900

Means 12-digit Line Number

Level Four - Means

Level Three - CSI

3 *Description (Concrete-In-Place, etc.)*

Each line item is described in detail. Sub-items and additional sizes are indented beneath the appropriate line items. The first line or two after the main item (in boldface) may contain descriptive information that pertains to all line items beneath this boldface listing.

4 *Reference Number Information*

R03310 -010 You'll see reference numbers shown in bold rectangles at the beginning of some sections. These refer to related items in the Reference Section, visually identified by a vertical gray bar on the edge of pages.

The relation may be: (1) an estimating procedure that should be read before estimating, (2) an alternate pricing method, or (3) technical information.

The "R" designates the Reference Section. The numbers refer to the MasterFormat classification system.

It is strongly recommended that you review all reference numbers that appear within the section in which you are working.

Note: Not all reference numbers appear in all Means publications.

03300 | Cast-In-Place Concrete 1

03310 Structural Concrete			CREW	DAILY OUTPUT	LABOR-HOURS	UNIT	2004 BARE COSTS MAT.	LABOR	EQUIP.	TOTAL	TOTAL INCL O&P	
240	0010	**CONCRETE IN PLACE** Including forms (4 uses), reinforcing R03310 -010										240
	0050	steel and finishing unless otherwise indicated										
	3800	Footings, spread under 1 C.Y.	C-14C	38.07	2.942	C.Y.	105	93	.68	198.68	272	
	3850	Over 5 C.Y.		81.04	1.382		96.50	43.50	.32	140.32	179	
	3900	Footings, strip, 18″ x 9″, unreinforced		40	2.800		91	88.50	.64	180.14	248	
	3920	18″ x 9″, reinforced		35	3.200		105	101	.74	206.74	284	
	3925	20″ x 10″, unreinforced		45	2.489		88	78.50	.57	167.07	229	
	3930	20″ x 10″, reinforced		40	2.800		99.50	88.5[illegible]	.64	188.64	257	
	3935	24″ x 12″, unreinforced		55	2.036		87	64.5[illegible]	.47	151.97	[illegible]	
	3940	24″ x 12″, reinforced		48	2.333		98.50	74	.54	173.04	[illegible]	
	3945	36″ x 12″, unreinforced		70	1.600		83.50	50.50	.37	134.37	176	
	3950	36″ x 12″, reinforced		[illegible]	[illegible]867		94	59	.43	153.43	201	
	4000	Foundation mat, under 10 C.Y.		38.[illegible]	[illegible].896		129	91.50	.67	221.17	295	
	4050	Over 20 C.Y.		56.40	1.986		115	63	.46	178.46	232	
	4200	Grade walls, 8″ thick, 8′ high	C-14D	45.83	4.364		126	143	16.30	285.30	395	
	4250	14′ high		27.26	7.337		158	241	27.50	426.50	605	

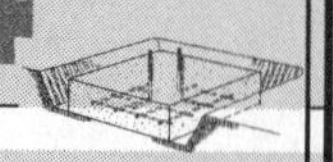

2 3 4 5 6 7 8 9

5 *Crew (C-14C)*

The "Crew" column designates the typical trade or crew used to install the item. If an installation can be accomplished by one trade and requires no power equipment, that trade and the number of workers are listed (for example, "2 Carpenters"). If an installation requires a composite crew, a crew code designation is listed (for example, "C-14C"). You'll find full details on all composite crews in the Crew Listings.

- For a complete list of all trades utilized in this book and their abbreviations, see the inside back cover.

Crews

Crew No.	Bare Costs		Incl. Subs O & P		Cost Per Labor-Hour	
Crew C-14C	Hr.	Daily	Hr.	Daily	Bare Costs	Incl. O&P
1 Carpenter Foreman (out)	$35.00	$280.00	$58.05	$464.40	$31.63	$52.57
6 Carpenters	33.00	1584.00	54.70	2625.60		
2 Rodmen (reinf.)	37.10	593.60	63.85	1021.60		
4 Laborers	26.00	832.00	43.10	1379.20		
1 Cement Finisher	31.55	252.40	49.60	396.80		
1 Gas Engine Vibrator		26.00		28.60	.23	.26
112 L.H., Daily Totals		$3568.00		$5916.20	$31.86	$52.83

6 *Productivity: Daily Output (40)/Labor-Hours (2.80)*

The "Daily Output" represents the typical number of units the designated crew will install in a normal 8-hour day. To find out the number of days the given crew would require to complete the installation, divide your quantity by the daily output. For example:

Quantity	÷	Daily Output	=	Duration
100 C.Y.	÷	40/ Crew Day	=	2.50 Crew Days

The "Labor-Hours" figure represents the number of labor-hours required to install one unit of work. To find out the number of labor-hours required for your particular task, multiply the quantity of the item times the number of labor-hours shown. For example:

Quantity	x	Productivity Rate	=	Duration
100 C.Y.	x	2.800 Labor-Hours/ C.Y.	=	280 Labor-Hours

7 *Unit (C.Y.)*

The abbreviated designation indicates the unit of measure upon which the price, production, and crew are based (C.Y. = Cubic Yard). For a complete listing of abbreviations refer to the Abbreviations Listing in the Reference Section of this book.

8 *Bare Costs:*

Mat. (Bare Material Cost) (91)

The unit material cost is the "bare" material cost with no overhead and profit included. *Costs shown reflect national average material prices for January of the current year and include delivery to the job site. No sales taxes are included.*

Labor (88.50)

The unit labor cost is derived by multiplying bare labor-hour costs for Crew C-14C by labor-hour units. The bare labor-hour cost is found in the Crew Section under C-14C. (If a trade is listed, the hourly labor cost—the wage rate—is found on the inside back cover.)

Labor-Hour Cost Crew C-14C	x	Labor-Hour Units	=	Labor
$31.63	x	2.80	=	$88.56

Equip. (Equipment) (.64)

Equipment costs for each crew are listed in the description of each crew. Tools or equipment whose value justifies purchase or ownership by a contractor are considered overhead as shown on the inside back cover. The unit equipment cost is derived by multiplying the bare equipment hourly cost by the labor-hour units.

Equipment Cost Crew C-14C	x	Labor-Hour Units	=	Equip.
.23	x	2.80	=	.64

Total (180.14)

The total of the bare costs is the arithmetic total of the three previous columns: mat., labor, and equip.

Material	+	Labor	+	Equip.	=	Total
$91	+	$88.50	+	$.64	=	$180.14

9 *Total Costs Including O&P*

This figure is the sum of the bare material cost plus 10% for profit; the bare labor cost plus total overhead and profit (per the inside back cover or, if a crew is listed, from the crew listings); and the bare equipment cost plus 10% for profit.

Material is Bare Material Cost + 10% = 91 + 9.10	=	$100.10
Labor for Crew C-14C = Labor-Hour Cost (52.57) x Labor-Hour Units (2.80)	=	$147.20
Equip. is Bare Equip. Cost + 10% = .64 + .06	=	$ .70
Total	=	$248

Division 1 General Requirements

Estimating Tips

The General Requirements of any contract are very important to both the bidder and the owner. These lay the ground rules under which the contract will be executed and have a significant influence on the cost of operations. Therefore, it is extremely important to thoroughly read and understand the General Requirements both before preparing an estimate and when the estimate is complete, to ascertain that nothing in the contract is overlooked. Caution should be exercised when applying items listed in Division 1 to an estimate. Many of the items are included in the unit prices listed in the other divisions such as mark-ups on labor and company overhead.

01200 Price & Payment Procedures

- When estimating historic preservation projects (depending on the condition of the existing structure and the owner's requirements), a 15-20% contingency or allowance is recommended, regardless of the stage of the drawings.

01300 Administrative Requirements

- Before determining a final cost estimate, it is a good practice to review all the items listed in subdivision 01300 to make final adjustments for items that may need customizing to specific job conditions.
- Historic preservation projects may require specialty labor and methods, as well as extra time to protect existing materials that must be preserved and/or restored. Some additional expenses may be incurred in architectural fees for facility surveys and other special inspections and analyses.

01330 Submittal Procedures

- Requirements for initial and periodic submittals can represent a significant cost to the General Requirements of a job. Thoroughly check the submittal specifications when estimating a project to determine any costs that should be included.

01400 Quality Requirements

- All projects will require some degree of Quality Control. This cost is not included in the unit cost of construction listed in each division. Depending upon the terms of the contract, the various costs of inspection and testing can be the responsibility of either the owner or the contractor. Be sure to include the required costs in your estimate.

01500 Temporary Facilities & Controls

- Barricades, access roads, safety nets, scaffolding, security and many more requirements for the execution of a safe project are elements of direct cost. These costs can easily be overlooked when preparing an estimate. When looking through the major classifications of this subdivision, determine which items apply to each division in your estimate.

01590 Equipment Rental

- This subdivision contains transportation, handling, storage, protection and product options and substitutions. Listed in this cost manual are average equipment rental rates for all types of equipment. This is useful information when estimating the time and materials requirement of any particular operation in order to establish a unit or total cost.
- A good rule of thumb is that weekly rental is 3 times daily rental and that monthly rental is 3 times weekly rental.
- The figures in the column for Crew Equipment Cost represent the rental rate used in determining the daily cost of equipment in a crew. It is calculated by dividing the weekly rate by 5 days and adding the hourly operating cost times 8 hours.

01770 Closeout Procedures

- When preparing an estimate, read the specifications to determine the requirements for Contract Closeout thoroughly. Final cleaning, record documentation, operation and maintenance data, warranties and bonds, and spare parts and maintenance materials can all be elements of cost for the completion of a contract. Do not overlook these in your estimate.

01830 Operations & Maintenance

- If maintenance and repair are included in your contract, they require special attention. To estimate the cost to remove and replace any unit usually requires a site visit to determine the accessibility and the specific difficulty at that location. Obstructions, dust control, safety, and often overtime hours must be considered when preparing your estimate.

Reference Numbers

Reference numbers are shown in bold squares at the beginning of some major classifications. These numbers refer to related items in the Reference Section. The reference information may be an estimating procedure, an alternate pricing method or technical information.

Note: Not all subdivisions listed here necessarily appear in this publication.

01100 | Summary

		01103 | Models & Renderings	CREW	DAILY OUTPUT	LABOR-HOURS	UNIT	2004 BARE COSTS MAT.	LABOR	EQUIP.	TOTAL	TOTAL INCL O&P	
500	0010	**RENDERINGS** Color, matted, 20" x 30", eye level,										500
	0020	1 building, minimum				Ea.	1,750			1,750	1,925	
	0050	Average					2,500			2,500	2,750	
	0100	Maximum					4,000			4,000	4,400	
	1000	5 buildings, minimum					3,500			3,500	3,850	
	1100	Maximum				↓	7,000			7,000	7,700	

		01107 | Professional Consultant	CREW	DAILY OUTPUT	LABOR-HOURS	UNIT	2004 BARE COSTS MAT.	LABOR	EQUIP.	TOTAL	TOTAL INCL O&P	
100	0011	**ARCHITECTURAL FEES** R01107-010										100
	0020	For new construction										
	0060	Minimum				Project					4.90%	
	0090	Maximum									16%	
	0100	For alteration work, to $500,000, add to fee									50%	
	0150	Over $500,000, add to fee				↓					25%	
200	0011	**CONSTRUCTION MANAGEMENT FEES**										200
	0060	For work to $10,000				Project					10%	
	0070	To $25,000									9%	
	0090	To $100,000									6%	
	0100	To $500,000									5%	
	0110	To $1,000,000				↓					4%	
300	0010	**ENGINEERING FEES** R01107-030										300
	0020	Educational planning consultant, minimum				Project					.50%	
	0100	Maximum				"					2.50%	
	0400	Elevator & conveying systems, minimum				Contrct					2.50%	
	0500	Maximum									5%	
	1000	Mechanical (plumbing & HVAC), minimum									4.10%	
	1100	Maximum				↓					10.10%	
	1200	Structural, minimum				Project					1%	
	1300	Maximum				"					2.50%	
700	0010	**SURVEYING** Conventional, topographical, minimum	A-7	3.30	7.273	Acre	16.30	259	18.05	293.35	460	700
	0100	Maximum	A-8	.60	53.333		49	1,850	99	1,998	3,175	
	0300	Lot location and lines, minimum, for large quantities	A-7	2	12		25.50	430	30	485.50	750	
	0320	Average	"	1.25	19.200		46	685	47.50	778.50	1,200	
	0400	Maximum, for small quantities	A-8	1	32	↓	73.50	1,100	59.50	1,233	1,950	
	1100	Crew for layout of building, trenching or pipe laying, 2 person crew	A-6	1	16	Day		585	59.50	644.50	1,000	
	1200	3 person crew	A-7	1	24			855	59.50	914.50	1,450	
	1400	Crew for roadway layout, 4 person crew	A-8	1	32	↓		1,100	59.50	1,159.50	1,875	

01200 | Price & Payment Procedures

		01250 | Contract Modification Procedures	CREW	DAILY OUTPUT	LABOR-HOURS	UNIT	2004 BARE COSTS MAT.	LABOR	EQUIP.	TOTAL	TOTAL INCL O&P	
200	0010	**CONTINGENCIES** for estimate at conceptual stage				Project					25%	200
	0050	Schematic stage									20%	
	0100	Preliminary working drawing stage (Design Dev.)									15%	
	0150	Final working drawing stage				↓					8%	
400	0010	**FACTORS** Cost adjustments R01250-010										400
	0100	Add to construction costs for particular job requirements										
	0500	Cut & patch to match existing construction, add, minimum				Costs	2%	3%				
	0550	Maximum				↓	5%	9%				

01200 | Price & Payment Procedures

01250 | Contract Modification Procedures

			CREW	DAILY OUTPUT	LABOR-HOURS	UNIT	2004 BARE COSTS MAT.	LABOR	EQUIP.	TOTAL	TOTAL INCL O&P	
400	0800	Dust protection, add, minimum R01250-010				Costs	1%	2%				400
	0850	Maximum					4%	11%				
	1100	Equipment usage curtailment, add, minimum					1%	1%				
	1150	Maximum					3%	10%				
	1400	Material handling & storage limitation, add, minimum					1%	1%				
	1450	Maximum					6%	7%				
	1700	Protection of existing work, add, minimum					2%	2%				
	1750	Maximum					5%	7%				
	2000	Shift work requirements, add, minimum						5%				
	2050	Maximum						30%				
	2300	Temporary shoring and bracing, add, minimum					2%	5%				
	2350	Maximum					5%	12%				
	2400	Work inside prisons and high security areas, add, minimum						30%				
	2450	Maximum						50%				
500	0010	**JOB CONDITIONS** Modifications to total										500
	0020	project cost summaries										
	0100	Economic conditions, favorable, deduct				Project					2%	
	0200	Unfavorable, add									5%	
	0300	Hoisting conditions, favorable, deduct									2%	
	0400	Unfavorable, add									5%	
	0500	General Contractor management, experienced, deduct									2%	
	0600	Inexperienced, add									10%	
	0700	Labor availability, surplus, deduct									1%	
	0800	Shortage, add									10%	
	0900	Material storage area, available, deduct									1%	
	1000	Not available, add									2%	
	1100	Subcontractor availability, surplus, deduct									5%	
	1200	Shortage, add									12%	
	1300	Work space, available, deduct									2%	
	1400	Not available, add									5%	

01290 | Payment Procedures

			CREW	DAILY OUTPUT	LABOR-HOURS	UNIT	MAT.	LABOR	EQUIP.	TOTAL	TOTAL INCL O&P	
800	0010	**TAXES** Sales tax, State, average R01100-090				%	4.65%					800
	0050	Maximum					7%					
	0200	Social Security, on first $87,000 of wages R01100-100						7.65%				
	0300	Unemployment, MA, combined Federal and State, minimum						2.10%				
	0350	Average						6.20%				
	0400	Maximum						8%				

01300 | Administrative Requirements

01310 | Project Management/Coordination

			CREW	DAILY OUTPUT	LABOR-HOURS	UNIT	2004 BARE COSTS MAT.	LABOR	EQUIP.	TOTAL	TOTAL INCL O&P	
150	0010	**PERMITS** Rule of thumb, most cities, minimum				Job					.50%	150
	0100	Maximum				"					2%	
350	0010	**INSURANCE** Builders risk, standard, minimum R01100-040				Job					.22%	350
	0050	Maximum									.59%	
	0200	All-risk type, minimum R01100-060									.25%	
	0250	Maximum									.62%	
	0400	Contractor's equipment floater, minimum				Value					.50%	
	0450	Maximum				"					1.50%	

01310 | Project Management/Coordination

		01310 Project Management/Coordination	CREW	DAILY OUTPUT	LABOR-HOURS	UNIT	2004 BARE COSTS MAT.	LABOR	EQUIP.	TOTAL	TOTAL INCL O&P	
350	0600	Public liability, average R01100-040				Job					1.55%	350
	0810	Workers' compensation & employer's liability										
	2000	Range of 35 trades in 50 states, excl. wrecking, min. R01100-060				Payroll		2.50%				
	2100	Average						16.20%				
	2200	Maximum						110.10%				
400	0010	**MAIN OFFICE EXPENSE** Average for General Contractors										400
	0020	As a percentage of their annual volume										
	0030	Annual volume to $50,000, minimum				% Vol.				20%		
	0040	Maximum								30%		
	0060	To $100,000, minimum								17%		
	0070	Maximum								22%		
	0080	To $250,000, minimum								16%		
	0090	Maximum								19%		
	0110	To $500,000, minimum								14%		
	0120	Maximum								16%		
	0130	To $1,000,000, minimum								8%		
	0140	Maximum								10%		
500	0010	**MARK-UP** For General Contractors for change										500
	0100	of scope of job as bid										
	0200	Extra work, by subcontractors, add				%					10%	
	0250	By General Contractor, add									15%	
	0400	Omitted work, by subcontractors, deduct all but									5%	
	0450	By General Contractor, deduct all but									7.50%	
	0600	Overtime work, by subcontractors, add									15%	
	0650	By General Contractor, add									10%	
	1000	Installing contractors, on his own labor, minimum						53.70%				
	1100	Maximum						97.30%				
600	0010	**OVERHEAD** As percent of direct costs, minimum R01100-050				%				5%		600
	0050	Average								16%		
	0100	Maximum R01100-070								30%		
620	0010	**OVERHEAD & PROFIT** Allowance to add to items in this R01100-050										620
	0020	book that do not include Subs O&P, average				%				30%		
	0100	Allowance to add to items in this book that										
	0110	do include Subs O&P, minimum				%					5%	
	0150	Average									10%	
	0200	Maximum									15%	
	0290	Typical, by size of project, under $50,000								40%		
	0310	$50,000 to $100,000								35%		
	0320	$100,000 to $500,000								25%		
	0330	$500,000 to $1,000,000								20%		
700	0011	**FIELD PERSONNEL**										700
	0020											
	0180	Project manager, minimum				Week		1,375		1,375	2,150	
	0200	Average						1,575		1,575	2,450	
	0220	Maximum						1,775		1,775	2,750	
	0240	Superintendent, minimum						1,325		1,325	2,050	
	0260	Average						1,450		1,450	2,250	
	0280	Maximum						1,675		1,675	2,600	

01400 | Quality Requirements

		01450 | Quality Control	CREW	DAILY OUTPUT	LABOR-HOURS	UNIT	2004 BARE COSTS MAT.	LABOR	EQUIP.	TOTAL	TOTAL INCL O&P	
500	0010	**TESTING** and Inspectional Services										500
	1800	Compressive test, cylinder, delivered to lab, ASTM C 39				Ea.					13	
	1900	Picked up by lab, minimum									15	
	1950	Average									20	
	2000	Maximum									30	
	2200	Compressive strength, cores (not incl. drilling), ASTM C 42									40	
	2300	Patching core holes									24	
	4730	Soil testing										
	4735	Soil density, nuclear method, ASTM D2922				Ea.					38.67	
	4740	Sand cone method ASTM D1556									30.17	
	4750	Moisture content, ASTM D 2216									10	
	4780	Permeability test, double ring infiltrometer									550	
	4800	Permeability, var. or constant head, undist., ASTM D 2434									250	
	4850	Recompacted									275	
	4900	Proctor compaction, 4″ standard mold, ASTM D 698									135	
	4950	6″ modified mold									75	
	5100	Shear tests, triaxial, minimum									450	
	5150	Maximum									600	
	5550	Technician for inspection, per day, earthwork									220	
	5650	Bolting									280	
	5750	Roofing									255	
	5790	Welding									270	
	5820	Non-destructive testing, dye penetrant				Day					330	
	5840	Magnetic particle									330	
	5860	Radiography									495	
	5880	Ultrasonic									340	
	6000	Welding certification, minimum				Ea.					100	
	6100	Maximum				″					275	
	7000	Underground storage tank										
	7500	Volumetric tightness test ,<=12,000 gal				Ea.					350	
	7510	<=30,000 gal				″					605	
	7600	Vadose zone (soil gas) sampling, 10-40 samples, min.				Day					1,500	
	7610	Maximum				″					2,500	
	7700	Ground water monitoring incl. drilling 3 wells, min.				Total					5,000	
	7710	Maximum				″					7,000	
	8000	X-ray concrete slabs				Ea.					200	

01500 | Temporary Facilities & Controls

		01510 | Temporary Utilities	CREW	DAILY OUTPUT	LABOR-HOURS	UNIT	2004 BARE COSTS MAT.	LABOR	EQUIP.	TOTAL	TOTAL INCL O&P	
800	0010	**TEMPORARY UTILITIES**										800
	0100	Heat, incl. fuel and operation, per week, 12 hrs. per day	1 Skwk	100	.080	CSF Flr	5.20	2.69		7.89	10.15	
	0200	24 hrs. per day	″	60	.133		7.85	4.49		12.34	15.95	
	0350	Lighting, incl. service lamps, wiring & outlets, minimum	1 Elec	34	.235		2.11	9.25		11.36	16.55	
	0360	Maximum	″	17	.471		4.60	18.55		23.15	33.50	
	0400	Power for temp lighting only, per month, min/month 6.6 KWH								.75	1.18	
	0450	Maximum/month 23.6 KWH								2.85	3.14	
	0600	Power for job duration incl. elevator, etc., minimum								47	51.70	
	0650	Maximum								110	121	

01520 | Construction Facilities

			CREW	DAILY OUTPUT	LABOR-HOURS	UNIT	2004 BARE COSTS MAT.	LABOR	EQUIP.	TOTAL	TOTAL INCL O&P	
500	0010	**OFFICE** Trailer, furnished, no hookups, 20′ x 8′, buy	2 Skwk	1	16	Ea.	6,125	540		6,665	7,625	500
	0250	Rent per month					153			153	169	
	0300	32′ x 8′, buy	2 Skwk	.70	22.857		9,225	770		9,995	11,400	
	0350	Rent per month					164			164	180	
	0400	50′ x 10′, buy	2 Skwk	.60	26.667		16,800	895		17,695	20,000	
	0450	Rent per month					260			260	286	
	0500	50′ x 12′, buy	2 Skwk	.50	32		20,000	1,075		21,075	23,800	
	0550	Rent per month					286			286	315	
	0700	For air conditioning, rent per month, add				↓	39.50			39.50	43.50	
	0800	For delivery, add per mile				Mile	1.53			1.53	1.68	
	1200	Storage boxes, 20′ x 8′, buy	2 Skwk	1.80	8.889	Ea.	3,050	299		3,349	3,875	
	1250	Rent per month					74			74	81.50	
	1300	40′ x 8′, buy	2 Skwk	1.40	11.429		4,100	385		4,485	5,125	
	1350	Rent per month				↓	105			105	115	

01530 | Temporary Construction

			CREW	DAILY OUTPUT	LABOR-HOURS	UNIT	MAT.	LABOR	EQUIP.	TOTAL	TOTAL INCL O&P	
700	0010	**PROTECTION** Stair tread, 2″ x 12″ planks, 1 use	1 Carp	75	.107	Tread	4.42	3.52		7.94	10.70	700
	0100	Exterior plywood, 1/2″ thick, 1 use		65	.123		1.18	4.06		5.24	8.05	
	0200	3/4″ thick, 1 use	↓	60	.133	↓	1.72	4.40		6.12	9.20	
900	0010	**WINTER PROTECTION** Reinforced plastic on wood										900
	0100	framing to close openings	2 Clab	750	.021	S.F.	.38	.55		.93	1.34	
	0200	Tarpaulins hung over scaffolding, 8 uses, not incl. scaffolding		1,500	.011		.18	.28		.46	.66	
	0250	Tarpaulin polyester reinf. w/ integral fastening system 11 mils thick		1,600	.010		.78	.26		1.04	1.29	
	0300	Prefab fiberglass panels, steel frame, 8 uses	↓	1,200	.013	↓	.74	.35		1.09	1.38	

01540 | Construction Aids

			CREW	DAILY OUTPUT	LABOR-HOURS	UNIT	MAT.	LABOR	EQUIP.	TOTAL	TOTAL INCL O&P	
550	0010	**PUMP STAGING**, Aluminum R01540-200										550
	0200	24′ long pole section, buy				Ea.	355			355	395	
	0300	18′ long pole section, buy					276			276	305	
	0400	12′ long pole section, buy					187			187	206	
	0500	6′ long pole section, buy					99.50			99.50	110	
	0600	6′ long splice joint section, buy					73.50			73.50	80.50	
	0700	Pump jack					119			119	131	
	0900	Foldable brace					51.50			51.50	56.50	
	1000	Workbench/back safety rail support					63			63	69	
	1100	Scaffolding planks/workbench, 14″ wide x 24′ long					580			580	640	
	1200	Plank end safety rail					196			196	215	
	1250	Safety net, 22′ long				↓	285			285	315	
	1300	System in place, 50′ working height, per use based on 50 uses	2 Carp	84.80	.189	C.S.F.	5.30	6.25		11.55	16.10	
	1400	100 uses		84.80	.189		2.64	6.25		8.89	13.20	
	1500	150 uses	↓	84.80	.189	↓	1.77	6.25		8.02	12.25	
750	0010	**SCAFFOLDING** R01540-100										750
	0015	Steel tubular, reg, rent/mo, no plank, incl erect or dismantle										
	0090	Building exterior, wall face, 1 to 5 stories, 6′-4″ x 5′ frames	3 Carp	24	1	C.S.F.	24.50	33		57.50	81	
	0200	6 to 12 stories	4 Carp	21.20	1.509		24.50	50		74.50	109	
	0310	13 to 20 stories	5 Carp	20	2		24.50	66		90.50	136	
	0460	Building interior, wall face area, up to 16′ high	3 Carp	25	.960		24.50	31.50		56	79	
	0560	16′ to 40′ high		23	1.043	↓	24.50	34.50		59	83.50	
	0800	Building interior floor area, up to 30′ high	↓	312	.077	C.C.F.	2.57	2.54		5.11	7.05	
	0900	Over 30′ high	4 Carp	275	.116	″	2.57	3.84		6.41	9.20	
	0910	Steel tubular, heavy duty shoring, buy										
	0920	Frames 5′ high 2′ wide				Ea.	75			75	82.50	
	0925	5′ high 4′ wide				↓	85			85	93.50	

1 GENERAL REQUIREMENTS

		01540 | Construction Aids	CREW	DAILY OUTPUT	LABOR-HOURS	UNIT	2004 BARE COSTS MAT.	LABOR	EQUIP.	TOTAL	TOTAL INCL O&P	
750	0930	6′ high 2′ wide R01540-100				Ea.	86			86	94.50	750
	0935	6′ high 4′ wide					101			101	111	
	0940	Accessories										
	0945	Cross braces				Ea.	16			16	17.60	
	0950	U-head, 8″ x 8″					17.50			17.50	19.25	
	0955	J-head, 4″ x 8″					12.80			12.80	14.10	
	0960	Base plate, 8″ x 8″					14.20			14.20	15.60	
	0965	Leveling jack					30.50			30.50	33.50	
	1000	Steel tubular, regular, buy										
	1100	Frames 3′ high 5′ wide				Ea.	58			58	64	
	1150	5′ high 5′ wide					67			67	73.50	
	1200	6′-4″ high 5′ wide					84			84	92.50	
	1350	7′-6″ high 6′ wide					145			145	160	
	1500	Accessories cross braces					15			15	16.50	
	1550	Guardrail post					15			15	16.50	
	1600	Guardrail 7′ section					7.25			7.25	8	
	1650	Screw jacks & plates					24			24	26.50	
	1700	Sidearm brackets					28			28	31	
	1750	8″ casters					33			33	36.50	
	1800	Plank 2″ x 10″ x 16′-0″					42.50			42.50	47	
	1900	Stairway section					245			245	270	
	1910	Stairway starter bar					29			29	32	
	1920	Stairway inside handrail					53			53	58.50	
	1930	Stairway outside handrail					73			73	80.50	
	1940	Walk-thru frame guardrail					37			37	40.50	
	2000	Steel tubular, regular, rent/mo.										
	2100	Frames 3′ high 5′ wide				Ea.	3.75			3.75	4.13	
	2150	5′ high 5′ wide					3.75			3.75	4.13	
	2200	6′-4″ high 5′ wide					3.75			3.75	4.13	
	2250	7′-6″ high 6′ wide					7			7	7.70	
	2500	Accessories, cross braces					.60			.60	.66	
	2550	Guardrail post					1			1	1.10	
	2600	Guardrail 7′ section					.75			.75	.83	
	2650	Screw jacks & plates					1.50			1.50	1.65	
	2700	Sidearm brackets					1.50			1.50	1.65	
	2750	8″ casters					6			6	6.60	
	2800	Outrigger for rolling tower					3			3	3.30	
	2850	Plank 2″ x 10″ x 16′-0″					5			5	5.50	
	2900	Stairway section					10			10	11	
	2910	Stairway starter bar					.10			.10	.11	
	2920	Stairway inside handrail					5			5	5.50	
	2930	Stairway outside handrail					5			5	5.50	
	2940	Walk-thru frame guardrail					2			2	2.20	
	3000	Steel tubular, heavy duty shoring, rent/mo.										
	3250	5′ high 2′ & 4′ wide				Ea.	5			5	5.50	
	3300	6′ high 2′ & 4′ wide					5			5	5.50	
	3500	Accessories, cross braces					1			1	1.10	
	3600	U - head, 8″ x 8″					1			1	1.10	
	3650	J - head, 4″ x 8″					1			1	1.10	
	3700	Base plate, 8″ x 8″					1			1	1.10	
	3750	Leveling jack					2			2	2.20	
	5700	Planks, 2x10x16′-0″, labor only, erect or remove to 50′ H	3 Carp	144	.167			5.50		5.50	9.10	
	5800	Over 50′ high	4 Carp	160	.200			6.60		6.60	10.95	
755	0010	**SCAFFOLDING SPECIALTIES**										755
	1200	Sidewalk bridge, heavy duty steel posts & beams, including										

	01540 | Construction Aids		CREW	DAILY OUTPUT	LABOR-HOURS	UNIT	2004 BARE COSTS MAT.	LABOR	EQUIP.	TOTAL	TOTAL INCL O&P	
755	1210	parapet protection & waterproofing										755
	1220	8′ to 10′ wide, 2 posts	3 Carp	15	1.600	L.F.	28.50	53		81.50	119	
	1230	3 posts	"	10	2.400	"	44.50	79		123.50	180	
	1500	Sidewalk bridge using tubular steel										
	1512	scaffold frames, including planking	3 Carp	55	.436	L.F.	4.72	14.40		19.12	29	
	1600	For 2 uses per month, deduct from all above					50%					
	1700	For 1 use every 2 months, add to all above					100%					
	1900	Catwalks, 20″ wide, no guardrails, 7′ span, buy				Ea.	125			125	138	
	2000	10′ span, buy				"	160			160	176	
	2800	Hand winch-operated masons scaffolding, no plank										
	2810	plank moving not required										
	2900	98′ long, 10′-6″ high, buy				Ea.	26,200			26,200	28,900	
	3000	Rent per month					1,050			1,050	1,150	
	3100	28′-6″ high, buy					32,200			32,200	35,400	
	3200	Rent per month					1,275			1,275	1,425	
	3400	196′ long, 28′-6″ high, buy					62,000			62,000	68,500	
	3500	Rent per month					2,475			2,475	2,725	
	3600	64′-6″ high, buy					85,000			85,000	93,500	
	3700	Rent per month					3,400			3,400	3,750	
	3720	Putlog, standard, 8′ span, with hangers, buy					61			61	67	
	3730	Rent per month					10			10	11	
	3750	12′ span, buy					92			92	101	
	3755	Rent per month					15			15	16.50	
	3760	Trussed type, 16′ span, buy					210			210	231	
	3770	Rent per month					20			20	22	
	3790	22′ span, buy					252			252	277	
	3795	Rent per month					30			30	33	
	4000	7 step					510			510	560	
	4100	Rolling towers, buy, 5′ wide, 7′ long, 10′ high					1,150			1,150	1,250	
	4200	For 5′ high added sections, to buy, add					190			190	209	
	4300	Complete incl. wheels, railings, outriggers,										
	4350	21′ high, to buy				Ea.	1,925			1,925	2,125	
	4400	Rent/month				"	145			145	159	
760	0010	**STAGING AIDS** and fall protection equipment										760
	0100	Sidewall staging bracket, tubular, buy				Ea.	29			29	32	
	0110	Cost each per day, based on 250 days use				Day	.12			.12	.13	
	0200	Guard post, buy				Ea.	15			15	16.50	
	0210	Cost each per day, based on 250 days use				Day	.06			.06	.07	
	0300	End guard chains, buy per pair				Pair	25			25	27.50	
	0310	Cost per set per day, based on 250 days use				Day	.12			.12	.13	
	1010	Cost each per day, based on 250 days use				"	.03			.03	.03	
	1100	Wood bracket, buy				Ea.	13.15			13.15	14.45	
	1110	Cost each per day, based on 250 days use				Day	.05			.05	.06	
	2010	Cost per pair per day, based on 250 days use				"	.33			.33	.36	
	2100	Steel siderail jack, buy per pair				Pair	63			63	69.50	
	2110	Cost per pair per day, based on 250 days use				Day	.25			.25	.28	
	3010	Cost each per day, based on 250 days use				"	.17			.17	.19	
	3100	Aluminum scaffolding plank, 20″ wide x 24′ long, buy				Ea.	690			690	760	
	3110	Cost each per day, based on 250 days use				Day	2.76			2.76	3.04	
	4000	Nylon full body harness, lanyard and rope grab				Ea.	207			207	228	
	4010	Cost each per day, based on 250 days use				Day	.83			.83	.91	
	4100	Rope for safety line, 5/8″ x 100′ nylon, buy				Ea.	41			41	45	
	4110	Cost each per day, based on 250 days use				Day	.16			.16	.18	
	4200	Permanent U-Bolt roof anchor, buy				Ea.	31			31	34	
	4300	Temporary (one use) roof ridge anchor, buy				"	24			24	26.50	

	01540 | Construction Aids		CREW	DAILY OUTPUT	LABOR-HOURS	UNIT	2004 BARE COSTS MAT.	LABOR	EQUIP.	TOTAL	TOTAL INCL O&P	
760	5000	Installation (setup and removal) of staging aids										760
	5010	Sidewall staging bracket	2 Carp	64	.250	Ea.		8.25		8.25	13.70	
	5020	Guard post with 2 wood rails	″	64	.250			8.25		8.25	13.70	
	5030	End guard chains, set	1 Carp	64	.125			4.13		4.13	6.85	
	5100	Roof shingling bracket		96	.083			2.75		2.75	4.56	
	5200	Ladder jack		64	.125			4.13		4.13	6.85	
	5300	Wood plank, 2x10x16′	2 Carp	80	.200			6.60		6.60	10.95	
	5310	Aluminum scaffold plank, 20″ x 24′	″	40	.400			13.20		13.20	22	
	5410	Safety rope	1 Carp	40	.200			6.60		6.60	10.95	
	5420	Permanent U-Bolt roof anchor (install only)	2 Carp	40	.400			13.20		13.20	22	
	5430	Temporary roof ridge anchor (install only)	1 Carp	64	.125			4.13		4.13	6.85	
780	0010	**SWING STAGING**, 500 lb cap., 2′ wide to 24′ long, hand operated hoist										780
	0020	steel cable type, with 60′ cables, buy				Ea.	4,250			4,250	4,675	
	0030	Rent per month				″	425			425	470	
	0600	Lightweight (not for masons) 24′ long for 150′ height.										
	0610	manual type, buy				Ea.	4,500			4,500	4,950	
	0620	Rent per month					450			450	495	
	0700	Powered, electric or air, to 150′ high, buy					15,900			15,900	17,500	
	0710	Rent per month					1,125			1,125	1,225	
	0780	To 300′ high, buy					18,700			18,700	20,500	
	0800	Rent per month					1,300			1,300	1,450	
	1000	Bosun's chair or work basket 3′ x 3.5′, to 300′ high, electric, buy					7,025			7,025	7,725	
	1010	Rent per month					490			490	540	
	2200	Move swing staging (setup and remove)	E-4	2	16	Move		600	37.50	637.50	1,175	
800	0010	**TARPAULINS** Cotton duck, 10 oz. to 13.13 oz. per S.Y., minimum				S.F.	.48			.48	.53	800
	0050	Maximum					.57			.57	.63	
	0100	Polyvinyl coated nylon, 14 oz. to 18 oz., minimum					.47			.47	.52	
	0150	Maximum					.67			.67	.74	
	0200	Reinforced polyethylene 3 mils thick, white					.11			.11	.12	
	0300	4 mils thick, white, clear or black					.14			.14	.15	
	0400	5.5 mils thick, clear					.19			.19	.21	
	0500	White, fire retardant					.18			.18	.20	
	0600	7.5 mils, oil resistant, fire retardant					.19			.19	.21	
	0700	8.5 mils, black					.24			.24	.26	
	0720	Steel reinforced polyethylene, 4 mils thick					.52			.52	.57	
	0730	Polyester reinforced w/ integral fastening system 11 mils thick					1.05			1.05	1.16	
	0740	Mylar polyester, non-reinforced, 7 mils thick					1.15			1.15	1.27	
820	0010	**SMALL TOOLS** As % of contractor's work, minimum R01100-050				Total					.50%	820
	0100	Maximum				″					2%	

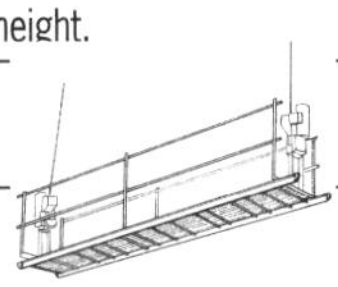

	01550 | Vehicular Access & Parking		CREW	DAILY OUTPUT	LABOR-HOURS	UNIT	MAT.	LABOR	EQUIP.	TOTAL	TOTAL INCL O&P	
700	0010	**ROADS AND SIDEWALKS** Temporary										700
	2200	Sidewalks, 2″ x 12″ planks, 2 uses	1 Carp	350	.023	S.F.	.74	.75		1.49	2.06	
	2300	Exterior plywood, 2 uses, 1/2″ thick		750	.011		.20	.35		.55	.80	
	2400	5/8″ thick		650	.012		.23	.41		.64	.92	
	2500	3/4″ thick		600	.013		.29	.44		.73	1.05	

	01560 | Barriers & Enclosures		CREW	DAILY OUTPUT	LABOR-HOURS	UNIT	MAT.	LABOR	EQUIP.	TOTAL	TOTAL INCL O&P	
100	0010	**BARRICADES** 5′ high, 3 rail @ 2″ x 8″, fixed	2 Carp	30	.533	L.F.	11	17.60		28.60	41	100
	0150	Movable	″	20	.800	″	11	26.50		37.50	56	
	0300	Stock units, 6′ high, 8′ wide, plain, buy				Ea.	435			435	480	
	0350	With reflective tape, buy				″	525			525	580	

01560 | Barriers & Enclosures

		01560 Barriers & Enclosures	CREW	DAILY OUTPUT	LABOR-HOURS	UNIT	2004 BARE COSTS MAT.	LABOR	EQUIP.	TOTAL	TOTAL INCL O&P	
100	0400	Break-a-way 3" PVC pipe barricade										100
	0410	with 3 ea. 1' x 4' reflectorized panels, buy				Ea.	305			305	335	
	0500	Plywood with steel legs, 32" wide					72			72	79	
	0600	Telescoping Christmas tree, 9' high, 5 flags, buy					122			122	134	
	0800	Traffic cones, PVC, 18" high					6			6	6.60	
	0850	28" high					18.25			18.25	20	
	0900	Barrels, 55 gal., with flasher	1 Clab	96	.083		52	2.17		54.17	60.50	
	1000	Guardrail, wooden, 3' high, 1" x 6", on 2" x 4" posts	2 Carp	200	.080	L.F.	.78	2.64		3.42	5.25	
	1100	2" x 6", on 4" x 4" posts	"	165	.097		1.74	3.20		4.94	7.20	
	1200	Portable metal with base pads, buy					15.50			15.50	17.05	
	1250	Typical installation, assume 10 reuses	2 Carp	600	.027		1.60	.88		2.48	3.22	
	1300	Barricade tape, polyethelyne, 7 mil, 3" wide x 500' long roll				Ea.	25			25	27.50	
250	0010	**TEMPORARY FENCING** Chain link, 11 ga, 5' high	2 Clab	400	.040	L.F.	3.58	1.04		4.62	5.65	250
	0100	6' high		300	.053		3.45	1.39		4.84	6.10	
	0200	Rented chain link, 6' high, to 1000' (up to 12 mo.)		400	.040		2.26	1.04		3.30	4.21	
	0250	Over 1000' (up to 12 mo.)		300	.053		1.64	1.39		3.03	4.10	
	0350	Plywood, painted, 2" x 4" frame, 4' high	A-4	135	.178		3.62	5.65		9.27	13.30	
	0400	4" x 4" frame, 8' high	"	110	.218		7.70	6.95		14.65	19.90	
	0500	Wire mesh on 4" x 4" posts, 4' high	2 Carp	100	.160		6.85	5.30		12.15	16.25	
	0550	8' high	"	80	.200		10.25	6.60		16.85	22	
400	0010	**TEMPORARY CONSTRUCTION** See also division 01530										400
800	0010	**WATCHMAN** Service, monthly basis, uniformed person, minimum				Hr.					8.20	800
	0100	Maximum									14.85	
	0200	Person and command dog, minimum									10.80	
	0300	Maximum									16.05	
	0500	Sentry dog, leased, with job patrol (yard dog), 1 dog				Week					210	
	0600	2 dogs				"					290	
	0800	Purchase, trained sentry dog, minimum				Ea.					850	
	0900	Maximum				"					2,000	

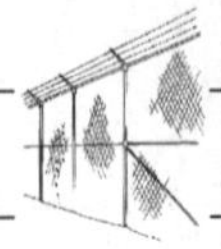

01580 | Project Signs

		01580 Project Signs	CREW	DAILY OUTPUT	LABOR-HOURS	UNIT	MAT.	LABOR	EQUIP.	TOTAL	TOTAL INCL O&P	
700	0010	**SIGNS** Hi-intensity reflectorized, no posts, buy				S.F.	16.35			16.35	18	700

01590 | Equipment Rental

			UNIT	HOURLY OPER. COST	RENT PER DAY	RENT PER WEEK	RENT PER MONTH	CREW EQUIPMENT COST/DAY	
100	0010	**CONCRETE EQUIPMENT RENTAL** R01590-100							100
	0100	without operators							
	0150	For batch plant, see div. 01590-500 R03310-090							
	0200	Bucket, concrete lightweight, 1/2 C.Y.	Ea.	.50	15.35	46	138	13.20	
	0300	1 C.Y.		.55	18.65	56	168	15.60	
	0400	1-1/2 C.Y.		.70	25.50	77	231	21	
	0500	2 C.Y.		.80	30.50	92	276	24.80	
	0580	8 C.Y.		4.25	202	605	1,825	155	
	0600	Cart, concrete, self propelled, operator walking, 10 C.F.		1.85	56.50	170	510	48.80	
	0700	Operator riding, 18 C.F.		2.85	86.50	260	780	74.80	
	0800	Conveyer for concrete, portable, gas, 16" wide, 26' long		6.15	117	350	1,050	119.20	
	0900	46' long		6.55	143	430	1,300	138.40	
	1000	56' long		6.65	153	460	1,375	145.20	
	1100	Core drill, electric, 2-1/2 H.P., 1" to 8" bit diameter		1.53	58.50	176	530	47.45	
	1150	11 HP, 8" to 18" cores		6.73	87	260.80	780	106	
	1200	Finisher, concrete floor, gas, riding trowel, 48" diameter		3.55	86.50	260	780	80.40	
	1300	Gas, manual, 3 blade, 36" trowel		.80	25.50	76	228	21.60	
	1400	4 blade, 48" trowel		1.15	30	90	270	27.20	
	1500	Float, hand-operated (Bull float) 48" wide		.08	13.35	40	120	8.65	
	1570	Curb builder, 14 H.P., gas, single screw		8.50	150	450	1,350	158	
	1590	Double screw		9.25	190	570	1,700	188	
	1600	Grinder, concrete and terrazzo, electric, floor		1.82	75	225	675	59.55	
	1700	Wall grinder		.91	37.50	113	340	29.90	
	1800	Mixer, powered, mortar and concrete, gas, 6 C.F., 18 H.P.		4.50	90	270	810	90	
	1900	10 C.F., 25 H.P.		5.40	103	310	930	105.20	
	2000	16 C.F.		5.70	127	380	1,150	121.60	
	2100	Concrete, stationary, tilt drum, 2 C.Y.		5	202	605	1,825	161	
	2120	Pump, concrete, truck mounted 4" line 80' boom		21.15	915	2,750	8,250	719.20	
	2140	5" line, 110' boom		28.60	1,275	3,800	11,400	988.80	
	2160	Mud jack, 50 C.F. per hr.		4.91	119	358	1,075	110.90	
	2180	225 C.F. per hr.		7.70	159	476.80	1,425	156.95	
	2190	Shotcrete pump rig, 12 CY/hr		10.80	235	705	2,125	227.40	
	2600	Saw, concrete, manual, gas, 18 H.P.		2.85	33.50	100	300	42.80	
	2650	Self-propelled, gas, 30 H.P.		5.40	81.50	245	735	92.20	
	2700	Vibrators, concrete, electric, 60 cycle, 2 H.P.		.38	15	45	135	12.05	
	2800	3 H.P.		.58	21.50	65	195	17.65	
	2900	Gas engine, 5 H.P.		.75	24	72	216	20.40	
	3000	8 H.P.		1.05	29.50	88	264	26	
	3050	Vibrating screed, gas engine, 8HP		1.53	42.50	127	380	37.65	
	3100	Concrete transit mixer, hydraulic drive							
	3120	6 x 4, 250 H.P., 8 C.Y., rear discharge		32.85	630	1,885	5,650	639.80	
	3200	Front discharge		36.20	680	2,045	6,125	698.60	
	3300	6 x 6, 285 H.P., 12 C.Y., rear discharge		35.10	640	1,915	5,750	663.80	
	3400	Front discharge		36.90	690	2,070	6,200	709.20	
200	0010	**EARTHWORK EQUIPMENT RENTAL** Without operators R01590-100							200
	0040	Aggregate spreader, push type 8' to 12' wide	Ea.	1.80	45	135	405	41.40	
	0045	Tailgate type, 8' wide R02315-450	"	1.75	31.50	95	285	33	
	0050	Augers for truck or trailer mounting, vertical drilling							
	0055	Fence post auger, truck mounted	Ea.	8.30	485	1,455	4,375	357.40	
	0060	4" to 36" diam., 54 H.P., gas, 10' spindle travel		28.35	645	1,940	5,825	614.80	
	0070	14' spindle travel		32.25	805	2,410	7,225	740	
	0075	Auger, truck mounted, vertical drilling, to 25' depth		138.75	2,950	8,835	26,500	2,877	
	0080	Auger, horizontal boring machine, 12" to 36" diameter, 45 H.P.		15.45	193	580	1,750	239.60	
	0090	12" to 48" diameter, 65 H.P.		21.70	345	1,040	3,125	381.60	
	0100	Excavator, diesel hydraulic, crawler mounted, 1/2 C.Y. cap.		14.25	350	1,045	3,125	323	
	0120	5/8 C.Y. capacity		17.50	465	1,390	4,175	418	
	0140	3/4 C.Y. capacity		20.35	495	1,480	4,450	458.80	
	0150	1 C.Y. capacity		22.30	515	1,545	4,625	487.40	

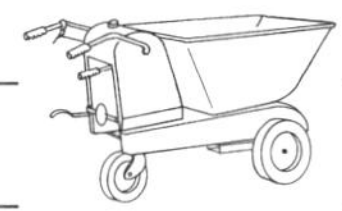

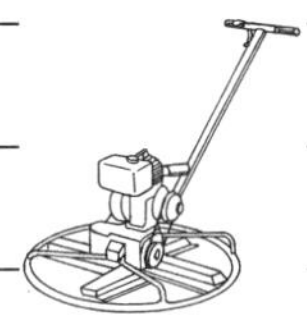

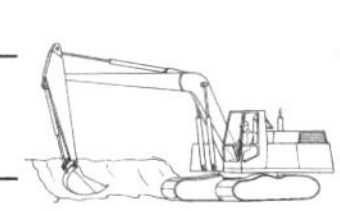

01590 | Equipment Rental

200

R01590-100

R02315-450

01590	Equipment Rental	UNIT	HOURLY OPER. COST	RENT PER DAY	RENT PER WEEK	RENT PER MONTH	CREW EQUIPMENT COST/DAY
0200	1-1/2 C.Y. capacity	Ea.	29.30	750	2,245	6,725	683.40
0300	2 C.Y. capacity		38.15	970	2,910	8,725	887.20
0320	2-1/2 C.Y. capacity		49.75	1,300	3,915	11,700	1,181
0340	3-1/2 C.Y. capacity		84.80	2,150	6,435	19,300	1,965
0341	Attachments						
0342	Bucket thumbs		2.40	208	625	1,875	144.20
0345	Grapples		1	220	658.80	1,975	139.75
0350	Gradall type, truck mounted, 3 ton @ 15' radius, 5/8 C.Y.		35.55	900	2,695	8,075	823.40
0370	1 C.Y. capacity		41.10	1,050	3,150	9,450	958.80
0400	Backhoe-loader, 40 to 45 H.P., 5/8 C.Y. capacity		7.55	180	540	1,625	168.40
0450	45 H.P. to 60 H.P., 3/4 C.Y. capacity		9.10	225	675	2,025	207.80
0460	80 H.P., 1-1/4 C.Y. capacity		11.10	237	710	2,125	230.80
0470	112 H.P., 1-1/2 C.Y. capacity		16.35	390	1,170	3,500	364.80
0480	Attachments						
0482	Compactor, 20,000 lb		4.25	117	350	1,050	104
0485	Hydraulic hammer, 750 ft-lbs		1.95	68.50	205	615	56.60
0486	Hydraulic hammer, 1200 ft-lbs		4	133	400	1,200	112
0500	Brush chipper, gas engine, 6″ cutter head, 35 H.P.		5.90	93.50	280	840	103.20
0550	12″ cutter head, 130 H.P.		9.35	150	450	1,350	164.80
0600	15″ cutter head, 165 H.P.		13.40	158	475	1,425	202.20
0750	Bucket, clamshell, general purpose, 3/8 C.Y.		.95	33.50	100	300	27.60
0800	1/2 C.Y.		1.05	41.50	125	375	33.40
0850	3/4 C.Y.		1.20	51.50	155	465	40.60
0900	1 C.Y.		1.25	56.50	170	510	44
0950	1-1/2 C.Y.		1.95	75	225	675	60.60
1000	2 C.Y.		2.05	85	255	765	67.40
1010	Bucket, dragline, medium duty, 1/2 C.Y.		.55	22.50	67	201	17.80
1020	3/4 C.Y.		.55	23.50	71	213	18.60
1030	1 C.Y.		.60	25.50	77	231	20.20
1040	1-1/2 C.Y.		.90	38.50	115	345	30.20
1050	2 C.Y.		1	43.50	130	390	34
1070	3 C.Y.		1.45	60	180	540	47.60
1200	Compactor, roller, 2 drum, 2000 lb., operator walking		6.10	137	410	1,225	130.80
1250	Rammer compactor, gas, 1000 lb. blow		1.55	38.50	115	345	35.40
1300	Vibratory plate, gas, 18″ plate, 3000 lb. blow		1.45	28.50	85	255	28.60
1350	21″ plate, 5000 lb. blow		1.60	47.50	143	430	41.40
1370	Curb builder/extruder, 14 H.P., gas, single screw		8.50	152	456	1,375	159.20
1390	Double screw		9.25	191	572	1,725	188.40
1500	Disc harrow attachment, for tractor		.35	59	177	530	38.20
1750	Extractor, piling, see lines 2500 to 2750						
1810	Feller buncher, shearing & accumulating trees, 100 H.P.	Ea.	19.80	455	1,370	4,100	432.40
1860	Grader, self-propelled, 25,000 lb.		16.25	410	1,235	3,700	377
1910	30,000 lb.		18.25	475	1,430	4,300	432
1920	40,000 lb.		28.20	735	2,205	6,625	666.60
1930	55,000 lb.		37.70	1,025	3,060	9,175	913.60
1950	Hammer, pavement demo., hyd., gas, self-prop., 1000 to 1250 lb.		20.15	390	1,170	3,500	395.20
2000	Diesel 1300 to 1500 lb.		27.80	590	1,765	5,300	575.40
2050	Pile driving hammer, steam or air, 4150 ft.-lb. @ 225 BPM		6.35	275	825	2,475	215.80
2100	8750 ft.-lb. @ 145 BPM		8.25	450	1,350	4,050	336
2150	15,000 ft.-lb. @ 60 BPM		8.60	485	1,455	4,375	359.80
2200	24,450 ft.-lb. @ 111 BPM		11.35	535	1,605	4,825	411.80
2250	Leads, 15,000 ft.-lb. hammers	L.F.	.03	2.10	6.29	18.85	1.50
2300	24,450 ft.-lb. hammers and heavier	″	.05	3	9	27	2.20
2350	Diesel type hammer, 22,400 ft.-lb.	Ea.	24	620	1,860	5,575	564
2400	41,300 ft.-lb.		32.20	675	2,025	6,075	662.60
2450	141,000 ft.-lb.		60.90	1,475	4,430	13,300	1,373
2500	Vib. elec. hammer/extractor, 200 KW diesel generator, 34 H.P.		26.25	655	1,965	5,900	603
2550	80 H.P.		44.80	960	2,885	8,650	935.40

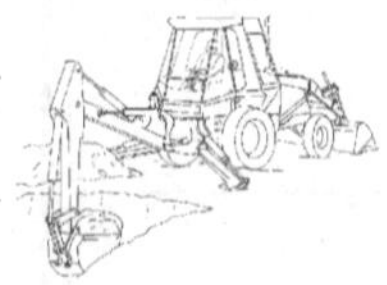

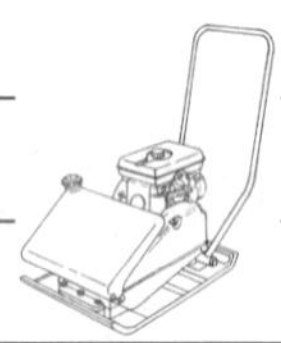

Important: See the Reference Section for critical supporting data - Reference Nos., Crews, & City Cost Indexes

01590 | Equipment Rental

		Description	UNIT	HOURLY OPER. COST	RENT PER DAY	RENT PER WEEK	RENT PER MONTH	CREW EQUIPMENT COST/DAY	
200	2600	150 H.P. R01590-100	Ea.	64	1,475	4,455	13,400	1,403	200
	2700	Extractor, steam or air, 700 ft.-lb.		14.50	345	1,040	3,125	324	
	2750	1000 ft.-lb. R03310-090		16.60	450	1,350	4,050	402.80	
	2800	Log chipper, up to 22" diam, 600 H.P.		41.93	1,450	4,360	13,100	1,207	
	2850	Logger, for skidding & stacking logs, 150 H.P.		33.80	790	2,375	7,125	745.40	
	2900	Rake, spring tooth, with tractor		7.04	214	643	1,925	184.90	
	3000	Roller, tandem, gas, 3 to 5 ton		4.90	115	345	1,025	108.20	
	3050	Diesel, 8 to 12 ton		6.95	213	640	1,925	183.60	
	3100	Towed type vibratory compactor, diesel, 50 HP, 72" smooth drum		31.15	600	1,805	5,425	610.20	
	3150	Sheepsfoot, double 60" x 60"		2.35	98.50	295	885	77.80	
	3170	Landfill compactor, 220 HP		43.05	1,175	3,505	10,500	1,045	
	3200	Pneumatic tire diesel roller, 12 ton		7.95	300	905	2,725	244.60	
	3250	21 to 25 ton		13	550	1,650	4,950	434	
	3300	Sheepsfoot roller, self-propelled, 4 wheel, 130 H.P.		31.95	845	2,535	7,600	762.60	
	3320	300 H.P.		43.50	1,200	3,590	10,800	1,066	
	3350	Vibratory steel drum & pneumatic tire, diesel, 18,000 lb.		16.45	320	960	2,875	323.60	
	3400	29,000 lb.		24.10	405	1,220	3,650	436.80	
	3410	Rotary mower, brush, 60", with tractor		10.40	232	695	2,075	222.20	
	3450	Scrapers, towed type, 9 to 12 C.Y. capacity		4.65	221	664	2,000	170	
	3500	12 to 17 C.Y. capacity		1.51	295	885.20	2,650	189.10	
	3550	Scrapers, self-propelled, 4 x 4 drive, 2 engine, 14 C.Y. capacity		76.50	1,425	4,280	12,800	1,468	
	3600	2 engine, 24 C.Y. capacity		116.80	2,275	6,815	20,400	2,297	
	3640	32 - 44 C.Y. capacity		139.15	2,650	7,935	23,800	2,700	
	3650	Self-loading, 11 C.Y. capacity		39.80	810	2,435	7,300	805.40	
	3700	22 C.Y. capacity		75.20	1,650	4,975	14,900	1,597	
	3710	Screening plant 110 hp. w / 5' x 10'screen		20.35	375	1,125	3,375	387.80	
	3720	5' x 16' screen	▼	22.40	475	1,425	4,275	464.20	
	3850	Shovels, see Cranes division 01590-600							
	3860	Shovel/backhoe bucket, 1/2 C.Y.	Ea.	1.65	53.50	160	480	45.20	
	3870	3/4 C.Y.		1.75	60	180	540	50	
	3880	1 C.Y.		1.85	70	210	630	56.80	
	3890	1-1/2 C.Y.		1.95	83.50	250	750	65.60	
	3910	3 C.Y.		2.30	118	355	1,075	89.40	
	3950	Stump chipper, 18" deep, 30 H.P.		4.65	83.50	250	750	87.20	
	4110	Tractor, crawler, with bulldozer, torque converter, diesel 75 H.P.		14.40	310	925	2,775	300.20	
	4150	105 H.P.		19.25	465	1,400	4,200	434	
	4200	140 H.P.		23.90	585	1,755	5,275	542.20	
	4260	200 H.P.		35.55	965	2,895	8,675	863.40	
	4310	300 H.P.		46	1,225	3,655	11,000	1,099	
	4360	410 H.P.		64.80	1,600	4,775	14,300	1,473	
	4370	500 H.P.		83.45	2,075	6,205	18,600	1,909	
	4380	700 H.P.		128	3,375	10,115	30,300	3,047	
	4400	Loader, crawler, torque conv., diesel, 1-1/2 C.Y., 80 H.P.		13.45	320	965	2,900	300.60	
	4450	1-1/2 to 1-3/4 C.Y., 95 H.P.		15.55	385	1,150	3,450	354.40	
	4510	1-3/4 to 2-1/4 C.Y., 130 H.P.		21.25	615	1,840	5,525	538	
	4530	2-1/2 to 3-1/4 C.Y., 190 H.P.		32	840	2,525	7,575	761	
	4560	3-1/2 to 5 C.Y., 275 H.P.		42.85	1,200	3,590	10,800	1,061	
	4610	Tractor loader, wheel, torque conv., 4 x 4, 1 to 1-1/4 C.Y., 65 H.P.		9	200	600	1,800	192	
	4620	1-1/2 to 1-3/4 C.Y., 80 H.P.		11.15	245	735	2,200	236.20	
	4650	1-3/4 to 2 C.Y., 100 H.P.		12.35	278	835	2,500	265.80	
	4710	2-1/2 to 3-1/2 C.Y., 130 H.P.		13.45	315	940	2,825	295.60	
	4730	3 to 4-1/2 C.Y., 170 H.P.		19.05	475	1,430	4,300	438.40	
	4760	5-1/4 to 5-3/4 C.Y., 270 H.P.		32.10	695	2,090	6,275	674.80	
	4810	7 to 8 C.Y., 375 H.P.		57.75	1,275	3,830	11,500	1,228	
	4870	12-1/2 C.Y., 690 H.P.		82.45	2,025	6,085	18,300	1,877	
	4880	Wheeled, skid steer, 10 C.F., 30 H.P. gas		7.40	140	420	1,250	143.20	
	4890	1 C.Y., 78 H.P., diesel	▼	8.85	188	565	1,700	183.80	
	4891	Attachments for all skid steer loaders							

01590 | Equipment Rental

				UNIT	HOURLY OPER. COST	RENT PER DAY	RENT PER WEEK	RENT PER MONTH	CREW EQUIPMENT COST/DAY	
200	4892	Auger	R01590 -100	Ea.	.39	65	195	585	42.10	200
	4893	Backhoe			.65	108	325	975	70.20	
	4894	Broom	R02315 -450		.63	106	317	950	68.45	
	4895	Forks			.22	36.50	109	325	23.55	
	4896	Grapple			.50	82.50	248	745	53.60	
	4897	Concrete hammer			.96	161	482	1,450	104.10	
	4898	Tree spade			.93	155	464	1,400	100.25	
	4899	Trencher			.68	113	340	1,025	73.45	
	4900	Trencher, chain, boom type, gas, operator walking, 12 H.P.			2.30	68.50	205	615	59.40	
	4910	Operator riding, 40 H.P.			7.70	245	735	2,200	208.60	
	5000	Wheel type, diesel, 4' deep, 12" wide			41.30	730	2,185	6,550	767.40	
	5100	Diesel, 6' deep, 20" wide			59.45	1,425	4,250	12,800	1,326	
	5150	Ladder type, diesel, 5' deep, 8" wide			21.05	650	1,950	5,850	558.40	
	5200	Diesel, 8' deep, 16" wide			52.05	1,500	4,470	13,400	1,310	
	5210	Tree spade, self-propelled			9.60	267	800	2,400	236.80	
	5250	Truck, dump, tandem, 12 ton payload			20.25	272	815	2,450	325	
	5300	Three axle dump, 16 ton payload			27.95	420	1,265	3,800	476.60	
	5350	Dump trailer only, rear dump, 16-1/2 C.Y.			4.15	117	350	1,050	103.20	
	5400	20 C.Y.			4.55	133	400	1,200	116.40	
	5450	Flatbed, single axle, 1-1/2 ton rating			11	56.50	170	510	122	
	5500	3 ton rating			14.05	80	240	720	160.40	
	5550	Off highway rear dump, 25 ton capacity			40.55	995	2,980	8,950	920.40	
	5600	35 ton capacity			41.45	1,025	3,075	9,225	946.60	
	5610	50 ton capacity			53.40	1,325	3,995	12,000	1,226	
	5620	65 ton capacity			57.20	1,425	4,260	12,800	1,310	
	5630	100 ton capacity			73.40	1,800	5,380	16,100	1,663	
	6000	Vibratory plow, 25 H.P., walking		↓	3.90	58.50	175	525	66.20	
400	0010	**GENERAL EQUIPMENT RENTAL** Without operators								400
	0150	Aerial lift, scissor type, to 15' high, 1000 lb. cap., electric		Ea.	2.20	43.50	130	390	43.60	
	0160	To 25' high, 2000 lb. capacity			2.60	63.50	190	570	58.80	
	0170	Telescoping boom to 40' high, 500 lb. capacity, gas			10.90	272	815	2,450	250.20	
	0180	To 45' high, 500 lb. capacity			11.70	315	940	2,825	281.60	
	0190	To 60' high, 600 lb. capacity			13.65	415	1,250	3,750	359.20	
	0195	Air compressor, portable, 6.5 CFM, electric			.31	17.65	53	159	13.10	
	0196	gasoline			.35	26.50	79	237	18.60	
	0200	Air compressor, portable, gas engine, 60 C.F.M.			4.10	32.50	98	294	52.40	
	0300	160 C.F.M.			7	43.50	130	390	82	
	0400	Diesel engine, rotary screw, 250 C.F.M.			7.95	107	320	960	127.60	
	0500	365 C.F.M.			9.95	130	390	1,175	157.60	
	0550	450 C.F.M.			13.40	158	475	1,425	202.20	
	0600	600 C.F.M.			19.50	217	650	1,950	286	
	0700	750 C.F.M.			21	228	685	2,050	305	
	0800	For silenced models, small sizes, add			3%	5%	5%	5%		
	0900	Large sizes, add		↓	5%	7%	7%	7%		
	0920	Air tools and accessories								
	0930	Breaker, pavement, 60 lb.		Ea.	.35	14	42	126	11.20	
	0940	80 lb.			.35	17	51	153	13	
	0950	Drills, hand (jackhammer) 65 lb.			.45	14.35	43	129	12.20	
	0960	Track or wagon, swing boom, 4" drifter			33.90	605	1,815	5,450	634.20	
	0970	5" drifter			46.05	765	2,295	6,875	827.40	
	0975	Track mounted quarry drill, 6" diameter drill			48.65	825	2,480	7,450	885.20	
	0980	Dust control per drill			.79	12.35	37	111	13.70	
	0990	Hammer, chipping, 12 lb.			.40	21.50	64	192	16	
	1000	Hose, air with couplings, 50' long, 3/4" diameter			.03	5.35	16	48	3.45	
	1100	1" diameter			.03	5.35	16	48	3.45	
	1200	1-1/2" diameter			.05	7.65	23	69	5	
	1300	2" diameter		↓	.10	16.65	50	150	10.80	

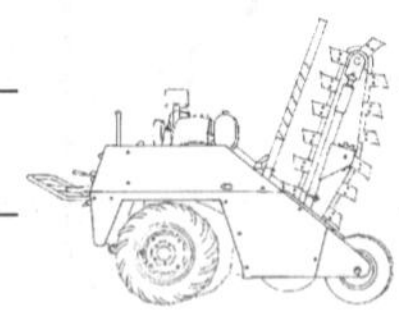

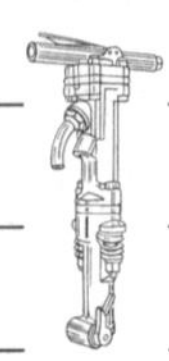

01590 | Equipment Rental

			UNIT	HOURLY OPER. COST	RENT PER DAY	RENT PER WEEK	RENT PER MONTH	CREW EQUIPMENT COST/DAY	
400	1400	2-1/2″ diameter	Ea.	.12	19.65	59	177	12.75	400
	1410	3″ diameter		.16	27.50	82	246	17.70	
	1450	Drill, steel, 7/8″ x 2′		.05	6.65	20	60	4.40	
	1460	7/8″ x 6′		.06	8	24	72	5.30	
	1520	Moil points		.03	4.67	14	42	3.05	
	1525	Pneumatic nailer w/accessories		.41	27	81	243	19.50	
	1530	Sheeting driver for 60 lb. breaker		.10	6.95	20.80	62.50	4.95	
	1540	For 90 lb. breaker		.15	10	30	90	7.20	
	1550	Spade, 25 lb.		.35	6.65	20	60	6.80	
	1560	Tamper, single, 35 lb.		.49	32.50	98	294	23.50	
	1570	Triple, 140 lb.		.74	49	147	440	35.30	
	1580	Wrenches, impact, air powered, up to 3/4″ bolt		.20	10	30	90	7.60	
	1590	Up to 1-1/4″ bolt		.30	20	60	180	14.40	
	1600	Barricades, barrels, reflectorized, 1 to 50 barrels		.02	3.10	9.30	28	2	
	1610	100 to 200 barrels		.01	2.33	7	21	1.50	
	1620	Barrels with flashers, 1 to 50 barrels		.02	3.77	11.30	34	2.40	
	1630	100 to 200 barrels		.02	3	9	27	1.95	
	1640	Barrels with steady burn type C lights		.03	5	15	45	3.25	
	1650	Illuminated board, trailer mounted, with generator		.65	117	350	1,050	75.20	
	1670	Portable barricade, stock, with flashers, 1 to 6 units		.02	3.77	11.30	34	2.40	
	1680	25 to 50 units		.02	3.50	10.50	31.50	2.25	
	1690	Butt fusion machine, electric		21.70	435	1,300	3,900	433.60	
	1695	Electro fusion machine		8.60	173	520	1,550	172.80	
	1700	Carts, brick, hand powered, 1000 lb. capacity		.23	38.50	115	345	24.85	
	1800	Gas engine, 1500 lb., 7-1/2′ lift		2.80	95	285	855	79.40	
	1822	Dehumidifier, medium, 6 Lb/Hr, 150 CFM		.68	41.50	124	370	30.25	
	1824	Large, 18 Lb/Hr, 600 CFM		1.36	82.50	248	745	60.50	
	1830	Distributor, asphalt, trailer mtd, 2000 gal., 38 H.P. diesel		7	253	760	2,275	208	
	1840	3000 gal., 38 H.P. diesel		8.30	290	870	2,600	240.40	
	1850	Drill, rotary hammer, electric, 1-1/2″ diameter		.40	24.50	74	222	18	
	1860	Carbide bit for above		.03	5.35	16	48	3.45	
	1865	Rotary, crawler, 250 HP		81.10	1,675	5,055	15,200	1,660	
	1870	Emulsion sprayer, 65 gal., 5 H.P. gas engine		1.83	74	222	665	59.05	
	1880	200 gal., 5 H.P. engine	↓	4.75	123	370	1,100	112	
	1900	Fencing, see division 01560-250 & 02820-000							
	1920	Floodlight, mercury vapor, or quartz, on tripod							
	1930	1000 watt	Ea.	.30	11.65	35	105	9.40	
	1940	2000 watt		.52	21.50	65	195	17.15	
	1950	Floodlights, trailer mounted with generator, 1 - 300 watt light		2.45	65	195	585	58.60	
	1960	2 - 1000 watt lights		3.30	108	325	975	91.40	
	2000	4 - 300 watt lights		2.75	76.50	230	690	68	
	2020	Forklift, wheeled, for brick, 18′, 3000 lb., 2 wheel drive, gas		13.15	188	565	1,700	218.20	
	2040	28′, 4000 lb., 4 wheel drive, diesel		11.35	248	745	2,225	239.80	
	2050	For rough terrain, 8000 lb., 16′ lift, 68 HP		15	370	1,110	3,325	342	
	2060	For plant, 4 T. capacity, 80 H.P., 2 wheel drive, gas		7.50	95	285	855	117	
	2080	10 T. capacity, 120 H.P., 2 wheel drive, diesel		11.25	172	515	1,550	193	
	2100	Generator, electric, gas engine, 1.5 KW to 3 KW		1.65	21.50	64	192	26	
	2200	5 KW		2.25	31.50	95	285	37	
	2300	10 KW		3.70	68.50	205	615	70.60	
	2400	25 KW		7.10	92.50	278	835	112.40	
	2500	Diesel engine, 20 KW		5.95	63.50	190	570	85.60	
	2600	50 KW		11.15	70	210	630	131.20	
	2700	100 KW		16.45	80	240	720	179.60	
	2800	250 KW		46.45	143	430	1,300	457.60	
	2850	Hammer, hydraulic, for mounting on boom, to 500 ft.-lb.		1.75	63.50	190	570	52	
	2860	1000 ft.-lb.		3.10	102	305	915	85.80	
	2900	Heaters, space, oil or electric, 50 MBH		.92	12.35	37	111	14.75	
	3000	100 MBH	↓	1.66	16.65	50	150	23.30	

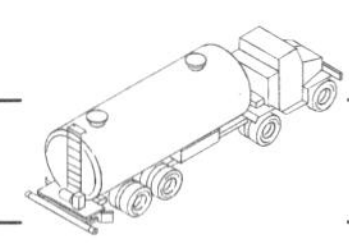

01590 | Equipment Rental

			UNIT	HOURLY OPER. COST	RENT PER DAY	RENT PER WEEK	RENT PER MONTH	CREW EQUIPMENT COST/DAY	
400	3100	300 MBH	Ea.	5.33	35	105	315	63.65	400
	3150	500 MBH		10.70	50	150	450	115.60	
	3200	Hose, water, suction with coupling, 20' long, 2" diameter		.02	5.65	17	51	3.55	
	3210	3" diameter		.03	8.65	26	78	5.45	
	3220	4" diameter		.04	11.65	35	105	7.30	
	3230	6" diameter		.09	23	69	207	14.50	
	3240	8" diameter		.26	43.50	130	390	28.10	
	3250	Discharge hose with coupling, 50' long, 2" diameter		.02	5	15	45	3.15	
	3260	3" diameter		.02	6	18	54	3.75	
	3270	4" diameter		.03	8.35	25	75	5.25	
	3280	6" diameter		.06	19	57	171	11.90	
	3290	8" diameter		.36	59.50	178	535	38.50	
	3295	Insulation blower		.11	7.35	22	66	5.30	
	3300	Ladders, extension type, 16' to 36' long		.16	26	78	234	16.90	
	3400	40' to 60' long		.19	31	93	279	20.10	
	3405	Lance for cutting concrete		2.84	109	327	980	88.10	
	3407	Lawn mower, rotary, 22", 5HP		.98	27.50	83	249	24.45	
	3408	48" self propelled		2.44	85.50	257	770	70.90	
	3410	Level, laser type, transit		1.24	82.50	248	745	59.50	
	3430	For pipe laying, manual leveling		.71	47.50	142	425	34.10	
	3440	Rotary beacon		.92	61	183	550	43.95	
	3460	Builders level with tripod and rod		.09	14.65	44	132	9.50	
	3500	Light towers, towable, with diesel generator, 2000 watt		2.75	76.50	230	690	68	
	3600	4000 watt		3.30	108	325	975	91.40	
	3700	Mixer, powered, plaster and mortar, 6 C.F., 7 H.P.		1.05	36.50	110	330	30.40	
	3800	10 C.F., 9 H.P.		1.30	53.50	160	480	42.40	
	3850	Nailer, pneumatic		.41	27	81	243	19.50	
	3900	Paint sprayers complete, 8 CFM		.69	45.50	137	410	32.90	
	4000	17 CFM		1.10	73.50	220	660	52.80	
	4020	Pavers, bituminous, rubber tires, 8' wide, 52 H.P., gas		25.75	770	2,315	6,950	669	
	4030	8' wide, 64 H.P., diesel		38.40	11.20	33.65	101	313.95	
	4050	Crawler, 10' wide, 78 H.P., gas		44.95	1,325	3,940	11,800	1,148	
	4060	10' wide, 87 H.P., diesel		58.05	1,650	4,965	14,900	1,457	
	4070	Concrete paver, 12' to 24' wide, 250 H.P.		51.75	1,225	3,710	11,100	1,156	
	4080	Placer-spreader-trimmer, 24' wide, 300 H.P.		64.05	1,775	5,340	16,000	1,580	
	4100	Pump, centrifugal gas pump, 1-1/2", 4 MGPH		2.50	35	105	315	41	
	4200	2", 8 MGPH		3.15	43.50	130	390	51.20	
	4300	3", 15 MGPH		3.35	45	135	405	53.80	
	4400	6", 90 MGPH		15.60	175	525	1,575	229.80	
	4500	Submersible electric pump, 1-1/4", 55 GPM		.35	20.50	62	186	15.20	
	4600	1-1/2", 83 GPM		.41	23.50	70	210	17.30	
	4700	2", 120 GPM		.57	29.50	88	264	22.15	
	4800	3", 300 GPM		.97	38.50	115	345	30.75	
	4900	4", 560 GPM		6.24	124	372	1,125	124.30	
	5000	6", 1590 GPM		9.05	185	555	1,675	183.40	
	5100	Diaphragm pump, gas, single, 1-1/2" diameter		.74	36	108	325	27.50	
	5200	2" diameter		2.40	45	135	405	46.20	
	5300	3" diameter		2.45	48.50	145	435	48.60	
	5400	Double, 4" diameter		4.30	88.50	265	795	87.40	
	5500	Trash pump, self-priming, gas, 2" diameter		2.90	35	105	315	44.20	
	5600	Diesel, 4" diameter		4.35	80	240	720	82.80	
	5650	Diesel, 6" diameter		9	128	385	1,150	149	
	5655	Grout Pump	↓	3.90	38.50	116	350	54.40	
	5660	Rollers, see division 01590-200							
	5700	Salamanders, L.P. gas fired, 100,000 B.T.U.	Ea.	1.66	10.65	32	96	19.70	
	5705	50,000 BTU		1.25	7.65	23	69	14.60	
	5720	Sandblaster, portable, open top, 3 C.F. capacity		.40	20.50	62	186	15.60	
	5730	6 C.F. capacity	↓	.65	30	90	270	23.20	

01590 | Equipment Rental

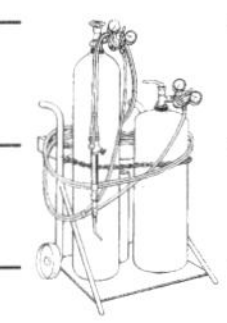

			UNIT	HOURLY OPER. COST	RENT PER DAY	RENT PER WEEK	RENT PER MONTH	CREW EQUIPMENT COST/DAY	
400	5740	Accessories for above	Ea.	.11	18	54	162	11.70	400
	5750	Sander, floor		.70	16.35	49	147	15.40	
	5760	Edger		.60	20	60	180	16.80	
	5800	Saw, chain, gas engine, 18" long		1.15	16	48	144	18.80	
	5900	36" long		.55	48.50	145	435	33.40	
	5950	60" long		.55	50	150	450	34.40	
	6000	Masonry, table mounted, 14" diameter, 5 H.P.		1.30	56	168	505	44	
	6050	Portable cut-off, 8 H.P.		1.20	25.50	76	228	24.80	
	6100	Circular, hand held, electric, 7-1/4" diameter	Ea.	.20	10	30	90	7.60	
	6200	12" diameter		.27	14	42	126	10.55	
	6250	Wall saw, w/hydraulic power, 10 H.P		2.08	97.50	292.40	875	75.10	
	6275	Shot blaster, walk behind, 20" wide		1.04	445	1,330	4,000	274.30	
	6300	Steam cleaner, 100 gallons per hour		2.20	63.50	190	570	55.60	
	6310	200 gallons per hour		2.90	78.50	235	705	70.20	
	6340	Tar Kettle/Pot, 400 gallon		2.69	51.50	155	465	52.50	
	6350	Torch, cutting, acetylene-oxygen, 150' hose		1.50	13.35	40	120	20	
	6360	Hourly operating cost includes tips and gas		8.10				64.80	
	6410	Toilet, portable chemical		.10	17.35	52	156	11.20	
	6420	Recycle flush type		.13	21.50	64	192	13.85	
	6430	Toilet, fresh water flush, garden hose,		.14	24	72	216	15.50	
	6440	Hoisted, non-flush, for high rise		.13	21	63	189	13.65	
	6450	Toilet, trailers, minimum		.22	36	108	325	23.35	
	6460	Maximum		.65	108	324	970	70	
	6465	Tractor, farm with attachment		9.30	225	675	2,025	209.40	
	6470	Trailer, office, see division 01520-500							
	6500	Trailers, platform, flush deck, 2 axle, 25 ton capacity	Ea.	4.20	91.50	275	825	88.60	
	6600	40 ton capacity		5.45	128	385	1,150	120.60	
	6700	3 axle, 50 ton capacity		5.90	142	425	1,275	132.20	
	6800	75 ton capacity		7.40	185	555	1,675	170.20	
	6810	Trailer mounted cable reel for H.V. line work		4.38	209	626	1,875	160.25	
	6820	Trailer mounted cable tensioning rig		8.61	410	1,230	3,700	314.90	
	6830	Cable pulling rig		54.46	2,325	6,980	20,900	1,832	
	6850	Trailer, storage, see division 01520-500							
	6900	Water tank, engine driven discharge, 5000 gallons	Ea.	5.50	123	370	1,100	118	
	6925	10,000 gallons		7.65	175	525	1,575	166.20	
	6950	Water truck, off highway, 6000 gallons		49.05	715	2,150	6,450	822.40	
	7010	Tram car for H.V. line work, powered, 2 conductor		5.68	113	340	1,025	113.45	
	7020	Transit (builder's level) with tripod		.09	14.65	44	132	9.50	
	7030	Trench box, 3000 lbs. 6'x8'		.42	70	210	630	45.35	
	7040	7200 lbs. 6'x20'		.82	136	409	1,225	88.35	
	7050	8000 lbs., 8' x 16'		.87	145	436	1,300	94.15	
	7060	9500 lbs., 8'x20'		1.17	195	584	1,750	126.15	
	7065	11,000 lbs., 8'x24'		1.31	218	654	1,950	141.30	
	7070	12,000 lbs., 10' x 20'		1.63	272	817	2,450	176.45	
	7100	Truck, pickup, 3/4 ton, 2 wheel drive		5.35	55	165	495	75.80	
	7200	4 wheel drive		5.50	63.50	190	570	82	
	7250	Crew carrier, 9 passenger		5.53	108	324.40	975	109.10	
	7290	Tool van, 24,000 G.V.W.		9.09	94.50	283.20	850	129.35	
	7300	Tractor, 4 x 2, 30 ton capacity, 195 H.P.		13.90	173	520	1,550	215.20	
	7410	250 H.P.		18.90	272	815	2,450	314.20	
	7500	6 x 2, 40 ton capacity, 240 H.P.		17.50	272	815	2,450	303	
	7600	6 x 4, 45 ton capacity, 240 H.P.		21.45	267	800	2,400	331.60	
	7620	Vacuum truck, hazardous material, 2500 gallon		6.72	315	943	2,825	242.35	
	7625	5,000 gallon		9.09	420	1,256.80	3,775	324.10	
	7640	Tractor, with A frame, boom and winch, 225 H.P.		14.30	233	700	2,100	254.40	
	7650	Vacuum, H.E.P.A., 16 gal., wet/dry		.27	24	72	216	16.55	
	7655	55 gal, wet/dry		.60	36	108	325	26.40	
400	7660	Water tank, portable		1	9.35	28	84	13.60	400

01590 | Equipment Rental

			UNIT	HOURLY OPER. COST	RENT PER DAY	RENT PER WEEK	RENT PER MONTH	CREW EQUIPMENT COST/DAY	
400	7690	Large production vacuum loader, 3150 CFM	Ea.	15.63	630	1,890	5,675	503.05	400
	7700	Welder, electric, 200 amp		3.74	57	171	515	64.10	
	7800	300 amp		5.22	60	180	540	77.75	
	7900	Gas engine, 200 amp		5.30	35.50	106	320	63.60	
	8000	300 amp		6.20	42.50	128	385	75.20	
	8100	Wheelbarrow, any size		.06	10.35	31	93	6.70	
	8200	Wrecking ball, 4000 lb.		1.85	68.50	205	615	55.80	
500	0010	**HIGHWAY EQUIPMENT RENTAL**							500
	0050	Asphalt batch plant, portable drum mixer, 100 ton/hr.	Ea.	55.10	1,350	4,020	12,100	1,245	
	0060	200 ton/hr.		61.10	1,400	4,215	12,600	1,332	
	0070	300 ton/hr.		71.20	1,675	4,990	15,000	1,568	
	0100	Backhoe attachment, long stick, up to 185 HP, 10.5′ long		.30	19.65	59	177	14.20	
	0140	Up to 250 HP, 12′ long		.32	21.50	64	192	15.35	
	0180	Over 250 HP, 15′ long		.42	27.50	83	249	19.95	
	0200	Special dipper arm, up to 100 HP, 32′ long		.86	57.50	172	515	41.30	
	0240	Over 100 HP, 33′ long		1.08	72	216	650	51.85	
	0300	Concrete batch plant, portable, electric, 200 CY/Hr		12.67	645	1,940	5,825	489.35	
	0500	Grader attachment, ripper/scarifier, rear mounted							
	0520	Up to 135 HP	Ea.	2.70	58.50	175	525	56.60	
	0540	Up to 180 HP		3.20	75	225	675	70.60	
	0580	Up to 250 HP		3.55	86.50	260	780	80.40	
	0700	Pvmt. removal bucket, for hyd. excavator, up to 90 HP		1.30	41.50	125	375	35.40	
	0740	Up to 200 HP		1.50	63.50	190	570	50	
	0780	Over 200 HP		1.65	76.50	230	690	59.20	
	0900	Aggregate spreader, self-propelled, 187 HP		37.15	790	2,370	7,100	771.20	
	1000	Chemical spreader, 3 C.Y.		2.20	80.50	242	725	66	
	1900	Hammermill, traveling, 250 HP		38.85	1,675	5,020	15,100	1,315	
	2000	Horizontal borer, 3″ diam, 13 HP gas driven		3.85	53.50	160	480	62.80	
	2200	Hydromulchers, gas power, 3000 gal., for truck mounting		10.50	187	560	1,675	196	
	2400	Joint & crack cleaner, walk behind, 25 HP		2.05	46.50	140	420	44.40	
	2500	Filler, trailer mounted, 400 gal., 20 HP		6	182	545	1,625	157	
	3000	Paint striper, self propelled, double line, 30 HP		5.15	155	465	1,400	134.20	
	3200	Post drivers, 6″ I-Beam frame, for truck mounting		8.15	415	1,250	3,750	315.20	
	3400	Road sweeper, self propelled, 8′ wide, 90 HP		23.50	385	1,160	3,475	420	
	4000	Road mixer, self-propelled, 130 HP		28.35	590	1,770	5,300	580.80	
	4100	310 HP		53.85	1,975	5,895	17,700	1,610	
	4200	Cold mix paver, incl pug mill and bitumen tank,							
	4220	165 HP	Ea.	65.05	1,925	5,770	17,300	1,674	
	4250	Paver, asphalt, wheel or crawler, 130 H.P., diesel		58.05	1,650	4,965	14,900	1,457	
	4300	Paver, road widener, gas 1′ to 6′, 67 HP		28.70	635	1,900	5,700	609.60	
	4400	Diesel, 2′ to 14′, 88 HP		38.80	995	2,980	8,950	906.40	
	4600	Slipform pavers, curb and gutter, 2 track, 75 HP		22.95	650	1,950	5,850	573.60	
	4700	4 track, 165 HP		31.05	730	2,195	6,575	687.40	
	4800	Median barrier, 215 HP		31.50	755	2,260	6,775	704	
	4901	Trailer, low bed, 75 ton capacity		7.85	182	545	1,625	171.80	
	5000	Road planer, walk behind, 10″ cutting width, 10 HP		2	26	78	234	31.60	
	5100	Self propelled, 12″ cutting width, 64 HP		5.30	288	865	2,600	215.40	
	5200	Pavement profiler, 4′ to 6′ wide, 450 HP		141.40	2,800	8,380	25,100	2,807	
	5300	8′ to 10′ wide, 750 HP		225.75	4,250	12,725	38,200	4,351	
	5400	Roadway plate, steel, 1″x8′x20′		.06	9.65	29	87	6.30	
	5600	Stabilizer,self-propelled, 150 HP		25.25	560	1,685	5,050	539	
	5700	310 HP		42.50	1,200	3,605	10,800	1,061	
	5800	Striper, thermal, truck mounted 120 gal. paint, 150H.P.		32.40	485	1,450	4,350	549.20	
	6000	Tar kettle, 330 gal., trailer mounted		2.37	36.50	110	330	40.95	
	7000	Tunnel locomotive, diesel, 8 to 12 ton		20.45	560	1,675	5,025	498.60	
	7005	Electric, 10 ton		20.45	635	1,905	5,725	544.60	
	7010	Muck cars, 1/2 C.Y. capacity		1.50	20.50	62	186	24.40	

Important: See the Reference Section for critical supporting data - Reference Nos., Crews, & City Cost Indexes

01590 | Equipment Rental

			UNIT	HOURLY OPER. COST	RENT PER DAY	RENT PER WEEK	RENT PER MONTH	CREW EQUIPMENT COST/DAY	
500	7020	1 C.Y. capacity	Ea.	1.70	29	87	261	31	500
	7030	2 C.Y. capacity		1.80	33.50	100	300	34.40	
	7040	Side dump, 2 C.Y. capacity		2	41.50	125	375	41	
	7050	3 C.Y. capacity		2.70	48.50	145	435	50.60	
	7060	5 C.Y. capacity		3.80	61.50	185	555	67.40	
	7100	Ventilating blower for tunnel, 7-1/2 H.P.		1.25	40	120	360	34	
	7110	10 H.P.		1.44	41.50	125	375	36.50	
	7120	20 H.P.		2.33	48.50	145	435	47.65	
	7140	40 H.P.		4.07	73.50	220	660	76.55	
	7160	60 H.P.		6.16	110	330	990	115.30	
	7175	75 H.P.		7.89	148	445	1,325	152.10	
	7180	200 H.P.		17.60	205	615	1,850	263.80	
	7800	Windrow loader, elevating	↓	32.70	905	2,720	8,150	805.60	
600	0010	**LIFTING AND HOISTING EQUIPMENT RENTAL** R02315-450							600
	0100	without operators							
	0120	Aerial lift truck, 2 person, to 80' R01590-100	Ea.	17.85	600	1,795	5,375	501.80	
	0140	Boom work platform, 40' snorkel		9.25	205	615	1,850	197	
	0150	Crane, flatbed mntd, 3 ton cap.		11.60	182	545	1,625	201.80	
	0200	Crane, climbing, 106' jib, 6000 lb. capacity, 410 FPM		41.95	1,325	3,990	12,000	1,134	
	0300	101' jib, 10,250 lb. capacity, 270 FPM	↓	47.25	1,675	5,050	15,200	1,388	
	0400	Tower, static, 130' high, 106' jib,							
	0500	6200 lb. capacity at 400 FPM	Ea.	45.05	1,525	4,610	13,800	1,282	
	0600	Crawler mounted, lattice boom, 1/2 C.Y., 15 tons at 12' radius		20.11	485	1,450	4,350	450.90	
	0700	3/4 C.Y., 20 tons at 12' radius		26.81	650	1,950	5,850	604.50	
	0800	1 C.Y., 25 tons at 12' radius		35.75	845	2,530	7,600	792	
	0900	1-1/2 C.Y., 40 tons at 12' radius		39.90	995	2,980	8,950	915.20	
	1000	2 C.Y., 50 tons at 12' radius		49.90	1,425	4,305	12,900	1,260	
	1100	3 C.Y., 75 tons at 12' radius		47.75	1,400	4,175	12,500	1,217	
	1200	100 ton capacity, 60' boom		61.05	1,850	5,570	16,700	1,602	
	1300	165 ton capacity, 60' boom		90.10	2,400	7,235	21,700	2,168	
	1400	200 ton capacity, 70' boom		95.85	2,625	7,855	23,600	2,338	
	1500	350 ton capacity, 80' boom		138	3,750	11,215	33,600	3,347	
	1600	Truck mounted, lattice boom, 6 x 4, 20 tons at 10' radius		20.93	800	2,400	7,200	647.45	
	1700	25 tons at 10' radius		22.32	855	2,560	7,675	690.55	
	1800	8 x 4, 30 tons at 10' radius		28.67	905	2,720	8,150	773.35	
	1900	40 tons at 12' radius		24.56	960	2,880	8,650	772.50	
	2000	8 x 4, 60 tons at 15' radius		30.91	1,100	3,290	9,875	905.30	
	2050	82 tons at 15' radius		45.40	1,600	4,770	14,300	1,317	
	2100	90 tons at 15' radius		46.24	1,750	5,230	15,700	1,416	
	2200	115 tons at 15' radius		51.10	1,950	5,815	17,400	1,572	
	2300	150 tons at 18' radius		47.45	2,050	6,150	18,500	1,610	
	2350	165 tons at 18' radius		73.35	2,400	7,235	21,700	2,034	
	2400	Truck mounted, hydraulic, 12 ton capacity		31	605	1,820	5,450	612	
	2500	25 ton capacity		31.15	625	1,875	5,625	624.20	
	2550	33 ton capacity		32	655	1,965	5,900	649	
	2560	40 ton capacity		30.40	645	1,940	5,825	631.20	
	2600	55 ton capacity		44.75	920	2,765	8,300	911	
	2700	80 ton capacity		54	980	2,940	8,825	1,020	
	2720	100 ton capacity		78.90	2,600	7,770	23,300	2,185	
	2740	120 ton capacity		82.60	2,825	8,460	25,400	2,353	
	2760	150 ton capacity		100.45	3,550	10,625	31,900	2,929	
	2800	Self-propelled, 4 x 4, with telescoping boom, 5 ton		14.60	335	1,005	3,025	317.80	
	2900	12-1/2 ton capacity		21.90	510	1,535	4,600	482.20	
	3000	15 ton capacity		23.85	600	1,800	5,400	550.80	
	3050	20 ton capacity		24.80	625	1,880	5,650	574.40	
	3100	25 ton capacity		25.90	615	1,840	5,525	575.20	
	3150	40 ton capacity	↓	41.65	880	2,640	7,925	861.20	

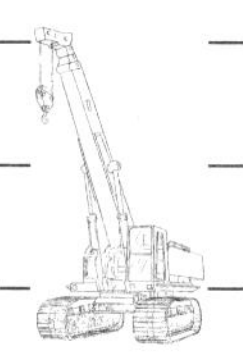

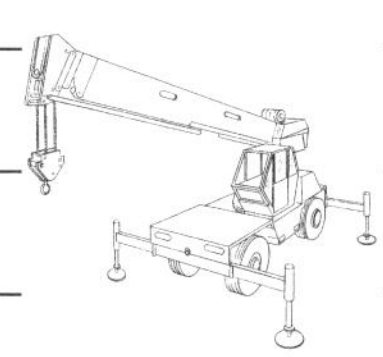

GENERAL REQUIREMENTS 1

01590 | Equipment Rental

			Unit	Hourly Oper. Cost	Rent Per Day	Rent Per Week	Rent Per Month	Crew Equipment Cost/Day	
600	3200	Derricks, guy, 20 ton capacity, 60' boom, 75' mast (R02315-450)	Ea.	12.33	325	976	2,925	293.85	600
	3300	100' boom, 115' mast		20.01	560	1,680	5,050	496.10	
	3400	Stiffleg, 20 ton capacity, 70' boom, 37' mast (R01590-100)		14.25	415	1,250	3,750	364	
	3500	100' boom, 47' mast		22.53	680	2,040	6,125	588.25	
	3550	Helicopter, small, lift to 1250 lbs. maximum, w/pilot		66.23	2,625	7,890	23,700	2,108	
	3600	Hoists, chain type, overhead, manual, 3/4 ton		.10	3	9	27	2.60	
	3900	10 ton		.55	16	48	144	14	
	4000	Hoist and tower, 5000 lb. cap., portable electric, 40' high		4.14	188	563	1,700	145.70	
	4100	For each added 10' section, add		.09	14.65	44	132	9.50	
	4200	Hoist and single tubular tower, 5000 lb. electric, 100' high		5.58	262	785	2,350	201.65	
	4300	For each added 6'-6″ section, add		.15	24.50	74	222	16	
	4400	Hoist and double tubular tower, 5000 lb., 100' high		5.98	288	865	2,600	220.85	
	4500	For each added 6'-6″ section, add		.17	27.50	83	249	17.95	
	4550	Hoist and tower, mast type, 6000 lb., 100' high		6.47	299	897	2,700	231.15	
	4570	For each added 10' section, add		.11	18	54	162	11.70	
	4600	Hoist and tower, personnel, electric, 2000 lb., 100' @ 125 FPM		13.27	795	2,390	7,175	584.15	
	4700	3000 lb., 100' @ 200 FPM		15.15	900	2,700	8,100	661.20	
	4800	3000 lb., 150' @ 300 FPM		16.80	1,000	3,030	9,100	740.40	
	4900	4000 lb., 100' @ 300 FPM		17.43	1,025	3,090	9,275	757.45	
	5000	6000 lb., 100' @ 275 FPM		18.84	1,075	3,240	9,725	798.70	
	5100	For added heights up to 500', add	L.F.	.01	1.67	5	15	1.10	
	5200	Jacks, hydraulic, 20 ton	Ea.	.05	8	24	72	5.20	
	5500	100 ton	″	.30	23	69	207	16.20	
	6000	Jacks, hydraulic, climbing with 50' jackrods							
	6010	and control consoles, minimum 3 mo. rental							
	6100	30 ton capacity	Ea.	1.62	108	323	970	77.55	
	6150	For each added 10' jackrod section, add		.05	3.33	10	30	2.40	
	6300	50 ton capacity		2.60	173	520	1,550	124.80	
	6350	For each added 10' jackrod section, add		.06	4	12	36	2.90	
	6500	125 ton capacity		6.80	455	1,360	4,075	326.40	
	6550	For each added 10' jackrod section, add		.47	31	93	279	22.35	
	6600	Cable jack, 10 ton capacity with 200' cable		1.35	90	270	810	64.80	
	6650	For each added 50' of cable, add		.15	9.65	29	87	7	
700	0010	**WELLPOINT EQUIPMENT RENTAL** See also division 02240							700
	0020	Based on 2 months rental							
	0100	Combination jetting & wellpoint pump, 60 H.P. diesel	Ea.	8.91	267	801	2,400	231.50	
	0200	High pressure gas jet pump, 200 H.P., 300 psi	″	15.79	228	684	2,050	263.10	
	0300	Discharge pipe, 8″ diameter	L.F.	.01	.43	1.29	3.87	.35	
	0350	12″ diameter		.01	.64	1.92	5.75	.45	
	0400	Header pipe, flows up to 150 G.P.M., 4″ diameter		.01	.39	1.18	3.54	.30	
	0500	400 G.P.M., 6″ diameter		.01	.46	1.39	4.17	.35	
	0600	800 G.P.M., 8″ diameter		.01	.64	1.92	5.75	.45	
	0700	1500 G.P.M., 10″ diameter		.01	.67	2.02	6.05	.50	
	0800	2500 G.P.M., 12″ diameter		.02	1.27	3.81	11.45	.90	
	0900	4500 G.P.M., 16″ diameter		.02	1.63	4.88	14.65	1.15	
	0950	For quick coupling aluminum and plastic pipe, add		.03	1.68	5.05	15.15	1.25	
	1100	Wellpoint, 25' long, with fittings & riser pipe, 1-1/2″ or 2″ diameter	Ea.	.05	3.36	10.08	30	2.40	
	1200	Wellpoint pump, diesel powered, 4″ diameter, 20 H.P.		4.33	154	462	1,375	127.05	
	1300	6″ diameter, 30 H.P.		5.66	191	573	1,725	159.90	
	1400	8″ suction, 40 H.P.		7.70	262	785	2,350	218.60	
	1500	10″ suction, 75 H.P.		10.55	305	918	2,750	268	
	1600	12″ suction, 100 H.P.		15.79	490	1,470	4,400	420.30	
	1700	12″ suction, 175 H.P.		20.97	540	1,620	4,850	491.75	
800	0010	**MARINE EQUIPMENT RENTAL**							800
	0200	Barge, 400 Ton, 30' wide x 90' long	Ea.	16.15	240	720	2,150	273.20	
	0240	800 Ton, 45' wide x 90' long		26.65	345	1,030	3,100	419.20	
	2000	Tugboat, diesel, 100 HP		16.95	168	505	1,525	236.60	

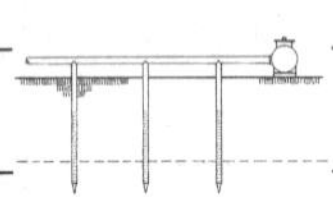

01590 | Equipment Rental

			UNIT	HOURLY OPER. COST	RENT PER DAY	RENT PER WEEK	RENT PER MONTH	CREW EQUIPMENT COST/DAY	
800	2040	250 HP	Ea.	31.45	315	945	2,825	440.60	800
	2080	380 HP	↓	73.10	925	2,780	8,350	1,141	

01700 | Execution Requirements

	01740 Cleaning		Crew	Daily Output	Labor-Hours	Unit	2004 Bare Costs Mat.	Labor	Equip.	Total	Total Incl O&P	
500	0010	**CLEANING UP** After job completion, allow, minimum				Job					.30%	500
	0040	Maximum				"					1%	
	0052	Cleanup of floor area, continuous, per day, during const.	A-5	16	1.125	M.S.F.	1.70	29	1.90	32.60	52.50	
	0100	Final by GC at end of job	"	11.50	1.565	"	2.65	40.50	2.65	45.80	73	

01800 | Facility Operation

	01810 Commissioning		Crew	Daily Output	Labor-Hours	Unit	2004 Bare Costs Mat.	Labor	Equip.	Total	Total Incl O&P	
100	0010	**COMMISSIONING** Including documentation of design intent										100
	0100	performance verification, O&M, training, min				Project					.50%	
	0150	Maximum				"					.75%	

	01840 Moving Equipment		Crew	Daily Output	Labor-Hours	Unit	2004 Bare Costs Mat.	Labor	Equip.	Total	Total Incl O&P	
100	0010	**MOVING EQUIPMENT,** Remove and reset, 100′ distance,										100
	0020	No obstructions, no assembly or leveling unless noted										
	0100	Annealing furnace, 24′ overall	B-67	4	4	Ea.		133	60	193	276	
	0200	Annealing oven, small		14	1.143			38	17.15	55.15	79	
	0240	Very large		1	16			530	240	770	1,100	
	0400	Band saw, small		12	1.333			44.50	20	64.50	92	
	0440	Large		8	2			66.50	30	96.50	138	
	0500	Blue print copy machine		7	2.286			76	34.50	110.50	158	
	0600	Bonding mill, 6″		7	2.286			76	34.50	110.50	158	
	0620	12″		6	2.667			88.50	40	128.50	184	
	0640	18″		4	4			133	60	193	276	
	0660	24″		2	8			266	120	386	550	
	0700	Boring machine (jig)	B-68	7	3.429			115	34.50	149.50	220	
	0800	Bridgeport mill, standard	B-67	14	1.143			38	17.15	55.15	79	
	1000	Calibrator, 6 unit	"	14	1.143			38	17.15	55.15	79	
	1100	Comparitor, bench top	2 Clab	14	1.143			29.50		29.50	49.50	
	1140	Floor mounted	B-67	7	2.286			76	34.50	110.50	158	
	1200	Computer, desk top	2 Clab	25	.640			16.65		16.65	27.50	
	1300	Copy machine	"	25	.640			16.65		16.65	27.50	
	1500	Deflasher	B-67	14	1.143			38	17.15	55.15	79	
	1600	Degreaser, small		14	1.143			38	17.15	55.15	79	
	1640	Large 24′ overall		1	16			530	240	770	1,100	
	1700	Desk with chair	2 Clab	25	.640			16.65		16.65	27.50	
	1800	Dial press	B-67	7	2.286			76	34.50	110.50	158	
	1900	Drafting table	2 Clab	14	1.143			29.50		29.50	49.50	
	2000	Drill press, bench top	"	14	1.143			29.50		29.50	49.50	
	2040	Floor mounted	B-67	14	1.143			38	17.15	55.15	79	
	2080	Industrial radial	"	7	2.286			76	34.50	110.50	158	
	2100	Dust collector, portable	2 Clab	25	.640			16.65		16.65	27.50	
	2140	Stationary, small	B-67	7	2.286			76	34.50	110.50	158	
	2180	Stationary, large	"	2	8			266	120	386	550	
	2300	Electric discharge machine	B-68	7	3.429			115	34.50	149.50	220	
	2400	Environmental chamber walls, including assembly	4 Clab	18	1.778	L.F.		46		46	76.50	
	2600	File cabinet	2 Clab	25	.640	Ea.		16.65		16.65	27.50	
	2800	Grinder/sander, pedestal mount	B-67	14	1.143			38	17.15	55.15	79	
	3000	Hack saw, power	2 Clab	24	.667			17.35		17.35	28.50	

01840 | Moving Equipment

			CREW	DAILY OUTPUT	LABOR-HOURS	UNIT	2004 BARE COSTS MAT.	LABOR	EQUIP.	TOTAL	TOTAL INCL O&P	
100	3100	Hydraulic press	B-67	14	1.143	Ea.		38	17.15	55.15	79	100
	3500	Laminar flow tables	"	14	1.143			38	17.15	55.15	79	
	3600	Lathe, bench	2 Clab	14	1.143			29.50		29.50	49.50	
	3640	6"	B-67	14	1.143			38	17.15	55.15	79	
	3680	10"		13	1.231			41	18.45	59.45	85	
	3720	12"		12	1.333			44.50	20	64.50	92	
	4000	Milling machine		8	2			66.50	30	96.50	138	
	4100	Molding press, 25 ton		5	3.200			106	48	154	221	
	4140	60 ton		4	4			133	60	193	276	
	4180	100 ton		2	8			266	120	386	550	
	4220	150 ton		1.50	10.667			355	160	515	735	
	4260	200 ton		1	16			530	240	770	1,100	
	4300	300 ton		.75	21.333			710	320	1,030	1,475	
	4700	Oil pot stand		14	1.143			38	17.15	55.15	79	
	5000	Press, 10 ton		14	1.143			38	17.15	55.15	79	
	5040	15 ton		12	1.333			44.50	20	64.50	92	
	5080	20 ton		10	1.600			53	24	77	111	
	5120	30 ton		8	2			66.50	30	96.50	138	
	5160	45 ton		6	2.667			88.50	40	128.50	184	
	5200	60 ton		4	4			133	60	193	276	
	5240	75 ton		2.50	6.400			213	96	309	440	
	5280	100 ton		2	8			266	120	386	550	
	5500	Raised floor, including assembly	2 Carp	250	.064	S.F.		2.11		2.11	3.50	
	5600	Rolling mill, 6"	B-67	7	2.286	Ea.		76	34.50	110.50	158	
	5640	9"		6	2.667			88.50	40	128.50	184	
	5680	12"		4	4			133	60	193	276	
	5720	13"		3.50	4.571			152	68.50	220.50	315	
	5760	18"		2	8			266	120	386	550	
	5800	25"		1	16			530	240	770	1,100	
	6000	Sander, floor stand		14	1.143			38	17.15	55.15	79	
	6100	Screw machine		7	2.286			76	34.50	110.50	158	
	6200	Shaper, 16"		14	1.143			38	17.15	55.15	79	
	6300	Shear, power assist		4	4			133	60	193	276	
	6400	Slitter, 6"		14	1.143			38	17.15	55.15	79	
	6440	8"		13	1.231			41	18.45	59.45	85	
	6480	10"		12	1.333			44.50	20	64.50	92	
	6520	12"		11	1.455			48.50	22	70.50	100	
	6560	16"		10	1.600			53	24	77	111	
	6600	20"		8	2			66.50	30	96.50	138	
	6640	24"		6	2.667			88.50	40	128.50	184	
	6800	Snag and tap machine		7	2.286			76	34.50	110.50	158	
	6900	Solder machine (auto)		7	2.286			76	34.50	110.50	158	
	7000	Storage cabinet metal, small	2 Clab	36	.444			11.55		11.55	19.15	
	7040	Large		25	.640			16.65		16.65	27.50	
	7100	Storage rack open, small		14	1.143			29.50		29.50	49.50	
	7140	Large		7	2.286			59.50		59.50	98.50	
	7200	Surface bench, small	B-67	14	1.143			38	17.15	55.15	79	
	7240	Large	"	5	3.200			106	48	154	221	
	7300	Surface grinder, large wet	B-68	5	4.800			161	48	209	305	
	7500	Time check machine	2 Clab	14	1.143			29.50		29.50	49.50	
	8000	Welder, 30 KVA (bench)		14	1.143			29.50		29.50	49.50	
	8100	Work bench with chair		25	.640			16.65		16.65	27.50	

Division Notes

	CREW	DAILY OUTPUT	LABOR-HOURS	UNIT	2004 BARE COSTS				TOTAL INCL O&P
					MAT.	LABOR	EQUIP.	TOTAL	

Division 2
Site Construction

Estimating Tips

02200 Site Preparation

- If possible visit the site and take an inventory of the type, quantity and size of the trees. Certain trees may have a landscape resale value or firewood value. Stump disposal can be very expensive, particularly if they cannot be buried at the site. Consider using a bulldozer in lieu of hand cutting trees.
- Estimators should visit the site to determine the need for haul road, access, storage of materials, and security considerations. When estimating for access roads on unstable soil, consider using a geotextile stabilization fabric. It can greatly reduce the quantity of crushed stone or gravel. Sites of limited size and access can cause cost overruns due to lost productivity. Theft and damage is another consideration if the location is isolated. A temporary fence or security guards may be required. Investigate the site thoroughly.

02210 Subsurface Investigation

In preparing estimates on structures involving earthwork or foundations, all information concerning soil characteristics should be obtained. Look particularly for hazardous waste, evidence of prior dumping of debris, and previous stream beds.

02220 Selective Demolition

The costs shown for selective demolition do not include rubbish handling or disposal. These items should be estimated separately using Means data or other sources.

- Historic preservation often requires that the contractor remove materials from the existing structure, rehab them and replace them. The estimator must be aware of any related measures and precautions that must be taken when doing selective demolition, and cutting and patching. Requirements may include special handling and storage, as well as security.
- In addition to Section 02220, you can find selective demolition items in each division. Example: Roofing demolition is in division 7.

02300 Earthwork

- Estimating the actual cost of performing earthwork requires careful consideration of the variables involved. This includes items such as type of soil, whether or not water will be encountered, dewatering, whether or not banks need bracing, disposal of excavated earth, length of haul to fill or spoil sites, etc. If the project has large quantities of cut or fill, consider raising or lowering the site to reduce costs while paying close attention to the effect on site drainage and utilities if doing this.
- If the project has large quantities of fill, creating a borrow pit on the site can significantly lower the costs.
- It is very important to consider what time of year the project is scheduled for completion. Bad weather can create large cost overruns from dewatering, site repair and lost productivity from cold weather.

02500 Utility Services
02600 Drainage & Containment

- Never assume that the water, sewer and drainage lines will go in at the early stages of the project. Consider the site access needs before dividing the site in half with open trenches, loose pipe, and machinery obstructions. Always inspect the site to establish that the site drawings are complete. Check off all existing utilities on your drawing as you locate them. If you find any discrepancies, mark up the site plan for further research. Differing site conditions can be very costly if discovered later in the project.
- See also Section 02955 for restoration of pipe where removal/replacement may be undesirable.

02700 Bases, Ballasts, Pavements/Appurtenances

- When estimating paving, keep in mind the project schedule. If an asphaltic paving project is in a colder climate and runs through to the spring, consider placing the base course in the autumn, then topping it in the spring just prior to completion. This could save considerable costs in spring repair. Keep in mind that prices for asphalt and concrete are generally higher in the cold seasons.
- See also Sections 02960/02965.

02900 Planting

- The timing of planting and guarantee specifications often dictate the costs for establishing tree and shrub growth and a stand of grass or ground cover. Establish the work performance schedule to coincide with the local planting season. Maintenance and growth guarantees can add from 20% to 100% to the total landscaping cost. The cost to replace trees and shrubs can be as high as 5% of the total cost depending on the planting zone, soil conditions and time of year.

02960 & 02965 Flexible Pavement Surfacing Recovery

- Recycling of asphalt pavement is becoming very popular and is an alternative to removal and replacement of asphalt pavement. It can be a good value engineering proposal if removed pavement can be recycled either at the site or another site that is reasonably close to the project site.

Reference Numbers

Reference numbers are shown in bold squares at the beginning of some major classifications. These numbers refer to related items in the Reference Section. The reference information may be an estimating procedure, an alternate pricing method or technical information.

Note: Not all subdivisions listed here necessarily appear in this publication.

02050 | Basic Site Materials & Methods

	02055 | Soils		CREW	DAILY OUTPUT	LABOR-HOURS	UNIT	2004 BARE COSTS MAT.	LABOR	EQUIP.	TOTAL	TOTAL INCL O&P	
150	0010	**BORROW**										150
	0020	And spread, with 200 H.P. dozer, no compaction										
	0200	Common borrow	B-15	600	.047	C.Y.	5.30	1.33	3.03	9.66	11.25	
	0700	Screened loam		600	.047		20	1.33	3.03	24.36	27.50	
	0800	Topsoil, weed free		600	.047		20	1.33	3.03	24.36	27.50	
	0900	For 5 mile haul, add	B-34B	200	.040			1.06	2.38	3.44	4.34	

	02060 | Aggregate		CREW	DAILY OUTPUT	LABOR-HOURS	UNIT	MAT.	LABOR	EQUIP.	TOTAL	TOTAL INCL O&P	
150	0010	**BORROW**										150
	0020	And spread, with 200 H.P. dozer, no compaction										
	0100	Bank run gravel	B-15	600	.047	C.Y.	15.40	1.33	3.03	19.76	22.50	
	0300	Crushed stone, (1.40 tons per CY), 1-1/2"		600	.047		24.50	1.33	3.03	28.86	32.50	
	0320	3/4"		600	.047		27	1.33	3.03	31.36	35	
	0340	1/2"		600	.047		17.75	1.33	3.03	22.11	25	
	0360	3/8"		600	.047		17.90	1.33	3.03	22.26	25	
	0400	Sand, washed, concrete		600	.047		25	1.33	3.03	29.36	33	
	0500	Dead or bank sand		600	.047		3.93	1.33	3.03	8.29	9.80	
	0600	Select structural fill		600	.047		7.50	1.33	3.03	11.86	13.70	
	0900	For 5 mile haul, add	B-34B	200	.040			1.06	2.38	3.44	4.34	

	02080 | Utility Materials		CREW	DAILY OUTPUT	LABOR-HOURS	UNIT	MAT.	LABOR	EQUIP.	TOTAL	TOTAL INCL O&P	
400	0010	**UTILITY BOXES** Precast concrete, 6" thick										400
	0050	5' x 10' x 6' high, I.D.	B-13	2	28	Ea.	1,450	785	310	2,545	3,225	
	0350	Hand hole, precast concrete, 1-1/2" thick										
	0400	1'-0" x 2'-0" x 1'-9", I.D., light duty	B-1	4	6	Ea.	248	160		408	540	
	0450	4'-6" x 3'-2" x 2'-0", O.D., heavy duty	B-6	3	8	"	760	224	69.50	1,053.50	1,275	
600	0010	**UTILITY ACCESSORIES** R02315-400										600
	0400	Underground tape, detectable, reinforced, alum. foil core, 2"	1 Clab	150	.053	C.L.F.	1.43	1.39		2.82	3.87	
	0500	6"		140	.057	"	3.58	1.49		5.07	6.40	
	9000	Minimum labor/equipment charge		4	2	Job		52		52	86	

02100 | Site Remediation

	02110 | Hazmat Removal & Handling		CREW	DAILY OUTPUT	LABOR-HOURS	UNIT	2004 BARE COSTS MAT.	LABOR	EQUIP.	TOTAL	TOTAL INCL O&P	
300	0010	**HAZARDOUS WASTE CLEANUP/PICKUP/DISPOSAL**										300
	0100	For contractor equipment, i.e. dozer,										
	0110	Front end loader, dump truck, etc., see div. 01590-200										
	1000	Solid pickup										
	1100	55 gal. drums				Ea.					220	
	1120	Bulk material, minimum				Ton					165	
	1130	Maximum				"					550	
	1200	Transportation to disposal site										
	1220	Truckload = 80 drums or 25 C.Y. or 18 tons										
	1260	Minimum				Mile					2.50	
	1270	Maximum				"					4.40	
	3000	Liquid pickup, vacuum truck, stainless steel tank										
	3100	Minimum charge, 4 hours										
	3110	1 compartment, 2200 gallon				Hr.					110	

02100 | Site Remediation

	02110 | Hazmat Removal & Handling		CREW	DAILY OUTPUT	LABOR-HOURS	UNIT	2004 BARE COSTS MAT.	LABOR	EQUIP.	TOTAL	TOTAL INCL O&P	
300	3120	2 compartment, 5000 gallon				Hr.					110	300
	3400	Transportation in 6900 gallon bulk truck				Mile					4.75	
	3410	In teflon lined truck				"					5.50	
	5000	Heavy sludge or dry vacuumable material				Hr.					110	
	6000	Dumpsite disposal charge, minimum				Ton					110	
	6020	Maximum				"					440	

	02115 | Underground Storage Tank Removal		CREW	DAILY OUTPUT	LABOR-HOURS	UNIT	MAT.	LABOR	EQUIP.	TOTAL	TOTAL INCL O&P	
200	0010	REMOVAL OF UNDERGROUND STORAGE TANKS R02115-200										200
	0011	Petroleum storage tanks, non-leaking										
	0100	Excavate & load onto trailer										
	0110	3000 gal. to 5000 gal. tank	B-14	4	12	Ea.		330	52	382	595	
	0120	6000 gal to 8000 gal tank	B-3A	3	13.333			365	228	593	850	
	0130	9000 gal to 12000 gal tank	"	2	20			550	340	890	1,275	
	0190	Known leaking tank add				%				100%	100%	
	0200	Remove sludge, water and remaining product from tank bottom										
	0201	of tank with vacuum truck										
	0300	3000 gal to 5000 gal tank	A-13	5	1.600	Ea.		51.50	101	152.50	192	
	0310	6000 gal to 8000 gal tank		4	2			64.50	126	190.50	239	
	0320	9000 gal to 12000 gal tank		3	2.667			85.50	168	253.50	320	
	0390	Dispose of sludge off-site, average				Gal.					4.40	
	0400	Insert inert solid CO2 "dry ice" into tank										
	0401	For cleaning/transporting tanks (1.5 lbs./100 gal. cap)	1 Clab	500	.016	Lb.	1.34	.42		1.76	2.16	
	1020	Haul tank to certified salvage dump, 100 miles round trip										
	1023	3000 gal. to 5000 gal. tank				Ea.				550	690	
	1026	6000 gal. to 8000 gal. tank								650	825	
	1029	9,000 gal. to 12,000 gal. tank								875	1,100	
	1100	Disposal of contaminated soil to landfill										
	1110	Minimum				C.Y.					110	
	1111	Maximum				"					330	
	1120	Disposal of contaminated soil to										
	1121	bituminous concrete batch plant										
	1130	Minimum				C.Y.					55	
	1131	Maximum				"					110	
	2010	Decontamination of soil on site incl poly tarp on top/bottom										
	2011	Soil containment berm, and chemical treatment										
	2020	Minimum	B-11C	100	.160	C.Y.	5.60	4.77	2.08	12.45	16.15	
	2021	Maximum	"	100	.160		7.25	4.77	2.08	14.10	18	
	2050	Disposal of decontaminated soil, minimum									66	
	2055	Maximum									135	

02200 | Site Preparation

	02210 | Subsurface Investigation		CREW	DAILY OUTPUT	LABOR-HOURS	UNIT	2004 BARE COSTS MAT.	LABOR	EQUIP.	TOTAL	TOTAL INCL O&P	
120	0010	BORING AND EXPLORATORY DRILLING										120
	0020	Borings, initial field stake out & determination of elevations	A-6	1	16	Day		585	59.50	644.50	1,000	
	0100	Drawings showing boring details				Total		185		185	270	
	0200	Report and recommendations from P.E.						415		415	595	
	0300	Mobilization and demobilization, minimum	B-55	4	6			155	194	349	470	
	0350	For over 100 miles, per added mile		450	.053	Mile		1.38	1.72	3.10	4.16	

02210 | Subsurface Investigation

		02210 Subsurface Investigation	CREW	DAILY OUTPUT	LABOR-HOURS	UNIT	2004 BARE COSTS MAT.	LABOR	EQUIP.	TOTAL	TOTAL INCL O&P	
120	0600	Auger holes in earth, no samples, 2-1/2" diameter	B-55	78.60	.305	L.F.		7.90	9.85	17.75	24	120
	0650	4" diameter		67.50	.356			9.20	11.50	20.70	28	
	0800	Cased borings in earth, with samples, 2-1/2" diameter		55.50	.432		14.20	11.20	13.95	39.35	49.50	
	0850	4" diameter		32.60	.736		22.50	19.05	24	65.55	82	
	1000	Drilling in rock, "BX" core, no sampling	B-56	34.90	.458			13.35	27	40.35	51	
	1050	With casing & sampling		31.70	.505		14.20	14.70	29.50	58.40	71.50	
	1200	"NX" core, no sampling		25.92	.617			17.95	36	53.95	69	
	1250	With casing and sampling		25	.640		17.35	18.60	37.50	73.45	90.50	
	1400	Drill rig and crew with truck mounted auger	B-55	1	24	Day		620	775	1,395	1,875	
	1450	With crawler type drill	B-56	1	16	"		465	940	1,405	1,775	
	1500	For inner city borings add, minimum									10%	
	1510	Maximum									20%	
200	0010	**CORE DRILLING** Reinforced concrete slab, up to 6" thick slab										200
	0020	Including bit, layout and set up										
	0100	1" diameter core	B-89A	28	.571	Ea.	2.40	17.05	3.79	23.24	35	
	0150	Each added inch thick, add		300	.053		.43	1.59	.35	2.37	3.48	
	0300	3" diameter core		23	.696		5.35	21	4.61	30.96	45	
	0350	Each added inch thick, add		186	.086		.96	2.57	.57	4.10	5.90	
	0500	4" diameter core		19	.842		5.35	25	5.60	35.95	53.50	
	0550	Each added inch thick, add		170	.094		1.21	2.81	.62	4.64	6.65	
	0700	6" diameter core		14	1.143		8.80	34	7.60	50.40	74	
	0750	Each added inch thick, add		140	.114		1.49	3.41	.76	5.66	8.05	
	0900	8" diameter core		11	1.455		12	43.50	9.65	65.15	95.50	
	0950	Each added inch thick, add		95	.168		2.02	5	1.12	8.14	11.70	
	1100	10" diameter core		10	1.600		16	47.50	10.60	74.10	108	
	1150	Each added inch thick, add		80	.200		2.65	5.95	1.33	9.93	14.20	
	1300	12" diameter core		9	1.778		19.20	53	11.80	84	121	
	1350	Each added inch thick, add		68	.235		3.18	7	1.56	11.74	16.75	
	1500	14" diameter core		7	2.286		23.50	68	15.15	106.65	154	
	1550	Each added inch thick, add		55	.291		4.04	8.70	1.93	14.67	21	
	1700	18" diameter core		4	4		30	119	26.50	175.50	259	
	1750	Each added inch thick, add		28	.571		5.30	17.05	3.79	26.14	38	
	1760	For horizontal holes, add to above								30%	30%	
	1770	Prestressed hollow core plank, 6" thick										
	1780	1" diameter core	B-89A	52	.308	Ea.	1.59	9.20	2.04	12.83	19.10	
	1790	Each added inch thick, add		350	.046		.28	1.36	.30	1.94	2.88	
	1800	3" diameter core		50	.320		3.51	9.55	2.12	15.18	22	
	1810	Each added inch thick, add		240	.067		.58	1.99	.44	3.01	4.40	
	1820	4" diameter core		48	.333		4.67	9.95	2.21	16.83	24	
	1830	Each added inch thick, add		216	.074		.81	2.21	.49	3.51	5.05	
	1840	6" diameter core		44	.364		5.80	10.85	2.41	19.06	27	
	1850	Each added inch thick, add		175	.091		.96	2.73	.61	4.30	6.20	
	1860	8" diameter core		32	.500		7.75	14.90	3.32	25.97	36.50	
	1870	Each added inch thick, add		118	.136		1.34	4.04	.90	6.28	9.10	
	1880	10" diameter core		28	.571		10.45	17.05	3.79	31.29	43.50	
	1890	Each added inch thick, add		99	.162		1.44	4.82	1.07	7.33	10.70	
	1900	12" diameter core		22	.727		12.75	21.50	4.82	39.07	55	
	1910	Each added inch thick, add		85	.188		2.12	5.60	1.25	8.97	12.95	
	1950	Minimum charge for above, 3" diameter core		7	2.286	Total		68	15.15	83.15	129	
	2000	4" diameter core		6.80	2.353			70	15.60	85.60	132	
	2050	6" diameter core		6	2.667			79.50	17.70	97.20	150	
	2100	8" diameter core		5.50	2.909			87	19.30	106.30	164	
	2150	10" diameter core		4.75	3.368			100	22.50	122.50	190	
	2200	12" diameter core		3.90	4.103			122	27	149	231	
	2250	14" diameter core		3.38	4.734			141	31.50	172.50	267	
	2300	18" diameter core		3.15	5.079			152	33.50	185.50	286	

	02210 | Subsurface Investigation		CREW	DAILY OUTPUT	LABOR-HOURS	UNIT	2004 BARE COSTS MAT.	LABOR	EQUIP.	TOTAL	TOTAL INCL O&P	
200	3010	Bits for core drill, diamond, premium, 1″ diameter				Ea.	107			107	118	200
	3020	3″ diameter					264			264	291	
	3040	4″ diameter					294			294	325	
	3050	6″ diameter					470			470	520	
	3080	8″ diameter					645			645	710	
	3120	12″ diameter					1,025			1,025	1,125	
	3180	18″ diameter					2,100			2,100	2,300	
	3240	24″ diameter					2,800			2,800	3,075	

	02220 | Site Demolition		CREW	DAILY OUTPUT	LABOR-HOURS	UNIT	MAT.	LABOR	EQUIP.	TOTAL	TOTAL INCL O&P	
110	0010	**BUILDING DEMOLITION** Large urban projects, incl. 20 Mi. haul										110
	0012	No foundation or dump fees, C.F. is volume of building standing, steel	B-8	21,500	.003	C.F.		.09	.11	.20	.26	
	0050	Concrete		15,300	.004			.12	.15	.27	.36	
	0080	Masonry		20,100	.003			.09	.12	.21	.28	
	0100	Mixture of types, average		20,100	.003			.09	.12	.21	.28	
	0500	Small bldgs, or single bldgs, no salvage included, steel	B-3	14,800	.003			.09	.12	.21	.28	
	0600	Concrete		11,300	.004			.12	.15	.27	.36	
	0650	Masonry		14,800	.003			.09	.12	.21	.28	
	0700	Wood		14,800	.003			.09	.12	.21	.28	
	1000	Single family, one story house, wood, minimum				Ea.				2,525	2,975	
	1020	Maximum								4,400	5,275	
	1200	Two family, two story house, wood, minimum								3,300	3,950	
	1220	Maximum								6,375	7,700	
	1300	Three family, three story house, wood, minimum								4,400	5,275	
	1320	Maximum								7,700	9,250	
	1400	Gutting building, see division 02225-400										
	5000	For buildings with no interior walls, deduct				Ea.				50%		
130	0010	**BLDG. FOOTINGS AND FOUNDATIONS DEMOLITION**										130
	0200	Floors, concrete slab on grade,										
	0240	4″ thick, plain concrete	B-9C	500	.080	S.F.		2.11	.32	2.43	3.85	
	0280	Reinforced, wire mesh		470	.085			2.25	.34	2.59	4.09	
	0300	Rods		400	.100			2.64	.40	3.04	4.82	
	0400	6″ thick, plain concrete		375	.107			2.82	.43	3.25	5.15	
	0420	Reinforced, wire mesh		340	.118			3.11	.47	3.58	5.65	
	0440	Rods		300	.133			3.52	.53	4.05	6.45	
	1000	Footings, concrete, 1′ thick, 2′ wide	B-5	300	.187	L.F.		5.30	3.07	8.37	12.05	
	1080	1′-6″ thick, 2′ wide		250	.224			6.40	3.68	10.08	14.45	
	1120	3′ wide		200	.280			7.95	4.60	12.55	18.05	
	1140	2′ thick, 3′ wide		175	.320			9.10	5.25	14.35	20.50	
	1200	Average reinforcing, add								10%	10%	
	1220	Heavy reinforcing, add								20%	20%	
	2000	Walls, block, 4″ thick	1 Clab	180	.044	S.F.		1.16		1.16	1.92	
	2040	6″ thick		170	.047			1.22		1.22	2.03	
	2080	8″ thick		150	.053			1.39		1.39	2.30	
	2100	12″ thick		150	.053			1.39		1.39	2.30	
	2200	For horizontal reinforcing, add								10%	10%	
	2220	For vertical reinforcing, add								20%	20%	
	2400	Concrete, plain concrete, 6″ thick	B-9	160	.250			6.60	1	7.60	12.05	
	2420	8″ thick		140	.286			7.55	1.14	8.69	13.75	
	2440	10″ thick		120	.333			8.80	1.33	10.13	16.05	
	2500	12″ thick		100	.400			10.55	1.60	12.15	19.25	
	2600	For average reinforcing, add								10%	10%	
	2620	For heavy reinforcing, add								20%	20%	
	9000	Minimum labor/equipment charge	A-1	2	4	Job		104	21.50	125.50	196	

	02220	Site Demolition	CREW	DAILY OUTPUT	LABOR-HOURS	UNIT	2004 BARE COSTS MAT.	LABOR	EQUIP.	TOTAL	TOTAL INCL O&P
210	0010	**MINOR BUILDING DECONSTRUCTION** For salvage									
	0225	Avg 2 story, pre 1970 house, 1400 SF, labor for all salv mat, min	6 Clab	25	1.920	SF Flr.		50		50	83
	0230	Maximum	"	15	3.200	"		83		83	138
	0235	Salvage of carpet, tackless	2 Clab	2,400	.007	S.F.		.17		.17	.29
	0240	Wood floors, incl denailing and packaging	3 Clab	270	.089	"		2.31		2.31	3.83
	0260	Wood doors and trim, standard	1 Clab	16	.500	Ea.		13		13	21.50
	0280	Base or cove mouldings, incl denailing and packaging		500	.016	L.F.		.42		.42	.69
	0320	Closet shelving and trim, incl denailing and packaging		18	.444	Set		11.55		11.55	19.15
	0340	Kitchen cabinets, uppers and lowers, prefab type	2 Clab	24	.667	L.F.		17.35		17.35	28.50
	0360	Kitchen cabinets, uppers and lowers, built-in		12	1.333	"		34.50		34.50	57.50
	0370	Bath fixt, incl toilet, tub, vanity, and med cabinet		3	5.333	Set		139		139	230
220	0010	**FENCING DEMOLITION**									
	1600	Fencing, barbed wire, 3 strand	2 Clab	430	.037	L.F.		.97		.97	1.60
	1650	5 strand	"	280	.057			1.49		1.49	2.46
	1700	Chain link, posts & fabric, remove only, 8′ to 10′ high	B-6	445	.054			1.51	.47	1.98	2.97
230	0010	**HYDRODEMOLITION**									
	0015	Hydrodemolition, concrete pavement, 4000 PSI, 2″ depth	B-5	500	.112	S.F.		3.19	1.84	5.03	7.25
	0120	4″ depth		450	.124			3.54	2.05	5.59	8.05
	0130	6″ depth		400	.140			3.99	2.30	6.29	9.05
	0410	6000 PSI, 2″ depth		410	.137			3.89	2.25	6.14	8.80
	0420	4″ depth		350	.160			4.56	2.63	7.19	10.35
	0430	6″ depth		300	.187			5.30	3.07	8.37	12.05
	0510	8000 PSI, 2″ depth		330	.170			4.83	2.79	7.62	10.95
	0520	4″ depth		280	.200			5.70	3.29	8.99	12.90
	0530	6″ depth		240	.233			6.65	3.84	10.49	15.05
240	0010	**MINOR SITE DEMOLITION**									
	0015	Minor site demolition, no hauling, abandon catch basin or manhole	B-6	7	3.429	Ea.		96	29.50	125.50	189
	0020	Remove existing catch basin or manhole, masonry		4	6			168	52	220	330
	0030	Catch basin or manhole frames and covers, stored		13	1.846			52	16	68	102
	0040	Remove and reset		7	3.429			96	29.50	125.50	189
	0100	Roadside delineators, remove only	B-80	175	.183			5.10	2.96	8.06	11.55
	0110	Remove and reset	"	100	.320			8.95	5.20	14.15	20.50
	0400	Minimum labor/equipment charge	B-6	4	6	Job		168	52	220	330
	0800	Guiderail, corrugated steel, remove only	B-80A	100	.240	L.F.		6.25	1.60	7.85	12.10
	0850	Remove and reset	"	40	.600	"		15.60	4.01	19.61	30.50
	0860	Guide posts, remove only	B-80B	120	.267	Ea.		7.35	1.68	9.03	13.85
	0870	Remove and reset	B-55	50	.480	"		12.45	15.50	27.95	37.50
	0890	Minimum labor/equipment charge	2 Clab	4	4	Job		104		104	172
	0900	Hydrants, fire, remove only	B-21A	5	8	Ea.		256	96.50	352.50	515
	0950	Remove and reset	"	2	20	"		640	241	881	1,300
	0990	Minimum labor/equipment charge	2 Plum	2	8	Job		315		315	490
	1000	Masonry walls, block or tile, solid, remove	B-5	1,800	.031	C.F.		.89	.51	1.40	2
	1100	Cavity wall		2,200	.025			.72	.42	1.14	1.64
	1200	Brick, solid		900	.062			1.77	1.02	2.79	4.02
	1300	With block back-up		1,130	.050			1.41	.81	2.22	3.20
	1400	Stone, with mortar		900	.062			1.77	1.02	2.79	4.02
	1500	Dry set		1,500	.037			1.06	.61	1.67	2.41
	1600	Median barrier, precast concrete, remove and store	B-3	430	.112	L.F.		3.10	3.99	7.09	9.45
	1610	Remove and reset	"	390	.123	"		3.42	4.40	7.82	10.40
	1650	Minimum labor/equipment charge	A-1	4	2	Job		52	10.70	62.70	98
	2900	Pipe removal, sewer/water, no excavation, 12″ diameter	B-6	175	.137	L.F.		3.85	1.19	5.04	7.55
	2930	15″ diameter		150	.160			4.49	1.39	5.88	8.80
	2960	24″ diameter		120	.200			5.60	1.73	7.33	11
	3000	36″ diameter		90	.267			7.50	2.31	9.81	14.70
	3200	Steel, welded connections, 4″ diameter		160	.150			4.21	1.30	5.51	8.30

	02220 | Site Demolition		CREW	DAILY OUTPUT	LABOR-HOURS	UNIT	2004 BARE COSTS MAT.	LABOR	EQUIP.	TOTAL	TOTAL INCL O&P
240	3300	10″ diameter	B-6	80	.300	L.F.		8.40	2.60	11	16.55
	3390	Minimum labor/equipment charge	↓	3	8	Job		224	69.50	293.50	440
	3500	Railroad track removal, ties and track	B-13	330	.170	L.F.		4.75	1.89	6.64	9.85
	3600	Ballast	B-14	500	.096	C.Y.		2.63	.42	3.05	4.77
	3700	Remove and re-install, ties & track using new bolts & spikes		50	.960	L.F.		26.50	4.16	30.66	47.50
	3800	Turnouts using new bolts and spikes		1	48	Ea.		1,325	208	1,533	2,375
	3890	Minimum labor/equipment charge	↓	5	9.600	Job		263	41.50	304.50	475
	4000	Sidewalk removal, bituminous, 2-1/2″ thick	B-6	325	.074	S.Y.		2.07	.64	2.71	4.07
	4050	Brick, set in mortar		185	.130			3.64	1.12	4.76	7.15
	4100	Concrete, plain, 4″		160	.150			4.21	1.30	5.51	8.30
	4200	Mesh reinforced	↓	150	.160	↓		4.49	1.39	5.88	8.80
	4290	Minimum labor/equipment charge	B-39	12	4	Job		109	13.30	122.30	195
250	0010	**DEMOLISH, REMOVE PAVEMENT AND CURB**									
	5010	Pavement removal, bituminous roads, 3″ thick	B-38	690	.058	S.Y.		1.69	1.17	2.86	4.03
	5050	4″ to 6″ thick		420	.095			2.78	1.92	4.70	6.60
	5100	Bituminous driveways		640	.063			1.82	1.26	3.08	4.34
	5200	Concrete to 6″ thick, hydraulic hammer, mesh reinforced		255	.157			4.57	3.17	7.74	10.90
	5300	Rod reinforced		200	.200	↓		5.85	4.04	9.89	13.90
	5400	Concrete, 7″ to 24″ thick, plain		33	1.212	C.Y.		35.50	24.50	60	84.50
	5500	Reinforced		24	1.667	″		48.50	33.50	82	116
	5590	Minimum labor/equipment charge	↓	6	6.667	Job		194	135	329	465
	5600	With hand held air equipment, bituminous, to 6″ thick	B-39	1,900	.025	S.F.		.69	.08	.77	1.22
	5700	Concrete to 6″ thick, no reinforcing		1,600	.030			.82	.10	.92	1.46
	5800	Mesh reinforced		1,400	.034			.94	.11	1.05	1.67
	5900	Rod reinforced	↓	765	.063	↓		1.72	.21	1.93	3.05
	5990	Minimum labor/equipment charge	B-38	6	6.667	Job		194	135	329	465
	6000	Curbs, concrete, plain	B-6	360	.067	L.F.		1.87	.58	2.45	3.67
	6100	Reinforced		275	.087			2.45	.76	3.21	4.81
	6200	Granite		360	.067			1.87	.58	2.45	3.67
	6300	Bituminous		528	.045	↓		1.27	.39	1.66	2.50
	6390	Minimum labor/equipment charge	↓	6	4	Job		112	34.50	146.50	220
310	0010	**SELECTIVE DEMOLITION, CUTOUT**									
	0020	Concrete, elev. slab, light reinforcement, under 6 CF	B-9C	65	.615	C.F.		16.25	2.46	18.71	29.50
	0050	Light reinforcing, over 6 C.F.	″	75	.533	″		14.10	2.13	16.23	26
	0200	Slab on grade to 6″ thick, not reinforced, under 8 S.F.	B-9	85	.471	S.F.		12.40	1.88	14.28	22.50
	0250	Not reinforced, over 8 S.F.		175	.229	″		6.05	.91	6.96	11
	0600	Walls, not reinforced, under 6 C.F.		60	.667	C.F.		17.60	2.67	20.27	32
	0650	Not reinforced, over 6 C.F.	↓	65	.615			16.25	2.46	18.71	29.50
	1000	Concrete, elevated slab, bar reinforced, under 6 C.F.	B-9C	45	.889			23.50	3.56	27.06	43
	1050	Bar reinforced, over 6 C.F.	″	50	.800	↓		21	3.20	24.20	38.50
	1200	Slab on grade to 6″ thick, bar reinforced, under 8 S.F.	B-9	75	.533	S.F.		14.10	2.13	16.23	26
	1250	Bar reinforced, over 8 S.F.	″	105	.381	″		10.05	1.52	11.57	18.35
	1400	Walls, bar reinforced, under 6 C.F.	B-9C	50	.800	C.F.		21	3.20	24.20	38.50
	1450	Bar reinforced, over 6 C.F.	″	55	.727	″		19.20	2.91	22.11	35
	2000	Brick, to 4 S.F. opening, not including toothing									
	2040	4″ thick	B-9C	30	1.333	Ea.		35	5.35	40.35	64.50
	2060	8″ thick		18	2.222			58.50	8.90	67.40	107
	2080	12″ thick		10	4			106	16	122	193
	2400	Concrete block, to 4 S.F. opening, 2″ thick		35	1.143			30	4.57	34.57	55
	2420	4″ thick		30	1.333			35	5.35	40.35	64.50
	2440	8″ thick		27	1.481			39	5.95	44.95	71.50
	2460	12″ thick	↓	24	1.667			44	6.65	50.65	80.50
	2600	Gypsum block, to 4 S.F. opening, 2″ thick	B-9	80	.500			13.20	2	15.20	24
	2620	4″ thick		70	.571			15.10	2.29	17.39	27.50
	2640	8″ thick	↓	55	.727	↓		19.20	2.91	22.11	35

	02220 | Site Demolition		CREW	DAILY OUTPUT	LABOR-HOURS	UNIT	2004 BARE COSTS MAT.	LABOR	EQUIP.	TOTAL	TOTAL INCL O&P	
310	2800	Terra cotta, to 4 S.F. opening, 4" thick	B-9	70	.571	Ea.		15.10	2.29	17.39	27.50	310
	2840	8" thick		65	.615			16.25	2.46	18.71	29.50	
	2880	12" thick		50	.800			21	3.20	24.20	38.50	
	4000	For toothing masonry, see Division 04910-800										
	6000	Walls, interior, not including re-framing,										
	6010	openings to 5 S.F.										
	6100	Drywall to 5/8" thick	1 Clab	24	.333	Ea.		8.65		8.65	14.35	
	6200	Paneling to 3/4" thick		20	.400			10.40		10.40	17.25	
	6300	Plaster, on gypsum lath		20	.400			10.40		10.40	17.25	
	6340	On wire lath		14	.571			14.85		14.85	24.50	
	7000	Wood frame, not including re-framing, openings to 5 S.F.										
	7200	Floors, sheathing and flooring to 2" thick	1 Clab	5	1.600	Ea.		41.50		41.50	69	
	7310	Roofs, sheathing to 1" thick, not including roofing		6	1.333			34.50		34.50	57.50	
	7410	Walls, sheathing to 1" thick, not including siding		7	1.143			29.50		29.50	49.50	
	8500	Minimum labor/equipment charge		4	2	Job		52		52	86	
320	0010	**SELECTIVE DEMOLITION, DISPOSAL ONLY**										320
	0015	Urban bldg w/salvage value allowed										
	0020	Including loading and 5 mile haul to dump										
	0200	Steel frame	B-3	430	.112	C.Y.		3.10	3.99	7.09	9.45	
	0300	Concrete frame		365	.132			3.65	4.70	8.35	11.10	
	0400	Masonry construction		445	.108			2.99	3.85	6.84	9.10	
	0500	Wood frame		247	.194			5.40	6.95	12.35	16.45	
330	0010	**SELECTIVE DEMOLITION, DUMP CHARGES**										330
	0020	Dump charges, typical urban city, tipping fees only										
	0100	Building construction materials				Ton					70	
	0200	Trees, brush, lumber									50	
	0300	Rubbish only									60	
	0500	Reclamation station, usual charge									85	
340	0010	**SELECTIVE DEMOLITION, GUTTING**										340
	0020	Building interior, including disposal, dumpter fees not included										
	0500	Residential building										
	0560	Minimum	B-16	400	.080	SF Flr.		2.13	1.19	3.32	4.82	
	0580	Maximum	"	360	.089	"		2.37	1.32	3.69	5.35	
	0900	Commercial building										
	1000	Minimum	B-16	350	.091	SF Flr.		2.43	1.36	3.79	5.50	
	1020	Maximum		250	.128	"		3.41	1.91	5.32	7.70	
	3000	Minimum labor/equipment charge		4	8	Job		213	119	332	480	
350	0010	**SELECTIVE DEMOLITION, RUBBISH HANDLING**										350
	0020	The following are to be added to the demolition prices										
	0400	Chute, circular, prefabricated steel, 18" diameter	B-1	40	.600	L.F.	28	16		44	57.50	
	0440	30" diameter	"	30	.800	"	37.50	21.50		59	76.50	
	0725	Dumpster, weekly rental, 1 dump/week, 20 C.Y. capacity (8 Tons)				Week					440	
	0800	30 C.Y. capacity (10 Tons)									665	
	0840	40 C.Y. capacity (13 Tons)									805	
	0900	Alternate pricing for dumpsters										
	0910	Delivery, average for all sizes				Ea.					52	
	0920	Haul, average for all sizes									130	
	0930	Rent per day, average for all sizes									3.10	
	0940	Rent per month, average for all sizes									31	
	0950	Disposal fee per ton, average for all sizes				Ton					47	
	1000	Dust partition, 6 mil polyethylene, 1" x 3" frame	2 Carp	2,000	.008	S.F.	.16	.26		.42	.61	
	1080	2" x 4" frame	"	2,000	.008	"	.26	.26		.52	.73	
	2000	Load, haul, and dump, 50' haul	2 Clab	24	.667	C.Y.		17.35		17.35	28.50	

		02220 | Site Demolition	CREW	DAILY OUTPUT	LABOR-HOURS	UNIT	2004 BARE COSTS MAT.	LABOR	EQUIP.	TOTAL	TOTAL INCL O&P	
350	2040	100′ haul	2 Clab	16.50	.970	C.Y.		25		25	42	350
	2080	Over 100′ haul, add per 100 L.F.		35.50	.451			11.70		11.70	19.45	
	2120	In elevators, per 10 floors, add		140	.114			2.97		2.97	4.93	
	3000	Loading & trucking, including 2 mile haul, chute loaded	B-16	45	.711			18.90	10.60	29.50	42.50	
	3040	Hand loading truck, 50′ haul	″	48	.667			17.75	9.95	27.70	40.50	
	3080	Machine loading truck	B-17	120	.267			7.35	4.44	11.79	16.90	
	3120	Wheeled 50′ and ramp dump loaded	2 Clab	24	.667			17.35		17.35	28.50	
	5000	Haul, per mile, up to 8 C.Y. truck	B-34B	1,165	.007			.18	.41	.59	.75	
	5100	Over 8 C.Y. truck	″	1,550	.005			.14	.31	.45	.56	
360	0010	**SELECTIVE DEMOLITION, SAW CUTTING**										360
	0015	Asphalt, up to 3″ deep	B-89	1,050	.015	L.F.	.26	.44	.25	.95	1.27	
	0020	Each additional inch of depth		1,800	.009		.06	.26	.15	.47	.64	
	0400	Concrete slabs, mesh reinforcing, up to 3″ deep		980	.016		.35	.47	.27	1.09	1.44	
	0420	Each additional inch of depth		1,600	.010		.12	.29	.17	.58	.77	
	0800	Concrete walls, hydraulic saw, plain, per inch of depth	B-89B	250	.064		.32	1.85	1.71	3.88	5.20	
	0820	Rod reinforcing, per inch of depth		150	.107		.44	3.09	2.86	6.39	8.55	
	1200	Masonry walls, hydraulic saw, brick, per inch of depth		300	.053		.32	1.54	1.43	3.29	4.38	
	1220	Block walls, solid, per inch of depth		250	.064		.33	1.85	1.71	3.89	5.20	
	2000	Brick or masonry w/hand held saw, per inch of depth	A-1	125	.064		.26	1.66	.34	2.26	3.43	
	5000	Wood sheathing to 1″ thick, on walls	1 Carp	200	.040			1.32		1.32	2.19	
	5020	On roof	″	250	.032			1.06		1.06	1.75	
	9000	Minimum labor/equipment charge	A-1	2	4	Job		104	21.50	125.50	196	
	9950	See also div. 02210-200 core drilling										
370	0010	**SELECTIVE DEMOLITION, TORCH CUTTING**										370
	0020	Steel, 1″ thick plate	1 Clab	360	.022	L.F.	.18	.58		.76	1.16	
	0040	1″ diameter bar	″	210	.038	Ea.		.99		.99	1.64	
	1000	Oxygen lance cutting, reinforced concrete walls										
	1040	12″ to 16″ thick walls	1 Clab	10	.800	L.F.		21		21	34.50	
	1080	24″ thick walls	″	6	1.333	″		34.50		34.50	57.50	
	1090	Minimum labor/equipment charge	E-25	2	4	Job		157	42.50	199.50	340	
	1100	See also division 05090-920										
		02250 | Shoring & Underpinning										
100	0010	**GROUTING, PRESSURE**										100
	0020	Grouting, pressure, cement & sand, 1:1 mix, minimum	B-61	124	.323	Bag	8.90	8.90	1.96	19.76	26.50	
	0100	Maximum		51	.784	″	8.90	21.50	4.77	35.17	50.50	
	0200	Cement and sand, 1:1 mix, minimum		250	.160	C.F.	17.80	4.42	.97	23.19	28	
	0300	Maximum		100	.400		26.50	11.05	2.43	39.98	50.50	
	0400	Epoxy cement grout, minimum		137	.292		126	8.05	1.78	135.83	154	
	0500	Maximum		57	.702		126	19.40	4.27	149.67	176	
	0600	Structural epoxy grout				Gal.	49.50			49.50	54.50	
	0700	Alternate pricing method: (Add for materials)										
	0710	5 person crew and equipment	B-61	1	40	Day		1,100	243	1,343	2,075	
400	0010	**SHEET PILING** R02250-400										400
	0020	Sht. piling stl., not incl. wales, 22 psf, 15′ excav., left in place	B-40	10.81	5.920	Ton	815	193	214	1,222	1,450	
	0100	Drive, extract & salvage R02250-400		6	10.667	″	360	350	385	1,095	1,400	
	1200	15′ deep excavation, 22 psf, left in place		983	.065	S.F.	9.50	2.12	2.36	13.98	16.55	
	1300	Drive, extract & salvage		545	.117	″	4.07	3.83	4.25	12.15	15.50	
	2100	Rent steel sheet piling and wales, first month				Ton	194			194	214	
	2200	Per added month				″	19.40			19.40	21.50	
	3900	Wood, solid sheeting, incl. wales, braces and spacers,										
	3910	drive, extract & salvage, 8′ deep excavation	B-31	330	.121	S.F.	1.67	3.37	.44	5.48	7.90	
	4520	Left in place, 8′ deep, 55 S.F./hr.		440	.091	″	3	2.53	.33	5.86	7.85	
	4990	Minimum labor/equipment charge		2	20	Job		555	72.50	627.50	1,000	
	5000	For treated lumber add cost of treatment to lumber										

02200 | Site Preparation

	02250 | Shoring & Underpinning		CREW	DAILY OUTPUT	LABOR-HOURS	UNIT	2004 BARE COSTS MAT.	LABOR	EQUIP.	TOTAL	TOTAL INCL O&P	
400	5010	See division 06070-400 R02250-400										400
500	0010	**SHORING**										500
	0020	Shoring, existing building, with timber, no salvage allowance	B-51	2.20	21.818	M.B.F.	735	575	55.50	1,365.50	1,825	
	1000	On cribbing with 35 ton screw jacks, per box and jack		3.60	13.333	Jack	44.50	350	34	428.50	665	
	1090	Minimum labor/equipment charge		2	24	Ea.		630	61	691	1,125	
	1100	Masonry openings in walls, see div. 02225-320										
800	0010	**UNDERPINNING FOUNDATIONS** Including excavation,										800
	0020	forming, reinforcing, concrete and equipment										
	0100	5′ to 16′ below grade, 100 to 500 C.Y.	B-52	2.30	24.348	C.Y.	203	740	165	1,108	1,625	
	0200	Over 500 C.Y.		2.50	22.400		182	680	152	1,014	1,500	
	0400	16′ to 25′ below grade, 100 to 500 C.Y.		2	28		223	850	190	1,263	1,850	
	0500	Over 500 C.Y.		2.10	26.667		211	810	181	1,202	1,750	
	0700	26′ to 40′ below grade, 100 to 500 C.Y.		1.60	35		243	1,075	238	1,556	2,275	
	0800	Over 500 C.Y.		1.80	31.111		223	945	211	1,379	2,025	
	0900	For under 50 C.Y., add					10%	40%				

02300 | Earthwork

	02305 | Equipment		CREW	DAILY OUTPUT	LABOR-HOURS	UNIT	2004 BARE COSTS MAT.	LABOR	EQUIP.	TOTAL	TOTAL INCL O&P	
250	0010	**MOBILIZATION OR DEMOB.** (One or the other, unless noted) R01590-100										250
	0015	Up to 25 mi haul dist (50 mi round trip for mob/demob crew)										
	0020	Dozer, loader, backhoe, excav., grader, paver, roller, 70 to 150 H.P.	B-34N	4	2	Ea.		53	111	164	209	
	0100	Above 150 HP	B-34K	3	2.667			70.50	168	238.50	300	
	0300	Scraper, towed type (incl. tractor), 6 C.Y. capacity		3	2.667			70.50	168	238.50	300	
	0400	10 C.Y.		2.50	3.200			84.50	201	285.50	360	
	0600	Self-propelled scraper, 15 C.Y.		2.50	3.200			84.50	201	285.50	360	
	0700	24 C.Y.		2	4			106	252	358	450	
	0900	Shovel or dragline, 3/4 C.Y.		3.60	2.222			59	140	199	250	
	1000	1-1/2 C.Y.		3	2.667			70.50	168	238.50	300	
	1100	Small equipment, placed in rear of, or towed by pickup truck	A-3A	8	1			25.50	10.25	35.75	53.50	
	1150	Equip up to 70 HP, on flatbed trailer behind pickup truck	A-3D	4	2			51.50	42.50	94	131	
	2000	Crane, truck-mounted, up to 75 ton (costs incl both mob & demob)	1 EQHV	3.60	2.222			77.50		77.50	122	
	2100	Crane, truck-mounted, over 75 ton	A-3E	2.50	6.400			196	33	229	350	
	2200	Crawler-mounted, up to 75 ton	A-3F	2	8			245	278	523	695	
	2300	Over 75 ton	A-3G	1.50	10.667			325	390	715	950	
	2500	For each additional 5 miles haul distance, add						10%	10%			
	3000	For large pieces of equipment, allow for assembly/knockdown										
	3100	For mob/demob of micro-tunneling equip, see section 02441-400										
	3200	For mob/demob of pile driving equip, see section 02455-900										
	3300	For mob/demob of caisson drilling equip, see section 02465-800										

02315 | Excavation and Fill

			CREW	DAILY OUTPUT	LABOR-HOURS	UNIT	2004 BARE COSTS MAT.	LABOR	EQUIP.	TOTAL	TOTAL INCL O&P	
110	0010	**BACKFILL, GENERAL** R02315-300										110
	0015	By hand, no compaction, light soil	1 Clab	14	.571	C.Y.		14.85		14.85	24.50	
	0100	Heavy soil		11	.727			18.90		18.90	31.50	
	0300	Compaction in 6″ layers, hand tamp, add to above		20.60	.388			10.10		10.10	16.75	
	0400	Roller compaction operator walking, add	B-10A	100	.120			3.73	1.31	5.04	7.40	
	0500	Air tamp, add	B-9D	190	.211			5.55	.97	6.52	10.25	
	0600	Vibrating plate, add	A-1D	60	.133			3.47	.48	3.95	6.25	
	0800	Compaction in 12″ layers, hand tamp, add to above	1 Clab	34	.235			6.10		6.10	10.15	
	1000	Air tamp, add	B-9	285	.140			3.71	.56	4.27	6.75	
	1100	Vibrating plate, add	A-1E	90	.089			2.31	.46	2.77	4.34	
	1200	Trench, dozer, no compaction, 60 HP	B-10L	425	.028			.88	.71	1.59	2.18	
	1300	Dozer backfilling, bulk, up to 300′ haul, no compaction	B-10B	1,200	.010			.31	.72	1.03	1.29	
	1400	Air tamped	B-11B	240	.067			1.94	.80	2.74	4.01	
	1900	Dozer backfilling, trench, up to 300′ haul, no compaction	B-10B	900	.013			.41	.96	1.37	1.72	
	2000	Air tamped	B-11B	80	.200	E.C.Y.		5.80	2.39	8.19	12.05	
	2350	Spreading in 8″ layers, small dozer	B-10B	1,060	.011	C.Y.		.35	.81	1.16	1.46	
	2450	Compacting with vibrating plate, 8″ lifts	A-1D	73	.110	″		2.85	.39	3.24	5.15	
120	0010	**BACKFILL, STRUCTURAL** Dozer or F.E. loader										120
	0020	From existing stockpile, no compaction										
	2000	75 H.P., 50′ haul, sand & gravel	B-10L	1,100	.011	C.Y.		.34	.27	.61	.84	
	2020	Common earth		975	.012			.38	.31	.69	.95	
	2040	Clay		850	.014			.44	.35	.79	1.09	
	2400	300′ haul, sand & gravel		370	.032			1.01	.81	1.82	2.50	
	2420	Common earth		330	.036			1.13	.91	2.04	2.81	
	2440	Clay		290	.041			1.29	1.04	2.33	3.20	
	3000	105 H.P., 50′ haul, sand & gravel	B-10W	1,350	.009			.28	.32	.60	.79	
	3020	Common earth		1,225	.010			.30	.35	.65	.88	
	3040	Clay		1,100	.011			.34	.39	.73	.97	
	3300	300′ haul, sand & gravel		465	.026			.80	.93	1.73	2.31	
	3320	Common earth		415	.029			.90	1.05	1.95	2.59	
	3340	Clay		370	.032			1.01	1.17	2.18	2.90	
310	0010	**COMPACTION, GENERAL**										310
	5000	Riding, vibrating roller, 6″ lifts, 2 passes	B-10Y	3,000	.004	C.Y.		.12	.11	.23	.32	
	5020	3 passes		2,300	.005			.16	.14	.30	.41	
	5040	4 passes		1,900	.006			.20	.17	.37	.50	
	5060	12″ lifts, 2 passes		5,200	.002			.07	.06	.13	.18	
	5080	3 passes		3,500	.003			.11	.09	.20	.27	
	5100	4 passes		2,600	.005			.14	.12	.26	.37	
	5600	Sheepsfoot or wobbly wheel roller, 6″ lifts, 2 passes	B-10G	2,400	.005			.16	.32	.48	.60	
	5620	3 passes		1,735	.007			.22	.44	.66	.82	
	5640	4 passes		1,300	.009			.29	.59	.88	1.11	
	5680	12″ lifts, 2 passes		5,200	.002			.07	.15	.22	.27	
	5700	3 passes		3,500	.003			.11	.22	.33	.41	
	5720	4 passes		2,600	.005			.14	.29	.43	.55	
	7000	Walk behind, vibrating plate 18″ wide, 6″ lifts, 2 passes	A-1D	200	.040			1.04	.14	1.18	1.88	
	7020	3 passes		185	.043			1.12	.15	1.27	2.03	
	7040	4 passes		140	.057			1.49	.20	1.69	2.68	
	7200	12″ lifts, 2 passes	A-1E	560	.014			.37	.07	.44	.70	
	7220	3 passes		375	.021			.55	.11	.66	1.04	
	7240	4 passes		280	.029			.74	.15	.89	1.39	
	7500	Vibrating roller 24″ wide, 6″ lifts, 2 passes	B-10A	420	.029			.89	.31	1.20	1.76	
	7520	3 passes		280	.043			1.33	.47	1.80	2.64	
	7540	4 passes		210	.057			1.78	.62	2.40	3.53	
	7600	12″ lifts, 2 passes		840	.014			.44	.16	.60	.88	
	7620	3 passes		560	.021			.67	.23	.90	1.33	

02315 | Excavation and Fill

				DAILY	LABOR-		2004 BARE COSTS				TOTAL	
			CREW	OUTPUT	HOURS	UNIT	MAT.	LABOR	EQUIP.	TOTAL	INCL O&P	
310	7640	4 passes	B-10A	420	.029	C.Y.		.89	.31	1.20	1.76	310
	8000	Rammer tamper, 6″ to 11″, 4″ lifts, 2 passes	A-1F	130	.062			1.60	.27	1.87	2.95	
	8050	3 passes		97	.082			2.14	.37	2.51	3.95	
	8100	4 passes		65	.123			3.20	.55	3.75	5.90	
	8200	8″ lifts, 2 passes		260	.031			.80	.14	.94	1.48	
	8250	3 passes		195	.041			1.07	.18	1.25	1.97	
	8300	4 passes		130	.062			1.60	.27	1.87	2.95	
	8400	13″ to 18″, 4″ lifts, 2 passes	A-1G	390	.021			.53	.09	.62	.98	
	8450	3 passes		290	.028			.72	.12	.84	1.32	
	8500	4 passes		195	.041			1.07	.18	1.25	1.97	
	8600	8″ lifts, 2 passes		780	.010			.27	.05	.32	.49	
	8650	3 passes		585	.014			.36	.06	.42	.66	
	8700	4 passes		390	.021			.53	.09	.62	.98	
	9000	Water, 3000 gal. truck, 3 mile haul	B-45	1,888	.008		.20	.25	.29	.74	.95	
	9010	6 mile haul		1,444	.011		.20	.33	.38	.91	1.17	
	9020	12 mile haul		1,000	.016		.20	.48	.55	1.23	1.60	
	9030	6000 gal. wagon, 3 mile haul	B-59	2,000	.004		.20	.11	.17	.48	.57	
	9040	6 mile haul	″	1,600	.005		.20	.13	.21	.54	.66	
320	0010	**COMPACTION, STRUCTURAL** R02315 -300										320
	0020	Steel wheel tandem roller, 5 tons	B-10E	8	1.500	Hr.		46.50	13.55	60.05	89.50	
	0050	Air tamp, 6″ to 8″ lifts, common fill	B-9	250	.160	C.Y.		4.22	.64	4.86	7.70	
	0060	Select fill	″	300	.133			3.52	.53	4.05	6.45	
	0600	Vibratory plate, 8″ lifts, common fill	A-1D	200	.040			1.04	.14	1.18	1.88	
	0700	Select fill	″	216	.037			.96	.13	1.09	1.75	
	9000	Minimum labor/equipment charge	1 Clab	4	2	Job		52		52	86	
424	0010	**EXCAVATING, BULK BANK MEASURE** Common earth piled R02315 -400										424
	0020	For loading onto trucks, add								15%	15%	
	0200	Backhoe, hydraulic, crawler mtd., 1 C.Y. cap. = 75 C.Y./hr. R02315 -450	B-12A	600	.027	C.Y.		.85	.81	1.66	2.24	
	1200	Front end loader, track mtd., 1-1/2 C.Y. cap. = 70 C.Y./hr.	B-10N	560	.021			.67	.54	1.21	1.66	
	1500	Wheel mounted, 3/4 C.Y. cap. = 45 C.Y./hr.	B-10R	360	.033			1.04	.53	1.57	2.25	
	9000	Minimum labor/equipment charge	B-10L	2	6	Job		187	150	337	465	
432	0010	**EXCAVATING, BULK, DOZER** Open site										432
	2000	80 H.P., 50′ haul, sand & gravel	B-10L	460	.026	C.Y.		.81	.65	1.46	2.02	
	2200	150′ haul, sand & gravel		230	.052			1.62	1.31	2.93	4.03	
	2400	300′ haul, sand & gravel		120	.100			3.11	2.50	5.61	7.70	
	3000	105 H.P., 50′ haul, sand & gravel	B-10W	700	.017			.53	.62	1.15	1.53	
	3200	150′ haul, sand & gravel		310	.039			1.20	1.40	2.60	3.47	
	3300	300′ haul, sand & gravel		140	.086			2.67	3.10	5.77	7.65	
	4000	200 H.P., 50′ haul, sand & gravel	B-10B	1,400	.009			.27	.62	.89	1.11	
	4200	150′ haul, sand & gravel		595	.020			.63	1.45	2.08	2.60	
	4400	300′ haul, sand & gravel		310	.039			1.20	2.79	3.99	4.99	
	5040	Clay	B-10M	1,025	.012			.36	1.07	1.43	1.76	
	5400	300′ haul, sand & gravel	″	470	.026			.79	2.34	3.13	3.84	
462	0010	**EXCAVATION, STRUCTURAL**										462
	0015	Hand, pits to 6′ deep, sandy soil	1 Clab	8	1	C.Y.		26		26	43	
	0100	Heavy soil or clay		4	2			52		52	86	
	0300	Pits 6′ to 12′ deep, sandy soil		5	1.600			41.50		41.50	69	
	0500	Heavy soil or clay		3	2.667			69.50		69.50	115	
	0700	Pits 12′ to 18′ deep, sandy soil		4	2			52		52	86	
	0900	Heavy soil or clay		2	4			104		104	172	
	1100	Hand loading trucks from stock pile, sandy soil		12	.667			17.35		17.35	28.50	
	1300	Heavy soil or clay		8	1			26		26	43	
	1500	For wet or muck hand excavation, add to above				%				50%	50%	
	6000	Machine excavation, for spread and mat footings, elevator pits,										
	6001	and small building foundations										

	02315 | Excavation and Fill		CREW	DAILY OUTPUT	LABOR-HOURS	UNIT	2004 BARE COSTS MAT.	LABOR	EQUIP.	TOTAL	TOTAL INCL O&P	
462	6030	Common earth, hydraulic backhoe, 1/2 C.Y. bucket	B-12E	55	.291	C.Y.		9.30	5.85	15.15	21	462
	6035	3/4 C.Y. bucket	B-12F	90	.178			5.70	5.10	10.80	14.55	
	6040	1 C.Y. bucket	B-12A	108	.148			4.74	4.51	9.25	12.40	
	6050	1-1/2 C.Y. bucket	B-12B	144	.111			3.56	4.75	8.31	10.80	
	6060	2 C.Y. bucket	B-12C	200	.080			2.56	4.44	7	8.90	
	6070	Sand and gravel, 3/4 C.Y. bucket	B-12F	100	.160			5.10	4.59	9.69	13.10	
	6080	1 C.Y. bucket	B-12A	120	.133			4.27	4.06	8.33	11.15	
	6090	1-1/2 C.Y. bucket	B-12B	160	.100			3.20	4.27	7.47	9.75	
	6100	2 C.Y. bucket	B-12C	220	.073			2.33	4.03	6.36	8.10	
	6110	Clay, till, or blasted rock, 3/4 C.Y. bucket	B-12F	80	.200			6.40	5.75	12.15	16.40	
	6120	1 C.Y. bucket	B-12A	95	.168			5.40	5.15	10.55	14.15	
	6130	1-1/2 C.Y. bucket	B-12B	130	.123			3.94	5.25	9.19	12	
	6140	2 C.Y. bucket	B-12C	175	.091	↓		2.93	5.05	7.98	10.20	
	9000	Minimum labor/equipment charge	1 Clab	4	2	Job		52		52	86	
490	0010	**HAULING**, excavated or borrow, loose cubic yards										490
	0012	no loading included, highway haulers										
	0020	6 C.Y. dump truck, 1/4 mile round trip, 5.0 loads/hr.	B-34A	195	.041	C.Y.		1.09	1.67	2.76	3.59	
	0200	4 mile round trip, 1.8 loads/hr.	″	70	.114			3.02	4.64	7.66	10	
	0310	12 C.Y. dump truck, 1/4 mile round trip 3.7 loads/hr.	B-34B	288	.028			.73	1.66	2.39	3.01	
	0500	4 mile round trip, 1.6 loads/hr.	″	125	.064			1.69	3.81	5.50	6.95	
	0600	16.5 C.Y. dump trailer, 1 mile round trip, 2.6 loads/hr.	B-34C	280	.029			.76	1.45	2.21	2.83	
	1100	4 mile round trip, 1.6 loads/hr.	″	172	.047			1.23	2.36	3.59	4.60	
	1150	20 C.Y. dump trailer, 1 mile round trip, 2.5 loads/hr.	B-34D	325	.025			.65	1.29	1.94	2.48	
	1240	4 mile round trip, 1.5 loads/hr.	″	195	.041			1.09	2.15	3.24	4.13	
	1300	Hauling in medium traffic, add								20%	20%	
	1400	Heavy traffic, add								30%	30%	
	1600	Grading at dump, or embankment if required, by dozer	B-10B	1,000	.012	↓		.37	.86	1.23	1.55	
	1800	Spotter at fill or cut, if required	1 Clab	8	1	Hr.		26		26	43	
	4700	Highway hauling beyond 20 miles, per loaded mile, minimum				Mile					1	
	4750	Maximum				″					2	
520	0010	**FILL**, spread dumped material, no compaction										520
	0020	By dozer, no compaction	B-10B	1,000	.012	C.Y.		.37	.86	1.23	1.55	
	0100	By hand	1 Clab	12	.667	″		17.35		17.35	28.50	
	9000	Minimum labor/equipment charge	″	4	2	Job		52		52	86	
610	0010	**EXCAVATING, TRENCH** or continuous footing, common earth										610
	0020	No sheeting or dewatering included										
	1400	By hand with pick and shovel 2′ to 6′ deep, light soil	1 Clab	8	1	C.Y.		26		26	43	
	1500	Heavy soil	″	4	2			52		52	86	
	1700	For tamping backfilled trenches, air tamp, add	A-1G	100	.080			2.08	.35	2.43	3.84	
	1900	Vibrating plate, add	B-18	180	.133	↓		3.56	.23	3.79	6.15	
	2100	Trim sides and bottom for concrete pours, common earth		1,500	.016	S.F.		.43	.03	.46	.74	
	2300	Hardpan	↓	600	.040	″		1.07	.07	1.14	1.85	
	9000	Minimum labor/equipment charge	1 Clab	4	2	Job		52		52	86	
620	0010	**EXCAVATING, UTILITY TRENCH** Common earth										620
	0050	Trenching with chain trencher, 12 H.P., operator walking										
	0100	4″ wide trench, 12″ deep	B-53	800	.010	L.F.		.32	.07	.39	.59	
	0150	18″ deep		750	.011			.34	.08	.42	.63	
	0200	24″ deep		700	.011			.37	.08	.45	.67	
	0300	6″ wide trench, 12″ deep		650	.012			.40	.09	.49	.72	
	0350	18″ deep		600	.013			.43	.10	.53	.79	
	0400	24″ deep		550	.015			.47	.11	.58	.86	
	0450	36″ deep		450	.018			.57	.13	.70	1.05	
	0600	8″ wide trench, 12″ deep		475	.017			.54	.13	.67	.99	
	0650	18″ deep		400	.020			.64	.15	.79	1.17	
	0700	24″ deep	↓	350	.023	↓		.73	.17	.90	1.35	

02315 | Excavation and Fill

		02315 Excavation and Fill	CREW	DAILY OUTPUT	LABOR-HOURS	UNIT	2004 BARE COSTS MAT.	LABOR	EQUIP.	TOTAL	TOTAL INCL O&P	
620	0750	36″ deep	B-53	300	.027	L.F.		.86	.20	1.06	1.57	620
	0900	Minimum labor/equipment charge		2	4	Job		129	29.50	158.50	236	
	1000	Backfill by hand including compaction, add										
	1050	4″ wide trench, 12″ deep	A-1G	800	.010	L.F.		.26	.04	.30	.48	
	1100	18″ deep		530	.015			.39	.07	.46	.72	
	1150	24″ deep		400	.020			.52	.09	.61	.96	
	1300	6″ wide trench, 12″ deep		540	.015			.39	.07	.46	.71	
	1350	18″ deep		405	.020			.51	.09	.60	.95	
	1400	24″ deep		270	.030			.77	.13	.90	1.42	
	1450	36″ deep		180	.044			1.16	.20	1.36	2.14	
	1600	8″ wide trench, 12″ deep		400	.020			.52	.09	.61	.96	
	1650	18″ deep		265	.030			.78	.13	.91	1.45	
	1700	24″ deep		200	.040			1.04	.18	1.22	1.91	
	1750	36″ deep		135	.059			1.54	.26	1.80	2.84	
	2000	Chain trencher, 40 H.P. operator riding										
	2050	6″ wide trench and backfill, 12″ deep	B-54	1,200	.007	L.F.		.21	.17	.38	.53	
	2100	18″ deep		1,000	.008			.26	.21	.47	.64	
	2150	24″ deep		975	.008			.26	.21	.47	.66	
	2200	36″ deep		900	.009			.29	.23	.52	.71	
	2250	48″ deep		750	.011			.34	.28	.62	.85	
	2300	60″ deep		650	.012			.40	.32	.72	.97	
	2400	8″ wide trench and backfill, 12″ deep		1,000	.008			.26	.21	.47	.64	
	2450	18″ deep		950	.008			.27	.22	.49	.67	
	2500	24″ deep		900	.009			.29	.23	.52	.71	
	2550	36″ deep		800	.010			.32	.26	.58	.80	
	2600	48″ deep		650	.012			.40	.32	.72	.97	
	2700	12″ wide trench and backfill, 12″ deep		975	.008			.26	.21	.47	.66	
	2750	18″ deep		860	.009			.30	.24	.54	.74	
	2800	24″ deep		800	.010			.32	.26	.58	.80	
	2850	36″ deep		725	.011			.35	.29	.64	.88	
	3000	16″ wide trench and backfill, 12″ deep		835	.010			.31	.25	.56	.76	
	3050	18″ deep		750	.011			.34	.28	.62	.85	
	3100	24″ deep		700	.011			.37	.30	.67	.91	
	3200	Compaction with vibratory plate, add								50%	50%	
	5100	Hand excavate and trim for pipe bells after trench excavation										
	5200	8″ pipe	1 Clab	155	.052	L.F.		1.34		1.34	2.22	
	5300	18″ pipe	″	130	.062	″		1.60		1.60	2.65	
	9000	Minimum labor/equipment charge	A-1G	4	2	Job		52	8.85	60.85	96	
640	0010	**UTILITY BEDDING** For pipe and conduit, not incl. compaction										640
	0050	Crushed or screened bank run gravel	B-6	150	.160	C.Y.	19.25	4.49	1.39	25.13	30	
	0100	Crushed stone 3/4″ to 1/2″		150	.160		24.50	4.49	1.39	30.38	36	
	0200	Sand, dead or bank		150	.160		3.93	4.49	1.39	9.81	13.15	
	0500	Compacting bedding in trench	A-1D	90	.089			2.31	.32	2.63	4.18	

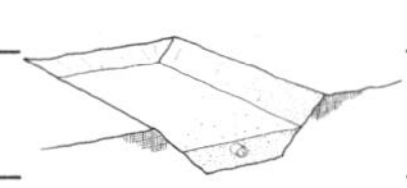

02360 | Soil Treatment

		02360 Soil Treatment	CREW	DAILY OUTPUT	LABOR-HOURS	UNIT	MAT.	LABOR	EQUIP.	TOTAL	TOTAL INCL O&P	
200	0010	**TERMITE PRETREATMENT**										200
	0020	Slab and walls, residential	1 Skwk	1,200	.007	SF Flr.	.26	.22		.48	.66	
	0100	Commercial, minimum		2,496	.003		.28	.11		.39	.49	
	0200	Maximum		1,645	.005		.42	.16		.58	.73	
	0390	Minimum labor/equipment charge		4	2	Job		67.50		67.50	110	
	0400	Insecticides for termite control, minimum		14.20	.563	Gal.	11.10	18.95		30.05	43.50	
	0500	Maximum		11	.727	″	19	24.50		43.50	61	

02300 | Earthwork

		02370 | Erosion & Sedimentation Control	CREW	DAILY OUTPUT	LABOR-HOURS	UNIT	2004 BARE COSTS MAT.	LABOR	EQUIP.	TOTAL	TOTAL INCL O&P	
700	0010	**SYNTHETIC EROSION CONTROL**										700
	0020	Jute mesh, 100 SY per roll, 4' wide, stapled	B-80A	2,400	.010	S.Y.	.66	.26	.07	.99	1.23	
	0100	Plastic netting, stapled, 2" x 1" mesh, 20 mil	B-1	2,500	.010		.60	.26		.86	1.08	
	0200	Polypropylene mesh, stapled, 6.5 oz./S.Y.		2,500	.010		1.24	.26		1.50	1.78	
	0300	Tobacco netting, or jute mesh #2, stapled		2,500	.010		.07	.26		.33	.50	
	1000	Silt fence, polypropylene, 3' high, ideal conditions	2 Clab	1,600	.010	L.F.	.30	.26		.56	.76	
	1100	Adverse conditions	"	950	.017	"	.30	.44		.74	1.06	
	1200	Place and remove hay bales	A-2	3	8	Ton	52	207	40.50	299.50	440	
	1250	Hay bales, staked	"	2,500	.010	L.F.	2.08	.25	.05	2.38	2.75	

02400 | Tunneling, Boring & Jacking

		02441 | Microtunneling	CREW	DAILY OUTPUT	LABOR-HOURS	UNIT	2004 BARE COSTS MAT.	LABOR	EQUIP.	TOTAL	TOTAL INCL O&P	
400	0010	**MICROTUNNELING** Not including excavation, backfill, shoring,										400
	0020	or dewatering, average 50'/day, slurry method										
	0100	24" to 48" outside diameter, minimum				L.F.					640	
	0110	Adverse conditions, add				%					50%	
	1000	Rent microtunneling machine, average monthly lease				Month					85,500	
	1010	Operating technician				Day					640	
	1100	Mobilization and demobilization, minimum				Job					42,800	
	1110	Maximum				"					430,000	
		02445 | Boring or Jacking Conduits										
300	0010	**HORIZONTAL BORING** Casing only, 100' minimum,										300
	0020	not incl. jacking pits or dewatering										
	0100	Roadwork, 1/2" thick wall, 24" diameter casing	B-42	20	3.200	L.F.	51.50	93.50	54	199	273	
	0200	36" diameter		16	4		82	117	67.50	266.50	360	
	0300	48" diameter		15	4.267		120	124	72	316	420	
	0500	Railroad work, 24" diameter		15	4.267		51.50	124	72	247.50	345	
	0600	36" diameter		14	4.571		82	133	77	292	395	
	0700	48" diameter		12	5.333		120	155	90	365	490	
	0900	For ledge, add								155	190	
	1000	Small diameter boring, 3", sandy soil	B-82	900	.018		14.20	.52	.07	14.79	16.55	
	1040	Rocky soil	"	500	.032		14.20	.93	.13	15.26	17.30	

02450 | Foundation & Load Bearing Elements

		02455 | Driven Piles	CREW	DAILY OUTPUT	LABOR-HOURS	UNIT	2004 BARE COSTS MAT.	LABOR	EQUIP.	TOTAL	TOTAL INCL O&P	
100	0010	**CAST IN PLACE CONCRETE PILES**, 200 piles, 60' long										100
	0020	unless specified otherwise, not incl. pile caps or mobilization										
	0800	Cast in place friction pile, 50' long, fluted,										
	0810	tapered steel, 4000 psi concrete, no reinforcing										

02450 | Foundation & Load Bearing Elements

02455 | Driven Piles

			CREW	DAILY OUTPUT	LABOR-HOURS	UNIT	2004 BARE COSTS MAT.	LABOR	EQUIP.	TOTAL	TOTAL INCL O&P
100	0900	12" diameter, 7 ga.	B-19	600	.107	V.L.F.	13.05	3.48	2.62	19.15	23
	1200	18" diameter, 7 ga.	"	480	.133	"	19.60	4.35	3.27	27.22	32.50
	1300	End bearing, fluted, constant diameter,									
	1320	4000 psi concrete, no reinforcing									
	1340	12" diameter, 7 ga.	B-19	600	.107	V.L.F.	13.65	3.48	2.62	19.75	23.50
	1400	18" diameter, 7 ga.	"	480	.133	"	22	4.35	3.27	29.62	35
450	0010	**PRESTRESSED CONCRETE PILES**, 200 piles									
	0020	unless specified otherwise, not incl. pile caps or mobilization									
	3100	Precast, prestressed, 40' long, 10" thick, square	B-19	700	.091	V.L.F.	7.55	2.98	2.24	12.77	15.70
	3200	12" thick, square		680	.094		9.50	3.07	2.31	14.88	18.10
	3400	14" thick, square		600	.107		11.20	3.48	2.62	17.30	21
	4000	18" thick, square	B-19A	520	.123		21.50	4.02	3.87	29.39	34.50
600	0010	**STEEL PILES**, Not including mobilization or demobilization									
	0100	Step tapered, round, concrete filled									
	0110	8" tip, 60 ton capacity, 30' depth	B-19	760	.084	V.L.F.	5.30	2.75	2.06	10.11	12.65
	0120	60' depth		740	.086		6	2.82	2.12	10.94	13.60
	0250	"H" Sections, 50' long, HP8 x 36		640	.100		8.65	3.26	2.45	14.36	17.60
	1300	HP14 X 102	B-19A	510	.125		25	4.09	3.94	33.03	38.50
	2600	Pipe piles, 50' lg. 8" diam., 29 lb. per L.F., no concrete	B-19	500	.128		9.85	4.18	3.14	17.17	21
	2700	Concrete filled		460	.139		10.60	4.54	3.41	18.55	23
	3500	14" diameter, 46 lb. per L.F., no concrete		430	.149		16.30	4.86	3.65	24.81	30
	3600	Concrete filled		355	.180		17.85	5.90	4.42	28.17	34
	4100	18" diameter, 59 lb. per L.F., no concrete		355	.180		23.50	5.90	4.42	33.82	40.50
	4200	Concrete filled		310	.206		24	6.75	5.05	35.80	43
650	0010	**TIMBER PILES**, Friction or end bearing, not including (R02455-900)									
	0050	mobilization or demobilization									
	0100	Untreated piles, up to 30' long, 12" butts, 8" points	B-19	625	.102	V.L.F.	6	3.34	2.51	11.85	14.90
	0200	30' to 39' long, 12" butts, 8" points	"	700	.091	"	6	2.98	2.24	11.22	14
	0800	Treated piles, 12 lb. per C.F.,									
	0810	friction or end bearing, ASTM class B									
	1000	Up to 30' long, 12" butts, 8" points	B-19	625	.102	V.L.F.	8.85	3.34	2.51	14.70	18.05
	1100	30' to 39' long, 12" butts, 8" points	"	700	.091	"	8.65	2.98	2.24	13.87	16.95
800	0011	**PILING SPECIAL COSTS** pile caps, see Division 03310-240									
	0500	Cutoffs, concrete piles, plain	1 Pile	5.50	1.455	Ea.		46.50		46.50	79.50
	0600	With steel thin shell, add		38	.211			6.75		6.75	11.50
	0700	Steel pile or "H" piles		19	.421			13.50		13.50	23
	0800	Wood piles		38	.211			6.75		6.75	11.50
	1000	Testing, any type piles, test load is twice the design load									
	1050	50 ton design load, 100 ton test				Ea.				15,000	16,050
	1100	100 ton design load, 200 ton test								19,250	20,300
	1200	200 ton design load, 400 ton test								26,200	28,800
900	0010	**MOBILIZATION**									
	0020	Set up & remove, air compressor, 600 C.F.M.	A-5	3.30	5.455	Ea.		142	9.20	151.20	244
	0100	1200 C.F.M.	"	2.20	8.182			212	13.85	225.85	365
	0200	Crane, with pile leads and pile hammer, 75 ton	B-19	.60	106			3,475	2,625	6,100	8,625
	0300	150 ton	"	.36	177			5,800	4,350	10,150	14,400

02465 | Bored Piles

			CREW	DAILY OUTPUT	LABOR-HOURS	UNIT	MAT.	LABOR	EQUIP.	TOTAL	TOTAL INCL O&P
800	0010	**DRILLED CAISSONS** Incl. excav., concrete, 50 lbs. reinf. per C.Y., not									
	0020	incl. mobilization, boulder removal, disposal									
	0100	Open style, machine drilled, to 50' deep, in stable ground, no (R02465-600)									
	0110	casings or ground water, 18" diam., 0.065 C.Y./L.F.	B-43	200	.240	V.L.F.	5.60	6.80	14.40	26.80	33
	0200	24" diameter, 0.116 C.Y./L.F.		190	.253	"	10	7.15	15.15	32.30	39.50
	0210	4' bell diameter , add		20	2.400	Ea.	113	68	144	325	395

02450 | Foundation & Load Bearing Elements

02465	Bored Piles		CREW	DAILY OUTPUT	LABOR-HOURS	UNIT	2004 BARE COSTS MAT.	LABOR	EQUIP.	TOTAL	TOTAL INCL O&P	
800	0500	48″ diameter, 0.465 C.Y./L.F. R02465 -600	B-43	100	.480	V.L.F.	40	13.60	29	82.60	97.50	800
	0510	9′ bell diameter , add	↓	2	24	Ea.	905	680	1,450	3,035	3,675	
	1210	Open style, machine drilled, to 25′ deep, in wet ground, 										
	1220	pulled casing and pumping										
	1400	24″ diameter, 0.116 C.Y./L.F.	B-48	125	.448	V.L.F.	10	12.95	25	47.95	59.50	
	1410	4′ bell diameter , add	″	19.80	2.828	Ea.	113	81.50	158	352.50	430	
	1700	48″ diameter, 0.465 C.Y./L.F.	B-49	55	1.600	V.L.F.	40	48	68.50	156.50	197	
	1710	9′ bell diameter , add	″	3.30	26.667	Ea.	905	800	1,150	2,855	3,550	
	2310	Open style, machine drilled, to 25′ deep, in soft rocks and										
	2320	medium hard shales										
	2500	24″ diameter, 0.116 C.Y./L.F.	B-49	30	2.933	V.L.F.	10	88	125	223	292	
	2510	4′ bell diameter, add		10.90	8.073	Ea.	113	242	345	700	900	
	2800	48″ diameter, 0.465 C.Y./L.F.		10	8.800	V.L.F.	40	264	375	679	890	
	2810	9′ bell diameter , add		1.10	80	Ea.	680	2,400	3,425	6,505	8,400	
	3600	For rock excavation, sockets, add, minimum		120	.733	C.F.		22	31.50	53.50	70.50	
	3650	Average		95	.926			28	39.50	67.50	88.50	
	3700	Maximum	↓	48	1.833	↓		55	78.50	133.50	176	
	3900	For 50′ to 100′ deep, add				V.L.F.				7%	7%	
	4000	For 100′ to 150′ deep, add								25%	25%	
	4100	For 150′ to 200′ deep, add				↓				30%	30%	
	4200	For casings left in place, add				Lb.	.63			.63	.69	
	4300	For other than 50 lb. reinf. per C.Y., add or deduct				″	.68			.68	.75	
	4400	For steel "I" beam cores, add	B-49	8.30	10.602	Ton	1,025	320	455	1,800	2,150	
	4500	Load and haul excess excavation, 2 miles	B-34B	178	.045	C.Y.		1.19	2.68	3.87	4.87	
	4600	For mobilization, 50 mile radius, rig to 36″	B-43	2	24	Ea.		680	1,450	2,130	2,675	
	4650	Rig to 84″	B-48	1.75	32			925	1,800	2,725	3,475	
	4700	For low headroom, add								50%		
	4750	For difficult access, add				↓				25%		

2 SITE CONSTRUCTION

02500 | Utility Services

02510	Water Distribution		CREW	DAILY OUTPUT	LABOR-HOURS	UNIT	2004 BARE COSTS MAT.	LABOR	EQUIP.	TOTAL	TOTAL INCL O&P	
730	0010	**WATER SUPPLY, DUCTILE IRON PIPE** cement lined R02510 -800										730
	0020	Not including excavation or backfill										
	2000	Pipe, class 50 water piping, 18′ lengths R15100 -050										
	2020	Mechanical joint, 4″ diameter	B-21A	200	.200	L.F.	9.60	6.40	2.41	18.41	23.50	
	3000	Tyton, Push-on joint, 4″ diameter		400	.100		7.20	3.20	1.21	11.61	14.40	
	3020	6″ diameter	↓	333.33	.120	↓	8.20	3.84	1.45	13.49	16.75	
	4000	Drill and tap pressurized main (labor only)										
	4100	6″ main, 1″ to 2″ service	Q-1	3	5.333	Ea.		190		190	295	
	4150	8″ main, 1″ to 2″ service	″	2.75	5.818	″		207		207	320	
	4500	Tap and insert gate valve										
	4600	8″ main, 4″ branch	B-21	3.20	8.750	Ea.		263	49.50	312.50	485	
	4651	Piping, drill, tap & insert gate valve, 8″ main, 6″ branch		2.70	10.370			310	59	369	575	
	4800	12″ main, 6″ branch	↓	2.35	11.915			360	67.50	427.50	660	
	8020	6″ diameter	B-20A	12.80	2.500		171	78.50		249.50	315	
	8200	Wye or tee, 4″ diameter		10.67	2.999		150	94		244	315	
	8220	6″ diameter		8.53	3.751		188	118		306	395	
	8398	45° bends, 4″ diameter		16	2		132	62.50		194.50	246	
	8450	Decreaser, 6″ x 4″ diameter	↓	14.22	2.250	↓	102	70.50		172.50	226	

	02510 | Water Distribution		CREW	DAILY OUTPUT	LABOR-HOURS	UNIT	2004 BARE COSTS MAT.	LABOR	EQUIP.	TOTAL	TOTAL INCL O&P	
730	8460	8" x 6" diameter (R02510-800)	B-20A	11.64	2.749	Ea.	209	86		295	365	730
	8550	Butterfly valves with boxes, cast iron										
	8560	4" diameter (R15100-050)	B-20	6	4	Ea.	360	117		477	590	
	9600	Steel sleeve and tap, 4" diameter		3	8		400	234		634	820	
	9620	6" diameter		2	12		470	350		820	1,100	

	02520 | Wells		CREW	DAILY OUTPUT	LABOR-HOURS	UNIT	MAT.	LABOR	EQUIP.	TOTAL	TOTAL INCL O&P	
510	0010	**WELLS & ACCESSORIES**, domestic (R15100-050)										510
	0100	Drilled, 4" to 6" diameter	B-23	120	.333	L.F.		8.80	25.50	34.30	42.50	
	0200	8" diameter	"	95.20	.420	"		11.10	32	43.10	53.50	
	1400	Remove & reset pump, minimum	B-21	4	7	Ea.		210	40	250	390	
	1420	Maximum	"	2	14	"		420	79.50	499.50	780	
	1500	Pumps, installed in wells to 100' deep, 4" submersible										
	1510	1/2 H.P.	Q-1	3.22	4.969	Ea.	320	177		497	625	
	1520	3/4 H.P.		2.66	6.015		375	214		589	750	
	1600	1 H.P.		2.29	6.987		395	249		644	820	
	1700	1-1/2 H.P.	Q-22	1.60	10		1,025	355	385	1,765	2,100	
	1800	2 H.P.		1.33	12.030		1,200	430	460	2,090	2,500	
	1900	3 H.P.		1.14	14.035		1,375	500	535	2,410	2,900	
	2000	5 H.P.		1.14	14.035		1,875	500	535	2,910	3,450	
	2050	Remove and install motor only, 4 H.P.		1.14	14.035		670	500	535	1,705	2,100	
	5000	Wells to 180 ft. deep, 4" submersible, 1 HP	B-21	1.10	25.455		1,725	765	145	2,635	3,300	
	5500	2 HP		1.10	25.455		2,075	765	145	2,985	3,675	
	6000	3 HP		1	28		2,825	840	159	3,824	4,650	
	7000	5 HP		.90	31.111		3,425	935	177	4,537	5,475	
	9000	Minimum labor/equipment charge		1.80	15.556	Job		465	88.50	553.50	860	

	02530 | Sanitary Sewerage		CREW	DAILY OUTPUT	LABOR-HOURS	UNIT	MAT.	LABOR	EQUIP.	TOTAL	TOTAL INCL O&P	
730	0010	**SEWAGE COLLECTION, CONCRETE PIPE** (R02510-810)										730
	0020	Not including excavation or backfill										
	1000	Non-reinforced pipe, extra strength, B&S or T&G joints										
	1010	6" diameter	B-14	265.04	.181	L.F.	4.05	4.95	.78	9.78	13.45	
	1020	8" diameter		224	.214		4.46	5.85	.93	11.24	15.50	
	1040	12" diameter		200	.240		6.10	6.55	1.04	13.69	18.65	
	2000	Reinforced culvert, class 3, no gaskets										
	2010	12" diameter	B-14	210	.229	L.F.	10	6.25	.99	17.24	22.50	
	2020	15" diameter	"	175	.274	"	12.85	7.50	1.19	21.54	27.50	
780	0010	**SEWAGE COLLECTION, POLYVINYL CHLORIDE PIPE**										780
	0020	Not including excavation or backfill										
	2000	10' lengths, S.D.R. 35, B&S, 4" diameter	B-20	375	.064	L.F.	1.75	1.87		3.62	5	
	2040	6" diameter		350	.069		3.13	2		5.13	6.75	
	2080	8" diameter		335	.072		5.25	2.09		7.34	9.25	
	2120	10" diameter	B-21	330	.085		7.95	2.55	.48	10.98	13.45	
	4000	Piping, DWV PVC,no exc/bkfill, 10' L, Sch 40, 4" dia	B-20	375	.064		1.71	1.87		3.58	4.97	
	4010	6" dia		350	.069		3.65	2		5.65	7.30	
	4020	8" dia		335	.072		11.15	2.09		13.24	15.70	

	02540 | Septic Tank Systems		CREW	DAILY OUTPUT	LABOR-HOURS	UNIT	MAT.	LABOR	EQUIP.	TOTAL	TOTAL INCL O&P	
400	0010	**SEPTIC TANKS**										400
	0015	Septic tanks, not incl exc or piping, precast, 1,000 gal	B-21	8	3.500	Ea.	530	105	19.90	654.90	780	
	0020	1,250 gallon		8	3.500		740	105	19.90	864.90	1,000	
	0060	1,500 gallon		7	4		875	120	22.50	1,017.50	1,175	

	02540 | Septic Tank Systems		CREW	DAILY OUTPUT	LABOR-HOURS	UNIT	2004 BARE COSTS MAT.	LABOR	EQUIP.	TOTAL	TOTAL INCL O&P	
400	0100	2,000 gallon	B-21	5	5.600	Ea.	1,050	168	32	1,250	1,450	400
	0140	2,500 gallon		5	5.600		1,400	168	32	1,600	1,850	
	0180	4,000 gallon		4	7		4,100	210	40	4,350	4,900	
	0220	5,000 gal., 4 piece	B-13	3	18.667		6,575	525	208	7,308	8,300	
	0300	15,000 gallon, 4 piece	B-13B	1.70	32.941		13,000	920	535	14,455	16,400	
	0400	25,000 gallon, 4 piece		1.10	50.909		27,500	1,425	830	29,755	33,400	
	0500	40,000 gallon, 4 piece		.80	70		34,400	1,950	1,150	37,500	42,300	
	0520	50,000 gallon, 5 piece	B-13C	.60	93.333		39,500	2,625	2,675	44,800	50,500	
	0540	75,000 gallon, cast in place	C-14C	.25	448		48,100	14,200	103	62,403	76,500	
	0560	100,000 gallon	"	.15	746		59,500	23,600	172	83,272	105,000	
	0600	High density polyethylene, 1,000 gallon	B-21	6	4.667		850	140	26.50	1,016.50	1,200	
	0700	1,500 gallon		4	7		1,100	210	40	1,350	1,600	
	0900	Galley, 4′ x 4′ x 4′		16	1.750		196	52.50	9.95	258.45	310	
	1000	Distribution boxes, concrete, 7 outlets	2 Clab	16	1		99	26		125	152	
	1100	9 outlets	"	8	2		268	52		320	380	
	1150	Leaching field chambers, 13′ x 3′-7″ x 1′-4″, standard	B-13	16	3.500		535	98	39	672	795	
	1200	Heavy duty, 8′ x 4′ x 1′-6″		14	4		370	112	44.50	526.50	640	
	1300	13′ x 3′-9″ x 1′-6″		12	4.667		1,300	131	52	1,483	1,725	
	1350	20′ x 4′ x 1′-6″		5	11.200		865	315	125	1,305	1,600	
	1420	Leaching pit, 6′, dia, 3′ deep complete					590			590	645	
	1600	Leaching pit, 6′-6″ diameter, 6′ deep	B-21	5	5.600		565	168	32	765	935	
	1620	8′ deep		4	7		665	210	40	915	1,125	
	1700	8′ diameter, H-20 load, 6′ deep		4	7		920	210	40	1,170	1,400	
	1720	8′ deep		3	9.333		1,125	280	53	1,458	1,775	
	2000	Velocity reducing pit, precast conc., 6′ diameter, 3′ deep		4.70	5.957		283	179	34	496	640	
	2200	Excavation for septic tank, 3/4 C.Y. backhoe	B-12F	145	.110	C.Y.		3.53	3.16	6.69	9.05	
	2400	4′ trench for disposal field, 3/4 C.Y. backhoe	"	335	.048	L.F.		1.53	1.37	2.90	3.92	
	2600	Gravel fill, run of bank	B-6	150	.160	C.Y.	15.40	4.49	1.39	21.28	26	
	2800	Crushed stone, 3/4″	"	150	.160	"	22.50	4.49	1.39	28.38	33.50	

	02550 | Piped Energy Distribution		CREW	DAILY OUTPUT	LABOR-HOURS	UNIT	MAT.	LABOR	EQUIP.	TOTAL	TOTAL INCL O&P	
450	0010	GAS STATION PRODUCT LINE										450
	0020	Primary containment pipe, fiberglass-reinforced										
	0030	Plastic pipe 15′ & 30′ lengths										
	0040	2″ diameter	Q-6	425	.056	L.F.	3.17	2.10		5.27	6.75	
	0050	3″ diameter		400	.060		4.15	2.23		6.38	8	
	0060	4″ diameter		375	.064		5.35	2.37		7.72	9.60	
	0100	Fittings										
	0110	Elbows, 90° & 45°, bell-ends, 2″	Q-6	24	1	Ea.	32	37		69	92.50	
	0120	3″ diameter		22	1.091		33.50	40.50		74	100	
	0130	4″ diameter		20	1.200		44.50	44.50		89	118	
	0200	Tees, bell ends, 2″		21	1.143		39	42.50		81.50	109	
	0210	3″ diameter		18	1.333		39	49.50		88.50	120	
	0220	4″ diameter		15	1.600		53.50	59.50		113	151	
	0230	Flanges bell ends, 2″		24	1		12.85	37		49.85	71.50	
	0240	3″ diameter		22	1.091		16.20	40.50		56.70	81	
	0250	4″ diameter		20	1.200		22	44.50		66.50	93.50	
	0260	Sleeve couplings, 2″		21	1.143		8.20	42.50		50.70	75	
	0270	3″ diameter		18	1.333		11.60	49.50		61.10	89.50	
	0280	4″ diameter		15	1.600		16.05	59.50		75.55	110	
	0290	Threaded adapters 2″		21	1.143		10.75	42.50		53.25	78	
	0300	3″ diameter		18	1.333		18.85	49.50		68.35	97.50	
	0310	4″ diameter		15	1.600		25.50	59.50		85	120	
	0320	Reducers, 2″		27	.889		14.30	33		47.30	67	
	0330	3″ diameter		22	1.091		16.55	40.50		57.05	81.50	

02550 | Piped Energy Distribution

		02550 Piped Energy Distribution	CREW	DAILY OUTPUT	LABOR-HOURS	UNIT	2004 BARE COSTS MAT.	LABOR	EQUIP.	TOTAL	TOTAL INCL O&P	
450	0340	4″ diameter	Q-6	20	1.200	Ea.	21.50	44.50		66	92.50	450
	1010	Gas station product line for secondary containment (double wall)										
	1100	Fiberglass reinforced plastic pipe 25′ lengths										
	1120	Pipe, plain end, 3″	Q-6	375	.064	L.F.	5.45	2.37		7.82	9.70	
	1130	4″ diameter		350	.069		8.95	2.54		11.49	13.75	
	1140	5″ diameter		325	.074		11.60	2.74		14.34	17	
	1150	6″ diameter		300	.080		11.95	2.97		14.92	17.75	
	1200	Fittings										
	1230	Elbows, 90° & 45°, 3″	Q-6	18	1.333	Ea.	39	49.50		88.50	120	
	1240	4″ diameter		16	1.500		68.50	55.50		124	162	
	1250	5″ diameter		14	1.714		158	63.50		221.50	273	
	1260	6″ diameter		12	2		160	74		234	291	
	1270	Tees, 3″		15	1.600		58	59.50		117.50	156	
	1280	4″ diameter		12	2		85.50	74		159.50	209	
	1290	5″ diameter		9	2.667		173	99		272	345	
	1300	6″ diameter		6	4		180	148		328	430	
	1310	Couplings, 3″		18	1.333		27.50	49.50		77	107	
	1320	4″ diameter		16	1.500		71	55.50		126.50	165	
	1330	5″ diameter		14	1.714		148	63.50		211.50	261	
	1340	6″ diameter		12	2		153	74		227	283	
	1350	Cross-over nipples, 3″		18	1.333		6.30	49.50		55.80	83.50	
	1360	4″ diameter		16	1.500		7.35	55.50		62.85	94.50	
	1370	5″ diameter		14	1.714		10.95	63.50		74.45	111	
	1380	6″ diameter		12	2		11.45	74		85.45	128	
	1400	Telescoping, reducers, concentric 4″ x 3″		18	1.333		21	49.50		70.50	99.50	
	1410	5″ x 4″		17	1.412		54.50	52.50		107	141	
	1420	6″ x 5″		16	1.500		131	55.50		186.50	231	
464	0010	**PIPING, GAS SERVICE & DISTRIBUTION, POLYETHYLENE**										464
	0020	not including excavation or backfill										
	1000	60 psi coils, comp cplg @ 100′, 1/2″ diameter, SDR 9.3	B-20A	608	.053	L.F.	.47	1.65		2.12	3.15	
	1040	1-1/4″ diameter, SDR 11		544	.059		.86	1.84		2.70	3.88	
	1100	2″ diameter, SDR 11		488	.066		1.07	2.05		3.12	4.45	
	1160	3″ diameter, SDR 11		408	.078		2.24	2.46		4.70	6.40	
	1500	60 PSI 40′ joints with coupling, 3″ diameter, SDR 11	B-21A	408	.098		2.24	3.14	1.18	6.56	8.75	
	1540	4″ diameter, SDR 11		352	.114		5.10	3.64	1.37	10.11	12.95	
	1600	6″ diameter, SDR 11		328	.122		15.95	3.90	1.47	21.32	25.50	
	1640	8″ diameter, SDR 11		272	.147		21.50	4.71	1.77	27.98	33.50	
	9000	Minimum labor/equipment charge	B-20	2	12	Job		350		350	580	

02580 | Elec/Communication Structures

		02580 Elec/Communication Structures	CREW	DAILY OUTPUT	LABOR-HOURS	UNIT	MAT.	LABOR	EQUIP.	TOTAL	TOTAL INCL O&P	
300	0010	**UNDERGROUND STRUCTURES**, communication and power										300
	9990	Minimum labor/equipment charge	1 Elec	3.50	2.286	Job		90		90	138	
420	0010	**ELECTRIC & TELEPHONE UNDERGROUND**, Not including excavation										420
	0200	backfill and cast in place concrete										
	4200	Underground duct, banks ready for concrete fill, min. of 7.5″										
	4400	between conduits, ctr. to ctr.(for wire & cable see div. 16120)										
	4600	2 @ 2″ diameter	2 Elec	240	.067	L.F.	1.40	2.63		4.03	5.60	
	4800	4 @ 2″ diameter		120	.133		2.80	5.25		8.05	11.15	
	5600	4 @ 4″ diameter		80	.200		6.05	7.90		13.95	18.75	
	6200	Rigid galvanized steel, 2 @ 2″ diameter		180	.089		10.20	3.50		13.70	16.60	
	6400	4 @ 2″ diameter		90	.178		20.50	7		27.50	33.50	
	7400	4 @ 4″ diameter		34	.471		65.50	18.55		84.05	101	

02600 | Drainage & Containment

02630 | Storm Drainage

		02630 Storm Drainage	CREW	DAILY OUTPUT	LABOR-HOURS	UNIT	2004 BARE COSTS MAT.	LABOR	EQUIP.	TOTAL	TOTAL INCL O&P	
110	0010	**CATCH BASIN GRATES AND FRAMES n**ot including footing, excavation										110
	1600	Frames & covers, C.I., 24″ square, 500 lb.	B-6	7.80	3.077	Ea.	231	86.50	26.50	344	425	
	1700	26″ D shape, 600 lb.	″	7	3.429	″	380	96	29.50	505.50	610	
	3320	Frames and covers, existing, raised for paving, 2″, including										
	3340	row of brick, concrete collar, up to 12″ wide frame	B-6	18	1.333	Ea.	35.50	37.50	11.55	84.55	113	
120	0010	**REMOVE AND REPLACE CATCH BASIN COVER**										120
	0023	Remove catch basin cover	1 Clab	80	.100	Ea.		2.60		2.60	4.31	
	0033	Re-place catch basin cover	″	80	.100	″		2.60		2.60	4.31	
400	0010	**STORM DRAINAGE MANHOLES, FRAMES & COVERS** not including										400
	0020	footing, excavation, backfill (See line items for frame & cover)										
	0050	Brick, 4′ inside diameter, 4′ deep	D-1	1	16	Ea.	290	480		770	1,100	
	0100	6′ deep		.70	22.857		405	685		1,090	1,550	
	0150	8′ deep		.50	32		520	960		1,480	2,125	
	0200	For depths over 8′, add		4	4	V.L.F.	116	120		236	320	
	1110	Precast, 4′ I.D., 4′ deep	B-22	4.10	7.317	Ea.	450	222	58	730	920	
	1120	6′ deep		3	10		580	305	79.50	964.50	1,225	
	1130	8′ deep		2	15		675	455	119	1,249	1,625	
	1140	For depths over 8′, add		16	1.875	V.L.F.	95	57	14.90	166.90	214	
510	0010	**PIPING, STORM DRAINAGE, CORRUGATED METAL** R15100-050										510
	0020	Not including excavation or backfill										
	2000	Corrugated metal pipe, galvanized and coated										
	2020	Bituminous coated with paved invert, 20′ lengths										
	2040	8″ diameter, 16 ga.	B-14	330	.145	L.F.	6.90	3.98	.63	11.51	14.80	
	2080	12″ diameter, 16 ga.	″	210	.229	″	9.90	6.25	.99	17.14	22	

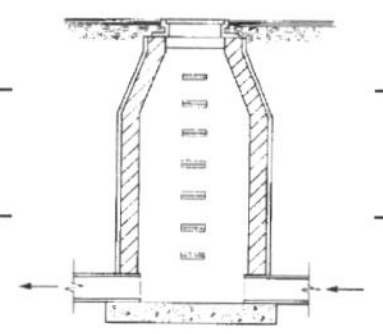

02700 | Bases, Ballasts, Pavements & Appurtenances

02720 | Unbound Base Courses & Ballasts

		02720 Unbound Base Courses & Ballasts	CREW	DAILY OUTPUT	LABOR-HOURS	UNIT	2004 BARE COSTS MAT.	LABOR	EQUIP.	TOTAL	TOTAL INCL O&P	
200	0010	**AGGREGRATE BASE COURSE** For roadways and large paved areas										200
	0050	Crushed 3/4″ stone base, compacted, 3″ deep	B-36C	5,200	.008	S.Y.	2.60	.24	.46	3.30	3.75	
	0100	6″ deep		5,000	.008		5.20	.25	.48	5.93	6.70	
	0200	9″ deep		4,600	.009		7.80	.27	.53	8.60	9.60	
	0300	12″ deep		4,200	.010		10.45	.30	.58	11.33	12.55	
	0301	Crushed 1-1/2″ stone base, compacted to 4″ deep	B-36B	6,000	.011		3.97	.32	.46	4.75	5.40	
	0302	6″ deep		5,400	.012		5.95	.36	.51	6.82	7.70	
	0303	8″ deep		4,500	.014		7.95	.43	.62	9	10.10	
	0304	12″ deep		3,800	.017		11.90	.51	.73	13.14	14.70	
	0310	Minimum labor/equipment charge	B-36	1	40	Job		1,175	1,100	2,275	3,100	
	0350	Bank run gravel, spread and compacted										
	0370	6″ deep	B-32	6,000	.005	S.Y.	3.30	.17	.25	3.72	4.17	
	0390	9″ deep		4,900	.007		4.95	.21	.30	5.46	6.10	
	0400	12″ deep		4,200	.008		6.60	.24	.35	7.19	8.05	
	6900	For small and irregular areas, add						50%	50%			
	6990	Minimum labor/equipment charge	1 Clab	3	2.667	Job		69.50		69.50	115	
	7000	Prepare and roll sub-base, small areas to 2500 S.Y.	B-32A	1,500	.016	S.Y.		.50	.58	1.08	1.44	
	8000	Large areas over 2500 S.Y.	B-32	3,700	.009	″		.27	.40	.67	.88	
	9000	Minimum labor/equipment charge	1 Clab	4	2	Job		52		52	86	

		02740 | Flexible Pavement	Crew	Daily Output	Labor-Hours	Unit	2004 Bare Costs Mat.	Labor	Equip.	Total	Total Incl O&P	
310	0010	**ASPHALTIC CONCRETE PAVEMENT, HIGHWAYS** (R02065-300)										310
	0020	and large paved areas										
	0080	Binder course, 1-1/2" thick	B-25	7,725	.011	S.Y.	2.16	.32	.24	2.72	3.17	
	0120	2" thick		6,345	.014		2.88	.39	.30	3.57	4.13	
	0160	3" thick		4,905	.018		4.26	.51	.38	5.15	5.95	
	0300	Wearing course, 1" thick 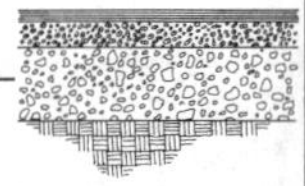	B-25B	10,575	.009		1.59	.26	.20	2.05	2.39	
	0340	1-1/2" thick		7,725	.012		2.42	.36	.27	3.05	3.53	
	0380	2" thick		6,345	.015		3.25	.43	.33	4.01	4.65	
	0420	2-1/2" thick		5,480	.018		4.01	.50	.38	4.89	5.65	
	0460	3" thick		4,900	.020		4.78	.56	.42	5.76	6.60	
	0500	Open graded friction course	B-25C	5,000	.010		1.58	.28	.33	2.19	2.55	
	0800	Alternate method of figuring paving costs										
	0810	Binder course, 1-1/2" thick	B-25	630	.140	Ton	28.50	3.95	2.99	35.44	40.50	
	0811	2" thick		690	.128		28.50	3.61	2.73	34.84	40	
	0812	3" thick		800	.110		28.50	3.11	2.36	33.97	38.50	
	0813	4" thick		850	.104		28.50	2.93	2.22	33.65	38	
	0850	Wearing course, 1" thick	B-25B	575	.167		29.50	4.80	3.60	37.90	44.50	
	0851	1-1/2" thick		630	.152		29.50	4.38	3.28	37.16	43	
	0852	2" thick		690	.139		29.50	4	3	36.50	42.50	
	0853	2-1/2" thick		745	.129		29.50	3.70	2.78	35.98	41.50	
	0854	3" thick		800	.120		29.50	3.45	2.59	35.54	41	
	1000	Pavement replacement over trench, 2" thick	B-37	90	.533	S.Y.	3.31	14.60	1.20	19.11	29	
	1050	4" thick		70	.686		6.55	18.75	1.54	26.84	40	
	1080	6" thick		55	.873		10.40	24	1.96	36.36	52.50	
315	0011	**ASPHALTIC CONCRETE PAVEMENT, LOTS & DRIVEWAYS**										315
	0020	6" stone base, 2" binder course, 1" topping	B-25C	9,000	.005	S.F.	1.30	.15	.18	1.63	1.88	
	0300	Binder course, 1-1/2" thick		35,000	.001		.29	.04	.05	.38	.43	
	0400	2" thick		25,000	.002		.37	.06	.07	.50	.57	
	0500	3" thick		15,000	.003		.57	.09	.11	.77	.90	
	0600	4" thick		10,800	.004		.75	.13	.15	1.03	1.21	
	0800	Sand finish course, 3/4" thick		41,000	.001		.18	.03	.04	.25	.30	
	0900	1" thick		34,000	.001		.22	.04	.05	.31	.36	
	1000	Fill pot holes, hot mix, 2" thick	B-16	4,200	.008		.40	.20	.11	.71	.89	
	1100	4" thick		3,500	.009		.58	.24	.14	.96	1.19	
	1120	6" thick		3,100	.010		.78	.27	.15	1.20	1.48	
	1140	Cold patch, 2" thick	B-51	3,000	.016		.48	.42	.04	.94	1.25	
	1160	4" thick		2,700	.018		.90	.47	.05	1.42	1.82	
	1180	6" thick		1,900	.025		1.41	.66	.06	2.13	2.72	
		02750 | Rigid Pavement										
300	0010	**PLAIN CEMENT CONCRETE PAVEMENT** (R02065-300)										300
	0015	Including joints, finishing and curing										
	0020	Fixed form, 12' pass, unreinforced, 6" thick	B-26	3,000	.029	S.Y.	18	.85	.67	19.52	22	
	0100	8" thick		2,750	.032		25	.93	.73	26.66	30	
	0400	12" thick		1,800	.049		33.50	1.42	1.12	36.04	40.50	
	0505	Minimum labor/equipment charge	1 Cefi	1	8	Job		252		252	395	
	0700	Finishing, broom finish small areas	2 Cefi	120	.133	S.Y.		4.21		4.21	6.60	
	1000	Curing, with sprayed membrane by hand	2 Clab	1,500	.011	"	.47	.28		.75	.97	
		02760 | Paving Specialties										
300	0010	**PAINTED TRAFFIC LINES AND MARKINGS** (R02065-300)										300
	0020	Acrylic waterborne, white or yellow, 4" wide	B-78	20,000	.002	L.F.	.15	.06	.02	.23	.29	
	0200	6" wide		11,000	.004		.14	.11	.03	.28	.38	
	0500	8" wide		10,000	.005		.20	.13	.04	.37	.46	
	0600	12" wide		4,000	.012		.36	.32	.09	.77	1.02	
	0620	Arrows or gore lines		2,300	.021	S.F.	.56	.55	.16	1.27	1.71	

	02760 | Paving Specialties		CREW	DAILY OUTPUT	LABOR-HOURS	UNIT	2004 BARE COSTS MAT.	LABOR	EQUIP.	TOTAL	TOTAL INCL O&P	
300	0640	Temporary paint, white or yellow R02065-300	B-78	15,000	.003	L.F.	.17	.08	.02	.27	.36	300
	0660	Removal	1 Clab	300	.027			.69		.69	1.15	
	0680	Temporary tape	2 Clab	1,500	.011		1.69	.28		1.97	2.32	
	0710	Thermoplastic, white or yellow, 4" wide	B-79	15,000	.003		.64	.07	.06	.77	.88	
	0730	6" wide		14,000	.003		.93	.08	.06	1.07	1.21	
	0740	8" wide		12,000	.003		1.25	.09	.07	1.41	1.59	
	0750	12" wide		6,000	.007		1.86	.18	.14	2.18	2.49	
	0760	Arrows		660	.061	S.F.	1.71	1.60	1.30	4.61	5.95	
	0770	Gore lines		2,500	.016		1.13	.42	.34	1.89	2.32	
	0780	Letters		660	.061		1.42	1.60	1.30	4.32	5.65	
500	0010	PAVEMENT MARKINGS R02065-300										500
	0790	Layout of pavement marking	A-2	25,000	.001	L.F.		.02		.02	.05	
	0800	Parking stall, paint, white	B-78	440	.109	Stall	3.03	2.87	.84	6.74	9	
	1000	Street letters and numbers	"	1,600	.030	S.F.	.58	.79	.23	1.60	2.19	

	02770 | Curbs and Gutters		CREW	DAILY OUTPUT	LABOR-HOURS	UNIT	MAT.	LABOR	EQUIP.	TOTAL	TOTAL INCL O&P	
100	0010	BITUMINOUS CONCRETE CURBS										100
	0012	Curbs, asphaltic, machine formed, 8" wide, 6" high, 40 L.F./ton	B-27	1,000	.032	L.F.	.59	.85	.19	1.63	2.27	
	0100	8" wide, 8" high, 30 L.F. per ton		900	.036		.68	.94	.21	1.83	2.54	
	0150	Asphaltic berm, 12" W, 3"-6" H, 35 L.F./ton, before pavement		700	.046		.91	1.21	.27	2.39	3.31	
	0200	12" W, 1-1/2" to 4" H, 60 L.F. per ton, laid with pavement	B-2	1,050	.038		.55	1.01		1.56	2.28	
300	0010	CEMENT CONCRETE CURBS										300
	0300	Concrete, wood forms, 6" x 18", straight	C-2A	500	.096	L.F.	2.10	3.07		5.17	7.35	
	0400	6" x 18", radius	"	200	.240		2.20	7.65		9.85	15	
	0415	Machine formed, 6" x 18", straight	B-69A	2,000	.024		3.44	.68	.29	4.41	5.20	
	0416	6" x 18", radius	"	900	.053		3.59	1.52	.64	5.75	7.10	
	0550	Precast, 6" x 18", straight	B-29	700	.080		7	2.24	1.18	10.42	12.65	
	0600	6" x 18", radius	"	325	.172		8	4.82	2.53	15.35	19.45	
500	0010	STONE CURBS										500
	1000	Granite, split face, straight, 5" x 16"	D-13	500	.096	L.F.	9.35	3.03	.96	13.34	16.20	
	1100	6" x 18"	"	450	.107		12.25	3.37	1.07	16.69	20	
	1300	Radius curbing, 6" x 18", over 10' radius	B-29	260	.215		15	6.05	3.17	24.22	30	
	1400	Corners, 2' radius		80	.700	Ea.	50.50	19.60	10.30	80.40	99	
	1600	Edging, 4-1/2" x 12", straight		300	.187	L.F.	4.67	5.25	2.74	12.66	16.65	
	1800	Curb inlets, (guttermouth) straight		41	1.366	Ea.	112	38	20	170	208	
	2000	Indian granite (belgian block)										
	2100	Jumbo, 10-1/2" x 7-1/2" x 4", grey	D-1	150	.107	L.F.	1.54	3.19		4.73	6.90	
	2150	Pink		150	.107		2.06	3.19		5.25	7.45	
	2200	Regular, 9" x 4-1/2" x 4-1/2", grey		160	.100		1.42	2.99		4.41	6.40	
	2250	Pink		160	.100		2	2.99		4.99	7.05	
	2300	Cubes, 4" x 4" x 4", grey		175	.091		1.40	2.74		4.14	6	
	2350	Pink		175	.091		1.48	2.74		4.22	6.05	
	2400	6" x 6" x 6", pink		155	.103		3.60	3.09		6.69	8.95	
	2500	Alternate pricing method for indian granite										
	2550	Jumbo, 10-1/2" x 7-1/2" x 4" (30lb), grey				Ton	88			88	97	
	2600	Pink					120			120	132	
	2650	Regular, 9" x 4-1/2" x 4-1/2" (20lb), grey					100			100	110	
	2700	Pink					140			140	154	
	2750	Cubes, 4" x 4" x 4" (5lb), grey					170			170	187	
	2800	Pink					190			190	209	
	2850	6" x 6" x 6" (25lb), pink					140			140	154	
	2900	For pallets, add					16.50			16.50	18.15	

02700 | Bases, Ballasts, Pavements & Appurtenances

		02775 | Sidewalks	CREW	DAILY OUTPUT	LABOR-HOURS	UNIT	2004 BARE COSTS MAT.	LABOR	EQUIP.	TOTAL	TOTAL INCL O&P	
275	0010	**SIDEWALKS, DRIVEWAYS, & PATIOS** No base										275
	0020	Asphaltic concrete, 2" thick	B-37	720	.067	S.Y.	3.35	1.82	.15	5.32	6.85	
	0100	2-1/2" thick	"	660	.073	"	4.24	1.99	.16	6.39	8.10	
	0110	Bedding for brick or stone, mortar, 1" thick	D-1	300	.053	S.F.	.35	1.60		1.95	2.97	
	0120	2" thick	"	200	.080		.87	2.39		3.26	4.85	
	0130	Sand, 2" thick	B-18	8,000	.003		.15	.08	.01	.24	.30	
	0140	4" thick	"	4,000	.006		.29	.16	.01	.46	.60	
	0300	Concrete, 3000 psi, CIP, 6 x 6 - W1.4 x W1.4 mesh,										
	0310	broomed finish, no base, 4" thick	B-24	600	.040	S.F.	1.21	1.21		2.42	3.30	
	0350	5" thick		545	.044		1.61	1.33		2.94	3.93	
	0400	6" thick		510	.047		1.88	1.42		3.30	4.38	
	0450	For bank run gravel base, 4" thick, add	B-18	2,500	.010		.39	.26	.02	.67	.87	
	0520	8" thick, add	"	1,600	.015		.80	.40	.03	1.23	1.57	
	0550	Exposed aggregate finish, add to above, minimum	B-24	1,875	.013		.10	.39		.49	.74	
	0600	Maximum	"	455	.053		.33	1.59		1.92	2.95	
	0950	Concrete tree grate, 5' square	B-6	25	.960	Ea.	305	27	8.30	340.30	390	
	0960	Cast iron tree grate with frame, 2 piece, round, 5' diameter		25	.960		835	27	8.30	870.30	975	
	0980	Square, 5' side		25	.960		865	27	8.30	900.30	1,000	
	1700	Redwood, prefabricated, 4' x 4' sections	2 Carp	316	.051	S.F.	8.10	1.67		9.77	11.65	
	1750	Redwood planks, 1" thick, on sleepers	"	240	.067	"	5.65	2.20		7.85	9.90	
	2250	Stone dust, 4" thick	B-62	900	.027	S.Y.	2.74	.75	.16	3.65	4.42	
	9000	Minimum labor/equipment charge	D-1	2	8	Job		239		239	390	
		02780 | Unit Pavers										
100	0010	**ASPHALT BLOCKS**										100
	0020	Asphalt blocks, 6" x 12" x 1-1/4", W/bed & neopr. adhesive	D-1	135	.119	S.F.	4.02	3.55		7.57	10.15	
	0100	3" thick		130	.123		5.65	3.68		9.33	12.20	
	0300	Hexagonal tile, 8" wide, 1-1/4" thick		135	.119		4.02	3.55		7.57	10.15	
	0400	2" thick		130	.123		5.65	3.68		9.33	12.20	
	0500	Square, 8" x 8", 1-1/4" thick		135	.119		4.02	3.55		7.57	10.15	
	0600	2" thick		130	.123		5.65	3.68		9.33	12.20	
	9000	Minimum labor/equipment charge	1 Bric	2	4	Job		137		137	222	
200	0010	**BRICK PAVING** 4" x 8" x 1-1/2", without joints (4.5 brick/S.F.)	D-1	110	.145	S.F.	2.20	4.35		6.55	9.50	200
	0100	Grouted, 3/8" joint (3.9 brick/S.F.)		90	.178		2.55	5.30		7.85	11.45	
	0200	4" x 8" x 2-1/4", without joints (4.5 bricks/S.F.)		110	.145		2.82	4.35		7.17	10.15	
	0300	Grouted, 3/8" joint (3.9 brick/S.F.)		90	.178		2.60	5.30		7.90	11.50	
	0500	Bedding, asphalt, 3/4" thick	B-25	5,130	.017		.32	.48	.37	1.17	1.54	
	0540	Course washed sand bed, 1" thick	B-18	5,000	.005		.16	.13	.01	.30	.40	
	0580	Mortar, 1" thick	D-1	300	.053		.29	1.60		1.89	2.91	
	0620	2" thick		200	.080		.58	2.39		2.97	4.53	
	1500	Brick on 1" thick sand bed laid flat, 4.5 per S.F.		100	.160		2.38	4.79		7.17	10.35	
	2000	Brick pavers, laid on edge, 7.2 per S.F.		70	.229		2.16	6.85		9.01	13.45	
	2500	For 4" thick concrete bed and joints, add		595	.027		.96	.80		1.76	2.37	
	2800	For steam cleaning, add	A-1H	950	.008		.05	.22	.06	.33	.48	
	9000	Minimum labor/equipment charge	1 Bric	2	4	Job		137		137	222	
600	0010	**STONE PAVERS**										600
	1300	Slate, natural cleft, irregular, 3/4" thick	D-1	92	.174	S.F.	5.15	5.20		10.35	14.15	
	1350	Random rectangular, gauged, 1/2" thick		105	.152		11.15	4.56		15.71	19.70	
	1400	Random rectangular, butt joint, gauged, 1/4" thick		150	.107		12	3.19		15.19	18.45	
	1450	For sand rubbed finish, add					5.60			5.60	6.15	
	1550	Granite blocks, 3-1/2" x 3-1/2" x 3-1/2"	D-1	92	.174		6.25	5.20		11.45	15.35	
	1600	4" to 12" long, 3" to 5" wide, 3" to 5" thick	"	98	.163		5.20	4.89		10.09	13.70	
700	0010	**STEPS** Incl. excav., borrow & concrete base, where applicable										700
	0100	Brick steps	B-24	35	.686	LF Riser	8.35	20.50		28.85	42.50	

	02780 | Unit Pavers		CREW	DAILY OUTPUT	LABOR-HOURS	UNIT	2004 BARE COSTS MAT.	LABOR	EQUIP.	TOTAL	TOTAL INCL O&P	
700	0200	Railroad ties	2 Clab	25	.640	LF Riser	2.83	16.65		19.48	30.50	700
	0300	Bluestone treads, 12″ x 2″ or 12″ x 1-1/2″	B-24	30	.800	↓	21	24		45	63	
	0490	Minimum labor/equipment charge	D-1	2	8	Job		239		239	390	
	0500	Concrete, cast in place, see division 03310-240										
	0600	Precast concrete, see division 03480-800										

	02785 | Flexible Pavement Coating		CREW	DAILY OUTPUT	LABOR-HOURS	UNIT	MAT.	LABOR	EQUIP.	TOTAL	TOTAL INCL O&P	
250	0010	**FOG SEAL**										250
	0012	Sealcoating, 2 coat coal tar pitch emulsion over 10,000 SY	B-45	5,000	.003	S.Y.	.44	.10	.11	.65	.75	
	0030	1000 to 10,000 S.Y.	"	3,000	.005		.44	.16	.18	.78	.94	
	0100	Under 1000 S.Y.	B-1	1,050	.023		.44	.61		1.05	1.49	
	0300	Petroleum resistant, over 10,000 S.Y.	B-45	5,000	.003		.54	.10	.11	.75	.86	
	0320	1000 to 10,000 S.Y.	"	3,000	.005		.54	.16	.18	.88	1.05	
	0400	Under 1000 S.Y.	B-1	1,050	.023		.54	.61		1.15	1.60	
	0600	Non-skid pavement renewal, over 10,000 S.Y.	B-45	5,000	.003		.64	.10	.11	.85	.97	
	0620	1000 to 10,000 S.Y.	"	3,000	.005		.64	.16	.18	.98	1.16	
	0700	Under 1000 S.Y.	B-1	1,050	.023		.64	.61		1.25	1.71	
	0800	Prepare and clean surface for above	A-2	8,545	.003	↓		.07	.01	.08	.14	
	1000	Hand seal asphalt curbing	B-1	4,420	.005	L.F.	.32	.14		.46	.59	
	1900	Asphalt surface treatment, single course, small area										
	1901	0.30 gal/S.Y. asphalt material, 20#/S.Y. aggregate	B-91	5,000	.013	S.Y.	.75	.39	.29	1.43	1.77	
	1910	Roadway or large area		10,000	.006		.69	.19	.14	1.02	1.23	
	1950	Asphalt surface treatment, dbl. course for small area		3,000	.021		1.39	.64	.48	2.51	3.09	
	1960	Roadway or large area		6,000	.011		1.25	.32	.24	1.81	2.16	
	1980	Asphalt surface treatment,single course, for shoulders	↓	7,500	.009	↓	.80	.26	.19	1.25	1.50	
300	0010	**LATEX MODIFIED EMULSION**										300
	3600	Waterproofing, membrane, tar and fabric, small area	B-63	233	.172	S.Y.	5.15	4.67	.61	10.43	14	
	3640	Large area		1,435	.028		4.73	.76	.10	5.59	6.55	
	3680	Preformed rubberized asphalt, small area		100	.400		6.95	10.90	1.43	19.28	27	
	3720	Large area	↓	367	.109		6.35	2.97	.39	9.71	12.25	
	3780	Rubberized asphalt (latex) seal	B-45	5,000	.003	↓	1	.10	.11	1.21	1.37	
	9000	Minimum labor/equipment charge	1 Clab	2	4	Job		104		104	172	
400	0010	**SAND SEAL**										400
	2080	Sand sealing, sharp sand, asphalt emulsion, small area	B-91	10,000	.006	S.Y.	.54	.19	.14	.87	1.06	
	2120	Roadway or large area	"	18,000	.004	"	.46	.11	.08	.65	.77	
	3000	Sealing random cracks, min 1/2″ wide, to 1-1/2″, 1,000 L.F.	B-77	2,800	.014	L.F.	.43	.38	.13	.94	1.23	
	3040	10,000 L.F.		4,000	.010	"	.33	.26	.09	.68	.89	
	3080	Alternate method, 1,000 L.F.		200	.200	Gal.	6.50	5.25	1.81	13.56	17.85	
	3120	10,000 L.F.	↓	325	.123	"	5.25	3.24	1.11	9.60	12.30	
	3200	Multi-cracks (flooding), 1 coat, small area	B-92	460	.070	S.Y.	1.95	1.84	.65	4.44	5.90	
	3240	Large area		2,850	.011		1.74	.30	.10	2.14	2.52	
	3280	2 coat, small area		230	.139		6.45	3.69	1.30	11.44	14.60	
	3320	Large area		1,425	.022	↓	5.85	.60	.21	6.66	7.60	
	3360	Alternate method, small area		115	.278	Gal.	6.70	7.35	2.59	16.64	22.50	
	3400	Large area	↓	715	.045	"	6.30	1.19	.42	7.91	9.40	
500	0010	**SLURRY SEAL**										500
	0100	Slurry seal, type I, 8 lbs agg./S.Y., 1 coat, small or irregular area	B-90	2,800	.023	S.Y.	.74	.64	.65	2.03	2.56	
	0150	Roadway or large area		10,000	.006		.74	.18	.18	1.10	1.30	
	0200	Type II, 12 lbs aggregate/S.Y., 2 coats,small or irregular area		2,000	.032		1.48	.89	.91	3.28	4.08	
	0250	Roadway or large area		8,000	.008		1.48	.22	.23	1.93	2.24	
	0300	Type III, 20 lbs aggregate/S.Y., 2 coats, small or irregular area		1,800	.036		1.75	.99	1.01	3.75	4.64	
	0350	Roadway or large area		6,000	.011		1.75	.30	.30	2.35	2.73	
	0400	Slurry seal, thermoplastic coal-tar, type I, small or irregular area	↓	2,400	.027	↓	1.55	.74	.76	3.05	3.75	

For expanded coverage of these items see *Means Heavy Construction Cost Data 2004*

02700 | Bases, Ballasts, Pavements & Appurtenances

02785 | Flexible Pavement Coating

		02785 Flexible Pavement Coating	CREW	DAILY OUTPUT	LABOR-HOURS	UNIT	2004 BARE COSTS MAT.	LABOR	EQUIP.	TOTAL	TOTAL INCL O&P	
500	0450	Roadway or large area	B-90	8,000	.008	S.Y.	1.55	.22	.23	2	2.32	500
	0500	Type II, small or irregular area		2,400	.027		1.98	.74	.76	3.48	4.22	
	0550	Roadway or large area		7,800	.008		1.98	.23	.23	2.44	2.81	
	0600	Average mobilization cost				Ea.				3,750	3,750	
600	0010	SURFACE TREATMENT										600
	3000	Pavement overlay, polypropylene										
	3040	6 oz. per S.Y., ideal conditions	B-63	10,000	.004	S.Y.	1.07	.11	.01	1.19	1.38	
	3080	Adverse conditions		1,000	.040		1.43	1.09	.14	2.66	3.52	
	3120	4 oz. per S.Y., ideal conditions		10,000	.004		.70	.11	.01	.82	.97	
	3160	Adverse conditions		1,000	.040		.90	1.09	.14	2.13	2.93	
	3200	Tack coat, emulsion, .05 gal per S.Y., 1000 S.Y	B-45	2,500	.006		.20	.19	.22	.61	.77	
	3240	10,000 S.Y.		10,000	.002		.16	.05	.06	.27	.32	
	3270	.10 gal per S.Y., 1000 S.Y.		2,500	.006		.38	.19	.22	.79	.96	
	3275	.10 gal per S.Y., 10,000 S.Y.		10,000	.002		.38	.05	.06	.49	.55	
	3280	.15 gal per S.Y., 1000 S.Y.		2,500	.006		.55	.19	.22	.96	1.16	
	3320	10,000 S.Y.		10,000	.002		.44	.05	.06	.55	.62	

02800 | Site Improvements and Amenities

02810 | Irrigation System

		02810 Irrigation System	CREW	DAILY OUTPUT	LABOR-HOURS	UNIT	2004 BARE COSTS MAT.	LABOR	EQUIP.	TOTAL	TOTAL INCL O&P	
300	0010	**SPRINKLER IRRIGATION SYSTEM** For lawns										300
	0100	Golf course with fully automatic system	C-17	.05	1,600	9 holes	79,000	54,500		133,500	175,500	
	0200	24′ diam. head at 15′ O.C incl. piping, auto oper., minimum	B-20	70	.343	Head	17.35	10		27.35	35.50	
	0300	Maximum		40	.600		40	17.55		57.55	73	
	0500	60′ diam. head at 40′ O.C. incl. piping, auto oper., minimum		28	.857		52.50	25		77.50	99.50	
	0600	Maximum		23	1.043		147	30.50		177.50	213	
	0800	Residential system, custom, 1″ supply		2,000	.012	S.F.	.26	.35		.61	.87	
	0900	1-1/2″ supply		1,800	.013	″	.29	.39		.68	.96	

02820 | Fences & Gates

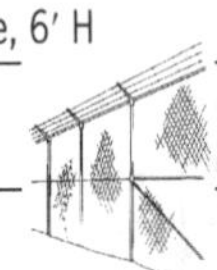

		02820 Fences & Gates	CREW	DAILY OUTPUT	LABOR-HOURS	UNIT	2004 BARE COSTS MAT.	LABOR	EQUIP.	TOTAL	TOTAL INCL O&P	
130	0010	**FENCE, CHAIN LINK INDUSTRIAL,** schedule 40										130
	0020	3 strands barb wire, 2″ post @ 10′ O.C., set in concrete, 6′ H										
	0200	9 ga. wire, galv. steel	B-80	240	.133	L.F.	9.20	3.73	2.16	15.09	18.55	
	0300	Aluminized steel		240	.133		11.80	3.73	2.16	17.69	21.50	
	0500	6 ga. wire, galv. steel		240	.133		14.35	3.73	2.16	20.24	24	
	0600	Aluminized steel		240	.133		16.30	3.73	2.16	22.19	26.50	
	0800	6 ga. wire, 6′ high but omit barbed wire, galv. steel		250	.128		13.95	3.58	2.07	19.60	23.50	
	0900	Aluminized steel		250	.128		19.15	3.58	2.07	24.80	29	
	0920	8′ H, 6 ga. wire, 2-1/2″ line post, galv. steel		180	.178		22	4.97	2.88	29.85	36	
	0940	Aluminized steel		180	.178		27	4.97	2.88	34.85	41.50	
	1100	Add for corner posts, 3″ diam., galv. steel		40	.800	Ea.	63	22.50	12.95	98.45	120	
	1200	Aluminized steel		40	.800		75.50	22.50	12.95	110.95	134	
	1300	Add for braces, galv. steel		80	.400		17.20	11.20	6.45	34.85	44.50	
	1350	Aluminized steel		80	.400		23	11.20	6.45	40.65	51	
	1400	Gate for 6′ high fence, 1-5/8″ frame, 3′ wide, galv. steel		10	3.200		101	89.50	52	242.50	315	
	1500	Aluminized steel		10	3.200		124	89.50	52	265.50	340	
	2000	5′-0″ high fence, 9 ga., no barbed wire, 2″ line post,										
	2010	10′ O.C., 1-5/8″ top rail										
	2100	Galvanized steel	B-80	300	.107	L.F.	8.05	2.98	1.73	12.76	15.60	
	2200	Aluminized steel		300	.107	″	9.25	2.98	1.73	13.96	16.95	

Important: See the Reference Section for critical supporting data - Reference Nos., Crews, & City Cost Indexes

	02820 | Fences & Gates		CREW	DAILY OUTPUT	LABOR-HOURS	UNIT	2004 BARE COSTS MAT.	LABOR	EQUIP.	TOTAL	TOTAL INCL O&P	
130	2400	Gate, 4′ wide, 5′ high, 2″ frame, galv. steel	B-80	10	3.200	Ea.	115	89.50	52	256.50	330	130
	2500	Aluminized steel		10	3.200	″	126	89.50	52	267.50	340	
	3100	Overhead slide gate, chain link, 6′ high, to 18′ wide		38	.842	L.F.	101	23.50	13.65	138.15	165	
	3110	Cantilever type		48	.667		43.50	18.65	10.80	72.95	90.50	
	3120	8′ high		24	1.333		63	37.50	21.50	122	154	
	3130	10′ high		18	1.778		74.50	49.50	29	153	195	
	3150	Tennis courts, 11 ga. wire, 1-3/4″ mesh, 2-1/2″ line posts										
	3170	1-5/8″ top rail, 3″ corner and gate posts										
	3190	10′ high, galvanized steel	B-80	190	.168	L.F.	11.45	4.71	2.73	18.89	23.50	
	3210	Vinyl covered 9 ga. wire		190	.168	″	12.65	4.71	2.73	20.09	24.50	
	3240	Corner posts for above, 3″ diameter, 10′ high		30	1.067	Ea.	120	30	17.25	167.25	201	
	3300	Residential, 11 ga. wire, 1-5/8″ line post @ 10′ O.C.										
	3310	1-3/8″ top rail										
	3350	4′ high, galvanized steel	B-80	475	.067	L.F.	6.30	1.88	1.09	9.27	11.20	
	3400	Aluminized		475	.067	″	6.35	1.88	1.09	9.32	11.25	
	3600	Gate, 3′ wide, 1-3/8″ frame, galv. steel		10	3.200	Ea.	51.50	89.50	52	193	260	
	3700	Aluminized		10	3.200	″	57.50	89.50	52	199	266	
	3900	3′ high, galvanized steel		620	.052	L.F.	3.74	1.44	.84	6.02	7.40	
	4000	Aluminized		620	.052	″	4.59	1.44	.84	6.87	8.30	
	4200	Gate, 3′ wide, 1-3/8″ frame, galv. steel		12	2.667	Ea.	46	74.50	43	163.50	219	
	4300	Aluminized		12	2.667	″	51.50	74.50	43	169	226	
	7001	Snow fence on steel posts 10′ O.C., 4′ high	B-1	500	.048	L.F.	1.69	1.28		2.97	3.98	
	8000	Components only, fabric, galvanized, 4′ high	B-80	585	.055		2.46	1.53	.89	4.88	6.15	
	8040	6′ high		430	.074		3.03	2.08	1.20	6.31	8.05	
	8200	Posts, line post, 4′ high		34	.941	Ea.	8.35	26.50	15.25	50.10	69	
	8240	6′ high		26	1.231	″	9.25	34.50	19.90	63.65	88	
	8400	Top rails, 1-3/8″ O.D.		1,000	.032	L.F.	1.18	.89	.52	2.59	3.33	
	8440	1-5/8″ O.D.		1,000	.032	″	1.32	.89	.52	2.73	3.48	
	9000	Minimum labor/equipment charge		2	16	Job		445	259	704	1,025	
410	0010	**FENCES, MISC. METAL**										410
	0012	Chicken wire, posts @ 4′, 1″ mesh, 4′ high	B-80	410	.078	L.F.	1.25	2.18	1.26	4.69	6.30	
	0100	2″ mesh, 6′ high		350	.091		1.13	2.56	1.48	5.17	7.05	
	0200	Galv. steel, 12 ga., 2″ x 4″ mesh, posts 5′ O.C., 3′ high		300	.107		1.70	2.98	1.73	6.41	8.60	
	0300	5′ high		300	.107		2.27	2.98	1.73	6.98	9.25	
	0400	14 ga., 1″ x 2″ mesh, 3′ high		300	.107		1.81	2.98	1.73	6.52	8.75	
	0500	5′ high		300	.107		2.50	2.98	1.73	7.21	9.50	
	1000	Kennel fencing, 1-1/2″ mesh, 6′ long, 3′-6″ wide, 6′-2″ high	2 Clab	4	4	Ea.	284	104		388	480	
	1050	12′ long		4	4		340	104		444	545	
	1200	Top covers, 1-1/2″ mesh, 6′ long		15	1.067		57.50	27.50		85	110	
	1250	12′ long		12	1.333		92.50	34.50		127	160	
	1300	For kennel doors, see division 08344-350										
	4500	Security fence, prison grade, set in concrete, 12′ high	B-80	25	1.280	L.F.	23.50	36	20.50	80	107	
	4600	16′ high		20	1.600		28	44.50	26	98.50	133	
	5300	Tubular picket, steel, 6′ sections, 1-9/16″ posts, 4′ high		300	.107		17.65	2.98	1.73	22.36	26	
	5400	2″ posts, 5′ high		240	.133		24.50	3.73	2.16	30.39	35.50	
	5600	2″ posts, 6′ high		200	.160		27.50	4.47	2.59	34.56	40.50	
	5700	Staggered picket 1-9/16″ posts, 4′ high		300	.107		15.90	2.98	1.73	20.61	24.50	
	5800	2″ posts, 5′ high		240	.133		26	3.73	2.16	31.89	37.50	
	5900	2″ posts, 6′ high		200	.160		27	4.47	2.59	34.06	40	
	6200	Gates, 4′ high, 3′ wide	B-1	10	2.400	Ea.	153	64		217	275	
	6300	5′ high, 3′ wide		10	2.400		199	64		263	325	
	6400	6′ high, 3′ wide		10	2.400		205	64		269	330	
	6500	4′ wide		10	2.400		239	64		303	370	
520	0010	**FENCE, WOOD RAIL** Picket, No. 2 cedar, Gothic, 2 rail, 3′ high	B-1	160	.150	L.F.	5	4		9	12.15	520
	0050	Gate, 3′-6″ wide		9	2.667	Ea.	43	71		114	166	

				DAILY	LABOR-		2004 BARE COSTS				TOTAL	
		02820 | Fences & Gates	CREW	OUTPUT	HOURS	UNIT	MAT.	LABOR	EQUIP.	TOTAL	INCL O&P	
520	0600	Open rail, rustic, No. 1 cedar, 2 rail, 3′ high	B-1	160	.150	L.F.	4.48	4		8.48	11.60	520
	0650	Gate, 3′ wide		9	2.667	Ea.	50	71		121	173	
	1200	Stockade, No. 2 cedar, treated wood rails, 6′ high		160	.150	L.F.	6.25	4		10.25	13.50	
	1250	Gate, 3′ wide		9	2.667	Ea.	51	71		122	175	
	3300	Board, shadow box, 1″ x 6″, treated pine, 6′ high		160	.150	L.F.	9.20	4		13.20	16.80	
	3400	No. 1 cedar, 6′ high		150	.160		18.20	4.27		22.47	27	
	3900	Basket weave, No. 1 cedar, 6′ high		160	.150		18	4		22	26.50	
	3950	Gate, 3′-6″ wide		8	3	Ea.	115	80		195	260	
	4000	Treated pine, 6′ high		150	.160	L.F.	9.75	4.27		14.02	17.80	
	4200	Gate, 3′-6″ wide		9	2.667	Ea.	54	71		125	177	
	8000	Posts only, 4′ high		40	.600		13.35	16		29.35	41	
	8040	6′ high		30	.800		18.45	21.50		39.95	56	
	9000	Minimum labor/equipment charge	1 Clab	2	4	Job		104		104	172	
	9015	Fence rail, made from recycled plastic, various colors, 2 rail	B-1	150	.160	L.F.	4.06	4.27		8.33	11.50	
	9018	3 rail		150	.160		5.40	4.27		9.67	13	
	9020	4 rail		150	.160		6.55	4.27		10.82	14.25	
	9030	Fence pole, made from recycled plastic, various colors, 7′		96	.250	Ea.	15.60	6.65		22.25	28	
	9040	Stockade fence, made from recycled plastic, various colors, 4′ high		160	.150	L.F.	27	4		31	36	
	9050	6′ high		160	.150	″	33.50	4		37.50	43	
	9060	6′ pole		96	.250	Ea.	16.65	6.65		23.30	29.50	
	9070	9′ pole		96	.250	″	25	6.65		31.65	38.50	
	9080	Picket fence, made from recycled plastic, various colors, 3′ high		160	.150	L.F.	27	4		31	36	
	9090	4′ high		160	.150	″	33.50	4		37.50	43	
	9100	3′ high gate		8	3	Ea.	52	80		132	190	
	9110	4′ high gate		8	3		60.50	80		140.50	200	
	9120	5′ high pole		96	.250		9.35	6.65		16	21.50	
	9130	6′ high pole		96	.250		11.45	6.65		18.10	23.50	
	9140	Pole cap only					3.12			3.12	3.43	
	9150	Keeper pins only					.21			.21	.23	
		02830 | Retaining Walls										
800	0010	**STONE RETAINING WALLS**										800
	0015	Including excavation, concrete footing and										
	0020	stone 3′ below grade. Price is exposed face area.										
	0200	Decorative random stone, to 6′ high, 1′-6″ thick, dry set	D-1	35	.457	S.F.	8	13.70		21.70	31	
	0300	Mortar set		40	.400		9.70	11.95		21.65	30	
	0500	Cut stone, to 6′ high, 1′-6″ thick, dry set		35	.457		12.10	13.70		25.80	35.50	
	0600	Mortar set		40	.400		14.20	11.95		26.15	35	
	0800	Retaining wall, random stone, 6′ to 10′ high, 2′ thick, dry set		45	.356		10	10.65		20.65	28.50	
	0900	Mortar set		50	.320		12.10	9.60		21.70	29	
	1100	Cut stone, 6′ to 10′ high, 2′ thick, dry set		45	.356		15.25	10.65		25.90	34	
	1200	Mortar set		50	.320		16.55	9.60		26.15	34	
	9000	Minimum labor/equipment charge		2	8	Job		239		239	390	
		02840 | Walk/Road/Parking Appurtenances										
200	0010	**GUIDE/GUARD RAIL** Corrugated stl, galv. stl posts, 6′-3″ O.C.	B-80	850	.038	L.F.	12.15	1.05	.61	13.81	15.80	200
	0200	End sections, galvanized, flared		50	.640	Ea.	48.50	17.90	10.35	76.75	94	
	0300	Wrap around end		50	.640	″	73	17.90	10.35	101.25	120	
	0400	Timber guide rail, 4″ x 8″ with 6″ x 8″ wood posts, treated		960	.033	L.F.	15.75	.93	.54	17.22	19.40	
400	0010	**FENDERS**										400
	0015	Bumper rails for garages, 12 Ga. rail, 6″ wide, with steel										
	0020	posts 12′-6″ O.C., minimum	E-4	190	.168	L.F.	9.75	6.35	.40	16.50	23	
	0030	Average		165	.194		12.15	7.30	.46	19.91	27.50	
	0100	Maximum		140	.229		14.60	8.60	.54	23.74	32.50	
	9000	Minimum labor/equipment charge		2	16	Job		600	37.50	637.50	1,175	

Important: See the Reference Section for critical supporting data - Reference Nos., Crews, & City Cost Indexes

	02840 | Walk/Road/Parking Appurtenances		CREW	DAILY OUTPUT	LABOR-HOURS	UNIT	2004 BARE COSTS MAT.	LABOR	EQUIP.	TOTAL	TOTAL INCL O&P	
800	0010	**PARKING BUMPERS**										800
	0020	Parking barriers, timber w/saddles, treated type										
	0100	4″ x 4″ for cars	B-2	520	.077	L.F.	2.64	2.03		4.67	6.30	
	0200	6″ x 6″ for trucks		520	.077	″	5.65	2.03		7.68	9.60	
	1000	Wheel stops, precast concrete incl. dowels, 6″ x 10″ x 6′-0″		120	.333	Ea.	32	8.80		40.80	50	
	1100	8″ x 13″ x 6′-0″		120	.333		37	8.80		45.80	55.50	
	1300	Pipe bollards, conc filled/paint, 8′ L x 4′ D hole, 6″ diam.	B-6	20	1.200		193	33.50	10.40	236.90	278	
	1400	8″ diam.		15	1.600		293	45	13.85	351.85	410	
	1500	12″ diam.		12	2		380	56	17.30	453.30	530	
	1540	Bollards, precast concrete, free standing, round, 15″ Dia x 36″ H		14	1.714		102	48	14.85	164.85	206	
	1550	16″ Dia x 30″ H		14	1.714		102	48	14.85	164.85	206	
	1560	18″ Dia x 34″ H		14	1.714		209	48	14.85	271.85	325	
	1570	18″ Dia x 72″ H		14	1.714		255	48	14.85	317.85	375	
	1572	26″ Dia x 18″ H		12	2		255	56	17.30	328.30	390	
	1574	26″ Dia x 30″ H		12	2		375	56	17.30	448.30	525	
	1578	26″ Dia x 50″ H		12	2		510	56	17.30	583.30	670	
	1580	Square, 22″ x 32″ H		12	2		270	56	17.30	343.30	405	
	1585	22″ x 41″ H		12	2		480	56	17.30	553.30	635	
	1590	Hexagon, 26″ x 41″ H		10	2.400		500	67.50	21	588.50	680	
	1592	Bollards, steel, 3′ H, retractable, incl hydraulic controls, min		4	6		17,900	168	52	18,120	19,900	
	1594	Max		2	12		19,900	335	104	20,339	22,600	
	1600	Bollards, recycled plastic, 5″ x 5″ x 5′ L		20	1.200		37.50	33.50	10.40	81.40	107	
	1610	Wheel stops, recycled plastic, yellow, 4″ x 6″ x 6′ L		20	1.200		37.50	33.50	10.40	81.40	107	
	1620	Speed bumps, recycled plastic, yellow, 3″H x 10″ W x 6′ L		20	1.200		81.50	33.50	10.40	125.40	156	
	1630	3″H x 10″ W x 9′ L		20	1.200		112	33.50	10.40	155.90	189	

	02870 | Site Furnishings		CREW	DAILY OUTPUT	LABOR-HOURS	UNIT	MAT.	LABOR	EQUIP.	TOTAL	TOTAL INCL O&P	
310	0010	**BENCHES**										310
	0012	Seating, benches, park, precast conc, w/backs, wood rails, 4′ long	2 Clab	5	3.200	Ea.	370	83		453	545	
	0100	8′ long		4	4		645	104		749	880	
	0300	Fiberglass, without back, one piece, 4′ long		10	1.600		405	41.50		446.50	520	
	0400	8′ long		7	2.286		840	59.50		899.50	1,025	
	0500	Steel barstock pedestals w/backs, 2″ x 3″ wood rails, 4′ long		10	1.600		770	41.50		811.50	920	
	0510	8′ long		7	2.286		910	59.50		969.50	1,100	
	0520	3″ x 8″ wood plank, 4′ long		10	1.600		775	41.50		816.50	925	
	0530	8′ long		7	2.286		810	59.50		869.50	990	
	0540	Backless, 4″ x 4″ wood plank, 4′ square		10	1.600		755	41.50		796.50	900	
	0550	8′ long		7	2.286		715	59.50		774.50	890	
	0600	Aluminum pedestals, with backs, aluminum slats, 8′ long		8	2		380	52		432	505	
	0610	15′ long		5	3.200		360	83		443	540	
	0620	Portable, aluminum slats, 8′ long		8	2		320	52		372	440	
	0630	15′ long		5	3.200		470	83		553	660	
	0800	Cast iron pedestals, back & arms, wood slats, 4′ long		8	2		300	52		352	415	
	0820	8′ long		5	3.200		825	83		908	1,050	
	0840	Backless, wood slats, 4′ long		8	2		505	52		557	640	
	0860	8′ long		5	3.200		545	83		628	740	
	1700	Steel frame, fir seat, 10′ long		10	1.600		153	41.50		194.50	237	
	9000	Minimum labor/equipment charge		2	8	Job		208		208	345	

	02890 | Traffic Signs & Signals		CREW	DAILY OUTPUT	LABOR-HOURS	UNIT	MAT.	LABOR	EQUIP.	TOTAL	TOTAL INCL O&P	
100	0010	**SIGNS** Stock, 24″ x 24″, no posts, .080″ alum. reflectorized	B-80	70	.457	Ea.	51.50	12.80	7.40	71.70	85.50	100
	0100	High intensity		70	.457		51.50	12.80	7.40	71.70	85.50	
	0300	30″ x 30″, reflectorized		70	.457		71	12.80	7.40	91.20	107	
	0400	High intensity		70	.457		67	12.80	7.40	87.20	103	

02800 | Site Improvements and Amenities

		02890 | Traffic Signs & Signals	CREW	DAILY OUTPUT	LABOR-HOURS	UNIT	2004 BARE COSTS MAT.	LABOR	EQUIP.	TOTAL	TOTAL INCL O&P	
100	0600	Guide and directional signs, 12" x 18", reflectorized	B-80	70	.457	Ea.	31.50	12.80	7.40	51.70	64	100
	0700	High intensity		70	.457		29.50	12.80	7.40	49.70	61.50	
	0900	18" x 24", stock signs, reflectorized		70	.457		34	12.80	7.40	54.20	66.50	
	1000	High intensity		70	.457		34	12.80	7.40	54.20	66.50	
	1200	24" x 24", stock signs, reflectorized		70	.457		42	12.80	7.40	62.20	75	
	1300	High intensity		70	.457		42	12.80	7.40	62.20	75	
	1500	Add to above for steel posts, galvanized, 10'-0" upright, bolted		200	.160		14.35	4.47	2.59	21.41	26	
	1600	12'-0" upright, bolted		140	.229		17.40	6.40	3.70	27.50	33.50	
	1800	Highway road signs, aluminum, over 20 S. F., reflectorized		350	.091	S.F.	12.55	2.56	1.48	16.59	19.60	
	2000	High intensity		350	.091		15.70	2.56	1.48	19.74	23	
	2200	Highway, suspended over road, 80 S.F. min., reflectorized		165	.194		15.70	5.40	3.14	24.24	29.50	
	2300	High intensity		165	.194		21	5.40	3.14	29.54	35.50	

02900 | Planting

		02905 | Plants, Planting, Transplanting	CREW	DAILY OUTPUT	LABOR-HOURS	UNIT	2004 BARE COSTS MAT.	LABOR	EQUIP.	TOTAL	TOTAL INCL O&P	
725	0010	PLANTING R02930-900										725
	0012	Moving shrubs on site, 12" ball	B-62	28	.857	Ea.		24	5.10	29.10	44.50	
	0100	24" ball	"	22	1.091			30.50	6.50	37	57	
	0300	Moving trees on site, 36" ball	B-6	3.75	6.400			180	55.50	235.50	355	
	0400	60" ball	"	1	24			675	208	883	1,325	
925	0010	TREE REMOVAL										925
	0100	Dig & lace, shrubs, broadleaf evergreen, 18"-24"	B-1	55	.436	Ea.		11.65		11.65	19.30	
	0200	2'-3'	"	35	.686			18.30		18.30	30.50	
	0300	3'-4'	B-6	30	.800			22.50	6.95	29.45	44	
	0400	4'-5'	"	20	1.200			33.50	10.40	43.90	66	
	1000	Deciduous, 12"-15"	B-1	110	.218			5.80		5.80	9.65	
	1100	18"-24"		65	.369			9.85		9.85	16.30	
	1200	2'-3'		55	.436			11.65		11.65	19.30	
	1300	3'-4'	B-6	50	.480			13.45	4.16	17.61	26.50	
	2000	Evergreeen, 18"-24"	B-1	55	.436			11.65		11.65	19.30	
	2100	2'-0" to 2'-6"		50	.480			12.80		12.80	21	
	2200	2'-6" to 3'-0"		35	.686			18.30		18.30	30.50	
	2300	3'-0" to 3'-6"		20	1.200			32		32	53	
	3000	Trees, deciduous, small, 2'-3'		55	.436			11.65		11.65	19.30	
	3100	3'-4'	B-6	50	.480			13.45	4.16	17.61	26.50	
	3200	4'-5'		35	.686			19.25	5.95	25.20	38	
	3300	5'-6'		30	.800			22.50	6.95	29.45	44	
	4000	Shade, 5'-6'		50	.480			13.45	4.16	17.61	26.50	
	4100	6'-8'		35	.686			19.25	5.95	25.20	38	
	4200	8'-10'		25	.960			27	8.30	35.30	53	
	4300	2" caliper		12	2			56	17.30	73.30	110	
	5000	Evergreen, 4'-5'		35	.686			19.25	5.95	25.20	38	
	5100	5'-6'		25	.960			27	8.30	35.30	53	
	5200	6'-7'		19	1.263			35.50	10.95	46.45	69.50	
	5300	7'-8'		15	1.600			45	13.85	58.85	88.50	
	5400	8'-10'		11	2.182			61	18.90	79.90	121	

02910 | Plant Preparation

		02910 Plant Preparation	CREW	DAILY OUTPUT	LABOR-HOURS	UNIT	2004 BARE COSTS MAT.	LABOR	EQUIP.	TOTAL	TOTAL INCL O&P	
710	0010	LAWN BED PREPARATION										710
	0100	Rake topsoil, site material, harley rock rake, ideal	B-6	33	.727	M.S.F.		20.50	6.30	26.80	40	
	0200	Adverse	″	7	3.429			96	29.50	125.50	189	
	0300	Screened loam, york rake and finish, ideal	B-62	24	1			28	5.95	33.95	52	
	0400	Adverse	″	20	1.200			33.50	7.15	40.65	62.50	
	1000	Remove topsoil & stock pile on site, 75 HP dozer, 6″ deep, 50′ haul	B-10L	30	.400			12.45	10	22.45	31	
	1050	300′ haul		6.10	1.967			61	49	110	152	
	1100	12″ deep, 50′ haul		15.50	.774			24	19.35	43.35	60	
	1150	300′ haul		3.10	3.871			120	97	217	300	
	1200	200 HP dozer, 6″ deep, 50′ haul	B-10B	125	.096			2.99	6.90	9.89	12.35	
	1250	300′ haul		30.70	.391			12.15	28	40.15	50.50	
	1300	12″ deep, 50′ haul		62	.194			6	13.95	19.95	25	
	1350	300′ haul		15.40	.779			24	56	80	101	
	1400	Alternate method, 75 HP dozer, 50′ haul	B-10L	860	.014	C.Y.		.43	.35	.78	1.07	
	1450	300′ haul	″	114	.105			3.27	2.63	5.90	8.15	
	1500	200 HP dozer, 50′ haul	B-10B	2,660	.005			.14	.32	.46	.58	
	1600	300′ haul	″	570	.021			.65	1.51	2.16	2.72	
	1800	Rolling topsoil, hand push roller	1 Clab	3,200	.002	S.F.		.07		.07	.11	
	1850	Tractor drawn roller	B-66	10,666	.001	″		.02	.02	.04	.06	
	2000	Root raking and loading, residential, no boulders	B-6	53.30	.450	M.S.F.		12.65	3.90	16.55	25	
	2100	With boulders		32	.750			21	6.50	27.50	41	
	2200	Municipal, no boulders		200	.120			3.37	1.04	4.41	6.60	
	2300	With boulders		120	.200			5.60	1.73	7.33	11	
	2400	Large commercial, no boulders	B-10B	400	.030			.93	2.16	3.09	3.86	
	2500	With boulders	″	240	.050			1.56	3.60	5.16	6.45	
	3000	Scarify subsoil, residential, skid steer loader w/scarifiers, 50 HP	B-66	32	.250			8.05	5.25	13.30	18.45	
	3050	Municipal, skid steer loader w/scarifiers, 50 HP	″	120	.067			2.14	1.40	3.54	4.92	
	3100	Large commercial, 75 HP, dozer w/scarifier	B-10L	240	.050			1.56	1.25	2.81	3.87	
	3500	Screen topsoil from stockpile, vibrating screen, wet material (organic)	B-10P	200	.060	C.Y.		1.87	3.81	5.68	7.15	
	3550	Dry material	″	300	.040			1.24	2.54	3.78	4.78	
	3600	Mixing with conditioners, manure and peat	B-10R	550	.022			.68	.35	1.03	1.47	
	3650	Mobilization add for 2 days or less operation	B-34K	3	2.667	Job		70.50	168	238.50	300	
	3800	Spread conditioned topsoil, 6″ deep, by hand	B-1	360	.067	S.Y.	4.49	1.78		6.27	7.90	
	3850	300 HP dozer	B-10M	27	.444	M.S.F.	485	13.80	40.50	539.30	600	
	4000	Spread soil conditioners, alum. sulfate, 1#/S.Y., hand push spreader	1 Clab	17,500	.001	S.Y.	13.65	.01		13.66	15	
	4050	Tractor spreader	B-66	700	.011	M.S.F.	1,525	.37	.24	1,525.61	1,675	
	4100	Fertilizer, 0.2#/S.Y., push spreader	1 Clab	17,500	.001	S.Y.	.07	.01		.08	.10	
	4150	Tractor spreader	B-66	700	.011	M.S.F.	7.80	.37	.24	8.41	9.40	
	4200	Ground limestone, 1#/S.Y., push spreader	1 Clab	17,500	.001	S.Y.	.08	.01		.09	.11	
	4250	Tractor spreader	B-66	700	.011	M.S.F.	8.90	.37	.24	9.51	10.65	
	4300	Lusoil, 3#/S.Y., push spreader	1 Clab	17,500	.001	S.Y.	.42	.01		.43	.48	
	4350	Tractor spreader	B-66	700	.011	M.S.F.	46.50	.37	.24	47.11	52.50	
	4400	Manure, 18#/S.Y., push spreader	1 Clab	2,500	.003	S.Y.	2.49	.08		2.57	2.88	
	4450	Tractor spreader	B-66	280	.029	M.S.F.	277	.92	.60	278.52	305	
	4500	Perlite, 1″ deep, push spreader	1 Clab	17,500	.001	S.Y.	7.70	.01		7.71	8.45	
	4550	Tractor spreader	B-66	700	.011	M.S.F.	855	.37	.24	855.61	940	
	4600	Vermiculite, push spreader	1 Clab	17,500	.001	S.Y.	2.34	.01		2.35	2.59	
	4650	Tractor spreader	B-66	700	.011	M.S.F.	260	.37	.24	260.61	287	
	5000	Spread topsoil, skid steer loader and hand dress	B-62	270	.089	C.Y.	20	2.49	.53	23.02	26.50	
	5100	Articulated loader and hand dress	B-100	320	.037		20	1.17	1.68	22.85	25.50	
	5200	Articulated loader and 75HP dozer	B-10M	500	.024		20	.75	2.20	22.95	25.50	
	5300	Road grader and hand dress	B-11L	1,000	.016		20	.48	.43	20.91	23.50	
	6000	Tilling topsoil, 20 HP tractor, disk harrow, 2″ deep	B-66	450	.018	M.S.F.		.57	.37	.94	1.31	
	6050	4″ deep		360	.022			.71	.47	1.18	1.64	
	6100	6″ deep		270	.030			.95	.62	1.57	2.19	
	6150	26″ rototiller, 2″ deep	A-1J	1,250	.006	S.Y.		.17	.06	.23	.34	

02900 | Planting

02910 | Plant Preparation

			CREW	DAILY OUTPUT	LABOR-HOURS	UNIT	2004 BARE COSTS MAT.	LABOR	EQUIP.	TOTAL	TOTAL INCL O&P	
710	6200	4″ deep	A-1J	1,000	.008	S.Y.		.21	.07	.28	.42	710
	6250	6″ deep		750	.011			.28	.09	.37	.56	
720	0010	**PLANT BED PREPARATION, SHRUB & TREE**										720
	0100	Backfill planting pit, by hand, on site topsoil	2 Clab	18	.889	C.Y.		23		23	38.50	
	0200	Prepared planting mix	"	24	.667			17.35		17.35	28.50	
	0300	Skid steer loader, on site topsoil	B-62	340	.071			1.98	.42	2.40	3.68	
	0400	Prepared planting mix	"	410	.059			1.64	.35	1.99	3.05	
	1000	Excavate planting pit, by hand, sandy soil	2 Clab	16	1			26		26	43	
	1100	Heavy soil or clay	"	8	2			52		52	86	
	1200	1/2 C.Y. backhoe, sandy soil	B-11C	150	.107			3.18	1.39	4.57	6.65	
	1300	Heavy soil or clay	"	115	.139			4.15	1.81	5.96	8.70	
	2000	Mix planting soil, incl. loam, manure, peat, by hand	2 Clab	60	.267		35.50	6.95		42.45	50.50	
	2100	Skid steer loader	B-62	150	.160		35.50	4.49	.96	40.95	47.50	
	3000	Pile sod, skid steer loader	"	2,800	.009	S.Y.		.24	.05	.29	.45	
	3100	By hand	2 Clab	400	.040			1.04		1.04	1.72	
	4000	Remove sod, F.E. loader	B-10S	2,000	.006			.19	.12	.31	.43	
	4100	Sod cutter	B-12K	3,200	.005			.16	.30	.46	.58	
	4200	By hand	2 Clab	240	.067			1.73		1.73	2.87	
810	0010	**LOAM & TOPSOIL**										810
	0700	Furnish and place, truck dumped, screened, 4″ deep	B-10S	1,300	.009	S.Y.	2.53	.29	.18	3	3.44	
	0800	6″ deep		820	.015	"	3.23	.46	.29	3.98	4.61	
	0810	Minimum labor/equipment charge		2	6	Ea.		187	118	305	430	
	0900	Fine grading and seeding, incl. lime, fertilizer & seed,										
	1000	With equipment	B-14	1,000	.048	S.Y.	.34	1.31	.21	1.86	2.76	

02915 | Shrub and Tree Transplanting

			CREW	DAILY OUTPUT	LABOR-HOURS	UNIT	MAT.	LABOR	EQUIP.	TOTAL	TOTAL INCL O&P	
200	0010	**GROUND COVER** Plants, pachysandra, in prepared beds	B-1	15	1.600	C	25	42.50		67.50	98	200
	0200	Vinca minor, 1 yr, bare root		12	2	"	26	53.50		79.50	117	
	0600	Stone chips, in 50 lb. bags, Georgia marble		520	.046	Bag	2.35	1.23		3.58	4.63	
	0700	Onyx gemstone		260	.092		17	2.46		19.46	23	
	0800	Quartz		260	.092		6.35	2.46		8.81	11.10	
	0900	Pea gravel, truckload lots		28	.857	Ton	23.50	23		46.50	63.50	

02920 | Lawns & Grasses

			CREW	DAILY OUTPUT	LABOR-HOURS	UNIT	MAT.	LABOR	EQUIP.	TOTAL	TOTAL INCL O&P	
310	0010	**SEEDING, GENERAL** R02920-500										310
	0020	Mechanical seeding, 215 lb./acre	B-66	1.50	5.333	Acre	500	171	112	783	945	
	0100	44 lb./M.S.Y.	"	2,500	.003	S.Y.	.15	.10	.07	.32	.40	
	0600	Limestone hand push spreader, 50 lbs. per M.S.F.	1 Clab	180	.044	M.S.F.	3.31	1.16		4.47	5.55	
	9000	Minimum labor/equipment charge	"	4	2	Job		52		52	86	
400	0010	**SODDING**										400
	0020	Sodding, 1″ deep, bluegrass sod, on level ground, over 8 MSF	B-63	22	1.818	M.S.F.	217	49.50	6.50	273	325	
	0200	4 M.S.F.		17	2.353		243	64	8.40	315.40	380	
	0300	1000 S.F.		13.50	2.963		265	80.50	10.60	356.10	435	
	0500	Sloped ground, over 8 M.S.F.		6	6.667		217	182	24	423	565	
	0600	4 M.S.F.		5	8		243	218	28.50	489.50	655	
	0700	1000 S.F.		4	10		265	272	36	573	775	
	1000	Bent grass sod, on level ground, over 6 M.S.F.		20	2		485	54.50	7.15	546.65	630	
	1100	3 M.S.F.		18	2.222		540	60.50	7.95	608.45	705	
	1200	Sodding 1000 S.F. or less		14	2.857		615	78	10.25	703.25	820	
	1500	Sloped ground, over 6 M.S.F.		15	2.667		485	72.50	9.55	567.05	665	
	1600	3 M.S.F.		13.50	2.963		540	80.50	10.60	631.10	740	
	1700	1000 S.F.		12	3.333		615	91	11.95	717.95	840	

02930 | Exterior Plants

		02930 Exterior Plants	CREW	DAILY OUTPUT	LABOR-HOURS	UNIT	MAT.	LABOR	EQUIP.	TOTAL	TOTAL INCL O&P	
310	0010	**SHRUBS AND TREES** Evergreen, in prepared beds, B & B										310
	0100	Arborvitae pyramidal, 4'-5'	B-17	30	1.067	Ea.	42.50	29.50	17.75	89.75	115	
	0150	Globe, 12"-15"	B-1	96	.250		10.70	6.65		17.35	23	
	0300	Cedar, blue, 8'-10'	B-17	18	1.778		118	49	29.50	196.50	242	
	0500	Hemlock, canadian, 2-1/2'-3'	B-1	36	.667		17.40	17.80		35.20	48.50	
	0550	Holly, Savannah, 8' - 10' H		9.68	2.479		530	66		596	690	
	0600	Juniper, andorra, 18"-24"		80	.300		15.60	8		23.60	30.50	
	0620	Wiltoni, 15"-18"		80	.300		12	8		20	26.50	
	0640	Skyrocket, 4-1/2'-5'	B-17	55	.582		47	16.10	9.70	72.80	88	
	0660	Blue pfitzer, 2'-2-1/2'	B-1	44	.545		20.50	14.55		35.05	47	
	0680	Ketleerie, 2-1/2'-3'		50	.480		31	12.80		43.80	55.50	
	0700	Pine, black, 2-1/2'-3'		50	.480		33	12.80		45.80	57	
	0720	Mugo, 18"-24"		60	.400		33.50	10.65		44.15	54.50	
	0740	White, 4'-5'	B-17	75	.427		48	11.80	7.10	66.90	79.50	
	0800	Spruce, blue, 18"-24"	B-1	60	.400		32	10.65		42.65	53	
	0840	Norway, 4'-5'	B-17	75	.427		61	11.80	7.10	79.90	94	
	0900	Yew, denisforma, 12"-15"	B-1	60	.400		22	10.65		32.65	41.50	
	1000	Capitata, 18"-24"		30	.800		18.95	21.50		40.45	56.50	
	1100	Hicksi, 2'-2-1/2'		30	.800		28	21.50		49.50	66.50	
320	0010	**SHRUBS** Broadleaf evergreen, planted in prepared beds										320
	0100	Andromeda, 15"-18", container	B-1	96	.250	Ea.	17.15	6.65		23.80	30	
	0200	Azalea, 15" - 18", container		96	.250		22	6.65		28.65	35.50	
	0300	Barberry, 9"-12", container		130	.185		9.75	4.92		14.67	18.85	
	0400	Boxwood, 15"-18", B & B		96	.250		21	6.65		27.65	34	
	0500	Euonymus, emerald gaiety, 12" to 15", container		115	.209		14.30	5.55		19.85	25	
	0600	Holly, 15"-18", B & B		96	.250		15.35	6.65		22	28	
	0900	Mount laurel, 18" - 24", B & B		80	.300		48.50	8		56.50	67	
	1000	Paxistema, 9 - 12" high		130	.185		15.75	4.92		20.67	25.50	
	1100	Rhododendron, 18"-24", container		48	.500		27	13.35		40.35	51.50	
	1200	Rosemary, 1 gal container		600	.040		57.50	1.07		58.57	65.50	
	2000	Deciduous, amelanchier, 2'-3', B & B		57	.421		78	11.25		89.25	105	
	2100	Azalea, 15"-18", B & B		96	.250		20.50	6.65		27.15	33.50	
	2300	Bayberry, 2'-3', B & B		57	.421		24	11.25		35.25	45	
	2600	Cotoneaster, 15"-18", B & B		80	.300		14.25	8		22.25	29	
	2800	Dogwood, 3'-4', B & B	B-17	40	.800		22.50	22	13.30	57.80	75.50	
	2900	Euonymus, alatus compacta, 15" to 18", container	B-1	80	.300		18.75	8		26.75	34	
	3200	Forsythia, 2'-3', container	"	60	.400		16.50	10.65		27.15	36	
	3300	Hibiscus, 3'-4', B & B	B-17	75	.427		12.90	11.80	7.10	31.80	41	
	3400	Honeysuckle, 3'-4', B & B	B-1	60	.400		17.10	10.65		27.75	36.50	
	3500	Hydrangea, 2'-3', B & B	"	57	.421		20.50	11.25		31.75	41	
	3600	Lilac, 3'-4', B & B	B-17	40	.800		25	22	13.30	60.30	78	
	3900	Privet, bare root, 18"-24"	B-1	80	.300		10.25	8		18.25	24.50	
	4100	Quince, 2'-3', B & B	"	57	.421		18.95	11.25		30.20	39.50	
	4200	Russian olive, 3'-4', B & B	B-17	75	.427		20.50	11.80	7.10	39.40	49.50	
	4400	Spirea, 3'-4', B & B	B-1	70	.343		27.50	9.15		36.65	45	
	4500	Viburnum, 3'-4', B & B	B-17	40	.800		19.95	22	13.30	55.25	72.50	
410	0010	**TREES** Deciduous, in prep. beds, balled & burlapped (B&B) R02930-900										410
	0100	Ash, 2" caliper	B-17	8	4	Ea.	110	111	66.50	287.50	375	
	0200	Beech, 5'-6'		50	.640		215	17.70	10.65	243.35	278	
	0300	Birch, 6'-8', 3 stems		20	1.600		126	44	26.50	196.50	240	
	0500	Crabapple, 6'-8'		20	1.600		155	44	26.50	225.50	272	
	0600	Dogwood, 4'-5'		40	.800		66	22	13.30	101.30	123	
	0700	Eastern redbud 4'-5'		40	.800		120	22	13.30	155.30	183	
	0800	Elm, 8'-10'		20	1.600		113	44	26.50	183.50	226	

(MAT., LABOR, EQUIP., TOTAL are 2004 BARE COSTS)

2 SITE CONSTRUCTION

		02930 | Exterior Plants	CREW	DAILY OUTPUT	LABOR-HOURS	UNIT	MAT.	LABOR	EQUIP.	TOTAL	TOTAL INCL O&P	
							2004 BARE COSTS					
410	0900	Ginkgo, 6′-7′ R02930-900	B-17	24	1.333	Ea.	159	37	22	218	260	410
	1000	Hawthorn, 8′-10′, 1″ caliper		20	1.600		128	44	26.50	198.50	243	
	1100	Honeylocust, 10′-12′, 1-1/2″ caliper		10	3.200		136	88.50	53.50	278	350	
	1300	Larch, 8′		32	1		90	27.50	16.65	134.15	162	
	1400	Linden, 8′-10′, 1″ caliper		20	1.600		114	44	26.50	184.50	227	
	1500	Magnolia, 4′-5′		20	1.600		61	44	26.50	131.50	169	
	1600	Maple, red, 8′-10′, 1-1/2″ caliper		10	3.200		152	88.50	53.50	294	370	
	1700	Mountain ash, 8′-10′, 1″ caliper		16	2		147	55.50	33.50	236	289	
	1800	Oak, 2-1/2″-3″ caliper		6	5.333		229	147	89	465	590	
	2100	Planetree, 9′-11′, 1-1/4″ caliper		10	3.200		99.50	88.50	53.50	241.50	310	
	2200	Plum, 6′-8′, 1″ caliper		20	1.600		93.50	44	26.50	164	205	
	2300	Poplar, 9′-11′, 1-1/4″ caliper		10	3.200		46	88.50	53.50	188	253	
	2500	Sumac, 2′-3′		75	.427		23	11.80	7.10	41.90	52.50	
	2700	Tulip, 5′-6′		40	.800		47.50	22	13.30	82.80	103	
	2800	Willow, 6′-8′, 1″ caliper		20	1.600		59	44	26.50	129.50	167	
	9000	Minimum labor/equipment charge	1 Clab	4	2	Job		52		52	86	
		02945 | Planting Accessories										
120	0010	**EDGING**										120
	0050	Aluminum alloy, including stakes, 1/8″ x 4″, mill finish	B-1	390	.062	L.F.	1.92	1.64		3.56	4.83	
	0051	Black paint		390	.062		2.23	1.64		3.87	5.15	
	0052	Black anodized		390	.062		2.57	1.64		4.21	5.55	
	0100	Brick, set horizontally, 1-1/2 bricks per L.F.	D-1	370	.043		.93	1.29		2.22	3.12	
	0150	Set vertically, 3 bricks per L.F.	″	135	.119		2.64	3.55		6.19	8.65	
	0200	Corrugated aluminum, roll, 4″ wide	1 Carp	650	.012		.32	.41		.73	1.02	
	0250	6″ wide	″	550	.015		.40	.48		.88	1.24	
	0600	Railroad ties, 6″ x 8″	2 Carp	170	.094		2.44	3.11		5.55	7.85	
	0650	7″ x 9″		136	.118		2.71	3.88		6.59	9.45	
	0750	2″ x 4″		330	.048		2.79	1.60		4.39	5.70	
	0800	Steel edge strips, incl. stakes, 1/4″ x 5″	B-1	390	.062		2.91	1.64		4.55	5.90	
	0850	3/16″ x 4″	″	390	.062		2.30	1.64		3.94	5.25	
	9000	Minimum labor/equipment charge	1 Carp	4	2	Job		66		66	109	
300	0010	**PLANTERS** Concrete, sandblasted, precast, 48″ diameter, 24″ high	2 Clab	15	1.067	Ea.	530	27.50		557.50	630	300
	0100	Fluted, precast, 7′ diameter, 36″ high		10	1.600		870	41.50		911.50	1,025	
	0300	Fiberglass, circular, 36″ diameter, 24″ high		15	1.067		365	27.50		392.50	450	
	0400	60″ diameter, 24″ high		10	1.600		620	41.50		661.50	750	
	9000	Minimum labor/equipment charge	1 Clab	2	4	Job		104		104	172	
510	0010	**TREE GUYING**										510
	0015	Tree guying Including stakes, guy wire and wrap										
	0100	Less than 3″ caliper, 2 stakes	2 Clab	35	.457	Ea.	15.10	11.90		27	36.50	
	0200	3″ to 4″ caliper, 3 stakes	″	21	.762	″	17.65	19.80		37.45	52.50	
	1000	Including arrowhead anchor, cable, turnbuckles and wrap										
	1100	Less than 3″ caliper, 3″ anchors	2 Clab	20	.800	Ea.	46.50	21		67.50	85.50	
	1200	3″ to 6″ caliper, 4″ anchors		15	1.067		66.50	27.50		94	120	
	1300	6″ caliper, 6″ anchors		12	1.333		82.50	34.50		117	148	
	1400	8″ caliper, 8″ anchors		9	1.778		94.50	46		140.50	181	

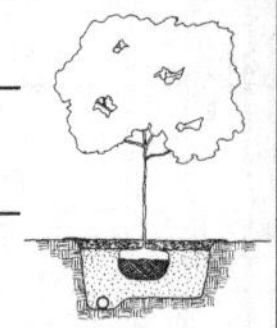

		02955 | Restoration of Underground Piping	CREW	DAILY OUTPUT	LABOR-HOURS	UNIT	2004 BARE COSTS MAT.	LABOR	EQUIP.	TOTAL	TOTAL INCL O&P	
230	0010	**LINING PIPE** with cement, incl. bypass and cleaning										230
	0020	Less than 10,000 L.F., urban, 6" to 10"	C-17E	130	.615	L.F.	6.55	21	.60	28.15	42.50	
	0050	10" to 12"		125	.640		8.05	22	.62	30.67	45	
	0070	12" to 16"		115	.696		8.30	23.50	.67	32.47	48.50	
	0100	16" to 20"		95	.842		9.70	28.50	.82	39.02	58.50	
	0200	24" to 36"		90	.889		10.45	30.50	.86	41.81	62	
	0300	48" to 72"		80	1		16.65	34	.97	51.62	75	
	0500	Rural, 6" to 10"		180	.444		6.55	15.15	.43	22.13	32	
	0550	10" to 12"		175	.457		8.05	15.55	.44	24.04	35	
	0570	12" to 16"		160	.500		8.40	17.05	.49	25.94	38	
	0600	16" to 20"		135	.593		8.85	20	.57	29.42	43.50	
	0700	24" to 36"		125	.640		10.60	22	.62	33.22	48	
	0800	48" to 72"		100	.800		16.65	27	.78	44.43	63.50	
	1000	Greater than 10,000 L.F., urban, 6" to 10"		160	.500		6.55	17.05	.49	24.09	35.50	
	1050	10" to 12"		155	.516		7.95	17.55	.50	26	38	
	1070	12" to 16"		140	.571		8.30	19.45	.55	28.30	41.50	
	1100	16" to 20"		120	.667		8.85	22.50	.65	32	47.50	
	1200	24" to 36"		115	.696		10.60	23.50	.67	34.77	51	
	1300	48" to 72"		95	.842		16.65	28.50	.82	45.97	66	
	1500	Rural, 6" to 10"		215	.372		6.55	12.65	.36	19.56	28	
	1550	10" to 12"		210	.381		8.05	12.95	.37	21.37	30.50	
	1570	12" to 16"		185	.432		8.30	14.70	.42	23.42	33.50	
	1600	16" to 20"		150	.533		8.85	18.15	.52	27.52	40	
	1700	24" to 36"		140	.571		10.60	19.45	.55	30.60	44.50	
	1800	48" to 72"		120	.667		16.75	22.50	.65	39.90	56	
		02960 | Flex. Pavement Surfacing Recovery										
100	0010	**PAVEMENT MILLING AND PAVEMENT COLD PLANING**										100
	5200	Cold planing & cleaning, 1" to 3" asphalt pavmt., over 25,000 S.Y.	B-71	6,000	.009	S.Y.		.28	.83	1.11	1.37	
	5280	5,000 S.Y. to 10,000 S.Y.	"	4,000	.014	"		.41	1.25	1.66	2.05	
	5300	Asphalt pavement removal from conc. base, no haul										
	5320	Rip, load & sweep 1" to 3"	B-70	8,000	.007	S.Y.		.21	.14	.35	.49	
	5330	3" to 6" deep	"	5,000	.011			.33	.23	.56	.79	
	5340	Profile grooving, asphalt pavement load & sweep, 1" deep	B-71	12,500	.004			.13	.40	.53	.65	
	5350	3" deep		9,000	.006			.18	.56	.74	.91	
	5360	6" deep		5,000	.011			.33	1	1.33	1.64	
200	0010	**COLD IN-PLACE RECYCLED BITUMINOUS PAVEMENT COURSES**										200
	5000	Reclamation, pulverizing and blending with existing base										
	5040	Aggregate base, 4" thick pavement, over 15,000 S.Y.	B-73	2,400	.027	S.Y.		.83	1.59	2.42	3.08	
	5080	5,000 S.Y. to 15,000 S.Y.		2,200	.029			.90	1.74	2.64	3.36	
	5120	8" thick pavement, over 15,000 S.Y.		2,200	.029			.90	1.74	2.64	3.36	
	5160	5,000 S.Y. to 15,000 S.Y.		2,000	.032			.99	1.91	2.90	3.69	
	5180	Add for mobilization and demobilization for crew B-73				Ea.				1,700	2,150	
400	0010	**HOT IN-PLACE RECYCLED BITUMINOUS PAVEMENT COURSES**										400
	5500	Recycle asphalt pavement at site										
	5520	Remove, rejuvenate and spread 4" deep	B-72	2,500	.026	S.Y.	2.14	.77	3.36	6.27	7.30	
	5521	6" deep	"	2,000	.032	"	3.21	.96	4.20	8.37	9.70	
		02990 | Structure Moving										
300	0010	**MOVING BUILDINGS** One day move, up to 24' wide										300
	0020	Reset on new foundation, patch & hook-up, average move				Total					9,300	
	0040	Wood or steel frame bldg., based on ground floor area	B-4	185	.259	S.F.		6.85	1.82	8.67	13.30	
	0060	Masonry bldg., based on ground floor area	"	137	.350			9.25	2.45	11.70	18	

	02990 | Structure Moving		CREW	DAILY OUTPUT	LABOR-HOURS	UNIT	2004 BARE COSTS MAT.	LABOR	EQUIP.	TOTAL	TOTAL INCL O&P	
300	0200	For 24′ to 42′ wide, add				S.F.					15%	300
	0220	For each additional day on road, add	B-4	1	48	Day		1,275	335	1,610	2,475	
	0240	Construct new basement, move building, 1 day										
	0300	move, patch & hook-up, based on ground floor area	B-3	155	.310	S.F.	6	8.60	11.05	25.65	33	

For information about Means Estimating Seminars, see yellow pages 12 and 13 in back of book

Division 3
Concrete

Estimating Tips

General

- Carefully check all the plans and specifications. Concrete often appears on drawings other than structural drawings, including mechanical and electrical drawings for equipment pads. The cost of cutting and patching is often difficult to estimate. See Subdivision 02220 for demolition costs.
- Always obtain concrete prices from suppliers near the job site. A volume discount can often be negotiated depending upon competition in the area. Remember to add for waste, particularly for slabs and footings on grade.

03100 Concrete Forms & Accessories

- A primary cost for concrete construction is forming. Most jobs today are constructed with prefabricated forms. The selection of the forms best suited for the job and the total square feet of forms required for efficient concrete forming and placing are key elements in estimating concrete construction. Enough forms must be available for erection to make efficient use of the concrete placing equipment and crew.
- Concrete accessories for forming and placing depend upon the systems used. Study the plans and specifications to assure that all special accessory requirements have been included in the cost estimate such as anchor bolts, inserts and hangers.

03200 Concrete Reinforcement

- Ascertain that the reinforcing steel supplier has included all accessories, cutting, bending and an allowance for lapping, splicing and waste. A good rule of thumb is 10% for lapping, splicing and waste. Also, 10% waste should be allowed for welded wire fabric.

03300 Cast-in-Place Concrete

- When estimating structural concrete, pay particular attention to requirements for concrete additives, curing methods and surface treatments. Special consideration for climate, hot or cold, must be included in your estimate. Be sure to include requirements for concrete placing equipment and concrete finishing.

03400 Precast Concrete
03500 Cementitious Decks & Toppings

- The cost of hauling precast concrete structural members is often an important factor. For this reason, it is important to get a quote from the nearest supplier. It may become economically feasible to set up precasting beds on the site if the hauling costs are prohibitive.

Reference Numbers

Reference numbers are shown in bold squares at the beginning of some major classifications. These numbers refer to related items in the Reference Section. The reference information may be an estimating procedure, an alternate pricing method or technical information.

Note: Not all subdivisions listed here necessarily appear in this publication.

03050 | Basic Concrete Materials & Methods

03055 | Selective Demolition

				DAILY	LABOR-		2004 BARE COSTS				TOTAL	
			CREW	OUTPUT	HOURS	UNIT	MAT.	LABOR	EQUIP.	TOTAL	INCL O&P	
110	0010	SELECTIVE DEMOLITION, CONCRETE FRAMING										110
	1020	Concrete, average reinforcing, beams, 8" x 10"	B-9	120	.333	L.F.		8.80	1.33	10.13	16.05	
	1040	10" x 12"		110	.364			9.60	1.45	11.05	17.50	
	1060	12" x 14"		90	.444			11.75	1.78	13.53	21.50	
	1200	Columns, 8" x 8"		120	.333			8.80	1.33	10.13	16.05	
	1240	10" x 10"		120	.333			8.80	1.33	10.13	16.05	
	1280	12" x 12"		110	.364			9.60	1.45	11.05	17.50	
	1320	14" x 14"		100	.400			10.55	1.60	12.15	19.25	
	1400	Girders, 14" x 16"		55	.727			19.20	2.91	22.11	35	
	1440	16" x 18"		40	1			26.50	4	30.50	48.50	
	1600	Slabs, elevated, 6" thick		600	.067	S.F.		1.76	.27	2.03	3.21	
	1640	8" thick		450	.089			2.35	.36	2.71	4.28	
	1680	10" thick		360	.111			2.93	.44	3.37	5.35	
	1900	Add for heavy reinforcement									25%	
	1910	Minimum labor/equipment charge	B-9	2	20	Job		530	80	610	965	

03060 | Basic Concrete Materials

			CREW	DAILY OUTPUT	LABOR-HOURS	UNIT	MAT.	LABOR	EQUIP.	TOTAL	TOTAL INCL O&P	
100	0010	CONCRETE ADMIXTURES & SURFACE TREATMENTS										100
	0040	Abrasives, aluminum oxide, over 20 tons				Lb.	1.02			1.02	1.12	
	0070	Under 1 ton					1.09			1.09	1.20	
	0100	Silicon carbide, black, over 20 tons					1.42			1.42	1.56	
	0120	Under 1 ton					1.51			1.51	1.66	
	0200	Air entraining agent, .7 to 1.5 oz. per bag, 55 gallon lots				Gal.	9.15			9.15	10.05	
	0220	5 gallon lots					9.45			9.45	10.40	
	0300	Bonding agent, acrylic latex, 250 S.F. per gallon					20.50			20.50	22.50	
	0320	Epoxy resin, 80 S.F. per gallon					45.50			45.50	50	
	0400	Calcium chloride, 50 lb. bags				Ton	355			355	390	
	0420	Less than truckload lots				Bag	30			30	33	
	0500	Carbon black, liquid, 2 to 8 lbs. per bag of cement				Lb.	3.64			3.64	4	
	0600	Colors, integral, 2 to 10 lb. per bag of cement, minimum					2.18			2.18	2.40	
	0610	Average					2.60			2.60	2.86	
	0620	Maximum					2.98			2.98	3.28	
	0700	Curing compound, solvent based, 400 S.F./gal, 55 gal. lots				Gal.	8.50			8.50	9.35	
	0720	5 gallon lots					10.05			10.05	11.05	
	0800	Water based, 250 S.F./gal, 55 gallon lots					4.12			4.12	4.53	
	0820	5 gallon lots					5.15			5.15	5.65	
	0900	Dustproofing compound, (200-600 S.F./gal.), 55 gallon lots					3.91			3.91	4.30	
	0920	5 gallon lots					4.53			4.53	4.98	
	1000	Epoxy dustproof coating, colors, (300-400 S.F. per coat),										
	1010	or transparent, (400-600 S.F. per coat)				Gal.	24.50			24.50	27	
	1100	Hardeners, metallic, 55 lb. bags, natural (grey)				Lb.	.59			.59	.65	
	1200	Colors, average					1.03			1.03	1.13	
	1300	Non-metallic, 55 lb. bags, natural (grey), minimum					.36			.36	.40	
	1310	Maximum					.41			.41	.45	
	1320	Non-metallic, colors, mininum					.47			.47	.52	
	1340	Maximum					.75			.75	.83	
	1400	Non-metallic, non-slip, 100 lb. bags, minimum					.47			.47	.52	
	1420	Maximum					.94			.94	1.03	
	1500	Solution type, (300 to 400 S.F. per gallon)				Gal.	6.20			6.20	6.80	
	1600	Sealer, hardener and dustproofer, epoxy, 150 S.F., minimum					7.05			7.05	7.75	
	1620	Maximum					15.45			15.45	17	
	1630	Solvent based, 200 S.F., minimum					8.55			8.55	9.40	
	1640	Maximum					15.85			15.85	17.45	
	1650	Water based, 250 S.F., minimum					10.05			10.05	11.05	
	1660	Maximum					11.60			11.60	12.75	

03050 | Basic Concrete Materials & Methods

03060 | Basic Concrete Materials

		03060 Basic Concrete Materials	CREW	DAILY OUTPUT	LABOR-HOURS	UNIT	2004 BARE COSTS MAT.	LABOR	EQUIP.	TOTAL	TOTAL INCL O&P	
100	1700	Colors (300-400 S.F. per gallon)				Gal.	41.50			41.50	46	100
	1750											
	1800	Set accelerator for below freezing, 1 to 1-1/2 gal. per C.Y.				Gal.	5.55			5.55	6.10	
	1900	Set retarder, 2 to 4 fl. oz. per bag of cement				″	11.40			11.40	12.55	
	2000	Waterproofing, integral 1 lb. per bag of cement				Lb.	.83			.83	.91	
	2100	Powdered metallic, 40 lbs. per 100 S.F., minimum					1.04			1.04	1.14	
	2120	Maximum					2.08			2.08	2.29	
870	0010	**WINTER PROTECTION** For heated ready mix, add, minimum				C.Y.	4.25			4.25	4.68	870
	0050	Maximum				″	5.25			5.25	5.80	
	0100	Protecting concrete and temporary heat, add, minimum	2 Clab	6,000	.003	S.F.	.25	.07		.32	.40	
	0150	Maximum, see also division 01510-800	″	2,000	.008	″	.75	.21		.96	1.17	
	0200	Temporary shelter for slab on grade, wood frame and polyethylene										
	0201	sheeting, minimum	2 Carp	10	1.600	M.S.F.	237	53		290	350	
	0210	Maximum	″	3	5.333	″	285	176		461	605	
	0300	See also Division 03390-200										

03100 | Concrete Forms & Accessories

03110 | Structural C.I.P. Forms

		03110 Structural C.I.P. Forms	CREW	DAILY OUTPUT	LABOR-HOURS	UNIT	2004 BARE COSTS MAT.	LABOR	EQUIP.	TOTAL	TOTAL INCL O&P	
405	0010	**FORMS IN PLACE, BEAMS AND GIRDERS** R03110-040										405
	0020	See also Elevated Slabs, division 03310-240										
	0500	Exterior spandrel, job-built plywood, 12″ wide, 1 use R03110-050	C-2	225	.213	SFCA	2.11	6.85		8.96	13.70	
	0650	4 use		310	.155		.69	4.98		5.67	9	
	1000	18″ wide, 1 use		250	.192		1.90	6.20		8.10	12.35	
	1150	4 use		315	.152		.62	4.90		5.52	8.85	
	1500	24″ wide, 1 use		265	.181		1.73	5.85		7.58	11.55	
	1650	4 use		325	.148		.56	4.75		5.31	8.50	
	2000	Interior beam, job-built plywood, 12″ wide, 1 use		300	.160		2.26	5.15		7.41	11.05	
	2150	4 use		377	.127		.74	4.10		4.84	7.60	
	2500	24″ wide, 1 use		320	.150		1.78	4.83		6.61	9.95	
	2650	4 use		395	.122		.57	3.91		4.48	7.15	
	3000	Encasing steel beam, hung, job-built plywood, 1 use		325	.148		2.15	4.75		6.90	10.25	
	3150	4 use		430	.112		.70	3.59		4.29	6.70	
	9000	Minimum labor/equipment charge	2 Carp	2	8	Job		264		264	440	
410	0010	**FORMS IN PLACE, COLUMNS** R03110-040										410
	0020											
	0500	Round fiberglass, 4 use per mo., rent, 12″ diameter R03110-060	C-1	160	.200	L.F.	6.10	6.25		12.35	17.05	
	0550	16″ diameter		150	.213		7.35	6.65		14	19.15	
	0600	18″ diameter		140	.229		8.10	7.15		15.25	21	
	0650	24″ diameter		135	.237		10.10	7.40		17.50	23.50	
	0700	28″ diameter		130	.246		11.25	7.70		18.95	25	
	0800	30″ diameter		125	.256		11.75	8		19.75	26	
	0850	36″ diameter		120	.267		15.65	8.35		24	31	
	1500	Round fiber tube, 1 use, 8″ diameter		155	.206		1.46	6.45		7.91	12.30	
	1550	10″ diameter		155	.206		1.83	6.45		8.28	12.70	
	1600	12″ diameter		150	.213		2.10	6.65		8.75	13.35	
	1650	14″ diameter		145	.221		2.95	6.90		9.85	14.70	
	1700	16″ diameter		140	.229		3.48	7.15		10.63	15.70	

03100 | Concrete Forms & Accessories

03110 | Structural C.I.P. Forms

		03110 Structural C.I.P. Forms		Crew	Daily Output	Labor-Hours	Unit	2004 Bare Costs Mat.	Labor	Equip.	Total	Total Incl O&P	
410	1750	20″ diameter	R03110-040	C-1	135	.237	L.F.	5.35	7.40		12.75	18.20	410
	1800	24″ diameter			130	.246		6.80	7.70		14.50	20.50	
	1850	30″ diameter	R03110-060		125	.256		9.90	8		17.90	24	
	1900	36″ diameter			115	.278		12.45	8.70		21.15	28	
	1950	42″ diameter			100	.320		30	10		40	49.50	
	2000	48″ diameter			85	.376		38	11.75		49.75	61.50	
	2200	For seamless type, add						15%					
	3000	Round, steel, 4 use per mo., rent, regular duty, 12″ diameter		C-1	145	.221	L.F.	10.45	6.90		17.35	23	
	3050	16″ diameter			125	.256		11.75	8		19.75	26	
	3100	Heavy duty, 20″ diameter			105	.305		12.90	9.50		22.40	30	
	3150	24″ diameter			85	.376		14.15	11.75		25.90	35	
	3200	30″ diameter			70	.457		16.25	14.30		30.55	41.50	
	3250	36″ diameter			60	.533		17.40	16.65		34.05	46.50	
	3300	48″ diameter			50	.640		24	20		44	59.50	
	3350	60″ diameter			45	.711		32	22		54	72	
	4000	Column capitals, steel, 4 uses/mo., 24″ col, 4′ cap diameter			12	2.667	Ea.	187	83.50		270.50	345	
	4050	5′ cap diameter			11	2.909		213	91		304	385	
	4100	6′ cap diameter			10	3.200		289	100		389	485	
	4150	7′ cap diameter			9	3.556		355	111		466	575	
	4500	For second and succeeding months, deduct						50%					
	5000	Job-built plywood, 8″ x 8″ columns, 1 use		C-1	165	.194	SFCA	1.53	6.05		7.58	11.75	
	5050	2 use			195	.164		.88	5.15		6.03	9.45	
	5100	3 use			210	.152		.61	4.76		5.37	8.55	
	5150	4 use			215	.149		.50	4.65		5.15	8.25	
	5500	12″ x 12″ columns, 1 use			180	.178		1.53	5.55		7.08	10.90	
	5550	2 use			210	.152		.84	4.76		5.60	8.85	
	5600	3 use			220	.145		.61	4.55		5.16	8.20	
	5650	4 use			225	.142		.50	4.44		4.94	7.90	
	6000	16″ x 16″ columns, 1 use			185	.173		1.57	5.40		6.97	10.65	
	6050	2 use			215	.149		.83	4.65		5.48	8.60	
	6100	3 use			230	.139		.63	4.35		4.98	7.90	
	6150	4 use			235	.136		.51	4.26		4.77	7.60	
	6500	24″ x 24″ columns, 1 use			190	.168		1.80	5.25		7.05	10.70	
	6550	2 use			216	.148		.99	4.63		5.62	8.75	
	6600	3 use			230	.139		.72	4.35		5.07	8	
	6650	4 use			238	.134		.59	4.20		4.79	7.60	
	7000	36″ x 36″ columns, 1 use			200	.160		1.58	5		6.58	10.05	
	7050	2 use			230	.139		.89	4.35		5.24	8.15	
	7100	3 use			245	.131		.63	4.08		4.71	7.45	
	7150	4 use			250	.128		.51	4		4.51	7.20	
	7500	Steel framed plywood, 4 use per mo., rent, 8″ x 8″			340	.094		2.85	2.94		5.79	8	
	7550	10″ x 10″			350	.091		2.40	2.86		5.26	7.40	
	7600	12″ x 12″			370	.086		2.58	2.70		5.28	7.30	
	7650	16″ x 16″			400	.080		2.74	2.50		5.24	7.15	
	7700	20″ x 20″			420	.076		1.53	2.38		3.91	5.65	
	7750	24″ x 24″			440	.073		1.43	2.27		3.70	5.35	
	7755	30″ x 30″			440	.073		1.35	2.27		3.62	5.25	
	9000	Minimum labor/equipment charge		2 Carp	2	8	Job		264		264	440	
420	0010	**FORMS IN PLACE, ELEVATED SLABS**	R03110-020										420
	1000	Flat plate, job-built plywood, to 15′ high, 1 use		C-2	470	.102	S.F.	2.50	3.29		5.79	8.20	
	1150	4 use	R03110-050		560	.086		.81	2.76		3.57	5.45	
	1500	15′ to 20′ high ceilings, 4 use			495	.097		1.04	3.12		4.16	6.30	
	2000	Flat slab, drop panels, job-built plywood, to 15′ high, 1 use			449	.107		2.83	3.44		6.27	8.80	
	2150	4 use			544	.088		.92	2.84		3.76	5.70	
	2250	15′ to 20′ high ceilings, 4 use			480	.100		1.37	3.22		4.59	6.85	
	3500	Floor slab, with 20″ metal pans, 1 use			415	.116		3.97	3.72		7.69	10.50	

	03110 | Structural C.I.P. Forms		CREW	DAILY OUTPUT	LABOR-HOURS	UNIT	2004 BARE COSTS MAT.	LABOR	EQUIP.	TOTAL	TOTAL INCL O&P	
420	3650	4 use	C-2	500	.096	S.F.	2.05	3.09		5.14	7.35	420
	4500	With 30" fiberglass domes, 1 use (R03110-020)		405	.119		4.18	3.81		7.99	10.90	
	4550	4 use		470	.102		2.26	3.29		5.55	7.95	
	5000	Box out for slab openings, over 16" deep, 1 use (R03110-050)		190	.253	SFCA	3.28	8.15		11.43	17.05	
	5050	2 use		240	.200	"	1.80	6.45		8.25	12.65	
	5500	Shallow slab box outs, to 10 S.F.		42	1.143	Ea.	9.35	37		46.35	71.50	
	5550	Over 10 S.F. (use perimeter)		600	.080	L.F.	1.25	2.57		3.82	5.65	
	6000	Bulkhead forms for slab, with keyway, 1 use, 2 piece		500	.096		1.51	3.09		4.60	6.75	
	6100	3 piece (see also edge forms)		460	.104		1.98	3.36		5.34	7.75	
	6200	Bulkhead forms for slab, w/keyway expanded metal										
	6210	In lieu of 2 piece form	C-1	1,100	.029	L.F.	.76	.91		1.67	2.35	
	6215	In lieu of 3 piece form		960	.033		.76	1.04		1.80	2.57	
	6220	6" high, 4 uses		1,100	.029		.92	.91		1.83	2.52	
	6500	Curb forms, wood, 6" to 12" high, on elevated slabs, 1 use		180	.178	SFCA	1.24	5.55		6.79	10.55	
	6550	2 use		205	.156		1.08	4.88		5.96	9.30	
	6600	3 use		220	.145		.79	4.55		5.34	8.40	
	6650	4 use		225	.142		.64	4.44		5.08	8.05	
	7000	Edge forms to 6" high, on elevated slab, 4 use		500	.064	L.F.	.35	2		2.35	3.70	
	7500	Depressed area forms to 12" high, 4 use		300	.107		.73	3.33		4.06	6.35	
	7550	12" to 24" high, 4 use		175	.183		.99	5.70		6.69	10.55	
	8000	Perimeter deck and rail for elevated slabs, straight		90	.356		8.60	11.10		19.70	28	
	8050	Curved		65	.492		11.80	15.40		27.20	38.50	
	8500	Void forms, round fiber, 3" diameter		450	.071		1.20	2.22		3.42	5	
	8650	8" diameter		375	.085		3.73	2.67		6.40	8.50	
	9000	Minimum labor/equipment charge	2 Carp	2	8	Job		264		264	440	
425	0010	**FORMS IN PLACE, EQUIPMENT FOUNDATIONS** job built (R03110-050)										425
	0020	1 use	C-2	160	.300	SFCA	2.52	9.65		12.17	18.75	
	0050	2 use		190	.253		1.39	8.15		9.54	14.95	
	0100	3 use		200	.240		1.01	7.70		8.71	13.90	
	0150	4 use		205	.234		.82	7.55		8.37	13.40	
	9000	Minimum labor/equipment charge (R03110-060)	1 Carp	3	2.667	Job		88		88	146	
430	0010	**FORMS IN PLACE, FOOTINGS** Continuous wall, plywood, 1 use (R03110-050)	C-1	375	.085	SFCA	2.20	2.67		4.87	6.85	430
	0150	4 use	"	485	.066	"	.72	2.06		2.78	4.21	
	1500	Keyway, 4 use, tapered wood, 2" x 4" (R03110-060)	1 Carp	530	.015	L.F.	.17	.50		.67	1.01	
	1550	2" x 6"	"	500	.016	"	.26	.53		.79	1.17	
	3000	Pile cap, square or rectangular, job-built plywood, 1 use	C-1	290	.110	SFCA	1.75	3.45		5.20	7.65	
	3150	4 use		383	.084		.57	2.61		3.18	4.96	
	5000	Spread footings, job-built lumber, 1 use		305	.105		1.61	3.28		4.89	7.20	
	5150	4 use		414	.077		.52	2.42		2.94	4.57	
	9000	Minimum labor/equipment charge	1 Carp	3	2.667	Job		88		88	146	
435	0010	**FORMS IN PLACE, GRADE BEAM** Job-built plywood, 1 use	C-2	530	.091	SFCA	1.50	2.91		4.41	6.50	435
	0150	4 use	"	605	.079	"	.49	2.55		3.04	4.77	
	9000	Minimum labor/equipment charge	2 Carp	2	8	Job		264		264	440	
445	0010	**FORMS IN PLACE, SLAB ON GRADE**										445
	1000	Bulkhead forms with keyway, wood, 1 use, 2 piece	C-1	510	.063	L.F.	.93	1.96		2.89	4.27	
	1400	Bulkhead forms w/keyway, 1 piece expanded metal, left in place										
	1410	In lieu of 2 piece form	C-1	1,375	.023	L.F.	1.28	.73		2.01	2.62	
	1420	In lieu of 3 piece form		1,200	.027		1.28	.83		2.11	2.79	
	1430	In lieu of 4 piece form		1,050	.030		1.28	.95		2.23	2.99	
	2000	Curb forms, wood, 6" to 12" high, on grade, 1 use		215	.149	SFCA	1.63	4.65		6.28	9.50	
	2150	4 use		275	.116	"	.53	3.64		4.17	6.65	
	3000	Edge forms, wood, 4 use, on grade, to 6" high		600	.053	L.F.	.40	1.67		2.07	3.20	
	3050	7" to 12" high		435	.074	SFCA	.85	2.30		3.15	4.75	

		03110 | Structural C.I.P. Forms	CREW	DAILY OUTPUT	LABOR-HOURS	UNIT	2004 BARE COSTS MAT.	LABOR	EQUIP.	TOTAL	TOTAL INCL O&P	
445	3500	For depressed slabs, 4 use, to 12″ high	C-1	300	.107	L.F.	.60	3.33		3.93	6.20	445
	3550	To 24″ high		175	.183		.80	5.70		6.50	10.35	
	4000	For slab blockouts, to 12″ high, 1 use		200	.160		.67	5		5.67	9.05	
	4050	To 24″ high, 1 use		120	.267		.85	8.35		9.20	14.75	
	4100	Plastic (extruded), to 6″ high, multiple use, on grade		800	.040		3.73	1.25		4.98	6.15	
	5000	Screed, 24 ga. metal key joint, see Div 03150-250										
	5020	Wood, incl. wood stakes, 1″ x 3″	C-1	900	.036	L.F.	.50	1.11		1.61	2.39	
	5050	2″ x 4″		900	.036	″	1.38	1.11		2.49	3.36	
	6000	Trench forms in floor, wood, 1 use		160	.200	SFCA	2.09	6.25		8.34	12.65	
	6150	4 use		185	.173	″	.66	5.40		6.06	9.70	
	9000	Minimum labor/equipment charge	1 Carp	2	4	Job		132		132	219	
455	0010	**FORMS IN PLACE, WALLS** R03110-010										455
	0100	Box out for wall openings, to 16″ thick, to 10 S.F.	C-2	24	2	Ea.	17.35	64.50		81.85	126	
	0150	Over 10 S.F. (use perimeter) R03110-050	″	280	.171	L.F.	1.55	5.50		7.05	10.85	
	0250	Brick shelf, 4″ w, add to wall forms, use wall area abv shelf										
	0260	1 use R03110-060	C-2	240	.200	SFCA	1.73	6.45		8.18	12.55	
	0350	4 use		300	.160	″	.69	5.15		5.84	9.30	
	0500	Bulkhead, with keyway, 1 use, 2 piece		265	.181	L.F.	2.71	5.85		8.56	12.65	
	0550	3 piece		175	.274	″	3.41	8.80		12.21	18.40	
	0600	Bulkhead w/keyway, 1 piece expanded metal, left in place										
	0610	In lieu of 2 piece form	C-1	800	.040	L.F.	1.31	1.25		2.56	3.51	
	0620	In lieu of 3 piece form	″	525	.061	″	1.31	1.90		3.21	4.60	
	0700	Buttress, to 8′ high, 1 use	C-2	350	.137	SFCA	2.96	4.41		7.37	10.55	
	0850	4 use		480	.100	″	.98	3.22		4.20	6.40	
	1000	Corbel or haunch, to 12″ wide, add to wall forms, 1 use		150	.320	L.F.	1.52	10.30		11.82	18.70	
	1150	4 use		180	.267	″	.49	8.60		9.09	14.75	
	2000	Wall, below grade, job-built plywood, to 8′ high, 1 use		300	.160	SFCA	1.76	5.15		6.91	10.50	
	2150	4 use		435	.110		.68	3.55		4.23	6.65	
	2400	Over 8′ to 16′ high, 1 use		280	.171		3.70	5.50		9.20	13.20	
	2420	2 use		345	.139		1.58	4.48		6.06	9.15	
	2430	3 use		375	.128		1.32	4.12		5.44	8.30	
	2440	4 use		395	.122		1.19	3.91		5.10	7.80	
	2445	Exterior wall, 8′ to 16′ high, 1 use		280	.171		1.66	5.50		7.16	11	
	2550	4 use		395	.122		.54	3.91		4.45	7.10	
	2700	Over 16′ high, 1 use		235	.204		1.88	6.55		8.43	12.95	
	2850	4 use		330	.145		.61	4.68		5.29	8.40	
	3000	For architectural finish, add		1,820	.026		.68	.85		1.53	2.16	
	3500	Polystyrene (expanded) wall forms										
	3510	To 8′ high, 1 use, left in place	1 Carp	295	.027	SFCA	1.77	.90		2.67	3.43	
	4000	Radial, smooth curved, job-built plywood, 1 use	C-2	245	.196		1.64	6.30		7.94	12.25	
	4150	4 use		335	.143		.53	4.61		5.14	8.25	
	4200	Below grade, job-built plywood, 1 use		225	.213		2.52	6.85		9.37	14.15	
	4210	2 use		225	.213		1.39	6.85		8.24	12.95	
	4220	3 use		225	.213		1.14	6.85		7.99	12.65	
	4230	4 use		225	.213		.82	6.85		7.67	12.30	
	4600	Retaining wall, battered, job-built plywood, to 8′ high, 1 use		300	.160		1.38	5.15		6.53	10.05	
	4750	4 use		390	.123		.41	3.96		4.37	7	
	4900	Over 8′ to 16′ high, 1 use		240	.200		1.51	6.45		7.96	12.30	
	5050	4 use		320	.150		.49	4.83		5.32	8.55	
	5750	Liners for forms (add to wall forms), A.B.S. plastic										
	5800	Aged wood, 4″ wide, 1 use	1 Carp	250	.032	SFCA	5.45	1.06		6.51	7.75	
	5820	2 use		400	.020		2.94	.66		3.60	4.32	
	5840	4 use		750	.011		3.76	.35		4.11	4.72	
	5900	Fractured rope rib, 1 use		250	.032		2.69	1.06		3.75	4.71	
	6000	4 use		750	.011		5.10	.35		5.45	6.20	

Important: See the Reference Section for critical supporting data - Reference Nos., Crews, & City Cost Indexes

03110 | Structural C.I.P. Forms

		03110 Structural C.I.P. Forms	CREW	DAILY OUTPUT	LABOR-HOURS	UNIT	2004 BARE COSTS MAT.	LABOR	EQUIP.	TOTAL	TOTAL INCL O&P	
455	6100	Ribbed look, 1/2" & 3/4" deep, 1 use (R03110-010)	1 Carp	300	.027	SFCA	4.47	.88		5.35	6.40	455
	6200	4 use		800	.010		4.23	.33		4.56	5.20	
	6300	Rustic brick pattern, 1 use (R03110-050)		250	.032		2.69	1.06		3.75	4.71	
	6400	4 use		750	.011		5.10	.35		5.45	6.20	
	6500	Striated, random, 3/8" x 3/8" deep, 1 use (R03110-060)		300	.027		2.69	.88		3.57	4.42	
	6600	4 use		800	.010		5.10	.33		5.43	6.15	
	7500	Lintel or sill forms, 1 use		30	.267		2.19	8.80		10.99	17	
	7560	4 use	↓	37	.216		.71	7.15		7.86	12.65	
	7800	Modular prefabricated plywood, to 8' high, 1 use	C-2	1,180	.041		1.54	1.31		2.85	3.86	
	7860	4 use		1,260	.038		.51	1.23		1.74	2.59	
	8000	To 16' high, 1 use		715	.067		1.99	2.16		4.15	5.75	
	8060	4 use		790	.061		.66	1.95		2.61	3.96	
	8100	Over 16' high, 1 use		715	.067		2.39	2.16		4.55	6.20	
	8160	4 use		790	.061		.80	1.95		2.75	4.12	
	8600	Pilasters, 1 use		270	.178		2.09	5.70		7.79	11.80	
	8660	4 use	↓	385	.125	↓	.68	4.01		4.69	7.40	
	9475	For elevated walls, add						10%				
	9480	For battered walls, 1 side battered, add					10%	10%				
	9485	For battered walls, 2 sides battered, add					15%	15%				
	9900	Minimum labor/equipment charge	2 Carp	2	8	Job		264		264	440	
500	0010	**GAS STATION FORMS** Curb fascia, with template,										500
	0050	12 ga. steel, left in place, 9" high	1 Carp	50	.160	L.F.	7.90	5.30		13.20	17.45	
	1000	Sign or light bases, 18" diameter, 9" high		9	.889	Ea.	43	29.50		72.50	95.50	
	1050	30" diameter, 13" high		8	1	"	70.50	33		103.50	132	
	1990	Minimum labor/equipment charge	↓	2	4	Job		132		132	219	
	2000	Island forms, 10' long, 9" high, 3'- 6" wide	C-1	10	3.200	Ea.	189	100		289	375	
	2050	4' wide		9	3.556		201	111		312	405	
	2500	20' long, 9" high, 4' wide		6	5.333		325	167		492	635	
	2550	5' wide	↓	5	6.400	↓	340	200		540	705	
	9000	Minimum labor/equipment charge	1 Carp	3	2.667	Job		88		88	146	
800	0010	**SCAFFOLDING** See division 01540-750										800

03150 | Concrete Accessories

		03150 Concrete Accessories	CREW	DAILY OUTPUT	LABOR-HOURS	UNIT	MAT.	LABOR	EQUIP.	TOTAL	TOTAL INCL O&P	
085	0012	**ANCHOR BOLTS** See division 04080-070										085
250	0010	**EXPANSION JOINT** Keyed, cold, 24 ga, incl. stakes, 3-1/2" high	1 Carp	200	.040	L.F.	.56	1.32		1.88	2.81	250
	0050	4-1/2" high		200	.040		.68	1.32		2	2.94	
	0100	5-1/2" high		195	.041		.76	1.35		2.11	3.08	
	0150	7-1/2" high	↓	190	.042		.92	1.39		2.31	3.31	
	0300	Poured asphalt, plain, 1/2" x 1"	1 Clab	450	.018		.35	.46		.81	1.16	
	0350	1" x 2"		400	.020		1.24	.52		1.76	2.22	
	0500	Neoprene, liquid, cold applied, 1/2" x 1"		450	.018		1.60	.46		2.06	2.53	
	0550	1" x 2"		400	.020		6.25	.52		6.77	7.75	
	0700	Polyurethane, poured, 2 part, 1/2" x 1"		400	.020		2	.52		2.52	3.06	
	0750	1" x 2"		350	.023		7.30	.59		7.89	9.05	
	0900	Rubberized asphalt, hot or cold applied, 1/2" x 1"		450	.018		.43	.46		.89	1.24	
	0950	1" x 2"		400	.020		1.20	.52		1.72	2.18	
	1100	Hot applied, fuel resistant, 1/2" x 1"		450	.018		.75	.46		1.21	1.60	
	1150	1" x 2"	↓	400	.020		1.53	.52		2.05	2.54	
	2000	Premolded, bituminous fiber, 1/2" x 6"	1 Carp	375	.021		.38	.70		1.08	1.59	
	2050	1" x 12"		300	.027		1.34	.88		2.22	2.93	
	2250	Cork with resin binder, 1/2" x 6"		375	.021		1.49	.70		2.19	2.81	
	2300	1" x 12"		300	.027		5.35	.88		6.23	7.35	
	2500	Neoprene sponge, closed cell, 1/2" x 6"		375	.021		1.28	.70		1.98	2.58	
	2550	1" x 12"	↓	300	.027	↓	5.90	.88		6.78	7.95	

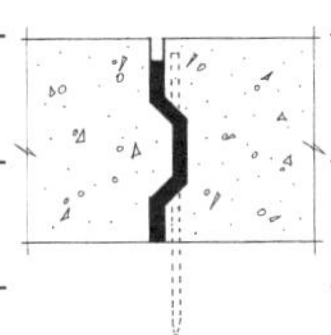

	03150 | Concrete Accessories		Crew	Daily Output	Labor-Hours	Unit	2004 Bare Costs Mat.	Labor	Equip.	Total	Total Incl O&P	
250	2750	Polyethylene foam, 1/2" x 6"	1 Carp	375	.021	L.F.	.48	.70		1.18	1.70	250
	2800	1" x 12"		300	.027		1.37	.88		2.25	2.97	
	3000	Polyethylene backer rod, 3/8" diameter		460	.017		.04	.57		.61	.99	
	3050	3/4" diameter		460	.017		.06	.57		.63	1.02	
	3100	1" diameter		460	.017		.07	.57		.64	1.03	
	3500	Polyurethane foam, with polybutylene, 1/2" x 1/2"		475	.017		.60	.56		1.16	1.58	
	3550	1" x 1"		450	.018		1.65	.59		2.24	2.79	
	3750	Polyurethane foam, regular, closed cell, 1/2" x 6"		375	.021		1.16	.70		1.86	2.45	
	3800	1" x 12"		300	.027		1.36	.88		2.24	2.96	
	4000	Polyvinyl chloride foam, closed cell, 1/2" x 6"		375	.021		1.96	.70		2.66	3.33	
	4050	1" x 12"		300	.027		5.45	.88		6.33	7.45	
	4250	Rubber, gray sponge, 1/2" x 6"		375	.021		2.44	.70		3.14	3.85	
	4300	1" x 12"		300	.027		9.55	.88		10.43	11.95	
	4500	Lead wool for joints, 1 ton lots				Lb.	1.93			1.93	2.12	
	5000	For installation in walls, add						75%				
	5250	For installation in boxouts, add						25%				
600	0010	**SHORES** Erect and strip, by hand, horizontal members										600
	0500	Aluminum joists and stringers	2 Carp	60	.267	Ea.		8.80		8.80	14.60	
	0600	Steel, adjustable beams		45	.356			11.75		11.75	19.45	
	0700	Wood joists		50	.320			10.55		10.55	17.50	
	0800	Wood stringers		30	.533			17.60		17.60	29	
	1000	Vertical members to 10' high		55	.291			9.60		9.60	15.90	
	1050	To 13' high		50	.320			10.55		10.55	17.50	
	1100	To 16' high		45	.356			11.75		11.75	19.45	
	1500	Reshoring		1,400	.011	S.F.	.21	.38		.59	.86	
	1600	Flying truss system	C-17D	9,600	.009	SFCA		.30	.05	.35	.55	
	1760	Horizontal, aluminum joists, 6' to 30' spans, buy				L.F.	26			26	28.50	
	1770	Aluminum stringers, 12' & 16' spans				"	35.50			35.50	39	
	1810	Horizontal, steel beam, adjustable, 4' to 7' span				Ea.	120			120	132	
	1830	6' to 10' span					154			154	169	
	1920	9' to 15' span					286			286	315	
	1940	12' to 20' span					335			335	365	
	1970	Steel stringer, 6' to 15' span				L.F.	8.60			8.60	9.45	
	3000	Rent for job duration, aluminum, first month				SF Flr.	.31			.31	.34	
	3050	Steel				"	.22			.22	.24	
	3500	Vertical, adjustable steel, 5'-7" to 9'-6" high, 10,000# cap., buy				Ea.	63			63	69.50	
	3550	7'-3" to 12'-10" high, 7800# capacity					74.50			74.50	82	
	3600	8'-10" to 12'-4" high, 10,000# capacity					86.50			86.50	95	
	3650	8'-10" to 16'-1" high, 3800# capacity					92			92	101	
	4000	Frame shoring systems, aluminum, 10,000# per leg,										
	4050	6' wide, 5' & 6' high				Ea.	182			182	200	
	4100	5' to 7' post with base, jack screw & top plate				Set	56			56	61.50	
	5010	Steel, 10,000# per leg										
	5040	2' & 4' wide, 3', 4', 5' & 6' high				Ea.	101			101	111	
	5250	6' extension tube with adjusting collar					95			95	104	
	5550	Base plate					10.20			10.20	11.20	
	5600	12" adjustable leg					41			41	45	
	5650	Top plate					22.50			22.50	24.50	
660	0010	**STAIR TREAD INSERTS** Cast iron, abrasive, 3" wide	1 Carp	90	.089	L.F.	6.10	2.93		9.03	11.60	660
	0020	4" wide		80	.100		7.45	3.30		10.75	13.65	
	0040	6" wide		75	.107		9.20	3.52		12.72	15.95	
	0050	9" wide		70	.114		13.60	3.77		17.37	21	
	0100	12" wide		65	.123		20.50	4.06		24.56	29.50	
	0300	Cast aluminum, compared to cast iron, deduct					10%					
	0500	Extruded aluminum safety tread, 3" wide	1 Carp	75	.107		5.10	3.52		8.62	11.45	
	0550	4" wide		75	.107		6.90	3.52		10.42	13.40	

03100 | Concrete Forms & Accessories

03150 | Concrete Accessories

							2004 Bare Costs				Total	
			Crew	Daily Output	Labor-Hours	Unit	Mat.	Labor	Equip.	Total	Incl O&P	
660	0600	6″ wide	1 Carp	75	.107	L.F.	11.55	3.52		15.07	18.55	660
	0650	9″ wide to resurface stairs		70	.114		16.25	3.77		20.02	24	
	1700	Cement filled pan type, plain	1 Cefi	115	.070	S.F.	2.55	2.19		4.74	6.25	
	1750	Non-slip	″	100	.080	″	3.83	2.52		6.35	8.20	
860	0010	**WATERSTOP** PVC, ribbed 3/16″ thick, 4″ wide	1 Carp	155	.052	L.F.	.76	1.70		2.46	3.66	860
	0050	6″ wide		145	.055		1.30	1.82		3.12	4.45	
	0500	Ribbed, PVC, with center bulb, 9″ wide, 3/16″ thick		135	.059		1.95	1.96		3.91	5.40	
	0550	3/8″ thick		130	.062		2.95	2.03		4.98	6.60	

03200 | Concrete Reinforcement

03210 | Reinforcing Steel

							2004 Bare Costs				Total	
			Crew	Daily Output	Labor-Hours	Unit	Mat.	Labor	Equip.	Total	Incl O&P	
600	0010	**REINFORCING IN PLACE** A615 Grade 60, incl. reinf. access.										600
	0102	Beams & Girders, #3 to #7	4 Rodm	3,200	.010	Lb.	.30	.37		.67	.97	
	0152	#8 to #18		5,400	.006		.29	.22		.51	.70	
	0202	Columns, #3 to #7		3,000	.011		.30	.40		.70	1.01	
	0252	#8 to #18		4,600	.007		.29	.26		.55	.76	
	0402	Elevated slabs, #4 to #7		5,800	.006		.31	.20		.51	.69	
	0502	Footings, #4 to #7		4,200	.008		.30	.28		.58	.82	
	0552	#8 to #18		7,200	.004		.29	.16		.45	.60	
	0602	Slab on grade, #3 to #7		4,200	.008		.29	.28		.57	.80	
	0702	Walls, #3 to #7		6,000	.005		.30	.20		.50	.67	
	0752	#8 to #18		8,000	.004		.29	.15		.44	.58	
	2000	Unloading & sorting, add to above	C-5	100	.560	Ton		20	6.25	26.25	41	
	2200	Crane cost for handling, add to above, minimum		135	.415			14.90	4.63	19.53	30	
	2210	Average		92	.609			22	6.80	28.80	44.50	
	2220	Maximum		35	1.600			57.50	17.85	75.35	117	
	2400	Dowels, 2 feet long, deformed, #3	2 Rodm	520	.031	Ea.	.24	1.14		1.38	2.22	
	2410	#4		480	.033		.42	1.24		1.66	2.59	
	2420	#5		435	.037		.66	1.36		2.02	3.08	
	2430	#6		360	.044		.94	1.65		2.59	3.88	
	2450	Longer and heavier dowels		725	.022	Lb.	.41	.82		1.23	1.86	
	2500	Smooth dowels, 12″ long, 1/4″ or 3/8″ diameter		140	.114	Ea.	.62	4.24		4.86	8	
	2520	5/8″ diameter		125	.128		1.09	4.75		5.84	9.35	
	2530	3/4″ diameter		110	.145		1.35	5.40		6.75	10.80	
	2700	Dowel caps, 5″ long, 1/2″ to 3/4″ diameter		800	.020		.14	.74		.88	1.44	
	2720	1-1/4″ diameter		750	.021		.36	.79		1.15	1.76	
	9000	Minimum labor/equipment charge	1 Rodm	4	2	Job		74		74	128	

03220 | Welded Wire Fabric

			Crew	Daily Output	Labor-Hours	Unit	Mat.	Labor	Equip.	Total	Total Incl O&P	
200	0010	**WELDED WIRE FABRIC** ASTM A185										200
	0050	Sheets										
	0100	6 x 6 - W1.4 x W1.4 (10 x 10) 21 lb. per C.S.F.	2 Rodm	35	.457	C.S.F.	7.30	16.95		24.25	37	
	0200	6 x 6 - W2.1 x W2.1 (8 x 8) 30 lb. per C.S.F.		31	.516		9.40	19.15		28.55	43.50	
	0300	6 x 6 - W2.9 x W2.9 (6 x 6) 42 lb. per C.S.F.		29	.552		12.50	20.50		33	49	
	0400	6 x 6 - W4 x W4 (4 x 4) 58 lb. per C.S.F.		27	.593		17.90	22		39.90	57.50	
	0500	4 x 4 - W1.4 x W1.4 (10 x 10) 31 lb. per C.S.F.		31	.516		12.35	19.15		31.50	46.50	
	0600	4 x 4 - W2.1 x W2.1 (8 x 8) 44 lb. per C.S.F.		29	.552		14.70	20.50		35.20	51	
	0650	4 x 4 - W2.9 x W2.9 (6 x 6) 61 lb. per C.S.F.		27	.593		20	22		42	60	
	0700	4 x 4 - W4 x W4 (4 x 4) 85 lb. per C.S.F.		25	.640		26	23.50		49.50	69.50	

03200 | Concrete Reinforcement

	03220 | Welded Wire Fabric		Crew	Daily Output	Labor-Hours	Unit	2004 Bare Costs Mat.	Labor	Equip.	Total	Total Incl O&P
200	0750	Rolls									
	0800	2 x 2 - #14 galv., 21 lb/C.S.F., beam & column wrap	2 Rodm	6.50	2.462	C.S.F.	17.95	91.50		109.45	177
	0900	2 x 2 - #12 galv. for gunite reinforcing	″	6.50	2.462	″	20	91.50		111.50	179
	9000	Minimum labor/equipment charge	1 Rodm	4	2	Job		74		74	128

	03240 | Fibrous Reinforcing		Crew	Daily Output	Labor-Hours	Unit	Mat.	Labor	Equip.	Total	Total Incl O&P
300	0010	**FIBROUS REINFORCING**									
	0100	Synthetic fibers, add to concrete				Lb.	3.85			3.85	4.24
	0110	1-1/2 lb. per C.Y.				C.Y.	5.95			5.95	6.55
	0150	Steel fibers, add to concrete				Lb.	.44			.44	.48
	0155	25 lb. per C.Y.				C.Y.	11			11	12.10
	0160	50 lb. per C.Y.					22			22	24
	0170	75 lb. per C.Y.					34			34	37.50
	0180	100 lb. per C.Y.					44			44	48.50

03300 | Cast-In-Place Concrete

	03310 | Structural Concrete		Crew	Daily Output	Labor-Hours	Unit	2004 Bare Costs Mat.	Labor	Equip.	Total	Total Incl O&P
220	0010	**CONCRETE, READY MIX** Normal weight (R03310 -020)									
	0020	2000 psi				C.Y.	69.50			69.50	76.50
	0100	2500 psi (R03310 -060)					71			71	78
	0150	3000 psi					72.50			72.50	80
	0200	3500 psi					74			74	81.50
	0300	4000 psi					76.50			76.50	84
	0350	4500 psi					78.50			78.50	86
	0400	5000 psi					81			81	89
	0411	6000 psi					92.50			92.50	102
	0412	8000 psi					151			151	166
	0413	10,000 psi					214			214	235
	0414	12,000 psi					258			258	284
	1000	For high early strength cement, add					10%				
	2000	For all lightweight aggregate, add					45%				
	3000	For integral colors, 2500 psi, 5 bag mix									
	3100	Red, yellow or brown, 1.8 lb. per bag, add				C.Y.	19.60			19.60	21.50
	3200	9.4 lb. per bag, add					102			102	113
	3400	Black, 1.8 lb. per bag, add					23.50			23.50	25.50
	3500	7.5 lb. per bag, add					97.50			97.50	107
	3700	Green, 1.8 lb. per bag, add					27			27	29.50
	3800	7.5 lb. per bag, add					112			112	123
240	0010	**CONCRETE IN PLACE** Including forms (4 uses), reinforcing (R03310 -010)									
	0050	steel and finishing unless otherwise indicated									
	0300	Beams, 5 kip per L.F., 10′ span (R03310 -100)	C-14A	15.62	12.804	C.Y.	233	425	48	706	1,025
	0350	25′ span	″	18.55	10.782		214	355	40	609	875
	0500	Chimney foundations, industrial, minimum (R04210 -055)	C-14C	32.22	3.476		145	110	.80	255.80	345
	0510	Maximum	″	23.71	4.724		168	149	1.09	318.09	435
	0700	Columns, square, 12″ x 12″, minimum reinforcing	C-14A	11.96	16.722		212	555	62.50	829.50	1,225
	0720	Average reinforcing		10.13	19.743		310	655	73.50	1,038.50	1,525
	0740	Maximum reinforcing		9.03	22.148		390	735	82.50	1,207.50	1,750
	0800	16″ x 16″, minimum reinforcing		16.22	12.330		197	410	46	653	945

03310 | Structural Concrete

				CREW	DAILY OUTPUT	LABOR-HOURS	UNIT	2004 BARE COSTS MAT.	LABOR	EQUIP.	TOTAL	TOTAL INCL O&P	
240	0820	Average reinforcing	R03310-010	C-14A	12.57	15.911	C.Y.	310	525	59.50	894.50	1,275	240
	0840	Maximum reinforcing			10.25	19.512		425	645	73	1,143	1,625	
	0900	24" x 24", minimum reinforcing	R03310-100		23.66	8.453		166	280	31.50	477.50	685	
	0920	Average reinforcing			17.71	11.293		258	375	42	675	950	
	0940	Maximum reinforcing	R04210-055		14.15	14.134		350	470	52.50	872.50	1,225	
	1000	36" x 36", minimum reinforcing			33.69	5.936		153	197	22	372	520	
	1020	Average reinforcing			23.32	8.576		240	284	32	556	775	
	1040	Maximum reinforcing			17.82	11.223		325	370	42	737	1,025	
	1200	16" diameter, minimum reinforcing			31.49	6.351		188	210	23.50	421.50	585	
	1220	Average reinforcing			19.12	10.460		320	345	39	704	970	
	1240	Maximum reinforcing			13.77	14.524		445	480	54	979	1,350	
	1300	20" diameter, minimum reinforcing			41.04	4.873		181	162	18.20	361.20	490	
	1320	Average reinforcing			24.05	8.316		295	276	31	602	820	
	1340	Maximum reinforcing			17.01	11.758		410	390	44	844	1,150	
	1400	24" diameter, minimum reinforcing			51.85	3.857		176	128	14.40	318.40	425	
	1420	Average reinforcing			27.06	7.391		293	245	27.50	565.50	755	
	1440	Maximum reinforcing			18.29	10.935		410	360	41	811	1,100	
	1500	36" diameter, minimum reinforcing			75.04	2.665		165	88.50	9.95	263.45	340	
	1520	Average reinforcing			37.49	5.335		253	177	19.90	449.90	595	
	1540	Maximum reinforcing		↓	22.84	8.757		365	290	32.50	687.50	920	
	1900	Elevated slabs, flat slab, 125 psf Sup. Load, 20' span		C-14B	38.45	5.410		159	179	19.35	357.35	495	
	1950	30' span			50.99	4.079		145	135	14.60	294.60	400	
	2100	Flat plate, 125 psf Sup. Load, 15' span			30.24	6.878		169	228	24.50	421.50	595	
	2150	25' span			49.60	4.194		139	139	15	293	400	
	2300	Waffle const., 30" domes, 125 psf Sup. Load, 20' span			37.07	5.611		280	186	20	486	640	
	2350	30' span			44.07	4.720		247	156	16.90	419.90	550	
	2500	One way joists, 30" pans, 125 psf Sup. Load, 15' span			27.38	7.597		380	251	27	658	865	
	2550	25' span			31.15	6.677		345	221	24	590	765	
	2700	One way beam & slab, 125 psf Sup. Load, 15' span			20.59	10.102		194	335	36	565	810	
	2750	25' span			28.36	7.334		176	243	26.50	445.50	630	
	2900	Two way beam & slab, 125 psf Sup. Load, 15' span			24.04	8.652		182	286	31	499	710	
	2950	25' span		↓	35.87	5.799	↓	154	192	21	367	510	
	3100	Elevated slabs including finish, not											
	3110	including forms or reinforcing											
	3150	Regular concrete, 4" slab		C-8	2,613	.021	S.F.	.97	.62	.28	1.87	2.37	
	3200	6" slab			2,585	.022		1.51	.63	.28	2.42	2.98	
	3250	2-1/2" thick floor fill			2,685	.021		.66	.60	.27	1.53	2	
	3300	Lightweight, 110# per C.F., 2-1/2" thick floor fill			2,585	.022		.93	.63	.28	1.84	2.34	
	3400	Cellular concrete, 1-5/8" fill, under 5000 S.F.			2,000	.028		.54	.81	.36	1.71	2.30	
	3450	Over 10,000 S.F.			2,200	.025		.43	.74	.33	1.50	2.02	
	3500	Add per floor for 3 to 6 stories high			31,800	.002			.05	.02	.07	.10	
	3520	For 7 to 20 stories high		↓	21,200	.003	↓		.08	.03	.11	.16	
	3800	Footings, spread under 1 C.Y.		C-14C	38.07	2.942	C.Y.	105	93	.68	198.68	272	
	3850	Over 5 C.Y.	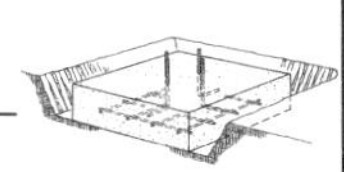		81.04	1.382		96.50	43.50	.32	140.32	179	
	3900	Footings, strip, 18" x 9", unreinforced			40	2.800		91	88.50	.64	180.14	248	
	3920	18" x 9", reinforced			35	3.200		105	101	.74	206.74	284	
	3925	20" x 10", unreinforced			45	2.489		88	78.50	.57	167.07	229	
	3930	20" x 10", reinforced			40	2.800		99.50	88.50	.64	188.64	257	
	3935	24" x 12", unreinforced			55	2.036		87	64.50	.47	151.97	204	
	3940	24" x 12", reinforced			48	2.333		98.50	74	.54	173.04	233	
	3945	36" x 12", unreinforced			70	1.600		83.50	50.50	.37	134.37	176	
	3950	36" x 12", reinforced			60	1.867		94	59	.43	153.43	201	
	4000	Foundation mat, under 10 C.Y.			38.67	2.896		129	91.50	.67	221.17	295	
	4050	Over 20 C.Y.		↓	56.40	1.986		115	63	.46	178.46	232	
	4200	Grade walls, 8" thick, 8' high		C-14D	45.83	4.364		126	143	16.30	285.30	395	
	4250	14' high		↓	27.26	7.337	↓	158	241	27.50	426.50	605	

	03310 | Structural Concrete		Crew	Daily Output	Labor-Hours	Unit	2004 Bare Costs Mat.	Labor	Equip.	Total	Total Incl O&P	
240	4260 12″ thick, 8′ high	R03310-010	C-14D	64.32	3.109	C.Y.	112	102	11.60	225.60	305	240
	4270 14′ high			40.01	4.999		126	164	18.65	308.65	430	
	4300 15″ thick, 8′ high	R03310-100		80.02	2.499		106	82	9.30	197.30	262	
	4350 12′ high			51.26	3.902		112	128	14.55	254.55	350	
	4500 18′ high	R04210-055		48.85	4.094		122	134	15.25	271.25	375	
	4520 Handicap access ramp, railing both sides, 3′ wide		C-14H	14.58	3.292	L.F.	149	107	1.78	257.78	345	
	4525 5′ wide			12.22	3.928		156	128	2.12	286.12	385	
	4530 With 6″ curb and rails both sides, 3′ wide			8.55	5.614		155	183	3.03	341.03	480	
	4535 5′ wide			7.31	6.566		159	214	3.55	376.55	535	
	4650 Slab on grade, not including finish, 4″ thick		C-14E	60.75	1.449	C.Y.	88.50	47.50	.43	136.43	177	
	4700 6″ thick		″	92	.957	″	85	31	.29	116.29	146	
	4751 Slab on grade, incl. troweled finish, not incl. forms											
	4760 or reinforcing, over 10,000 S.F., 4″ thick		C-14F	3,425	.021	S.F.	.96	.63	.01	1.60	2.07	
	4820 6″ thick			3,350	.021		1.41	.64	.01	2.06	2.59	
	4840 8″ thick			3,184	.023		1.93	.68	.01	2.62	3.21	
	4900 12″ thick			2,734	.026		2.89	.79	.01	3.69	4.45	
	4950 15″ thick			2,505	.029		3.64	.86	.01	4.51	5.40	
	5000 Slab on grade, incl. textured finish, not incl. forms											
	5001 or reinforcing, 4″ thick		C-14G	2,873	.019	S.F.	.94	.57	.01	1.52	1.97	
	5010 6″ thick			2,590	.022		1.47	.64	.01	2.12	2.65	
	5020 8″ thick			2,320	.024		1.92	.71	.01	2.64	3.26	
	5200 Lift slab in place above the foundation, incl. forms,											
	5210 reinforcing, concrete and columns, minimum		C-14B	2,113	.098	S.F.	5.10	3.26	.35	8.71	11.40	
	5250 Average			1,650	.126		5.75	4.17	.45	10.37	13.70	
	5300 Maximum			1,500	.139		6.50	4.59	.50	11.59	15.30	
	5500 Lightweight, ready mix, including screed finish only,											
	5510 not including forms or reinforcing											
	5550 1:4 for structural roof decks		C-14B	260	.800	C.Y.	105	26.50	2.86	134.36	163	
	5600 1:6 for ground slab with radiant heat		C-14F	92	.783		100	23.50	.28	123.78	148	
	5650 1:3:2 with sand aggregate, roof deck		C-14B	260	.800		104	26.50	2.86	133.36	161	
	5700 Ground slab		C-14F	107	.673		104	20	.24	124.24	146	
	5900 Pile caps, incl. forms and reinf., sq. or rect., under 5 C.Y.		C-14C	54.14	2.069		89	65.50	.48	154.98	208	
	5950 Over 10 C.Y.			75	1.493		92	47	.34	139.34	180	
	6000 Triangular or hexagonal, under 5 C.Y.			53	2.113		86.50	67	.49	153.99	207	
	6050 Over 10 C.Y.			85	1.318		94.50	41.50	.30	136.30	174	
	6200 Retaining walls, gravity, 4′ high see division 02830-100		C-14D	66.20	3.021		97.50	99	11.25	207.75	283	
	6250 10′ high			125	1.600		86	52.50	5.95	144.45	188	
	6300 Cantilever, level backfill loading, 8′ high			70	2.857		109	94	10.65	213.65	287	
	6350 16′ high			91	2.198		103	72	8.20	183.20	241	
	6800 Stairs, not including safety treads, free standing, 3′-6″ wide		C-14H	83	.578	LF Nose	6.90	18.85	.31	26.06	39	
	6850 Cast on ground			125	.384	″	4.85	12.50	.21	17.56	26	
	7000 Stair landings, free standing			200	.240	S.F.	2.65	7.85	.13	10.63	16	
	7050 Cast on ground			475	.101	″	1.53	3.30	.05	4.88	7.20	
	9000 Minimum labor/equipment charge		2 Carp	1	16	Job		530		530	875	
700	0010 **PLACING CONCRETE** and vibrating, including labor & equipment	R03310-090										700
	0050 Beams, elevated, small beams, pumped		C-20	60	1.067	C.Y.		30	12.85	42.85	62.50	
	0100 With crane and bucket		C-7	45	1.600			45	21.50	66.50	97	
	0200 Large beams, pumped		C-20	90	.711			19.85	8.55	28.40	42	
	0250 With crane and bucket		C-7	65	1.108			31	15.05	46.05	67	
	0400 Columns, square or round, 12″ thick, pumped		C-20	60	1.067			30	12.85	42.85	62.50	
	0450 With crane and bucket		C-7	40	1.800			50.50	24.50	75	109	
	0600 18″ thick, pumped		C-20	90	.711			19.85	8.55	28.40	42	
	0650 With crane and bucket		C-7	55	1.309			36.50	17.80	54.30	79	
	0800 24″ thick, pumped		C-20	92	.696			19.40	8.40	27.80	40.50	
	0850 With crane and bucket		C-7	70	1.029			29	14	43	62.50	
	1000 36″ thick, pumped		C-20	140	.457			12.75	5.50	18.25	27	

	03310 | Structural Concrete	CREW	DAILY OUTPUT	LABOR-HOURS	UNIT	MAT.	LABOR	EQUIP.	TOTAL	TOTAL INCL O&P	
700	1050 With crane and bucket R03310-090	C-7	100	.720	C.Y.		20	9.80	29.80	44	700
	1400 Elevated slabs, less than 6" thick, pumped	C-20	140	.457			12.75	5.50	18.25	27	
	1450 With crane and bucket	C-7	95	.758			21.50	10.30	31.80	46	
	1500 6" to 10" thick, pumped	C-20	160	.400			11.15	4.82	15.97	23.50	
	1550 With crane and bucket	C-7	110	.655			18.35	8.90	27.25	40	
	1600 Slabs over 10" thick, pumped	C-20	180	.356			9.90	4.28	14.18	21	
	1650 With crane and bucket	C-7	130	.554			15.55	7.55	23.10	34	
	1900 Footings, continuous, shallow, direct chute	C-6	120	.400			10.90	.43	11.33	18.40	
	1950 Pumped	C-20	150	.427			11.90	5.15	17.05	25	
	2000 With crane and bucket	C-7	90	.800			22.50	10.85	33.35	48.50	
	2100 Footings, continuous, deep, direct chute	C-6	140	.343			9.35	.37	9.72	15.75	
	2150 Pumped	C-20	160	.400			11.15	4.82	15.97	23.50	
	2200 With crane and bucket	C-7	110	.655			18.35	8.90	27.25	40	
	2400 Footings, spread, under 1 C.Y., direct chute	C-6	55	.873			24	.94	24.94	40	
	2450 Pumped	C-20	65	.985			27.50	11.85	39.35	58	
	2500 With crane and bucket	C-7	45	1.600			45	21.50	66.50	97	
	2600 Over 5 C.Y., direct chute	C-6	120	.400			10.90	.43	11.33	18.40	
	2650 Pumped	C-20	150	.427			11.90	5.15	17.05	25	
	2700 With crane and bucket	C-7	100	.720			20	9.80	29.80	44	
	2900 Foundation mats, over 20 C.Y., direct chute	C-6	350	.137			3.74	.15	3.89	6.30	
	2950 Pumped	C-20	400	.160			4.46	1.93	6.39	9.40	
	3000 With crane and bucket	C-7	300	.240			6.75	3.26	10.01	14.55	
	3200 Grade beams, direct chute	C-6	150	.320			8.70	.35	9.05	14.70	
	3250 Pumped	C-20	180	.356			9.90	4.28	14.18	21	
	3300 With crane and bucket	C-7	120	.600			16.80	8.15	24.95	36.50	
	3500 High rise, for more than 5 stories, pumped, add per story	C-20	2,100	.030			.85	.37	1.22	1.79	
	3510 With crane and bucket, add per story	C-7	2,100	.034			.96	.47	1.43	2.07	
	3700 Pile caps, under 5 C.Y., direct chute	C-6	90	.533			14.55	.58	15.13	24.50	
	3750 Pumped	C-20	110	.582			16.25	7	23.25	34	
	3800 With crane and bucket	C-7	80	.900			25	12.25	37.25	54.50	
	3850 Pile cap, 5 C.Y. to 10 C.Y., direct chute	C-6	175	.274			7.50	.30	7.80	12.60	
	3900 Pumped	C-20	200	.320			8.95	3.86	12.81	18.85	
	3950 With crane and bucket	C-7	150	.480			13.45	6.50	19.95	29	
	4000 Over 10 C.Y., direct chute	C-6	215	.223			6.10	.24	6.34	10.25	
	4050 Pumped	C-20	240	.267			7.45	3.21	10.66	15.70	
	4100 With crane and bucket	C-7	185	.389			10.90	5.30	16.20	23.50	
	4300 Slab on grade, 4" thick, direct chute	C-6	110	.436			11.90	.47	12.37	20	
	4350 Pumped	C-20	130	.492			13.75	5.95	19.70	29	
	4400 With crane and bucket	C-7	110	.655			18.35	8.90	27.25	40	
	4600 Over 6" thick, direct chute	C-6	165	.291			7.95	.31	8.26	13.35	
	4650 Pumped	C-20	185	.346			9.65	4.17	13.82	20.50	
	4700 With crane and bucket	C-7	145	.497			13.90	6.75	20.65	30	
	4900 Walls, 8" thick, direct chute	C-6	90	.533			14.55	.58	15.13	24.50	
	4950 Pumped	C-20	100	.640			17.85	7.70	25.55	37.50	
	5000 With crane and bucket	C-7	80	.900			25	12.25	37.25	54.50	
	5050 12" thick, direct chute	C-6	100	.480			13.10	.52	13.62	22	
	5100 Pumped	C-20	110	.582			16.25	7	23.25	34	
	5200 With crane and bucket	C-7	90	.800			22.50	10.85	33.35	48.50	
	5300 15" thick, direct chute	C-6	105	.457			12.45	.49	12.94	21	
	5350 Pumped	C-20	120	.533			14.90	6.45	21.35	31.50	
	5400 With crane and bucket	C-7	95	.758	↓		21.50	10.30	31.80	46	
	5600 Wheeled concrete dumping, add to placing costs above										
	5610 Walking cart, 50' haul, add	C-18	32	.281	C.Y.		7.35	1.52	8.87	13.95	
	5620 150' haul, add		24	.375			9.85	2.03	11.88	18.55	
	5700 250' haul, add	↓	18	.500			13.10	2.71	15.81	24.50	
	5800 Riding cart, 50' haul, add	C-19	80	.112	↓		2.95	.93	3.88	5.90	

Table header: 2004 BARE COSTS (MAT., LABOR, EQUIP., TOTAL)

3 CONCRETE

03300 | Cast-In-Place Concrete

03310 | Structural Concrete

		03310 Structural Concrete	CREW	DAILY OUTPUT	LABOR-HOURS	UNIT	MAT.	LABOR	EQUIP.	TOTAL	TOTAL INCL O&P	
							2004 BARE COSTS					
700	5810	150′ haul, add (R03310-090)	C-19	60	.150	C.Y.		3.93	1.25	5.18	7.85	700
	5900	250′ haul, add		45	.200			5.25	1.66	6.91	10.55	
	9000	Minimum labor/equipment charge	C-6	2	24	Job		655	26	681	1,100	

03350 | Concrete Finishing

		03350 Concrete Finishing	CREW	DAILY OUTPUT	LABOR-HOURS	UNIT	MAT.	LABOR	EQUIP.	TOTAL	TOTAL INCL O&P	
300	0010	**FINISHING FLOORS** Monolithic, screed finish	1 Cefi	900	.009	S.F.		.28		.28	.44	300
	0100	Screed and bull float (darby) finish		725	.011			.35		.35	.55	
	0150	Screed, float, and broom finish		630	.013			.40		.40	.63	
	0200	Screed, float, and hand trowel		600	.013			.42		.42	.66	
	0250	Machine trowel		550	.015			.46		.46	.72	
	0370	Minimum labor/equipment charge		4	2	Job		63		63	99	
	0380											
	0400	Integral topping and finish, using 1:1:2 mix, 3/16″ thick	C-10B	1,000	.040	S.F.	.06	1.13	.16	1.35	2.08	
	0450	1/2″ thick		950	.042		.16	1.19	.17	1.52	2.28	
	0500	3/4″ thick		850	.047		.25	1.33	.19	1.77	2.64	
	0600	1″ thick		750	.053		.33	1.51	.21	2.05	3.03	
	0800	Granolithic topping, laid after, 1:1:1-1/2 mix, 1/2″ thick		590	.068		.18	1.91	.27	2.36	3.59	
	0820	3/4″ thick		580	.069		.27	1.95	.28	2.50	3.74	
	0850	1″ thick		575	.070		.35	1.96	.28	2.59	3.88	
	0950	2″ thick		500	.080		.71	2.26	.32	3.29	4.79	
	9100	Minimum labor/equipment charge	C-10	2	12	Job		355		355	570	
325	0010	**CONTROL JOINT**, concrete floor slab										325
	0100	Sawcut in green concrete										
	0120	1″ depth	C-27	2,000	.008	L.F.		.25	.05	.30	.45	
	0140	1-1/2″ depth		1,800	.009			.28	.05	.33	.50	
	0160	2″ depth		1,600	.010			.32	.06	.38	.56	
	0200	Clean out control joint of debris	C-28	6,000	.001			.04		.04	.07	
	0300	Joint sealant										
	0320	Backer rod, polyethylene, 1/4″ diameter	1 Cefi	460	.017	L.F.	.02	.55		.57	.88	
	0340	Sealant, polyurethane										
	0360	1/4″ x 1/4″ (308 LF/Gal)	1 Cefi	270	.030	L.F.	.15	.93		1.08	1.64	
	0380	1/4″ x 1/2″ (154 LF/Gal)	″	255	.031	″	.31	.99		1.30	1.90	
350	0010	**FINISHING WALLS** Break ties and patch voids	1 Cefi	540	.015	S.F.	.03	.47		.50	.76	350
	0050	Burlap rub with grout	″	450	.018		.03	.56		.59	.91	
	0300	Bush hammer, green concrete	B-39	1,000	.048		.03	1.31	.16	1.50	2.37	
	0350	Cured concrete	″	650	.074		.03	2.02	.25	2.30	3.62	
	0700	Sandblast, light penetration	C-10	1,100	.022		.22	.65		.87	1.27	
	0750	Heavy penetration		375	.064		.44	1.90		2.34	3.52	
	9000	Minimum labor/equipment charge		2	12	Job		355		355	570	
600	0010	**SLAB TEXTURE STAMPING**, buy										600
	0020	Approx. 3 S.F.- 5 S.F. each, minimum				Ea.	44			44	48.50	
	0030	Average				″	48.50			48.50	53	
	0120	Per S.F. of tool, average				S.F.	53			53	58	
	0200	Commonly used chemicals for texture systems										
	0210	Hardeners w/colors average				S.F.	.44			.44	.48	
	0220	Release agents w/colors, average					.18			.18	.20	
	0225	Clear, average					.14			.14	.15	
	0230	Sealers, clear, average					.12			.12	.13	
	0240	Colors, average					.15			.15	.17	

03370 | Specially Placed Concrete

		03370 Specially Placed Concrete	CREW	DAILY OUTPUT	LABOR-HOURS	UNIT	MAT.	LABOR	EQUIP.	TOTAL	TOTAL INCL O&P	
300	0010	**GUNITE**, dry mix										300
	0020	Applied in 1″ layers, no mesh included	C-8	2,000	.028	S.F.	.98	.81	.36	2.15	2.79	

03300 | Cast-In-Place Concrete

03370 | Specially Placed Concrete

		03370 Specially Placed Concrete	CREW	DAILY OUTPUT	LABOR-HOURS	UNIT	2004 BARE COSTS MAT.	LABOR	EQUIP.	TOTAL	TOTAL INCL O&P	
300	0300	Typical in place, including mesh, 2″ thick, minimum	C-16	1,000	.072	S.F.	1.89	2.22	.72	4.83	6.50	300
	0350	Maximum		500	.144		2.93	4.43	1.44	8.80	12.10	
	0500	4″ thick, minimum		750	.096		2.80	2.95	.96	6.71	9	
	0550	Maximum	↓	350	.206		4.35	6.35	2.06	12.76	17.45	
	0900	Prepare old walls, no scaffolding, minimum	C-10	1,000	.024		.66	.71		1.37	1.87	
	0950	Maximum	″	275	.087		2.04	2.59		4.63	6.40	
	1100	For high finish requirement or close tolerance, add, minimum						50%				
	1150	Maximum				↓		110%				
	9000	Minimum labor/equipment charge	C-10	1	24	Job		715		715	1,150	

03390 | Concrete Curing

		03390 Concrete Curing	CREW	DAILY OUTPUT	LABOR-HOURS	UNIT	MAT.	LABOR	EQUIP.	TOTAL	TOTAL INCL O&P	
200	0010	**CURING** Burlap, 4 uses assumed, 7.5 oz.	2 Clab	55	.291	C.S.F.	6.85	7.55		14.40	20	200
	0100	10 oz.		55	.291		11.15	7.55		18.70	25	
	0200	Waterproof curing paper, 2 ply, reinforced		70	.229		5.95	5.95		11.90	16.40	
	0300	Sprayed membrane curing compound	↓	95	.168	↓	5.20	4.38		9.58	12.95	
	0710	Electrically, heated pads, 15 watts/S.F., 20 uses, minimum				S.F.	.16			.16	.18	
	0800	Maximum				″	.27			.27	.30	
	9000	Minimum labor/equipment charge	1 Clab	5	1.600	Job		41.50		41.50	69	

03400 | Precast Concrete

03410 | Plant-Precast Structural Concrete

		03410 Plant-Precast Structural Concrete	CREW	DAILY OUTPUT	LABOR-HOURS	UNIT	2004 BARE COSTS MAT.	LABOR	EQUIP.	TOTAL	TOTAL INCL O&P	
100	0011	**BEAMS**, "L" shaped, 20′ span, 12″ x 20″	C-11	32	2.250	Ea.	1,425	81.50	50.50	1,557	1,775	100
	1000	Inverted tee beams, add to above, small beams				L.F.	15%					
	1050	Large beams				″	5.85			5.85	6.40	
	1200	Rectangular, 20′ span, 12″ x 20″	C-11	32	2.250	Ea.	975	81.50	50.50	1,107	1,275	
	1250	18″ x 36″		24	3		1,800	109	67	1,976	2,250	
	1300	24″ x 44″		22	3.273		2,575	119	73	2,767	3,125	
	1400	30′ span, 12″ x 36″		24	3		2,275	109	67	2,451	2,775	
	1450	18″ x 44″		20	3.600		3,200	130	80.50	3,410.50	3,850	
	1500	24″ x 52″		16	4.500		4,550	163	101	4,814	5,400	
	1600	40′ span, 12″ x 52″		20	3.600		4,225	130	80.50	4,435.50	4,975	
	1650	18″ x 52″		16	4.500		5,150	163	101	5,414	6,075	
	1700	24″ x 52″		12	6		6,300	217	134	6,651	7,475	
	2000	"T" shaped, 20′ span, 12″ x 20″		32	2.250		1,700	81.50	50.50	1,832	2,050	
	2050	18″ x 36″		24	3		2,700	109	67	2,876	3,225	
	2100	24″ x 44″		22	3.273		3,775	119	73	3,967	4,450	
	2200	30′ span, 12″ x 36″		24	3		3,825	109	67	4,001	4,475	
	2250	18″ x 44″		20	3.600		5,225	130	80.50	5,435.50	6,075	
	2300	24″ x 52″		16	4.500		5,400	163	101	5,664	6,350	
	2500	40′ span, 12″ x 52″		20	3.600		7,200	130	80.50	7,410.50	8,250	
	2550	18″ x 52″		16	4.500		7,875	163	101	8,139	9,075	
	2600	24″ x 52″	↓	12	6	↓	9,625	217	134	9,976	11,100	
210	0010	**COLUMNS** Rectangular to 12′ high, small columns	C-11	120	.600	L.F.	71	21.50	13.40	105.90	132	210
	0050	Large columns	″	96	.750	″	113	27	16.75	156.75	191	
620	0010	**SLABS** Prestressed roof/floor members, grouted, solid, 4″ thick (R03410-030)	C-11	2,400	.030	S.F.	3.82	1.09	.67	5.58	6.90	620
	0050	6″ thick	↓	2,800	.026	↓	4.67	.93	.57	6.17	7.45	

03400 | Precast Concrete

03410 | Plant-Precast Structural Concrete

			CREW	DAILY OUTPUT	LABOR-HOURS	UNIT	2004 BARE COSTS MAT.	LABOR	EQUIP.	TOTAL	TOTAL INCL O&P	
620	0100	Hollow, 8″ thick R03410-030	C-11	3,200	.023	S.F.	5.10	.82	.50	6.42	7.60	620
	0150	10″ thick		3,600	.020		4.55	.72	.45	5.72	6.80	
	0200	12″ thick		4,000	.018		5.65	.65	.40	6.70	7.80	

03450 | Plant-Precast Architectural Concrete

			CREW	DAILY OUTPUT	LABOR-HOURS	UNIT	MAT.	LABOR	EQUIP.	TOTAL	TOTAL INCL O&P	
850	0011	**WALL PANELS**										850
	2200	Fiberglass reinforced cement with urethane core										
	2210	R20, 8′ x 8′, minimum	E-2	750	.075	S.F.	8.70	2.69	1.89	13.28	16.50	
	2220	Maximum	″	600	.093	″	16.30	3.36	2.36	22.02	26.50	
855	0010	**PRECAST WINDOW SILLS**										855
	0600	Precast concrete, 4″ tapers to 3″, 9″ wide	D-1	70	.229	L.F.	9.70	6.85		16.55	22	
	0650	11″ wide		60	.267		13.60	8		21.60	28	
	0700	13″ wide, 3 1/2″ tapers to 2 1/2″, 12″ wall		50	.320		13.50	9.60		23.10	30.50	

03480 | Precast Concrete Specialties

			CREW	DAILY OUTPUT	LABOR-HOURS	UNIT	MAT.	LABOR	EQUIP.	TOTAL	TOTAL INCL O&P	
400	0010	**LINTELS**										400
	0800	Precast concrete, 4″ wide, 8″ high, to 5′ long	D-10	28	1.429	Ea.	24	44.50	17.25	85.75	117	
	0850	5′-12′ long		24	1.667		68	52	20	140	181	
	1000	6″ wide, 8″ high, to 5′ long		26	1.538		29	48	18.55	95.55	130	
	1050	5′-12′ long		22	1.818		84	57	22	163	209	
	1200	8″ wide, 8″ high, to 5′ long		24	1.667		35	52	20	107	145	
	1250	5′-12′ long		20	2		90.50	62.50	24	177	227	
	1400	10″ wide, 8″ high, to 14′ long		18	2.222		149	69.50	27	245.50	305	
	1450	12″ wide, 8″ high, to 19′ long		16	2.500		195	78.50	30	303.50	375	
800	0010	**STAIRS**, Precast concrete treads on steel stringers, 3′ wide	C-12	75	.640	Riser	56.50	21	8.15	85.65	106	800
	0300	Front entrance, 5′ wide with 48″ platform, 2 risers		16	3	Flight	289	97.50	38.50	425	520	
	0350	5 risers		12	4		330	130	51	511	635	
	0500	6′ wide, 2 risers		15	3.200		330	104	41	475	580	
	0550	5 risers		11	4.364		355	142	55.50	552.50	690	
	1200	Basement entrance stairs, steel bulkhead doors, minimum	B-51	22	2.182		485	57.50	5.55	548.05	630	
	1250	Maximum	″	11	4.364		840	115	11.10	966.10	1,125	

03500 | Cementitious Decks & Underlayments

03510 | Cementitious Roof Deck

			CREW	DAILY OUTPUT	LABOR-HOURS	UNIT	2004 BARE COSTS MAT.	LABOR	EQUIP.	TOTAL	TOTAL INCL O&P	
200	0010	**WOOD FIBER** Lightweight cement system										200
	0050	Plank, beveled, 1″ thick	2 Carp	1,000	.016	S.F.	1.77	.53		2.30	2.83	
	0100	Plank, T & G, 1-1/2″ thick		975	.016		1.89	.54		2.43	2.98	
	0150	2″ thick		950	.017		2.52	.56		3.08	3.69	
	0200	2-1/2″ thick		925	.017		2.75	.57		3.32	3.98	
	0300	3-1/2″ thick		875	.018		4.63	.60		5.23	6.10	
	0350	4″ thick		850	.019		5.15	.62		5.77	6.75	
	1000	Bulb tee, sub-purlin and grout, 6′ span, add	E-1	5,000	.005		1.72	.17	.02	1.91	2.22	
	1100	8′ span	″	4,200	.006		1.77	.21	.02	2	2.33	

Important: See the Reference Section for critical supporting data - Reference Nos., Crews, & City Cost Indexes

03500 | Cementitious Decks & Underlayments

		03520 | Lightweight Concrete Roof Insulation	CREW	DAILY OUTPUT	LABOR-HOURS	UNIT	2004 BARE COSTS MAT.	LABOR	EQUIP.	TOTAL	TOTAL INCL O&P	
250	0010	**INSULATING ROOF FILL**, lightweight cellular concrete R03520-010										250
	0020	Portland cement and foaming agent	C-8	50	1.120	C.Y.	73.50	32.50	14.40	120.40	149	
	0100	Poured vermiculite or perlite, field mix,										
	0110	1:6 field mix	C-8	50	1.120	C.Y.	90.50	32.50	14.40	137.40	168	
	0200	Ready mix, 1:6 mix, roof fill, 2" thick	↓	10,000	.006	S.F.	.50	.16	.07	.73	.89	
	0250	3" thick	↓	7,700	.007	"	.75	.21	.09	1.05	1.27	

CONCRETE 3

03900 | Concrete Restoration & Cleaning

		03920 | Concrete Resurfacing	CREW	DAILY OUTPUT	LABOR-HOURS	UNIT	2004 BARE COSTS MAT.	LABOR	EQUIP.	TOTAL	TOTAL INCL O&P	
600	0010	**PATCHING CONCRETE**										600
	0100	Floors, 1/4" thick, small areas, regular grout	1 Cefi	170	.047	S.F.	.07	1.48		1.55	2.41	
	0150	Epoxy grout	"	100	.080	"	4.05	2.52		6.57	8.45	
	0300	Slab on Grade, cut outs, up to 50 C.F.	2 Cefi	50	.320	C.F.	5.35	10.10		15.45	22	
	2000	Walls, including chipping, cleaning and epoxy grout										
	2100	Minimum	1 Cefi	65	.123	S.F.	3.29	3.88		7.17	9.70	
	2150	Average	↓	50	.160	↓	6.60	5.05		11.65	15.20	
	2200	Maximum	↓	40	.200	↓	13.15	6.30		19.45	24.50	
	2510	Underlayment, P.C based self-leveling, 4100 psi, pumped, 1/4"	C-8	20,000	.003	↓	1.39	.08	.04	1.51	1.70	
	2520	1/2"	↓	19,000	.003	↓	2.50	.09	.04	2.63	2.93	
	2530	3/4"	↓	18,000	.003	↓	3.89	.09	.04	4.02	4.47	
	2540	1"	↓	17,000	.003	↓	5.30	.10	.04	5.44	6	
	2550	1-1/2"	↓	15,000	.004	↓	8.05	.11	.05	8.21	9.05	
	2560	Hand mix, 1/2"	C-18	4,000	.002	↓	2.50	.06	.01	2.57	2.86	
	2610	Topping, P.C. based self-level/dry 6100 psi, pumped, 1/4"	C-8	20,000	.003	↓	1.96	.08	.04	2.08	2.33	
	2620	1/2"	↓	19,000	.003	↓	3.53	.09	.04	3.66	4.06	
	2630	3/4"	↓	18,000	.003	↓	5.50	.09	.04	5.63	6.25	
	2660	1"	↓	17,000	.003	↓	7.45	.10	.04	7.59	8.40	
	2670	1-1/2"	↓	15,000	.004	↓	11.35	.11	.05	11.51	12.70	
	2680	Hand mix, 1/2"	C-18	4,000	.002	↓	3.53	.06	.01	3.60	3.99	
	9000	Minimum labor/equipment charge	1 Cefi	4.50	1.778	Job		56		56	88	

		03930 | Concrete Rehabilitation	CREW	DAILY OUTPUT	LABOR-HOURS	UNIT	MAT.	LABOR	EQUIP.	TOTAL	TOTAL INCL O&P	
300	0010	**CRACK REPAIR**, including chipping, sand blasting and cleaning										300
	0100	Epoxy injection, 1/8" wide, 12" deep	B-9	80	.500	L.F.	4.66	13.20	2	19.86	29.50	
	0110	1/4" wide, 12" deep	↓	60	.667	↓	9.30	17.60	2.67	29.57	42	
	0200	Latex injection, 1/8" wide, 12" deep	↓	100	.400	↓	1.28	10.55	1.60	13.43	20.50	
	0210	1/4" wide, 12" deep	↓	75	.533	↓	2.57	14.10	2.13	18.80	28.50	

Division Notes

		CREW	DAILY OUTPUT	LABOR-HOURS	UNIT	2004 BARE COSTS MAT.	LABOR	EQUIP.	TOTAL	TOTAL INCL O&P

Division 4 Masonry

Estimating Tips

04050 Basic Masonry Materials & Methods

- The terms *mortar* and *grout* are often used interchangeably, and incorrectly. Mortar is used to bed masonry units, seal the entry of air and moisture, provide architectural appearance, and allow for size variations in the units. Grout is used primarily in reinforced masonry construction and is used to bond the masonry to the reinforcing steel. Common mortar types are M(2500 psi), S(1800 psi), N(750 psi), and O(350 psi), and conform to ASTM C270. Grout is either fine or coarse, conforms to ASTM C476, and in-place strengths generally exceed 2500 psi. Mortar and grout are different components of masonry construction and are placed by entirely different methods. An estimator should be aware of their unique uses and costs.
- Waste, specifically the loss/droppings of mortar and the breakage of brick and block, is included in all masonry assemblies in this division. A factor of 25% is added for mortar and 3% for brick and concrete masonry units.
- Scaffolding or staging is not included in any of the Division 4 costs. Refer to section 01540 for scaffolding and staging costs.

04800 Masonry Assemblies

- The most common types of unit masonry are brick and concrete masonry. The major classifications of brick are building brick (ASTM C62), facing brick (ASTM C216) and glazed brick, fire brick and pavers. Many varieties of texture and appearance can exist within these classifications, and the estimator would be wise to check local custom and availability within the project area. On repair and remodeling jobs, matching the existing brick may be the most important criteria.
- Brick and concrete block are priced by the piece and then converted into a price per square foot of wall. Openings less than two square feet are generally ignored by the estimator because any savings in units used is offset by the cutting and trimming required.
- It is often difficult and expensive to find and purchase small lots of historic brick. Costs can vary widely. Many design issues affect costs, selection of mortar mix, and repairs or replacement of masonry materials. Cleaning techniques must be reflected in the estimate.
- All masonry walls, whether interior or exterior, require bracing. The cost of bracing walls during construction should be included by the estimator and this bracing must remain in place until permanent bracing is complete. Permanent bracing of masonry walls is accomplished by masonry itself, in the form of pilasters or abutting wall corners, or by anchoring the walls to the structural frame. Accessories in the form of anchors, anchor slots and ties are used, but their supply and installation can be by different trades. For instance, anchor slots on spandrel beams and columns are supplied and welded in place by the steel fabricator, but the ties from the slots into the masonry are installed by the bricklayer. Regardless of the installation method the estimator must be certain that these accessories are accounted for in pricing.

Reference Numbers

Reference numbers are shown in bold squares at the beginning of some major classifications. These numbers refer to related items in the Reference Section. The reference information may be an estimating procedure, an alternate pricing method or technical information.

Note: Not all subdivisions listed here necessarily appear in this publication.

		04055 | Selective Demolition	CREW	DAILY OUTPUT	LABOR-HOURS	UNIT	2004 BARE COSTS MAT.	LABOR	EQUIP.	TOTAL	TOTAL INCL O&P
110	0010	**SELECTIVE DEMOLITION, MASONRY**									
	1000	Chimney, 16″ x 16″, soft old mortar	1 Clab	24	.333	V.L.F.		8.65		8.65	14.35
	1020	Hard mortar		18	.444			11.55		11.55	19.15
	1080	20″ x 20″, soft old mortar		12	.667			17.35		17.35	28.50
	1100	Hard mortar		10	.800			21		21	34.50
	1140	20″ x 32″, soft old mortar		10	.800			21		21	34.50
	1160	Hard mortar		8	1			26		26	43
	1200	48″ x 48″, soft old mortar		5	1.600			41.50		41.50	69
	1220	Hard mortar		4	2			52		52	86
	2000	Columns, 8″ x 8″, soft old mortar		48	.167			4.33		4.33	7.20
	2020	Hard mortar		40	.200			5.20		5.20	8.60
	2060	16″ x 16″, soft old mortar		16	.500			13		13	21.50
	2100	Hard mortar		14	.571			14.85		14.85	24.50
	2140	24″ x 24″, soft old mortar		8	1			26		26	43
	2160	Hard mortar		6	1.333			34.50		34.50	57.50
	2200	36″ x 36″, soft old mortar		4	2			52		52	86
	2220	Hard mortar		3	2.667			69.50		69.50	115
	3000	Copings, precast or masonry, to 8″ wide									
	3020	Soft old mortar	1 Clab	180	.044	L.F.		1.16		1.16	1.92
	3040	Hard mortar	″	160	.050	″		1.30		1.30	2.16
	3100	To 12″ wide									
	3120	Soft old mortar	1 Clab	160	.050	L.F.		1.30		1.30	2.16
	3140	Hard mortar	″	140	.057	″		1.49		1.49	2.46
	4000	Fireplace, brick, 30″ x 24″ opening									
	4020	Soft old mortar	1 Clab	2	4	Ea.		104		104	172
	4040	Hard mortar		1.25	6.400			166		166	276
	4100	Stone, soft old mortar		1.50	5.333			139		139	230
	4120	Hard mortar		1	8			208		208	345
	5000	Veneers, brick, soft old mortar		140	.057	S.F.		1.49		1.49	2.46
	5020	Hard mortar		125	.064			1.66		1.66	2.76
	5100	Granite and marble, 2″ thick		180	.044			1.16		1.16	1.92
	5120	4″ thick		170	.047			1.22		1.22	2.03
	5140	Stone, 4″ thick		180	.044			1.16		1.16	1.92
	5160	8″ thick		175	.046			1.19		1.19	1.97
	5400	Alternate pricing method, stone, 4″ thick		60	.133	C.F.		3.47		3.47	5.75
	5420	8″ thick		85	.094	″		2.45		2.45	4.06
	9000	Minimum labor/equipment charge		2	4	Job		104		104	172
		04060 | Masonry Mortar									
200	0010	**CEMENT** Gypsum 80 lb. bag, T.L. lots				Bag	11.75			11.75	12.95
	0050	L.T.L. lots					12.15			12.15	13.35
	0100	Masonry, 70 lb. bag, T.L. lots					6.15			6.15	6.75
	0150	L.T.L. lots					5.80			5.80	6.35
	0200	White, 70 lb. bag, T.L. lots					16.35			16.35	18
	0250	L.T.L. lots					17.25			17.25	19
400	0010	**LIME** Masons, hydrated, 50 lb. bag, T.L. lots				Bag	5.60			5.60	6.15
	0050	L.T.L. lots					6			6	6.60
	0200	Finish, double hydrated, 50 lb. bag, T.L. lots					7			7	7.70
	0250	L.T.L. lots					7.75			7.75	8.55
500	0010	**MORTAR**									
	0100	Type M, 1:1:6 mix	1 Brhe	143	.056	C.F.	3.54	1.43		4.97	6.20
	0200	Type N, 1:3 mix		143	.056		3.21	1.43		4.64	5.85
	0300	Type O, 1:3 mix		143	.056		3.28	1.43		4.71	5.95
	0400	Type PM, 1:1:6 mix, 2500 psi		143	.056		3.48	1.43		4.91	6.15
	0500	Type S, 1/2:1:4 mix		143	.056		3.69	1.43		5.12	6.40

04060 | Masonry Mortar

			CREW	DAILY OUTPUT	LABOR-HOURS	UNIT	2004 BARE COSTS MAT.	LABOR	EQUIP.	TOTAL	TOTAL INCL O&P	
500	2000	With portland cement and lime										500
	2100	Type M, 1:1/4:3 mix	1 Brhe	143	.056	C.F.	4.16	1.43		5.59	6.90	
	2200	Type N, 1:1:6 mix, 750 psi		143	.056		3.41	1.43		4.84	6.05	
	2300	Type O, 1:2:9 mix (Pointing Mortar)		143	.056		3.31	1.43		4.74	5.95	
	2400	Type PL, 1:1/2:4 mix, 2500 psi		143	.056		3.78	1.43		5.21	6.50	
	2500	Type K, 1:3:12 mix, 75 psi		143	.056		3.27	1.43		4.70	5.90	
	2600	Type S, 1:1/2:4 mix, 1800 psi		143	.056		3.78	1.43		5.21	6.50	
	2650	Pre-mixed, type S or N					4.07			4.07	4.48	
	2700	Mortar for glass block	1 Brhe	143	.056		7.15	1.43		8.58	10.20	
	2800	Gypsum cement mortar					6.25			6.25	6.85	
	2900	Mortar for Fire Brick, 80 lb. bag, T.L. Lots				Bag	13.50			13.50	14.85	

04070 | Masonry Grout

			CREW	DAILY OUTPUT	LABOR-HOURS	UNIT	MAT.	LABOR	EQUIP.	TOTAL	TOTAL INCL O&P	
420	0010	**GROUTING** Bond bms. & lintels, 8″ dp., pumped, not incl. block										420
	0020	8″ thick, 0.2 C.F. per L.F.	D-4	1,400	.023	L.F.	.74	.67	.10	1.51	2.01	
	0050	10″ thick, 0.25 C.F. per L.F.		1,200	.027		.83	.78	.12	1.73	2.30	
	0060	12″ thick, 0.3 C.F. per L.F.		1,040	.031		.99	.90	.13	2.02	2.70	
	0200	Concrete block cores, solid, 4″ thk., by hand, 0.067 C.F./S.F. of wall	D-8	1,100	.036	S.F.	.22	1.12		1.34	2.06	
	0210	6″ thick, pumped, 0.175 C.F. per S.F.	D-4	720	.044		.58	1.31	.19	2.08	2.95	
	0250	8″ thick, pumped, 0.258 C.F. per S.F.		680	.047		.85	1.38	.20	2.43	3.39	
	0300	10″ thick, pumped, 0.340 C.F. per S.F.		660	.048		1.12	1.43	.21	2.76	3.77	
	0350	12″ thick, pumped, 0.422 C.F. per S.F.		640	.050		1.39	1.47	.22	3.08	4.14	
	0500	Cavity walls, 2″ space, pumped, 0.167 C.F./S.F. of wall		1,700	.019		.55	.55	.08	1.18	1.59	
	0550	3″ space, 0.250 C.F./S.F.		1,200	.027		.83	.78	.12	1.73	2.30	
	0600	4″ space, 0.333 C.F. per S.F.		1,150	.028		1.10	.82	.12	2.04	2.66	
	0700	6″ space, 0.500 C.F. per S.F.		800	.040		1.65	1.18	.17	3	3.90	
	0800	Door frames, 3′ x 7′ opening, 2.5 C.F. per opening		60	.533	Opng.	8.25	15.70	2.30	26.25	37	
	0850	6′ x 7′ opening, 3.5 C.F. per opening		45	.711	″	11.55	21	3.07	35.62	49.50	
	2000	Grout, C476, for bond beams, lintels and CMU cores		350	.091	C.F.	3.31	2.69	.40	6.40	8.40	
	9000	Minimum labor/equipment charge	1 Bric	2	4	Job		137		137	222	

04080 | Anchorage & Reinforcement

			CREW	DAILY OUTPUT	LABOR-HOURS	UNIT	MAT.	LABOR	EQUIP.	TOTAL	TOTAL INCL O&P	
070	0010	**ANCHOR BOLTS** Hooked type with nut and washer, 1/2″ diam., 8″ long	1 Bric	200	.040	Ea.	.52	1.37		1.89	2.79	070
	0030	12″ long		190	.042		.98	1.44		2.42	3.42	
	0040	5/8″ diameter, 8″ long		180	.044		.80	1.52		2.32	3.35	
	0050	12″ long		170	.047		.88	1.61		2.49	3.59	
	0060	3/4″ diameter, 8″ long		160	.050		1.16	1.71		2.87	4.06	
	0070	12″ long		150	.053		1.45	1.83		3.28	4.57	
200	0010	**REINFORCING** Steel bars A615, placed horiz., #3 & #4 bars	1 Bric	450	.018	Lb.	.29	.61		.90	1.30	200
	0020	#5 & #6 bars		800	.010		.29	.34		.63	.87	
	0050	Placed vertical, #3 & #4 bars		350	.023		.29	.78		1.07	1.58	
	0060	#5 & #6 bars		650	.012		.29	.42		.71	.99	
	0500	Joint reinforcing, ladder type, mill std galvanized										
	0600	9 ga. sides, 9 ga. ties, 4″ wall	1 Bric	30	.267	C.L.F.	6.80	9.15		15.95	22.50	
	0650	6″ wall		30	.267		7.70	9.15		16.85	23.50	
	0700	8″ wall		25	.320		8	10.95		18.95	26.50	
	0750	10″ wall		20	.400		8.50	13.70		22.20	31.50	
	0800	12″ wall		20	.400		9.40	13.70		23.10	32.50	
	1000	Truss type										
	1100	9 ga. sides, 9 ga. ties, 4″ wall	1 Bric	30	.267	C.L.F.	10.25	9.15		19.40	26	
	1150	6″ wall		30	.267		10.35	9.15		19.50	26.50	
	1200	8″ wall		25	.320		10.80	10.95		21.75	29.50	
	1250	10″ wall		20	.400		11.30	13.70		25	34.50	
	1300	12″ wall		20	.400		11.95	13.70		25.65	35	

04080 | Anchorage & Reinforcement

		04080 Anchorage & Reinforcement	CREW	DAILY OUTPUT	LABOR-HOURS	UNIT	2004 BARE COSTS MAT.	LABOR	EQUIP.	TOTAL	TOTAL INCL O&P	
200	1500	3/16" sides, 9 ga. ties, 4" wall	1 Bric	30	.267	C.L.F.	14.25	9.15		23.40	30.50	200
	1550	6" wall		30	.267		14.45	9.15		23.60	31	
	1600	8" wall		25	.320		14.85	10.95		25.80	34	
	1650	10" wall		20	.400		14.90	13.70		28.60	38.50	
	1700	12" wall		20	.400		15.90	13.70		29.60	39.50	
	2000	3/16" sides, 3/16" ties, 4" wall		30	.267		18.10	9.15		27.25	35	
	2050	6" wall		30	.267		18.30	9.15		27.45	35	
	2100	8" wall		25	.320		19.05	10.95		30	39	
	2150	10" wall		20	.400		20	13.70		33.70	44	
	2200	12" wall	↓	20	.400	↓	20.50	13.70		34.20	45	
	2500	Cavity truss type, galvanized										
	2600	9 ga. sides, 9 ga. ties, 4" wall	1 Bric	25	.320	C.L.F.	17.55	10.95		28.50	37	
	2650	6" wall		25	.320		17.90	10.95		28.85	37.50	
	2700	8" wall		20	.400		19.45	13.70		33.15	43.50	
	2750	10" wall		15	.533		21.50	18.25		39.75	53.50	
	2800	12" wall		15	.533		27	18.25		45.25	59.50	
	3000	3/16" sides, 9 ga. ties, 4" wall		25	.320		23.50	10.95		34.45	44	
	3050	6" wall		25	.320		24	10.95		34.95	44	
	3100	8" wall		20	.400		26	13.70		39.70	50.50	
	3150	10" wall		15	.533		28	18.25		46.25	60	
	3200	12" wall	↓	15	.533	↓	33.50	18.25		51.75	66	
	3500	For hot dip galvanizing, add					80%					
650	0010	**WALL TIES** To brick veneer, galv., corrugated, 7/8" x 7", 22 Ga.	1 Bric	10.50	.762	C	4.47	26		30.47	47.50	650
	0100	24 Ga.		10.50	.762		4	26		30	47	
	0150	16 Ga.		10.50	.762		13.45	26		39.45	57.50	
	0200	Buck anchors, galv., corrugated, 16 gauge, 2" bend, 8" x 2"		10.50	.762		93	26		119	145	
	0250	8" x 3"	↓	10.50	.762	↓	99	26		125	152	
	0300	Adjustable, rectangular, 4-1/8" wide										
	0350	Anchor and tie, 3/16" wire, mill galv.										
	0400	2-3/4" eye, 3-1/4" tie	1 Bric	1.05	7.619	M	340	261		601	800	
	0500	4-3/4" tie		1.05	7.619		375	261		636	835	
	0520	5-1/2" tie		1.05	7.619		390	261		651	850	
	0550	4-3/4" eye, 3-1/4" tie		1.05	7.619		380	261		641	845	
	0570	4-3/4" tie		1.05	7.619		415	261		676	880	
	0580	5-1/2" tie		1.05	7.619	↓	430	261		691	895	
	0600	Cavity wall, Z type, galvanized, 6" long, 1/4" diameter		10.50	.762	C	20	26		46	64.50	
	0650	3/16" diameter		10.50	.762		13.60	26		39.60	57.50	
	0800	8" long, 1/4" diameter		10.50	.762		24	26		50	69	
	0850	3/16" diameter		10.50	.762		10.95	26		36.95	54.50	
	1000	Rectangular type, galvanized, 1/4" diameter, 2" x 6"		10.50	.762		25.50	26		51.50	71	
	1050	4" x 6"		10.50	.762		29	26		55	74	
	1100	3/16" diameter, 2" x 6"		10.50	.762		16.95	26		42.95	61	
	1150	4" x 6"	↓	10.50	.762	↓	19.40	26		45.40	64	
	1200	Mesh wall tie, 1/2" mesh, hot dip galvanized										
	1400	16 gauge, 12" long, 3" wide	1 Bric	9	.889	C	71	30.50		101.50	128	
	1420	6" wide		9	.889		105	30.50		135.50	165	
	1440	12" wide		8.50	.941		167	32		199	236	
	1500	Rigid partition anchors, plain, 8" long, 1" x 1/8"		10.50	.762		48	26		74	95.50	
	1550	1" x 1/4"		10.50	.762		94	26		120	146	
	1580	1-1/2" x 1/8"		10.50	.762		66.50	26		92.50	116	
	1600	1-1/2" x 1/4"		10.50	.762		158	26		184	216	
	1650	2" x 1/8"		10.50	.762		83	26		109	134	
	1700	2" x 1/4"	↓	10.50	.762	↓	225	26		251	291	
	2000	Column flange ties, wire, galvanized										
	2300	3/16" diameter, up to 3" wide	1 Bric	10.50	.762	C	48	26		74	95.50	
	2350	To 5" wide	↓	10.50	.762	↓	51.50	26		77.50	99	

	04080 | Anchorage & Reinforcement		CREW	DAILY OUTPUT	LABOR-HOURS	UNIT	2004 BARE COSTS MAT.	LABOR	EQUIP.	TOTAL	TOTAL INCL O&P	
650	2400	To 7″ wide	1 Bric	10.50	.762	C	53.50	26		79.50	102	650
	2600	To 9″ wide		10.50	.762		57	26		83	105	
	2650	1/4″ diameter, up to 3″ wide		10.50	.762		66.50	26		92.50	116	
	2700	To 5″ wide		10.50	.762		72.50	26		98.50	123	
	2800	To 7″ wide		10.50	.762		76	26		102	126	
	2850	To 9″ wide	↓	10.50	.762		82	26		108	133	
	2900	For hot dip galvanized, add				↓	35%					
	4000	Channel slots, 1-3/8″ x 1/2″ x 8″										
	4100	12 gauge, plain	1 Bric	10.50	.762	C	112	26		138	166	
	4150	16 gauge, galvanized	″	10.50	.762	″	69.50	26		95.50	119	
	4200	Channel slot anchors										
	4300	16 gauge, galvanized, 1-1/4″ x 3-1/2″				C	24			24	26.50	
	4350	1-1/4″ x 5-1/2″					29			29	32	
	4400	1-1/4″ x 7-1/2″					33.50			33.50	37	
	4500	1/8″ plain, 1-1/4″ x 3-1/2″					47			47	51.50	
	4550	1-1/4″ x 5-1/2″					57.50			57.50	63.50	
	4600	1-1/4″ x 7-1/2″					68			68	75	
	4700	For corrugation, add					32			32	35	
	4750	For hot dip galvanized, add				↓	35%					
	5000	Dowels										
	5100	Plain, 1/4″ diameter, 3″ long				C	17			17	18.70	
	5150	4″ long					19.20			19.20	21	
	5200	6″ long					23.50			23.50	26	
	5300	3/8″ diameter, 3″ long					23			23	25.50	
	5350	4″ long					28.50			28.50	31.50	
	5400	6″ long					39.50			39.50	43.50	
	5500	1/2″ diameter, 3″ long					34			34	37.50	
	5550	4″ long					42			42	46	
	5600	6″ long					56			56	61.50	
	5700	5/8″ diameter, 3″ long					47			47	51.50	
	5750	4″ long					59			59	65	
	5800	6″ long					83			83	91.50	
	6000	3/4″ diameter, 3″ long					60			60	66	
	6100	4″ long					78			78	86	
	6150	6″ long					114			114	125	
	6300	For hot dip galvanized, add				↓	35%					

	04090 | Masonry Accessories		CREW	DAILY OUTPUT	LABOR-HOURS	UNIT	MAT.	LABOR	EQUIP.	TOTAL	TOTAL INCL O&P	
170	0010	**CONTROL JOINT** Rubber, 4″ and wider wall	1 Bric	400	.020	L.F.	1.60	.69		2.29	2.87	170
	0050	PVC, 4″ wall		400	.020		.97	.69		1.66	2.18	
	0100	Rubber, 6″ wall		320	.025		2.04	.86		2.90	3.63	
	0120	PVC, 6″ wall		320	.025		1.21	.86		2.07	2.72	
	0140	Rubber, 8″ and wider wall		280	.029		2.63	.98		3.61	4.48	
	0160	PVC, 8″ wall		280	.029		1.41	.98		2.39	3.14	
	0180	12″ wall	↓	240	.033	↓	2.99	1.14		4.13	5.15	
860	0010	**VENT BOX** Extruded aluminum, 4″ deep, 2-3/8″ x 8-1/8″	1 Bric	30	.267	Ea.	24	9.15		33.15	41.50	860
	0050	5″ x 8-1/8″		25	.320		32.50	10.95		43.45	53.50	
	0100	2-1/4″ x 25″		25	.320		64	10.95		74.95	88.50	
	0150	5″ x 16-1/2″		22	.364		47.50	12.45		59.95	72.50	
	0200	6″ x 16-1/2″		22	.364		69	12.45		81.45	96	
	0250	7-3/4″ x 16-1/2″	↓	20	.400		71	13.70		84.70	100	
	0400	For baked enamel finish, add					35%					
	0500	For cast aluminum, painted, add					60%					
	1000	Stainless steel ventilators, 6″ x 6″	1 Bric	25	.320		92	10.95		102.95	119	
	1050	8″ x 8″	↓	24	.333	↓	97	11.40		108.40	126	

For expanded coverage of these items see *Means Concrete & Masonry Cost Data 2004*

04050 | Basic Masonry Materials and Methods

04090 | Masonry Accessories

		04090 Masonry Accessories	CREW	DAILY OUTPUT	LABOR-HOURS	UNIT	2004 BARE COSTS MAT.	LABOR	EQUIP.	TOTAL	TOTAL INCL O&P	
860	1100	12" x 12"	1 Bric	23	.348	Ea.	112	11.90		123.90	142	860
	1150	12" x 6"		24	.333		98	11.40		109.40	127	
	1200	Foundation block vent, galv., 1-1/4" thk, 8" high, 16" long, no damper	↓	30	.267		18	9.15		27.15	34.50	
	1250	For damper, add				↓	6			6	6.60	

04200 | Masonry Units

04210 | Clay Masonry Units

		04210 Clay Masonry Units	CREW	DAILY OUTPUT	LABOR-HOURS	UNIT	2004 BARE COSTS MAT.	LABOR	EQUIP.	TOTAL	TOTAL INCL O&P	
100	0010	**COMMON BUILDING BRICK** C62, TL lots, material only (R04210-120)										100
	0020	Standard, minimum				M	270			270	297	
	0050	Average (select)	↓			"	325			325	360	
300	0010	**FACE BRICK** C216, TL lots, material only (R04210-120)										300
	0300	Standard modular, 4" x 2-2/3" x 8", minimum				M	355			355	395	
	0350	Maximum (R04210-500)					465			465	510	
	0450	Economy, 4" x 4" x 8", minimum					690			690	760	
	0500	Maximum					550			550	605	
	0510	Economy, 4" x 4" x 12", minimum					690			690	760	
	0520	Maximum					550			550	605	
	0550	Jumbo, 6" x 4" x 12", minimum					1,125			1,125	1,225	
	0600	Maximum					1,425			1,425	1,575	
	0610	Jumbo, 8" x 4" x 12", minimum					1,125			1,125	1,225	
	0620	Maximum					1,425			1,425	1,575	
	0650	Norwegian, 4" x 3-1/5" x 12", minimum					720			720	790	
	0700	Maximum					715			715	785	
	0710	Norwegian, 6" x 3-1/5" x 12", minimum					720			720	790	
	0720	Maximum					715			715	785	
	0850	Standard glazed, plain colors, 4" x 2-2/3" x 8", minimum					1,100			1,100	1,225	
	0900	Maximum					1,125			1,125	1,225	
	1000	Deep trim shades, 4" x 2-2/3" x 8", minimum					1,150			1,150	1,250	
	1050	Maximum					1,275			1,275	1,400	
	1080	Jumbo utility, 4" x 4" x 12"					1,025			1,025	1,125	
	1120	4" x 8" x 8"					1,075			1,075	1,175	
	1140	4" x 8" x 16"					2,250			2,250	2,475	
	1260	Engineer, 4" x 3-1/5" x 8", minimum					455			455	500	
	1270	Maximum					485			485	535	
	1350	King, 4" x 2-3/4" x 10", minimum					430			430	470	
	1360	Maximum					460			460	505	
	1400	Norman, 4" x 2-3/4" x 12"					430			430	470	
	1450	Roman, 4" x 2" x 12"					430			430	470	
	1500	SCR, 6" x 2-2/3" x 12"					430			430	470	
	1550	Double, 4" x 5-1/3" x 8"					430			430	470	
	1600	Triple, 4" x 5-1/3" x 12"					430			430	470	
	1770	Standard modular, double glazed, 4" x 2-2/3" x 8"					970			970	1,075	
	1850	Jumbo, colored glazed ceramic, 6" x 4" x 12"					1,425			1,425	1,575	
	2050	Jumbo utility, glazed, 4" x 4" x 12"					1,075			1,075	1,175	
	2100	4" x 8" x 8"					1,475			1,475	1,625	
	2150	4" x 16" x 8"					2,550			2,550	2,800	
	2170	For less than truck load lots, add					10			10	11	
	2180	For buff or gray brick, add	↓			↓	15			15	16.50	

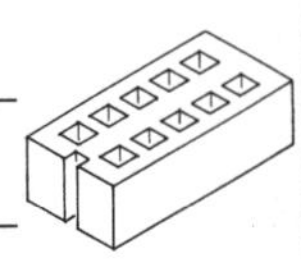

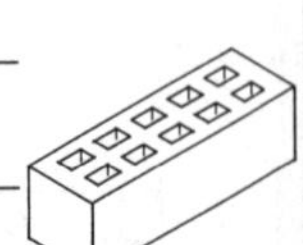

Important: See the Reference Section for critical supporting data - Reference Nos., Crews, & City Cost Indexes

04200 | Masonry Units

04210 | Clay Masonry Units

		04210 Clay Masonry Units	CREW	DAILY OUTPUT	LABOR-HOURS	UNIT	2004 BARE COSTS MAT.	LABOR	EQUIP.	TOTAL	TOTAL INCL O&P	
300	3050	Used brick, minimum (R04210-120)				M	300			300	330	300
	3100	Maximum					390			390	430	
	3150	Add for brick to match existing work, minimum (R04210-500)					5%					
	3200	Maximum				↓	50%					

04500 | Refractories

04580 | Refractory Brick

		04580 Refractory Brick	CREW	DAILY OUTPUT	LABOR-HOURS	UNIT	2004 BARE COSTS MAT.	LABOR	EQUIP.	TOTAL	TOTAL INCL O&P	
260	0010	**FIRE CLAY** Gray, high duty, 100 lb. bag				Bag	42			42	46	260
	0050	100 lb. drum, premixed (400 brick per drum)				Drum	52			52	57	

04700 | Simulated Masonry

04710 | Simulated Brick

		04710 Simulated Brick	CREW	DAILY OUTPUT	LABOR-HOURS	UNIT	2004 BARE COSTS MAT.	LABOR	EQUIP.	TOTAL	TOTAL INCL O&P	
600	0010	**SIMULATED BRICK** Aluminum, baked on colors	1 Carp	200	.040	S.F.	2.50	1.32		3.82	4.94	600
	0050	Fiberglass panels		200	.040		2.75	1.32		4.07	5.20	
	0100	Urethane pieces cemented in mastic		150	.053		4.75	1.76		6.51	8.15	
	0150	Vinyl siding panels	↓	200	.040	↓	1.90	1.32		3.22	4.28	

04800 | Masonry Assemblies

04810 | Unit Masonry Assemblies

		04810 Unit Masonry Assemblies	CREW	DAILY OUTPUT	LABOR-HOURS	UNIT	2004 BARE COSTS MAT.	LABOR	EQUIP.	TOTAL	TOTAL INCL O&P	
050	0010	**AUTOCLAVED AERATED CONCRETE BLOCK** Scaffolding not incl										050
	0050	Solid, 4" x 12" x 24", incl mortar	D-8	600	.067	S.F.	1.81	2.05		3.86	5.30	
	0060	6" x 12" x 24"		600	.067		2.42	2.05		4.47	6	
	0070	8" x 8" x 24"		575	.070		3.04	2.14		5.18	6.85	
	0080	10" x 12" x 24"		575	.070		3.96	2.14		6.10	7.85	
	0090	12" x 12" x 24"	↓	550	.073	↓	4.61	2.24		6.85	8.75	
100	0010	**BRICK VENEER** Scaffolding not included, truck load lots (R04210-120)										100
	0015	Material costs incl. 3% brick and 25% mortar waste										
	2000	Standard, sel. common, 4" x 2-2/3" x 8", (6.75/S.F.) (R04210-500)	D-8	230	.174	S.F.	2.76	5.35		8.11	11.75	
	2020	Standard, red, 4" x 2-2/3" x 8", running bond (6.75/SF)		220	.182		2.76	5.60		8.36	12.15	
	2050	Full header every 6th course (7.88/S.F.)		185	.216		3.21	6.65		9.86	14.35	
	2100	English, full header every 2nd course (10.13/S.F.)		140	.286		4.12	8.80		12.92	18.85	
	2150	Flemish, alternate header every course (9.00/S.F.)		150	.267		3.67	8.20		11.87	17.40	
	2200	Flemish, alt. header every 6th course (7.13/S.F.)		205	.195		2.91	6		8.91	12.95	
	2250	Full headers throughout (13.50/S.F.)		105	.381		5.50	11.75		17.25	25	
	2300	Rowlock course (13.50/S.F.)	↓	100	.400	↓	5.50	12.30		17.80	26	

	04810 | Unit Masonry Assemblies		CREW	DAILY OUTPUT	LABOR-HOURS	UNIT	2004 BARE COSTS MAT.	LABOR	EQUIP.	TOTAL	TOTAL INCL O&P	
100	2350	Rowlock stretcher (4.50/S.F.)	D-8	310	.129	S.F.	1.85	3.97		5.82	8.50	100
	2400	Soldier course (6.75/S.F.)		200	.200		2.76	6.15		8.91	13.05	
	2450	Sailor course (4.50/S.F.) R04210-120		290	.138		1.85	4.25		6.10	8.95	
	2600	Buff or gray face, running bond, (6.75/S.F.)		220	.182		2.93	5.60		8.53	12.30	
	2700	Glazed face brick, running bond R04210-500		210	.190		7.95	5.85		13.80	18.25	
	2750	Full header every 6th course (7.88/S.F.)		170	.235		9.30	7.25		16.55	22	
	3000	Jumbo, 6" x 4" x 12" running bond (3.00/S.F.)		435	.092		3.61	2.83		6.44	8.55	
	3050	Norman, 4" x 2-2/3" x 12" running bond, (4.5/S.F.)		320	.125		3.83	3.85		7.68	10.45	
	3100	Norwegian, 4" x 3-1/5" x 12" (3.75/S.F.)		375	.107		2.94	3.28		6.22	8.60	
	3150	Economy, 4" x 4" x 8" (4.50/S.F.)		310	.129		3.39	3.97		7.36	10.20	
	3200	Engineer, 4" x 3-1/5" x 8" (5.63/S.F.)		260	.154		2.86	4.74		7.60	10.85	
	3250	Roman, 4" x 2" x 12" (6.00/S.F.)		250	.160		4.62	4.93		9.55	13.10	
	3300	SCR, 6" x 2-2/3" x 12" (4.50/S.F.)		310	.129		4.23	3.97		8.20	11.10	
	3350	Utility, 4" x 4" x 12" (3.00/S.F.)		450	.089		3.31	2.74		6.05	8.10	
	3400	For cavity wall construction, add						15%				
	3450	For stacked bond, add						10%				
	3500	For interior veneer construction, add						15%				
	3550	For curved walls, add						30%				
	9000	Minimum labor/equipment charge	D-1	2	8	Job		239		239	390	
160	0010	**CHIMNEY** See Div. 03310-240 for foundation, add to prices below										160
	0100	Brick, 16" x 16", 8" flue, scaff. not incl.	D-1	18.20	.879	V.L.F.	14.90	26.50		41.40	59	
	0150	16" x 20" with one 8" x 12" flue		16	1		23.50	30		53.50	74.50	
	0200	16" x 24" with two 8" x 8" flues		14	1.143		34.50	34		68.50	93.50	
	0250	20" x 20" with one 12" x 12" flue		13.70	1.168		26	35		61	85	
	0300	20" x 24" with two 8" x 12" flues		12	1.333		38.50	40		78.50	108	
	0350	20" x 32" with two 12" x 12" flues		10	1.600		45	48		93	127	
170	0010	**COLUMNS** Face brick, includes mortar, scaffolding not included										170
	0050	8" x 8", 9 brick per course	D-1	56	.286	V.L.F.	3.60	8.55		12.15	17.85	
	0100	12" x 8", 13.5 brick		37	.432		5.40	12.95		18.35	27	
	0200	12" x 12", 20 brick		25	.640		8	19.15		27.15	40	
	0300	16" x 12", 27 brick		19	.842		10.80	25		35.80	53	
	0400	16" x 16", 36 brick		14	1.143		14.40	34		48.40	71.50	
	0500	20" x 16", 45 brick		11	1.455		18	43.50		61.50	90.50	
	0600	20" x 20", 56 brick		9	1.778		22.50	53		75.50	111	
	0700	24" x 20", 68 brick		7	2.286		27	68.50		95.50	141	
	0800	24" x 24", 81 brick		6	2.667		32.50	80		112.50	166	
	1000	36" x 36", 182 brick		3	5.333		73	160		233	340	
	9000	Minimum labor/equipment charge		2	8	Job		239		239	390	
172	0010	**CONCRETE BLOCK, BACK-UP,** C90, 2000 psi										172
	0020	Normal weight, 8" x 16" units, tooled joint 1 side										
	0050	Not-reinforced, 2000 psi, 2" thick	D-8	475	.084	S.F.	.82	2.59		3.41	5.10	
	0200	4" thick		460	.087		.97	2.68		3.65	5.40	
	0300	6" thick		440	.091		1.42	2.80		4.22	6.10	
	0350	8" thick		400	.100		1.54	3.08		4.62	6.70	
	0400	10" thick		330	.121		2.16	3.73		5.89	8.40	
	0450	12" thick	D-9	310	.155		2.22	4.63		6.85	9.95	
	1000	Reinforced, alternate courses, 4" thick	D-8	450	.089		1.05	2.74		3.79	5.60	
	1100	6" thick		430	.093		1.50	2.86		4.36	6.30	
	1150	8" thick		395	.101		1.62	3.12		4.74	6.85	
	1200	10" thick		320	.125		2.24	3.85		6.09	8.70	
	1250	12" thick	D-9	300	.160		2.31	4.79		7.10	10.30	
	9000	Minimum labor/equipment charge	D-1	2	8	Job		239		239	390	
175	0010	**CONCRETE BLOCK BOND BEAM** C90, 2000 psi										175
	0020	Not including grout or reinforcing										

	04810 Unit Masonry Assemblies	CREW	DAILY OUTPUT	LABOR-HOURS	UNIT	2004 BARE COSTS MAT.	LABOR	EQUIP.	TOTAL	TOTAL INCL O&P	
175	0100 Regular block, 8" high, 8" thick	D-8	565	.071	L.F.	1.62	2.18		3.80	5.30	175
	0150 12" thick	D-9	510	.094	"	2.24	2.82		5.06	7.05	
	2000 Including grout and 2 #5 bars										
	2100 Regular block, 8" high, 8" thick	D-8	300	.133	L.F.	3.01	4.11		7.12	9.95	
	2150 12" thick	D-9	250	.192	"	4.06	5.75		9.81	13.80	
	9000 Minimum labor/equipment charge	D-1	2	8	Job		239		239	390	
182	0010 **CONCRETE BLOCK, DECORATIVE** C90, 2000 psi										182
	1000 Fluted high strength										
	1100 8" x 16" x 4" thick, flutes 1 side,	D-8	345	.116	S.F.	2.73	3.57		6.30	8.80	
	1150 Flutes 2 sides		335	.119		3.35	3.68		7.03	9.65	
	1200 8" thick		300	.133		4.32	4.11		8.43	11.40	
	1250 For special colors, add					.28			.28	.31	
	1400 Deep grooved, smooth face										
	1450 8" x 16" x 4" thick	D-8	345	.116	S.F.	1.74	3.57		5.31	7.70	
	1500 8" thick	"	300	.133	"	2.96	4.11		7.07	9.90	
	1600										
	4000 Slump block										
	4100 4" face height x 16" x 4" thick	D-1	165	.097	S.F.	2.79	2.90		5.69	7.80	
	4150 6" thick		160	.100		4.01	2.99		7	9.25	
	4200 8" thick		155	.103		4.48	3.09		7.57	9.90	
	4250 10" thick		140	.114		8.10	3.42		11.52	14.45	
	4300 12" thick		130	.123		8.55	3.68		12.23	15.40	
	5000 Split rib profile units, 1" deep ribs, 8 ribs										
	5100 8" x 16" x 4" thick	D-8	345	.116	S.F.	2.09	3.57		5.66	8.10	
	5150 6" thick		325	.123		2.41	3.79		6.20	8.80	
	5200 8" thick		300	.133		2.77	4.11		6.88	9.70	
	5250 12" thick	D-9	275	.175		3.27	5.20		8.47	12.10	
	9000 Minimum labor/equipment charge	D-1	2	8	Job		239		239	390	
184	0010 **CONCRETE BLOCK, EXTERIOR** C90, 2000 psi										184
	0020 Reinforced alt courses, tooled joints 2 sides										
	0100 Normal weight, 8" x 16" x 6" thick	D-8	395	.101	S.F.	1.65	3.12		4.77	6.85	
	0200 8" thick		360	.111		2.46	3.42		5.88	8.25	
	0250 10" thick		290	.138		2.93	4.25		7.18	10.10	
	0300 12" thick	D-9	250	.192		2.99	5.75		8.74	12.65	
	9000 Minimum labor/equipment charge	D-1	2	8	Job		239		239	390	
186	0010 **CONCRETE BLOCK FOUNDATION WALL** C90/C145										186
	0050 Normal-weight, cut joints, horiz joint reinf, no vert reinf										
	0200 Hollow, 8" x 16" x 6" thick	D-8	455	.088	S.F.	1.63	2.71		4.34	6.20	
	0250 8" thick		425	.094		1.76	2.90		4.66	6.65	
	0300 10" thick		350	.114		2.38	3.52		5.90	8.30	
	0350 12" thick	D-9	300	.160		2.46	4.79		7.25	10.45	
	1000 Reinforced, #4 vert @ 48"										
	1100 Hollow, 8" x 16" block, 4" thick	D-8	455	.088	S.F.	1.57	2.71		4.28	6.10	
	1125 6" thick		445	.090		2.27	2.77		5.04	7	
	1150 8" thick		415	.096		2.67	2.97		5.64	7.75	
	1200 10" thick		340	.118		3.56	3.62		7.18	9.80	
	1250 12" thick	D-9	290	.166		3.91	4.95		8.86	12.35	
	9000 Minimum labor/equipment charge	D-1	2	8	Job		239		239	390	
188	0010 **CONCRETE BLOCK INSULATION INSERTS**										188
	0100 Inserts, styrofoam, plant installed, add to block prices										
	0200 8" x 16" units, 6" thick				S.F.	.85			.85	.94	
	0250 8" thick					.85			.85	.94	
	0300 10" thick					1			1	1.10	
	0350 12" thick					1.05			1.05	1.16	

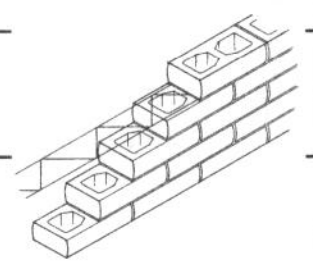

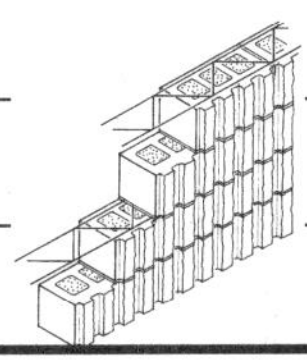

04810 | Unit Masonry Assemblies

			CREW	DAILY OUTPUT	LABOR-HOURS	UNIT	2004 BARE COSTS MAT.	LABOR	EQUIP.	TOTAL	TOTAL INCL O&P	
188	0500	8″ x 8″ units, 8″ thick				S.F.	.70			.70	.77	188
	0550	12″ thick				↓	.85			.85	.94	
189	0010	**CONCRETE BLOCK, INTERLOCKING**										189
	0100	Not including grout or reinforcing										
	0200	8″ x 16″ units, 2,000 psi, 8″ thick	D-1	245	.065	S.F.	1.73	1.95		3.68	5.10	
	0300	12″ thick	″	220	.073		2.58	2.18		4.76	6.35	
	0400	Including grout & reinforcing, 8″ thick	D-4	245	.131		5.30	3.84	.56	9.70	12.60	
	0450	12″ thick	″	220	.145	↓	6.25	4.28	.63	11.16	14.45	
	9000	Minimum labor/equipment charge	D-1	2	8	Job		239		239	390	
190	0010	**CONCRETE BLOCK, LINTELS** C90, normal weight										190
	0100	Including grout and horizontal reinforcing										
	0200	8″ x 8″ x 8″, 1 #4 bar	D-4	300	.107	L.F.	3.57	3.14	.46	7.17	9.50	
	0250	2 #4 bars		295	.108		3.69	3.19	.47	7.35	9.75	
	1000	12″ x 8″ x 8″, 1 #4 bar		275	.116		5.05	3.42	.50	8.97	11.60	
	1150	2 #5 bars	↓	270	.119	↓	5.30	3.48	.51	9.29	12	
	9000	Minimum labor/equipment charge	D-1	2	8	Job		239		239	390	
210	0010	**CONCRETE BLOCK, PARTITIONS**, scaffolding not included										210
	1000	Lightweight block, tooled joints, 2 sides, hollow										
	1100	Not reinforced, 8″ x 16″ x 4″ thick	D-8	440	.091	S.F.	1.11	2.80		3.91	5.75	
	1150	6″ thick		410	.098		1.51	3		4.51	6.55	
	1200	8″ thick		385	.104		1.86	3.20		5.06	7.25	
	1250	10″ thick	↓	370	.108		2.45	3.33		5.78	8.10	
	1300	12″ thick	D-9	350	.137		2.87	4.10		6.97	9.80	
	1500	Reinforced alternate courses, 4″ thick	D-8	435	.092		1.24	2.83		4.07	5.95	
	1600	6″ thick		405	.099		1.56	3.04		4.60	6.65	
	1650	8″ thick		380	.105		1.92	3.24		5.16	7.35	
	1700	10″ thick	↓	365	.110		2.51	3.37		5.88	8.25	
	1750	12″ thick	D-9	345	.139	↓	2.95	4.16		7.11	10	
	4000	Regular block, tooled joints, 2 sides, hollow										
	4100	Not reinforced, 8″ x 16″ x 4″ thick	D-8	430	.093	S.F.	.93	2.86		3.79	5.65	
	4150	6″ thick		400	.100		1.38	3.08		4.46	6.50	
	4200	8″ thick		375	.107		1.50	3.28		4.78	7	
	4250	10″ thick	↓	360	.111		2.12	3.42		5.54	7.90	
	4300	12″ thick	D-9	340	.141		2.19	4.23		6.42	9.25	
	4500	Reinforced alternate courses, 8″ x 16″ x 4″ thick	D-8	425	.094		1.01	2.90		3.91	5.80	
	4550	6″ thick		395	.101		1.46	3.12		4.58	6.65	
	4600	8″ thick		370	.108		1.58	3.33		4.91	7.15	
	4650	10″ thick	↓	355	.113		2.21	3.47		5.68	8.10	
	4700	12″ thick	D-9	335	.143	↓	2.27	4.29		6.56	9.45	
	9000	Minimum labor/equipment charge	D-1	2	8	Job		239		239	390	
225	0010	**CONCRETE BRICK** C55, grade N, type I										225
	0100	Regular, 4 x 2-1/4 x 8	D-8	220	.182	Ea.	.33	5.60		5.93	9.45	
	0125	Rusticated, 4 x 2-1/4 x 8		220	.182		.37	5.60		5.97	9.50	
	0150	Frog, 4 x 2-1/4 x 8		220	.182		.36	5.60		5.96	9.50	
	0200	Double, 4 x 4-7/8 x 8	↓	180	.222	↓	.58	6.85		7.43	11.75	
250	0010	**COPING** Stock units										250
	0050	Precast concrete, 10″ wide, 4″ tapers to 3-1/2″, 8″ wall	D-1	75	.213	L.F.	17.55	6.40		23.95	29.50	
	0100	12″ wide, 3-1/2″ tapers to 3″, 10″ wall		70	.229		10.50	6.85		17.35	22.50	
	0150	16″ wide, 4″ tapers to 3-1/2″, 14″ wall		60	.267		14.50	8		22.50	29	
	0300	Limestone for 12″ wall, 4″ thick		90	.178		13.50	5.30		18.80	23.50	
	0350	6″ thick		80	.200		15.75	6		21.75	27	
	0500	Marble, to 4″ thick, no wash, 9″ wide		90	.178		19.25	5.30		24.55	29.50	
	0550	12″ wide	↓	80	.200	↓	29	6		35	41.50	

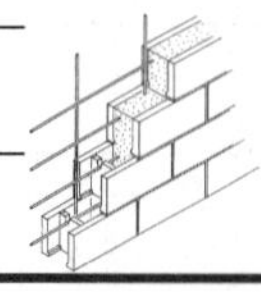

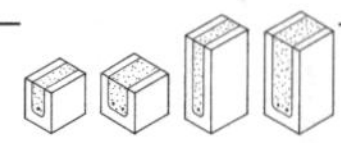

	04810	Unit Masonry Assemblies	CREW	DAILY OUTPUT	LABOR-HOURS	UNIT	2004 BARE COSTS MAT.	LABOR	EQUIP.	TOTAL	TOTAL INCL O&P	
250	0700	Terra cotta, 9″ wide	D-1	90	.178	L.F.	4.75	5.30		10.05	13.90	250
	0750	12″ wide		80	.200		7.80	6		13.80	18.30	
	0800	Aluminum, for 12″ wall		80	.200		11.25	6		17.25	22	
	9000	Minimum labor/equipment charge		2	8	Job		239		239	390	
260	0010	**CORNICES** Brick, on existing building										260
	0110	Face bricks, 12 brick/S.F., minimum	D-1	30	.533	SF Face	5	15.95		20.95	31.50	
	0150	15 brick/S.F., maximum		23	.696	″	6	21		27	40.50	
	9000	Minimum labor/equipment charge		1.50	10.667	Job		320		320	520	
325	0010	**GLASS BLOCK**										325
	0100	Plain, 4″ thick, under 1,000 S.F., 6″ x 6″	D-8	115	.348	S.F.	15.05	10.70		25.75	34	
	0150	8″ x 8″		160	.250		8.65	7.70		16.35	22	
	0160	end block		160	.250		26.50	7.70		34.20	41.50	
	0170	90 deg corner		160	.250		27	7.70		34.70	42	
	0180	45 deg corner		160	.250		12.95	7.70		20.65	27	
	0200	12″ x 12″		175	.229		11.40	7.05		18.45	24	
	0210	4″ x 8″		160	.250		7.65	7.70		15.35	21	
	0220	6″ x 8″		160	.250		9.40	7.70		17.10	23	
	0300	1,000 to 5,000 S.F., 6″ x 6″		135	.296		15	9.10		24.10	31.50	
	0350	8″ x 8″		190	.211		9.55	6.50		16.05	21	
	0400	12″ x 12″		215	.186		11.95	5.75		17.70	22.50	
	0410	4″ x 8″		215	.186		3.66	5.75		9.41	13.35	
	0420	6″ x 8″		215	.186		4.10	5.75		9.85	13.80	
	0700	For solar reflective blocks, add					100%					
	1000	Thinline, plain, 3-1/8″ thick, under 1,000 S.F., 6″ x 6″	D-8	115	.348	S.F.	13.05	10.70		23.75	32	
	1050	8″ x 8″		160	.250		7.70	7.70		15.40	21	
	1400	For cleaning block after installation (both sides), add		1,000	.040		.10	1.23		1.33	2.11	
	9000	Minimum labor/equipment charge	D-1	2	8	Job		239		239	390	
350	0010	**GLAZED CONCRETE BLOCK** C744										350
	0100	Single face, 8″ x 16″ units, 2″ thick	D-8	360	.111	S.F.	5.80	3.42		9.22	11.95	
	0200	4″ thick		345	.116		5.90	3.57		9.47	12.30	
	0250	6″ thick 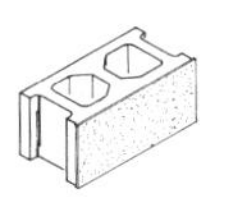		330	.121		6.25	3.73		9.98	12.95	
	0300	8″ thick		310	.129		7.10	3.97		11.07	14.30	
	0350	10″ thick		295	.136		7.85	4.17		12.02	15.45	
	0400	12″ thick	D-9	280	.171		8.45	5.15		13.60	17.65	
	0700	Double face, 8″ x 16″ units, 4″ thick	D-8	340	.118		9.55	3.62		13.17	16.40	
	0750	6″ thick		320	.125		10.15	3.85		14	17.40	
	0800	8″ thick		300	.133		10.55	4.11		14.66	18.25	
	1500	Cove base, 8″ x 16″, 2″ thick		315	.127	L.F.	5.85	3.91		9.76	12.80	
	1550	4″ thick		285	.140		5.65	4.32		9.97	13.20	
	1600	6″ thick		265	.151		5.90	4.65		10.55	14	
	1650	8″ thick		245	.163		6.20	5.05		11.25	14.95	
	9000	Minimum labor/equipment charge	D-1	2	8	Job		239		239	390	
500	0010	**TERRA COTTA** Coping, split type, not glazed, 9″ wide	D-1	90	.178	L.F.	5.95	5.30		11.25	15.20	500
	0100	13″ wide		80	.200		9.50	6		15.50	20	
	0200	Split type, glazed, 9″ wide		90	.178		11.40	5.30		16.70	21	
	0250	13″ wide		80	.200		15.25	6		21.25	26.50	
	0500	Partition or back-up blocks, scored, in C.L. lots										
	0700	Non-load bearing 12″ x 12″, 3″ thick, special order	D-8	550	.073	S.F.	15.80	2.24		18.04	21	
	0850	8″ thick		400	.100		6.35	3.08		9.43	12	
	1000	Load bearing, 12″ x 12″, 4″ thick, in walls		500	.080		4.64	2.46		7.10	9.10	
	1400	8″ thick, in walls		400	.100		6.35	3.08		9.43	12	
	9000	Minimum labor/equipment charge	D-1	2	8	Job		239		239	390	
650	0016	**WALLS** R04210-120										650
	0800	4″ wall, face, 4″ x 2-2/3″ x 8″	D-8	215	.186	S.F.	2.72	5.75		8.47	12.30	

04800 | Masonry Assemblies

04810 | Unit Masonry Assemblies

		04810 Unit Masonry Assemblies	CREW	DAILY OUTPUT	LABOR-HOURS	UNIT	2004 BARE COSTS MAT.	LABOR	EQUIP.	TOTAL	TOTAL INCL O&P
650	0850	4" thick, as back up, 6.75 bricks per S.F. (R04210-120)	D-8	240	.167	S.F.	2.12	5.15		7.27	10.70
	0900	8" thick wall, 13.50 brick per S.F.		135	.296		4.34	9.10		13.44	19.60
	1000	12" thick wall, 20.25 bricks per S.F. (R04210-500)		95	.421		6.55	12.95		19.50	28
	1050	16" thick wall, 27.00 bricks per S.F.		75	.533		8.80	16.40		25.20	36
	1200	Reinforced, 4" x 2-2/3" x 8" , 4" wall		205	.195		2.13	6		8.13	12.10
	1250	8" thick wall, 13.50 brick per S.F.		130	.308		4.35	9.45		13.80	20
	1300	12" thick wall, 20.25 bricks per S.F.		90	.444		6.55	13.70		20.25	29
	1350	16" thick wall, 27.00 bricks per S.F.		70	.571		8.80	17.60		26.40	38
	9000	Minimum labor/equipment charge	D-1	2	8	Job		239		239	390
700	0010	**STRUCTURAL BRICK** C652, Grade SW, incl mortar, scaffolding not incl									
	0100	Standard unit, 4-5/8" x 2-3/4" x 9-5/8"	D-8	245	.163	S.F.	4.03	5.05		9.08	12.60
	0120	Bond beam		225	.178		7.75	5.45		13.20	17.40
	0140	V cut bond beam		225	.178		7.95	5.45		13.40	17.60
	0160	Stretcher quoin, 5-5/8" x 2-3/4" x 9-5/8"		245	.163		7.65	5.05		12.70	16.55
	0180	Corner quoin		245	.163		8.55	5.05		13.60	17.60
	0200	Corner, 45 deg, 4-5/8" x 2-3/4" x 10-7/16"		235	.170		8.90	5.25		14.15	18.30
750	0010	**STRUCTURAL FACING TILE** Scaffolding not incl, standard colors									
	0020	6T series, 5-1/3" x 12", 2.3 pieces per S.F., glazed 1 side, 2" thick	D-8	225	.178	S.F.	5.85	5.45		11.30	15.35
	0100	4" thick		220	.182		6.85	5.60		12.45	16.60
	0150	Glazed 2 sides		195	.205		11.50	6.30		17.80	23
	0250	6" thick		210	.190		10.60	5.85		16.45	21
	0300	Glazed 2 sides		185	.216		13.20	6.65		19.85	25.50
	0400	8" thick		180	.222		13.25	6.85		20.10	25.50
	9000	Minimum labor/equipment charge	D-1	2	8	Job		239		239	390

04850 | Stone Assemblies

		04850 Stone Assemblies	CREW	DAILY OUTPUT	LABOR-HOURS	UNIT	MAT.	LABOR	EQUIP.	TOTAL	TOTAL INCL O&P
100	0010	**BLUESTONE** Cut to size									
	0500	Sills, natural cleft, 10" wide to 6' long, 1-1/2" thick	D-11	70	.343	L.F.	11	11		22	30
	0550	2" thick		63	.381		12.50	12.20		24.70	33.50
	0600	Smooth finish, 1-1/2" thick		70	.343		16	11		27	35.50
	0650	2" thick		63	.381		19	12.20		31.20	41
	0800	Thermal finish, 1-1/2" thick		70	.343		17.50	11		28.50	37
	0850	2" thick		63	.381		19	12.20		31.20	41
	1000	Stair treads, natural cleft, 12" wide, 6' long, 1-1/2" thick	D-10	115	.348		15.75	10.90	4.19	30.84	39.50
	1050	2" thick		105	.381		16.55	11.90	4.59	33.04	42.50
	1100	Smooth finish, 1-1/2" thick		115	.348		19.05	10.90	4.19	34.14	43
	1150	2" thick		105	.381		19.05	11.90	4.59	35.54	45.50
	1300	Thermal finish		115	.348		24	10.90	4.19	39.09	48
	1350	2" thick		105	.381		27	11.90	4.59	43.49	54
	9000	Minimum labor/equipment charge	D-1	2.50	6.400	Job		192		192	310
300	0010	**GRANITE** Cut to size									
	0050	Veneer, polished face, 3/4" to 1-1/2" thick									
	0150	Low price, gray, light gray, etc.	D-10	130	.308	S.F.	20	9.65	3.71	33.36	41.50
	0180	Medium price, pink, brown, etc.		130	.308		22	9.65	3.71	35.36	43.50
	0220	High price, red, black, etc.		130	.308		33.50	9.65	3.71	46.86	56.50
	2500	Steps, copings, etc., finished on more than one surface									
	2550	Minimum	D-10	50	.800	C.F.	75	25	9.65	109.65	134
	2600	Maximum	"	50	.800	"	120	25	9.65	154.65	183
	3500	Curbing, city street type, See Division 02770-225									
	3800	Radius curbs, over 5' radius, add				L.F.	50%				
	3850	Under 5' radius, add				"	100%				
	9000	Minimum labor/equipment charge	D-1	2	8	Job		239		239	390
400	0010	**LIMESTONE,** Cut to size									
	0020	Veneer facing panels									

	04850 | Stone Assemblies		CREW	DAILY OUTPUT	LABOR-HOURS	UNIT	2004 BARE COSTS MAT.	LABOR	EQUIP.	TOTAL	TOTAL INCL O&P
400	0750	5″ thick, 5′ x 14′ panels	D-10	275	.145	S.F.	21.50	4.55	1.75	27.80	33
	1000	Sugarcube finish, 2″ Thick, 3′ x 5′ panels		275	.145		8.35	4.55	1.75	14.65	18.45
	1050	3″ Thick, 4′ x 9′ panels		275	.145		12.50	4.55	1.75	18.80	23
	1200	4″ Thick, 5′ x 11′ panels		275	.145		16.35	4.55	1.75	22.65	27.50
	1400	Sugarcube, textured finish, 4-1/2″ thick, 5′ x 12′		275	.145		21	4.55	1.75	27.30	32.50
	1450	5″ thick, 5′ x 14′ panels		275	.145		21.50	4.55	1.75	27.80	33
	2000	Coping, sugarcube finish, top & 2 sides		30	1.333	C.F.	37	41.50	16.10	94.60	126
	2100	Sills, lintels, jambs, trim, stops, sugarcube finish, average		20	2		55	62.50	24	141.50	188
	2150	Detailed		20	2		55	62.50	24	141.50	188
	2300	Steps, extra hard, 14″ wide, 6″ rise		50	.800	L.F.	19.50	25	9.65	54.15	72.50
	3000	Quoins, plain finish, 6″x12″x12″	D-12	25	1.280	Ea.	50	39		89	118
	3050	6″x16″x24″	″	25	1.280	″	66.50	39		105.50	137
	9000	Minimum labor/equipment charge	D-1	2	8	Job		239		239	390
500	0011	**MARBLE**, ashlar, split face, 4″ + or - thick, random									
	0040	lengths 1′ to 4′ & heights 2″ to 7-1/2″, average	D-8	175	.229	S.F.	14.15	7.05		21.20	27
	0100	Base, polished, 3/4″ or 7/8″ thick, polished, 6″ high	D-10	65	.615	L.F.	12.85	19.25	7.40	39.50	53.50
	0300	Carvings or bas relief, from templates, average		80	.500	S.F.	115	15.65	6.05	136.70	158
	0350	Maximum		80	.500	″	268	15.65	6.05	289.70	325
	1000	Facing, polished finish, cut to size, 3/4″ to 7/8″ thick									
	1050	Average	D-10	130	.308	S.F.	18.85	9.65	3.71	32.21	40.50
	1100	Maximum	″	130	.308	″	43.50	9.65	3.71	56.86	67.50
	2500	Flooring, polished tiles, 12″ x 12″ x 3/8″ thick									
	2510	Thin set, average	D-11	90	.267	S.F.	8.75	8.55		17.30	23.50
	2600	Maximum		90	.267		31.50	8.55		40.05	49
	2700	Mortar bed, average		65	.369		8.90	11.85		20.75	29
	2740	Maximum		65	.369		29	11.85		40.85	51
	3500	Thresholds, 3′ long, 7/8″ thick, 4″ to 5″ wide, plain	D-12	24	1.333	Ea.	13.60	40.50		54.10	81
	3550	Beveled		24	1.333	″	15.85	40.50		56.35	83.50
	3700	Window stools, polished, 7/8″ thick, 5″ wide		85	.376	L.F.	12.75	11.45		24.20	32.50
	9000	Minimum labor/equipment charge	D-1	2	8	Job		239		239	390
600	0011	**ROUGH STONE WALL**, Dry									
	0100	Random fieldstone, under 18″ thick	D-12	60	.533	C.F.	7.45	16.25		23.70	34.50
	0150	Over 18″ thick	″	63	.508	″	8.90	15.45		24.35	35
	9000	Minimum labor/equipment charge	D-1	2	8	Job		239		239	390
700	0011	**SANDSTONE OR BROWNSTONE**									
	0100	Sawed face veneer, 2-1/2″ thick, to 2′ x 4′ panels	D-10	130	.308	S.F.	16.65	9.65	3.71	30.01	38
	0150	4′ thick, to 3′-6″ x 8′panels		100	.400		16.65	12.50	4.82	33.97	43.50
	0300	Split face, random sizes		100	.400		9.65	12.50	4.82	26.97	36
	9000	Minimum labor/equipment charge	D-1	2.50	6.400	Job		192		192	310
800	0010	**SLATE** Pennsylvania, blue gray to gray black; Vermont,									
	0050	Unfading green, mottled green & purple, gray & purple									
	0100	Virginia, blue black									
	3100	Stair landings, 1″ thick, black, clear	D-1	65	.246	S.F.	12	7.35		19.35	25
	3200	Ribbon	″	65	.246	″	11.50	7.35		18.85	24.50
	3500	Stair treads, sand finish, 1″ thick x 12″ wide									
	3550	Under 3 L.F.	D-10	85	.471	L.F.	14.50	14.75	5.70	34.95	46
	3600	3 L.F. to 6 L.F.	″	120	.333	″	15.50	10.45	4.02	29.97	38.50
	3700	Ribbon, sand finish, 1″ thick x 12″ wide									
	3750	To 6 L.F.	D-10	120	.333	L.F.	10.25	10.45	4.02	24.72	32.50
	4000	Stools or sills, sand finish, 1″ thick, 6″ wide	D-12	160	.200		7.50	6.10		13.60	18.15
	4200	10″ wide		90	.356		11.25	10.80		22.05	30
	4400	2″ thick, 6″ wide		140	.229		12	6.95		18.95	24.50
	4600	10″ wide		90	.356		18	10.80		28.80	37.50

For expanded coverage of these items see *Means Concrete & Masonry Cost Data 2004*

04800 | Masonry Assemblies

04850 | Stone Assemblies

			CREW	DAILY OUTPUT	LABOR-HOURS	UNIT	2004 BARE COSTS MAT.	LABOR	EQUIP.	TOTAL	TOTAL INCL O&P
800	4800	For lengths over 3′, add				L.F.	25%				
	9000	Minimum labor/equipment charge	D-1	2.50	6.400	Job		192		192	310
900	0010	**WINDOW SILL** Bluestone, thermal top, 10″ wide, 1-1/2″ thick		85	.188	S.F.	13.50	5.65		19.15	24
	0050	2″ thick		75	.213	″	15.75	6.40		22.15	27.50
	0100	Cut stone, 5″ x 8″ plain		48	.333	L.F.	10.20	10		20.20	27.50
	0200	Face brick on edge, brick, 8″ wide		80	.200		2.15	6		8.15	12.05
	0400	Marble, 9″ wide, 1″ thick		85	.188		7.50	5.65		13.15	17.40
	0900	Slate, colored, unfading, honed, 12″ wide, 1″ thick		85	.188		15.25	5.65		20.90	26
	0950	2″ thick		70	.229		21.50	6.85		28.35	34.50
	9000	Minimum labor/equipment charge	1 Bric	2	4	Job		137		137	222

04880 | Masonry Fireplaces

			CREW	DAILY OUTPUT	LABOR-HOURS	UNIT	MAT.	LABOR	EQUIP.	TOTAL	TOTAL INCL O&P
600	0010	**FIREPLACE** For prefabricated fireplace, see div. 10305-100									
	0100	Brick fireplace, not incl. foundations or chimneys									
	0110	30″ x 29″ opening, incl. chamber, plain brickwork	D-1	.40	40	Ea.	380	1,200		1,580	2,375
	0200	Fireplace box only (110 brick)	″	2	8	″	125	239		364	530
	0300	For elaborate brickwork and details, add					35%	35%			
	0400	For hearth, brick & stone, add	D-1	2	8	Ea.	140	239		379	545
	0410	For steel angle, damper, cleanouts, add		4	4		98	120		218	300
	0600	Plain brickwork, incl. metal circulator		.50	32		725	960		1,685	2,350
	0800	Face brick only, standard size, 8″ x 2-2/3″ x 4″		.30	53.333	M	380	1,600		1,980	3,025
	0900	Stone fireplace, fieldstone, add				SF Face	10			10	11
	1000	Cut stone, add				″	11			11	12.10
	9000	Minimum labor/equipment charge	D-1	2	8	Job		239		239	390

04900 | Masonry Restoration and Cleaning

04910 | Unit Masonry Restoration

			CREW	DAILY OUTPUT	LABOR-HOURS	UNIT	2004 BARE COSTS MAT.	LABOR	EQUIP.	TOTAL	TOTAL INCL O&P
200	0010	**CAULKING MASONRY** 1/2″ x 1/2″ joint									
	0050	Re-caulk only, oil base	1 Bric	225	.036	L.F.	.22	1.22		1.44	2.22
	0100	Acrylic latex		205	.039		.24	1.34		1.58	2.44
	0200	Polyurethane		200	.040		.32	1.37		1.69	2.57
	0300	Silicone		195	.041		.47	1.41		1.88	2.80
	1000	Cut out and re-caulk, oil base		145	.055		.22	1.89		2.11	3.31
	1050	Acrylic latex		130	.062		.24	2.11		2.35	3.69
	1100	Polyurethane		125	.064		.32	2.19		2.51	3.91
	1150	Silicone		120	.067		.47	2.28		2.75	4.23
	9000	Minimum labor/equipment charge		4	2	Job		68.50		68.50	111
600	0010	**NEEDLE BEAM MASONRY** Incl. wood shoring 10′ x 10′ opening									
	0400	Block, concrete, 8″ thick	B-9	7.10	5.634	Ea.	33.50	149	22.50	205	310
	0420	12″ thick		6.70	5.970		45.50	158	24	227.50	340
	0800	Brick, 4″ thick with 8″ backup block		5.70	7.018		45.50	185	28	258.50	385
	1000	Brick, solid, 8″ thick		6.20	6.452		33.50	170	26	229.50	350
	1040	12″ thick		4.90	8.163		45.50	216	32.50	294	440
	1080	16″ thick		4.50	8.889		70.50	235	35.50	341	505
	2000	Add for additional floors of shoring	B-1	6	4		33.50	107		140.50	214
	9000	Minimum labor/equipment charge	″	2	12	Job		320		320	530
720	0010	**POINTING MASONRY**									
	0300	Cut and repoint brick, hard mortar, running bond	1 Bric	80	.100	S.F.	.27	3.43		3.70	5.85

	04910 | Unit Masonry Restoration		CREW	DAILY OUTPUT	LABOR-HOURS	UNIT	2004 BARE COSTS MAT.	LABOR	EQUIP.	TOTAL	TOTAL INCL O&P	
720	0320	Common bond	1 Bric	77	.104	S.F.	.27	3.56		3.83	6.10	720
	0360	Flemish bond		70	.114		.28	3.91		4.19	6.65	
	0400	English bond		65	.123		.28	4.22		4.50	7.15	
	0600	Soft old mortar, running bond		100	.080		.27	2.74		3.01	4.75	
	0620	Common bond		96	.083		.27	2.85		3.12	4.92	
	0640	Flemish bond		90	.089		.28	3.04		3.32	5.25	
	0680	English bond		82	.098		.28	3.34		3.62	5.70	
	0700	Stonework, hard mortar		140	.057	L.F.	.35	1.96		2.31	3.57	
	0720	Soft old mortar		160	.050	"	.35	1.71		2.06	3.17	
	1000	Repoint, mask and grout method, running bond		95	.084	S.F.	.35	2.88		3.23	5.05	
	1020	Common bond		90	.089		.35	3.04		3.39	5.35	
	1040	Flemish bond		86	.093		.35	3.19		3.54	5.55	
	1060	English bond		77	.104		.35	3.56		3.91	6.20	
	2000	Scrub coat, sand grout on walls, minimum		120	.067		2.80	2.28		5.08	6.80	
	2020	Maximum		98	.082		2.01	2.80		4.81	6.75	
	9000	Minimum labor/equipment charge		3	2.667	Job		91.50		91.50	148	
800	0010	**TOOTHING MASONRY**										800
	0500	Brickwork, soft old mortar	1 Clab	40	.200	V.L.F.		5.20		5.20	8.60	
	0520	Hard mortar		30	.267			6.95		6.95	11.50	
	0700	Blockwork, soft old mortar		70	.114			2.97		2.97	4.93	
	0720	Hard mortar		50	.160			4.16		4.16	6.90	
	9000	Minimum labor/equipment charge		4	2	Job		52		52	86	

	04930 | Unit Masonry Cleaning		CREW	DAILY OUTPUT	LABOR-HOURS	UNIT	MAT.	LABOR	EQUIP.	TOTAL	TOTAL INCL O&P	
220	0010	**MASONRY CLEANING**										220
	0200	Chemical cleaning, new construction, brush and wash, minimum	D-1	1,000	.016	S.F.	.04	.48		.52	.82	
	0220	Average		800	.020		.05	.60		.65	1.03	
	0240	Maximum		600	.027		.07	.80		.87	1.38	
	0260	Light restoration, minimum		800	.020		.07	.60		.67	1.04	
	0270	Average		400	.040		.10	1.20		1.30	2.05	
	0280	Maximum		330	.048		.13	1.45		1.58	2.51	
	0300	Heavy restoration, minimum		600	.027		.07	.80		.87	1.38	
	0310	Average		400	.040		.11	1.20		1.31	2.06	
	0320	Maximum		250	.064		.15	1.92		2.07	3.27	
	0400	High pressure water only, minimum	B-9	2,000	.020			.53	.08	.61	.97	
	0420	Average		1,500	.027			.70	.11	.81	1.29	
	0440	Maximum		1,000	.040			1.06	.16	1.22	1.93	
	0800	High pressure water and chemical, minimum		1,800	.022		.08	.59	.09	.76	1.16	
	0820	Average		1,200	.033		.12	.88	.13	1.13	1.74	
	0840	Maximum		800	.050		.16	1.32	.20	1.68	2.59	
	1200	Sandblast, wet system, minimum		1,750	.023		.15	.60	.09	.84	1.26	
	1220	Average		1,100	.036		.22	.96	.15	1.33	1.99	
	1240	Maximum		700	.057		.29	1.51	.23	2.03	3.07	
	1400	Dry system, minimum		2,500	.016		.15	.42	.06	.63	.93	
	1420	Average		1,750	.023		.22	.60	.09	.91	1.34	
	1440	Maximum		1,000	.040		.29	1.06	.16	1.51	2.25	
	1800	For walnut shells, add					.38			.38	.42	
	1820	For corn chips, add					.38			.38	.42	
	2000	Steam cleaning, minimum	B-9	3,000	.013			.35	.05	.40	.64	
	2020	Average		2,500	.016			.42	.06	.48	.77	
	2040	Maximum		1,500	.027			.70	.11	.81	1.29	
	4000	Add for masking doors and windows									.80	
	4200	Add for pedestrian protection				Job					10%	
	4400	Add for wire cut face brick				S.F.	.13			.13	.15	
	9000	Minimum labor/equipment charge	D-4	2	16	Job		470	69	539	830	

For information about Means Estimating Seminars, see yellow pages 12 and 13 in back of book

Division Notes

		CREW	DAILY OUTPUT	LABOR-HOURS	UNIT	2004 BARE COSTS MAT.	LABOR	EQUIP.	TOTAL	TOTAL INCL O&P

Division 5 Metals

Estimating Tips

05050 Basic Metal Materials & Methods

- Nuts, bolts, washers, connection angles and plates can add a significant amount to both the tonnage of a structural steel job as well as the estimated cost. As a rule of thumb add 10% to the total weight to account for these accessories.
- Type 2 steel construction, commonly referred to as "simple construction," consists generally of field bolted connections with lateral bracing supplied by other elements of the building, such as masonry walls or x-bracing. The estimator should be aware, however, that shop connections may be accomplished by welding or bolting. The method may be particular to the fabrication shop and may have an impact on the estimated cost.

05200 Metal Joists

- In any given project the total weight of open web steel joists is determined by the loads to be supported and the design. However, economies can be realized in minimizing the amount of labor used to place the joists. This is done by maximizing the joist spacing and therefore minimizing the number of joists required to be installed on the job. Certain spacings and locations may be required by the design, but in other cases maximizing the spacing and keeping it as uniform as possible will keep the costs down.

05300 Metal Deck

- The takeoff and estimating of metal deck involves more than simply the area of the floor or roof and the type of deck specified or shown on the drawings. Many different sizes and types of openings may exist. Small openings for individual pipes or conduits may be drilled after the floor/roof is installed, but larger openings may require special deck lengths as well as reinforcing or structural support. The estimator should determine who will be supplying this reinforcing. Additionally, some deck terminations are part of the deck package, such as screed angles and pour stops, and others will be part of the steel contract, such as angles attached to structural members and cast-in-place angles and plates. The estimator must ensure that all pieces are accounted for in the complete estimate.

05500 Metal Fabrications

- The most economical steel stairs are those that use common materials, standard details and most importantly, a uniform and relatively simple method of field assembly. Commonly available A36 channels and plates are very good choices for the main stringers of the stairs, as are angles and tees for the carrier members. Risers and treads are usually made by specialty shops, and it is most economical to use a typical detail in as many places as possible. The stairs should be pre-assembled and shipped directly to the site. The field connections should be simple and straightforward to be accomplished efficiently and with a minimum of equipment and labor.

Reference Numbers

Reference numbers are shown in bold squares at the beginning of some major classifications. These numbers refer to related items in the Reference Section. The reference information may be an estimating procedure, an alternate pricing method or technical information.

Note: Not all subdivisions listed here necessarily appear in this publication.

		05060 | Selective Demolition	CREW	DAILY OUTPUT	LABOR-HOURS	UNIT	2004 BARE COSTS MAT.	LABOR	EQUIP.	TOTAL	TOTAL INCL O&P	
110	0010	**SELECTIVE DEMOLITION, STEEL FRAMING**										110
	2000	Steel framing, beams, 4" x 6"	B-13	500	.112	L.F.		3.14	1.25	4.39	6.45	
	2020	4" x 8"		400	.140			3.92	1.56	5.48	8.10	
	2080	8" x 12"		250	.224			6.25	2.50	8.75	13	
	2200	Columns, 6" x 6"		400	.140			3.92	1.56	5.48	8.10	
	2240	8" x 8"		350	.160			4.48	1.78	6.26	9.25	
	2280	10" x 10"		320	.175			4.90	1.95	6.85	10.15	
	2400	Girders, 10" x 12"		225	.249			6.95	2.78	9.73	14.40	
	2440	10" x 14"		200	.280			7.85	3.12	10.97	16.25	
	2480	10" x 16"		165	.339			9.50	3.78	13.28	19.65	
	2520	10" x 24"		125	.448			12.55	5	17.55	26	
	2950	Minimum labor/equipment charge		2	28	Job		785	310	1,095	1,625	
		05090 | Metal Fastenings										
300	0010	**CHEMICAL ANCHORS**, Includes layout & drilling										300
	1430	Chemical anchor, w/rod & epoxy cartridge, 3/4" diam. x 9-1/2" long	B-89A	27	.593	Ea.	10.25	17.70	3.93	31.88	44.50	
	1435	1" diameter x 11-3/4" long		24	.667		19.65	19.90	4.42	43.97	59	
	1440	1-1/4" diameter x 14" long		21	.762		37.50	22.50	5.05	65.05	84	
	1445	Concrete anchor, w/rod & epoxy cartridge, 1-3/4" diameter x 15" long		20	.800		70.50	24	5.30	99.80	122	
	1450	18" long		17	.941		84.50	28	6.25	118.75	146	
	1455	2" diameter x 18" long		16	1		108	30	6.65	144.65	175	
	1460	24" long		15	1.067		141	32	7.05	180.05	215	
340	0010	**DRILLING** For anchors, up to 4" deep, incl. bit and layout										340
	0050	in concrete or brick walls and floors, no anchor										
	0100	Holes, 1/4" diameter	1 Carp	75	.107	Ea.	.08	3.52		3.60	5.95	
	0150	For each additional inch of depth, add		430	.019		.02	.61		.63	1.04	
	0200	3/8" diameter		63	.127		.07	4.19		4.26	7.05	
	0250	For each additional inch of depth, add		340	.024		.02	.78		.80	1.31	
	0300	1/2" diameter		50	.160		.07	5.30		5.37	8.85	
	0350	For each additional inch of depth, add		250	.032		.02	1.06		1.08	1.77	
	0400	5/8" diameter		48	.167		.13	5.50		5.63	9.25	
	0450	For each additional inch of depth, add		240	.033		.03	1.10		1.13	1.86	
	0500	3/4" diameter		45	.178		.16	5.85		6.01	9.90	
	0550	For each additional inch of depth, add		220	.036		.04	1.20		1.24	2.03	
	0600	7/8" diameter		43	.186		.19	6.15		6.34	10.40	
	0650	For each additional inch of depth, add		210	.038		.05	1.26		1.31	2.13	
	0700	1" diameter		40	.200		.22	6.60		6.82	11.20	
	0750	For each additional inch of depth, add		190	.042		.06	1.39		1.45	2.36	
	0800	1-1/4" diameter		38	.211		.31	6.95		7.26	11.85	
	0850	For each additional inch of depth, add		180	.044		.08	1.47		1.55	2.52	
	0900	1-1/2" diameter		35	.229		.48	7.55		8.03	13.05	
	0950	For each additional inch of depth, add		165	.048		.12	1.60		1.72	2.78	
	1000	For ceiling installations, add						40%				
	1100	Drilling & layout for drywall/plaster walls, up to 1" deep, no anchor										
	1200	Holes, 1/4" diameter	1 Carp	150	.053	Ea.	.01	1.76		1.77	2.93	
	1300	3/8" diameter		140	.057		.01	1.89		1.90	3.14	
	1400	1/2" diameter		130	.062		.01	2.03		2.04	3.38	
	1500	3/4" diameter		120	.067		.02	2.20		2.22	3.67	
	1600	1" diameter		110	.073		.03	2.40		2.43	4.01	
	1700	1-1/4" diameter		100	.080		.04	2.64		2.68	4.42	
	1800	1-1/2" diameter		90	.089		.06	2.93		2.99	4.93	
	1900	For ceiling installations, add						40%				
	1910	Drilling & layout for steel, up to 1/4" deep, no anchor										
	1920	Holes, 1/4" diameter	1 Sswk	112	.071	Ea.	.10	2.65		2.75	5.05	
	1925	For each additional 1/4" depth, add		336	.024		.10	.88		.98	1.76	
	1930	3/8" diameter		104	.077		.12	2.86		2.98	5.45	

05090 | Metal Fastenings

		05090 | Metal Fastenings	Crew	Daily Output	Labor-Hours	Unit	2004 Bare Costs				Total Incl O&P	
							Mat.	Labor	Equip.	Total		
340	1935	For each additional 1/4" depth, add	1 Sswk	312	.026	Ea.	.12	.95		1.07	1.90	340
	1940	1/2" diameter		96	.083		.13	3.10		3.23	5.90	
	1945	For each additional 1/4" depth, add		288	.028		.13	1.03		1.16	2.06	
	1950	5/8" diameter		88	.091		.21	3.38		3.59	6.55	
	1955	For each additional 1/4" depth, add		264	.030		.21	1.13		1.34	2.33	
	1960	3/4" diameter		80	.100		.24	3.72		3.96	7.15	
	1965	For each additional 1/4" depth, add		240	.033		.24	1.24		1.48	2.56	
	1970	7/8" diameter		72	.111		.28	4.13		4.41	8	
	1975	For each additional 1/4" depth, add		216	.037		.28	1.38		1.66	2.87	
	1980	1" diameter		64	.125		.32	4.64		4.96	9	
	1985	For each additional 1/4" depth, add		192	.042		.32	1.55		1.87	3.23	
	1990	For drilling up, add						40%				
	2000	Minimum labor/equipment charge	1 Carp	4	2	Job		66		66	109	
380	0010	**EXPANSION ANCHORS** & shields										380
	0100	Bolt anchors for concrete, brick or stone, no layout and drilling										
	0200	Expansion shields, zinc, 1/4" diameter, 1-5/16" long, single	1 Carp	90	.089	Ea.	.96	2.93		3.89	5.90	
	0300	1-3/8" long, double		85	.094		1.06	3.11		4.17	6.30	
	0400	3/8" diameter, 1-1/2" long, single 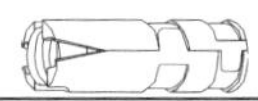		85	.094		1.58	3.11		4.69	6.90	
	0500	2" long, double		80	.100		1.95	3.30		5.25	7.60	
	0600	1/2" diameter, 2-1/16" long, single		80	.100		2.62	3.30		5.92	8.35	
	0700	2-1/2" long, double		75	.107		2.52	3.52		6.04	8.60	
	0800	5/8" diameter, 2-5/8" long, single 		75	.107		3.74	3.52		7.26	9.95	
	0900	2-3/4" long, double		70	.114		3.74	3.77		7.51	10.35	
	1000	3/4" diameter, 2-3/4" long, single		70	.114		5.55	3.77		9.32	12.35	
	1100	3-15/16" long, double		65	.123		7.40	4.06		11.46	14.90	
	1500	Self drilling anchor, snap-off, for 1/4" diameter bolt		26	.308		.72	10.15		10.87	17.65	
	1600	3/8" diameter bolt		23	.348		1.04	11.50		12.54	20	
	1700	1/2" diameter bolt 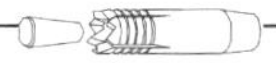		20	.400		1.60	13.20		14.80	24	
	1800	5/8" diameter bolt		18	.444		2.68	14.65		17.33	27.50	
	1900	3/4" diameter bolt		16	.500		4.50	16.50		21	32.50	
	2100	Hollow wall anchors for gypsum wall board, plaster or tile										
	2300	1/8" diameter, short	1 Carp	160	.050	Ea.	.23	1.65		1.88	2.99	
	2400	Long 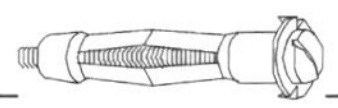		160	.050		.27	1.65		1.92	3.04	
	2500	3/16" diameter, short		150	.053		.48	1.76		2.24	3.45	
	2600	Long		150	.053		.52	1.76		2.28	3.49	
	2700	1/4" diameter, short		140	.057		.59	1.89		2.48	3.78	
	2800	Long		140	.057		.66	1.89		2.55	3.86	
	3000	Toggle bolts, bright steel, 1/8" diameter, 2" long		85	.094		.21	3.11		3.32	5.40	
	3100	4" long		80	.100		.32	3.30		3.62	5.80	
	3200	3/16" diameter, 3" long 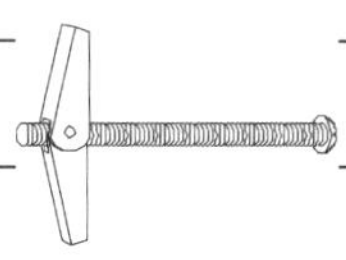		80	.100		.36	3.30		3.66	5.85	
	3300	6" long		75	.107		.50	3.52		4.02	6.40	
	3400	1/4" diameter, 3" long		75	.107		.41	3.52		3.93	6.30	
	3500	6" long		70	.114		.57	3.77		4.34	6.90	
	3600	3/8" diameter, 3" long		70	.114		.78	3.77		4.55	7.10	
	3700	6" long		60	.133		1.37	4.40		5.77	8.80	
	3800	1/2" diameter, 4" long		60	.133		2.03	4.40		6.43	9.55	
	3900	6" long		50	.160		3.35	5.30		8.65	12.45	
	4000	Nailing anchors										
	4100	Nylon nailing anchor, 1/4" diameter, 1" long	1 Carp	3.20	2.500	C	16.80	82.50		99.30	156	
	4200	1-1/2" long		2.80	2.857		21.50	94.50		116	180	
	4300	2" long 		2.40	3.333		36	110		146	222	
	4400	Metal nailing anchor, 1/4" diameter, 1" long		3.20	2.500		25	82.50		107.50	165	
	4500	1-1/2" long		2.80	2.857		34	94.50		128.50	194	
	4600	2" long		2.40	3.333		43.50	110		153.50	230	
	8000	Wedge anchors, not including layout or drilling										

	05090 | Metal Fastenings		CREW	DAILY OUTPUT	LABOR-HOURS	UNIT	2004 BARE COSTS MAT.	LABOR	EQUIP.	TOTAL	TOTAL INCL O&P	
380	8050	Carbon steel, 1/4" diameter, 1-3/4" long	1 Carp	150	.053	Ea.	.36	1.76		2.12	3.32	380
	8100	3 1/4" long		145	.055		.48	1.82		2.30	3.55	
	8150	3/8" diameter, 2-1/4" long 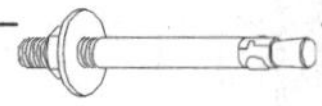		150	.053		.55	1.76		2.31	3.52	
	8200	5" long		145	.055		.96	1.82		2.78	4.08	
	8250	1/2" diameter, 2-3/4" long		140	.057		.84	1.89		2.73	4.05	
	8300	7" long		130	.062		1.43	2.03		3.46	4.94	
	8350	5/8" diameter, 3-1/2" long		130	.062		1.65	2.03		3.68	5.20	
	8400	8-1/2" long		115	.070		3.52	2.30		5.82	7.70	
	8450	3/4" diameter, 4-1/4" long		115	.070		1.99	2.30		4.29	6	
	8500	10" long		100	.080		4.53	2.64		7.17	9.35	
	8550	1" diameter, 6" long		100	.080		6.70	2.64		9.34	11.75	
	8575	9" long		80	.100		8.70	3.30		12	15	
	8600	12" long		80	.100		9.35	3.30		12.65	15.75	
	8650	1-1/4" diameter, 9" long		70	.114		12.15	3.77		15.92	19.65	
	8700	12" long	↓	60	.133	↓	15.60	4.40		20	24.50	
	8750	For type 303 stainless steel, add					350%					
	8800	For type 316 stainless steel, add					450%					
	9000	Minimum labor/equipment charge	1 Carp	4	2	Job		66		66	109	
460	0010	**LAG SCREWS**										460
	0020	Steel, 1/4" diameter, 2" long	1 Carp	200	.040	Ea.	.07	1.32		1.39	2.27	
	0200	1/2" diameter, 3" long		130	.062		.33	2.03		2.36	3.73	
	0300	5/8" diameter, 3" long	↓	120	.067	↓	.65	2.20		2.85	4.37	
540	0010	**MACHINERY ANCHORS**, heavy duty, incl. sleeve, floating base nut,										540
	0020	lower stud & coupling nut, fiber plug, connecting stud, washer & nut.										
	0030	For flush mounted embedment in poured concrete heavy equip. pads.										
	0200	Material only, 1/2" diameter stud & bolt				Ea.	46			46	50.50	
	0300	5/8" diameter					51			51	56	
	0500	3/4" diameter					58.50			58.50	64.50	
	0600	7/8" diameter					64			64	70.50	
	0800	1" diameter					67			67	74	
	0900	1-1/4" diameter				↓	89.50			89.50	98	
580	0010	**POWDER ACTUATED** Tools & fasteners										580
	0020	Stud driver, .22 caliber, buy, minimum				Ea.	280			280	310	
	0100	Maximum				"	450			450	495	
	0300	Powder charges for above, low velocity				C	15.80			15.80	17.40	
	0400	Standard velocity					22.50			22.50	24.50	
	0600	Drive pins & studs, 1/4" & 3/8" diam., to 3" long, minimum					10.30			10.30	11.35	
	0700	Maximum				↓	40.50			40.50	44.50	
	0800	Pneumatic stud driver for 1/8" diameter studs				Ea.	1,950			1,950	2,150	
	0900	Drive pins for above, 1/2" to 3/4" long				M	425			425	465	
860	0010	**WELD STUDS**										860
	0020	1/4" diameter, 2-11/16" long	E-10	1,030	.016	Ea.	.21	.59	.23	1.03	1.58	
	0100	4-1/8" long		1,030	.016		.20	.59	.23	1.02	1.57	
	0200	3/8" diameter, 4-1/8" long		1,030	.016		.23	.59	.23	1.05	1.60	
	0300	6-1/8" long	↓	1,030	.016	↓	.30	.59	.23	1.12	1.68	
	9000	Minimum labor/equipment charge	1 Sswk	2	4	Job		149		149	277	
900	0010	**WELDING STRUCTURAL** R05090-520										900
	0020	Field welding, 1/8" E6011, cost per welder, no oper. engr	E-14	8	1	Hr.	3	39	9.40	51.40	86.50	
	0200	With 1/2 operating engineer	E-13	8	1.500		3	55	9.40	67.40	112	
	0300	With 1 operating engineer	E-12	8	2	↓	3	71.50	9.40	83.90	138	
	0500	With no operating engineer, 2# weld rod per ton	E-14	8	1	Ton	3	39	9.40	51.40	86.50	
	0600	8# E6011 per ton	"	2	4		12	157	37.50	206.50	345	
	0800	With one operating engineer per welder, 2# E6011 per ton	E-12	8	2		3	71.50	9.40	83.90	138	
	0900	8# E6011 per ton	"	2	8	↓	12	285	37.50	334.50	550	

 Important: See the Reference Section for critical supporting data - Reference Nos., Crews, & City Cost Indexes

05050 | Basic Metal Materials & Methods

05090 | Metal Fastenings

			CREW	DAILY OUTPUT	LABOR-HOURS	UNIT	2004 BARE COSTS MAT.	LABOR	EQUIP.	TOTAL	TOTAL INCL O&P	
900	1200	Continuous fillet, stick welding, incl. equipment R05090-520										900
	1300	Single pass, 1/8" thick, 0.1#/L.F.	E-14	150	.053	L.F.	.15	2.09	.50	2.74	4.60	
	1400	3/16" thick, 0.2#/L.F.		75	.107		.30	4.18	1	5.48	9.25	
	1500	1/4" thick, 0.3#/L.F.		50	.160		.45	6.25	1.50	8.20	13.80	
	1610	5/16" thick, 0.4#/L.F.		38	.211		.60	8.25	1.98	10.83	18.20	
	1800	3 passes, 3/8" thick, 0.5#/L.F.		30	.267		.75	10.45	2.51	13.71	23	
	2010	4 passes, 1/2" thick, 0.7#/L.F.		22	.364		1.05	14.25	3.42	18.72	31.50	
	2200	5 to 6 passes, 3/4" thick, 1.3#/L.F.		12	.667		1.95	26	6.25	34.20	57.50	
	2400	8 to 11 passes, 1" thick, 2.4#/L.F.		6	1.333		3.60	52	12.55	68.15	115	
	2600	For all position welding, add, minimum						20%				
	2700	Maximum						300%				
	2900	For semi-automatic welding, deduct, minimum						5%				
	3000	Maximum						15%				
	4000	Cleaning and welding plates, bars, or rods										
	4010	to existing beams, columns, or trusses	E-14	12	.667	L.F.	.75	26	6.25	33	56	
	9000	Minimum labor/equipment charge	"	4	2	Job		78.50	18.80	97.30	167	
920	0010	**STEEL CUTTING**										920
	0020	Hand burning, incl. preparation, torch cutting & grinding, no staging										
	0100	Steel to 1/2" thick	E-25	320	.025	L.F.		.98	.27	1.25	2.11	
	0150	3/4" thick		260	.031			1.20	.33	1.53	2.60	
	0200	1" thick		200	.040			1.57	.42	1.99	3.39	
	9000	Minimum labor/equipment charge		2	4	Job		157	42.50	199.50	340	

05100 | Structural Metal Framing

05120 | Structural Steel

			CREW	DAILY OUTPUT	LABOR-HOURS	UNIT	2004 BARE COSTS MAT.	LABOR	EQUIP.	TOTAL	TOTAL INCL O&P	
180	0010	**CANOPY FRAMING**										180
	0020	6" and 8" members	E-4	3,000	.011	Lb.	.77	.40	.03	1.20	1.62	
	9000	Minimum labor/equipment charge	1 Sswk	1	8	Job		297		297	555	
220	0010	**CEILING SUPPORTS**										220
	1000	Entrance door/folding partition supports	E-4	60	.533	L.F.	12.75	20	1.25	34	53	
	1100	Linear accelerator door supports		14	2.286		58	86	5.35	149.35	230	
	1200	Lintels or shelf angles, hung, exterior hot dipped galv.		267	.120		8.70	4.51	.28	13.49	18.25	
	1250	Two coats primer paint instead of galv.		267	.120		7.55	4.51	.28	12.34	17	
	1400	Monitor support, ceiling hung, expansion bolted		4	8	Ea.	202	300	18.80	520.80	805	
	1450	Hung from pre-set inserts		6	5.333		217	201	12.55	430.55	630	
	1600	Motor supports for overhead doors		4	8		103	300	18.80	421.80	695	
	1700	Partition support for heavy folding partitions, without pocket		24	1.333	L.F.	29	50	3.13	82.13	129	
	1750	Supports at pocket only		12	2.667		58	100	6.25	164.25	258	
	2000	Rolling grilles & fire door supports		34	.941		25	35.50	2.21	62.71	96	
	2100	Spider-leg light supports, expansion bolted to ceiling slab		8	4	Ea.	83	151	9.40	243.40	380	
	2150	Hung from pre-set inserts		12	2.667	"	89.50	100	6.25	195.75	292	
	2400	Toilet partition support		36	.889	L.F.	29	33.50	2.09	64.59	97	
	2500	X-ray travel gantry support		12	2.667	"	99.50	100	6.25	205.75	305	
260	0010	**COLUMNS, STRUCTURAL** R05120-210										260
	0020	Shop fab'd for 100-ton, 1-2 story project, bolted conn's.										
	0800	Steel, concrete filled, extra strong pipe, 3-1/2" diameter	E-2	660	.085	L.F.	21	3.05	2.15	26.20	31	
	0830	4" diameter		780	.072		23.50	2.58	1.82	27.90	32.50	

5 METALS

	05120 | Structural Steel		CREW	DAILY OUTPUT	LABOR-HOURS	UNIT	2004 BARE COSTS MAT.	LABOR	EQUIP.	TOTAL	TOTAL INCL O&P	
260	0890	5" diameter R05120-210	E-2	1,020	.055	L.F.	28	1.97	1.39	31.36	35.50	260
	1500	Steel pipe, extra strong, no concrete, 3" to 5" diameter		16,000	.003	Lb.	.64	.13	.09	.86	1.03	
	3300	Structural tubing, square, A500GrB, 4" to 6" square, light section		11,270	.005	"	.64	.18	.13	.95	1.16	
	8090	For projects 75 to 99 tons, add				All	10%					
	8092	50 to 74 tons, add					20%					
	8094	25 to 49 tons, add					30%	10%				
	8096	10 to 24 tons, add					50%	25%				
	8098	2 to 9 tons, add					75%	50%				
	8099	Less than 2 tons, add					100%	100%				
	9000	Minimum labor/equipment charge	1 Sswk	1	8	Job		297		297	555	
300	0010	**CURB EDGING**										300
	0020	Steel angle w/anchors, on forms, 1" x 1", 0.8#/L.F.	E-4	350	.091	L.F.	1.17	3.44	.21	4.82	7.95	
	0300	4" x 4" angles, 8.2#/L.F.		275	.116	"	7.25	4.38	.27	11.90	16.40	
	9000	Minimum labor/equipment charge		4	8	Job		300	18.80	318.80	580	
440	0010	**LIGHTWEIGHT FRAMING**										440
	0200	For load-bearing steel studs see division 05410-400										
	0400	Angle framing, field fabricated, 4" and larger	E-3	440	.055	Lb.	.37	2.06	.17	2.60	4.44	
	0450	Less than 4" angles		265	.091		.38	3.43	.28	4.09	7.15	
	0600	Channel framing, field fabricated, 8" and larger		500	.048		.38	1.82	.15	2.35	3.97	
	0650	Less than 8" channels		335	.072		.38	2.71	.22	3.31	5.70	
	1000	Continuous slotted channel framing system, shop fab, min	2 Sswk	2,400	.007		1.98	.25		2.23	2.63	
	1200	Maximum	"	1,600	.010		2.23	.37		2.60	3.14	
	1250	Plate & bar stock for reinforcing beams and trusses					.70			.70	.77	
	1300	Cross bracing, rods, shop fabricated, 3/4" diameter	E-3	700	.034		.77	1.30	.11	2.18	3.37	
	1310	7/8" diameter		850	.028		.77	1.07	.09	1.93	2.93	
	1320	1" diameter		1,000	.024		.77	.91	.08	1.76	2.61	
	1330	Angle, 5" x 5" x 3/8"		2,800	.009		.77	.32	.03	1.12	1.47	
	1350	Hanging lintels, shop fabricated, average		850	.028		.77	1.07	.09	1.93	2.93	
	1380	Roof frames, shop fabricated, 3'-0" square, 5' span	E-2	4,200	.013		.77	.48	.34	1.59	2.07	
	1400	Tie rod, not upset, 1-1/2" to 4" diameter, with turnbuckle	2 Sswk	800	.020		.83	.74		1.57	2.29	
	1420	No turnbuckle		700	.023		.80	.85		1.65	2.46	
	1500	Upset, 1-3/4" to 4" diameter, with turnbuckle		800	.020		.83	.74		1.57	2.29	
	1520	No turnbuckle		700	.023		.80	.85		1.65	2.46	
	9000	Minimum labor/equipment charge		2	8	Job		297		297	555	
480	0010	**LINTELS**										480
	0020	Plain steel angles, under 500 lb.	1 Bric	550	.015	Lb.	.49	.50		.99	1.35	
	0100	500 to 1000 lb.		640	.013		.48	.43		.91	1.23	
	0200	1,000 to 2,000 lb.		640	.013		.47	.43		.90	1.21	
	0300	2,000 to 4,000 lb.		640	.013		.45	.43		.88	1.20	
	0500	For built-up angles and plates, add to above					.16			.16	.18	
	0700	For engineering, add to above					.06			.06	.07	
	0900	For galvanizing, add to above, under 500 lb.					.27			.27	.30	
	0950	500 to 2,000 lb.					.24			.24	.27	
	1000	Over 2,000 lb.					.20			.20	.22	
	2000	Steel angles, 3-1/2" x 3", 1/4" thick, 2'-6" long	1 Bric	47	.170	Ea.	6.90	5.85		12.75	17	
	2100	4'-6" long		26	.308		12.40	10.55		22.95	31	
	2500	3-1/2" x 3-1/2" x 5/16", 5'-0" long		18	.444		18.35	15.20		33.55	44.50	
	2600	4" x 3-1/2", 1/4" thick, 5'-0" long		21	.381		15.80	13.05		28.85	38.50	
	2700	9'-0" long		12	.667		28.50	23		51.50	68.50	
	2800	4" x 3-1/2" x 5/16", 7'-0" long		12	.667		27.50	23		50.50	67	
	2900	5" x 3-1/2" x 5/16", 10'-0" long		8	1		44.50	34.50		79	105	
	3500	For precast concrete lintels, see div. 03480-400										
	9000	Minimum labor/equipment charge	1 Bric	4	2	Job		68.50		68.50	111	
520	0010	**PIPE SUPPORT FRAMING**										520
	0020	Under 10#/L.F.	E-4	3,900	.008	Lb.	.85	.31	.02	1.18	1.54	

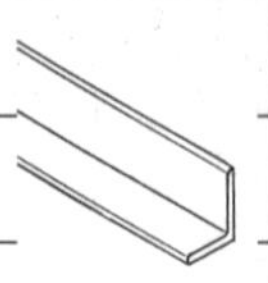

05100 | Structural Metal Framing

		05120 | Structural Steel	CREW	DAILY OUTPUT	LABOR-HOURS	UNIT	2004 BARE COSTS MAT.	LABOR	EQUIP.	TOTAL	TOTAL INCL O&P	
520	0200	10.1 to 15#/L.F.	E-4	4,300	.007	Lb.	.84	.28	.02	1.14	1.47	520
	0400	15.1 to 20#/L.F.		4,800	.007		.83	.25	.02	1.10	1.40	
	0600	Over 20#/L.F.		5,400	.006		.82	.22	.01	1.05	1.34	
680	0010	**STRUCTURAL STEEL PROJECTS** R05120-210										680
	0020	Shop fab'd for 100-ton, 1-2 story project, bolted conn's.										
	0700	Offices, hospitals, etc., steel bearing, 1 to 2 stories R05120-215	E-5	10.30	7.767	Ton	1,275	284	145	1,704	2,075	
	1300	Industrial bldgs., 1 story, beams & girders, steel bearing		12.90	6.202		1,275	226	116	1,617	1,925	
	1400	Masonry bearing		10	8		1,275	292	149	1,716	2,100	
	1600	1 story with roof trusses, steel bearing		10.60	7.547		1,500	276	141	1,917	2,300	
	1700	Masonry bearing		8.30	9.639		1,500	350	180	2,030	2,500	
	5390	For projects 75 to 99 tons, add					10%					
	5392	50 to 74 tons, add					20%					
	5394	25 to 49 tons, add					30%	10%				
	5396	10 to 24 tons, add					50%	25%				
	5398	2 to 9 tons, add					75%	50%				
	5399	Less than 2 tons, add					100%	100%				
720	0010	**STRUCTURAL STEEL** Bolted, incl. fabrication R05120-210										720
	0020	Shop fab'd for 100-ton, 1-2 story project, bolted conn's.										
	0050	Beams, W 6 x 9 R05120-215	E-2	720	.078	L.F.	6.90	2.80	1.97	11.67	14.70	
	0100	W 8 x 10		720	.078		7.65	2.80	1.97	12.42	15.55	
	0150	W 10 x 15		720	.078		11.50	2.80	1.97	16.27	19.75	
	0200	Columns, W 6 x 15		540	.104		12.45	3.73	2.62	18.80	23	
	0250	W 8 x 31		540	.104		25.50	3.73	2.62	31.85	38	
	0500	Girders, W 12 x 22		900	.062		16.85	2.24	1.57	20.66	24	
	0550	W 14 x 26		900	.062		19.90	2.24	1.57	23.71	27.50	
	0600	W 16 x 31		900	.062		23.50	2.24	1.57	27.31	31.50	
	0700	Joists (bar joists, H or K series), span to 30′	E-7	30,000	.003	Lb.	.46	.10	.05	.61	.75	
	0750	Span to 50′	"	20,000	.004	"	.45	.15	.08	.68	.85	
	7990	For projects 75 to 99 tons, add				All	10%					
	7992	50 to 75 tons, add					20%					
	7994	25 to 49 tons, add					30%	10%				
	7996	10 to 24 tons, add					50%	25%				
	7998	2 to 9 tons, add					75%	50%				
	7999	Less than 2 tons, add					100%	100%				
	9000	Minimum labor/equipment charge	E-2	2	28	Job		1,000	710	1,710	2,575	

05200 | Metal Joists

		05210 | Steel Joists	CREW	DAILY OUTPUT	LABOR-HOURS	UNIT	2004 BARE COSTS MAT.	LABOR	EQUIP.	TOTAL	TOTAL INCL O&P	
600	0010	**OPEN WEB JOISTS**, 40-ton job lots										600
	0020	K series, horizontal bridging, spans up to 30′, minimum	E-7	15	5.333	Ton	750	195	104	1,049	1,300	
	0050	Average		12	6.667		840	243	131	1,214	1,500	
	0080	Maximum		9	8.889		1,000	325	174	1,499	1,875	
	0410	Span 30′ to 50′, minimum		17	4.706		735	172	92	999	1,225	
	0440	Average		17	4.706		825	172	92	1,089	1,325	
	0460	Maximum		10	8		875	292	157	1,324	1,650	
	1010	CS series, horizontal bridging										

05200 | Metal Joists

05210 | Steel Joists

			CREW	DAILY OUTPUT	LABOR-HOURS	UNIT	2004 BARE COSTS MAT.	LABOR	EQUIP.	TOTAL	TOTAL INCL O&P	
600	1020	Spans to 30′, minimum	E-7	15	5.333	Ton	775	195	104	1,074	1,325	600
	1040	Average		12	6.667		860	243	131	1,234	1,525	
	1060	Maximum		9	8.889		1,025	325	174	1,524	1,900	
	2000	LH series, bolted cross bridging										
	2020	Spans to 96′, minimum	E-7	16	5	Ton	845	183	98	1,126	1,375	
	2040	Average		13	6.154		930	225	120	1,275	1,575	
	2080	Maximum		11	7.273		1,100	266	142	1,508	1,825	
	3010	DLH series, bolted cross bridging										
	3020	Spans to 144′ (shipped in 2 pieces), minimum	E-7	16	5	Ton	920	183	98	1,201	1,450	
	3040	Average		13	6.154		990	225	120	1,335	1,650	
	3100	Maximum		11	7.273		1,200	266	142	1,608	1,950	
	4010	SLH series, bolted cross bridging										
	4020	Spans to 200′, minimum	E-7	16	5	Ton	905	183	98	1,186	1,425	
	4040	Average		13	6.154		1,025	225	120	1,370	1,675	
	4060	Maximum		11	7.273		1,200	266	142	1,608	1,950	
	6100	For less than 40-ton job lots										
	6102	For 30 to 39 tons, add					10%					
	6104	20 to 29 tons, add					20%					
	6106	10 to 19 tons, add					30%					
	6107	5 to 9 tons, add					50%	25%				
	6108	1 to 4 tons, add					75%	50%				
	6109	Less than 1 ton, add					100%	100%				

05300 | Metal Deck

05310 | Steel Deck

			CREW	DAILY OUTPUT	LABOR-HOURS	UNIT	2004 BARE COSTS MAT.	LABOR	EQUIP.	TOTAL	TOTAL INCL O&P	
300	0010	**METAL DECKING** Steel decking										300
	0200	Cellular units, galvanized, 2″ deep, 20-20 gauge, over 15 squares	E-4	1,460	.022	S.F.	3.34	.83	.05	4.22	5.25	
	0400	3″ deep, galvanized, 20-20 gauge		1,375	.023		3.67	.88	.05	4.60	5.75	
	1000	4-1/2″ deep, galvanized, 20-18 gauge		1,100	.029		5.15	1.10	.07	6.32	7.75	
	1500	For acoustical deck, add					15%					
	1900	For multi-story or congested site, add						50%				
	2100	Open type, galv., 1-1/2″ deep wide rib, 22 gauge, under 50 squares	E-4	4,500	.007	S.F.	.84	.27	.02	1.13	1.44	
	2400	Over 500 squares		5,100	.006		.60	.24	.01	.85	1.12	
	5200	Non-cellular composite deck, galv., 2″ deep, 22 gauge		3,860	.008		.80	.31	.02	1.13	1.48	
	5300	20 gauge		3,600	.009		.88	.33	.02	1.23	1.61	
	5400	18 gauge 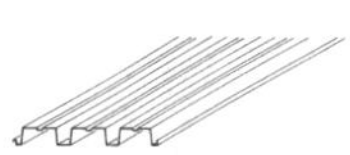		3,380	.009		1.13	.36	.02	1.51	1.92	
	5500	16 gauge		3,200	.010		1.41	.38	.02	1.81	2.28	
	5700	3″ deep, galv., 22 gauge		3,200	.010		.87	.38	.02	1.27	1.69	
	5800	20 gauge		3,000	.011		.97	.40	.03	1.40	1.84	
	5900	18 gauge		2,850	.011		1.20	.42	.03	1.65	2.14	
	6000	16 gauge		2,700	.012		1.60	.45	.03	2.08	2.62	
	6100	Slab form, steel, 28 gauge, 9/16″ deep, uncoated		4,000	.008		.55	.30	.02	.87	1.19	
	6200	Galvanized		4,000	.008		.49	.30	.02	.81	1.12	
	9000	Minimum labor/equipment charge	1 Sswk	1	8	Job		297		297	555	

	05410 | Load-Bearing Metal Studs		CREW	DAILY OUTPUT	LABOR-HOURS	UNIT	2004 BARE COSTS MAT.	LABOR	EQUIP.	TOTAL	TOTAL INCL O&P	
100	0010	**BRACING**, shear wall X-bracing, per 10′ x 10′ bay, one face										100
	0120	Metal strap, 20 ga x 4″ wide	2 Carp	18	.889	Ea.	12.60	29.50		42.10	62.50	
	0130	6″ wide		18	.889		19.80	29.50		49.30	70.50	
	0160	18 ga x 4″ wide		16	1		18.45	33		51.45	75	
	0170	6″ wide		16	1		27.50	33		60.50	84.50	
	0410	Continuous strap bracing, per horizontal row on both faces										
	0420	Metal strap, 20 ga x 2″ wide, studs 12″ O.C.	1 Carp	7	1.143	C.L.F.	33	37.50		70.50	99	
	0430	16″ O.C.		8	1		33	33		66	91	
	0440	24″ O.C.		10	.800		33	26.50		59.50	80.50	
	0450	18 ga x 2″ wide, studs 12″ O.C.		6	1.333		46	44		90	124	
	0460	16″ O.C.		7	1.143		46	37.50		83.50	114	
	0470	24″ O.C.		8	1		46	33		79	106	
120	0010	**BRIDGING**, solid between studs w/ 1-1/4″ leg track, per stud bay										120
	0200	Studs 12″ O.C., 18 ga x 2-1/2″ wide	1 Carp	125	.064	Ea.	.55	2.11		2.66	4.11	
	0210	3-5/8″ wide		120	.067		.67	2.20		2.87	4.39	
	0220	4″ wide		120	.067		.71	2.20		2.91	4.43	
	0230	6″ wide		115	.070		.93	2.30		3.23	4.83	
	0240	8″ wide		110	.073		1.17	2.40		3.57	5.25	
	0300	16 ga x 2-1/2″ wide		115	.070		.69	2.30		2.99	4.57	
	0310	3-5/8″ wide		110	.073		.84	2.40		3.24	4.90	
	0320	4″ wide		110	.073		.90	2.40		3.30	4.97	
	0330	6″ wide		105	.076		1.15	2.51		3.66	5.45	
	0340	8″ wide		100	.080		1.47	2.64		4.11	6	
	1200	Studs 16″ O.C., 18 ga x 2-1/2″ wide		125	.064		.71	2.11		2.82	4.28	
	1210	3-5/8″ wide		120	.067		.86	2.20		3.06	4.59	
	1220	4″ wide		120	.067		.91	2.20		3.11	4.65	
	1230	6″ wide		115	.070		1.19	2.30		3.49	5.10	
	1240	8″ wide		110	.073		1.50	2.40		3.90	5.65	
	1300	16 ga x 2-1/2″ wide		115	.070		.89	2.30		3.19	4.79	
	1310	3-5/8″ wide		110	.073		1.07	2.40		3.47	5.15	
	1320	4″ wide		110	.073		1.16	2.40		3.56	5.25	
	1330	6″ wide		105	.076		1.47	2.51		3.98	5.80	
	1340	8″ wide		100	.080		1.88	2.64		4.52	6.45	
	2200	Studs 24″ O.C., 18 ga x 2-1/2″ wide		125	.064		1.03	2.11		3.14	4.63	
	2210	3-5/8″ wide		120	.067		1.24	2.20		3.44	5	
	2220	4″ wide		120	.067		1.31	2.20		3.51	5.10	
	2230	6″ wide		115	.070		1.72	2.30		4.02	5.70	
	2240	8″ wide		110	.073		2.17	2.40		4.57	6.35	
	2300	16 ga x 2-1/2″ wide		115	.070		1.29	2.30		3.59	5.25	
	2310	3-5/8″ wide		110	.073		1.55	2.40		3.95	5.70	
	2320	4″ wide		110	.073		1.67	2.40		4.07	5.80	
	2330	6″ wide		105	.076		2.12	2.51		4.63	6.50	
	2340	8″ wide		100	.080		2.72	2.64		5.36	7.35	
	3000	Continuous bridging, per row										
	3100	16 ga x 1-1/2″ channel thru studs 12″ O.C.	1 Carp	6	1.333	C.L.F.	31	44		75	108	
	3110	16″ O.C.		7	1.143		31	37.50		68.50	97	
	3120	24″ O.C.		8.80	.909		31	30		61	84	
	4100	2″ x 2″ angle x 18 ga, studs 12″ O.C.		7	1.143		45	37.50		82.50	112	
	4110	16″ O.C.		9	.889		45	29.50		74.50	98	
	4120	24″ O.C.		12	.667		45	22		67	86	
	4200	16 ga, studs 12″ O.C.		5	1.600		57	53		110	151	
	4210	16″ O.C.		7	1.143		57	37.50		94.50	126	
	4220	24″ O.C.		10	.800		57	26.50		83.50	107	
300	0010	**FRAMING,** boxed headers/beams										300
	0200	Double, 18 ga x 6″ deep	2 Carp	220	.073	L.F.	3.20	2.40		5.60	7.50	

	05410 | Load-Bearing Metal Studs		CREW	DAILY OUTPUT	LABOR-HOURS	UNIT	2004 BARE COSTS MAT.	LABOR	EQUIP.	TOTAL	TOTAL INCL O&P	
300	0210	8″ deep	2 Carp	210	.076	L.F.	3.56	2.51		6.07	8.10	300
	0220	10″ deep		200	.080		4.32	2.64		6.96	9.15	
	0230	12 ″ deep		190	.084		4.74	2.78		7.52	9.80	
	0300	16 ga x 8″ deep		180	.089		4.08	2.93		7.01	9.35	
	0310	10″ deep		170	.094		4.90	3.11		8.01	10.55	
	0320	12 ″ deep		160	.100		5.35	3.30		8.65	11.30	
	0400	14 ga x 10″ deep		140	.114		5.70	3.77		9.47	12.50	
	0410	12 ″ deep		130	.123		6.25	4.06		10.31	13.65	
	1210	Triple, 18 ga x 8″ deep		170	.094		5.15	3.11		8.26	10.85	
	1220	10″ deep		165	.097		6.20	3.20		9.40	12.10	
	1230	12 ″ deep		160	.100		6.80	3.30		10.10	12.95	
	1300	16 ga x 8″ deep		145	.110		5.95	3.64		9.59	12.60	
	1310	10″ deep		140	.114		7.05	3.77		10.82	14	
	1320	12 ″ deep		135	.119		7.70	3.91		11.61	15	
	1400	14 ga x 10″ deep		115	.139		7.80	4.59		12.39	16.15	
	1410	12 ″ deep	↓	110	.145	↓	8.60	4.80		13.40	17.45	
400	0010	**FRAMING, STUD WALLS** w/ top & bottom track, no openings,										400
	0020	headers, beams, bridging or bracing										
	4100	8′ high walls, 18 ga x 2-1/2″ wide, studs 12″ O.C.	2 Carp	54	.296	L.F.	5.35	9.80		15.15	22	
	4110	16″ O.C.		77	.208		4.29	6.85		11.14	16.05	
	4120	24″ O.C.		107	.150		3.21	4.93		8.14	11.75	
	4130	3-5/8″ wide, studs 12″ O.C.		53	.302		6.40	9.95		16.35	23.50	
	4140	16″ O.C.		76	.211		5.10	6.95		12.05	17.15	
	4150	24″ O.C.		105	.152		3.83	5.05		8.88	12.55	
	4160	4″ wide, studs 12″ O.C.		52	.308		6.70	10.15		16.85	24.50	
	4170	16″ O.C.		74	.216		5.35	7.15		12.50	17.75	
	4180	24″ O.C.		103	.155		4.03	5.15		9.18	12.95	
	4190	6″ wide, studs 12″ O.C.		51	.314		8.50	10.35		18.85	26.50	
	4200	16″ O.C.		73	.219		6.80	7.25		14.05	19.50	
	4210	24″ O.C.		101	.158		5.10	5.25		10.35	14.30	
	4220	8″ wide, studs 12″ O.C.		50	.320		10.40	10.55		20.95	29	
	4230	16″ O.C.		72	.222		8.35	7.35		15.70	21.50	
	4240	24″ O.C.		100	.160		6.30	5.30		11.60	15.70	
	4300	16 ga x 2-1/2″ wide, studs 12″ O.C.		47	.340		6.35	11.25		17.60	25.50	
	4310	16″ O.C.		68	.235		5	7.75		12.75	18.35	
	4320	24″ O.C.		94	.170		3.69	5.60		9.29	13.35	
	4330	3-5/8″ wide, studs 12″ O.C.		46	.348		7.60	11.50		19.10	27.50	
	4340	16″ O.C.		66	.242		6	8		14	19.85	
	4350	24″ O.C.		92	.174		4.43	5.75		10.18	14.40	
	4360	4″ wide, studs 12″ O.C.		45	.356		8	11.75		19.75	28.50	
	4370	16″ O.C.		65	.246		6.35	8.10		14.45	20.50	
	4380	24″ O.C.		90	.178		4.67	5.85		10.52	14.85	
	4390	6″ wide, studs 12″ O.C.		44	.364		10	12		22	31	
	4400	16″ O.C.		64	.250		7.95	8.25		16.20	22.50	
	4410	24″ O.C.		88	.182		5.90	6		11.90	16.40	
	4420	8″ wide, studs 12″ O.C.		43	.372		12.40	12.30		24.70	34	
	4430	16″ O.C.		63	.254		9.85	8.40		18.25	25	
	4440	24″ O.C.		86	.186		7.30	6.15		13.45	18.25	
	5100	10′ high walls, 18 ga x 2-1/2″ wide, studs 12″ O.C.		54	.296		6.45	9.80		16.25	23.50	
	5110	16″ O.C.		77	.208		5.10	6.85		11.95	16.95	
	5120	24″ O.C.		107	.150		3.75	4.93		8.68	12.35	
	5130	3-5/8″ wide, studs 12″ O.C.		53	.302		7.65	9.95		17.60	25	
	5140	16″ O.C.		76	.211		6.05	6.95		13	18.20	
	5150	24″ O.C.		105	.152		4.47	5.05		9.52	13.25	
	5160	4″ wide, studs 12″ O.C.		52	.308		8.05	10.15		18.20	25.50	
	5170	16″ O.C.	↓	74	.216	↓	6.35	7.15		13.50	18.85	

05400 | Cold Formed Metal Framing

05410 | Load-Bearing Metal Studs

			CREW	DAILY OUTPUT	LABOR-HOURS	UNIT	2004 BARE COSTS MAT.	LABOR	EQUIP.	TOTAL	TOTAL INCL O&P	
400	5180	24" O.C.	2 Carp	103	.155	L.F.	4.70	5.15		9.85	13.65	400
	5190	6" wide, studs 12" O.C.		51	.314		10.15	10.35		20.50	28.50	
	5200	16" O.C.		73	.219		8.05	7.25		15.30	21	
	5210	24" O.C.		101	.158		5.95	5.25		11.20	15.20	
	5220	8" wide, studs 12" O.C.		50	.320		12.45	10.55		23	31	
	5230	16" O.C.		72	.222		9.90	7.35		17.25	23	
	5240	24" O.C.		100	.160		7.35	5.30		12.65	16.80	
	5300	16 ga x 2-1/2" wide, studs 12" O.C.		47	.340		7.65	11.25		18.90	27	
	5310	16" O.C.		68	.235		6	7.75		13.75	19.45	
	5320	24" O.C.		94	.170		4.35	5.60		9.95	14.10	
	5330	3-5/8" wide, studs 12" O.C.		46	.348		9.15	11.50		20.65	29	
	5340	16" O.C.		66	.242		7.20	8		15.20	21	
	5350	24" O.C.		92	.174		5.20	5.75		10.95	15.25	
	5360	4" wide, studs 12" O.C.		45	.356		9.65	11.75		21.40	30	
	5370	16" O.C.		65	.246		7.55	8.10		15.65	22	
	5380	24" O.C.		90	.178		5.50	5.85		11.35	15.75	
	5390	6" wide, studs 12" O.C.		44	.364		12.05	12		24.05	33	
	5400	16" O.C.		64	.250		9.50	8.25		17.75	24	
	5410	24" O.C.		88	.182		6.90	6		12.90	17.55	
	5420	8" wide, studs 12" O.C.		43	.372		14.95	12.30		27.25	37	
	5430	16" O.C.		63	.254		11.75	8.40		20.15	27	
	5440	24" O.C.		86	.186		8.60	6.15		14.75	19.65	
	6190	12' high walls, 18 ga x 6" wide, studs 12" O.C.		41	.390		11.85	12.90		24.75	34.50	
	6200	16" O.C.		58	.276		9.30	9.10		18.40	25.50	
	6210	24" O.C.		81	.198		6.80	6.50		13.30	18.30	
	6220	8" wide, studs 12" O.C.		40	.400		14.45	13.20		27.65	38	
	6230	16" O.C.		57	.281		11.40	9.25		20.65	28	
	6240	24" O.C.		80	.200		8.35	6.60		14.95	20	
	6390	16 ga x 6" wide, studs 12" O.C.		35	.457		14.10	15.10		29.20	40.50	
	6400	16" O.C.		51	.314		11.05	10.35		21.40	29.50	
	6410	24" O.C.		70	.229		7.95	7.55		15.50	21.50	
	6420	8" wide, studs 12" O.C.		34	.471		17.45	15.55		33	44.50	
	6430	16" O.C.		50	.320		13.65	10.55		24.20	32.50	
	6440	24" O.C.		69	.232		9.85	7.65		17.50	23.50	
	6530	14 ga x 3-5/8" wide, studs 12" O.C.		34	.471		13.45	15.55		29	40.50	
	6540	16" O.C.		48	.333		10.50	11		21.50	30	
	6550	24" O.C.		65	.246		7.55	8.10		15.65	22	
	6560	4" wide, studs 12" O.C.		33	.485		14.20	16		30.20	42	
	6570	16" O.C.		47	.340		11.10	11.25		22.35	31	
	6580	24" O.C.		64	.250		7.95	8.25		16.20	22.50	
	6730	12 ga x 3-5/8" wide, studs 12" O.C.		31	.516		18.65	17.05		35.70	48.50	
	6740	16" O.C.		43	.372		14.35	12.30		26.65	36.50	
	6750	24" O.C.		59	.271		10.10	8.95		19.05	26	
	6760	4" wide, studs 12" O.C.		30	.533		19.95	17.60		37.55	51	
	6770	16" O.C.		42	.381		15.40	12.55		27.95	38	
	6780	24" O.C.		58	.276		10.85	9.10		19.95	27	
	7390	16' high walls, 16 ga x 6" wide, studs 12" O.C.		33	.485		18.25	16		34.25	46.50	
	7400	16" O.C.		48	.333		14.10	11		25.10	34	
	7410	24" O.C.		67	.239		10	7.90		17.90	24	
	7420	8" wide, studs 12" O.C.		32	.500		22.50	16.50		39	52.50	
	7430	16" O.C.		47	.340		17.45	11.25		28.70	38	
	7440	24" O.C.		66	.242		12.40	8		20.40	27	
	7560	14 ga x 4" wide, studs 12" O.C.		31	.516		18.35	17.05		35.40	48	
	7570	16" O.C.		45	.356		14.20	11.75		25.95	35	
	7580	24" O.C.		61	.262		10.05	8.65		18.70	25.50	
400	7590	6" wide, studs 12" O.C.	↓	30	.533	↓	23	17.60		40.60	54.50	

05410 | Load-Bearing Metal Studs

		05410 Load-Bearing Metal Studs	CREW	DAILY OUTPUT	LABOR-HOURS	UNIT	2004 BARE COSTS MAT.	LABOR	EQUIP.	TOTAL	TOTAL INCL O&P	
400	7600	16" O.C.	2 Carp	44	.364	L.F.	17.90	12		29.90	39.50	400
	7610	24" O.C.		60	.267		12.65	8.80		21.45	28.50	
	7760	12 ga x 4" wide, studs 12" O.C.		29	.552		26	18.20		44.20	58.50	
	7770	16" O.C.		40	.400		19.95	13.20		33.15	44	
	7780	24" O.C.		55	.291		13.90	9.60		23.50	31	
	7790	6" wide, studs 12" O.C.		28	.571		33	18.85		51.85	67.50	
	7800	16" O.C.		39	.410		25	13.55		38.55	50	
	7810	24" O.C.		54	.296		17.55	9.80		27.35	35.50	
	8590	20' high walls, 14 ga x 6" wide, studs 12" O.C.		29	.552		28.50	18.20		46.70	61	
	8600	16" O.C.		42	.381		22	12.55		34.55	45	
	8610	24" O.C.		57	.281		15.30	9.25		24.55	32	
	8620	8" wide, studs 12" O.C.		28	.571		35	18.85		53.85	70	
	8630	16" O.C.		41	.390		27	12.90		39.90	51	
	8640	24" O.C.		56	.286		18.80	9.45		28.25	36	
	8790	12 ga x 6" wide, studs 12" O.C.		27	.593		40.50	19.55		60.05	77	
	8800	16" O.C.		37	.432		31	14.25		45.25	57.50	
	8810	24" O.C.		51	.314		21.50	10.35		31.85	40.50	
	8820	8" wide, studs 12" O.C.		26	.615		49.50	20.50		70	88	
	8830	16" O.C.		36	.444		37.50	14.65		52.15	66	
	8840	24" O.C.		50	.320	↓	26	10.55		36.55	46	
	9000	Minimum labor/equipment charge	↓	4	4	Job		132		132	219	

05420 | Cold-Formed Metal Joists

		05420 Cold-Formed Metal Joists	CREW	DAILY OUTPUT	LABOR-HOURS	UNIT	MAT.	LABOR	EQUIP.	TOTAL	TOTAL INCL O&P	
100	0010	**BRACING**, continuous, per row, top & bottom										100
	0120	Flat strap, 20 ga x 2" wide, joists at 12" O.C.	1 Carp	4.67	1.713	C.L.F.	34.50	56.50		91	132	
	0130	16" O.C.		5.33	1.501		33.50	49.50		83	119	
	0140	24" O.C.		6.66	1.201		32	39.50		71.50	101	
	0150	18 ga x 2" wide, joists at 12" O.C.		4	2		46	66		112	160	
	0160	16" O.C.		4.67	1.713		45	56.50		101.50	143	
	0170	24" O.C.	↓	5.33	1.501	↓	44	49.50		93.50	131	
120	0010	**BRIDGING**, solid between joists w/ 1-1/4" leg track, per joist bay										120
	0230	Joists 12" O.C., 18 ga track x 6" wide	1 Carp	80	.100	Ea.	.93	3.30		4.23	6.45	
	0240	8" wide		75	.107		1.17	3.52		4.69	7.15	
	0250	10" wide		70	.114		1.44	3.77		5.21	7.85	
	0260	12" wide		65	.123		1.67	4.06		5.73	8.60	
	0330	16 ga track x 6" wide		70	.114		1.15	3.77		4.92	7.50	
	0340	8" wide		65	.123		1.47	4.06		5.53	8.35	
	0350	10" wide		60	.133		1.80	4.40		6.20	9.30	
	0360	12" wide		55	.145		2.08	4.80		6.88	10.25	
	0440	14 ga track x 8" wide		60	.133		1.85	4.40		6.25	9.35	
	0450	10" wide		55	.145		2.28	4.80		7.08	10.45	
	0460	12" wide		50	.160		2.63	5.30		7.93	11.65	
	0550	12 ga track x 10" wide		45	.178		3.35	5.85		9.20	13.40	
	0560	12" wide		40	.200		3.77	6.60		10.37	15.10	
	1230	16" O.C., 18 ga track x 6" wide		80	.100		1.19	3.30		4.49	6.75	
	1240	8" wide		75	.107		1.50	3.52		5.02	7.50	
	1250	10" wide		70	.114		1.85	3.77		5.62	8.30	
	1260	12" wide		65	.123		2.15	4.06		6.21	9.10	
	1330	16 ga track x 6" wide		70	.114		1.47	3.77		5.24	7.85	
	1340	8" wide		65	.123		1.88	4.06		5.94	8.80	
	1350	10" wide		60	.133		2.31	4.40		6.71	9.85	
	1360	12" wide		55	.145		2.67	4.80		7.47	10.90	
	1440	14 ga track x 8" wide		60	.133		2.38	4.40		6.78	9.90	
	1450	10" wide	↓	55	.145	↓	2.92	4.80		7.72	11.15	

	05420 | Cold-Formed Metal Joists		CREW	DAILY OUTPUT	LABOR-HOURS	UNIT	2004 BARE COSTS MAT.	LABOR	EQUIP.	TOTAL	TOTAL INCL O&P	
120	1460	12" wide	1 Carp	50	.160	Ea.	3.37	5.30		8.67	12.45	120
	1550	12 ga track x 10" wide		45	.178		4.29	5.85		10.14	14.40	
	1560	12" wide		40	.200		4.83	6.60		11.43	16.25	
	2230	24" O.C., 18 ga track x 6" wide		80	.100		1.72	3.30		5.02	7.35	
	2240	8" wide		75	.107		2.17	3.52		5.69	8.25	
	2250	10" wide		70	.114		2.67	3.77		6.44	9.20	
	2260	12" wide		65	.123		3.10	4.06		7.16	10.15	
	2330	16 ga track x 6" wide		70	.114		2.12	3.77		5.89	8.60	
	2340	8" wide		65	.123		2.72	4.06		6.78	9.75	
	2350	10" wide		60	.133		3.34	4.40		7.74	11	
	2360	12" wide		55	.145		3.87	4.80		8.67	12.20	
	2440	14 ga track x 8" wide		60	.133		3.44	4.40		7.84	11.10	
	2450	10" wide		55	.145		4.22	4.80		9.02	12.60	
	2460	12" wide		50	.160		4.87	5.30		10.17	14.10	
	2550	12 ga track x 10" wide		45	.178		6.20	5.85		12.05	16.55	
	2560	12" wide	↓	40	.200	↓	7	6.60		13.60	18.65	
200	0010	**FRAMING, BAND JOIST** (track) fastened to bearing wall										200
	0220	18 ga track x 6" deep	2 Carp	1,000	.016	L.F.	.76	.53		1.29	1.71	
	0230	8" deep		920	.017		.96	.57		1.53	2	
	0240	10" deep		860	.019		1.18	.61		1.79	2.31	
	0320	16 ga track x 6" deep		900	.018		.93	.59		1.52	2	
	0330	8" deep		840	.019		1.20	.63		1.83	2.36	
	0340	10" deep		780	.021		1.47	.68		2.15	2.74	
	0350	12" deep		740	.022		1.70	.71		2.41	3.05	
	0430	14 ga track x 8" deep		750	.021		1.51	.70		2.21	2.83	
	0440	10" deep		720	.022		1.86	.73		2.59	3.26	
	0450	12" deep		700	.023		2.14	.75		2.89	3.61	
	0540	12 ga track x 10" deep		670	.024		2.73	.79		3.52	4.31	
	0550	12" deep	↓	650	.025	↓	3.08	.81		3.89	4.73	
300	0010	**FRAMING, BOXED HEADERS/BEAMS**										300
	0200	Double, 18 ga x 6" deep	2 Carp	220	.073	L.F.	3.20	2.40		5.60	7.50	
	0210	8" deep		210	.076		3.56	2.51		6.07	8.10	
	0220	10" deep		200	.080		4.32	2.64		6.96	9.15	
	0230	12" deep		190	.084		4.74	2.78		7.52	9.80	
	0300	16 ga x 8" deep		180	.089		4.08	2.93		7.01	9.35	
	0310	10" deep		170	.094		4.90	3.11		8.01	10.55	
	0320	12" deep		160	.100		5.35	3.30		8.65	11.30	
	0400	14 ga x 10" deep		140	.114		5.70	3.77		9.47	12.50	
	0410	12" deep		130	.123		6.25	4.06		10.31	13.65	
	0500	12 ga x 10" deep		110	.145		7.55	4.80		12.35	16.25	
	0510	12" deep		100	.160		8.35	5.30		13.65	17.95	
	1210	Triple, 18 ga x 8" deep		170	.094		5.15	3.11		8.26	10.85	
	1220	10" deep		165	.097		6.20	3.20		9.40	12.10	
	1230	12" deep		160	.100		6.80	3.30		10.10	12.95	
	1300	16 ga x 8" deep		145	.110		5.95	3.64		9.59	12.60	
	1310	10" deep		140	.114		7.05	3.77		10.82	14	
	1320	12" deep		135	.119		7.70	3.91		11.61	15	
	1400	14 ga x 10" deep		115	.139		8.25	4.59		12.84	16.70	
	1410	12" deep		110	.145		9.10	4.80		13.90	17.95	
	1500	12 ga x 10" deep		90	.178		11.05	5.85		16.90	22	
	1510	12" deep	↓	85	.188	↓	12.25	6.20		18.45	24	
410	0010	**FRAMING, JOISTS**, no band joists (track), web stiffeners, headers,										410
	0020	beams, bridging or bracing										
	0030	Joists (2" flange) and fasteners, materials only										
	0220	18 ga x 6" deep				L.F.	.99			.99	1.09	

05420 | Cold-Formed Metal Joists

		05420 Cold-Formed Metal Joists	CREW	DAILY OUTPUT	LABOR-HOURS	UNIT	2004 BARE COSTS MAT.	LABOR	EQUIP.	TOTAL	TOTAL INCL O&P	
410	0230	8″ deep				L.F.	1.18			1.18	1.29	410
	0240	10″ deep					1.39			1.39	1.52	
	0320	16 ga x 6″ deep					1.22			1.22	1.34	
	0330	8″ deep					1.45			1.45	1.59	
	0340	10″ deep					1.69			1.69	1.86	
	0350	12″ deep					1.92			1.92	2.11	
	0430	14 ga x 8″ deep					1.84			1.84	2.02	
	0440	10″ deep					2.11			2.11	2.32	
	0450	12″ deep					2.40			2.40	2.64	
	0540	12 ga x 10″ deep					3.09			3.09	3.40	
	0550	12″ deep				▼	3.51			3.51	3.86	
	1010	Installation of joists to band joists, beams & headers, labor only										
	1220	18 ga x 6″ deep	2 Carp	110	.145	Ea.		4.80		4.80	7.95	
	1230	8″ deep		90	.178			5.85		5.85	9.70	
	1240	10″ deep		80	.200			6.60		6.60	10.95	
	1320	16 ga x 6″ deep		95	.168			5.55		5.55	9.20	
	1330	8″ deep		70	.229			7.55		7.55	12.50	
	1340	10″ deep		60	.267			8.80		8.80	14.60	
	1350	12″ deep		55	.291			9.60		9.60	15.90	
	1430	14 ga x 8″ deep		65	.246			8.10		8.10	13.45	
	1440	10″ deep		45	.356			11.75		11.75	19.45	
	1450	12″ deep		35	.457			15.10		15.10	25	
	1540	12 ga x 10″ deep		40	.400			13.20		13.20	22	
	1550	12″ deep		30	.533	▼		17.60		17.60	29	
	9000	Minimum labor/equipment charge	▼	4	4	Job		132		132	219	
500	0010	**FRAMING, WEB STIFFENERS** at joist bearing, fabricated from										500
	0020	stud piece (1-5/8″ flange) to stiffen joist (2″ flange)										
	2120	For 6″ deep joist, with 18 ga x 2-1/2″ stud	1 Carp	120	.067	Ea.	1.19	2.20		3.39	4.96	
	2130	3-5/8″ stud		110	.073		1.31	2.40		3.71	5.40	
	2140	4″ stud		105	.076		1.27	2.51		3.78	5.55	
	2150	6″ stud		100	.080		1.39	2.64		4.03	5.90	
	2160	8″ stud		95	.084		1.43	2.78		4.21	6.20	
	2220	8″ deep joist, with 2-1/2″ stud		120	.067		1.30	2.20		3.50	5.10	
	2230	3-5/8″ stud		110	.073		1.42	2.40		3.82	5.55	
	2240	4″ stud		105	.076		1.39	2.51		3.90	5.70	
	2250	6″ stud		100	.080		1.52	2.64		4.16	6.05	
	2260	8″ stud		95	.084		1.64	2.78		4.42	6.40	
	2320	10″ deep joist, with 2-1/2″ stud		110	.073		1.84	2.40		4.24	6	
	2330	3-5/8″ stud		100	.080		2.02	2.64		4.66	6.60	
	2340	4″ stud		95	.084		2	2.78		4.78	6.80	
	2350	6″ stud		90	.089		2.16	2.93		5.09	7.25	
	2360	8″ stud		85	.094		2.20	3.11		5.31	7.55	
	2420	12″ deep joist, with 2-1/2″ stud		110	.073		1.94	2.40		4.34	6.10	
	2430	3-5/8″ stud		100	.080		2.11	2.64		4.75	6.70	
	2440	4″ stud		95	.084		2.08	2.78		4.86	6.90	
	2450	6″ stud		90	.089		2.27	2.93		5.20	7.35	
	2460	8″ stud		85	.094		2.45	3.11		5.56	7.85	
	3130	For 6″ deep joist, with 16 ga x 3-5/8″ stud		100	.080		1.38	2.64		4.02	5.90	
	3140	4″ stud		95	.084		1.37	2.78		4.15	6.10	
	3150	6″ stud		90	.089		1.49	2.93		4.42	6.50	
	3160	8″ stud		85	.094		1.59	3.11		4.70	6.90	
	3230	8″ deep joist, with 3-5/8″ stud		100	.080		1.53	2.64		4.17	6.05	
	3240	4″ stud		95	.084		1.50	2.78		4.28	6.25	
	3250	6″ stud		90	.089		1.66	2.93		4.59	6.70	
	3260	8″ stud	▼	85	.094	▼	1.79	3.11		4.90	7.10	

05400 | Cold Formed Metal Framing

05420 | Cold-Formed Metal Joists

			CREW	DAILY OUTPUT	LABOR-HOURS	UNIT	2004 BARE COSTS MAT.	LABOR	EQUIP.	TOTAL	TOTAL INCL O&P	
500	3330	10″ deep joist, with 3-5/8″ stud	1 Carp	85	.094	Ea.	2.10	3.11		5.21	7.45	500
	3340	4″ stud		80	.100		2.14	3.30		5.44	7.80	
	3350	6″ stud		75	.107		2.31	3.52		5.83	8.40	
	3360	8″ stud		70	.114		2.42	3.77		6.19	8.90	
	3430	12″ deep joist, with 3-5/8″ stud		85	.094		2.29	3.11		5.40	7.65	
	3440	4″ stud		80	.100		2.24	3.30		5.54	7.90	
	3450	6″ stud		75	.107		2.47	3.52		5.99	8.55	
	3460	8″ stud		70	.114		2.67	3.77		6.44	9.20	
	4230	For 8″ deep joist, with 14 ga x 3-5/8″ stud		90	.089		1.99	2.93		4.92	7.05	
	4240	4″ stud		85	.094		2.02	3.11		5.13	7.35	
	4250	6″ stud		80	.100		2.19	3.30		5.49	7.85	
	4260	8″ stud		75	.107		2.36	3.52		5.88	8.45	
	4330	10″ deep joist, with 3-5/8″ stud		75	.107		2.79	3.52		6.31	8.90	
	4340	4″ stud		70	.114		2.76	3.77		6.53	9.30	
	4350	6″ stud		65	.123		3.04	4.06		7.10	10.10	
	4360	8″ stud		60	.133		3.19	4.40		7.59	10.80	
	4430	12″ deep joist, with 3-5/8″ stud		75	.107		2.97	3.52		6.49	9.10	
	4440	4″ stud		70	.114		3.02	3.77		6.79	9.55	
	4450	6″ stud		65	.123		3.28	4.06		7.34	10.35	
	4460	8″ stud		60	.133		3.52	4.40		7.92	11.15	
	5330	For 10″ deep joist, with 12 ga x 3-5/8″ stud		65	.123		2.95	4.06		7.01	10	
	5340	4″ stud		60	.133		3.03	4.40		7.43	10.65	
	5350	6″ stud		55	.145		3.35	4.80		8.15	11.65	
	5360	8″ stud		50	.160		3.67	5.30		8.97	12.80	
	5430	12″ deep joist, with 3-5/8″ stud		65	.123		3.27	4.06		7.33	10.35	
	5440	4″ stud		60	.133		3.19	4.40		7.59	10.80	
	5450	6″ stud		55	.145		3.65	4.80		8.45	11.95	
	5460	8″ stud		50	.160		4.19	5.30		9.49	13.35	

05425 | Cold-Formed Roof Framing

			CREW	DAILY OUTPUT	LABOR-HOURS	UNIT	MAT.	LABOR	EQUIP.	TOTAL	TOTAL INCL O&P	
100	0010	**FRAMING, BRACING**										100
	0020	Continuous bracing, per row										
	0100	16 ga x 1-1/2″ channel thru rafters/trusses @ 16″ O.C.	1 Carp	4.50	1.778	C.L.F.	31	58.50		89.50	132	
	0120	24″ O.C.		6	1.333		31	44		75	108	
	0300	2″ x 2″ angle x 18 ga, rafters/trusses @ 16″ O.C.		6	1.333		45	44		89	123	
	0320	24″ O.C.		8	1		45	33		78	104	
	0400	16 ga, rafters/trusses @ 16″ O.C.		4.50	1.778		57	58.50		115.50	160	
	0420	24″ O.C.		6.50	1.231		57	40.50		97.50	131	
200	0010	**FRAMING, BRIDGING**										200
	0020	Solid, between rafters w/ 1-1/4″ leg track, per rafter bay										
	1200	Rafters 16″ O.C., 18 ga x 4″ deep	1 Carp	60	.133	Ea.	.91	4.40		5.31	8.30	
	1210	6″ deep		57	.140		1.19	4.63		5.82	9	
	1220	8″ deep		55	.145		1.50	4.80		6.30	9.60	
	1230	10″ deep		52	.154		1.85	5.10		6.95	10.45	
	1240	12″ deep		50	.160		2.15	5.30		7.45	11.10	
	2200	24″ O.C., 18 ga x 4″ deep		60	.133		1.31	4.40		5.71	8.75	
	2210	6″ deep		57	.140		1.72	4.63		6.35	9.60	
	2220	8″ deep		55	.145		2.17	4.80		6.97	10.35	
	2230	10″ deep		52	.154		2.67	5.10		7.77	11.35	
	2240	12″ deep		50	.160		3.10	5.30		8.40	12.15	
500	0010	**FRAMING, PARAPETS**										500
	0100	3′ high installed on 1st story, 18 ga x 4″ wide studs, 12″ O.C.	2 Carp	100	.160	L.F.	3.36	5.30		8.66	12.45	
	0110	16″ O.C.		150	.107		2.86	3.52		6.38	9	
	0120	24″ O.C.		200	.080		2.35	2.64		4.99	6.95	
	0200	6″ wide studs, 12″ O.C.		100	.160		4.28	5.30		9.58	13.45	
	0210	16″ O.C.		150	.107		3.65	3.52		7.17	9.85	

05425 | Cold-Formed Roof Framing

			CREW	DAILY OUTPUT	LABOR-HOURS	UNIT	2004 BARE COSTS MAT.	LABOR	EQUIP.	TOTAL	TOTAL INCL O&P	
500	0220	24" O.C.	2 Carp	200	.080	L.F.	3.02	2.64		5.66	7.70	500
	1100	Installed on 2nd story, 18 ga x 4" wide studs, 12" O.C.		95	.168		3.36	5.55		8.91	12.90	
	1110	16" O.C.		145	.110		2.86	3.64		6.50	9.20	
	1120	24" O.C.		190	.084		2.35	2.78		5.13	7.20	
	1200	6" wide studs, 12" O.C.		95	.168		4.28	5.55		9.83	13.90	
	1210	16" O.C.		145	.110		3.65	3.64		7.29	10.05	
	1220	24" O.C.		190	.084		3.02	2.78		5.80	7.95	
	2100	Installed on gable, 18 ga x 4" wide studs, 12" O.C.		85	.188		3.36	6.20		9.56	14	
	2110	16" O.C.		130	.123		2.86	4.06		6.92	9.90	
	2120	24" O.C.		170	.094		2.35	3.11		5.46	7.75	
	2200	6" wide studs, 12" O.C.		85	.188		4.28	6.20		10.48	15	
	2210	16" O.C.		130	.123		3.65	4.06		7.71	10.75	
	2220	24" O.C.	↓	170	.094	↓	3.02	3.11		6.13	8.50	
550	0010	**FRAMING, ROOF RAFTERS**										550
	0100	Boxed ridge beam, double, 18 ga x 6" deep	2 Carp	160	.100	L.F.	3.20	3.30		6.50	8.95	
	0110	8" deep		150	.107		3.56	3.52		7.08	9.75	
	0120	10" deep		140	.114		4.32	3.77		8.09	11	
	0130	12" deep		130	.123		4.74	4.06		8.80	11.95	
	0200	16 ga x 6" deep		150	.107		3.64	3.52		7.16	9.85	
	0210	8" deep		140	.114		4.08	3.77		7.85	10.75	
	0220	10" deep		130	.123		4.90	4.06		8.96	12.15	
	0230	12" deep	↓	120	.133		5.35	4.40		9.75	13.15	
	1100	Rafters, 2" flange, material only, 18 ga x 6" deep					.99			.99	1.09	
	1110	8" deep					1.18			1.18	1.29	
	1120	10" deep					1.39			1.39	1.52	
	1130	12" deep					1.61			1.61	1.77	
	1200	16 ga x 6" deep					1.22			1.22	1.34	
	1210	8" deep					1.45			1.45	1.59	
	1220	10" deep					1.69			1.69	1.86	
	1230	12" deep				↓	1.92			1.92	2.11	
	2100	Installation only, ordinary rafter to 4:12 pitch, 18 ga x 6" deep	2 Carp	35	.457	Ea.		15.10		15.10	25	
	2110	8" deep		30	.533			17.60		17.60	29	
	2120	10" deep		25	.640			21		21	35	
	2130	12" deep		20	.800			26.50		26.50	44	
	2200	16 ga x 6" deep		30	.533			17.60		17.60	29	
	2210	8" deep		25	.640			21		21	35	
	2220	10" deep		20	.800			26.50		26.50	44	
	2230	12" deep	↓	15	1.067	↓		35		35	58.50	
	8100	Add to labor, ordinary rafters on steep roofs						25%				
	8110	Dormers & complex roofs						50%				
	8200	Hip & valley rafters to 4:12 pitch						25%				
	8210	Steep roofs						50%				
	8220	Dormers & complex roofs						75%				
	8300	Hip & valley jack rafters to 4:12 pitch						50%				
	8310	Steep roofs						75%				
	8320	Dormers & complex roofs						100%				
	9000	Minimum labor/equipment charge	2 Carp	4	4	Job		132		132	219	
600	0010	**FRAMING, ROOF TRUSSES**										600
	0020	Fabrication of trusses on ground, Fink (W) or King Post, to 4:12 pitch										
	0120	18 ga x 4" chords, 16' span	2 Carp	12	1.333	Ea.	37.50	44		81.50	115	
	0130	20' span		11	1.455		47	48		95	131	
	0140	24' span		11	1.455		56.50	48		104.50	142	
	0150	28' span		10	1.600		65.50	53		118.50	160	
	0160	32' span		10	1.600		75	53		128	170	
	0250	6" chords, 28' span	↓	9	1.778	↓	82.50	58.50		141	188	

Important: See the Reference Section for critical supporting data - Reference Nos., Crews, & City Cost Indexes

	05425 | Cold-Formed Roof Framing		CREW	DAILY OUTPUT	LABOR-HOURS	UNIT	2004 BARE COSTS MAT.	LABOR	EQUIP.	TOTAL	TOTAL INCL O&P	
600	0260	32′ span	2 Carp	9	1.778	Ea.	94	58.50		152.50	200	600
	0270	36′ span		8	2		106	66		172	225	
	0280	40′ span		8	2		118	66		184	238	
	1120	5:12 to 8:12 pitch, 18 ga x 4″ chords, 16′ span		10	1.600		43	53		96	135	
	1130	20′ span		9	1.778		53.50	58.50		112	156	
	1140	24′ span		9	1.778		64.50	58.50		123	168	
	1150	28′ span		8	2		75	66		141	192	
	1160	32′ span		8	2		86	66		152	204	
	1250	6″ chords, 28′ span		7	2.286		94	75.50		169.50	228	
	1260	32′ span		7	2.286		108	75.50		183.50	243	
	1270	36′ span		6	2.667		121	88		209	279	
	1280	40′ span		6	2.667		134	88		222	294	
	2120	9:12 to 12:12 pitch, 18 ga x 4″ chords, 16′ span		8	2		53.50	66		119.50	168	
	2130	20′ span		7	2.286		67	75.50		142.50	199	
	2140	24′ span		7	2.286		80.50	75.50		156	214	
	2150	28′ span		6	2.667		94	88		182	249	
	2160	32′ span		6	2.667		107	88		195	264	
	2250	6″ chords, 28′ span		5	3.200		118	106		224	305	
	2260	32′ span		5	3.200		134	106		240	325	
	2270	36′ span		4	4		151	132		283	385	
	2280	40′ span		4	4		168	132		300	405	
	4900	Minimum labor/equipment charge		4	4	Job		132		132	219	
	5120	Erection only of roof trusses, to 4:12 pitch, 16′ span	F-6	48	.833	Ea.		25.50	12.75	38.25	55.50	
	5130	20′ span		46	.870			26.50	13.30	39.80	58	
	5140	24′ span		44	.909			28	13.90	41.90	61	
	5150	28′ span		42	.952			29	14.55	43.55	63.50	
	5160	32′ span		40	1			30.50	15.30	45.80	67	
	5170	36′ span		38	1.053			32	16.10	48.10	70	
	5180	40′ span		36	1.111			34	17	51	74	
	5220	5:12 to 8:12 pitch, 16′ span		42	.952			29	14.55	43.55	63.50	
	5230	20′ span		40	1			30.50	15.30	45.80	67	
	5240	24′ span		38	1.053			32	16.10	48.10	70	
	5250	28′ span		36	1.111			34	17	51	74	
	5260	32′ span		34	1.176			36	18	54	79	
	5270	36′ span		32	1.250			38	19.15	57.15	83.50	
	5280	40′ span		30	1.333			41	20.50	61.50	89.50	
	5320	9:12 to 12:12 pitch, 16′ span		36	1.111			34	17	51	74	
	5330	20′ span		34	1.176			36	18	54	79	
	5340	24′ span		32	1.250			38	19.15	57.15	83.50	
	5350	28′ span		30	1.333			41	20.50	61.50	89.50	
	5360	32′ span		28	1.429			43.50	22	65.50	95.50	
	5370	36′ span		26	1.538			47	23.50	70.50	103	
	5380	40′ span		24	1.667			51	25.50	76.50	112	
	9000	Minimum labor/equipment charge		2	20	Job		610	305	915	1,325	
650	0010	**FRAMING, SOFFITS & CANOPIES**										650
	0130	Continuous ledger track @ wall, studs @ 16″ O.C., 18 ga x 4″ wide	2 Carp	535	.030	L.F.	.61	.99		1.60	2.31	
	0140	6″ wide		500	.032		.79	1.06		1.85	2.62	
	0150	8″ wide		465	.034		1	1.14		2.14	2.98	
	0160	10″ wide		430	.037		1.23	1.23		2.46	3.40	
	0230	Studs @ 24″ O.C., 18 ga x 4″ wide		800	.020		.58	.66		1.24	1.73	
	0240	6″ wide		750	.021		.76	.70		1.46	2	
	0250	8″ wide		700	.023		.96	.75		1.71	2.30	
	0260	10″ wide		650	.025		1.18	.81		1.99	2.64	
	1000	Horizontal soffit and canopy members, material only										
	1030	1-5/8″ flange studs, 18 ga x 4″ deep				L.F.	.80			.80	.88	
	1040	6″ deep					1.01			1.01	1.11	

METALS 5

05400 | Cold Formed Metal Framing

	05425 | Cold-Formed Roof Framing		CREW	DAILY OUTPUT	LABOR-HOURS	UNIT	2004 BARE COSTS MAT.	LABOR	EQUIP.	TOTAL	TOTAL INCL O&P	
650	1050	8″ deep				L.F.	1.22			1.22	1.35	650
	1140	2″ flange joists, 18 ga x 6″ deep					1.13			1.13	1.24	
	1150	8″ deep					1.34			1.34	1.48	
	1160	10″ deep					1.58			1.58	1.74	
	4030	Installation only, 18 ga, 1-5/8″ flange x 4″ deep	2 Carp	130	.123	Ea.		4.06		4.06	6.75	
	4040	6″ deep		110	.145			4.80		4.80	7.95	
	4050	8″ deep		90	.178			5.85		5.85	9.70	
	4140	2″ flange, 18 ga x 6″ deep		110	.145			4.80		4.80	7.95	
	4150	8″ deep		90	.178			5.85		5.85	9.70	
	4160	10″ deep		80	.200			6.60		6.60	10.95	
	6010	Clips to attach facia to rafter tails, 2″ x 2″ x 18 ga angle	1 Carp	120	.067		.53	2.20		2.73	4.24	
	6020	16 ga angle	″	100	.080		.68	2.64		3.32	5.10	
	9000	Minimum labor/equipment charge	2 Carp	4	4	Job		132		132	219	

05500 | Metal Fabrications

	05514 | Ladders		CREW	DAILY OUTPUT	LABOR-HOURS	UNIT	2004 BARE COSTS MAT.	LABOR	EQUIP.	TOTAL	TOTAL INCL O&P	
500	0010	**LADDER**										500
	0020	Steel, 20″ wide, bolted to concrete, with cage	E-4	50	.640	V.L.F.	56.50	24	1.50	82	109	
	0100	Without cage		85	.376		26	14.15	.88	41.03	56.50	
	0300	Aluminum, bolted to concrete, with cage		50	.640		83	24	1.50	108.50	138	
	0400	Without cage		85	.376		48	14.15	.88	63.03	80.50	
	1350	Alternating tread stair, 56/68°, steel, standard paint color	2 Sswk	50	.320		143	11.90		154.90	180	
	1360	Non-standard paint color		50	.320		162	11.90		173.90	201	
	1370	Galvanized steel		50	.320		162	11.90		173.90	200	
	1380	Stainless steel		50	.320		241	11.90		252.90	287	
	1390	68°, aluminum		50	.320		176	11.90		187.90	216	
	9000	Minimum labor/equipment charge	E-4	2	16	Job		600	37.50	637.50	1,175	
	05517 | Metal Stairs											
300	0010	**FIRE ESCAPE**										300
	0200	2′ wide balcony, 1″ x 1/4″ bars 1-1/2″ O.C.	1 Sswk	5	1.600	L.F.	38	59.50		97.50	153	
	0400	1st story cantilevered stair, standard	″	.09	88.889	Ea.	1,575	3,300		4,875	7,875	
	0500	Cable counterweight				″	1,475			1,475	1,625	
	0700	Platform & fixed stair, 36″ x 40″	1 Sswk	.17	47.059	Flight	700	1,750		2,450	4,025	
	0900	For 3′-6″ wide escapes, add to above					100%	150%				
350	0010	**FIRE ESCAPE STAIRS**										350
	0020	One story, disappearing, stainless steel	2 Sswk	20	.800	V.L.F.	152	29.50		181.50	223	
	0100	Portable ladder				Ea.	50.50			50.50	55.50	
700	0010	**STAIR** Steel, safety nosing, steel stringers										700
	0020	Grating tread and pipe railing, 3′-6″ wide	E-4	35	.914	Riser	111	34.50	2.15	147.65	189	
	0100	4′-0″ wide		30	1.067		145	40	2.51	187.51	237	
	0200	Cement fill metal pan, picket rail, 3′-6″ wide		35	.914		167	34.50	2.15	203.65	250	
	0300	4′-0″ wide		30	1.067		189	40	2.51	231.51	286	
	0350	Wall rail, both sides, 3′-6″ wide		53	.604		128	22.50	1.42	151.92	184	
	0500	Checkered plate tread, industrial, 3′-6″ wide		28	1.143		111	43	2.69	156.69	206	
	0550	Circular, for tanks, 3′-0″ wide		33	.970		123	36.50	2.28	161.78	206	

05517 | Metal Stairs

			CREW	DAILY OUTPUT	LABOR-HOURS	UNIT	2004 BARE COSTS MAT.	LABOR	EQUIP.	TOTAL	TOTAL INCL O&P	
700	0600	For isolated stairs, add				Riser		100%				700
	0800	Custom steel stairs, 3'-6" wide, minimum	E-4	35	.914		167	34.50	2.15	203.65	250	
	0810	Average		30	1.067		223	40	2.51	265.51	325	
	0900	Maximum		20	1.600		279	60	3.76	342.76	420	
	1100	For 4' wide stairs, add					5%	5%				
	1300	For 5' wide stairs, add					10%	10%				
	1500	Landing, steel pan, conventional	E-4	160	.200	S.F.	22.50	7.55	.47	30.52	39	
	1810	Spiral aluminum, 5'-0" diameter, stock units		45	.711	Riser	202	27	1.67	230.67	274	
	1820	Custom units		45	.711		380	27	1.67	408.67	470	
	1900	Spiral, cast iron, 4'-0" diameter, ornamental, minimum		45	.711		180	27	1.67	208.67	250	
	1920	Maximum		25	1.280		246	48	3.01	297.01	365	
	2000	Spiral, steel, industrial checkered plate, 4' diameter		45	.711		180	27	1.67	208.67	250	
	2200	Stock units, 6'-0" diameter		40	.800		218	30	1.88	249.88	298	
	3110	Spiral steel, stock units, primed, flat metal tread, 3'-6" dia	2 Carp	1.60	10	Flight	880	330		1,210	1,500	
	3120	4'-0" dia		1.45	11.034		1,000	365		1,365	1,700	
	3130	4'-6" dia		1.35	11.852		1,100	390		1,490	1,875	
	3140	5'-0" dia		1.25	12.800		1,200	420		1,620	2,025	
	3210	Galvanized, 3'-6" dia		1.60	10		1,550	330		1,880	2,250	
	3220	4'-0" dia		1.45	11.034		1,725	365		2,090	2,500	
	3230	4'-6" dia		1.35	11.852		1,875	390		2,265	2,725	
	3240	5'-0" dia		1.25	12.800		2,025	420		2,445	2,925	
	3310	Checkered plate tread, 3'-6" dia		1.45	11.034		1,075	365		1,440	1,800	
	3320	4'-0" dia		1.35	11.852		1,225	390		1,615	2,000	
	3330	4'-6" dia		1.25	12.800		1,350	420		1,770	2,175	
	3340	5'-0" dia		1.15	13.913		1,450	460		1,910	2,325	
	3410	Galvanized, 3'-6" dia		1.45	11.034		1,750	365		2,115	2,525	
	3420	4'-0" dia		1.35	11.852		1,975	390		2,365	2,825	
	3430	4'-6" dia		1.25	12.800		2,150	420		2,570	3,050	
	3440	5'-0" dia		1.15	13.913		2,300	460		2,760	3,300	
	3510	Red oak tread on flat metal, 3'-6" dia		1.35	11.852		1,475	390		1,865	2,250	
	3520	4'-0" dia		1.25	12.800		1,625	420		2,045	2,500	
	3530	4'-6" dia		1.15	13.913		1,775	460		2,235	2,700	
	3540	5'-0" dia		1.05	15.238		1,900	505		2,405	2,925	
	3900	Industrial ships ladder, 3' W, grating treads, 2 line pipe rail	E-4	30	1.067	Riser	72.50	40	2.51	115.01	157	
	4000	Aluminum		30	1.067	"	111	40	2.51	153.51	201	
	9000	Minimum labor/equipment charge		2	16	Job		600	37.50	637.50	1,175	

05520 | Handrails & Railings

			CREW	DAILY OUTPUT	LABOR-HOURS	UNIT	MAT.	LABOR	EQUIP.	TOTAL	TOTAL INCL O&P	
700	0010	**RAILING, PIPE**										700
	0020	Aluminum, 2 rail, satin finish, 1-1/4" diameter	E-4	160	.200	L.F.	13.90	7.55	.47	21.92	30	
	0030	Clear anodized		160	.200		17.20	7.55	.47	25.22	33.50	
	0040	Dark anodized		160	.200		19.45	7.55	.47	27.47	36	
	0080	1-1/2" diameter, satin finish		160	.200		16.65	7.55	.47	24.67	33	
	0090	Clear anodized		160	.200		18.60	7.55	.47	26.62	35	
	0100	Dark anodized		160	.200		20.50	7.55	.47	28.52	37	
	0140	Aluminum, 3 rail, 1-1/4" diam., satin finish		137	.234		21.50	8.80	.55	30.85	40.50	
	0150	Clear anodized		137	.234		26.50	8.80	.55	35.85	46.50	
	0160	Dark anodized		137	.234		29.50	8.80	.55	38.85	49.50	
	0200	1-1/2" diameter, satin finish		137	.234		25.50	8.80	.55	34.85	45	
	0210	Clear anodized		137	.234		29	8.80	.55	38.35	49	
	0220	Dark anodized		137	.234		32	8.80	.55	41.35	52	
	0500	Steel, 2 rail, on stairs, primed, 1-1/4" diameter		160	.200		10.45	7.55	.47	18.47	26	
	0520	1-1/2" diameter		160	.200		11.45	7.55	.47	19.47	27	
	0540	Galvanized, 1-1/4" diameter		160	.200		14.45	7.55	.47	22.47	30.50	
	0560	1-1/2" diameter		160	.200		16.20	7.55	.47	24.22	32.50	
	0580	Steel, 3 rail, primed, 1-1/4" diameter		137	.234		15.55	8.80	.55	24.90	34	

		05520 | Handrails & Railings	CREW	DAILY OUTPUT	LABOR-HOURS	UNIT	2004 BARE COSTS MAT.	LABOR	EQUIP.	TOTAL	TOTAL INCL O&P	
700	0600	1-1/2" diameter	E-4	137	.234	L.F.	16.50	8.80	.55	25.85	35	700
	0620	Galvanized, 1-1/4" diameter		137	.234		22	8.80	.55	31.35	41	
	0640	1-1/2" diameter		137	.234		25.50	8.80	.55	34.85	45	
	0700	Stainless steel, 2 rail, 1-1/4" diam. #4 finish		137	.234		37.50	8.80	.55	46.85	58.50	
	0720	High polish		137	.234		60.50	8.80	.55	69.85	83.50	
	0740	Mirror polish		137	.234		76	8.80	.55	85.35	100	
	0760	Stainless steel, 3 rail, 1-1/2" diam., #4 finish		120	.267		56.50	10.05	.63	67.18	82	
	0770	High polish		120	.267		93.50	10.05	.63	104.18	122	
	0780	Mirror finish		120	.267		114	10.05	.63	124.68	145	
	0900	Wall rail, alum. pipe, 1-1/4" diam., satin finish		213	.150		7.95	5.65	.35	13.95	19.70	
	0905	Clear anodized		213	.150		9.70	5.65	.35	15.70	21.50	
	0910	Dark anodized		213	.150		11.75	5.65	.35	17.75	24	
	0915	1-1/2" diameter, satin finish		213	.150		8.80	5.65	.35	14.80	20.50	
	0920	Clear anodized		213	.150		11.10	5.65	.35	17.10	23	
	0925	Dark anodized		213	.150		13.70	5.65	.35	19.70	26	
	0930	Steel pipe, 1-1/4" diameter, primed		213	.150		6.35	5.65	.35	12.35	17.90	
	0935	Galvanized		213	.150		9.15	5.65	.35	15.15	21	
	0940	1-1/2" diameter		176	.182		6.50	6.85	.43	13.78	20.50	
	0945	Galvanized		213	.150		9.20	5.65	.35	15.20	21	
	0955	Stainless steel pipe, 1-1/2" diam., #4 finish		107	.299		30	11.25	.70	41.95	55	
	0960	High polish		107	.299		61	11.25	.70	72.95	89.50	
	0965	Mirror polish		107	.299		72	11.25	.70	83.95	101	
	9000	Minimum labor/equipment charge	1 Sswk	2	4	Job		149		149	277	
780	0010	**RAILINGS, INDUSTRIAL** Welded										780
	0020	2 rail, 3'-6" high, 1-1/2" pipe	E-4	255	.125	L.F.	15.60	4.72	.29	20.61	26.50	
	0200	For 4" high kick plate, 10 gauge, add					3.24			3.24	3.56	
	0500	For curved rails, add					30%	30%				
	9000	Minimum labor/equipment charge	1 Sswk	2	4	Job		149		149	277	
		05560 | Metal Castings										
200	0010	**CONSTRUCTION CASTINGS**										200
	0020	Manhole covers and frames see Division 02630-110										
	0100	Column bases, cast iron, 16" x 16", approx. 65 lbs.	E-4	46	.696	Ea.	89	26	1.63	116.63	149	
	0200	32" x 32", approx. 256 lbs.		23	1.391	"	330	52.50	3.27	385.77	465	
	0600	Miscellaneous C.I. castings, light sections, less than 150 lbs		3,200	.010	Lb.	1.37	.38	.02	1.77	2.24	
	1300	Special low volume items		3,200	.010	"	2.39	.38	.02	2.79	3.36	
		05580 | Formed Metal Fabrications										
200	0010	**ALUMINUM COLUMNS**										200
	0020	Aluminum, extruded, stock units, no cap or base, 6" diameter	E-4	240	.133	L.F.	7.85	5	.31	13.16	18.35	
	0100	8" diameter	"	170	.188		10.60	7.10	.44	18.14	25.50	
	0500	For square columns, add to column prices above					50%					
600	0010	**LAMP POSTS**										600
	0020	Aluminum, 7' high, stock units, post only	1 Carp	16	.500	Ea.	32	16.50		48.50	62.50	
	0100	Mild steel, plain		16	.500	"	28	16.50		44.50	58.50	
	9000	Minimum labor/equipment charge		4	2	Job		66		66	109	
900	0010	**WINDOW GUARDS**										900
	0015	Expanded metal, steel angle frame, permanent	E-4	350	.091	S.F.	16.30	3.44	.21	19.95	24.50	
	0025	Steel bars, 1/2" x 1/2", spaced 5" O.C.	"	290	.110	"	11.30	4.15	.26	15.71	20.50	
	0030	Hinge mounted, add				Opng.	32.50			32.50	36	
	0040	Removable type, add				"	20.50			20.50	23	
	0050	For galvanized guards, add				S.F.	35%					
	0070	For pivoted or projected type, add					105%	40%				
	0100	Mild steel, stock units, economy	E-4	405	.079		4.42	2.97	.19	7.58	10.60	

05500 | Metal Fabrications

05580 | Formed Metal Fabrications

			CREW	DAILY OUTPUT	LABOR-HOURS	UNIT	2004 BARE COSTS MAT.	LABOR	EQUIP.	TOTAL	TOTAL INCL O&P	
900	0200	Deluxe	E-4	405	.079	S.F.	9.10	2.97	.19	12.26	15.75	900
	0400	Woven wire, stock units, 3/8″ channel frame, 3′ x 5′ opening		40	.800	Opng.	119	30	1.88	150.88	189	
	0500	4′ x 6′ opening		38	.842		191	31.50	1.98	224.48	271	
	0800	Basket guards for above, add					164			164	180	
	1000	Swinging guards for above, add					56			56	61.50	
	9000	Minimum labor/equipment charge	1 Sswk	2	4	Job		149		149	277	

05700 | Ornamental Metal

05720 | Ornamental Handrails & Railings

			CREW	DAILY OUTPUT	LABOR-HOURS	UNIT	2004 BARE COSTS MAT.	LABOR	EQUIP.	TOTAL	TOTAL INCL O&P	
700	0010	**RAILINGS, ORNAMENTAL**										700
	0020	Aluminum, bronze or stainless, minimum	1 Sswk	24	.333	L.F.	25	12.40		37.40	50.50	
	0100	Maximum		9	.889		216	33		249	300	
	0200	Aluminum ornamental rail, minimum		15	.533		25	19.80		44.80	64.50	
	0300	Maximum		8	1		75	37		112	152	
	0400	Hand-forged wrought iron, minimum		12	.667		71	25		96	124	
	0500	Maximum		8	1		213	37		250	305	
	0600	Composite metal/wood/glass, minimum		6	1.333		150	49.50		199.50	257	
	0700	Maximum		5	1.600		299	59.50		358.50	440	
	9000	Minimum labor/equipment charge		2	4	Job		149		149	277	

05800 | Expansion Control

05810 | Exp. Joint Cover Assemblies

			CREW	DAILY OUTPUT	LABOR-HOURS	UNIT	2004 BARE COSTS MAT.	LABOR	EQUIP.	TOTAL	TOTAL INCL O&P	
350	0010	**EXPANSION JOINT ASSEMBLIES** Custom units										350
	0200	Floor cover assemblies, 1″ space, aluminum	1 Sswk	38	.211	L.F.	14.70	7.80		22.50	31	
	0300	Bronze		38	.211		28.50	7.80		36.30	46	
	0500	2″ space, aluminum		38	.211		17.80	7.80		25.60	34	
	0600	Bronze		38	.211		30.50	7.80		38.30	48	
	0800	Wall and ceiling assemblies, 1″ space, aluminum		38	.211		8.80	7.80		16.60	24	
	0900	Bronze		38	.211		26.50	7.80		34.30	44	
	1100	2″ space, aluminum		38	.211		14.90	7.80		22.70	31	
	1200	Bronze		38	.211		33	7.80		40.80	50.50	
	1400	Floor to wall assemblies, 1″ space, aluminum		38	.211		13.40	7.80		21.20	29.50	
	1500	Bronze or stainless		38	.211		32.50	7.80		40.30	50	
	1700	Gym floor angle covers, aluminum, 3″ x 3″ angle		46	.174		11.60	6.45		18.05	25	
	1800	3″ x 4″ angle		46	.174		13.70	6.45		20.15	27	
	2000	Roof closures, aluminum, flat roof, low profile, 1″ space		57	.140		27	5.20		32.20	39	
	2100	High profile		57	.140		33	5.20		38.20	45.50	
	2300	Roof to wall, low profile, 1″ space		57	.140		14.80	5.20		20	26	
	2400	High profile		57	.140		19.05	5.20		24.25	30.50	
	9000	Minimum labor/equipment charge		2	4	Job		149		149	277	

Division Notes

	CREW	DAILY OUTPUT	LABOR-HOURS	UNIT	2004 BARE COSTS MAT.	LABOR	EQUIP.	TOTAL	TOTAL INCL O&P

Division 6 Wood & Plastics

Estimating Tips

06050 Basic Wood & Plastic Materials & Methods

- Common to any wood framed structure are the accessory connector items such as screws, nails, adhesives, hangers, connector plates, straps, angles and holdowns. For typical wood framed buildings, such as residential projects, the aggregate total for these items can be significant, especially in areas where seismic loading is a concern. For floor and wall framing, the material cost is based on 10 to 25 lbs. per MBF. Holdowns, hangers and other connectors should be taken off by the piece.

06100 Rough Carpentry

- Lumber is a traded commodity and therefore sensitive to supply and demand in the marketplace. Even in "budgetary" estimating of wood framed projects, it is advisable to call local suppliers for the latest market pricing.
- Common quantity units for wood framed projects are "thousand board feet" (MBF). A board foot is a volume of wood, 1″ x 1′ x 1′, or 144 cubic inches. Board foot quantities are generally calculated using nominal material dimensions—dressed sizes are ignored. Board foot per lineal foot of any stick of lumber can be calculated by dividing the nominal cross sectional area by 12. As an example, 2,000 lineal feet of 2 x 12 equates to 4 MBF by dividing the nominal area, 2 x 12, by 12, which equals 2, and multiplying by 2,000 to give 4,000 board feet. This simple rule applies to all nominal dimensioned lumber.
- Waste is an issue of concern at the quantity takeoff for any area of construction. Framing lumber is sold in even foot lengths, i.e., 10′, 12′, 14′, 16′, and depending on spans, wall heights and the grade of lumber, waste is inevitable. A rule of thumb for lumber waste is 5% to 10% depending on material quality and the complexity of the framing.
- Wood in various forms and shapes is used in many projects, even where the main structural framing is steel, concrete or masonry. Plywood as a back-up partition material and 2x boards used as blocking and cant strips around roof edges are two common examples. The estimator should ensure that the costs of all wood materials are included in the final estimate.

06200 Finish Carpentry

- It is necessary to consider the grade of workmanship when estimating labor costs for erecting millwork and interior finish. In practice, there are three grades: premium, custom and economy. The Means daily output for base and case moldings is in the range of 200 to 250 L.F. per carpenter per day. This is appropriate for most average custom grade projects. For premium projects an adjustment to productivity of 25% to 50% should be made depending on the complexity of the job.

Reference Numbers

Reference numbers are shown in bold squares at the beginning of some major classifications. These numbers refer to related items in the Reference Section. The reference information may be an estimating procedure, an alternate pricing method or technical information.

Note: Not all subdivisions listed here necessarily appear in this publication.

	06052 | Selective Demolition		CREW	DAILY OUTPUT	LABOR-HOURS	UNIT	2004 BARE COSTS MAT.	LABOR	EQUIP.	TOTAL	TOTAL INCL O&P	
110	0010	**SELECTIVE DEMOLITION, WOOD FRAMING**										110
	3000	Wood framing, beams, 6″ x 8″	B-2	275	.145	L.F.		3.84		3.84	6.35	
	3040	6″ x 10″		220	.182			4.80		4.80	7.95	
	3080	6″ x 12″		185	.216			5.70		5.70	9.45	
	3120	8″ x 12″		140	.286			7.55		7.55	12.50	
	3160	10″ x 12″		110	.364			9.60		9.60	15.90	
	3400	Fascia boards, 1″ x 6″	1 Clab	500	.016			.42		.42	.69	
	3440	1″ x 8″		450	.018			.46		.46	.77	
	3480	1″ x 10″		400	.020			.52		.52	.86	
	3520	For trim boards, see division 02225-620										
	3800	Headers over openings, 2 @ 2″ x 6″	1 Clab	110	.073	L.F.		1.89		1.89	3.13	
	3840	2 @ 2″ x 8″		100	.080			2.08		2.08	3.45	
	3880	2 @ 2″ x 10″		90	.089			2.31		2.31	3.83	
	4200	Joists, 2″ x 4″	2 Clab	1,000	.016			.42		.42	.69	
	4230	Joists, 2″ x 6″		970	.016			.43		.43	.71	
	4240	2″ x 8″		940	.017			.44		.44	.73	
	4250	2″ x 10″		910	.018			.46		.46	.76	
	4280	2″ x 12″		880	.018			.47		.47	.78	
	5400	Posts, 4″ x 4″		800	.020			.52		.52	.86	
	5440	6″ x 6″		400	.040			1.04		1.04	1.72	
	5480	8″ x 8″		300	.053			1.39		1.39	2.30	
	5500	10″ x 10″		240	.067			1.73		1.73	2.87	
	5800	Rafters, ordinary, 2″ x 6″		850	.019			.49		.49	.81	
	5840	2″ x 8″		837	.019			.50		.50	.82	
	5900	Hip & valley, 2″ x 6″		500	.032			.83		.83	1.38	
	5940	2″ x 8″		420	.038			.99		.99	1.64	
	6200	Stairs and stringers, minimum		40	.400	Riser		10.40		10.40	17.25	
	6240	Maximum		26	.615	″		16		16	26.50	
	6600	Studs, 2″ x 4″		2,000	.008	L.F.		.21		.21	.34	
	6640	2″ x 6″		1,600	.010	″		.26		.26	.43	
	9000	Minimum labor/equipment charge	1 Clab	4	2	Job		52		52	86	
	9500	See Div. 02220-350 for rubbish handling										
120	0010	**SELECTIVE DEMOLITION, MILLWORK AND TRIM** R02220-510										120
	1000	Cabinets, wood, base cabinets	2 Clab	80	.200	L.F.		5.20		5.20	8.60	
	1020	Wall cabinets	″	80	.200	″		5.20		5.20	8.60	
	1060	Remove and reset, base cabinets	2 Carp	18	.889	Ea.		29.50		29.50	48.50	
	1070	Wall cabinets	″	20	.800	″		26.50		26.50	44	
	1100	Steel, painted, base cabinets	2 Clab	60	.267	L.F.		6.95		6.95	11.50	
	1120	Wall cabinets		60	.267	″		6.95		6.95	11.50	
	1200	Casework, large area		320	.050	S.F.		1.30		1.30	2.16	
	1220	Selective		200	.080	″		2.08		2.08	3.45	
	1500	Counter top, minimum		200	.080	L.F.		2.08		2.08	3.45	
	1510	Maximum		120	.133			3.47		3.47	5.75	
	1550	Remove and reset, minimum	2 Carp	50	.320			10.55		10.55	17.50	
	1560	Maximum	″	40	.400			13.20		13.20	22	
	2000	Paneling, 4′ x 8′ sheets, 1/4″ thick	2 Clab	2,000	.008	S.F.		.21		.21	.34	
	2100	Boards, 1″ x 4″		700	.023			.59		.59	.99	
	2120	1″ x 6″		750	.021			.55		.55	.92	
	2140	1″ x 8″		800	.020			.52		.52	.86	
	3000	Trim, baseboard, to 6″ wide		1,200	.013	L.F.		.35		.35	.57	
	3040	12″ wide		1,000	.016			.42		.42	.69	
	3080	Remove and reset, minimum	2 Carp	400	.040			1.32		1.32	2.19	
	3090	Maximum	″	300	.053			1.76		1.76	2.92	
	3100	Ceiling trim	2 Clab	1,000	.016			.42		.42	.69	
	3120	Chair rail		1,200	.013			.35		.35	.57	
	3140	Railings with balusters		240	.067			1.73		1.73	2.87	

		06052 | Selective Demolition	CREW	DAILY OUTPUT	LABOR-HOURS	UNIT	2004 BARE COSTS MAT.	LABOR	EQUIP.	TOTAL	TOTAL INCL O&P	
120	3160	Wainscoting R02220-510	2 Clab	700	.023	S.F.		.59		.59	.99	120
	9000	Minimum labor/equipment charge	1 Clab	4	2	Job		52		52	86	
		06070 | Lumber Treatment										
400	0011	**LUMBER TREATMENT**										400
	0400	Fire retardant, wet				M.B.F.	291			291	320	
	0500	KDAT					277			277	305	
	0700	Salt treated, water borne, .40 lb. retention					133			133	146	
	0800	Oil borne, 8 lb. retention					156			156	171	
	1000	Kiln dried lumber, 1" & 2" thick, softwoods					88.50			88.50	97.50	
	1100	Hardwoods					94.50			94.50	104	
	1500	For small size 1" stock, add				↓	11.95			11.95	13.15	
	1700	For full size rough lumber, add					20%					
600	0010	**PLYWOOD TREATMENT** Fire retardant, 1/4" thick				M.S.F.	222			222	244	600
	0030	3/8" thick					244			244	268	
	0050	1/2" thick					261			261	287	
	0070	5/8" thick					277			277	305	
	0100	3/4" thick					305			305	335	
	0200	For KDAT, add					66.50			66.50	73	
	0500	Salt treated water borne, .25 lb., wet, 1/4" thick					122			122	134	
	0530	3/8" thick					127			127	140	
	0550	1/2" thick					133			133	146	
	0570	5/8" thick					145			145	159	
	0600	3/4" thick					150			150	165	
	0800	For KDAT add					66.50			66.50	73	
	0900	For .40 lb., per C.F. retention, add					55.50			55.50	61	
	1000	For certification stamp, add				↓	32.50			32.50	36	
		06090 | Wood & Plastic Fastenings										
600	0010	**NAILS** Prices of material only, based on 50# box purchase, copper, plain				Lb.	4.66			4.66	5.15	600
	0400	Stainless steel, plain					5.25			5.25	5.80	
	0500	Box, 3d to 20d, bright					1.19			1.19	1.31	
	0520	Galvanized					1.43			1.43	1.57	
	0600	Common, 3d to 60d, plain 					.78			.78	.86	
	0700	Galvanized					.95			.95	1.05	
	0800	Aluminum					3.38			3.38	3.72	
	1000	Annular or spiral thread, 4d to 60d, plain					1.60			1.60	1.76	
	1200	Galvanized					1.70			1.70	1.87	
	1400	Drywall nails, plain					.70			.70	.77	
	1600	Galvanized					1.49			1.49	1.64	
	1800	Finish nails, 4d to 10d, plain 					.81			.81	.89	
	2000	Galvanized					1.12			1.12	1.23	
	2100	Aluminum					4.10			4.10	4.51	
	2300	Flooring nails, hardened steel, 2d to 10d, plain 					1.37			1.37	1.51	
	2400	Galvanized					2.21			2.21	2.43	
	2500	Gypsum lath nails, 1-1/8", 13 ga. flathead, blued					1.36			1.36	1.50	
	2600	Masonry nails, hardened steel, 3/4" to 3" long, plain					1.41			1.41	1.55	
	2700	Galvanized 					1.74			1.74	1.91	
	2900	Roofing nails, threaded, galvanized					1.32			1.32	1.45	
	3100	Aluminum					4.79			4.79	5.25	
	3300	Compressed lead head, threaded, galvanized					1.47			1.47	1.62	
	3600	Siding nails, plain shank, galvanized					1.36			1.36	1.50	
	3800	Aluminum					4.10			4.10	4.51	
	5000	Add to prices above for cement coating					.07			.07	.08	
	5200	Zinc or tin plating				↓	.12			.12	.13	

	06090 | Wood & Plastic Fastenings		Crew	Daily Output	Labor-Hours	Unit	2004 Bare Costs Mat.	Labor	Equip.	Total	Total Incl O&P	
600	5500	Vinyl coated sinkers, 8d to 16d				Lb.	.56			.56	.62	600
650	0010	**NAILS** mat. only, for pneumatic tools, framing, per carton of 5000, 2"				Ea.	37			37	41	650
	0100	2-3/8"					42.50			42.50	46.50	
	0200	Per carton of 4000, 3"					37.50			37.50	41.50	
	0300	3-1/4"					40			40	44	
	0400	Per carton of 5000, 2-3/8", galv.					57.50			57.50	63.50	
	0500	Per carton of 4000, 3", galv.					65			65	71.50	
	0600	3-1/4", galv.					80.50			80.50	88.50	
	0700	Roofing, per carton of 7200, 1"					35			35	38.50	
	0800	1-1/4"					32.50			32.50	36	
	0900	1-1/2"					37.50			37.50	41.50	
	1000	1-3/4"				↓	45.50			45.50	50	
750	0010	**WOOD SCREWS** #8, 1" long, steel				C	3.36			3.36	3.70	750
	0100	Brass					11.20			11.20	12.35	
	0600	#10, 2" long, steel					7.65			7.65	8.45	
	0700	Brass					40.50			40.50	44.50	
	1500	#12, 3" long, steel				↓	16.20			16.20	17.85	
800	0010	**TIMBER CONNECTORS** Add up cost of each part for total										800
	0020	cost of connection										
	0100	Connector plates, steel, with bolts, straight	2 Carp	75	.213	Ea.	18.60	7.05		25.65	32	
	0110	Tee		50	.320		27.50	10.55		38.05	47.50	
	0120	T- Strap, 14 gauge12" x 8" x 2"		50	.320		27.50	10.55		38.05	47.50	
	0150	Anchor plate, 7 gauge, 9" x 7"	↓	75	.213		18.60	7.05		25.65	32	
	0200	Bolts, machine, sq. hd. with nut & washer, 1/2" diameter, 4" long	1 Carp	140	.057		.30	1.89		2.19	3.46	
	0300	7-1/2" long		130	.062		.53	2.03		2.56	3.95	
	0500	3/4" diameter, 7-1/2" long		130	.062		1.45	2.03		3.48	4.97	
	0610	Machine bolts, w/ nut, washer, 3/4" dia, 15" L, HD's & beam hangers		95	.084	↓	2.66	2.78		5.44	7.55	
	0800	Drilling bolt holes in timber, 1/2" diameter		450	.018	Inch		.59		.59	.97	
	0900	1" diameter		350	.023	"		.75		.75	1.25	
	1100	Framing anchors, 2 or 3 dimensional, 10 gauge, no nails incl.		175	.046	Ea.	.46	1.51		1.97	3.01	
	1150	Framing anchors, 18 gauge, 4 1/2" x 2 3/4"		175	.046		.46	1.51		1.97	3.01	
	1160	Framing anchors, 18 gauge, 4 1/2" x 3"		175	.046		.46	1.51		1.97	3.01	
	1170	Clip anchors plates, 18 gauge, 12" x 1 1/8"		175	.046		.46	1.51		1.97	3.01	
	1250	Holdowns, 3 gauge base, 10 gauge body		8	1		15.45	33		48.45	71.50	
	1260	Holdowns, 7 gauge 11 1/16" x 3 1/4"		8	1		15.45	33		48.45	71.50	
	1270	Holdowns, 7 gauge 14 3/8" x 3 1/8"		8	1		15.45	33		48.45	71.50	
	1275	Holdowns, 12 gauge 8" x 2 1/2"		8	1		15.45	33		48.45	71.50	
	1300	Joist and beam hangers, 18 ga. galv., for 2" x 4" joist		175	.046		.57	1.51		2.08	3.13	
	1400	2" x 6" to 2" x 10" joist		165	.048		.49	1.60		2.09	3.19	
	1600	16 ga. galv., 3" x 6" to 3" x 10" joist		160	.050		2.61	1.65		4.26	5.60	
	1700	3" x 10" to 3" x 14" joist		160	.050		3.03	1.65		4.68	6.05	
	1800	4" x 6" to 4" x 10" joist		155	.052		2.30	1.70		4	5.35	
	1900	4" x 10" to 4" x 14" joist		155	.052		3.11	1.70		4.81	6.25	
	2000	Two-2" x 6" to two-2" x 10" joists		150	.053		2.50	1.76		4.26	5.65	
	2100	Two-2" x 10" to two-2" x 14" joists		150	.053		2.50	1.76		4.26	5.65	
	2300	3/16" thick, 6" x 8" joist		145	.055		5.45	1.82		7.27	9	
	2400	6" x 10" joist		140	.057		6.45	1.89		8.34	10.20	
	2500	6" x 12" joist		135	.059		7.75	1.96		9.71	11.75	
	2700	1/4" thick, 6" x 14" joist	↓	130	.062		9.60	2.03		11.63	13.90	
	2900	Plywood clips, extruded aluminum H clip, for 3/4" panels					.15			.15	.17	
	3000	Galvanized 18 ga. back-up clip					.14			.14	.15	
	3200	Post framing, 16 ga. galv. for 4" x 4" base, 2 piece	1 Carp	130	.062		5.40	2.03		7.43	9.30	
	3300	Cap	↓	130	.062	↓	2.64	2.03		4.67	6.25	

06050 | Basic Wood / Plastic Materials / Methods

06090 | Wood & Plastic Fastenings

		06090 Wood & Plastic Fastenings	CREW	DAILY OUTPUT	LABOR-HOURS	UNIT	2004 BARE COSTS MAT.	LABOR	EQUIP.	TOTAL	TOTAL INCL O&P	
800	3500	Rafter anchors, 18 ga. galv., 1-1/2" wide, 5-1/4" long	1 Carp	145	.055	Ea.	.46	1.82		2.28	3.53	800
	3600	10-3/4" long		145	.055		.90	1.82		2.72	4.01	
	3800	Shear plates, 2-5/8" diameter		120	.067		1.61	2.20		3.81	5.40	
	3900	4" diameter		115	.070		3.71	2.30		6.01	7.90	
	4000	Sill anchors, embedded in concrete or block, 18-5/8" long		115	.070		1.09	2.30		3.39	5	
	4100	Spike grids, 4" x 4", flat or curved		120	.067		.41	2.20		2.61	4.10	
	4400	Split rings, 2-1/2" diameter		120	.067		1.31	2.20		3.51	5.10	
	4500	4" diameter		110	.073		2.02	2.40		4.42	6.20	
	4550	Tie plate, 20 gauge, 7" x 3 1/8"		110	.073		2.02	2.40		4.42	6.20	
	4560	Tie plate, 20 gauge, 5" x 4 1/8"		110	.073		2.02	2.40		4.42	6.20	
	4575	Twist straps, 18 gauge, 12" x 1 1/4"		110	.073		2.02	2.40		4.42	6.20	
	4580	Twist straps, 18 gauge, 16" x 1 1/4"		110	.073		2.02	2.40		4.42	6.20	
	4600	Strap ties, 20 ga., 2 -1/16" wide, 12 13/16" long		180	.044		1.18	1.47		2.65	3.73	
	4700	Strap ties, 16 ga., 1-3/8" wide, 12" long		180	.044		1.18	1.47		2.65	3.73	
	4800	24" long		160	.050		1.62	1.65		3.27	4.52	
	5000	Toothed rings, 2-5/8" or 4" diameter		90	.089		1.12	2.93		4.05	6.10	
	5200	Truss plates, nailed, 20 gauge, up to 32' span		17	.471	Truss	8.05	15.55		23.60	34.50	
	5400	Washers, 2" x 2" x 1/8"				Ea.	.26			.26	.29	
	5500	3" x 3" x 3/16"				"	.66			.66	.73	
	9000	Minimum labor/equipment charge	1 Carp	4	2	Job		66		66	109	
825	0010	**ROUGH HARDWARE** Average % of carpentry material, minimum					.50%					825
	0200	Maximum					1.50%					
850	0010	**BRACING**										850
	0300	Let-in, "T" shaped, 22 ga. galv. steel, studs at 16" O.C.	1 Carp	5.80	1.379	C.L.F.	43	45.50		88.50	123	
	0400	Studs at 24" O.C.		6	1.333		43	44		87	121	
	0500	16 ga. galv. steel straps, studs at 16" O.C.		6	1.333		64	44		108	144	
	0600	Studs at 24" O.C.		6.20	1.290		64	42.50		106.50	141	

06100 | Rough Carpentry

06110 | Wood Framing

		06110 Wood Framing	CREW	DAILY OUTPUT	LABOR-HOURS	UNIT	2004 BARE COSTS MAT.	LABOR	EQUIP.	TOTAL	TOTAL INCL O&P	
100	0010	**BLOCKING**										100
	2600	Miscellaneous, to wood construction										
	2620	2" x 4"	1 Carp	.17	47.059	M.B.F.	465	1,550		2,015	3,075	
	2625	Pneumatic nailed		.21	38.095		465	1,250		1,715	2,575	
	2660	2" x 8"		.27	29.630		600	980		1,580	2,275	
	2665	Pneumatic nailed		.33	24.242		600	800		1,400	1,975	
	2720	To steel construction										
	2740	2" x 4"	1 Carp	.14	57.143	M.B.F.	465	1,875		2,340	3,625	
	2780	2" x 8"		.21	38.095	"	600	1,250		1,850	2,725	
	9000	Minimum labor/equipment charge		4	2	Job		66		66	109	
150	0012	**BRACING** Let-in, with 1" x 6" boards, studs @ 16" O.C.	1 Carp	150	.053	L.F.	.38	1.76		2.14	3.33	150
	0202	Studs @ 24" O.C.	"	230	.035	"	.38	1.15		1.53	2.31	
200	0012	**BRIDGING** Wood, for joists 16" O.C., 1" x 3"	1 Carp	130	.062	Pr.	.31	2.03		2.34	3.71	200
	0017	Pneumatic nailed		170	.047		.31	1.55		1.86	2.91	
	0102	2" x 3" bridging		130	.062		.36	2.03		2.39	3.77	
	0107	Pneumatic nailed		170	.047		.36	1.55		1.91	2.97	
	0302	Steel, galvanized, 18 ga., for 2" x 10" joists at 12" O.C.		130	.062		.87	2.03		2.90	4.33	
	0402	24" O.C.		140	.057		1.59	1.89		3.48	4.88	

	06110 | Wood Framing		Crew	Daily Output	Labor-Hours	Unit	2004 Bare Costs Mat.	Labor	Equip.	Total	Total Incl O&P	
200	0902	Compression type, 16" O.C., 2" x 8" joists	1 Carp	200	.040	Pr.	1.12	1.32		2.44	3.42	200
	1002	2" x 12" joists		200	.040		1.12	1.32		2.44	3.42	
505	0010	**FRAMING, BEAMS & GIRDERS** R06100-010										505
	1002	Single, 2" x 6"	2 Carp	700	.023	L.F.	.52	.75		1.27	1.82	
	1007	Pneumatic nailed R06110-030		812	.020		.52	.65		1.17	1.65	
	1022	2" x 8"		650	.025		.80	.81		1.61	2.23	
	1027	Pneumatic nailed		754	.021		.80	.70		1.50	2.04	
	1042	2" x 10"		600	.027		1.08	.88		1.96	2.65	
	1047	Pneumatic nailed		696	.023		1.08	.76		1.84	2.45	
	1062	2" x 12"		550	.029		1.47	.96		2.43	3.21	
	1067	Pneumatic nailed		638	.025		1.47	.83		2.30	2.99	
	1082	2" x 14"		500	.032		1.77	1.06		2.83	3.70	
	1087	Pneumatic nailed		580	.028		1.77	.91		2.68	3.46	
	1102	3" x 8"		550	.029		2.19	.96		3.15	4	
	1122	3" x 10"		500	.032		2.73	1.06		3.79	4.76	
	1142	3" x 12"		450	.036		3.28	1.17		4.45	5.55	
	1162	3" x 14"		400	.040		3.83	1.32		5.15	6.40	
	1170	4" x 6"	F-3	1,100	.036		2.51	1.21	.56	4.28	5.35	
	1182	4" x 8"		1,000	.040		3.34	1.33	.61	5.28	6.55	
	1202	4" x 10"		950	.042		4.18	1.40	.64	6.22	7.60	
	1222	4" x 12"		900	.044		5	1.48	.68	7.16	8.70	
	1242	4" x 14"		850	.047		5.85	1.57	.72	8.14	9.80	
	2002	Double, 2" x 6"	2 Carp	625	.026		1.04	.84		1.88	2.55	
	2007	Pneumatic nailed		725	.022		1.04	.73		1.77	2.36	
	2022	2" x 8"		575	.028		1.60	.92		2.52	3.28	
	2027	Pneumatic nailed		667	.024		1.60	.79		2.39	3.07	
	2042	2" x 10"		550	.029		2.16	.96		3.12	3.97	
	2047	Pneumatic nailed		638	.025		2.16	.83		2.99	3.75	
	2062	2" x 12"		525	.030		2.95	1.01		3.96	4.91	
	2067	Pneumatic nailed		610	.026		2.95	.87		3.82	4.67	
	2082	2" x 14"		475	.034		3.54	1.11		4.65	5.75	
	2087	Pneumatic nailed		551	.029		3.54	.96		4.50	5.50	
	3002	Triple, 2" x 6"		550	.029		1.56	.96		2.52	3.31	
	3007	Pneumatic nailed		638	.025		1.56	.83		2.39	3.09	
	3022	2" x 8"		525	.030		2.40	1.01		3.41	4.31	
	3027	Pneumatic nailed		609	.026		2.40	.87		3.27	4.08	
	3042	2" x 10"		500	.032		3.24	1.06		4.30	5.30	
	3047	Pneumatic nailed		580	.028		3.24	.91		4.15	5.05	
	3062	2" x 12"		475	.034		4.42	1.11		5.53	6.70	
	3067	Pneumatic nailed		551	.029		4.42	.96		5.38	6.45	
	3082	2" x 14"		450	.036		5.30	1.17		6.47	7.80	
	3087	Pneumatic nailed		522	.031		5.30	1.01		6.31	7.55	
	9000	Minimum labor/equipment charge	1 Carp	2	4	Job		132		132	219	
510	0010	**FRAMING, CEILINGS**										510
	6002	Suspended, 2" x 3"	2 Carp	1,000	.016	L.F.	.24	.53		.77	1.15	
	6052	2" x 4"		900	.018		.31	.59		.90	1.31	
	6102	2" x 6"		800	.020		.52	.66		1.18	1.66	
	6152	2" x 8"		650	.025		.80	.81		1.61	2.23	
	9000	Minimum labor/equipment charge	1 Carp	4	2	Job		66		66	109	
515	0010	**FRAMING, COLUMNS**										515
	0100	4" x 4"	2 Carp	390	.041	L.F.	1.16	1.35		2.51	3.52	
	0150	4" x 6"		275	.058		2.51	1.92		4.43	5.95	
	0200	4" x 8"		220	.073		3.34	2.40		5.74	7.65	
	0250	6" x 6"		215	.074		5.05	2.46		7.51	9.60	
	0300	6" x 8"		175	.091		7.10	3.02		10.12	12.80	

	06110 | Wood Framing		CREW	DAILY OUTPUT	LABOR-HOURS	UNIT	2004 BARE COSTS MAT.	LABOR	EQUIP.	TOTAL	TOTAL INCL O&P	
515	0350	6″ x 10″	2 Carp	150	.107	L.F.	9.20	3.52		12.72	15.95	515
	9000	Minimum labor/equipment charge	1 Carp	2	4	Job		132		132	219	
520	0010	**FRAMING, HEAVY** Mill timber, beams, single 6″ x 10″ R06100-010	2 Carp	1.10	14.545	M.B.F.	2,400	480		2,880	3,425	520
	0100	Single 8″ x 16″		1.20	13.333	″	2,675	440		3,115	3,675	
	0202	Built from 2″ lumber, multiple 2″ x 14″		900	.018	B.F.	.76	.59		1.35	1.81	
	0212	Built from 3″ lumber, multiple 3″ x 6″		700	.023		1.09	.75		1.84	2.45	
	0222	Multiple 3″ x 8″		800	.020		1.09	.66		1.75	2.29	
	0232	Multiple 3″ x 10″		900	.018		1.09	.59		1.68	2.17	
	0242	Multiple 3″ x 12″		1,000	.016		1.09	.53		1.62	2.08	
	0252	Built from 4″ lumber, multiple 4″ x 6″		800	.020		1.25	.66		1.91	2.47	
	0262	Multiple 4″ x 8″		900	.018		1.25	.59		1.84	2.35	
	0272	Multiple 4″ x 10″		1,000	.016		1.25	.53		1.78	2.26	
	0281											
	0282	Multiple 4″ x 12″	2 Carp	1,100	.015	B.F.	1.25	.48		1.73	2.18	
	0292	Columns, structural grade, 1500f, 4″ x 4″		450	.036	L.F.	2.23	1.17		3.40	4.40	
	0302	6″ x 6″		225	.071		6.50	2.35		8.85	11.05	
	0402	8″ x 8″		240	.067		11.80	2.20		14	16.60	
	0502	10″ x 10″		90	.178		19.30	5.85		25.15	30.50	
	0602	12″ x 12″		70	.229		28.50	7.55		36.05	43.50	
	0802	Floor planks, 2″ thick, T & G, 2″ x 6″		1,050	.015	B.F.	1.72	.50		2.22	2.72	
	0902	2″ x 10″		1,100	.015		.82	.48		1.30	1.70	
	1102	3″ thick, 3″ x 6″		1,050	.015		1.24	.50		1.74	2.19	
	1202	3″ x 10″		1,100	.015		1.24	.48		1.72	2.16	
	1402	Girders, structural grade, 12″ x 12″		800	.020		1.78	.66		2.44	3.04	
	1502	10″ x 16″		1,000	.016		1.73	.53		2.26	2.79	
	2050	Roof planks, see division 06150-600										
	2302	Roof purlins, 4″ thick, structural grade	2 Carp	1,050	.015	B.F.	1.21	.50		1.71	2.16	
	2502	Roof trusses, add timber connectors, division 06090-800	″	450	.036	″	1.18	1.17		2.35	3.25	
	9000	Minimum labor/equipment charge	1 Carp	2	4	Job		132		132	219	
530	0010	**FRAMING, JOISTS** R06100-010										530
	2002	Joists, 2″ x 4″	2 Carp	1,250	.013	L.F.	.31	.42		.73	1.04	
	2007	Pneumatic nailed R06110-030		1,438	.011		.31	.37		.68	.95	
	2100	2″ x 6″		1,250	.013		.52	.42		.94	1.27	
	2105	Pneumatic nailed		1,438	.011		.52	.37		.89	1.18	
	2152	2″ x 8″		1,100	.015		.80	.48		1.28	1.68	
	2157	Pneumatic nailed		1,265	.013		.80	.42		1.22	1.57	
	2202	2″ x 10″		900	.018		1.08	.59		1.67	2.16	
	2207	Pneumatic nailed		1,035	.015		1.08	.51		1.59	2.04	
	2252	2″ x 12″		875	.018		1.47	.60		2.07	2.62	
	2257	Pneumatic nailed		1,006	.016		1.47	.52		1.99	2.49	
	2302	2″ x 14″		770	.021		1.77	.69		2.46	3.09	
	2307	Pneumatic nailed		886	.018		1.77	.60		2.37	2.94	
	2352	3″ x 6″		925	.017		1.64	.57		2.21	2.75	
	2402	3″ x 10″		780	.021		2.73	.68		3.41	4.13	
	2452	3″ x 12″		600	.027		3.28	.88		4.16	5.05	
	2502	4″ x 6″		800	.020		2.51	.66		3.17	3.85	
	2552	4″ x 10″		600	.027		4.18	.88		5.06	6.05	
	2602	4″ x 12″		450	.036		5	1.17		6.17	7.45	
	2607	Sister joist, 2″ x 6″		800	.020		.52	.66		1.18	1.66	
	2608	Pneumatic nailed		960	.017		.52	.55		1.07	1.48	
	2612	2″ x 8″		640	.025		.80	.83		1.63	2.25	
	2613	Pneumatic nailed		768	.021		.80	.69		1.49	2.02	
	2617	2″ x 10″		535	.030		1.08	.99		2.07	2.83	
	2618	Pneumatic nailed		642	.025		1.08	.82		1.90	2.55	
	2622	2″ x 12″		455	.035		1.47	1.16		2.63	3.54	

06110 | Wood Framing

		06110 Wood Framing	CREW	DAILY OUTPUT	LABOR-HOURS	UNIT	2004 BARE COSTS MAT.	LABOR	EQUIP.	TOTAL	TOTAL INCL O&P	
530	2627	Pneumatic nailed R06100-010	2 Carp	546	.029	L.F.	1.47	.97		2.44	3.22	530
	3000	Composite wood joist 9-1/2" deep		.90	17.778	M.L.F.	1,600	585		2,185	2,725	
	3010	11-1/2" deep R06110-030		.88	18.182		1,700	600		2,300	2,875	
	3020	14" deep		.82	19.512		1,875	645		2,520	3,125	
	3030	16" deep		.78	20.513		2,550	675		3,225	3,925	
	4000	Open web joist 12" deep		.88	18.182		1,675	600		2,275	2,850	
	4010	14" deep		.82	19.512		1,950	645		2,595	3,225	
	4020	16" deep		.78	20.513		2,025	675		2,700	3,350	
	4030	18" deep		.74	21.622		2,050	715		2,765	3,425	
	6000	Composite rim joist, 1-1/4" x 9-1/2"		90	.178		1,975	5.85		1,980.85	2,150	
	6010	1-1/4" x 11-1/2"		.88	18.182		2,225	600		2,825	3,450	
	6020	1-1/4" x 14-1/2"		.82	19.512		2,650	645		3,295	4,000	
	6030	1-1/4" x 16-1/2"		.78	20.513		2,875	675		3,550	4,275	
	9000	Minimum labor/equipment charge	1 Carp	4	2	Job		66		66	109	
545	0010	**FRAMING, MISCELLANEOUS**										545
	2002	Firestops, 2" x 4"	2 Carp	780	.021	L.F.	.31	.68		.99	1.46	
	2007	Pneumatic nailed		952	.017		.31	.55		.86	1.26	
	2102	2" x 6"		600	.027		.52	.88		1.40	2.03	
	2107	Pneumatic nailed		732	.022		.52	.72		1.24	1.77	
	5002	Nailers, treated, wood construction, 2" x 4"		800	.020		.42	.66		1.08	1.55	
	5007	Pneumatic nailed		960	.017		.42	.55		.97	1.37	
	5102	2" x 6"		750	.021		.76	.70		1.46	2.01	
	5107	Pneumatic nailed		900	.018		.76	.59		1.35	1.81	
	5122	2" x 8"		700	.023		.95	.75		1.70	2.30	
	5127	Pneumatic nailed		840	.019		.95	.63		1.58	2.09	
	5202	Steel construction, 2" x 4"		750	.021		.42	.70		1.12	1.63	
	5222	2" x 6"		700	.023		.76	.75		1.51	2.09	
	5242	2" x 8"		650	.025		.95	.81		1.76	2.40	
	7002	Rough bucks, treated, for doors or windows, 2" x 6"		400	.040		.76	1.32		2.08	3.03	
	7007	Pneumatic nailed		480	.033		.76	1.10		1.86	2.66	
	7102	2" x 8"		380	.042		.95	1.39		2.34	3.35	
	7107	Pneumatic nailed		456	.035		.95	1.16		2.11	2.97	
	8000	Stair stringers, 2" x 10"		130	.123		1.08	4.06		5.14	7.95	
	8100	2" x 12"		130	.123		1.47	4.06		5.53	8.35	
	8150	3" x 10"		125	.128		2.73	4.22		6.95	10	
	8200	3" x 12"		125	.128		3.28	4.22		7.50	10.60	
	8870	Composite LSL, 1-1/4" x 11-1/2"		130	.123		2.23	4.06		6.29	9.20	
	8880	1-1/4" x 14-1/2"		130	.123		2.65	4.06		6.71	9.65	
	9000	Minimum labor/equipment charge	1 Carp	4	2	Job		66		66	109	
550	0010	**PARTITIONS** Wood stud with single bottom plate and										550
	0020	double top plate, no waste, std. & better lumber										
	0182	2" x 4" studs, 8' high, studs 12" O.C.	2 Carp	80	.200	L.F.	3.74	6.60		10.34	15.05	
	0187	12" O.C., pneumatic nailed		96	.167		3.74	5.50		9.24	13.20	
	0202	16" O.C.		100	.160		3.06	5.30		8.36	12.10	
	0207	16" O.C., pneumatic nailed		120	.133		3.06	4.40		7.46	10.65	
	0302	24" O.C.		125	.128		2.38	4.22		6.60	9.60	
	0307	24" O.C., pneumatic nailed		150	.107		2.38	3.52		5.90	8.45	
	0382	10' high, studs 12" O.C.		80	.200		4.42	6.60		11.02	15.80	
	0387	12" O.C., pneumatic nailed		96	.167		4.42	5.50		9.92	13.95	
	0402	16" O.C.		100	.160		3.57	5.30		8.87	12.70	
	0407	16" O.C., pneumatic nailed		120	.133		3.57	4.40		7.97	11.25	
	0502	24" O.C.		125	.128		2.72	4.22		6.94	10	
	0507	24" O.C., pneumatic nailed		150	.107		2.72	3.52		6.24	8.85	
	0582	12' high, studs 12" O.C.		65	.246		5.10	8.10		13.20	19.05	
	0587	12" O.C., pneumatic nailed		78	.205		5.10	6.75		11.85	16.80	

06110 | Wood Framing

			CREW	DAILY OUTPUT	LABOR-HOURS	UNIT	2004 BARE COSTS MAT.	LABOR	EQUIP.	TOTAL	TOTAL INCL O&P	
550	0602	16" O.C.	2 Carp	80	.200	L.F.	4.08	6.60		10.68	15.45	550
	0607	16" O.C., pneumatic nailed		96	.167		4.08	5.50		9.58	13.60	
	0701	24" O.C.		100	.160		3.06	5.30		8.36	12.10	
	0702											
	0706	24" O.C., pneumatic nailed	2 Carp	120	.133	L.F.	3.06	4.40		7.46	10.65	
	0782	2" x 6" studs, 8' high, studs 12" O.C.		70	.229		6.30	7.55		13.85	19.45	
	0787	12" O.C., pneumatic nailed		84	.190		6.30	6.30		12.60	17.35	
	0802	16" O.C.		90	.178		5.15	5.85		11	15.40	
	0807	16" O.C., pneumatic nailed		108	.148		5.15	4.89		10.04	13.80	
	0902	24" O.C.		115	.139		4.01	4.59		8.60	12	
	0907	24" O.C., pneumatic nailed		138	.116		4.01	3.83		7.84	10.75	
	0982	10' high, studs 12" O.C.		70	.229		7.45	7.55		15	20.50	
	0987	12" O.C., pneumatic nailed		84	.190		7.45	6.30		13.75	18.60	
	1002	16" O.C.		90	.178		6	5.85		11.85	16.30	
	1007	16" O.C., pneumatic nailed		108	.148		6	4.89		10.89	14.70	
	1102	24" O.C.		115	.139		4.59	4.59		9.18	12.65	
	1107	24" O.C., pneumatic nailed		138	.116		4.59	3.83		8.42	11.40	
	1182	12' high, studs 12" O.C.		55	.291		8.60	9.60		18.20	25.50	
	1187	12" O.C., pneumatic nailed		66	.242		8.60	8		16.60	22.50	
	1202	16" O.C.		70	.229		6.90	7.55		14.45	20	
	1207	16" O.C., pneumatic nailed		84	.190		6.90	6.30		13.20	17.95	
	1302	24" O.C.		90	.178		5.15	5.85		11	15.40	
	1307	24" O.C., pneumatic nailed		108	.148		5.15	4.89		10.04	13.80	
	1402	For horizontal blocking, 2" x 4", add		600	.027		.34	.88		1.22	1.83	
	1502	2" x 6", add		600	.027		.57	.88		1.45	2.09	
	1600	For openings, add		250	.064			2.11		2.11	3.50	
	1702	Headers for above openings, material only, add				B.F.	.66			.66	.73	
	9000	Minimum labor/equipment charge	1 Carp	4	2	Job		66		66	109	
555	0010	**FRAMING, ROOFS** R06110-030										555
	2000	Fascia boards, 2" x 8"	2 Carp	225	.071	L.F.	.80	2.35		3.15	4.77	
	2100	2" x 10"		180	.089		1.08	2.93		4.01	6.05	
	5000	Rafters, to 4 in 12 pitch, 2" x 6", ordinary		1,000	.016		.52	.53		1.05	1.45	
	5060	2" x 8", ordinary		950	.017		.80	.56		1.36	1.80	
	5250	Composite rafter, 9-1/2" deep		575	.028		1.59	.92		2.51	3.27	
	5260	11-1/2" deep		575	.028		1.70	.92		2.62	3.39	
	5300	Hip and valley rafters, 2" x 6", ordinary		760	.021		.52	.69		1.21	1.72	
	5360	2" x 8", ordinary		720	.022		.80	.73		1.53	2.10	
	5540	Hip and valley jacks, 2" x 6", ordinary		600	.027		.52	.88		1.40	2.03	
	5600	2" x 8", ordinary		490	.033		.80	1.08		1.88	2.67	
	5761	For slopes steeper than 4 in 12, add						30%				
	5770	For dormers or complex roofs, add						50%				
	5780	Rafter tie, 1" x 4", #3	2 Carp	800	.020	L.F.	.34	.66		1	1.46	
	5800	Ridge board, #2 or better, 1" x 6"		600	.027		.91	.88		1.79	2.46	
	5820	1" x 8"		550	.029		1.22	.96		2.18	2.93	
	5840	1" x 10"		500	.032		1.52	1.06		2.58	3.42	
	5860	2" x 6"		500	.032		.52	1.06		1.58	2.32	
	5880	2" x 8"		450	.036		.80	1.17		1.97	2.83	
	5900	2" x 10"		400	.040		1.08	1.32		2.40	3.38	
	5920	Roof cants, split, 4" x 4"		650	.025		1.16	.81		1.97	2.63	
	5940	6" x 6"		600	.027		5.05	.88		5.93	7	
	5960	Roof curbs, untreated, 2" x 6"		520	.031		.52	1.02		1.54	2.25	
	5980	2" x 12"		400	.040		1.47	1.32		2.79	3.81	
	6000	Sister rafters, 2" x 6"		800	.020		.52	.66		1.18	1.66	
	6020	2" x 8"		640	.025		.80	.83		1.63	2.25	
	6040	2" x 10"		535	.030		1.08	.99		2.07	2.83	
	6060	2" x 12"		455	.035		1.47	1.16		2.63	3.54	

WOOD & PLASTICS 6

		06110 | Wood Framing	CREW	DAILY OUTPUT	LABOR-HOURS	UNIT	2004 BARE COSTS MAT.	LABOR	EQUIP.	TOTAL	TOTAL INCL O&P	
555	9000	Minimum labor/equipment charge (R06110-030)	1 Carp	4	2	Job		66		66	109	555
560	0010	**FRAMING, SILLS**										560
	2002	Ledgers, nailed, 2″ x 4″	2 Carp	755	.021	L.F.	.31	.70		1.01	1.50	
	2052	2″ x 6″		600	.027		.52	.88		1.40	2.03	
	2102	Bolted, not including bolts, 3″ x 6″		325	.049		1.64	1.62		3.26	4.49	
	2152	3″ x 12″		233	.069		3.28	2.27		5.55	7.35	
	2602	Mud sills, redwood, construction grade, 2″ x 4″		895	.018		2.79	.59		3.38	4.05	
	2622	2″ x 6″		780	.021		4.18	.68		4.86	5.70	
	4002	Sills, 2″ x 4″		600	.027		.31	.88		1.19	1.80	
	4052	2″ x 6″		550	.029		.52	.96		1.48	2.16	
	4082	2″ x 8″		500	.032		.80	1.06		1.86	2.63	
	4202	Treated, 2″ x 4″		550	.029		.42	.96		1.38	2.05	
	4222	2″ x 6″		500	.032		.76	1.06		1.82	2.59	
	4242	2″ x 8″		450	.036		.95	1.17		2.12	3	
	4402	4″ x 4″		450	.036		1.29	1.17		2.46	3.36	
	4422	4″ x 6″		350	.046		2.07	1.51		3.58	4.77	
	4462	4″ x 8″		300	.053		2.77	1.76		4.53	5.95	
	4481	4″ x 10″		260	.062		3.18	2.03		5.21	6.85	
	9000	Minimum labor/equipment charge	1 Carp	4	2	Job		66		66	109	
565	0010	**FRAMING, SLEEPERS**										565
	0100	On concrete, treated, 1″ x 2″	2 Carp	2,350	.007	L.F.	.09	.22		.31	.47	
	0150	1″ x 3″		2,000	.008		.17	.26		.43	.63	
	0200	2″ x 4″		1,500	.011		.42	.35		.77	1.04	
	0250	2″ x 6″		1,300	.012		.76	.41		1.17	1.51	
	9000	Minimum labor/equipment charge	1 Carp	4	2	Job		66		66	109	
570	0010	**FRAMING, SOFFITS & CANOPIES**										570
	1002	Canopy or soffit framing , 1″ x 4″	2 Carp	900	.018	L.F.	.60	.59		1.19	1.63	
	1042	1″ x 8″		750	.021		1.22	.70		1.92	2.51	
	1102	2″ x 4″		620	.026		.31	.85		1.16	1.75	
	1142	2″ x 8″		500	.032		.80	1.06		1.86	2.63	
	1202	3″ x 4″		500	.032		1.09	1.06		2.15	2.95	
	1242	3″ x 10″		300	.053		2.73	1.76		4.49	5.95	
	9000	Minimum labor/equipment charge	1 Carp	4	2	Job		66		66	109	
575	0010	**FRAMING, TREATED LUMBER**										575
	0020	Water-borne salt, C.C.A., A.C.A., wet, .40 P.C.F. retention										
	0100	2″ x 4″				M.B.F.	630			630	690	
	0110	2″ x 6″					765			765	840	
	0120	2″ x 8″					715			715	785	
	0130	2″ x 10″					710			710	780	
	0140	2″ x 12″					905			905	995	
	0200	4″ x 4″					965			965	1,050	
	0210	4″ x 6″					1,025			1,025	1,125	
	0220	4″ x 8″					1,050			1,050	1,150	
	0250	Add for .60 P.C.F. retention					40%					
	0260	Add for 2.5 P.C.F. retention					200%					
	0270	Add for K.D.A.T.					20%					
590	0010	**FRAMING, WALLS** (R06100-010)										590
	0100	Door buck, studs, header & access, 8′h 2″ x 4″ wall, 3′w	1 Carp	32	.250	Ea.	13.30	8.25		21.55	28.50	
	0110	4′ wide (R06110-030)		32	.250		14.35	8.25		22.60	29.50	
	0120	5′ wide		32	.250		18.30	8.25		26.55	33.50	
	0130	6′ wide		32	.250		19.90	8.25		28.15	35.50	
	0140	8′ wide		30	.267		27.50	8.80		36.30	45	
	0150	10′ wide		30	.267		40	8.80		48.80	58.50	
	0160	12′ wide		30	.267		53.50	8.80		62.30	73	

	06110 | Wood Framing	CREW	DAILY OUTPUT	LABOR-HOURS	UNIT	2004 BARE COSTS MAT.	LABOR	EQUIP.	TOTAL	TOTAL INCL O&P	
590	0170 2″ x 6″ wall, 3′ wide R06100-010	1 Carp	32	.250	Ea.	20	8.25		28.25	35.50	590
	0180 4′ wide		32	.250		21	8.25		29.25	36.50	
	0190 5′ wide R06110-030		32	.250		25	8.25		33.25	41	
	0200 6′ wide		32	.250		26.50	8.25		34.75	43	
	0210 8′ wide		30	.267		34.50	8.80		43.30	52.50	
	0220 10′ wide		30	.267		47	8.80		55.80	66	
	0230 12′ wide		30	.267		60	8.80		68.80	80.50	
	0240 Window buck, studs, header & access, 8′ high 2″ x 4″ wall, 2′ wide		24	.333		13.75	11		24.75	33.50	
	0250 3′ wide		24	.333		16.40	11		27.40	36.50	
	0260 4′ wide		24	.333		18.05	11		29.05	38	
	0270 5′ wide		24	.333		22	11		33	42.50	
	0280 6′ wide		24	.333		24.50	11		35.50	45.50	
	0300 8′ wide		22	.364		34	12		46	57.50	
	0310 10′ wide		22	.364		47.50	12		59.50	72.50	
	0320 12′ wide		22	.364		62.50	12		74.50	89	
	0330 2″ x 6″ wall, 2′ wide		24	.333		22	11		33	43	
	0340 3′ wide		24	.333		25.50	11		36.50	46.50	
	0350 4′ wide		24	.333		27	11		38	48.50	
	0360 5′ wide		24	.333		31.50	11		42.50	53	
	0370 6′ wide		24	.333		34.50	11		45.50	56.50	
	0380 7′ wide		24	.333		41.50	11		52.50	64.50	
	0390 8′ wide		22	.364		45.50	12		57.50	70	
	0400 10′ wide		22	.364		59.50	12		71.50	85.50	
	0410 12′ wide	↓	22	.364	↓	75.50	12		87.50	103	
	2002 Headers over openings, 2″ x 6″	2 Carp	360	.044	L.F.	.52	1.47		1.99	3	
	2007 2″ x 6″, pneumatic nailed		432	.037		.52	1.22		1.74	2.60	
	2052 2″ x 8″		340	.047		.80	1.55		2.35	3.45	
	2057 2″ x 8″, pneumatic nailed		408	.039		.80	1.29		2.09	3.03	
	2100 2″ x 10″		320	.050		1.08	1.65		2.73	3.93	
	2105 2″ x 10″, pneumatic nailed		384	.042		1.08	1.38		2.46	3.47	
	2152 2″ x 12″		300	.053		1.47	1.76		3.23	4.54	
	2157 2″ x 12″, pneumatic nailed		360	.044		1.47	1.47		2.94	4.05	
	2202 4″ x 12″		190	.084		5	2.78		7.78	10.10	
	2207 4″ x 12″, pneumatic nailed		228	.070		5	2.32		7.32	9.35	
	2252 6″ x 12″		140	.114		10.85	3.77		14.62	18.15	
	3000 Radius, 2″ x 6″		270	.059		.81	1.96		2.77	4.14	
	3010 2″ x 6″, pneumatic nailed		324	.049		.81	1.63		2.44	3.60	
	3020 2″ x 8″		255	.063		1.25	2.07		3.32	4.81	
	3030 2″ x 8″, pneumatic nailed		296	.054		1.25	1.78		3.03	4.34	
	3040 2″ x 10″		240	.067		1.69	2.20		3.89	5.50	
	3050 2″ x 10″, pneumatic nailed		285	.056		1.69	1.85		3.54	4.93	
	3060 2″ x 12″		225	.071		2.30	2.35		4.65	6.40	
	3070 2″ x 12″, pneumatic nailed		270	.059		2.30	1.96		4.26	5.75	
	5002 Plates, untreated, 2″ x 3″		850	.019		.24	.62		.86	1.30	
	5007 2″ x 3″, pneumatic nailed		1,020	.016		.24	.52		.76	1.13	
	5022 2″ x 4″		800	.020		.31	.66		.97	1.43	
	5027 2″ x 4″, pneumatic nailed		960	.017		.31	.55		.86	1.25	
	5041 2″ x 6″		750	.021		.52	.70		1.22	1.74	
	5046 2″ x 6″, pneumatic nailed		900	.018		.52	.59		1.11	1.54	
	5122 Studs, 8′ high wall, 2″ x 3″		1,200	.013		.24	.44		.68	1	
	5127 2″ x 3″, pneumatic nailed		1,440	.011		.24	.37		.61	.88	
	5142 2″ x 4″		1,100	.015		.31	.48		.79	1.14	
	5147 2″ x 4″, pneumatic nailed		1,320	.012		.31	.40		.71	1	
	5162 2″ x 6″		1,000	.016		.52	.53		1.05	1.45	
	5167 2″ x 6″, pneumatic nailed		1,200	.013		.52	.44		.96	1.30	
	5182 3″ x 4″	↓	800	.020	↓	1.09	.66		1.75	2.29	

		06110 | Wood Framing	CREW	DAILY OUTPUT	LABOR-HOURS	UNIT	2004 BARE COSTS MAT.	LABOR	EQUIP.	TOTAL	TOTAL INCL O&P	
590	5187	3" x 4", pneumatic nailed R06100-010	2 Carp	960	.017	L.F.	1.09	.55		1.64	2.11	590
	8200	For 12' high walls, deduct						5%				
	8220	For stub wall, 6' high, add R06110-030						20%				
	8240	3' high, add						40%				
	8250	For second story & above, add						5%				
	8300	For dormer & gable, add						15%				
	9000	Minimum labor/equipment charge	1 Carp	4	2	Job		66		66	109	
600	0010	**FURRING** Wood strips, 1" x 2", on walls, on wood	1 Carp	550	.015	L.F.	.19	.48		.67	1.01	600
	0017	Pneumatic nailed		710	.011		.19	.37		.56	.83	
	0302	On masonry		495	.016		.19	.53		.72	1.09	
	0402	On concrete		260	.031		.19	1.02		1.21	1.89	
	0602	1" x 3", on walls, on wood		550	.015		.21	.48		.69	1.03	
	0607	Pneumatic nailed		710	.011		.21	.37		.58	.85	
	0702	On masonry		495	.016		.21	.53		.74	1.11	
	0802	On concrete		260	.031		.21	1.02		1.23	1.91	
	0852	On ceilings, on wood		350	.023		.21	.75		.96	1.48	
	0857	Pneumatic nailed		450	.018		.21	.59		.80	1.20	
	0902	On masonry		320	.025		.21	.83		1.04	1.60	
	0952	On concrete		210	.038		.21	1.26		1.47	2.31	
	9000	Minimum labor/equipment charge		4	2	Job		66		66	109	
700	0010	**GROUNDS** For casework, 1" x 2" wood strips, on wood	1 Carp	330	.024	L.F.	.19	.80		.99	1.54	700
	0102	On masonry		285	.028		.19	.93		1.12	1.75	
	0202	On concrete		250	.032		.19	1.06		1.25	1.96	
	0402	For plaster, 3/4" deep, on wood		450	.018		.19	.59		.78	1.18	
	0502	On masonry		225	.036		.19	1.17		1.36	2.16	
	0602	On concrete		175	.046		.19	1.51		1.70	2.71	
	0702	On metal lath		200	.040		.19	1.32		1.51	2.40	
	9000	Minimum labor/equipment charge		4	2	Job		66		66	109	
		06120 | Structural Panels										
900	0010	**STRUCTURAL INSULATED PANELS**										900
	0100	Structural insul. panels, 7/16" OSB both faces, EPS insul, 3-5/8" T	F-3	2,075	.019	S.F.	2.35	.64	.30	3.29	3.97	
	0110	5-5/8" thick		1,725	.023		2.60	.77	.35	3.72	4.52	
	0120	7-3/8" thick		1,425	.028		2.85	.94	.43	4.22	5.15	
	0130	9-3/8" thick		1,125	.036		3.10	1.19	.54	4.83	5.95	
	0140	7/16" OSB one face, EPS insul, 3-5/8" thick		2,175	.018		1.35	.61	.28	2.24	2.81	
	0150	5-5/8" thick		1,825	.022		1.60	.73	.34	2.67	3.33	
	0160	7-3/8" thick		1,525	.026		1.85	.88	.40	3.13	3.92	
	0170	9-3/8" thick		1,225	.033		2.10	1.09	.50	3.69	4.65	
	0180	11-3/8" thick		925	.043		2.35	1.44	.66	4.45	5.70	
	0190	7/16" OSB - 1/2" GWB faces , EPS insul, 3-5/8" T		2,075	.019		2.40	.64	.30	3.34	4.02	
	0200	5-5/8" thick		1,725	.023		2.65	.77	.35	3.77	4.58	
	0210	7-3/8" thick		1,425	.028		2.90	.94	.43	4.27	5.20	
	0220	9-3/8" thick		1,125	.036		3.15	1.19	.54	4.88	6	
	0230	11-3/8" thick		825	.048		3.40	1.62	.74	5.76	7.20	
	0240	7/16" OSB - 1/2" MRGWB faces , EPS insul, 3-5/8" T		2,075	.019		2.65	.64	.30	3.59	4.30	
	0250	5-5/8" thick		1,725	.023		2.90	.77	.35	4.02	4.85	
	0300	For 1/2" GWB added to OSB skin, add					.60			.60	.66	
	0310	For 1/2" MRGWB added to OSB skin, add					.75			.75	.83	
	0320	For one T1-11 skin, add to OSB-OSB					.60			.60	.66	
	0330	For one 19/32" CDX skin, add to OSB-OSB					.36			.36	.40	

	06150 | Wood Decking	CREW	DAILY OUTPUT	LABOR-HOURS	UNIT	2004 BARE COSTS MAT.	LABOR	EQUIP.	TOTAL	TOTAL INCL O&P	
600	0010 **ROOF DECKS**										600
	0020 For laminated decks, see division 06170-550										
	0400 Cedar planks, 3″ thick	2 Carp	320	.050	S.F.	5.95	1.65		7.60	9.25	
	0500 4″ thick		250	.064		8	2.11		10.11	12.30	
	0702 Douglas fir, 3″ thick		320	.050		2.22	1.65		3.87	5.20	
	0802 4″ thick		250	.064		2.97	2.11		5.08	6.75	
	1002 Hemlock, 3″ thick		320	.050		2.22	1.65		3.87	5.20	
	1102 4″ thick		250	.064		2.96	2.11		5.07	6.75	
	1302 Western white spruce, 3″ thick		320	.050		2.14	1.65		3.79	5.10	
	1402 4″ thick		250	.064		2.85	2.11		4.96	6.65	
	9000 Minimum labor/equipment charge	1 Carp	2	4	Job		132		132	219	

	06160 | Sheathing	CREW	DAILY OUTPUT	LABOR-HOURS	UNIT	MAT.	LABOR	EQUIP.	TOTAL	TOTAL INCL O&P	
800	0010 **SHEATHING** Plywood on roof, CDX R06160-020										800
	0032 5/16″ thick	2 Carp	1,600	.010	S.F.	.42	.33		.75	1.01	
	0037 Pneumatic nailed R06110-030		1,952	.008		.42	.27		.69	.91	
	0052 3/8″ thick		1,525	.010		.42	.35		.77	1.03	
	0057 Pneumatic nailed		1,860	.009		.42	.28		.70	.93	
	0102 1/2″ thick		1,400	.011		.39	.38		.77	1.06	
	0103 Pneumatic nailed		1,708	.009		.39	.31		.70	.94	
	0202 5/8″ thick		1,300	.012		.45	.41		.86	1.17	
	0207 Pneumatic nailed		1,586	.010		.45	.33		.78	1.05	
	0302 3/4″ thick		1,200	.013		.57	.44		1.01	1.36	
	0307 Pneumatic nailed		1,464	.011		.57	.36		.93	1.23	
	0502 Plywood on walls with exterior CDX, 3/8″ thick		1,200	.013		.42	.44		.86	1.19	
	0507 Pneumatic nailed		1,488	.011		.42	.35		.77	1.05	
	0602 1/2″ thick		1,125	.014		.39	.47		.86	1.21	
	0607 Pneumatic nailed		1,395	.011		.39	.38		.77	1.06	
	0702 5/8″ thick		1,050	.015		.45	.50		.95	1.33	
	0707 Pneumatic nailed		1,302	.012		.45	.41		.86	1.17	
	0802 3/4″ thick		975	.016		.57	.54		1.11	1.53	
	0807 Pneumatic nailed		1,209	.013		.57	.44		1.01	1.35	
	1000 For shear wall construction, add						20%				
	1200 For structural 1 exterior plywood, add				S.F.	10%					
	1402 With boards, on roof 1″ x 6″ boards, laid horizontal	2 Carp	725	.022		.82	.73		1.55	2.11	
	1502 Laid diagonal		650	.025		.82	.81		1.63	2.25	
	1702 1″ x 8″ boards, laid horizontal		875	.018		.80	.60		1.40	1.88	
	1802 Laid diagonal		725	.022		.80	.73		1.53	2.09	
	2000 For steep roofs, add						40%				
	2200 For dormers, hips and valleys, add					5%	50%				
	2402 Boards on walls, 1″ x 6″ boards, laid regular	2 Carp	650	.025		.82	.81		1.63	2.25	
	2502 Laid diagonal		585	.027		.82	.90		1.72	2.40	
	2702 1″ x 8″ boards, laid regular		765	.021		.80	.69		1.49	2.02	
	2802 Laid diagonal		650	.025		.80	.81		1.61	2.23	
	2852 Gypsum, weatherproof, 1/2″ thick		1,050	.015		.28	.50		.78	1.14	
	2902 Sealed, 4/10″ thick		1,100	.015		.44	.48		.92	1.28	
	3000 Wood fiber, regular, no vapor barrier, 1/2″ thick		1,200	.013		.53	.44		.97	1.31	
	3100 5/8″ thick		1,200	.013		.71	.44		1.15	1.51	
	3300 No vapor barrier, in colors, 1/2″ thick		1,200	.013		.77	.44		1.21	1.58	
	3400 5/8″ thick		1,200	.013		.95	.44		1.39	1.78	
	3600 With vapor barrier one side, white, 1/2″ thick		1,200	.013		.54	.44		.98	1.32	
	3700 Vapor barrier 2 sides, 1/2″ thick		1,200	.013		.75	.44		1.19	1.56	
	3800 Asphalt impregnated, 25/32″ thick		1,200	.013		.34	.44		.78	1.10	
	3850 Intermediate, 1/2″ thick		1,200	.013		.27	.44		.71	1.03	
	9000 Minimum labor/equipment charge	1 Carp	2	4	Job		132		132	219	

6 WOOD & PLASTICS

		06160 | Sheathing	CREW	DAILY OUTPUT	LABOR-HOURS	UNIT	2004 BARE COSTS MAT.	LABOR	EQUIP.	TOTAL	TOTAL INCL O&P	
850	0010	**SUBFLOOR** Plywood, CDX, 1/2" thick (R06160-020)	2 Carp	1,500	.011	SF Flr.	.39	.35		.74	1.01	850
	0017	Pneumatic nailed		1,860	.009	"	.39	.28		.67	.90	
	0102	5/8" thick		1,350	.012	SF Flr.	.45	.39		.84	1.15	
	0107	Pneumatic nailed		1,674	.010		.45	.32		.77	1.02	
	0202	3/4" thick		1,250	.013		.57	.42		.99	1.33	
	0207	Pneumatic nailed		1,550	.010		.57	.34		.91	1.19	
	0302	1-1/8" thick, 2-4-1 including underlayment		1,050	.015		1.80	.50		2.30	2.81	
	0502	With boards, 1" x 10" S4S, laid regular		1,100	.015		1.06	.48		1.54	1.96	
	0602	Laid diagonal		900	.018		1.06	.59		1.65	2.13	
	0802	1" x 8" S4S, laid regular		1,000	.016		.80	.53		1.33	1.76	
	0902	Laid diagonal		850	.019		.80	.62		1.42	1.91	
	9000	Minimum labor/equipment charge	1 Carp	4	2	Job		66		66	109	
900	0011	**UNDERLAYMENT** Plywood, underlayment grade, 3/8" thick (R06160-020)	2 Carp	1,500	.011	SF Flr.	.53	.35		.88	1.16	900
	0017	Pneumatic nailed	"	1,860	.009	"	.53	.28		.81	1.05	
	0102	1/2" thick	2 Carp	1,450	.011	SF Flr.	.62	.36		.98	1.28	
	0107	Pneumatic nailed		1,798	.009		.62	.29		.91	1.17	
	0202	5/8" thick		1,400	.011		.72	.38		1.10	1.42	
	0207	Pneumatic nailed		1,736	.009		.72	.30		1.02	1.29	
	0302	3/4" thick		1,300	.012		.83	.41		1.24	1.58	
	0306	Pneumatic nailed		1,612	.010		.83	.33		1.16	1.45	
	0502	Particle board, 3/8" thick		1,500	.011		.37	.35		.72	.99	
	0507	Pneumatic nailed		1,860	.009		.37	.28		.65	.88	
	0602	1/2" thick		1,450	.011		.39	.36		.75	1.03	
	0607	Pneumatic nailed		1,798	.009		.39	.29		.68	.92	
	0802	5/8" thick		1,400	.011		.49	.38		.87	1.17	
	0807	Pneumatic nailed		1,736	.009		.49	.30		.79	1.04	
	0902	3/4" thick		1,300	.012		.54	.41		.95	1.26	
	0907	Pneumatic nailed		1,612	.010		.54	.33		.87	1.13	
	1102	Hardboard, underlayment grade, 4' x 4', .215" thick		1,500	.011		.39	.35		.74	1.01	
	9000	Minimum labor/equipment charge	1 Carp	4	2	Job		66		66	109	
		06170 | Prefabricated Structural Wood										
550	0010	**LAMINATED ROOF DECK** Pine or hemlock, 3" thick	2 Carp	425	.038	S.F.	2.80	1.24		4.04	5.15	550
	0100	4" thick		325	.049		3.72	1.62		5.34	6.80	
	0300	Cedar, 3" thick		425	.038		3.42	1.24		4.66	5.80	
	0400	4" thick		325	.049		4.35	1.62		5.97	7.50	
	0600	Fir, 3" thick		425	.038		2.85	1.24		4.09	5.20	
	0700	4" thick		325	.049		3.57	1.62		5.19	6.60	
	9000	Minimum labor/equipment charge	1 Carp	3	2.667	Job		88		88	146	
600	0010	**STRUCTURAL JOISTS** Fabricated "I" joists with wood flanges,										600
	0100	Plywood webs, incl. bridging & blocking, panels 24" O.C.										
	1200	15' to 24' span, 50 psf live load	F-5	2,400	.013	SF Flr.	1.81	.45		2.26	2.73	
	1300	55 psf live load		2,250	.014		1.94	.48		2.42	2.92	
	1400	24' to 30' span, 45 psf live load		2,600	.012		2.13	.41		2.54	3.02	
	1500	55 psf live load		2,400	.013		2.90	.45		3.35	3.93	
	1600	Tubular steel open webs, 45 psf, 24" O.C., 40' span	F-3	6,250	.006		1.80	.21	.10	2.11	2.44	
	1700	55' span		7,750	.005		1.75	.17	.08	2	2.30	
	1800	70' span		9,250	.004		2.27	.14	.07	2.48	2.81	
	1900	85 psf live load, 26' span		2,300	.017		2.11	.58	.27	2.96	3.56	
980	0010	**ROOF TRUSSES** (R06170-100)										980
	0020	For timber connectors, see div. 06090-800										
	0100	Fink (W) or King post type, 2'-0" O.C.										
	0200	Metal plate connected, 4 in 12 slope										
	0210	24' to 29' span	F-3	3,000	.013	SF Flr.	1.46	.44	.20	2.10	2.56	
	0300	30' to 43' span		3,000	.013		1.62	.44	.20	2.26	2.73	

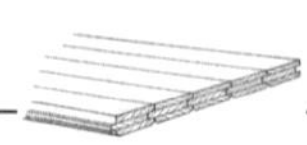

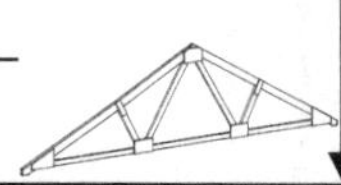

Important: See the Reference Section for critical supporting data - Reference Nos., Crews, & City Cost Indexes

		06170 | Prefabricated Structural Wood	CREW	DAILY OUTPUT	LABOR-HOURS	UNIT	2004 BARE COSTS MAT.	LABOR	EQUIP.	TOTAL	TOTAL INCL O&P	
980	0400	44′ to 60′ span R06170-100	F-3	3,000	.013	SF Flr.	1.79	.44	.20	2.43	2.92	980
	0700	Glued and nailed, add					50%					

		06180 | Glued-Laminated Construction	CREW	DAILY OUTPUT	LABOR-HOURS	UNIT	MAT.	LABOR	EQUIP.	TOTAL	TOTAL INCL O&P	
400	0010	**LAMINATED FRAMING** Not including decking										400
	0020	30 lb., short term live load, 15 lb. dead load										
	0200	Straight roof beams, 20′ clear span, beams 8′ O.C.	F-3	2,560	.016	SF Flr.	1.58	.52	.24	2.34	2.86	
	0300	Beams 16′ O.C.		3,200	.013		1.14	.42	.19	1.75	2.14	
	0500	40′ clear span, beams 8′ O.C.		3,200	.013		3.03	.42	.19	3.64	4.22	
	0600	Beams 16′ O.C.		3,840	.010		2.47	.35	.16	2.98	3.47	
	0800	60′ clear span, beams 8′ O.C.	F-4	2,880	.017		5.20	.54	.32	6.06	6.95	
	0900	Beams 16′ O.C.	″	3,840	.013		3.87	.41	.24	4.52	5.20	
	1100	Tudor arches, 30′ to 40′ clear span, frames 8′ O.C.	F-3	1,680	.024		6.80	.79	.36	7.95	9.15	
	1200	Frames 16′ O.C.	″	2,240	.018		5.30	.60	.27	6.17	7.15	
	1400	50′ to 60′ clear span, frames 8′ O.C.	F-4	2,200	.022		7.30	.71	.41	8.42	9.65	
	1500	Frames 16′ O.C.		2,640	.018		6.20	.59	.35	7.14	8.20	
	1700	Radial arches, 60′ clear span, frames 8′ O.C.		1,920	.025		6.85	.82	.47	8.14	9.35	
	1800	Frames 16′ O.C.		2,880	.017		5.25	.54	.32	6.11	7.05	
	2000	100′ clear span, frames 8′ O.C.		1,600	.030		7.10	.98	.57	8.65	10.05	
	2100	Frames 16′ O.C.		2,400	.020		6.20	.65	.38	7.23	8.35	
	2300	120′ clear span, frames 8′ O.C.		1,440	.033		9.40	1.09	.63	11.12	12.85	
	2400	Frames 16′ O.C.		1,920	.025		8.60	.82	.47	9.89	11.30	
	2600	Bowstring trusses, 20′ O.C., 40′ clear span	F-3	2,400	.017		4.24	.56	.26	5.06	5.85	
	2700	60′ clear span	F-4	3,600	.013		3.81	.44	.25	4.50	5.20	
	2800	100′ clear span		4,000	.012		5.40	.39	.23	6.02	6.85	
	2900	120′ clear span		3,600	.013		5.80	.44	.25	6.49	7.35	
	3100	For premium appearance, add to S.F. prices					5%					
	3300	For industrial type, deduct					15%					
	3500	For stain and varnish, add					5%					
	3900	For 3/4″ laminations, add to straight					25%					
	4100	Add to curved					15%					
	4300	Alternate pricing method: (use nominal footage of										
	4310	components). Straight beams, camber less than 6″	F-3	3.50	11.429	M.B.F.	2,350	380	175	2,905	3,400	
	4400	Columns, including hardware		2	20		2,525	665	305	3,495	4,200	
	4600	Curved members, radius over 32′		2.50	16		2,575	535	245	3,355	4,000	
	4700	Radius 10′ to 32′		3	13.333		2,550	445	204	3,199	3,775	
	4900	For complicated shapes, add maximum					100%					
	5100	For pressure treating, add to straight					35%					
	5200	Add to curved					45%					
	6000	Laminated veneer members, southern pine or western species										
	6050	1-3/4″ wide x 5-1/2″ deep	2 Carp	480	.033	L.F.	2.69	1.10		3.79	4.78	
	6100	9-1/2″ deep		480	.033		3.35	1.10		4.45	5.50	
	6150	14″ deep		450	.036		4.99	1.17		6.16	7.45	
	6200	18″ deep		450	.036		6.75	1.17		7.92	9.40	
	6300	Parallel strand members, southern pine or western species										
	6350	1-3/4″ wide x 9-1/4″ deep	2 Carp	480	.033	L.F.	3.17	1.10		4.27	5.30	
	6400	11-1/4″ deep		450	.036		3.90	1.17		5.07	6.25	
	6450	14″ deep		400	.040		4.65	1.32		5.97	7.30	
	6500	3-1/2″ wide x 9-1/4″ deep		480	.033		7.70	1.10		8.80	10.30	
	6550	11-1/4″ deep		450	.036		9.55	1.17		10.72	12.45	
	6600	14″ deep		400	.040		11.30	1.32		12.62	14.65	
	6650	7″ wide x 9-1/4″ deep		450	.036		16.10	1.17		17.27	19.65	
	6700	11-1/4″ deep		420	.038		20	1.26		21.26	24	
	6750	14″ deep		400	.040		24	1.32		25.32	28.50	
	9000	Minimum labor/equipment charge	F-3	2.50	16	Job		535	245	780	1,150	

	06220 | Millwork		CREW	DAILY OUTPUT	LABOR-HOURS	UNIT	2004 BARE COSTS MAT.	LABOR	EQUIP.	TOTAL	TOTAL INCL O&P	
200	0010	**MOLDINGS, BASE**										200
	0500	Base, stock pine, 9/16" x 3-1/2"	1 Carp	240	.033	L.F.	1.22	1.10		2.32	3.16	
	0550	9/16" x 4-1/2"		200	.040		1.54	1.32		2.86	3.88	
	0561	Base shoe, oak, 3/4" x 1"		240	.033		1.04	1.10		2.14	2.96	
	9000	Minimum labor/equipment charge		4	2	Job		66		66	109	
400	0010	**MOLDINGS, CASINGS**										400
	0090	Apron, stock pine, 5/8" x 2"	1 Carp	250	.032	L.F.	.99	1.06		2.05	2.84	
	0110	5/8" x 3-1/2"		220	.036		1.44	1.20		2.64	3.57	
	0300	Band, stock pine, 11/16" x 1-1/8"		270	.030		.54	.98		1.52	2.21	
	0350	11/16" x 1-3/4"		250	.032		.86	1.06		1.92	2.70	
	0700	Casing, stock pine, 11/16" x 2-1/2"		240	.033		.85	1.10		1.95	2.76	
	0750	11/16" x 3-1/2"		215	.037		1.33	1.23		2.56	3.50	
	9000	Minimum labor/equipment charge		4	2	Job		66		66	109	
450	0010	**MOLDINGS, CEILINGS**										450
	0600	Bed, stock pine, 9/16" x 1-3/4"	1 Carp	270	.030	L.F.	.62	.98		1.60	2.30	
	0650	9/16" x 2"		240	.033		.76	1.10		1.86	2.66	
	1200	Cornice molding, stock pine, 9/16" x 1-3/4"		330	.024		.67	.80		1.47	2.07	
	1300	9/16" x 2-1/4"		300	.027		.89	.88		1.77	2.44	
	2400	Cove scotia, stock pine, 9/16" x 1-3/4"		270	.030		.59	.98		1.57	2.27	
	2500	11/16" x 2-3/4"		255	.031		1.25	1.04		2.29	3.10	
	2600	Crown, stock pine, 9/16" x 3-5/8"		250	.032		1.67	1.06		2.73	3.59	
	2700	11/16" x 4-5/8"		220	.036		3.60	1.20		4.80	5.95	
	9000	Minimum labor/equipment charge		4	2	Job		66		66	109	
500	0010	**MOLDINGS, EXTERIOR**										500
	1500	Cornice, boards, pine, 1" x 2"	1 Carp	330	.024	L.F.	.32	.80		1.12	1.68	
	1700	1" x 6"		250	.032		1.15	1.06		2.21	3.02	
	2000	1" x 12"		180	.044		1.72	1.47		3.19	4.32	
	2200	Three piece, built-up, pine, minimum		80	.100		2.13	3.30		5.43	7.80	
	2300	Maximum		65	.123		4.71	4.06		8.77	11.95	
	3000	Corner board, sterling pine, 1" x 4"		200	.040		.66	1.32		1.98	2.92	
	3100	1" x 6"		200	.040		.97	1.32		2.29	3.26	
	3350	Fascia, sterling pine, 1" x 6"		250	.032		.97	1.06		2.03	2.82	
	3370	1" x 8"		225	.036		1.51	1.17		2.68	3.61	
	3400	Trim, exterior, sterling pine, back band		250	.032		.66	1.06		1.72	2.48	
	3500	Casing		250	.032		1.75	1.06		2.81	3.68	
	3600	Crown		250	.032		1.68	1.06		2.74	3.60	
	3700	Porch rail with balusters		22	.364		13.70	12		25.70	35	
	3800	Screen		395	.020		1.06	.67		1.73	2.28	
	3850											
	4100	Verge board, sterling pine, 1" x 4"	1 Carp	200	.040	L.F.	.64	1.32		1.96	2.89	
	4200	1" x 6"		200	.040		.96	1.32		2.28	3.25	
	4300	2" x 6"		165	.048		1.56	1.60		3.16	4.37	
	4400	2" x 8"		165	.048		2.07	1.60		3.67	4.93	
	4700	For redwood trim, add					200%					
	9000	Minimum labor/equipment charge	1 Carp	4	2	Job		66		66	109	
700	0010	**MOLDINGS, TRIM**										700
	0200	Astragal, stock pine, 11/16" x 1-3/4"	1 Carp	255	.031	L.F.	.96	1.04		2	2.78	
	0250	1-5/16" x 2-3/16"		240	.033		2.94	1.10		4.04	5.05	
	0800	Chair rail, stock pine, 5/8" x 2-1/2"		270	.030		.85	.98		1.83	2.56	
	0900	5/8" x 3-1/2"		240	.033		1.41	1.10		2.51	3.37	
	1000	Closet pole, stock pine, 1-1/8" diameter		200	.040		.80	1.32		2.12	3.07	
	1100	Fir, 1-5/8" diameter		200	.040		1.19	1.32		2.51	3.50	
	3300	Half round, stock pine, 1/4" x 1/2"		270	.030		.16	.98		1.14	1.80	
	3350	1/2" x 1"		255	.031		.39	1.04		1.43	2.15	
	3400	Handrail, fir, single piece, stock, hardware not included										

Important: See the Reference Section for critical supporting data - Reference Nos., Crews, & City Cost Indexes

	06220 | Millwork		CREW	DAILY OUTPUT	LABOR-HOURS	UNIT	2004 BARE COSTS MAT.	LABOR	EQUIP.	TOTAL	TOTAL INCL O&P	
700	3450	1-1/2″ x 1-3/4″	1 Carp	80	.100	L.F.	1.22	3.30		4.52	6.80	700
	3470	Pine, 1-1/2″ x 1-3/4″		80	.100		1.07	3.30		4.37	6.65	
	3500	1-1/2″ x 2-1/2″		76	.105		1.28	3.47		4.75	7.15	
	3600	Lattice, stock pine, 1/4″ x 1-1/8″		270	.030		.23	.98		1.21	1.87	
	3700	1/4″ x 1-3/4″		250	.032		.35	1.06		1.41	2.14	
	3800	Miscellaneous, custom, pine, 1″ x 1″		270	.030		.31	.98		1.29	1.96	
	3900	1″ x 3″		240	.033		.63	1.10		1.73	2.51	
	4100	Birch or oak, nominal 1″ x 1″		240	.033		.49	1.10		1.59	2.36	
	4200	Nominal 1″ x 3″		215	.037		1.66	1.23		2.89	3.87	
	4400	Walnut, nominal 1″ x 1″		215	.037		.80	1.23		2.03	2.92	
	4500	Nominal 1″ x 3″		200	.040		2.39	1.32		3.71	4.82	
	4700	Teak, nominal 1″ x 1″		215	.037		1.13	1.23		2.36	3.28	
	4800	Nominal 1″ x 3″		200	.040		3.22	1.32		4.54	5.75	
	4900	Quarter round, stock pine, 1/4″ x 1/4″		275	.029		.15	.96		1.11	1.76	
	4950	3/4″ x 3/4″		255	.031		.36	1.04		1.40	2.12	
	5600	Wainscot moldings, 1-1/8″ x 9/16″, 2′ high, minimum		76	.105	S.F.	8.75	3.47		12.22	15.35	
	5700	Maximum		65	.123	″	18.05	4.06		22.11	26.50	
	9000	Minimum labor/equipment charge		4	2	Job		66		66	109	
800	0010	**MOLDINGS, WINDOW AND DOOR**										800
	2800	Door moldings, stock, decorative, 1-1/8″ wide, plain	1 Carp	17	.471	Set	29	15.55		44.55	57.50	
	2900	Detailed		17	.471	″	76.50	15.55		92.05	110	
	2960	Clear pine door jamb, no stops, 11/16″ x 4-9/16″		240	.033	L.F.	2.53	1.10		3.63	4.60	
	3150	Door trim set, 1 head and 2 sides, pine, 2-1/2 wide		5.90	1.356	Opng.	14.45	45		59.45	90	
	3170	3-1/2″ wide		5.30	1.509	″	22.50	50		72.50	108	
	3250	Glass beads, stock pine, 3/8″ x 1/2″		275	.029	L.F.	.36	.96		1.32	1.99	
	3270	3/8″ x 7/8″		270	.030		.40	.98		1.38	2.06	
	4850	Parting bead, stock pine, 3/8″ x 3/4″		275	.029		.26	.96		1.22	1.88	
	4870	1/2″ x 3/4″		255	.031		.31	1.04		1.35	2.06	
	5000	Stool caps, stock pine, 11/16″ x 3-1/2″		200	.040		1.53	1.32		2.85	3.87	
	5100	1-1/16″ x 3-1/4″		150	.053		2.67	1.76		4.43	5.85	
	5300	Threshold, oak, 3′ long, inside, 5/8″ x 3-5/8″		32	.250	Ea.	7.60	8.25		15.85	22	
	5400	Outside, 1-1/2″ x 7-5/8″		16	.500	″	33	16.50		49.50	64	
	5900	Window trim sets, including casings, header, stops,										
	5910	stool and apron, 2-1/2″ wide, minimum	1 Carp	13	.615	Opng.	19.90	20.50		40.40	55.50	
	5950	Average		10	.800		23.50	26.50		50	69.50	
	6000	Maximum		6	1.333		55	44		99	134	
	9000	Minimum labor/equipment charge		4	2	Job		66		66	109	
900	0005	**SOFFITS** R06110 -030										900
	0200	Soffits, pine, 1″ x 4″	2 Carp	420	.038	L.F.	.34	1.26		1.60	2.45	
	0210	1″ x 6″		420	.038		.38	1.26		1.64	2.49	
	0220	1″ x 8″		420	.038		.50	1.26		1.76	2.63	
	0230	1″ x 10″		400	.040		.84	1.32		2.16	3.11	
	0240	1″ x 12″		400	.040		.96	1.32		2.28	3.24	
	0250	STK cedar, 1″ x 4″		420	.038		.50	1.26		1.76	2.63	
	0260	1″ x 6″		420	.038		.81	1.26		2.07	2.97	
	0270	1″ x 8″		420	.038		1.06	1.26		2.32	3.25	
	0280	1″ x 10″		400	.040		1.27	1.32		2.59	3.59	
	0290	1″ x 12″		400	.040		1.56	1.32		2.88	3.91	
	1000	Exterior AC plywood, 1/4″ thick		420	.038	S.F.	.45	1.26		1.71	2.58	
	1050	3/8″ thick		420	.038		.53	1.26		1.79	2.66	
	1100	1/2″ thick		420	.038		.62	1.26		1.88	2.76	
	1150	Polyvinyl chloride, white, solid	1 Carp	230	.035		.66	1.15		1.81	2.63	
	1160	Perforated	″	230	.035		.66	1.15		1.81	2.63	
	9000	Minimum labor/equipment charge	2 Carp	5	3.200	Job		106		106	175	

06250 | Prefinished Paneling

		06250 Prefinished Paneling	CREW	DAILY OUTPUT	LABOR-HOURS	UNIT	2004 BARE COSTS MAT.	LABOR	EQUIP.	TOTAL	TOTAL INCL O&P	
200	0010	**PANELING, HARDBOARD**										200
	0050	Not incl. furring or trim, hardboard, tempered, 1/8″ thick	2 Carp	500	.032	S.F.	.30	1.06		1.36	2.08	
	0100	1/4″ thick		500	.032		.38	1.06		1.44	2.17	
	0300	Tempered pegboard, 1/8″ thick		500	.032		.37	1.06		1.43	2.16	
	0400	1/4″ thick		500	.032		.42	1.06		1.48	2.21	
	0600	Untempered hardboard, natural finish, 1/8″ thick		500	.032		.33	1.06		1.39	2.11	
	0700	1/4″ thick		500	.032		.32	1.06		1.38	2.10	
	0900	Untempered pegboard, 1/8″ thick		500	.032		.33	1.06		1.39	2.11	
	1000	1/4″ thick		500	.032		.37	1.06		1.43	2.16	
	1200	Plastic faced hardboard, 1/8″ thick		500	.032		.54	1.06		1.60	2.34	
	1300	1/4″ thick		500	.032		.72	1.06		1.78	2.54	
	1500	Plastic faced pegboard, 1/8″ thick		500	.032		.51	1.06		1.57	2.31	
	1600	1/4″ thick		500	.032		.63	1.06		1.69	2.44	
	1800	Wood grained, plain or grooved, 1/4″ thick, minimum		500	.032		.48	1.06		1.54	2.28	
	1900	Maximum		425	.038		1.01	1.24		2.25	3.17	
	2100	Moldings for hardboard, wood or aluminum, minimum		500	.032	L.F.	.33	1.06		1.39	2.11	
	2200	Maximum		425	.038	″	.91	1.24		2.15	3.06	
	9000	Minimum labor/equipment charge	1 Carp	2	4	Job		132		132	219	
500	0010	**PANELING, PLYWOOD** R06160-020										500
	2400	Plywood, prefinished, 1/4″ thick, 4′ x 8′ sheets										
	2410	with vertical grooves. Birch faced, minimum	2 Carp	500	.032	S.F.	.79	1.06		1.85	2.62	
	2420	Average		420	.038		1.21	1.26		2.47	3.41	
	2430	Maximum		350	.046		1.76	1.51		3.27	4.44	
	2600	Mahogany, African		400	.040		2.25	1.32		3.57	4.67	
	2700	Philippine (Lauan)		500	.032		.97	1.06		2.03	2.82	
	2900	Oak or Cherry, minimum		500	.032		1.89	1.06		2.95	3.83	
	3000	Maximum		400	.040		2.90	1.32		4.22	5.40	
	3200	Rosewood		320	.050		4.11	1.65		5.76	7.25	
	3400	Teak		400	.040		2.90	1.32		4.22	5.40	
	3600	Chestnut		375	.043		4.28	1.41		5.69	7.05	
	3800	Pecan		400	.040		1.85	1.32		3.17	4.23	
	3900	Walnut, minimum		500	.032		2.47	1.06		3.53	4.47	
	3950	Maximum		400	.040		4.68	1.32		6	7.35	
	4000	Plywood, prefinished, 3/4″ thick, stock grades, minimum		320	.050		1.12	1.65		2.77	3.97	
	4100	Maximum		224	.071		4.83	2.36		7.19	9.20	
	4300	Architectural grade, minimum		224	.071		3.57	2.36		5.93	7.85	
	4400	Maximum		160	.100		5.45	3.30		8.75	11.45	
	4600	Plywood, "A" face, birch, V.C., 1/2″ thick, natural		450	.036		1.69	1.17		2.86	3.81	
	4700	Select		450	.036		1.85	1.17		3.02	3.99	
	4900	Veneer core, 3/4″ thick, natural		320	.050		1.78	1.65		3.43	4.70	
	5000	Select		320	.050		2.01	1.65		3.66	4.95	
	5200	Lumber core, 3/4″ thick, natural		320	.050		2.68	1.65		4.33	5.70	
	5500	Plywood, knotty pine, 1/4″ thick, A2 grade		450	.036		1.46	1.17		2.63	3.56	
	5600	A3 grade		450	.036		1.85	1.17		3.02	3.99	
	5800	3/4″ thick, veneer core, A2 grade		320	.050		1.90	1.65		3.55	4.83	
	5900	A3 grade		320	.050		2.13	1.65		3.78	5.10	
	6100	Aromatic cedar, 1/4″ thick, plywood		400	.040		1.87	1.32		3.19	4.25	
	6200	1/4″ thick, particle board		400	.040		.91	1.32		2.23	3.19	
	9000	Minimum labor/equipment charge	1 Carp	2	4	Job		132		132	219	

06260 | Board Paneling

		06260 Board Paneling	CREW	DAILY OUTPUT	LABOR-HOURS	UNIT	MAT.	LABOR	EQUIP.	TOTAL	TOTAL INCL O&P	
400	0010	**PANELING, BOARDS**										400
	6400	Wood board paneling, 3/4″ thick, knotty pine	2 Carp	300	.053	S.F.	1.26	1.76		3.02	4.31	
	6500	Rough sawn cedar		300	.053		1.61	1.76		3.37	4.69	
	6700	Redwood, clear, 1″ x 4″ boards		300	.053		3.76	1.76		5.52	7.05	

6 WOOD & PLASTICS

06200 | Finish Carpentry

	06260 Board Paneling		Crew	Daily Output	Labor-Hours	Unit	2004 Bare Costs Mat.	Labor	Equip.	Total	Total Incl O&P	
400	6900	Aromatic cedar, closet lining, boards	2 Carp	275	.058	S.F.	2.90	1.92		4.82	6.35	400
	9000	Minimum labor/equipment charge	1 Carp	2	4	Job		132		132	219	

	06270 Closet/Utility Wood Shelving		Crew	Daily Output	Labor-Hours	Unit	Mat.	Labor	Equip.	Total	Total Incl O&P	
200	0010	**SHELVING** Pine, clear grade, no edge band, 1″ x 8″	1 Carp	115	.070	L.F.	1.73	2.30		4.03	5.70	200
	0100	1″ x 10″		110	.073		2.21	2.40		4.61	6.40	
	0200	1″ x 12″		105	.076		2.97	2.51		5.48	7.45	
	0600	Plywood, 3/4″ thick with lumber edge, 12″ wide		75	.107		.31	3.52		3.83	6.20	
	0700	24″ wide		70	.114		1.97	3.77		5.74	8.40	
	0900	Bookcase, clear grade pine, shelves 12″ O.C., 8″ deep, /SF shelf		70	.114	S.F.	5.60	3.77		9.37	12.45	
	1000	12″ deep shelves		65	.123	″	9.65	4.06		13.71	17.35	
	1200	Adjustable closet rod and shelf, 12″ wide, 3′ long		20	.400	Ea.	8.50	13.20		21.70	31.50	
	1300	8′ long		15	.533	″	20.50	17.60		38.10	51.50	
	1500	Prefinished shelves with supports, stock, 8″ wide		75	.107	L.F.	3.68	3.52		7.20	9.90	
	1600	10″ wide		70	.114	″	4.09	3.77		7.86	10.75	
	9000	Minimum labor/equipment charge		4	2	Job		66		66	109	

06400 | Architectural Woodwork

	06410 Custom Cabinets		Crew	Daily Output	Labor-Hours	Unit	2004 Bare Costs Mat.	Labor	Equip.	Total	Total Incl O&P	
100	0010	**CABINETS** Corner china cabinets, stock pine,										100
	0020	80″ high, unfinished, minimum	2 Carp	6.60	2.424	Ea.	435	80		515	615	
	0100	Maximum	″	4.40	3.636	″	960	120		1,080	1,250	
	0300	Built-in drawer units, pine, 18″ deep, 32″ high, unfinished										
	0400	Minimum	2 Carp	53	.302	L.F.	115	9.95		124.95	144	
	0500	Maximum	″	40	.400	″	140	13.20		153.20	176	
	0700	Kitchen base cabinets, hardwood, not incl. counter tops,										
	0710	24″ deep, 35″ high, prefinished										
	0800	One top drawer, one door below, 12″ wide	2 Carp	24.80	.645	Ea.	127	21.50		148.50	175	
	0820	15″ wide		24	.667		173	22		195	227	
	0840	18″ wide		23.30	.687		188	22.50		210.50	245	
	0860	21″ wide		22.70	.705		196	23.50		219.50	255	
	0880	24″ wide		22.30	.717		225	23.50		248.50	288	
	1000	Four drawers, 12″ wide		24.80	.645		300	21.50		321.50	365	
	1020	15″ wide		24	.667		231	22		253	291	
	1040	18″ wide		23.30	.687		257	22.50		279.50	320	
	1060	24″ wide		22.30	.717		280	23.50		303.50	350	
	1200	Two top drawers, two doors below, 27″ wide		22	.727		250	24		274	315	
	1220	30″ wide		21.40	.748		268	24.50		292.50	335	
	1240	33″ wide		20.90	.766		277	25.50		302.50	345	
	1260	36″ wide		20.30	.788		288	26		314	360	
	1280	42″ wide		19.80	.808		310	26.50		336.50	385	
	1300	48″ wide		18.90	.847		330	28		358	405	
	1500	Range or sink base, two doors below, 30″ wide		21.40	.748		221	24.50		245.50	284	
	1520	33″ wide		20.90	.766		237	25.50		262.50	305	
	1540	36″ wide		20.30	.788		248	26		274	315	
	1560	42″ wide		19.80	.808		265	26.50		291.50	335	
	1580	48″ wide		18.90	.847		278	28		306	350	
	1800	For sink front units, deduct					48			48	53	
	2000	Corner base cabinets, 36″ wide, standard	2 Carp	18	.889		365	29.50		394.50	450	

	06410 Custom Cabinets		CREW	DAILY OUTPUT	LABOR-HOURS	UNIT	2004 BARE COSTS MAT.	LABOR	EQUIP.	TOTAL	TOTAL INCL O&P	
100	2100	Lazy Susan with revolving door	2 Carp	16.50	.970	Ea.	355	32		387	445	100
	4000	Kitchen wall cabinets, hardwood, 12" deep with two doors										
	4050	12" high, 30" wide	2 Carp	24.80	.645	Ea.	139	21.50		160.50	189	
	4100	36" wide		24	.667		162	22		184	215	
	4400	15" high, 30" wide		24	.667		146	22		168	197	
	4420	33" wide		23.30	.687		163	22.50		185.50	217	
	4440	36" wide		22.70	.705		165	23.50		188.50	220	
	4450	42" wide		22.70	.705		191	23.50		214.50	249	
	4700	24" high, 30" wide		23.30	.687		181	22.50		203.50	237	
	4720	36" wide		22.70	.705		200	23.50		223.50	259	
	4740	42" wide		22.30	.717		221	23.50		244.50	283	
	5000	30" high, one door, 12" wide		22	.727		122	24		146	174	
	5020	15" wide		21.40	.748		138	24.50		162.50	193	
	5040	18" wide		20.90	.766		151	25.50		176.50	208	
	5060	24" wide		20.30	.788		170	26		196	230	
	5300	Two doors, 27" wide		19.80	.808		211	26.50		237.50	276	
	5320	30" wide		19.30	.829		204	27.50		231.50	270	
	5340	36" wide		18.80	.851		233	28		261	305	
	5360	42" wide		18.50	.865		254	28.50		282.50	325	
	5380	48" wide		18.40	.870		286	28.50		314.50	365	
	6000	Corner wall, 30" high, 24" wide		18	.889		139	29.50		168.50	202	
	6050	30" wide		17.20	.930		166	30.50		196.50	234	
	6100	36" wide		16.50	.970		180	32		212	251	
	6500	Revolving Lazy Susan		15.20	1.053		274	34.50		308.50	360	
	7000	Broom cabinet, 84" high, 24" deep, 18" wide		10	1.600		375	53		428	500	
	7500	Oven cabinets, 84" high, 24" deep, 27" wide		8	2	↓	540	66		606	705	
	7750	Valance board trim	↓	396	.040	L.F.	7.75	1.33		9.08	10.75	
	9000	For deluxe models of all cabinets, add					40%					
	9500	For custom built in place, add					25%	10%				
	9550	Rule of thumb, kitchen cabinets not including										
	9560	appliances & counter top, minimum	2 Carp	30	.533	L.F.	88	17.60		105.60	126	
	9600	Maximum	"	25	.640	"	230	21		251	288	
	9700	Minimum labor/equipment charge	1 Carp	3	2.667	Job		88		88	146	
210	0010	**CASEWORK, FRAMES**										210
	0050	Base cabinets, counter storage, 36" high, one bay										
	0100	18" wide	1 Carp	2.70	2.963	Ea.	98.50	98		196.50	270	
	0400	Two bay, 36" wide		2.20	3.636		150	120		270	365	
	1100	Three bay, 54" wide		1.50	5.333		179	176		355	490	
	2800	Book cases, one bay, 7' high, 18" wide		2.40	3.333		116	110		226	310	
	3500	Two bay, 36" wide		1.60	5		168	165		333	460	
	4100	Three bay, 54" wide		1.20	6.667		278	220		498	670	
	5100	Coat racks, one bay, 7' high, 24" wide		4.50	1.778		116	58.50		174.50	224	
	5300	Two bay, 48" wide		2.75	2.909		161	96		257	335	
	5800	Three bay, 72" wide		2.10	3.810		237	126		363	470	
	6100	Wall mounted cabinet, one bay, 24" high, 18" wide		3.60	2.222		63.50	73.50		137	192	
	6800	Two bay, 36" wide		2.20	3.636		93	120		213	300	
	7400	Three bay, 54" wide		1.70	4.706		116	155		271	385	
	8400	30" high, one bay, 18" wide		3.60	2.222		69.50	73.50		143	198	
	9000	Two bay, 36" wide		2.15	3.721		92	123		215	305	
	9400	Three bay, 54" wide		1.60	5		115	165		280	400	
	9800	Wardrobe, 7' high, single, 24" wide		2.70	2.963		127	98		225	300	
	9880	Partition & adjustable shelves, 48" wide		1.70	4.706		162	155		317	435	
	9950	Partition, adjustable shelves & drawers, 48" wide		1.40	5.714	↓	243	189		432	585	
	9970	Minimum labor/equipment charge	↓	4	2	Job		66		66	109	

	06410	Custom Cabinets	CREW	DAILY OUTPUT	LABOR-HOURS	UNIT	2004 BARE COSTS MAT.	LABOR	EQUIP.	TOTAL	TOTAL INCL O&P	
220	0010	**CABINET DOORS**										220
	2000	Glass panel, hardwood frame										
	2200	12" wide, 18" high	1 Carp	34	.235	Ea.	22	7.75		29.75	37.50	
	2600	30" high		32	.250		25.50	8.25		33.75	41.50	
	4450	18" wide, 18" high		32	.250		24	8.25		32.25	39.50	
	4550	30" high	↓	29	.276	↓	26.50	9.10		35.60	44	
	5000	Hardwood, raised panel										
	5100	12" wide, 18" high	1 Carp	16	.500	Ea.	33.50	16.50		50	64.50	
	5200	30" high		15	.533		40	17.60		57.60	73	
	5500	18" wide, 18" high		15	.533		37	17.60		54.60	69.50	
	5600	30" high	↓	14	.571	↓	45	18.85		63.85	81	
	6000	Plastic laminate on particle board										
	6100	12" wide, 18" high	1 Carp	25	.320	Ea.	15.20	10.55		25.75	34	
	6140	30" high		23	.348		24.50	11.50		36	46	
	6500	18" wide, 18" high		24	.333		22	11		33	43	
	6600	30" high	↓	22	.364	↓	37.50	12		49.50	61	
	7000	Plywood, with edge band										
	7010	12" wide, 18" high	1 Carp	27	.296	Ea.	20.50	9.80		30.30	38.50	
	7120	30" high		25	.320		35	10.55		45.55	56	
	7650	18" wide, 18" high		26	.308		30	10.15		40.15	50	
	7750	30" high		24	.333	↓	51.50	11		62.50	75	
	9000	Minimum labor/equipment charge	↓	4	2	Job		66		66	109	
230	0010	**CABINET HARDWARE**										230
	1000	Catches, minimum	1 Carp	235	.034	Ea.	.75	1.12		1.87	2.69	
	1040	Maximum	"	80	.100	"	4.65	3.30		7.95	10.55	
	2000	Door/drawer pulls, handles										
	2200	Handles and pulls, projecting, metal, minimum	1 Carp	160	.050	Ea.	3.48	1.65		5.13	6.55	
	2240	Maximum		68	.118		6.40	3.88		10.28	13.45	
	2300	Wood, minimum		160	.050		3.48	1.65		5.13	6.55	
	2340	Maximum		68	.118		6.40	3.88		10.28	13.45	
	2600	Flush, metal, minimum		160	.050		3.48	1.65		5.13	6.55	
	2640	Maximum		68	.118	↓	6.40	3.88		10.28	13.45	
	3000	Drawer tracks/glides, minimum		48	.167	Pr.	5.90	5.50		11.40	15.60	
	3040	Maximum		24	.333		17.15	11		28.15	37	
	4000	Cabinet hinges, minimum		160	.050		2	1.65		3.65	4.94	
	4040	Maximum	↓	68	.118	↓	7	3.88		10.88	14.15	
240	0010	**DRAWERS**										240
	0100	Solid hardwood front										
	1000	4" high, 12" wide	1 Carp	17	.471	Ea.	17.30	15.55		32.85	44.50	
	1200	18" wide	"	16	.500	"	24	16.50		40.50	54	
	2800	Plastic laminate on particle board front										
	3000	4" high, 12" wide	1 Carp	17	.471	Ea.	22.50	15.55		38.05	50.50	
	3200	18" wide	"	16	.500	"	26.50	16.50		43	56.50	
	5400	Plywood, flush panel front										
	6000	4" high, 12" wide	1 Carp	17	.471	Ea.	23.50	15.55		39.05	51.50	
	6200	18" wide		16	.500	"	28.50	16.50		45	58.50	
	9000	Minimum labor/equipment charge	↓	4	2	Job		66		66	109	
400	0010	**VANITIES**										400
	8000	Vanity bases, 2 doors, 30" high, 21" deep, 24" wide	2 Carp	20	.800	Ea.	169	26.50		195.50	230	
	8050	30" wide		16	1		194	33		227	269	
	8100	36" wide		13.33	1.200		259	39.50		298.50	350	
	8150	48" wide	↓	11.43	1.400		310	46		356	415	
	9000	For deluxe models of all vanities, add to above					40%					
	9500	For custom built in place, add to above				↓	25%	10%				

	06415 | Countertops		Crew	Daily Output	Labor-Hours	Unit	2004 Bare Costs Mat.	Labor	Equip.	Total	Total Incl O&P	
100	0010	**COUNTER TOP** Stock, plastic lam., 24" wide w/backsplash, min.	1 Carp	30	.267	L.F.	8.30	8.80		17.10	24	100
	0100	Maximum		25	.320		15.85	10.55		26.40	35	
	0300	Custom plastic, 7/8" thick, aluminum molding, no splash		30	.267		17	8.80		25.80	33.50	
	0400	Cove splash		30	.267		22	8.80		30.80	39	
	0600	1-1/4" thick, no splash		28	.286		20	9.45		29.45	37.50	
	0700	Square splash		28	.286		25	9.45		34.45	43	
	0900	Square edge, plastic face, 7/8" thick, no splash		30	.267		21.50	8.80		30.30	38	
	1000	With splash		30	.267		28	8.80		36.80	45	
	1200	For stainless channel edge, 7/8" thick, add					2.27			2.27	2.50	
	1300	1-1/4" thick, add					2.66			2.66	2.93	
	1500	For solid color suede finish, add					2.12			2.12	2.33	
	1700	For end splash, add				Ea.	13.35			13.35	14.70	
	1900	For cut outs, standard, add, minimum	1 Carp	32	.250		2.71	8.25		10.96	16.70	
	2000	Maximum		8	1		3.23	33		36.23	58	
	2100	Postformed, including backsplash and front edge		30	.267	L.F.	8.80	8.80		17.60	24.50	
	2110	Mitred, add		12	.667	Ea.		22		22	36.50	
	2200	Built-in place, 25" wide, plastic laminate		25	.320	L.F.	11.70	10.55		22.25	30.50	
	2300	Ceramic tile mosaic		25	.320		25.50	10.55		36.05	45.50	
	2500	Marble, stock, with splash, 1/2" thick, minimum	1 Bric	17	.471		31	16.10		47.10	60	
	2700	3/4" thick, maximum	"	13	.615		78.50	21		99.50	121	
	2900	Maple, solid, laminated, 1-1/2" thick, no splash	1 Carp	28	.286		52.50	9.45		61.95	73	
	3000	With square splash		28	.286		62	9.45		71.45	84	
	3200	Stainless steel		24	.333	S.F.	94	11		105	121	
	3400	Recessed cutting block with trim, 16" x 20" x 1"		8	1	Ea.	61	33		94	122	
	9000	Minimum labor/equipment charge		3.75	2.133	Job		70.50		70.50	117	

	06430 | Stairs & Railings		Crew	Daily Output	Labor-Hours	Unit	Mat.	Labor	Equip.	Total	Total Incl O&P	
500	0010	**RAILING** Custom design, architectural grade, hardwood, minimum	1 Carp	38	.211	L.F.	5.60	6.95		12.55	17.65	500
	0100	Maximum		30	.267		45.50	8.80		54.30	64.50	
	0300	Stock interior railing with spindles 6" O.C., 4' long		40	.200		29	6.60		35.60	42.50	
	0400	8' long		48	.167		27	5.50		32.50	38.50	
	9000	Minimum labor/equipment charge		3	2.667	Job		88		88	146	
620	0011	**STAIRS, PREFABRICATED**										620
	0100	Box stairs, prefabricated, 3'-0" wide										
	0110	Oak treads, up to 14 risers	2 Carp	39	.410	Riser	71.50	13.55		85.05	101	
	0600	With pine treads for carpet, up to 14 risers	"	39	.410	"	46	13.55		59.55	73	
	1100	For 4' wide stairs, add				Flight	25%					
	1550	Stairs, prefabricated stair handrail with balusters	1 Carp	30	.267	L.F.	49	8.80		57.80	68	
	1700	Basement stairs, prefabricated, pine treads										
	1710	Pine risers, 3' wide, up to 14 risers	2 Carp	52	.308	Riser	46	10.15		56.15	67.50	
	4000	Residential, wood, oak treads, prefabricated		1.50	10.667	Flight	930	350		1,280	1,600	
	4200	Built in place		.44	36.364	"	1,675	1,200		2,875	3,850	
	4400	Spiral, oak, 4'-6" diameter, unfinished, prefabricated,										
	4500	incl. railing, 9' high	2 Carp	1.50	10.667	Flight	4,400	350		4,750	5,425	
	9000	Minimum labor/equipment charge	"	3	5.333	Job		176		176	292	
630	0010	**STAIR PARTS** Balusters, turned, 30" high, pine, minimum	1 Carp	28	.286	Ea.	4.10	9.45		13.55	20	630
	0100	Maximum		26	.308	"	19	10.15		29.15	38	
	0300	30" high birch balusters, minimum		28	.286	Ea.	6.30	9.45		15.75	22.50	
	0400	Maximum		26	.308		25.50	10.15		35.65	45	
	0600	42" high, pine balusters, minimum		27	.296		5.30	9.80		15.10	22	
	0700	Maximum		25	.320		27.50	10.55		38.05	47.50	
	0900	42" high birch balusters, minimum		27	.296		5.30	9.80		15.10	22	
	1000	Maximum		25	.320		37	10.55		47.55	58	

	06430 | Stairs & Railings		CREW	DAILY OUTPUT	LABOR-HOURS	UNIT	2004 BARE COSTS MAT.	LABOR	EQUIP.	TOTAL	TOTAL INCL O&P
630	1050	Baluster, stock pine, 1-1/4" x 1-1/4"	1 Carp	240	.033	L.F.	1.55	1.10		2.65	3.53
	1100	1-3/4" x 1-3/4"		220	.036	"	7.60	1.20		8.80	10.35
	1200	Newels, 3-1/4" wide, starting, minimum		7	1.143	Ea.	37	37.50		74.50	104
	1300	Maximum		6	1.333		425	44		469	545
	1500	Landing, minimum		5	1.600		98	53		151	196
	1600	Maximum		4	2		460	66		526	615
	1800	Railings, oak, built-up, minimum		60	.133	L.F.	30	4.40		34.40	40.50
	1900	Maximum		55	.145		43.50	4.80		48.30	56
	2100	Add for sub rail		110	.073		5	2.40		7.40	9.50
	2300	Risers, beech, 3/4" x 7-1/2" high		64	.125		5.80	4.13		9.93	13.20
	2400	Fir, 3/4" x 7-1/2" high		64	.125		1.60	4.13		5.73	8.60
	2600	Oak, 3/4" x 7-1/2" high		64	.125		6	4.13		10.13	13.45
	2800	Pine, 3/4" x 7-1/2" high		66	.121		3.05	4		7.05	10
	2850	Skirt board, pine, 1" x 10"		55	.145		2.90	4.80		7.70	11.15
	2900	1" x 12"		52	.154		3.45	5.10		8.55	12.20
	3000	Treads, oak, 1-1/4" x 10" wide, 3' long		18	.444	Ea.	74	14.65		88.65	106
	3100	4' long, oak		17	.471		98.50	15.55		114.05	134
	3300	1-1/4" x 11-1/2" wide, 3' long, oak		18	.444		26	14.65		40.65	53
	3400	6' long, oak		14	.571		155	18.85		173.85	202
	3600	Beech treads, add					40%				
	3800	For mitered return nosings, add				L.F.	3.01			3.01	3.31
	9000	Minimum labor/equipment charge	1 Carp	3	2.667	Job		88		88	146
	06440 | Wood Ornaments										
150	0010	**BEAMS, DECORATIVE** Rough sawn cedar, non-load bearing, 4" x 4"	2 Carp	180	.089	L.F.	1.35	2.93		4.28	6.35
	0100	4" x 6"		170	.094		2.60	3.11		5.71	8
	0200	4" x 8"		160	.100		3.34	3.30		6.64	9.10
	0300	4" x 10"		150	.107		4.64	3.52		8.16	10.95
	0400	4" x 12"		140	.114		5.60	3.77		9.37	12.45
	0500	8" x 8"		130	.123		7.85	4.06		11.91	15.40
	1100	Beam connector plates see div. 06090-800									
	9000	Minimum labor/equipment charge	1 Carp	3	2.667	Job		88		88	146
350	0010	**GRILLES** and panels, hardwood, sanded									
	0020	2' x 4' to 4' x 8', custom designs, unfinished, minimum	1 Carp	38	.211	S.F.	12.10	6.95		19.05	25
	0050	Average		30	.267		26.50	8.80		35.30	43.50
	0100	Maximum		19	.421		40.50	13.90		54.40	67.50
	0300	As above, but prefinished, minimum		38	.211		12.10	6.95		19.05	25
	0400	Maximum		19	.421		45.50	13.90		59.40	73
	9000	Minimum labor/equipment charge		2	4	Job		132		132	219
400	0010	**LOUVERS** Redwood, 2'-0" diameter, full circle	1 Carp	16	.500	Ea.	129	16.50		145.50	170
	0100	Half circle		16	.500	"	124	16.50		140.50	164
	0200	Octagonal 		16	.500	Ea.	99	16.50		115.50	137
	0300	Triangular, 5/12 pitch, 5'-0" at base		16	.500	"	209	16.50		225.50	258
	9000	Minimum labor/equipment charge		3.50	2.286	Job		75.50		75.50	125
500	0010	**FIREPLACE MANTELS** 6" molding, 6' x 3'-6" opening, minimum	1 Carp	5	1.600	Opng.	135	53		188	237
	0100	Maximum		5	1.600		168	53		221	273
	0300	Prefabricated pine, colonial type, stock, deluxe		2	4		2,125	132		2,257	2,550
	0400	Economy		3	2.667		365	88		453	545
	9000	Minimum labor/equipment charge		3	2.667	Job		88		88	146
550	0010	**FIREPLACE MANTEL BEAMS** Rough texture wood, 4" x 8"	1 Carp	36	.222	L.F.	5	7.35		12.35	17.65
	0100	4" x 10"		35	.229	"	6	7.55		13.55	19.10
	0300	Laminated hardwood, 2-1/4" x 10-1/2" wide, 6' long		5	1.600	Ea.	95.50	53		148.50	193
	0400	8' long		5	1.600	"	133	53		186	234

	06440 | Wood Ornaments		CREW	DAILY OUTPUT	LABOR-HOURS	UNIT	2004 BARE COSTS MAT.	LABOR	EQUIP.	TOTAL	TOTAL INCL O&P	
550	0600	Brackets for above, rough sawn	1 Carp	12	.667	Pr.	8.80	22		30.80	46	550
	0700	Laminated		12	.667	"	13.30	22		35.30	51	
	9000	Minimum labor/equipment charge		4	2	Job		66		66	109	
700	0010	**COLUMNS**										700
	0050	Aluminum, round colonial, 6" diameter	2 Carp	80	.200	V.L.F.	18.05	6.60		24.65	31	
	0100	8" diameter		62.25	.257		24	8.50		32.50	40.50	
	0200	10" diameter		55	.291		29.50	9.60		39.10	48.50	
	0250	Fir, stock units, hollow round, 6" diameter		80	.200		16.20	6.60		22.80	29	
	0300	8" diameter		80	.200		19	6.60		25.60	32	
	0350	10" diameter		70	.229		24.50	7.55		32.05	39.50	
	0400	Solid turned, to 8' high, 3-1/2" diameter		80	.200		7.45	6.60		14.05	19.10	
	0500	4-1/2" diameter		75	.213		10.60	7.05		17.65	23.50	
	0600	5-1/2" diameter		70	.229		14.85	7.55		22.40	29	
	0800	Square columns, built-up, 5" x 5"		65	.246		13.80	8.10		21.90	28.50	
	0900	Solid, 3-1/2" x 3-1/2"		130	.123		6.35	4.06		10.41	13.75	
	1600	Hemlock, tapered, T & G, 12" diam, 10' high		100	.160		32	5.30		37.30	44	
	1700	16' high		65	.246		56	8.10		64.10	75.50	
	1900	10' high, 14" diameter		100	.160		81.50	5.30		86.80	98.50	
	2000	18' high		65	.246		77.50	8.10		85.60	98.50	
	2200	18" diameter, 12' high		65	.246		109	8.10		117.10	133	
	2300	20' high		50	.320		105	10.55		115.55	134	
	2500	20" diameter, 14' high		40	.400		129	13.20		142.20	164	
	2600	20' high		35	.457		134	15.10		149.10	172	
	2800	For flat pilasters, deduct					33%					
	3000	For splitting into halves, add				Ea.	63.50			63.50	70	
	4000	Rough sawn cedar posts, 4" x 4"	2 Carp	250	.064	V.L.F.	2.58	2.11		4.69	6.35	
	4100	4" x 6"		235	.068		3.84	2.25		6.09	7.95	
	4200	6" x 6"		220	.073		5.80	2.40		8.20	10.35	
	4300	8" x 8"		200	.080		5.95	2.64		8.59	10.95	
	9000	Minimum labor/equipment charge	1 Carp	3	2.667	Job		88		88	146	
	06445 | Simulated Wood Ornaments											
100	0010	**MILLWORK, HIGH DENSITY POLYMER**										100
	0100	Base, 9/16" x 3-3/16"	1 Carp	230	.035	L.F.	1.31	1.15		2.46	3.34	
	0200	Casing, fluted, 5/8" x 3-1/4"		215	.037		3.66	1.23		4.89	6.05	
	0300	Chair rail, 9/16" x 2-1/4"		260	.031		1.91	1.02		2.93	3.78	
	0600	Cove, 13/16" x 3-3/4"		260	.031		3.83	1.02		4.85	5.90	
	0700	Crown, 3/4" x 3-13/16"		260	.031		5.40	1.02		6.42	7.65	
	0800	Half round, 15/16" x 2"		240	.033		17.10	1.10		18.20	20.50	
	06470 | Screen, Blinds & Shutters											
100	0010	**SHUTTERS, EXTERIOR** Aluminum, louvered, 1'-4" wide, 3'-0" long	1 Carp	10	.800	Pr.	42	26.50		68.50	90	100
	0400	6'-8" long		9	.889		84	29.50		113.50	141	
	1000	Pine, louvered, primed, each 1'-2" wide, 3'-3" long		10	.800		79.50	26.50		106	132	
	1100	4'-7" long		10	.800		107	26.50		133.50	162	
	1250	Each 1'-4" wide, 3'-0" long		10	.800		81	26.50		107.50	134	
	1350	5'-3" long		10	.800		122	26.50		148.50	178	
	1500	Each 1'-6" wide, 3'-3" long		10	.800		84.50	26.50		111	137	
	1600	4'-7" long		10	.800		119	26.50		145.50	174	
	1610	Door blinds, 6'-9" long 1'-3" wide		9	.889		148	29.50		177.50	212	
	1615	1' - 6" wide		9	.889		163	29.50		192.50	228	
	1620	Hemlock, louvered, 1'-2" wide, 5'-7" long		10	.800		130	26.50		156.50	187	
	1630	Each 1'-4" wide, 2'-2" long		10	.800		81	26.50		107.50	134	

06400 | Architectural Woodwork

06470 | Screen, Blinds & Shutters

		06470 Screen, Blinds & Shutters	CREW	DAILY OUTPUT	LABOR-HOURS	UNIT	2004 BARE COSTS MAT.	LABOR	EQUIP.	TOTAL	TOTAL INCL O&P	
100	1640	3′-0″ long	1 Carp	10	.800	Pr.	81	26.50		107.50	134	100
	1650	3′-3″ long		10	.800		87.50	26.50		114	141	
	1660	3′-11″ long		10	.800		99.50	26.50		126	153	
	1670	4′-3″ long		10	.800		97	26.50		123.50	151	
	1680	5′-3″ long		10	.800		121	26.50		147.50	177	
	1690	5′-11″ long		10	.800		137	26.50		163.50	195	
	1700	Door blinds, 6′-9″ long, each 1′-3″ wide		9	.889		138	29.50		167.50	200	
	1710	1′-6″ wide		9	.889		148	29.50		177.50	212	
	1720	Hemlock, solid raised panel, each 1′-4″ wide, 3′-3″ long		10	.800		131	26.50		157.50	188	
	1730	3′-11″ long		10	.800		154	26.50		180.50	213	
	1740	4′-3″ long		10	.800		166	26.50		192.50	227	
	1750	4′-7″ long		10	.800		179	26.50		205.50	241	
	1760	4′-11″ long		10	.800		195	26.50		221.50	258	
	1770	5′-11″ long		10	.800		222	26.50		248.50	288	
	1800	Door blinds, 6′-9″ long, each 1′-3″ wide		9	.889		250	29.50		279.50	325	
	1900	1′-6″ wide		9	.889		272	29.50		301.50	350	
	2500	Polystyrene, solid raised panel, each 1′-4″ wide, 3′-3″ long		10	.800		55	26.50		81.50	105	
	2700	4′-7″ long		10	.800		66.50	26.50		93	117	
	3500	Polystyrene, solid raised panel, each 3′-3″ wide, 3′-0″ long		10	.800		166	26.50		192.50	226	
	3600	3′-11″ long		10	.800		177	26.50		203.50	238	
	3700	4′-7″ long		10	.800		209	26.50		235.50	274	
	3800	5′-3″ long		10	.800		231	26.50		257.50	299	
	3900	6′-8″ long		9	.889		261	29.50		290.50	335	
	4500	Polystyrene, louvered, each 1′-2″ wide, 3′-3″ long		10	.800		40	26.50		66.50	88	
	4600	4′-7″ long		10	.800		49.50	26.50		76	98.50	
	4750	5′-3″ long		10	.800		53	26.50		79.50	102	
	4850	6′-8″ long		9	.889		87	29.50		116.50	144	
	6000	Vinyl, louvered, each 1′-2″ x 4′-7″ long		10	.800		52	26.50		78.50	101	
	6200	Each 1′-4″ x 6′-8″ long		9	.889		79	29.50		108.50	136	
	9000	Minimum labor/equipment charge		4	2	Job		66		66	109	
200	0010	**SHUTTERS, INTERIOR** Wood, louvered,				Set	105			105	116	200
	0200	Two panel, 27″ wide, 36″ high	1 Carp	5	1.600	″	105	53		158	204	
	0300	33″ wide, 36″ high		5	1.600	Set	125	53		178	226	
	0500	47″ wide, 36″ high		5	1.600		182	53		235	288	
	1000	Four panel, 27″ wide, 36″ high		5	1.600		241	53		294	355	
	1100	33″ wide, 36″ high		5	1.600		241	53		294	355	
	1300	47″ wide, 36″ high		5	1.600		320	53		373	445	

06600 | Plastic Fabrications

06620 | Non-Structural Plastics

		06620 Non-Structural Plastics	CREW	DAILY OUTPUT	LABOR-HOURS	UNIT	2004 BARE COSTS MAT.	LABOR	EQUIP.	TOTAL	TOTAL INCL O&P	
810	0010	**SOLID SURFACE COUNTERTOPS**, Acrylic polymer										810
	0020	Pricing for orders of 100 L.F. or greater										
	0100	25″ wide, solid colors	2 Carp	28	.571	L.F.	43.50	18.85		62.35	79.50	
	0200	Patterned colors		28	.571		55.50	18.85		74.35	92.50	
	0300	Premium patterned colors		28	.571		69.50	18.85		88.35	108	
	0400	With silicone attached 4″ backsplash, solid colors		27	.593		48	19.55		67.55	85.50	
	0500	Patterned colors		27	.593		61	19.55		80.55	99.50	
	0600	Premium patterned colors		27	.593		75.50	19.55		95.05	116	

	06620 | Non-Structural Plastics	CREW	DAILY OUTPUT	LABOR-HOURS	UNIT	2004 BARE COSTS MAT.	LABOR	EQUIP.	TOTAL	TOTAL INCL O&P	
810	0700 With hard seam attached 4" backsplash, solid colors	2 Carp	23	.696	L.F.	48	23		71	91	810
	0800 Patterned colors		23	.696		61	23		84	105	
	0900 Premium patterned colors	↓	23	.696	↓	75.50	23		98.50	122	
	1000 Pricing for order of 51 - 99 L.F.										
	1100 25" wide, solid colors	2 Carp	24	.667	L.F.	50.50	22		72.50	92	
	1200 Patterned colors		24	.667		64	22		86	107	
	1300 Premium patterned colors		24	.667		79.50	22		101.50	124	
	1400 With silicone attached 4" backsplash, solid colors		23	.696		55	23		78	98.50	
	1500 Patterned colors		23	.696		70	23		93	115	
	1600 Premium patterned colors		23	.696		87	23		110	134	
	1700 With hard seam attached 4" backsplash, solid colors		20	.800		55	26.50		81.50	105	
	1800 Patterned colors		20	.800		70	26.50		96.50	121	
	1900 Premium patterned colors	↓	20	.800	↓	87	26.50		113.50	140	
	2000 Pricing for order of 1 - 50 L.F.										
	2100 25" wide, solid colors	2 Carp	20	.800	L.F.	59	26.50		85.50	109	
	2200 Patterned colors		20	.800		75	26.50		101.50	127	
	2300 Premium patterned colors		20	.800		93.50	26.50		120	147	
	2400 With silicone attached 4" backsplash, solid colors		19	.842		65	28		93	118	
	2500 Patterned colors		19	.842		82	28		110	137	
	2600 Premium patterned colors		19	.842		102	28		130	158	
	2700 With hard seam attached 4" backsplash, solid colors		4	4		65	132		197	291	
	2800 Patterned colors		15	1.067		82	35		117	149	
	2900 Premium patterned colors	↓	15	1.067	↓	102	35		137	171	
	3000 Sinks, pricing for order of 100 or greater units										
	3100 Single bowl, hard seamed, solid colors, 13" x 17"	1 Carp	3	2.667	Ea.	295	88		383	470	
	3200 10" x 15"		7	1.143		137	37.50		174.50	213	
	3300 Cutouts for sinks	↓	8	1	↓		33		33	54.50	
	3400 Sinks, pricing for order of 51 - 99 units										
	3500 Single bowl, hard seamed, solid colors, 13" x 17"	1 Carp	2.55	3.137	Ea.	340	104		444	545	
	3600 10" x 15"		6	1.333		157	44		201	246	
	3700 Cutouts for sinks	↓	7	1.143	↓		37.50		37.50	62.50	
	3800 Sinks, pricing for order of 1 - 50 units										
	3900 Single bowl, hard seamed, solid colors, 13" x 17"	1 Carp	2	4	Ea.	400	132		532	660	
	4000 10" x 15"		4.55	1.758		184	58		242	299	
	4100 Cutouts for sinks		5.25	1.524			50.50		50.50	83.50	
	4200 Cooktop cutouts, pricing for 100 or greater units		4	2		21.50	66		87.50	133	
	4300 51 - 99 units		3.40	2.353		25	77.50		102.50	157	
	4400 1 - 50 units	↓	3	2.667	↓	29	88		117	178	
850	0010 **VANITY TOPS**										850
	0015 Solid surface, center bowl, 17" x 19"	1 Carp	12	.667	Ea.	168	22		190	222	
	0020 19" x 25"		12	.667		204	22		226	261	
	0030 19" x 31"		12	.667		247	22		269	310	
	0040 19" x 37"		12	.667		288	22		310	350	
	0050 22" x 25"		10	.800		229	26.50		255.50	296	
	0060 22" x 31"		10	.800		267	26.50		293.50	340	
	0070 22" x 37"		10	.800		310	26.50		336.50	385	
	0080 22" x 43"		10	.800		355	26.50		381.50	435	
	0090 22" x 49"		10	.800		395	26.50		421.50	475	
	0110 22" x 55"		8	1		445	33		478	545	
	0120 22" x 61"		8	1		510	33		543	615	
	0220 Double bowl, 22" x 61"		8	1		575	33		608	685	
	0230 Double bowl, 22" x 73"	↓	8	1	↓	625	33		658	745	
	0240 For aggregate colors, add					35%					
	0250 For faucets and fittings see 15410-300										

Division 7
Thermal & Moisture Protection

Estimating Tips

07100 Dampproofing & Waterproofing

- Be sure of the job specifications before pricing this subdivision. The difference in cost between waterproofing and dampproofing can be great. Waterproofing will hold back standing water. Dampproofing prevents the transmission of water vapor. Also included in this section are vapor retarding membranes.

07200 Thermal Protection

- Insulation and fireproofing products are measured by area, thickness, volume or R value. Specifications may only give what the specific R value should be in a certain situation. The estimator may need to choose the type of insulation to meet that R value.

07300 Shingles, Roof Tiles & Roof Coverings
07400 Roofing & Siding Panels

- Many roofing and siding products are bought and sold by the square. One square is equal to an area that measures 100 square feet.

This simple change in unit of measure could create a large error if the estimator is not observant. Accessories necessary for a complete installation must be figured into any calculations for both material and labor.

07500 Membrane Roofing
07600 Flashing & Sheet Metal
07700 Roof Specialties & Accessories

- The items in these subdivisions compose a roofing system. No one component completes the installation and all must be estimated. Built-up or single ply membrane roofing systems are made up of many products and installation trades. Wood blocking at roof perimeters or penetrations, parapet coverings, reglets, roof drains, gutters, downspouts, sheet metal flashing, skylights, smoke vents or roof hatches all need to be considered along with the roofing material. Several different installation trades will need to work together on the roofing system. Inherent difficulties in the scheduling and coordination of various trades must be accounted for when estimating labor costs.

07900 Joint Sealers

- To complete the weather-tight shell the sealants and caulkings must be estimated. Where different materials meet—at expansion joints, at flashing penetrations, and at hundreds of other locations throughout a construction project—they provide another line of defense against water penetration. Often, an entire system is based on the proper location and placement of caulking or sealants. The detail drawings that are included as part of a set of architectural plans, show typical locations for these materials. When caulking or sealants are shown at typical locations, this means the estimator must include them for all the locations where this detail is applicable. Be careful to keep different types of sealants separate, and remember to consider backer rods and primers if necessary.

Reference Numbers

Reference numbers are shown in bold squares at the beginning of some major classifications. These numbers refer to related items in the Reference Section. The reference information may be an estimating procedure, an alternate pricing method or technical information.

Note: Not all subdivisions listed here necessarily appear in this publication.

07050 | Basic Thermal & Moisture Protection Materials & Methods

07060 | Selective Demolition

		Description	CREW	DAILY OUTPUT	LABOR-HOURS	UNIT	2004 BARE COSTS MAT.	LABOR	EQUIP.	TOTAL	TOTAL INCL O&P	
110	0010	**SELECTIVE DEMOLITION, ROOFING AND SIDING** R02220-510										110
	1000	Deck, roof, concrete plank	B-13	1,680	.033	S.F.		.93	.37	1.30	1.93	
	1100	Gypsum plank		3,900	.014			.40	.16	.56	.84	
	1150	Metal decking		3,500	.016			.45	.18	.63	.93	
	1200	Wood, boards, tongue and groove, 2" x 6"	2 Clab	960	.017			.43		.43	.72	
	1220	2" x 10"		1,040	.015			.40		.40	.66	
	1280	Standard planks, 1" x 6"		1,080	.015			.39		.39	.64	
	1320	1" x 8"		1,160	.014			.36		.36	.59	
	1340	1" x 12"		1,200	.013			.35		.35	.57	
	1350	Plywood, to 1" thick		2,000	.008			.21		.21	.34	
	2000	Gutters, aluminum or wood, edge hung	1 Clab	240	.033	L.F.		.87		.87	1.44	
	2100	Built-in		100	.080	"		2.08		2.08	3.45	
	2500	Roof accessories, plumbing vent flashing		14	.571	Ea.		14.85		14.85	24.50	
	2600	Adjustable metal chimney flashing		9	.889	"		23		23	38.50	
	2650	Coping, sheet metal, up to 12" wide		240	.033	L.F.		.87		.87	1.44	
	2660	Concrete, up to 12" wide	2 Clab	160	.100	"		2.60		2.60	4.31	
	3000	Roofing, built-up, 5 ply roof, no gravel	B-2	1,600	.025	S.F.		.66		.66	1.09	
	3001	Including gravel		890	.045			1.19		1.19	1.97	
	3100	Gravel removal, minimum		5,000	.008			.21		.21	.35	
	3120	Maximum		2,000	.020			.53		.53	.88	
	3400	Roof insulation board, up to 2" thick		3,900	.010			.27		.27	.45	
	3450	Roll roofing, cold adhesive	1 Clab	12	.667	Sq.		17.35		17.35	28.50	
	4000	Shingles, asphalt strip, 1 layer	B-2	3,500	.011	S.F.		.30		.30	.50	
	4100	Slate		2,500	.016			.42		.42	.70	
	4300	Wood		2,200	.018			.48		.48	.80	
	4500	Skylight to 10 S.F.	1 Clab	8	1	Ea.		26		26	43	
	5000	Siding, metal, horizontal		444	.018	S.F.		.47		.47	.78	
	5020	Vertical		400	.020			.52		.52	.86	
	5200	Wood, boards, vertical		400	.020			.52		.52	.86	
	5220	Clapboards, horizontal		380	.021			.55		.55	.91	
	5240	Shingles		350	.023			.59		.59	.99	
	5260	Textured plywood		725	.011			.29		.29	.48	
	9000	Minimum labor/equipment charge		2	4	Job		104		104	172	

07100 | Dampproofing and Waterproofing

07110 | Dampproofing

		Description	CREW	DAILY OUTPUT	LABOR-HOURS	UNIT	2004 BARE COSTS MAT.	LABOR	EQUIP.	TOTAL	TOTAL INCL O&P	
100	0010	**BITUMINOUS ASPHALT COATING** For foundation										100
	0030	Brushed on, below grade, 1 coat	1 Rofc	665	.012	S.F.	.07	.34		.41	.69	
	0100	2 coat		500	.016		.10	.46		.56	.93	
	0300	Sprayed on, below grade, 1 coat, 25.6 S.F./gal.		830	.010		.07	.27		.34	.57	
	0400	2 coat, 20.5 S.F./gal.		500	.016		.14	.46		.60	.98	
	0600	Troweled on, asphalt with fibers, 1/16" thick		500	.016		.16	.46		.62	1	
	0700	1/8" thick		400	.020		.29	.57		.86	1.34	
	1000	1/2" thick		350	.023		.94	.65		1.59	2.19	
	9000	Minimum labor/equipment charge		3	2.667	Job		76		76	136	

	07130 | Sheet Waterproofing		CREW	DAILY OUTPUT	LABOR-HOURS	UNIT	2004 BARE COSTS MAT.	LABOR	EQUIP.	TOTAL	TOTAL INCL O&P	
200	0010	**ELASTOMERIC WATERPROOFING**										200
	0050	Acrylic rubber, fluid applied, 20 mils thick	3 Rofc	1,000	.024	S.F.	1.57	.68		2.25	2.95	
	0060	50 mil, reinforced, stucco texture	″	600	.040		2.53	1.14		3.67	4.82	
	0090	EPDM, plain, 45 mils thick	2 Rofc	580	.028		.29	.78		1.07	1.73	
	0100	60 mils thick		570	.028		.39	.80		1.19	1.86	
	0300	Nylon reinforced sheets, 45 mils thick		580	.028		.79	.78		1.57	2.28	
	0400	60 mils thick		570	.028		1.15	.80		1.95	2.70	
	0600	Vulcanizing splicing tape for above, 2″ wide				C.L.F.	33.50			33.50	36.50	
	0700	4″ wide				″	67			67	73.50	
	0900	Adhesive, bonding, 60 SF per gal				Gal.	13.60			13.60	14.95	
	1000	Splicing, 75 SF per gal				″	21			21	23	
	1200	Neoprene sheets, plain, 45 mils thick	2 Rofc	580	.028	S.F.	.88	.78		1.66	2.38	
	1300	60 mils thick		570	.028		1.42	.80		2.22	2.99	
	1500	Nylon reinforced, 45 mils thick		580	.028		1.15	.78		1.93	2.68	
	1600	60 mils thick		570	.028		1.29	.80		2.09	2.85	
	1800	120 mils thick		500	.032		2.25	.91		3.16	4.11	
	1900	Adhesive, splicing, 150 S.F. per gal. per coat				Gal.	18.15			18.15	19.95	
	2100	Fiberglass reinforced, fluid applied, 1/8″ thick	2 Rofc	500	.032	S.F.	1.50	.91		2.41	3.28	
	2200	Polyethylene and rubberized asphalt sheets, 1/8″ thick		550	.029		.54	.83		1.37	2.07	
	2210	Asphaltic hardboard protection board, 1/8″ thick		500	.032		.30	.91		1.21	1.96	
	2220	Asphaltic hardboard protection board, 1/4″ thick		450	.036		.52	1.01		1.53	2.38	
	2400	Polyvinyl chloride sheets, plain, 10 mils thick		580	.028		.14	.78		.92	1.56	
	2500	20 mils thick		570	.028		.22	.80		1.02	1.67	
	2700	30 mils thick		560	.029		.30	.81		1.11	1.79	
	3000	Adhesives, trowel grade, 40-100 SF per gal				Gal.	20			20	22	
	3100	Brush grade, 100-250 SF per gal.				″	20			20	22	
	3300	Bitumen modified polyurethane, fluid applied, 55 mils thick	2 Rofc	665	.024	S.F.	.65	.68		1.33	1.95	
	3600	Vinyl plastic, sprayed on, 25 to 40 mils thick		475	.034	″	.98	.96		1.94	2.80	
	9000	Minimum labor/equipment charge		2	8	Job		228		228	410	
500	0010	**MEMBRANE WATERPROOFING** On slabs, 1 ply, felt	G-1	3,000	.019	S.F.	.19	.50	.10	.79	1.21	500
	0015	Membrane waterproofing on walls, 1 ply, felt, mopped		3,000	.019		.19	.50	.10	.79	1.21	
	0300	2 ply, felt		2,500	.022		.37	.60	.12	1.09	1.61	
	0400	Glass fiber fabric		1,650	.034		.45	.90	.18	1.53	2.31	
	0600	3 ply, felt		2,100	.027		.56	.71	.14	1.41	2.05	
	0700	Glass fiber fabric		1,550	.036		.60	.96	.19	1.75	2.60	
	1000	For adhered 1/4″ EPS protection board, add	2 Rofc	3,500	.005		.14	.13		.27	.38	
	1050	3/8″ thick, add		3,500	.005		.16	.13		.29	.41	
	1060	1/2″ thick, add		3,500	.005		.18	.13		.31	.43	
	1070	Fiberglass fabric, black, 20/10 mesh		116	.138	Sq.	9.95	3.92		13.87	18	
	1080	White, 20/10 mesh		116	.138	″	10.25	3.92		14.17	18.35	
	1100	1/16″ urethane, troweled		200	.080	S.F.	.66	2.28		2.94	4.81	
	1200	Roller applied		120	.133	″	.60	3.79		4.39	7.45	
	9000	Minimum labor/equipment charge		2	8	Job		228		228	410	
	07160 | Cement. & Reactive Waterproofing											
150	0010	**CEMENTITIOUS WATERPROOFING** One coat cement base										150
	0020	1/8″ application, sprayed on	G-2	1,000	.024	S.F.	1.50	.65	.11	2.26	2.84	
	0030	2 coat, cementitious/metallic slurry, troweled, 1/4″ thick	1 Cefi	2.48	3.226	C.S.F.	35	102		137	199	
	0040	3 coat, 3/8″ thick		1.84	4.348		52.50	137		189.50	274	
	0050	4 coat, 1/2″ thick		1.20	6.667		70	210		280	405	
	07190 | Water Repellents											
700	0010	**RUBBER COATING** Water base liquid, roller applied	2 Rofc	7,000	.002	S.F.	.55	.07		.62	.73	700
	0200	Silicone or stearate, sprayed on CMU, 1 coat	1 Rofc	4,000	.002		.27	.06		.33	.40	

07100 | Dampproofing and Waterproofing

	07190 Water Repellents		Crew	Daily Output	Labor-Hours	Unit	2004 Bare Costs Mat.	Labor	Equip.	Total	Total Incl O&P	
700	0300	2 coats	1 Rofc	3,000	.003	S.F.	.54	.08		.62	.74	700
	9000	Minimum labor/equipment charge		3	2.667	Job		76		76	136	

07200 | Thermal Protection

	07210 Building Insulation		Crew	Daily Output	Labor-Hours	Unit	2004 Bare Costs Mat.	Labor	Equip.	Total	Total Incl O&P	
150	0010	**BLOWN-IN INSULATION** Ceilings, with open access										150
	0020	Cellulose, 3-1/2″ thick, R13	G-4	5,000	.005	S.F.	.13	.13	.04	.30	.39	
	0030	5-3/16″ thick, R19		3,800	.006		.21	.17	.05	.43	.57	
	0050	6-1/2″ thick, R22		3,000	.008		.26	.21	.07	.54	.71	
	0100	8-11/16″ thick, R30		2,600	.009		.34	.25	.08	.67	.87	
	1000	Fiberglass, 5″ thick, R11		3,800	.006		.16	.17	.05	.38	.52	
	1050	6″ thick, R13		3,000	.008		.22	.21	.07	.50	.66	
	1100	8-1/2″ thick, R19		2,200	.011		.31	.29	.09	.69	.92	
	1200	10″ thick, R22		1,800	.013		.36	.36	.11	.83	1.11	
	1300	12″ thick, R26		1,500	.016		.43	.43	.14	1	1.33	
	2000	Mineral wool, 4″ thick, R12		3,500	.007		.18	.18	.06	.42	.56	
	2050	6″ thick, R17		2,500	.010		.20	.26	.08	.54	.73	
	2100	9″ thick, R23		1,750	.014		.29	.37	.12	.78	1.06	
	2500	Wall installation, incl. drilling & patching from outside, two 1″										
	2510	diam. holes @ 16″ O.C., top & mid-point of wall, add to above										
	2700	For masonry	G-4	415	.058	S.F.	.06	1.54	.49	2.09	3.17	
	2800	For wood siding		840	.029		.06	.76	.24	1.06	1.60	
	2900	For stucco/plaster		665	.036		.06	.96	.31	1.33	2.01	
	9000	Minimum labor/equipment charge		4	6	Job		160	51	211	320	
350	0010	**FLOOR INSULATION, NONRIGID** Including										350
	0020	spring type wire fasteners										
	2000	Fiberglass, blankets or batts, paper or foil backing										
	2100	1 side, 3-1/2″ thick, R11	1 Carp	700	.011	S.F.	.28	.38		.66	.94	
	2150	6″ thick, R19		600	.013		.36	.44		.80	1.13	
	2200	8-1/2″ thick, R30		550	.015		.65	.48		1.13	1.52	
	9000	Minimum labor/equipment charge		4	2	Job		66		66	109	
500	0010	**POURED INSULATION** Cellulose fiber, R3.8 per inch	1 Carp	200	.040	C.F.	.48	1.32		1.80	2.72	500
	0040	Ceramic type (perlite), R3.2 per inch		200	.040	″	1.65	1.32		2.97	4.01	
	0080	Fiberglass wool, R4 per inch		200	.040	C.F.	.37	1.32		1.69	2.60	
	0100	Mineral wool, R3 per inch		200	.040		.33	1.32		1.65	2.55	
	0300	Polystyrene, R4 per inch		200	.040		2.23	1.32		3.55	4.64	
	0400	Vermiculite or perlite, R2.7 per inch		200	.040		1.65	1.32		2.97	4.01	
	9000	Minimum labor/equipment charge		4	2	Job		66		66	109	
550	0010	**MASONRY INSULATION** Vermiculite or perlite, poured										550
	0100	In cores of concrete block, 4″ thick wall, .115 CF/SF	D-1	4,800	.003	S.F.	.19	.10		.29	.37	
	0200	6″ thick wall, .175 CF/SF		3,000	.005		.29	.16		.45	.58	
	0300	8″ thick wall, .258 CF/SF		2,400	.007		.43	.20		.63	.79	
	0400	10″ thick wall, .340 CF/SF		1,850	.009		.56	.26		.82	1.04	
	0500	12″ thick wall, .422 CF/SF		1,200	.013		.70	.40		1.10	1.42	
	0550	For sand fill, deduct from above					70%					
	0600	Poured cavity wall, vermiculite or perlite, water repellant	D-1	250	.064	C.F.	1.65	1.92		3.57	4.93	

07210 | Building Insulation

		07210 Building Insulation	CREW	DAILY OUTPUT	LABOR-HOURS	UNIT	2004 BARE COSTS MAT.	LABOR	EQUIP.	TOTAL	TOTAL INCL O&P	
550	0700	Foamed in place, urethane in 2-5/8" cavity	G-2	1,035	.023	S.F.	.40	.63	.11	1.14	1.59	550
	0800	For each 1" added thickness, add	"	2,372	.010	"	.12	.28	.05	.45	.63	
600	0010	**PERIMETER INSULATION**										600
	0600	Polystyrene, expanded, 1" thick, R4	1 Carp	680	.012	S.F.	.19	.39		.58	.85	
	0700	2" thick, R8		675	.012	"	.37	.39		.76	1.06	
	9000	Minimum labor/equipment charge		4	2	Job		66		66	109	
700	0012	**REFLECTIVE INSULATION**, aluminum foil on reinforced scrim	1 Carp	1,900	.004	S.F.	.14	.14		.28	.38	700
	0102	Reinforced with woven polyolefin		1,900	.004		.17	.14		.31	.42	
	0502	With single bubble air space, R8.8		1,500	.005		.27	.18		.45	.59	
	0602	With double bubble air space, R9.8		1,500	.005		.29	.18		.47	.61	
	9000	Minimum labor/equipment charge		4	2	Job		66		66	109	
800	0010	**SPRAYED** Fibrous/cementitious, finished wall, 1" thick, R3.7	G-2	2,050	.012	S.F.	.22	.32	.05	.59	.82	800
	0100	Attic, 5.2" thick, R19	"	1,550	.015	"	.36	.42	.07	.85	1.17	
	0300	Foam type, incl. preparation										
	0600	3 #/CF, 1" thick, R3.8	G-2	770	.031	S.F.	.49	.85	.14	1.48	2.09	
	0700	2" thick, R7.5	"	475	.051	"	.98	1.38	.23	2.59	3.59	
	1000	Coating, 50-70 mils, elastomeric aromatic urethane										
	1100	Finish coat, 5 mils, elastomeric aliphatic, 200 S.F./gal.				Gal.	40.50			40.50	44.50	
	9000	Minimum labor/equipment charge	G-2	2	12	Job		325	55.50	380.50	595	
900	0010	**WALL INSULATION, RIGID**										900
	0040	Fiberglass, 1.5#/CF, unfaced, 1" thick, R4.1	1 Carp	1,000	.008	S.F.	.29	.26		.55	.76	
	0060	1-1/2" thick, R6.2		1,000	.008		.41	.26		.67	.89	
	0080	2" thick, R8.3		1,000	.008		.46	.26		.72	.95	
	0120	3" thick, R12.4		800	.010		.57	.33		.90	1.18	
	0370	3#/CF, unfaced, 1" thick, R4.3		1,000	.008		.36	.26		.62	.84	
	0390	1-1/2" thick, R6.5		1,000	.008		.69	.26		.95	1.20	
	0400	2" thick, R8.7		890	.009		.83	.30		1.13	1.40	
	0420	2-1/2" thick, R10.9		800	.010		1.02	.33		1.35	1.67	
	0440	3" thick, R13		800	.010		1.21	.33		1.54	1.88	
	0520	Foil faced, 1" thick, R4.3		1,000	.008		.80	.26		1.06	1.32	
	0540	1-1/2" thick, R6.5		1,000	.008		1.08	.26		1.34	1.63	
	0560	2" thick, R8.7		890	.009		1.35	.30		1.65	1.98	
	0580	2-1/2" thick, R10.9		800	.010		1.60	.33		1.93	2.31	
	0600	3" thick, R13		800	.010		1.74	.33		2.07	2.46	
	0670	6#/CF, unfaced, 1" thick, R4.3		1,000	.008		.77	.26		1.03	1.29	
	0690	1-1/2" thick, R6.5		890	.009		1.19	.30		1.49	1.80	
	0700	2" thick, R8.7		800	.010		1.68	.33		2.01	2.40	
	0721	2-1/2" thick, R10.9		800	.010		1.84	.33		2.17	2.57	
	0741	3" thick, R13		730	.011		2.20	.36		2.56	3.02	
	0821	Foil faced, 1" thick, R4.3		1,000	.008		1.09	.26		1.35	1.64	
	0840	1-1/2" thick, R6.5		890	.009		1.57	.30		1.87	2.22	
	0850	2" thick, R8.7		800	.010		2.05	.33		2.38	2.81	
	0880	2-1/2" thick, R10.9		800	.010		2.46	.33		2.79	3.26	
	0900	3" thick, R13		730	.011		2.94	.36		3.30	3.83	
	1500	Foamglass, 1-1/2" thick, R4.5		800	.010		1.30	.33		1.63	1.98	
	1550	3" thick, R9		730	.011		2.93	.36		3.29	3.82	
	1600	Isocyanurate, 4' x 8' sheet, foil faced, both sides										
	1610	1/2" thick, R3.9	1 Carp	800	.010	S.F.	.29	.33		.62	.87	
	1620	5/8" thick, R4.5		800	.010		.30	.33		.63	.88	
	1630	3/4" thick, R5.4		800	.010		.24	.33		.57	.81	
	1640	1" thick, R7.2		800	.010		.33	.33		.66	.91	
	1650	1-1/2" thick, R10.8		730	.011		.35	.36		.71	.99	
	1660	2" thick, R14.4		730	.011		.46	.36		.82	1.11	
	1670	3" thick, R21.6		730	.011		1.11	.36		1.47	1.82	
	1680	4" thick, R28.8		730	.011		1.37	.36		1.73	2.11	

07210 | Building Insulation

		07210 Building Insulation	CREW	DAILY OUTPUT	LABOR-HOURS	UNIT	2004 BARE COSTS MAT.	LABOR	EQUIP.	TOTAL	TOTAL INCL O&P	
900	1700	Perlite, 1" thick, R2.77	1 Carp	800	.010	S.F.	.26	.33		.59	.84	900
	1750	2" thick, R5.55		730	.011		.50	.36		.86	1.15	
	1900	Extruded polystyrene, 25 PSI compressive strength, 1" thick, R5		800	.010		.36	.33		.69	.95	
	1940	2" thick R10		730	.011		.71	.36		1.07	1.38	
	1960	3" thick, R15		730	.011		.98	.36		1.34	1.68	
	2100	Expanded polystyrene, 1" thick, R3.85		800	.010		.16	.33		.49	.73	
	2120	2" thick, R7.69		730	.011		.40	.36		.76	1.04	
	2140	3" thick, R11.49		730	.011		.52	.36		.88	1.17	
	9000	Minimum labor/equipment charge		4	2	Job		66		66	109	
950	0010	**WALL OR CEILING INSUL., NON-RIGID**										950
	0040	Fiberglass, kraft faced, batts or blankets										
	0060	3-1/2" thick, R11, 11" wide	1 Carp	1,150	.007	S.F.	.23	.23		.46	.63	
	0080	15" wide		1,600	.005		.23	.17		.40	.52	
	0100	23" wide		1,600	.005		.23	.17		.40	.52	
	0140	6" thick, R19, 11" wide		1,000	.008		.31	.26		.57	.78	
	0160	15" wide		1,350	.006		.31	.20		.51	.66	
	0180	23" wide		1,600	.005		.31	.17		.48	.61	
	0200	9" thick, R30, 15" wide		1,150	.007		.60	.23		.83	1.04	
	0220	23" wide		1,350	.006		.60	.20		.80	.98	
	0240	12" thick, R38, 15" wide		1,000	.008		.76	.26		1.02	1.28	
	0260	23" wide		1,350	.006		.76	.20		.96	1.16	
	0400	Fiberglass, foil faced, batts or blankets										
	0420	3-1/2" thick, R11, 15" wide	1 Carp	1,600	.005	S.F.	.34	.17		.51	.64	
	0440	23" wide		1,600	.005		.34	.17		.51	.64	
	0460	6" thick, R19, 15" wide		1,350	.006		.41	.20		.61	.77	
	0480	23" wide		1,600	.005		.41	.17		.58	.72	
	0500	9" thick, R30, 15" wide		1,150	.007		.71	.23		.94	1.16	
	0550	23" wide		1,350	.006		.71	.20		.91	1.10	
	0800	Fiberglass, unfaced, batts or blankets										
	0820	3-1/2" thick, R11, 15" wide	1 Carp	1,350	.006	S.F.	.21	.20		.41	.55	
	0830	23" wide		1,600	.005		.21	.17		.38	.50	
	0860	6" thick, R19, 15" wide		1,150	.007		.36	.23		.59	.78	
	0880	23" wide		1,350	.006		.34	.20		.54	.69	
	0900	9" thick, R30, 15" wide		1,000	.008		.60	.26		.86	1.10	
	0920	23" wide		1,150	.007		.60	.23		.83	1.04	
	0940	12" thick, R38, 15" wide		1,000	.008		.76	.26		1.02	1.28	
	0960	23" wide		1,150	.007		.76	.23		.99	1.22	
	1300	Mineral fiber batts, kraft faced										
	1320	3-1/2" thick, R12	1 Carp	1,600	.005	S.F.	.26	.17		.43	.56	
	1340	6" thick, R19		1,600	.005		.34	.17		.51	.64	
	1380	10" thick, R30		1,350	.006		.52	.20		.72	.89	
	1850	Friction fit wire insulation supports, 16" O.C.		960	.008	Ea.	.05	.27		.32	.52	
	1900	For foil backing, add				S.F.	.04			.04	.04	
	9000	Minimum labor/equipment charge	1 Carp	4	2	Job		66		66	109	

07220 | Roof and Deck Insulation

		07220 Roof and Deck Insulation	CREW	DAILY OUTPUT	LABOR-HOURS	UNIT	MAT.	LABOR	EQUIP.	TOTAL	TOTAL INCL O&P	
700	0010	**ROOF DECK INSULATION**										700
	0020	Fiberboard low density, 1/2" thick R1.39	1 Rofc	1,000	.008	S.F.	.19	.23		.42	.62	
	0030	1" thick R2.78		800	.010		.34	.28		.62	.88	
	0080	1 1/2" thick R4.17		800	.010		.50	.28		.78	1.06	
	0100	2" thick R5.56		800	.010		.68	.28		.96	1.26	
	0110	Fiberboard high density, 1/2" thick R1.3		1,000	.008		.20	.23		.43	.63	
	0120	1" thick R2.5		800	.010		.36	.28		.64	.91	
	0130	1-1/2" thick R3.8		800	.010		.59	.28		.87	1.16	

07220 Roof and Deck Insulation		CREW	DAILY OUTPUT	LABOR-HOURS	UNIT	2004 BARE COSTS MAT.	LABOR	EQUIP.	TOTAL	TOTAL INCL O&P
700	0200 Fiberglass, 3/4" thick R2.78	1 Rofc	1,000	.008	S.F.	.46	.23		.69	.92
	0400 15/16" thick R3.70		1,000	.008		.61	.23		.84	1.08
	0460 1-1/16" thick R4.17		1,000	.008		.76	.23		.99	1.25
	0600 1-5/16" thick R5.26		1,000	.008		1.05	.23		1.28	1.57
	0650 2-1/16" thick R8.33		800	.010		1.12	.28		1.40	1.74
	0700 2-7/16" thick R10		800	.010		1.28	.28		1.56	1.92
	1500 Foamglass, 1-1/2" thick R4.5		800	.010		1.26	.28		1.54	1.90
	1530 3" thick R9		700	.011	↓	2.57	.33		2.90	3.41
	1600 Tapered for drainage		600	.013	B.F.	1.10	.38		1.48	1.89
	1650 Perlite, 1/2" thick R1.32		1,050	.008	S.F.	.27	.22		.49	.69
	1655 3/4" thick R2.08		800	.010		.29	.28		.57	.83
	1660 1" thick R2.78		800	.010		.30	.28		.58	.84
	1670 1-1/2" thick R4.17		800	.010		.39	.28		.67	.94
	1680 2" thick R5.56		700	.011		.61	.33		.94	1.25
	1685 2-1/2" thick R6.67		700	.011	↓	.77	.33		1.10	1.43
	1690 Tapered for drainage		800	.010	B.F.	.59	.28		.87	1.16
	1700 Polyisocyanurate, 2#/CF density, 3/4" thick, R5.1		1,500	.005	S.F.	.28	.15		.43	.58
	1705 1" thick R7.14		1,400	.006		.34	.16		.50	.66
	1715 1-1/2" thick R10.87		1,250	.006		.37	.18		.55	.74
	1725 2" thick R14.29		1,100	.007		.48	.21		.69	.90
	1735 2-1/2" thick R16.67		1,050	.008		.50	.22		.72	.94
	1745 3" thick R21.74		1,000	.008		.72	.23		.95	1.20
	1755 3-1/2" thick R25		1,000	.008	↓	.74	.23		.97	1.22
	1765 Tapered for drainage	↓	1,400	.006	B.F.	.38	.16		.54	.71
	1900 Extruded Polystyrene									
	1910 15 PSI compressive strength, 1" thick, R5	1 Rofc	1,500	.005	S.F.	.23	.15		.38	.52
	1920 2" thick, R10		1,250	.006		.35	.18		.53	.72
	1930 3" thick R15		1,000	.008		.68	.23		.91	1.16
	1932 4" thick R20		1,000	.008	↓	1.07	.23		1.30	1.59
	1934 Tapered for drainage		1,500	.005	B.F.	.35	.15		.50	.66
	1940 25 PSI compressive strength, 1" thick R5		1,500	.005	S.F.	.42	.15		.57	.73
	1942 2" thick R10		1,250	.006		.82	.18		1	1.23
	1944 3" thick R15		1,000	.008		1.24	.23		1.47	1.77
	1946 4" thick R20		1,000	.008	↓	1.19	.23		1.42	1.72
	1948 Tapered for drainage		1,500	.005	B.F.	.40	.15		.55	.71
	1950 40 psi compressive strength, 1" thick R5		1,500	.005	S.F.	.35	.15		.50	.66
	1952 2" thick R10		1,250	.006		.69	.18		.87	1.09
	1954 3" thick R15		1,000	.008		1.01	.23		1.24	1.52
	1956 4" thick R20		1,000	.008	↓	1.35	.23		1.58	1.90
	1958 Tapered for drainage		1,400	.006	B.F.	.50	.16		.66	.84
	1960 60 PSI compressive strength, 1" thick R5		1,450	.006	S.F.	.42	.16		.58	.74
	1962 2" thick R10		1,200	.007		.75	.19		.94	1.17
	1964 3" thick R15		975	.008		1.12	.23		1.35	1.65
	1966 4" thick R20		950	.008	↓	1.56	.24		1.80	2.15
	1968 Tapered for drainage		1,400	.006	B.F.	.61	.16		.77	.96
	2010 Expanded polystyrene, 1#/CF density, 3/4" thick R2.89		1,500	.005	S.F.	.19	.15		.34	.48
	2020 1" thick R3.85		1,500	.005		.19	.15		.34	.48
	2100 2" thick R7.69		1,250	.006		.37	.18		.55	.74
	2110 3" thick R11.49		1,250	.006		.59	.18		.77	.98
	2120 4" thick R15.38		1,200	.007		.57	.19		.76	.97
	2130 5" thick R19.23		1,150	.007		.71	.20		.91	1.13
	2140 6" thick R23.26		1,150	.007	↓	.83	.20		1.03	1.26
	2150 Tapered for drainage	↓	1,500	.005	B.F.	.34	.15		.49	.64
	2400 Composites with 2" EPS									
	2410 1" fiberboard	1 Rofc	950	.008	S.F.	.78	.24		1.02	1.29
	2420 7/16" oriented strand board	↓	800	.010	↓	.93	.28		1.21	1.53

	07220 | Roof and Deck Insulation		CREW	DAILY OUTPUT	LABOR-HOURS	UNIT	2004 BARE COSTS MAT.	LABOR	EQUIP.	TOTAL	TOTAL INCL O&P	
700	2430	1/2" plywood	1 Rofc	800	.010	S.F.	1	.28		1.28	1.61	700
	2440	1" perlite		800	.010		.82	.28		1.10	1.41	
	2450	Composites with 1 1/2" polyisocyanurate										
	2460	1" fiberboard	1 Rofc	800	.010	S.F.	.84	.28		1.12	1.43	
	2470	1" perlite		850	.009		.88	.27		1.15	1.45	
	2480	7/16" oriented strand board		800	.010		1.01	.28		1.29	1.62	
	9000	Minimum labor/equipment charge		3.25	2.462	Job		70		70	125	

	07240 | Ext. Insulation Finish Systems (EIFS)		CREW	DAILY OUTPUT	LABOR-HOURS	UNIT	MAT.	LABOR	EQUIP.	TOTAL	TOTAL INCL O&P	
100	0010	**EXTERIOR INSULATION FINISH SYSTEM**										100
	0095	Field applied, 1" EPS insulation	J-1	295	.136	S.F.	1.83	3.84	.31	5.98	8.55	
	0100	With 1/2" cement board sheathing		220	.182		2.59	5.15	.41	8.15	11.65	
	0105	2" EPS insulation		295	.136		2.07	3.84	.31	6.22	8.80	
	0110	With 1/2" cement board sheathing		220	.182		2.83	5.15	.41	8.39	11.90	
	0115	3" EPS insulation		295	.136		2.19	3.84	.31	6.34	8.95	
	0120	With 1/2" cement board sheathing		220	.182		2.95	5.15	.41	8.51	12.05	
	0125	4" EPS insulation		295	.136		2.47	3.84	.31	6.62	9.25	
	0130	With 1/2" cement board sheathing		220	.182		3.98	5.15	.41	9.54	13.20	
	0140	Premium finish add		1,265	.032		.28	.89	.07	1.24	1.84	
	0150	Heavy duty reinforcement add		914	.044		1.68	1.24	.10	3.02	3.96	
	0160	2.5#/S.Y. metal lath substrate add	1 Lath	75	.107	S.Y.	2.16	3.26		5.42	7.55	
	0170	3.4#/S.Y. metal lath substrate add	"	75	.107	"	2.25	3.26		5.51	7.65	
	0180	Color or texture change,	J-1	1,265	.032	S.F.	.73	.89	.07	1.69	2.33	
	0190	With substrate leveling base coat	1 Plas	530	.015		.73	.45		1.18	1.53	
	0210	With substrate sealing base coat	1 Pord	1,224	.007		.07	.19		.26	.39	
	0220	Prefab. panels, with hat channels 2 1/2" x 3/4", 2" EPS insul.	L-5	1,800	.031		7.70	1.15	.35	9.20	10.95	
	0240	3" EPS insulation		1,800	.031		8.55	1.15	.35	10.05	11.90	
	0250	4" EPS insulation		1,800	.031		9.55	1.15	.35	11.05	13	
	0260	1-1/2" profile lightweight metal backing, 1" EPS insulation		1,800	.031		11.75	1.15	.35	13.25	15.45	
	0270	2" EPS insulation		1,800	.031		13.10	1.15	.35	14.60	16.95	
	0280	3" EPS insulation		1,800	.031		14.65	1.15	.35	16.15	18.60	
	0290	4" EPS insulation		1,800	.031		16.35	1.15	.35	17.85	20.50	
	0310	6"-16 Ga. steel stud back-up, 1" EPS insulation		1,800	.031		13.55	1.15	.35	15.05	17.40	
	0320	2" EPS insulation		1,800	.031		15.15	1.15	.35	16.65	19.15	
	0330	3" EPS insulation		1,800	.031		16.85	1.15	.35	18.35	21	
	0340	4" EPS insulation		1,800	.031		18.85	1.15	.35	20.35	23	
	0350	Premium finish add	J-1	1,265	.032		.27	.89	.07	1.23	1.83	
	0360	Heavy duty reinforcement add	"	914	.044		1.65	1.24	.10	2.99	3.93	
	0370	V groove shape in panel face				L.F.	.52			.52	.57	
	0380	U groove shape in panel face				"	.69			.69	.76	
	0433	Crack repair, acrylic rubber, fluid applied, 20 mils thick	1 Plas	350	.023	S.F.	1.32	.68		2	2.56	
	0437	50 mils thick, reinforced	"	200	.040	"	2.46	1.20		3.66	4.65	
	0440	For higher than one story, add						25%				

	07260 | Vapor Retarders		CREW	DAILY OUTPUT	LABOR-HOURS	UNIT	MAT.	LABOR	EQUIP.	TOTAL	TOTAL INCL O&P	
100	0010	**BUILDING PAPER** Aluminum and kraft laminated, foil 1 side	1 Carp	37	.216	Sq.	3.64	7.15		10.79	15.85	100
	0100	Foil 2 sides		37	.216		6.40	7.15		13.55	18.85	
	0300	Asphalt, two ply, 30#, for subfloors		19	.421		14.25	13.90		28.15	38.50	
	0400	Asphalt felt sheathing paper, 15#		37	.216		2.68	7.15		9.83	14.80	
	0450	Housewrap, exterior, spun bonded polypropylene										
	0470	Small roll	1 Carp	3,800	.002	S.F.	.16	.07		.23	.30	
	0480	Large roll	"	4,000	.002	"	.10	.07		.17	.22	
	0500	Material only, 3' x 111.1' roll				Ea.	55			55	60.50	
	0520	9' x 111.1' roll				"	100			100	110	
	0600	Polyethylene vapor barrier, standard, .002" thick	1 Carp	37	.216	Sq.	.96	7.15		8.11	12.90	

Important: See the Reference Section for critical supporting data - Reference Nos., Crews, & City Cost Indexes

07200 | Thermal Protection

	07260 Vapor Retarders		CREW	DAILY OUTPUT	LABOR-HOURS	UNIT	2004 BARE COSTS MAT.	LABOR	EQUIP.	TOTAL	TOTAL INCL O&P	
100	0700	.004" thick	1 Carp	37	.216	Sq.	1.87	7.15		9.02	13.90	100
	0900	.006" thick		37	.216		2.75	7.15		9.90	14.90	
	1200	.010" thick		37	.216		5.10	7.15		12.25	17.45	
	1300	Clear reinforced, fire retardant, .008" thick		37	.216		8.55	7.15		15.70	21.50	
	1350	Cross laminated type, .003" thick		37	.216		6.65	7.15		13.80	19.15	
	1400	.004" thick		37	.216		7.40	7.15		14.55	20	
	1500	Red rosin paper, 5 sq rolls, 4 lb per square		37	.216		1.58	7.15		8.73	13.60	
	1600	5 lbs. per square		37	.216		2.04	7.15		9.19	14.10	
	1800	Reinf. waterproof, .002" polyethylene backing, 1 side		37	.216		5	7.15		12.15	17.35	
	1900	2 sides	↓	37	.216		6.65	7.15		13.80	19.15	
	2100	Roof deck vapor barrier, class 1 metal decks	1 Rofc	37	.216		10.30	6.15		16.45	22.50	
	2200	For all other decks	"	37	.216		7.60	6.15		13.75	19.35	
	2400	Waterproofed kraft with sisal or fiberglass fibers, minimum	1 Carp	37	.216		5.40	7.15		12.55	17.80	
	2500	Maximum		37	.216	↓	13.50	7.15		20.65	26.50	
	9950	Minimum labor/equipment charge	↓	4	2	Job		66		66	109	

07300 | Shingles, Roof Tiles and Roof Coverings

	07310 Shingles		CREW	DAILY OUTPUT	LABOR-HOURS	UNIT	2004 BARE COSTS MAT.	LABOR	EQUIP.	TOTAL	TOTAL INCL O&P	
050	0010	**ALUMINUM** Shingles, mill finish, .019" thick	1 Carp	5	1.600	Sq.	145	53		198	248	050
	0100	.020" thick	"	5	1.600		149	53		202	251	
	0300	For colors, add				↓	15.15			15.15	16.65	
	0600	Ridge cap, .024" thick	1 Carp	170	.047	L.F.	1.72	1.55		3.27	4.46	
	0700	End wall flashing, .024" thick		170	.047		1.29	1.55		2.84	3.99	
	0900	Valley section, .024" thick		170	.047		2.18	1.55		3.73	4.97	
	1000	Starter strip, .024" thick		400	.020		1.15	.66		1.81	2.36	
	1200	Side wall flashing, .024" thick		170	.047	↓	1.29	1.55		2.84	3.99	
	9000	Minimum labor/equipment charge	↓	3	2.667	Job		88		88	146	
100	0010	**ASPHALT SHINGLES**										100
	0100	Standard strip shingles										
	0150	Inorganic, class A, 210-235 lb/sq	1 Rofc	5.50	1.455	Sq.	31	41.50		72.50	108	
	0155	Pneumatic nailed		7	1.143		31	32.50		63.50	92	
	0200	Organic, class C, 235-240 lb/sq		5	1.600		38	45.50		83.50	124	
	0205	Pneumatic nailed	↓	6.25	1.280	↓	38	36.50		74.50	107	
	0250	Standard, laminated multi-layered shingles										
	0300	Class A, 240-260 lb/sq	1 Rofc	4.50	1.778	Sq.	42.50	50.50		93	137	
	0305	Pneumatic nailed		5.63	1.422		42.50	40.50		83	119	
	0350	Class C, 260-300 lb/square, 4 bundles/square		4	2		49	57		106	156	
	0355	Pneumatic nailed	↓	5	1.600	↓	49	45.50		94.50	136	
	0400	Premium, laminated multi-layered shingles										
	0450	Class A, 260-300 lb, 4 bundles/sq	1 Rofc	3.50	2.286	Sq.	50.50	65		115.50	172	
	0455	Pneumatic nailed		4.37	1.831		50.50	52		102.50	150	
	0500	Class C, 300-385 lb/square, 5 bundles/square		3	2.667		65.50	76		141.50	208	
	0505	Pneumatic nailed		3.75	2.133		65.50	60.50		126	181	
	0800	#15 felt underlayment		64	.125		2.68	3.56		6.24	9.30	
	0825	#30 felt underlayment		58	.138		7.10	3.92		11.02	14.90	
	0850	Self adhering polyethylene and rubberized asphalt underlayment		22	.364	↓	38	10.35		48.35	60.50	
	0900	Ridge shingles	↓	330	.024	L.F.	.72	.69		1.41	2.03	

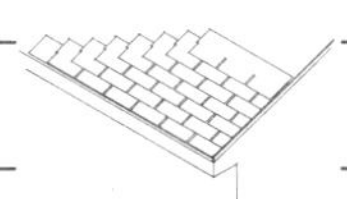

7 THERMAL & MOISTURE PROTECTION

	07310 | Shingles		CREW	DAILY OUTPUT	LABOR-HOURS	UNIT	2004 BARE COSTS MAT.	LABOR	EQUIP.	TOTAL	TOTAL INCL O&P	
100	0905	Pneumatic nailed	1 Rofc	412.50	.019	L.F.	.72	.55		1.27	1.78	100
	1000	For steep roofs (7 to 12 pitch or greater), add						50%				
	9000	Minimum labor/equipment charge	1 Rofc	3	2.667	Job		76		76	136	
500	0010	**FIBER CEMENT** shingles, 16″ x 9.35″, 500 lb per square	1 Rofc	2.20	3.636	Sq.	244	103		347	455	500
	0200	Shakes, 16″ x 9.35″, 550 lb per square		2.20	3.636	″	221	103		324	430	
	0300	Hip & ridge, 4.75 x 14″		1	8	C.L.F.	600	228		828	1,075	
	0400	Hexagonal, 16″ x 16″		3	2.667	Sq.	165	76		241	320	
	0500	Square, 16″ x 16″		3	2.667		148	76		224	299	
	2000	For steep roofs (7/12 pitch or greater), add						50%				
	9000	Minimum labor/equipment charge	1 Rofc	3	2.667	Job		76		76	136	
800	0010	**SLATE**, Buckingham, Virginia, black										800
	0100	3/16″ - 1/4″ thick	1 Rots	1.75	4.571	Sq.	500	131		631	785	
	0200	1/4″ thick		1.75	4.571		570	131		701	860	
	0900	Pennsylvania black, Bangor, #1 clear		1.75	4.571		450	131		581	730	
	1200	Vermont, unfading, green, mottled green		1.75	4.571		345	131		476	610	
	1300	Semi-weathering green & gray		1.75	4.571		330	131		461	600	
	1400	Purple		1.75	4.571		370	131		501	640	
	1500	Black or gray		1.75	4.571		325	131		456	595	
	2500	Slate roof repair, extensive replacement		1	8		540	229		769	1,000	
	2600	Repair individual pieces, scattered		19	.421	Ea.	5.40	12.05		17.45	27.50	
	9000	Minimum labor/equipment charge		3	2.667	Job		76.50		76.50	137	
900	0010	**STEEL** Shingles, galvanized, 26 gauge	1 Rots	2.20	3.636	Sq.	152	104		256	355	900
	0200	24 gauge	″	2.20	3.636	″	160	104		264	365	
	0300	For colored galvanized shingles, add				Sq.	41.50			41.50	46	
	0500	For 1″ factory applied polystyrene insulation, add				″	30.50			30.50	33.50	
	9000	Minimum labor/equipment charge	1 Rots	3	2.667	Job		76.50		76.50	137	
980	0010	**WOOD** 16″ No. 1 red cedar shingles, 5″ exposure, on roof R06110-030	1 Carp	2.50	3.200	Sq.	163	106		269	355	980
	0015	Pneumatic nailed		3.25	2.462		163	81		244	315	
	0200	7-1/2″ exposure, on walls		2.05	3.902		108	129		237	330	
	0205	Pneumatic nailed		2.67	2.996		108	99		207	283	
	0300	18″ No. 1 red cedar perfections, 5-1/2″ exposure, on roof		2.75	2.909		160	96		256	335	
	0305	Pneumatic nailed		3.57	2.241		160	74		234	299	
	0600	Resquared, and rebutted, 5-1/2″ exposure, on roof		3	2.667		199	88		287	365	
	0605	Pneumatic nailed		3.90	2.051		199	67.50		266.50	330	
	0900	7-1/2″ exposure, on walls		2.45	3.265		146	108		254	340	
	0905	Pneumatic nailed		3.18	2.516		146	83		229	299	
	1000	Add to above for fire retardant shingles, 16″ long					30			30	33	
	1050	18″ long					28.50			28.50	31.50	
	1060	Preformed ridge shingles	1 Carp	400	.020	L.F.	1.65	.66		2.31	2.91	
	1100	Hand-split red cedar shakes, 1/2″ thick x 24″ long, 10″ exp. on roof		2.50	3.200	Sq.	138	106		244	325	
	1105	Pneumatic nailed		3.25	2.462		138	81		219	287	
	1110	3/4″ thick x 24″ long, 10″ exp. on roof		2.25	3.556		138	117		255	345	
	1115	Pneumatic nailed		2.92	2.740		138	90.50		228.50	300	
	1200	1/2″ thick, 18″ long, 8-1/2″ exp. on roof		2	4		97.50	132		229.50	325	
	1205	Pneumatic nailed		2.60	3.077		97.50	102		199.50	275	
	1210	3/4″ thick x 18″ long, 8 1/2″ exp. on roof		1.80	4.444		97.50	147		244.50	350	
	1215	Pneumatic nailed		2.34	3.419		97.50	113		210.50	294	
	1255	10″ exp. on walls		2	4		110	132		242	340	
	1260	10″ exposure on walls, pneumatic nailed		2.60	3.077		110	102		212	289	
	1700	Add to above for fire retardant shakes, 24″ long					30			30	33	
	1800	18″ long					30			30	33	
	1810	Ridge shakes	1 Carp	350	.023	L.F.	2.35	.75		3.10	3.84	

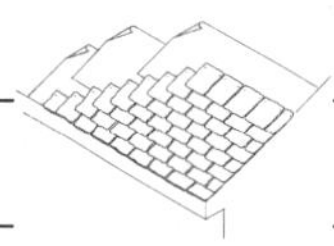

07300 | Shingles, Roof Tiles and Roof Coverings

	07310 Shingles		CREW	DAILY OUTPUT	LABOR-HOURS	UNIT	2004 BARE COSTS MAT.	LABOR	EQUIP.	TOTAL	TOTAL INCL O&P	
980	2000	White cedar shingles, 16" long, extras, 5" exposure, on roof (R06110-030)	1 Carp	2.40	3.333	Sq.	118	110		228	310	980
	2005	Pneumatic nailed		3.12	2.564		118	84.50		202.50	270	
	2050	5" exposure on walls		2	4		118	132		250	350	
	2055	Pneumatic nailed		2.60	3.077		118	102		220	298	
	2100	7-1/2" exposure, on walls		2	4		84.50	132		216.50	310	
	2105	Pneumatic nailed		2.60	3.077		84.50	102		186.50	261	
	2150	"B" grade, 5" exposure on walls		2	4		110	132		242	340	
	2155	Pneumatic nailed		2.60	3.077		110	102		212	289	
	2300	For 15# organic felt underlayment on roof, 1 layer, add		64	.125		2.68	4.13		6.81	9.80	
	2400	2 layers, add		32	.250		5.35	8.25		13.60	19.60	
	2600	For steep roofs (7/12 pitch or greater), add to above						50%				
	2700	Panelized systems, No.1 cedar shingles on 5/16" CDX plywood										
	2800	On walls, 8' strips, 7" or 14" exposure	2 Carp	700	.023	S.F.	3.20	.75		3.95	4.77	
	3500	On roofs, 8' strips, 7" or 14" exposure	1 Carp	3	2.667	Sq.	320	88		408	495	
	3505	Pneumatic nailed		4	2	"	320	66		386	460	
	9000	Minimum labor/equipment charge		3	2.667	Job		88		88	146	
	07320 Roof Tiles											
100	0010	**ALUMINUM** Tiles with accessories, .032" thick, mission tile	1 Carp	2.50	3.200	Sq.	440	106		546	660	100
	0200	Spanish tiles		3	2.667	"	380	88		468	560	
	9000	Minimum labor/equipment charge		3	2.667	Job		88		88	146	
200	0010	**CLAY TILE** ASTM C1167, GR 1, severe weathering, acces. incl.										200
	0200	Lanai tile or Classic tile, 158 pc per sq	1 Rots	1.65	4.848	Sq.	395	139		534	685	
	0300	Americana, 158 pc per sq, most colors		1.65	4.848		485	139		624	780	
	0350	Green, gray or brown		1.65	4.848		490	139		629	790	
	0400	Blue		1.65	4.848		490	139		629	790	
	0600	Spanish tile, 171 pc per sq, red		1.80	4.444		263	127		390	520	
	0800	Blend		1.80	4.444		420	127		547	690	
	0900	Glazed white		1.80	4.444		500	127		627	780	
	1100	Mission tile, 192 pc per sq, machine scored finish, red		1.15	6.957		635	199		834	1,050	
	1700	French tile, 133 pc per sq, smooth finish, red		1.35	5.926		575	170		745	935	
	1750	Blue or green		1.35	5.926		685	170		855	1,050	
	1800	Norman black 317 pc per sq		1	8		755	229		984	1,250	
	2200	Williamsburg tile, 158 pc per sq, aged cedar		1.35	5.926		490	170		660	845	
	2250	Gray or green		1.35	5.926		490	170		660	845	
	3000	For steep roofs (7/12 pitch or greater), add to above						50%				
	9000	Minimum labor/equipment charge	1 Rots	3	2.667	Job		76.50		76.50	137	
300	0010	**CONCRETE TILE** Including installation of accessories										300
	0050	Earthtone colors, nailed to wood deck	1 Rots	1.35	5.926	Sq.	90	170		260	405	
	0150	Blues		1.35	5.926		102	170		272	415	
	0200	Greens		1.35	5.926		102	170		272	415	
	0250	Premium colors		1.35	5.926		153	170		323	475	
	0500	Shakes, 13" x 16-1/2", 90 per sq, 950 lb per sq										
	0600	All colors, nailed to wood deck	1 Rots	1.50	5.333	Sq.	185	153		338	480	
	1500	Accessory pieces, ridge & hip, 10" x 16-1/2", 8 lbs. each				Ea.	2.25			2.25	2.48	
	1700	Rake, 6-1/2" x 16-3/4", 9 lbs. each					2.25			2.25	2.48	
	1800	Mansard hip, 10" x 16-1/2", 9.2 lbs. each					2.25			2.25	2.48	
	1900	Hip starter, 10" x 16-1/2", 10.5 lbs. each					9.50			9.50	10.45	
	2000	3 or 4 way apex, 10" each side, 11.5 lbs. each					10.25			10.25	11.30	
	9000	Minimum labor/equipment charge	1 Rots	3	2.667	Job		76.50		76.50	137	

07410 | Metal Roof and Wall Panels

		07410 Metal Roof and Wall Panels	CREW	DAILY OUTPUT	LABOR-HOURS	UNIT	2004 BARE COSTS MAT.	LABOR	EQUIP.	TOTAL	TOTAL INCL O&P	
100	0010	**ALUMINUM ROOFING** Corrugated or ribbed, .0155" thick, natural	G-3	1,200	.027	S.F.	.62	.86		1.48	2.07	100
	0300	Painted		1,200	.027		.89	.86		1.75	2.37	
	0400	Corrugated, .018" thick, on steel frame, natural finish		1,200	.027		.80	.86		1.66	2.27	
	0600	Painted		1,200	.027		.99	.86		1.85	2.48	
	0700	Corrugated, on steel frame, natural, .024" thick		1,200	.027		1.16	.86		2.02	2.67	
	0800	Painted, .024" thick		1,200	.027		1.40	.86		2.26	2.93	
	0900	.032" thick, natural		1,200	.027		1.35	.86		2.21	2.88	
	1200	painted		1,200	.027		1.73	.86		2.59	3.29	
	9000	Minimum labor/equipment charge	1 Rofc	3	2.667	Job		76		76	136	
500	0010	**MANSARD** Colored aluminum, with battens, .032" thick										500
	0600	Stock units, straight surfaces	1 Shee	115	.070	S.F.	2.13	2.70		4.83	6.60	
	0700	Concave or convex surfaces		75	.107	"	2.35	4.14		6.49	9.15	
	0800	For framing, to 5' high, add		115	.070	L.F.	2.35	2.70		5.05	6.85	
	0900	Soffits, to 1' wide		125	.064	S.F.	1.16	2.48		3.64	5.20	
	9000	Minimum labor/equipment charge		2.50	3.200	Job		124		124	197	
700	0010	**STEEL ROOFING** on steel frame, corrugated or ribbed, 30 ga galv	G-3	1,100	.029	S.F.	.77	.94		1.71	2.37	700
	0100	28 ga		1,050	.030	"	.81	.99		1.80	2.48	
	0300	26 ga		1,000	.032	S.F.	.97	1.04		2.01	2.74	
	0400	24 ga		950	.034		1.05	1.09		2.14	2.92	
	0600	Colored, 28 ga		1,050	.030		1.08	.99		2.07	2.78	
	0700	26 ga		1,000	.032		1.15	1.04		2.19	2.94	
	0710	Flat profile, 1-3/4" standing seams, 10" wide, standard finish, 26 ga		1,000	.032		2.37	1.04		3.41	4.28	
	0715	24 ga		950	.034		2.75	1.09		3.84	4.79	
	0720	22 ga		900	.036		3.39	1.15		4.54	5.60	
	0725	Zinc aluminum alloy finish, 26 ga		1,000	.032		1.85	1.04		2.89	3.71	
	0730	24 ga		950	.034		2.21	1.09		3.30	4.19	
	0735	22 ga		900	.036		2.54	1.15		3.69	4.65	
	0740	12" wide, standard finish, 26 ga		1,000	.032		2.36	1.04		3.40	4.27	
	0745	24 ga		950	.034		2.74	1.09		3.83	4.77	
	0750	Zinc aluminum alloy finish, 26 ga		1,000	.032		1.75	1.04		2.79	3.60	
	0755	24 ga		950	.034		2.09	1.09		3.18	4.06	
	0840	Flat profile, 1" x 3/8" batten, 12" wide, standard finish, 26 ga		1,000	.032		1.84	1.04		2.88	3.69	
	0845	24 ga		950	.034		2.16	1.09		3.25	4.14	
	0850	22 ga		900	.036		2.60	1.15		3.75	4.72	
	0855	Zinc aluminum alloy finish, 26 ga		1,000	.032		1.56	1.04		2.60	3.39	
	0860	24 ga		950	.034		1.73	1.09		2.82	3.66	
	0865	22 ga		900	.036		2.01	1.15		3.16	4.07	
	0870	16-1/2" wide, standard finish, 24 ga		950	.034		2.19	1.09		3.28	4.17	
	0875	22 ga		900	.036		2.46	1.15		3.61	4.57	
	0880	Zinc aluminum alloy finish, 24 ga		950	.034		1.65	1.09		2.74	3.58	
	0885	22 ga		900	.036		1.83	1.15		2.98	3.87	
	0890	Flat profile, 2" x 2" batten, 12" wide, standard finish, 26 ga		1,000	.032		2.11	1.04		3.15	3.99	
	0895	24 ga		950	.034		2.52	1.09		3.61	4.53	
	0900	22 ga		900	.036		3.09	1.15		4.24	5.25	
	0905	Zinc aluminum alloy finish, 26 ga		1,000	.032		1.74	1.04		2.78	3.58	
	0910	24 ga		950	.034		1.88	1.09		2.97	3.83	
	0915	22 ga		900	.036		2.32	1.15		3.47	4.41	
	0920	16-1/2" wide, standard finish, 24 ga		950	.034		2.34	1.09		3.43	4.33	
	0925	22 ga		900	.036		2.72	1.15		3.87	4.85	
	0930	Zinc aluminum alloy finish, 24 ga		950	.034		1.85	1.09		2.94	3.80	
	0935	22 ga		900	.036		2.11	1.15		3.26	4.18	
	9000	Minimum labor/equipment charge	1 Rofc	2	4	Job		114		114	204	

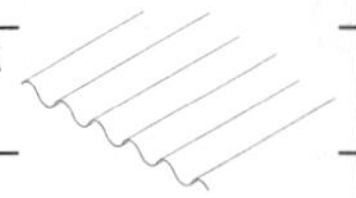

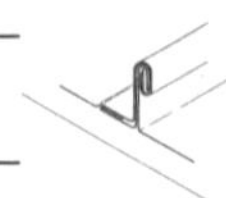

07400 | Roofing and Siding Panels

07420 | Plastic Roof and Wall Panels

		07420 Plastic Roof and Wall Panels	CREW	DAILY OUTPUT	LABOR-HOURS	UNIT	2004 BARE COSTS MAT.	LABOR	EQUIP.	TOTAL	TOTAL INCL O&P	
770	0010	**FIBERGLASS** Corrugated panels, roofing, 8 oz per SF	G-3	1,000	.032	S.F.	2.24	1.04		3.28	4.13	770
	0100	12 oz per SF		1,000	.032		3.23	1.04		4.27	5.20	
	0300	Corrugated siding, 6 oz per SF		880	.036		1.94	1.18		3.12	4.03	
	0400	8 oz per SF		880	.036		2.24	1.18		3.42	4.36	
	0500	Fire retardant		880	.036		3.09	1.18		4.27	5.30	
	0600	12 oz. siding, textured		880	.036		3.14	1.18		4.32	5.35	
	0700	Fire retardant		880	.036		4.14	1.18		5.32	6.45	
	0900	Flat panels, 6 oz per SF, clear or colors		880	.036		1.73	1.18		2.91	3.80	
	1100	Fire retardant, class A		880	.036		3	1.18		4.18	5.20	
	1300	8 oz per SF, clear or colors		880	.036		2.24	1.18		3.42	4.36	
	9000	Minimum labor/equipment charge	1 Rofc	2	4	Job		114		114	204	

07460 | Siding

		07460 Siding	CREW	DAILY OUTPUT	LABOR-HOURS	UNIT	MAT.	LABOR	EQUIP.	TOTAL	TOTAL INCL O&P	
100	0010	**ALUMINUM SIDING** .019" thick, on steel construction, natural	G-3	775	.041	S.F.	.66	1.34		2	2.89	100
	0100	Painted		775	.041		.86	1.34		2.20	3.11	
	0400	Farm type, .021" thick on steel frame, natural		775	.041		.77	1.34		2.11	3.01	
	0600	Painted		775	.041		.94	1.34		2.28	3.19	
	0700	Industrial type, corrugated, on steel, .024" thick, mill		775	.041		.97	1.34		2.31	3.23	
	0900	Painted		775	.041		1.17	1.34		2.51	3.45	
	1000	.032" thick, mill		775	.041		1.37	1.34		2.71	3.67	
	1200	Painted		775	.041		1.73	1.34		3.07	4.06	
	1300	V-Beam, on steel frame, .032" thick, mill		775	.041		1.59	1.34		2.93	3.91	
	1500	Painted		775	.041		1.89	1.34		3.23	4.24	
	1600	.040" thick, mill		775	.041		1.93	1.34		3.27	4.28	
	1800	Painted		775	.041		2.28	1.34		3.62	4.67	
	3800	Horizontal, colored clapboard, 8" wide, plain	2 Carp	515	.031		1.23	1.03		2.26	3.05	
	3810	Insulated		515	.031		1.39	1.03		2.42	3.23	
	3830	8" embossed, painted		515	.031		1.43	1.03		2.46	3.27	
	3840	Insulated		515	.031		1.57	1.03		2.60	3.43	
	3860	12" painted, smooth		600	.027		1.29	.88		2.17	2.88	
	3870	Insulated		600	.027		1.52	.88		2.40	3.13	
	3890	12" embossed, painted		600	.027		1.48	.88		2.36	3.09	
	3900	Insulated		515	.031		1.52	1.03		2.55	3.37	
	4000	Vertical board & batten, colored, non-insulated		515	.031		1.08	1.03		2.11	2.89	
	4200	For simulated wood design, add					.08			.08	.09	
	4300	Corners for above, outside	2 Carp	515	.031	V.L.F.	1.70	1.03		2.73	3.57	
	4500	Inside corners	"	515	.031	"	.96	1.03		1.99	2.76	
	4520	For simulated wood design, add				S.F.	.08			.08	.09	
	9000	Minimum labor/equipment charge	1 Carp	3	2.667	Job		88		88	146	
300	0010	**FASCIA** Aluminum, reverse board and batten,										300
	0100	.032" thick, colored, no furring included	1 Shee	145	.055	S.F.	2.14	2.14		4.28	5.75	
	0200	Residential type, aluminum	1 Carp	200	.040	L.F.	1.15	1.32		2.47	3.46	
	0220	Vinyl	"	200	.040	"	.80	1.32		2.12	3.07	
	0300	Steel, galv and enameled, stock, no furring, long panels	1 Shee	145	.055	S.F.	2.23	2.14		4.37	5.85	
	0600	Short panels		115	.070	"	3.38	2.70		6.08	8	
	9000	Minimum labor/equipment charge		4	2	Job		77.50		77.50	123	
500	0010	**FIBER CEMENT SIDING**										500
	0020	Lap siding, 5/16" thick, 6" wide, smooth texture	2 Carp	415	.039	S.F.	.98	1.27		2.25	3.19	
	0025	Woodgrain texture		415	.039		.98	1.27		2.25	3.19	
	0030	7-1/2" wide, smooth texture		425	.038		.88	1.24		2.12	3.03	
	0035	Woodgrain texture		425	.038		.88	1.24		2.12	3.03	
	0040	8" wide, smooth texture		425	.038		.87	1.24		2.11	3.02	
	0045	Roughsawn texture		425	.038		.87	1.24		2.11	3.02	
	0050	9-1/2" wide, smooth texture		440	.036		.84	1.20		2.04	2.92	

07460 | Siding

		07460 Siding	CREW	DAILY OUTPUT	LABOR-HOURS	UNIT	2004 BARE COSTS MAT.	LABOR	EQUIP.	TOTAL	TOTAL INCL O&P	
500	0055	Woodgrain texture	2 Carp	440	.036	S.F.	.84	1.20		2.04	2.92	500
	0060	12" wide, smooth texture		455	.035		.81	1.16		1.97	2.82	
	0065	Woodgrain texture		455	.035		.81	1.16		1.97	2.82	
	0070	Panel siding, 5/16" thick, smooth texture		750	.021		.73	.70		1.43	1.97	
	0075	Stucco texture		750	.021		.73	.70		1.43	1.97	
	0080	Grooved woodgrain texture		750	.021		.73	.70		1.43	1.97	
	0085	V - grooved woodgrain texture		750	.021	↓	.73	.70		1.43	1.97	
	0090	Wood starter strip	↓	400	.040	L.F.	.17	1.32		1.49	2.38	
600	0010	**VINYL SIDING** Solid PVC panels, 8" to 10" wide, plain	1 Carp	255	.031	S.F.	.69	1.04		1.73	2.48	600
	0100	with 3/8" insulation		255	.031		.76	1.04		1.80	2.56	
	0200	Soffit and fascia		205	.039	↓	1.45	1.29		2.74	3.73	
	0300	Window and door trim moldings		185	.043	L.F.	.36	1.43		1.79	2.77	
	0500	Corner posts, outside corner		205	.039		1.15	1.29		2.44	3.40	
	0600	Inside corner		205	.039	↓	.61	1.29		1.90	2.80	
	9000	Minimum labor/equipment charge	↓	3	2.667	Job		88		88	146	
750	0010	**SOFFIT** Aluminum, residential, stock units, .020" thick	1 Carp	210	.038	S.F.	1.01	1.26		2.27	3.19	750
	0100	Baked enamel on steel, 16 or 18 gauge		105	.076	"	3.74	2.51		6.25	8.30	
	0300	Polyvinyl chloride, white, solid		230	.035	S.F.	.66	1.15		1.81	2.63	
	0400	Perforated	↓	230	.035		.66	1.15		1.81	2.63	
	0500	For colors, add				↓	.07			.07	.08	
	9000	Minimum labor/equipment charge	1 Carp	3	2.667	Job		88		88	146	
800	0010	**STEEL SIDING**, Beveled, vinyl coated, 8" wide, including fasteners		265	.030	S.F.	1.15	1		2.15	2.92	800
	0050	10" wide		275	.029	"	1.23	.96		2.19	2.94	
	0060	Minimum labor/equipment charge	↓	3	2.667	Job		88		88	146	
	0070											
	0080	Galv, corrugated or ribbed, on steel frame, 30 gauge	G-3	800	.040	S.F.	.77	1.30		2.07	2.94	
	0100	28 gauge		795	.040		.81	1.30		2.11	2.99	
	0300	26 gauge		790	.041		.78	1.31		2.09	2.98	
	0400	24 gauge		785	.041		1.14	1.32		2.46	3.38	
	0600	22 gauge		770	.042		1.31	1.35		2.66	3.61	
	0700	Colored, corrugated/ribbed, on steel frame, 10 yr fnsh, 28 ga.		800	.040		1.20	1.30		2.50	3.41	
	0900	26 gauge		795	.040		1.07	1.30		2.37	3.28	
	1000	24 gauge		790	.041		1.26	1.31		2.57	3.51	
	1020	20 gauge	↓	785	.041	↓	1.53	1.32		2.85	3.81	
	9000	Minimum labor/equipment charge	1 Carp	3	2.667	Job		88		88	146	
900	0010	**WOOD SIDING, BOARDS**										900
	3200	Wood, cedar bevel, A grade, 1/2" x 6"	1 Carp	250	.032	S.F.	1.91	1.06		2.97	3.85	
	3300	1/2" x 8"		275	.029		2.46	.96		3.42	4.30	
	3500	3/4" x 10", clear grade		300	.027		3.54	.88		4.42	5.35	
	3600	"B" grade		300	.027		2.80	.88		3.68	4.54	
	3800	Cedar, rough sawn, 1" x 4", A grade, natural		240	.033		2.67	1.10		3.77	4.76	
	3900	Stained		240	.033		3.01	1.10		4.11	5.15	
	4100	1" x 12", board & batten, #3 & Btr., natural		260	.031		2.02	1.02		3.04	3.90	
	4200	Stained		260	.031		2.35	1.02		3.37	4.27	
	4400	1" x 8" channel siding, #3 & Btr., natural		250	.032		1.96	1.06		3.02	3.91	
	4500	Stained		250	.032		2.23	1.06		3.29	4.20	
	4700	Redwood, clear, beveled, vertical grain, 1/2" x 4"		200	.040		3.12	1.32		4.44	5.60	
	4750	1/2" x 6"		225	.036		2.62	1.17		3.79	4.83	
	4800	1/2" x 8"		250	.032		2.12	1.06		3.18	4.08	
	5000	3/4" x 10"		300	.027		3.46	.88		4.34	5.25	
	5200	Channel siding, 1" x 10", B grade	↓	285	.028		2.23	.93		3.16	3.99	
	5250	Redwood, T&G boards, B grade, 1" x 4"	2 Carp	300	.053		2.66	1.76		4.42	5.85	
	5270	1" x 8"	"	375	.043		2.29	1.41		3.70	4.85	
	5400	White pine, rough sawn, 1" x 8", natural	1 Carp	275	.029		.67	.96		1.63	2.33	
	5500	Stained	↓	275	.029	↓	.99	.96		1.95	2.68	

Important: See the Reference Section for critical supporting data - Reference Nos., Crews, & City Cost Indexes

07400 | Roofing and Siding Panels

07460 | Siding

		07460 Siding	CREW	DAILY OUTPUT	LABOR-HOURS	UNIT	2004 BARE COSTS MAT.	LABOR	EQUIP.	TOTAL	TOTAL INCL O&P	
900	9000	Minimum labor/equipment charge	1 Carp	2	4	Job		132		132	219	900
950	0010	**WOOD PRODUCT SIDING** R06160-020										950
	0030	Lap siding, hardboard, 7/16" x 8", primed										
	0051	Wood grain texture finish	L-2	552	.029	S.F.	1.04	.83		1.87	2.52	
	0100	Panels, 7/16" thick, smooth, textured or grooved, primed	2 Carp	700	.023		.73	.75		1.48	2.05	
	0200	Stained		700	.023		.88	.75		1.63	2.22	
	0700	Particle board, overlaid, 3/8" thick		750	.021		.63	.70		1.33	1.86	
	0900	Plywood, medium density overlaid, 3/8" thick		750	.021		.84	.70		1.54	2.09	
	1000	1/2" thick		700	.023		1.01	.75		1.76	2.36	
	1100	3/4" thick		650	.025		1.39	.81		2.20	2.88	
	1600	Texture 1-11, cedar, 5/8" thick, natural		675	.024		2.32	.78		3.10	3.85	
	1700	Factory stained		675	.024		2.21	.78		2.99	3.73	
	1900	Texture 1-11, fir, 5/8" thick, natural		675	.024		.93	.78		1.71	2.32	
	2000	Factory stained		675	.024		1.08	.78		1.86	2.49	
	2050	Texture 1-11, S.Y.P., 5/8" thick, natural		675	.024		.81	.78		1.59	2.19	
	2100	Factory stained		675	.024		.88	.78		1.66	2.27	
	2200	Rough sawn cedar, 3/8" thick, natural		675	.024		1.08	.78		1.86	2.49	
	2300	Factory stained		675	.024		1.20	.78		1.98	2.62	
	2500	Rough sawn fir, 3/8" thick, natural		675	.024		.58	.78		1.36	1.94	
	2600	Factory stained		675	.024		.65	.78		1.43	2.02	
	2800	Redwood, textured siding, 5/8" thick		675	.024		1.80	.78		2.58	3.28	
	3000	Polyvinyl chloride coated, 3/8" thick		750	.021		.88	.70		1.58	2.14	
	9000	Minimum labor/equipment charge	1 Carp	2	4	Job		132		132	219	

07500 | Membrane Roofing

07510 | Built-Up Bituminous Roofing

		07510 Built-Up Bituminous Roofing	CREW	DAILY OUTPUT	LABOR-HOURS	UNIT	2004 BARE COSTS MAT.	LABOR	EQUIP.	TOTAL	TOTAL INCL O&P	
050	0010	**ASPHALT** Coated felt, #30, 2 sq per roll, not mopped	1 Rofc	58	.138	Sq.	7.10	3.92		11.02	14.90	050
	0200	#15, 4 sq per roll, plain or perforated, not mopped		58	.138		2.68	3.92		6.60	10	
	0300	Roll roofing, smooth, #65		15	.533		6.35	15.15		21.50	34	
	0500	#90		15	.533		17.50	15.15		32.65	46.50	
	0520	Mineralized		15	.533		15.35	15.15		30.50	44	
	0540	D.C. (Double coverage), 19" selvage edge		10	.800		29	23		52	73	
	0580	Adhesive (lap cement)				Gal.	3.68			3.68	4.05	
	0600	Steep, flat or dead level asphalt, 10 ton lots, bulk				Ton	260			260	286	
	0800	Packaged				"	410			410	450	
	9000	Minimum labor/equipment charge	1 Rofc	4	2	Job		57		57	102	
300	0010	**BUILT-UP ROOFING**										300
	0120	Asphalt flood coat with gravel/slag surfacing, not including										
	0140	Insulation, flashing or wood nailers										
	0200	Asphalt base sheet, 3 plies #15 asphalt felt, mopped	G-1	22	2.545	Sq.	46.50	67.50	13.65	127.65	188	
	0350	On nailable decks		21	2.667		51.50	71	14.30	136.80	199	
	0500	4 plies #15 asphalt felt, mopped		20	2.800		66.50	74.50	15	156	223	
	0550	On nailable decks		19	2.947		59.50	78.50	15.80	153.80	223	
	0700	Coated glass base sheet, 2 plies glass (type IV), mopped		22	2.545		50	67.50	13.65	131.15	191	
	0850	3 plies glass, mopped		20	2.800		60	74.50	15	149.50	216	
	0950	On nailable decks		19	2.947		56.50	78.50	15.80	150.80	219	
	1100	4 plies glass fiber felt (type IV), mopped		20	2.800		73	74.50	15	162.50	230	
	1150	On nailable decks		19	2.947		66.50	78.50	15.80	160.80	230	

07510 | Built-Up Bituminous Roofing

		07510 Built-Up Bituminous Roofing	CREW	DAILY OUTPUT	LABOR-HOURS	UNIT	2004 BARE COSTS MAT.	LABOR	EQUIP.	TOTAL	TOTAL INCL O&P
300	1200	Coated & saturated base sheet, 3 plies #15 asph. felt, mopped	G-1	20	2.800	Sq.	53	74.50	15	142.50	208
	1250	On nailable decks		19	2.947		49.50	78.50	15.80	143.80	212
	1300	4 plies #15 asphalt felt, mopped		22	2.545		61.50	67.50	13.65	142.65	204
	2000	Asphalt flood coat, smooth surface									
	2200	Asphalt base sheet & 3 plies #15 asphalt felt, mopped	G-1	24	2.333	Sq.	51	62	12.50	125.50	181
	2400	On nailable decks		23	2.435		47.50	64.50	13.05	125.05	183
	2600	4 plies #15 asphalt felt, mopped		24	2.333		59.50	62	12.50	134	190
	2700	On nailable decks		23	2.435		56	64.50	13.05	133.55	192
	2900	Coated glass fiber base sheet, mopped, and 2 plies of									
	2910	glass fiber felt (type IV)	G-1	25	2.240	Sq.	46.50	59.50	12	118	171
	3100	On nailable decks		24	2.333		44	62	12.50	118.50	173
	3200	3 plies, mopped		23	2.435		56	64.50	13.05	133.55	192
	3300	On nailable decks		22	2.545		53	67.50	13.65	134.15	194
	3800	4 plies glass fiber felt (type IV), mopped		23	2.435		66	64.50	13.05	143.55	203
	3900	On nailable decks		22	2.545		62.50	67.50	13.65	143.65	205
	4000	Coated & saturated base sheet, 3 plies #15 asph. felt, mopped		24	2.333		49.50	62	12.50	124	179
	4200	On nailable decks		23	2.435		46	64.50	13.05	123.55	181
	4300	4 plies #15 organic felt, mopped		22	2.545		58	67.50	13.65	139.15	200
	4500	Coal tar pitch with gravel/slag surfacing									
	4600	4 plies #15 tarred felt, mopped	G-1	21	2.667	Sq.	101	71	14.30	186.30	254
	4800	3 plies glass fiber felt (type IV), mopped	"	19	2.947	"	83	78.50	15.80	177.30	249
	5000	Coated glass fiber base sheet, and 2 plies of									
	5010	glass fiber felt, (type IV), mopped	G-1	19	2.947	Sq.	84.50	78.50	15.80	178.80	250
	5300	On nailable decks		18	3.111		74	82.50	16.70	173.20	248
	5600	4 plies glass fiber felt (type IV), mopped		21	2.667		115	71	14.30	200.30	270
	5800	On nailable decks		20	2.800		105	74.50	15	194.50	266
400	0010	**CANTS** 4″ x 4″, treated timber, cut diagonally	1 Rofc	325	.025	L.F.	1.33	.70		2.03	2.71
	0100	Foamglass		325	.025		2.15	.70		2.85	3.62
	0300	Mineral or fiber, trapezoidal, 1″x 4″ x 48″		325	.025		.17	.70		.87	1.44
	0400	1-1/2″ x 5-5/8″ x 48″		325	.025		.29	.70		.99	1.57
	9000	Minimum labor/equipment charge		4	2	Job		57		57	102
700	0010	**FELT** Glass fibered, #15, no mopping	1 Rofc	58	.138	Sq.	4.11	3.92		8.03	11.55
	0300	Base sheet, #45, channel vented		58	.138		16.50	3.92		20.42	25
	0400	#50, coated		58	.138		8.95	3.92		12.87	16.85
	0500	Cap, mineral surfaced		58	.138		16.50	3.92		20.42	25
	0600	Flashing membrane, #65		16	.500		22.50	14.25		36.75	50.50
	0800	Coal tar fibered, #15, no mopping		58	.138		7.50	3.92		11.42	15.30
	0900	Asphalt felt, #15, 4 sq per roll, no mopping		58	.138		2.68	3.92		6.60	10
	1100	#30, 2 sq per roll		58	.138		7.10	3.92		11.02	14.90
	1200	Double coated, #33		58	.138		5.90	3.92		9.82	13.55
	1400	#40, base sheet		58	.138		8.05	3.92		11.97	15.90
	1450	Coated and saturated		58	.138		6.45	3.92		10.37	14.15
	1500	Tarred felt, organic, #15, 4 sq rolls		58	.138		7.80	3.92		11.72	15.65
	1550	#30, 2 sq roll		58	.138		15.60	3.92		19.52	24.50
	1700	Add for mopping above felts, per ply, asphalt, 24 lb per sq	G-1	192	.292		4.92	7.75	1.56	14.23	21
	1800	Coal tar mopping, 30 lb per sq		186	.301		8.80	8	1.61	18.41	26
	1900	Flood coat, with asphalt, 60 lb per sq		60	.933		12.30	25	5	42.30	63.50
	2000	With coal tar, 75 lb per sq		56	1		22	26.50	5.35	53.85	77.50
	9000	Minimum labor/equipment charge	1 Rofc	4	2	Job		57		57	102

07520 | Cold Applied Bituminous Roofing

		07520 Cold Applied Bituminous Roofing	CREW	DAILY OUTPUT	LABOR-HOURS	UNIT	MAT.	LABOR	EQUIP.	TOTAL	TOTAL INCL O&P
200	0010	**COLD APPLIED** 3-ply system (components listed below)	G-5	50	.800	Sq.		20.50	2.77	23.27	40
	0100	Spunbond poly. fabric, 1.35 oz/SY, 36″W, 10.8 Sq/roll				Ea.	123			123	135
	0200	49″ wide, 14.6 Sq./roll					169			169	186
	0300	2.10 oz./S.Y., 36″ wide, 10.8 Sq./roll					185			185	203

Important: See the Reference Section for critical supporting data - Reference Nos., Crews, & City Cost Indexes

07500 | Membrane Roofing

07520 | Cold Applied Bituminous Roofing

			CREW	DAILY OUTPUT	LABOR-HOURS	UNIT	2004 BARE COSTS MAT.	LABOR	EQUIP.	TOTAL	TOTAL INCL O&P
200	0400	49″ wide, 14.6 Sq./roll				Ea.	250			250	275
	0500	Base & finish coat, 3 gal./Sq., 5 gal./can				Gal.	3.25			3.25	3.58
	0600	Coating, ceramic granules, 1/2 Sq./bag				Ea.	11.50			11.50	12.65
	0700	Aluminum, 2 gal./Sq.				Gal.	9.25			9.25	10.20
	0800	Emulsion, fibered or non-fibered, 4 gal./Sq.				″	4.25			4.25	4.68

07530 | Elastomeric Membrane Roofing

			CREW	DAILY OUTPUT	LABOR-HOURS	UNIT	MAT.	LABOR	EQUIP.	TOTAL	TOTAL INCL O&P
350	0010	**ELASTOMERIC ROOFING**									
	0100	For Elastomeric waterproofing, see division 07130-200									
	0110	Acrylic rubber, fluid applied, 20 mils thick	G-5	2,000	.020	S.F.	1.80	.52	.07	2.39	2.99
	0120	50 mils, reinforced		1,200	.033		2.80	.86	.12	3.78	4.75
	0130	For walking surface, add		900	.044		.85	1.15	.15	2.15	3.17
	0300	Hypalon neoprene, fluid applied, 20 mil thick, not-reinforced	G-1	1,135	.049		2.05	1.31	.26	3.62	4.90
	0600	Non-woven polyester, reinforced		960	.058		2.07	1.55	.31	3.93	5.40
	0700	5 coat neoprene deck, 60 mil thick, under 10,000 SF		325	.172		4.56	4.58	.92	10.06	14.20
	0900	Over 10,000 SF		625	.090		4.25	2.38	.48	7.11	9.50
	1300	Vinyl plastic traffic deck, sprayed, 2 to 4 mils thick		625	.090		1.33	2.38	.48	4.19	6.25
	1500	Vinyl and neoprene membrane traffic deck		1,550	.036		1.41	.96	.19	2.56	3.48
	9000	Minimum labor/equipment charge	1 Rofc	2	4	Job		114		114	204
800	0010	**SINGLE-PLY MEMBRANE**									
	0800	Chlorosulfonated polyethylene-hypalon (CSPE), 45 mils,									
	0900	0.29 P.S.F., fully adhered	G-5	26	1.538	Sq.	123	40	5.30	168.30	213
	3500	Ethylene propylene diene monomer (EPDM), 45 mils, 0.28 P.S.F.									
	4810	45 mil, .28 PSF, membrane only				Sq.	29.50			29.50	32.50
	4820	60 mil, .40 PSF, membrane only				″	39.50			39.50	43.50
	4850	Seam tape for membrane, 4″ x 100′ roll				Ea.	67.50			67.50	74
	4900	Batten strips, 10′ sections					2.47			2.47	2.72
	4910	Cover tape for batten strips, 6″ x 100′ roll					127			127	140
	4930	Plate anchors				M	124			124	137
	4970	Adhesive for fully adhered systems, 60 S.F./gal.				Gal.	12.20			12.20	13.40
	7500	Polyisobutylene (PIB), 100 mils, 0.57 P.S.F.									
	7600	Loose-laid & ballasted with stone/gravel (10 P.S.F.)	G-5	51	.784	Sq.	124	20.50	2.71	147.21	176
	7700	Partially adhered with adhesive		35	1.143		156	29.50	3.95	189.45	228
	7800	Hot asphalt attachment		35	1.143		148	29.50	3.95	181.45	220
	7900	Fully adhered with contact cement		26	1.538		161	40	5.30	206.30	254

07540 | PVC Single Ply Membrane

			CREW	DAILY OUTPUT	LABOR-HOURS	UNIT	MAT.	LABOR	EQUIP.	TOTAL	TOTAL INCL O&P
800	0010	**POLYVINYL CHLORIDE ROOFING**									
	8200	Heat welded seams									
	8700	Reinforced, 48 mils, 0.33 P.S.F.									
	8750	Loose-laid & ballasted with stone/gravel (12 P.S.F.)	G-5	51	.784	Sq.	95	20.50	2.71	118.21	143
	8800	Mechanically attached		35	1.143		86	29.50	3.95	119.45	152
	8850	Fully adhered with adhesive		26	1.538		115	40	5.30	160.30	204
	8860	Reinforced, 60 mils, .40 P.S.F.									
	8870	Loose-laid & ballasted with stone/gravel (12 P.S.F.)	G-5	51	.784	Sq.	87.50	20.50	2.71	110.71	136
	8880	Mechanically attached		35	1.143		79	29.50	3.95	112.45	144
	8890	Fully adhered with adhesive		26	1.538		108	40	5.30	153.30	196

07550 | Modified Bit. Membrane Roofing

			CREW	DAILY OUTPUT	LABOR-HOURS	UNIT	MAT.	LABOR	EQUIP.	TOTAL	TOTAL INCL O&P
500	0010	**MODIFIED BITUMEN ROOFING** R07550-030									
	0020	Base sheet, #15 glass fiber felt, nailed to deck	1 Rofc	58	.138	Sq.	4.85	3.92		8.77	12.40
	0030	Spot mopped to deck	G-1	295	.190		6.55	5.05	1.02	12.62	17.40
	0040	Fully mopped to deck	″	192	.292		9.05	7.75	1.56	18.36	25.50
	0050	#15 organic felt, nailed to deck	1 Rofc	58	.138		3.42	3.92		7.34	10.80
	0060	Spot mopped to deck	G-1	295	.190		5.15	5.05	1.02	11.22	15.80

07500 | Membrane Roofing

07550 | Modified Bit. Membrane Roofing

			CREW	DAILY OUTPUT	LABOR-HOURS	UNIT	2004 BARE COSTS MAT.	LABOR	EQUIP.	TOTAL	TOTAL INCL O&P	
500	0070	Fully mopped to deck (R07550 -030)	G-1	192	.292	Sq.	7.60	7.75	1.56	16.91	24	500
	0080	SBS modified, granule surf cap sheet, polyester rein., mopped										
	1500	Glass fiber reinforced, mopped, 160 mils	G-1	2,000	.028	S.F.	.42	.74	.15	1.31	1.95	
	1600	Smooth surface cap sheet, mopped, 145 mils		2,100	.027		.42	.71	.14	1.27	1.89	
	1700	Smooth surface flashing, 145 mils		1,260	.044		.42	1.18	.24	1.84	2.84	
	1800	150 mils		1,260	.044		.41	1.18	.24	1.83	2.83	
	1900	Granular surface flashing, 150 mils		1,260	.044		.45	1.18	.24	1.87	2.88	
	2000	160 mils		1,260	.044		.63	1.18	.24	2.05	3.07	
	2100	APP mod., smooth surf. cap sheet, poly. reinf., torched, 160 mils	G-5	2,100	.019		.41	.49	.07	.97	1.40	
	2150	170 mils		2,100	.019		.46	.49	.07	1.02	1.46	
	2200	Granule surface cap sheet, poly. reinf., torched, 180 mils		2,000	.020		.50	.52	.07	1.09	1.56	
	2250	Smooth surface flashing, torched, 160 mils		1,260	.032		.41	.82	.11	1.34	2.04	
	2300	170 mils		1,260	.032		.46	.82	.11	1.39	2.10	
	2350	Granule surface flashing, torched, 180 mils		1,260	.032		.50	.82	.11	1.43	2.14	
	2400	Fibrated aluminum coating	1 Rofc	3,800	.002		.10	.06		.16	.22	

07580 | Roll Roofing

			CREW	DAILY OUTPUT	LABOR-HOURS	UNIT	MAT.	LABOR	EQUIP.	TOTAL	TOTAL INCL O&P	
200	0010	**ROLL ROOFING**										200
	0100	Asphalt, mineral surface										
	0200	1 ply #15 organic felt, 1 ply mineral surfaced										
	0300	Selvage roofing, lap 19", nailed & mopped	G-1	27	2.074	Sq.	39	55	11.10	105.10	154	
	0400	3 plies glass fiber felt (type IV), 1 ply mineral surfaced										
	0500	Selvage roofing, lapped 19", mopped	G-1	25	2.240	Sq.	61	59.50	12	132.50	187	
	0600	Coated glass fiber base sheet, 2 plies of glass fiber										
	0700	Felt (type IV), 1 ply mineral surfaced selvage										
	0800	Roofing, lapped 19", mopped	G-1	25	2.240	Sq.	66	59.50	12	137.50	193	
	0900	On nailable decks	"	24	2.333	"	61	62	12.50	135.50	192	
	1000	3 plies glass fiber felt (type III), 1 ply mineral surfaced										
	1100	Selvage roofing, lapped 19", mopped	G-1	25	2.240	Sq.	61	59.50	12	132.50	187	

07590 | Roof Maintenance and Repairs

			CREW	DAILY OUTPUT	LABOR-HOURS	UNIT	MAT.	LABOR	EQUIP.	TOTAL	TOTAL INCL O&P	
300	0010	**ROOF COATINGS** Asphalt				Gal.	2.95			2.95	3.25	300
	0200	Asphalt base, fibered aluminum coating					9.60			9.60	10.55	
	0300	Asphalt primer, 5 gallon					3.64			3.64	4	
	0600	Coal tar pitch, 200 lb. barrels				Ton	585			585	645	
	0700	Tar roof cement, 5 gal. lots				Gal.	6.05			6.05	6.65	
	0800	Glass fibered roof & patching cement, 5 gallon				"	4.43			4.43	4.87	
	0900	Reinforcing glass membrane, 450 S.F./roll				Ea.	45.50			45.50	50	
	1000	Neoprene roof coating, 5 gal, 2 gal/sq				Gal.	21.50			21.50	23.50	
	1100	Roof patch & flashing cement, 5 gallon					17.15			17.15	18.85	
	1200	Roof resaturant, glass fibered, 3 gal/sq					7.30			7.30	8.05	
	1300	Mineral rubber, 3 gal/sq					4.71			4.71	5.20	

07600 | Flashing and Sheet Metal

07610 | Sheet Metal Roofing

			CREW	DAILY OUTPUT	LABOR-HOURS	UNIT	2004 BARE COSTS MAT.	LABOR	EQUIP.	TOTAL	TOTAL INCL O&P	
300	0010	**COPPER ROOFING** Batten seam, over 10 sq, 16 oz, 130 lb/sq	1 Shee	1.10	7.273	Sq.	405	282		687	890	300
	0200	18 oz, 145 lb per sq		1	8		450	310		760	985	

07610 | Sheet Metal Roofing

		07610 Sheet Metal Roofing	CREW	DAILY OUTPUT	LABOR-HOURS	UNIT	2004 BARE COSTS MAT.	LABOR	EQUIP.	TOTAL	TOTAL INCL O&P	
300	0300	20 oz, 160 lb per sq	1 Shee	1	8	Sq.	500	310		810	1,050	300
	0400	Standing seam, over 10 squares, 16 oz, 125 lb per sq		1.30	6.154		390	239		629	805	
	0600	18 oz, 140 lb per sq		1.20	6.667		435	259		694	890	
	0700	20 oz, 150 lb per sq		1.10	7.273		465	282		747	960	
	0900	Flat seam, over 10 squares, 16 oz, 115 lb per sq		1.20	6.667		355	259		614	805	
	1000	20 oz, 145 lb per sq		1.10	7.273		450	282		732	940	
	1200	For abnormal conditions or small areas, add					25%	100%				
	1300	For lead-coated copper, add					25%					
	9000	Minimum labor/equipment charge	1 Shee	2	4	Job		155		155	246	
500	0010	**LEAD ROOFING** 5 lb. per SF, batten seam	1 Shee	1.20	6.667	Sq.	385	259		644	830	500
	0100	Flat seam		1.30	6.154	″	385	239		624	800	
	9000	Minimum labor/equipment charge		2	4	Job		155		155	246	
700	0010	**STAINLESS STEEL ROOFING** Type 304, batten seam, 28 gauge	1 Shee	1.20	6.667	Sq.	305	259		564	750	700
	0100	26 gauge	″	1.15	6.957		380	270		650	845	
	0200	For standing seam construction, deduct					2%					
	0500	For flat seam construction, deduct					3%					
	0800	For lead or terne coated stainless, 28 gauge, add					70			70	77	
	0900	For 26 gauge, add					92.50			92.50	102	
	9000	Minimum labor/equipment charge	1 Shee	2.75	2.909	Job		113		113	179	
900	0010	**ZINC** Copper alloy roofing, batten seam, .020″ thick	1 Shee	1.20	6.667	Sq.	520	259		779	980	900
	0100	.027″ thick		1.15	6.957		630	270		900	1,125	
	0300	.032″ thick		1.10	7.273		710	282		992	1,225	
	0400	.040″ thick		1.05	7.619		835	296		1,131	1,400	
	0600	For standing seam construction, deduct					2%					
	0700	For flat seam construction, deduct					3%					
	9000	Minimum labor/equipment charge	1 Shee	2.75	2.909	Job		113		113	179	

07620 | Sheet Metal Flashing and Trim

		07620 Sheet Metal Flashing and Trim	CREW	DAILY OUTPUT	LABOR-HOURS	UNIT	MAT.	LABOR	EQUIP.	TOTAL	TOTAL INCL O&P	
100	0010	**SHEET METAL CLADDING**										100
	0100	Aluminum, up to 6 bends, .032″ thick, window casing	1 Carp	180	.044	S.F.	.58	1.47		2.05	3.07	
	0200	Window sill		72	.111	L.F.	.58	3.67		4.25	6.75	
	0300	Door casing		180	.044	S.F.	.58	1.47		2.05	3.07	
	0400	Fascia		250	.032		.58	1.06		1.64	2.39	
	0500	Rake trim		225	.036		.58	1.17		1.75	2.59	
	0700	.024″ thick, window casing		180	.044		.90	1.47		2.37	3.42	
	0800	Window sill		72	.111	L.F.	.90	3.67		4.57	7.10	
	0900	Door casing		180	.044	S.F.	.90	1.47		2.37	3.42	
	1000	Fascia		250	.032		.90	1.06		1.96	2.74	
	1100	Rake trim		225	.036		.90	1.17		2.07	2.94	
	1200	Vinyl coated aluminum, up to 6 bends, window casing		180	.044		.61	1.47		2.08	3.10	
	1300	Window sill		72	.111	L.F.	.61	3.67		4.28	6.75	
	1400	Door casing		180	.044	S.F.	.61	1.47		2.08	3.10	
	1500	Fascia		250	.032		.61	1.06		1.67	2.42	
	1600	Rake trim		225	.036		.61	1.17		1.78	2.62	

07650 | Flexible Flashing

		07650 Flexible Flashing	CREW	DAILY OUTPUT	LABOR-HOURS	UNIT	MAT.	LABOR	EQUIP.	TOTAL	TOTAL INCL O&P	
600	0010	**FLASHING** Aluminum, mill finish, .013″ thick	1 Rofc	145	.055	S.F.	.35	1.57		1.92	3.20	600
	0030	.016″ thick		145	.055		.51	1.57		2.08	3.37	
	0060	.019″ thick		145	.055		.66	1.57		2.23	3.54	
	0100	.032″ thick		145	.055		1.06	1.57		2.63	3.98	
	0200	.040″ thick		145	.055		1.39	1.57		2.96	4.34	
	0300	.050″ thick		145	.055		1.84	1.57		3.41	4.83	

	07650 | Flexible Flashing		CREW	DAILY OUTPUT	LABOR-HOURS	UNIT	2004 BARE COSTS MAT.	LABOR	EQUIP.	TOTAL	TOTAL INCL O&P	
600	0325	Mill finish 5″ x 7″ step flashing, .016″ thick	1 Rofc	1,920	.004	Ea.	.10	.12		.22	.32	600
	0350	Mill finish 12″ x 12″ step flashing, .016″ thick		1,600	.005	″	.40	.14		.54	.69	
	0400	Painted finish, add				S.F.	.24			.24	.26	
	0500	Fabric-backed 2 sides, .004″ thick	1 Rofc	330	.024		.93	.69		1.62	2.26	
	0700	.005″ thick		330	.024		1.09	.69		1.78	2.44	
	0750	Mastic-backed, self adhesive		460	.017		2.26	.49		2.75	3.38	
	0800	Mastic-coated 2 sides, .004″ thick		330	.024		.93	.69		1.62	2.26	
	1000	.005″ thick		330	.024		1.09	.69		1.78	2.44	
	1100	.016″ thick		330	.024		1.20	.69		1.89	2.56	
	1300	Asphalt flashing cement, 5 gallon				Gal.	3.90			3.90	4.29	
	1600	Copper, 16 oz, sheets, under 1000 lbs.	1 Rofc	115	.070	S.F.	2.23	1.98		4.21	6	
	1700	Over 4000 lbs.		155	.052		2.91	1.47		4.38	5.85	
	1900	20 oz sheets, under 1000 lbs.		110	.073		3.90	2.07		5.97	8	
	2000	Over 4000 lbs.		145	.055		3.62	1.57		5.19	6.80	
	2200	24 oz sheets, under 1000 lbs.		105	.076		4.69	2.17		6.86	9.05	
	2300	Over 4000 lbs.		135	.059		4.34	1.69		6.03	7.80	
	2500	32 oz sheets, under 1000 lbs.		100	.080		6.20	2.28		8.48	10.95	
	2600	Over 4000 lbs.		130	.062		5.80	1.75		7.55	9.55	
	2700	W shape for valleys, 16 oz, 24″ wide		100	.080	L.F.	5.90	2.28		8.18	10.60	
	2800	Copper, paperbacked 1 side, 2 oz		330	.024	S.F.	.86	.69		1.55	2.19	
	2900	3 oz		330	.024		1.12	.69		1.81	2.47	
	3100	Paperbacked 2 sides, 2 oz		330	.024		.86	.69		1.55	2.19	
	3150	3 oz		330	.024		1.11	.69		1.80	2.46	
	3200	5 oz		330	.024		1.65	.69		2.34	3.06	
	3250	7 oz		330	.024		2.72	.69		3.41	4.23	
	3400	Mastic-backed 2 sides, copper, 2 oz		330	.024		1.02	.69		1.71	2.36	
	3500	3 oz		330	.024		1.26	.69		1.95	2.63	
	3700	5 oz		330	.024		1.85	.69		2.54	3.28	
	3800	Fabric-backed 2 sides, copper, 2 oz		330	.024		1.09	.69		1.78	2.44	
	4000	3 oz		330	.024		1.40	.69		2.09	2.78	
	4100	5 oz		330	.024		1.90	.69		2.59	3.33	
	4300	Copper-clad stainless steel, .015″ thick, under 500 lbs.		115	.070		3.16	1.98		5.14	7	
	4400	Over 2000 lbs.		155	.052		3.05	1.47		4.52	6	
	4600	.018″ thick, under 500 lbs.		100	.080		4.18	2.28		6.46	8.70	
	4700	Over 2000 lbs.		145	.055		3.06	1.57		4.63	6.20	
	4900	Fabric, asphalt-saturated cotton, specification grade		35	.229	S.Y.	1.93	6.50		8.43	13.75	
	5000	Utility grade		35	.229		1.22	6.50		7.72	13	
	5200	Open-mesh fabric, saturated, 40 oz per S.Y.		35	.229		1.35	6.50		7.85	13.15	
	5300	Close-mesh fabric, saturated, 17 oz per S.Y.		35	.229		1.42	6.50		7.92	13.20	
	5500	Fiberglass, resin-coated		35	.229		1.16	6.50		7.66	12.95	
	5600	Asphalt-coated, 40 oz per S.Y.		35	.229		7.90	6.50		14.40	20.50	
	5800	Lead, 2.5 lb. per SF, up to 12″ wide		135	.059	S.F.	2.80	1.69		4.49	6.10	
	5900	Over 12″ wide		135	.059		3.25	1.69		4.94	6.60	
	6100	Lead-coated copper, fabric-backed, 2 oz		330	.024		1.41	.69		2.10	2.79	
	6200	5 oz		330	.024		1.62	.69		2.31	3.02	
	6400	Mastic-backed 2 sides, 2 oz		330	.024		1.10	.69		1.79	2.45	
	6500	5 oz		330	.024		1.36	.69		2.05	2.74	
	6700	Paperbacked 1 side, 2 oz		330	.024		.95	.69		1.64	2.29	
	6800	3 oz		330	.024		1.12	.69		1.81	2.47	
	7000	Paperbacked 2 sides, 2 oz		330	.024		.98	.69		1.67	2.32	
	7100	5 oz		330	.024		1.59	.69		2.28	2.99	
	7300	Polyvinyl chloride, black, .010″ thick		285	.028		.14	.80		.94	1.58	
	7400	.020″ thick		285	.028		.22	.80		1.02	1.67	
	7600	.030″ thick		285	.028		.28	.80		1.08	1.74	
	7700	.056″ thick		285	.028		.69	.80		1.49	2.19	
	7900	Black or white for exposed roofs, .060″ thick		285	.028		1.55	.80		2.35	3.14	

07600 | Flashing and Sheet Metal

	07650 | Flexible Flashing	CREW	DAILY OUTPUT	LABOR-HOURS	UNIT	2004 BARE COSTS MAT.	LABOR	EQUIP.	TOTAL	TOTAL INCL O&P	
600	8060 PVC tape, 5" x 45 mils, for joint covers, 100 L.F./roll				Ea.	79.50			79.50	87	600
	8100 Rubber, butyl, 1/32" thick	1 Rofc	285	.028	S.F.	.70	.80		1.50	2.20	
	8200 1/16" thick		285	.028		1.05	.80		1.85	2.59	
	8300 Neoprene, cured, 1/16" thick		285	.028		1.48	.80		2.28	3.06	
	8400 1/8" thick		285	.028		2.99	.80		3.79	4.72	
	8500 Shower pan, bituminous membrane, 7 oz		155	.052		1.08	1.47		2.55	3.82	
	8550 3 ply copper and fabric, 3 oz		155	.052		1.63	1.47		3.10	4.42	
	8600 7 oz		155	.052		3.37	1.47		4.84	6.35	
	8650 Copper, 16 oz		100	.080		3.11	2.28		5.39	7.50	
	8700 Lead on copper and fabric, 5 oz		155	.052		1.62	1.47		3.09	4.41	
	8800 7 oz		155	.052		2.93	1.47		4.40	5.85	
	8900 Stainless steel sheets, 32 ga, .010" thick		155	.052		2.16	1.47		3.63	5	
	9000 28 ga, .015" thick		155	.052		2.68	1.47		4.15	5.60	
	9100 26 ga, .018" thick		155	.052		3.25	1.47		4.72	6.20	
	9200 24 ga, .025" thick	▼	155	.052	▼	4.22	1.47		5.69	7.25	
	9290 For mechanically keyed flashing, add					40%					
	9300 Stainless steel, paperbacked 2 sides, .005" thick	1 Rofc	330	.024	S.F.	1.90	.69		2.59	3.33	
	9320 Steel sheets, galvanized, 20 gauge		130	.062		.73	1.75		2.48	3.94	
	9340 30 gauge		160	.050		.31	1.42		1.73	2.89	
	9400 Terne coated stainless steel, .015" thick, 28 ga		155	.052		4.05	1.47		5.52	7.10	
	9500 .018" thick, 26 ga		155	.052		4.57	1.47		6.04	7.70	
	9600 Zinc and copper alloy (brass), .020" thick		155	.052		3.26	1.47		4.73	6.20	
	9700 .027" thick		155	.052		4.37	1.47		5.84	7.45	
	9800 .032" thick		155	.052		5.10	1.47		6.57	8.25	
	9900 .040" thick		155	.052	▼	6.20	1.47		7.67	9.50	
	9950 Minimum labor/equipment charge	▼	3	2.667	Job		76		76	136	

07700 | Roof Specialties and Accessories

	07710 | Manufactured Roof Specialties	CREW	DAILY OUTPUT	LABOR-HOURS	UNIT	2004 BARE COSTS MAT.	LABOR	EQUIP.	TOTAL	TOTAL INCL O&P	
400	0010 **DOWNSPOUTS** Aluminum 2" x 3", .020" thick, embossed	1 Shee	190	.042	L.F.	.65	1.63		2.28	3.31	400
	0100 Enameled		190	.042		.95	1.63		2.58	3.64	
	0300 Enameled, .024" thick, 2" x 3"		180	.044		1.12	1.72		2.84	3.96	
	0400 3" x 4"		140	.057		1.59	2.22		3.81	5.25	
	0600 Round, corrugated aluminum, 3" diameter, .020" thick		190	.042		.98	1.63		2.61	3.67	
	0700 4" diameter, .025" thick		140	.057	▼	1.60	2.22		3.82	5.25	
	0900 Wire strainer, round, 2" diameter		155	.052	Ea.	1.75	2		3.75	5.10	
	1000 4" diameter		155	.052		1.82	2		3.82	5.15	
	1200 Rectangular, perforated, 2" x 3"		145	.055		2.15	2.14		4.29	5.75	
	1300 3" x 4"		145	.055	▼	3.10	2.14		5.24	6.80	
	1500 Copper, round, 16 oz., stock, 2" diameter		190	.042	L.F.	4.53	1.63		6.16	7.55	
	1600 3" diameter		190	.042		4.51	1.63		6.14	7.55	
	1800 4" diameter		145	.055		4.65	2.14		6.79	8.50	
	1900 5" diameter		130	.062		6.40	2.39		8.79	10.85	
	2100 Rectangular, corrugated copper, stock, 2" x 3"		190	.042		3.71	1.63		5.34	6.65	
	2200 3" x 4"		145	.055		4.80	2.14		6.94	8.70	
	2400 Rectangular, plain copper, stock, 2" x 3"		190	.042		4.78	1.63		6.41	7.85	
	2500 3" x 4"		145	.055	▼	6.30	2.14		8.44	10.30	
	2700 Wire strainers, rectangular, 2" x 3"		145	.055	Ea.	2.80	2.14		4.94	6.45	
	2800 3" x 4"	▼	145	.055	▼	4.44	2.14		6.58	8.25	

07710	Manufactured Roof Specialties		CREW	DAILY OUTPUT	LABOR-HOURS	UNIT	2004 BARE COSTS MAT.	LABOR	EQUIP.	TOTAL	TOTAL INCL O&P
400	3000	Round, 2" diameter	1 Shee	145	.055	Ea.	2.63	2.14		4.77	6.30
	3100	3" diameter		145	.055		3.67	2.14		5.81	7.45
	3300	4" diameter		145	.055		5.70	2.14		7.84	9.65
	3400	5" diameter		115	.070		8.20	2.70		10.90	13.35
	3600	Lead-coated copper, round, stock, 2" diameter		190	.042	L.F.	5.35	1.63		6.98	8.50
	3700	3" diameter		190	.042		5.75	1.63		7.38	8.90
	3900	4" diameter		145	.055		7.60	2.14		9.74	11.80
	4000	5" diameter, corrugated		130	.062		10.50	2.39		12.89	15.35
	4200	6" diameter, corrugated		105	.076		10.95	2.96		13.91	16.75
	4300	Rectangular, corrugated, stock, 2" x 3"		190	.042		6.65	1.63		8.28	9.95
	4500	Plain, stock, 2" x 3"		190	.042		7.35	1.63		8.98	10.70
	4600	3" x 4"		145	.055		7.95	2.14		10.09	12.15
	4800	Steel, galvanized, round, corrugated, 2" or 3" diam, 28 ga		190	.042		.74	1.63		2.37	3.40
	4900	4" diameter, 28 gauge		145	.055		1.01	2.14		3.15	4.50
	5100	5" diameter, 28 gauge		130	.062		1.36	2.39		3.75	5.30
	5200	26 gauge		130	.062		1.36	2.39		3.75	5.30
	5400	6" diameter, 28 gauge		105	.076		1.97	2.96		4.93	6.85
	5500	26 gauge		105	.076		1.42	2.96		4.38	6.25
	5700	Rectangular, corrugated, 28 gauge, 2" x 3"		190	.042		.54	1.63		2.17	3.18
	5800	3" x 4"		145	.055		1.50	2.14		3.64	5.05
	6000	Rectangular, plain, 28 gauge, galvanized, 2" x 3"		190	.042		.65	1.63		2.28	3.31
	6100	3" x 4"		145	.055		1.23	2.14		3.37	4.74
	6300	Epoxy painted, 24 gauge, corrugated, 2" x 3"		190	.042		1.03	1.63		2.66	3.72
	6400	3" x 4"		145	.055		1.78	2.14		3.92	5.35
	6600	Wire strainers, rectangular, 2" x 3"		145	.055	Ea.	1.62	2.14		3.76	5.15
	6700	3" x 4"		145	.055		2.61	2.14		4.75	6.25
	6900	Round strainers, 2" or 3" diameter		145	.055		1.20	2.14		3.34	4.71
	7000	4" diameter		145	.055		1.42	2.14		3.56	4.95
	7200	5" diameter		145	.055		2.20	2.14		4.34	5.80
	7300	6" diameter		115	.070		2.63	2.70		5.33	7.15
	9000	Minimum labor/equipment charge		4	2	Job		77.50		77.50	123
450	0010	**DRIP EDGE**, aluminum, .016" thick, 5" wide, mill finish	1 Carp	400	.020	L.F.	.25	.66		.91	1.37
	0100	White finish		400	.020		.22	.66		.88	1.33
	0200	8" wide, mill finish		400	.020		.30	.66		.96	1.42
	0300	Ice belt, 28" wide, mill finish		100	.080		3.42	2.64		6.06	8.15
	0310	Vented, mill finish		400	.020		1.41	.66		2.07	2.64
	0320	Painted finish		400	.020		1.54	.66		2.20	2.78
	0400	Galvanized, 5" wide		400	.020		.22	.66		.88	1.33
	0500	8" wide, mill finish		400	.020		.33	.66		.99	1.45
	0510	Rake edge, aluminum, 1-1/2" x 1-1/2"		400	.020		.13	.66		.79	1.23
	0520	3-1/2" x 1-1/2"		400	.020		.19	.66		.85	1.30
	9000	Minimum labor/equipment charge		4	2	Job		66		66	109
500	0010	**ELBOWS** Aluminum, 2" x 3", embossed	1 Shee	100	.080	Ea.	.91	3.10		4.01	5.90
	0100	Enameled		100	.080		1.76	3.10		4.86	6.85
	0200	3" x 4", .025" thick, embossed		100	.080		3.20	3.10		6.30	8.45
	0300	Enameled		100	.080		3.20	3.10		6.30	8.45
	0400	Round corrugated, 3", embossed, .020" thick		100	.080		1.98	3.10		5.08	7.10
	0500	4", .025" thick		100	.080		2.99	3.10		6.09	8.20
	0600	Copper, 16 oz. round, 2" diameter		100	.080		11.15	3.10		14.25	17.20
	0700	3" diameter		100	.080		4.76	3.10		7.86	10.15
	0800	4" diameter		100	.080		7.80	3.10		10.90	13.50
	1000	2" x 3" corrugated		100	.080		5.10	3.10		8.20	10.55
	1100	3" x 4" corrugated		100	.080		6.45	3.10		9.55	11.95
	9000	Minimum labor/equipment charge		4	2	Job		77.50		77.50	123

	07710 | Manufactured Roof Specialties		CREW	DAILY OUTPUT	LABOR-HOURS	UNIT	2004 BARE COSTS MAT.	LABOR	EQUIP.	TOTAL	TOTAL INCL O&P	
550	0010	**GRAVEL STOP** Aluminum, .050″ thick, 4″ face height, mill finish	1 Shee	145	.055	L.F.	3.45	2.14		5.59	7.20	550
	0080	Duranodic finish		145	.055	″	3.69	2.14		5.83	7.45	
	0100	Painted		145	.055	L.F.	4.26	2.14		6.40	8.10	
	0300	6″ face height		135	.059		3.64	2.30		5.94	7.65	
	0350	Duranodic finish		135	.059		4.35	2.30		6.65	8.45	
	0400	Painted		135	.059		5.05	2.30		7.35	9.20	
	0600	8″ face height		125	.064		4.75	2.48		7.23	9.20	
	0650	Duranodic finish		125	.064		4.98	2.48		7.46	9.45	
	0700	Painted		125	.064		5.05	2.48		7.53	9.55	
	0900	12″ face height, .080 thick, 2 piece		100	.080		6.40	3.10		9.50	11.95	
	0950	Duranodic finish		100	.080		6.30	3.10		9.40	11.80	
	1000	Painted		100	.080		7.40	3.10		10.50	13.05	
	1500	Polyvinyl chloride, 6″ face height		135	.059		3.28	2.30		5.58	7.25	
	1600	9″ face height		125	.064		3.87	2.48		6.35	8.20	
	1800	Stainless steel, 24 ga., 6″ face height		135	.059		7.15	2.30		9.45	11.50	
	1900	12″ face height		100	.080		15	3.10		18.10	21.50	
	2100	20 ga., 6″ face height		135	.059		8.10	2.30		10.40	12.55	
	2200	12″ face height		100	.080		17.10	3.10		20.20	23.50	
	9000	Minimum labor/equipment charge		3.50	2.286	Job		88.50		88.50	140	
650	0010	**GUTTERS** Aluminum, stock units, 5″ box, .027″ thick, plain	1 Shee	120	.067	L.F.	1.22	2.59		3.81	5.45	650
	0100	Enameled		120	.067		1.09	2.59		3.68	5.30	
	0300	5″ box type, .032″ thick, plain		120	.067		1.22	2.59		3.81	5.45	
	0400	Enameled		120	.067		1.23	2.59		3.82	5.45	
	0600	5″ x 6″ combination fascia & gutter, .032″ thick, enameled		60	.133		3.52	5.15		8.67	12.05	
	0700	Copper, half round, 16 oz, stock units, 4″ wide		120	.067		4.01	2.59		6.60	8.50	
	0900	5″ wide		120	.067		5.15	2.59		7.74	9.75	
	1000	6″ wide		115	.070		6.20	2.70		8.90	11.10	
	1200	K type, 16 oz, stock, 4″ wide		120	.067		4.74	2.59		7.33	9.30	
	1300	5″ wide		120	.067		5.20	2.59		7.79	9.80	
	1500	Lead coated copper, half round, stock, 4″ wide		120	.067		6.50	2.59		9.09	11.25	
	1600	6″ wide		115	.070		9.95	2.70		12.65	15.25	
	1800	K type, stock, 4″ wide		120	.067		7.65	2.59		10.24	12.50	
	1900	5″ wide		120	.067		7.95	2.59		10.54	12.85	
	2100	Stainless steel, half round or box, stock, 4″ wide		120	.067		4.65	2.59		7.24	9.20	
	2200	5″ wide		120	.067		5	2.59		7.59	9.60	
	2400	Steel, galv, half round or box, 28 ga, 5″ wide, plain		120	.067		.95	2.59		3.54	5.15	
	2500	Enameled		120	.067		1.08	2.59		3.67	5.30	
	2700	26 ga, stock, 5″ wide		120	.067		.95	2.59		3.54	5.15	
	2800	6″ wide		120	.067		1.43	2.59		4.02	5.65	
	3000	Vinyl, O.G., 4″ wide	1 Carp	110	.073		.85	2.40		3.25	4.92	
	3100	5″ wide		110	.073		1	2.40		3.40	5.10	
	3200	4″ half round, stock units		110	.073		.68	2.40		3.08	4.73	
	3250	Joint connectors				Ea.	1.36			1.36	1.50	
	3300	Wood, clear treated cedar, fir or hemlock, 3″ x 4″	1 Carp	100	.080	L.F.	6.30	2.64		8.94	11.30	
	3400	4″ x 5″	″	100	.080	″	7.30	2.64		9.94	12.40	
	9000	Minimum labor/equipment charge	1 Shee	3.75	2.133	Job		83		83	131	
700	0010	**GUTTER GUARD** 6″ wide strip, aluminum mesh	1 Carp	500	.016	L.F.	.37	.53		.90	1.29	700
	0100	Vinyl mesh		500	.016	″	.40	.53		.93	1.32	
	9000	Minimum labor/equipment charge		4	2	Job		66		66	109	
750	0010	**REGLET** Aluminum, .025″ thick, in concrete parapet	1 Carp	225	.036	L.F.	.94	1.17		2.11	2.98	750
	0100	Copper, 10 oz.		225	.036		1.62	1.17		2.79	3.73	
	0300	16 oz.		225	.036		2.16	1.17		3.33	4.33	
	0400	Galvanized steel, 24 gauge		225	.036		.69	1.17		1.86	2.71	

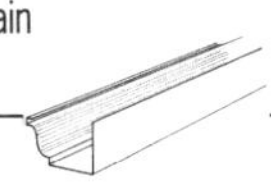

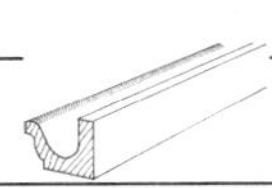

		07710 | Manufactured Roof Specialties	CREW	DAILY OUTPUT	LABOR-HOURS	UNIT	2004 BARE COSTS MAT.	LABOR	EQUIP.	TOTAL	TOTAL INCL O&P	
750	0600	Stainless steel, .020″ thick	1 Carp	225	.036	L.F.	1.60	1.17		2.77	3.71	750
	0700	Zinc and copper alloy, 20 oz.		225	.036		1.79	1.17		2.96	3.92	
	0900	Counter flashing for above, 12″ wide, .032″ aluminum	1 Shee	150	.053		1.21	2.07		3.28	4.61	
	1000	Copper, 10 oz.		150	.053		3.39	2.07		5.46	7	
	1200	16 oz.		150	.053		3.77	2.07		5.84	7.45	
	1300	Galvanized steel, .020″ thick		150	.053		.63	2.07		2.70	3.97	
	1500	Stainless steel, .020″ thick		150	.053		2.82	2.07		4.89	6.40	
	1600	Zinc and copper alloy, 20 oz.		150	.053		3.17	2.07		5.24	6.75	
	9000	Minimum labor/equipment charge	1 Carp	3	2.667	Job		88		88	146	
800	0010	**EXPANSION JOINT**										800
	0300	Butyl or neoprene center with foam insulation, metal flanges										
	0400	Aluminum, .032″ thick for openings to 2-1/2″	1 Rofc	165	.048	L.F.	7.55	1.38		8.93	10.75	
	0600	For joint openings to 3-1/2″		165	.048		8.80	1.38		10.18	12.15	
	0610	For joint openings to 5″		165	.048		10.60	1.38		11.98	14.10	
	0620	For joint openings to 8″		165	.048		16.75	1.38		18.13	21	
	0700	Copper, 16 oz. for openings to 2-1/2″		165	.048		10.60	1.38		11.98	14.10	
	0900	For joint openings to 3-1/2″		165	.048		12.20	1.38		13.58	15.85	
	0910	For joint openings to 5″		165	.048		14.30	1.38		15.68	18.15	
	0920	For joint openings to 8″		165	.048		21.50	1.38		22.88	26	
	1000	Galvanized steel, 26 ga. for openings to 2-1/2″		165	.048		6.40	1.38		7.78	9.45	
	1200	For joint openings to 3-1/2″		165	.048		7.55	1.38		8.93	10.75	
	1210	For joint openings to 5″		165	.048		9.45	1.38		10.83	12.85	
	1220	For joint openings to 8″		165	.048		16.05	1.38		17.43	20	
	1300	Lead-coated copper, 16 oz. for openings to 2-1/2″		165	.048		18.85	1.38		20.23	23.50	
	1500	For joint openings to 3-1/2″		165	.048		22.50	1.38		23.88	27	
	1600	Stainless steel, .018″, for openings to 2-1/2″		165	.048		9.45	1.38		10.83	12.85	
	1800	For joint openings to 3-1/2″		165	.048		10.75	1.38		12.13	14.30	
	1810	For joint openings to 5″		165	.048		13.30	1.38		14.68	17.10	
	1820	For joint openings to 8″		165	.048		20.50	1.38		21.88	25	
	1900	Neoprene, double-seal type with thick center, 4-1/2″ wide		125	.064		9.10	1.82		10.92	13.25	
	1950	Polyethylene bellows, with galv steel flat flanges		100	.080		3.52	2.28		5.80	7.95	
	1960	With galvanized angle flanges		100	.080		3.88	2.28		6.16	8.35	
	2000	Roof joint with extruded aluminum cover, 2″	1 Shee	115	.070		24.50	2.70		27.20	31	
	2100	Roof joint, plastic curbs, foam center, standard	1 Rofc	100	.080		8.85	2.28		11.13	13.85	
	2200	Large		100	.080		11.85	2.28		14.13	17.10	
	2300	Transitions, regular, minimum		10	.800	Ea.	76.50	23		99.50	125	
	2350	Maximum		4	2		97.50	57		154.50	209	
	2400	Large, minimum		9	.889		112	25.50		137.50	169	
	2450	Maximum		3	2.667		117	76		193	265	
	2500	Roof to wall joint with extruded aluminum cover	1 Shee	115	.070	L.F.	20.50	2.70		23.20	27	
	2650											
	2700	Wall joint, closed cell foam on PVC cover, 9″ wide	1 Rofc	125	.064	L.F.	3.16	1.82		4.98	6.75	
	2800	12″ wide	″	115	.070	″	3.57	1.98		5.55	7.45	
	9000	Minimum labor/equipment charge	1 Shee	3	2.667	Job		103		103	164	
		07720 | Roof Accessories										
480	0010	**PITCH POCKETS**										480
	0100	Adjustable, 4″ to 7″, welded corners, 4″ deep	1 Rofc	48	.167	Ea.	10.50	4.74		15.24	20	
	0200	Side extenders, 6″	″	240	.033	″	1.70	.95		2.65	3.57	
500	0010	**ROOF VENTS** Mushroom for built-up roofs, aluminum	1 Rofc	30	.267	Ea.	24	7.60		31.60	40	500
	0100	PVC, 6″ high		30	.267	″	27.50	7.60		35.10	44	
	9000	Minimum labor/equipment charge		2.75	2.909	Job		83		83	148	

07700 | Roof Specialties and Accessories

	07720 | Roof Accessories		Crew	Daily Output	Labor-Hours	Unit	2004 Bare Costs Mat.	Labor	Equip.	Total	Total Incl O&P	
550	0010	**RIDGE VENT**										550
	0100	Aluminum strips, mill finish	1 Rofc	160	.050	L.F.	1.20	1.42		2.62	3.87	
	0150	Painted finish		160	.050	"	2.12	1.42		3.54	4.89	
	0200	Connectors		48	.167	Ea.	1.88	4.74		6.62	10.55	
	0300	End caps		48	.167	"	.92	4.74		5.66	9.50	
	0400	Galvanized strips		160	.050	L.F.	2.07	1.42		3.49	4.83	
	0430	Molded polyethylene, shingles not included		160	.050	"	2.55	1.42		3.97	5.35	
	0440	End plugs		48	.167	Ea.	.92	4.74		5.66	9.50	
	0450	Flexible roll, shingles not included		160	.050	L.F.	1.99	1.42		3.41	4.74	
600	0010	**ROOF DRAINS**										600
	0020	For roof drains see division 15155-300										
700	0010	**ROOF HATCHES** With curb, 1″ fiberglass insulation, 2′-6″ x 3′-0″										700
	0500	Aluminum curb and cover	G-3	10	3.200	Ea.	430	104		534	635	
	0520	Galvanized steel curb and aluminum cover		10	3.200		350	104		454	550	
	0540	Galvanized steel curb and cover		10	3.200		385	104		489	590	
	0600	2′-6″ x 4′-6″, aluminum curb and cover		9	3.556		585	115		700	830	
	0800	Galvanized steel curb and aluminum cover		9	3.556		500	115		615	735	
	0900	Galvanized steel curb and cover		9	3.556		595	115		710	840	
	1100	4′ x 4′ aluminum curb and cover		8	4		610	130		740	880	
	1120	Galvanized steel curb and aluminum cover		8	4		500	130		630	760	
	1140	Galvanized steel curb and cover		8	4		580	130		710	845	
	1200	2′-6″ x 8′-0″, aluminum curb and cover		6.60	4.848		1,175	157		1,332	1,550	
	1400	Galvanized steel curb and aluminum cover		6.60	4.848		965	157		1,122	1,300	
	1500	Galvanized steel curb and cover		6.60	4.848		975	157		1,132	1,325	
	1800	For plexiglass panels, 2′-6″ x 3′-0″, add to above					345			345	380	
	9000	Minimum labor/equipment charge	2 Carp	2	8	Job		264		264	440	
800	0010	**WALKWAY** For built-up roofs, asphalt impregnated, 3′ x 6′ x 1/2″ thk	1 Rofc	400	.020	S.F.	1.08	.57		1.65	2.20	800
	0100	3′ x 3′ x 3/4″ thick		400	.020	"	1.50	.57		2.07	2.67	
	0600	100% recycled rubber, 3′ x 4′ x 3/8″		400	.020	L.F.	4	.57		4.57	5.40	
	0610	3′ x 4′ x 1/2″		400	.020		4.33	.57		4.90	5.80	
	0620	3′ x 4′ x 3/4″		400	.020		5.40	.57		5.97	6.95	
	9000	Minimum labor/equipment charge		2.75	2.909	Job		83		83	148	
850	0010	**SMOKE HATCHES** Unlabeled, not including hand winch operator										850
	0200	For 3′-0″ long, add to roof hatches from division 07720-700				Ea.	25%	5%				
	0250	For 4′-0″ long, add to roof hatches from division 07720-700					20%	5%				
	0300	For 8′-0″ long, add to roof hatches from division 07720-700					10%	5%				
860	0010	**SMOKE VENT,** insulated, 4′ x 4′										860
	0100	Aluminum cover and frame	G-3	13	2.462	Ea.	1,050	80		1,130	1,275	
	0200	Galvanized steel cover and frame		13	2.462		950	80		1,030	1,175	
	0300	4′ x 8′ aluminum cover and frame		8	4		1,425	130		1,555	1,775	
	0400	Galvanized steel cover and frame		8	4		1,250	130		1,380	1,575	
	9000	Minimum labor/equipment charge	2 Carp	2	8	Job		264		264	440	
870	0010	**VENTS, ONE-WAY** For insul. decks, 1 per M.S.F., plastic, min.	1 Rofc	40	.200	Ea.	12.50	5.70		18.20	24	870
	0100	Maximum		20	.400		28.50	11.40		39.90	52	
	0300	Aluminum		30	.267		12.50	7.60		20.10	27.50	
	0800	Polystyrene baffles, 12″ wide for 16″ O.C. rafter spacing	1 Carp	90	.089		.45	2.93		3.38	5.35	
	0900	For 24″ O.C. rafter spacing		110	.073		1.05	2.40		3.45	5.15	
	9000	Minimum labor/equipment charge		3	2.667	Job		88		88	146	

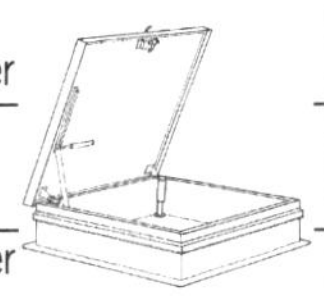

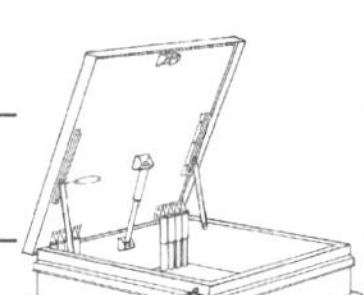

07800 | Fire and Smoke Protection

07812 | Cementitious Fireproofing

		07812 Cementitious Fireproofing	CREW	DAILY OUTPUT	LABOR-HOURS	UNIT	2004 BARE COSTS MAT.	LABOR	EQUIP.	TOTAL	TOTAL INCL O&P	
600	0010	**SPRAYED** Mineral fiber or cementitious for fireproofing,										600
	0050	not incl tamping or canvas protection										
	0100	1" thick, on flat plate steel	G-2	3,000	.008	S.F.	.42	.22	.04	.68	.87	
	0200	Flat decking		2,400	.010		.42	.27	.05	.74	.96	
	0400	Beams		1,500	.016		.42	.44	.07	.93	1.26	
	0500	Corrugated or fluted decks		1,250	.019		.64	.52	.09	1.25	1.65	
	0700	Columns, 1-1/8" thick		1,100	.022		.48	.59	.10	1.17	1.60	
	0800	2-3/16" thick		700	.034		.88	.93	.16	1.97	2.66	
	0850	For tamping, add						10%				
	0900	For canvas protection, add	G-2	5,000	.005	S.F.	.06	.13	.02	.21	.30	
	9000	Minimum labor/equipment charge	"	3	8	Job		218	37	255	395	

07840 | Firestopping

		07840 Firestopping	CREW	DAILY OUTPUT	LABOR-HOURS	UNIT	MAT.	LABOR	EQUIP.	TOTAL	TOTAL INCL O&P	
100	0010	**FIRESTOPPING** R07800-030										100
	0100	Metallic piping, non insulated										
	0110	Through walls, 2" diameter	1 Carp	16	.500	Ea.	9.75	16.50		26.25	38.50	
	0120	4" diameter		14	.571		14.90	18.85		33.75	48	
	0130	6" diameter		12	.667		20	22		42	58.50	
	0140	12" diameter		10	.800		35.50	26.50		62	83	
	0150	Through floors, 2" diameter		32	.250		5.95	8.25		14.20	20	
	0160	4" diameter		28	.286		8.50	9.45		17.95	25	
	0170	6" diameter		24	.333		11.20	11		22.20	30.50	
	0180	12" diameter		20	.400		18.90	13.20		32.10	43	
	0190	Metallic piping, insulated										
	0200	Through walls, 2" diameter	1 Carp	16	.500	Ea.	13.85	16.50		30.35	43	
	0210	4" diameter		14	.571		19	18.85		37.85	52.50	
	0220	6" diameter		12	.667		24	22		46	63	
	0230	12" diameter		10	.800		39.50	26.50		66	87.50	
	0240	Through floors, 2" diameter		32	.250		10	8.25		18.25	24.50	
	0250	4" diameter		28	.286		12.60	9.45		22.05	29.50	
	0260	6" diameter		24	.333		15.30	11		26.30	35	
	0270	12" diameter		20	.400		18.90	13.20		32.10	43	
	0280	Non metallic piping, non insulated										
	0290	Through walls, 2" diameter	1 Carp	12	.667	Ea.	40.50	22		62.50	81	
	0300	4" diameter		10	.800		50.50	26.50		77	99.50	
	0310	6" diameter		8	1		70.50	33		103.50	132	
	0330	Through floors, 2" diameter		16	.500		31.50	16.50		48	62	
	0340	4" diameter		6	1.333		39	44		83	116	
	0350	6" diameter		6	1.333		47	44		91	125	
	0370	Ductwork, insulated & non insulated, round										
	0380	Through walls, 6" diameter	1 Carp	12	.667	Ea.	20.50	22		42.50	59	
	0390	12" diameter		10	.800		41	26.50		67.50	89	
	0400	18" diameter		8	1		66.50	33		99.50	128	
	0410	Through floors, 6" diameter		16	.500		11.20	16.50		27.70	40	
	0420	12" diameter		14	.571		20.50	18.85		39.35	54	
	0430	18" diameter		12	.667		35.50	22		57.50	76	
	0440	Ductwork, insulated & non insulated, rectangular										
	0450	With stiffener/closure angle, through walls, 6" x 12"	1 Carp	8	1	Ea.	17	33		50	73	
	0460	12" x 24"		6	1.333		22.50	44		66.50	98	
	0470	24" x 48"		4	2		64.50	66		130.50	180	
	0480	With stiffener/closure angle, through floors, 6" x 12"		10	.800		9.20	26.50		35.70	54	
	0490	12" x 24"		8	1		16.55	33		49.55	72.50	
	0500	24" x 48"		6	1.333		32.50	44		76.50	109	
	0510	Multi trade openings										
	0520	Through walls, 6" x 12"	1 Carp	2	4	Ea.	35.50	132		167.50	259	

07800 | Fire and Smoke Protection

	07840 | Firestopping		CREW	DAILY OUTPUT	LABOR-HOURS	UNIT	2004 BARE COSTS MAT.	LABOR	EQUIP.	TOTAL	TOTAL INCL O&P	
100	0530	12″ x 24″ R07800-030	1 Carp	1	8	Ea.	144	264		408	600	100
	0540	24″ x 48″	2 Carp	1	16		575	530		1,105	1,500	
	0550	48″ x 96″	″	.75	21.333		2,300	705		3,005	3,725	
	0560	Through floors, 6″ x 12″	1 Carp	2	4		35.50	132		167.50	259	
	0570	12″ x 24″	″	1	8		144	264		408	600	
	0580	24″ x 48″	2 Carp	.75	21.333		575	705		1,280	1,800	
	0590	48″ x 96″	″	.50	32		2,300	1,050		3,350	4,300	
	0600	Structural penetrations, through walls										
	0610	Steel beams, W8 x 10	1 Carp	8	1	Ea.	22.50	33		55.50	79	
	0620	W12 x 14		6	1.333		35.50	44		79.50	113	
	0630	W21 x 44		5	1.600		71.50	53		124.50	166	
	0640	W36 x 135		3	2.667		173	88		261	335	
	0650	Bar joists, 18″ deep		6	1.333		32.50	44		76.50	109	
	0660	24″ deep		6	1.333		41	44		85	118	
	0670	36″ deep		5	1.600		61	53		114	155	
	0680	48″ deep		4	2		71.50	66		137.50	188	
	0690	Construction joints, floor slab at exterior wall										
	0700	Precast, brick, block or drywall exterior										
	0710	2″ wide joint	1 Carp	125	.064	L.F.	5.10	2.11		7.21	9.10	
	0720	4″ wide joint	″	75	.107	″	10.20	3.52		13.72	17.05	
	0730	Metal panel, glass or curtain wall exterior										
	0740	2″ wide joint	1 Carp	40	.200	L.F.	12.05	6.60		18.65	24.50	
	0750	4″ wide joint	″	25	.320	″	16.50	10.55		27.05	35.50	
	0760	Floor slab to drywall partition										
	0770	Flat joint	1 Carp	100	.080	L.F.	5	2.64		7.64	9.90	
	0780	Fluted joint		50	.160		10.20	5.30		15.50	19.95	
	0790	Etched fluted joint		75	.107		6.65	3.52		10.17	13.15	
	0800	Floor slab to concrete/masonry partition										
	0810	Flat joint	1 Carp	75	.107	L.F.	11.20	3.52		14.72	18.20	
	0820	Fluted joint	″	50	.160	″	13.25	5.30		18.55	23.50	
	0830	Concrete/CMU wall joints										
	0840	1″ wide	1 Carp	100	.080	L.F.	6.10	2.64		8.74	11.15	
	0850	2″ wide		75	.107		11.20	3.52		14.72	18.20	
	0860	4″ wide		50	.160		21.50	5.30		26.80	32.50	
	0870	Concrete/CMU floor joints										
	0880	1″ wide	1 Carp	200	.040	L.F.	3.06	1.32		4.38	5.55	
	0890	2″ wide		150	.053		5.60	1.76		7.36	9.05	
	0900	4″ wide		100	.080		10.70	2.64		13.34	16.20	

07900 | Joint Sealers

	07920 | Joint Sealants		CREW	DAILY OUTPUT	LABOR-HOURS	UNIT	2004 BARE COSTS MAT.	LABOR	EQUIP.	TOTAL	TOTAL INCL O&P	
800	0010	CAULKING AND SEALANTS										800
	0020	Acoustical sealant, elastomeric, cartridges				Ea.	2.05			2.05	2.26	
	0032	Backer rod, polyethylene, 1/4″ diameter	1 Bric	460	.017	L.F.	.02	.60		.62	.99	
	0052	1/2″ diameter		460	.017		.03	.60		.63	1.01	
	0072	3/4″ diameter		460	.017		.06	.60		.66	1.03	
	0092	1″ diameter		460	.017		.09	.60		.69	1.07	
	0100	Acrylic latex caulk, white										
	0200	11 fl. oz cartridge				Ea.	1.82			1.82	2	

THERMAL & MOISTURE PROTECTION 7

	07920 | Joint Sealants	CREW	DAILY OUTPUT	LABOR-HOURS	UNIT	2004 BARE COSTS MAT.	LABOR	EQUIP.	TOTAL	TOTAL INCL O&P	
800	0500 1/4" x 1/2"	1 Bric	248	.032	L.F.	.15	1.10		1.25	1.95	800
	0600 1/2" x 1/2"		250	.032		.30	1.10		1.40	2.11	
	0800 3/4" x 3/4"		230	.035		.67	1.19		1.86	2.67	
	0900 3/4" x 1"		200	.040		.89	1.37		2.26	3.20	
	1000 1" x 1"		180	.044		1.12	1.52		2.64	3.70	
	1400 Butyl based, bulk				Gal.	22			22	24.50	
	1500 Cartridges				"	27			27	29.50	
	1700 Bulk, in place 1/4" x 1/2", 154 L.F./gal.	1 Bric	230	.035	L.F.	.14	1.19		1.33	2.09	
	1800 1/2" x 1/2", 77 L.F./gal.	"	180	.044	"	.29	1.52		1.81	2.79	
	2000 Latex acrylic based, bulk				Gal.	23			23	25.50	
	2100 Cartridges				"	26			26	28.50	
	2200 Bulk in place, 1/4" x 1/2", 154 L.F./gal.	1 Bric	230	.035	L.F.	.15	1.19		1.34	2.10	
	2250										
	2300 Polysulfide compounds, 1 component, bulk				Gal.	44			44	48	
	2400 Cartridges				"	46.50			46.50	51.50	
	2600 1 or 2 component, in place, 1/4" x 1/4", 308 L.F./gal.	1 Bric	145	.055	L.F.	.14	1.89		2.03	3.23	
	2700 1/2" x 1/4", 154 L.F./gal.		135	.059		.28	2.03		2.31	3.60	
	2900 3/4" x 3/8", 68 L.F./gal.		130	.062		.64	2.11		2.75	4.13	
	3000 1" x 1/2", 38 L.F./gal.		130	.062		1.15	2.11		3.26	4.69	
	3200 Polyurethane, 1 or 2 component				Gal.	47.50			47.50	52.50	
	3300 Cartridges				"	46.50			46.50	51.50	
	3500 Bulk, in place, 1/4" x 1/4"	1 Bric	150	.053	L.F.	.15	1.83		1.98	3.14	
	3600 1/2" x 1/4"		145	.055		.31	1.89		2.20	3.41	
	3800 3/4" x 3/8", 68 L.F./gal.		130	.062		.70	2.11		2.81	4.19	
	3900 1" x 1/2"		110	.073		1.24	2.49		3.73	5.40	
	4100 Silicone rubber, bulk				Gal.	34.50			34.50	38	
	4200 Cartridges				"	40.50			40.50	44.50	
	4300 Bulk in place, 1/4" x 1/2", 154 L.F./gal.	1 Bric	235	.034	L.F.	.22	1.17		1.39	2.14	
	4350										
	4400 Neoprene gaskets, closed cell, adhesive, 1/8" x 3/8"	1 Bric	240	.033	L.F.	.20	1.14		1.34	2.07	
	4500 1/4" x 3/4"		215	.037		.48	1.27		1.75	2.60	
	4700 1/2" x 1"		200	.040		1.40	1.37		2.77	3.76	
	4800 3/4" x 1-1/2"		165	.048		2.91	1.66		4.57	5.90	
	5500 Resin epoxy coating, 2 component, heavy duty				Gal.	26			26	28.50	
	5802 Tapes, sealant, P.V.C. foam adhesive, 1/16" x 1/4"				L.F.	.05			.05	.05	
	5902 1/16" x 1/2"					.07			.07	.08	
	5952 1/16" x 1"					.11			.11	.12	
	6002 1/8" x 1/2"					.08			.08	.08	
	6200 Urethane foam, 2 component, handy pack, 1 C.F.				Ea.	27.50			27.50	30.50	
	6300 50.0 C.F. pack				C.F.	14.05			14.05	15.45	
	9000 Minimum labor/equipment charge	1 Bric	4	2	Job		68.50		68.50	111	

For information about Means Estimating Seminars, see yellow pages 12 and 13 in back of book

Division 8 Doors & Windows

Estimating Tips

08100 Metal Doors & Frames

- Most metal doors and frames look alike, but there may be significant differences among them. When estimating these items be sure to choose the line item that most closely compares to the specification or door schedule requirements regarding:
  - type of metal
  - metal gauge
  - door core material
  - fire rating
  - finish

08200 Wood & Plastic Doors

- Wood and plastic doors vary considerably in price. The primary determinant is the veneer material. Lauan, birch and oak are the most common veneers. Other variables include the following:
  - hollow or solid core
  - fire rating
  - flush or raised panel
  - finish
- If the specifications require compliance with AWI (Architectural Woodwork Institute) standards or acoustical standards, the cost of the door may increase substantially. All wood doors are priced pre-mortised for hinges and predrilled for cylindrical locksets.
- Frequently doors, frames, and windows are unique in old buildings. Specified replacement units could be stock, custom (similar to the original) or exact reproduction. The estimator should work closely with a window consultant to determine any extra costs that may be associated with the unusual installation requirements.

08300 Specialty Doors

- There are many varieties of special doors, and they are usually priced per each. Add frames, hardware or operators required for a complete installation.

08510 Steel Windows

- Most metal windows are delivered preglazed. However, some metal windows are priced without glass. Refer to 08800 Glazing for glass pricing. The grade C indicates commercial grade windows, usually ASTM C-35.

08550 Wood Windows

- All wood windows are priced preglazed. The two glazing options priced are single pane float glass and insulating glass 1/2" thick. Add the cost of screens and grills if required.

08700 Hardware

- Hardware costs add considerably to the cost of a door. The most efficient method to determine the hardware requirements for a project is to review the door schedule. This schedule, in conjunction with the specifications, is all you should need to take off the door hardware.
- Door hinges are priced by the pair, with most doors requiring 1-1/2 pairs per door. The hinge prices do not include installation labor because it is included in door installation. Hinges are classified according to the frequency of use.

08800 Glazing

- Different openings require different types of glass. The three most common types are:
  - float
  - tempered
  - insulating
- Most exterior windows are glazed with insulating glass. Entrance doors and window walls, where the glass is less than 18" from the floor, are generally glazed with tempered glass. Interior windows and some residential windows are glazed with float glass.

08900 Glazed Curtain Wall

- Glazed curtain walls consist of the metal tube framing and the glazing material. The cost data in this subdivision is presented for the metal tube framing alone or the composite wall. If your estimate requires a detailed takeoff of the framing, be sure to add the glazing cost.

Reference Numbers

Reference numbers are shown in bold squares at the beginning of some major classifications. These numbers refer to related items in the Reference Section. The reference information may be an estimating procedure, an alternate pricing method or technical information.

Note: Not all subdivisions listed here necessarily appear in this publication.

	08060 Selective Demolition	CREW	DAILY OUTPUT	LABOR-HOURS	UNIT	2004 BARE COSTS MAT.	LABOR	EQUIP.	TOTAL	TOTAL INCL O&P
110	0010 SELECTIVE DEMOLITION, DOORS R02220-510									
	0200 Doors, exterior, 1-3/4" thick, single, 3' x 7' high	1 Clab	16	.500	Ea.		13		13	21.50
	0220 Double, 6' x 7' high		12	.667			17.35		17.35	28.50
	0500 Interior, 1-3/8" thick, single, 3' x 7' high		20	.400			10.40		10.40	17.25
	0520 Double, 6' x 7' high		16	.500			13		13	21.50
	0700 Bi-folding, 3' x 6'-8" high		20	.400			10.40		10.40	17.25
	0720 6' x 6'-8" high		18	.444			11.55		11.55	19.15
	0900 Bi-passing, 3' x 6'-8" high		16	.500			13		13	21.50
	0940 6' x 6'-8" high	↓	14	.571			14.85		14.85	24.50
	1500 Remove and reset, minimum	1 Carp	8	1			33		33	54.50
	1520 Maximum		6	1.333			44		44	73
	2000 Frames, including trim, metal	↓	8	1			33		33	54.50
	2200 Wood	2 Carp	32	.500	↓		16.50		16.50	27.50
	2201 Alternate pricing method	1 Carp	200	.040	L.F.		1.32		1.32	2.19
	2950 Minimum labor/equipment charge	1 Clab	4	2	Job		52		52	86
	3000 Special doors, counter doors	2 Carp	6	2.667	Ea.		88		88	146
	3100 Double acting		10	1.600			53		53	87.50
	3200 Floor door (trap type)		8	2			66		66	109
	3300 Glass, sliding, including frames		12	1.333			44		44	73
	3400 Overhead, commercial, 12' x 12' high		4	4			132		132	219
	3440 20' x 16' high		3	5.333			176		176	292
	3500 Residential, 9' x 7' high		8	2			66		66	109
	3540 16' x 7' high		7	2.286			75.50		75.50	125
	3600 Remove and reset, minimum		4	4			132		132	219
	3620 Maximum		2.50	6.400			211		211	350
	3700 Roll-up grille		5	3.200			106		106	175
	3800 Revolving door		2	8			264		264	440
	3900 Storefront swing door	↓	3	5.333	↓		176		176	292
	9000 Minimum labor/equipment charge	1 Carp	4	2	Job		66		66	109
120	0010 SELECTIVE DEMOLITION, WINDOWS R02220-510									
	0200 Aluminum, including trim, to 12 S.F.	1 Clab	16	.500	Ea.		13		13	21.50
	0240 To 25 S.F.		11	.727			18.90		18.90	31.50
	0280 To 50 S.F.		5	1.600			41.50		41.50	69
	0320 Storm windows, to 12 S.F.		27	.296			7.70		7.70	12.75
	0360 To 25 S.F.		21	.381			9.90		9.90	16.40
	0400 To 50 S.F.		16	.500	↓		13		13	21.50
	0600 Glass, minimum		200	.040	S.F.		1.04		1.04	1.72
	0620 Maximum		150	.053	"		1.39		1.39	2.30
	1000 Steel, including trim, to 12 S.F.		13	.615	Ea.		16		16	26.50
	1020 To 25 S.F.		9	.889			23		23	38.50
	1040 To 50 S.F.		4	2			52		52	86
	2000 Wood, including trim, to 12 S.F.		22	.364			9.45		9.45	15.65
	2020 To 25 S.F.		18	.444			11.55		11.55	19.15
	2060 To 50 S.F.	↓	13	.615			16		16	26.50
	5020 Remove and reset window, minimum	1 Carp	6	1.333			44		44	73
	5040 Average		4	2			66		66	109
	5080 Maximum	↓	2	4	↓		132		132	219
	9000 Minimum labor/equipment charge	1 Clab	4	2	Job		52		52	86

08110 | Steel Doors and Frames

		08110 Steel Doors and Frames	Crew	Daily Output	Labor-Hours	Unit	2004 Bare Costs Mat.	2004 Bare Costs Labor	2004 Bare Costs Equip.	2004 Bare Costs Total	Total Incl O&P	
200	0010	**COMMERCIAL STEEL DOORS**										200
	0015	Flush, full panel, hollow core										
	0020	1-3/8" thick, 20 ga., 2'-0"x 6'-8"	2 Carp	20	.800	Ea.	159	26.50		185.50	218	
	0040	2'-8" x 6'-8"		18	.889		163	29.50		192.50	228	
	0060	3'-0" x 6'-8"		17	.941		166	31		197	234	
	0100	3'-0" x 7'-0"		17	.941		161	31		192	229	
	0120	For vision lite, add					48.50			48.50	53.50	
	0140	For narrow lite, add					48.50			48.50	53	
	0160	For bottom louver, add					88.50			88.50	97.50	
	0230	For baked enamel finish, add					30%	15%				
	0260	For galvanizing, add					15%					
	0320	Half glass, 20 ga., 2'-0" x 6'-8"	2 Carp	20	.800	Ea.	223	26.50		249.50	289	
	0340	2'-8" x 6'-8"		18	.889		228	29.50		257.50	299	
	0360	3'-0" x 6'-8"		17	.941		230	31		261	305	
	0400	3'-0" x 7'-0"		17	.941		230	31		261	305	
	0500	Hollow core, 1-3/4" thick, full panel, 20 ga., 2'-8" x 6'-8"		18	.889		173	29.50		202.50	239	
	0520	3'-0" x 6'-8"		17	.941		166	31		197	234	
	0640	3'-0" x 7'-0"		17	.941		181	31		212	251	
	0680	4'-0" x 7'-0"		15	1.067		259	35		294	345	
	0700	4'-0" x 8'-0"		13	1.231		310	40.50		350.50	410	
	1000	18 ga., 2'-8" x 6'-8"		17	.941		191	31		222	262	
	1020	3'-0" x 6'-8"		16	1		186	33		219	259	
	1120	3'-0" x 7'-0"		17	.941		201	31		232	273	
	1180	4'-0" x 7'-0"		14	1.143		259	37.50		296.50	350	
	1200	4'-0" x 8'-0"		17	.941		310	31		341	390	
	1230	Half glass, 20 ga., 2'-8" x 6'-8"		20	.800		230	26.50		256.50	297	
	1240	3'-0" x 6'-8"		18	.889		230	29.50		259.50	300	
	1260	3'-0" x 7'-0"		18	.889		238	29.50		267.50	310	
	1280	4'-0" x 7'-0"		16	1		293	33		326	375	
	1300	4'-0" x 8'-0"		13	1.231		370	40.50		410.50	480	
	1320	18 ga., 2'-8" x 6'-8"		18	.889		253	29.50		282.50	325	
	1340	3'-0" x 6'-8"		17	.941		250	31		281	325	
	1360	3'-0" x 7'-0"		17	.941		257	31		288	335	
	1380	4'-0" x 7'-0"		15	1.067		325	35		360	415	
	1400	4'-0" x 8'-0"		14	1.143		370	37.50		407.50	475	
	1720	Insulated, 1-3/4" thick, full panel, 18 ga., 3'-0" x 6'-8"		15	1.067		186	35		221	263	
	1740	2'-8" x 7'-0"		16	1		197	33		230	272	
	1760	3'-0" x 7'-0"		15	1.067		199	35		234	278	
	1800	4'-0" x 8'-0"		13	1.231		310	40.50		350.50	410	
	1820	Half glass, 18 ga., 3'-0" x 6'-8"		16	1		250	33		283	330	
	1840	2'-8" x 7'-0"		17	.941		261	31		292	340	
	1860	3'-0" x 7'-0"		16	1		280	33		313	365	
	1900	4'-0" x 8'-0"		14	1.143		370	37.50		407.50	475	
	9000	Minimum labor/equipment charge	1 Carp	4	2	Job		66		66	109	
250	0010	**DOOR FRAMES**										250
	0020	Steel channels with anchors and bar stops										
	0100	6" channel @ 8.2#/L.F., 3' x 7' door, weighs 150#	E-4	13	2.462	Ea.	115	92.50	5.80	213.30	305	
	0200	8" channel @ 11.5#/L.F., 6' x 8' door, weighs 275#		9	3.556		210	134	8.35	352.35	490	
	0300	8' x 12' door, weighs 400#		6.50	4.923		305	185	11.55	501.55	695	
	0800	For frames without bar stops, light sections, deduct					15%					
	0900	Heavy sections, deduct					10%					
	9000	Minimum labor/equipment charge	E-4	4	8	Job		300	18.80	318.80	580	
300	0010	**FIRE DOOR**										300
	0015	Steel, flush, "B" label, 90 minute										
	0020	Full panel, 20 ga., 2'-0" x 6'-8"	2 Carp	20	.800	Ea.	191	26.50		217.50	254	
	0040	2'-8" x 6'-8"		18	.889		198	29.50		227.50	267	

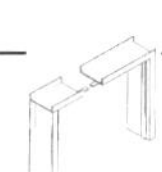

08110 | Steel Doors and Frames

		08110 Steel Doors and Frames	CREW	DAILY OUTPUT	LABOR-HOURS	UNIT	2004 BARE COSTS MAT.	LABOR	EQUIP.	TOTAL	TOTAL INCL O&P	
300	0060	3'-0" x 6'-8"	2 Carp	17	.941	Ea.	198	31		229	270	300
	0080	3'-0" x 7'-0"		17	.941		206	31		237	279	
	0140	18 ga., 3'-0" x 6'-8"		16	1		218	33		251	295	
	0160	2'-8" x 7'-0"		17	.941		231	31		262	305	
	0180	3'-0" x 7'-0"		16	1		226	33		259	305	
	0200	4'-0" x 7'-0"		15	1.067		292	35		327	380	
	0220	For "A" label, 3 hour, 18 ga., use same price as "B" label										
	0240	For vision lite, add				Ea.	47			47	51.50	
	0520	Flush, "B" label 90 min., composite, 20 ga., 2'-0" x 6'-8"	2 Carp	18	.889		252	29.50		281.50	325	
	0540	2'-8" x 6'-8"		17	.941		257	31		288	335	
	0560	3'-0" x 6'-8"		16	1		259	33		292	340	
	0580	3'-0" x 7'-0"		16	1		267	33		300	350	
	0640	Flush, "A" label 3 hour, composite, 18 ga., 3'-0" x 6'-8"		15	1.067		218	35		253	299	
	0660	2'-8" x 7'-0"		16	1		229	33		262	305	
	0680	3'-0" x 7'-0"		15	1.067		226	35		261	305	
	0700	4'-0" x 7'-0"		14	1.143		292	37.50		329.50	385	
	9000	Minimum labor/equipment charge	1 Carp	4	2	Job		66		66	109	
600	0010	**RESIDENTIAL STEEL DOOR**										600
	0020	Prehung, insulated, exterior										
	0030	Embossed, full panel, 2'-8" x 6'-8"	2 Carp	17	.941	Ea.	190	31		221	261	
	0040	3'-0" x 6'-8"		15	1.067		190	35		225	268	
	0060	3'-0" x 7'-0"		15	1.067		249	35		284	335	
	0070	5'-4" x 6'-8", double		8	2		375	66		441	525	
	0220	Half glass, 2'-8" x 6'-8"		17	.941		230	31		261	305	
	0240	3'-0" x 6'-8"		16	1		232	33		265	310	
	0260	3'-0" x 7'-0"		16	1		282	33		315	365	
	0270	5'-4" x 6'-8", double		8	2		480	66		546	640	
	0720	Raised plastic face, full panel, 2'-8" x 6'-8"		16	1		237	33		270	315	
	0740	3'-0" x 6'-8"		15	1.067		239	35		274	320	
	0760	3'-0" x 7'-0"		15	1.067		241	35		276	325	
	0780	5'-4" x 6'-8", double		8	2		445	66		511	600	
	0820	Half glass, 2'-8" x 6'-8"		17	.941		265	31		296	345	
	0840	3'-0" x 6'-8"		16	1		268	33		301	350	
	0860	3'-0" x 7'-0"		16	1		295	33		328	380	
	0880	5'-4" x 6'-8", double		8	2		595	66		661	765	
	1320	Flush face, full panel, 2'-6" x 6'-8"		16	1		192	33		225	266	
	1340	3'-0" x 6'-8"		15	1.067		195	35		230	274	
	1360	3'-0" x 7'-0"		15	1.067		247	35		282	330	
	1380	5'-4" x 6'-8", double		8	2		405	66		471	555	
	1420	Half glass, 2'-8" x 6'-8"		17	.941		241	31		272	320	
	1440	3'-0" x 6'-8"		16	1		245	33		278	325	
	1460	3'-0" x 7'-0"		16	1		284	33		317	365	
	1480	5'-4" x 6'-8", double		8	2		475	66		541	630	
	2300	Interior, residential, closet, bi-fold, 6'-8" x 2'-0" wide		16	1		129	33		162	197	
	2330	3'-0" wide		16	1		145	33		178	215	
	2360	4'-0" wide		15	1.067		220	35		255	300	
	2400	5'-0" wide		14	1.143		255	37.50		292.50	345	
	2420	6'-0" wide		13	1.231		286	40.50		326.50	385	
	9000	Minimum labor/equipment charge	1 Carp	4	2	Job		66		66	109	
820	0010	**STEEL FRAMES, KNOCK DOWN**										820
	0020	16 ga., up to 5-3/4" deep										
	0025	6'-8" high, 3'-0" wide, single	2 Carp	16	1	Ea.	75	33		108	137	
	0040	6'-0" wide, double		14	1.143		90.50	37.50		128	162	
	0100	7'-0" high, 3'-0" wide, single		16	1		77	33		110	139	
	0140	6'-0" wide, double		14	1.143		94.50	37.50		132	167	

		08110 | Steel Doors and Frames	CREW	DAILY OUTPUT	LABOR-HOURS	UNIT	2004 BARE COSTS MAT.	LABOR	EQUIP.	TOTAL	TOTAL INCL O&P	
820	1000	16 ga., up to 4-7/8" deep, 7'-0" H, 3'-0" W, single	2 Carp	16	1	Ea.	74	33		107	136	820
	1140	6'-0" wide, double		14	1.143		91	37.50		128.50	163	
	2800	14 ga., up to 3-7/8" deep, 7'-0" high, 3'-0" wide, single		16	1		76	33		109	138	
	2840	6'-0" wide, double		14	1.143		93	37.50		130.50	165	
	3600	5-3/4" deep, 7'-0" high, 4'-0" wide, single		15	1.067		65	35		100	130	
	3640	8'-0" wide, double		12	1.333		86	44		130	168	
	3700	8'-0" high, 4'-0" wide, single		15	1.067		86	35		121	153	
	3740	8'-0" wide, double		12	1.333		107	44		151	190	
	4000	6-3/4" deep, 7'-0" high, 4'-0" wide, single		15	1.067		90	35		125	158	
	4020	6'-0" wide, double		12	1.333		103	44		147	187	
	4040	8'-0", wide double		12	1.333		120	44		164	205	
	4100	8'-0" high, 4'-0" wide, single		15	1.067		102	35		137	171	
	4140	8'-0" wide, double		12	1.333		119	44		163	204	
	4400	8-3/4" deep, 7'-0" high, 4'-0" wide, single		15	1.067		95.50	35		130.50	164	
	4440	8'-0" wide, double		12	1.333		126	44		170	212	
	4500	8'-0" high, 4'-0" wide, single		15	1.067		112	35		147	182	
	4540	8'-0" wide, double		12	1.333		131	44		175	217	
	4900	For welded frames, add					28			28	31	
	5400	14 ga., "B" label, up to 5-3/4" deep, 7'-0" high, 4'-0" wide, single	2 Carp	15	1.067		99.50	35		134.50	168	
	5440	8'-0" wide, double		12	1.333		116	44		160	201	
	5800	6-3/4" deep, 7'-0" high, 4'-0" wide, single		15	1.067		88.50	35		123.50	156	
	5840	8'-0" wide, double		12	1.333		128	44		172	214	
	6200	8-3/4" deep, 7'-0" high, 4'-0" wide, single		15	1.067		108	35		143	177	
	6240	8'-0" wide, double		12	1.333		138	44		182	225	
	6300	For "A" label use same price as "B" label										
	6400	For baked enamel finish, add					30%	15%				
	6500	For galvanizing, add					15%					
	7900	Transom lite frames, fixed, add	2 Carp	155	.103	S.F.	27	3.41		30.41	35	
	8000	Movable, add	"	130	.123	"	32.50	4.06		36.56	42.50	
	9000	Minimum labor/equipment charge	1 Carp	4	2	Job		66		66	109	

		08180 | Metal Screen and Storm Doors	CREW	DAILY OUTPUT	LABOR-HOURS	UNIT	MAT.	LABOR	EQUIP.	TOTAL	TOTAL INCL O&P	
100	0010	**STORM DOORS & FRAMES** Aluminum, residential,										100
	0020	combination storm and screen										
	0400	Clear anodic coating, 6'-8" x 2'-6" wide	2 Carp	15	1.067	Ea.	152	35		187	226	
	0420	2'-8" wide		14	1.143		174	37.50		211.50	254	
	0440	3'-0" wide		14	1.143		174	37.50		211.50	254	
	0500	For 7' door height, add					5%					
	1000	Mill finish, 6'-8" x 2'-6" wide	2 Carp	15	1.067	Ea.	202	35		237	281	
	1020	2'-8" wide		14	1.143		202	37.50		239.50	285	
	1040	3'-0" wide		14	1.143		218	37.50		255.50	305	
	1100	For 7'-0" door, add					5%					
	1500	White painted, 6'-8" x 2'-6" wide	2 Carp	15	1.067		202	35		237	281	
	1520	2'-8" wide		14	1.143		204	37.50		241.50	288	
	1540	3'-0" wide		14	1.143		214	37.50		251.50	298	
	1600	For 7'-0" door, add					5%					
	2000	Wood door & screen, see division 08210-930										
	2020											
	9000	Minimum labor/equipment charge	1 Carp	4	2	Job		66		66	109	

	08210 | Wood Doors		Crew	Daily Output	Labor-Hours	Unit	2004 Bare Costs Mat.	Labor	Equip.	Total	Total Incl O&P	
450	0010	**KALAMEIN**										450
	0020	Interior, flush type, 3′ x 7′	2 Carp	4.30	3.721	Opng.	165	123		288	385	
	9000	Minimum labor/equipment charge	1 Carp	2	4	Job		132		132	219	
720	0010	**PRE-HUNG DOORS**										720
	0300	Exterior, wood, comb. storm & screen, 6′-9″ x 2′-6″ wide	2 Carp	15	1.067	Ea.	266	35		301	350	
	0320	2′-8″ wide		15	1.067		266	35		301	350	
	0340	3′-0″ wide		15	1.067		273	35		308	360	
	0360	For 7′-0″ high door, add					23			23	25.50	
	0370	For aluminum storm doors, see division 08280-800										
	1600	Entrance door, flush, birch, solid core										
	1620	4-5/8″ solid jamb, 1-3/4″ x 6′-8″ x 2′-8″ wide	2 Carp	16	1	Ea.	272	33		305	355	
	1640	3′-0″ wide	″	16	1		279	33		312	360	
	1680	For 7′-0″ high door, add					12.95			12.95	14.25	
	2000	Entrance door, colonial, 6 panel pine										
	2020	4-5/8″ solid jamb, 1-3/4″ x 6′-8″ x 2′-8″ wide	2 Carp	16	1	Ea.	445	33		478	545	
	2040	3′-0″ wide	″	16	1		470	33		503	575	
	2060	For 7′-0″ high door, add					35			35	38	
	2200	For 5-5/8″ solid jamb, add					22.50			22.50	24.50	
	2990											
	4000	Interior, passage door, 4-5/8″ solid jamb										
	4400	Lauan, flush, solid core, 1-3/8″ x 6′-8″ x 2′-6″ wide	2 Carp	20	.800	Ea.	167	26.50		193.50	228	
	4420	2′-8″ wide		20	.800		147	26.50		173.50	205	
	4440	3′-0″ wide		19	.842		151	28		179	213	
	4600	Hollow core, 1-3/8″ x 6′-8″ x 2′-6″ wide		20	.800		109	26.50		135.50	164	
	4620	2′-8″ wide		20	.800		110	26.50		136.50	166	
	4640	3′-0″ wide		19	.842		112	28		140	169	
	4700	For 7′-0″ high door, add					21.50			21.50	24	
	5000	Birch, flush, solid core, 1-3/8″ x 6′-8″ x 2′-6″ wide	2 Carp	20	.800		154	26.50		180.50	213	
	5020	2′-8″ wide		20	.800		174	26.50		200.50	235	
	5040	3′-0″ wide		19	.842		186	28		214	250	
	5200	Hollow core, 1-3/8″ x 6′-8″ x 2′-6″ wide		20	.800		127	26.50		153.50	184	
	5220	2′-8″ wide		20	.800		133	26.50		159.50	191	
	5240	3′-0″ wide		19	.842		133	28		161	192	
	5280	For 7′-0″ high door, add					15.45			15.45	17	
	5500	Hardboard paneled, 1-3/8″ x 6′-8″ x 2′-6″ wide	2 Carp	20	.800		128	26.50		154.50	185	
	5520	2′-8″ wide		20	.800		134	26.50		160.50	192	
	5540	3′-0″ wide		19	.842		132	28		160	192	
	6000	Pine paneled, 1-3/8″ x 6′-8″ x 2′-6″ wide		20	.800		220	26.50		246.50	286	
	6020	2′-8″ wide		20	.800		237	26.50		263.50	305	
	6040	3′-0″ wide		19	.842		246	28		274	315	
	6500	For 5-5/8″ solid jamb, add					10.80			10.80	11.85	
	6520	For split jamb, deduct					13.10			13.10	14.40	
	9000	Minimum labor/equipment charge	1 Carp	4	2	Job		66		66	109	
850	0010	**TIN CLAD**										850
	0020	3 ply, 6′ x 7′, double sliding, doors only	2 Carp	1	16	Opng.	560	530		1,090	1,500	
	1000	For electric operator, add	1 Elec	2	4	″	2,300	158		2,458	2,775	
	9000	Minimum labor/equipment charge	2 Carp	1	16	Job		530		530	875	
900	0010	**WOOD DOOR, ARCHITECTURAL**										900
	0015	Flush, int., 1-3/4″, 7 ply, hollow core,										
	0020	Lauan face, 2′-0″ x 6′-8″	2 Carp	17	.941	Ea.	27	31		58	81	
	0040	2′-6″ x 6′-8″		17	.941		31	31		62	85.50	
	0080	3′-0″ x 6′-8″		17	.941		37	31		68	92	
	0100	4′-0″ x 6′-8″		16	1		65	33		98	126	
	0120	Birch face, 2′-0″ x 6′-8″		17	.941		43.50	31		74.50	99.50	
	0140	2′-6″ x 6′-8″		17	.941		41.50	31		72.50	97.50	

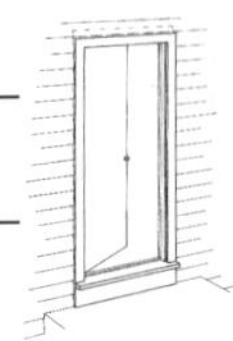

	08210 | Wood Doors	CREW	DAILY OUTPUT	LABOR-HOURS	UNIT	2004 BARE COSTS MAT.	LABOR	EQUIP.	TOTAL	TOTAL INCL O&P	
900	0180 3'-0" x 6'-8"	2 Carp	17	.941	Ea.	47	31		78	104	900
	0200 4'-0" x 6'-8"		16	1		88.50	33		121.50	152	
	0220 Oak face, 2'-0" x 6'-8"		17	.941		67.50	31		98.50	126	
	0240 2'-6" x 6'-8"		17	.941		72.50	31		103.50	131	
	0280 3'-0" x 6'-8"		17	.941		77.50	31		108.50	137	
	0300 4'-0" x 6'-8"		16	1		98	33		131	163	
	0320 Walnut face, 2'-0" x 6'-8"		17	.941		137	31		168	203	
	0340 2'-6" x 6'-8"		17	.941		140	31		171	206	
	0380 3'-0" x 6'-8"		17	.941		145	31		176	211	
	0400 4'-0" x 6'-8"	↓	16	1		164	33		197	236	
	0430 For 7'-0" high, add					13.90			13.90	15.30	
	0440 For 8'-0" high, add					19.45			19.45	21.50	
	0480 For prefinishing, clear, add					30.50			30.50	33.50	
	0500 For prefinishing, stain, add					41.50			41.50	46	
	1320 M.D. overlay on hardboard, 2'-0" x 6'-8"	2 Carp	17	.941		86	31		117	147	
	1340 2'-6" x 6'-8"		17	.941		86	31		117	147	
	1380 3'-0" x 6'-8"		17	.941		102	31		133	164	
	1400 4'-0" x 6'-8"	↓	16	1		139	33		172	208	
	1420 For 7'-0" high, add					7.65			7.65	8.40	
	1440 For 8'-0" high, add					20.50			20.50	22.50	
	1720 H.P. plastic laminate, 2'-0" x 6'-8"	2 Carp	16	1		203	33		236	278	
	1740 2'-6" x 6'-8"		16	1		203	33		236	278	
	1780 3'-0" x 6'-8"		15	1.067		232	35		267	315	
	1800 4'-0" x 6'-8"	↓	14	1.143		325	37.50		362.50	420	
	1820 For 7'-0" high, add					8.05			8.05	8.85	
	1840 For 8'-0" high, add					20.50			20.50	22.50	
	2020 5 ply particle core, lauan face, 2'-6" x 6'-8"	2 Carp	15	1.067		70	35		105	136	
	2040 3'-0" x 6'-8"		14	1.143		73	37.50		110.50	143	
	2080 3'-0" x 7'-0"		13	1.231		80	40.50		120.50	156	
	2100 4'-0" x 7'-0"		12	1.333		94.50	44		138.50	177	
	2120 Birch face, 2'-6" x 6'-8"		15	1.067		80	35		115	147	
	2140 3'-0" x 6'-8"		14	1.143		87.50	37.50		125	159	
	2180 3'-0" x 7'-0"		13	1.231		89.50	40.50		130	166	
	2200 4'-0" x 7'-0"		12	1.333		109	44		153	193	
	2220 Oak face, 2'-6" x 6'-8"		15	1.067		88	35		123	156	
	2240 3'-0" x 6'-8"		14	1.143		97	37.50		134.50	170	
	2280 3'-0" x 7'-0"		13	1.231		99.50	40.50		140	178	
	2300 4'-0" x 7'-0"		12	1.333		122	44		166	207	
	2320 Walnut face, 2'-0" x 6'-8"		15	1.067		98	35		133	167	
	2340 2'-6" x 6'-8"		14	1.143		112	37.50		149.50	186	
	2380 3'-0" x 6'-8"		13	1.231		126	40.50		166.50	206	
	2400 4'-0" x 6'-8"	↓	12	1.333		164	44		208	254	
	2440 For 8'-0" high, add					24			24	26.50	
	2460 For 8'-0" high walnut, add					13.15			13.15	14.45	
	2480 For solid wood core, add					29.50			29.50	32.50	
	2720 For prefinishing, clear, add					19.15			19.15	21	
	2740 For prefinishing, stain, add				↓	42.50			42.50	47	
	2750										
	3320 M.D. overlay on hardboard, 2'-6" x 6'-8"	2 Carp	14	1.143	Ea.	83.50	37.50		121	155	
	3340 3'-0" x 6'-8"		13	1.231		87.50	40.50		128	164	
	3380 3'-0" x 7'-0"		12	1.333		89	44		133	171	
	3400 4'-0" x 7'-0"	↓	10	1.600		109	53		162	208	
	3440 For 8'-0" height, add					24.50			24.50	27	
	3460 For solid wood core, add					31.50			31.50	35	
	3720 H.P. plastic laminate, 2'-6" x 6'-8"	2 Carp	13	1.231		122	40.50		162.50	203	
	3740 3'-0" x 6'-8"	↓	12	1.333	↓	139	44		183	225	

For expanded coverage of these items see *Means Interior Cost Data 2004*

	08210 | Wood Doors	CREW	DAILY OUTPUT	LABOR-HOURS	UNIT	2004 BARE COSTS MAT.	LABOR	EQUIP.	TOTAL	TOTAL INCL O&P	
900	3780 3'-0" x 7'-0"	2 Carp	11	1.455	Ea.	144	48		192	238	900
	3800 4'-0" x 7'-0"		8	2		175	66		241	300	
	3840 For 8'-0" height, add					24.50			24.50	27	
	3860 For solid wood core, add					30			30	32.50	
	4000 Exterior, flush, solid wood stave core, birch, 1-3/4" x 7'-0" x 2'-6"	2 Carp	15	1.067		152	35		187	226	
	4020 2'-8" wide		15	1.067		159	35		194	233	
	4040 3'-0" wide		14	1.143		167	37.50		204.50	247	
	4100 Oak faced 1-3/4" x 7'-0" x 2'-6" wide		15	1.067		167	35		202	243	
	4120 2'-8" wide		15	1.067		179	35		214	255	
	4140 3'-0" wide		14	1.143		190	37.50		227.50	272	
	4200 Walnut faced, 1-3/4" x 7'-0" x 2'-6" wide		15	1.067		245	35		280	330	
	4220 2'-8" wide		15	1.067		256	35		291	340	
	4240 3'-0" wide		14	1.143		267	37.50		304.50	355	
	4300 For 6'-8" high door, deduct from 7'-0" door					13.65			13.65	15	
	9000 Minimum labor/equipment charge	1 Carp	4	2	Job		66		66	109	
910	0010 **WOOD DOORS, DECORATOR**										910
	3000 Solid wood, 1-3/4" thick stile and rail										
	3020 Mahogany, 3'-0" x 7'-0", minimum	2 Carp	14	1.143	Ea.	855	37.50		892.50	1,000	
	3030 Maximum		10	1.600		1,075	53		1,128	1,300	
	3040 3'-6" x 8'-0", minimum		10	1.600		870	53		923	1,050	
	3050 Maximum		8	2		1,000	66		1,066	1,200	
	3100 Pine, 3'-0" x 7'-0", minimum		14	1.143		305	37.50		342.50	400	
	3110 Maximum		10	1.600		550	53		603	695	
	3120 3'-6" x 8'-0", minimum		10	1.600		605	53		658	760	
	3130 Maximum		8	2		1,000	66		1,066	1,200	
	3200 Red oak, 3'-0" x 7'-0", minimum		14	1.143		830	37.50		867.50	975	
	3210 Maximum		10	1.600		1,375	53		1,428	1,625	
	3220 3'-6" x 8'-0", minimum		10	1.600		555	53		608	700	
	3230 Maximum		8	2		1,600	66		1,666	1,850	
	4000 Hand carved door, mahogany										
	4020 3'-0" x 7'-0", minimum	2 Carp	14	1.143	Ea.	1,375	37.50		1,412.50	1,575	
	4030 Maximum		11	1.455		2,150	48		2,198	2,450	
	4040 3'-6" x 8'-0", minimum		10	1.600		1,375	53		1,428	1,625	
	4050 Maximum		8	2		2,450	66		2,516	2,775	
	4200 Red oak, 3'-0" x 7'-0", minimum		14	1.143		4,000	37.50		4,037.50	4,475	
	4210 Maximum		11	1.455		11,000	48		11,048	12,200	
	4220 3'-6" x 8'-0", minimum		10	1.600		4,500	53		4,553	5,050	
	4280 For 6'-8" high door, deduct from 7'-0" door					30			30	33	
	4400 For custom finish, add					310			310	340	
	4600 Side light, mahogany, 7'-0" x 1'-6" wide, minimum	2 Carp	18	.889		745	29.50		774.50	865	
	4610 Maximum		14	1.143		2,150	37.50		2,187.50	2,425	
	4620 8'-0" x 1'-6" wide, minimum		14	1.143		1,450	37.50		1,487.50	1,675	
	4630 Maximum		10	1.600		1,650	53		1,703	1,900	
	4640 Side light, oak, 7'-0" x 1'-6" wide, minimum		18	.889		840	29.50		869.50	975	
	4650 Maximum		14	1.143		1,575	37.50		1,612.50	1,825	
	4660 8'-0" x 1-6" wide, minimum		14	1.143		735	37.50		772.50	870	
	4670 Maximum		10	1.600		1,575	53		1,628	1,850	
	6520 Interior cafe doors, 2'-6" opening, stock, panel pine		16	1		168	33		201	240	
	6540 3'-0" opening		16	1		175	33		208	248	
	6550 Louvered pine										
	6560 2'-6" opening	2 Carp	16	1	Ea.	147	33		180	216	
	8000 3'-0" opening		16	1		157	33		190	228	
	8010 2'-6" opening, hardwood		16	1		260	33		293	340	
	8020 3'-0" opening		16	1		286	33		319	370	
	8800 Pre-hung doors, see division 08210-720										

	08210 | Wood Doors		CREW	DAILY OUTPUT	LABOR-HOURS	UNIT	2004 BARE COSTS MAT.	LABOR	EQUIP.	TOTAL	TOTAL INCL O&P	
910	9000	Minimum labor/equipment charge	1 Carp	4	2	Job		66		66	109	910
920	0010	WOOD DOORS, PANELED										920
	0020	Interior, six panel, hollow core, 1-3/8" thick										
	0040	Molded hardboard, 2'-0" x 6'-8"	2 Carp	17	.941	Ea.	44.50	31		75.50	100	
	0060	2'-6" x 6'-8"		17	.941		48	31		79	104	
	0070	2'-8" x 6'-8"		17	.941		50	31		81	107	
	0080	3'-0" x 6'-8"		17	.941		52.50	31		83.50	110	
	0140	Embossed print, molded hardboard, 2'-0" x 6'-8"		17	.941		48	31		79	104	
	0160	2'-6" x 6'-8"		17	.941		48	31		79	104	
	0180	3'-0" x 6'-8"		17	.941		52.50	31		83.50	110	
	0540	Six panel, solid, 1-3/8" thick, pine, 2'-0" x 6'-8"		15	1.067		116	35		151	186	
	0560	2'-6" x 6'-8"		14	1.143		130	37.50		167.50	206	
	0580	3'-0" x 6'-8"		13	1.231		150	40.50		190.50	232	
	1020	Two panel, bored rail, solid, 1-3/8" thick, pine, 1'-6" x 6'-8"		16	1		211	33		244	287	
	1040	2'-0" x 6'-8"		15	1.067		277	35		312	365	
	1060	2'-6" x 6'-8"		14	1.143		315	37.50		352.50	415	
	1340	Two panel, solid, 1-3/8" thick, fir, 2'-0" x 6'-8"		15	1.067		116	35		151	186	
	1360	2'-6" x 6'-8"		14	1.143		130	37.50		167.50	206	
	1380	3'-0" x 6'-8"		13	1.231		315	40.50		355.50	420	
	1740	Five panel, solid, 1-3/8" thick, fir, 2'-0" x 6'-8"		15	1.067		207	35		242	287	
	1760	2'-6" x 6'-8"		14	1.143		335	37.50		372.50	430	
	1780	3'-0" x 6'-8"	↓	13	1.231	↓	335	40.50		375.50	435	
	9000	Minimum labor/equipment charge	1 Carp	4	2	Job		66		66	109	
930	0010	WOOD DOORS, RESIDENTIAL										930
	0200	Exterior, combination storm & screen, pine										
	0260	2'-8" wide	2 Carp	10	1.600	Ea.	256	53		309	370	
	0280	3'-0" wide		9	1.778		261	58.50		319.50	385	
	0300	7'-1" x 3'-0" wide		9	1.778		279	58.50		337.50	400	
	0400	Full lite, 6'-9" x 2'-6" wide		11	1.455		266	48		314	370	
	0420	2'-8" wide		10	1.600		266	53		319	380	
	0440	3'-0" wide		9	1.778		273	58.50		331.50	395	
	0500	7'-1" x 3'-0" wide		9	1.778		295	58.50		353.50	420	
	0700	Dutch door, pine, 1-3/4" x 6'-8" x 2'-8" wide, minimum		12	1.333		595	44		639	730	
	0720	Maximum		10	1.600		630	53		683	780	
	0800	3'-0" wide, minimum		12	1.333		620	44		664	755	
	0820	Maximum		10	1.600		670	53		723	830	
	1000	Entrance door, colonial, 1-3/4" x 6'-8" x 2'-8" wide		16	1		330	33		363	415	
	1020	6 panel pine, 3'-0" wide		15	1.067		355	35		390	450	
	1100	8 panel pine, 2'-8" wide		16	1		585	33		618	700	
	1120	3'-0" wide	↓	15	1.067		500	35		535	610	
	1200	For tempered safety glass lites, add					24.50			24.50	27	
	1300	Flush, birch, solid core, 1-3/4" x 6'-8" x 2'-8" wide	2 Carp	16	1		83.50	33		116.50	146	
	1320	3'-0" wide		15	1.067		91.50	35		126.50	160	
	1350	7'-0" x 2'-8" wide		16	1		87	33		120	150	
	1360	3'-0" wide	↓	15	1.067		99.50	35		134.50	168	
	1380	For tempered safety glass lites, add				↓	51			51	56	
	1550	For handcarved door, see division 08210-910										
	2700	Interior, closet, bi-fold, w/hardware, no frame or trim incl.										
	2720	Flush, birch, 6'-6" or 6'-8" x 2'-6" wide	2 Carp	13	1.231	Ea.	45.50	40.50		86	118	
	2740	3'-0" wide		13	1.231		49.50	40.50		90	122	
	2760	4'-0" wide		12	1.333		90	44		134	172	
	2780	5'-0" wide		11	1.455		91	48		139	180	
	2800	6'-0" wide		10	1.600		98	53		151	196	
	2820	Flush, hardboard, primed, 6'-8" x 2'-6" wide		13	1.231		49	40.50		89.50	122	
	2840	3'-0" wide	↓	13	1.231	↓	53.50	40.50		94	126	

DOORS & WINDOWS 8

08210 | Wood Doors

			CREW	DAILY OUTPUT	LABOR-HOURS	UNIT	2004 BARE COSTS MAT.	LABOR	EQUIP.	TOTAL	TOTAL INCL O&P	
930	2860	4'-0" wide	2 Carp	12	1.333	Ea.	86.50	44		130.50	168	930
	2880	5'-0" wide		11	1.455		94	48		142	183	
	2900	6'-0" wide		10	1.600		111	53		164	210	
	3000	Raised panel pine, 6'-6" or 6'-8" x 2'-6" wide		13	1.231		151	40.50		191.50	234	
	3020	3'-0" wide		13	1.231		170	40.50		210.50	255	
	3040	4'-0" wide		12	1.333		218	44		262	315	
	3060	5'-0" wide		11	1.455		252	48		300	355	
	3080	6'-0" wide		10	1.600		283	53		336	400	
	3200	Louvered, pine 6'-6" or 6'-8" x 2'-6" wide		13	1.231		93.50	40.50		134	171	
	3220	3'-0" wide		13	1.231		147	40.50		187.50	230	
	3240	4'-0" wide		12	1.333		164	44		208	253	
	3260	5'-0" wide		11	1.455		185	48		233	283	
	3280	6'-0" wide		10	1.600		204	53		257	315	
	4400	Bi-passing closet, incl. hardware and frame, no trim incl.										
	4420	Flush, lauan, 6'-8" x 4'-0" wide	2 Carp	12	1.333	Opng.	156	44		200	245	
	4440	5'-0" wide		11	1.455		171	48		219	268	
	4460	6'-0" wide		10	1.600		183	53		236	289	
	4600	Flush, birch, 6'-8" x 4'-0" wide		12	1.333		188	44		232	280	
	4620	5'-0" wide		11	1.455		193	48		241	292	
	4640	6'-0" wide		10	1.600		226	53		279	335	
	4800	Louvered, pine, 6'-8" x 4'-0" wide		12	1.333		370	44		414	480	
	4820	5'-0" wide		11	1.455		360	48		408	475	
	4840	6'-0" wide		10	1.600		455	53		508	590	
	5000	Paneled, pine, 6'-8" x 4'-0" wide		12	1.333		355	44		399	465	
	5020	5'-0" wide		11	1.455		375	48		423	490	
	5040	6'-0" wide		10	1.600		440	53		493	570	
	6100	Folding accordion, closet, including track and frame										
	6120	Vinyl, 2 layer, stock (see also division 10651-100)	2 Carp	400	.040	S.F.	2.93	1.32		4.25	5.40	
	6140	Woven mahogany and vinyl, stock		400	.040		1.57	1.32		2.89	3.92	
	6160	Wood slats with vinyl overlay, stock		400	.040		9.45	1.32		10.77	12.60	
	6180	Economy vinyl, stock		400	.040		1.61	1.32		2.93	3.96	
	6200	Rigid PVC		400	.040		4.40	1.32		5.72	7.05	
	6220	For custom partition, add					25%	10%				
	7310	Passage doors, flush, no frame included										
	7320	Hardboard, hollow core, 1-3/8" x 6'-8" x 1'-6" wide	2 Carp	18	.889	Ea.	37.50	29.50		67	90	
	7330	2'-0" wide		18	.889		38	29.50		67.50	90	
	7340	2'-6" wide		18	.889		42	29.50		71.50	94.50	
	7350	2'-8" wide		18	.889		44	29.50		73.50	97	
	7360	3'-0" wide		17	.941		46.50	31		77.50	103	
	7420	Lauan, hollow core, 1-3/8" x 6'-8" x 1'-6" wide		18	.889		24	29.50		53.50	74.50	
	7440	2'-0" wide		18	.889		25.50	29.50		55	76.50	
	7450	2'-4" wide		18	.889		28.50	29.50		58	80	
	7460	2'-6" wide		18	.889		28.50	29.50		58	80	
	7480	2'-8" wide		18	.889		30	29.50		59.50	81.50	
	7500	3'-0" wide		17	.941		31.50	31		62.50	86.50	
	7700	Birch, hollow core, 1-3/8" x 6'-8" x 1'-6" wide		18	.889		30.50	29.50		60	82	
	7720	2'-0" wide		18	.889		37.50	29.50		67	89.50	
	7740	2'-6" wide		18	.889		41.50	29.50		71	94.50	
	7760	2'-8" wide		18	.889		43.50	29.50		73	96	
	7780	3'-0" wide		17	.941		47	31		78	104	
	8000	Pine louvered, 1-3/8" x 6'-8" x 1'-6" wide		19	.842		89.50	28		117.50	145	
	8020	2'-0" wide		18	.889		113	29.50		142.50	173	
	8040	2'-6" wide		18	.889		123	29.50		152.50	184	
	8060	2'-8" wide		18	.889		130	29.50		159.50	192	
	8080	3'-0" wide		17	.941		139	31		170	205	
	8090											

08210 | Wood Doors			CREW	DAILY OUTPUT	LABOR-HOURS	UNIT	2004 BARE COSTS MAT.	LABOR	EQUIP.	TOTAL	TOTAL INCL O&P	
930	8300	Pine paneled, 1-3/8" x 6'-8" x 1'-6" wide	2 Carp	19	.842	Ea.	101	28		129	157	930
	8320	2'-0" wide		18	.889		117	29.50		146.50	177	
	8330	2'-4" wide		18	.889		128	29.50		157.50	190	
	8340	2'-6" wide		18	.889		131	29.50		160.50	193	
	8360	2'-8" wide		18	.889		142	29.50		171.50	205	
	8380	3'-0" wide	↓	17	.941	↓	148	31		179	215	
	9900	Minimum labor/equipment charge	1 Carp	4	2	Job		66		66	109	
950	0010	**WOOD FIRE DOORS**										950
	0020	Particle core, 7 face plys, "B" label,										
	0040	1 hour, birch face, 1-3/4" x 2'-6" x 6'-8"	2 Carp	14	1.143	Ea.	268	37.50		305.50	360	
	0080	3'-0" x 6'-8"		13	1.231		277	40.50		317.50	375	
	0090	3'-0" x 7'-0"		12	1.333		289	44		333	395	
	0100	4'-0" x 7'-0"		12	1.333		385	44		429	500	
	0140	Oak face, 2'-6" x 6'-8"		14	1.143		268	37.50		305.50	360	
	0180	3'-0" x 6'-8"		13	1.231		278	40.50		318.50	375	
	0190	3'-0" x 7'-0"		12	1.333		290	44		334	395	
	0200	4'-0" x 7'-0"		12	1.333		380	44		424	490	
	0240	Walnut face, 2'-6" x 6'-8"		14	1.143		350	37.50		387.50	450	
	0280	3'-0" x 6'-8"		13	1.231		360	40.50		400.50	465	
	0290	3'-0" x 7'-0"		12	1.333		375	44		419	490	
	0300	4'-0" x 7'-0"		12	1.333		505	44		549	635	
	0440	M.D. overlay on hardboard, 2'-6" x 6'-8"		15	1.067		235	35		270	320	
	0480	3'-0" x 6'-8"		14	1.143		245	37.50		282.50	330	
	0490	3'-0" x 7'-0"		13	1.231		258	40.50		298.50	350	
	0500	4'-0" x 7'-0"		12	1.333		315	44		359	420	
	0540	H.P. plastic laminate, 2'-6" x 6'-8"		13	1.231		315	40.50		355.50	415	
	0590	3'-0" x 7'-0"		11	1.455		335	48		383	450	
	0600	4'-0" x 7'-0"		10	1.600		425	53		478	560	
	0740	90 minutes, birch face, 1-3/4" x 2'-6" x 6'-8"		14	1.143		236	37.50		273.50	325	
	0780	3'-0" x 6'-8"		13	1.231		245	40.50		285.50	335	
	0790	3'-0" x 7'-0"		12	1.333		292	44		336	395	
	0800	4'-0" x 7'-0"		12	1.333		350	44		394	465	
	0840	Oak face, 2'-6" x 6'-8"		14	1.143		241	37.50		278.50	330	
	0880	3'-0" x 6'-8"		13	1.231		251	40.50		291.50	345	
	0890	3'-0" x 7'-0"		12	1.333		263	44		307	360	
	0900	4'-0" x 7'-0"		12	1.333		365	44		409	475	
	0940	Walnut face, 2'-6" x 6'-8"		14	1.143		330	37.50		367.50	430	
	0980	3'-0" x 6'-8"		13	1.231		340	40.50		380.50	445	
	0990	3'-0" x 7'-0"		12	1.333		355	44		399	465	
	1000	4'-0" x 7'-0"		12	1.333		510	44		554	640	
	1140	M.D. overlay on hardboard, 2'-6" x 6'-8"		15	1.067		263	35		298	350	
	1180	3'-0" x 6'-8"		14	1.143		270	37.50		307.50	360	
	1190	3'-0" x 7'-0"		13	1.231		280	40.50		320.50	380	
	1200	4'-0" x 7'-0"	↓	12	1.333		380	44		424	495	
	1240	For 8'-0" height, add					48			48	53	
	1260	For 8'-0" height walnut, add					63.50			63.50	70	
	1340	H.P. plastic laminate, 2'-6" x 6'-8"	2 Carp	13	1.231		325	40.50		365.50	430	
	1380	3'-0" x 6'-8"		12	1.333		345	44		389	455	
	1390	3'-0" x 7'-0"		11	1.455		345	48		393	460	
	1400	4'-0" x 7'-0"	↓	10	1.600	↓	450	53		503	585	
	2200	Custom architectural "B" label, flush, 1-3/4" thick, birch,										
	2210	Solid core										
	2220	2'-6" x 7'-0"	2 Carp	15	1.067	Ea.	248	35		283	330	
	2260	3'-0" x 7'-0"		14	1.143		257	37.50		294.50	345	
	2300	4'-0" x 7'-0"	↓	13	1.231	↓	385	40.50		425.50	490	

	08210 | Wood Doors	CREW	DAILY OUTPUT	LABOR-HOURS	UNIT	2004 BARE COSTS MAT.	LABOR	EQUIP.	TOTAL	TOTAL INCL O&P	
950	2420 4'-0" x 8'-0"	2 Carp	11	1.455	Ea.	435	48		483	560	950
	2460 For 6'-8" high door, deduct from 7'-0" door					10.15			10.15	11.15	
	2480 For oak veneer, add					50%					
	2500 For walnut veneer, add				↓	75%					
	9000 Minimum labor/equipment charge	1 Carp	4	2	Job		66		66	109	
960	0010 WOOD FRAMES										960
	0400 Exterior frame, incl. ext. trim, pine, 5/4 x 4-9/16" deep	2 Carp	375	.043	L.F.	4.41	1.41		5.82	7.20	
	0420 5-3/16" deep		375	.043		7.35	1.41		8.76	10.45	
	0440 6-9/16" deep		375	.043		6.85	1.41		8.26	9.85	
	0600 Oak, 5/4 x 4-9/16" deep		350	.046		8.45	1.51		9.96	11.75	
	0620 5-3/16" deep		350	.046		9.50	1.51		11.01	12.95	
	0640 6-9/16" deep		350	.046		10.55	1.51		12.06	14.10	
	0800 Walnut, 5/4 x 4-9/16" deep		350	.046		9.90	1.51		11.41	13.40	
	0820 5-3/16" deep		350	.046		14.40	1.51		15.91	18.30	
	0840 6-9/16" deep		350	.046		16.95	1.51		18.46	21	
	1000 Sills, 8/4 x 8" deep, oak, no horns		100	.160		11.95	5.30		17.25	22	
	1020 2" horns		100	.160		13.30	5.30		18.60	23.50	
	1040 3" horns		100	.160		15.35	5.30		20.65	25.50	
	1100 8/4 x 10" deep, oak, no horns		90	.178		16.10	5.85		21.95	27.50	
	1120 2" horns		90	.178		17.95	5.85		23.80	29.50	
	1140 3" horns		90	.178	↓	19.55	5.85		25.40	31	
	2000 Exterior, colonial, frame & trim, 3' opng., in-swing, minimum		22	.727	Ea.	279	24		303	345	
	2010 Average		21	.762		415	25		440	495	
	2020 Maximum		20	.800		940	26.50		966.50	1,075	
	2100 5'-4" opening, in-swing, minimum		17	.941		315	31		346	395	
	2120 Maximum		15	1.067		940	35		975	1,075	
	2140 Out-swing, minimum		17	.941		325	31		356	405	
	2160 Maximum		15	1.067		975	35		1,010	1,125	
	2400 6'-0" opening, in-swing, minimum		16	1		300	33		333	390	
	2420 Maximum		10	1.600		975	53		1,028	1,175	
	2460 Out-swing, minimum		16	1		325	33		358	410	
	2480 Maximum		10	1.600	↓	1,200	53		1,253	1,425	
	2600 For two sidelights, add, minimum		30	.533	Opng.	310	17.60		327.60	375	
	2620 Maximum		20	.800	"	995	26.50		1,021.50	1,150	
	2700 Custom birch frame, 3'-0" opening		16	1	Ea.	179	33		212	252	
	2750 6'-0" opening		16	1		270	33		303	350	
	2900 Exterior, modern, plain trim, 3' opng., in-swing, minimum		26	.615		30	20.50		50.50	67	
	2920 Average		24	.667		36	22		58	76	
	2940 Maximum		22	.727	↓	44	24		68	88	
	3000 Interior frame, pine, 11/16" x 3-5/8" deep		375	.043	L.F.	3.99	1.41		5.40	6.70	
	3020 4-9/16" deep		375	.043		5.45	1.41		6.86	8.35	
	3200 Oak, 11/16" x 3-5/8" deep		350	.046		3.59	1.51		5.10	6.45	
	3220 4-9/16" deep		350	.046		3.87	1.51		5.38	6.75	
	3240 5-3/16" deep		350	.046		3.98	1.51		5.49	6.90	
	3400 Walnut, 11/16" x 3-5/8" deep		350	.046		6.05	1.51		7.56	9.15	
	3420 4-9/16" deep		350	.046		6.40	1.51		7.91	9.50	
	3440 5-3/16" deep		350	.046		6.60	1.51		8.11	9.80	
	3800 Threshold, oak, 5/8" x 3-5/8" deep		200	.080		2.33	2.64		4.97	6.95	
	3820 4-5/8" deep		190	.084		2.96	2.78		5.74	7.85	
	3840 5-5/8" deep	↓	180	.089	↓	5	2.93		7.93	10.35	
	4000 For casing see division 06220-400 & 06220-800										
	9000 Minimum labor/equipment charge	1 Carp	4	2	Job		66		66	109	

8 DOORS & WINDOWS

08200 | Wood and Plastic Doors

08260 Sliding Wood and Plastic Doors			CREW	DAILY OUTPUT	LABOR-HOURS	UNIT	2004 BARE COSTS MAT.	LABOR	EQUIP.	TOTAL	TOTAL INCL O&P	
700	0010	GLASS, SLIDING										700
	0012	Vinyl clad, 1" insul. glass, 6'-0" x 6'-10" high	2 Carp	4	4	Opng.	1,150	132		1,282	1,475	
	0030	6'-0" x 8'-0" high		4	4	Ea.	1,725	132		1,857	2,125	
	0100	8'-0" x 6'-10" high		4	4	Opng.	1,775	132		1,907	2,175	
	0500	3 leaf, 9'-0" x 6'-10" high		3	5.333		1,600	176		1,776	2,075	
	0600	12'-0" x 6'-10" high	↓	3	5.333	↓	2,000	176		2,176	2,500	
	9000	Minimum labor/equipment charge	1 Carp	4	2	Job		66		66	109	
900	0010	GLASS, SLIDING										900
	0020	Wood, 5/8" tempered insul. glass, 6' wide, premium	2 Carp	4	4	Ea.	985	132		1,117	1,300	
	0100	Economy		4	4		695	132		827	985	
	0150	8' wide, wood, premium		3	5.333		1,150	176		1,326	1,550	
	0200	Economy		3	5.333		805	176		981	1,175	
	0250	12' wide, wood, premium		2.50	6.400		2,700	211		2,911	3,325	
	0300	Economy	↓	2.50	6.400	↓	1,825	211		2,036	2,375	
	0350	Aluminum, 5/8" tempered insulated glass, 6' wide										
	0400	Premium	2 Carp	4	4	Ea.	1,375	132		1,507	1,725	
	0450	Economy		4	4		710	132		842	1,000	
	0500	8' wide, premium		3	5.333		1,300	176		1,476	1,725	
	0550	Economy		3	5.333		1,100	176		1,276	1,500	
	0600	12' wide, premium		2.50	6.400		2,150	211		2,361	2,725	
	0650	Economy	↓	2.50	6.400	↓	1,375	211		1,586	1,850	
	1000	Replacement doors, wood										
	1050	6' wide, premium	2 Carp	4	4	Ea.	635	132		767	915	
	9000	Minimum labor/equipment charge	1 Carp	2	4	Job		132		132	219	

08300 | Specialty Doors

08310 Access Doors and Panels			CREW	DAILY OUTPUT	LABOR-HOURS	UNIT	2004 BARE COSTS MAT.	LABOR	EQUIP.	TOTAL	TOTAL INCL O&P	
100	0010	ACCESS DOORS										100
	1000	Fire rated door with lock										
	1100	Metal, 12" x 12"	1 Carp	10	.800	Ea.	118	26.50		144.50	174	
	1150	18" x 18"		9	.889		154	29.50		183.50	218	
	1200	24" x 24"		9	.889		186	29.50		215.50	254	
	1250	24" x 36"		8	1		250	33		283	330	
	1300	24" x 48"		8	1		310	33		343	395	
	1350	36" x 36"		7.50	1.067		370	35		405	470	
	1400	48" x 48"		7.50	1.067		475	35		510	585	
	1600	Stainless steel, 12" x 12"		10	.800		211	26.50		237.50	276	
	1650	18" x 18"		9	.889		305	29.50		334.50	385	
	1700	24" x 24"		9	.889		375	29.50		404.50	465	
	1750	24" x 36"	↓	8	1	↓	480	33		513	585	
	2000	Flush door for finishing										
	2100	Metal 8" x 8"	1 Carp	10	.800	Ea.	34	26.50		60.50	81.50	
	2150	12" x 12"	"	10	.800	"	38	26.50		64.50	86	
	3000	Recessed door for acoustic tile										
	3100	Metal, 12" x 12"	1 Carp	4.50	1.778	Ea.	52	58.50		110.50	154	
	3150	12" x 24"		4.50	1.778		68	58.50		126.50	172	
	3200	24" x 24"	↓	4	2	↓	91	66		157	209	

8 DOORS & WINDOWS

	08310 Access Doors and Panels		CREW	DAILY OUTPUT	LABOR-HOURS	UNIT	2004 BARE COSTS MAT.	LABOR	EQUIP.	TOTAL	TOTAL INCL O&P	
100	3250	24″ x 36″	1 Carp	4	2	Ea.	117	66		183	238	100
	4000	Recessed door for drywall										
	4100	Metal 12″ x 12″	1 Carp	6	1.333	Ea.	58	44		102	137	
	4150	12″ x 24″		5.50	1.455		86	48		134	174	
	4200	24″ x 36″		5	1.600		137	53		190	239	
	6000	Standard door										
	6100	Metal, 8″ x 8″	1 Carp	10	.800	Ea.	30	26.50		56.50	77	
	6150	12″ x 12″		10	.800		34	26.50		60.50	81.50	
	6200	18″ x 18″		9	.889		47	29.50		76.50	100	
	6250	24″ x 24″		9	.889		61	29.50		90.50	116	
	6300	24″ x 36″		8	1		90	33		123	154	
	6350	36″ x 36″		8	1		110	33		143	176	
	6500	Stainless steel, 8″ x 8″		10	.800		60	26.50		86.50	110	
	6550	12″ x 12″		10	.800		80	26.50		106.50	132	
	6600	18″ x 18″		9	.889		148	29.50		177.50	212	
	6650	24″ x 24″		9	.889		194	29.50		223.50	262	
	7010	Aluminum cover	G-3	11	2.909		500	94.50		594.50	700	
	9000	Minimum labor/equipment charge	1 Carp	4	2	Job		66		66	109	
150	0010	**BULKHEAD CELLAR DOORS**										150
	0020	Steel, not incl. sides, 44″ x 62″	1 Carp	5.50	1.455	Ea.	204	48		252	305	
	0100	52″ x 73″		5.10	1.569		227	52		279	335	
	0500	With sides and foundation plates, 57″ x 45″ x 24″		4.70	1.702		266	56		322	385	
	0600	42″ x 49″ x 51″		4.30	1.860		320	61.50		381.50	455	
	9000	Minimum labor/equipment charge		2	4	Job		132		132	219	
300	0010	**FLOOR, COMMERCIAL**										300
	0020	Aluminum tile, steel frame, one leaf, 2′ x 2′ opng.	2 Sswk	3.50	4.571	Opng.	365	170		535	715	
	0050	3′-6″ x 3′-6″ opening		3.50	4.571		660	170		830	1,050	
	0500	Double leaf, 4′ x 4′ opening		3	5.333		965	198		1,163	1,425	
	0550	5′ x 5′ opening		3	5.333		1,425	198		1,623	1,950	
	9000	Minimum labor/equipment charge		2	8	Job		297		297	555	
350	0010	**FLOOR, INDUSTRIAL**										350
	0020	Steel 300 psf L.L., single leaf, 2′ x 2′, 175#	2 Sswk	6	2.667	Opng.	495	99		594	730	
	0050	3′ x 3′ opening, 300#		5.50	2.909		685	108		793	955	
	0300	Double leaf, 4′ x 4′ opening, 455#		5	3.200		1,025	119		1,144	1,350	
	0350	5′ x 5′ opening, 645#		4.50	3.556		1,350	132		1,482	1,725	
	1000	Aluminum, 300 psf L.L., single leaf, 2′ x 2′, 60#		6	2.667		485	99		584	720	
	1050	3′ x 3′ opening, 100#		5.50	2.909		760	108		868	1,025	
	1500	Double leaf, 4′ x 4′ opening, 160#		5	3.200		1,200	119		1,319	1,550	
	1550	5′ x 5′ opening, 235#		4.50	3.556		1,600	132		1,732	2,000	
	9000	Minimum labor/equipment charge		2	8	Job		297		297	555	
	08330	**Coiling Doors and Grilles**										
130	0010	**COUNTER DOORS**										130
	0020	Manual, incl. frm and hdwe, galv. stl., 4′ roll-up, 6′ long	2 Carp	2	8	Opng.	835	264		1,099	1,350	
	0300	Galvanized steel, UL label		1.80	8.889		995	293		1,288	1,575	
	0600	Stainless steel, 4′ high roll-up, 6′ long		2	8		1,250	264		1,514	1,825	
	0700	10′ long		1.80	8.889		2,000	293		2,293	2,675	
	2000	Aluminum, 4′ high, 4′ long		2.20	7.273		810	240		1,050	1,300	
	2020	6′ long		2	8		895	264		1,159	1,425	
	2040	8′ long		1.90	8.421		1,025	278		1,303	1,575	
	2060	10′ long		1.80	8.889		1,225	293		1,518	1,825	
	2080	14′ long		1.40	11.429		1,800	375		2,175	2,600	
	2100	6′ high, 4′ long		2	8		895	264		1,159	1,425	
	2120	6′ long		1.60	10		1,050	330		1,380	1,700	
	2140	10′ long		1.40	11.429		1,450	375		1,825	2,200	
	9000	Minimum labor/equipment charge	1 Carp	2	4	Job		132		132	219	

08330 | Coiling Doors and Grilles

		08330 Coiling Doors and Grilles	CREW	DAILY OUTPUT	LABOR-HOURS	UNIT	2004 BARE COSTS MAT.	LABOR	EQUIP.	TOTAL	TOTAL INCL O&P
640	0010	**COILING GRILLE**									
	2020	Aluminum, manual operated, mill finish	2 Sswk	82	.195	S.F.	22	7.25		29.25	37.50
	2040	Bronze anodized		82	.195	"	35	7.25		42.25	52
	2060	Steel, manual operated, 10′ x 10′ high		1	16	Opng.	1,975	595		2,570	3,250
	2080	15′ x 8′ high		.80	20	"	2,300	745		3,045	3,925
	3000	For safety edge bottom bar, electric, add				L.F.	40.50			40.50	44.50
	8000	For motor operation, add	2 Sswk	5	3.200	Opng.	990	119		1,109	1,325
	9000	Minimum labor/equipment charge	"	1	16	Job		595		595	1,100
720	0010	**ROLLING SERVICE DOORS** Steel, manual, 20 ga., incl. hardware									
	0120	8′ x 8′ high, class A fire door	2 Sswk	1.40	11.429	Ea.	960	425		1,385	1,850
	0130	12′ x 12′ high, standard		1.20	13.333		1,200	495		1,695	2,250
	0140	12′ x 12′ high, class A fire door		1	16		1,700	595		2,295	2,975
	0160	10′ x 20′ high, standard		.50	32		1,425	1,200		2,625	3,800
	0180	10′ x 20′ high, class A fire door		.40	40		2,550	1,475		4,025	5,575
	3000	For 18 ga. doors, add				S.F.	.72			.72	.79
	3300	For enamel finish, add				"	.86			.86	.95
	3600	For safety edge bottom bar, pneumatic, add				L.F.	11.30			11.30	12.40
	4000	For weatherstripping, extruded rubber, jambs, add					7.30			7.30	8.05
	4100	Hood, add					4.65			4.65	5.10
	4200	Sill, add					2.66			2.66	2.93
	4500	Motor operators, to 14′ x 14′ opening	2 Sswk	5	3.200	Ea.	705	119		824	995
	4700	For fire door, additional fusible link, add				"	13.30			13.30	14.60
	9000	Minimum labor/equipment charge	2 Sswk	1	16	Job		595		595	1,100

08340 | Special Function Doors

		08340 Special Function Doors	CREW	DAILY OUTPUT	LABOR-HOURS	UNIT	MAT.	LABOR	EQUIP.	TOTAL	TOTAL INCL O&P
100	0010	**COLD STORAGE**									
	0020	Single, 20 ga. galvanized steel									
	0300	Horizontal sliding, 5′ x 7′, manual operation, 2″ thick	2 Carp	2	8	Ea.	2,300	264		2,564	3,000
	0400	4″ thick		2	8		2,775	264		3,039	3,525
	0500	6″ thick		2	8		2,700	264		2,964	3,425
	0800	5′ x 7′, power operation, 2″ thick		1.90	8.421		4,500	278		4,778	5,400
	0900	4″ thick		1.90	8.421		4,575	278		4,853	5,475
	1000	6″ thick		1.90	8.421		5,200	278		5,478	6,175
	1300	9′ x 10′, manual operation, 2″ insulation		1.70	9.412		3,625	310		3,935	4,525
	1400	4″ insulation		1.70	9.412		3,750	310		4,060	4,650
	1500	6″ insulation		1.70	9.412		4,525	310		4,835	5,500
	1800	Power operation, 2″ insulation		1.60	10		6,250	330		6,580	7,425
	1900	4″ insulation		1.60	10		6,400	330		6,730	7,600
	2000	6″ insulation		1.70	9.412		7,250	310		7,560	8,500
	2300	For stainless steel face, add					20%				
	3000	Hinged, lightweight, 3′ x 7′-0″, galvanized 1 face, 2″ thick	2 Carp	2	8	Ea.	1,050	264		1,314	1,600
	3050	4″ thick		1.90	8.421		1,225	278		1,503	1,800
	3300	Aluminum doors, 3′ x 7′-0″, 4″ thick		1.90	8.421		1,075	278		1,353	1,625
	3350	6″ thick		1.40	11.429		1,900	375		2,275	2,725
	3600	Stainless steel, 3′ x 7′-0″, 4″ thick		1.90	8.421		1,375	278		1,653	1,975
	3650	6″ thick		1.40	11.429		2,300	375		2,675	3,175
	3900	Painted, 3′ x 7′-0″, 4″ thick		1.90	8.421		985	278		1,263	1,525
	3950	6″ thick		1.40	11.429		1,875	375		2,250	2,700
	5000	Bi-parting, electric operated									
	5010	6′ x 8′ opening, galv. faces, 4″ thick for cooler	2 Carp	.80	20	Opng.	5,875	660		6,535	7,575
	5050	For freezer, 4″ thick		.80	20		6,475	660		7,135	8,225
	5300	For door buck framing and door protection, add		2.50	6.400		480	211		691	880
	6000	Galvanized batten door, galvanized hinges, 4′ x 7′		2	8		1,450	264		1,714	2,050
	6050	6′ x 8′		1.80	8.889		1,975	293		2,268	2,650
	6500	Fire door, 3 hr., 6′ x 8′, single slide		.80	20		6,800	660		7,460	8,575

	08340 | Special Function Doors		Crew	Daily Output	Labor-Hours	Unit	2004 Bare Costs Mat.	Labor	Equip.	Total	Total Incl O&P	
100	6550	Double, bi-parting	2 Carp	.70	22.857	Opng.	9,450	755		10,205	11,700	100
	9000	Minimum labor/equipment charge	1 Carp	2	4	Job		132		132	219	
610	0010	**ACOUSTICAL DOORS**										610
	0020	Including framed seals, 3′ x 7′, wood, 27 STC rating	2 Carp	1.50	10.667	Ea.	340	350		690	955	
	0100	Steel, 40 STC rating		1.50	10.667		1,250	350		1,600	1,950	
	0200	45 STC rating		1.50	10.667		1,650	350		2,000	2,400	
	0300	48 STC rating		1.50	10.667		2,075	350		2,425	2,875	
	0400	52 STC rating		1.50	10.667		2,500	350		2,850	3,325	
	9000	Minimum labor/equipment charge	1 Carp	4	2	Job		66		66	109	
700	0010	**DOUBLE ACTING, SWING**										700
	0020	Including frame, closer, hardware and vision panel										
	1000	.063″ aluminum, 7′-0″ high, 4′-0″ wide	2 Carp	4.20	3.810	Pr.	1,300	126		1,426	1,625	
	1050	6′-8″ wide	″	4	4	″	2,300	132		2,432	2,750	
	2000	Solid core wood, 3/4″ thick, metal frame, stainless steel										
	2010	base plate, 7′ high opening, 4′ wide	2 Carp	4	4	Pr.	1,300	132		1,432	1,650	
	2050	7′ wide		3.80	4.211	″	2,300	139		2,439	2,750	
	9000	Minimum labor/equipment charge		2	8	Job		264		264	440	
710	0010	**GLASS DOOR, SWING**										710
	0020	Including hardware, 1/2″ thick, tempered, 3′ x 7′ opening	2 Glaz	2	8	Opng.	1,625	256		1,881	2,225	
	0100	6′ x 7′ opening		1.40	11.429	″	3,000	365		3,365	3,900	
	9000	Minimum labor/equipment charge		2	8	Job		256		256	415	
720	0010	**SHOCK ABSORBING DOORS**										720
	0020	Rigid, no frame, 1-1/2″ thick, 5′ x 7′	2 Sswk	1.90	8.421	Opng.	1,175	315		1,490	1,850	
	0100	8′ x 8′		1.80	8.889		1,650	330		1,980	2,425	
	0500	Flexible, no frame, insulated, .16″ thick, economy, 5′ x 7′		2	8		1,425	297		1,722	2,125	
	0600	Deluxe		1.90	8.421		2,175	315		2,490	2,950	
	1000	8′ x 8′ opening, economy		2	8		2,225	297		2,522	3,000	
	1100	Deluxe		1.90	8.421		2,850	315		3,165	3,725	
	9000	Minimum labor/equipment charge		2	8	Job		297		297	555	
	08360 | Overhead Doors											
550	0010	**OVERHEAD, COMMERCIAL** Frames not included										550
	1000	Stock, sectional, heavy duty, wood, 1-3/4″ thick, 8′ x 8′ high	2 Carp	2	8	Ea.	555	264		819	1,050	
	1200	12′ x 12′ high		1.50	10.667		1,200	350		1,550	1,900	
	1300	Chain hoist, 14′ x 14′ high		1.30	12.308		1,700	405		2,105	2,550	
	1600	20′ x 16′ high		.65	24.615		3,500	810		4,310	5,200	
	2100	For medium duty custom door, deduct					5%	5%				
	2150	For medium duty stock doors, deduct					10%	5%				
	2300	Fiberglass and aluminum, heavy duty, sectional, 12′ x 12′ high	2 Carp	1.50	10.667	Ea.	1,750	350		2,100	2,500	
	2450	Chain hoist, 20′ x 20′ high	″	.50	32		4,300	1,050		5,350	6,475	
	2900	For electric trolley operator, 1/3 H.P., to 12′ x 12′, add	1 Carp	2	4		650	132		782	935	
	2950	Over 12′ x 12′, 1/2 H.P., add	″	1	8		705	264		969	1,225	
	9000	Minimum labor/equipment charge	2 Carp	1.50	10.667	Job		350		350	585	
600	0010	**RESIDENTIAL GARAGE DOORS** Including hardware, no frame										600
	0050	Hinged, wood, custom, double door, 9′ x 7′	2 Carp	4	4	Ea.	345	132		477	600	
	0070	16′ x 7′		3	5.333		585	176		761	935	
	0200	Overhead, sectional, incl. hardware, fiberglass, 9′ x 7′, standard		5.28	3.030		555	100		655	775	
	0220	Deluxe		5.28	3.030		700	100		800	935	
	0300	16′ x 7′, standard		6	2.667		965	88		1,053	1,200	
	0320	Deluxe		6	2.667		1,200	88		1,288	1,475	
	0500	Hardboard, 9′ x 7′, standard		8	2		370	66		436	520	
	0520	Deluxe		8	2		480	66		546	635	
	0600	16′ x 7′, standard		6	2.667		700	88		788	915	
	0620	Deluxe		6	2.667		815	88		903	1,050	
	0700	Metal, 9′ x 7′, standard		5.28	3.030		435	100		535	640	

08360 | Overhead Doors

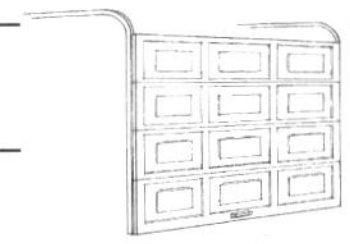

		08360 Overhead Doors	CREW	DAILY OUTPUT	LABOR-HOURS	UNIT	2004 BARE COSTS MAT.	LABOR	EQUIP.	TOTAL	TOTAL INCL O&P	
600	0720	Deluxe	2 Carp	8	2	Ea.	585	66		651	750	600
	0800	16' x 7', standard		3	5.333		550	176		726	895	
	0820	Deluxe		6	2.667		855	88		943	1,075	
	0900	Wood, 9' x 7', standard		8	2		440	66		506	595	
	0920	Deluxe		8	2		1,275	66		1,341	1,500	
	1000	16' x 7', standard		6	2.667		890	88		978	1,125	
	1020	Deluxe		6	2.667		1,850	88		1,938	2,200	
	1800	Door hardware, sectional	1 Carp	4	2		195	66		261	325	
	1810	Door tracks only		4	2		93	66		159	211	
	1820	One side only		7	1.143		63	37.50		100.50	132	
	3000	Swing-up, including hardware, fiberglass, 9' x 7', standard	2 Carp	8	2		610	66		676	780	
	3020	Deluxe		8	2		670	66		736	850	
	3100	16' x 7', standard		6	2.667		770	88		858	990	
	3120	Deluxe		6	2.667		830	88		918	1,050	
	3200	Hardboard, 9' x 7', standard		8	2		293	66		359	430	
	3220	Deluxe		8	2		390	66		456	535	
	3300	16' x 7', standard		6	2.667		410	88		498	595	
	3320	Deluxe		6	2.667		610	88		698	815	
	3400	Metal, 9' x 7', standard		8	2		320	66		386	465	
	3420	Deluxe		8	2		550	66		616	715	
	3500	16' x 7', standard		6	2.667		505	88		593	700	
	3520	Deluxe		6	2.667		810	88		898	1,025	
	3600	Wood, 9' x 7', standard		8	2		350	66		416	495	
	3620	Deluxe		8	2		600	66		666	770	
	3700	16' x 7', standard		6	2.667		610	88		698	815	
	3720	Deluxe		6	2.667		860	88		948	1,100	
	3900	Door hardware only, swing up	1 Carp	4	2		98.50	66		164.50	217	
	3920	One side only		7	1.143		57	37.50		94.50	126	
	4000	For electric operator, economy, add		8	1		254	33		287	335	
	4100	Deluxe, including remote control		8	1		370	33		403	465	
	4500	For transmitter/receiver control , add to operator				Total	80.50			80.50	88.50	
	4600	Transmitters, additional				"	28			28	31	
	6000	Replace section, on sectional door, fiberglass, 9' x 7'	1 Carp	4	2	Ea.	161	66		227	286	
	6020	16' x 7'		3.50	2.286		244	75.50		319.50	395	
	6200	Hardboard, 9' x 7'		4	2		85.50	66		151.50	203	
	6220	16' x 7'		3.50	2.286		164	75.50		239.50	305	
	6300	Metal, 9' x 7'		4	2		139	66		205	262	
	6320	16' x 7'		3.50	2.286		228	75.50		303.50	375	
	6500	Wood, 9' x 7'		4	2		84	66		150	202	
	6520	16' x 7'		3.50	2.286		163	75.50		238.50	305	
	9000	Minimum labor/equipment charge		2.50	3.200	Job		106		106	175	

08370 | Vertical Lift Doors

		08370 Vertical Lift Doors	CREW	DAILY OUTPUT	LABOR-HOURS	UNIT	MAT.	LABOR	EQUIP.	TOTAL	TOTAL INCL O&P	
950	0010	**VERTICAL LIFT DOORS**										950
	0020	Motorized, 14 ga. stl, incl., frm and ctrl pnl										
	0050	16' x 16' high	L-10	.50	48	Ea.	16,400	1,775	1,225	19,400	22,500	
	0100	10' x 20' high		1.30	18.462		21,600	685	470	22,755	25,500	
	0120	15' x 20' high		1.30	18.462		26,500	685	470	27,655	30,800	
	0140	20' x 20' high		1	24		31,100	890	610	32,600	36,500	
	0160	25' x 20' high		1	24		34,300	890	610	35,800	40,000	
	0170	32' x 24' high		.75	32		31,800	1,175	815	33,790	38,000	
	0180	20' x 25' high		1	24		35,700	890	610	37,200	41,600	
	0200	25' x 25' high		.70	34.286		40,300	1,275	875	42,450	47,500	
	0220	25' x 30' high		.70	34.286		47,200	1,275	875	49,350	55,000	
	0240	30' x 30' high		.70	34.286		54,500	1,275	875	56,650	63,000	

08300 | Specialty Doors

	08370 | Vertical Lift Doors		Crew	Daily Output	Labor-Hours	Unit	2004 Bare Costs Mat.	Labor	Equip.	Total	Total Incl O&P	
950	0260	35′ x 30′ high	L-10	.70	34.286	Ea.	59,500	1,275	875	61,650	68,500	950

08400 | Entrances and Storefronts

	08410 | Metal-Framed Storefronts		Crew	Daily Output	Labor-Hours	Unit	2004 Bare Costs Mat.	Labor	Equip.	Total	Total Incl O&P	
110	0010	**ALUMINUM-FRAMED STOREFRONTS**										110
	0020	Entrance, 3′ x 7′ opening, clear anodized finish	2 Sswk	7	2.286	Opng.	231	85		316	410	
	0100	Bronze finish		7	2.286		315	85		400	505	
	0500	6′ x 7′ opening, clear finish		6	2.667		288	99		387	500	
	0520	Bronze finish		6	2.667		325	99		424	545	
	1000	With 3′ high transoms, 3′ x 10′ opening, clear finish		6.50	2.462		345	91.50		436.50	550	
	1050	Bronze finish		6.50	2.462		385	91.50		476.50	590	
	1100	Black finish		6.50	2.462		450	91.50		541.50	665	
	1500	With 3′ high transoms, 6′ x 10′ opening, clear finish		5.50	2.909		420	108		528	660	
	1550	Bronze finish		5.50	2.909		450	108		558	695	
	1600	Black finish		5.50	2.909		535	108		643	790	
	9000	Minimum labor/equipment charge		4	4	Job		149		149	277	
120	0010	**ALUMINUM DOORS** Commercial entrance, no glazing										120
	0020	Incl. hinges, push/pull, deadlock, cyl., threshold										
	0800	Narrow stile, no glazing, standard hardware, pair of 2′-6″ x 7′-0″	2 Carp	1.70	9.412	Pr.	695	310		1,005	1,275	
	1000	3′-0″ x 7′-0″, single		3	5.333	Ea.	455	176		631	790	
	1200	Pair of 3′-0″ x 7′-0″		1.70	9.412	Pr.	915	310		1,225	1,525	
	1500	3′-6″ x 7′-0″, single		3	5.333	Ea.	490	176		666	830	
	2000	Medium stile, pair of 2′-6″ x 7′-0″		1.70	9.412	Pr.	855	310		1,165	1,450	
	2100	3′-0″ x 7′-0″, single		3	5.333	Ea.	600	176		776	950	
	2200	Pair of 3′-0″ x 7′-0″		1.70	9.412	Pr.	1,175	310		1,485	1,825	
	2300	3′-6″ x 7′-0″, single		5.33	3.002	Ea.	705	99		804	945	
	5000	Flush panel doors, pair of 2′-6″ x 7′-0″	2 Sswk	2	8	Pr.	895	297		1,192	1,550	
	5050	3′-0″ x 7′-0″, single		2.50	6.400	Ea.	445	238		683	935	
	5100	Pair of 3′-0″ x 7′-0″		2	8	Pr.	895	297		1,192	1,550	
	5150	3′-6″ x 7′-0″, single		2.50	6.400	Ea.	530	238		768	1,025	
130	0010	**ALUMINUM DOORS & FRAMES** Entrance, narrow stile, including										130
	0015	Standard hardware, clear finish, not incl. glass, 2′-6″ x 7′-0″ opng.	2 Sswk	2	8	Ea.	660	297		957	1,275	
	0020	3′-0″ x 7′-0″ opening		2	8		515	297		812	1,125	
	0030	3′-6″ x 7′-0″ opening		2	8		490	297		787	1,100	
	0100	3′-0″ x 10′-0″ opening, 3′ high transom		1.80	8.889		715	330		1,045	1,400	
	0200	3′-6″ x 10′-0″ opening, 3′ high transom		1.80	8.889		700	330		1,030	1,375	
	0280	5′-0″ x 7′-0″ opening		2	8		825	297		1,122	1,475	
	0300	6′-0″ x 7′-0″ opening		1.30	12.308	Pr.	795	455		1,250	1,725	
	0400	6′-0″ x 10′-0″ opening, 3′ high transom		1.10	14.545		650	540		1,190	1,725	
	0420	7′-0″ x 7′-0″ opening		1	16		815	595		1,410	2,000	
	0500	Wide stile, 2′-6″ x 7′-0″ opening		2	8	Ea.	665	297		962	1,275	
	0520	3′-0″ x 7′-0″ opening		2	8		675	297		972	1,300	
	0540	3′-6″ x 7′-0″ opening		2	8		720	297		1,017	1,350	
	0560	5′-0″ x 7′-0″ opening		2	8		1,175	297		1,472	1,850	
	0580	6′-0″ x 7′-0″ opening		1.30	12.308	Pr.	1,125	455		1,580	2,075	
	0600	7′-0″ x 7′-0″ opening		1	16	″	1,225	595		1,820	2,450	
	1100	For full vision doors, with 1/2″ glass, add				Leaf	55%					
	1200	For non-standard size, add					67%					

				DAILY	LABOR-		2004 BARE COSTS				TOTAL	
	08410 | Metal-Framed Storefronts		CREW	OUTPUT	HOURS	UNIT	MAT.	LABOR	EQUIP.	TOTAL	INCL O&P	
130	1300	Light bronze finish, add				Leaf	36%					130
	1400	Dark bronze finish, add					18%					
	1500	For black finish, add					36%					
	1600	Concealed panic device, add					940			940	1,025	
	1700	Electric striker release, add				Opng.	242			242	266	
	1800	Floor check, add				Leaf	720			720	790	
	1900	Concealed closer, add				″	475			475	525	
	2000	Flush 3′ x 7′ Insulated, 12″x 12″ lite, clear finish	2 Sswk	2	8	Ea.	900	297		1,197	1,550	
	9000	Minimum labor/equipment charge	2 Carp	4	4	Job		132		132	219	
140	0010	**STOREFRONT SYSTEMS** Aluminum frame, clear 3/8″ plate glass,										140
	0020	incl. 3′ x 7′ door with hardware (400 sq. ft. max. wall)										
	0500	Wall height to 12′ high, commercial grade	2 Glaz	150	.107	S.F.	12.60	3.42		16.02	19.35	
	0600	Institutional grade		130	.123		15.85	3.94		19.79	24	
	0700	Monumental grade		115	.139		24	4.46		28.46	33.50	
	1000	6′ x 7′ door with hardware, commercial grade		135	.119		12.90	3.80		16.70	20.50	
	1100	Institutional grade		115	.139		17.65	4.46		22.11	26.50	
	1200	Monumental grade		100	.160		32.50	5.15		37.65	44.50	
	1500	For bronze anodized finish, add					15%					
	1600	For black anodized finish, add					30%					
	1700	For stainless steel framing, add to monumental					75%					
	9000	Minimum labor/equipment charge	2 Glaz	1	16	Job		515		515	825	
300	0010	**STAINLESS STEEL AND GLASS** Entrance unit, narrow stiles										300
	0020	3′ x 7′ opening, including hardware, minimum	2 Sswk	1.60	10	Opng.	4,600	370		4,970	5,750	
	0050	Average		1.40	11.429		4,975	425		5,400	6,275	
	0100	Maximum		1.20	13.333		5,325	495		5,820	6,800	
	1000	For solid bronze entrance units, statuary finish, add					60%					
	1100	Without statuary finish, add					45%					
	2000	Balanced doors, 3′ x 7′, economy	2 Sswk	.90	17.778	Ea.	6,225	660		6,885	8,075	
	2100	Premium		.70	22.857	″	10,700	850		11,550	13,400	
	9000	Minimum labor/equipment charge		2	8	Job		297		297	555	
	08460 | Automatic Entrance Doors											
600	0010	**SLIDING ENTRANCE** 12′ x 7′-6″ opng., 5′ x 7′ door, 2 way traf.,										600
	0020	mat activated, panic pushout, incl. operator & hardware,										
	0030	not including glass or glazing	2 Glaz	.70	22.857	Opng.	5,900	735		6,635	7,675	
	9000	Minimum labor/equipment charge	″	.70	22.857	Job		735		735	1,175	
	08480 | Balanced Entrance Doors											
150	0010	**BALANCED DOORS**										150
	0020	Hardware & frame, alum. & glass, 3′ x 7′, econ.	2 Sswk	.90	17.778	Ea.	4,975	660		5,635	6,700	
	0150	Premium		.70	22.857	″	6,175	850		7,025	8,375	
	9000	Minimum labor/equipment charge		1	16	Job		595		595	1,100	
	08490 | Sliding Storefronts											
100	0010	**SLIDING PANELS**										100
	0020	Mall fronts, aluminum & glass, 15′ x 9′ high	2 Glaz	1.30	12.308	Opng.	2,325	395		2,720	3,175	
	0100	24′ x 9′ high		.70	22.857		3,350	735		4,085	4,875	
	0200	48′ x 9′ high, with fixed panels		.90	17.778		6,250	570		6,820	7,800	
	0500	For bronze finish, add					17%					
	9000	Minimum labor/equipment charge	2 Glaz	1	16	Job		515		515	825	

08510 | Steel Windows

			CREW	DAILY OUTPUT	LABOR-HOURS	UNIT	2004 BARE COSTS MAT.	LABOR	EQUIP.	TOTAL	TOTAL INCL O&P	
700	0010	**SCREENS**										700
	0020	For metal sash, aluminum or bronze mesh, flat screen	2 Sswk	1,200	.013	S.F.	3.30	.50		3.80	4.55	
	0500	Wicket screen, inside window		1,000	.016		5.05	.59		5.64	6.65	
	0800	Security screen, aluminum frame with stainless steel cloth		1,200	.013		17.80	.50		18.30	20.50	
	0900	Steel grate, painted, on steel frame		1,600	.010		9.55	.37		9.92	11.20	
	1000	For solar louvers, add		160	.100		18.45	3.72		22.17	27.50	
	4000	See also division 05580-900										
750	0010	**STEEL SASH** Custom units, glazing and trim not included										750
	0100	Casement, 100% vented	2 Sswk	200	.080	S.F.	38	2.97		40.97	47.50	
	0200	50% vented		200	.080		34.50	2.97		37.47	43.50	
	0300	Fixed		200	.080		24	2.97		26.97	31.50	
	1000	Projected, commercial, 40% vented		200	.080		39.50	2.97		42.47	49	
	1100	Intermediate, 50% vented		200	.080		43	2.97		45.97	52.50	
	1500	Industrial, horizontally pivoted		200	.080		41	2.97		43.97	50.50	
	1600	Fixed		200	.080		23.50	2.97		26.47	31	
	2000	Industrial security sash, 50% vented		200	.080		44	2.97		46.97	54	
	2100	Fixed		200	.080		36	2.97		38.97	45	
	2500	Picture window 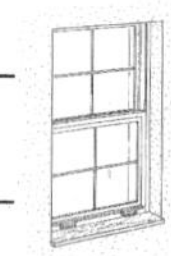		200	.080		22	2.97		24.97	30	
	3000	Double hung		200	.080		43	2.97		45.97	52.50	
	5000	Mullions for above, open interior face		240	.067	L.F.	7.65	2.48		10.13	13.05	
	5100	With interior cover		240	.067	"	12.70	2.48		15.18	18.55	
	5200	Single glazing for above, add	2 Glaz	200	.080	S.F.	4.92	2.56		7.48	9.55	
	6000	Double glazing for above, add		200	.080		9.80	2.56		12.36	14.95	
	6100	Triple glazing for above, add		85	.188		9.50	6.05		15.55	20	
	9000	Minimum labor/equipment charge	1 Sswk	2	4	Job		149		149	277	
770	0010	**STEEL WINDOWS** Stock, including frame, trim and insul. glass R08510-100										770
	1000	Custom units, double hung, 2'-8" x 4'-6" opening	2 Sswk	12	1.333	Ea.	545	49.50		594.50	690	
	1100	2'-4" x 3'-9" opening		12	1.333		450	49.50		499.50	585	
	1500	Commercial projected, 3'-9" x 5'-5" opening		10	1.600		955	59.50		1,014.50	1,150	
	1600	6'-9" x 4'-1" opening		7	2.286		1,250	85		1,335	1,525	
	2000	Intermediate projected, 2'-9" x 4'-1" opening		12	1.333		535	49.50		584.50	675	
	2100	4'-1" x 5'-5" opening		10	1.600		1,075	59.50		1,134.50	1,300	
	9000	Minimum labor/equipment charge	1 Sswk	3	2.667	Job		99		99	184	

08520 | Aluminum Windows

			CREW	DAILY OUTPUT	LABOR-HOURS	UNIT	MAT.	LABOR	EQUIP.	TOTAL	TOTAL INCL O&P	
100	0010	**ALUMINUM SASH**										100
	0020	Stock, grade C, glaze & trim not incl., casement	2 Sswk	200	.080	S.F.	26.50	2.97		29.47	34.50	
	0050	Double hung		200	.080		26.50	2.97		29.47	35	
	0100	Fixed casement		200	.080		10.60	2.97		13.57	17.20	
	0150	Picture window		200	.080		11.40	2.97		14.37	18.10	
	0200	Projected window		200	.080		24.50	2.97		27.47	32.50	
	0250	Single hung		200	.080		12.65	2.97		15.62	19.45	
	0300	Sliding		200	.080		16.35	2.97		19.32	23.50	
	1000	Mullions for above, tubular		240	.067	L.F.	4.16	2.48		6.64	9.20	
	2950	Single glazing for above, add	2 Glaz	200	.080	S.F.	5.80	2.56		8.36	10.55	
	3000	Double glazing for above, add		200	.080		9.35	2.56		11.91	14.45	
	3100	Triple glazing for above, add		85	.188		11	6.05		17.05	22	
	9000	Minimum labor/equipment charge	1 Sswk	2	4	Job		149		149	277	
120	0010	**ALUMINUM WINDOWS** Incl. frame and glazing, Commercial grade										120
	1000	Stock units, casement, 3'-1" x 3'-2" opening	2 Sswk	10	1.600	Ea.	284	59.50		343.50	425	
	1050	Add for storms					57			57	63	
	1600	Projected, with screen, 3'-1" x 3'-2" opening	2 Sswk	10	1.600		203	59.50		262.50	335	
	1700	Add for storms					54			54	59.50	
	2000	4'-5" x 5'-3" opening	2 Sswk	8	2		286	74.50		360.50	455	

		08520 | Aluminum Windows	CREW	DAILY OUTPUT	LABOR-HOURS	UNIT	2004 BARE COSTS MAT.	LABOR	EQUIP.	TOTAL	TOTAL INCL O&P	
120	2100	Add for storms				Ea.	74.50			74.50	82	120
	2500	Enamel finish windows, 3'-1" x 3'-2"	2 Sswk	10	1.600		182	59.50		241.50	310	
	2600	4'-5" x 5'-3"		8	2		273	74.50		347.50	440	
	3000	Single hung, 2' x 3' opening, enameled, standard glazed		10	1.600		136	59.50		195.50	261	
	3100	Insulating glass		10	1.600		165	59.50		224.50	293	
	3300	2'-8" x 6'-8" opening, standard glazed		8	2		290	74.50		364.50	460	
	3400	Insulating glass		8	2		375	74.50		449.50	550	
	3700	3'-4" x 5'-0" opening, standard glazed		9	1.778		187	66		253	330	
	3800	Insulating glass		9	1.778		263	66		329	410	
	4000	Sliding aluminum, 3' x 2' opening, standard glazed		10	1.600		154	59.50		213.50	280	
	4100	Insulating glass		10	1.600		170	59.50		229.50	298	
	4300	5' x 3' opening, standard glazed		9	1.778		195	66		261	335	
	4400	Insulating glass		9	1.778		273	66		339	425	
	4600	8' x 4' opening, standard glazed		6	2.667		281	99		380	495	
	4700	Insulating glass		6	2.667		450	99		549	680	
	5000	9' x 5' opening, standard glazed		4	4		425	149		574	740	
	5100	Insulating glass		4	4		680	149		829	1,025	
	5500	Sliding, with thermal barrier and screen, 6' x 4', 2 track		8	2		580	74.50		654.50	775	
	5700	4 track	↓	8	2		705	74.50		779.50	915	
	6000	For above units with bronze finish, add					12%					
	6200	For installation in concrete openings, add				↓	5%					
	6400											
	9000	Minimum labor/equipment charge	1 Sswk	3	2.667	Job		99		99	184	
500	0010	**JALOUSIES**										500
	0020	Aluminum incl. glazing & screens, stock, 1'-7" x 3'-2"	2 Sswk	10	1.600	Ea.	125	59.50		184.50	248	
	0100	2'-3" x 4'-0"		10	1.600		178	59.50		237.50	305	
	0200	3'-1" x 2'-0"		10	1.600		140	59.50		199.50	265	
	0300	3'-1" x 5'-3"		10	1.600		254	59.50		313.50	390	
	1000	Mullions for above, 2'-0" long		80	.200		9.60	7.45		17.05	24.50	
	1100	5'-3" long	↓	80	.200	↓	16.40	7.45		23.85	32	
	9000	Minimum labor/equipment charge	1 Sswk	3.50	2.286	Job		85		85	158	
		08550 | Wood Windows										
100	0010	**AWNING WINDOW** Including frame, screens and grills										100
	0100	Average quality, builders model, 34" x 22", double insulated glass	1 Carp	10	.800	Ea.	193	26.50		219.50	257	
	0200	Low E glass		10	.800		221	26.50		247.50	287	
	0300	40" x 28", double insulated glass		9	.889		264	29.50		293.50	340	
	0400	Low E Glass		9	.889		280	29.50		309.50	360	
	0500	48" x 36", double insulated glass		8	1		385	33		418	480	
	0600	Low E glass		8	1		405	33		438	500	
	1000	34" x 22"		10	.800		223	26.50		249.50	289	
	1100	40" x 22"		10	.800		243	26.50		269.50	310	
	1200	36" x 28"		9	.889		259	29.50		288.50	335	
	1300	36" x 36"		9	.889		288	29.50		317.50	365	
	1400	48" x 28"		8	1		310	33		343	395	
	1500	60" x 36"		8	1		450	33		483	550	
	2000	Metal clad, deluxe, double insulated glass, 34" x 22"		10	.800		208	26.50		234.50	273	
	2100	40" x 22"		10	.800		244	26.50		270.50	310	
	2200	36" x 25"		9	.889		226	29.50		255.50	298	
	2300	40" x 30"		9	.889		283	29.50		312.50	360	
	2400	48" x 28"		8	1		289	33		322	375	
	2500	60" x 36"		8	1	↓	310	33		343	395	
	9000	Minimum labor/equipment charge	↓	4	2	Job		66		66	109	
150	0010	**BOW-BAY WINDOW** Including frame, screens and grills,										150
	0020	end panels operable										

08550 | Wood Windows

		08550 Wood Windows	CREW	DAILY OUTPUT	LABOR-HOURS	UNIT	2004 BARE COSTS MAT.	LABOR	EQUIP.	TOTAL	TOTAL INCL O&P	
150	1000	Bow type, casement, wood, bldrs mdl, 8′ x 5′ dbl insltd glass, 4 panel	2 Carp	10	1.600	Ea.	960	53		1,013	1,150	150
	1050	Low E glass		10	1.600		1,175	53		1,228	1,375	
	1100	10′-0″ x 5′-0″ , double insulated glass, 6 panels		6	2.667		1,225	88		1,313	1,500	
	1200	Low E glass, 6 panels		6	2.667		1,300	88		1,388	1,575	
	1300	Vinyl clad, bldrs model, double insulated glass, 6′-0″ x 4′-0″, 3 panel		10	1.600		1,000	53		1,053	1,200	
	1340	9′-0″ x 4′-0″, 4 panel		8	2		1,325	66		1,391	1,550	
	1380	10′-0″ x 6′-0″, 5 panels		7	2.286		1,775	75.50		1,850.50	2,075	
	1420	12′-0″ x 6′-0″, 6 panels		6	2.667		1,800	88		1,888	2,125	
	1600	Metal clad, casement, bldrs mdl, 6′-0″ x 4′-0″, dbl insltd gls, 3 panels		10	1.600		830	53		883	1,000	
	1640	9′-0″ x 4′-0″ , 4 panels		8	2		1,175	66		1,241	1,375	
	1680	10′-0″ x 5′-0″, 5 panels		7	2.286		1,600	75.50		1,675.50	1,900	
	1720	12′-0″ x 6′-0″, 6 panels		6	2.667		2,250	88		2,338	2,625	
	2000	Bay window, casement, builders model, 8′ x 5′ dbl insul glass, 4 panels		10	1.600		1,425	53		1,478	1,650	
	2050	Low E glass,		10	1.600		1,725	53		1,778	1,975	
	2100	12′-0″ x 6′-0″ , double insulated glass, 6 panels		6	2.667		1,775	88		1,863	2,100	
	2200	Low E glass		6	2.667		1,825	88		1,913	2,175	
	2300	Vinyl clad, premium, double insulated glass, 8′-0″ x 5′-0″		10	1.600		1,150	53		1,203	1,350	
	2340	10′-0″ x 5′-0″ 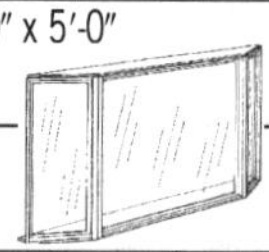		8	2		1,625	66		1,691	1,875	
	2380	10′-0″ x 6′-0″		7	2.286		1,700	75.50		1,775.50	2,000	
	2420	12′-0″ x 6′-0″		6	2.667		2,025	88		2,113	2,375	
	2600	Metal clad, deluxe, dbl insul. glass, 8′-0″ x 5′-0″ high, 4 panels		10	1.600		1,250	53		1,303	1,475	
	2640	10′-0″ x 5′-0″ high, 5 panels		8	2		1,350	66		1,416	1,575	
	2680	10′-0″ x 6′-0″ high, 5 panels		7	2.286		1,600	75.50		1,675.50	1,875	
	2720	12′-0″ x 6′-0″ high, 6 panels		6	2.667		2,200	88		2,288	2,575	
	3000	Double hung, bldrs. model, bay, 8′ x 4′ high,dbl insulated glass		10	1.600		995	53		1,048	1,200	
	3050	Low E glass		10	1.600		1,075	53		1,128	1,275	
	3100	9′-0″ x 5′-0″ high, doublel insulated glass		6	2.667		1,075	88		1,163	1,325	
	3200	Low E glass		6	2.667		1,125	88		1,213	1,400	
	3300	Vinyl clad, premium, double insulated glass, 7′-0″ x 4′-6″		10	1.600		1,025	53		1,078	1,225	
	3340	8′-0″ x 4′-6″		8	2		1,050	66		1,116	1,250	
	3380	8′-0″ x 5′-0″		7	2.286		1,100	75.50		1,175.50	1,325	
	3420	9′-0″ x 5′-0″		6	2.667		1,125	88		1,213	1,400	
	3600	Metal clad, deluxe, dbl insul. glass, 7′-0″ x 4′-0″ high		10	1.600		955	53		1,008	1,150	
	3640	8′-0″ x 4′-0″ high		8	2		990	66		1,056	1,175	
	3680	8′-0″ x 5′-0″ high		7	2.286		1,025	75.50		1,100.50	1,250	
	3720	9′-0″ x 5′-0″ high		6	2.667	↓	1,075	88		1,163	1,350	
	9000	Minimum labor/equipment charge	↓	2.50	6.400	Job		211		211	350	
200	0010	**CASEMENT WINDOW** Including frame, screen, and grills										200
	0100	Avg. quality, bldrs. model, 2′-0″ x 3′-0″ H, dbl. insulated glass	1 Carp	10	.800	Ea.	177	26.50		203.50	239	
	0150	Low E glass		10	.800		234	26.50		260.50	300	
	0200	2′-0″ x 4′-6″ high, double insulated glass 		9	.889		230	29.50		259.50	300	
	0250	Low E glass		9	.889		320	29.50		349.50	400	
	0300	2′-3″ x 6′-0″ high, double insulated glass		8	1		269	33		302	350	
	0350	Low E glass		8	1		325	33		358	415	
	0522	Vinyl clad, premium, double insulated glass, 2′-0″ x 3′-0″		10	.800		237	26.50		263.50	305	
	0524	2′-0″ x 4′-0″		9	.889		277	29.50		306.50	355	
	0525	2′-0″ x 5′-0″		8	1		315	33		348	405	
	0528	2′-0″ x 6′-0″		8	1		360	33		393	450	
	8100	Metal clad, deluxe, dbl. insul. glass, 2′-0″ x 3′-0″ high		10	.800		175	26.50		201.50	237	
	8120	2′-0″ x 4′-0″ high		9	.889		211	29.50		240.50	281	
	8140	2′-0″ x 5′-0″ high		8	1		240	33		273	320	
	8160	2′-0″ x 6′-0″ high	↓	8	1	↓	275	33		308	360	
	8200	For multiple leaf units, deduct for stationary sash										
	8220	2′ high				Ea.	18.20			18.20	20	
	8240	4′-6″ high				↓	21			21	23	

8 DOORS & WINDOWS

08500 | Windows

	08550 | Wood Windows		CREW	DAILY OUTPUT	LABOR-HOURS	UNIT	2004 BARE COSTS MAT.	LABOR	EQUIP.	TOTAL	TOTAL INCL O&P	
200	8260	6′ high				Ea.	28			28	31	200
	8300	For installation, add per leaf						15%				
	9000	Minimum labor/equipment charge	1 Carp	3	2.667	Job		88		88	146	
250	0010	**DOUBLE HUNG** Including frame, screens, and grills										250
	0100	Avg. quality, bldrs. model, 2′-0″ x 3′-0″ high, dbl insul. glass	1 Carp	10	.800	Ea.	183	26.50		209.50	246	
	0150	Low E glass		10	.800		183	26.50		209.50	246	
	0200	3′-0″ x 4′-0″ high, double insulated glass		9	.889		244	29.50		273.50	320	
	0250	Low E glass		9	.889		252	29.50		281.50	325	
	0300	4′-0″ x 4′-6″ high, double insulated glass		8	1		277	33		310	360	
	0350	Low E glass		8	1		300	33		333	385	
	1000	Vinyl clad, premium, double insulated glass, 2′-6″ x 3′-0″		10	.800		195	26.50		221.50	259	
	1100	3′-0″ x 3′-6″		10	.800		230	26.50		256.50	297	
	1200	3′-0″ x 4′-0″		9	.889		255	29.50		284.50	330	
	1300	3′-0″ x 4′-6″		9	.889		266	29.50		295.50	340	
	1400	3′-0″ x 5′-0″		8	1		278	33		311	360	
	1500	3′-6″ x 6′-0″		8	1		325	33		358	410	
	2000	Metal clad, deluxe, dbl. insul. glass, 2′-6″ x 3′-0″ high		10	.800		186	26.50		212.50	248	
	2100	3′-0″ x 3′-6″ high		10	.800		220	26.50		246.50	286	
	2200	3′-0″ x 4′-0″ high		9	.889		234	29.50		263.50	305	
	2300	3′-0″ x 4′-6″ high		9	.889		253	29.50		282.50	325	
	2400	3′-0″ x 5′-0″ high		8	1		270	33		303	350	
	2500	3′-6″ x 6′-0″ high		8	1		330	33		363	415	
	9000	Minimum labor/equipment charge		3	2.667	Job		88		88	146	
260	0010	**HALF ROUND WINDOW**, Vinyl clad, double insulated glass, including grill										260
	0800	14″ height x 24″ base	2 Carp	9	1.778	Ea.	325	58.50		383.50	455	
	1040	15″ height x 25″ base		8	2		330	66		396	470	
	1060	16″ height x 28″ base		7	2.286		360	75.50		435.50	520	
	1080	17″ height x 29″ base		7	2.286		375	75.50		450.50	535	
	2000	19″ height x 33″ base	1 Carp	6	1.333		400	44		444	515	
	2100	20″ height x 35″ base		6	1.333		445	44		489	565	
	2200	21″ height x 37″ base		6	1.333		425	44		469	545	
	2250	23″ height x 41″ base	2 Carp	6	2.667		465	88		553	655	
	2300	26″ height x 48″ base		6	2.667		480	88		568	675	
	2350	30″ height x 56″ base		6	2.667		565	88		653	765	
	3000	36″ height x 67″base	1 Carp	4	2		975	66		1,041	1,175	
	3040	38″ height x 71″ base	2 Carp	5	3.200		900	106		1,006	1,175	
	3050	40″ height x 75″ base	"	5	3.200		1,175	106		1,281	1,475	
	5000	Elliptical, 71″ x 16″	1 Carp	11	.727		740	24		764	855	
	5100	Elliptical, 95″ x 21″	"	10	.800		1,050	26.50		1,076.50	1,200	
670	0010	**PICTURE WINDOW** Including frame and grills										670
	0100	Average quality, bldrs. model, 3′-6″ x 4′-0″ high, dbl insulated glass	2 Carp	12	1.333	Ea.	260	44		304	360	
	0150	Low E glass		12	1.333		284	44		328	385	
	0200	4′-0″ x 4′-6″ high, double insulated glass		11	1.455		268	48		316	375	
	0250	Low E glass		11	1.455		294	48		342	405	
	0300	5′-0″ x 4′-0″ high, double insulated glass		11	1.455		330	48		378	445	
	0350	Low E glass		11	1.455		370	48		418	490	
	0400	6′-0″ x 4′-6″ high, double insulated glass		10	1.600		420	53		473	555	
	0450	Low E glass		10	1.600		475	53		528	610	
	1000	Vinyl clad, premium, dbl. insul. glass, 4′-0″ x 4′-0″		12	1.333		375	44		419	490	
	1100	4′-0″ x 6′-0″		11	1.455		645	48		693	790	
	1200	5′-0″ x 6′-0″		10	1.600		840	53		893	1,000	
	1300	6′-0″ x 6′-0″		10	1.600		855	53		908	1,025	
	2000	Metal clad, deluxe, dbl. insul. glass, 4′-0″ x 4′-0″ high		12	1.333		275	44		319	380	
	2100	4′-0″ x 6′-0″ high		11	1.455		405	48		453	525	
	2200	5′-0″ x 6′-0″ high		10	1.600		445	53		498	580	

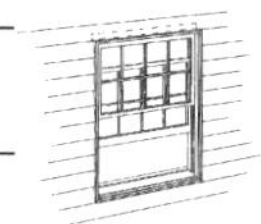

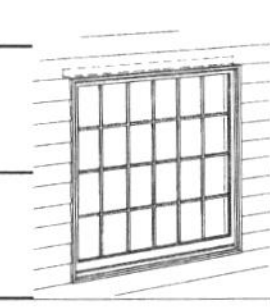

	08550 | Wood Windows		CREW	DAILY OUTPUT	LABOR-HOURS	UNIT	2004 BARE COSTS MAT.	LABOR	EQUIP.	TOTAL	TOTAL INCL O&P	
670	2300	6'-0" x 6'-0" high	2 Carp	10	1.600	Ea.	515	53		568	655	670
	9000	Minimum labor/equipment charge		2.75	5.818	Job		192		192	320	
750	0010	**SLIDING WINDOW** Including frame, screen, and grills										750
	0100	Average quality, bldrs. model, 3'-0" x 3'-0" high, double insulated	1 Carp	10	.800	Ea.	140	26.50		166.50	198	
	0120	Low E glass		10	.800		177	26.50		203.50	238	
	0200	4'-0" x 3'-6" high, double insulated		9	.889		166	29.50		195.50	232	
	0220	Low E glass		9	.889		208	29.50		237.50	278	
	0300	6'-0" x 5'-0" high, double insulated		8	1		305	33		338	390	
	0320	Low E glass		8	1		365	33		398	455	
	1000	Vinyl clad, premium, dbl. insulated glass, 3'-0" x 3'-0"		10	.800		515	26.50		541.50	610	
	1050	4'-0" x 3'-6"		9	.889		640	29.50		669.50	755	
	1100	5'-0" x 4'-0"		9	.889		770	29.50		799.50	900	
	1150	6'-0" x 5'-0"		8	1		945	33		978	1,075	
	2000	Metal clad, deluxe, double insulated glass, 3'-0" x 3'-0" high		10	.800		282	26.50		308.50	355	
	2050	4'-0" x 3'-6" high		9	.889		345	29.50		374.50	430	
	2100	5'-0" x 4'-0" high		9	.889		415	29.50		444.50	505	
	2150	6'-0" x 5'-0" high		8	1		640	33		673	760	
	9000	Minimum labor/equipment charge		3	2.667	Job		88		88	146	
760	0010	**TRANSOM WINDOWS**										760
	0050	Vinyl clad, premium, double insulated glass, 32" x 8"	1 Carp	16	.500	Ea.	144	16.50		160.50	186	
	0100	36" x 8"		16	.500		154	16.50		170.50	197	
	0110	36" x 12"		16	.500		173	16.50		189.50	218	
	0150	36" x 48"		12	.667		203	22		225	261	
	2000	Custom sizes, up to 350 sq. in.		12	.667		221	22		243	280	
	2100	351 to 750 sq. in.		12	.667		276	22		298	340	
	2200	751 to 1150 sq. in.		11	.727		340	24		364	415	
	2300	1151 to 1450 sq. in.		11	.727		390	24		414	470	
	2400	1451 to 1850 sq. in.	2 Carp	12	1.333		450	44		494	570	
	2500	1851 to 2250 sq. in.		12	1.333		510	44		554	635	
	2600	2251 to 2650 sq. in.		11	1.455		570	48		618	710	
	2700	2651 to 3050 sq. in.		11	1.455		585	48		633	720	
	2800	3051 to 3450 sq. in.		11	1.455		640	48		688	780	
	2900	3451 to 3850 sq. in.		10	1.600		690	53		743	850	
	3000	3851 to 4250 sq. in.		10	1.600		745	53		798	910	
	3100	4251 to 4650 sq. in.		10	1.600		760	53		813	925	
	3200	4651 to 5050 sq. in.		10	1.600		810	53		863	980	
	3300	5051 to 5450 sq. in.		9	1.778		860	58.50		918.50	1,050	
	3400	5451 to 5850 sq. in.		9	1.778		830	58.50		888.50	1,000	
	3600	6251 to 6650 sq. in.		8	2		1,025	66		1,091	1,225	
	3700	6651 to 7050 sq. in.		8	2		1,075	66		1,141	1,275	
	4960	36" x 30", round top		11	1.455		1,300	48		1,348	1,500	
	5000	36" x 114", round top		10	1.600		1,525	53		1,578	1,800	
	5200	48" x 72", round top		10	1.600		2,700	53		2,753	3,075	
	5400	66" x 90", round top		9	1.778		3,125	58.50		3,183.50	3,525	
	5600	70" x 108", round top		8	2		3,025	66		3,091	3,450	
780	0010	**TRAPEZOID WINDOWS**										780
	0900	20" base x 44" leg x 53" leg	2 Carp	13	1.231	Ea.	350	40.50		390.50	450	
	1000	24" base x 90" leg x 102" leg		8	2		580	66		646	745	
	3000	36" base x 0" leg x 22" leg		12	1.333		370	44		414	480	
	3010	36" base x 4" leg x 25" leg		13	1.231		390	40.50		430.50	500	
	3050	36" base x 26" leg x 48" leg		9	1.778		405	58.50		463.50	540	
	3100	36" base x 42" legs, 50" peak		9	1.778		470	58.50		528.50	615	
	3200	36" base x 60" leg x 81" leg		11	1.455		615	48		663	760	
	4320	44" base x 23" leg x 56" leg		11	1.455		485	48		533	615	
	4350	44" base x 59" leg x 92" leg		10	1.600		730	53		783	890	

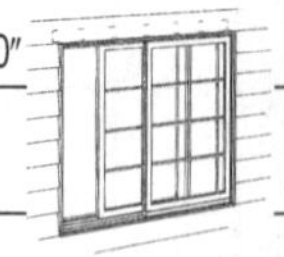

	08550 | Wood Windows		CREW	DAILY OUTPUT	LABOR-HOURS	UNIT	2004 BARE COSTS MAT.	LABOR	EQUIP.	TOTAL	TOTAL INCL O&P	
780	4500	46″ base x 15″ leg x 46″ leg	2 Carp	8	2	Ea.	365	66		431	515	780
	4550	46″ base x 16″ leg x 48″ leg		8	2		390	66		456	540	
	4600	46″ base x 50″ leg x 80″ leg		7	2.286		585	75.50		660.50	770	
	6600	66″ base x 12″ leg x 42″ leg		8	2		495	66		561	650	
	6650	66″ base x 12″ legs, 28″ peak		9	1.778		400	58.50		458.50	535	
	6700	68″ base x 3″ legs, 31″ peak		8	2		500	66		566	660	
800	0010	**WINDOW GRILLE OR MUNTIN** Snap-in type										800
	0020	Standard pattern interior grills										
	2000	Wood, awning window, glass size 28″ x 16″ high	1 Carp	30	.267	Ea.	18.55	8.80		27.35	35	
	2060	44″ x 24″ high		32	.250		27	8.25		35.25	43	
	2100	Casement, glass size, 20″ x 36″ high		30	.267		22.50	8.80		31.30	39.50	
	2180	20″ x 56″ high		32	.250		33	8.25		41.25	49.50	
	2200	Double hung, glass size, 16″ x 24″ high		24	.333	Set	40.50	11		51.50	63	
	2280	32″ x 32″ high		34	.235	″	113	7.75		120.75	137	
	2500	Picture, glass size, 48″ x 48″ high		30	.267	Ea.	129	8.80		137.80	157	
	2580	60″ x 68″ high		28	.286	″	97.50	9.45		106.95	123	
	2600	Sliding, glass size, 14″ x 36″ high		24	.333	Set	23	11		34	44	
	2680	36″ x 36″ high		22	.364	″	35	12		47	58	
	9000	Minimum labor/equipment charge		5	1.600	Job		53		53	87.50	
820	0010	**WOOD SASH** Including glazing but not including trim										820
	0050	Custom, 5′-0″ x 4′-0″, 1″ dbl. glazed, 3/16″ thick lites	2 Carp	3.20	5	Ea.	147	165		312	435	
	0100	1/4″ thick lites		5	3.200		151	106		257	340	
	0200	1″ thick, triple glazed		5	3.200		345	106		451	555	
	0300	7′-0″ x 4′-6″ high, 1″ double glazed, 3/16″ thick lites		4.30	3.721		350	123		473	590	
	0400	1/4″ thick lites		4.30	3.721		395	123		518	640	
	0500	1″ thick, triple glazed		4.30	3.721		455	123		578	705	
	0600	8′-6″ x 5′-0″ high, 1″ double glazed, 3/16″ thick lites		3.50	4.571		475	151		626	775	
	0700	1/4″ thick lites		3.50	4.571		520	151		671	820	
	0800	1″ thick, triple glazed		3.50	4.571		525	151		676	825	
	0900	Window frames only, based on perimeter length				L.F.	2.90			2.90	3.19	
	1200	Window sill, stock, per lineal foot					6.65			6.65	7.30	
	1250	Casing, stock					2.31			2.31	2.54	
	3000	Replacement sash, double hung, double glazing, to 12 S.F.	1 Carp	64	.125	S.F.	15.80	4.13		19.93	24.50	
	3100	12 S.F. to 20 S.F.		94	.085		16	2.81		18.81	22.50	
	3200	20 S.F. and over		106	.075		13.95	2.49		16.44	19.50	
	3800	Triple glazing for above, add					2.13			2.13	2.34	
	7000	Sash, single lite, 2′-0″ x 2′-0″ high	1 Carp	20	.400	Ea.	41	13.20		54.20	67	
	7050	2′-6″ x 2′-0″ high		19	.421		44	13.90		57.90	71.50	
	7100	2′-6″ x 2′-6″ high		18	.444		46.50	14.65		61.15	76	
	7150	3′-0″ x 2′-0″ high		17	.471		58.50	15.55		74.05	90	
	9000	Minimum labor/equipment charge		4	2	Job		66		66	109	
840	0010	**WOOD SCREENS**										840
	0020	Over 3 S.F., 3/4″ frames	2 Carp	375	.043	S.F.	3.34	1.41		4.75	6	
	0100	1-1/8″ frames		375	.043		5.95	1.41		7.36	8.90	
	0200	Rescreen wood frame		500	.032		.48	1.06		1.54	2.28	
	9000	Minimum labor/equipment charge	1 Carp	4	2	Job		66		66	109	
	08560 | Plastic Windows												
100	0010	**VINYL SINGLE HUNG WINDOWS**										100
	0100	Grids, low E, J fin, ext. jambs, 21″ x 53″	2 Carp	18	.889	Ea.	130	29.50		159.50	192	
	0110	21″ x 57″		17	.941		133	31		164	199	
	0120	21″ x 65″		16	1		139	33		172	208	
	0130	25″ x 41″		20	.800		123	26.50		149.50	179	
	0140	25″ x 49″		18	.889		136	29.50		165.50	198	

08560 | Plastic Windows

		08560 Plastic Windows	CREW	DAILY OUTPUT	LABOR-HOURS	UNIT	2004 BARE COSTS MAT.	LABOR	EQUIP.	TOTAL	TOTAL INCL O&P	
100	0150	25″ x 57″	2 Carp	17	.941	Ea.	139	31		170	205	100
	0160	25″ x 65″		16	1		145	33		178	214	
	0170	29″ x 41″		18	.889		131	29.50		160.50	193	
	0180	29″ x 53″		18	.889		140	29.50		169.50	203	
	0190	29″ x 57″		17	.941		143	31		174	210	
	0200	29″ x 65″		16	1		149	33		182	219	
	0210	33″ x 41″		20	.800		135	26.50		161.50	193	
	0220	33″ x 53″		18	.889		146	29.50		175.50	209	
	0230	33″ x 57″		17	.941		149	31		180	216	
	0240	33″ x 65″		16	1		155	33		188	225	
	0250	37″ x 41″		20	.800		143	26.50		169.50	201	
	0260	37″ x 53″		18	.889		153	29.50		182.50	217	
	0270	37″ x 57″		17	.941		156	31		187	224	
	0280	37″ x 65″		16	1		163	33		196	234	
200	0010	**VINYL DOUBLE HUNG WINDOWS**										200
	0100	Grids, low E, J fin, ext. jambs, 21″ x 53″	2 Carp	18	.889	Ea.	149	29.50		178.50	213	
	0102	21″ x 37″		18	.889		133	29.50		162.50	195	
	0104	21″ x 41″		18	.889		136	29.50		165.50	199	
	0106	21″ x 49″		18	.889		143	29.50		172.50	206	
	0110	21″ x 57″		17	.941		152	31		183	219	
	0120	21″ x 65″		16	1		158	33		191	229	
	0128	25″ x 37″		20	.800		140	26.50		166.50	198	
	0130	25″ x 41″		20	.800		143	26.50		169.50	202	
	0140	25″ x 49″		18	.889		148	29.50		177.50	211	
	0145	25″ x 53″		18	.889		154	29.50		183.50	218	
	0150	25″ x 57″		17	.941		154	31		185	222	
	0160	25″ x 65″		16	1		164	33		197	235	
	0162	25″ x 69″		16	1		170	33		203	242	
	0164	25″ x 77″		16	1		180	33		213	253	
	0168	29″ x 37″		18	.889		145	29.50		174.50	208	
	0170	29″ x 41″		18	.889		148	29.50		177.50	211	
	0172	29″ x 49″		18	.889		155	29.50		184.50	219	
	0180	29″ x 53″		18	.889		158	29.50		187.50	223	
	0190	29″ x 57″		17	.941		162	31		193	230	
	0200	29″ x 65″		16	1		168	33		201	240	
	0202	29″ x 69″		16	1		174	33		207	247	
	0205	29″ x 77″		16	1		184	33		217	258	
	0208	33″ x 37″		20	.800		149	26.50		175.50	207	
	0210	33″ x 41″		20	.800		152	26.50		178.50	211	
	0215	33″ x 49″		20	.800		160	26.50		186.50	220	
	0220	33″ x 53″		18	.889		163	29.50		192.50	229	
	0230	33″ x 57″		17	.941		167	31		198	236	
	0240	33″ x 65″		16	1		171	33		204	243	
	0242	33″ x 69″		16	1		181	33		214	255	
	0246	33″ x 77″		16	1		191	33		224	265	
	0250	37″ x 41″		20	.800		156	26.50		182.50	215	
	0255	37″ x 49″		20	.800		163	26.50		189.50	224	
	0260	37″ x 53″		18	.889		170	29.50		199.50	236	
	0270	37″ x 57″		17	.941		174	31		205	243	
	0280	37″ x 65″		16	1		178	33		211	251	
	0282	37″ x 69″		16	1		233	33		266	310	
	0286	37″ x 77″		16	1		244	33		277	325	
	0300	Solid vinyl, average quality, double insulated glass, 2′-0″ x 3′-0″	1 Carp	10	.800		127	26.50		153.50	183	
	0310	3′-0″ x 4′-0″		9	.889		156	29.50		185.50	220	
	0320	4′-0″ x 4′-6″		8	1		187	33		220	261	
	0330	Premium, double insulated glass, 2′-6″ x 3′-0″		10	.800		148	26.50		174.50	207	

Important: See the Reference Section for critical supporting data - Reference Nos., Crews, & City Cost Indexes

	08560 | Plastic Windows		CREW	DAILY OUTPUT	LABOR-HOURS	UNIT	2004 BARE COSTS MAT.	LABOR	EQUIP.	TOTAL	TOTAL INCL O&P	
200	0340	3′-0″ x 3′-6″	1 Carp	9	.889	Ea.	172	29.50		201.50	238	200
	0350	3′-0″ x 4″-0″		9	.889		183	29.50		212.50	250	
	0360	3′-0″ x 4′-6″		9	.889		187	29.50		216.50	255	
	0370	3′-0″ x 5′-0″		8	1		192	33		225	266	
	0380	3′-6″ x 6″-0″		8	1		211	33		244	287	
300	0010	VINYL CASEMENT WINDOWS										300
	0100	Grids, low E, J fin, ext. jambs, 1 lt, 21″ x 41″	2 Carp	20	.800	Ea.	193	26.50		219.50	256	
	0110	21″ x 47″		20	.800		210	26.50		236.50	275	
	0120	21″ x 53″		20	.800		227	26.50		253.50	294	
	0128	24″ x 35″		19	.842		185	28		213	250	
	0130	24″ x 41″		19	.842		201	28		229	267	
	0140	24″ x 47″		19	.842		218	28		246	286	
	0150	24″ x 53″		19	.842		235	28		263	305	
	0158	28″ x 35″		19	.842		197	28		225	263	
	0160	28″ x 41″		19	.842		213	28		241	280	
	0170	28″ x 47″		19	.842		230	28		258	299	
	0180	28″ x 53″		19	.842		254	28		282	325	
	0184	28″ x 59″		19	.842		258	28		286	330	
	0188	Two lites, 33″ x 35″		18	.889		315	29.50		344.50	400	
	0190	33″ x 41″		18	.889		340	29.50		369.50	425	
	0200	33″ x 47″		18	.889		365	29.50		394.50	450	
	0210	33″ x 53″		18	.889		390	29.50		419.50	480	
	0212	33″ x 59″		18	.889		415	29.50		444.50	505	
	0215	33″ x 72″		18	.889		430	29.50		459.50	520	
	0220	41″ x 41″		18	.889		370	29.50		399.50	460	
	0230	41″ x 47″		18	.889		395	29.50		424.50	485	
	0240	41″ x 53″		17	.941		420	31		451	515	
	0242	41″ x 59″		17	.941		445	31		476	540	
	0246	41″ x 72″		17	.941		465	31		496	560	
	0250	47″ x 41″		17	.941		375	31		406	460	
	0260	47″ x 47″		17	.941		400	31		431	490	
	0270	47″ x 53″		17	.941		425	31		456	515	
	0272	47″ x 59″		17	.941		465	31		496	560	
	0280	56″ x 41″		15	1.067		400	35		435	500	
	0290	56″ x 47″		15	1.067		425	35		460	525	
	0300	56″ x 53″		15	1.067		460	35		495	570	
	0302	56″ x 59″		15	1.067		480	35		515	590	
	0310	56″ x 72″		15	1.067		525	35		560	635	
	0340	Solid vinyl, premium, double insulated glass, 2′-0″ x 3′-0″ high	1 Carp	10	.800		170	26.50		196.50	231	
	0360	2′-0″ x 4′-0″ high		9	.889		209	29.50		238.50	279	
	0380	2′-0″ x 5′-0″ high		8	1		244	33		277	325	
400	0010	VINYL PICTURE WINDOWS										400
	0100	Grids, low E, J fin, ext. jambs, 33″ x 47″	2 Carp	12	1.333	Ea.	194	44		238	286	
	0110	35″ x 71″		12	1.333		205	44		249	298	
	0120	41″ x 47″		12	1.333		225	44		269	320	
	0130	41″ x 71″		12	1.333		244	44		288	340	
	0140	47″ x 47″		12	1.333		254	44		298	355	
	0150	47″ x 71″		11	1.455		267	48		315	375	
	0160	53″ x 47″		11	1.455		250	48		298	355	
	0170	53″ x 71″		11	1.455		261	48		309	365	
	0180	59″ x 47″		11	1.455		283	48		331	390	
	0190	59″ x 71″		11	1.455		305	48		353	415	
	0200	71″ x 47″		10	1.600		315	53		368	435	
	0210	71″ x 71″		10	1.600		330	53		383	455	

08560 | Plastic Windows

		08560 Plastic Windows	CREW	DAILY OUTPUT	LABOR-HOURS	UNIT	2004 BARE COSTS MAT.	LABOR	EQUIP.	TOTAL	TOTAL INCL O&P	
500	0010	**VINYL HALF ROUND WINDOW,** including grill, j fin, low E, ext. jambs										500
	0100	10″ height x 20″ base	2 Carp	9	1.778	Ea.	237	58.50		295.50	360	
	0110	15″ height x 30″ base		8	2		305	66		371	445	
	0120	17″ height x 34″ base		7	2.286		355	75.50		430.50	515	
	0130	19″ height x 38″ base		7	2.286		405	75.50		480.50	570	
	0140	20″ height x 40″ base	1 Carp	6	1.333		345	44		389	450	
	0150	24″ height x 48″ base		6	1.333		415	44		459	530	
	0160	25″ height x 50″ base		6	1.333		450	44		494	570	
	0170	30″ height x 60″ base	2 Carp	6	2.667		555	88		643	755	

08580 | Special Function Windows

		08580 Special Function Windows	CREW	DAILY OUTPUT	LABOR-HOURS	UNIT	MAT.	LABOR	EQUIP.	TOTAL	TOTAL INCL O&P	
900	0010	**STORM WINDOWS** Aluminum, residential										900
	0300	Basement, mill finish, incl. fiberglass screen										
	0320	1′-10″ x 1′-0″ high	2 Carp	30	.533	Ea.	28	17.60		45.60	60	
	0340	2′-9″ x 1′-6″ high		30	.533		30.50	17.60		48.10	62.50	
	0360	3′-4″ x 2′-0″ high		30	.533		37	17.60		54.60	69.50	
	1600	Double-hung, combination, storm & screen										
	1700	Custom, clear anodic coating, 2′-0″ x 3′-5″ high	2 Carp	30	.533	Ea.	73	17.60		90.60	109	
	1720	2′-6″ x 5′-0″ high		28	.571		97.50	18.85		116.35	139	
	1740	4′-0″ x 6′-0″ high		25	.640		206	21		227	262	
	1800	White painted, 2′-0″ x 3′-5″ high		30	.533		86.50	17.60		104.10	125	
	1820	2′-6″ x 5′-0″ high		28	.571		139	18.85		157.85	185	
	1840	4′-0″ x 6′-0″ high		25	.640		250	21		271	310	
	2000	Average quality, clear anodic coating, 2′-0″ x 3′-5″ high		30	.533		74	17.60		91.60	111	
	2020	2′-6″ x 5′-0″ high		28	.571		89.50	18.85		108.35	130	
	2040	4′-0″ x 6′-0″ high		25	.640		110	21		131	156	
	2400	White painted, 2′-0″ x 3′-5″ high		30	.533		72.50	17.60		90.10	109	
	2420	2′-6″ x 5′-0″ high		28	.571		80.50	18.85		99.35	120	
	2440	4′-0″ x 6′-0″ high		25	.640		88	21		109	132	
	2600	Mill finish, 2′-0″ x 3′-5″ high		30	.533		66.50	17.60		84.10	102	
	2620	2′-6″ x 5′-0″ high		28	.571		74	18.85		92.85	113	
	2640	4′-0″ x 6-8″ high		25	.640		83	21		104	126	
	4000	Picture window, storm, 1 lite, white or bronze finish										
	4020	4′-6″ x 4′-6″ high	2 Carp	25	.640	Ea.	111	21		132	157	
	4040	5′-8″ x 4′-6″ high		20	.800		126	26.50		152.50	183	
	4400	Mill finish, 4′-6″ x 4′-6″ high		25	.640		111	21		132	157	
	4420	5′-8″ x 4′-6″ high		20	.800		126	26.50		152.50	183	
	4600	3 lite, white or bronze finish										
	4620	4′-6″ x 4′-6″ high	2 Carp	25	.640	Ea.	135	21		156	184	
	4640	5′-8″ x 4′-6″ high		20	.800		150	26.50		176.50	210	
	4800	Mill finish, 4′-6″ x 4′-6″ high		25	.640		119	21		140	166	
	4820	5′-8″ x 4′-6″ high		20	.800		126	26.50		152.50	183	
	5001	Sliding glass door, storm window, 6′-0″ x 6′-8″, fixed	1 Glaz	1.60	5		297	160		457	585	
	5101	Operable	″	2.10	3.810		297	122		419	520	
	6000	Sliding window, storm, 2 lite, white or bronze finish										
	6020	3′-4″ x 2′-7″ high	2 Carp	28	.571	Ea.	92.50	18.85		111.35	134	
	6040	4′-4″ x 3′-3″ high		25	.640		126	21		147	174	
	6060	5′-4″ x 6′-0″ high		20	.800		202	26.50		228.50	267	
	6400	3 lite, white or bronze finish										
	6420	4′-4″ x 3′-3″ high	2 Carp	25	.640	Ea.	146	21		167	196	
	6440	5′-4″ x 6′-0″ high		20	.800		264	26.50		290.50	335	
	6460	6′-0″ x 6′-0″ high		18	.889		265	29.50		294.50	340	
	6800	Mill finish, 4′-4″ x 3′-3″ high		25	.640		126	21		147	174	
	6820	5′-4″ x 6′-0″ high		20	.800		265	26.50		291.50	335	
	6840	6′-0″ x 6-0″ high		18	.889		273	29.50		302.50	350	

Important: See the Reference Section for critical supporting data - Reference Nos., Crews, & City Cost Indexes

08580 | Special Function Windows

		08580 Special Function Windows	CREW	DAILY OUTPUT	LABOR-HOURS	UNIT	2004 BARE COSTS MAT.	LABOR	EQUIP.	TOTAL	TOTAL INCL O&P	
900	8000	PVC framed										900
	8100	Double-hung, combination, storm & screen										
	8120	2'-6" x 3'-5"	2 Carp	15	1.067	Ea.	58	35		93	122	
	8140	4' x 6'	"	12.50	1.280	"	99	42		141	179	
	8600	Single lite picture storm										
	8620	4'-6" x 4'-6"	2 Carp	12.50	1.280	Ea.	70	42		112	148	
	8640	5'-8" x 4'-6"	"	10	1.600	"	76.50	53		129.50	172	
	9000	Magnetic interior storm window										
	9100	3/16" plate glass	1 Glaz	107	.075	S.F.	3.98	2.40		6.38	8.25	
	9410	Minimum labor/equipment charge	1 Carp	4	2	Job		66		66	109	

08590 | Window Restoration & Replace

		08590 Window Restoration & Replace	CREW	DAILY OUTPUT	LABOR-HOURS	UNIT	MAT.	LABOR	EQUIP.	TOTAL	TOTAL INCL O&P	
600	0010	SOLID VINYL REPLACEMENT WINDOWS R08550-200										600
	0020	Double hung, insulated glass, up to 83 united inches	2 Carp	8	2	Ea.	196	66		262	325	
	0040	84 to 93		8	2		219	66		285	350	
	0060	94 to 101		6	2.667		249	88		337	420	
	0080	102 to 111		6	2.667		276	88		364	450	
	0100	112 to 120		6	2.667		310	88		398	485	
	0120	For each united inch over 120 , add		800	.020	Inch	3.28	.66		3.94	4.70	
	0140	Casement windows, one operating sash , 42 to 60 united inches		8	2	Ea.	170	66		236	296	
	0160	61 to 70		8	2		193	66		259	320	
	0180	71 to 80		8	2		210	66		276	340	
	0200	81 to 96		8	2		223	66		289	355	
	0220	Two operating sash, 58 to 78 united inches		8	2		340	66		406	485	
	0240	79 to 88		8	2		365	66		431	510	
	0260	89 to 98		8	2		395	66		461	545	
	0280	99 to 108		6	2.667		415	88		503	600	
	0300	109 to 121		6	2.667		445	88		533	635	
	0320	Three operating sash, 73 to 108 united inches		8	2		535	66		601	700	
	0340	109 to 118		8	2		565	66		631	730	
	0360	119 to 128		6	2.667		580	88		668	785	
	0380	129 to 138		6	2.667		620	88		708	825	
	0400	139 to 156		6	2.667		655	88		743	865	
	0420	Four operating sash, 98 to 118 united inches		8	2		770	66		836	955	
	0440	119 to 128		8	2		825	66		891	1,025	
	0460	129 to 138		6	2.667		875	88		963	1,100	
	0480	139 to 148		6	2.667		920	88		1,008	1,150	
	0500	149 to 168		6	2.667		980	88		1,068	1,225	
	0520	169 to 178		6	2.667		1,050	88		1,138	1,325	
	0540	For venting unit to fixed unit, deduct					15.50			15.50	17.05	
	0560	Fixed picture window, up to 63 united inches	2 Carp	8	2		123	66		189	244	
	0580	64 to 83		8	2		146	66		212	269	
	0600	84 to 101		8	2		187	66		253	315	
	0620	For each united inch over 101, add		900	.018	Inch	2.25	.59		2.84	3.45	
	0640	Picture window opt., low E glazing, up to 101 united inches				Ea.	17.50			17.50	19.25	
	0660	102 to 124					22.50			22.50	25	
	0680	124 and over					34			34	37.50	
	0700	Options, low E glazing, up to 101 united inches					8.75			8.75	9.65	
	0720	102 to 124					11.25			11.25	12.40	
	0740	124 and over					17			17	18.70	
	0760	Muntins, between glazing, square, per lite					1.65			1.65	1.82	
	0780	Diamond shape, per full or partial diamond					2.60			2.60	2.86	
	0800	Celluose fiber insulation, poured into sash balance cavity	1 Carp	36	.222	C.F.	.48	7.35		7.83	12.70	
	0820	Silicone caulking at perimeter	"	800	.010	L.F.	.13	.33		.46	.69	

08600 | Skylights

	08620	Unit Skylights	CREW	DAILY OUTPUT	LABOR-HOURS	UNIT	2004 BARE COSTS MAT.	LABOR	EQUIP.	TOTAL	TOTAL INCL O&P	
800	0010	**SKYLIGHT** Plastic domes, flush or curb mounted, ten or										800
	0100	more units, curb not included										
	0300	Nominal size under 10 S.F., double	G-3	130	.246	S.F.	18.30	8		26.30	33	
	0400	Single		160	.200		13.40	6.50		19.90	25	
	0600	10 S.F. to 20 S.F., double		315	.102		16.20	3.29		19.49	23	
	0700	Single		395	.081		8.25	2.62		10.87	13.30	
	0900	20 S.F. to 30 S.F., double		395	.081		14.70	2.62		17.32	20.50	
	1000	Single		465	.069		10.35	2.23		12.58	15	
	1200	30 S.F. to 65 S.F., double		465	.069		11.10	2.23		13.33	15.80	
	1300	Single		610	.052		14.10	1.70		15.80	18.25	
	1500	For insulated 4" curbs, double, add					25%					
	1600	Single, add					30%					
	1800	For integral insulated 9" curbs, double, add					30%					
	1900	Single, add					40%					
	2120	Ventilating insulated plexiglass dome with										
	2130	curb mounting, 36" x 36"	G-3	12	2.667	Ea.	370	86.50		456.50	545	
	2150	52" x 52"		12	2.667		555	86.50		641.50	750	
	2160	28" x 52"		10	3.200		430	104		534	640	
	2170	36" x 52"		10	3.200		465	104		569	680	
	2180	For electric opening system, add					277			277	305	
	2200	Field fabricated, factory type, aluminum and wire glass	G-3	120	.267	S.F.	14.25	8.65		22.90	29.50	
	2300	Insulated safety glass with aluminum frame		160	.200		83	6.50		89.50	102	
	2400	Sandwich panels, fiberglass, for walls, 1-9/16" thick, to 250 SF		200	.160		15.10	5.20		20.30	25	
	2500	250 SF and up		265	.121		13.50	3.91		17.41	21	
	2700	As above, but for roofs, 2-3/4" thick, to 250 SF		295	.108		22	3.51		25.51	29.50	
	2800	250 SF and up		330	.097		17.80	3.14		20.94	24.50	

08700 | Hardware

	08710	Door Hardware	CREW	DAILY OUTPUT	LABOR-HOURS	UNIT	2004 BARE COSTS MAT.	LABOR	EQUIP.	TOTAL	TOTAL INCL O&P	
100	0010	**AUTOMATIC OPENERS COMMERCIAL**										100
	0020	Pneumatic, incl opener, motion sens, control box, tubing, compressor										
	0050	For single swing door, per opening	2 Skwk	.80	20	Ea.	3,550	675		4,225	5,000	
	0100	Pair, per opening		.50	32	Opng.	5,900	1,075		6,975	8,250	
	1000	For single sliding door, per opening		.60	26.667		3,925	895		4,820	5,800	
	1300	Bi-parting pair		.50	32		5,925	1,075		7,000	8,275	
	1420	Electronic door opener incl motion sens, 12V control box, motor										
	1450	For single swing door, per opening	2 Skwk	.80	20	Opng.	3,200	675		3,875	4,600	
	1500	Pair, per opening		.50	32		5,200	1,075		6,275	7,475	
	1600	For single sliding door, per opening		.60	26.667		3,450	895		4,345	5,275	
	1700	Bi-parting pair		.50	32		5,300	1,075		6,375	7,575	
	1750	Handicap actuator buttons, 2, including 12V DC wiring, add	1 Carp	1.50	5.333	Pr.	355	176		531	685	
150	0010	**AVERAGE** Percentage for hardware, total job cost, minimum									.75%	150
	0050	Maximum									3.50%	
	0500	Total hardware for building, average distribution					85%	15%				
	1000	Door hardware, apartment, interior				Door	115			115	126	
	1500	Hospital bedroom, minimum					256			256	281	
	2000	Maximum					560			560	615	
	2250	School, single exterior, incl. lever, not incl. panic device					380			380	415	
	2500	Single interior, regular use, no lever included					252			252	277	

		08710 | Door Hardware	CREW	DAILY OUTPUT	LABOR-HOURS	UNIT	2004 BARE COSTS MAT.	LABOR	EQUIP.	TOTAL	TOTAL INCL O&P	
150	2600	Heavy use, incl. lever and closer				Door	440			440	485	150
	2850	Stairway, single interior					630			630	695	
	3100	Double exterior, with panic device				Pr.	890			890	975	
	3600	Toilet, public, single interior				Door	139			139	153	
300	0010	**DOOR CLOSER** Rack and pinion	1 Carp	6.50	1.231	Ea.	123	40.50		163.50	203	300
	0020	Adjustable backcheck, 3 way mount, all sizes, regular arm		6	1.333	"	125	44		169	210	
	0040	Hold open arm		6	1.333	Ea.	141	44		185	228	
	0100	Fusible link		6.50	1.231		110	40.50		150.50	189	
	0200	Non sized, regular arm		6	1.333		123	44		167	208	
	0240	Hold open arm		6	1.333		153	44		197	241	
	0400	4 way mount, non sized, regular arm		6	1.333		168	44		212	258	
	0440	Hold open arm		6	1.333		181	44		225	272	
	2000	Backcheck and adjustable power, hinge face mount										
	2010	All sizes, regular arm	1 Carp	6.50	1.231	Ea.	155	40.50		195.50	238	
	2040	Hold open arm		6.50	1.231		167	40.50		207.50	251	
	2400	Top jamb mount, all sizes, regular arm		6	1.333		155	44		199	243	
	2440	Hold open arm		6	1.333		167	44		211	256	
	2800	Top face mount, all sizes, regular arm		6.50	1.231		155	40.50		195.50	238	
	2840	Hold open arm		6.50	1.231		166	40.50		206.50	251	
	4000	Backcheck, overhead concealed, all sizes, regular arm		5.50	1.455		164	48		212	260	
	4040	Concealed arm		5	1.600		175	53		228	280	
	4400	Compact overhead, concealed, all sizes, regular arm		5.50	1.455		298	48		346	410	
	4440	Concealed arm		5	1.600		310	53		363	430	
	4800	Concealed in door, all sizes, regular arm		5.50	1.455		110	48		158	201	
	4840	Concealed arm		5	1.600		119	53		172	219	
	4900	Floor concealed, all sizes, single acting		2.20	3.636		140	120		260	355	
	4940	Double acting		2.20	3.636		181	120		301	400	
	5000	For cast aluminum cylinder, deduct					14.85			14.85	16.35	
	5040	For delayed action, add					26			26	28.50	
	5080	For fusible link arm, add					10.75			10.75	11.85	
	5120	For shock absorbing arm, add					32			32	35.50	
	5160	For spring power adjustment, add					25			25	27	
	6000	Closer-holder, hinge face mount, all sizes, exposed arm	1 Carp	6.50	1.231		114	40.50		154.50	193	
	7000	Electronic closer-holder, hinge facemount, concealed arm		5	1.600		173	53		226	279	
	7400	With built-in detector		5	1.600		520	53		573	665	
	9000	Minimum labor/equipment charge		4	2	Job		66		66	109	
320	0010	**DEADLOCKS** Mortise, heavy duty, outside key	1 Carp	9	.889	Ea.	118	29.50		147.50	178	320
	0020	Double cylinder		9	.889		130	29.50		159.50	192	
	0100	Medium duty, outside key		10	.800		91.50	26.50		118	144	
	0110	Double cylinder		10	.800		114	26.50		140.50	170	
	1000	Tubular, standard duty, outside key		10	.800		49.50	26.50		76	98	
	1010	Double cylinder		10	.800		63	26.50		89.50	114	
	1200	Night latch, outside key		10	.800		62	26.50		88.50	113	
340	0010	**DOORSTOPS** Holder and bumper, floor or wall	1 Carp	32	.250	Ea.	28	8.25		36.25	44.50	340
	1300	Wall bumper, 4" diameter, with rubber pad, aluminum		32	.250		8.50	8.25		16.75	23	
	1600	Door bumper, floor type, aluminum		32	.250		4.44	8.25		12.69	18.60	
	1900	Plunger type, door mounted		32	.250		24	8.25		32.25	39.50	
	9000	Minimum labor/equipment charge		6	1.333	Job		44		44	73	
400	0010	**ENTRANCE LOCKS** Cylinder, grip handle, deadlocking latch	1 Carp	9	.889	Ea.	108	29.50		137.50	168	400
	0020	Deadbolt		8	1		131	33		164	199	
	0100	Push and pull plate, dead bolt		8	1		125	33		158	192	
	0900	For handicapped lever, add					136			136	150	
500	0010	**HASP** Steel, 3" assembly	1 Carp	26	.308	Ea.	2.48	10.15		12.63	19.60	500
	0020	4-1/2"		13	.615		3.18	20.50		23.68	37	

		08710 | Door Hardware	CREW	DAILY OUTPUT	LABOR-HOURS	UNIT	2004 BARE COSTS MAT.	LABOR	EQUIP.	TOTAL	TOTAL INCL O&P	
500	0040	6″	1 Carp	12.50	.640	Ea.	4.81	21		25.81	40.50	500
520	0010	**HINGES** Full mortise, avg. freq., steel base, 4-1/2″ x 4-1/2″, USP R08700 -100				Pr.	19.35			19.35	21.50	520
	0100	5″ x 5″, USP					31.50			31.50	34.50	
	0200	6″ x 6″, USP					67			67	74	
	0400	Brass base, 4-1/2″ x 4-1/2″, US10					40			40	44	
	0500	5″ x 5″, US10					56.50			56.50	62.50	
	0600	6″ x 6″, US10					96.50			96.50	106	
	0800	Stainless steel base, 4-1/2″ x 4-1/2″, US32					61.50			61.50	67.50	
	0900	For non removable pin, add				Ea.	2.23			2.23	2.45	
	0910	For floating pin, driven tips, add					2.50			2.50	2.75	
	0930	For hospital type tip on pin, add					10.80			10.80	11.90	
	0940	For steeple type tip on pin, add					9.45			9.45	10.40	
	0950	Full mortise, high frequency, steel base, 3-1/2″ x 3-1/2″, US26D				Pr.	17.90			17.90	19.70	
	1000	4-1/2″ x 4-1/2″, USP					40.50			40.50	44.50	
	1100	5″ x 5″, USP					44.50			44.50	49	
	1200	6″ x 6″, USP					107			107	117	
	1400	Brass base, 3-1/2″ x 3-1/2″, US4					37			37	40.50	
	1430	4-1/2″ x 4-1/2″, US10					63.50			63.50	69.50	
	1500	5″ x 5″, US10					94			94	103	
	1600	6″ x 6″, US10					136			136	150	
	1800	Stainless steel base, 4-1/2″ x 4-1/2″, US32					100			100	110	
	1930	For hospital type tip on pin, add				Ea.	6.70			6.70	7.35	
	1950	Full mortise, low frequency, steel base, 3-1/2″ x 3-1/2″, US26D				Pr.	9.45			9.45	10.40	
	2000	4-1/2″ x 4-1/2″, USP					8.85			8.85	9.70	
	2100	5″ x 5″, USP					23.50			23.50	26	
	2200	6″ x 6″, USP					47.50			47.50	52	
	2300	4-1/2″ x 4-1/2″, US3					13.95			13.95	15.35	
	2310	5″ x 5″, US3					34.50			34.50	37.50	
	2400	Brass bass, 4-1/2″ x 4-1/2″, US10					33.50			33.50	36.50	
	2500	5″ x 5″, US10					50.50			50.50	55.50	
	2800	Stainless steel base, 4-1/2″ x 4-1/2″, US32					57.50			57.50	63	
550	0010	**KICK PLATE** 6″ high, for 3′ door, stainless steel	1 Carp	15	.533	Ea.	25.50	17.60		43.10	57	550
	0500	Bronze		15	.533		31.50	17.60		49.10	63.50	
	2000	Aluminum, .050, with 3 beveled edges, 10″ x 28″		15	.533		18	17.60		35.60	49	
	2010	10″ x 30″		15	.533		19.55	17.60		37.15	50.50	
	2020	10″ x 34″		15	.533		20.50	17.60		38.10	51.50	
	2040	10″ x 38″		15	.533		23	17.60		40.60	54	
	9000	Minimum labor/equipment charge		6	1.333	Job		44		44	73	
650	0010	**LOCKSET** Standard duty, cylindrical, with sectional trim										650
	0020	Non-keyed, passage	1 Carp	12	.667	Ea.	40	22		62	80	
	0100	Privacy		12	.667		48	22		70	89.50	
	0400	Keyed, single cylinder function		10	.800		68.50	26.50		95	120	
	0420	Hotel		8	1		95.50	33		128.50	160	
	0500	Lever handled, keyed, single cylinder function		10	.800		121	26.50		147.50	177	
	1000	Heavy duty with sectional trim, non-keyed, passages		12	.667		111	22		133	159	
	1100	Privacy		12	.667		140	22		162	191	
	1400	Keyed, single cylinder function		10	.800		166	26.50		192.50	226	
	1420	Hotel		8	1		247	33		280	325	
	1600	Communicating		10	.800		196	26.50		222.50	260	
	1690	For re-core cylinder, add					28			28	30.50	
	1700	Residential, interior door, minimum	1 Carp	16	.500		13.40	16.50		29.90	42	
	1720	Maximum		8	1		35.50	33		68.50	93.50	
	1800	Exterior, minimum		14	.571		30	18.85		48.85	64.50	
	1810	Average		8	1		63	33		96	124	

	08710 | Door Hardware		CREW	DAILY OUTPUT	LABOR-HOURS	UNIT	2004 BARE COSTS MAT.	LABOR	EQUIP.	TOTAL	TOTAL INCL O&P	
650	1820	Maximum	1 Carp	8	1	Ea.	125	33		158	193	650
	9000	Minimum labor/equipment charge		6	1.333	Job		44		44	73	
700	0010	**MORTISE LOCKSET** Comm., wrought knobs & full escutcheon trim										700
	0020	Non-keyed, passage, minimum	1 Carp	9	.889	Ea.	143	29.50		172.50	207	
	0030	Maximum		8	1		232	33		265	310	
	0040	Privacy, minimum		9	.889		153	29.50		182.50	217	
	0050	Maximum		8	1		250	33		283	330	
	0100	Keyed, office/entrance/apartment, minimum		8	1		175	33		208	248	
	0110	Maximum		7	1.143		300	37.50		337.50	395	
	0120	Single cylinder, typical, minimum		8	1		150	33		183	220	
	0130	Maximum		7	1.143		278	37.50		315.50	370	
	0200	Hotel, minimum		7	1.143		181	37.50		218.50	262	
	0210	Maximum		6	1.333		294	44		338	400	
	0300	Communication, double cylinder, minimum		8	1		181	33		214	254	
	0310	Maximum		7	1.143		234	37.50		271.50	320	
	1000	Wrought knobs and sectional trim, non-keyed, passage, minimum		10	.800		95	26.50		121.50	149	
	1010	Maximum		9	.889		188	29.50		217.50	255	
	1040	Privacy, minimum		10	.800		111	26.50		137.50	166	
	1050	Maximum		9	.889		200	29.50		229.50	269	
	1100	Keyed, entrance, office/apartment, minimum		9	.889		165	29.50		194.50	231	
	1110	Maximum		8	1		239	33		272	315	
	1120	Single cylinder, typical, minimum		9	.889		159	29.50		188.50	224	
	1130	Maximum		8	1		231	33		264	310	
	2000	Cast knobs and full escutcheon trim										
	2010	Non-keyed, passage, minimum	1 Carp	9	.889	Ea.	202	29.50		231.50	271	
	2020	Maximum		8	1		330	33		363	415	
	2040	Privacy, minimum		9	.889		243	29.50		272.50	315	
	2050	Maximum		8	1		355	33		388	445	
	2120	Keyed, single cylinder, typical, minimum		8	1		243	33		276	325	
	2130	Maximum		7	1.143		385	37.50		422.50	485	
	2200	Hotel, minimum		7	1.143		285	37.50		322.50	380	
	2210	Maximum		6	1.333		510	44		554	635	
	3000	Cast knob and sectional trim, non-keyed, passage, minimum		10	.800		157	26.50		183.50	217	
	3010	Maximum		10	.800		315	26.50		341.50	390	
	3040	Privacy, minimum		10	.800		179	26.50		205.50	241	
	3050	Maximum		10	.800		315	26.50		341.50	390	
	3100	Keyed, office/entrance/apartment, minimum		9	.889		203	29.50		232.50	272	
	3110	Maximum		9	.889		320	29.50		349.50	400	
	3120	Single cylinder, typical, minimum		9	.889		203	29.50		232.50	272	
	3130	Maximum		9	.889		395	29.50		424.50	485	
	3190	For re-core cylinder, add					27.50			27.50	30	
	3800	Cipher lockset	1 Carp	13	.615		645	20.50		665.50	745	
	3900	Keyless, pushbutton type										
	4000	Residential/light commercial, deadbolt, standard	1 Carp	9	.889	Ea.	91.50	29.50		121	150	
	4010	Heavy duty		9	.889		109	29.50		138.50	169	
	4020	Industrial, heavy duty, with deadbolt		9	.889		223	29.50		252.50	294	
	4030	Key override		9	.889		247	29.50		276.50	320	
	4040	Lever activated handle		9	.889		271	29.50		300.50	345	
	4050	Key override		9	.889		300	29.50		329.50	380	
	4060	Double sided pushbutton type		8	1		495	33		528	600	
	4070	Key override		8	1		535	33		568	640	
750	0010	**PANIC DEVICE** For rim locks, single door, exit only	1 Carp	6	1.333	Ea.	340	44		384	450	750
	0020	Outside key and pull		5	1.600		385	53		438	515	
	0200	Bar and vertical rod, exit only		5	1.600		490	53		543	630	
	0210	Outside key and pull		4	2		585	66		651	755	

DOORS & WINDOWS 8

For expanded coverage of these items see *Means Interior Cost Data 2004*

08710 | Door Hardware

		08710 Door Hardware	CREW	DAILY OUTPUT	LABOR-HOURS	UNIT	2004 BARE COSTS MAT.	LABOR	EQUIP.	TOTAL	TOTAL INCL O&P	
750	0400	Bar and concealed rod	1 Carp	4	2	Ea.	495	66		561	655	750
	0600	Touch bar, exit only		6	1.333		390	44		434	505	
	0610	Outside key and pull		5	1.600		470	53		523	610	
	0700	Touch bar and vertical rod, exit only		5	1.600		535	53		588	675	
	0710	Outside key and pull		4	2		620	66		686	795	
	1000	Mortise, bar, exit only		4	2		440	66		506	590	
	1600	Touch bar, exit only		4	2		500	66		566	660	
	2000	Narrow stile, rim mounted, bar, exit only		6	1.333		525	44		569	655	
	2010	Outside key and pull		5	1.600		570	53		623	720	
	2200	Bar and vertical rod, exit only		5	1.600		545	53		598	685	
	2210	Outside key and pull		4	2		545	66		611	705	
	2400	Bar and concealed rod, exit only		3	2.667		630	88		718	840	
	3000	Mortise, bar, exit only		4	2		455	66		521	610	
	3600	Touch bar, exit only		4	2		660	66		726	840	
	4000	Double doors, exit only		2	4	Pr.	685	132		817	970	
	4500	Exit & entrance		2	4	"	780	132		912	1,075	
	9000	Minimum labor/equipment charge		2.50	3.200	Job		106		106	175	
770	0010	**DOOR AND WINDOW ACCESSORIES**										770
	0050	Door closing coordinator, 36" (for paired openings up to 56")	1 Carp	8	1	Ea.	80.50	33		113.50	143	
	0060	48" (for paired openings up to 84")		8	1		86	33		119	149	
	0070	56" (for paired openings up to 96")		8	1		94.50	33		127.50	159	
780	0010	**PUSH-PULL PLATE**										780
	0100	Push plate, .050 thick, 4" x 16", aluminum	1 Carp	12	.667	Ea.	6.20	22		28.20	43.50	
	0500	Bronze		12	.667		15.70	22		37.70	54	
	1500	Pull handle and push bar, aluminum		11	.727		110	24		134	160	
	2000	Bronze		10	.800		142	26.50		168.50	200	
	3000	Push plate both sides, aluminum		14	.571		13.35	18.85		32.20	46	
	3500	Bronze		13	.615		33	20.50		53.50	70	
	4000	Door pull, designer style, cast aluminum, minimum		12	.667		58.50	22		80.50	101	
	5000	Maximum		8	1		293	33		326	380	
	6000	Cast bronze, minimum		12	.667		68	22		90	112	
	7000	Maximum		8	1		320	33		353	405	
	8000	Walnut, minimum		12	.667		52	22		74	94	
	9000	Maximum		8	1		292	33		325	375	
	9800	Minimum labor/equipment charge		5	1.600	Job		53		53	87.50	
800	0010	**SPECIAL HINGES**										800
	0015	Paumelle, high frequency										
	0020	Steel base, 6" x 4-1/2", US10				Pr.	116			116	127	
	0100	Bronze base, 5" x 4-1/2", US10					153			153	169	
	0200	Paumelle, average frequency, steel base, 4-1/2" x 3-1/2", US10					78.50			78.50	86	
	0400	Olive knuckle, low frequency, brass base, 6" x 4-1/2", US10					140			140	154	
	1000	Electric hinge with concealed conductor, average frequency										
	1010	Steel base, 4-1/2" x 4-1/2", US26D				Pr.	255			255	280	
	1100	Bronze base, 4-1/2" x 4-1/2", US26D				"	268			268	294	
	1200	Electric hinge with concealed conductor, high frequency										
	1210	Steel base, 4-1/2" x 4-1/2", US26D				Pr.	190			190	209	
	1600	Double weight, 800 lb., steel base, removable pin, 5" x 6", USP					117			117	129	
	1700	Steel base-welded pin, 5" x 6", USP					129			129	142	
	1800	Triple weight, 2000 lb., steel base, welded pin, 5" x 6", USP					135			135	148	
	2000	Pivot reinf., high frequency, steel base, 7-3/4" door plate, USP					155			155	170	
	2200	Bronze base, 7-3/4" door plate, US10					181			181	200	
	3000	Swing clear, full mortise, full or half surface, high frequency,										
	3010	Steel base, 5" high, USP				Pr.	129			129	142	
	3200	Swing clear, full mortise, average frequency										
	3210	Steel base, 4-1/2" high, USP				Pr.	102			102	113	

08700 | Hardware

	08710	Door Hardware	Crew	Daily Output	Labor-Hours	Unit	2004 Bare Costs Mat.	Labor	Equip.	Total	Total Incl O&P	
800	4000	Wide throw, average frequency, steel base, 4-1/2" x 6", USP				Pr.	78.50			78.50	86	800
	4200	High frequency, steel base, 4-1/2" x 6", USP					120			120	132	
	4600	Spring hinge, single acting, 6" flange, steel				Ea.	44			44	48.50	
	4700	Brass					77.50			77.50	85.50	
	4900	Double acting, 6" flange, steel					79.50			79.50	87.50	
	4950	Brass					130			130	143	
	5000	T-strap, galvanized, 4"				Pr.	15.50			15.50	17.05	
	5010	6"					24.50			24.50	27	
	5020	8"					37.50			37.50	41	
	9000	Continuous hinge, steel, full mortise, heavy duty	2 Carp	64	.250	L.F.	10.80	8.25		19.05	25.50	

	08720	Weatherstripping & Seals	Crew	Daily Output	Labor-Hours	Unit	Mat.	Labor	Equip.	Total	Total Incl O&P	
100	0010	**ASTRAGALS** One piece overlapping										100
	0400	Cadmium plated steel, flat, 3/16" x 2"	1 Carp	90	.089	L.F.	2.75	2.93		5.68	7.90	
	0600	Prime coated steel, flat, 1/8" x 3"		90	.089		3.82	2.93		6.75	9.05	
	0800	Stainless steel, flat, 3/32" x 1-5/8"		90	.089		14.30	2.93		17.23	20.50	
	1000	Aluminum, flat, 1/8" x 2"		90	.089		2.76	2.93		5.69	7.90	
	1200	Nail on, "T" extrusion		120	.067		.61	2.20		2.81	4.32	
	1300	Vinyl bulb insert		105	.076		.99	2.51		3.50	5.25	
	1600	Screw on, "T" extrusion		90	.089		4.02	2.93		6.95	9.30	
	1700	Vinyl insert		75	.107		2.71	3.52		6.23	8.85	
	2000	"L" extrusion, neoprene bulbs		75	.107		1.56	3.52		5.08	7.55	
	2100	Neoprene sponge insert		75	.107		4.76	3.52		8.28	11.10	
	2200	Magnetic		75	.107		7.85	3.52		11.37	14.45	
	2400	Spring hinged security seal, with cam		75	.107		5.05	3.52		8.57	11.40	
	2600	Spring loaded locking bolt, vinyl insert		45	.178		6.85	5.85		12.70	17.25	
	2800	Neoprene sponge strip, "Z" shaped, aluminum		60	.133		3.27	4.40		7.67	10.90	
	2900	Solid neoprene strip, nail on aluminum strip		90	.089		2.74	2.93		5.67	7.85	
	3000	One piece stile protection										
	3020	Neoprene fabric loop, nail on aluminum strips	1 Carp	60	.133	L.F.	.49	4.40		4.89	7.85	
	3110	Flush mounted aluminum extrusion, 1/2" x 1-1/4"		60	.133		2.60	4.40		7	10.15	
	3140	3/4" x 1-3/8"		60	.133		3.30	4.40		7.70	10.95	
	3160	1-1/8" x 1-3/4"		60	.133		5.25	4.40		9.65	13.10	
	3300	Mortise, 9/16" x 3/4"		60	.133		2.80	4.40		7.20	10.40	
	3320	13/16" x 1-3/8"		60	.133		3.03	4.40		7.43	10.65	
	3600	Spring bronze strip, nail on type		105	.076		2.38	2.51		4.89	6.80	
	3620	Screw on, with retainer		75	.107		1.95	3.52		5.47	8	
	3800	Flexible stainless steel housing, pile insert, 1/2" door		105	.076		5.40	2.51		7.91	10.10	
	3820	3/4" door		105	.076		6.10	2.51		8.61	10.85	
	4000	Extruded aluminum retainer, flush mount, pile insert		105	.076		1.89	2.51		4.40	6.25	
	4080	Mortise, felt insert		90	.089		3.36	2.93		6.29	8.55	
	4160	Mortise with spring, pile insert		90	.089		2.63	2.93		5.56	7.75	
	4400	Rigid vinyl retainer, mortise, pile insert		105	.076		1.88	2.51		4.39	6.25	
	4600	Wool pile filler strip, aluminum backing		105	.076		1.89	2.51		4.40	6.25	
	5000	Two piece overlapping astragal, extruded aluminum retainer										
	5010	Pile insert	1 Carp	60	.133	L.F.	2.45	4.40		6.85	10	
	5020	Vinyl bulb insert		60	.133		1.58	4.40		5.98	9.05	
	5040	Vinyl flap insert		60	.133		4.91	4.40		9.31	12.70	
	5060	Solid neoprene flap insert		60	.133		4.86	4.40		9.26	12.65	
	5080	Hypalon rubber flap insert		60	.133		4.97	4.40		9.37	12.75	
	5090	Snap on cover, pile insert		60	.133		5.70	4.40		10.10	13.55	
	5400	Magnetic aluminum, surface mounted		60	.133		19.20	4.40		23.60	28.50	
	5500	Interlocking aluminum, 5/8" x 1" neoprene bulb insert		45	.178		3.05	5.85		8.90	13.05	
	5600	Adjustable aluminum, 9/16" x 21/32", pile insert		45	.178		14.45	5.85		20.30	25.50	
	5790	For vinyl bulb, deduct					.37			.37	.41	
	5800	Magnetic, adjustable, 9/16" x 21/32"	1 Carp	45	.178		18.60	5.85		24.45	30	

For expanded coverage of these items see *Means Interior Cost Data 2004*

		08720 | Weatherstripping & Seals	CREW	DAILY OUTPUT	LABOR-HOURS	UNIT	2004 BARE COSTS MAT.	LABOR	EQUIP.	TOTAL	TOTAL INCL O&P	
100	6000	Two piece stile protection										100
	6010	Cloth backed rubber loop, 1" gap, nail on aluminum strips	1 Carp	45	.178	L.F.	3.12	5.85		8.97	13.15	
	6040	Screw on aluminum strips		45	.178		4.86	5.85		10.71	15.05	
	6100	1-1/2" gap, screw on aluminum extrusion		45	.178		4.37	5.85		10.22	14.50	
	6240	Vinyl fabric loop, slotted aluminum extrusion, 1" gap		45	.178		1.54	5.85		7.39	11.40	
	6300	1-1/4" gap	↓	45	.178	↓	4.60	5.85		10.45	14.75	
300	0010	**WEATHERSTRIPPING** Window, double hung, 3' x 5', zinc	1 Carp	7.20	1.111	Opng.	10.70	36.50		47.20	73	300
	0100	Bronze		7.20	1.111		21.50	36.50		58	84.50	
	0200	Vinyl V strip		7	1.143		3.65	37.50		41.15	66.50	
	0500	As above but heavy duty, zinc		4.60	1.739		13.80	57.50		71.30	110	
	0600	Bronze		4.60	1.739		24	57.50		81.50	122	
	1000	Doors, wood frame, interlocking, for 3' x 7' door, zinc		3	2.667		12.35	88		100.35	160	
	1100	Bronze		3	2.667		19.45	88		107.45	168	
	1300	6' x 7' opening, zinc		2	4		13.50	132		145.50	234	
	1400	Bronze	↓	2	4	↓	25.50	132		157.50	247	
	1700	Wood frame, spring type, bronze										
	1800	3' x 7' door	1 Carp	7.60	1.053	Opng.	15.80	34.50		50.30	75	
	1900	6' x 7' door	"	7	1.143	"	16.60	37.50		54.10	81	
	2200	Metal frame, spring type, bronze										
	2300	3' x 7' door	1 Carp	3	2.667	Opng.	26.50	88		114.50	175	
	2400	6' x 7' door	"	2.50	3.200	"	37	106		143	216	
	2500	For stainless steel, spring type, add					133%					
	2700	Metal frame, extruded sections, 3' x 7' door, aluminum	1 Carp	2	4	Opng.	36	132		168	259	
	2800	Bronze		2	4		90.50	132		222.50	320	
	3100	6' x 7' door, aluminum		1.20	6.667		45.50	220		265.50	415	
	3200	Bronze	↓	1.20	6.667	↓	107	220		327	485	
	3500	Threshold weatherstripping										
	3650	Door sweep, flush mounted, aluminum	1 Carp	25	.320	Ea.	10.65	10.55		21.20	29	
	3700	Vinyl		25	.320		12.55	10.55		23.10	31.50	
	5000	Garage door bottom weatherstrip, 12' aluminum, clear		14	.571		17	18.85		35.85	50	
	5010	Bronze		14	.571		65	18.85		83.85	103	
	5050	Bottom protection, 12' aluminum, clear		14	.571		20	18.85		38.85	53.50	
	5100	Bronze		14	.571	↓	80.50	18.85		99.35	120	
	9000	Minimum labor/equipment charge	↓	3	2.667	Job		88		88	146	
800	0010	**THRESHOLD** 3' long door saddles, aluminum	1 Carp	48	.167	L.F.	3.43	5.50		8.93	12.85	800
	0100	Aluminum, 8" wide, 1/2" thick		12	.667	Ea.	29	22		51	68.50	
	0500	Bronze ♿		60	.133	L.F.	27	4.40		31.40	37.50	
	0600	Bronze, panic threshold, 5" wide, 1/2" thick		12	.667	Ea.	57	22		79	99	
	0700	Rubber, 1/2" thick, 5-1/2" wide		20	.400		30	13.20		43.20	55	
	0800	2-3/4" wide		20	.400	↓	13.65	13.20		26.85	37	
	9000	Minimum labor/equipment charge	↓	4	2	Job		66		66	109	

		08750 | Window Hardware	CREW	DAILY OUTPUT	LABOR-HOURS	UNIT	MAT.	LABOR	EQUIP.	TOTAL	TOTAL INCL O&P	
400	0010	**WINDOW HARDWARE**										400
	1000	Handles, surface mounted, aluminum	1 Carp	24	.333	Ea.	1.84	11		12.84	20.50	
	1020	Brass		24	.333		2.14	11		13.14	20.50	
	1040	Chrome		24	.333		1.96	11		12.96	20.50	
	1500	Recessed, aluminum		12	.667		1.07	22		23.07	37.50	
	1520	Brass		12	.667		1.18	22		23.18	38	
	1540	Chrome		12	.667		1.12	22		23.12	37.50	
	2000	Latches, aluminum		20	.400		1.53	13.20		14.73	23.50	
	2020	Brass		20	.400		1.84	13.20		15.04	24	
	2040	Chrome		20	.400	↓	1.73	13.20		14.93	24	
	9000	Minimum labor/equipment charge	↓	6	1.333	Job		44		44	73	

08700 | Hardware

	08770 | Door/Window Accessories		CREW	DAILY OUTPUT	LABOR-HOURS	UNIT	2004 BARE COSTS MAT.	LABOR	EQUIP.	TOTAL	TOTAL INCL O&P
560	0010	**DOOR ACCESSORIES**									
	1000	Knockers, brass, standard	1 Carp	16	.500	Ea.	36	16.50		52.50	67
	1100	Deluxe		10	.800		111	26.50		137.50	166
	4000	Security chain, standard		18	.444		6.20	14.65		20.85	31.50
	4100	Deluxe		18	.444		37	14.65		51.65	65.50
	9000	Minimum labor/equipment charge		6	1.333	Job		44		44	73

08800 | Glazing

	08810 | Glass		CREW	DAILY OUTPUT	LABOR-HOURS	UNIT	2004 BARE COSTS MAT.	LABOR	EQUIP.	TOTAL	TOTAL INCL O&P
205	0010	**CURTAIN WALL** See division 08900									
260	0010	**FLOAT GLASS** 3/16" thick, clear, plain	2 Glaz	130	.123	S.F.	4.02	3.94		7.96	10.75
	0200	Tempered, clear		130	.123		4.77	3.94		8.71	11.60
	0300	Tinted		130	.123		6	3.94		9.94	12.95
	0600	1/4" thick, clear, plain		120	.133		4.83	4.27		9.10	12.20
	0700	Tinted		120	.133		4.76	4.27		9.03	12.15
	0800	Tempered, clear		120	.133		5.95	4.27		10.22	13.40
	0900	Tinted		120	.133		8.20	4.27		12.47	15.90
	1600	3/8" thick, clear, plain		75	.213		7.75	6.85		14.60	19.50
	1700	Tinted		75	.213		9.50	6.85		16.35	21.50
	1800	Tempered, clear		75	.213		11.95	6.85		18.80	24
	1900	Tinted		75	.213		14.70	6.85		21.55	27
	2200	1/2" thick, clear, plain		55	.291		15.35	9.30		24.65	32
	2300	Tinted		55	.291		16.65	9.30		25.95	33.50
	2400	Tempered, clear		55	.291		17.90	9.30		27.20	35
	2500	Tinted		55	.291		22	9.30		31.30	39.50
	2800	5/8" thick, clear, plain		45	.356		16.65	11.40		28.05	36.50
	2900	Tempered, clear		45	.356		19.05	11.40		30.45	39.50
	3200	3/4" thick, clear, plain		35	.457		21.50	14.65		36.15	47
	3300	Tempered, clear		35	.457		25	14.65		39.65	51
	3600	1" thick, clear, plain		30	.533		35.50	17.10		52.60	67
	8900	For low emissivity coating for 3/16" and 1/4" only, add to above					15%				
	9000	Minimum labor/equipment charge	1 Glaz	2	4	Job		128		128	207
270	0010	**FULL VISION** Window system with 3/4" glass mullions, 10' high	H-2	130	.185	S.F.	47.50	5.55		53.05	61
	0100	10' to 20' high, minimum		110	.218	"	50	6.55		56.55	65.50
	0150	Average		100	.240	S.F.	54	7.20		61.20	71
	0200	Maximum		80	.300	"	61	9		70	81.50
	9000	Minimum labor/equipment charge	1 Glaz	2	4	Job		128		128	207
300	0010	**GLAZING VARIABLES** R08810 -010									
	0500	For high rise glazing, exterior, add per S.F. per story				S.F.					.09
	0600	For glass replacement, add				"		100%			
	0700	For gasket settings, add				L.F.	3.57			3.57	3.93
	0900	For sloped glazing, add				S.F.		25%			
	2000	Fabrication, polished edges, 1/4" thick				Inch	.31			.31	.34
	2100	1/2" thick					.78			.78	.86
	2500	Mitered edges, 1/4" thick					.78			.78	.86
	2600	1/2" thick					1.26			1.26	1.39

08810 | Glass

		08810 Glass	Crew	Daily Output	Labor-Hours	Unit	2004 Bare Costs Mat.	2004 Bare Costs Labor	2004 Bare Costs Equip.	2004 Bare Costs Total	Total Incl O&P	
460	0010	**INSULATING GLASS** 2 lites 1/8" float, 1/2" thk, under 15 S.F. R08810-010										460
	0020	Clear	2 Glaz	95	.168	S.F.	6.85	5.40		12.25	16.25	
	0100	Tinted		95	.168		10.15	5.40		15.55	19.85	
	0200	2 lites 3/16" float, for 5/8" thk unit, 15 to 30 S.F., clear		90	.178		8.35	5.70		14.05	18.35	
	0400	1" thk, dbl. glazed, 1/4" float, 30-70 S.F., clear		75	.213		11.70	6.85		18.55	24	
	0500	Tinted		75	.213		14.30	6.85		21.15	26.50	
	2000	Both lites, light & heat reflective		85	.188		19.05	6.05		25.10	30.50	
	2500	Heat reflective, film inside, 1" thick unit, clear		85	.188		16.65	6.05		22.70	28	
	2600	Tinted		85	.188		17.95	6.05		24	29.50	
	3000	Film on weatherside, clear, 1/2" thick unit		95	.168		11.90	5.40		17.30	22	
	3100	5/8" thick unit		90	.178		14.65	5.70		20.35	25.50	
	3200	1" thick unit		85	.188		16.40	6.05		22.45	28	
	3350	Minimum	1 Glaz	50	.160		9	5.15		14.15	18.15	
	3360	Maximum	"	25	.320		9.75	10.25		20	27.50	
	3370	Reflective or tinted, add					2.49			2.49	2.74	
	9000	Minimum labor/equipment charge	1 Glaz	2	4	Job		128		128	207	
500	0010	**LAMINATED GLASS** Clear float, .03" vinyl, 1/4" thick	2 Glaz	90	.178	S.F.	8.15	5.70		13.85	18.15	500
	0100	3/8" thick		78	.205		13	6.55		19.55	25	
	0200	.06" vinyl, 1/2" thick		65	.246		15.20	7.90		23.10	29.50	
	1000	5/8" thick		90	.178		17.85	5.70		23.55	29	
	2000	Bullet-resisting, 1-3/16" thick, to 15 S.F.		16	1		48.50	32		80.50	105	
	2100	Over 15 S.F.		16	1		44	32		76	99.50	
	2500	2-1/4" thick, to 15 S.F.		12	1.333		58.50	42.50		101	134	
	2600	Over 15 S.F.		12	1.333		52	42.50		94.50	126	
675	0010	**REFLECTIVE GLASS** 1/4" float with fused metallic oxide, tinted	2 Glaz	115	.139	S.F.	9.90	4.46		14.36	18.10	675
	0500	1/4" float glass with reflective applied coating		115	.139		8.25	4.46		12.71	16.30	
	2000	Solar film on glass, not including glass, minimum		180	.089		3.99	2.85		6.84	9	
	2050	Maximum		225	.071		9.25	2.28		11.53	13.80	
850	0010	**WINDOW GLASS** Clear float, stops, putty bed, 1/8" thick	2 Glaz	480	.033	S.F.	3.31	1.07		4.38	5.35	850
	0500	3/16" thick, clear		480	.033		4.07	1.07		5.14	6.20	
	0600	Tinted		480	.033		4.59	1.07		5.66	6.75	
	0700	Tempered		480	.033		5.55	1.07		6.62	7.80	
	2000	Replace broken window lite, 1/8" glass (9 S.F. maximum)	1 Glaz	48	.167		3.72	5.35		9.07	12.70	
	2100	1/4" plate (16 S.F. maximum)	"	48	.167		4.10	5.35		9.45	13.10	
	9000	Minimum labor/equipment charge	2 Glaz	5	3.200	Job		103		103	165	
900	0010	**WIRE GLASS** 1/4" thick, rough obscure (chicken wire)	2 Glaz	135	.119	S.F.	10.90	3.80		14.70	18.10	900
	1000	Polished wire, 1/4" thick, diamond, clear		135	.119		13.90	3.80		17.70	21.50	
	1500	Pinstripe, obscure		135	.119		13.55	3.80		17.35	21	

08830 | Mirrors

		08830 Mirrors	Crew	Daily Output	Labor-Hours	Unit	Mat.	Labor	Equip.	Total	Total Incl O&P	
100	0010	**MIRRORS** No frames, wall type, 1/4" plate glass, polished edge										100
	0100	Up to 5 S.F.	2 Glaz	125	.128	S.F.	6.40	4.10		10.50	13.65	
	0200	Over 5 S.F.		160	.100		6.20	3.21		9.41	12	
	0500	Door type, 1/4" plate glass, up to 12 S.F.		160	.100		6.70	3.21		9.91	12.50	
	1000	Float glass, up to 10 S.F., 1/8" thick		160	.100		3.86	3.21		7.07	9.40	
	1100	3/16" thick		150	.107		4.48	3.42		7.90	10.45	
	1500	12" x 12" wall tiles, square edge, clear		195	.082		1.59	2.63		4.22	6	
	1600	Veined		195	.082		4.17	2.63		6.80	8.85	
	2000	1/4" thick, stock sizes, one way transparent		125	.128		14.45	4.10		18.55	22.50	
	2010	Bathroom, unframed, laminated		160	.100		10.80	3.21		14.01	17.05	
	2500	Tempered		160	.100		11.80	3.21		15.01	18.10	

Important: See the Reference Section for critical supporting data - Reference Nos., Crews, & City Cost Indexes

08840 | Plastic Glazing

		08840 Plastic Glazing	CREW	DAILY OUTPUT	LABOR-HOURS	UNIT	2004 BARE COSTS MAT.	LABOR	EQUIP.	TOTAL	TOTAL INCL O&P	
600	0010	**PLEXIGLASS ACRYLIC** Clear, masked, 1/8" thick, cut sheets	2 Glaz	170	.094	S.F.	3.32	3.02		6.34	8.50	600
	0200	Full sheets		195	.082		1.73	2.63		4.36	6.15	
	0500	1/4" thick, cut sheets		165	.097		5.85	3.11		8.96	11.45	
	0600	Full sheets		185	.086		3.18	2.77		5.95	7.95	
	0900	3/8" thick, cut sheets		155	.103		10.70	3.31		14.01	17.15	
	1000	Full sheets		180	.089		5.80	2.85		8.65	10.95	
	1300	1/2" thick, cut sheets		135	.119		12.35	3.80		16.15	19.70	
	1400	Full sheets		150	.107		11.95	3.42		15.37	18.65	
	1700	3/4" thick, cut sheets		115	.139		44	4.46		48.46	55	
	1800	Full sheets		130	.123		25.50	3.94		29.44	34.50	
	2100	1" thick, cut sheets		105	.152		49.50	4.88		54.38	62.50	
	2200	Full sheets		125	.128		30.50	4.10		34.60	40	
	3000	Colored, 1/8" thick, cut sheets		170	.094		10.20	3.02		13.22	16.05	
	3200	Full sheets		195	.082		6.60	2.63		9.23	11.50	
	3500	1/4" thick, cut sheets		165	.097		11.40	3.11		14.51	17.55	
	3600	Full sheets		185	.086		7.80	2.77		10.57	13	
	4000	Mirrors, untinted, cut sheets, 1/8" thick		185	.086		4.79	2.77		7.56	9.70	
	4200	1/4" thick		180	.089		7.70	2.85		10.55	13.05	
650	0010	**POLYCARBONATE** Clear, masked, cut sheets, 1/8" thick	2 Glaz	170	.094	S.F.	5.75	3.02		8.77	11.20	650
	0500	3/16" thick		165	.097		6.95	3.11		10.06	12.65	
	1000	1/4" thick		155	.103		7.70	3.31		11.01	13.80	
	1500	3/8" thick		150	.107		14.15	3.42		17.57	21	
	9000	Minimum labor/equipment charge	1 Glaz	2	4	Job		128		128	207	

08900 | Glazed Curtain Wall

08910 | Glazed Aluminum Curtain Wall

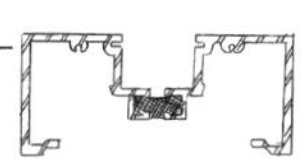

		08910 Glazed Aluminum Curtain Wall	CREW	DAILY OUTPUT	LABOR-HOURS	UNIT	2004 BARE COSTS MAT.	LABOR	EQUIP.	TOTAL	TOTAL INCL O&P	
700	0010	**TUBE FRAMING** For window walls and store fronts, aluminum, stock										700
	0050	Plain tube frame, mill finish, 1-3/4" x 1-3/4"	2 Glaz	103	.155	L.F.	6.35	4.98		11.33	15	
	0150	1-3/4" x 4"		98	.163		8.35	5.25		13.60	17.60	
	0200	1-3/4" x 4-1/2"		95	.168		9.30	5.40		14.70	18.90	
	0250	2" x 6"		89	.180		14	5.75		19.75	24.50	
	0350	4" x 4"		87	.184		13.70	5.90		19.60	24.50	
	0400	4-1/2" x 4-1/2"		85	.188		15.25	6.05		21.30	26.50	
	0450	Glass bead		240	.067		1.77	2.14		3.91	5.40	
	1000	Flush tube frame, mill finish, 1/4" glass, 1-3/4" x 4", open header		80	.200		8.20	6.40		14.60	19.35	
	1050	Open sill		82	.195		7.10	6.25		13.35	17.95	
	1100	Closed back header		83	.193		11.45	6.20		17.65	22.50	
	1150	Closed back sill		85	.188		10.90	6.05		16.95	21.50	
	1200	Vertical mullion, one piece		75	.213		12.10	6.85		18.95	24.50	
	1250	Two piece		73	.219		13	7		20	25.50	
	1300	90° or 180° vertical corner post		75	.213		20.50	6.85		27.35	33.50	
	1400	1-3/4" x 4-1/2", open header		80	.200		10	6.40		16.40	21.50	
	1450	Open sill		82	.195		8.25	6.25		14.50	19.15	
	1500	Closed back header		83	.193		12.30	6.20		18.50	23.50	
	1550	Closed back sill		85	.188		11.75	6.05		17.80	22.50	
	1600	Vertical mullion, one piece		75	.213		13.10	6.85		19.95	25.50	
	1650	Two piece		73	.219		13.85	7		20.85	26.50	
	1700	90° or 180° vertical corner post		75	.213		14.20	6.85		21.05	26.50	

08900 | Glazed Curtain Wall

8 DOORS & WINDOWS

08910 | Glazed Aluminum Curtain Wall

		08910 Glazed Aluminum Curtain Wall	Crew	Daily Output	Labor-Hours	Unit	2004 Bare Costs Mat.	Labor	Equip.	Total	Total Incl O&P	
700	2000	Flush tube frame, mill fin. for ins. glass, 2" x 4-1/2", open header	2 Glaz	75	.213	L.F.	11.15	6.85		18	23.50	700
	2050	Open sill		77	.208		9.75	6.65		16.40	21.50	
	2100	Closed back header		78	.205		12.15	6.55		18.70	24	
	2150	Closed back sill		80	.200		12	6.40		18.40	23.50	
	2200	Vertical mullion, one piece		70	.229		13.45	7.35		20.80	26.50	
	2250	Two piece		68	.235		14.40	7.55		21.95	28	
	2300	90° or 180° vertical corner post		70	.229		13.60	7.35		20.95	27	
	5000	Flush tube frame, mill fin., thermal brk., 2-1/4"x 4-1/2", open header		74	.216		12.25	6.95		19.20	24.50	
	5050	Open sill		75	.213		10.70	6.85		17.55	23	
	5100	Vertical mullion, one piece		69	.232		14.80	7.45		22.25	28.50	
	5150	Two piece		67	.239		15.80	7.65		23.45	30	
	5200	90° or 180° vertical corner post		69	.232		14.20	7.45		21.65	27.50	
	6980	Door stop (snap in)		380	.042		2.38	1.35		3.73	4.80	
	7000	For joints, 90°, clip type, add				Ea.	19.60			19.60	21.50	
	7050	Screw spline joint, add					15.35			15.35	16.85	
	7100	For joint other than 90°, add					32			32	35.50	
	8000	For bronze anodized aluminum, add					15%					
	8020	For black finish, add					27%					
	8050	For stainless steel materials, add					350%					
	8100	For monumental grade, add					50%					
	8150	For steel stiffener, add	2 Glaz	200	.080	L.F.	7.85	2.56		10.41	12.75	
	8200	For 2 to 5 stories, add per story				Story		5%				
	9000	Minimum labor/equipment charge	2 Glaz	2	8	Job		256		256	415	
900	0010	**WINDOW WALLS** Aluminum, stock, including glazing, minimum	H-2	160	.150	S.F.	27	4.50		31.50	37	900
	0050	Average		140	.171	"	33	5.15		38.15	44.50	
	0100	Maximum		110	.218	S.F.	103	6.55		109.55	124	
	0500	For translucent sandwich wall systems, see div. 07420-770										
	0850	Cost of the above walls depends on material,										
	0860	finish, repetition, and size of units.										
	0870	The larger the opening, the lower the S.F. cost										
	1200	Double glazed acoustical window wall for airports,										
	1220	including 1" thick glass with 2" x 4-1/2" tube frame	H-2	40	.600	S.F.	62.50	18		80.50	98	

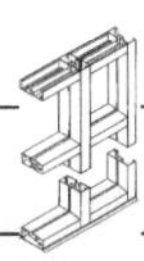

08950 | Translucent Wall & Roof Assemblies

		08950 Translucent Wall & Roof Assemblies	Crew	Daily Output	Labor-Hours	Unit	Mat.	Labor	Equip.	Total	Total Incl O&P	
100	0010	**SKYROOFS** Translucent panels, 2-3/4" thick, under 5000 S.F.	G-3	395	.081	SF Hor.	20.50	2.62		23.12	26.50	100
	0100	Over 5000 S.F.		465	.069		18.10	2.23		20.33	23.50	
	0300	Continuous vaulted, semi-circular, to 8' wide, double glazed		145	.221		45.50	7.15		52.65	61.50	
	0400	Single glazed		160	.200		30.50	6.50		37	44	
	0600	To 20' wide, single glazed		175	.183		34	5.90		39.90	47	
	0700	Over 20' wide, single glazed		200	.160		39	5.20		44.20	51.50	
	0900	Motorized opening type, single glazed, 1/3 opening		145	.221		41	7.15		48.15	56.50	
	1000	Full opening		130	.246		47	8		55	64.50	
	1200	Pyramid type units, self-supporting, to 30' clear opening,										
	1300	square or circular, single glazed, minimum	G-3	200	.160	SF Hor.	22	5.20		27.20	33	
	1310	Average		165	.194		31.50	6.30		37.80	44.50	
	1400	Maximum		130	.246		45	8		53	62.50	
	1500	Grid type, 4' to 10' modules, single glass glazed, minimum		200	.160		28.50	5.20		33.70	39.50	
	1550	Maximum		128	.250		46.50	8.10		54.60	64	
	1600	Preformed acrylic, minimum		300	.107		33.50	3.46		36.96	42.50	
	1650	Maximum		175	.183		47	5.90		52.90	61.50	
	9000	Minimum labor/equipment charge	2 Carp	8	2	Job		66		66	109	

Division 9 Finishes

Estimating Tips

General

- Room Finish Schedule: A complete set of plans should contain a room finish schedule. If one is not available, it would be well worth the time and effort to put one together. A room finish schedule should contain the room number, room name (for clarity), floor materials, base materials, wainscot materials, wainscot height, wall materials (for each wall), ceiling materials, ceiling height and special instructions.
- Surplus Finishes: Review the specifications to determine if there is any requirement to provide certain amounts of extra materials for the owner's maintenance department. In some cases the owner may require a substantial amount of materials, especially when it is a special order item or long lead time item.

09200 Plaster & Gypsum Board

- Lath is estimated by the square yard for both gypsum and metal lath, plus usually 5% allowance for waste. Furring, channels and accessories are measured by the linear foot. An extra foot should be allowed for each accessory miter or stop.
- Plaster is also estimated by the square yard. Deductions for openings vary by preference, from zero deduction to 50% of all openings over 2 feet in width. Some estimators deduct a percentage of the total yardage for openings. The estimator should allow one extra square foot for each linear foot of horizontal interior or exterior angle located below the ceiling level. Also, double the areas of small radius work.
- Each room should be measured, perimeter times maximum wall height. Floors and ceiling areas are equal to length times width.
- Drywall accessories, studs, track, and acoustical caulking are all measured by the linear foot. Drywall taping is figured by the square foot. Gypsum wallboard is estimated by the square foot. No material deductions should be made for door or window openings under 32 S.F. Coreboard can be obtained in a 1″ thickness for solid wall and shaft work. Additions should be made to price out the inside or outside corners.
- Different types of partition construction should be listed separately on the quantity sheets. There may be walls with studs of various widths, double studded, and similar or dissimilar surface materials. Shaft work is usually different construction from surrounding partitions requiring separate quantities and pricing of the work.

09300 Tile
09400 Terrazzo

- Tile and terrazzo areas are taken off on a square foot basis. Trim and base materials are measured by the linear foot. Accent tiles are listed per each. Two basic methods of installation are used. Mud set is approximately 30% more expensive than the thin set. In terrazzo work, be sure to include the linear footage of embedded decorative strips, grounds, machine rubbing and power cleanup.

09600 Flooring

- Wood flooring is available in strip, parquet, or block configuration. The latter two types are set in adhesives with quantities estimated by the square foot. The laying pattern will influence labor costs and material waste. In addition to the material and labor for laying wood floors, the estimator must make allowances for sanding and finishing these areas unless the flooring is prefinished.
- Most of the various types of flooring are all measured on a square foot basis. Base is measured by the linear foot. If adhesive materials are to be quantified, they are estimated at a specified coverage rate by the gallon depending upon the specified type and the manufacturer's recommendations.
- Sheet flooring is measured by the square yard. Roll widths vary, so consideration should be given to use the most economical width, as waste must be figured into the total quantity. Consider also the installation methods available, direct glue down or stretched.

09700 Wall Finishes

- Wall coverings are estimated by the square foot. The area to be covered is measured, length by height of wall above baseboards, to calculate the square footage of each wall. This figure is divided by the number of square feet in the single roll which is being used. Deduct, in full, the areas of openings such as doors and windows. Where a pattern match is required allow 25%-30% waste. One gallon of paste should be sufficient to hang 12 single rolls of light to medium weight paper.

09800 Acoustical Treatment

- Acoustical systems fall into several categories. The takeoff of these materials should be by the square foot of area with a 5% allowance for waste. Do not forget about scaffolding, if applicable, when estimating these systems.

09900 Paints & Coatings

- A major portion of the work in painting involves surface preparation. Be sure to include cleaning, sanding, filling and masking costs in the estimate.
- Painting is one area where bids vary to a greater extent than almost any other section of a project. This arises from the many methods of measuring surfaces to be painted. The estimator should check the plans and specifications carefully to be sure of the required number of coats.
- Protection of adjacent surfaces is not included in painting costs. When considering the method of paint application, an important factor is the amount of protection and masking required. These must be estimated separately and may be the determining factor in choosing the method of application.

Reference Numbers

Reference numbers are shown in bold squares at the beginning of some major classifications. These numbers refer to related items in the Reference Section. The reference information may be an estimating procedure, an alternate pricing method or technical information.

Note: Not all subdivisions listed here necessarily appear in this publication.

09060 | Selective Demolition

		09060 Selective Demolition	Crew	Daily Output	Labor-Hours	Unit	Mat.	Labor	Equip.	Total	Total Incl O&P	
							2004 Bare Costs					
110	0010	**SELECTIVE DEMOLITION, CEILINGS** R02220-510										110
	0200	Drywall, furred and nailed	2 Clab	800	.020	S.F.		.52		.52	.86	
	0220	On metal frame		760	.021			.55		.55	.91	
	0240	On suspension system, including system		720	.022			.58		.58	.96	
	1000	Plaster, lime and horse hair, on wood lath, incl. lath		700	.023			.59		.59	.99	
	1020	On metal lath		570	.028			.73		.73	1.21	
	1100	Gypsum, on gypsum lath		720	.022			.58		.58	.96	
	1120	On metal lath		500	.032			.83		.83	1.38	
	1200	Suspended ceiling, mineral fiber, 2'x2' or 2'x4'		1,500	.011			.28		.28	.46	
	1250	On suspension system, incl. system		1,200	.013			.35		.35	.57	
	1500	Tile, wood fiber, 12" x 12", glued		900	.018			.46		.46	.77	
	1540	Stapled		1,500	.011			.28		.28	.46	
	1580	On suspension system, incl. system		760	.021			.55		.55	.91	
	2000	Wood, tongue and groove, 1" x 4"		1,000	.016			.42		.42	.69	
	2040	1" x 8"		1,100	.015			.38		.38	.63	
	2400	Plywood or wood fiberboard, 4' x 8' sheets		1,200	.013			.35		.35	.57	
	9000	Minimum labor/equipment charge	1 Clab	2	4	Job		104		104	172	
120	0010	**SELECTIVE DEMOLITION, FLOORING** R02220-510										120
	0200	Brick with mortar	2 Clab	475	.034	S.F.		.88		.88	1.45	
	0400	Carpet, bonded, including surface scraping		2,000	.008			.21		.21	.34	
	0480	Tackless		9,000	.002			.05		.05	.08	
	0600	Composition, acrylic or epoxy		400	.040			1.04		1.04	1.72	
	0700	Concrete, scarify skin	A-1A	225	.036			1.20	1.22	2.42	3.30	
	0800	Resilient, sheet goods	2 Clab	1,400	.011			.30		.30	.49	
	0820	For gym floors		900	.018			.46		.46	.77	
	0900	Vinyl composition tile, 12" x 12"		1,000	.016			.42		.42	.69	
	2000	Tile, ceramic, thin set		675	.024			.62		.62	1.02	
	2020	Mud set		625	.026			.67		.67	1.10	
	2200	Marble, slate, thin set		675	.024			.62		.62	1.02	
	2220	Mud set		625	.026			.67		.67	1.10	
	2600	Terrazzo, thin set		450	.036			.92		.92	1.53	
	2620	Mud set		425	.038			.98		.98	1.62	
	2640	Cast in place		300	.053			1.39		1.39	2.30	
	3000	Wood, block, on end	1 Carp	400	.020			.66		.66	1.09	
	3200	Parquet		450	.018			.59		.59	.97	
	3400	Strip flooring, interior, 2-1/4" x 25/32" thick		325	.025			.81		.81	1.35	
	3500	Exterior, porch flooring, 1" x 4"		220	.036			1.20		1.20	1.99	
	3800	Subfloor, tongue and groove, 1" x 6"		325	.025			.81		.81	1.35	
	3820	1" x 8"		430	.019			.61		.61	1.02	
	3840	1" x 10"		520	.015			.51		.51	.84	
	4000	Plywood, nailed		600	.013			.44		.44	.73	
	4100	Glued and nailed		400	.020			.66		.66	1.09	
	8000	Remove flooring, bead blast, minimum	A-1A	1,000	.008			.27	.27	.54	.74	
	8100	Maximum		400	.020			.67	.69	1.36	1.85	
	8150	Mastic only		1,500	.005			.18	.18	.36	.49	
	9000	Minimum labor/equipment charge	1 Clab	4	2	Job		52		52	86	
130	0010	**SELECTIVE DEMOLITION, WALLS AND PARTITIONS** R02220-510										130
	0100	Brick, 4" to 12" thick	B-9C	220	.182	C.F.		4.80	.73	5.53	8.75	
	0200	Concrete block, 4" thick		1,000	.040	S.F.		1.06	.16	1.22	1.93	
	0280	8" thick		810	.049			1.30	.20	1.50	2.38	
	0300	Exterior stucco 1" thick over mesh	B-9	3,200	.013			.33	.05	.38	.61	
	1000	Drywall, nailed	1 Clab	1,000	.008			.21		.21	.34	
	1020	Glued and nailed		900	.009			.23		.23	.38	
	1500	Fiberboard, nailed		900	.009			.23		.23	.38	

09050 | Basic Material Finishes and Methods

	09060 | Selective Demolition		CREW	DAILY OUTPUT	LABOR-HOURS	UNIT	2004 BARE COSTS MAT.	LABOR	EQUIP.	TOTAL	TOTAL INCL O&P	
130	1520	Glued and nailed R02220-510	1 Clab	800	.010	S.F.		.26		.26	.43	130
	2000	Movable walls, metal, 5′ high		300	.027			.69		.69	1.15	
	2020	8′ high		400	.020			.52		.52	.86	
	2200	Metal or wood studs, finish 2 sides, fiberboard	B-1	520	.046			1.23		1.23	2.04	
	2250	Lath and plaster		260	.092			2.46		2.46	4.08	
	2300	Plasterboard (drywall)		520	.046			1.23		1.23	2.04	
	2350	Plywood		450	.053			1.42		1.42	2.36	
	3000	Plaster, lime and horsehair, on wood lath	1 Clab	400	.020			.52		.52	.86	
	3020	On metal lath		335	.024			.62		.62	1.03	
	3400	Gypsum or perlite, on gypsum lath		410	.020			.51		.51	.84	
	3420	On metal lath		300	.027			.69		.69	1.15	
	3800	Toilet partitions, slate or marble		5	1.600	Ea.		41.50		41.50	69	
	3820	Hollow metal		8	1	″		26		26	43	
	5000	Wallcovering, vinyl	1 Pape	700	.011	S.F.		.34		.34	.54	
	5040	Designer	″	480	.017	″		.49		.49	.78	
	9000	Minimum labor/equipment charge	1 Clab	4	2	Job		52		52	86	

FINISHES 9

09100 | Metal Support Assemblies

	09110 | Non-Load Bearing Wall Framing		CREW	DAILY OUTPUT	LABOR-HOURS	UNIT	2004 BARE COSTS MAT.	LABOR	EQUIP.	TOTAL	TOTAL INCL O&P	
100	0010	**METAL STUDS AND TRACK**										100
	1600	Non-load bearing, galv, 8′ high, 25 ga. 1-5/8″ wide, 16″ O.C.	1 Carp	619	.013	S.F.	.19	.43		.62	.92	
	1610	24″ O.C.		950	.008		.14	.28		.42	.62	
	1620	2-1/2″ wide, 16″ O.C.		613	.013		.21	.43		.64	.94	
	1630	24″ O.C.		938	.009		.16	.28		.44	.64	
	1640	3-5/8″ wide, 16″ O.C.		600	.013		.22	.44		.66	.98	
	1650	24″ O.C.		925	.009		.17	.29		.46	.66	
	1660	4″ wide, 16″ O.C.		594	.013		.25	.44		.69	1.02	
	1670	24″ O.C.		925	.009		.19	.29		.48	.68	
	1680	6″ wide, 16″ O.C.		588	.014		.36	.45		.81	1.13	
	1690	24″ O.C.		906	.009		.27	.29		.56	.77	
	1700	20 ga. studs, 1-5/8″ wide, 16″ O.C.		494	.016		.29	.53		.82	1.21	
	1710	24″ O.C.		763	.010		.22	.35		.57	.81	
	1720	2-1/2″ wide, 16″ O.C.		488	.016		.32	.54		.86	1.25	
	1730	24″ O.C.		750	.011		.24	.35		.59	.85	
	1740	3-5/8″ wide, 16″ O.C.		481	.017		.38	.55		.93	1.33	
	1750	24″ O.C.		738	.011		.29	.36		.65	.91	
	1760	4″ wide, 16″ O.C.		475	.017		.40	.56		.96	1.36	
	1770	24″ O.C.		738	.011		.30	.36		.66	.92	
	1780	6″ wide, 16″ O.C.		469	.017		.53	.56		1.09	1.51	
	1790	24″ O.C.		725	.011		.40	.36		.76	1.04	
	2000	Non-load bearing, galv, 10′ high, 25 ga. 1-5/8″ wide, 16″ O.C.		495	.016		.18	.53		.71	1.08	
	2100	24″ O.C.		760	.011		.13	.35		.48	.73	
	2200	2-1/2″ wide, 16″ O.C.		490	.016		.20	.54		.74	1.11	
	2250	24″ O.C.		750	.011		.15	.35		.50	.74	
	2300	3-5/8″ wide, 16″ O.C.		480	.017		.21	.55		.76	1.14	
	2350	24″ O.C.		740	.011		.16	.36		.52	.76	
	2400	4″ wide, 16″ O.C.		475	.017		.24	.56		.80	1.18	
	2450	24″ O.C.		740	.011		.18	.36		.54	.78	
	2500	6″ wide, 16″ O.C.		470	.017		.34	.56		.90	1.30	

09110 | Non-Load Bearing Wall Framing

			CREW	DAILY OUTPUT	LABOR-HOURS	UNIT	2004 BARE COSTS MAT.	LABOR	EQUIP.	TOTAL	TOTAL INCL O&P	
100	2550	24" O.C.	1 Carp	725	.011	S.F.	.25	.36		.61	.87	100
	2600	20 ga. studs, 1-5/8" wide, 16" O.C.		395	.020		.28	.67		.95	1.41	
	2650	24" O.C.		610	.013		.20	.43		.63	.95	
	2700	2-1/2" wide, 16" O.C.		390	.021		.30	.68		.98	1.45	
	2750	24" O.C.		600	.013		.23	.44		.67	.98	
	2800	3-5/8" wide, 16" O.C.		385	.021		.36	.69		1.05	1.54	
	2850	24" O.C.		590	.014		.27	.45		.72	1.04	
	2900	4" wide, 16" O.C.		380	.021		.38	.69		1.07	1.57	
	2950	24" O.C.		590	.014		.28	.45		.73	1.05	
	3000	6" wide, 16" O.C.		375	.021		.50	.70		1.20	1.72	
	3050	24" O.C.		580	.014		.37	.46		.83	1.16	
	3060	Non-load bearing, galv, 12' high, 25 ga. 1-5/8" wide, 16" O.C.		413	.019		.17	.64		.81	1.25	
	3070	24" O.C.		633	.013		.13	.42		.55	.83	
	3080	2-1/2" wide, 16" O.C.		408	.020		.19	.65		.84	1.28	
	3090	24" O.C.		625	.013		.14	.42		.56	.85	
	3100	3-5/8" wide, 16" O.C.		400	.020		.20	.66		.86	1.31	
	3110	24" O.C.		617	.013		.15	.43		.58	.87	
	3120	4" wide, 16" O.C.		396	.020		.23	.67		.90	1.35	
	3130	24" O.C.		617	.013		.17	.43		.60	.89	
	3140	6" wide, 16" O.C.		392	.020		.33	.67		1	1.48	
	3150	24" O.C.		604	.013		.24	.44		.68	.98	
	3160	20 ga. studs, 1-5/8" wide, 16" O.C.		329	.024		.26	.80		1.06	1.62	
	3170	24" O.C.		508	.016		.19	.52		.71	1.07	
	3180	2-1/2" wide, 16" O.C.		325	.025		.29	.81		1.10	1.67	
	3190	24" O.C.		500	.016		.21	.53		.74	1.11	
	3200	3-5/8" wide, 16" O.C.		321	.025		.35	.82		1.17	1.74	
	3210	24" O.C.		492	.016		.25	.54		.79	1.17	
	3220	4" wide, 16" O.C.		317	.025		.37	.83		1.20	1.78	
	3230	24" O.C.		492	.016		.27	.54		.81	1.18	
	3240	6" wide, 16" O.C.		313	.026		.48	.84		1.32	1.93	
	3250	24" O.C.		483	.017		.35	.55		.90	1.29	
	5000	Load bearing studs, see division 05410-400										
	9000	Minimum labor/equipment charge	1 Carp	4	2	Job		66		66	109	

09120 | Ceiling Suspension

			CREW	DAILY OUTPUT	LABOR-HOURS	UNIT	MAT.	LABOR	EQUIP.	TOTAL	TOTAL INCL O&P	
100	0010	**CEILING SUSPENSION SYSTEMS** For gypsum board or plaster										100
	8000	Suspended ceilings, including carriers										
	8200	1-1/2" carriers, 24" O.C. with:										
	8300	7/8" channels, 16" O.C.	1 Lath	165	.048	S.F.	.28	1.48		1.76	2.65	
	8320	24" O.C.		200	.040		.23	1.22		1.45	2.18	
	8400	1-5/8" channels, 16" O.C.		155	.052		.35	1.58		1.93	2.87	
	8420	24" O.C.		190	.042		.27	1.29		1.56	2.33	
	8600	2" carriers, 24" O.C. with:										
	8700	7/8" channels, 16" O.C.	1 Lath	155	.052	S.F.	.34	1.58		1.92	2.86	
	8720	24" O.C.		190	.042		.29	1.29		1.58	2.35	
	8800	1-5/8" channels, 16" O.C.		145	.055		.40	1.69		2.09	3.10	
	8820	24" O.C.		180	.044		.33	1.36		1.69	2.50	

09130 | Acoustical Suspension

			CREW	DAILY OUTPUT	LABOR-HOURS	UNIT	MAT.	LABOR	EQUIP.	TOTAL	TOTAL INCL O&P	
100	0010	**CEILING SUSPENSION SYSTEMS** For boards and tile										100
	0050	Class A suspension system, 15/16" T bar, 2' x 4' grid	1 Carp	800	.010	S.F.	.41	.33		.74	1	
	0300	2' x 2' grid	"	650	.012		.51	.41		.92	1.23	
	0350	For 9/16" grid, add					.13			.13	.14	
	0360	For fire rated grid, add					.07			.07	.08	
	0370	For colored grid, add					.15			.15	.17	

09100 | Metal Support Assemblies

	09130 | Acoustical Suspension	CREW	DAILY OUTPUT	LABOR-HOURS	UNIT	2004 BARE COSTS MAT.	LABOR	EQUIP.	TOTAL	TOTAL INCL O&P	
100	0400 Concealed Z bar suspension system, 12" module	1 Carp	520	.015	S.F.	.37	.51		.88	1.25	100
	0600 1-1/2" carrier channels, 4' O.C., add	"	470	.017	↓	.07	.56		.63	1.01	
	0700 Carrier channels for ceilings with										
	0900 recessed lighting fixtures, add	1 Carp	460	.017	S.F.	.13	.57		.70	1.10	
	9000 Minimum labor/equipment charge	"	5	1.600	Job		53		53	87.50	

09200 | Plaster & Gypsum Board

	09205 | Furring & Lathing	CREW	DAILY OUTPUT	LABOR-HOURS	UNIT	2004 BARE COSTS MAT.	LABOR	EQUIP.	TOTAL	TOTAL INCL O&P	
530	0010 **FURRING** Beams & columns, 7/8" galvanized channels,										530
	0030 12" O.C.	1 Lath	155	.052	S.F.	.18	1.58		1.76	2.69	
	0050 16" O.C.		170	.047		.15	1.44		1.59	2.43	
	0070 24" O.C.		185	.043		.10	1.32		1.42	2.20	
	0100 Ceilings, on steel, 7/8" channels, galvanized, 12" O.C.		210	.038		.16	1.16		1.32	2.02	
	0300 16" O.C.		290	.028		.15	.84		.99	1.49	
	0400 24" O.C.		420	.019		.10	.58		.68	1.03	
	0600 1-5/8" channels, galvanized, 12" O.C.		190	.042		.24	1.29		1.53	2.29	
	0700 16" O.C.		260	.031		.21	.94		1.15	1.71	
	0900 24" O.C.		390	.021		.14	.63		.77	1.15	
	1000 Walls, 7/8" channels, galvanized, 12" O.C.		235	.034		.16	1.04		1.20	1.82	
	1200 16" O.C.		265	.030		.15	.92		1.07	1.62	
	1300 24" O.C.		350	.023		.10	.70		.80	1.21	
	1500 1-5/8" channels, galvanized, 12" O.C.		210	.038		.24	1.16		1.40	2.10	
	1600 16" O.C.		240	.033		.21	1.02		1.23	1.84	
	1800 24" O.C.		305	.026	↓	.14	.80		.94	1.43	
	9000 Minimum labor/equipment charge	↓	4	2	Job		61		61	96.50	
540	0010 **GYPSUM LATH** Plain or perforated, nailed, 3/8" thick	1 Lath	85	.094	S.Y.	3.42	2.88		6.30	8.30	540
	0100 1/2" thick, nailed		80	.100	"	3.51	3.06		6.57	8.70	
	0300 Clipped to steel studs, 3/8" thick		75	.107	S.Y.	3.42	3.26		6.68	8.90	
	0400 1/2" thick		70	.114		3.51	3.49		7	9.35	
	0600 Firestop gypsum base, to steel studs, 3/8" thick		70	.114		3.51	3.49		7	9.35	
	0700 1/2" thick		65	.123		3.96	3.76		7.72	10.30	
	0900 Foil back, to steel studs, 3/8" thick		75	.107		3.60	3.26		6.86	9.10	
	1000 1/2" thick		70	.114		3.96	3.49		7.45	9.85	
	1500 For ceiling installations, add		216	.037			1.13		1.13	1.79	
	1600 For columns and beams, add		170	.047	↓		1.44		1.44	2.27	
	9000 Minimum labor/equipment charge	↓	4.25	1.882	Job		57.50		57.50	91	
560	0010 **METAL LATH** Diamond, expanded, 2.5 lb. per S.Y., painted				S.Y.	1.50			1.50	1.65	560
	0100 Galvanized, 2.5 lb. per S.Y.					1.67			1.67	1.84	
	0300 3.4 lb. per S.Y., painted					2.58			2.58	2.84	
	0400 Galvanized					1.79			1.79	1.97	
	0600 For 15# asphalt sheathing paper, add					.24			.24	.27	
	0900 Flat rib, 1/8" high, 2.75 lb., painted					2.25			2.25	2.48	
	1000 Foil backed					2.70			2.70	2.97	
	1200 3.4 lb. per S.Y., painted					2.42			2.42	2.66	
	1300 Galvanized					3.06			3.06	3.37	
	1500 For 15# asphalt sheating paper, add				↓	.24			.24	.27	

09205 | Furring & Lathing

		09205 Furring & Lathing	CREW	DAILY OUTPUT	LABOR-HOURS	UNIT	2004 BARE COSTS MAT.	LABOR	EQUIP.	TOTAL	TOTAL INCL O&P	
560	1800	High rib, 3/8" high, 3.4 lb. per S.Y., painted				S.Y.	3.26			3.26	3.59	560
	1900	Galvanized					2.72			2.72	2.99	
	2400	High rib, 3/4" high, painted, .60 lb. per S.F.				S.F.	.31			.31	.34	
	2500	.75 lb. per S.F.				"	.67			.67	.74	
	2800	Stucco mesh, painted, 3.6 lb.				S.Y.	1.90			1.90	2.09	
	3000	K-lath, perforated, absorbent paper, regular					2.38			2.38	2.62	
	3100	Heavy duty					2.81			2.81	3.09	
	3300	Waterproof, heavy duty, grade B backing					2.75			2.75	3.03	
	3400	Fire resistant backing					3.04			3.04	3.34	
	3600	2.5 lb. diamond painted, on wood framing, on walls	1 Lath	85	.094		1.50	2.88		4.38	6.20	
	3700	On ceilings		75	.107		1.50	3.26		4.76	6.80	
	3900	3.4 lb. diamond painted, on wood framing, on walls		80	.100		2.42	3.06		5.48	7.50	
	4000	On ceilings		70	.114		2.42	3.49		5.91	8.15	
	4200	3.4 lb. diamond painted, wired to steel framing		75	.107		2.42	3.26		5.68	7.80	
	4300	On ceilings		60	.133		2.42	4.07		6.49	9.10	
	4600	Cornices, wired to steel		35	.229		2.42	7		9.42	13.70	
	4800	Screwed to steel studs, 2.5 lb.		80	.100		1.50	3.06		4.56	6.50	
	4900	3.4 lb.		75	.107		2.58	3.26		5.84	8	
	5100	Rib lath, painted, wired to steel, on walls, 2.5 lb.		75	.107		2.25	3.26		5.51	7.65	
	5200	3.4 lb.		70	.114		3.26	3.49		6.75	9.10	
	5400	4.0 lb.		65	.123		3.35	3.76		7.11	9.65	
	5500	For self-furring lath, add					.06			.06	.07	
	5700	Suspended ceiling system, incl. 3.4 lb. diamond lath, painted	1 Lath	15	.533		8.80	16.30		25.10	35	
	5800	Galvanized	"	15	.533		9.10	16.30		25.40	35.50	
	6000	Hollow metal stud partitions, 3.4 lb. painted lath both sides										
	6010	Non-load bearing, 25 ga., w/rib lath 2-1/2" studs, 12" O.C.	1 Lath	20.30	.394	S.Y.	8.75	12.05		20.80	28.50	
	6300	16" O.C.		21.10	.379		8.25	11.60		19.85	27.50	
	6350	24" O.C.		22.70	.352		7.85	10.75		18.60	25.50	
	6400	3-5/8" studs, 16" O.C.		19.50	.410		8.45	12.55		21	29	
	6600	24" O.C.		20.40	.392		7.95	12		19.95	27.50	
	6700	4" studs, 16" O.C.		20.40	.392		8.65	12		20.65	28.50	
	6900	24" O.C.		21.60	.370		8.10	11.30		19.40	27	
	7000	6" studs, 16" O.C.		19.50	.410		9.60	12.55		22.15	30.50	
	7100	24" O.C.		21.10	.379		8.75	11.60		20.35	28	
	7200	L.B. partitions, 16 ga., w/rib lath, 2-1/2" studs, 16" O.C.		20	.400		8.20	12.20		20.40	28.50	
	7300	3-5/8" studs, 16 ga.		19.70	.406		9.30	12.40		21.70	30	
	7500	4" studs, 16 ga.		19.50	.410		9.65	12.55		22.20	30.50	
	7600	6" studs, 16 ga.		18.70	.428		11.35	13.05		24.40	33	
	9000	Minimum labor/equipment charge		4.25	1.882	Job		57.50		57.50	91	

09210 | Gypsum Plaster

		09210 Gypsum Plaster	CREW	DAILY OUTPUT	LABOR-HOURS	UNIT	MAT.	LABOR	EQUIP.	TOTAL	TOTAL INCL O&P	
100	0010	**GYPSUM PLASTER** 80# bag, less than 1 ton				Bag	14.05			14.05	15.45	100
	0300	2 coats, no lath included, on walls	J-1	105	.381	S.Y.	3.14	10.80	.86	14.80	22	
	0400	On ceilings		92	.435		3.14	12.30	.98	16.42	24.50	
	0900	3 coats, no lath included, on walls		87	.460		4.45	13	1.03	18.48	27	
	1000	On ceilings		78	.513		4.45	14.50	1.15	20.10	29.50	
	1600	For irregular or curved surfaces, add						30%				
	1800	For columns & beams, add						50%				
	9000	Minimum labor/equipment charge	1 Plas	1	8	Job		240		240	390	
500	0010	**PERLITE OR VERMICULITE PLASTER** 100 lb. bags Under 200 bags				Bag	13			13	14.30	500
	0300	2 coats, no lath included, on walls	J-1	92	.435	S.Y.	3.24	12.30	.98	16.52	24.50	
	0400	On ceilings		79	.506		3.24	14.30	1.14	18.68	28	
	0900	3 coats, no lath included, on walls		74	.541		5.40	15.30	1.22	21.92	32.50	
	1000	On ceilings		63	.635		5.40	17.95	1.43	24.78	36.50	
	1700	For irregular or curved surfaces, add to above						30%				

Important: See the Reference Section for critical supporting data - Reference Nos., Crews, & City Cost Indexes

09200 | Plaster & Gypsum Board

	09210 | Gypsum Plaster		CREW	DAILY OUTPUT	LABOR-HOURS	UNIT	2004 BARE COSTS MAT.	LABOR	EQUIP.	TOTAL	TOTAL INCL O&P	
500	1800	For columns and beams, add to above				S.Y.		50%				500
	1900	For soffits, add to ceiling prices				↓		40%				
	9000	Minimum labor/equipment charge	1 Plas	1	8	Job		240		240	390	
900	0010	**THIN COAT** Plaster, 1 coat veneer, not incl. lath	J-1	3,600	.011	S.F.	.07	.31	.03	.41	.61	900
	1000	In 50 lb. bags				Bag	8.95			8.95	9.85	
	09220 | Portland Cement Plaster											
200	0010	**STUCCO**, 3 coats 1" thick, float finish, with mesh, on wood frame	J-2	63	.762	S.Y.	3.95	22	1.43	27.38	41	200
	0100	On masonry construction, no mesh incl.	J-1	67	.597		2.05	16.90	1.34	20.29	31	
	0300	For trowel finish, add	1 Plas	170	.047			1.41		1.41	2.28	
	0400	For 3/4" thick, on masonry, deduct	J-1	880	.045		.56	1.29	.10	1.95	2.81	
	0600	For coloring and special finish, add, minimum		685	.058		.36	1.65	.13	2.14	3.21	
	0700	Maximum	↓	200	.200		1.26	5.65	.45	7.36	11.05	
	0900	For soffits, add	J-2	155	.310		1.94	8.90	.58	11.42	17.05	
	1000	Exterior stucco, with bonding agent, 3 coats, on walls, no mesh incl.	J-1	200	.200		3.25	5.65	.45	9.35	13.25	
	1200	Ceilings		180	.222		3.25	6.30	.50	10.05	14.35	
	1300	Beams		80	.500		3.25	14.15	1.13	18.53	28	
	1500	Columns	↓	100	.400	↓	3.25	11.30	.90	15.45	23	
	1550	Minimum labor/equipment charge	1 Plas	1	8	Job		240		240	390	
	1600	Mesh, painted, nailed to wood, 1.8 lb.	1 Lath	60	.133	S.Y.	3.09	4.07		7.16	9.85	
	1800	3.6 lb.		55	.145		1.90	4.44		6.34	9.10	
	1900	Wired to steel, painted, 1.8 lb.		53	.151		3.09	4.61		7.70	10.70	
	2100	3.6 lb.		50	.160	↓	1.90	4.89		6.79	9.80	
	9000	Minimum labor/equipment charge	↓	4	2	Job		61		61	96.50	
	09250 | Gypsum Board											
200	0010	**CEMENTITIOUS BACKERBOARD**										200
	0070	Cementitious backerboard, on floor, 3' x 4'x 1/2" sheets	2 Carp	525	.030	S.F.	1.04	1.01		2.05	2.82	
	0080	3' x 5' x 1/2" sheets		525	.030		1.04	1.01		2.05	2.82	
	0090	3' x 6' x 1/2" sheets		525	.030		.76	1.01		1.77	2.50	
	0100	3' x 4'x 5/8" sheets		525	.030		1.08	1.01		2.09	2.86	
	0110	3' x 5' x 5/8" sheets		525	.030		1.06	1.01		2.07	2.84	
	0120	3' x 6' x 5/8" sheets		525	.030		1.07	1.01		2.08	2.84	
	0150	On wall, 3' x 4'x 1/2" sheets		350	.046		1.04	1.51		2.55	3.65	
	0160	3' x 5' x 1/2" sheets		350	.046		1.04	1.51		2.55	3.65	
	0170	3' x 6' x 1/2" sheets		350	.046		.76	1.51		2.27	3.33	
	0180	3' x 4'x 5/8" sheets		350	.046		1.08	1.51		2.59	3.69	
	0190	3' x 5' x 5/8" sheets		350	.046		1.06	1.51		2.57	3.67	
	0200	3' x 6' x 5/8" sheets		350	.046		1.07	1.51		2.58	3.67	
	0250	On counter, 3' x 4'x 1/2" sheets		180	.089		1.04	2.93		3.97	6	
	0260	3' x 5' x 1/2" sheets		180	.089		1.04	2.93		3.97	6	
	0270	3' x 6' x 1/2" sheets		180	.089		.76	2.93		3.69	5.70	
	0300	3' x 4'x 5/8" sheets		180	.089		1.08	2.93		4.01	6.05	
	0310	3' x 5' x 5/8" sheets		180	.089		1.06	2.93		3.99	6.05	
	0320	3' x 6' x 5/8" sheets	↓	180	.089	↓	1.07	2.93		4	6.05	
300	0010	**BLUEBOARD** For use with thin coat										300
	0100	plaster application (see division 09210-900)										
	1000	3/8" thick, on walls or ceilings, standard, no finish included	2 Carp	1,900	.008	S.F.	.18	.28		.46	.66	
	1100	With thin coat plaster finish		875	.018		.25	.60		.85	1.27	
	1400	On beams, columns, or soffits, standard, no finish included		675	.024		.21	.78		.99	1.53	
	1450	With thin coat plaster finish		475	.034		.27	1.11		1.38	2.14	
	3000	1/2" thick, on walls or ceilings, standard, no finish included		1,900	.008		.20	.28		.48	.68	
	3100	With thin coat plaster finish	↓	875	.018	↓	.27	.60		.87	1.29	

	09250 | Gypsum Board		CREW	DAILY OUTPUT	LABOR-HOURS	UNIT	2004 BARE COSTS MAT.	LABOR	EQUIP.	TOTAL	TOTAL INCL O&P	
300	3300	Fire resistant, no finish included	2 Carp	1,900	.008	S.F.	.20	.28		.48	.68	300
	3400	With thin coat plaster finish		875	.018		.27	.60		.87	1.29	
	3450	On beams, columns, or soffits, standard, no finish included		675	.024		.23	.78		1.01	1.55	
	3500	With thin coat plaster finish		475	.034		.30	1.11		1.41	2.17	
	3700	Fire resistant, no finish included		675	.024		.23	.78		1.01	1.55	
	3800	With thin coat plaster finish		475	.034		.30	1.11		1.41	2.17	
	5000	5/8" thick, on walls or ceilings, fire resistant, no finish included		1,900	.008		.21	.28		.49	.69	
	5100	With thin coat plaster finish		875	.018		.28	.60		.88	1.30	
	5500	On beams, columns, or soffits, no finish included		675	.024		.24	.78		1.02	1.57	
	5600	With thin coat plaster finish		475	.034		.31	1.11		1.42	2.18	
	6000	For high ceilings, over 8' high, add		3,060	.005			.17		.17	.29	
	6500	For over 3 stories high, add per story	↓	6,100	.003	↓		.09		.09	.14	
	9000	Minimum labor/equipment charge	1 Carp	2	4	Job		132		132	219	
500	0010	**CEILINGS** Gypsum drywall, fire rated, finished										500
	0100	Screwed to grid, channel or joists, 1/2" thick	2 Carp	765	.021	S.F.	.20	.69		.89	1.36	
	0200	5/8" thick		765	.021		.23	.69		.92	1.39	
	0300	Over 8' high, 1/2" thick		615	.026		.20	.86		1.06	1.64	
	0400	5/8" thick	↓	615	.026	↓	.23	.86		1.09	1.67	
	0600	Grid suspension system, direct hung										
	0700	1-1/2" C.R.C., with 7/8" hi hat furring channel, 16" O.C.	2 Carp	600	.027	S.F.	.84	.88		1.72	2.38	
	0800	24" O.C.		900	.018		.77	.59		1.36	1.82	
	0900	3-5/8" C.R.C., with 7/8" hi hat furring channel, 16" O.C.		600	.027		.89	.88		1.77	2.44	
	1000	24" O.C.	↓	900	.018	↓	.78	.59		1.37	1.83	
700	0010	**DRYWALL** Gypsum plasterboard, nailed or screwed R09250-100										700
	0100	to studs unless otherwise noted										
	0150	3/8" thick, on walls, standard, no finish included	2 Carp	2,000	.008	S.F.	.20	.26		.46	.66	
	0200	On ceilings, standard, no finish included		1,800	.009		.20	.29		.49	.71	
	0250	On beams, columns, or soffits, no finish included		675	.024		.20	.78		.98	1.52	
	0300	1/2" thick, on walls, standard, no finish included		2,000	.008		.19	.26		.45	.65	
	0350	Taped and finished (level 4 finish)		965	.017		.23	.55		.78	1.16	
	0390	With compound skim coat (level 5 finish)		775	.021		.27	.68		.95	1.43	
	0400	Fire resistant, no finish included		2,000	.008		.20	.26		.46	.66	
	0450	Taped and finished (level 4 finish)		965	.017		.24	.55		.79	1.17	
	0490	With compound skim coat (level 5 finish)		775	.021		.28	.68		.96	1.44	
	0500	Water resistant, no finish included		2,000	.008		.24	.26		.50	.70	
	0550	Taped and finished (level 4 finish)		965	.017		.28	.55		.83	1.22	
	0590	With compound skim coat (level 5 finish)		775	.021		.32	.68		1	1.48	
	0600	Prefinished, vinyl, clipped to studs		900	.018		.56	.59		1.15	1.59	
	1000	On ceilings, standard, no finish included		1,800	.009		.19	.29		.48	.70	
	1050	Taped and finished (level 4 finish)		765	.021		.23	.69		.92	1.39	
	1090	With compound skim coat (level 5 finish)		610	.026		.27	.87		1.14	1.73	
	1100	Fire resistant, no finish included		1,800	.009		.20	.29		.49	.71	
	1150	Taped and finished (level 4 finish)		765	.021		.24	.69		.93	1.40	
	1195	With compound skim coat (level 5 finish)		610	.026		.28	.87		1.15	1.74	
	1200	Water resistant, no finish included		1,800	.009		.24	.29		.53	.75	
	1250	Taped and finished (level 4 finish)		765	.021		.28	.69		.97	1.45	
	1290	With compound skim coat (level 5 finish)		610	.026		.32	.87		1.19	1.78	
	1500	On beams, columns, or soffits, standard, no finish included		675	.024		.22	.78		1	1.54	
	1550	Taped and finished (level 4 finish)		475	.034		.23	1.11		1.34	2.09	
	1590	With compound skim coat (level 5 finish)		540	.030		.27	.98		1.25	1.92	
	1600	Fire resistant, no finish included		675	.024		.23	.78		1.01	1.55	
	1650	Taped and finished (level 4 finish)		475	.034		.24	1.11		1.35	2.10	
	1690	With compound skim coat (level 5 finish)		540	.030		.28	.98		1.26	1.93	
	1700	Water resistant, no finish included		675	.024		.28	.78		1.06	1.60	
	1750	Taped and finished (level 4 finish)	↓	475	.034	↓	.28	1.11		1.39	2.15	

09250 | Gypsum Board

		09250 Gypsum Board	CREW	DAILY OUTPUT	LABOR-HOURS	UNIT	2004 BARE COSTS MAT.	LABOR	EQUIP.	TOTAL	TOTAL INCL O&P	
700	1790	With compound skim coat (level 5 finish) R09250-100	2 Carp	540	.030	S.F.	.32	.98		1.30	1.97	700
	2000	5/8″ thick, on walls, standard, no finish included		2,000	.008		.22	.26		.48	.68	
	2050	Taped and finished (level 4 finish)		965	.017		.26	.55		.81	1.19	
	2090	With compound skim coat (level 5 finish)		775	.021		.30	.68		.98	1.46	
	2100	Fire resistant, no finish included		2,000	.008		.23	.26		.49	.69	
	2150	Taped and finished (level 4 finish)		965	.017		.27	.55		.82	1.20	
	2195	With compound skim coat (level 5 finish)		775	.021		.31	.68		.99	1.47	
	2200	Water resistant, no finish included		2,000	.008		.27	.26		.53	.74	
	2250	Taped and finished (level 4 finish)		965	.017		.31	.55		.86	1.25	
	2290	With compound skim coat (level 5 finish)		775	.021		.35	.68		1.03	1.52	
	2300	Prefinished, vinyl, clipped to studs		900	.018		.65	.59		1.24	1.69	
	3000	On ceilings, standard, no finish included		1,800	.009		.22	.29		.51	.73	
	3050	Taped and finished (level 4 finish)		765	.021		.26	.69		.95	1.42	
	3090	With compound skim coat (level 5 finish)		615	.026		.30	.86		1.16	1.75	
	3100	Fire resistant, no finish included		1,800	.009		.23	.29		.52	.74	
	3150	Taped and finished (level 4 finish)		765	.021		.27	.69		.96	1.43	
	3190	With compound skim coat (level 5 finish)		615	.026		.31	.86		1.17	1.76	
	3200	Water resistant, no finish included		1,800	.009		.27	.29		.56	.79	
	3250	Taped and finished (level 4 finish)		765	.021		.31	.69		1	1.48	
	3290	With compound skim coat (level 5 finish)		615	.026		.35	.86		1.21	1.81	
	3500	On beams, columns, or soffits, no finish included		675	.024		.25	.78		1.03	1.58	
	3550	Taped and finished (level 4 finish)		475	.034		.30	1.11		1.41	2.17	
	3590	With compound skim coat (level 5 finish)		380	.042		.35	1.39		1.74	2.68	
	3600	Fire resistant, no finish included		675	.024		.26	.78		1.04	1.59	
	3650	Taped and finished (level 4 finish)		475	.034		.31	1.11		1.42	2.18	
	3690	With compound skim coat (level 5 finish)		380	.042		.31	1.39		1.70	2.64	
	3700	Water resistant, no finish included		675	.024		.31	.78		1.09	1.64	
	3750	Taped and finished (level 4 finish)		475	.034		.35	1.11		1.46	2.23	
	3790	With compound skim coat (level 5 finish)		380	.042		.35	1.39		1.74	2.69	
	4000	Fireproofing, beams or columns, 2 layers, 1/2″ thick, incl finish		330	.048		.44	1.60		2.04	3.13	
	4050	5/8″ thick		300	.053		.54	1.76		2.30	3.51	
	4100	3 layers, 1/2″ thick		225	.071		.64	2.35		2.99	4.59	
	4150	5/8″ thick		210	.076		.80	2.51		3.31	5.05	
	5050	For 1″ thick coreboard on columns		480	.033		.45	1.10		1.55	2.32	
	5100	For foil-backed board, add					.10			.10	.11	
	5200	For work over 8′ high, add	2 Carp	3,060	.005			.17		.17	.29	
	5270	For textured spray, add	2 Lath	1,600	.010		.04	.31		.35	.52	
	5300	For over 3 stories high, add per story	2 Carp	6,100	.003			.09		.09	.14	
	5350	For finishing inner corners, add		950	.017	L.F.	.08	.56		.64	1.01	
	5355	For finishing outer corners, add		1,250	.013		.17	.42		.59	.88	
	5500	For acoustical sealant, add per bead	1 Carp	500	.016		.03	.53		.56	.91	
	5550	Sealant, 1 quart tube				Ea.	4.89			4.89	5.40	
	5600	Sound deadening board, 1/4″ gypsum	2 Carp	1,800	.009	S.F.	.19	.29		.48	.70	
	5650	1/2″ wood fiber	″	1,800	.009	″	.34	.29		.63	.86	
	9000	Minimum labor/equipment charge	1 Carp	2	4	Job		132		132	219	

09260 | Gypsum Board Systems

		09260 Gypsum Board Systems	CREW	DAILY OUTPUT	LABOR-HOURS	UNIT	MAT.	LABOR	EQUIP.	TOTAL	TOTAL INCL O&P	
100	0010	**PARTITION WALL** Stud wall, 8′ to 12′ high										100
	0050	1/2″, interior, gypsum drywall, standard, taped both sides										
	0500	Installed on and incl., 2″ x 4″ wood studs, 16″ O.C.	2 Carp	310	.052	S.F.	.80	1.70		2.50	3.70	
	1000	Metal studs, NLB, 25 ga., 16″ O.C., 3-5/8″ wide		350	.046		.67	1.51		2.18	3.23	
	1200	6″ wide		330	.048		.79	1.60		2.39	3.52	
	1400	Water resistant, on 2″ x 4″ wood studs, 16″ O.C.		310	.052		.90	1.70		2.60	3.81	
	1600	Metal studs, NLB, 25 ga., 16″ O.C., 3-5/8″ wide		350	.046		.77	1.51		2.28	3.34	
	1800	6″ wide		330	.048		.89	1.60		2.49	3.63	

09260 | Gypsum Board Systems

			CREW	DAILY OUTPUT	LABOR-HOURS	UNIT	2004 BARE COSTS MAT.	LABOR	EQUIP.	TOTAL	TOTAL INCL O&P	
100	2000	Fire res., 2 layers, 1-1/2 hr., on 2" x 4" wood studs, 16" O.C.	2 Carp	210	.076	S.F.	1.22	2.51		3.73	5.50	100
	2200	Metal studs, NLB, 25 ga., 16" O.C., 3-5/8" wide		250	.064		1.09	2.11		3.20	4.70	
	2400	6" wide		230	.070		1.21	2.30		3.51	5.15	
	2600	Fire & water res., 2 layers, 1-1/2 hr., 2"x4" studs, 16" O.C.		210	.076		1.22	2.51		3.73	5.50	
	2800	Metal studs, NLB, 25 ga., 16" O.C., 3-5/8" wide		250	.064		1.09	2.11		3.20	4.70	
	3000	6" wide	↓	230	.070	↓	1.21	2.30		3.51	5.15	
	3200	5/8", interior, gypsum drywall, standard, taped both sides										
	3400	Installed on and including 2" x 4" wood studs, 16" O.C.	2 Carp	300	.053	S.F.	.86	1.76		2.62	3.87	
	3600	24" O.C.		330	.048		.79	1.60		2.39	3.51	
	3800	Metal studs, NLB, 25 ga., 16" O.C., 3-5/8" wide		340	.047		.73	1.55		2.28	3.37	
	4000	6" wide		320	.050		.85	1.65		2.50	3.68	
	4200	24" O.C., 3-5/8" wide		360	.044		.67	1.47		2.14	3.17	
	4400	6" wide		340	.047		.76	1.55		2.31	3.41	
	4800	Water resistant, on 2" x 4" wood studs, 16" O.C.		300	.053		.96	1.76		2.72	3.98	
	5000	24" O.C.		330	.048		.89	1.60		2.49	3.62	
	5200	Metal studs, NLB, 25 ga. 16" O.C., 3-5/8" wide		340	.047		.83	1.55		2.38	3.48	
	5400	6" wide		320	.050		.95	1.65		2.60	3.79	
	5600	24" O.C., 3-5/8" wide		360	.044		.77	1.47		2.24	3.28	
	5800	6" wide		340	.047		.86	1.55		2.41	3.52	
	6000	Fire res., 2 layers, 2 hr., on 2" x 4" wood studs, 16" O.C.		205	.078		1.27	2.58		3.85	5.65	
	6200	24" O.C.		235	.068		1.27	2.25		3.52	5.10	
	6400	Metal studs, NLB, 25 ga., 16" O.C., 3-5/8" wide		245	.065		1.23	2.16		3.39	4.92	
	6600	6" wide		225	.071		1.33	2.35		3.68	5.35	
	6800	24" O.C., 3-5/8" wide		265	.060		1.15	1.99		3.14	4.57	
	7000	6" wide		245	.065		1.24	2.16		3.40	4.94	
	7200	Fire & water res., 2 layers, 2 hr., 2" x 4" studs, 16" O.C.		205	.078		1.34	2.58		3.92	5.75	
	7400	24" O.C.		235	.068		1.27	2.25		3.52	5.10	
	7600	Metal studs, NLB, 25 ga., 16" O.C., 3-5/8" wide		245	.065		1.21	2.16		3.37	4.90	
	7800	6" wide		225	.071		1.33	2.35		3.68	5.35	
	8000	24" O.C., 3-5/8" wide		265	.060		1.15	1.99		3.14	4.57	
	8200	6" wide	↓	245	.065	↓	1.24	2.16		3.40	4.94	
	8600	1/2" blueboard, mesh tape both sides										
	8620	Installed on and including 2" x 4" wood studs, 16" O.C.	2 Carp	300	.053	S.F.	.86	1.76		2.62	3.87	
	8640	Metal studs, NLB, 25 ga., 16" O.C., 3-5/8" wide		340	.047		.73	1.55		2.28	3.37	
	8660	6" wide	↓	320	.050	↓	.85	1.65		2.50	3.68	
	9000	Exterior, 1/2" gypsum sheathing, 1/2" gypsum finished, interior,										
	9100	including foil faced insulation, metal studs, 20 ga.										
	9200	16" O.C., 3-5/8" wide	2 Carp	290	.055	S.F.	1.21	1.82		3.03	4.35	
	9400	6" wide	"	270	.059	"	1.35	1.96		3.31	4.72	
800	0010	**SHAFT WALL** Cavity type on 25 ga. J track & C-H studs, 24" O.C.										800
	0030	1" thick coreboard wall liner on shaft side										
	0040	2-hour assembly with double layer										
	0060	5/8" fire rated gypsum board on room side	2 Carp	220	.073	S.F.	.89	2.40		3.29	4.96	
	0100	3-hour assembly with triple layer										
	0300	5/8" fire rated gypsum board on room side	2 Carp	180	.089	S.F.	1.11	2.93		4.04	6.10	
	0400	4-hour assembly, 1" coreboard, 5/8" fire rated gypsum board										
	0600	and 3/4" galv. metal furring channels, 24" O.C., with										
	0700	Double layer 5/8" fire rated gypsum board on room side	2 Carp	110	.145	S.F.	.99	4.80		5.79	9.05	
	0900	For taping & finishing, add per side	1 Carp	1,050	.008	"	.04	.25		.29	.46	
	1000	For insulation, see div. 07210-000										

09270 | Drywall Accessories

			CREW	DAILY OUTPUT	LABOR-HOURS	UNIT	MAT.	LABOR	EQUIP.	TOTAL	TOTAL INCL O&P	
100	0010	**ACCESSORIES, DRYWALL** Casing bead, galvanized steel	1 Carp	2.90	2.759	C.L.F.	16.55	91		107.55	169	100
	0100	Vinyl	↓	3	2.667	↓	15.85	88		103.85	163	

Important: See the Reference Section for critical supporting data - Reference Nos., Crews, & City Cost Indexes

09200 | Plaster & Gypsum Board

		09270 | Drywall Accessories	CREW	DAILY OUTPUT	LABOR-HOURS	UNIT	2004 BARE COSTS MAT.	LABOR	EQUIP.	TOTAL	TOTAL INCL O&P	
100	0300	Corner bead, galvanized steel, 1" x 1"	1 Carp	4	2	C.L.F.	10.35	66		76.35	120	100
	0400	Corner bead, galvanized steel, 1-1/4" x 1-1/4"		3.50	2.286		9.50	75.50		85	135	
	0600	Vinyl corner bead		4	2		20.50	66		86.50	132	
	0900	Furring channel, galv. steel, 7/8" deep, standard		2.60	3.077		17.40	102		119.40	187	
	1000	Resilient		2.55	3.137		18.50	104		122.50	193	
	1100	J trim, galvanized steel, 1/2" wide		3	2.667		14.60	88		102.60	162	
	1120	5/8" wide		2.95	2.712		14.95	89.50		104.45	164	
	1500	Z stud, galvanized steel, 1-1/2" wide		2.60	3.077		31.50	102		133.50	203	
	9000	Minimum labor/equipment charge		3	2.667	Job		88		88	146	

		09280 | Gypsum Wallboard Repairs	CREW	DAILY OUTPUT	LABOR-HOURS	UNIT	MAT.	LABOR	EQUIP.	TOTAL	TOTAL INCL O&P	
100	0010	**GYPSUM WALLBOARD REPAIRS**										100
	0100	Fill and sand, pin / nail holes	1 Carp	960	.008	Ea.		.27		.27	.46	
	0110	Screw head pops		480	.017			.55		.55	.91	
	0120	Dents, up to 2" square		48	.167		.01	5.50		5.51	9.10	
	0130	2" to 4" square		24	.333		.03	11		11.03	18.30	
	0140	Cut square, patch, sand and finish, holes, up to 2" square		12	.667		.02	22		22.02	36.50	
	0150	2" to 4" square		11	.727		.06	24		24.06	40	
	0160	4" to 8" square		10	.800		.14	26.50		26.64	44	
	0170	8" to 12" square		8	1		.28	33		33.28	55	
	9000	Minimum labor/equipment charge		2	4	Job		132		132	219	

FINISHES 9

09300 | Tile

		09310 | Ceramic Tile	CREW	DAILY OUTPUT	LABOR-HOURS	UNIT	2004 BARE COSTS MAT.	LABOR	EQUIP.	TOTAL	TOTAL INCL O&P	
100	0010	**CERAMIC TILE**										100
	0600	Cove base, 4-1/4" x 4-1/4" high, mud set	D-7	91	.176	L.F.	3.07	4.92		7.99	11.10	
	0700	Thin set		128	.125		3.09	3.50		6.59	8.90	
	0900	6" x 4-1/4" high, mud set		100	.160		2.77	4.48		7.25	10.05	
	1000	Thin set		137	.117		2.89	3.27		6.16	8.30	
	1200	Sanitary cove base, 6" x 4-1/4" high, mud set		93	.172		3.56	4.81		8.37	11.45	
	1300	Thin set		124	.129		3.62	3.61		7.23	9.65	
	1500	6" x 6" high, mud set		84	.190		3.28	5.35		8.63	11.95	
	1600	Thin set		117	.137		3.32	3.83		7.15	9.65	
	2400	Bullnose trim, 4-1/4" x 4-1/4", mud set		82	.195		2.66	5.45		8.11	11.50	
	2500	Thin set		128	.125		2.60	3.50		6.10	8.35	
	2700	6" x 4-1/4" bullnose trim, mud set		84	.190		2.02	5.35		7.37	10.55	
	2800	Thin set		124	.129		2.02	3.61		5.63	7.85	
	3000	Floors, natural clay, random or uniform, thin set, color group 1		183	.087	S.F.	3.62	2.45		6.07	7.80	
	3100	Color group 2		183	.087		3.90	2.45		6.35	8.15	
	3300	Porcelain type, 1 color, color group 2, 1" x 1"		183	.087		4.19	2.45		6.64	8.45	
	3310	2" x 2" or 2" x 1", thin set		190	.084		4.01	2.36		6.37	8.10	
	3350	For random blend, 2 colors, add					.77			.77	.85	
	3360	4 colors, add					1.10			1.10	1.21	
	4300	Specialty tile, 4-1/4" x 4-1/4" x 1/2", decorator finish	D-7	183	.087		8.95	2.45		11.40	13.70	
	4500	Add for epoxy grout, 1/16" joint, 1" x 1" tile		800	.020		.54	.56		1.10	1.47	
	4600	2" x 2" tile		820	.020		.50	.55		1.05	1.41	
	4800	Pregrouted sheets, walls, 4-1/4" x 4-1/4", 6" x 4-1/4"										
	4810	and 8-1/2" x 4-1/4", 4 S.F. sheets, silicone grout	D-7	240	.067	S.F.	4.15	1.87		6.02	7.50	

09310 | Ceramic Tile

			CREW	DAILY OUTPUT	LABOR-HOURS	UNIT	2004 BARE COSTS MAT.	LABOR	EQUIP.	TOTAL	TOTAL INCL O&P	
100	5100	Floors, unglazed, 2 S.F. sheets,										100
	5110	urethane adhesive	D-7	180	.089	S.F.	4.13	2.49		6.62	8.45	
	5400	Walls, interior, thin set, 4-1/4" x 4-1/4" tile		190	.084		2.09	2.36		4.45	6	
	5500	6" x 4-1/4" tile		190	.084		2.30	2.36		4.66	6.25	
	5700	8-1/2" x 4-1/4" tile		190	.084		3.25	2.36		5.61	7.30	
	5800	6" x 6" tile		200	.080		2.63	2.24		4.87	6.40	
	5810	8" x 8" tile		225	.071		3.16	1.99		5.15	6.60	
	5820	12" x 12" tile		300	.053		2.98	1.49		4.47	5.60	
	5830	16" x 16" tile		500	.032		3.23	.90		4.13	4.95	
	6000	Decorated wall tile, 4-1/4" x 4-1/4", minimum		270	.059		3.50	1.66		5.16	6.45	
	6100	Maximum		180	.089		39	2.49		41.49	46.50	
	6600	Crystalline glazed, 4-1/4" x 4-1/4", mud set, plain		100	.160		3.25	4.48		7.73	10.60	
	6700	4-1/4" x 4-1/4", scored tile		100	.160		4.05	4.48		8.53	11.45	
	6900	6"x 6" plain		93	.172		3.25	4.81		8.06	11.15	
	7000	For epoxy grout, 1/16" joints, 4-1/4" tile, add		800	.020		.33	.56		.89	1.24	
	7200	For tile set in dry mortar, add		1,735	.009			.26		.26	.40	
	7300	For tile set in portland cement mortar, add		290	.055	▼		1.54		1.54	2.42	
	9500	Minimum labor/equipment charge	▼	3.25	4.923	Job		138		138	216	

09330 | Quarry Tile

			CREW	DAILY OUTPUT	LABOR-HOURS	UNIT	MAT.	LABOR	EQUIP.	TOTAL	TOTAL INCL O&P	
100	0010	**QUARRY TILE** Base, cove or sanitary, 2" or 5" high, mud set										100
	0100	1/2" thick	D-7	110	.145	L.F.	3.85	4.07		7.92	10.65	
	0300	Bullnose trim, red, mud set, 6" x 6" x 1/2" thick		120	.133		3.81	3.73		7.54	10.05	
	0400	4" x 4" x 1/2" thick		110	.145		4.30	4.07		8.37	11.15	
	0600	4" x 8" x 1/2" thick, using 8" as edge		130	.123	▼	3.78	3.44		7.22	9.55	
	0700	Floors, mud set, 1,000 S.F. lots, red, 4" x 4" x 1/2" thick		120	.133	S.F.	3.81	3.73		7.54	10.05	
	0900	6" x 6" x 1/2" thick		140	.114		2.96	3.20		6.16	8.25	
	1000	4" x 8" x 1/2" thick	▼	130	.123		3.81	3.44		7.25	9.60	
	1300	For waxed coating, add					.61			.61	.67	
	1500	For colors other than green, add					.36			.36	.40	
	1600	For abrasive surface, add					.43			.43	.47	
	1800	Brown tile, imported, 6" x 6" x 3/4"	D-7	120	.133		4.53	3.73		8.26	10.85	
	1900	8" x 8" x 1"		110	.145		5.15	4.07		9.22	12.05	
	2100	For thin set mortar application, deduct		700	.023			.64		.64	1	
	2700	Stair tread, 6" x 6" x 3/4", plain		50	.320		4.20	8.95		13.15	18.65	
	2800	Abrasive		47	.340		4.68	9.55		14.23	20	
	3000	Wainscot, 6" x 6" x 1/2", thin set, red		105	.152		3.49	4.26		7.75	10.55	
	3100	Colors other than green		105	.152	▼	3.89	4.26		8.15	11	
	3300	Window sill, 6" wide, 3/4" thick		90	.178	L.F.	4.47	4.97		9.44	12.70	
	3400	Corners		80	.200	Ea.	4.91	5.60		10.51	14.20	
	9000	Minimum labor/equipment charge	▼	3.25	4.923	Job		138		138	216	

09370 | Metal Tile

			CREW	DAILY OUTPUT	LABOR-HOURS	UNIT	MAT.	LABOR	EQUIP.	TOTAL	TOTAL INCL O&P	
100	0010	**METAL TILE** 4' x 4' sheet, 24 ga., tile pattern, nailed										100
	0200	Stainless steel	2 Carp	512	.031	S.F.	22	1.03		23.03	26	
	0400	Aluminized steel	"	512	.031	"	11.90	1.03		12.93	14.80	
	9000	Minimum labor/equipment charge	1 Carp	4	2	Job		66		66	109	

09400 | Terrazzo

		09410 | Portland Cement Terrazzo	CREW	DAILY OUTPUT	LABOR-HOURS	UNIT	2004 BARE COSTS MAT.	LABOR	EQUIP.	TOTAL	TOTAL INCL O&P	
100	0010	**PORTLAND CEMENT TERRAZZO**, cast-in-place										100
	0020	Cove base, 6″ high, 16ga. zinc	1 Mstz	20	.400	L.F.	2.85	12.65		15.50	23	
	0100	Curb, 6″ high and 6″ wide		6	1.333		4.57	42		46.57	71	
	0300	Divider strip for floors, 14 ga., 1-1/4″ deep, zinc		375	.021		1.04	.67		1.71	2.20	
	0400	Brass		375	.021		1.84	.67		2.51	3.08	
	0600	Heavy top strip 1/4″ thick, 1-1/4″ deep, zinc		300	.027		1.45	.84		2.29	2.92	
	1200	For thin set floors, 16 ga., 1/2″ x 1/2″, zinc		350	.023		.63	.72		1.35	1.82	
	1500	Floor, bonded to concrete, 1-3/4″ thick, gray cement	J-3	130	.123	S.F.	2.44	3.50	1.39	7.33	9.70	
	1600	White cement, mud set		130	.123		2.78	3.50	1.39	7.67	10.10	
	1800	Not bonded, 3″ total thickness, gray cement		115	.139		3.05	3.96	1.58	8.59	11.30	
	1900	White cement, mud set		115	.139		3.33	3.96	1.58	8.87	11.60	
	9000	Minimum labor/equipment charge	1 Mstz	1	8	Job		253		253	395	

		09420 | Precast Terrazzo	CREW	DAILY OUTPUT	LABOR-HOURS	UNIT	MAT.	LABOR	EQUIP.	TOTAL	TOTAL INCL O&P	
900	0010	**TERRAZZO, PRECAST** Base, 6″ high, straight	1 Mstz	35	.229	L.F.	9.10	7.20		16.30	21.50	900
	0100	Cove		30	.267		10.15	8.45		18.60	24.50	
	0300	8″ high base, straight		30	.267		9.10	8.45		17.55	23.50	
	0400	Cove		25	.320		13.40	10.10		23.50	30.50	
	0600	For white cement, add					.37			.37	.41	
	0700	For 16 ga. zinc toe strip, add					1.34			1.34	1.47	
	0900	Curbs, 4″ x 4″ high	1 Mstz	19	.421		25.50	13.30		38.80	49.50	
	1000	8″ x 8″ high	″	15	.533		30	16.85		46.85	59.50	
	1200	Floor tiles, non-slip, 1″ thick, 12″ x 12″	D-1	29	.552	S.F.	15.40	16.50		31.90	44	
	1300	1-1/4″ thick, 12″ x 12″		29	.552		17.05	16.50		33.55	46	
	1500	16″ x 16″		23	.696		18.55	21		39.55	54.50	
	1600	1-1/2″ thick, 16″ x 16″		21	.762		16.95	23		39.95	55.50	
	1800	For Venetian terrazzo, add					5.10			5.10	5.60	
	1900	For white cement, add					.48			.48	.53	
	2400	Stair treads, 1-1/2″ thick, non-slip, three line pattern	2 Mstz	70	.229	L.F.	34.50	7.20		41.70	49	
	2500	Nosing and two lines		70	.229		34.50	7.20		41.70	49	
	2700	2″ thick treads, straight		60	.267		36	8.45		44.45	53	
	2800	Curved		50	.320		47	10.10		57.10	67.50	
	3000	Stair risers, 1″ thick, to 6″ high, straight sections		60	.267		8.20	8.45		16.65	22.50	
	3100	Cove		50	.320		12.30	10.10		22.40	29.50	
	3300	Curved, 1″ thick, to 6″ high, vertical		48	.333		16.75	10.55		27.30	35	
	3400	Cove		38	.421		32	13.30		45.30	56	
	3600	Stair tread and riser, single piece, straight, minimum		60	.267		43.50	8.45		51.95	61.50	
	3700	Maximum		40	.400		56.50	12.65		69.15	82	
	3900	Curved tread and riser, minimum		40	.400		61	12.65		73.65	87	
	4000	Maximum		32	.500		77	15.80		92.80	110	
	4200	Stair stringers, notched, 1″ thick		25	.640		25	20		45	59	
	4300	2″ thick		22	.727		29.50	23		52.50	68.50	
	4500	Stair landings, structural, non-slip, 1-1/2″ thick		85	.188	S.F.	27	5.95		32.95	39.50	
	4600	3″ thick		75	.213		38.50	6.75		45.25	53	
	4800	Wainscot, 12″ x 12″ x 1″ tiles	1 Mstz	12	.667		5.45	21		26.45	39	
	4900	16″ x 16″ x 1-1/2″ tiles	″	8	1		11.95	31.50		43.45	62.50	
	9500	Minimum labor/equipment charge	1 Tilf	2	4	Job		126		126	198	

		09450 | Cast-in-Place Terrazzo	CREW	DAILY OUTPUT	LABOR-HOURS	UNIT	MAT.	LABOR	EQUIP.	TOTAL	TOTAL INCL O&P	
200	0010	**TILE OR TERRAZZO BASE** Scratch coat only	1 Mstz	150	.053	S.F.	.31	1.69		2	2.99	200
	0500	Scratch and brown coat only		75	.107	″	.56	3.37		3.93	5.90	
	9000	Minimum labor/equipment charge		1	8	Job		253		253	395	

		09510 Acoustical Ceilings	CREW	DAILY OUTPUT	LABOR-HOURS	UNIT	2004 BARE COSTS MAT.	LABOR	EQUIP.	TOTAL	TOTAL INCL O&P	
700	0010	**SUSPENDED ACOUSTIC CEILING TILES**, Not including										700
	0100	suspension system										
	0300	Fiberglass boards, film faced, 2' x 2' or 2' x 4', 5/8" thick	1 Carp	625	.013	S.F.	.51	.42		.93	1.26	
	0400	3/4" thick		600	.013		1.14	.44		1.58	1.98	
	0500	3" thick, thermal, R11		450	.018		1.26	.59		1.85	2.36	
	0600	Glass cloth faced fiberglass, 3/4" thick		500	.016		1.62	.53		2.15	2.66	
	0700	1" thick		485	.016		1.79	.54		2.33	2.87	
	0820	1-1/2" thick, nubby face		475	.017		2.23	.56		2.79	3.37	
	1110	Mineral fiber tile, lay-in, 2' x 2' or 2' x 4', 5/8" thick, fine texture		625	.013		.43	.42		.85	1.17	
	1115	Rough textured		625	.013		1.06	.42		1.48	1.87	
	1125	3/4" thick, fine textured		600	.013		1.17	.44		1.61	2.02	
	1130	Rough textured		600	.013		1.47	.44		1.91	2.35	
	1135	Fissured		600	.013		1.74	.44		2.18	2.64	
	1150	Tegular, 5/8" thick, fine textured		470	.017		1.04	.56		1.60	2.07	
	1155	Rough textured		470	.017		1.35	.56		1.91	2.42	
	1165	3/4" thick, fine textured		450	.018		1.47	.59		2.06	2.59	
	1170	Rough textured		450	.018		1.66	.59		2.25	2.80	
	1175	Fissured		450	.018		2.59	.59		3.18	3.82	
	1180	For aluminum face, add					4.72			4.72	5.20	
	1185	For plastic film face, add					.70			.70	.77	
	1190	For fire rating, add					.35			.35	.39	
	1300	Mirror faced panels, 15/16" thick, 2' x 2'	1 Carp	500	.016		9.80	.53		10.33	11.70	
	1900	Eggcrate, acrylic, 1/2" x 1/2" x 1/2" cubes		500	.016		1.42	.53		1.95	2.44	
	2100	Polystyrene eggcrate, 3/8" x 3/8" x 1/2" cubes		510	.016		1.19	.52		1.71	2.17	
	2200	1/2" x 1/2" x 1/2" cubes		500	.016		1.60	.53		2.13	2.64	
	2400	Luminous panels, prismatic, acrylic		400	.020		1.73	.66		2.39	2.99	
	2500	Polystyrene		400	.020		.88	.66		1.54	2.06	
	2700	Flat white acrylic		400	.020		3.01	.66		3.67	4.40	
	2800	Polystyrene		400	.020		2.06	.66		2.72	3.36	
	3000	Drop pan, white, acrylic		400	.020		4.41	.66		5.07	5.95	
	3100	Polystyrene		400	.020		3.69	.66		4.35	5.15	
	3600	Perforated aluminum sheets, .024" thick, corrugated, painted		490	.016		1.76	.54		2.30	2.83	
	3700	Plain		500	.016		3.07	.53		3.60	4.26	
	3720	Mineral fiber, 24" x 24" or 48", reveal edge, painted, 5/8" thick		600	.013		.98	.44		1.42	1.81	
	3740	3/4" thick		575	.014		1.60	.46		2.06	2.52	
	9000	Minimum labor/equipment charge		4	2	Job		66		66	109	
760	0010	**SUSPENDED CEILINGS, COMPLETE** Including standard										760
	0100	suspension system but not incl. 1-1/2" carrier channels										
	0600	Fiberglass ceiling board, 2' x 4' x 5/8", plain faced,	1 Carp	500	.016	S.F.	.92	.53		1.45	1.89	
	0700	Offices, 2' x 4' x 3/4"		380	.021		1.55	.69		2.24	2.85	
	0800	Mineral fiber, on 15/16" T bar susp. 2' x 2' x 3/4" lay-in board		345	.023		1.68	.77		2.45	3.12	
	0810	2' x 4' x 5/8" tile		380	.021		.84	.69		1.53	2.07	
	0820	Tegular, 2' x 2' x 5/8" tile on 9/16" grid		250	.032		1.68	1.06		2.74	3.60	
	0830	2' x 4' x 3/4" tile		275	.029		2.01	.96		2.97	3.80	
	0900	Luminous panels, prismatic, acrylic		255	.031		2.14	1.04		3.18	4.07	
	1200	Metal pan with acoustic pad, steel		75	.107		3.22	3.52		6.74	9.40	
	1300	Painted aluminum		75	.107		2.17	3.52		5.69	8.25	
	1500	Aluminum, degreased finish		75	.107		3.74	3.52		7.26	9.95	
	1600	Stainless steel		75	.107		7.10	3.52		10.62	13.65	
	1800	Tile, Z bar suspension, 5/8" mineral fiber tile		150	.053		1.55	1.76		3.31	4.63	
	1900	3/4" mineral fiber tile		150	.053		1.65	1.76		3.41	4.74	
	2402	For strip lighting, see division 16510-440										
	2500	For rooms under 500 S.F., add				S.F.		25%				
	9000	Minimum labor/equipment charge	1 Carp	2	4	Job		132		132	219	
900	0010	**CEILING TILE**, Stapled or cemented										900
	0100	12" x 12" or 12" x 24", not including furring										

09500 | Ceilings

09510 | Acoustical Ceilings

			CREW	DAILY OUTPUT	LABOR-HOURS	UNIT	2004 BARE COSTS MAT.	LABOR	EQUIP.	TOTAL	TOTAL INCL O&P	
900	0600	Mineral fiber, vinyl coated, 5/8" thick	1 Carp	1,000	.008	S.F.	.89	.26		1.15	1.42	900
	0700	3/4" thick		1,000	.008		1.30	.26		1.56	1.87	
	0900	Fire rated, 3/4" thick, plain faced		1,000	.008		.85	.26		1.11	1.38	
	1000	Plastic coated face		1,000	.008		1.04	.26		1.30	1.58	
	1200	Aluminum faced, 5/8" thick, plain		1,000	.008		1.12	.26		1.38	1.67	
	3000	Wood fiber tile, 1/2" thick		400	.020		.76	.66		1.42	1.93	
	3100	3/4" thick		400	.020		1.03	.66		1.69	2.22	
	3300	For flameproofing, add					.09			.09	.10	
	3400	For sculptured 3 dimensional, add					.25			.25	.28	
	3900	For ceiling primer, add					.12			.12	.13	
	4000	For ceiling cement, add					.33			.33	.36	
	9000	Minimum labor/equipment charge	1 Carp	4	2	Job		66		66	109	

09600 | Flooring

09620 | Specialty Flooring

			CREW	DAILY OUTPUT	LABOR-HOURS	UNIT	2004 BARE COSTS MAT.	LABOR	EQUIP.	TOTAL	TOTAL INCL O&P	
100	0010	**ATHLETIC FLOORING**										100
	3700	Polyethylene, in rolls, no base incl., landscape surfaces	1 Tilf	275	.029	S.F.	2.45	.92		3.37	4.14	
	3800	Nylon action surface, 1/8" thick		275	.029		2.63	.92		3.55	4.33	
	3900	1/4" thick		275	.029		3.79	.92		4.71	5.60	
	4000	3/8" thick		275	.029		4.76	.92		5.68	6.70	
	5500	Polyvinyl chloride, sheet goods for gyms, 1/4" thick		80	.100		3.85	3.15		7	9.20	
	5600	3/8" thick		60	.133		4.34	4.20		8.54	11.35	

09631 | Brick Flooring

			CREW	DAILY OUTPUT	LABOR-HOURS	UNIT	2004 BARE COSTS MAT.	LABOR	EQUIP.	TOTAL	TOTAL INCL O&P	
100	0010	**BRICK FLOORING**										100
	0020	Acid proof shales, red, 8" x 3-3/4" x 1-1/4" thick	D-7	.43	37.209	M	695	1,050		1,745	2,400	
	0050	2-1/4" thick	D-1	.40	40		755	1,200		1,955	2,775	
	0200	Acid proof clay brick, 8" x 3-3/4" x 2-1/4" thick	"	.40	40		755	1,200		1,955	2,775	
	0260	Cast ceramic, pressed, 4" x 8" x 1/2", unglazed	D-7	100	.160	S.F.	4.94	4.48		9.42	12.45	
	0270	Glazed		100	.160		6.60	4.48		11.08	14.25	
	0280	Hand molded flooring, 4" x 8" x 3/4", unglazed		95	.168		6.55	4.71		11.26	14.60	
	0290	Glazed		95	.168		8.20	4.71		12.91	16.40	
	0300	8" hexagonal, 3/4" thick, unglazed		85	.188		7.15	5.25		12.40	16.15	
	0310	Glazed		85	.188		12.95	5.25		18.20	22.50	
	0400	Heavy duty industrial, cement mortar bed, 2" thick, not incl. brick	D-1	80	.200		.67	6		6.67	10.45	
	0450	Acid proof joints, 1/4" wide	"	65	.246		1.13	7.35		8.48	13.20	
	0500	Pavers, 8" x 4", 1" to 1-1/4" thick, red	D-7	95	.168		2.88	4.71		7.59	10.55	
	0510	Ironspot	"	95	.168		4.07	4.71		8.78	11.90	
	0540	1-3/8" to 1-3/4" thick, red	D-1	95	.168		2.78	5.05		7.83	11.25	
	0560	Ironspot		95	.168		4.02	5.05		9.07	12.60	
	0580	2-1/4" thick, red		90	.178		2.83	5.30		8.13	11.75	
	0590	Ironspot		90	.178		4.38	5.30		9.68	13.45	
	0800	For sidewalks and patios with pavers, see division 02780-200										
	0870	For epoxy joints, add	D-1	600	.027	S.F.	2.15	.80		2.95	3.67	
	0880	For Furan underlayment, add	"	600	.027		1.78	.80		2.58	3.26	
	0890	For waxed surface, steam cleaned, add	D-5	1,000	.008		.15	.27		.42	.61	
	9000	Minimum labor/equipment charge	1 Bric	2	4	Job		137		137	222	

09600 | Flooring

		09635 | Marble Flooring	Crew	Daily Output	Labor-Hours	Unit	2004 Bare Costs Mat.	Labor	Equip.	Total	Total Incl O&P	
100	0010	**MARBLE** Thin gauge tile, 12″ x 6″, 3/8″, White Carara	D-7	60	.267	S.F.	8.75	7.45		16.20	21.50	100
	0100	Travertine		60	.267		9.65	7.45		17.10	22.50	
	0200	12″ x 12″ x 3/8″, thin set, floors		60	.267		6.70	7.45		14.15	19.05	
	0300	On walls		52	.308		8.90	8.60		17.50	23.50	
	9000	Minimum labor/equipment charge		3	5.333	Job		149		149	234	

		09637 | Stone Flooring	Crew	Daily Output	Labor-Hours	Unit	Mat.	Labor	Equip.	Total	Total Incl O&P	
100	0010	**SLATE TILE** Vermont, 6″ x 6″ x 1/4″ thick, thin set	D-7	180	.089	S.F.	4.14	2.49		6.63	8.45	100
	9000	Minimum labor/equipment charge	″	3	5.333	Job		149		149	234	

		09643 | Wood Block Flooring	Crew	Daily Output	Labor-Hours	Unit	Mat.	Labor	Equip.	Total	Total Incl O&P	
100	0010	**WOOD BLOCK FLOORING** End grain flooring, coated, 2″ thick	1 Carp	295	.027	S.F.	2.98	.90		3.88	4.76	100
	0400	Natural finish, 1″ thick, fir		125	.064		3.08	2.11		5.19	6.90	
	0600	1-1/2″ thick, pine		125	.064		3.03	2.11		5.14	6.85	
	0700	2″ thick, pine		125	.064		3.40	2.11		5.51	7.25	
	9000	Minimum labor/equipment charge		2	4	Job		132		132	219	

		09647 | Wood Parquet Flooring	Crew	Daily Output	Labor-Hours	Unit	Mat.	Labor	Equip.	Total	Total Incl O&P	
100	0010	**WOOD PARQUET** flooring										100
	5200	Parquetry, standard, 5/16″ thick, not incl. finish, oak, minimum	1 Carp	160	.050	S.F.	2.85	1.65		4.50	5.90	
	5300	Maximum		100	.080		5.20	2.64		7.84	10.15	
	5500	Teak, minimum		160	.050		4.40	1.65		6.05	7.60	
	5600	Maximum		100	.080		7.70	2.64		10.34	12.85	
	5650	13/16″ thick, select grade oak, minimum		160	.050		8.50	1.65		10.15	12.10	
	5700	Maximum		100	.080		12.90	2.64		15.54	18.55	
	5800	Custom parquetry, including finish, minimum		100	.080		14.20	2.64		16.84	20	
	5900	Maximum		50	.160		18.80	5.30		24.10	29.50	
	6700	Parquetry, prefinished white oak, 5/16″ thick, minimum		160	.050		3.50	1.65		5.15	6.60	
	6800	Maximum		100	.080		7.15	2.64		9.79	12.30	
	7000	Walnut or teak, parquetry, minimum		160	.050		4.77	1.65		6.42	8	
	7100	Maximum		100	.080		8.30	2.64		10.94	13.55	
	7200	Acrylic wood parquet blocks, 12″ x 12″ x 5/16″,										
	7210	irradiated, set in epoxy	1 Carp	160	.050	S.F.	6.95	1.65		8.60	10.40	

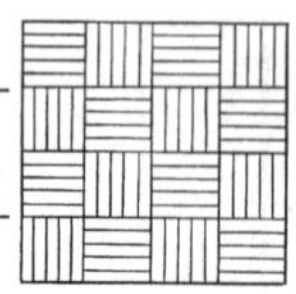

		09648 | Wood Strip Flooring	Crew	Daily Output	Labor-Hours	Unit	Mat.	Labor	Equip.	Total	Total Incl O&P	
100	0010	**WOOD** Fir, vertical grain, 1″ x 4″, not incl. finish, B & better (R06110-030)	1 Carp	255	.031	S.F.	2.44	1.04		3.48	4.40	100
	0100	C grade & better		255	.031		2.29	1.04		3.33	4.24	
	0300	Flat grain, 1″ x 4″, not incl. finish, B & better		255	.031		2.79	1.04		3.83	4.79	
	0400	C & better		255	.031		2.68	1.04		3.72	4.67	
	4000	Maple, strip, 25/32″ x 2-1/4″, not incl. finish, select		170	.047		4.65	1.55		6.20	7.65	
	4100	#2 & better		170	.047		2.81	1.55		4.36	5.65	
	4300	33/32″ x 3-1/4″, not incl. finish, #1 grade		170	.047		3.45	1.55		5	6.35	
	4400	#2 & better		170	.047		3.07	1.55		4.62	5.95	
	4600	Oak, white or red, 25/32″ x 2-1/4″, not incl. finish										
	4700	#1 common	1 Carp	170	.047	S.F.	2.86	1.55		4.41	5.70	
	4900	Select quartered, 2-1/4″ wide		170	.047		3.60	1.55		5.15	6.55	
	5000	Clear		170	.047		3.72	1.55		5.27	6.65	
	6100	Prefinished, white oak, prime grade, 2-1/4″ wide		170	.047		6.20	1.55		7.75	9.35	
	6200	3-1/4″ wide		185	.043		7.95	1.43		9.38	11.10	
	6400	Ranch plank		145	.055		7.70	1.82		9.52	11.45	
	6500	Hardwood blocks, 9″ x 9″, 25/32″ thick		160	.050		5.15	1.65		6.80	8.40	
	7400	Yellow pine, 3/4″ x 3-1/8″, T & G, C & better, not incl. finish		200	.040		2.28	1.32		3.60	4.70	
	7500	Refinish wood floor, sand, 2 cts poly, wax, soft wood, min.	1 Clab	400	.020		.69	.52		1.21	1.62	

09600 | Flooring

		09648 | Wood Strip Flooring	CREW	DAILY OUTPUT	LABOR-HOURS	UNIT	2004 BARE COSTS MAT.	LABOR	EQUIP.	TOTAL	TOTAL INCL O&P	
100	7600	Hard wood, max (R06110 -030)	1 Clab	130	.062	S.F.	1.04	1.60		2.64	3.79	100
	7800	Sanding and finishing, 2 coats polyurethane		295	.027		.69	.71		1.40	1.93	
	7900	Subfloor and underlayment, see division 06160										
	8015	Transition molding, 2 1/4" wide, 5' long	1 Carp	19.20	.417	Ea.	12.65	13.75		26.40	37	
	8300	Floating floor, wood composition strip, complete.	1 Clab	133	.060	S.F.	3.36	1.56		4.92	6.30	
	8310	Floating floor components, T & G wood composite strips					3.02			3.02	3.33	
	8320	Film					.14			.14	.15	
	8330	Foam					.17			.17	.18	
	8340	Adhesive					.07			.07	.08	
	8350	Installation kit					.16			.16	.18	
	8360	Trim, 2" wide x 3' long				L.F.	2.05			2.05	2.26	
	8370	Reducer moulding				"	4.15			4.15	4.57	
	9000	Minimum labor/equipment charge	1 Carp	2	4	Job		132		132	219	
		09651 | Resilient Base & Access.										
100	0010	**STAIR TREADS AND RISERS** See index for materials other										100
	0100	than rubber and vinyl										
	0300	Rubber, molded tread, 12" wide, 5/16" thick, black	1 Tilf	115	.070	L.F.	8.90	2.19		11.09	13.25	
	0400	Colors		115	.070		8.25	2.19		10.44	12.50	
	0600	1/4" thick, black		115	.070		7.95	2.19		10.14	12.20	
	0700	Colors		115	.070		8.25	2.19		10.44	12.50	
	0900	Grip strip safety tread, colors, 5/16" thick		115	.070		10.95	2.19		13.14	15.50	
	1000	3/16" thick 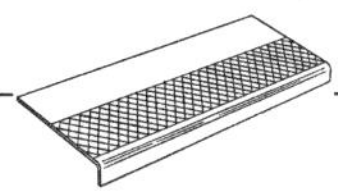		120	.067		7.75	2.10		9.85	11.80	
	1200	Landings, smooth sheet rubber, 1/8" thick		120	.067	S.F.	3.71	2.10		5.81	7.35	
	1300	3/16" thick		120	.067	"	4.77	2.10		6.87	8.55	
	1500	Nosings, 3" wide, 3/16" thick, black		140	.057	L.F.	2.67	1.80		4.47	5.75	
	1600	Colors		140	.057		2.60	1.80		4.40	5.70	
	1800	Risers, 7" high, 1/8" thick, flat		250	.032		3.14	1.01		4.15	5.05	
	1900	Coved		250	.032		2.83	1.01		3.84	4.69	
	2100	Vinyl, molded tread, 12" wide, colors, 1/8" thick		115	.070		3.13	2.19		5.32	6.90	
	2200	1/4" thick		115	.070		5.05	2.19		7.24	9	
	2300	Landing material, 1/8" thick 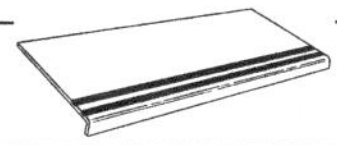		200	.040	S.F.	3.45	1.26		4.71	5.80	
	2400	Riser, 7" high, 1/8" thick, coved		175	.046	L.F.	2.02	1.44		3.46	4.48	
	2500	Tread and riser combined, 1/8" thick		80	.100	"	5.85	3.15		9	11.35	
	9000	Minimum labor/equipment charge		3	2.667	Job		84		84	132	
200	0010	**RESILIENT BASE**										200
	0800	Base, cove, rubber or vinyl, .080" thick										
	1100	Standard colors, 2-1/2" high	1 Tilf	315	.025	L.F.	.44	.80		1.24	1.73	
	1150	4" high		315	.025		.49	.80		1.29	1.79	
	1200	6" high		315	.025		.80	.80		1.60	2.13	
	1450	1/8" thick, standard colors, 2-1/2" high		315	.025		.51	.80		1.31	1.81	
	1500	4" high		315	.025		.67	.80		1.47	1.99	
	1550	6" high		315	.025		.88	.80		1.68	2.22	
	1600	Corners, 2-1/2" high		315	.025	Ea.	1.11	.80		1.91	2.47	
	1630	4" high		315	.025		1.16	.80		1.96	2.53	
	1660	6" high		315	.025		1.51	.80		2.31	2.91	
		09653 | Resilient Sheet Flooring										
100	0010	**RESILIENT SHEET FLOORING**										100
	5900	Rubber, sheet goods, 36" wide, 1/8" thick	1 Tilf	120	.067	S.F.	3.37	2.10		5.47	7	
	5950	3/16" thick		100	.080		4.79	2.52		7.31	9.20	
	6000	1/4" thick		90	.089		5.55	2.80		8.35	10.50	
	8000	Vinyl sheet goods, backed, .065" thick, minimum		250	.032		1.97	1.01		2.98	3.75	
	8050	Maximum		200	.040		2.52	1.26		3.78	4.75	

09653 | Resilient Sheet Flooring

		09653 Resilient Sheet Flooring	CREW	DAILY OUTPUT	LABOR-HOURS	UNIT	2004 BARE COSTS MAT.	LABOR	EQUIP.	TOTAL	TOTAL INCL O&P	
100	8100	.080" thick, minimum	1 Tilf	230	.035	S.F.	2.12	1.10		3.22	4.05	100
	8150	Maximum		200	.040		3.03	1.26		4.29	5.30	
	8200	.125" thick, minimum		230	.035		2.39	1.10		3.49	4.35	
	8250	Maximum		200	.040		3.79	1.26		5.05	6.15	
	8700	Adhesive cement, 1 gallon does 200 to 300 S.F.				Gal.	15.55			15.55	17.10	
	8800	Asphalt primer, 1 gallon per 300 S.F.					9.20			9.20	10.10	
	8900	Emulsion, 1 gallon per 140 S.F.					11.65			11.65	12.85	
	8950	Latex underlayment, liquid, fortified					31.50			31.50	34.50	

09658 | Resilient Tile Flooring

		09658 Resilient Tile Flooring	CREW	DAILY OUTPUT	LABOR-HOURS	UNIT	MAT.	LABOR	EQUIP.	TOTAL	TOTAL INCL O&P	
100	0010	**RESILIENT TILE FLOORING**										100
	2200	Cork tile, standard finish, 1/8" thick	1 Tilf	315	.025	S.F.	3.57	.80		4.37	5.20	
	2250	3/16" thick		315	.025		3.56	.80		4.36	5.15	
	2300	5/16" thick		315	.025		4.78	.80		5.58	6.50	
	2350	1/2" thick		315	.025		5.50	.80		6.30	7.30	
	2500	Urethane finish, 1/8" thick		315	.025		4.50	.80		5.30	6.20	
	2550	3/16" thick		315	.025		4.88	.80		5.68	6.60	
	2600	5/16" thick		315	.025		6.05	.80		6.85	7.90	
	2650	1/2" thick		315	.025		8.50	.80		9.30	10.60	
	6050	Tile, marbleized colors, 12" x 12", 1/8" thick		400	.020		3.85	.63		4.48	5.25	
	6100	3/16" thick		400	.020		5.20	.63		5.83	6.70	
	6300	Special tile, plain colors, 1/8" thick		400	.020		4.18	.63		4.81	5.60	
	6350	3/16" thick		400	.020		5.65	.63		6.28	7.20	
	7000	Vinyl composition tile, 12" x 12", 1/16" thick		500	.016		.75	.50		1.25	1.62	
	7050	Embossed		500	.016		.95	.50		1.45	1.84	
	7100	Marbleized		500	.016		.95	.50		1.45	1.84	
	7150	Solid		500	.016		1.06	.50		1.56	1.96	
	7200	3/32" thick, embossed		500	.016		.96	.50		1.46	1.85	
	7250	Marbleized		500	.016		1.07	.50		1.57	1.97	
	7300	Solid		500	.016		1.56	.50		2.06	2.51	
	7350	1/8" thick, marbleized		500	.016		1.02	.50		1.52	1.91	
	7400	Solid		500	.016		1.91	.50		2.41	2.89	
	7450	Conductive		500	.016		3.58	.50		4.08	4.73	
	7500	Vinyl tile, 12" x 12", .050" thick, minimum		500	.016		1.68	.50		2.18	2.64	
	7550	Maximum		500	.016		3.29	.50		3.79	4.41	
	7600	1/8" thick, minimum		500	.016		2.12	.50		2.62	3.12	
	7650	Solid colors		500	.016		4.57	.50		5.07	5.85	
	7700	Marbleized or Travertine pattern		500	.016		3.39	.50		3.89	4.52	
	7750	Florentine pattern		500	.016		3.90	.50		4.40	5.10	
	7800	Maximum		500	.016		8	.50		8.50	9.60	
	9500	Minimum labor/equipment charge		4	2	Job		63		63	99	

09662 | Static Control Flooring

		09662 Static Control Flooring	CREW	DAILY OUTPUT	LABOR-HOURS	UNIT	MAT.	LABOR	EQUIP.	TOTAL	TOTAL INCL O&P	
100	0010	**CONDUCTIVE RESILIENT FLOORING**										100
	1700	Conductive flooring, rubber tile, 1/8" thick	1 Tilf	315	.025	S.F.	2.74	.80		3.54	4.26	
	1800	Homogeneous vinyl tile, 1/8" thick	"	315	.025	"	3.73	.80		4.53	5.35	

09673 | Composition Flooring

		09673 Composition Flooring	CREW	DAILY OUTPUT	LABOR-HOURS	UNIT	MAT.	LABOR	EQUIP.	TOTAL	TOTAL INCL O&P	
100	0010	**COMPOSITION FLOORING** Acrylic, 1/4" thick	C-6	520	.092	S.F.	1.24	2.52	.10	3.86	5.60	100
	0600	Epoxy, with colored quartz chips, broadcast, minimum		675	.071		2.19	1.94	.08	4.21	5.65	
	0700	Maximum		490	.098		2.66	2.67	.11	5.44	7.45	
	0900	Trowelled, minimum		560	.086		2.83	2.34	.09	5.26	7.05	
	1000	Maximum		480	.100		4.12	2.73	.11	6.96	9.10	
	1200	Heavy duty epoxy topping, 1/4" thick,										

		09673 Composition Flooring	CREW	DAILY OUTPUT	LABOR-HOURS	UNIT	2004 BARE COSTS MAT.	LABOR	EQUIP.	TOTAL	TOTAL INCL O&P	
100	1300	500 to 1,000 S.F.	C-6	420	.114	S.F.	4.89	3.12	.12	8.13	10.65	100
	1500	1,000 to 2,000 S.F.		450	.107		3.87	2.91	.12	6.90	9.15	
	1600	Over 10,000 S.F.		480	.100		3.58	2.73	.11	6.42	8.55	
	1800	Epoxy terrazzo, 1/4" thick, chemical resistant, minimum	J-3	200	.080		5.20	2.28	.91	8.39	10.25	
	1900	Maximum	"	150	.107		7.95	3.03	1.21	12.19	14.85	

		09680 Carpet	CREW	DAILY OUTPUT	LABOR-HOURS	UNIT	MAT.	LABOR	EQUIP.	TOTAL	TOTAL INCL O&P	
600	0010	**CARPET PAD**, commercial grade										600
	9000	Sponge rubber pad, minimum	1 Tilf	150	.053	S.Y.	3.18	1.68		4.86	6.15	
	9100	Maximum		150	.053		8.25	1.68		9.93	11.75	
	9200	Felt pad, minimum		150	.053		3.47	1.68		5.15	6.45	
	9300	Maximum		150	.053		6.40	1.68		8.08	9.70	
	9400	Bonded urethane pad, minimum		150	.053		3.76	1.68		5.44	6.75	
	9500	Maximum		150	.053		6.45	1.68		8.13	9.75	
	9600	Prime urethane pad, minimum		150	.053		2.16	1.68		3.84	5	
	9700	Maximum		150	.053		3.99	1.68		5.67	7	
800	0010	**CARPET** Commercial grades, direct cement										800
	0700	Nylon, level loop, 26 oz., light to medium traffic	1 Tilf	75	.107	S.Y.	14.95	3.36		18.31	21.50	
	0720	28 oz., light to medium traffic		75	.107		16.15	3.36		19.51	23	
	0900	32 oz., medium traffic		75	.107		21	3.36		24.36	29	
	1100	40 oz., medium to heavy traffic		75	.107		31.50	3.36		34.86	40	
	2920	Nylon plush, 30 oz., medium traffic		57	.140		15.65	4.42		20.07	24	
	3000	36 oz., medium traffic		75	.107		19.70	3.36		23.06	27	
	3100	42 oz., medium to heavy traffic		70	.114		20.50	3.60		24.10	28	
	3200	46 oz., medium to heavy traffic		70	.114		27	3.60		30.60	35.50	
	3300	54 oz., heavy traffic		70	.114		30.50	3.60		34.10	39	
	4500	50 oz., medium to heavy traffic		75	.107		69.50	3.36		72.86	82	
	4700	Patterned, 32 oz., medium to heavy traffic		70	.114		68.50	3.60		72.10	81	
	4900	48 oz., heavy traffic		70	.114		69.50	3.60		73.10	82	
	5000	For less than full roll, add					25%					
	5100	For small rooms, less than 12' wide, add						25%				
	5200	For large open areas (no cuts), deduct						25%				
	5600	For bound carpet baseboard, add	1 Tilf	300	.027	L.F.	1.09	.84		1.93	2.52	
	5610	For stairs, not incl. price of carpet, add	"	30	.267	Riser		8.40		8.40	13.15	
	5620	For borders and patterns, add to labor						18%				
	8950	For tackless, stretched installation, add padding to above										
	9850	For "branded" fiber, add				S.Y.	25%					
	9900	Carpet cleaning machine, rent				Day					35	
	9910	Minimum labor/equipment charge	1 Tilf	3	2.667	Job		84		84	132	
900	0010	**CARPET TILE**										900
	0100	Tufted nylon, 18" x 18", hard back, 20 oz.	1 Tilf	150	.053	S.Y.	19.65	1.68		21.33	24	
	0110	26 oz.		150	.053		33.50	1.68		35.18	39.50	
	0200	Cushion back, 20 oz.		150	.053		24.50	1.68		26.18	29.50	
	0210	26 oz.		150	.053		38.50	1.68		40.18	45	

09720 | Wall Coverings

			CREW	DAILY OUTPUT	LABOR-HOURS	UNIT	2004 BARE COSTS MAT.	LABOR	EQUIP.	TOTAL	TOTAL INCL O&P	
100	0010	**WALL COVERING** Including sizing, add 10%-30% waste at takeoff										100
	0050	Aluminum foil	1 Pape	275	.029	S.F.	.83	.85		1.68	2.28	
	0100	Copper sheets, .025″ thick, vinyl backing		240	.033		4.44	.98		5.42	6.45	
	0300	Phenolic backing		240	.033		5.75	.98		6.73	7.90	
	0600	Cork tiles, light or dark, 12″ x 12″ x 3/16″		240	.033		2.60	.98		3.58	4.43	
	0700	5/16″ thick		235	.034		2.95	1		3.95	4.85	
	0900	1/4″ basketweave		240	.033		4.53	.98		5.51	6.55	
	1000	1/2″ natural, non-directional pattern		240	.033		6.05	.98		7.03	8.20	
	1100	3/4″ natural, non-directional pattern		240	.033		9.85	.98		10.83	12.40	
	1200	Granular surface, 12″ x 36″, 1/2″ thick		385	.021		.98	.61		1.59	2.06	
	1300	1″ thick		370	.022		1.27	.63		1.90	2.42	
	1500	Polyurethane coated, 12″ x 12″ x 3/16″ thick		240	.033		3.06	.98		4.04	4.94	
	1600	5/16″ thick		235	.034		4.35	1		5.35	6.40	
	1800	Cork wallpaper, paperbacked, natural		480	.017		1.74	.49		2.23	2.69	
	1900	Colors		480	.017		2.15	.49		2.64	3.15	
	2100	Flexible wood veneer, 1/32″ thick, plain woods		100	.080		1.83	2.35		4.18	5.75	
	2200	Exotic woods		95	.084		2.78	2.47		5.25	7	
	2400	Gypsum-based, fabric-backed, fire										
	2500	resistant for masonry walls, minimum, 21 oz./S.Y.	1 Pape	800	.010	S.F.	.68	.29		.97	1.22	
	2600	Average		720	.011		1	.33		1.33	1.62	
	2700	Maximum, (small quantities)		640	.013		1.11	.37		1.48	1.81	
	2750	Acrylic, modified, semi-rigid PVC, .028″ thick	2 Carp	330	.048		.92	1.60		2.52	3.66	
	2800	.040″ thick	″	320	.050		1.21	1.65		2.86	4.07	
	3000	Vinyl wall covering, fabric-backed, lightweight, (12-15 oz./S.Y.)	1 Pape	640	.013		.58	.37		.95	1.23	
	3300	Medium weight, type 2, (20-24 oz./S.Y.)		480	.017		.72	.49		1.21	1.57	
	3400	Heavy weight, type 3, (28 oz./S.Y.)		435	.018		1.16	.54		1.70	2.14	
	3600	Adhesive, 5 gal. lots, (18SY/Gal.)				Gal.	8.95			8.95	9.80	
	3700	Wallpaper, average workmanship, solid pattern, low cost paper	1 Pape	640	.013	S.F.	.29	.37		.66	.91	
	3900	basic patterns (matching required), avg. cost paper		535	.015		.64	.44		1.08	1.40	
	4000	Paper at $85per double roll, quality workmanship		435	.018		1.52	.54		2.06	2.53	
	4200	Grass cloths with lining paper, minimum		400	.020		.63	.59		1.22	1.63	
	4300	Maximum		350	.023		2.03	.67		2.70	3.30	
	5990	Wallpaper removal, 1 layer, minimum		800	.010		.06	.29		.35	.54	
	6000	Wallpaper removal, 3 layer, maximum		400	.020		.11	.59		.70	1.06	
	9000	Minimum labor/equipment charge		2	4	Job		117		117	188	

09770 | Special Wall Surfaces

			CREW	DAILY OUTPUT	LABOR-HOURS	UNIT	MAT.	LABOR	EQUIP.	TOTAL	TOTAL INCL O&P	
400	0010	**FIBERGLASS REINFORCED PLASTIC** panels, .090″ thick, on walls										400
	0020	Adhesive mounted, embossed surface	2 Carp	640	.025	S.F.	1.25	.83		2.08	2.75	
	0030	Smooth surface		640	.025		1.47	.83		2.30	2.99	
	0040	Fire rated, embossed surface		640	.025		1.64	.83		2.47	3.17	
	0050	Nylon rivet mounted, on drywall, embossed surface		480	.033		1.16	1.10		2.26	3.10	
	0060	Smooth surface		480	.033		1.37	1.10		2.47	3.33	
	0070	Fire rated, embossed surface		480	.033		1.53	1.10		2.63	3.50	
	0080	On masonry, embossed surface		320	.050		1.16	1.65		2.81	4.02	
	0090	Smooth surface		320	.050		1.47	1.65		3.12	4.36	
	0100	Fire rated, embossed surface		320	.050		1.55	1.65		3.20	4.45	
	0110	Nylon rivet and adhesive mounted, on drywall, embossed surface		240	.067		1.37	2.20		3.57	5.15	
	0120	Smooth surface		240	.067		1.55	2.20		3.75	5.35	
	0130	Fire rated, embossed surface		240	.067		1.73	2.20		3.93	5.55	
	0140	On masonry, embossed surface		190	.084		1.37	2.78		4.15	6.10	
	0150	Smooth surface		190	.084		1.55	2.78		4.33	6.30	
	0160	Fire rated, embossed surface		190	.084		1.73	2.78		4.51	6.50	
	0170	For moldings add	1 Carp	250	.032	L.F.	.23	1.06		1.29	2	
	0180	On ceilings, for lay in grid system, embossed surface		400	.020	S.F.	1.25	.66		1.91	2.47	

 Important: See the Reference Section for critical supporting data - Reference Nos., Crews, & City Cost Indexes

09700 | Wall Finishes

		09770 | Special Wall Surfaces	Crew	Daily Output	Labor-Hours	Unit	2004 Bare Costs Mat.	Labor	Equip.	Total	Total Incl O&P	
400	0190	Smooth surface	1 Carp	400	.020	S.F.	1.47	.66		2.13	2.71	400
	0200	Fire rated, embossed surface		400	.020		1.64	.66		2.30	2.89	
700	0010	**PANEL SYSTEM**										700
	0100	Raised panel, eng. wood core w/ wood veneer, std., paint grade	2 Carp	300	.053	S.F.	6.90	1.76		8.66	10.50	
	0110	Oak veneer		300	.053		10.60	1.76		12.36	14.60	
	0120	Maple veneer		300	.053		17.90	1.76		19.66	22.50	
	0130	Cherry veneer		300	.053		21.50	1.76		23.26	27	
	0300	Class I fire rated, paint grade		300	.053		16.55	1.76		18.31	21	
	0310	Oak veneer		300	.053		28	1.76		29.76	34	
	0320	Maple veneer		300	.053		46.50	1.76		48.26	54	
	0330	Cherry veneer		300	.053		54	1.76		55.76	62.50	
	5000	For prefinished paneling, see division 06250-500 & 06250-200										
750	0010	**SLATWALL PANELS AND ACCESSORIES**										750
	0100	Slatwall panel, 4′ x 8′ x 3/4″ T, MDF, paint grade	1 Carp	500	.016	S.F.	1.20	.53		1.73	2.20	
	0110	Melamine finish		500	.016		1.82	.53		2.35	2.88	
	0120	High pressure plastic laminate finish		500	.016		2.90	.53		3.43	4.07	
	0130	Aluminum channel inserts, add					2.18			2.18	2.40	
	0200	Accessories, corner forms, 8′ L				L.F.	3.10			3.10	3.41	
	0210	T-connector, 8′ L					4.50			4.50	4.95	
	0220	J-mold, 8′ L					1.15			1.15	1.27	
	0230	Edge cap, 8′ L					.75			.75	.83	
	0240	Finish end cap, 8′ L					2.75			2.75	3.03	
	0300	Display hook, 4″ L				Ea.	.85			.85	.94	
	0310	6″ L					.90			.90	.99	
	0320	8″ L					1			1	1.10	
	0330	10″ L					1.10			1.10	1.21	
	0340	12″ L					1.20			1.20	1.32	
	0350	Acrylic, 4″ L					.70			.70	.77	
	0360	6″ L					.80			.80	.88	
	0370	8″ L					.85			.85	.94	
	0380	10″ L					.95			.95	1.05	
	0400	Waterfall hanger, metal, 12″ - 16″					5.40			5.40	5.95	
	0410	Acrylic					7.70			7.70	8.45	
	0500	Shelf bracket, metal, 8″					4.20			4.20	4.62	
	0510	10″					4.50			4.50	4.95	
	0520	12″					4.80			4.80	5.30	
	0530	14″					5.40			5.40	5.95	
	0540	16″					6			6	6.60	
	0550	Acrylic, 8″					2.50			2.50	2.75	
	0560	10″					2.80			2.80	3.08	
	0570	12″					3.10			3.10	3.41	
	0580	14″					3.50			3.50	3.85	
	0600	Shelf, acrylic, 12″ x 16″ x 1/4″					21			21	23	
	0610	12″ x 24″ x 1/4″					35			35	38.50	

09800 | Acoustical Treatment

		09820 | Acoustical Insul/Sealants	Crew	Daily Output	Labor-Hours	Unit	2004 Bare Costs Mat.	Labor	Equip.	Total	Total Incl O&P	
500	0010	**SOUND ATTENUATION** Blanket, 1″ thick	1 Carp	925	.009	S.F.	.24	.29		.53	.73	500
	0500	1-1/2″ thick		920	.009		.23	.29		.52	.73	

09800 | Acoustical Treatment

	09820 | Acoustical Insul/Sealants		CREW	DAILY OUTPUT	LABOR-HOURS	UNIT	2004 BARE COSTS MAT.	LABOR	EQUIP.	TOTAL	TOTAL INCL O&P	
500	1000	2″ thick	1 Carp	915	.009	S.F.	.29	.29		.58	.80	500
	1500	3″ thick	↓	910	.009		.39	.29		.68	.91	
	3400	Urethane plastic foam, open cell, on wall, 2″ thick	2 Carp	2,050	.008		2.56	.26		2.82	3.25	
	3500	3″ thick		1,550	.010		3.40	.34		3.74	4.30	
	3600	4″ thick		1,050	.015		4.77	.50		5.27	6.10	
	3700	On ceiling, 2″ thick		1,700	.009		2.55	.31		2.86	3.32	
	3800	3″ thick		1,300	.012		3.40	.41		3.81	4.41	
	3900	4″ thick	↓	900	.018	↓	4.77	.59		5.36	6.20	
	4000	Nylon matting 0.4″ thick, with carbon black spinerette										
	4010	plus polyester fabric, on floor	J-4	4,000	.004	S.F.	1.90	.11		2.01	2.27	
	4200	Fiberglass reinf. backer board underlayment, 7/16″ thick, on floor	″	800	.020	″	1.59	.56		2.15	2.63	
	9000	Minimum labor/equipment charge	1 Carp	5	1.600	Job		53		53	87.50	

	09840 | Acoustical Wall Treatment		CREW	DAILY OUTPUT	LABOR-HOURS	UNIT	MAT.	LABOR	EQUIP.	TOTAL	TOTAL INCL O&P	
100	0010	**SOUND ABSORBING PANELS** Perforated steel facing, painted with										100
	0100	fiberglass or mineral filler, no backs, 2-1/4″ thick, modular										
	0200	space units, ceiling or wall hung, white or colored	1 Carp	100	.080	S.F.	8.90	2.64		11.54	14.20	
	0300	Fiberboard sound deadening panels, 1/2″ thick	″	600	.013	″	.29	.44		.73	1.05	
	0500	Fiberglass panels, 4′ x 8′ x 1″ thick, with										
	0600	glass cloth face for walls, cemented	1 Carp	155	.052	S.F.	5.85	1.70		7.55	9.25	
	0700	1-1/2″ thick, dacron covered, inner aluminum frame,										
	0710	wall mounted	1 Carp	300	.027	S.F.	7.35	.88		8.23	9.55	
	0900	Mineral fiberboard panels, fabric covered, 30″x 108″,										
	1000	3/4″ thick, concealed spline, wall mounted	1 Carp	150	.053	S.F.	5.30	1.76		7.06	8.75	
	9000	Minimum labor/equipment charge	″	4	2	Job		66		66	109	

09900 | Paints & Coatings

	09910 | Paints		CREW	DAILY OUTPUT	LABOR-HOURS	UNIT	2004 BARE COSTS MAT.	LABOR	EQUIP.	TOTAL	TOTAL INCL O&P	
100	0010	**CABINETS AND CASEWORK**										100
	1000	Primer coat, oil base, brushwork	1 Pord	650	.012	S.F.	.05	.36		.41	.63	
	2000	Paint, oil base, brushwork, 1 coat		650	.012		.06	.36		.42	.65	
	3000	Stain, brushwork, wipe off		650	.012		.05	.36		.41	.63	
	4000	Shellac, 1 coat, brushwork		650	.012		.06	.36		.42	.64	
	4500	Varnish, 3 coats, brushwork, sand after 1st coat	↓	325	.025		.17	.73		.90	1.36	
	5000	For latex paint, deduct				↓	10%					
300	0010	**DOORS AND WINDOWS, EXTERIOR**										300
	0100	Door frames & trim, only										
	0110	Brushwork, primer	1 Pord	512	.016	L.F.	.05	.46		.51	.80	
	0120	Finish coat, exterior latex		512	.016		.06	.46		.52	.80	
	0130	Primer & 1 coat, exterior latex		300	.027	↓	.11	.79		.90	1.38	
	0135	2 coats, exterior latex, both sides		15	.533	Ea.	5.10	15.80		20.90	31	
	0140	Primer & 2 coats, exterior latex	↓	265	.030	L.F.	.16	.89		1.05	1.61	
	0150	Doors, flush, both sides, incl. frame & trim										
	0160	Roll & brush, primer	1 Pord	10	.800	Ea.	3.82	23.50		27.32	42	
	0170	Finish coat, exterior latex		10	.800		4.35	23.50		27.85	43	
	0180	Primer & 1 coat, exterior latex		7	1.143		8.15	34		42.15	63	
	0190	Primer & 2 coats, exterior latex	↓	5	1.600	↓	12.50	47.50		60	90	

	09910 | Paints	CREW	DAILY OUTPUT	LABOR-HOURS	UNIT	2004 BARE COSTS MAT.	LABOR	EQUIP.	TOTAL	TOTAL INCL O&P	
300	0200 Brushwork, stain, sealer & 2 coats polyurethane	1 Pord	4	2	Ea.	15.55	59		74.55	112	300
	0210 Doors, French, both sides, 10-15 lite, incl. frame & trim										
	0220 Brushwork, primer	1 Pord	6	1.333	Ea.	1.91	39.50		41.41	65	
	0230 Finish coat, exterior latex		6	1.333		2.17	39.50		41.67	65.50	
	0240 Primer & 1 coat, exterior latex		3	2.667		4.09	79		83.09	130	
	0250 Primer & 2 coats, exterior latex		2	4		6.15	118		124.15	197	
	0260 Brushwork, stain, sealer & 2 coats polyurethane	↓	2.50	3.200	↓	5.60	94.50		100.10	158	
	0270 Doors, louvered, both sides, incl. frame & trim										
	0280 Brushwork, primer	1 Pord	7	1.143	Ea.	3.82	34		37.82	58	
	0290 Finish coat, exterior latex		7	1.143		4.35	34		38.35	59	
	0300 Primer & 1 coat, exterior latex		4	2		8.15	59		67.15	104	
	0310 Primer & 2 coats, exterior latex		3	2.667		12.25	79		91.25	140	
	0320 Brushwork, stain, sealer & 2 coats polyurethane	↓	4.50	1.778	↓	15.55	52.50		68.05	102	
	0330 Doors, panel, both sides, incl. frame & trim										
	0340 Roll & brush, primer	1 Pord	6	1.333	Ea.	3.82	39.50		43.32	67	
	0350 Finish coat, exterior latex		6	1.333		4.35	39.50		43.85	68	
	0360 Primer & 1 coat, exterior latex		3	2.667		8.15	79		87.15	135	
	0370 Primer & 2 coats, exterior latex		2.50	3.200		12.25	94.50		106.75	166	
	0380 Brushwork, stain, sealer & 2 coats polyurethane	↓	3	2.667	↓	15.55	79		94.55	143	
	0400 Windows, per ext. side, based on 15 SF										
	0410 1 to 6 lite										
	0420 Brushwork, primer	1 Pord	13	.615	Ea.	.75	18.20		18.95	30	
	0430 Finish coat, exterior latex		13	.615		.86	18.20		19.06	30	
	0440 Primer & 1 coat, exterior latex		8	1		1.61	29.50		31.11	49.50	
	0450 Primer & 2 coats, exterior latex		6	1.333		2.42	39.50		41.92	65.50	
	0460 Stain, sealer & 1 coat varnish	↓	7	1.143	↓	2.20	34		36.20	56.50	
	0470 7 to 10 lite										
	0480 Brushwork, primer	1 Pord	11	.727	Ea.	.75	21.50		22.25	35.50	
	0490 Finish coat, exterior latex		11	.727		.86	21.50		22.36	35.50	
	0500 Primer & 1 coat, exterior latex		7	1.143		1.61	34		35.61	56	
	0510 Primer & 2 coats, exterior latex		5	1.600		2.42	47.50		49.92	78.50	
	0520 Stain, sealer & 1 coat varnish	↓	6	1.333	↓	2.20	39.50		41.70	65.50	
	0530 12 lite										
	0540 Brushwork, primer	1 Pord	10	.800	Ea.	.75	23.50		24.25	39	
	0550 Finish coat, exterior latex		10	.800		.86	23.50		24.36	39	
	0560 Primer & 1 coat, exterior latex		6	1.333		1.61	39.50		41.11	65	
	0570 Primer & 2 coats, exterior latex		5	1.600		2.42	47.50		49.92	78.50	
	0580 Stain, sealer & 1 coat varnish	↓	6	1.333		2.20	39.50		41.70	65.50	
	0590 For oil base paint, add				↓	10%					
310	0010 **DOORS & WINDOWS, INTERIOR LATEX**										310
	0100 Doors flush, both sides, incl. frame & trim										
	0110 Roll & brush, primer	1 Pord	10	.800	Ea.	3.33	23.50		26.83	41.50	
	0120 Finish coat, latex		10	.800		3.60	23.50		27.10	42	
	0130 Primer & 1 coat latex		7	1.143		6.90	34		40.90	61.50	
	0140 Primer & 2 coats latex		5	1.600		10.30	47.50		57.80	87.50	
	0160 Spray, both sides, primer		20	.400		3.50	11.85		15.35	23	
	0170 Finish coat, latex		20	.400		3.77	11.85		15.62	23	
	0180 Primer & 1 coat latex		11	.727		7.30	21.50		28.80	42.50	
	0190 Primer & 2 coats latex	↓	8	1	↓	10.90	29.50		40.40	59.50	
	0200 Doors, French, both sides, 10-15 lite, incl. frame & trim										
	0210 Roll & brush, primer	1 Pord	6	1.333	Ea.	1.66	39.50		41.16	65	
	0220 Finish coat, latex		6	1.333		1.80	39.50		41.30	65	
	0230 Primer & 1 coat latex		3	2.667		3.46	79		82.46	130	
	0240 Primer & 2 coats latex	↓	2	4	↓	5.15	118		123.15	196	
	0260 Doors, louvered, both sides, incl. frame & trim										

09900 | Paints & Coatings

09910 | Paints

				DAILY	LABOR-		2004 BARE COSTS				TOTAL	
			CREW	OUTPUT	HOURS	UNIT	MAT.	LABOR	EQUIP.	TOTAL	INCL O&P	
310	0270	Roll & brush, primer	1 Pord	7	1.143	Ea.	3.33	34		37.33	57.50	310
	0280	Finish coat, latex		7	1.143		3.60	34		37.60	58	
	0290	Primer & 1 coat, latex		4	2		6.75	59		65.75	102	
	0300	Primer & 2 coats, latex		3	2.667		10.50	79		89.50	138	
	0320	Spray, both sides, primer		20	.400		3.50	11.85		15.35	23	
	0330	Finish coat, latex		20	.400		3.77	11.85		15.62	23	
	0340	Primer & 1 coat, latex		11	.727		7.30	21.50		28.80	42.50	
	0350	Primer & 2 coats, latex		8	1		11.15	29.50		40.65	60	
	0360	Doors, panel, both sides, incl. frame & trim										
	0370	Roll & brush, primer	1 Pord	6	1.333	Ea.	3.50	39.50		43	67	
	0380	Finish coat, latex		6	1.333		3.60	39.50		43.10	67	
	0390	Primer & 1 coat, latex		3	2.667		6.90	79		85.90	134	
	0400	Primer & 2 coats, latex		2.50	3.200		10.50	94.50		105	164	
	0420	Spray, both sides, primer		10	.800		3.50	23.50		27	42	
	0430	Finish coat, latex		10	.800		3.77	23.50		27.27	42	
	0440	Primer & 1 coat, latex		5	1.600		7.30	47.50		54.80	84	
	0450	Primer & 2 coats, latex		4	2		11.15	59		70.15	107	
	0460	Windows, per interior side, based on 15 SF										
	0470	1 to 6 lite										
	0480	Brushwork, primer	1 Pord	13	.615	Ea.	.66	18.20		18.86	29.50	
	0490	Finish coat, enamel		13	.615		.71	18.20		18.91	30	
	0500	Primer & 1 coat enamel		8	1		1.37	29.50		30.87	49	
	0510	Primer & 2 coats enamel		6	1.333		2.08	39.50		41.58	65.50	
	0530	7 to 10 lite										
	0540	Brushwork, primer	1 Pord	11	.727	Ea.	.66	21.50		22.16	35	
	0550	Finish coat, enamel		11	.727		.71	21.50		22.21	35.50	
	0560	Primer & 1 coat enamel		7	1.143		1.37	34		35.37	55.50	
	0570	Primer & 2 coats enamel		5	1.600		2.08	47.50		49.58	78.50	
	0590	12 lite										
	0600	Brushwork, primer	1 Pord	10	.800	Ea.	.66	23.50		24.16	38.50	
	0610	Finish coat, enamel		10	.800		.71	23.50		24.21	39	
	0620	Primer & 1 coat enamel		6	1.333		1.37	39.50		40.87	64.50	
	0630	Primer & 2 coats enamel		5	1.600		2.08	47.50		49.58	78.50	
	0650	For oil base paint, add					10%					
320	0010	**DOORS AND WINDOWS, INTERIOR ALKYD (OIL BASE)**										320
	0500	Flush door & frame, 3′ x 7′, oil, primer, brushwork	1 Pord	10	.800	Ea.	2.09	23.50		25.59	40.50	
	1000	Paint, 1 coat		10	.800		2	23.50		25.50	40	
	1400	Stain, brushwork, wipe off		18	.444		1	13.15		14.15	22	
	1600	Shellac, 1 coat, brushwork		25	.320		1.16	9.45		10.61	16.45	
	1800	Varnish, 3 coats, brushwork, sand after 1st coat		9	.889		3.53	26.50		30.03	46	
	2000	Panel door & frame, 3′ x 7′, oil, primer, brushwork		6	1.333		1.80	39.50		41.30	65	
	2200	Paint, 1 coat		6	1.333		2	39.50		41.50	65	
	2600	Stain, brushwork, panel door, 3′ x 7′, not incl. frame		16	.500		1	14.80		15.80	24.50	
	2800	Shellac, 1 coat, brushwork		22	.364		1.16	10.75		11.91	18.55	
	3000	Varnish, 3 coats, brushwork, sand after 1st coat		7.50	1.067		3.53	31.50		35.03	54.50	
	4400	Windows, including frame and trim, per side										
	4600	Colonial type, 6/6 lites, 2′ x 3′, oil, primer, brushwork	1 Pord	14	.571	Ea.	.28	16.90		17.18	27.50	
	5800	Paint, 1 coat		14	.571		.32	16.90		17.22	27.50	
	6200	3′ x 5′ opening, 6/6 lites, primer coat, brushwork		12	.667		.71	19.75		20.46	32.50	
	6400	Paint, 1 coat		12	.667		.79	19.75		20.54	32.50	
	6800	4′ x 8′ opening, 6/6 lites, primer coat, brushwork		8	1		1.52	29.50		31.02	49	
	7000	Paint, 1 coat		8	1		1.69	29.50		31.19	49.50	
	8000	Single lite type, 2′ x 3′, oil base, primer coat, brushwork		33	.242		.28	7.20		7.48	11.80	
	8200	Paint, 1 coat		33	.242		.32	7.20		7.52	11.85	
	8600	3′ x 5′ opening, primer coat, brushwork		20	.400		.71	11.85		12.56	19.75	
	8800	Paint, 1 coat		20	.400		.79	11.85		12.64	19.80	

Important: See the Reference Section for critical supporting data - Reference Nos., Crews, & City Cost Indexes

	09910 | Paints		CREW	DAILY OUTPUT	LABOR-HOURS	UNIT	2004 BARE COSTS MAT.	LABOR	EQUIP.	TOTAL	TOTAL INCL O&P	
320	9200	4′ x 8′ opening, primer coat, brushwork	1 Pord	14	.571	Ea.	1.52	16.90		18.42	28.50	320
	9400	Paint, 1 coat	↓	14	.571	↓	1.69	16.90		18.59	29	
400	0010	**FENCES**										400
	0100	Chain link or wire metal, one side, water base										
	0110	Roll & brush, first coat	1 Pord	960	.008	S.F.	.06	.25		.31	.45	
	0120	Second coat		1,280	.006		.05	.19		.24	.36	
	0130	Spray, first coat		2,275	.004		.06	.10		.16	.23	
	0140	Second coat	↓	2,600	.003	↓	.06	.09		.15	.21	
	0150	Picket, water base										
	0160	Roll & brush, first coat	1 Pord	865	.009	S.F.	.06	.27		.33	.51	
	0170	Second coat		1,050	.008		.06	.23		.29	.43	
	0180	Spray, first coat		2,275	.004		.06	.10		.16	.24	
	0190	Second coat	↓	2,600	.003	↓	.06	.09		.15	.22	
	0200	Stockade, water base										
	0210	Roll & brush, first coat	1 Pord	1,040	.008	S.F.	.06	.23		.29	.43	
	0220	Second coat		1,200	.007		.06	.20		.26	.39	
	0230	Spray, first coat		2,275	.004		.06	.10		.16	.24	
	0240	Second coat	↓	2,600	.003	↓	.06	.09		.15	.22	
500	0010	**FLOORS, INTERIOR**										500
	0100	Concrete										
	0120	1st coat	1 Pord	975	.008	S.F.	.11	.24		.35	.51	
	0130	2nd coat		1,150	.007		.07	.21		.28	.41	
	0140	3rd coat	↓	1,300	.006	↓	.06	.18		.24	.35	
	0150	Roll, latex, block filler										
	0160	1st coat	1 Pord	2,600	.003	S.F.	.15	.09		.24	.31	
	0170	2nd coat		3,250	.002		.09	.07		.16	.22	
	0180	3rd coat	↓	3,900	.002	↓	.06	.06		.12	.17	
	0190	Spray, latex, block filler										
	0200	1st coat	1 Pord	2,600	.003	S.F.	.13	.09		.22	.29	
	0210	2nd coat		3,250	.002		.07	.07		.14	.20	
	0220	3rd coat	↓	3,900	.002	↓	.05	.06		.11	.16	
620	0010	**MISCELLANEOUS, EXTERIOR**										620
	0015	For painting metals, see Div. 05950-650										
	0100	Railing, ext., decorative wood, incl. cap & baluster										
	0110	newels & spindles @ 12″ O.C.										
	0120	Brushwork, stain, sand, seal & varnish										
	0130	First coat	1 Pord	90	.089	L.F.	.44	2.63		3.07	4.69	
	0140	Second coat	″	120	.067	″	.44	1.97		2.41	3.64	
	0150	Rough sawn wood, 42″ high, 2″x2″ verticals, 6″ O.C.										
	0160	Brushwork, stain, each coat	1 Pord	90	.089	L.F.	.14	2.63		2.77	4.37	
	0170	Wrought iron, 1″ rail, 1/2″ sq. verticals										
	0180	Brushwork, zinc chromate, 60″ high, bars 6″ O.C.										
	0190	Primer	1 Pord	130	.062	L.F.	.49	1.82		2.31	3.46	
	0200	Finish coat		130	.062		.15	1.82		1.97	3.09	
	0210	Additional coat	↓	190	.042	↓	.18	1.25		1.43	2.20	
	0220	Shutters or blinds, single panel, 2′x4′, paint all sides										
	0230	Brushwork, primer	1 Pord	20	.400	Ea.	.55	11.85		12.40	19.55	
	0240	Finish coat, exterior latex		20	.400		.46	11.85		12.31	19.45	
	0250	Primer & 1 coat, exterior latex		13	.615		.89	18.20		19.09	30	
	0260	Spray, primer		35	.229		.80	6.75		7.55	11.75	
	0270	Finish coat, exterior latex		35	.229		.97	6.75		7.72	11.90	
	0280	Primer & 1 coat, exterior latex	↓	20	.400	↓	.87	11.85		12.72	19.90	
	0290	For louvered shutters, add				S.F.	10%					
	0300	Stair stringers, exterior, metal										
	0310	Roll & brush, zinc chromate, to 14″, each coat	1 Pord	320	.025	L.F.	.05	.74		.79	1.25	

	09910 | Paints		CREW	DAILY OUTPUT	LABOR-HOURS	UNIT	2004 BARE COSTS MAT.	LABOR	EQUIP.	TOTAL	TOTAL INCL O&P	
620	0320	Rough sawn wood, 4″ x 12″										620
	0330	Roll & brush, exterior latex, each coat	1 Pord	215	.037	L.F.	.07	1.10		1.17	1.83	
	0340	Trellis/lattice, 2″x2″ @ 3″ O.C. with 2″x8″ supports										
	0350	Spray, latex, per side, each coat	1 Pord	475	.017	S.F.	.07	.50		.57	.87	
	0450	Decking, Ext., sealer, alkyd, brushwork, sealer coat		1,140	.007		.05	.21		.26	.39	
	0460	1st coat		1,140	.007		.05	.21		.26	.39	
	0470	2nd coat		1,300	.006		.04	.18		.22	.33	
	0500	Paint, alkyd, brushwork, primer coat		1,140	.007		.07	.21		.28	.41	
	0510	1st coat		1,140	.007		.07	.21		.28	.40	
	0520	2nd coat		1,300	.006		.05	.18		.23	.34	
	0600	Sand paint, alkyd, brushwork, 1 coat		150	.053		.09	1.58		1.67	2.62	
630	0010	**MISCELLANEOUS, INTERIOR**										630
	2400	Floors, conc./wood, oil base, primer/sealer coat, brushwork	2 Pord	1,950	.008	S.F.	.06	.24		.30	.45	
	2450	Roller		5,200	.003		.06	.09		.15	.22	
	2600	Spray		6,000	.003		.06	.08		.14	.20	
	2650	Paint 1 coat, brushwork		1,950	.008		.05	.24		.29	.45	
	2800	Roller		5,200	.003		.05	.09		.14	.21	
	2850	Spray		6,000	.003		.06	.08		.14	.19	
	3000	Stain, wood floor, brushwork, 1 coat		4,550	.004		.05	.10		.15	.22	
	3200	Roller		5,200	.003		.05	.09		.14	.21	
	3250	Spray		6,000	.003		.05	.08		.13	.18	
	3400	Varnish, wood floor, brushwork		4,550	.004		.06	.10		.16	.23	
	3450	Roller		5,200	.003		.06	.09		.15	.21	
	3600	Spray		6,000	.003		.06	.08		.14	.20	
	3650	For dust proofing or anti skid, see division 03350-300										
	3800	Grilles, per side, oil base, primer coat, brushwork	1 Pord	520	.015	S.F.	.09	.46		.55	.83	
	3850	Spray		1,140	.007		.10	.21		.31	.44	
	3920	Paint 2 coats, brushwork		325	.025		.20	.73		.93	1.40	
	3940	Spray		650	.012		.23	.36		.59	.84	
	5000	Pipe, to 4″ diameter, primer or sealer coat, oil base, brushwork	2 Pord	1,250	.013	L.F.	.06	.38		.44	.68	
	5100	Spray		2,165	.007		.06	.22		.28	.41	
	5350	Paint 2 coats, brushwork		775	.021		.11	.61		.72	1.10	
	5400	Spray		1,240	.013		.12	.38		.50	.75	
	6300	To 16″ diameter, primer or sealer coat, brushwork		310	.052		.24	1.53		1.77	2.71	
	6450	Spray		540	.030		.26	.88		1.14	1.68	
	6500	Paint 2 coats, brushwork		195	.082		.45	2.43		2.88	4.38	
	6550	Spray		310	.052		.50	1.53		2.03	3	
	7000	Trim, wood, incl. puttying, under 6″ wide										
	7200	Primer coat, oil base, brushwork	1 Pord	650	.012	L.F.	.02	.36		.38	.61	
	7250	Paint, 1 coat, brushwork		650	.012		.03	.36		.39	.61	
	7450	3 coats		325	.025		.08	.73		.81	1.25	
	7500	Over 6″ wide, primer coat, brushwork		650	.012		.05	.36		.41	.63	
	7550	Paint, 1 coat, brushwork		650	.012		.05	.36		.41	.64	
	7650	3 coats		325	.025		.15	.73		.88	1.34	
	8000	Cornice, simple design, primer coat, oil base, brushwork		650	.012	S.F.	.05	.36		.41	.63	
	8250	Paint, 1 coat		650	.012		.05	.36		.41	.64	
	8350	Ornate design, primer coat		350	.023		.05	.68		.73	1.13	
	8400	Paint, 1 coat		350	.023		.05	.68		.73	1.14	
	8600	Balustrades, primer coat, oil base, brushwork		520	.015		.05	.46		.51	.78	
	8800											
	8900	Trusses and wood frames, primer coat, oil base, brushwork	1 Pord	800	.010	S.F.	.05	.30		.35	.52	
	8950	Spray		1,200	.007		.05	.20		.25	.37	
	9220	Paint 2 coats, brushwork		500	.016		.10	.47		.57	.87	
	9240	Spray		600	.013		.11	.39		.50	.76	
	9260	Stain, brushwork, wipe off		600	.013		.05	.39		.44	.68	

Important: See the Reference Section for critical supporting data - Reference Nos., Crews, & City Cost Indexes

		09910 | Paints	CREW	DAILY OUTPUT	LABOR-HOURS	UNIT	2004 BARE COSTS MAT.	LABOR	EQUIP.	TOTAL	TOTAL INCL O&P	
630	9280	Varnish, 3 coats, brushwork	1 Pord	275	.029	S.F.	.17	.86		1.03	1.57	630
	9350	For latex paint, deduct					10%					
700	0010	**SIDING EXTERIOR**, Alkyd (oil base)										700
	0450	Steel siding, oil base, paint 1 coat, brushwork	2 Pord	2,015	.008	S.F.	.05	.24		.29	.44	
	0500	Spray		4,550	.004		.08	.10		.18	.26	
	0800	Paint 2 coats, brushwork		1,300	.012		.11	.36		.47	.70	
	1000	Spray		4,550	.004		.14	.10		.24	.32	
	1200	Stucco, rough, oil base, paint 2 coats, brushwork		1,300	.012		.11	.36		.47	.70	
	1400	Roller		1,625	.010		.11	.29		.40	.60	
	1600	Spray		2,925	.005		.12	.16		.28	.39	
	1800	Texture 1-11 or clapboard, oil base, primer coat, brushwork		1,300	.012		.09	.36		.45	.68	
	2000	Spray		4,550	.004		.09	.10		.19	.27	
	2400	Paint 2 coats, brushwork		810	.020		.16	.58		.74	1.11	
	2600	Spray		2,600	.006		.18	.18		.36	.48	
	3400	Stain 2 coats, brushwork		950	.017		.10	.50		.60	.90	
	4000	Spray		3,050	.005		.11	.16		.27	.37	
	4200	Wood shingles, oil base primer coat, brushwork		1,300	.012		.08	.36		.44	.67	
	4400	Spray		3,900	.004		.08	.12		.20	.28	
	5000	Paint 2 coats, brushwork		810	.020		.13	.58		.71	1.08	
	5200	Spray		2,275	.007		.12	.21		.33	.47	
	6500	Stain 2 coats, brushwork		950	.017		.10	.50		.60	.90	
	7000	Spray		2,660	.006		.13	.18		.31	.43	
	8000	For latex paint, deduct					10%					
	8100	For work over 12′ H, from pipe scaffolding, add						15%				
	8200	For work over 12′ H, from extension ladder, add						25%				
	8300	For work over 12′ H, from swing staging, add						35%				
710	0010	**SIDING, MISC.**										710
	0100	Aluminum siding										
	0110	Brushwork, primer	2 Pord	2,275	.007	S.F.	.05	.21		.26	.39	
	0120	Finish coat, exterior latex		2,275	.007		.04	.21		.25	.37	
	0130	Primer & 1 coat exterior latex		1,300	.012		.10	.36		.46	.69	
	0140	Primer & 2 coats exterior latex		975	.016		.13	.49		.62	.93	
	0150	Mineral Fiber shingles										
	0160	Brushwork, primer	2 Pord	1,495	.011	S.F.	.09	.32		.41	.61	
	0170	Finish coat, industrial enamel		1,495	.011		.10	.32		.42	.62	
	0180	Primer & 1 coat enamel		810	.020		.19	.58		.77	1.15	
	0190	Primer & 2 coats enamel		540	.030		.29	.88		1.17	1.72	
	0200	Roll, primer		1,625	.010		.10	.29		.39	.58	
	0210	Finish coat, industrial enamel		1,625	.010		.11	.29		.40	.59	
	0220	Primer & 1 coat enamel		975	.016		.21	.49		.70	1.01	
	0230	Primer & 2 coats enamel		650	.025		.32	.73		1.05	1.52	
	0240	Spray, primer		3,900	.004		.08	.12		.20	.28	
	0250	Finish coat, industrial enamel		3,900	.004		.09	.12		.21	.29	
	0260	Primer & 1 coat enamel		2,275	.007		.17	.21		.38	.51	
	0270	Primer & 2 coats enamel		1,625	.010		.26	.29		.55	.75	
	0280	Waterproof sealer, first coat		4,485	.004		.06	.11		.17	.24	
	0290	Second coat		5,235	.003		.06	.09		.15	.22	
	0300	Rough wood incl. shingles, shakes or rough sawn siding										
	0310	Brushwork, primer	2 Pord	1,280	.013	S.F.	.11	.37		.48	.71	
	0320	Finish coat, exterior latex		1,280	.013		.07	.37		.44	.67	
	0330	Primer & 1 coat exterior latex		960	.017		.18	.49		.67	.99	
	0340	Primer & 2 coats exterior latex		700	.023		.25	.68		.93	1.35	
	0350	Roll, primer		2,925	.005		.15	.16		.31	.42	
	0360	Finish coat, exterior latex		2,925	.005		.08	.16		.24	.35	
	0370	Primer & 1 coat exterior latex		1,790	.009		.23	.26		.49	.67	
	0380	Primer & 2 coats exterior latex		1,300	.012		.31	.36		.67	.92	

	09910 | Paints	CREW	DAILY OUTPUT	LABOR-HOURS	UNIT	2004 BARE COSTS MAT.	LABOR	EQUIP.	TOTAL	TOTAL INCL O&P	
710	0390 Spray, primer	2 Pord	3,900	.004	S.F.	.13	.12		.25	.33	710
	0400 Finish coat, exterior latex		3,900	.004		.06	.12		.18	.26	
	0410 Primer & 1 coat exterior latex		2,600	.006		.19	.18		.37	.50	
	0420 Primer & 2 coats exterior latex		2,080	.008		.25	.23		.48	.64	
	0430 Waterproof sealer, first coat		4,485	.004		.12	.11		.23	.30	
	0440 Second coat	↓	4,485	.004	↓	.06	.11		.17	.24	
	0450 Smooth wood incl. butt, T&G, beveled, drop or B&B siding										
	0460 Brushwork, primer	2 Pord	2,325	.007	S.F.	.08	.20		.28	.42	
	0470 Finish coat, exterior latex		1,280	.013		.07	.37		.44	.67	
	0480 Primer & 1 coat exterior latex		800	.020		.15	.59		.74	1.11	
	0490 Primer & 2 coats exterior latex		630	.025		.22	.75		.97	1.44	
	0500 Roll, primer		2,275	.007		.09	.21		.30	.43	
	0510 Finish coat, exterior latex		2,275	.007		.07	.21		.28	.41	
	0520 Primer & 1 coat exterior latex		1,300	.012		.16	.36		.52	.76	
	0530 Primer & 2 coats exterior latex		975	.016		.24	.49		.73	1.04	
	0540 Spray, primer		4,550	.004		.07	.10		.17	.25	
	0550 Finish coat, exterior latex		4,550	.004		.06	.10		.16	.24	
	0560 Primer & 1 coat exterior latex		2,600	.006		.13	.18		.31	.44	
	0570 Primer & 2 coats exterior latex		1,950	.008		.20	.24		.44	.61	
	0580 Waterproof sealer, first coat		5,230	.003		.06	.09		.15	.22	
	0590 Second coat	↓	5,980	.003		.06	.08		.14	.20	
	0600 For oil base paint, add				↓	10%					
800	0010 **TRIM, EXTERIOR**										800
	0100 Door frames & trim (see Doors, interior or exterior)										
	0110 Fascia, latex paint, one coat coverage										
	0120 1" x 4", brushwork	1 Pord	640	.013	L.F.	.02	.37		.39	.61	
	0130 Roll		1,280	.006		.02	.19		.21	.32	
	0140 Spray		2,080	.004		.01	.11		.12	.20	
	0150 1" x 6" to 1" x 10", brushwork		640	.013		.06	.37		.43	.66	
	0160 Roll		1,230	.007		.06	.19		.25	.38	
	0170 Spray		2,100	.004		.05	.11		.16	.23	
	0180 1" x 12", brushwork		640	.013		.06	.37		.43	.66	
	0190 Roll		1,050	.008		.06	.23		.29	.43	
	0200 Spray	↓	2,200	.004	↓	.05	.11		.16	.22	
	0210 Gutters & downspouts, metal, zinc chromate paint										
	0220 Brushwork, gutters, 5", first coat	1 Pord	640	.013	L.F.	.05	.37		.42	.65	
	0230 Second coat		960	.008		.05	.25		.30	.45	
	0240 Third coat		1,280	.006		.04	.19		.23	.35	
	0250 Downspouts, 4", first coat		640	.013		.05	.37		.42	.65	
	0260 Second coat		960	.008		.05	.25		.30	.45	
	0270 Third coat	↓	1,280	.006	↓	.04	.19		.23	.35	
	0280 Gutters & downspouts, wood										
	0290 Brushwork, gutters, 5", primer	1 Pord	640	.013	L.F.	.05	.37		.42	.65	
	0300 Finish coat, exterior latex		640	.013		.05	.37		.42	.65	
	0310 Primer & 1 coat exterior latex		400	.020		.11	.59		.70	1.07	
	0320 Primer & 2 coats exterior latex		325	.025		.16	.73		.89	1.35	
	0330 Downspouts, 4", primer		640	.013		.05	.37		.42	.65	
	0340 Finish coat, exterior latex		640	.013		.05	.37		.42	.65	
	0350 Primer & 1 coat exterior latex		400	.020		.11	.59		.70	1.07	
	0360 Primer & 2 coats exterior latex	↓	325	.025	↓	.08	.73		.81	1.26	
	0370 Molding, exterior, up to 14" wide										
	0380 Brushwork, primer	1 Pord	640	.013	L.F.	.06	.37		.43	.66	
	0390 Finish coat, exterior latex		640	.013		.06	.37		.43	.66	
	0400 Primer & 1 coat exterior latex		400	.020		.13	.59		.72	1.09	
	0410 Primer & 2 coats exterior latex		315	.025		.13	.75		.88	1.34	
	0420 Stain & fill	↓	1,050	.008	↓	.06	.23		.29	.42	

		09910 | Paints	CREW	DAILY OUTPUT	LABOR-HOURS	UNIT	2004 BARE COSTS MAT.	LABOR	EQUIP.	TOTAL	TOTAL INCL O&P	
800	0430	Shellac	1 Pord	1,850	.004	L.F.	.07	.13		.20	.27	800
	0440	Varnish		1,275	.006		.07	.19		.26	.37	
910	0350	**WALLS, MASONRY (CMU), EXTERIOR**										910
	0360	Concrete masonry units (CMU), smooth surface										
	0370	Brushwork, latex, first coat	1 Pord	640	.013	S.F.	.03	.37		.40	.63	
	0380	Second coat		960	.008		.03	.25		.28	.42	
	0390	Waterproof sealer, first coat		736	.011		.07	.32		.39	.60	
	0400	Second coat		1,104	.007		.05	.21		.26	.39	
	0410	Roll, latex, paint, first coat		1,465	.005		.04	.16		.20	.31	
	0420	Second coat		1,790	.004		.03	.13		.16	.24	
	0430	Waterproof sealer, first coat		1,680	.005		.07	.14		.21	.31	
	0440	Second coat		2,060	.004		.05	.11		.16	.23	
	0450	Spray, latex, paint, first coat		1,950	.004		.03	.12		.15	.23	
	0460	Second coat		2,600	.003		.03	.09		.12	.18	
	0470	Waterproof sealer, first coat		2,245	.004		.09	.11		.20	.26	
	0480	Second coat		2,990	.003		.03	.08		.11	.17	
	0490	Concrete masonry unit (CMU), porous										
	0500	Brushwork, latex, first coat	1 Pord	640	.013	S.F.	.07	.37		.44	.67	
	0510	Second coat		960	.008		.03	.25		.28	.43	
	0520	Waterproof sealer, first coat		736	.011		.09	.32		.41	.61	
	0530	Second coat		1,104	.007		.04	.21		.25	.39	
	0540	Roll latex, first coat		1,465	.005		.05	.16		.21	.32	
	0550	Second coat		1,790	.004		.03	.13		.16	.25	
	0560	Waterproof sealer, first coat		1,680	.005		.09	.14		.23	.32	
	0570	Second coat		2,060	.004		.05	.11		.16	.23	
	0580	Spray latex, first coat		1,950	.004		.04	.12		.16	.23	
	0590	Second coat		2,600	.003		.03	.09		.12	.18	
	0600	Waterproof sealer, first coat		2,245	.004		.07	.11		.18	.25	
	0610	Second coat		2,990	.003		.04	.08		.12	.17	
920	0010	**WALLS AND CEILINGS**, Interior										920
	0100	Concrete, dry wall or plaster, oil base, primer or sealer coat										
	0200	Smooth finish, brushwork	1 Pord	1,150	.007	S.F.	.05	.21		.26	.38	
	0240	Roller		2,040	.004		.04	.12		.16	.24	
	0300	Sand finish, brushwork		975	.008		.04	.24		.28	.44	
	0340	Roller		1,150	.007		.05	.21		.26	.38	
	0380	Spray		2,275	.004		.04	.10		.14	.21	
	0800	Paint 2 coats, smooth finish, brushwork		680	.012		.09	.35		.44	.66	
	0840	Roller		800	.010		.09	.30		.39	.57	
	0880	Spray		1,625	.005		.08	.15		.23	.32	
	0900	Sand finish, brushwork		605	.013		.09	.39		.48	.73	
	0940	Roller		1,020	.008		.09	.23		.32	.47	
	0980	Spray		1,700	.005		.08	.14		.22	.31	
	1600	Glaze coating, 5 coats, spray, clear		900	.009		.60	.26		.86	1.08	
	1640	Multicolor		900	.009		.83	.26		1.09	1.33	
	1700	For latex paint, deduct					10%					
	1800	For ceiling installations, add						25%				
	2000	Masonry or concrete block, oil base, primer or sealer coat										
	2100	Smooth finish, brushwork	1 Pord	1,224	.007	S.F.	.05	.19		.24	.36	
	2180	Spray		2,400	.003		.07	.10		.17	.23	
	2200	Sand finish, brushwork		1,089	.007		.07	.22		.29	.43	
	2280	Spray		2,400	.003		.07	.10		.17	.23	
	2800	Paint 2 coats, smooth finish, brushwork		756	.011		.15	.31		.46	.66	
	2880	Spray		1,360	.006		.14	.17		.31	.43	
	2900	Sand finish, brushwork		672	.012		.15	.35		.50	.72	
	2980	Spray		1,360	.006		.14	.17		.31	.43	

	09910 Paints		CREW	DAILY OUTPUT	LABOR-HOURS	UNIT	2004 BARE COSTS MAT.	LABOR	EQUIP.	TOTAL	TOTAL INCL O&P	
920	3600	Glaze coating, 5 coats, spray, clear	1 Pord	900	.009	S.F.	.60	.26		.86	1.08	920
	3620	Multicolor		900	.009		.83	.26		1.09	1.33	
	4000	Block filler, 1 coat, brushwork		425	.019		.12	.56		.68	1.02	
	4100	Silicone, water repellent, 2 coats, spray		2,000	.004		.25	.12		.37	.46	
	4120	For latex paint, deduct					10%					
	8200	For work 8 - 15′ H, add						10%				
	8300	For work over 15′ H, add						20%				
940	0010	**DRY FALL PAINTING** R09910-220										940
	0100	Walls										
	0200	Wallboard and smooth plaster, one coat, brush	1 Pord	910	.009	S.F.	.04	.26		.30	.47	
	0210	Roll		1,560	.005		.04	.15		.19	.29	
	0220	Spray		2,600	.003		.04	.09		.13	.20	
	0230	Two coats, brush		520	.015		.08	.46		.54	.82	
	0240	Roll		877	.009		.08	.27		.35	.52	
	0250	Spray		1,560	.005		.08	.15		.23	.33	
	0260	Concrete or textured plaster, one coat, brush		747	.011		.04	.32		.36	.56	
	0270	Roll		1,300	.006		.04	.18		.22	.34	
	0280	Spray		1,560	.005		.04	.15		.19	.29	
	0290	Two coats, brush		422	.019		.08	.56		.64	.99	
	0300	Roll		747	.011		.08	.32		.40	.60	
	0310	Spray		1,300	.006		.08	.18		.26	.38	
	0320	Concrete block, one coat, brush		747	.011		.04	.32		.36	.56	
	0330	Roll		1,300	.006		.04	.18		.22	.34	
	0340	Spray		1,560	.005		.04	.15		.19	.29	
	0350	Two coats, brush		422	.019		.08	.56		.64	.99	
	0360	Roll		747	.011		.08	.32		.40	.60	
	0370	Spray		1,300	.006		.08	.18		.26	.38	
	0380	Wood, one coat, brush		747	.011		.04	.32		.36	.56	
	0390	Roll		1,300	.006		.04	.18		.22	.34	
	0400	Spray		877	.009		.04	.27		.31	.48	
	0410	Two coats, brush		487	.016		.08	.49		.57	.87	
	0420	Roll		747	.011		.08	.32		.40	.60	
	0430	Spray		650	.012		.08	.36		.44	.67	
	0440	Ceilings										
	0450	Wallboard and smooth plaster, one coat, brush	1 Pord	600	.013	S.F.	.04	.39		.43	.68	
	0460	Roll		1,040	.008		.04	.23		.27	.41	
	0470	Spray		1,560	.005		.04	.15		.19	.29	
	0480	Two coats, brush		341	.023		.08	.69		.77	1.20	
	0490	Roll		650	.012		.08	.36		.44	.67	
	0500	Spray		1,300	.006		.08	.18		.26	.38	
	0510	Concrete or textured plaster, one coat, brush		487	.016		.04	.49		.53	.83	
	0520	Roll		877	.009		.04	.27		.31	.48	
	0530	Spray		1,560	.005		.04	.15		.19	.29	
	0540	Two coats, brush		276	.029		.08	.86		.94	1.46	
	0550	Roll		520	.015		.08	.46		.54	.82	
	0560	Spray		1,300	.006		.08	.18		.26	.38	
	0570	Structural steel, bar joists or metal deck, one coat, spray		1,560	.005		.04	.15		.19	.29	
	0580	Two coats, spray		1,040	.008		.08	.23		.31	.45	
	09930 Stains/Transp. Finishes											
100	0010	**VARNISH** 1 coat + sealer, on wood trim, no sanding included	1 Pord	400	.020	S.F.	.06	.59		.65	1.02	100
	0100	Hardwood floors, 2 coats, no sanding included, roller		1,890	.004	"	.12	.13		.25	.33	
	9000	Minimum labor/equipment charge		4	2	Job		59		59	95	

Important: See the Reference Section for critical supporting data - Reference Nos., Crews, & City Cost Indexes

		09963 Glazed Coatings	CREW	DAILY OUTPUT	LABOR-HOURS	UNIT	2004 BARE COSTS MAT.	LABOR	EQUIP.	TOTAL	TOTAL INCL O&P	
200	0010	**WALL COATINGS**										200
	0100	Acrylic glazed coatings, minimum	1 Pord	525	.015	S.F.	.25	.45		.70	1	
	0200	Maximum		305	.026		.52	.78		1.30	1.81	
	0300	Epoxy coatings, minimum		525	.015		.32	.45		.77	1.07	
	0400	Maximum		170	.047		.99	1.39		2.38	3.32	
	2400	Sprayed perlite or vermiculite, 1/16″ thick, minimum		2,935	.003		.21	.08		.29	.36	
	2500	Maximum		640	.013		.60	.37		.97	1.25	
	2700	Vinyl plastic wall coating, minimum		735	.011		.27	.32		.59	.82	
	2800	Maximum		240	.033		.66	.99		1.65	2.31	
	3000	Urethane on smooth surface, 2 coats, minimum		1,135	.007		.20	.21		.41	.55	
	3100	Maximum		665	.012		.45	.36		.81	1.07	

		09990 Paint Restoration	CREW	DAILY OUTPUT	LABOR-HOURS	UNIT	MAT.	LABOR	EQUIP.	TOTAL	TOTAL INCL O&P	
500	0010	**SCRAPE AFTER FIRE DAMAGE**										500
	0050	Boards, 1″ x 4″	1 Pord	336	.024	L.F.		.70		.70	1.13	
	0060	1″ x 6″		260	.031			.91		.91	1.46	
	0070	1″ x 8″		207	.039			1.14		1.14	1.83	
	0080	1″ x 10″		174	.046			1.36		1.36	2.18	
	0500	Framing, 2″ x 4″		265	.030			.89		.89	1.43	
	0510	2″ x 6″		221	.036			1.07		1.07	1.72	
	0520	2″ x 8″		190	.042			1.25		1.25	2	
	0530	2″ x 10″		165	.048			1.44		1.44	2.30	
	0540	2″ x 12″		144	.056			1.64		1.64	2.63	
	1000	Heavy framing, 3″ x 4″		226	.035			1.05		1.05	1.68	
	1010	4″ x 4″		210	.038			1.13		1.13	1.81	
	1020	4″ x 6″		191	.042			1.24		1.24	1.99	
	1030	4″ x 8″		165	.048			1.44		1.44	2.30	
	1040	4″ x 10″		144	.056			1.64		1.64	2.63	
	1060	4″ x 12″		131	.061			1.81		1.81	2.89	
	2900	For sealing, minimum		825	.010	S.F.	.12	.29		.41	.59	
	2920	Maximum		460	.017	″	.25	.51		.76	1.10	
	3000	For sandblasting, see division 04930-220										
	3020											
	9000	Minimum labor/equipment charge	1 Pord	3	2.667	Job		79		79	126	
800	0010	**SANDING** and puttying interior trim, compared to										800
	0100	Painting 1 coat, on quality work				L.F.		100%				
	0300	Medium work						50%				
	0400	Industrial grade						25%				
	0500	Surface protection, placement and removal										
	0510	Basic drop cloths	1 Pord	6,400	.001	S.F.		.04		.04	.06	
	0520	Masking with paper		800	.010		.03	.30		.33	.50	
	0530	Volume cover up (using plastic sheathing, or building paper)		16,000	.001			.01		.01	.02	
900	0010	**SURFACE PREPARATION, EXTERIOR**										900
	0015	Doors, per side, not incl. frames or trim										
	0020	Scrape & sand										
	0030	Wood, flush	1 Pord	616	.013	S.F.		.38		.38	.62	
	0040	Wood, detail		496	.016			.48		.48	.76	
	0050	Wood, louvered		280	.029			.85		.85	1.35	
	0060	Wood, overhead		616	.013			.38		.38	.62	
	0070	Wire brush										
	0080	Metal, flush	1 Pord	640	.013	S.F.		.37		.37	.59	
	0090	Metal, detail		520	.015			.46		.46	.73	
	0100	Metal, louvered		360	.022			.66		.66	1.05	
	0110	Metal or fibr., overhead		640	.013			.37		.37	.59	

		09990 | Paint Restoration	CREW	DAILY OUTPUT	LABOR-HOURS	UNIT	2004 BARE COSTS				TOTAL
							MAT.	LABOR	EQUIP.	TOTAL	INCL O&P
900	0120	Metal, roll up	1 Pord	560	.014	S.F.		.42		.42	.68
	0130	Metal, bulkhead		640	.013			.37		.37	.59
	0140	Power wash, based on 2500 lb. operating pressure									
	0150	Metal, flush	B-9	2,240	.018	S.F.		.47	.07	.54	.86
	0160	Metal, detail		2,120	.019			.50	.08	.58	.91
	0170	Metal, louvered		2,000	.020			.53	.08	.61	.97
	0180	Metal or fibr., overhead		2,400	.017			.44	.07	.51	.80
	0190	Metal, roll up		2,400	.017			.44	.07	.51	.80
	0200	Metal, bulkhead		2,200	.018			.48	.07	.55	.88
	0400	Windows, per side, not incl. trim									
	0410	Scrape & sand									
	0420	Wood, 1-2 lite	1 Pord	320	.025	S.F.		.74		.74	1.19
	0430	Wood, 3-6 lite		280	.029			.85		.85	1.35
	0440	Wood, 7-10 lite		240	.033			.99		.99	1.58
	0450	Wood, 12 lite		200	.040			1.18		1.18	1.90
	0460	Wood, Bay / Bow		320	.025			.74		.74	1.19
	0470	Wire brush									
	0480	Metal, 1-2 lite	1 Pord	480	.017	S.F.		.49		.49	.79
	0490	Metal, 3-6 lite		400	.020			.59		.59	.95
	0500	Metal, Bay / Bow		480	.017			.49		.49	.79
	0510	Power wash, based on 2500 lb. operating pressure									
	0520	1-2 lite	B-9	4,400	.009	S.F.		.24	.04	.28	.44
	0530	3-6 lite		4,320	.009			.24	.04	.28	.45
	0540	7-10 lite		4,240	.009			.25	.04	.29	.45
	0550	12 lite		4,160	.010			.25	.04	.29	.46
	0560	Bay / Bow		4,400	.009			.24	.04	.28	.44
	0600	Siding, scrape and sand, light=10-30%, med.=30-70%									
	0610	Heavy=70-100%, % of surface to sand									
	0650	Texture 1-11, light	1 Pord	480	.017	S.F.		.49		.49	.79
	0660	Med.		440	.018			.54		.54	.86
	0670	Heavy		360	.022			.66		.66	1.05
	0680	Wood shingles, shakes, light		440	.018			.54		.54	.86
	0690	Med.		360	.022			.66		.66	1.05
	0700	Heavy		280	.029			.85		.85	1.35
	0710	Clapboard, light		520	.015			.46		.46	.73
	0720	Med.		480	.017			.49		.49	.79
	0730	Heavy		400	.020			.59		.59	.95
	0740	Wire brush									
	0750	Aluminum, light	1 Pord	600	.013	S.F.		.39		.39	.63
	0760	Med.		520	.015			.46		.46	.73
	0770	Heavy		440	.018			.54		.54	.86
	0780	Pressure wash, based on 2500 lb.. operating pressure									
	0790	Stucco	B-9	3,080	.013	S.F.		.34	.05	.39	.63
	0800	Aluminum or vinyl		3,200	.013			.33	.05	.38	.61
	0810	Siding, masonry, brick & block		2,400	.017			.44	.07	.51	.80
	1300	Miscellaneous, wire brush									
	1310	Metal, pedestrian gate	1 Pord	100	.080	S.F.		2.37		2.37	3.79
	8000	For Chemical Washing, see Division 04930									
	8010	For Steam Cleaning, see Division 04930									
910	0010	**SURFACE PREPARATION, INTERIOR**									
	0020	Doors									
	0030	Scrape & sand									
	0040	Wood, flush	1 Pord	616	.013	S.F.		.38		.38	.62
	0050	Wood, detail		496	.016			.48		.48	.76
	0060	Wood, louvered		280	.029			.85		.85	1.35

	09990 | Paint Restoration		CREW	DAILY OUTPUT	LABOR-HOURS	UNIT	2004 BARE COSTS MAT.	LABOR	EQUIP.	TOTAL	TOTAL INCL O&P	
910	0070	Wire brush										910
	0080	Metal, flush	1 Pord	640	.013	S.F.		.37		.37	.59	
	0090	Metal, detail		520	.015			.46		.46	.73	
	0100	Metal, louvered		360	.022			.66		.66	1.05	
	0110	Hand wash										
	0120	Wood, flush	1 Pord	2,160	.004	S.F.		.11		.11	.18	
	0130	Wood, detailed		2,000	.004			.12		.12	.19	
	0140	Wood, louvered		1,360	.006			.17		.17	.28	
	0150	Metal, flush		2,160	.004			.11		.11	.18	
	0160	Metal, detail		2,000	.004			.12		.12	.19	
	0170	Metal, louvered		1,360	.006			.17		.17	.28	
	0400	Windows, per side, not incl. trim										
	0410	Scrape & sand										
	0420	Wood, 1-2 lite	1 Pord	360	.022	S.F.		.66		.66	1.05	
	0430	Wood, 3-6 lite		320	.025			.74		.74	1.19	
	0440	Wood, 7-10 lite		280	.029			.85		.85	1.35	
	0450	Wood, 12 lite		240	.033			.99		.99	1.58	
	0460	Wood, Bay / Bow		360	.022			.66		.66	1.05	
	0470	Wire brush										
	0480	Metal, 1-2 lite	1 Pord	520	.015	S.F.		.46		.46	.73	
	0490	Metal, 3-6 lite		440	.018			.54		.54	.86	
	0500	Metal, Bay / Bow		520	.015			.46		.46	.73	
	0600	Walls, sanding, light=10-30%										
	0610	Med.=30-70%, heavy=70-100%, % of surface to sand										
	0650	Walls, sand										
	0660	Drywall, gypsum, plaster, light	1 Pord	3,077	.003	S.F.		.08		.08	.12	
	0670	Drywall, gypsum, plaster, med.		2,160	.004			.11		.11	.18	
	0680	Drywall, gypsum, plaster, heavy		923	.009			.26		.26	.41	
	0690	Wood, T&G, light		2,400	.003			.10		.10	.16	
	0700	Wood, T&G, med.		1,600	.005			.15		.15	.24	
	0710	Wood, T&G, heavy		800	.010			.30		.30	.47	
	0720	Walls, wash										
	0730	Drywall, gypsum, plaster	1 Pord	3,200	.002	S.F.		.07		.07	.12	
	0740	Wood, T&G		3,200	.002			.07		.07	.12	
	0750	Masonry, brick & block, smooth		2,800	.003			.08		.08	.14	
	0760	Masonry, brick & block, coarse		2,000	.004			.12		.12	.19	
	8000	For Chemical Washing, see Division 04930										
	8010	For Steam Cleaning, see Division 04930										
	8020	For Sand Blasting, see Division 03350-350 and 05910-500										

For information about Means Estimating Seminars, see yellow pages 12 and 13 in back of book

Division Notes

		CREW	DAILY OUTPUT	LABOR-HOURS	UNIT	2004 BARE COSTS MAT.	LABOR	EQUIP.	TOTAL	TOTAL INCL O&P

Division 10 Specialties

Estimating Tips

General

- The items in this division are usually priced per square foot or each.
- Many items in Division 10 require some type of support system or special anchors that are not usually furnished with the item. The required anchors must be added to the estimate in the appropriate division.
- Some items in Division 10, such as lockers, may require assembly before installation. Verify the amount of assembly required. Assembly can often exceed installation time.

10150 Compartments & Cubicles

- Support angles and blocking are not included in the installation of toilet compartments, shower/dressing compartments or cubicles. Appropriate line items from Divisions 5 or 6 may need to be added to support the installations.
- Toilet partitions are priced by the stall. A stall consists of a side wall, pilaster and door with hardware. Toilet tissue holders and grab bars are extra.

10600 Partitions

- The required acoustical rating of a folding partition can have a significant impact on costs. Verify the sound transmission coefficient rating of the panel priced to the specification requirements.

10800 Toilet/Bath/Laundry Accessories

- Grab bar installation does not include supplemental blocking or backing to support the required load. When grab bars are installed at an existing facility provisions must be made to attach the grab bars to solid structure.

Reference Numbers

Reference numbers are shown in bold squares at the beginning of some major classifications. These numbers refer to related items in the Reference Section. The reference information may be an estimating procedure, an alternate pricing method or technical information.

Note: Not all subdivisions listed here necessarily appear in this publication.

10100 | Visual Display Boards

		10110 | Chalkboards	CREW	DAILY OUTPUT	LABOR-HOURS	UNIT	2004 BARE COSTS MAT.	LABOR	EQUIP.	TOTAL	TOTAL INCL O&P	
240	0010	**CHALKBOARDS** Porcelain enamel steel										240
	0100	Freestanding, reversible										
	0120	Economy, wood frame, 4′ x 6′										
	0140	Chalkboard both sides				Ea.	315			315	345	
	0160	Chalkboard one side, cork other side				"	445			445	490	
	0200	Standard, lightweight satin finished aluminum, 4′ x 6′										
	0220	Chalkboard both sides				Ea.	405			405	445	
	0240	Chalkboard one side, cork other side				"	480			480	530	
	0300	Deluxe, heavy duty extruded aluminum, 4′ x 6′										
	0320	Chalkboard both sides				Ea.	1,150			1,150	1,275	
	0340	Chalkboard one side, cork other side				"	1,150			1,150	1,275	
	3900	Wall hung										
	4000	Aluminum frame and chalktrough										
	4300	3′ x 5′	2 Carp	15	1.067	Ea.	211	35		246	291	
	4600	4′ x 12′	"	13	1.231	"	465	40.50		505.50	580	
	4700	Wood frame and chalktrough										
	4800	3′ x 4′	2 Carp	16	1	Ea.	102	33		135	167	
	5300	4′ x 8′	"	13	1.231	"	158	40.50		198.50	242	
	5400	Liquid chalk, white porcelain enamel, wall hung										
	5420	Deluxe units, aluminum trim and chalktrough										
	5450	4′ x 4′	2 Carp	16	1	Ea.	197	33		230	271	
	5550	4′ x 12′	"	12	1.333	"	480	44		524	600	
	5700	Wood trim and chalktrough										
	5900	4′ x 4′	2 Carp	16	1	Ea.	430	33		463	530	
	6200	4′ x 8′	↓	14	1.143	"	600	37.50		637.50	725	
	9000	Minimum labor/equipment charge	↓	3	5.333	Job		176		176	292	

		10120 | Tack & Visual Aid Boards	CREW	DAILY OUTPUT	LABOR-HOURS	UNIT	MAT.	LABOR	EQUIP.	TOTAL	TOTAL INCL O&P	
940	0011	**BULLETIN BOARD**										940
	2120	Prefabricated, 1/4″ cork, 3′ x 5′ with aluminum frame	2 Carp	16	1	Ea.	110	33		143	176	
	2140	Wood frame	"	16	1	"	142	33		175	212	
	2300	Glass enclosed cabinets, alum., cork panel, hinged doors										
	2600	4′ x 7′, 3 door	2 Carp	10	1.600	Ea.	1,350	53		1,403	1,600	
	9000	Minimum labor/equipment charge	"	4	4	Job		132		132	219	

10150 | Compartments & Cubicles

		10155 | Metal Toilet Compartments	CREW	DAILY OUTPUT	LABOR-HOURS	UNIT	2004 BARE COSTS MAT.	LABOR	EQUIP.	TOTAL	TOTAL INCL O&P	
100	0010	**TOILET PARTITIONS, METAL**										100
	0110	Cubicles, ceiling hung										
	0200	Painted metal	2 Carp	4	4	Ea.	430	132		562	690	
	0500	Stainless steel	"	4	4	↓	1,075	132		1,207	1,400	
	0600	For handicap units, incl. 52″ grab bars, add				↓	320			320	355	
	0900	Floor and ceiling anchored										
	1000	Painted metal	2 Carp	5	3.200	Ea.	390	106		496	605	
	1300	Stainless steel	"	5	3.200	↓	1,225	106		1,331	1,525	
	1400	For handicap units, incl. 52″ grab bars, add				↓	273			273	300	
	1610	Floor mounted										

10 SPECIALTIES

10150 | Compartments & Cubicles

10160 | Metal Toilet Compartments

		10160 | Metal Toilet Compartments	CREW	DAILY OUTPUT	LABOR-HOURS	UNIT	2004 BARE COSTS MAT.	LABOR	EQUIP.	TOTAL	TOTAL INCL O&P	
100	1700	Painted metal	2 Carp	7	2.286	Ea.	390	75.50		465.50	555	100
	2000	Stainless steel	"	7	2.286		1,300	75.50		1,375.50	1,575	
	2100	For handicap units, incl. 52″ grab bars, add					273			273	300	
	2200	For juvenile units, deduct					37.50			37.50	41	
	2450	Floor mounted, headrail braced										
	2500	Painted metal	2 Carp	6	2.667	Ea.	390	88		478	575	
	2800	Stainless steel	"	6	2.667		1,225	88		1,313	1,500	
	2900	For handicap units, incl. 52″ grab bars, add					273			273	300	
	3000	Wall hung partitions, painted metal	2 Carp	7	2.286		495	75.50		570.50	670	
	3300	Stainless steel	"	7	2.286		1,200	75.50		1,275.50	1,450	
	3400	For handicap units, incl. 52″ grab bars, add					273			273	300	
	4000	Screens, entrance, floor mounted, 58″ high, 48″ wide										
	4200	Painted metal	2 Carp	15	1.067	Ea.	177	35		212	254	
	4500	Stainless steel	"	15	1.067	"	650	35		685	775	
	4650	Urinal screen, 18″ wide										
	4700	Painted metal	2 Carp	8	2	Ea.	147	66		213	271	
	5000	Stainless steel	"	8	2	"	465	66		531	625	
	5100	Floor mounted, head rail braced										
	5300	Painted metal	2 Carp	8	2	Ea.	162	66		228	287	
	5600	Stainless steel	"	8	2	"	545	66		611	710	
	5750	Pilaster, flush										
	5800	Painted metal	2 Carp	10	1.600	Ea.	238	53		291	350	
	6100	Stainless steel		10	1.600		410	53		463	540	
	6300	Urinal screen, post braced, painted metal		10	1.600		245	53		298	360	
	6600	Stainless steel		10	1.600		410	53		463	540	
	6700	Wall hung, bracket supported										
	6800	Painted metal	2 Carp	10	1.600	Ea.	247	53		300	360	
	7100	Stainless steel		10	1.600		375	53		428	500	
	7400	Flange supported, painted metal		10	1.600		182	53		235	288	
	7700	Stainless steel		10	1.600		435	53		488	565	
	7800	Wedge type, painted metal		10	1.600		215	53		268	325	
	8100	Stainless steel		10	1.600		445	53		498	580	
	9000	Minimum labor/equipment charge	1 Carp	2.50	3.200	Job		106		106	175	

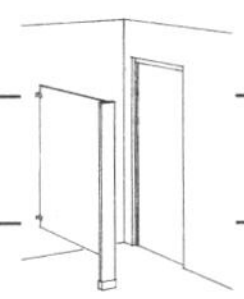

10165 | Plastic Laminate Toilet Compartment

		10165 | Plastic Laminate Toilet Compartment	CREW	DAILY OUTPUT	LABOR-HOURS	UNIT	MAT.	LABOR	EQUIP.	TOTAL	TOTAL INCL O&P	
100	0010	**TOILET PARTITIONS, PLASTIC LAMINATE**										100
	0110	Cubicles, ceiling hung										
	0300	Plastic laminate on particle board	2 Carp	4	4	Ea.	575	132		707	855	
	0600	For handicap units, incl. 52″ grab bars, add				"	320			320	355	
	0900	Floor and ceiling anchored										
	1100	Plastic laminate on particle board	2 Carp	5	3.200	Ea.	630	106		736	870	
	1400	For handicap units, incl. 52″ grab bars, add				"	273			273	300	
	1610	Floor mounted										
	1800	Plastic laminate on particle board	2 Carp	7	2.286	Ea.	535	75.50		610.50	715	
	2450	Floor mounted, headrail braced										
	2600	Plastic laminate on particle board	2 Carp	6	2.667	Ea.	735	88		823	950	
	3400	For handicap units, incl. 52″ grab bars, add					273			273	300	
	4300	Entrance screen, floor mtd., plas. lam., 58″ high, 48″ wide	2 Carp	15	1.067		415	35		450	515	
	4800	Urinal screen, 18″ wide, ceiling braced, plastic laminate		8	2		279	66		345	415	
	5400	Floor mounted, headrail braced		8	2		255	66		321	390	
	5900	Pilaster, flush, plastic laminate		10	1.600		345	53		398	470	
	6400	Post braced, plastic laminate		10	1.600		345	53		398	470	
	6700	Wall hung, bracket supported										
	6900	Plastic laminate on particle board	2 Carp	10	1.600	Ea.	131	53		184	232	
	7450	Flange supported										

10 SPECIALTIES

	10165 | Plastic Laminate Toilet Compartment		CREW	DAILY OUTPUT	LABOR-HOURS	UNIT	2004 BARE COSTS MAT.	LABOR	EQUIP.	TOTAL	TOTAL INCL O&P	
100	7500	Plastic laminate on particle board	2 Carp	10	1.600	Ea.	375	53		428	505	100

	10170 | Plastic Toilet Compartments		CREW	DAILY OUTPUT	LABOR-HOURS	UNIT	MAT.	LABOR	EQUIP.	TOTAL	TOTAL INCL O&P	
100	0010	**TOILET PARTITIONS, PLASTIC**										100
	0110	Cubicles, ceiling hung										
	0250	Phenolic	2 Carp	4	4	Ea.	820	132		952	1,125	
	0600	For handicap units, incl. 52″ grab bars, add				″	320			320	355	
	0900	Floor and ceiling anchored										
	1050	Phenolic	2 Carp	5	3.200	Ea.	865	106		971	1,125	
	1400	For handicap units, incl. 52″ grab bars, add				″	273			273	300	
	1610	Floor mounted										
	1750	Phenolic	2 Carp	7	2.286	Ea.	875	75.50		950.50	1,075	
	2100	For handicap units, incl. 52″ grab bars, add					273			273	300	
	2200	For juvenile units, deduct					37.50			37.50	41	
	2450	Floor mounted, headrail braced										
	2550	Phenolic	2 Carp	6	2.667	Ea.	840	88		928	1,075	

	10180 | Stone Toilet Compartments		CREW	DAILY OUTPUT	LABOR-HOURS	UNIT	MAT.	LABOR	EQUIP.	TOTAL	TOTAL INCL O&P	
100	0010	**TOILET PARTITIONS, STONE**										100
	0100	Cubicles, ceiling hung, marble	2 Marb	2	8	Ea.	1,425	256		1,681	2,000	
	0600	For handicap units, incl. 52″ grab bars, add					320			320	355	
	0800	Floor & ceiling anchored, marble	2 Marb	2.50	6.400		1,550	204		1,754	2,050	
	1400	For handicap units, incl. 52″ grab bars, add					273			273	300	
	1600	Floor mounted, marble	2 Marb	3	5.333		920	170		1,090	1,275	
	2400	Floor mounted, headrail braced, marble	″	3	5.333		875	170		1,045	1,250	
	2900	For handicap units, incl. 52″ grab bars, add					273			273	300	
	4100	Entrance screen, floor mounted marble, 58″ high, 48″ wide	2 Marb	9	1.778		575	57		632	725	
	4600	Urinal screen, 18″ wide, ceiling braced, marble	D-1	6	2.667		575	80		655	765	
	5100	Floor mounted, head rail braced										
	5200	Marble	D-1	6	2.667	Ea.	505	80		585	685	
	5700	Pilaster, flush, marble		9	1.778		645	53		698	790	
	6200	Post braced, marble		9	1.778		635	53		688	785	
	9000	Minimum labor/equipment charge	1 Carp	2.50	3.200	Job		106		106	175	

	10185 | Shower/Dressing Compartments		CREW	DAILY OUTPUT	LABOR-HOURS	UNIT	MAT.	LABOR	EQUIP.	TOTAL	TOTAL INCL O&P	
100	0010	**PARTITIONS, SHOWER** Floor mounted, no plumbing										100
	0100	Cabinet, incl. base, no door, painted steel, 1″ thick walls	2 Shee	5	3.200	Ea.	690	124		814	950	
	0300	With door, fiberglass		4.50	3.556		560	138		698	840	
	0600	Galvanized and painted steel, 1″ thick walls		5	3.200		730	124		854	1,000	
	0800	Stall, 1″ thick wall, no base, enameled steel		5	3.200		795	124		919	1,075	
	1500	Circular fiberglass, cabinet 36″ diameter,		4	4		575	155		730	880	
	1700	One piece, 36″ diameter, less door		4	4		485	155		640	780	
	1800	With door		3.50	4.571		800	177		977	1,150	
	4100	Shower doors, economy plastic, 24″ wide	1 Shee	9	.889		98.50	34.50		133	163	
	4200	Tempered glass door, economy		8	1		164	39		203	242	
	4700	Deluxe, tempered glass, chrome on brass frame, minimum		8	1		244	39		283	330	
	4800	Maximum		1	8		665	310		975	1,225	
	9990	Minimum labor/equipment charge		2.50	3.200	Job		124		124	197	

Important: See the Reference Section for critical supporting data - Reference Nos., Crews, & City Cost Indexes

10200 | Louvers & Vents

10210 | Wall Louvers

		10210 Wall Louvers	CREW	DAILY OUTPUT	LABOR-HOURS	UNIT	2004 BARE COSTS MAT.	LABOR	EQUIP.	TOTAL	TOTAL INCL O&P	
800	0010	**LOUVERS** Aluminum with screen, residential, 8″ x 8″	1 Carp	38	.211	Ea.	7.65	6.95		14.60	19.90	800
	0100	12″ x 12″		38	.211		8.45	6.95		15.40	21	
	0200	12″ x 18″		35	.229		12.30	7.55		19.85	26	
	0250	14″ x 24″		30	.267		15.95	8.80		24.75	32	
	0300	18″ x 24″		27	.296		18.55	9.80		28.35	36.50	
	0500	24″ x 30″		24	.333		24	11		35	44.50	
	0700	Triangle, adjustable, small		20	.400		21	13.20		34.20	45	
	0800	Large		15	.533		40	17.60		57.60	73	
	2100	Midget, aluminum, 3/4″ deep, 1″ diameter		85	.094		.67	3.11		3.78	5.90	
	2150	3″ diameter		60	.133		1.39	4.40		5.79	8.85	
	2200	4″ diameter		50	.160		2.16	5.30		7.46	11.15	
	2250	6″ diameter		30	.267		2.58	8.80		11.38	17.45	
	2300	Ridge vent strip, mill finish	1 Shee	155	.052	L.F.	2.30	2		4.30	5.70	
	2400	Under eaves vent, aluminum, mill finish, 16″ x 4″	1 Carp	48	.167	Ea.	1.61	5.50		7.11	10.85	
	2500	16″ x 8″		48	.167		1.79	5.50		7.29	11.05	
	7000	Vinyl gable vent, 8″ x 8″		38	.211		8.95	6.95		15.90	21.50	
	7020	12″ x 12″		38	.211		18.40	6.95		25.35	31.50	
	7080	12″ x 18″		35	.229		23.50	7.55		31.05	38.50	
	7200	18″ x 24″		30	.267		28	8.80		36.80	45	
	9000	Minimum labor/equipment charge		2.50	3.200	Job		106		106	175	

10260 | Wall & Corner Guards

10265 | Wall & Corner Guards

		10265 Wall & Corner Guards	CREW	DAILY OUTPUT	LABOR-HOURS	UNIT	2004 BARE COSTS MAT.	LABOR	EQUIP.	TOTAL	TOTAL INCL O&P	
200	0010	**CORNER GUARDS** Steel angle w/anchors, 1″ x 1″ x 1/4″, 1.5#/L.F.	2 Carp	160	.100	L.F.	4.78	3.30		8.08	10.70	200
	0100	2″ x 2″ x 1/4″ angles, 3.2#/L.F.	″	150	.107	″	9.30	3.52		12.82	16.10	
	9000	Minimum labor/equipment charge	1 Carp	2	4	Job		132		132	219	

10270 | Access Flooring

10275 | Access Flooring

		10275 Access Flooring	CREW	DAILY OUTPUT	LABOR-HOURS	UNIT	2004 BARE COSTS MAT.	LABOR	EQUIP.	TOTAL	TOTAL INCL O&P	
150	0010	**PEDESTAL ACCESS FLOORS** Computer room application, metal										150
	0020	Particle board or steel panels, no covering, under 6,000 S.F.	2 Carp	400	.040	S.F.	10.60	1.32		11.92	13.90	
	0300	Metal covered, over 6,000 S.F.		450	.036		7.55	1.17		8.72	10.25	
	0400	Aluminum, 24″ panels		500	.032		25.50	1.06		26.56	30.50	
	0600	For carpet covering, add					5.45			5.45	5.95	
	0700	For vinyl floor covering, add					5.50			5.50	6.05	
	0900	For high pressure laminate covering, add					3.67			3.67	4.04	
	0910	For snap on stringer system, add	2 Carp	1,000	.016		1.25	.53		1.78	2.26	
	0950	Office applications, to 8″ high, steel panels,										
	0960	no covering, over 6,000 S.F.	2 Carp	500	.032	S.F.	8	1.06		9.06	10.55	
	1050	Pedestals, 6″ to 12″	″	85	.188	Ea.	7.70	6.20		13.90	18.80	

10305 | Manufactured Fireplaces

		10305 Manufactured Fireplaces	CREW	DAILY OUTPUT	LABOR-HOURS	UNIT	2004 BARE COSTS MAT.	LABOR	EQUIP.	TOTAL	TOTAL INCL O&P	
100	0010	**FIREPLACE, PREFABRICATED** Free standing or wall hung										100
	0100	with hood & screen, minimum	1 Carp	1.30	6.154	Ea.	1,050	203		1,253	1,475	
	0150	Average		1	8		1,250	264		1,514	1,825	
	0200	Maximum		.90	8.889		3,075	293		3,368	3,850	
	0500	Chimney dbl. wall, all stainless, over 8′-6″, 7″ diam., add		33	.242	V.L.F.	47.50	8		55.50	66	
	0600	10″ diameter, add		32	.250		50.50	8.25		58.75	69	
	0700	12″ diameter, add		31	.258		66	8.50		74.50	86.50	
	0800	14″ diameter, add		30	.267		83.50	8.80		92.30	107	
	1000	Simulated brick chimney top, 4′ high, 16″ x 16″		10	.800	Ea.	180	26.50		206.50	242	
	1100	24″ x 24″		7	1.143	″	335	37.50		372.50	435	
	1500	Simulated logs, gas fired, 40,000 BTU, 2′ long, minimum		7	1.143	Set	440	37.50		477.50	550	
	1600	Maximum		6	1.333		615	44		659	750	
	1700	Electric, 1,500 BTU, 1′-6″ long, minimum		7	1.143		124	37.50		161.50	200	
	1800	11,500 BTU, maximum		6	1.333		269	44		313	370	
	2000	Fireplace, built-in, 36″ hearth, radiant		1.30	6.154	Ea.	540	203		743	925	
	2100	Recirculating, small fan		1	8		770	264		1,034	1,300	
	2150	Large fan		.90	8.889		1,425	293		1,718	2,050	
	2200	42″ hearth, radiant		1.20	6.667		685	220		905	1,125	
	2300	Recirculating, small fan		.90	8.889		900	293		1,193	1,475	
	2350	Large fan		.80	10		1,725	330		2,055	2,450	
	2400	48″ hearth, radiant		1.10	7.273		1,275	240		1,515	1,800	
	2500	Recirculating, small fan		.80	10		1,575	330		1,905	2,300	
	2550	Large fan		.70	11.429		2,450	375		2,825	3,325	
	3000	See through, including doors		.80	10		2,025	330		2,355	2,775	
	3200	Corner (2 wall)		1	8		1,000	264		1,264	1,550	

10310 | Fireplace Specialties & Accessories

		10310 Fireplace Specialties & Accessories	CREW	DAILY OUTPUT	LABOR-HOURS	UNIT	2004 BARE COSTS MAT.	LABOR	EQUIP.	TOTAL	TOTAL INCL O&P	
100	0010	**FIREPLACE ACCESSORIES** Chimney screens, galv., 13″ x 13″ flue	1 Bric	8	1	Ea.	33.50	34.50		68	92.50	100
	0050	Galv., 24″ x 24″ flue		5	1.600		100	55		155	199	
	0200	Stainless steel, 13″ x 13″ flue		8	1		264	34.50		298.50	345	
	0250	20″ x 20″ flue		5	1.600		360	55		415	485	
	0400	Cleanout doors and frames, cast iron, 8″ x 8″		12	.667		29	23		52	69	
	0450	12″ x 12″		10	.800		33.50	27.50		61	81	
	0500	18″ x 24″		8	1		105	34.50		139.50	171	
	0550	Cast iron frame, steel door, 24″ x 30″		5	1.600		227	55		282	340	
	0800	Damper, rotary control, steel, 30″ opening		6	1.333		64	45.50		109.50	144	
	0850	Cast iron, 30″ opening		6	1.333		71	45.50		116.50	152	
	0880	36″ opening		6	1.333		77	45.50		122.50	159	
	0900	48″ opening		6	1.333		110	45.50		155.50	195	
	0920	60″ opening		6	1.333		247	45.50		292.50	345	
	1000	84″ opening, special order		5	1.600		635	55		690	790	
	1050	96″ opening, special order		4	2		645	68.50		713.50	820	
	1200	Steel plate, poker control, 60″ opening		8	1		225	34.50		259.50	305	
	1250	84″ opening, special opening		5	1.600		410	55		465	540	
	1400	"Universal" type, chain operated, 32″ x 20″ opening		8	1		156	34.50		190.50	227	
	1450	48″ x 24″ opening		5	1.600		261	55		316	375	
	1600	Dutch Oven door and frame, cast iron, 12″ x 15″ opening		13	.615		92	21		113	135	
	1650	Copper plated, 12″ x 15″ opening		13	.615		177	21		198	228	
	1800	Fireplace forms, no accessories, 32″ opening		3	2.667		525	91.50		616.50	730	
	1900	36″ opening		2.50	3.200		635	110		745	880	
	2000	40″ opening		2	4		765	137		902	1,050	
	2100	78″ opening		1.50	5.333		1,100	183		1,283	1,500	
	2400	Squirrel and bird screens, galvanized, 8″ x 8″ flue		16	.500		36	17.15		53.15	68	
	2450	13″ x 13″ flue		12	.667		41.50	23		64.50	82.50	
	9000	Minimum labor/equipment charge		3.50	2.286	Job		78.50		78.50	127	

10300 | Fireplaces & Stoves

		10320 | Stoves	CREW	DAILY OUTPUT	LABOR-HOURS	UNIT	2004 BARE COSTS MAT.	LABOR	EQUIP.	TOTAL	TOTAL INCL O&P	
100	0010	**WOODBURNING STOVES** Cast iron, minimum	2 Carp	1.30	12.308	Ea.	775	405		1,180	1,525	100
	0020	Average		1	16		1,150	530		1,680	2,150	
	0030	Maximum	↓	.80	20		1,875	660		2,535	3,175	
	0050	For gas log lighter, add				↓	39			39	43	

10340 | Manufactured Exterior Specialties

		10342 | Cupolas	CREW	DAILY OUTPUT	LABOR-HOURS	UNIT	2004 BARE COSTS MAT.	LABOR	EQUIP.	TOTAL	TOTAL INCL O&P	
100	0010	**CUPOLA** Stock units, pine, painted, 18″ sq., 28″ high, alum. roof	1 Carp	4.10	1.951	Ea.	141	64.50		205.50	262	100
	0100	Copper roof		3.80	2.105		143	69.50		212.50	273	
	0300	23″ square, 33″ high, aluminum roof		3.70	2.162		236	71.50		307.50	380	
	0400	Copper roof		3.30	2.424		238	80		318	395	
	0600	30″ square, 37″ high, aluminum roof		3.70	2.162		360	71.50		431.50	515	
	0700	Copper roof		3.30	2.424		370	80		450	545	
	0900	Hexagonal, 31″ wide, 46″ high, copper roof		4	2		540	66		606	705	
	1000	36″ wide, 50″ high, copper roof	↓	3.50	2.286		575	75.50		650.50	755	
	1200	For deluxe stock units, add to above					25%					
	1400	For custom built units, add to above				↓	50%	50%				
	9000	Minimum labor/equipment charge	1 Carp	2.75	2.909	Job		96		96	159	

		10344 | Weathervanes	CREW	DAILY OUTPUT	LABOR-HOURS	UNIT	MAT.	LABOR	EQUIP.	TOTAL	TOTAL INCL O&P	
800	0010	**WEATHERVANES**										800
	0020	Residential types, minimum	1 Carp	8	1	Ea.	40	33		73	98.50	
	0100	Maximum		2	4	″	800	132		932	1,100	
	9000	Minimum labor/equipment charge	↓	4	2	Job		66		66	109	

10400 | Identification Devices

		10410 | Directories	CREW	DAILY OUTPUT	LABOR-HOURS	UNIT	2004 BARE COSTS MAT.	LABOR	EQUIP.	TOTAL	TOTAL INCL O&P	
100	0010	**DIRECTORY BOARDS**										100
	0050	Plastic, glass covered, 30″ x 20″	2 Carp	3	5.333	Ea.	234	176		410	550	
	0100	36″ x 48″		2	8		720	264		984	1,225	
	0900	Outdoor, weatherproof, black plastic, 36″ x 24″		2	8		670	264		934	1,175	
	1000	36″ x 36″	↓	1.50	10.667	↓	775	350		1,125	1,425	
	9000	Minimum labor/equipment charge	1 Carp	1	8	Job		264		264	440	

		10430 | Exterior Signage	CREW	DAILY OUTPUT	LABOR-HOURS	UNIT	MAT.	LABOR	EQUIP.	TOTAL	TOTAL INCL O&P	
200	0012	**PLAQUES**										200
	3910	20″ x 30″, up to 450 letters, cast alum.	2 Carp	4	4	Ea.	735	132		867	1,025	
	4000	Cast bronze		4	4		1,075	132		1,207	1,400	
	4200	30″ x 36″, up to 900 letters cast aluminum		3	5.333		1,725	176		1,901	2,200	
	4300	Cast bronze	↓	3	5.333		2,225	176		2,401	2,750	
	5100	Exit signs, 24 ga. alum., 14″ x 12″ surface mounted	1 Carp	30	.267	↓	26.50	8.80		35.30	43.50	

10400 | Identification Devices

	10430 | Exterior Signage		CREW	DAILY OUTPUT	LABOR-HOURS	UNIT	2004 BARE COSTS MAT.	LABOR	EQUIP.	TOTAL	TOTAL INCL O&P	
200	5200	10″ x 7″	1 Carp	20	.400	Ea.	10.40	13.20		23.60	33.50	200
	6400	Replacement sign faces, 6″ or 8″	1 Clab	50	.160	↓	22	4.16		26.16	31	
	9000	Minimum labor/equipment charge	1 Carp	4	2	Job		66		66	109	

10500 | Lockers

	10505 | Metal Lockers		CREW	DAILY OUTPUT	LABOR-HOURS	UNIT	2004 BARE COSTS MAT.	LABOR	EQUIP.	TOTAL	TOTAL INCL O&P	
500	0011	**LOCKERS** Steel, baked enamel										500
	0110	Single tier box locker, 12″ x 15″ x 72″	1 Shee	8	1	Ea.	148	39		187	225	
	0120	18″ x 15″ x 72″		8	1		160	39		199	237	
	0130	12″ x 18″ x 72″		8	1		152	39		191	230	
	0140	18″ x 18″ x 72″		8	1		172	39		211	252	
	0410	Double tier, 12″ x 15″ x 36″		21	.381		120	14.80		134.80	156	
	0420	18″ x 15″ x 36″		21	.381		159	14.80		173.80	199	
	0430	12″ x 18″ x 36″		21	.381		126	14.80		140.80	163	
	0440	18″ x 18″ x 36″		21	.381		188	14.80		202.80	231	
	0500	Two person, 18″ x 15″ x 72″		8	1		197	39		236	279	
	0510	18″ x 18″ x 72″		8	1		225	39		264	310	
	0520	Duplex, 15″ x 15″ x 72″		8	1		220	39		259	305	
	0530	15″ x 21″ x 72″		8	1		252	39		291	340	
	1100	Wire meshed wardrobe, floor. mtd., open front varsity type	↓	7.50	1.067	↓	158	41.50		199.50	240	
	2400	16-person locker unit with clothing rack										
	2500	72 wide x 15″ deep x 72″ high	1 Shee	8	1	Ea.	410	39		449	510	
	2550	18″ deep		8	1	″	360	39		399	460	
	3250	Rack w/ 24 wire mesh baskets		1.50	5.333	Set	284	207		491	645	
	3260	30 baskets		1.25	6.400		223	248		471	640	
	3270	36 baskets		.95	8.421		296	325		621	840	
	3280	42 baskets	↓	.80	10	↓	500	390		890	1,175	
	3600	For hanger rods, add				Ea.	1.70			1.70	1.87	
	9000	Minimum labor/equipment charge	1 Shee	2.50	3.200	Job		124		124	197	

10520 | Fire Protection Specialties

	10525 | Fire Prot. Specialties		CREW	DAILY OUTPUT	LABOR-HOURS	UNIT	2004 BARE COSTS MAT.	LABOR	EQUIP.	TOTAL	TOTAL INCL O&P	
200	0010	**FIRE EQUIPMENT CABINETS** Not equipped, 20 ga. steel box,										200
	0040	recessed, D.S. glass in door, box size given										
	1000	Portable extinguisher, single, 8″ x 12″ x 27″, alum. door & frame	Q-12	8	2	Ea.	82.50	70.50		153	201	
	1100	Steel door and frame	″	8	2	″	59	70.50		129.50	175	
	3000	Hose rack assy., 1-1/2″ valve & 100′ hose, 24″ x 40″ x 5-1/2″										
	3200	Steel door and frame	Q-12	6	2.667	Ea.	127	94		221	287	
	4000	Hose rack assy., 2-1/2″ x 1-1/2″ valve, 100′ hose, 24″ x 40″ x 8″										
	4200	Steel door and frame	Q-12	6	2.667	Ea.	138	94		232	299	

Important: See the Reference Section for critical supporting data - Reference Nos., Crews, & City Cost Indexes

10520 | Fire Protection Specialties

10525 | Fire Prot. Specialties

		Description	CREW	DAILY OUTPUT	LABOR-HOURS	UNIT	2004 BARE COSTS MAT.	LABOR	EQUIP.	TOTAL	TOTAL INCL O&P	
200	5000	Hose rack assy., 2-1/2" x 1-1/2" valve, 100' hose										200
	5010	and extinguisher, 30" x 40" x 8"										
	5200	Steel door and frame	Q-12	5	3.200	Ea.	146	113		259	340	
300	0010	**FIRE EXTINGUISHERS**										300
	0120	CO2, portable with swivel horn, 5 lb.				Ea.	101			101	111	
	0140	With hose and "H" horn, 10 lb.				"	150			150	165	
	1000	Dry chemical, pressurized										
	1040	Standard type, portable, painted, 2-1/2 lb.				Ea.	27.50			27.50	30.50	
	1080	10 lb.					67			67	73.50	
	1100	20 lb.					90			90	99	
	1120	30 lb.					157			157	173	
	2000	ABC all purpose type, portable, 2-1/2 lb.					27.50			27.50	30.50	
	2080	9-1/2 lb.					60			60	66	

10530 | Protective Covers

10535 | Awnings & Canopies

		Description	CREW	DAILY OUTPUT	LABOR-HOURS	UNIT	2004 BARE COSTS MAT.	LABOR	EQUIP.	TOTAL	TOTAL INCL O&P	
050	0010	**AWNINGS, FABRIC**										050
	0020	Including acrylic canvas and frame, standard design										
	0100	Door and window, slope, 3' high, 4' wide	1 Carp	4.50	1.778	Ea.	510	58.50		568.50	655	
	0110	6' wide		3.50	2.286		655	75.50		730.50	845	
	0120	8' wide		3	2.667		800	88		888	1,025	
	0200	Quarter round convex, 4' wide		3	2.667		790	88		878	1,025	
	0210	6' wide		2.25	3.556		1,025	117		1,142	1,325	
	0220	8' wide		1.80	4.444		1,250	147		1,397	1,625	
	0300	Dome, 4' wide		7.50	1.067		305	35		340	395	
	0310	6' wide		3.50	2.286		685	75.50		760.50	880	
	0320	8' wide		2	4		1,225	132		1,357	1,575	
	0350	Elongated dome, 4' wide		1.33	6.015		1,150	199		1,349	1,575	
	0360	6' wide		1.11	7.207		1,375	238		1,613	1,900	
	0370	8' wide		1	8		1,600	264		1,864	2,200	
	1000	Entry or walkway, peak, 12' long, 4' wide	2 Carp	.90	17.778		3,600	585		4,185	4,950	
	1010	6' wide		.60	26.667		5,575	880		6,455	7,575	
	1020	8' wide		.40	40		7,675	1,325		9,000	10,700	
	1100	Radius with dome end, 4' wide		1.10	14.545		2,750	480		3,230	3,825	
	1110	6' wide		.70	22.857		4,400	755		5,155	6,100	
	1120	8' wide		.50	32		6,250	1,050		7,300	8,625	
	2000	Retractable lateral arm awning, manual										
	2010	To 12' wide, 8' - 6" projection	2 Carp	1.70	9.412	Ea.	815	310		1,125	1,425	
	2020	To 14' wide, 8' - 6" projection		1.10	14.545		950	480		1,430	1,850	
	2030	To 19' wide, 8' - 6" projection		.85	18.824		1,300	620		1,920	2,450	
	2040	To 24' wide, 8' - 6" projection		.67	23.881		1,625	790		2,415	3,100	
	2050	Motor for above, add	1 Carp	2.67	3		700	99		799	935	
	3000	Patio/deck canopy with frame										
	3010	12' wide, 12' projection	2 Carp	2	8	Ea.	1,150	264		1,414	1,725	
	3020	16' wide, 14' projection	"	1.20	13.333		1,800	440		2,240	2,700	
	9000	For fire retardant canvas, add					7%					
	9010	For lettering or graphics, add					35%					
	9020	For painted or coated acrylic canvas, deduct					8%					

		10535 | Awnings & Canopies	CREW	DAILY OUTPUT	LABOR-HOURS	UNIT	2004 BARE COSTS MAT.	LABOR	EQUIP.	TOTAL	TOTAL INCL O&P	
050	9030	For translucent or opaque vinyl canvas, add				Ea.	10%					050
	9040	For 6 or more units, deduct				↓	20%	15%				
100	0010	**CANOPIES, RESIDENTIAL** Prefabricated										100
	0500	Carport, free standing, baked enamel, alum., .032″, 40 psf										
	0520	16′ x 8′, 4 posts	2 Carp	3	5.333	Ea.	2,775	176		2,951	3,350	
	0600	20′ x 10′, 6 posts	″	2	8		2,900	264		3,164	3,650	
	1000	Door canopies, extruded alum., .032″, 42″ projection, 4′ wide	1 Carp	8	1		335	33		368	425	
	1020	6′ wide	″	6	1.333		410	44		454	525	
	1040	8′ wide	2 Carp	9	1.778		525	58.50		583.50	670	
	1060	10′ wide		7	2.286		615	75.50		690.50	800	
	1080	12′ wide	↓	5	3.200		725	106		831	975	
	1200	54″ projection, 4′ wide	1 Carp	8	1		430	33		463	530	
	1220	6′ wide	″	6	1.333		550	44		594	680	
	1240	8′ wide	2 Carp	9	1.778		725	58.50		783.50	890	
	1260	10′ wide		7	2.286		820	75.50		895.50	1,025	
	1280	12′ wide	↓	5	3.200		930	106		1,036	1,200	
	1300	Painted, add					20%					
	1310	Bronze anodized, add					50%					
	3000	Window awnings, aluminum, window 3′ high, 4′ wide	1 Carp	10	.800		224	26.50		250.50	290	
	3020	6′ wide	″	8	1		261	33		294	340	
	3040	9′ wide	2 Carp	9	1.778		425	58.50		483.50	560	
	3060	12′ wide	″	5	3.200		580	106		686	810	
	3100	Window, 4′ high, 4′ wide	1 Carp	10	.800		274	26.50		300.50	345	
	3120	6′ wide	″	8	1		370	33		403	460	
	3140	9′ wide	2 Carp	9	1.778		500	58.50		558.50	645	
	3160	12′ wide	″	5	3.200		645	106		751	885	
	3200	Window, 6′ high, 4′ wide	1 Carp	10	.800		415	26.50		441.50	500	
	3220	6′ wide	″	8	1		570	33		603	685	
	3240	9′ wide	2 Carp	9	1.778		785	58.50		843.50	955	
	3260	12′ wide	″	5	3.200		1,075	106		1,181	1,350	
	3400	Roll-up aluminum, 2′-6″ wide	1 Carp	14	.571		91.50	18.85		110.35	133	
	3420	3′ wide		12	.667		110	22		132	158	
	3440	4′ wide		10	.800		141	26.50		167.50	199	
	3460	6′ wide	↓	8	1		174	33		207	247	
	3480	9′ wide	2 Carp	9	1.778		253	58.50		311.50	375	
	3500	12′ wide	″	5	3.200	↓	310	106		416	515	
	3600	Window awnings, canvas, 24″ drop, 3′ wide	1 Carp	30	.267	L.F.	36	8.80		44.80	54	
	3620	4′ wide		40	.200		33	6.60		39.60	47.50	
	3700	30″ drop, 3′ wide		30	.267		51.50	8.80		60.30	71	
	3720	4′ wide		40	.200		44.50	6.60		51.10	59.50	
	3740	5′ wide		45	.178		40	5.85		45.85	53.50	
	3760	6′ wide		48	.167		37	5.50		42.50	49.50	
	3780	8′ wide		48	.167		30.50	5.50		36	42.50	
	3800	10′ wide		50	.160	↓	28	5.30		33.30	40	
	3900	Repair canvas, minimum		8	1	Ea.		33		33	54.50	
	3920	Maximum		4	2	″		66		66	109	
	9000	Minimum labor/equipment charge	↓	3	2.667	Job		88		88	146	
200	0010	**CANOPIES** Wall hung, .032″, aluminum, prefinished, 8′ x 10′	K-2	1.30	18.462	Ea.	1,525	630	123	2,278	2,925	200
	0300	8′ x 20′		1.10	21.818		3,075	740	146	3,961	4,850	
	1000	12′ x 20′	↓	1	24	↓	3,925	815	160	4,900	5,975	
	1050											
	2300	Aluminum entrance canopies, flat soffit, .032″										
	2500	3′-6″ x 4′-0″, clear anodized	2 Carp	4	4	Ea.	645	132		777	930	
	9000	Minimum labor/equipment charge	″	2	8	Job		264		264	440	

Important: See the Reference Section for critical supporting data - Reference Nos., Crews, & City Cost Indexes

10550 | Postal Specialties

		10555 | Mail Delivery Systems	CREW	DAILY OUTPUT	LABOR-HOURS	UNIT	2004 BARE COSTS MAT.	LABOR	EQUIP.	TOTAL	TOTAL INCL O&P	
600	0010	**MAIL BOXES** Horiz., key lock, 5"H x 6"W x 15"D, alum., rear load	1 Carp	34	.235	Ea.	31.50	7.75		39.25	47.50	600
	0100	Front loading		34	.235		35.50	7.75		43.25	52	
	0200	Double, 5"H x 12"W x 15"D, rear loading		26	.308		60	10.15		70.15	83	
	0300	Front loading		26	.308		63	10.15		73.15	86.50	
	0500	Quadruple, 10"H x 12"W x 15"D, rear loading		20	.400		112	13.20		125.20	145	
	0600	Front loading		20	.400		107	13.20		120.20	140	
	1600	Vault type, horizontal, for apartments, 4" x 5"		34	.235		36.50	7.75		44.25	53	
	1700	Alphabetical directories, 120 names		10	.800		111	26.50		137.50	166	
	1900	Letter slot, residential		20	.400		56.50	13.20		69.70	84	
	2000	Post office type		8	1		220	33		253	297	
	9000	Minimum labor/equipment charge		5	1.600	Job		53		53	87.50	
700	0010	**MAIL CHUTES** Aluminum & glass, 14-1/4" wide, 4-5/8" deep	2 Shee	4	4	Floor	645	155		800	955	700
	0100	8-5/8" deep		3.80	4.211	"	720	163		883	1,050	
	0600	Lobby collection boxes, aluminum		5	3.200	Ea.	1,800	124		1,924	2,175	
	0700	Bronze or stainless		4.50	3.556	"	2,300	138		2,438	2,750	

10600 | Partitions

		10610 | Folding Gates	CREW	DAILY OUTPUT	LABOR-HOURS	UNIT	2004 BARE COSTS MAT.	LABOR	EQUIP.	TOTAL	TOTAL INCL O&P	
100	0010	**SECURITY GATES** For roll up type, see division 08330-130										100
	0300	Scissors type folding gate, ptd. steel, single, 6' high, 5-1/2'wide	2 Sswk	4	4	Opng.	127	149		276	415	
	0350	6-1/2' wide		4	4		136	149		285	425	
	0400	7-1/2' wide		4	4		168	149		317	460	
	0600	Double gate, 7-1/2' high, 8' wide		2.50	6.400		216	238		454	680	
	0650	10' wide		2.50	6.400		248	238		486	720	
	0700	12' wide		2	8		335	297		632	925	
	0750	14' wide		2	8		390	297		687	980	
	0900	Door gate, folding steel, 4' wide, 61" high		4	4		64	149		213	345	
	1000	71" high		4	4		66.50	149		215.50	350	
	1200	81" high		4	4		72	149		221	355	
	1300	Window gates, 2' to 4' wide, 31" high		4	4		35	149		184	315	
	1500	55" high		3.75	4.267		48	159		207	350	
	1600	79" high		3.50	4.571		63	170		233	385	

		10615 | Demountable Partitions	CREW	DAILY OUTPUT	LABOR-HOURS	UNIT	MAT.	LABOR	EQUIP.	TOTAL	TOTAL INCL O&P	
100	0010	**PARTITIONS, MOVABLE OFFICE** Demountable, add for doors										100
	0100	Do not deduct door openings from total L.F.										
	0900	Demountable gypsum system on 2" to 2-1/2"										
	1000	steel studs, 9' high, 3" to 3-3/4" thick										
	1200	Vinyl clad gypsum	2 Carp	48	.333	L.F.	39.50	11		50.50	62	
	1300	Fabric clad gypsum		44	.364		98.50	12		110.50	128	
	1500	Steel clad gypsum		40	.400		107	13.20		120.20	139	
	1600	1.75 system, aluminum framing, vinyl clad hardboard,										
	1800	paper honeycomb core panel, 1-3/4" to 2-1/2" thick										
	1900	9' high	2 Carp	48	.333	L.F.	66.50	11		77.50	91.50	
	2100	7' high		60	.267		59.50	8.80		68.30	80	
	2200	5' high		80	.200		50.50	6.60		57.10	66.50	
	2250	Unitized gypsum system										
	2300	Unitized panel, 9' high, 2" to 2-1/2" thick										

10600 | Partitions

10615 | Demountable Partitions

			CREW	DAILY OUTPUT	LABOR-HOURS	UNIT	2004 BARE COSTS MAT.	LABOR	EQUIP.	TOTAL	TOTAL INCL O&P	
100	2350	Vinyl clad gypsum	2 Carp	48	.333	L.F.	85.50	11		96.50	112	100
	2400	Fabric clad gypsum	"	44	.364	"	141	12		153	175	
	2500	Unitized mineral fiber system										
	2510	Unitized panel, 9′ high, 2-1/4″ thick, aluminum frame										
	2550	Vinyl clad mineral fiber	2 Carp	48	.333	L.F.	85	11		96	112	
	2600	Fabric clad mineral fiber	"	44	.364	"	127	12		139	159	
	2800	Movable steel walls, modular system										
	2900	Unitized panels, 9′ high, 48″ wide										
	3100	Baked enamel, pre-finished	2 Carp	60	.267	L.F.	96.50	8.80		105.30	121	
	3200	Fabric clad steel	"	56	.286	"	139	9.45		148.45	169	
	5500	For acoustical partitions, add, minimum				S.F.	1.63			1.63	1.79	
	5550	Maximum				"	7.60			7.60	8.35	
	5700	For doors, see Div. 08100 & 08200										
	5800	For door hardware, see div. 08700										
	6100	In-plant modular office system, w/prehung hollow core door										
	6200	3″ thick polystyrene core panels										
	6250	12′ x 12′, 2 wall	2 Clab	3.80	4.211	Ea.	2,875	109		2,984	3,350	
	6300	4 wall	↓	1.90	8.421	↓	4,000	219		4,219	4,775	
	6350	16′ x 16′, 2 wall	↓	3.60	4.444	↓	4,500	116		4,616	5,150	
	6400	4 wall	↓	1.80	8.889	↓	5,775	231		6,006	6,725	
	9000	Minimum labor/equipment charge	2 Carp	3	5.333	Job		176		176	292	

10630 | Port. Partitions/Screens/Panels

			CREW	DAILY OUTPUT	LABOR-HOURS	UNIT	MAT.	LABOR	EQUIP.	TOTAL	TOTAL INCL O&P	
100	0010	**PARTITIONS, PORTABLE** Divider panels, free standing, fiber core										100
	0020	Fabric face straight										
	0100	3′-0″ long, 4′-0″ high	2 Carp	100	.160	L.F.	99	5.30		104.30	118	
	0200	5′-0″ high	↓	90	.178	↓	100	5.85		105.85	120	
	0500	6′-0″ high	↓	75	.213	↓	113	7.05		120.05	136	
	0900	5′-0″ long, 4′-0″ high	↓	175	.091	↓	75	3.02		78.02	87	
	1000	5′-0″ high	↓	150	.107	↓	92	3.52		95.52	107	
	1500	6″-0″ high	↓	125	.128	↓	88	4.22		92.22	104	
	1600	6′-0″ long, 5′-0″ high	↓	162	.099	↓	73.50	3.26		76.76	86.50	
	3100	Curved, 3′-0″ long, 5′-0″ high	↓	90	.178	↓	89.50	5.85		95.35	108	
	3150	6′-0″ high	↓	75	.213	↓	103	7.05		110.05	126	
	3200	Economical panels, fabric face, 4′-0″ long, 5′-0″ high	↓	132	.121	↓	33.50	4		37.50	43.50	
	3250	6′-0″ high	↓	112	.143	↓	37.50	4.71		42.21	49.50	
	3300	5′-0″ long, 5′-0″ high	↓	150	.107	↓	28	3.52		31.52	37	
	3350	6′-0″ high	↓	125	.128	↓	33.50	4.22		37.72	43.50	
	3380	3′-0″ curved, 5′-0″ high	↓	90	.178	↓	89.50	5.85		95.35	108	
	3390	6′-0″ high	↓	75	.213	↓	103	7.05		110.05	126	
	3450	Acoustical panels, 60 to 90 NRC, 3′-0″ long, 5′-0″ high	↓	90	.178	↓	88.50	5.85		94.35	107	
	3550	6′-0″ high	↓	75	.213	↓	85	7.05		92.05	105	
	3600	5′-0″ long, 5′-0″ high	↓	150	.107	↓	62	3.52		65.52	74	
	3650	6′-0″ high	↓	125	.128	↓	68.50	4.22		72.72	82.50	
	3700	6′-0″ long, 5′-0″ high	↓	162	.099	↓	60	3.26		63.26	71.50	
	3750	6′-0″ high	↓	138	.116	↓	60	3.83		63.83	72.50	
	3800	Economy acoustical panels, 40 NRC, 4′-0″ long, 5′-0″ high	↓	132	.121	↓	33.50	4		37.50	43.50	
	3850	6′-0″ high	↓	112	.143	↓	38.50	4.71		43.21	50.50	
	3900	5′-0″ long, 6′-0″ high	↓	125	.128	↓	33.50	4.22		37.72	43.50	
	3950	6′-0″ long, 5′-0″ high	↓	162	.099	↓	27.50	3.26		30.76	36	
	9000	Minimum labor/equipment charge	↓	3	5.333	Job		176		176	292	

10651 | Accordion Folding Partitions

			CREW	DAILY OUTPUT	LABOR-HOURS	UNIT	MAT.	LABOR	EQUIP.	TOTAL	TOTAL INCL O&P	
100	0010	**PARTITIONS, FOLDING ACCORDION**										100
	0100	Vinyl covered, over 150 S.F., frame not included										

10600 | Partitions

	10651 Accordion Folding Partitions		CREW	DAILY OUTPUT	LABOR-HOURS	UNIT	2004 BARE COSTS MAT.	LABOR	EQUIP.	TOTAL	TOTAL INCL O&P
100	0300	Residential, 1.25 lb. per S.F., 8′ maximum height	2 Carp	300	.053	S.F.	15.45	1.76		17.21	19.90
	0400	Commercial, 1.75 lb. per S.F., 8′ maximum height		225	.071		15.45	2.35		17.80	21
	0900	Acoustical, 3 lb. per S.F., 17′ maximum height		100	.160		22.50	5.30		27.80	33.50
	1200	5 lb. per S.F., 20′ maximum height		95	.168		31	5.55		36.55	43
	1500	Vinyl clad wood or steel, electric operation, 5.0 psf		160	.100		40	3.30		43.30	49.50
	1900	Wood, non-acoustic, birch or mahogany, to 10′ high		300	.053		20.50	1.76		22.26	25.50
	9000	Minimum labor/equipment charge		4	4	Job		132		132	219

	10653 Folding Panel Partitions		CREW	DAILY OUTPUT	LABOR-HOURS	UNIT	MAT.	LABOR	EQUIP.	TOTAL	TOTAL INCL O&P
200	0010	**PARTITIONS, FOLDING LEAF** Acoustic, wood									
	0100	Vinyl faced, to 18′ high, 6 psf, minimum	2 Carp	60	.267	S.F.	41	8.80		49.80	59.50
	0150	Average		45	.356		48.50	11.75		60.25	73
	0200	Maximum		30	.533		63	17.60		80.60	98
	0400	Formica or hardwood finish, minimum		60	.267		42	8.80		50.80	60.50
	0500	Maximum		30	.533		45	17.60		62.60	78.50
	0600	Wood, low acoustical type, 4.5 psf, to 14′ high		50	.320		30.50	10.55		41.05	51
	9000	Minimum labor/equipment charge		4	4	Job		132		132	219

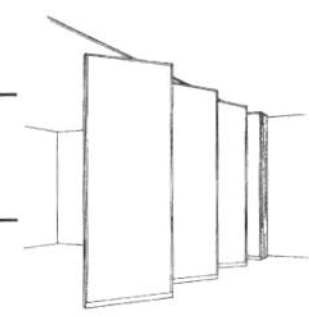

10670 | Storage Shelving

	10674 Storage Shelving		CREW	DAILY OUTPUT	LABOR-HOURS	UNIT	2004 BARE COSTS MAT.	LABOR	EQUIP.	TOTAL	TOTAL INCL O&P
500	0010	**SHELVING** Metal, industrial, cross-braced, 3′ wide, 12″ deep	1 Sswk	175	.046	SF Shlf	5.90	1.70		7.60	9.65
	0100	24″ deep		330	.024		4.50	.90		5.40	6.65
	2200	Wide span, 1600 lb. capacity per shelf, 6′ wide, 24″ deep		380	.021		7.35	.78		8.13	9.55
	2400	36″ deep		440	.018		6.45	.68		7.13	8.30
	9000	Minimum labor/equipment charge	1 Carp	4	2	Job		66		66	109

10750 | Telephone Specialties

	10755 Telephone Enclosures		CREW	DAILY OUTPUT	LABOR-HOURS	UNIT	2004 BARE COSTS MAT.	LABOR	EQUIP.	TOTAL	TOTAL INCL O&P
400	0010	**TELEPHONE ENCLOSURE**									
	0300	Shelf type, wall hung, minimum	2 Carp	5	3.200	Ea.	940	106		1,046	1,200
	0400	Maximum		5	3.200		2,425	106		2,531	2,850
	0600	Booth type, painted steel, indoor or outdoor, minimum		1.50	10.667		2,975	350		3,325	3,850
	0700	Maximum (stainless steel)		1.50	10.667		9,900	350		10,250	11,500
	1900	Outdoor, drive-up type, wall mounted		4	4		795	132		927	1,100
	2000	Post mounted, stainless steel posts		3	5.333		1,225	176		1,401	1,650
	9000	Minimum labor/equipment charge		4	4	Job		132		132	219

	10810 | Toilet Accessories		Crew	Daily Output	Labor-Hours	Unit	2004 Bare Costs Mat.	Labor	Equip.	Total	Total Incl O&P	
100	0010	COMMERCIAL TOILET ACCESSORIES										100
	0200	Curtain rod, stainless steel, 5′ long, 1″ diameter	1 Carp	13	.615	Ea.	30.50	20.50		51	67	
	0300	1-1/4″ diameter		13	.615		29	20.50		49.50	65	
	0400	Diaper changing station, horizontal, wall mounted, plastic	↓	10	.800	↓	183	26.50		209.50	245	
	0500	Dispenser units, combined soap & towel dispensers,										
	0510	mirror and shelf, flush mounted	1 Carp	10	.800	Ea.	330	26.50		356.50	410	
	0600	Towel dispenser and waste receptacle,										
	0610	18 gallon capacity	1 Carp	10	.800	Ea.	256	26.50		282.50	325	
	0800	Grab bar, straight, 1-1/4″ diameter, stainless steel, 18″ long		24	.333		21	11		32	41.50	
	0900	24″ long		23	.348		23.50	11.50		35	45	
	1000	30″ long		22	.364		24.50	12		36.50	47	
	1100	36″ long		20	.400		26.50	13.20		39.70	51	
	1200	1-1/2″ diameter, 24″ long		23	.348		46	11.50		57.50	69.50	
	1300	36″ long		20	.400		51.50	13.20		64.70	78.50	
	1310	42″ long		18	.444		48	14.65		62.65	77.50	
	1500	Tub bar, 1-1/4″ diameter, 24″ x 36″		14	.571		78.50	18.85		97.35	118	
	1600	Plus vertical arm		12	.667		86.50	22		108.50	132	
	1900	End tub bar, 1″ diameter, 90° angle, 16″ x 32″		12	.667		118	22		140	167	
	2300	Hand dryer, surface mounted, electric, 115 volt, 20 amp		4	2		485	66		551	640	
	2400	230 volt, 10 amp		4	2		485	66		551	640	
	2600	Hat and coat strip, stainless steel, 4 hook, 36″ long		24	.333		48	11		59	71.50	
	2700	6 hook, 60″ long		20	.400		84	13.20		97.20	115	
	3000	Mirror, with stainless steel 3/4″ square frame, 18″ x 24″		20	.400		58.50	13.20		71.70	86.50	
	3100	36″ x 24″		15	.533		101	17.60		118.60	140	
	3200	48″ x 24″		10	.800		137	26.50		163.50	195	
	3300	72″ x 24″		6	1.333		196	44		240	289	
	3500	With 5″ stainless steel shelf, 18″ x 24″		20	.400		90.50	13.20		103.70	122	
	3600	36″ x 24″		15	.533		126	17.60		143.60	168	
	3700	48″ x 24″		10	.800		149	26.50		175.50	208	
	3800	72″ x 24″		6	1.333		255	44		299	355	
	4100	Mop holder strip, stainless steel, 5 holders, 48″ long		20	.400		61	13.20		74.20	89	
	4200	Napkin/tampon dispenser, recessed		15	.533		310	17.60		327.60	375	
	4300	Robe hook, single, regular		36	.222		4.76	7.35		12.11	17.40	
	4400	Heavy duty, concealed mounting		36	.222		10.80	7.35		18.15	24	
	4600	Soap dispenser, chrome, surface mounted, liquid		20	.400		41.50	13.20		54.70	67.50	
	4700	Powder		20	.400		41.50	13.20		54.70	67.50	
	5000	Recessed stainless steel, liquid		10	.800		135	26.50		161.50	193	
	5100	Powder		10	.800		174	26.50		200.50	235	
	5300	Soap tank, stainless steel, 1 gallon		10	.800		159	26.50		185.50	219	
	5400	5 gallon		5	1.600		219	53		272	330	
	5600	Shelf, stainless steel, 5″ wide, 18 ga., 24″ long		24	.333		38.50	11		49.50	60.50	
	5700	48″ long		16	.500		57.50	16.50		74	90.50	
	5800	8″ wide shelf, 18 ga., 24″ long		22	.364		46.50	12		58.50	71	
	5900	48″ long		14	.571		81	18.85		99.85	121	
	6000	Toilet seat cover dispenser, stainless steel, recessed		20	.400		88.50	13.20		101.70	120	
	6050	Surface mounted		15	.533		23.50	17.60		41.10	55	
	6100	Toilet tissue dispenser, surface mounted, SS, single roll		30	.267		11.15	8.80		19.95	27	
	6200	Double roll	↓	24	.333		15.55	11		26.55	35.50	
	6290	Toilet seat	1 Plum	40	.200		19.35	7.90		27.25	34	
	6400	Towel bar, stainless steel, 18″ long	1 Carp	23	.348		30	11.50		41.50	52	
	6500	30″ long		21	.381		51	12.55		63.55	77	
	6700	Towel dispenser, stainless steel, surface mounted		16	.500		36.50	16.50		53	67.50	
	6800	Flush mounted, recessed		10	.800		170	26.50		196.50	231	
	7000	Towel holder, hotel type, 2 guest size		20	.400		15.40	13.20		28.60	39	
	7200	Towel shelf, stainless steel, 24″ long, 8″ wide		20	.400		47.50	13.20		60.70	74	
	7400	Tumbler holder, tumbler only	↓	30	.267	↓	25	8.80		33.80	42	

Important: See the Reference Section for critical supporting data - Reference Nos., Crews, & City Cost Indexes

		10810 Toilet Accessories	CREW	DAILY OUTPUT	LABOR-HOURS	UNIT	2004 BARE COSTS MAT.	LABOR	EQUIP.	TOTAL	TOTAL INCL O&P	
100	7500	Soap, tumbler & toothbrush	1 Carp	30	.267	Ea.	22.50	8.80		31.30	39.50	100
	7700	Wall urn ash receiver, surface mount, 11" long		12	.667		99.50	22		121.50	146	
	8000	Waste receptacles, stainless steel, with top, 13 gallon		10	.800		161	26.50		187.50	221	
	8100	36 gallon		8	1		289	33		322	375	
	9000	Minimum labor/equipment charge		5	1.600	Job		53		53	87.50	

		10820 Bath Accessories	CREW	DAILY OUTPUT	LABOR-HOURS	UNIT	MAT.	LABOR	EQUIP.	TOTAL	TOTAL INCL O&P	
400	0010	**MEDICINE CABINETS** With mirror, st. st. frame, 16" x 22", unlighted	1 Carp	14	.571	Ea.	69.50	18.85		88.35	108	400
	0100	Wood frame		14	.571		96.50	18.85		115.35	138	
	0300	Sliding mirror doors, 20" x 16" x 4-3/4", unlighted		7	1.143		86.50	37.50		124	158	
	0400	24" x 19" x 8-1/2", lighted		5	1.600		136	53		189	238	
	0600	Triple door, 30" x 32", unlighted, plywood body		7	1.143		214	37.50		251.50	299	
	0700	Steel body		7	1.143		282	37.50		319.50	375	
	0900	Oak door, wood body, beveled mirror, single door		7	1.143		125	37.50		162.50	201	
	1000	Double door		6	1.333		320	44		364	425	
	1200	Hotel cabinets, stainless, with lower shelf, unlighted		10	.800		173	26.50		199.50	234	
	1300	Lighted		5	1.600		256	53		309	370	
	9000	Minimum labor/equipment charge		4	2	Job		66		66	109	

Division Notes

		CREW	DAILY OUTPUT	LABOR-HOURS	UNIT	2004 BARE COSTS MAT.	LABOR	EQUIP.	TOTAL	TOTAL INCL O&P

Division 11
Equipment

Estimating Tips

General

- The items in this division are usually priced per square foot or each. Many of these items are purchased by the owner for installation by the contractor. Check the specifications for responsibilities, and include time for receiving, storage, installation and mechanical and electrical hook-ups in the appropriate divisions.
- Many items in Division 11 require some type of support system that is not usually furnished with the item. Examples of these systems include blocking for the attachment of casework and support angles for ceiling hung projection screens. The required blocking or supports must be added to the estimate in the appropriate division.
- Some items in Division 11 may require assembly or electrical hook-ups. Verify the amount of assembly required or the need for a hard electrical connection and add the appropriate costs.

Reference Numbers

Reference numbers are shown in bold squares at the beginning of some major classifications. These numbers refer to related items in the Reference Section. The reference information may be an estimating procedure, an alternate pricing method or technical information.

Note: Not all subdivisions listed here necessarily appear in this publication.

11 EQUIPMENT

11010 | Maintenance Equipment

	11013 Floor/Wall Cleaning Equipment		CREW	DAILY OUTPUT	LABOR-HOURS	UNIT	2004 BARE COSTS MAT.	LABOR	EQUIP.	TOTAL	TOTAL INCL O&P	
800	0010	VACUUM CLEANING										800
	0020	Central, 3 inlet, residential	1 Skwk	.90	8.889	Total	590	299		889	1,150	
	0200	Commercial		.70	11.429		1,125	385		1,510	1,850	
	0400	5 inlet system, residential		.50	16		895	540		1,435	1,875	
	0600	7 inlet system, commercial		.40	20		1,000	675		1,675	2,200	
	4010	Rule of thumb: First 1200 S.F., installed									1,125	
	4020	For each additional S.F., add				S.F.					.18	

11020 | Security & Vault Equipment

	11021 Safes and Vault Doors		CREW	DAILY OUTPUT	LABOR-HOURS	UNIT	2004 BARE COSTS MAT.	LABOR	EQUIP.	TOTAL	TOTAL INCL O&P	
600	0010	SAFE										600
	0015	Office, 4 hr. rating, 30" x 18" x 18" inside				Ea.	2,925			2,925	3,200	
	0800	Money, "B" label, 9" x 14" x 14"					450			450	495	
	0900	Tool resistive, 24" x 24" x 20"					2,525			2,525	2,775	
	1050	Tool and torch resistive, 24" x 24" x 20"					6,400			6,400	7,050	

11030 | Teller & Service Equipment

	11038 Bank Equipment		CREW	DAILY OUTPUT	LABOR-HOURS	UNIT	2004 BARE COSTS MAT.	LABOR	EQUIP.	TOTAL	TOTAL INCL O&P	
150	0010	BANK EQUIPMENT										150
	0020	Alarm system, police	2 Elec	1.60	10	Ea.	3,650	395		4,045	4,625	
	0400	Bullet resistant teller window, 44" x 60"	1 Glaz	.60	13.333		2,625	425		3,050	3,575	
	0500	48" x 60"	"	.60	13.333		3,300	425		3,725	4,325	
	3000	Counters for banks, frontal only	2 Carp	1	16	Station	1,375	530		1,905	2,375	
	3100	Complete with steel undercounter		.50	32	"	2,675	1,050		3,725	4,675	
	5400	Partitions, bullet-resistant, 1-3/16" glass, 8' high		10	1.600	L.F.	148	53		201	251	
	5600	Pass thru, bullet-res. window, painted steel, 24" x 36"	2 Sswk	1.60	10	Ea.	1,675	370		2,045	2,550	

11040 | Ecclesiastical Equipment

	11041 Ecclesiastical Equipment		CREW	DAILY OUTPUT	LABOR-HOURS	UNIT	2004 BARE COSTS MAT.	LABOR	EQUIP.	TOTAL	TOTAL INCL O&P	
250	0010	CHURCH EQUIPMENT										250
	0020	Altar, wood, custom design, plain	1 Carp	1.40	5.714	Ea.	1,725	189		1,914	2,225	
	0050	Deluxe	"	.20	40	"	8,300	1,325		9,625	11,300	
	0150	Baptistry, fiberglass, 3'-6" deep, x 13'-7" long,										

11040 | Ecclesiastical Equipment

	11041 | Ecclesiastical Equipment	CREW	DAILY OUTPUT	LABOR-HOURS	UNIT	2004 BARE COSTS MAT.	LABOR	EQUIP.	TOTAL	TOTAL INCL O&P	
250	0160 steps at both ends, incl. plumbing, minimum	L-8	1	20	Ea.	2,500	685		3,185	3,875	250
	0200 Maximum	"	.70	28.571		4,925	980		5,905	7,000	
	0250 Add for filter, heater and lights					1,075			1,075	1,175	
	1500 Pews, bench type, hardwood, minimum	1 Carp	20	.400	L.F.	60	13.20		73.20	88	
	1550 Maximum	"	15	.533		120	17.60		137.60	161	
	1570 For kneeler, add					13.25			13.25	14.60	
	4000 Steeples, translucent fiberglass, 30" square, 15' high	F-3	2	20	Ea.	3,275	665	305	4,245	5,025	
	4150 25' high	"	1.80	22.222		3,825	740	340	4,905	5,800	
	4600 Aluminum, baked finish, 14' high, 16" square					1,600			1,600	1,750	
	4640 35' high, 8' base					16,300			16,300	17,900	
	4680 152' high, custom					360,500			360,500	396,500	
	5000 Wall cross, aluminum, extruded, 2" x 2" section	1 Carp	34	.235	L.F.	35	7.75		42.75	51.50	
	5150 4" x 4" section		29	.276		50	9.10		59.10	70	
	5300 Bronze, extruded, 1" x 2" section		31	.258		69	8.50		77.50	89.50	
	5350 2-1/2" x 2-1/2" section		34	.235		104	7.75		111.75	128	

11050 | Library Equipment

	11051 | Library Equipment	CREW	DAILY OUTPUT	LABOR-HOURS	UNIT	2004 BARE COSTS MAT.	LABOR	EQUIP.	TOTAL	TOTAL INCL O&P	
400	0010 **LIBRARY EQUIPMENT**										400
	0020 Bookshelf, mtl, 90" high, 10" shelf, dbl face	1 Carp	11.50	.696	L.F.	167	23		190	221	
	0300 Single face	"	12	.667	"	133	22		155	184	
	0600 For 8" shelving, subtract from above					10%					
	0800 For 42" high with countertop, subtract from above					20%					
	2500 Carrels, hardwood, 36" x 24", minimum	1 Carp	5	1.600	Ea.	615	53		668	765	
	2650 Maximum		4	2		790	66		856	975	
	2850 Metal, minimum		5	1.600		219	53		272	330	
	3000 Maximum		4	2		530	66		596	690	

11060 | Theater & Stage Equipment

	11063 | Stage Equipment	CREW	DAILY OUTPUT	LABOR-HOURS	UNIT	2004 BARE COSTS MAT.	LABOR	EQUIP.	TOTAL	TOTAL INCL O&P	
600	0010 **STAGE EQUIPMENT**										600
	0050 Control boards with dimmers and breakers, minimum	1 Elec	1	8	Ea.	9,050	315		9,365	10,500	
	0150 Maximum	"	.20	40	"	76,000	1,575		77,575	86,500	
	0160										
	0500 Curtain track, straight, light duty	2 Carp	20	.800	L.F.	18.75	26.50		45.25	64.50	
	0700 Curved sections		12	1.333	"	112	44		156	196	
	1000 Curtains, velour, medium weight		600	.027	S.F.	6.25	.88		7.13	8.30	
	5000 Stages, portable with steps, folding legs, stock, 8" high				SF Stg.	15.65			15.65	17.20	
	5100 16" high					20			20	22	
	5200 32" high					23.50			23.50	25.50	

EQUIPMENT 11

11060 | Theater & Stage Equipment

	11063 | Stage Equipment		CREW	DAILY OUTPUT	LABOR-HOURS	UNIT	2004 BARE COSTS MAT.	LABOR	EQUIP.	TOTAL	TOTAL INCL O&P	
600	5300	40″ high				SF Stg.	47			47	51.50	600

11100 | Mercantile Equipment

	11102 | Barber Shop Equipment		CREW	DAILY OUTPUT	LABOR-HOURS	UNIT	2004 BARE COSTS MAT.	LABOR	EQUIP.	TOTAL	TOTAL INCL O&P	
150	0010	BARBER EQUIPMENT										150
	0020	Chair, hydraulic, movable, minimum	1 Carp	24	.333	Ea.	420	11		431	480	
	0050	Maximum	″	16	.500		2,750	16.50		2,766.50	3,050	
	0200	Wall hung styling station with mirrors, minimum	L-2	8	2		375	57.50		432.50	505	
	0300	Maximum	″	4	4		1,825	115		1,940	2,225	
	0500	Sink, hair washing basin, rough plumbing not incl.	1 Plum	8	1		278	39.50		317.50	365	
	1000	Sterilizer, liquid solution for tools					128			128	140	
	1100	Total equipment, rule of thumb, per chair, minimum	L-8	1	20		1,425	685		2,110	2,700	
	1150	Maximum	″	1	20	↓	4,075	685		4,760	5,625	

	11103 | Cash Register/Checking		CREW	DAILY OUTPUT	LABOR-HOURS	UNIT	MAT.	LABOR	EQUIP.	TOTAL	TOTAL INCL O&P	
200	0010	CHECKOUT COUNTER										200
	0020	Supermarket conveyor, single belt	2 Clab	10	1.600	Ea.	2,125	41.50		2,166.50	2,400	
	0100	Double belt, power take-away		9	1.778		3,575	46		3,621	4,000	
	0400	Double belt, power take-away, incl. side scanning		7	2.286		4,425	59.50		4,484.50	4,975	
	0800	Warehouse or bulk type	↓	6	2.667	↓	5,200	69.50		5,269.50	5,850	
	1000	Scanning system, 2 lanes, w/registers, scan gun & memory				System	13,000			13,000	14,300	
	1100	10 lanes, single processor, full scan, with scales				″	123,500			123,500	135,500	
	2000	Register, restaurant, minimum				Ea.	525			525	580	
	2100	Maximum					2,300			2,300	2,550	
	2150	Store, minimum					525			525	580	
	2200	Maximum				↓	2,300			2,300	2,550	

	11104 | Display Cases & Systems		CREW	DAILY OUTPUT	LABOR-HOURS	UNIT	MAT.	LABOR	EQUIP.	TOTAL	TOTAL INCL O&P	
700	0010	REFRIGERATED FOOD CASES										700
	0030	Dairy, multi-deck, 12′ long	Q-5	3	5.333	Ea.	7,800	191		7,991	8,875	
	0100	For rear sliding doors, add					1,075			1,075	1,175	
	0200	Delicatessen case, service deli, 12′ long, single deck	Q-5	3.90	4.103		5,250	147		5,397	6,000	
	0300	Multi-deck, 18 S.F. shelf display		3	5.333		6,425	191		6,616	7,375	
	0400	Freezer, self-contained, chest-type, 30 C.F.		3.90	4.103		3,825	147		3,972	4,425	
	0500	Glass door, upright, 78 C.F.		3.30	4.848		7,300	173		7,473	8,300	
	0600	Frozen food, chest type, 12′ long		3.30	4.848		5,275	173		5,448	6,075	
	0700	Glass door, reach-in, 5 door		3	5.333		10,100	191		10,291	11,400	
	0800	Island case, 12′ long, single deck		3.30	4.848		5,975	173		6,148	6,825	
	0900	Multi-deck		3	5.333		12,600	191		12,791	14,200	
	1000	Meat case, 12′ long, single deck		3.30	4.848		4,325	173		4,498	5,050	
	1050	Multi-deck		3.10	5.161		7,450	185		7,635	8,475	
	1100	Produce, 12′ long, single deck		3.30	4.848		5,725	173		5,898	6,575	
	1200	Multi-deck	↓	3.10	5.161	↓	6,375	185		6,560	7,300	

11110 | Commercial Laundry & Dry Cleaning Equipment

	11119 | Laundry Cleaning		CREW	DAILY OUTPUT	LABOR-HOURS	UNIT	2004 BARE COSTS MAT.	LABOR	EQUIP.	TOTAL	TOTAL INCL O&P	
450	0010	**LAUNDRY EQUIPMENT** Not incl. rough-in										450
	0500	Dryers, gas fired residential, 16 lb. capacity, average	1 Plum	3	2.667	Ea.	545	106		651	765	
	1000	Commercial, 30 lb. capacity, coin operated, single		3	2.667		2,500	106		2,606	2,925	
	1100	Double stacked		2	4		5,350	158		5,508	6,150	
	1500	Industrial, 30 lb. capacity		2	4		2,125	158		2,283	2,600	
	1600	50 lb. capacity		1.70	4.706		2,800	186		2,986	3,375	
	2000	Dry cleaners, electric, 20 lb. capacity	L-1	.20	80		29,300	3,150		32,450	37,200	
	2050	25 lb. capacity		.17	94.118		38,200	3,725		41,925	47,800	
	2100	30 lb. capacity		.15	106		40,200	4,225		44,425	50,500	
	2150	60 lb. capacity		.09	177		63,000	7,025		70,025	80,500	
	3500	Folders, blankets & sheets, minimum	1 Elec	.17	47.059		26,800	1,850		28,650	32,400	
	3700	King size with automatic stacker		.10	80		46,400	3,150		49,550	56,000	
	3800	For conveyor delivery, add		.45	17.778		5,525	700		6,225	7,150	
	4500	Ironers, institutional, 110", single roll		.20	40		25,400	1,575		26,975	30,400	
	4700	Lint collector, ductwork not included, 8,000 to 10,000 C.F.M.	Q-10	.30	80		7,275	2,900		10,175	12,600	
	5000	Washers, residential, 4 cycle, average	1 Plum	3	2.667		630	106		736	855	
	5300	Commercial, coin operated, average	"	3	2.667		1,000	106		1,106	1,275	
	6000	Combination washer/extractor, 20 lb. capacity	L-6	1.50	8		3,625	315		3,940	4,475	
	6100	30 lb. capacity		.80	15		7,000	595		7,595	8,600	
	6200	50 lb. capacity		.68	17.647		8,700	700		9,400	10,700	
	6300	75 lb. capacity		.30	40		18,000	1,575		19,575	22,200	
	6350	125 lb. capacity		.16	75		21,800	2,975		24,775	28,500	

11130 | Audio-Visual Equipment

	11136 | Projection Screens		CREW	DAILY OUTPUT	LABOR-HOURS	UNIT	2004 BARE COSTS MAT.	LABOR	EQUIP.	TOTAL	TOTAL INCL O&P	
500	0010	**PROJECTION SCREENS** Wall or ceiling hung, matte white										500
	0100	Manually operated, economy	2 Carp	500	.032	S.F.	4.75	1.06		5.81	7	
	0300	Intermediate		450	.036		5.55	1.17		6.72	8.05	
	0400	Deluxe		400	.040		7.70	1.32		9.02	10.65	
	9000	Minimum labor/equipment charge		3	5.333	Job		176		176	292	
600	0011	**MOVIE EQUIPMENT**										600
	0020	Changeover, minimum				Ea.	380			380	415	
	3000	Projection screens, rigid, in wall, acrylic, 1/4" thick	2 Glaz	195	.082	S.F.	34	2.63		36.63	41.50	
	3100	1/2" thick	"	130	.123	"	39.50	3.94		43.44	50	
	3700	Sound systems, incl. amplifier, mono, minimum	1 Elec	.90	8.889	Ea.	2,700	350		3,050	3,500	
	3800	Dolby/Super Sound, maximum		.40	20		14,700	790		15,490	17,400	
	4100	Dual system, 2 channel, front surround, minimum		.70	11.429		3,775	450		4,225	4,850	
	4200	Dolby/Super Sound, 4 channel, maximum		.40	20		13,500	790		14,290	16,000	
	5700	Seating, painted steel, upholstered, minimum	2 Carp	35	.457		108	15.10		123.10	143	
	5800	Maximum	"	28	.571		345	18.85		363.85	410	

11140 | Vehicle Service Equipment

		11141 | Service Station Equipment	CREW	DAILY OUTPUT	LABOR-HOURS	UNIT	2004 BARE COSTS MAT.	LABOR	EQUIP.	TOTAL	TOTAL INCL O&P	
100	0010	**COMPRESSED AIR EQUIPMENT**										100
	0030	Compressors, electric, 1-1/2 H.P., standard controls	L-4	1.50	16	Ea.	292	490		782	1,125	
	0550	Dual controls	↓	1.50	16	↓	470	490		960	1,325	
	0600	5 H.P., 115/230 volt, standard controls		1	24		1,575	735		2,310	2,925	
	0650	Dual controls	↓	1	24	↓	1,675	735		2,410	3,050	
200	0010	**FUEL DISPENSING EQUIPMENT**										200
	1100	Product dispenser with vapor recovery for 6 nozzles, installed, not										
	1110	including piping to storage tanks				Ea.	15,400			15,400	16,900	
300	0010	**LUBRICATION EQUIPMENT**										300
	3000	Lube equipment, 3 reel type, with pumps, not including piping	L-4	.50	48	Set	6,150	1,475		7,625	9,175	
400	0010	**SPRAY PAINTING EQUIPMENT**										400
	4000	Spray painting booth, 26′ long, complete	L-4	.40	60	Ea.	12,700	1,825		14,525	16,900	

11150 | Parking Control Equipment

		11156 | Parking Equipment	CREW	DAILY OUTPUT	LABOR-HOURS	UNIT	2004 BARE COSTS MAT.	LABOR	EQUIP.	TOTAL	TOTAL INCL O&P	
600	0010	**PARKING EQUIPMENT**										600
	5000	Barrier gate with programmable controller	2 Elec	3	5.333	Ea.	3,050	210		3,260	3,675	
	5020	Industrial	"	3	5.333		4,150	210		4,360	4,900	
	5100	Card reader	1 Elec	2	4		1,675	158		1,833	2,075	
	5120	Proximity with customer display	2 Elec	1	16		5,050	630		5,680	6,525	
	5200	Cashier booth, average	B-22	1	30		8,800	910	239	9,949	11,400	
	5300	Collector station, pay on foot	2 Elec	.20	80		100,500	3,150		103,650	115,500	
	5320	Credit card only		.50	32		18,500	1,250		19,750	22,400	
	5500	Exit verifier	↓	1	16		16,000	630		16,630	18,600	
	5600	Fee computer	1 Elec	1.50	5.333		12,200	210		12,410	13,700	
	5700	Full sign, 4″ letters	"	2	4		1,100	158		1,258	1,475	
	5800	Inductive loop	2 Elec	4	4		155	158		313	410	
	5900	Ticket spitter with time/date stamp, standard		2	8		5,800	315		6,115	6,850	
	5920	Mag stripe encoding	↓	2	8		17,000	315		17,315	19,200	
	5950	Vehicle detector, microprocessor based	1 Elec	3	2.667		355	105		460	550	
	6000	Parking control software, minimum		.50	16		20,100	630		20,730	23,100	
	6020	Maximum	↓	.20	40	↓	83,500	1,575		85,075	94,500	

11160 | Loading Dock Equipment

		11161 | Loading Dock Equipment	CREW	DAILY OUTPUT	LABOR-HOURS	UNIT	2004 BARE COSTS MAT.	LABOR	EQUIP.	TOTAL	TOTAL INCL O&P	
200	0010	**DOCK BUMPERS** Bolts not included										200
	0012	2″ x 6″ to 4″ x 8″, average	1 Carp	300	.027	B.F.	.53	.88		1.41	2.04	
400	0010	**LOADING DOCK**										400
	0020	Bumpers, rubber blocks 4-1/2″ thk, 10″ H, 14″ long	1 Carp	26	.308	Ea.	43.50	10.15		53.65	65	

11160 | Loading Dock Equipment

	11161 | Loading Dock Equipment	CREW	DAILY OUTPUT	LABOR-HOURS	UNIT	2004 BARE COSTS MAT.	LABOR	EQUIP.	TOTAL	TOTAL INCL O&P	
400	0200 24″ long	1 Carp	22	.364	Ea.	74	12		86	101	400
	0300 36″ long		17	.471		79	15.55		94.55	113	
	0500 12″ high, 14″ long	↓	25	.320		78.50	10.55		89.05	104	
	2200 Dock boards, heavy duty, 60″ x 60″, aluminum, 5,000 lb. cap.					1,125			1,125	1,250	
	4200 Platform lifter, 6′ x 6′, portable, 3,000 lb. capacity					7,150			7,150	7,850	
	6200 Shelters, fabric, for truck or train, scissor arms, minimum	1 Carp	1	8		1,175	264		1,439	1,725	
	6300 Maximum	″	.50	16	↓	1,700	530		2,230	2,750	

11170 | Solid Waste Handling Equipment

	11179 | Waste Handling Equipment	CREW	DAILY OUTPUT	LABOR-HOURS	UNIT	2004 BARE COSTS MAT.	LABOR	EQUIP.	TOTAL	TOTAL INCL O&P	
150	0010 WASTE HANDLING										150
	0020 Compactors, 115 volt, 250#/hr., chute fed	L-4	1	24	Ea.	8,600	735		9,335	10,700	
	0100 Hand fed		2.40	10		6,200	305		6,505	7,325	
	1000 Heavy duty industrial compactor, 0.5 C.Y. capacity		1	24		5,525	735		6,260	7,275	
	1050 1.0 C.Y. capacity	↓	1	24		8,500	735		9,235	10,600	
	1400 For handling hazardous waste materials, 55 gallon drum packer, std.					13,000			13,000	14,300	
	1410 55 gallon drum packer w/HEPA filter					16,300			16,300	17,900	
	1420 55 gallon drum packer w/charcoal & HEPA filter					21,700			21,700	23,900	
	1430 All of the above made explosion proof, add				↓	9,925			9,925	10,900	

11190 | Detention Equipment

	11191 | Detention Equipment	CREW	DAILY OUTPUT	LABOR-HOURS	UNIT	2004 BARE COSTS MAT.	LABOR	EQUIP.	TOTAL	TOTAL INCL O&P	
150	0010 DETENTION EQUIPMENT										150
	0020										
	2000 Cells, prefab., 5′ to 6′ wide, 7′ to 8′ high, 7′ to 8′ deep,										
	2010 bar front, cot, not incl. plumbing	E-4	1.50	21.333	Ea.	7,625	805	50	8,480	9,925	

11400 | Food Service Equipment

	11405 | Food Storage Equipment	CREW	DAILY OUTPUT	LABOR-HOURS	UNIT	2004 BARE COSTS MAT.	LABOR	EQUIP.	TOTAL	TOTAL INCL O&P	
110	0010 FOOD STORAGE EQUIPMENT										110
	2350 Cooler, reach-in, beverage, 6′ long	Q-1	6	2.667	Ea.	3,200	95		3,295	3,675	
	4300 Freezers, reach-in, 44 C.F.		4	4		8,000	143		8,143	9,025	
	4500 68 C.F.	↓	3	5.333	↓	9,150	190		9,340	10,400	

		11405 | Food Storage Equipment	CREW	DAILY OUTPUT	LABOR-HOURS	UNIT	2004 BARE COSTS MAT.	LABOR	EQUIP.	TOTAL	TOTAL INCL O&P
110	4600	Freezer, pre-fab, 8′ x 8′ w/refrigeration	2 Carp	.45	35.556	Ea.	7,425	1,175		8,600	10,100
	4620	8′ x 12′		.35	45.714		9,125	1,500		10,625	12,500
	4640	8′ x 16′		.25	64		11,500	2,100		13,600	16,200
	4660	8′ x 20′		.17	94.118		14,100	3,100		17,200	20,700
	4680	Reach-in, 1 compartment	Q-1	4	4		1,975	143		2,118	2,400
	4700	2 compartment		3	5.333		4,025	190		4,215	4,725
	8300	Refrigerators, reach-in type, 44 C.F.		5	3.200		4,500	114		4,614	5,125
	8310	With glass doors, 68 C.F.		4	4		6,925	143		7,068	7,850
	8320	Refrigerator, reach-in, 1 compartment	R-18	7.80	3.333		1,475	101		1,576	1,775
	8330	2 compartment		6.20	4.194		1,800	127		1,927	2,175
	8340	3 compartment		5.60	4.643		2,650	141		2,791	3,150
	8350	Pre-fab, with refrigeration, 8′ x 8′	2 Carp	.45	35.556		4,750	1,175		5,925	7,150
	8360	8′ x 12′		.35	45.714		6,025	1,500		7,525	9,150
	8370	8′ x 16′		.25	64		7,925	2,100		10,025	12,200
	8380	8′ x 20′		.17	94.118		9,700	3,100		12,800	15,900
	8390	Pass-thru/roll-in, 1 compartment	R-18	7.80	3.333		3,125	101		3,226	3,600
	8400	2 compartment		6.24	4.167		4,475	126		4,601	5,125
	8410	3 compartment		5.60	4.643		5,650	141		5,791	6,450
	8420	Walk-in, alum, door & floor only, no refrig, 6′ x 6′ x 7′-6″	2 Carp	1.40	11.429		6,125	375		6,500	7,375
	8430	10′ x 6′ x 7′-6″		.55	29.091		8,800	960		9,760	11,300
	8440	12′ x 14′ x 7′-6″		.25	64		12,200	2,100		14,300	16,900
	8450	12′ x 20′ x 7′-6″		.17	94.118		15,000	3,100		18,100	21,700
	8460	Refrigerated cabinets, mobile					1,875			1,875	2,075
	8470	Refrigerator/freezer, reach-in, 1 compartment	R-18	5.60	4.643		3,975	141		4,116	4,600
	8480	2 compartment	"	4.80	5.417		5,750	164		5,914	6,575
	8600	Stainless steel shelving, louvered 4-tier, 20″ x 3′	1 Clab	6	1.333		890	34.50		924.50	1,050
	8605	20″ x 4′		6	1.333		1,275	34.50		1,309.50	1,450
	8610	20″ x 6′		6	1.333		1,825	34.50		1,859.50	2,075
	8615	24″ x 3′		6	1.333		1,150	34.50		1,184.50	1,325
	8620	24″ x 4′		6	1.333		1,350	34.50		1,384.50	1,525
	8625	24″ x 6′		6	1.333		1,950	34.50		1,984.50	2,200
	8630	Flat 4-tier, 20″ x 3′		6	1.333		1,025	34.50		1,059.50	1,175
	8635	20″ x 4′		6	1.333		1,225	34.50		1,259.50	1,400
	8640	20″ x 5′		6	1.333		1,425	34.50		1,459.50	1,600
	8645	24″ x 3′		6	1.333		1,075	34.50		1,109.50	1,250
	8650	24″ x 4′		6	1.333		1,250	34.50		1,284.50	1,425
	8655	24″ x 6′		6	1.333		2,650	34.50		2,684.50	2,975
	8700	Galvanized shelving, louvered 4-tier, 20″ x 3′		6	1.333		450	34.50		484.50	555
	8705	20″ x 4′		6	1.333		505	34.50		539.50	615
	8710	20″ x 6′		6	1.333		525	34.50		559.50	635
	8715	24″ x 3′		6	1.333		410	34.50		444.50	515
	8720	24″ x 4′		6	1.333		540	34.50		574.50	655
	8725	24″ x 6′		6	1.333		770	34.50		804.50	910
	8730	Flat 4-tier, 20″ x 3′		6	1.333		320	34.50		354.50	415
	8735	20″ x 4′		6	1.333		360	34.50		394.50	460
	8740	20″ x 6′		6	1.333		650	34.50		684.50	775
	8745	24″ x 3′		6	1.333		315	34.50		349.50	405
	8750	24″ x 4′		6	1.333		480	34.50		514.50	590
	8755	24″ x 6′		6	1.333		690	34.50		724.50	820
	8760	Stainless steel dunnage rack, 24″ x 3′		8	1		625	26		651	730
	8765	24″ x 4′		8	1		670	26		696	785
	8770	Galvanized dunnage rack, 24″ x 3′		8	1		104	26		130	157
	8775	24″ x 4′		8	1		132	26		158	189
800	0010	**WINE CELLAR**, refrigerated, Redwood interior, carpeted, walk-in type									
	0020	6′-8″ high, including racks									

11 EQUIPMENT

Important: See the Reference Section for critical supporting data - Reference Nos., Crews, & City Cost Indexes

11400 | Food Service Equipment

		11405 | Food Storage Equipment	CREW	DAILY OUTPUT	LABOR-HOURS	UNIT	2004 BARE COSTS MAT.	LABOR	EQUIP.	TOTAL	TOTAL INCL O&P	
800	0200	80 ″W x 48″D for 900 bottles	2 Carp	1.50	10.667	Ea.	2,650	350		3,000	3,500	800
	0250	80″ W x 72″ D for 1300 bottles		1.33	12.030		3,500	395		3,895	4,500	
	0300	80″ W x 94″ D for 1900 bottles		1.17	13.675		4,550	450		5,000	5,750	
	0400	80″ W x 124″ D for 2500 bottles		1	16		5,500	530		6,030	6,925	
		11410 | Food Preparation Equipment										
110	0010	FOOD PREPARATION EQUIPMENT										110
	1700	Choppers, 5 pounds	R-18	7	3.714	Ea.	2,025	113		2,138	2,400	
	1720	16 pounds		5	5.200		2,475	158		2,633	2,975	
	1740	35 to 40 pounds		4	6.500		4,275	197		4,472	5,025	
	1840	Coffee brewer, 5 burners	1 Plum	3	2.667		1,050	106		1,156	1,325	
	1850	Coffee urn, twin 6 gallon urns		2	4		6,625	158		6,783	7,550	
	1860	Single, 3 gallon		3	2.667		4,850	106		4,956	5,500	
	3000	Fast food equipment, total package, minimum	6 Skwk	.08	600		129,000	20,200		149,200	175,000	
	3100	Maximum	"	.07	685		175,500	23,100		198,600	230,500	
	3800	Food mixers, 20 quarts	L-7	7	4		2,750	128		2,878	3,200	
	3850	40 quarts		5.40	5.185		6,800	165		6,965	7,750	
	3900	60 quarts		5	5.600		10,000	179		10,179	11,300	
	4040	80 quarts		3.90	7.179		12,100	229		12,329	13,700	
	4080	130 quarts		2.20	12.727		16,200	405		16,605	18,600	
	4100	Floor type, 20 quarts		15	1.867		2,800	59.50		2,859.50	3,175	
	4120	60 quarts		14	2		7,575	64		7,639	8,425	
	4140	80 quarts		12	2.333		10,800	74.50		10,874.50	12,000	
	4160	140 quarts		8.60	3.256		21,500	104		21,604	23,800	
	6700	Peelers, small	R-18	8	3.250		1,800	98.50		1,898.50	2,150	
	6720	Large	"	6	4.333		3,350	131		3,481	3,900	
	6800	Pulper/extractor, close coupled, 5 HP	1 Plum	1.90	4.211		3,000	167		3,167	3,550	
	8580	Slicer with table	R-18	9	2.889		3,675	87.50		3,762.50	4,200	
		11415 | Food Delivery Carts and Conveyors										
110	0010	FOOD DELIVERY CARTS										110
	1650	Cabinet, heated, 1 compartment, reach-in	R-18	5.60	4.643	Ea.	2,050	141		2,191	2,475	
	1655	Pass-thru roll-in		5.60	4.643		3,425	141		3,566	4,000	
	1660	2 compartment, reach-in		4.80	5.417		4,375	164		4,539	5,050	
	1670	Mobile					2,525			2,525	2,775	
	6850	Mobile rack w/pan slide					885			885	970	
	9180	Tray and silver dispenser, mobile	1 Clab	16	.500		705	13		718	795	
		11420 | Food Cooking Equipment										
110	0010	COOKING EQUIPMENT										110
	0020	Bake oven, gas, one section	Q-1	8	2	Ea.	3,925	71.50		3,996.50	4,400	
	0300	Two sections		7	2.286		7,900	81.50		7,981.50	8,800	
	0600	Three sections		6	2.667		11,600	95		11,695	12,900	
	0900	Electric convection, single deck	L-7	4	7		4,150	223		4,373	4,925	
	1300	Broiler, without oven, standard	Q-1	8	2		3,775	71.50		3,846.50	4,250	
	1550	Infra-red	L-7	4	7		5,950	223		6,173	6,925	
	4750	Fryer, with twin baskets, modular model	Q-1	7	2.286		1,725	81.50		1,806.50	2,025	
	5000	Floor model on 6″ legs	"	5	3.200		1,475	114		1,589	1,800	
	5100	Extra single basket, large					87			87	96	
	5300	3′ Long SST griddle, 24″ plate dp w/4″ legs, elec, 208V 3Ph.	Q-1	7	2.286		1,850	81.50		1,931.50	2,150	
	5550	4′ long	"	6	2.667		1,875	95		1,970	2,200	
	6200	Iced tea brewer	1 Plum	3.44	2.326		580	92		672	785	
	6350	20 Gal kettle w/steam jacket, tilting w/positive lock, S/S	L-7	7	4		6,075	128		6,203	6,875	
	6600	60 gallons	"	6	4.667		6,650	149		6,799	7,550	
	6900	Range, restaurant type, 6 burners and 1 standard oven, 36″ wide	Q-1	7	2.286		1,625	81.50		1,706.50	1,900	

11400 | Food Service Equipment

11420 | Food Cooking Equipment

				DAILY	LABOR-		2004 BARE COSTS				TOTAL
			CREW	OUTPUT	HOURS	UNIT	MAT.	LABOR	EQUIP.	TOTAL	INCL O&P
110	6950	Convection	Q-1	7	2.286	Ea.	2,350	81.50		2,431.50	2,700
	7150	2 standard ovens, 24" griddle, 60" wide		6	2.667		4,100	95		4,195	4,675
	7200	1 standard, 1 convection oven		6	2.667		5,000	95		5,095	5,650
	7450	Heavy duty, single 34" standard oven, open top		5	3.200		3,650	114		3,764	4,200
	7500	Convection oven		5	3.200		5,125	114		5,239	5,825
	7700	Griddle top		6	2.667		4,450	95		4,545	5,050
	7750	Convection oven		6	2.667		5,950	95		6,045	6,675
	8850	Steamer, electric 27 KW	L-7	7	4		8,425	128		8,553	9,475
	9100	Electric, 10 KW or gas 100,000 BTU	"	5	5.600		4,000	179		4,179	4,700
	9150	Toaster, conveyor type, 16-22 slices per minute					945			945	1,050
	9160	Pop-up, 2 slot					525			525	580
	9200	For deluxe models of above equipment, add					75%				
	9400	Rule of thumb: Equipment cost based									
	9410	on kitchen work area									
	9420	Office buildings, minimum	L-7	77	.364	S.F.	56.50	11.60		68.10	81
	9450	Maximum		58	.483		95	15.40		110.40	130
	9550	Public eating facilities, minimum		77	.364		74	11.60		85.60	100
	9600	Maximum		46	.609		120	19.40		139.40	164
	9750	Hospitals, minimum		58	.483		76	15.40		91.40	109
	9800	Maximum		39	.718		127	23		150	178

11425 | Hood and Ventilation Equipment

			CREW	DAILY OUTPUT	LABOR-HOURS	UNIT	MAT.	LABOR	EQUIP.	TOTAL	TOTAL INCL O&P
110	0010	**HOOD & VENTILATION EQUIPMENT**									
	2970	Exhaust hood, sst, gutter on all sides, 4' x 4' x 2'	1 Carp	1.80	4.444	Ea.	3,025	147		3,172	3,575
	2980	4'x 4' x 7'	"	1.60	5		4,675	165		4,840	5,425
	7950	Hood fire protection system, minimum	Q-1	3	5.333		3,150	190		3,340	3,750
	8050	Maximum	"	1	16		23,400	570		23,970	26,700

11430 | Food Dispensing Equipment

			CREW	DAILY OUTPUT	LABOR-HOURS	UNIT	MAT.	LABOR	EQUIP.	TOTAL	TOTAL INCL O&P
110	0010	**FOOD DISPENSING EQUIPMENT**									
	1050	Butter pat dispenser	1 Clab	13	.615	Ea.	535	16		551	615
	1100	Bread dispenser, counter top		13	.615		535	16		551	615
	1900	Cup and glass dispenser, drop in		4	2		980	52		1,032	1,150
	1920	Disposable cup, drop in		16	.500		293	13		306	340
	2650	Dish dispenser, drop in, 12"		11	.727		1,125	18.90		1,143.90	1,250
	2660	Mobile		10	.800		1,875	21		1,896	2,075
	3300	Food warmer, counter, 1.2 KW					565			565	620
	3550	1.6 KW					1,125			1,125	1,225
	3600	Well, hot food, built-in, rectangular, 12" x 20"	R-30	10	2.600		335	81.50		416.50	495
	3610	Circular, 7 qt		10	2.600		305	81.50		386.50	465
	3620	Refrigerated, 2 compartments		10	2.600		1,525	81.50		1,606.50	1,800
	3630	3 compartments		9	2.889		1,700	90.50		1,790.50	2,025
	3640	4 compartments		8	3.250		2,375	102		2,477	2,775
	4720	Frost cold plate		9	2.889		11,800	90.50		11,890.50	13,000
	5700	Hot chocolate dispenser	1 Plum	4	2		795	79		874	995
	6250	Jet spray dispenser	R-18	4.50	5.778		2,425	175		2,600	2,925
	6300	Juice dispenser, concentrate	"	4.50	5.778		900	175		1,075	1,275
	6690	Milk dispenser, bulk, 2 flavor	R-30	8	3.250		1,075	102		1,177	1,375
	6695	3 flavor	"	8	3.250		1,475	102		1,577	1,800
	8800	Serving counter, straight	1 Carp	40	.200	L.F.	570	6.60		576.60	635
	8820	Curved section	"	30	.267	"	710	8.80		718.80	795
	8825	Solid surface, see section 06620-810									
	8830	Soft serve ice cream machine, medium	R-18	11	2.364	Ea.	7,800	71.50		7,871.50	8,700
	8840	Large	"	9	2.889	"	11,600	87.50		11,687.50	12,800

11400 | Food Service Equipment

11435 | Ice Machines

			CREW	DAILY OUTPUT	LABOR-HOURS	UNIT	2004 BARE COSTS MAT.	LABOR	EQUIP.	TOTAL	TOTAL INCL O&P	
110	0010	**ICE MACHINES**										110
	5800	Ice cube maker, 50 pounds per day	Q-1	6	2.667	Ea.	1,275	95		1,370	1,550	
	5900	250 pounds per day		1.20	13.333		1,675	475		2,150	2,575	
	6050	500 pounds per day		4	4		2,800	143		2,943	3,325	
	6060	With bin		1.20	13.333		3,250	475		3,725	4,300	
	6090	1000 pounds per day, with bin		1	16		6,125	570		6,695	7,625	
	6100	Ice flakers, 300 pounds per day		1.60	10		2,825	355		3,180	3,650	
	6120	600 pounds per day		.95	16.842		3,675	600		4,275	4,950	
	6130	1000 pounds per day		.75	21.333		4,800	760		5,560	6,450	
	6140	2000 pounds per day		.65	24.615		13,600	880		14,480	16,400	
	6160	Ice storage bin, 500 pound capacity	Q-5	1	16		1,175	570		1,745	2,200	
	6180	1000 pound	″	.56	28.571		2,825	1,025		3,850	4,675	

11440 | Cleaning and Disposal Equipment

			CREW	DAILY OUTPUT	LABOR-HOURS	UNIT	MAT.	LABOR	EQUIP.	TOTAL	TOTAL INCL O&P	
110	0010	**CLEANING & DISPOSAL EQUIPMENT**										110
	2700	Dishwasher, commercial, rack type										
	2720	10 to 12 racks per hour	Q-1	3.20	5	Ea.	2,900	178		3,078	3,475	
	2750	Semi-automatic 38 to 50 racks per hour	″	1.30	12.308		6,000	440		6,440	7,275	
	2800	Automatic, 190 to 230 racks per hour	L-6	.35	34.286		8,200	1,350		9,550	11,100	
	2820	235 to 275 racks per hour		.25	48		19,000	1,900		20,900	23,800	
	2840	8,750 to 12,500 dishes per hour		.10	120		41,200	4,750		45,950	52,500	
	2950	Dishwasher hood, canopy type	L-3A	10	1.200	L.F.	430	43.50		473.50	545	
	2960	Pant leg type	″	2.50	4.800	Ea.	6,050	174		6,224	6,925	
	5200	Garbage disposal 1.5 HP, 100 GPH	L-1	4.80	3.333		1,425	132		1,557	1,775	
	5210	3 HP, 120 GPH		4.60	3.478		2,500	137		2,637	2,950	
	5220	5 HP, 250 GPH		4.50	3.556		3,250	140		3,390	3,800	
	6750	Pot sink, 3 compartment	1 Plum	7.25	1.103	L.F.	545	43.50		588.50	670	
	6760	Pot washer, small		1.60	5	Ea.	15,800	198		15,998	17,700	
	6770	Large		1.20	6.667		35,200	264		35,464	39,200	
	9170	Trash compactor, small, up to 125 lb. compacted weight	L-4	4	6		16,000	184		16,184	17,900	
	9175	Large, up to 175 lb. compacted weight	″	3	8		19,500	245		19,745	21,900	

EQUIPMENT 11

11450 | Residential Equipment

11454 | Residential Appliances

			CREW	DAILY OUTPUT	LABOR-HOURS	UNIT	2004 BARE COSTS MAT.	LABOR	EQUIP.	TOTAL	TOTAL INCL O&P	
500	0010	**RESIDENTIAL APPLIANCES**										500
	0020	Cooking range, 30″ free standing, 1 oven, minimum	2 Clab	10	1.600	Ea.	241	41.50		282.50	335	
	0050	Maximum	″	4	4		1,450	104		1,554	1,750	
	0350	Built-in, 30″ wide, 1 oven, minimum	1 Elec	6	1.333		445	52.50		497.50	570	
	0400	Maximum	2 Carp	2	8		1,325	264		1,589	1,900	
	0900	Counter top cook tops, 4 burner, standard, minimum	1 Elec	6	1.333		176	52.50		228.50	274	
	0950	Maximum		3	2.667		440	105		545	645	
	1250	Microwave oven, minimum		4	2		75.50	79		154.50	204	
	1300	Maximum		2	4		390	158		548	670	
	1750	Compactor, residential size, 4 to 1 compaction, minimum	1 Carp	5	1.600		415	53		468	545	
	1800	Maximum	″	3	2.667		470	88		558	660	
	2750	Dishwasher, built-in, 2 cycles, minimum	L-1	4	4		245	158		403	515	
	2800	Maximum		2	8		289	315		604	810	
	3300	Garbage disposer, sink type, minimum		10	1.600		42.50	63		105.50	144	
	3350	Maximum		10	1.600		145	63		208	258	
	4150	Hood for range, 2 speed, vented, 30″ wide, minimum	L-3	5	3.200		37.50	115		152.50	226	

11454 | Residential Appliances

			CREW	DAILY OUTPUT	LABOR-HOURS	UNIT	2004 BARE COSTS MAT.	LABOR	EQUIP.	TOTAL	TOTAL INCL O&P	
500	4200	Maximum	L-3	3	5.333	Ea.	595	192		787	965	500
	4300	42″ wide, minimum		5	3.200		225	115		340	430	
	4330	Custom		5	3.200		625	115		740	870	
	4350	Maximum		3	5.333		760	192		952	1,150	
	4400	Ventless hood, 2 speed, 30″ wide, minimum	1 Elec	8	1		37.50	39.50		77	102	
	4450	Maximum		7	1.143		129	45		174	211	
	4510	36″ wide, minimum		7	1.143		39	45		84	112	
	4550	Maximum		6	1.333		285	52.50		337.50	395	
	4580	42″ wide, minimum		6	1.333		201	52.50		253.50	300	
	4600	Maximum		5	1.600		269	63		332	395	
	4650	For vented 1 speed, deduct from maximum					39			39	43	
	4700											
	5380	Oven, built in, standard	1 Elec	4	2	Ea.	385	79		464	540	
	5390	Deluxe	″	2	4		1,625	158		1,783	2,025	
	5500	Refrigerator, no frost, 10 C.F. to 12 C.F. minimum	2 Clab	10	1.600		440	41.50		481.50	555	
	5600	Maximum	″	6	2.667		680	69.50		749.50	865	
	6400	Sump pump cellar drainer, pedestal, 1/3 H.P., molded PVC base	1 Plum	3	2.667		87	106		193	260	
	6450	Solid brass	″	2	4		179	158		337	445	
	6460	Sump pump, see also division 15440-940										
	7350	Water softener, automatic, to 30 grains per gallon	2 Plum	5	3.200	Ea.	430	127		557	670	
	7400	To 100 grains per gallon	″	4	4	″	600	158		758	905	
550	0010	**DISAPPEARING STAIRWAY** No trim included										550
	0020	One piece, yellow pine, 8′-0″ ceiling	2 Carp	4	4	Ea.	890	132		1,022	1,200	
	0030	9′-0″ ceiling		4	4		900	132		1,032	1,200	
	0040	10′-0″ ceiling		3	5.333		950	176		1,126	1,350	
	0050	11′-0″ ceiling		3	5.333		1,125	176		1,301	1,525	
	0060	12′-0″ ceiling		3	5.333		1,150	176		1,326	1,575	
	0100	Custom grade, pine, 8′-6″ ceiling, minimum	1 Carp	4	2		99.50	66		165.50	218	
	0150	Average		3.50	2.286		100	75.50		175.50	235	
	0200	Maximum		3	2.667		165	88		253	330	
	0250											
	0500	Heavy duty, pivoted, from 7′-7″ to 12′-10″ floor to floor	1 Carp	3	2.667	Ea.	325	88		413	505	
	0600	16′-0″ ceiling		2	4		1,075	132		1,207	1,425	
	0800	Economy folding, pine, 8′-6″ ceiling		4	2		90.50	66		156.50	209	
	0900	9′-6″ ceiling		4	2		98.50	66		164.50	217	
	1000	Fire escape, galvanized steel, 8′-0″ to 10′-4″ ceiling	2 Carp	1	16		1,150	530		1,680	2,150	
	1010	10′-6″ to 13′-6″ ceiling		1	16		1,450	530		1,980	2,475	
	1100	Automatic electric, aluminum, floor to floor height, 8′ to 9′		1	16		5,675	530		6,205	7,125	
	1500	11′ to 12′		.90	17.778		5,725	585		6,310	7,275	
	1700	14′ to 15′		.70	22.857		6,500	755		7,255	8,400	
	9000	Minimum labor/equipment charge	1 Carp	2	4	Job		132		132	219	

11460 | Unit Kitchens

			CREW	DAILY OUTPUT	LABOR-HOURS	UNIT	MAT.	LABOR	EQUIP.	TOTAL	TOTAL INCL O&P	
100	0010	**UNIT KITCHENS**										100
	1500	Combination range, refrigerator and sink, 30″ wide, minimum	L-1	2	8	Ea.	730	315		1,045	1,300	
	1550	Maximum	″	1	16	″	1,450	630		2,080	2,575	
	1640	Combination range, refrigerator, sink, microwave										
	1660	oven and ice maker	L-1	.80	20	Ea.	4,025	790		4,815	5,650	

11470 | Darkroom Equipment

	11471 | Darkroom Processing		CREW	DAILY OUTPUT	LABOR-HOURS	UNIT	2004 BARE COSTS MAT.	LABOR	EQUIP.	TOTAL	TOTAL INCL O&P
700	0010	**DARKROOM EQUIPMENT**									
	0020	Developing sink, 5″ deep, 24″ x 48″	Q-1	2	8	Ea.	4,100	285		4,385	4,975
	0050	48″ x 52″		1.70	9.412		4,175	335		4,510	5,100
	0200	10″ deep, 24″ x 48″		1.70	9.412		5,125	335		5,460	6,150
	0250	24″ x 108″		1.50	10.667		1,850	380		2,230	2,650
	0500	Dryers, dehumidified filtered air, 36″ x 25″ x 68″ high	L-7	6	4.667		4,000	149		4,149	4,650
	0550	48″ x 25″ x 68″ high		5	5.600		7,375	179		7,554	8,425
	2000	Processors, automatic, color print, minimum		4	7		10,400	223		10,623	11,800
	2050	Maximum		.60	46.667		10,600	1,500		12,100	14,000
	2300	Black and white print, minimum		2	14		7,875	445		8,320	9,375
	2350	Maximum		.80	35		51,500	1,125		52,625	58,500
	2600	Manual processor, 16″ x 20″ maximum print size		2	14		7,925	445		8,370	9,450
	2650	20″ x 24″ maximum print size		1	28		7,450	895		8,345	9,650
	3000	Viewing lites, 20″ x 24″		6	4.667		325	149		474	600
	3100	20″ x 24″ with color correction		6	4.667		460	149		609	750
	3500	Washers, round, minimum sheet 11″ x 14″	Q-1	2	8		2,525	285		2,810	3,225
	3550	Maximum sheet 20″ x 24″		1	16		2,975	570		3,545	4,150
	3800	Square, minimum sheet 20″ x 24″		1	16		2,650	570		3,220	3,800
	3900	Maximum sheet 50″ x 56″		.80	20		4,150	715		4,865	5,650
	4500	Combination tank sink, tray sink, washers, with									
	4510	dry side tables, average	Q-1	.45	35.556	Ea.	7,875	1,275		9,150	10,600

	11472 | Revolving Darkroom Doors		CREW	DAILY OUTPUT	LABOR-HOURS	UNIT	MAT.	LABOR	EQUIP.	TOTAL	TOTAL INCL O&P
370	0010	**DARKROOM DOORS**									
	0015	Revolving, standard, 2 way, 36″ diameter	2 Carp	3.10	5.161	Opng.	1,675	170		1,845	2,125
	0020	41″ diameter		3.10	5.161		1,825	170		1,995	2,275
	0050	3 way, 51″ diameter		1.40	11.429		2,275	375		2,650	3,125
	1000	4 way, 49″ diameter		1.40	11.429		2,700	375		3,075	3,600
	2000	Hinged safety, 2 way, 41″ diameter		2.30	6.957		2,175	230		2,405	2,775
	2500	3 way, 51″ diameter		1.40	11.429		2,750	375		3,125	3,650
	3000	Pop out safety, 2 way, 41″ diameter		3.10	5.161		2,675	170		2,845	3,225
	4000	3 way, 51″ diameter		1.40	11.429		2,675	375		3,050	3,575
	5000	Wheelchair-type, pop out, 51″ diameter		1.40	11.429		2,675	375		3,050	3,575
	5020	72″ diameter		.90	17.778		5,550	585		6,135	7,100

11480 | Athletic Recreational & Therapeutic Equipment

	11484 | Exercise Equipment		CREW	DAILY OUTPUT	LABOR-HOURS	UNIT	2004 BARE COSTS MAT.	LABOR	EQUIP.	TOTAL	TOTAL INCL O&P
400	0010	**HEALTH CLUB EQUIPMENT**									
	0020	Abdominal rack, 2 board capacity				Ea.	395			395	435
	0050	Abdominal board, upholstered					445			445	490
	0200	Bicycle trainer, minimum					690			690	760
	0300	Deluxe, electric					3,525			3,525	3,875
	0400	Bar bell set, chrome plated steel, 25 lbs.					207			207	228
	0420	100 lbs.					315			315	345
	0450	200 lbs.					605			605	665
	0500	Weight plates, cast iron, per lb.				Lb.	4.24			4.24	4.66
	0520	Storage rack, 10 station				Ea.	670			670	735

11480 | Athletic Recreational & Therapeutic Equipment

11484 | Exercise Equipment

			CREW	DAILY OUTPUT	LABOR-HOURS	UNIT	2004 BARE COSTS MAT.	LABOR	EQUIP.	TOTAL	TOTAL INCL O&P	
400	0600	Circuit training apparatus, 12 machines minimum	2 Clab	1.25	12.800	Set	23,300	335		23,635	26,300	400
	0700	Average		1	16		28,700	415		29,115	32,200	
	0800	Maximum	↓	.75	21.333		34,000	555		34,555	38,300	
	0820	Dumbbell set, cast iron, with rack and 5 pair				↓	650			650	715	
	0900	Squat racks	2 Clab	5	3.200	Ea.	645	83		728	850	
	1600	For saunas, see Div. 13035-800										
	1640	For steam baths, see Div. 13035-940										

11486 | Gymnasium Equipment

			CREW	DAILY OUTPUT	LABOR-HOURS	UNIT	MAT.	LABOR	EQUIP.	TOTAL	TOTAL INCL O&P	
700	0010	SCHOOL EQUIPMENT										700
	0200	For exterior equipment, see Div. 02880										
	0201	For chairs & desks, see Div. 12560-200										
	0300	For chalkboards & bulletin boards, see Div. 10120-940 & 10110-240										
	0400	For lockers, see Div. 10505-500										
	1000	Basketball backstops, wall mtd., 6′ extended, fixed, minimum	L-2	1	16	Ea.	965	460		1,425	1,800	
	1100	Maximum		1	16		1,425	460		1,885	2,325	
	1200	Swing up, minimum		1	16		1,150	460		1,610	2,000	
	1250	Maximum		1	16		2,050	460		2,510	3,025	
	1300	Portable, manual, heavy duty, spring operated		1.90	8.421		9,625	242		9,867	11,000	
	1400	Ceiling suspended, stationary, minimum		.78	20.513		1,725	590		2,315	2,850	
	1450	Fold up, with accessories, maximum	↓	.40	40		5,175	1,150		6,325	7,575	
	1600	For electrically operated, add	1 Elec	1	8	↓	1,700	315		2,015	2,350	
	2000	Benches, folding, in wall, 14′ table, 2 benches	L-4	2	12	Set	520	365		885	1,175	
	3000	Bleachers, telescoping, manual to 15 tier, minimum	F-5	65	.492	Seat	52	16.50		68.50	84.50	
	3100	Maximum		60	.533		78	17.85		95.85	116	
	3300	16 to 20 tier, minimum		60	.533		125	17.85		142.85	167	
	3400	Maximum		55	.582		156	19.50		175.50	205	
	3600	21 to 30 tier, minimum		50	.640		130	21.50		151.50	179	
	3700	Maximum	↓	40	.800		156	27		183	217	
	3900	For integral power operation, add, minimum	2 Elec	300	.053		26	2.10		28.10	31.50	
	4000	Maximum	″	250	.064	↓	41.50	2.52		44.02	50	
	4100	Boxing ring, elevated, 22′ x 22′	L-4	.10	240	Ea.	8,150	7,350		15,500	21,000	
	4110	For cellular plastic foam padding, add		.10	240		2,750	7,350		10,100	15,000	
	4120	Floor level, including posts and ropes only, 20′ x 20′		.80	30		1,475	920		2,395	3,125	
	4130	Canvas, 30′ x 30′	↓	5	4.800		1,000	147		1,147	1,350	
	4150	Exercise equipment, bicycle trainer					365			365	405	
	4180	Chinning bar, adjustable, wall mounted	1 Carp	5	1.600		260	53		313	375	
	4200	Exercise ladder, 16′ x 1′-7″, suspended	L-2	3	5.333		1,125	153		1,278	1,500	
	4210	High bar, floor plate attached	1 Carp	4	2		980	66		1,046	1,175	
	4240	Parallel bars, adjustable		4	2		2,150	66		2,216	2,475	
	4270	Uneven parallel bars, adjustable	↓	4	2	↓	2,575	66		2,641	2,925	
	4280	Wall mounted, adjustable	L-2	1.50	10.667	Set	720	305		1,025	1,300	
	4300	Rope, ceiling mounted, 18′ long	1 Carp	3.66	2.186	Ea.	192	72		264	330	
	4330	Side horse, vaulting		5	1.600		1,550	53		1,603	1,800	
	4360	Treadmill, motorized, deluxe, training type	↓	5	1.600		4,925	53		4,978	5,525	
	4390	Weight lifting multi-station, minimum	2 Clab	1	16	↓	2,325	415		2,740	3,250	
	4500	Gym divider curtain, mesh top, vinyl bottom, manual	L-4	500	.048	S.F.	6.85	1.47		8.32	9.95	
	4700	Electric roll up	L-7	400	.070		10.05	2.23		12.28	14.70	
	5500	Gym mats, 2″ thick, naugahyde covered					3.20			3.20	3.52	
	5600	Vinyl/nylon covered					4.81			4.81	5.30	
	5800	Wall pads, 1-1/2″ thick	2 Carp	640	.025		5.25	.83		6.08	7.15	
	6000	Wrestling mats, 1″ thick, heavy duty				↓	4.29			4.29	4.72	
	7000	Scoreboards, baseball, minimum	R-3	1.30	15.385	Ea.	2,625	595	122	3,342	3,950	
	7200	Maximum		.05	400		14,200	15,500	3,175	32,875	43,100	
	7300	Football, minimum	↓	.86	23.256	↓	3,600	900	185	4,685	5,575	

11480 | Athletic Recreational & Therapeutic Equipment

	11486 | Gymnasium Equipment		CREW	DAILY OUTPUT	LABOR-HOURS	UNIT	2004 BARE COSTS MAT.	LABOR	EQUIP.	TOTAL	TOTAL INCL O&P	
700	7400	Maximum	R-3	.20	100	Ea.	12,300	3,875	795	16,970	20,400	700
	7500	Basketball (one side), minimum	↓	2.07	9.662	↓	2,000	375	77	2,452	2,875	
	7600	Maximum		.30	66.667		12,400	2,575	530	15,505	18,300	
	7700	Hockey-basketball (four sides), minimum		.25	80		5,650	3,100	635	9,385	11,700	
	7800	Maximum	↓	.15	133	↓	22,300	5,150	1,050	28,500	33,700	

11500 | Industrial & Process Equipment

	11520 | Industrial Equipment		CREW	DAILY OUTPUT	LABOR-HOURS	UNIT	2004 BARE COSTS MAT.	LABOR	EQUIP.	TOTAL	TOTAL INCL O&P	
850	0010	VOCATIONAL SHOP EQUIPMENT										850
	0020	Benches, work, wood, average	2 Carp	5	3.200	Ea.	390	106		496	605	
	0100	Metal, average	↓	5	3.200	↓	370	106		476	580	
	0400	Combination belt & disc sander, 6″		4	4		720	132		852	1,000	
	0700	Drill press, floor mounted, 12″, 1/2 H.P.	↓	4	4		335	132		467	590	
	0800	Dust collector, not incl. ductwork, 6″ diameter	1 Shee	1.10	7.273		2,725	282		3,007	3,450	
	1000	Grinders, double wheel, 1/2 H.P.	2 Carp	5	3.200		226	106		332	425	
	1300	Jointer, 4″, 3/4 H.P.	↓	4	4		845	132		977	1,150	
	1600	Kilns, 16 C.F., to 2000°		4	4		1,950	132		2,082	2,350	
	1900	Lathe, woodworking, 10″, 1/2 H.P.		4	4		630	132		762	910	
	2200	Planer, 13″ x 6″		4	4		1,225	132		1,357	1,575	
	2500	Potter's wheel, motorized		4	4		990	132		1,122	1,325	
	2800	Saws, band, 14″, 3/4 H.P.		4	4		760	132		892	1,050	
	3100	Metal cutting band saw, 14″		4	4		1,025	132		1,157	1,350	
	3400	Radial arm saw, 10″, 2 H.P.		4	4		665	132		797	950	
	3700	Scroll saw, 24″		4	4		1,400	132		1,532	1,750	
	4000	Table saw, 10″, 3 H.P.		4	4		1,525	132		1,657	1,900	
	4300	Welder AC arc, 30 amp capacity	↓	4	4	↓	2,450	132		2,582	2,925	

11600 | Laboratory Equipment

	11620 | Laboratory Equipment		CREW	DAILY OUTPUT	LABOR-HOURS	UNIT	2004 BARE COSTS MAT.	LABOR	EQUIP.	TOTAL	TOTAL INCL O&P	
110	0010	LABORATORY EQUIPMENT										110
	1400	Safety equipment, eye wash, hand held				Ea.	294			294	325	
	1450	Deluge shower				″	565			565	625	
	8000	Alternate pricing method: as percent of lab furniture										
	8050	Installation, not incl. plumbing & duct work				% Furn.					22%	
	8100	Plumbing, final connections, simple system				↓					10%	
	8110	Moderately complex system									15%	
	8120	Complex system									20%	
	8150	Electrical, simple system									10%	
	8160	Moderately complex system									20%	
	8170	Complex system				↓					35%	

Division Notes

	CREW	DAILY OUTPUT	LABOR-HOURS	UNIT	2004 BARE COSTS MAT.	LABOR	EQUIP.	TOTAL	TOTAL INCL O&P

Division 12 Furnishings

Estimating Tips

General

- The items in this division are usually priced per square foot or each. Most of these items are purchased by the owner and placed by the supplier. Do not assume the items in Division 12 will be purchased and installed by the supplier. Check the specifications for responsibilities and include receiving, storage, installation and mechanical and electrical hook-ups in the appropriate divisions.
- Some items in this division require some type of support system that is not usually furnished with the item. Examples of these systems include blocking for the attachment of casework and heavy drapery rods. The required blocking must be added to the estimate in the appropriate division.

Reference Numbers

Reference numbers are shown in bold squares at the beginning of some major classifications. These numbers refer to related items in the Reference Section. The reference information may be an estimating procedure, an alternate pricing method or technical information.

Note: Not all subdivisions listed here necessarily appear in this publication.

12300 | Manufactured Casework

		12350 | Specialty Casework	CREW	DAILY OUTPUT	LABOR-HOURS	UNIT	2004 BARE COSTS MAT.	LABOR	EQUIP.	TOTAL	TOTAL INCL O&P	
500	0010	**LABORATORY CASEWORK**										500
	0020	Cabinets, base, door units, metal	2 Carp	18	.889	L.F.	140	29.50		169.50	204	
	0300	Drawer units	↓	18	.889	↓	305	29.50		334.50	385	
	0700	Tall storage cabinets, open, 7′ high		20	.800		305	26.50		331.50	380	
	0900	With glazed doors		20	.800		375	26.50		401.50	455	
	1300	Wall cabinets, metal, 12-1/2″ deep, open		20	.800		94.50	26.50		121	148	
	1500	With doors		20	.800	↓	197	26.50		223.50	260	
	1550	Counter tops, not incl. base cabinets, acidproof, minimum		82	.195	S.F.	23	6.45		29.45	36	
	1600	Maximum		70	.229	↓	32	7.55		39.55	47.50	
	1650	Stainless steel	↓	82	.195	↓	73	6.45		79.45	90.50	

12400 | Furnishings & Accessories

		12492 | Blinds and Shades	CREW	DAILY OUTPUT	LABOR-HOURS	UNIT	2004 BARE COSTS MAT.	LABOR	EQUIP.	TOTAL	TOTAL INCL O&P	
100	0010	**BLINDS, INTERIOR**										100
	0020	Horizontal, 1″ aluminum slats, solid color, stock	1 Carp	590	.014	S.F.	2.82	.45		3.27	3.84	
	3000	Wood folding panels with movable louvers, 7″ x 20″ each	↓	17	.471	Pr.	42	15.55		57.55	72	
	4000	Fixed louver type, stock units, 8″ x 20″ each		17	.471	↓	63	15.55		78.55	95	
	4450	18″ x 40″ each	↓	17	.471	↓	131	15.55		146.55	170	
600	0011	**SHADES** Basswood roll-up, stain finish, 3/8″ slats	1 Carp	300	.027	S.F.	9.85	.88		10.73	12.25	600
	0030	Double layered, heat reflective	↓	685	.012	↓	6.90	.39		7.29	8.25	
	0250	7/8″ slats		300	.027		9.30	.88		10.18	11.65	
	0950	Mylar, single layer, non-heat reflective		685	.012		4.21	.39		4.60	5.25	
	1050	Double layered, heat reflective		685	.012		7.15	.39		7.54	8.50	
	1150	Triple layered, heat reflective		685	.012		8.10	.39		8.49	9.55	
	5000	Thermal, roll up, R-4		44	.182		8.45	6		14.45	19.25	
	5030	R-10.7	↓	44	.182	↓	9.65	6		15.65	20.50	
	5050	Magnetic clips, set of 20				Set	18.05			18.05	19.85	

12500 | Furniture

		12510 | Office Furniture	CREW	DAILY OUTPUT	LABOR-HOURS	UNIT	2004 BARE COSTS MAT.	LABOR	EQUIP.	TOTAL	TOTAL INCL O&P	
600	0010	**OFFICE CASE GOODS**										600
	0020	Desks, 29″ high, double pedestal, 30″ x 60″, metal, minimum				Ea.	295			295	325	
	0030	Maximum				″	1,050			1,050	1,150	
		12540 | Hospitality Furniture										
500	0010	**FURNITURE, HOTEL**										500
	0020	Standard quality set, minimum				Room	1,525			1,525	1,675	
	0200	Maximum				″	7,900			7,900	8,675	

		12540 | Hospitality Furniture	CREW	DAILY OUTPUT	LABOR-HOURS	UNIT	2004 BARE COSTS MAT.	LABOR	EQUIP.	TOTAL	TOTAL INCL O&P	
700	0010	**FURNITURE, RESTAURANT**										700
	0020	Bars, built-in, front bar	1 Carp	5	1.600	L.F.	204	53		257	310	
	0200	Back bar		5	1.600	"	148	53		201	251	
	0500	Booth unit, molded plastic, stub wall and 2 seats, minimum		2	4	Set	278	132		410	525	
	0600	Maximum		1.50	5.333	"	875	176		1,051	1,250	
	0800	Booth seat, upholstered, foursome, single (end) minimum		5	1.600	Ea.	735	53		788	895	
	0900	Maximum		4	2		1,325	66		1,391	1,550	
	1000	Foursome, double, minimum		4	2		735	66		801	915	
	1100	Maximum		3	2.667		1,325	88		1,413	1,600	
	1300	Circle booth, upholstered, 1/4 circle, minimum		3	2.667		480	88		568	675	
	1400	Maximum		2	4		1,100	132		1,232	1,425	
	1500	3/4 circle, minimum		1.50	5.333		1,425	176		1,601	1,875	
	1600	Maximum	▼	1	8	▼	3,050	264		3,314	3,800	
		12560 | Institutional Furniture										
310	0010	**DORMITORY FURNITURE** Beds, free standing, minimum				Ea.	294			294	325	310
	0100	Maximum				"	635			635	700	
	2050	Rule of thumb: Total cost for furniture, minimum				Student					1,975	
	2150	Maximum				"					3,850	
400	0010	**FURNITURE, HOSPITAL**										400
	0020	Beds, manual, minimum				Ea.	865			865	950	
	0100	Maximum				"	2,375			2,375	2,625	
	1100	Patient wall systems, not incl. plumbing, minimum				Room	870			870	960	
	1200	Maximum				"	1,600			1,600	1,775	

For information about Means Estimating Seminars, see yellow pages 12 and 13 in back of book

Division Notes

		CREW	DAILY OUTPUT	LABOR-HOURS	UNIT	2004 BARE COSTS MAT.	LABOR	EQUIP.	TOTAL	TOTAL INCL O&P

Division 13
Special Construction

Estimating Tips

General

- The items and systems in this division are usually estimated, purchased, supplied and installed as a unit by one or more subcontractors. The estimator must ensure that all parties are operating from the same set of specifications and assumptions and that all necessary items are estimated and will be provided. Many times the complex items and systems are covered but the more common ones such as excavation or a crane are overlooked for the very reason that everyone assumes nobody could miss them. The estimator should be the central focus and be able to ensure that all systems are complete.
- Another area where problems can develop in this division is at the interface between systems. The estimator must ensure, for instance, that anchor bolts, nuts and washers are estimated and included for the air-supported structures and pre-engineered buildings to be bolted to their foundations. Utility supply is a common area where essential items or pieces of equipment can be missed or overlooked due to the fact that each subcontractor may feel it is the others' responsibility. The estimator should also be aware of certain items which may be supplied as part of a package but installed by others, and ensure that the installing contractor's estimate includes the cost of installation. Conversely, the estimator must also ensure that items are not costed by two different subcontractors, resulting in an inflated overall estimate.

13120 Pre-Engineered Structures

- The foundations and floor slab, as well as rough mechanical and electrical, should be estimated, as this work is required for the assembly and erection of the structure. Generally, as noted in the book, the pre-engineered building comes as a shell and additional features must be included by the estimator. Here again, the estimator must have a clear understanding of the scope of each portion of the work and all the necessary interfaces.

13200 Storage Tanks

- The prices in this subdivision for above and below ground storage tanks do not include foundations or hold-down slabs. The estimator should refer to Divisions 2 and 3 for foundation system pricing. In addition to the foundations, required tank accessories such as tank gauges, leak detection devices, and additional manholes and piping must be added to the tank prices.

Reference Numbers

Reference numbers are shown in bold squares at the beginning of some major classifications. These numbers refer to related items in the Reference Section. The reference information may be an estimating procedure, an alternate pricing method or technical information.

Note: Not all subdivisions listed here necessarily appear in this publication.

13030 | Special Purpose Rooms

		13035 | Special Purpose Rooms	CREW	DAILY OUTPUT	LABOR-HOURS	UNIT	2004 BARE COSTS MAT.	LABOR	EQUIP.	TOTAL	TOTAL INCL O&P	
700	0010	**REFRIGERATION** Curbs, 12″ high, 4″ thick, concrete	2 Carp	58	.276	L.F.	3.02	9.10		12.12	18.40	700
	6300	Rule of thumb for complete units, w/o doors & refrigeration, cooler		146	.110	SF Flr.	88	3.62		91.62	103	
	6400	Freezer	↓	109.60	.146	″	104	4.82		108.82	122	
800	0010	**SAUNA** Prefabricated, incl. heater & controls, 7′ high, 6′ x 4′, C/C	L-7	2.20	12.727	Ea.	3,450	405		3,855	4,475	800
	1700	Door only, cedar, 2′x6′, with tempered insulated glass window	2 Carp	3.40	4.706		455	155		610	760	
	1800	Prehung, incl. jambs, pulls & hardware	″	12	1.333		465	44		509	590	
	2500	Heaters only (incl. above), wall mounted, to 200 C.F.				↓	450			450	495	
	4480	For additional equipment, see div. 11484-400										
900	0010	**SPORT COURT** Floors, No. 2 & better maple, 25/32″ thick				SF Flr.					6.65	900
	0300	Squash, regulation court in existing building, minimum				Court					15,500	
	0400	Maximum				″					29,000	
	0450	Rule of thumb for components:										
	0470	Walls	3 Carp	.15	160	Court	10,800	5,275		16,075	20,700	
	0500	Floor	″	.25	96		4,850	3,175		8,025	10,600	
	0550	Lighting	2 Elec	.60	26.667	↓	1,525	1,050		2,575	3,300	
940	0010	**STEAM BATH** Heater, timer & head, single, to 140 C.F.	1 Plum	1.20	6.667	Ea.	930	264		1,194	1,425	940
	0500	To 300 C.F.	″	1.10	7.273		1,050	288		1,338	1,600	
	2000	Multiple, motels, apts., 2 baths, w/ blow-down assm., 500 C.F.	Q-1	1.30	12.308		3,575	440		4,015	4,600	
	2500	4 baths	″	.70	22.857	↓	3,875	815		4,690	5,525	

13080 | Sound, Vibration & Seismic Control

		13081 | Sound Control	CREW	DAILY OUTPUT	LABOR-HOURS	UNIT	2004 BARE COSTS MAT.	LABOR	EQUIP.	TOTAL	TOTAL INCL O&P	
100	0010	**ACOUSTICAL** Enclosure, 4″ thick wall and ceiling panels										100
	0020	8# per S.F., up to 12′ span	3 Carp	72	.333	SF Surf	25	11		36	46	
	0300	Better quality panels, 10.5# per S.F.		64	.375		28.50	12.40		40.90	51.50	
	0400	Reverb-chamber, 4″ thick, parallel walls	↓	60	.400	↓	35.50	13.20		48.70	61	

13090 | Radiation Protection

		13091 | X-Ray/Radio Freq Protection	CREW	DAILY OUTPUT	LABOR-HOURS	UNIT	2004 BARE COSTS MAT.	LABOR	EQUIP.	TOTAL	TOTAL INCL O&P	
600	0010	**SHIELDING LEAD**										600
	0050											
	0300	Lead sheets, 1/16″ thick	2 Lath	135	.119	S.F.	4.35	3.62		7.97	10.50	
	0400	1/8″ thick		120	.133		8.90	4.07		12.97	16.25	
	0500	Lead shielding, 1/4″ thick		135	.119		15.35	3.62		18.97	22.50	
	0550	1/2″ thick	↓	120	.133	↓	30	4.07		34.07	39.50	
	0600	Lead glass, 1/4″ thick, 2.0 mm LE, 12″ x 16″	2 Glaz	13	1.231	Ea.	247	39.50		286.50	335	
	0700	24″ x 36″	″	8	2	″	950	64		1,014	1,150	
	1200	X-ray protection, average radiography or fluoroscopy										
	1210	room, up to 300 S.F. floor, 1/16″ lead, minimum	2 Lath	.25	64	Total	4,850	1,950		6,800	8,450	

Important: See the Reference Section for critical supporting data - Reference Nos., Crews, & City Cost Indexes

13090 | Radiation Protection

13091 | X-Ray/Radio Freq Protection

	Line	Description	Crew	Daily Output	Labor-Hours	Unit	2004 Bare Costs Mat.	Labor	Equip.	Total	Total Incl O&P	
600	1500	Maximum, 7'-0" walls	2 Lath	.15	106	Total	6,025	3,250		9,275	11,800	600
	1600	Deep therapy X-ray room, 250 KV capacity,										
	1800	up to 300 S.F. floor, 1/4" lead, minimum	2 Lath	.08	200	Total	16,400	6,100		22,500	27,700	
	1900	Maximum, 7'-0" walls	↓	.06	266	"	21,100	8,150		29,250	36,100	
	1999	Minimum labor/equipment charge	↓	4.50	3.556	Job		109		109	172	
	2000	X-ray viewing panels, clear lead plastic										
	2010	7 mm thick, 0.3 mm LE, 2.3 lbs/S.F.	H-3	139	.115	S.F.	123	3.26		126.26	141	
	2020	12 mm thick, 0.5 mm LE, 3.9 lbs/S.F.		82	.195		167	5.50		172.50	192	
	2030	18 mm thick, 0.8mm LE, 5.9 lbs/S.F.		54	.296		181	8.40		189.40	213	
	2040	22 mm thick, 1.0 mm LE, 7.2 lbs/S.F.		44	.364		185	10.30		195.30	220	
	2050	35 mm thick, 1.5 mm LE, 11.5 lbs/S.F.		28	.571		207	16.15		223.15	255	
	2060	46 mm thick, 2.0 mm LE, 15.0 lbs/S.F.	↓	21	.762	↓	273	21.50		294.50	335	
	2090	For panels 12 S.F. to 48 S.F., add crating charge				Ea.					50	
	4000	X-ray barriers, modular, panels mounted within framework for										
	4002	attaching to floor, wall or ceiling, upper portion is clear lead										
	4005	plastic window panels 48"H, lower portion is opaque leaded										
	4008	steel panels 36"H, structural supports not incl.										
	4010	1-section barrier, 36"W x 84"H overall										
	4020	0.5 mm LE panels	H-3	6.40	2.500	Ea.	2,775	71		2,846	3,175	
	4030	0.8 mm LE panels		6.40	2.500		2,950	71		3,021	3,375	
	4040	1.0 mm LE panels		5.33	3.002		3,050	85		3,135	3,500	
	4050	1.5 mm LE panels	↓	5.33	3.002	↓	3,250	85		3,335	3,725	
	4060	2-section barrier, 72"W x 84"H overall										
	4070	0.5 mm LE panels	H-3	4	4	Ea.	5,750	113		5,863	6,500	
	4080	0.8 mm LE panels		4	4		6,075	113		6,188	6,875	
	4090	1.0 mm LE panels		3.56	4.494		6,200	127		6,327	7,025	
	5000	1.5 mm LE panels	↓	3.20	5	↓	6,725	142		6,867	7,625	
	5010	3-section barrier, 108"W x 84"H overall										
	5020	0.5 mm LE panels	H-3	3.20	5	Ea.	8,225	142		8,367	9,275	
	5030	0.8 mm LE panels		3.20	5		8,625	142		8,767	9,725	
	5040	1.0 mm LE panels		2.67	5.993		8,850	170		9,020	10,000	
	5050	1.5 mm LE panels	↓	2.46	6.504	↓	9,600	184		9,784	10,900	
	7000	X-ray barriers, mobile, mounted within framework w/casters on										
	7005	bottom, clear lead plastic window panels on upper portion,										
	7010	opaque on lower, 30"W x 75"H overall, incl. framework										
	7020	24"H upper w/0.5 mm LE, 48"H lower w/0.8 mm LE	1 Carp	16	.500	Ea.	1,900	16.50		1,916.50	2,100	
	7030	48"W x 75"H overall, incl. framework										
	7040	36"H upper w/0.5 mm LE, 36"H lower w/0.8 mm LE	1 Carp	16	.500	Ea.	3,325	16.50		3,341.50	3,700	
	7050	36"H upper w/1.0 mm LE, 36"H lower w/1.5 mm LE	"	16	.500	"	4,025	16.50		4,041.50	4,475	
	7060	72"W x 75"H overall, incl. framework										
	7070	36"H upper w/0.5 mm LE, 36"H lower w/0.8 mm LE	1 Carp	16	.500	Ea.	3,950	16.50		3,966.50	4,375	
	7080	36"H upper w/1.0 mm LE, 36"H lower w/1.5 mm LE	"	16	.500	"	5,000	16.50		5,016.50	5,525	

13120 | Pre-Engineered Structures

13128 | Pre-Engineered Structures

	Line	Description	Crew	Daily Output	Labor-Hours	Unit	2004 Bare Costs Mat.	Labor	Equip.	Total	Total Incl O&P	
540	0010	**GREENHOUSE** Shell only, stock units, not incl. 2' stub walls,										540
	0020	foundation, floors, heat or compartments										
	0300	Residential type, free standing, 8'-6" long x 7'-6" wide	2 Carp	59	.271	SF Flr.	37	8.95		45.95	56	
	0400	10'-6" wide	↓	85	.188	↓	28.50	6.20		34.70	42	

13120 | Pre-Engineered Structures

	13128 | Pre-Engineered Structures	CREW	DAILY OUTPUT	LABOR-HOURS	UNIT	2004 BARE COSTS MAT.	LABOR	EQUIP.	TOTAL	TOTAL INCL O&P	
540	0600 13'-6" wide	2 Carp	108	.148	SF Flr.	25.50	4.89		30.39	36	540
	0700 17'-0" wide		160	.100		28.50	3.30		31.80	37	
	0900 Lean-to type, 3'-10" wide		34	.471		33	15.55		48.55	61.50	
	1000 6'-10" wide		58	.276		25.50	9.10		34.60	43	
	1050 8'-0" wide	↓	60	.267	↓	23.50	8.80		32.30	40	
	1060										
	1100 Wall mounted, to existing window, 3' x 3'	1 Carp	4	2	Ea.	355	66		421	500	
	1120 4' x 5'	"	3	2.667	"	530	88		618	730	
	3900 For cooling, add, minimum				SF Flr.	2.23			2.23	2.45	
	4000 Maximum					5.50			5.50	6.05	
	4200 For heaters, 13.6 MBH, add					4.24			4.24	4.66	
	4300 60 MBH, add				↓	1.59			1.59	1.75	
	4500 For benches, 2' x 3'-6", add				SF Hor.	18.95			18.95	21	
	4600 3' x 10', add				S.F.	10.30			10.30	11.30	
	4800 For controls, add, minimum				Total	1,900			1,900	2,100	
	4900 Maximum				"	11,300			11,300	12,500	
	5100 For humidification equipment, add				M.C.F.	4.88			4.88	5.35	
	5200 For vinyl shading, add				S.F.	1.03			1.03	1.13	
600	0010 **KIOSKS** Round, 5' diameter, 8' high, 1/4" fiberglass wall				Ea.	5,500			5,500	6,050	600
	0100 1" insulated double wall, fiberglass				"	6,250			6,250	6,875	
	0500 Rectangular, 5' x 9', 7'-6" high, 1/4" fiberglass wall				Ea.	8,000			8,000	8,800	
	0600 1" insulated double wall, fiberglass				"	9,500			9,500	10,500	
880	0010 **SWIMMING POOL ENCLOSURE** Translucent, free standing,										880
	0020 not including foundations, heat or light										
	0200 Economy, minimum	2 Carp	200	.080	SF Hor.	11.35	2.64		13.99	16.85	
	0300 Maximum		100	.160		30	5.30		35.30	42	
	0400 Deluxe, minimum		100	.160		34	5.30		39.30	46.50	
	0600 Maximum	↓	70	.229	↓	163	7.55		170.55	192	

13150 | Swimming Pools

	13151 | Swimming Pools	CREW	DAILY OUTPUT	LABOR-HOURS	UNIT	2004 BARE COSTS MAT.	LABOR	EQUIP.	TOTAL	TOTAL INCL O&P	
200	0011 **SWIMMING POOLS,** Outdoor, incl. equip. & houses, minimum				SF Surf					36	200
	0300 Maximum									68	
	0400 Residential, incl. equipment, permanent type, minimum									12	
	0700 Maximum									25	
	0900 Municipal, including equipment only, over 5000 S.F., minimum									25	
	1000 Maximum									51	
	1300 Motel or apt., incl. equipment only, under 5000 S.F., minimum									22	
	1400 Maximum				↓					34	

Important: See the Reference Section for critical supporting data - Reference Nos., Crews, & City Cost Indexes

	13281	Hazardous Material Remediation	CREW	DAILY OUTPUT	LABOR-HOURS	UNIT	2004 BARE COSTS MAT.	LABOR	EQUIP.	TOTAL	TOTAL INCL O&P	
120	0010	**BULK ASBESTOS REMOVAL**										120
	0020	Includes disposable tools and 2 suits and 1respirator filter/day/worker										
	0100	Beams, W 10 x 19	A-9	235	.272	L.F.	.83	9.80		10.63	16.85	
	0110	W 12 x 22		210	.305		.93	11		11.93	18.90	
	0120	W 14 x 26		180	.356		1.09	12.80		13.89	22	
	0130	W 16 x 31		160	.400		1.23	14.40		15.63	25	
	0140	W 18 x 40		140	.457		1.40	16.50		17.90	28.50	
	0150	W 24 x 55		110	.582		1.78	21		22.78	36	
	0160	W 30 x 108		85	.753		2.31	27		29.31	46.50	
	0170	W 36 x 150		72	.889		2.72	32		34.72	55	
	0200	Boiler insulation		480	.133	S.F.	.47	4.81		5.28	8.30	
	0210	With metal lath add				%				50%		
	0300	Boiler breeching or flue insulation	A-9	520	.123	S.F.	.38	4.44		4.82	7.60	
	0310	For active boiler, add				%				100%		
	0400	Duct or AHU insulation	A-10B	440	.073	S.F.	.22	2.63		2.85	4.52	
	0500	Duct vibration isolation joints, up to 24 Sq. In. duct	A-9	56	1.143	Ea.	3.50	41		44.50	71	
	0520	25 Sq. In. to 48 Sq. In. duct		48	1.333		4.08	48		52.08	82.50	
	0530	49 Sq. In. to 76 Sq. In. duct		40	1.600		4.90	57.50		62.40	99.50	
	0600	Pipe insulation, air cell type, up to 4" diameter pipe		900	.071	L.F.	.22	2.56		2.78	4.41	
	0610	4" to 8" diameter pipe		800	.080		.25	2.88		3.13	4.96	
	0620	10" to 12" diameter pipe		700	.091		.28	3.30		3.58	5.65	
	0630	14" to 16" diameter pipe		550	.116		.36	4.20		4.56	7.20	
	0650	Over 16" diameter pipe		650	.098	S.F.	.30	3.55		3.85	6.10	
	0700	With glove bag up to 3" diameter pipe		100	.640	L.F.	3.15	23		26.15	41	
	1000	Pipe fitting insulation up to 4" diameter pipe		320	.200	Ea.	.61	7.20		7.81	12.35	
	1100	6" to 8" diameter pipe		304	.211		.64	7.60		8.24	13.05	
	1110	10" to 12" diameter pipe		192	.333		1.02	12		13.02	20.50	
	1120	14" to 16" diameter pipe		128	.500		1.53	18.05		19.58	31	
	1130	Over 16" diameter pipe		176	.364	S.F.	1.11	13.10		14.21	22.50	
	1200	With glove bag, up to 8" diameter pipe		40	1.600	Ea.	6.55	57.50		64.05	101	
	2000	Scrape foam fireproofing from flat surface		2,400	.027	S.F.	.08	.96		1.04	1.65	
	2100	Irregular surfaces		1,200	.053		.16	1.92		2.08	3.31	
	3000	Remove cementitious material from flat surface		1,800	.036		.11	1.28		1.39	2.20	
	3100	Irregular surface		1,400	.046		.14	1.65		1.79	2.83	
	4000	Scrape acoustical coating/fireproofing, from ceiling		3,200	.020		.06	.72		.78	1.24	
	5000	Remove VAT from floor by hand		2,400	.027		.08	.96		1.04	1.65	
	5100	By machine	A-11	4,800	.013		.04	.48	.01	.53	.83	
	5150	For 2 layers, add				%				50%		
	6000	Remove contaminated soil from crawl space by hand	A-9	400	.160	C.F.	.49	5.75		6.24	9.95	
	6100	With large production vacuum loader	A-12	700	.091	"	.28	3.30	.72	4.30	6.45	
	7000	Radiator backing, not including radiator removal	A-9	1,200	.053	S.F.	.16	1.92		2.08	3.31	
	8000	Cement-asbestos transite board	2 Asbe	1,000	.016		.13	.58		.71	1.08	
	8100	Transite shingle siding	A-10B	750	.043		.21	1.54		1.75	2.73	
	8200	Shingle roofing	"	2,000	.016		.07	.58		.65	1.02	
	8250	Built-up, no gravel, non-friable	B-2	1,400	.029		.07	.75		.82	1.33	
	8300	Asbestos millboard	2 Asbe	1,000	.016		.08	.58		.66	1.03	
	9000	For type C (supplied air) respirator equipment, add				%					10%	
125	0010	**ASBESTOS ABATEMENT WORK AREA** Containment and preparation.										125
	0100	Pre-cleaning, HEPA vacuum and wet wipe, flat surfaces	A-10	12,000	.005	S.F.	.02	.19		.21	.33	
	0200	Protect carpeted area, 2 layers 6 mil poly on 3/4" plywood	"	1,000	.064		1.50	2.31		3.81	5.40	
	0300	Separation barrier, 2" x 4" @ 16", 1/2" plywood ea. side, 8' high	2 Carp	400	.040		1.25	1.32		2.57	3.57	
	0310	12' high		320	.050		1.40	1.65		3.05	4.28	
	0320	16' high		200	.080		1.50	2.64		4.14	6.05	
	0400	Personnel decontam. chamber, 2" x 4" @ 16", 3/4" ply ea. side		280	.057		2.50	1.89		4.39	5.90	
	0450	Waste decontam. chamber, 2" x 4" studs @ 16", 3/4" ply ea. side		360	.044		3	1.47		4.47	5.75	

13280 | Hazardous Material Remediation

	13281	Hazardous Material Remediation	CREW	DAILY OUTPUT	LABOR-HOURS	UNIT	2004 BARE COSTS MAT.	LABOR	EQUIP.	TOTAL	TOTAL INCL O&P	
125	0500	Cover surfaces with polyethelene sheeting										125
	0501	Including glue and tape										
	0550	Floors, each layer, 6 mil	A-10	8,000	.008	S.F.	.10	.29		.39	.58	
	0551	4 mil		9,000	.007		.06	.26		.32	.49	
	0560	Walls, each layer, 6 mil		6,000	.011		.09	.38		.47	.73	
	0561	4 mil		7,000	.009		.07	.33		.40	.62	
	0570	For heights above 12′, add						20%				
	0575	For heights above 20′, add						30%				
	0580	For fire retardant poly, add					100%					
	0590	For large open areas, deduct					10%	20%				
	0600	Seal floor penetrations with foam firestop to 36 Sq. In.	2 Carp	200	.080	Ea.	6.25	2.64		8.89	11.30	
	0610	36 Sq. In. to 72 Sq. In.		125	.128		12.50	4.22		16.72	21	
	0615	72 Sq. In. to 144 Sq. In.		80	.200		25	6.60		31.60	38.50	
	0620	Wall penetrations, to 36 square inches		180	.089		6.25	2.93		9.18	11.75	
	0630	36 Sq. In. to 72 Sq. In.		100	.160		12.50	5.30		17.80	22.50	
	0640	72 Sq. In. to 144 Sq. In.		60	.267		25	8.80		33.80	42	
	0800	Caulk seams with latex	1 Carp	230	.035	L.F.	.15	1.15		1.30	2.07	
	0900	Set up neg. air machine, 1-2k C.F.M. /25 M.C.F. volume	1 Asbe	4.30	1.860	Ea.		67		67	109	
130	0010	**DEMOLITION IN ASBESTOS CONTAMINATED AREA**										130
	0200	Ceiling, including suspension system, plaster and lath	A-9	2,100	.030	S.F.	.09	1.10		1.19	1.89	
	0210	Finished plaster, leaving wire lath		585	.109		.34	3.95		4.29	6.75	
	0220	Suspended acoustical tile		3,500	.018		.06	.66		.72	1.13	
	0230	Concealed tile grid system		3,000	.021		.07	.77		.84	1.32	
	0240	Metal pan grid system		1,500	.043		.13	1.54		1.67	2.64	
	0250	Gypsum board		2,500	.026		.08	.92		1	1.59	
	0260	Lighting fixtures up to 2′ x 4′		72	.889	Ea.	2.72	32		34.72	55	
	0400	Partitions, non load bearing										
	0410	Plaster, lath, and studs	A-9	690	.093	S.F.	.83	3.34		4.17	6.35	
	0450	Gypsum board and studs	"	1,390	.046	"	.14	1.66		1.80	2.86	
	9000	For type C (supplied air) respirator equipment, add				%					10%	
135	0010	**ASBESTOS ABATEMENT EQUIPMENT** and supplies, buy R13281-120										135
	0200	Air filtration device, 2000 C.F.M.				Ea.	2,500			2,500	2,750	
	0250	Large volume air sampling pump, minimum					365			365	400	
	0260	Maximum					680			680	745	
	0300	Airless sprayer unit, 2 gun					2,000			2,000	2,200	
	0350	Light stand, 500 watt					250			250	275	
	0400	Personal respirators										
	0410	Negative pressure, 1/2 face, dual operation, min.				Ea.	22.50			22.50	25	
	0420	Maximum					24			24	26.50	
	0450	P.A.P.R., full face, minimum					400			400	440	
	0460	Maximum					700			700	770	
	0470	Supplied air, full face, incl. air line, minimum					450			450	495	
	0480	Maximum					600			600	660	
	0500	Personnel sampling pump, minimum					450			450	495	
	0510	Maximum					750			750	825	
	1500	Power panel, 20 unit, incl. G.F.I.					1,800			1,800	1,975	
	1600	Shower unit, including pump and filters					1,125			1,125	1,250	
	1700	Supplied air system (type C)					10,000			10,000	11,000	
	1750	Vacuum cleaner, HEPA, 16 gal., stainless steel, wet/dry					1,000			1,000	1,100	
	1760	55 gallon					2,200			2,200	2,425	
	1800	Vacuum loader, 9-18 ton/hr					90,000			90,000	99,000	
	1900	Water atomizer unit, including 55 gal. drum					230			230	253	
	2000	Worker protection, whole body, foot, head cover & gloves, plastic					6.50			6.50	7.15	
	2500	Respirator, single use					10.50			10.50	11.55	
	2550	Cartridge for respirator					11.50			11.50	12.65	
	2570	Glove bag, 7 mil, 50″ x 64″					8.50			8.50	9.35	

	13281 | Hazardous Material Remediation		Crew	Daily Output	Labor-Hours	Unit	2004 Bare Costs Mat.	Labor	Equip.	Total	Total Incl O&P	
135	2580	10 mil, 44″ x 60″ (R13281-120)				Ea.	8.40			8.40	9.25	135
	3000	HEPA vacuum for work area, minimun					1,000			1,000	1,100	
	3050	Maximum					2,775			2,775	3,050	
	6000	Disposable polyethelene bags, 6 mil, 3 C.F.					1.15			1.15	1.27	
	6300	Disposable fiber drums, 3 C.F.					6.50			6.50	7.15	
	6400	Pressure sensitive caution lables, 3″ x 5″					1.64			1.64	1.80	
	6450	11″ x 17″					6.50			6.50	7.15	
	6500	Negative air machine, 1800 C.F.M.					775			775	855	
140	0010	**DECONTAMINATION CONTAINMENT AREA DEMOLITION** and clean-up										140
	0100	Spray exposed substrate with surfactant (bridging)										
	0200	Flat surfaces	A-9	6,000	.011	S.F.	.35	.38		.73	1.02	
	0250	Irregular surfaces		4,000	.016	″	.30	.58		.88	1.27	
	0300	Pipes, beams, and columns		2,000	.032	L.F.	.55	1.15		1.70	2.49	
	1000	Spray encapsulate polyethelene sheeting		8,000	.008	S.F.	.30	.29		.59	.80	
	1100	Roll down polyethelene sheeting		8,000	.008	″		.29		.29	.47	
	1500	Bag polyethelene sheeting		400	.160	Ea.	.75	5.75		6.50	10.25	
	2000	Fine clean exposed substrate, with nylon brush		2,400	.027	S.F.		.96		.96	1.56	
	2500	Wet wipe substrate		4,800	.013			.48		.48	.78	
	2600	Vacuum surfaces, fine brush		6,400	.010			.36		.36	.59	
	3000	Structural demolition										
	3100	Wood stud walls	A-9	2,800	.023	S.F.		.82		.82	1.34	
	3500	Window manifolds, not incl. window replacement		4,200	.015			.55		.55	.89	
	3600	Plywood carpet protection		2,000	.032			1.15		1.15	1.88	
	4000	Remove custom decontamination facility	A-10A	8	3	Ea.	15	109		124	193	
	4100	Remove portable decontamination facility	3 Asbe	12	2	″	12.50	72		84.50	131	
	5000	HEPA vacuum, shampoo carpeting	A-9	4,800	.013	S.F.	.05	.48		.53	.84	
	9000	Final cleaning of protected surfaces	A-10A	8,000	.003	″		.11		.11	.18	
145	0010	**OSHA TESTING**										145
	0100	Certified technician, minimum				Day					300	
	0110	Maximum				″					500	
	0200	Personal sampling, PCM analysis, NIOSH 7400, minimum	1 Asbe	8	1	Ea.	2.75	36		38.75	61.50	
	0210	Maximum	″	4	2	″	3	72		75	120	
	0300	Industrial hygenist, minimum				Day					400	
	0310	Maximum				″					550	
	1000	Cleaned area samples	1 Asbe	8	1	Ea.	2.63	36		38.63	61.50	
	1100	PCM air sample analysis, NIOSH 7400, minimum		8	1		30	36		66	91.50	
	1110	Maximum		4	2		3.09	72		75.09	120	
	1200	TEM air sample analysis, NIOSH 7402, minimum								100	125	
	1210	Maximum								400	500	
150	0010	**ENCAPSULATION WITH SEALANTS**										150
	0100	Ceilings and walls, minimum	A-9	21,000	.003	S.F.	.26	.11		.37	.47	
	0110	Maximum		10,600	.006		.41	.22		.63	.80	
	0200	Columns and beams, minimum		13,300	.005		.26	.17		.43	.57	
	0210	Maximum		5,325	.012		.46	.43		.89	1.21	
	0300	Pipes to 12″ diameter including minor repairs, minimum		800	.080	L.F.	.36	2.88		3.24	5.10	
	0310	Maximum		400	.160	″	1.02	5.75		6.77	10.50	
155	0010	**WASTE PACKAGING, HANDLING, & DISPOSAL**										155
	0100	Collect and bag bulk material, 3 C.F. bags, by hand	A-9	400	.160	Ea.	1.15	5.75		6.90	10.65	
	0200	Large production vacuum loader	A-12	880	.073		.80	2.62	.57	3.99	5.75	
	1000	Double bag and decontaminate	A-9	960	.067		2.30	2.40		4.70	6.45	
	2000	Containerize bagged material in drums, per 3 C.F. drum	″	800	.080		6.50	2.88		9.38	11.85	
	3000	Cart bags 50′ to dumpster	2 Asbe	400	.040			1.44		1.44	2.34	
	5000	Disposal charges, not including haul, minimum				C.Y.					50	
	5020	Maximum				″					175	

		13281 | Hazardous Material Remediation	CREW	DAILY OUTPUT	LABOR-HOURS	UNIT	2004 BARE COSTS MAT.	LABOR	EQUIP.	TOTAL	TOTAL INCL O&P	
155	5100	Remove refrigerant from system	1 Plum	40	.200	Lb.		7.90		7.90	12.30	155
	9000	For type C (supplied air) respirator equipment, add				%					10%	
440	0010	**REMOVAL** Existing lead paint, by chemicals, per application R13281-460										440
	0020	See also, Div. 13280, Haz. Mat'l. Abatement										
	0050	Baseboard, to 6" wide	1 Pord	64	.125	L.F.	1.49	3.70		5.19	7.60	
	0070	To 12" wide		32	.250	"	2.94	7.40		10.34	15.10	
	0200	Balustrades, one side		28	.286	S.F.	3.33	8.45		11.78	17.20	
	1400	Cabinets, simple design		32	.250		2.92	7.40		10.32	15.05	
	1420	Ornate design		25	.320		3.75	9.45		13.20	19.30	
	1600	Cornice, simple design		60	.133		1.57	3.95		5.52	8.05	
	1620	Ornate design		20	.400		4.62	11.85		16.47	24	
	2800	Doors, one side, flush		84	.095		1.13	2.82		3.95	5.75	
	2820	Two panel		80	.100		1.17	2.96		4.13	6.05	
	2840	Four panel		45	.178		2.07	5.25		7.32	10.75	
	2880	For trim, one side, add		64	.125	L.F.	1.49	3.70		5.19	7.60	
	3000	Fence, picket, one side		30	.267	S.F.	3.13	7.90		11.03	16.10	
	3200	Grilles, one side, simple design		30	.267		3.13	7.90		11.03	16.10	
	3220	Ornate design		25	.320		3.75	9.45		13.20	19.30	
	4400	Pipes, to 4" diameter		90	.089	L.F.	1.07	2.63		3.70	5.40	
	4420	To 8" diameter		50	.160		1.86	4.74		6.60	9.65	
	4440	To 12" diameter		36	.222		2.61	6.60		9.21	13.40	
	4460	To 16" diameter		20	.400		4.65	11.85		16.50	24	
	4500	For hangers, add		40	.200	Ea.	2.33	5.90		8.23	12.05	
	4800	Siding		90	.089	S.F.	1.07	2.63		3.70	5.40	
	5000	Trusses, open		55	.145	SF Face	1.71	4.31		6.02	8.80	
	6200	Windows, one side only, double hung, 1/1 light, 24" x 48" high		4	2	Ea.	23.50	59		82.50	121	
	6220	30" x 60" high		3	2.667		31.50	79		110.50	161	
	6240	36" x 72" high		2.50	3.200		37.50	94.50		132	194	
	6280	40" x 80" high		2	4		47	118		165	242	
	6400	Colonial window, 6/6 light, 24" x 48" high		2	4		47	118		165	242	
	6420	30" x 60" high		1.50	5.333		62.50	158		220.50	320	
	6440	36" x 72" high		1	8		94	237		331	485	
	6480	40" x 80" high		1	8		94	237		331	485	
	6600	8/8 light, 24" x 48" high		2	4		47	118		165	242	
	6620	40" x 80" high		1	8		94	237		331	485	
	6800	12/12 light, 24" x 48" high		1	8		94	237		331	485	
	6820	40" x 80" high		.75	10.667		125	315		440	645	
	6840	Window frame & trim items, included in pricing above										
	9000	Minimum labor/equipment charge	1 Pord	3	2.667	Job		79		79	126	
460	0010	**LEAD PAINT ENCAPSULATION**, water based polymer coating, 14 mil DFT										460
	0020	Interior, brushwork, trim, under 6"	1 Pord	240	.033	L.F.	2.20	.99		3.19	4	
	0030	6" to 12" wide		180	.044		2.93	1.32		4.25	5.35	
	0040	Balustrades		300	.027		1.77	.79		2.56	3.21	
	0050	Pipe to 4" diameter		500	.016		1.06	.47		1.53	1.93	
	0060	To 8" diameter		375	.021		1.40	.63		2.03	2.55	
	0070	To 12" diameter		250	.032		2.11	.95		3.06	3.84	
	0080	To 16" diameter		170	.047		3.10	1.39		4.49	5.65	
	0090	Cabinets, ornate design		200	.040	S.F.	2.65	1.18		3.83	4.82	
	0100	Simple design		250	.032	"	2.11	.95		3.06	3.84	
	0110	Doors, 3'x 7', both sides, incl. frame & trim										
	0120	Flush	1 Pord	6	1.333	Ea.	27	39.50		66.50	92.50	
	0130	French, 10-15 lite		3	2.667		5.40	79		84.40	132	
	0140	Panel		4	2		32.50	59		91.50	131	
	0150	Louvered		2.75	2.909		29.50	86		115.50	171	
	0160	Windows, per interior side, per 15 S.F.										

13280 | Hazardous Material Remediation

	13281 | Hazardous Material Remediation	CREW	DAILY OUTPUT	LABOR-HOURS	UNIT	2004 BARE COSTS MAT.	LABOR	EQUIP.	TOTAL	TOTAL INCL O&P	
460	0170 1 to 6 lite	1 Pord	14	.571	Ea.	18.65	16.90		35.55	47.50	460
	0180 7 to 10 lite		7.50	1.067		20.50	31.50		52	73	
	0190 12 lite		5.75	1.391		27.50	41		68.50	96.50	
	0200 Radiators		8	1	↓	66	29.50		95.50	120	
	0210 Grilles, vents		275	.029	S.F.	1.92	.86		2.78	3.49	
	0220 Walls, roller, drywall or plaster		1,000	.008		.53	.24		.77	.96	
	0230 With spunbonded reinforcing fabric		720	.011		.60	.33		.93	1.19	
	0240 Wood		800	.010		.66	.30		.96	1.20	
	0250 Ceilings, roller, drywall or plaster		900	.009		.60	.26		.86	1.08	
	0260 Wood		700	.011	↓	.75	.34		1.09	1.37	
	0270 Exterior, brushwork, gutters and downspouts		300	.027	L.F.	1.77	.79		2.56	3.21	
	0280 Columns		400	.020	S.F.	1.31	.59		1.90	2.39	
	0290 Spray, siding	↓	600	.013	"	.88	.39		1.27	1.60	
	0300 Miscellaneous										
	0310 Electrical conduit, brushwork, to 2" diameter	1 Pord	500	.016	L.F.	1.06	.47		1.53	1.93	
	0320 Brick, block or concrete, spray		500	.016	S.F.	1.06	.47		1.53	1.93	
	0330 Steel, flat surfaces and tanks to 12"		500	.016		1.06	.47		1.53	1.93	
	0340 Beams, brushwork		400	.020		1.31	.59		1.90	2.39	
	0350 Trusses	↓	400	.020	↓	1.31	.59		1.90	2.39	
750	0010 **MOLD ABATEMENT METHODS**										750
	0020 Initial inspection, average 3 bedroom home				Total					250	
	0030 Initial inspection, average 5 bedroom home									350	
	0040 Air sample test, each									250	
	0050 Swab sample test, each									150	
	0060 Tape sample test, each									150	
	0070 After remediation air test, each									250	
	0080 Mold abatement plan, average 3 bedroom home									1,250	
	0090 Mold abatement plan, average 5 bedroom home									1,500	
	0100 Packup & removal of contents, average 3 bedroom home, excl storage									10,000	
	0110 Packup & removal of contents, average 5 bedroom home, excl storage				↓					25,000	
	0120 For demolition in mold contaminated areas, see Div. 13281-130										
	0130 For personal protection equipment, see Div. 13281-135										

13600 | Solar and Wind Energy Equipment

	13630 | Solar Collector Components	CREW	DAILY OUTPUT	LABOR-HOURS	UNIT	2004 BARE COSTS MAT.	LABOR	EQUIP.	TOTAL	TOTAL INCL O&P	
200	0010 **SOLAR ENERGY**										200
	0020 System/Package prices, not including connecting										
	0030 pipe, insulation, or special heating/plumbing fixtures										
	0500 Hot water, standard package, low temperature										
	0580 2 collectors, circulator, fittings, 120 gal. tank	Q-1	.40	40	Ea.	1,925	1,425		3,350	4,325	
	2250 Controller, liquid temperature	1 Plum	5	1.600	"	87.50	63.50		151	194	
	2300 Circulators, air										
	2310 Blowers										
	2520 Space & DHW system, less duct work	Q-9	.50	32	Ea.	1,450	1,125		2,575	3,375	
	2580 8" diameter, 150 CFM		16	1		38	35		73	97.50	
	2660 Shutter/damper		12	1.333		38	46.50		84.50	116	
	2670 Shutter motor	↓	16	1	↓	90	35		125	155	

	13630 | Solar Collector Components		CREW	DAILY OUTPUT	LABOR-HOURS	UNIT	2004 BARE COSTS MAT.	LABOR	EQUIP.	TOTAL	TOTAL INCL O&P	
200	2800	Circulators, liquid, 1/25 HP, 5.3 GPM	Q-1	14	1.143	Ea.	143	40.50		183.50	220	200
	2820	1/20 HP, 17 GPM	″	12	1.333		150	47.50		197.50	239	
	3000	Collector panels, air with aluminum absorber plate										
	3010	Wall or roof mount										
	3040	Flat black, plastic glazing										
	3080	4′ x 8′	Q-9	6	2.667	Ea.	615	93		708	825	
	3300	Collector panels, liquid with copper absorber plate										
	3320	Black chrome, tempered glass glazing										
	3330	Alum. frame, 4′ x 8′, 5/32″ single glazing	Q-1	9.50	1.684	Ea.	545	60		605	695	
	3450	Flat black, alum. frame, 3′ x 8′		9	1.778		430	63.50		493.50	575	
	3500	4′ x 8.5′		5.50	2.909		490	104		594	700	
	3600	Liquid, full wetted, plastic, alum. frame, 3′ x 10′		5	3.200		195	114		309	390	
	3650	Collector panel mounting, flat roof or ground rack		7	2.286		45	81.50		126.50	176	
	3670	Roof clamps		70	.229	Set	1.61	8.15		9.76	14.40	
	3700	Roof strap, teflon	1 Plum	205	.039	L.F.	9.50	1.55		11.05	12.85	
	3900	Differential controller with two sensors										
	3930	Thermostat, hard wired	1 Plum	8	1	Ea.	87.50	39.50		127	158	
	4050	Pool valve system	″	2.50	3.200	″	215	127		342	430	
	4150	Sensors										
	4220	Freeze prevention	1 Plum	32	.250	Ea.	22	9.90		31.90	39.50	
	4260											
	4300	Heat exchanger										
	4315	includes coil, blower, circulator										
	4316	and controller for DHW and space hot air										
	4380	70 MBH	Q-1	3.50	4.571	Ea.	325	163		488	610	
	4400	80 MBH	″	3	5.333	″	430	190		620	770	
	4580	Fluid to fluid package includes two circulating pumps										
	4590	expansion tank, check valve, relief valve										
	4600	controller, high temperature cutoff and sensors	Q-1	2.50	6.400	Ea.	695	228		923	1,125	
	4650	Heat transfer fluid										
	4700	Propylene glycol, inhibited anti-freeze	1 Plum	28	.286	Gal.	8.80	11.30		20.10	27.50	
	4800	Solar storage tanks, knocked down										
	5020	7′ x 10′-6″ = 459 C.F./3000 gallons	Q-10	.80	30	Ea.	10,400	1,075		11,475	13,100	
	5210	7′ x 14′ = 613 C.F./4000 gallons		.60	40		12,400	1,450		13,850	16,000	
	5230	10′-6″ x 14′ = 919 C.F./6000 gallons		.40	60		15,100	2,175		17,275	20,100	
	5250	14′ x 17′-6″ = 1531 C.F./10,000 gallons	Q-11	.30	106		19,800	3,950		23,750	28,100	
	7000	Solar control valves and vents										
	7070	Air eliminator, automatic 3/4″ size	1 Plum	32	.250	Ea.	27.50	9.90		37.40	46	
	7090	Air vent, automatic, 1/8″ fitting		32	.250		8.90	9.90		18.80	25	
	7100	Manual, 1/8″ NPT		32	.250		1.95	9.90		11.85	17.50	
	7120	Backflow preventer, 1/2″ pipe size		16	.500		58.50	19.80		78.30	94.50	
	7130	3/4″ pipe size		16	.500		60	19.80		79.80	96.50	
	7150	Balancing valve, 3/4″ pipe size		20	.400		25	15.85		40.85	52	
	7180	Draindown valve, 1/2″ copper tube		9	.889		202	35		237	277	
	7200	Flow control valve, 1/2″ pipe size		22	.364		56.50	14.40		70.90	84.50	
	7220	Expansion tank, up to 5 gal.		32	.250		59.50	9.90		69.40	81	
	7400	Pressure gauge, 2″ dial		32	.250		21.50	9.90		31.40	39	
	7450	Relief valve, temp. and pressure 3/4″ pipe size		30	.267		8.75	10.55		19.30	26	
	7500	Solenoid valve, normally closed										
	7750	Vacuum relief valve, 3/4″ pipe size	1 Plum	32	.250	Ea.	25.50	9.90		35.40	43.50	
	7800	Thermometers										
	8250	Water storage tank with heat exchanger and electric element										
	8270	66 gal. with 2″ x 2 lb. density insulation	1 Plum	1.60	5	Ea.	650	198		848	1,025	
	8300	80 gal. with 2″ x 2 lb. density insulation	″	1.60	5	″	745	198		943	1,125	
	8500	Water storage module, plastic										
	8600	Tubular, 12″ diameter, 4′ high	1 Carp	48	.167	Ea.	78	5.50		83.50	94.50	

Important: See the Reference Section for critical supporting data - Reference Nos., Crews, & City Cost Indexes

13600 | Solar and Wind Energy Equipment

	13630 | Solar Collector Components	CREW	DAILY OUTPUT	LABOR-HOURS	UNIT	2004 BARE COSTS MAT.	LABOR	EQUIP.	TOTAL	TOTAL INCL O&P	
200	8610 12″ diameter, 8′ high	1 Carp	40	.200	Ea.	119	6.60		125.60	142	200
	8650 Cap, 12″ diameter				↓	12.30			12.30	13.55	
	9000 Minimum labor/equipment charge	1 Plum	2	4	Job		158		158	246	

13700 | Security Access and Surveillance

	13710 | Security Access	CREW	DAILY OUTPUT	LABOR-HOURS	UNIT	2004 BARE COSTS MAT.	LABOR	EQUIP.	TOTAL	TOTAL INCL O&P	
300	0010 **ACCESS CONTROL**										300
	0020 Card type, 1 time zone, minimum				Ea.	310			310	340	
	0040 Maximum					1,050			1,050	1,150	
	0060 3 time zones, minimum					765			765	840	
	0080 Maximum				↓	1,750			1,750	1,925	
	0100 System with printer, and control console, 3 zones				Total	8,550			8,550	9,400	
	0120 6 zones				″	11,300			11,300	12,400	
	0140 For each door, minimum, add				Ea.	1,250			1,250	1,375	
	0160 Maximum, add				″	1,875			1,875	2,050	

	13720 | Detection & Alarm	CREW	DAILY OUTPUT	LABOR-HOURS	UNIT	MAT.	LABOR	EQUIP.	TOTAL	TOTAL INCL O&P	
065	0010 **DETECTION SYSTEMS**, not including wires & conduits										065
	0100 Burglar alarm, battery operated, mechanical trigger	1 Elec	4	2	Ea.	249	79		328	395	
	0200 Electrical trigger		4	2		297	79		376	445	
	0400 For outside key control, add		8	1		70.50	39.50		110	138	
	0600 For remote signaling circuitry, add		8	1		112	39.50		151.50	184	
	0800 Card reader, flush type, standard		2.70	2.963		835	117		952	1,100	
	1000 Multi-code		2.70	2.963		1,075	117		1,192	1,350	
	1200 Door switches, hinge switch		5.30	1.509		52.50	59.50		112	150	
	1400 Magnetic switch		5.30	1.509		62	59.50		121.50	160	
	1600 Exit control locks, horn alarm		4	2		310	79		389	460	
	1800 Flashing light alarm		4	2		350	79		429	505	
	2000 Indicating panels, 1 channel	↓	2.70	2.963		330	117		447	545	
	2200 10 channel	2 Elec	3.20	5		1,125	197		1,322	1,550	
	2400 20 channel		2	8		2,200	315		2,515	2,900	
	2600 40 channel	↓	1.14	14.035		4,000	555		4,555	5,250	
	2800 Ultrasonic motion detector, 12 volt	1 Elec	2.30	3.478		206	137		343	440	
	3000 Infrared photoelectric detector		2.30	3.478		170	137		307	400	
	3200 Passive infrared detector		2.30	3.478		254	137		391	490	
	3400 Glass break alarm switch		8	1		42.50	39.50		82	108	
	3420 Switchmats, 30″ x 5′		5.30	1.509		76	59.50		135.50	175	
	3440 30″ x 25′		4	2		182	79		261	320	
	3460 Police connect panel		4	2		219	79		298	360	
	3480 Telephone dialer		5.30	1.509		345	59.50		404.50	470	
	3500 Alarm bell		4	2		69.50	79		148.50	198	
	3520 Siren		4	2		131	79		210	265	
	3540 Microwave detector, 10′ to 200′		2	4		600	158		758	900	
	3560 10′ to 350′	↓	2	4	↓	1,750	158		1,908	2,175	
	3594 Fire, alarm control panel										
	3600 4 zone	2 Elec	2	8	Ea.	920	315		1,235	1,475	
	3800 8 zone	↓	1	16	↓	1,400	630		2,030	2,525	

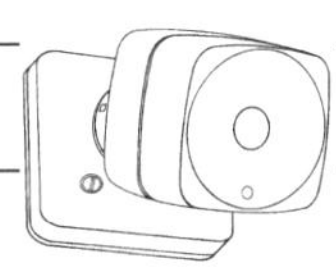

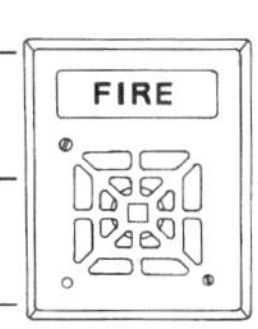

	13720 | Detection & Alarm		CREW	DAILY OUTPUT	LABOR-HOURS	UNIT	2004 BARE COSTS MAT.	LABOR	EQUIP.	TOTAL	TOTAL INCL O&P	
065	4000	12 zone	2 Elec	.67	23.988	Ea.	1,825	945		2,770	3,450	065
	4200	Battery and rack	1 Elec	4	2		690	79		769	880	
	4400	Automatic charger		8	1		445	39.50		484.50	550	
	4600	Signal bell		8	1		49.50	39.50		89	115	
	4800	Trouble buzzer or manual station		8	1		37	39.50		76.50	101	
	5000	Detector, rate of rise		8	1		33.50	39.50		73	97.50	
	5100	Fixed temperature		8	1		28	39.50		67.50	91.50	
	5200	Smoke detector, ceiling type		6.20	1.290		75	51		126	161	
	5400	Duct type		3.20	2.500		250	98.50		348.50	425	
	5600	Strobe and horn		5.30	1.509		95	59.50		154.50	197	
	5800	Fire alarm horn		6.70	1.194		36.50	47		83.50	113	
	6000	Door holder, electro-magnetic		4	2		77.50	79		156.50	207	
	6200	Combination holder and closer		3.20	2.500		430	98.50		528.50	625	
	6400	Code transmitter		4	2		690	79		769	880	
	6600	Drill switch		8	1		86.50	39.50		126	156	
	6800	Master box		2.70	2.963		3,100	117		3,217	3,575	
	7000	Break glass station		8	1		50	39.50		89.50	116	
	7800	Remote annunciator, 8 zone lamp	↓	1.80	4.444		175	175		350	460	
	8000	12 zone lamp	2 Elec	2.60	6.154		300	242		542	705	
	8200	16 zone lamp	"	2.20	7.273		300	287		587	770	
	8400	Standpipe or sprinkler alarm, alarm device	1 Elec	8	1		125	39.50		164.50	199	
	8600	Actuating device		8	1	↓	290	39.50		329.50	380	
	9410	Minimum labor/equipment charge	↓	4	2	Job		79		79	121	

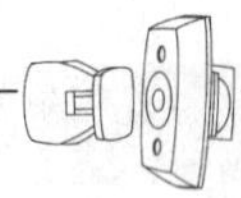

13800 | Building Automation & Control

	13834 | Electric/Electronic Control		CREW	DAILY OUTPUT	LABOR-HOURS	UNIT	2004 BARE COSTS MAT.	LABOR	EQUIP.	TOTAL	TOTAL INCL O&P	
200	0010	CONTROL SYSTEMS, ELECTRONIC										200
	9000	Minimum labor/equipment charge	1 Plum	8	1	Job		39.50		39.50	61.50	
	13838 | Pneumatic/Electric Controls											
200	0010	CONTROL COMPONENTS										200
	5000	Thermostats										
	5030	Manual	1 Shee	8	1	Ea.	27	39		66	91.50	
	5040	1 set back, electric, timed		8	1		81	39		120	151	
	5050	2 set back, electric, timed		8	1		178	39		217	258	
	5200	24 hour, automatic, clock	↓	8	1	↓	124	39		163	198	
	6000	Valves, motorized zone										
	6100	Sweat connections, 1/2" C x C	1 Stpi	20	.400	Ea.	67.50	15.90		83.40	99	
	6110	3/4" C x C		20	.400		72.50	15.90		88.40	104	
	6120	1" C x C	↓	19	.421	↓	84.50	16.75		101.25	119	
	9000	Minimum labor/equipment charge	1 Plum	4	2	Job		79		79	123	

		13910 | Basic Fire Protection Matl/Methd	CREW	DAILY OUTPUT	LABOR-HOURS	UNIT	2004 BARE COSTS MAT.	LABOR	EQUIP.	TOTAL	TOTAL INCL O&P	
400	0010	FIRE HOSE AND EQUIPMENT										400
	2200	Hose, less couplings										
	2260	Synthetic jacket, lined, 300 lb. test, 1-1/2" diameter	Q-12	2,600	.006	L.F.	1.59	.22		1.81	2.09	
	2280	2-1/2" diameter	"	2,200	.007	"	2.65	.26		2.91	3.32	
	2600	Hose rack, swinging, for 1-1/2" diameter hose,										
	2620	Enameled steel, 50' & 75' lengths of hose	Q-12	20	.800	Ea.	32.50	28.50		61	79.50	
	3750	Hydrants, wall, w/caps, single, flush, polished brass										
	3800	2-1/2" x 2-1/2"	Q-12	5	3.200	Ea.	117	113		230	305	
	4350	Double, projecting, polished brass										
	4400	2-1/2" x 2-1/2" x 4"	Q-12	5	3.200	Ea.	143	113		256	335	
	5600	Nozzles, brass										
	5620	Adjustable fog, 3/4" booster line				Ea.	69			69	76	
	5640	1-1/2" leader line				↓	89.50			89.50	98.50	
	5660	2-1/2" direct connection				↓	175			175	192	
	7140	Standpipe connections, wall, w/plugs & chains										
	7280	Double, flush, polished brass										
	7300	2-1/2" x 2-1/2" x 4"	Q-12	5	3.200	Ea.	292	113		405	495	
	9900	Minimum labor/equipment charge	1 Plum	2	4	Job		158		158	246	
800	0010	FIRE VALVES										800
	3000	Gate, hose, wheel handle, N.R.S., rough brass, 1-1/2"	1 Spri	12	.667	Ea.	73.50	26		99.50	122	
	3040	2-1/2", 300 lb.	↓	7	1.143	"	100	45		145	180	
	9990	Minimum labor/equipment charge	↓	4	2	Job		78.50		78.50	123	

		13930 | Wet-Pipe Fire Supp. Sprinklers	CREW	DAILY OUTPUT	LABOR-HOURS	UNIT	MAT.	LABOR	EQUIP.	TOTAL	TOTAL INCL O&P	
400	0010	SPRINKLER SYSTEM COMPONENTS										400
	0020											
	0800	Air compressor for dry pipe system, automatic, complete										
	0820	280 gal. system capacity, 3/4 HP	1 Spri	1.30	6.154	Ea.	720	242		962	1,175	
	1220	Water motor, complete with gong	"	4	2	"	185	78.50		263.50	325	
	1800	Firecycle system, controls, includes panel,										
	1820	batteries, solenoid valves and pressure switches	Q-13	1	32	Ea.	8,625	1,200		9,825	11,400	
	1980	Detector	1 Spri	16	.500	"	315	19.65		334.65	380	
	2600	Sprinkler heads, not including supply piping										
	2640	Dry, pendent, 1/2" orifice, 3/4" or 1" NPT										
	2660	1/2" to 6" length	1 Spri	14	.571	Ea.	35	22.50		57.50	73.50	
	2700	15-1/4" to 18" length	↓	14	.571	↓	51.50	22.50		74	91.50	
	3600	Foam-water, pendent or upright, 1/2" NPT	↓	12	.667	↓	54	26		80	101	
	3700	Standard spray, pendent or upright, brass, 135° to 286°F										
	3740	1/2" NPT, 1/2" orifice	1 Spri	16	.500	Ea.	5	19.65		24.65	36	
	3790	For riser & feeder piping, see div. 15107										
	6500	Check, swing, C.I. body, brass fittings, auto. ball drip										
	6520	4" size	Q-12	3	5.333	Ea.	160	188		348	470	
	8200	Dry pipe valve, incl. trim and gauges, 3" size	↓	2	8	↓	1,225	283		1,508	1,800	
	8220	4" size	↓	1	16	↓	1,275	565		1,840	2,275	
	9990	Minimum labor/equipment charge	1 Spri	3	2.667	Job		105		105	164	

Division Notes

		CREW	DAILY OUTPUT	LABOR-HOURS	UNIT	2004 BARE COSTS MAT.	LABOR	EQUIP.	TOTAL	TOTAL INCL O&P

Division 14
Conveying Systems

Estimating Tips

General

- Many products in Division 14 will require some type of support or blocking for installation not included with the item itself. Examples are supports for conveyors or tube systems, attachment points for lifts, and footings for hoists or cranes. Add these supports in the appropriate division.

14100 Dumbwaiters
14200 Elevators

- Dumbwaiters and elevators are estimated and purchased in a method similar to buying a car. The manufacturer has a base unit with standard features. Added to this base unit price will be whatever options the owner or specifications require. Increased load capacity, additional vertical travel, additional stops, higher speed, and cab finish options are items to be considered. When developing an estimate for dumbwaiters and elevators, remember that some items needed by the installers may have to be included as part of the general contract.

  Examples are:
  - shaftway
  - rail support brackets
  - machine room
  - electrical supply
  - sill angles
  - electrical connections
  - pits
  - roof penthouses
  - pit ladders

  Check the job specifications and drawings before pricing.
- Installation of elevators and handicapped lifts in historic structures can require significant additional costs. The associated structural requirements may involve cutting into and repairing finishes, mouldings, flooring, etc. The estimator must account for these special conditions.

14300 Escalators & Moving Walks

- Escalators and moving walks are specialty items installed by specialty contractors. There are numerous options associated with these items. For specific options contact a manufacturer or contractor. In a method similar to estimating dumbwaiters and elevators, you should verify the extent of general contract work and add items as necessary.

14400 Lifts
14500 Material Handling
14600 Hoists & Cranes

- Products such as correspondence lifts, conveyors, chutes, pneumatic tube systems, material handling cranes and hoists as well as other items specified in this subdivision may require trained installers. The general contractor might not have any choice as to who will perform the installation or when it will be performed. Long lead times are often required for these products, making early decisions in scheduling necessary.

Reference Numbers

Reference numbers are shown in bold squares at the beginning of some major classifications. These numbers refer to related items in the Reference Section. The reference information may be an estimating procedure, an alternate pricing method or technical information.

Note: Not all subdivisions listed here necessarily appear in this publication.

14100 | Dumbwaiters

		14110 | Manual Dumbwaiters	CREW	DAILY OUTPUT	LABOR-HOURS	UNIT	2004 BARE COSTS MAT.	LABOR	EQUIP.	TOTAL	TOTAL INCL O&P	
400	0010	**DUMBWAITERS** 2 stop, hand, minimum	2 Elev	.75	21.333	Ea.	2,275	885		3,160	3,875	400
	0100	Maximum		.50	32	"	5,200	1,325		6,525	7,750	
	0300	For each additional stop, add	↓	.75	21.333	Stop	825	885		1,710	2,275	

		14120 | Electric Dumbwaiters	CREW	DAILY OUTPUT	LABOR-HOURS	UNIT	MAT.	LABOR	EQUIP.	TOTAL	TOTAL INCL O&P	
400	0010	**DUMBWAITERS** 2 stop, electric, minimum	2 Elev	.13	123	Ea.	5,775	5,125		10,900	14,300	400
	0100	Maximum		.11	145	"	17,300	6,050		23,350	28,400	
	0600	For each additional stop, add	↓	.54	29.630	Stop	2,550	1,225		3,775	4,700	
	0750	45 lb capacity	2 Elec	.20	80	Ea.	6,750	3,150		9,900	12,300	

14200 | Elevators

		14210 | Electric Traction Elevators	CREW	DAILY OUTPUT	LABOR-HOURS	UNIT	2004 BARE COSTS MAT.	LABOR	EQUIP.	TOTAL	TOTAL INCL O&P	
100	0012	**ELEVATOR SYSTEMS**										100
	7000	Residential, cab type, 1 floor, 2 stop, minimum	2 Elev	.20	80	Ea.	8,075	3,325		11,400	14,100	
	7100	Maximum		.10	160		13,700	6,650		20,350	25,300	
	7200	2 floor, 3 stop, minimum		.12	133		12,000	5,550		17,550	21,800	
	7300	Maximum	↓	.06	266	↓	19,600	11,100		30,700	38,600	
200	0010	**ELEVATORS** R14200-100										200
	0020	For multi-story buildings, housing project, minimum				% total					2.50%	
	0100	Maximum R14200-200									4.50%	
	0300	Office building, minimum									2.50%	
	0400	Maximum R14200-300				↓					10%	
	0425	Electric freight, base unit, 4000 lb, 200 fpm, 4 stop, std. fin.	2 Elev	.05	320	Ea.	68,000	13,300		81,300	95,500	
	0450	For 5000 lb capacity, add R14200-400					4,875			4,875	5,375	
	0500	For 6000 lb capacity, add					8,600			8,600	9,475	
	0525	For 7000 lb capacity, add					11,500			11,500	12,700	
	0550	For 8000 lb capacity, add					16,000			16,000	17,700	
	0575	For 10000 lb capacity, add					19,300			19,300	21,300	
	0600	For 12000 lb capacity, add					23,600			23,600	26,000	
	0625	For 16000 lb capacity, add					28,400			28,400	31,200	
	0650	For 20000 lb capacity, add					31,300			31,300	34,400	
	0675	For increased speed, 250 fpm, add					9,350			9,350	10,300	
	0700	300 fpm, geared electric, add					11,700			11,700	12,800	
	0725	350 fpm, geared electric, add					13,800			13,800	15,100	
	0750	400 fpm, geared electric, add					15,600			15,600	17,100	
	0775	500 fpm, gearless electric, add					19,600			19,600	21,500	
	0800	600 fpm, gearless electric, add					21,600			21,600	23,800	
	0825	700 fpm, gearless electric, add					25,200			25,200	27,700	
	0850	800 fpm, gearless electric, add					28,000			28,000	30,800	
	0875	For class "B" loading, add					1,700			1,700	1,875	
	0900	For class "C-1" loading, add					4,200			4,200	4,600	
	0925	For class "C-2" loading, add					5,000			5,000	5,500	
	0950	For class "C-3" loading, add				↓	6,875			6,875	7,550	
	0975	For travel over 40 V.L.F., add	2 Elev	7.25	2.207	V.L.F.	99.50	92		191.50	252	
	1000	For number of stops over 4, add		.27	59.259	Stop	1,775	2,475		4,250	5,750	
	1625	Electric pass., base unit, 2000 lb, 200 fpm, 4 stop, std. fin.	↓	.05	320	Ea.	65,500	13,300		78,800	92,500	
	1650	For 2500 lb capacity, add ↓				↓	2,675			2,675	2,925	

14200 | Elevators

		14210 | Electric Traction Elevators	CREW	DAILY OUTPUT	LABOR-HOURS	UNIT	2004 BARE COSTS MAT.	LABOR	EQUIP.	TOTAL	TOTAL INCL O&P	
200	1675	For 3000 lb capacity, add R14200-100				Ea.	4,075			4,075	4,475	200
	1700	For 3500 lb capacity, add					5,425			5,425	5,975	
	1725	For 4000 lb capacity, add R14200-200					5,925			5,925	6,525	
	1750	For 4500 lb capacity, add					7,800			7,800	8,575	
	1775	For 5000 lb capacity, add R14200-300					9,925			9,925	10,900	
	1800	For increased speed, 250 fpm, geared electric, add					2,200			2,200	2,425	
	1825	300 fpm, geared electric, add R14200-400					4,525			4,525	4,975	
	1850	350 fpm, geared electric, add					5,350			5,350	5,900	
	1875	400 fpm, geared electric, add					7,600			7,600	8,375	
	1900	500 fpm, gearless electric, add					35,800			35,800	39,300	
	1925	600 fpm, gearless electric, add					37,800			37,800	41,500	
	1950	700 fpm, gearless electric, add					41,300			41,300	45,500	
	1975	800 fpm, gearless electric, add					45,500			45,500	50,000	
	2000	For travel over 40 V.L.F., add	2 Elev	7.25	2.207	V.L.F.	99.50	92		191.50	252	
	2025	For number of stops over 4, add		.27	59.259	Stop	2,225	2,475		4,700	6,250	
	2400	Electric hospital, base unit, 4000 lb, 200 fpm, 4 stop, std fin.		.05	320	Ea.	70,000	13,300		83,300	97,500	
	2425	For 4500 lb capacity, add					4,825			4,825	5,325	
	2450	For 5000 lb capacity, add					6,325			6,325	6,950	
	2475	For increased speed, 250 fpm, geared electric, add					2,350			2,350	2,600	
	2500	300 fpm, geared electric, add					4,475			4,475	4,925	
	2525	350 fpm, geared electric, add					5,475			5,475	6,025	
	2550	400 fpm, geared electric, add					7,625			7,625	8,375	
	2575	500 fpm, gearless electric, add					34,200			34,200	37,600	
	2600	600 fpm, gearless electric, add					37,800			37,800	41,500	
	2625	700 fpm, gearless electric, add					41,200			41,200	45,300	
	2650	800 fpm, gearless electric, add					45,500			45,500	50,000	
	2675	For travel over 40 V.L.F., add	2 Elev	7.25	2.207	V.L.F.	99.50	92		191.50	252	
	2700	For number of stops over 4, add	″	.27	59.259	Stop	2,775	2,475		5,250	6,850	
		14240 | Hydraulic Elevators										
200	0010	HYDRAULIC ELEVATORS										200
	1025	Hydraulic freight, base unit, 2000 lb, 50 fpm, 2 stop, std. fin.	2 Elev	.10	160	Ea.	34,400	6,650		41,050	48,100	
	1050	For 2500 lb capacity, add					2,475			2,475	2,700	
	1075	For 3000 lb capacity, add					3,900			3,900	4,300	
	1100	For 3500 lb capacity, add					5,925			5,925	6,500	
	1125	For 4000 lb capacity, add					6,325			6,325	6,975	
	1150	For 4500 lb capacity, add					7,350			7,350	8,075	
	1175	For 5000 lb capacity, add					10,000			10,000	11,000	
	1200	For 6000 lb capacity, add					10,600			10,600	11,600	
	1225	For 7000 lb capacity, add					16,100			16,100	17,800	
	1250	For 8000 lb capacity, add					18,100			18,100	19,900	
	1275	For 10000 lb capacity, add					19,200			19,200	21,200	
	1300	For 12000 lb capacity, add					23,200			23,200	25,600	
	1325	For 16000 lb capacity, add					30,300			30,300	33,300	
	1350	For 20000 lb capacity, add					33,600			33,600	37,000	
	1375	For increased speed, 100 fpm, add					720			720	790	
	1400	125 fpm, add					1,350			1,350	1,500	
	1425	150 fpm, add					2,575			2,575	2,825	
	1450	175 fpm, add					3,950			3,950	4,350	
	1475	For class "B" loading, add					1,650			1,650	1,825	
	1500	For class "C-1" loading, add					4,150			4,150	4,550	
	1525	For class "C-2" loading, add					4,975			4,975	5,450	
	1550	For class "C-3" loading, add					6,825			6,825	7,500	
	1575	For travel over 20 V.L.F., add	2 Elev	7.25	2.207	V.L.F.	330	92		422	500	
	1600	For number of stops over 2, add		.27	59.259	Stop	465	2,475		2,940	4,325	
	2050	Hyd. pass., base unit, 1500 lb, 100 fpm, 2 stop, std. fin.		.10	160	Ea.	28,600	6,650		35,250	41,800	

	14240 | Hydraulic Elevators		CREW	DAILY OUTPUT	LABOR-HOURS	UNIT	2004 BARE COSTS MAT.	LABOR	EQUIP.	TOTAL	TOTAL INCL O&P	
200	2075	For 2000 lb capacity, add				Ea.	910			910	1,000	200
	2100	For 2500 lb capacity, add				↓	1,875			1,875	2,050	
	2125	For 3000 lb capacity, add					3,400			3,400	3,725	
	2150	For 3500 lb capacity, add					5,025			5,025	5,550	
	2175	For 4000 lb capacity, add					6,050			6,050	6,675	
	2200	For 4500 lb capacity, add					7,300			7,300	8,025	
	2225	For 5000 lb capacity, add					11,000			11,000	12,100	
	2250	For increased speed, 125 fpm, add					1,075			1,075	1,200	
	2275	150 fpm, add					2,150			2,150	2,375	
	2300	175 fpm, add					3,475			3,475	3,825	
	2325	200 fpm, add				↓	5,075			5,075	5,575	
	2350	For travel over 12 V.L.F., add	2 Elev	7.25	2.207	V.L.F.	330	92		422	500	
	2375	For number of stops over 2, add	↓	.27	59.259	Stop	485	2,475		2,960	4,325	
	2725	Hydraulic hospital, base unit, 4000 lb, 100 fpm, 2 stop, std. fin.	↓	.10	160	Ea.	39,600	6,650		46,250	54,000	
	2750	For 4000 lb capacity, add				↓	4,775			4,775	5,275	
	2775	For 4500 lb capacity, add					5,625			5,625	6,175	
	2800	For 5000 lb capacity, add					8,175			8,175	9,000	
	2825	For increased speed, 125 fpm, add					1,325			1,325	1,475	
	2850	150 fpm, add					2,250			2,250	2,475	
	2875	175 fpm, add					3,600			3,600	3,975	
	2900	200 fpm, add				↓	5,275			5,275	5,800	
	2925	For travel over 12 V.L.F., add	2 Elev	7.25	2.207	V.L.F.	219	92		311	380	
	2950	For number of stops over 2, add	"	.27	59.259	Stop	2,775	2,475		5,250	6,850	

	14270 | Custom Elevator Cabs		CREW	DAILY OUTPUT	LABOR-HOURS	UNIT	MAT.	LABOR	EQUIP.	TOTAL	TOTAL INCL O&P	
200	0010	**CAB FINISHES**										200
	3325	Passenger elevator cab finishes (based on 3500 lb cab size)										
	3350	Acrylic panel ceiling				Ea.	415			415	455	
	3375	Aluminum eggcrate ceiling				↓	475			475	525	
	3400	Stainless steel doors					1,525			1,525	1,675	
	3425	Carpet flooring					345			345	380	
	3450	Epoxy flooring					283			283	310	
	3475	Quarry tile flooring					435			435	480	
	3500	Slate flooring					610			610	670	
	3525	Textured rubber flooring					111			111	123	
	3550	Stainless steel walls					2,475			2,475	2,725	
	3575	Stainless steel returns at door				↓	520			520	575	
	4450	Hospital elevator cab finishes (based on 3500# cab size)										
	4475	Aluminum eggcrate ceiling				Ea.	475			475	525	
	4500	Stainless steel doors				↓	1,525			1,525	1,675	
	4525	Epoxy flooring					283			283	310	
	4550	Quarry tile flooring					435			435	480	
	4575	Textured rubber flooring					111			111	123	
	4600	Stainless steel walls					2,475			2,475	2,725	
	4625	Stainless steel returns at door				↓	520			520	575	

	14280 | Elevator Equipment and Controls		CREW	DAILY OUTPUT	LABOR-HOURS	UNIT	MAT.	LABOR	EQUIP.	TOTAL	TOTAL INCL O&P	
200	0010	**ELEVATOR CONTROLS AND DOORS**										200
	2975	Passenger elevator options										
	3000	2 car group automatic controls	2 Elev	.66	24.242	Ea.	2,275	1,000		3,275	4,050	
	3025	3 car group automatic controls	↓	.44	36.364	↓	3,425	1,525		4,950	6,100	
	3050	4 car group automatic controls		.33	48.485		5,675	2,025		7,700	9,350	
	3075	5 car group automatic controls		.26	61.538		7,725	2,550		10,275	12,500	
	3100	6 car group automatic controls		.22	72.727		11,700	3,025		14,725	17,500	
	3125	Intercom service	↓	3	5.333	↓	300	222		522	680	

Important: See the Reference Section for critical supporting data - Reference Nos., Crews, & City Cost Indexes

	14280 | Elevator Equipment and Controls		CREW	DAILY OUTPUT	LABOR-HOURS	UNIT	2004 BARE COSTS MAT.	LABOR	EQUIP.	TOTAL	TOTAL INCL O&P	
200	3150	Duplex car selective collective	2 Elev	.66	24.242	Ea.	2,500	1,000		3,500	4,300	200
	3175	Center opening 1 speed doors		2	8		1,300	335		1,635	1,975	
	3200	Center opening 2 speed doors		2	8		1,575	335		1,910	2,250	
	3225	Rear opening doors (opposite front)		2	8		3,625	335		3,960	4,525	
	3250	Side opening 2 speed doors		2	8		4,950	335		5,285	5,975	
	3275	Automatic emergency power switching		.66	24.242		815	1,000		1,815	2,450	
	3300	Manual emergency power switching		8	2		315	83		398	475	
	3625	Hall finishes, stainless steel doors					865			865	950	
	3650	Stainless steel frames					865			865	950	
	3675	12 month maintenance contract									2,640	
	3700	Signal devices, hall lanterns	2 Elev	8	2		320	83		403	480	
	3725	Position indicators, up to 3		9.40	1.702		223	71		294	355	
	3750	Position indicators, per each over 3		32	.500		61.50	21		82.50	99.50	
	3775	High speed heavy duty door opener					1,575			1,575	1,725	
	3800	Variable voltage, O.H. gearless machine, min.	2 Elev	.16	100		23,000	4,150		27,150	31,700	
	3815	Maximum		.07	228		51,000	9,500		60,500	70,500	
	3825	Basement installed geared machine		.33	48.485		11,400	2,025		13,425	15,600	
	3850	Freight elevator options										
	3875	Doors, bi-parting	2 Elev	.66	24.242	Ea.	3,900	1,000		4,900	5,850	
	3900	Power operated door and gate	"	.66	24.242		15,800	1,000		16,800	19,000	
	3925	Finishes, steel plate floor					685			685	755	
	3950	14 ga. 1/4" x 4' steel plate walls					1,675			1,675	1,850	
	3975	12 month maintenance contract									1,992	
	4000	Signal devices, hall lanterns	2 Elev	8	2		320	83		403	480	
	4025	Position indicators, up to 3		9.40	1.702		223	71		294	355	
	4050	Position indicators, per each over 3		32	.500		61.50	21		82.50	99.50	
	4075	Variable voltage basement installed geared machine		.66	24.242		12,700	1,000		13,700	15,500	
	4100	Hospital elevator options										
	4125	2 car group automatic controls	2 Elev	.66	24.242	Ea.	2,275	1,000		3,275	4,050	
	4150	3 car group automatic controls		.44	36.364		3,425	1,525		4,950	6,075	
	4175	4 car group automatic controls		.33	48.485		5,675	2,025		7,700	9,375	
	4200	5 car group automatic controls		.26	61.538		7,750	2,550		10,300	12,500	
	4225	6 car group automatic controls		.22	72.727		11,700	3,025		14,725	17,600	
	4250	Intercom service		3	5.333		300	222		522	680	
	4275	Duplex car selective collective		.66	24.242		2,500	1,000		3,500	4,300	
	4300	Center opening 1 speed doors		2	8		1,300	335		1,635	1,975	
	4325	Center opening 2 speed doors		2	8		1,725	335		2,060	2,425	
	4350	Rear opening doors (opposite front)		2	8		3,625	335		3,960	4,525	
	4375	Side opening 2 speed doors		2	8		5,525	335		5,860	6,600	
	4400	Automatic emergency power switching		.66	24.242		815	1,000		1,815	2,450	
	4425	Manual emergency power switching		8	2		315	83		398	475	
	4675	Hall finishes, stainless steel doors					865			865	950	
	4700	Stainless steel frames					865			865	950	
	4725	12 month maintenance contract									3,985	
	4750	Signal devices, hall lanterns	2 Elev	8	2		320	83		403	480	
	4775	Position indicators, up to 3		9.40	1.702		223	71		294	355	
	4800	Position indicators, per each over 3		32	.500		61.50	21		82.50	99.50	
	4825	High speed heavy duty door opener					1,575			1,575	1,725	
	4850	Variable voltage, O.H. gearless machine, min.	2 Elev	.16	100		23,000	4,150		27,150	31,700	
	4865	Maximum		.07	228		51,000	9,500		60,500	70,500	
	4875	Basement installed geared machine		.33	48.485		11,400	2,025		13,425	15,600	
	5000	Drilling for piston, casing included, 18" diameter	B-48	80	.700	V.L.F.	30	20	39	89	109	

14400 | Lifts

		14420 | Wheelchair Lifts	CREW	DAILY OUTPUT	LABOR-HOURS	UNIT	2004 BARE COSTS MAT.	LABOR	EQUIP.	TOTAL	TOTAL INCL O&P	
100	0010	**CHAIR / WHEELCHAIR LIFT**										100
	7700	Stair climber (chair lift), single seat, minimum	2 Elev	1	16	Ea.	3,925	665		4,590	5,325	
	7800	Maximum		.20	80		5,375	3,325		8,700	11,100	
	8000	Wheelchair lift, minimum		1	16		5,375	665		6,040	6,925	
	8500	Maximum		.50	32		12,700	1,325		14,025	16,100	
	8700	Stair lift, minimum		1	16		10,600	665		11,265	12,700	
	8900	Maximum		.20	80		16,800	3,325		20,125	23,700	
		14450 | Vehicle Lifts										
700	0010	**HYDRAULIC LIFT AUTO/TRUCK**										700
	0200	Double post, 9000 lb frame	L-4	1.15	20.870	Ea.	5,225	640		5,865	6,800	
	0500	Four post, 50000 lb frame		.18	133		56,500	4,075		60,575	68,500	
	2200	Hoists, single post, 8,000# capacity, swivel arms		.40	60		3,600	1,825		5,425	6,950	
	2400	Two posts, adjustable frames, 11,000# capacity		.25	96		4,625	2,950		7,575	9,900	
	2500	24,000# capacity		.15	160		6,150	4,900		11,050	14,800	
	2700	7,500# capacity, frame supports		.50	48		5,125	1,475		6,600	8,050	
	2800	Four post, roll on ramp		.50	48		4,625	1,475		6,100	7,475	
	2810	Hydraulic lifts, above ground, 2 post, clear floor, 6000 lb cap		2.67	8.989		4,825	275		5,100	5,750	
	2815	9000 lb capacity		2.29	10.480		11,400	320		11,720	13,100	
	2820	15,000 lb capacity		2	12		12,900	365		13,265	14,800	
	2825	30,000 lb capacity		1.60	15		28,500	460		28,960	32,200	
	2830	4 post, ramp style, 25,000 lb capacity		2	12		11,200	365		11,565	12,900	
	2835	35,000 lb capacity		1	24		52,000	735		52,735	58,500	
	2840	50,000 lb capacity		1	24		58,000	735		58,735	65,000	
	2845	75,000 lb capacity		1	24		67,500	735		68,235	75,500	
	2850	For drive thru tracks, add, minimum					720			720	790	
	2855	Maximum					1,225			1,225	1,350	
	2860	Ramp extensions, 3′(set of 2)					590			590	650	
	2865	Rolling jack platform					2,050			2,050	2,250	
	2870	Elec/hyd jacking beam					5,475			5,475	6,025	
	2880	Scissor lift, portable, 6000 lb capacity					5,375			5,375	5,925	
		14460 | Correspondence & Parcel Lifts										
200	0010	**CORRESPONDENCE LIFT**, 1 floor 2 stop, 25 lb capacity, electric	2 Elev	.20	80	Ea.	4,825	3,325		8,150	10,500	200
	0100	Hand, 5 lb capacity	″	.20	80	″	1,825	3,325		5,150	7,150	

14500 | Material Transport

		14560 | Chutes	CREW	DAILY OUTPUT	LABOR-HOURS	UNIT	2004 BARE COSTS MAT.	LABOR	EQUIP.	TOTAL	TOTAL INCL O&P	
250	0010	**CHUTES** Linen or refuse, incl. sprinklers, 12′ floor height										250
	0020											
	0050	Aluminized steel, 16 ga., 18″ diameter	2 Shee	3.50	4.571	Floor	775	177		952	1,125	
	0100	24″ diameter		3.20	5		845	194		1,039	1,225	
	0400	Galvanized steel, 16 ga., 18″ diameter		3.50	4.571		700	177		877	1,050	
	0500	24″ diameter		3.20	5		790	194		984	1,175	
	0800	Stainless steel, 18″ diameter		3.50	4.571		1,600	177		1,777	2,025	
	0900	24″ diameter		3.20	5		1,700	194		1,894	2,175	
	9000	Minimum labor/equipment charge	1 Shee	1	8	Job		310		310	490	

14500 | Material Transport

14580 | Pneumatic Tube Systems

			CREW	DAILY OUTPUT	LABOR-HOURS	UNIT	2004 BARE COSTS MAT.	LABOR	EQUIP.	TOTAL	TOTAL INCL O&P	
800	0010	**PNEUMATIC TUBE SYSTEM** Single tube, 2 stations, blower										800
	0020	100′ long, stock										
	0100	3″ diameter	2 Stpi	.12	133	Total	4,800	5,300		10,100	13,500	
	0300	4″ diameter	"	.09	177	"	5,400	7,075		12,475	17,000	
	0400	Twin tube, two stations or more, conventional system										
	0700	3″ round	2 Stpi	46	.348	L.F.	11.10	13.85		24.95	33.50	
	1050	Add for blower		2	8	System	3,775	320		4,095	4,675	
	1110	Plus for each round station, add		7.50	2.133	Ea.	425	85		510	600	
	1200	Alternate pricing method: base cost, minimum		.75	21.333	Total	4,650	850		5,500	6,425	
	1300	Maximum		.25	64	"	9,200	2,550		11,750	14,100	
	1500	Plus total system length, add, minimum		93.40	.171	L.F.	5.95	6.80		12.75	17.10	
	1600	Maximum	▼	37.60	.426	"	17.65	16.90		34.55	45.50	

14600 | Hoists and Cranes

14630 | Bridge Cranes

			CREW	DAILY OUTPUT	LABOR-HOURS	UNIT	2004 BARE COSTS MAT.	LABOR	EQUIP.	TOTAL	TOTAL INCL O&P	
300	0010	**CRANE RAIL** Box beam bridge, no equipment included	E-4	3,400	.009	Lb.	.87	.35	.02	1.24	1.64	300
	0210	Running track only, 104 lb per yard, 20′ piece	"	160	.200	L.F.	15.90	7.55	.47	23.92	32	
700	0010	**OVERHEAD BRIDGE CRANES**										700
	0100	1 girder, 20′ span, 3 ton	M-3	1	34	Ea.	15,900	1,200	143	17,243	19,500	
	0125	5 ton		1	34		17,400	1,200	143	18,743	21,200	
	0150	7.5 ton		1	34		20,800	1,200	143	22,143	24,800	
	0175	10 ton		.80	42.500		27,400	1,500	179	29,079	32,700	
	0200	15 ton		.80	42.500		35,300	1,500	179	36,979	41,300	
	0225	30′ span, 3 ton		1	34		16,600	1,200	143	17,943	20,200	
	0250	5 ton		1	34		18,200	1,200	143	19,543	22,000	
	0275	7.5 ton		1	34		21,800	1,200	143	23,143	26,000	
	0300	10 ton		.80	42.500		28,400	1,500	179	30,079	33,700	
	0325	15 ton	▼	.80	42.500		36,800	1,500	179	38,479	43,000	
	0350	2 girder, 40′ span, 3 ton	M-4	.50	72		27,400	2,525	430	30,355	34,600	
	0375	5 ton		.50	72		28,700	2,525	430	31,655	36,000	
	0400	7.5 ton		.50	72		31,400	2,525	430	34,355	39,100	
	0425	10 ton		.40	90		36,700	3,175	540	40,415	45,900	
	0450	15 ton		.40	90		49,900	3,175	540	53,615	60,500	
	0475	25 ton		.30	120		59,000	4,225	720	63,945	72,500	
	0500	50′ span, 3 ton		.50	72		31,200	2,525	430	34,155	38,800	
	0525	5 ton		.50	72		32,400	2,525	430	35,355	40,100	
	0550	7.5 ton		.50	72		34,700	2,525	430	37,655	42,700	
	0575	10 ton		.40	90		40,000	3,175	540	43,715	49,500	
	0600	15 ton		.40	90		52,500	3,175	540	56,215	63,000	
	0625	25 ton	▼	.30	120	▼	62,000	4,225	720	66,945	75,500	

Division Notes

		CREW	DAILY OUTPUT	LABOR-HOURS	UNIT	2004 BARE COSTS MAT.	LABOR	EQUIP.	TOTAL	TOTAL INCL O&P

Division 15 Mechanical

Estimating Tips

15100 Building Services Piping

This subdivision is primarily basic pipe and related materials. The pipe may be used by any of the mechanical disciplines, i.e., plumbing, fire protection, heating, and air conditioning.

- The piping section lists the add to labor for elevated pipe installation. These adds apply to all elevated pipe, fittings, valves, insulation, etc., that are placed above 10′ high. CAUTION: the correct percentage may vary for the same pipe. For example, the percentage add for the basic pipe installation should be based on the maximum height that the craftsman must install for that particular section. If the pipe is to be located 14′ above the floor but it is suspended on threaded rod from beams, the bottom flange of which is 18′ high (4′ rods), then the height is actually 18′ and the add is 20%. The pipe coverer, however, does not have to go above the 14′ and so his add should be 10%.
- Most pipe is priced first as straight pipe with a joint (coupling, weld, etc.) every 10′ and a hanger usually every 10′. There are exceptions with hanger spacing such as: for cast iron pipe (5′) and plastic pipe (3 per 10′). Following each type of pipe there are several lines listing sizes and the amount to be subtracted to delete couplings and hangers. This is for pipe that is to be buried or supported together on trapeze hangers. The reason that the couplings are deleted is that these runs are usually long and frequently longer lengths of pipe are used. By deleting the couplings the estimator is expected to look up and add back the correct reduced number of couplings.
- When preparing an estimate it may be necessary to approximate the fittings. Fittings usually run between 25% and 50% of the cost of the pipe. The lower percentage is for simpler runs, and the higher number is for complex areas like mechanical rooms.
- For historic restoration projects, the systems must be as invisible as possible, and pathways must be sought for pipes, conduits, and ductwork. While installations in accessible spaces (such as basements and attics) are relatively straightforward to estimate, labor costs may be more difficult to determine when delivery systems must be concealed.

15400 Plumbing Fixtures & Equipment

- Plumbing fixture costs usually require two lines, the fixture itself and its "rough-in, supply and waste".
- In the Assemblies Section (Plumbing D2010) for the desired fixture, the System Components Group in the center of the page shows the fixture itself on the first line while the rest of the list (fittings, pipe, tubing, etc.) will total up to what we refer to in the Unit Price section as "Rough-in, supply, waste and vent". Note that for most fixtures we allow a nominal 5′ of tubing to reach from the fixture to a main or riser.
- Remember that gas and oil fired units need venting.

15500 Heat Generation Equipment

- When estimating the cost of an HVAC system, check to see who is responsible for providing and installing the temperature control system. It is possible to overlook controls, assuming that they would be included in the electrical estimate.
- When looking up a boiler be careful on specified capacity. Some manufacturers rate their products on output while others use input.
- Include HVAC insulation for pipe, boiler and duct (wrap and liner).
- Be careful when looking up mechanical items to get the correct pressure rating and connection type (thread, weld, flange).

15700 Heating/Ventilation/ Air Conditioning Equipment

- Combination heating and cooling units are sized by the air conditioning requirements. (See Reference No. R15710-020 for preliminary sizing guide.)
- A ton of air conditioning is nominally 400 CFM.
- Rectangular duct is taken off by the linear foot for each size, but its cost is usually estimated by the pound. Remember that SMACNA standards now base duct on internal pressure.
- Prefabricated duct is estimated and purchased like pipe: straight sections and fittings.
- Note that cranes or other lifting equipment are not included on any lines in Division 15. For example, if a crane is required to lift a heavy piece of pipe into place high above a gym floor, or to put a rooftop unit on the roof of a four-story building, etc., it must be added. Due to the potential for extreme variation—from nothing additional required, to a major crane or helicopter—we feel that including a nominal amount for "lifting contingency" would be useless and detract from the accuracy of the estimate. When using equipment rental from Means do not forget to include the cost of the operator(s).

Reference Numbers

Reference numbers are shown in bold squares at the beginning of some major classifications. These numbers refer to related items in the Reference Section. The reference information may be an estimating procedure, an alternate pricing method or technical information.

Note: Not all subdivisions listed here necessarily appear in this publication.

		15051 | Mechanical General	CREW	DAILY OUTPUT	LABOR-HOURS	UNIT	MAT.	LABOR	EQUIP.	TOTAL	TOTAL INCL O&P	
							2004 BARE COSTS					
700	0010	**PIPING** See also divisions 02500 & 02600 for site work										700
	1000	Add to labor for elevated installation										
	1080	10′ to 15′ high						10%				
	1100	15′ to 20′ high						20%				
	1120	20′ to 25′ high						25%				
	1140	25′ to 30′ high						35%				
	1160	30′ to 35′ high						40%				
	1180	35′ to 40′ high						50%				
	1200	Over 40′ high						55%				
		15055 | Selective Mech Demolition										
300	0010	**HVAC DEMOLITION** R02220-510										300
	0100	Air conditioner, split unit, 3 ton	Q-5	2	8	Ea.		286		286	445	
	0150	Package unit, 3 ton R15050-720	Q-6	3	8			297		297	460	
	0300	Boiler, electric	Q-19	2	12			445		445	685	
	0340	Gas or oil, steel, under 150 MBH	Q-6	3	8			297		297	460	
	0380	Over 150 MBH	″	2	12			445		445	690	
	1000	Ductwork, 4″ high, 8″ wide	1 Clab	200	.040	L.F.		1.04		1.04	1.72	
	1020	10″ wide		190	.042			1.09		1.09	1.81	
	1040	14″ wide		180	.044			1.16		1.16	1.92	
	1100	6″ high, 8″ wide		165	.048			1.26		1.26	2.09	
	1120	12″ wide		150	.053			1.39		1.39	2.30	
	1140	18″ wide		135	.059			1.54		1.54	2.55	
	1200	10″ high, 12″ wide		125	.064			1.66		1.66	2.76	
	1220	18″ wide		115	.070			1.81		1.81	3	
	1240	24″ wide		110	.073			1.89		1.89	3.13	
	1300	12″-14″ high, 16″-18″ wide		85	.094			2.45		2.45	4.06	
	1320	24″ wide		75	.107			2.77		2.77	4.60	
	1340	48″ wide		71	.113			2.93		2.93	4.86	
	1400	18″ high, 24″ wide		67	.119			3.10		3.10	5.15	
	1420	36″ wide		63	.127			3.30		3.30	5.45	
	1440	48″ wide		59	.136			3.53		3.53	5.85	
	1500	30″ high, 36″ wide		56	.143			3.71		3.71	6.15	
	1520	48″ wide		53	.151			3.92		3.92	6.50	
	1540	72″ wide		50	.160			4.16		4.16	6.90	
	1550	Duct heater, electric strip	1 Elec	8	1	Ea.		39.50		39.50	60.50	
	1850	Minimum labor/equipment charge	1 Clab	3	2.667	Job		69.50		69.50	115	
	2200	Furnace, electric	Q-20	2	10	Ea.		360		360	565	
	2300	Gas or oil, under 120 MBH	Q-9	4	4			140		140	221	
	2340	Over 120 MBH	″	3	5.333			186		186	295	
	2800	Heat pump, package unit, 3 ton	Q-5	2.40	6.667			239		239	370	
	2840	Split unit, 3 ton		2	8			286		286	445	
	3000	Mechanical equipment, light items. Unit is weight, not cooling.		.90	17.778	Ton		635		635	985	
	3600	Heavy items		1.10	14.545	″		520		520	805	
	9000	Minimum labor/equipment charge	Q-6	3	8	Job		297		297	460	
600	0010	**PLUMBING DEMOLITION** R15050-720										600
	1020	Fixtures, including 10′ piping										
	1100	Bath tubs, cast iron	1 Plum	4	2	Ea.		79		79	123	
	1120	Fiberglass		6	1.333			53		53	82	
	1140	Steel		5	1.600			63.50		63.50	98	
	1200	Lavatory, wall hung		10	.800			31.50		31.50	49	
	1220	Counter top		8	1			39.50		39.50	61.50	
	1300	Sink, steel or cast iron, single		8	1			39.50		39.50	61.50	
	1320	Double		7	1.143			45.50		45.50	70	
	1400	Water closet, floor mounted		8	1			39.50		39.50	61.50	

		15055 | Selective Mech Demolition	CREW	DAILY OUTPUT	LABOR-HOURS	UNIT	MAT. (2004 BARE COSTS)	LABOR (2004 BARE COSTS)	EQUIP. (2004 BARE COSTS)	TOTAL (2004 BARE COSTS)	TOTAL INCL O&P	
600	1420	Wall mounted (R15050-720)	1 Plum	7	1.143	Ea.		45.50		45.50	70	600
	1500	Urinal, floor mounted		4	2			79		79	123	
	1520	Wall mounted		7	1.143			45.50		45.50	70	
	1600	Water fountains, free standing		8	1			39.50		39.50	61.50	
	1620	Recessed		6	1.333			53		53	82	
	2000	Piping, metal, to 1-1/2" diameter		200	.040	L.F.		1.58		1.58	2.46	
	2050	2" to 3-1/2" diameter		150	.053			2.11		2.11	3.27	
	2100	4" to 6" diameter	2 Plum	100	.160			6.35		6.35	9.80	
	2150	8" to 14" diameter		60	.267			10.55		10.55	16.35	
	2153	16" diameter		55	.291			11.50		11.50	17.85	
	2240	Toilet partitions, see division 02225-840										
	2250	Water heater, 40 gal.	1 Plum	6	1.333	Ea.		53		53	82	
	6000	Remove and reset fixtures, minimum		6	1.333			53		53	82	
	6100	Maximum		4	2			79		79	123	
	9000	Minimum labor/equipment charge		2	4	Job		158		158	246	

		15080 | Mechanical Insulation	CREW	DAILY OUTPUT	LABOR-HOURS	UNIT	MAT.	LABOR	EQUIP.	TOTAL	TOTAL INCL O&P	
200	0010	**DUCT INSULATION**										200
	0100	Rule of thumb, as a percentage of total mechanical costs				Job				10%		
	0110	Insulation req'd is based on the surface size/area to be covered										
	3000	Ductwork										
	3020	Blanket type, fiberglass, flexible										
	3030	Fire resistant liner, black coating one side										
	3050	1/2" thick, 2 lb. density	Q-14	380	.042	S.F.	.36	1.36		1.72	2.62	
	3060	1" thick, 1-1/2 lb. density	"	350	.046	"	.48	1.48		1.96	2.94	
	3140	FRK vapor barrier wrap, .75 lb. density										
	3160	1" thick	Q-14	350	.046	S.F.	.34	1.48		1.82	2.78	
	3170	1-1/2" thick		320	.050		.35	1.62		1.97	3.02	
	3180	2" thick		300	.053		.36	1.73		2.09	3.21	
	3190	3" thick		260	.062		.53	1.99		2.52	3.82	
	3200	4" thick		242	.066		.70	2.14		2.84	4.25	
	3490	Board type, fiberglass liner, 3 lb. density										
	3500	Fire resistant, black pigmented, 1 side										
	3520	1" thick	Q-14	150	.107	S.F.	1.44	3.46		4.90	7.20	
	3540	1-1/2" thick	"	130	.123	"	1.76	3.99		5.75	8.45	
	9600	Minimum labor/equipment charge	1 Stpi	4	2	Job		79.50		79.50	123	
400	0010	**EQUIPMENT INSULATION**										400
	0100	Rule of thumb, as a percentage of total mechanical costs				Job				10%		
	0110	Insulation req'd is based on the surface size/area to be covered										
	2900	Domestic water heater wrap kit										
	2920	1-1/2" with vinyl jacket, 20-60 gal.	1 Plum	8	1	Ea.	16.60	39.50		56.10	80	
	2925	50 to 80 gallons	"	8	1	"	25.50	39.50		65	89.50	
600	0010	**PIPING INSULATION**										600
	0100	Rule of thumb, as a percentage of total mechanical costs				Job				10%		
	0110	Insulation req'd is based on the surface size/area to be covered										
	2930	Insulated protectors, (ADA)										
	2935	For exposed piping under sinks or lavatories.										
	2940	Vinyl coated foam, velcro tabs										
	2945	P Trap, 1-1/4" or 1-1/2"	1 Plum	32	.250	Ea.	17.15	9.90		27.05	34	
	2960	Valve and supply cover										
	2965	1/2", 3/8", and 7/16" pipe size	1 Plum	32	.250	Ea.	17.15	9.90		27.05	34	
	2970	Extension drain cover										
	2975	1-1/4", or 1-1/2" pipe size	1 Plum	32	.250	Ea.	17.55	9.90		27.45	34.50	
	2980	Tailpiece offset (wheelchair)										

	15080 | Mechanical Insulation	CREW	DAILY OUTPUT	LABOR-HOURS	UNIT	2004 BARE COSTS MAT.	LABOR	EQUIP.	TOTAL	TOTAL INCL O&P	
600	2985 1-1/4" pipe size	1 Plum	32	.250	Ea.	20	9.90		29.90	37.50	600
	4000 Pipe covering (price copper tube one size less than IPS)										
	6120 2" wall, 1/2" iron pipe size	Q-14	145	.110	L.F.	3.22	3.58		6.80	9.35	
	6210 4" iron pipe size	"	125	.128	"	5.85	4.15		10	13.20	
	6600 Fiberglass, with all service jacket										
	6840 1" wall, 1/2" iron pipe size	Q-14	240	.067	L.F.	.69	2.16		2.85	4.27	
	6860 3/4" iron pipe size		230	.070		.79	2.25		3.04	4.53	
	6870 1" iron pipe size		220	.073		.80	2.36		3.16	4.71	
	6880 1-1/4" iron pipe size		210	.076		.92	2.47		3.39	5	
	6890 1-1/2" iron pipe size		210	.076		.98	2.47		3.45	5.10	
	6900 2" iron pipe size		200	.080		1.08	2.59		3.67	5.40	
	7080 1-1/2" wall, 1/2" iron pipe size		230	.070		1.41	2.25		3.66	5.20	
	7140 2" iron pipe size		190	.084		1.91	2.73		4.64	6.55	
	7160 3" iron pipe size	↓	170	.094	↓	2.16	3.05		5.21	7.35	
	7879 Rubber tubing, flexible closed cell foam										
	8100 1/2" wall, 1/4" iron pipe size	1 Asbe	90	.089	L.F.	.34	3.20		3.54	5.55	
	8120 3/8" iron pipe size		90	.089		.36	3.20		3.56	5.60	
	8130 1/2" iron pipe size		89	.090		.54	3.24		3.78	5.85	
	8140 3/4" iron pipe size		89	.090		.56	3.24		3.80	5.85	
	8150 1" iron pipe size		88	.091		.66	3.27		3.93	6.05	
	8170 1-1/2" iron pipe size		87	.092		.98	3.31		4.29	6.50	
	8180 2" iron pipe size		86	.093		1.61	3.35		4.96	7.20	
	8200 3" iron pipe size		85	.094		1.72	3.39		5.11	7.40	
	8220 4" iron pipe size		80	.100		2.73	3.60		6.33	8.85	
	8300 3/4" wall, 1/4" iron pipe size		90	.089		.67	3.20		3.87	5.95	
	8330 1/2" iron pipe size		89	.090		.81	3.24		4.05	6.15	
	8340 3/4" iron pipe size		89	.090		.97	3.24		4.21	6.30	
	8350 1" iron pipe size		88	.091		1.18	3.27		4.45	6.60	
	8360 1-1/4" iron pipe size		87	.092		1.36	3.31		4.67	6.90	
	8370 1-1/2" iron pipe size		87	.092		2.07	3.31		5.38	7.70	
	8380 2" iron pipe size		86	.093		2.43	3.35		5.78	8.10	
	8444 1" wall, 1/2" iron pipe size		86	.093		1.53	3.35		4.88	7.15	
	8445 3/4" iron pipe size		84	.095		1.90	3.43		5.33	7.65	
	8446 1" iron pipe size		84	.095		2.30	3.43		5.73	8.10	
	8447 1-1/4" iron pipe size		82	.098		2.68	3.51		6.19	8.65	
	8448 1-1/2" iron pipe size		82	.098		3.02	3.51		6.53	9	
	8449 2" iron pipe size		80	.100		4.70	3.60		8.30	11	
	8450 2-1/2" iron pipe size	↓	80	.100	↓	6.15	3.60		9.75	12.60	
	8456 Rubber insulation tape, 1/8" x 2" x 30'				Ea.	10.80			10.80	11.90	
	9600 Minimum labor/equipment charge	1 Plum	4	2	Job		79		79	123	

15 MECHANICAL

15100 | Building Services Piping

	15106 | Glass Pipe & Fittings	CREW	DAILY OUTPUT	LABOR-HOURS	UNIT	2004 BARE COSTS MAT.	LABOR	EQUIP.	TOTAL	TOTAL INCL O&P	
120	0010 **PIPE, GLASS** Borosilicate, couplings & hangers 10' O.C.										120
	0020 Drainage										
	1100 1-1/2" diameter	Q-1	52	.308	L.F.	8.10	10.95		19.05	26	
	1120 2" diameter		44	.364		10.90	12.95		23.85	32	
	1140 3" diameter		39	.410		14.55	14.65		29.20	38.50	
	1160 4" diameter	↓	30	.533	↓	26	19		45	58	

Important: See the Reference Section for critical supporting data - Reference Nos., Crews, & City Cost Indexes

	15106 | Glass Pipe & Fittings		Crew	Daily Output	Labor-Hours	Unit	2004 Bare Costs Mat.	Labor	Equip.	Total	Total Incl O&P
120	1180	6″ diameter	Q-1	26	.615	L.F.	47.50	22		69.50	86
	9000	Minimum labor/equipment charge	1 Plum	4	2	Job		79		79	123
160	0010	**PIPE, GLASS, FITTINGS**									
	0020	Drainage, beaded ends									
	0040	Coupling & labor required at joints not incl. in fitting									
	0050	price. Add 1 per joint for installed price									
	0070	90° Bend or sweep, 1-1/2″				Ea.	19.10			19.10	21
	0090	2″					24			24	26.50
	0100	3″					40			40	44
	0110	4″					63.50			63.50	70
	0120	6″ (sweep only)					194			194	213
	0350	Tee, single sanitary, 1-1/2″					31			31	34
	0370	2″					31			31	34
	0380	3″					46			46	50.50
	0390	4″					82.50			82.50	90.50
	0400	6″					171			171	188
	0500	Coupling, stainless steel, TFE seal ring									
	0520	1-1/2″	Q-1	32	.500	Ea.	11.65	17.85		29.50	40.50
	0530	2″		30	.533		16.70	19		35.70	48
	0540	3″		25	.640		22.50	23		45.50	60
	0550	4″		23	.696		38.50	25		63.50	81
	0560	6″		20	.800		86.50	28.50		115	140
	9000	Minimum labor/equipment charge	1 Plum	4	2	Job		79		79	123

	15107 | Metal Pipe & Fittings		Crew	Daily Output	Labor-Hours	Unit	Mat.	Labor	Equip.	Total	Total Incl O&P
220	0010	**PIPE, BRASS** Plain end,									
	0900	Field threaded, coupling & clevis hanger 10′ O.C.									
	0920	Regular weight									
	1120	1/2″ diameter	1 Plum	48	.167	L.F.	4.30	6.60		10.90	15
	1140	3/4″ diameter		46	.174		5.80	6.90		12.70	17.10
	1160	1″ diameter		43	.186		8.35	7.35		15.70	20.50
	1180	1-1/4″ diameter	Q-1	72	.222		12.60	7.90		20.50	26
	1200	1-1/2″ diameter		65	.246		14.85	8.80		23.65	30
	1220	2″ diameter		53	.302		20	10.75		30.75	38.50
	9000	Minimum labor/equipment charge	1 Plum	4	2	Job		79		79	123
260	0010	**PIPE, BRASS, FITTINGS** Rough bronze, threaded									
	0020										
	1000	Standard wt., 90° Elbow									
	1100	1/2″	1 Plum	12	.667	Ea.	7.35	26.50		33.85	49
	1120	3/4″		11	.727		9.80	29		38.80	55.50
	1140	1″		10	.800		15.95	31.50		47.45	66.50
	1160	1-1/4″	Q-1	17	.941		26	33.50		59.50	80.50
	1180	1-1/2″	″	16	1		32	35.50		67.50	90.50
	1500	45° Elbow, 1/8″	1 Plum	13	.615		9	24.50		33.50	48
	1580	1/2″		12	.667		9	26.50		35.50	51
	1600	3/4″		11	.727		12.80	29		41.80	58.50
	1620	1″		10	.800		22	31.50		53.50	73
	1640	1-1/4″	Q-1	17	.941		34.50	33.50		68	90
	1660	1-1/2″	″	16	1		43.50	35.50		79	104
	2000	Tee, 1/8″	1 Plum	9	.889		8.60	35		43.60	64
	2080	1/2″		8	1		8.60	39.50		48.10	71
	2100	3/4″		7	1.143		12.25	45.50		57.75	83.50
	2120	1″		6	1.333		22	53		75	107
	2140	1-1/4″	Q-1	10	1.600		38	57		95	131
	2160	1-1/2″	″	9	1.778		43	63.50		106.50	146

		15107 | Metal Pipe & Fittings	CREW	DAILY OUTPUT	LABOR-HOURS	UNIT	2004 BARE COSTS MAT.	LABOR	EQUIP.	TOTAL	TOTAL INCL O&P	
260	2500	Coupling, 1/8"	1 Plum	26	.308	Ea.	6.15	12.20		18.35	25.50	260
	2580	1/2"		15	.533		6.15	21		27.15	40	
	2600	3/4"		14	.571		8.60	22.50		31.10	44.50	
	2620	1"		13	.615		14.70	24.50		39.20	54	
	2640	1-1/4"	Q-1	22	.727		24.50	26		50.50	67	
	2660	1-1/2"		20	.800		32	28.50		60.50	79	
	2680	2"		18	.889		53	31.50		84.50	107	
	9000	Minimum labor/equipment charge	1 Plum	4	2	Job		79		79	123	
320	0010	**PIPE, CAST IRON** Soil, on hangers 5' O.C. R15100-050										320
	0020	Single hub, service wt., lead & oakum joints 10' O.C.										
	2120	2" diameter	Q-1	63	.254	L.F.	4.69	9.05		13.74	19.20	
	2140	3" diameter		60	.267		6.50	9.50		16	22	
	2160	4" diameter		55	.291		8.30	10.35		18.65	25	
	2180	5" diameter	Q-2	76	.316		11.30	11.65		22.95	30.50	
	2200	6" diameter	"	73	.329		13.85	12.15		26	34	
	2220	8" diameter 	Q-3	59	.542		21.50	20.50		42	55.50	
	4000	No hub, couplings 10' O.C.										
	4100	1-1/2" diameter	Q-1	71	.225	L.F.	5.35	8.05		13.40	18.35	
	4120	2" diameter		67	.239		5.50	8.50		14	19.25	
	4140	3" diameter		64	.250		7.35	8.90		16.25	22	
	4160	4" diameter		58	.276		9.30	9.85		19.15	25.50	
	4180	5" diameter	Q-2	83	.289		13.15	10.70		23.85	31	
	9000	Minimum labor/equipment charge	1 Plum	4	2	Job		79		79	123	
360	0010	**PIPE, CAST IRON, FITTINGS** Soil R15100-050										360
	0040	Hub and spigot, service weight, lead & oakum joints										
	0080	1/4 bend, 2"	Q-1	16	1	Ea.	9.05	35.50		44.55	65.50	
	0120	3"		14	1.143		12	40.50		52.50	76	
	0140	4"		13	1.231		18.80	44		62.80	88.50	
	0160	5"	Q-2	18	1.333		26.50	49.50		76	106	
	0180	6"	"	17	1.412		32.50	52		84.50	117	
	0200	8"	Q-3	11	2.909		98.50	110		208.50	279	
	0340	1/8 bend, 2"	Q-1	16	1		6.40	35.50		41.90	62.50	
	0350	3"		14	1.143		10.10	40.50		50.60	74	
	0360	4"		13	1.231		14.70	44		58.70	84	
	0380	5"	Q-2	18	1.333		20.50	49.50		70	99	
	0400	6"	"	17	1.412		25	52		77	109	
	0420	8"	Q-3	11	2.909		74.50	110		184.50	252	
	0500	Sanitary tee, 2"	Q-1	10	1.600		11.10	57		68.10	101	
	0540	3"		9	1.778		20.50	63.50		84	121	
	0620	4"		8	2		25	71.50		96.50	139	
	0700	5"	Q-2	12	2		49.50	74		123.50	170	
	0800	6"	"	11	2.182		56	80.50		136.50	187	
	0880	8"	Q-3	7	4.571		149	173		322	430	
	0954	10" x 6"	"	8	4		215	151		366	470	
	5990	No hub										
	6000	Cplg. & labor required at joints not incl. in fitting										
	6010	price. Add 1 coupling per joint for installed price										
	6020	1/4 Bend, 1-1/2"				Ea.	5.10			5.10	5.60	
	6060	2"					5.50			5.50	6.05	
	6080	3"					7.65			7.65	8.45	
	6120	4"					11			11	12.10	
	6184	1/4 Bend, long sweep, 1-1/2"					11.90			11.90	13.05	
	6186	2"					11.90			11.90	13.05	
	6188	3"					14.15			14.15	15.60	
	6189	4"					22.50			22.50	25	

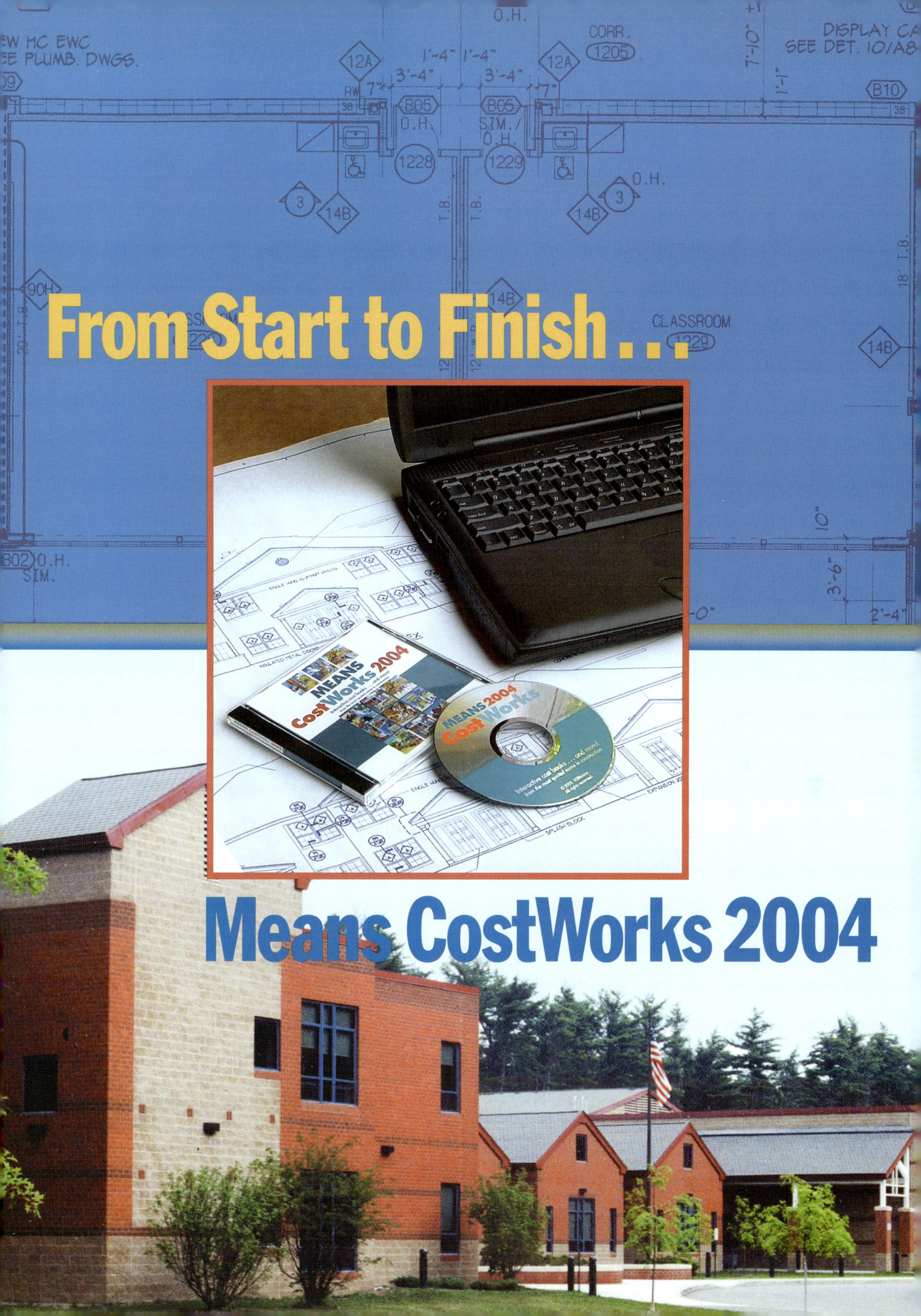
From Start to Finish . . .
Means CostWorks 2004

From the planning stages through the punch list . . .

Means CostWorks is the answer!

As building requirements grow more and more complex, the estimating and budgeting process has never been more demanding. Tighter deadlines and more stringent cost controls mean you must find ways to work more efficiently. Let **Means CostWorks 2004 CD** help you. Quickly and easily perform in-depth cost research and produce reports that make for successful projects.

Purchase your titles separately or in value priced packages. Plus, with your purchase you receive access to regular cost updates and additional valuable resources on the RSMeans Web site absolutely free!

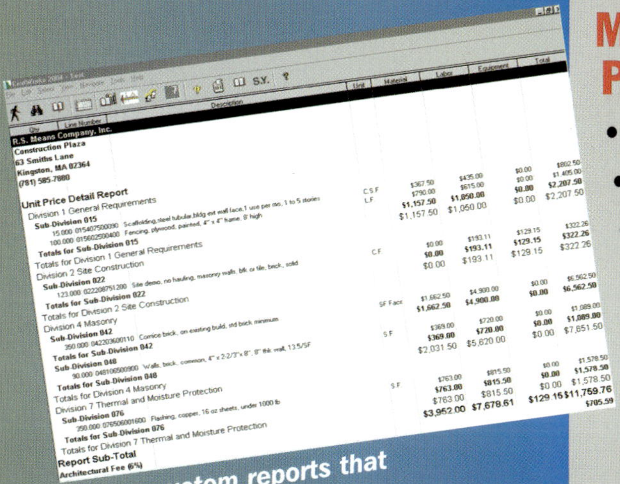

Design custom reports that meet your specifications

Means CostWorks Estimator 2004 . . . Powerful, efficient, and easy to use!

- Create customized estimates and detailed cost reports
- Create user defined columns, rows, and formulas
- Add subtotals, create reusable estimates, and merge multiple CostLists
- Quickly adjust or add your own costs
- Create custom templates and XML documents

Harness the power of Means Data . . . Order CostWorks Estimator today!

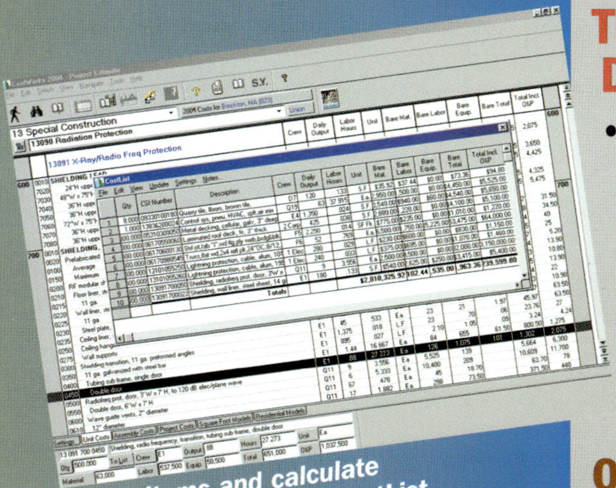

Select items and calculate project costs with the CostList

The same Cost Data you know and trust . . . Delivered the way you want it!

- Localize costs to your geographic area for better accuracy
- Easily search and organize detailed cost data
- Determine best-cost solutions for each project
- Change and calculate results automatically
- Receive free online cost factor updates every three months

Quick and easy . . . Get this powerful tool before you estimate your next project!

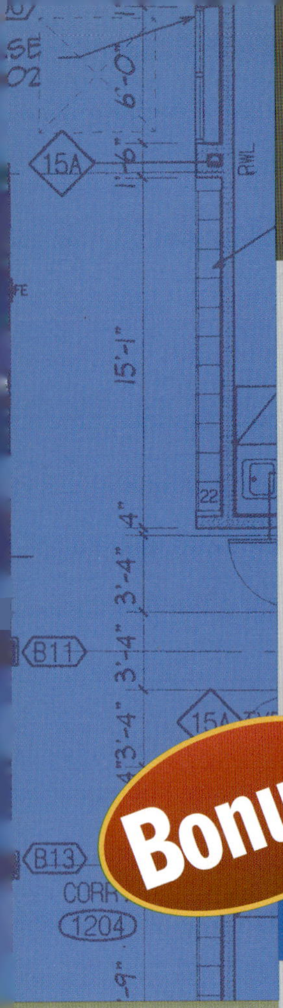

15107 | Metal Pipe & Fittings

			CREW	DAILY OUTPUT	LABOR-HOURS	UNIT	2004 BARE COSTS MAT.	LABOR	EQUIP.	TOTAL	TOTAL INCL O&P	
360	6190	5″ R15100-050				Ea.	41.50			41.50	45.50	360
	6191	6″					50.50			50.50	56	
	6192	8″					123			123	136	
	6193	10″					221			221	243	
	6200	1/8 Bend, 1-1/2″					4.22			4.22	4.64	
	6210	2″					4.69			4.69	5.15	
	6212	3″					6.35			6.35	6.95	
	6214	4″					8.05			8.05	8.85	
	6380	Sanitary Tee, tapped, 1-1/2″					9.30			9.30	10.25	
	6382	2″ x 1-1/2″					8.45			8.45	9.30	
	6384	2″					9.35			9.35	10.30	
	6386	3″ x 2″					13.05			13.05	14.40	
	6388	3″					24			24	26.50	
	6390	4″ x 1-1/2″					11.70			11.70	12.85	
	6392	4″ x 2″					13.15			13.15	14.45	
	6394	6″ x 1-1/2″					27			27	30	
	6396	6″ x 2″					27.50			27.50	30.50	
	6459	Sanitary Tee, 1-1/2″					7			7	7.70	
	6460	2″					7.65			7.65	8.45	
	6470	3″					9.30			9.30	10.25	
	6472	4″				▼	14.40			14.40	15.85	
	8000	Coupling, standard (by CISPI Mfrs.)										
	8020	1-1/2″	Q-1	48	.333	Ea.	4.84	11.90		16.74	24	
	8040	2″		44	.364		4.84	12.95		17.79	25.50	
	8080	3″		38	.421		5.75	15		20.75	30	
	8120	4″	▼	33	.485	▼	6.80	17.30		24.10	34.50	
	9000	Minimum labor/equipment charge	1 Plum	4	2	Job		79		79	123	
420	0010	**PIPE, COPPER** Solder joints										420
	1000	Type K tubing, couplings & clevis hangers 10′ O.C.										
	1100	1/4″ diameter	1 Plum	84	.095	L.F.	.99	3.77		4.76	6.95	
	1200	1″ diameter		66	.121		2.27	4.80		7.07	9.95	
	1260	2″ diameter	▼	40	.200	▼	5.30	7.90		13.20	18.15	
	2000	Type L tubing, couplings & hangers 10′ O.C.										
	2100	1/4″ diameter	1 Plum	88	.091	L.F.	.73	3.60		4.33	6.40	
	2120	3/8″ diameter		84	.095		.90	3.77		4.67	6.85	
	2140	1/2″ diameter		81	.099		1.01	3.91		4.92	7.15	
	2160	5/8″ diameter		79	.101		1.35	4.01		5.36	7.70	
	2180	3/4″ diameter		76	.105		1.42	4.17		5.59	8	
	2200	1″ diameter		68	.118		1.87	4.66		6.53	9.25	
	2220	1-1/4″ diameter		58	.138		2.44	5.45		7.89	11.15	
	2240	1-1/2″ diameter		52	.154		3.02	6.10		9.12	12.75	
	2260	2″ diameter	▼	42	.190		4.55	7.55		12.10	16.70	
	2280	2-1/2″ diameter	Q-1	62	.258		6.90	9.20		16.10	22	
	2300	3″ diameter		56	.286		9.25	10.20		19.45	26	
	2320	3-1/2″ diameter		43	.372		12.65	13.25		25.90	34.50	
	2340	4″ diameter		39	.410		15.50	14.65		30.15	39.50	
	2360	5″ diameter	▼	34	.471		40.50	16.80		57.30	70.50	
	2380	6″ diameter	Q-2	40	.600		50.50	22		72.50	90	
	2410	For other than full hard temper, add				▼	21%					
	2590	For silver solder, add						15%				
	3000	Type M tubing, couplings & hangers 10′ O.C.										
	3140	1/2″ diameter	1 Plum	84	.095	L.F.	.86	3.77		4.63	6.80	
	3180	3/4″ diameter		78	.103		1.22	4.06		5.28	7.65	
	3200	1″ diameter	▼	70	.114	▼	1.96	4.53		6.49	9.15	
	4000	Type DWV tubing, couplings & hangers 10′ O.C.										

15107 | Metal Pipe & Fittings

		15107 Metal Pipe & Fittings	CREW	DAILY OUTPUT	LABOR-HOURS	UNIT	2004 BARE COSTS MAT.	LABOR	EQUIP.	TOTAL	TOTAL INCL O&P	
420	4100	1-1/4" diameter	1 Plum	60	.133	L.F.	2.08	5.30		7.38	10.50	420
	4120	1-1/2" diameter		54	.148		2.53	5.85		8.38	11.90	
	4140	2" diameter		44	.182		3.29	7.20		10.49	14.75	
	4160	3" diameter	Q-1	58	.276		5.90	9.85		15.75	22	
	4180	4" diameter		40	.400		10.10	14.25		24.35	33	
	4200	5" diameter		36	.444		28.50	15.85		44.35	55.50	
	4220	6" diameter	Q-2	42	.571		40.50	21		61.50	77.50	
	9000	Minimum labor/equipment charge	1 Plum	4	2	Job		79		79	123	
460	0010	**PIPE, COPPER, FITTINGS** Wrought unless otherwise noted										460
	0040	Solder joints, copper x copper										
	0070	90° elbow, 1/4"	1 Plum	22	.364	Ea.	.94	14.40		15.34	23.50	
	0100	1/2"		20	.400		.30	15.85		16.15	25	
	0120	3/4"		19	.421		.67	16.65		17.32	26.50	
	0130	1"		16	.500		1.65	19.80		21.45	32.50	
	0250	45° elbow, 1/4"		22	.364		1.86	14.40		16.26	24.50	
	0270	3/8"		22	.364		1.52	14.40		15.92	24	
	0280	1/2"		20	.400		.55	15.85		16.40	25	
	0290	5/8"		19	.421		2.89	16.65		19.54	29	
	0300	3/4"		19	.421		.96	16.65		17.61	27	
	0310	1"		16	.500		3.45	19.80		23.25	34.50	
	0320	1-1/4"		15	.533		3.45	21		24.45	37	
	0330	1-1/2"		13	.615		4.14	24.50		28.64	42.50	
	0340	2"		11	.727		6.95	29		35.95	52	
	0350	2-1/2"	Q-1	13	1.231		14.75	44		58.75	84.50	
	0360	3"		13	1.231		22	44		66	92	
	0370	3-1/2"		10	1.600		39	57		96	132	
	0380	4"		9	1.778		46.50	63.50		110	150	
	0390	5"		6	2.667		181	95		276	345	
	0400	6"	Q-2	9	2.667		271	98.50		369.50	450	
	0450	Tee, 1/4"	1 Plum	14	.571		2.10	22.50		24.60	37.50	
	0470	3/8"		14	.571		1.61	22.50		24.11	37	
	0480	1/2"		13	.615		.51	24.50		25.01	38.50	
	0490	5/8"		12	.667		3.48	26.50		29.98	45	
	0500	3/4"		12	.667		1.24	26.50		27.74	42.50	
	0510	1"		10	.800		3.81	31.50		35.31	53	
	0520	1-1/4"		9	.889		6.10	35		41.10	61	
	0530	1-1/2"		8	1		8.45	39.50		47.95	71	
	0540	2"		7	1.143		13.20	45.50		58.70	84.50	
	0550	2-1/2"	Q-1	8	2		26.50	71.50		98	141	
	0560	3"		7	2.286		40.50	81.50		122	171	
	0570	3-1/2"		6	2.667		118	95		213	277	
	0580	4"		5	3.200		98.50	114		212.50	285	
	0590	5"		4	4		305	143		448	555	
	0600	6"	Q-2	6	4		420	148		568	690	
	0612	Tee, reducing on the outlet, 1/4"	1 Plum	15	.533		3.37	21		24.37	36.50	
	0613	3/8"		15	.533		3.17	21		24.17	36.50	
	0614	1/2"		14	.571		2.78	22.50		25.28	38	
	0615	5/8"		13	.615		5.60	24.50		30.10	44	
	0616	3/4"		12	.667		1.20	26.50		27.70	42.50	
	0617	1"		11	.727		4.03	29		33.03	49	
	0618	1-1/4"		10	.800		5.95	31.50		37.45	55.50	
	0619	1-1/2"		9	.889		6.30	35		41.30	61.50	
	0620	2"		8	1		10.30	39.50		49.80	73	
	0621	2-1/2"	Q-1	9	1.778		24.50	63.50		88	126	
	0622	3"		8	2		34	71.50		105.50	149	
	0623	4"		6	2.667		66.50	95		161.50	220	

15107 | Metal Pipe & Fittings

		15107 Metal Pipe & Fittings	CREW	DAILY OUTPUT	LABOR-HOURS	UNIT	2004 BARE COSTS MAT.	LABOR	EQUIP.	TOTAL	TOTAL INCL O&P	
460	0624	5"	Q-1	5	3.200	Ea.	305	114		419	510	460
	0625	6"	Q-2	7	3.429		420	127		547	655	
	0626	8"	"	6	4		1,700	148		1,848	2,100	
	0630	Tee, reducing on the run, 1/4"	1 Plum	15	.533		4.22	21		25.22	37.50	
	0631	3/8"		15	.533		5.65	21		26.65	39	
	0632	1/2"		14	.571		5.05	22.50		27.55	40.50	
	0633	5/8"		13	.615		5.65	24.50		30.15	44	
	0634	3/4"		12	.667		2.88	26.50		29.38	44	
	0635	1"		11	.727		4.80	29		33.80	50	
	0636	1-1/4"		10	.800		7.60	31.50		39.10	57.50	
	0637	1-1/2"		9	.889		13.85	35		48.85	70	
	0638	2"		8	1		18.45	39.50		57.95	82	
	0639	2-1/2"	Q-1	9	1.778		44.50	63.50		108	147	
	0640	3"		8	2		62.50	71.50		134	180	
	0641	4"		6	2.667		132	95		227	292	
	0642	5"		5	3.200		305	114		419	510	
	0643	6"	Q-2	7	3.429		420	127		547	655	
	0644	8"	"	6	4		1,700	148		1,848	2,100	
	0650	Coupling, 1/4"	1 Plum	24	.333		.23	13.20		13.43	21	
	0670	3/8"		24	.333		.31	13.20		13.51	21	
	0680	1/2"		22	.364		.23	14.40		14.63	23	
	0690	5/8"		21	.381		.71	15.10		15.81	24.50	
	0700	3/4"		21	.381		.45	15.10		15.55	24	
	0710	1"		18	.444		.91	17.60		18.51	28.50	
	0715	1-1/4"		17	.471		1.71	18.65		20.36	31	
	0716	1-1/2"		15	.533		2.25	21		23.25	35.50	
	0718	2"		13	.615		3.76	24.50		28.26	42	
	0721	2-1/2"	Q-1	15	1.067		8	38		46	68	
	0722	3"		13	1.231		12	44		56	81	
	0724	3-1/2"		8	2		23	71.50		94.50	137	
	0726	4"		7	2.286		24	81.50		105.50	153	
	0728	5"		6	2.667		56	95		151	209	
	0731	6"	Q-2	8	3		92.50	111		203.50	274	
	2000	DWV, solder joints, copper x copper										
	2030	90° Elbow, 1-1/4"	1 Plum	13	.615	Ea.	3.06	24.50		27.56	41.50	
	2050	1-1/2"		12	.667		4.12	26.50		30.62	45.50	
	2070	2"		10	.800		5.95	31.50		37.45	55.50	
	2090	3"	Q-1	10	1.600		14.75	57		71.75	105	
	2100	4"	"	9	1.778		71	63.50		134.50	177	
	2250	Tee, Sanitary, 1-1/4"	1 Plum	9	.889		6.05	35		41.05	61	
	2270	1-1/2"		8	1		7.50	39.50		47	70	
	2290	2"		7	1.143		8.75	45.50		54.25	79.50	
	2310	3"	Q-1	7	2.286		32	81.50		113.50	161	
	2330	4"	"	6	2.667		81	95		176	236	
	2400	Coupling, 1-1/4"	1 Plum	14	.571		1.43	22.50		23.93	36.50	
	2420	1-1/2"		13	.615		1.78	24.50		26.28	40	
	2440	2"		11	.727		2.46	29		31.46	47	
	2460	3"	Q-1	11	1.455		4.77	52		56.77	86	
	2480	4"	"	10	1.600		15.20	57		72.20	105	
	9000	Minimum labor/equipment charge	1 Plum	4	2	Job		79		79	123	
500	0010	**PIPE, CORROSION RESISTANT** No couplings or hangers										500
	0020	Iron alloy, drain, mechanical joint										
	1000	1-1/2" diameter	Q-1	70	.229	L.F.	22.50	8.15		30.65	37.50	
	1100	2" diameter		66	.242		26	8.65		34.65	42	
	1120	3" diameter		60	.267		36	9.50		45.50	55	
	1140	4" diameter		52	.308		46.50	10.95		57.45	68	

15107 | Metal Pipe & Fittings

			CREW	DAILY OUTPUT	LABOR-HOURS	UNIT	2004 BARE COSTS MAT.	LABOR	EQUIP.	TOTAL	TOTAL INCL O&P	
500	2980	Plastic, epoxy, fiberglass filament wound										500
	3000	2″ diameter	Q-1	62	.258	L.F.	5.20	9.20		14.40	19.95	
	3100	3″ diameter		51	.314		7.10	11.20		18.30	25	
	3120	4″ diameter		45	.356		9.40	12.70		22.10	30	
	3160	8″ diameter	Q-2	38	.632		19.80	23.50		43.30	58	
	3200	12″ diameter	″	28	.857		43	31.50		74.50	96.50	
	9800	Minimum labor/equipment charge	1 Plum	4	2	Job		79		79	123	
560	0010	**PIPE, CORROSION RESISTANT, FITTINGS**										560
	0030	Iron alloy										
	0050	Mechanical joint										
	0060	1/4 Bend, 1-1/2″	Q-1	12	1.333	Ea.	28	47.50		75.50	105	
	0080	2″		10	1.600		36.50	57		93.50	129	
	0090	3″		9	1.778		51	63.50		114.50	155	
	0100	4″		8	2		70.50	71.50		142	189	
	0160	Tee and Y, sanitary, straight										
	0170	1-1/2″	Q-1	8	2	Ea.	46	71.50		117.50	162	
	0180	2″		7	2.286		53.50	81.50		135	185	
	0190	3″		6	2.667		70.50	95		165.50	225	
	0200	4″		5	3.200		140	114		254	330	
	0360	Coupling, 1-1/2″		14	1.143		26.50	40.50		67	92	
	0380	2″		12	1.333		30.50	47.50		78	107	
	0390	3″		11	1.455		32.50	52		84.50	116	
	0400	4″		10	1.600		36.50	57		93.50	129	
	3000	Epoxy, filament wound										
	3030	Quick-lock joint										
	3040	90° Elbow, 2″	Q-1	28	.571	Ea.	58.50	20.50		79	96	
	3060	3″		16	1		67.50	35.50		103	130	
	3070	4″		13	1.231		92	44		136	169	
	3190	Tee, 2″		19	.842		140	30		170	200	
	3200	3″		11	1.455		169	52		221	267	
	3210	4″		9	1.778		203	63.50		266.50	320	
	9000	Minimum labor/equipment charge	1 Plum	4	2	Job		79		79	123	
620	0010	**PIPE, STEEL** R15100-050										620
	0020	All pipe sizes are to Spec. A-53 unless noted otherwise										
	0050	Schedule 40, threaded, with couplings, and clevis type										
	0060	hangers sized for covering, 10′ O.C.										
	0540	Black, 1/4″ diameter	1 Plum	66	.121	L.F.	1.28	4.80		6.08	8.85	
	0560	1/2″ diameter		63	.127		1.22	5.05		6.27	9.15	
	0570	3/4″ diameter		61	.131		1.42	5.20		6.62	9.60	
	0580	1″ diameter		53	.151		2.04	6		8.04	11.50	
	0590	1-1/4″ diameter	Q-1	89	.180		2.48	6.40		8.88	12.70	
	0600	1-1/2″ diameter 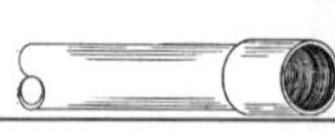		80	.200		2.83	7.15		9.98	14.15	
	0610	2″ diameter		64	.250		3.64	8.90		12.54	17.80	
	0620	2-1/2″ diameter		50	.320		6.20	11.40		17.60	24.50	
	0630	3″ diameter		43	.372		7.95	13.25		21.20	29.50	
	0640	3-1/2″ diameter		40	.400		10.90	14.25		25.15	34	
	0650	4″ diameter		36	.444		12.60	15.85		28.45	38.50	
	0670	6″ diameter	Q-2	31	.774		24.50	28.50		53	71.50	
	0680	8″ diameter	″	27	.889		36	33		69	90.50	
	1290	Galvanized, 1/4″ diameter	1 Plum	66	.121		1.93	4.80		6.73	9.60	
	1310	1/2″ diameter		63	.127		1.37	5.05		6.42	9.30	
	1320	3/4″ diameter		61	.131		1.56	5.20		6.76	9.75	
	1330	1″ diameter		53	.151		2.20	6		8.20	11.65	
	1350	1-1/2″ diameter	Q-1	80	.200		3.07	7.15		10.22	14.40	

15107 | Metal Pipe & Fittings

		15107 Metal Pipe & Fittings	CREW	DAILY OUTPUT	LABOR-HOURS	UNIT	2004 BARE COSTS MAT.	LABOR	EQUIP.	TOTAL	TOTAL INCL O&P	
620	1360	2″ diameter R15100-050	Q-1	64	.250	L.F.	4.06	8.90		12.96	18.25	620
	1370	2-1/2″ diameter		50	.320		6.60	11.40		18	25	
	1380	3″ diameter		43	.372		8.40	13.25		21.65	30	
	1400	4″ diameter		36	.444		13.55	15.85		29.40	39.50	
	1420	6″ diameter	Q-2	31	.774		23.50	28.50		52	70.50	
	1430	8″ diameter	″	27	.889		37	33		70	91.50	
	2000	Welded, sch. 40, on yoke & roll hangers, sized for covering,										
	2040	Black, 1″ diameter	Q-15	93	.172	L.F.	2.13	6.15	.84	9.12	12.75	
	2070	2″ diameter		61	.262		3.34	9.35	1.27	13.96	19.60	
	2090	3″ diameter		43	.372		5	13.25	1.81	20.06	28	
	2110	4″ diameter		37	.432		10.35	15.40	2.10	27.85	37.50	
	2120	5″ diameter		32	.500		12.30	17.85	2.43	32.58	43.50	
	2130	6″ diameter	Q-16	36	.667		15.45	24.50	2.16	42.11	57.50	
	9990	Minimum labor/equipment charge	1 Plum	3	2.667	Job		106		106	164	
640	0010	**PIPE, STEEL, FITTINGS** Threaded R15100-050										640
	0020	Cast Iron										
	0040	Standard weight, black										
	0060	90° Elbow, straight										
	0070	1/4″	1 Plum	16	.500	Ea.	3.54	19.80		23.34	34.50	
	0080	3/8″		16	.500		5.10	19.80		24.90	36	
	0090	1/2″		15	.533		2.24	21		23.24	35.50	
	0100	3/4″		14	.571		2.34	22.50		24.84	37.50	
	0110	1″		13	.615		2.77	24.50		27.27	41	
	0130	1-1/2″	Q-1	20	.800		5.45	28.50		33.95	50	
	0140	2″		18	.889		8.50	31.50		40	58.50	
	0150	2-1/2″		14	1.143		20.50	40.50		61	85.50	
	0160	3″		10	1.600		33.50	57		90.50	125	
	0180	4″		6	2.667		62	95		157	215	
	0200	6″	Q-2	7	3.429		156	127		283	370	
	0210	8″	″	6	4		325	148		473	585	
	0500	Tee, straight										
	0510	1/4″	1 Plum	10	.800	Ea.	5.05	31.50		36.55	54.50	
	0520	3/8″		10	.800		5.40	31.50		36.90	55	
	0540	3/4″		9	.889		4.08	35		39.08	59	
	0550	1″		8	1		3.62	39.50		43.12	65.50	
	0570	1-1/2″	Q-1	13	1.231		8.60	44		52.60	77.50	
	0580	2″		11	1.455		12	52		64	93.50	
	0590	2-1/2″		9	1.778		31	63.50		94.50	133	
	0600	3″		6	2.667		48	95		143	200	
	0620	4″		4	4		93	143		236	325	
	0640	6″	Q-2	4	6		219	222		441	585	
	0650	8″	″	3	8		495	296		791	1,000	
	0700	Standard weight, galvanized cast iron										
	0720	90° Elbow, straight										
	0730	1/4″	1 Plum	16	.500	Ea.	6.10	19.80		25.90	37	
	0740	3/8″		16	.500		6.10	19.80		25.90	37	
	0750	1/2″		15	.533		6.40	21		27.40	40	
	0760	3/4″		14	.571		6.85	22.50		29.35	42.50	
	0770	1″		13	.615		7.90	24.50		32.40	46.50	
	0790	1-1/2″	Q-1	20	.800		16.85	28.50		45.35	62.50	
	0800	2″		18	.889		25	31.50		56.50	76.50	
	0810	2-1/2″		14	1.143		51	40.50		91.50	120	
	0820	3″		10	1.600		78	57		135	174	
	0840	4″		6	2.667		143	95		238	305	
	0860	6″	Q-2	7	3.429		320	127		447	545	
	0870	8″	″	6	4		650	148		798	945	

15107 | Metal Pipe & Fittings

			CREW	DAILY OUTPUT	LABOR-HOURS	UNIT	2004 BARE COSTS MAT.	LABOR	EQUIP.	TOTAL	TOTAL INCL O&P	
640	1100	Tee, straight R15100-050										640
	1110	1/4"	1 Plum	10	.800	Ea.	7.80	31.50		39.30	57.50	
	1120	3/8"		10	.800		7.80	31.50		39.30	57.50	
	1140	3/4"		9	.889		9.85	35		44.85	65.50	
	1150	1"	↓	8	1		10.65	39.50		50.15	73.50	
	1170	1-1/2"	Q-1	13	1.231		24.50	44		68.50	95	
	1180	2"		11	1.455		31	52		83	115	
	1190	2-1/2"		9	1.778		66.50	63.50		130	172	
	1200	3"		6	2.667		172	95		267	335	
	1220	4"	↓	4	4		198	143		341	440	
	1240	6"	Q-2	4	6		299	222		521	675	
	1250	8"	"	3	8	↓	680	296		976	1,200	
	5000	Malleable iron, 150 lb.										
	5020	Black										
	5040	90° elbow, straight										
	5060	1/4"	1 Plum	16	.500	Ea.	1.79	19.80		21.59	32.50	
	5070	3/8"		16	.500		1.79	19.80		21.59	32.50	
	5090	3/4"		14	.571		1.57	22.50		24.07	36.50	
	5100	1"	↓	13	.615		2.73	24.50		27.23	41	
	5120	1-1/2"	Q-1	20	.800		5.90	28.50		34.40	50.50	
	5130	2"		18	.889		10.20	31.50		41.70	60	
	5140	2-1/2"		14	1.143		22.50	40.50		63	88	
	5150	3"		10	1.600		33.50	57		90.50	125	
	5170	4"	↓	6	2.667		71.50	95		166.50	226	
	5190	6"	Q-2	7	3.429	↓	220	127		347	440	
	5450	Tee, straight										
	5470	1/4"	1 Plum	10	.800	Ea.	2.73	31.50		34.23	52	
	5480	3/8"		10	.800		2.73	31.50		34.23	52	
	5500	3/4"		9	.889		2.50	35		37.50	57.50	
	5510	1"	↓	8	1		3.82	39.50		43.32	65.50	
	5520	1-1/4"	Q-1	14	1.143		6.90	40.50		47.40	70.50	
	5530	1-1/2"		13	1.231		8.60	44		52.60	77.50	
	5540	2"		11	1.455		14.65	52		66.65	96.50	
	5550	2-1/2"		9	1.778		31.50	63.50		95	134	
	5560	3"		6	2.667		46.50	95		141.50	199	
	5570	3-1/2"		5	3.200		113	114		227	300	
	5580	4"	↓	4	4		113	143		256	345	
	5600	6"	Q-2	4	6	↓	291	222		513	665	
	5650	Coupling										
	5670	1/4"	1 Plum	19	.421	Ea.	2.32	16.65		18.97	28.50	
	5680	3/8"		19	.421		2.32	16.65		18.97	28.50	
	5690	1/2"		19	.421		1.79	16.65		18.44	28	
	5700	3/4"		18	.444		2	17.60		19.60	29.50	
	5710	1"	↓	15	.533		3.15	21		24.15	36.50	
	5730	1-1/2"	Q-1	24	.667		5.50	24		29.50	43	
	5740	2"		21	.762		8.15	27		35.15	51	
	5750	2-1/2"		18	.889		22.50	31.50		54	74	
	5760	3"		14	1.143		30.50	40.50		71	96.50	
	5780	4"	↓	10	1.600		61.50	57		118.50	156	
	5800	6"	Q-2	8	3		127	111		238	310	
	6000	For galvanized elbows, tees, and couplings add				↓	20%					
	9990	Minimum labor/equipment charge	1 Plum	4	2	Job		79		79	123	
660	0010	**PIPE, STEEL, FITTINGS** Flanged, welded and special type										660
	3000	Weld joint, butt, carbon steel, standard weight										
	3040	90° elbow, long radius										
	3050	1/2" pipe size	Q-15	16	1	Ea.	14.10	35.50	4.86	54.46	76.50	

Important: See the Reference Section for critical supporting data - Reference Nos., Crews, & City Cost Indexes

15107 | Metal Pipe & Fittings

			CREW	DAILY OUTPUT	LABOR-HOURS	UNIT	2004 BARE COSTS				TOTAL INCL O&P	
							MAT.	LABOR	EQUIP.	TOTAL		
660	3060	3/4" pipe size	Q-15	16	1	Ea.	14.10	35.50	4.86	54.46	76.50	660
	3070	1" pipe size		16	1		7.05	35.50	4.86	47.41	68.50	
	3100	2" pipe size		10	1.600		7.05	57	7.80	71.85	105	
	3120	3" pipe size		7	2.286		10.90	81.50	11.10	103.50	150	
	3130	4" pipe size		5	3.200		18.55	114	15.55	148.10	215	
	3136	5" pipe size		4	4		39.50	143	19.45	201.95	286	
	3140	6" pipe size	Q-16	5	4.800		44	177	15.55	236.55	340	
	3350	Tee, straight										
	3360	1/2" pipe size	Q-15	10	1.600	Ea.	33	57	7.80	97.80	134	
	3370	3/4" pipe size		10	1.600		33	57	7.80	97.80	134	
	3380	1" pipe size		10	1.600		18.55	57	7.80	83.35	118	
	3410	2" pipe size		6	2.667		18.55	95	12.95	126.50	182	
	3430	3" pipe size		4	4		25.50	143	19.45	187.95	271	
	3440	4" pipe size		3	5.333		36.50	190	26	252.50	365	
	3446	5" pipe size		2.50	6.400		66	228	31	325	460	
	3450	6" pipe size	Q-16	3	8		66	296	26	388	560	
	9990	Minimum labor/equipment charge	Q-15	3	5.333	Job		190	26	216	325	
690	0010	**PIPE, GROOVED-JOINT STEEL FITTINGS & VALVES**										690
	0012	Fittings are ductile iron. Steel fittings noted.										
	0020	Pipe includes coupling & clevis type hanger 10' O.C.										
	1000	Schedule 40, black										
	1040	3/4" diameter	1 Plum	71	.113	L.F.	1.72	4.46		6.18	8.80	
	1050	1" diameter		63	.127		1.64	5.05		6.69	9.60	
	1060	1-1/4" diameter		58	.138		2.04	5.45		7.49	10.70	
	1070	1-1/2" diameter		51	.157		2.30	6.20		8.50	12.20	
	1080	2" diameter 		40	.200		2.73	7.90		10.63	15.30	
	1090	2-1/2" diameter	Q-1	57	.281		3.90	10		13.90	19.80	
	1100	3" diameter		50	.320		4.81	11.40		16.21	23	
	1110	4" diameter		45	.356		9.90	12.70		22.60	30.50	
	1120	5" diameter		37	.432		11.75	15.40		27.15	37	
	4000	Elbow, 90° or 45°, painted										
	4030	3/4" diameter	1 Plum	50	.160	Ea.	16.95	6.35		23.30	28.50	
	4040	1" diameter		50	.160		9	6.35		15.35	19.70	
	4050	1-1/4" diameter		40	.200		9	7.90		16.90	22	
	4060	1-1/2" diameter		33	.242		9	9.60		18.60	25	
	4070	2" diameter		25	.320		9	12.65		21.65	29.50	
	4080	2-1/2" diameter	Q-1	40	.400		9	14.25		23.25	32	
	4100	4" diameter		25	.640		17.40	23		40.40	54.50	
	4110	5" diameter		20	.800		42	28.50		70.50	90.50	
	4250	For galvanized elbows, add					26%					
	4690	Tee, painted										
	4700	3/4" diameter	1 Plum	38	.211	Ea.	18.20	8.35		26.55	33	
	4740	1" diameter		33	.242		13.85	9.60		23.45	30	
	4750	1-1/4" diameter		27	.296		13.85	11.75		25.60	33.50	
	4760	1-1/2" diameter		22	.364		13.85	14.40		28.25	38	
	4770	2" diameter		17	.471		13.85	18.65		32.50	44.50	
	4780	2-1/2" diameter	Q-1	27	.593		13.85	21		34.85	48.50	
	4800	4" diameter		17	.941		29.50	33.50		63	84.50	
	4810	5" diameter		13	1.231		69.50	44		113.50	145	
	4900	For galvanized tees, add					24%					
	9990	Minimum labor/equipment charge	1 Plum	4	2	Job		79		79	123	

15108 | Plastic Pipe & Fittings

520	0010	**PIPE, PLASTIC**										520
	0020	Fiberglass reinforced, couplings 10' O.C., hangers 3 per 10'										

MECHANICAL 15

15108 | Plastic Pipe & Fittings

		15108 Plastic Pipe & Fittings	CREW	DAILY OUTPUT	LABOR-HOURS	UNIT	2004 BARE COSTS MAT.	LABOR	EQUIP.	TOTAL	TOTAL INCL O&P	
520	0240	2″ diameter	Q-1	58	.276	L.F.	9.75	9.85		19.60	26	520
	0280	4″ diameter		47	.340		16.70	12.15		28.85	37	
	0300	6″ diameter		38	.421		25.50	15		40.50	51.50	
	1800	PVC, couplings 10′ O.C., hangers 3 per 10′										
	1820	Schedule 40										
	1860	1/2″ diameter	1 Plum	54	.148	L.F.	1.41	5.85		7.26	10.65	
	1870	3/4″ diameter		51	.157		1.49	6.20		7.69	11.30	
	1880	1″ diameter		46	.174		1.66	6.90		8.56	12.55	
	1890	1-1/4″ diameter		42	.190		1.86	7.55		9.41	13.75	
	1900	1-1/2″ diameter		36	.222		1.97	8.80		10.77	15.80	
	1910	2″ diameter	Q-1	59	.271		2.30	9.65		11.95	17.55	
	1920	2-1/2″ diameter		56	.286		3	10.20		13.20	19.10	
	1930	3″ diameter		53	.302		3.77	10.75		14.52	21	
	1940	4″ diameter		48	.333		4.76	11.90		16.66	23.50	
	1950	5″ diameter		43	.372		6.15	13.25		19.40	27.50	
	1960	6″ diameter		39	.410		8.10	14.65		22.75	31.50	
	4100	DWV type, schedule 40, couplings 10′ O.C., hangers 3 per 10′										
	4120	ABS										
	4140	1-1/4″ diameter	1 Plum	42	.190	L.F.	1.57	7.55		9.12	13.45	
	4150	1-1/2″ diameter	″	36	.222		1.60	8.80		10.40	15.40	
	4160	2″ diameter	Q-1	59	.271		1.71	9.65		11.36	16.90	
	4400	PVC										
	4410	1-1/4″ diameter	1 Plum	42	.190	L.F.	1.70	7.55		9.25	13.55	
	4460	2″ diameter	Q-1	59	.271		1.90	9.65		11.55	17.10	
	4470	3″ diameter		53	.302		3.19	10.75		13.94	20	
	4480	4″ diameter		48	.333		4.04	11.90		15.94	23	
	4490	6″ diameter		39	.410		7.10	14.65		21.75	30.50	
	5360	CPVC, couplings 10′ O.C., hangers 3 per 10′										
	5380	Schedule 40										
	5460	1/2″ diameter	1 Plum	54	.148	L.F.	2.60	5.85		8.45	11.95	
	5470	3/4″ diameter		51	.157		3.29	6.20		9.49	13.25	
	5480	1″ diameter		46	.174		3.97	6.90		10.87	15.05	
	5490	1-1/4″ diameter		42	.190		4.56	7.55		12.11	16.70	
	5500	1-1/2″ diameter		36	.222		5.05	8.80		13.85	19.20	
	5510	2″ diameter	Q-1	59	.271		6.10	9.65		15.75	22	
	5520	2-1/2″ diameter		56	.286		9.65	10.20		19.85	26.50	
	5530	3″ diameter		53	.302		11.75	10.75		22.50	29.50	
	5560											
	9900	Minimum labor/equipment charge	1 Plum	4	2	Job		79		79	123	
560	0010	**PIPE, PLASTIC, FITTINGS**										560
	2700	PVC (white), schedule 40, socket joints										
	2760	90° elbow, 1/2″	1 Plum	33.30	.240	Ea.	.29	9.50		9.79	15.05	
	2810	2″	Q-1	36.40	.440	″	1.68	15.65		17.33	26.50	
	4500	DWV, ABS, non pressure, socket joints										
	4540	1/4 Bend, 1-1/4″	1 Plum	20.20	.396	Ea.	2.94	15.70		18.64	27.50	
	4560	1-1/2″	″	18.20	.440		1.03	17.40		18.43	28	
	4570	2″	Q-1	33.10	.483		1.55	17.25		18.80	28	
	4800	Tee, sanitary										
	4820	1-1/4″	1 Plum	13.50	.593	Ea.	1.85	23.50		25.35	38.50	
	4830	1-1/2″	″	12.10	.661		1.56	26		27.56	42	
	4840	2″	Q-1	20	.800		2.49	28.50		30.99	46.50	
	5000	DWV, PVC, schedule 40, socket joints										
	5040	1/4 bend, 1-1/4″	1 Plum	20.20	.396	Ea.	1.46	15.70		17.16	26	
	5060	1-1/2″	″	18.20	.440		1.03	17.40		18.43	28	
	5070	2″	Q-1	33.10	.483		1.78	17.25		19.03	28.50	

15108 | Plastic Pipe & Fittings

			CREW	DAILY OUTPUT	LABOR-HOURS	UNIT	2004 BARE COSTS MAT.	LABOR	EQUIP.	TOTAL	TOTAL INCL O&P	
560	5080	3"	Q-1	20.80	.769	Ea.	4.37	27.50		31.87	47.50	560
	5090	4"		16.50	.970		6.95	34.50		41.45	61	
	5100	6"	↓	10.10	1.584		33.50	56.50		90	124	
	5105	8"	Q-2	9.30	2.581		52	95.50		147.50	205	
	5110	1/4 bend, long sweep, 1-1/2"	1 Plum	18.20	.440		1.94	17.40		19.34	29	
	5112	2"	Q-1	33.10	.483		1.76	17.25		19.01	28.50	
	5114	3"		20.80	.769		4.35	27.50		31.85	47.50	
	5116	4"	↓	16.50	.970		8.10	34.50		42.60	62.50	
	5215	8"	Q-2	9.30	2.581		56.50	95.50		152	210	
	5250	Tee, sanitary 1-1/4"	1 Plum	13.50	.593		2.13	23.50		25.63	39	
	5254	1-1/2"	"	12.10	.661		1.17	26		27.17	42	
	5255	2"	Q-1	20	.800		1.90	28.50		30.40	46	
	5256	3"		13.90	1.151		4	41		45	68	
	5257	4"		11	1.455		7.35	52		59.35	88.50	
	5259	6"	↓	6.70	2.388		38	85		123	174	
	5261	8"	Q-2	6.20	3.871		114	143		257	345	
	5264	2" x 1-1/2"	Q-1	22	.727		4.21	26		30.21	44.50	
	5266	3" x 1-1/2"		15.50	1.032		2.67	37		39.67	60	
	5268	4" x 3"		12.10	1.322		12.05	47		59.05	86.50	
	5271	6" x 4"	↓	6.90	2.319		58	82.50		140.50	192	
	5314	Combination Y & 1/8 bend, 1-1/2"	1 Plum	12.10	.661		2.91	26		28.91	43.50	
	5315	2"	Q-1	20	.800		3.76	28.50		32.26	48	
	5317	3"		13.90	1.151		6.65	41		47.65	71	
	5318	4"	↓	11	1.455	↓	12.90	52		64.90	94.50	
	5324	Combination Y & 1/8 bend, reducing										
	5325	2" x 2" x 1-1/2"	Q-1	22	.727	Ea.	4.17	26		30.17	44.50	
	5327	3" x 3" x 1-1/2"		15.50	1.032		7.25	37		44.25	65	
	5328	3" x 3" x 2"		15.30	1.046		5.05	37.50		42.55	63.50	
	5329	4" x 4" x 2"	↓	12.20	1.311		10.10	47		57.10	83.50	
	5331	Wye, 1-1/4"	1 Plum	13.50	.593		3.21	23.50		26.71	40	
	5332	1-1/2"	"	12.10	.661		1.96	26		27.96	42.50	
	5333	2"	Q-1	20	.800		2.30	28.50		30.80	46.50	
	5334	3"		13.90	1.151		5.90	41		46.90	70	
	5335	4"		11	1.455		9.05	52		61.05	90.50	
	5336	6"	↓	6.70	2.388		35.50	85		120.50	171	
	5337	8"	Q-2	6.20	3.871		64	143		207	293	
	5341	2" x 1-1/2"	Q-1	22	.727		3.21	26		29.21	43.50	
	5342	3" x 1-1/2"		15.50	1.032		4.25	37		41.25	61.50	
	5343	4" x 3"		12.10	1.322		7.20	47		54.20	81	
	5344	6" x 4"	↓	6.90	2.319		28.50	82.50		111	159	
	5345	8" x 6"	Q-2	6.40	3.750		48.50	139		187.50	269	
	5347	Double wye, 1-1/2"	1 Plum	9.10	.879		3.87	35		38.87	58.50	
	5348	2"	Q-1	16.60	.964		4.97	34.50		39.47	59	
	5349	3"		10.40	1.538		12.85	55		67.85	99	
	5350	4"		8.25	1.939		26	69		95	136	
	5354	2" x 1-1/2"		16.80	.952		4.54	34		38.54	57.50	
	5355	3" x 2"		10.60	1.509		9.55	54		63.55	94	
	5356	4" x 3"		8.45	1.893		20.50	67.50		88	128	
	5357	6" x 4"		7.25	2.207		43	78.50		121.50	169	
	5410	Reducer bushing, 2" x 1-1/4"		36.50	.438		.69	15.65		16.34	25	
	5412	3" x 1-1/2"		27.30	.586		3.02	21		24.02	36	
	5414	4" x 2"		18.20	.879		6.35	31.50		37.85	55.50	
	5416	6" x 4"	↓	11.10	1.441		18.20	51.50		69.70	99.50	
	5418	8" x 6"	Q-2	10.20	2.353	↓	36.50	87		123.50	175	
	5500	CPVC, Schedule 80, threaded joints										
	5540	90° Elbow, 1/4"	1 Plum	32	.250	Ea.	7.05	9.90		16.95	23	

	15108 | Plastic Pipe & Fittings		CREW	DAILY OUTPUT	LABOR-HOURS	UNIT	2004 BARE COSTS MAT.	LABOR	EQUIP.	TOTAL	TOTAL INCL O&P	
560	5560	1/2″	1 Plum	30.30	.264	Ea.	4.10	10.45		14.55	20.50	560
	5570	3/4″		26	.308		6.15	12.20		18.35	25.50	
	5580	1″		22.70	.352		8.60	13.95		22.55	31	
	5590	1-1/4″		20.20	.396		16.55	15.70		32.25	43	
	5600	1-1/2″		18.20	.440		17.85	17.40		35.25	46.50	
	5610	2″	Q-1	33.10	.483		24	17.25		41.25	53	
	5620	2-1/2″		24.20	.661		74	23.50		97.50	118	
	5630	3″		20.80	.769		79.50	27.50		107	130	
	6000	Coupling, 1/4″	1 Plum	32	.250		9	9.90		18.90	25.50	
	6020	1/2″		30.30	.264		7.40	10.45		17.85	24.50	
	6030	3/4″		26	.308		11.95	12.20		24.15	32	
	6040	1″		22.70	.352		13.55	13.95		27.50	36.50	
	6050	1-1/4″		20.20	.396		14.40	15.70		30.10	40.50	
	6060	1-1/2″		18.20	.440		15.45	17.40		32.85	44	
	6070	2″	Q-1	33.10	.483		18.25	17.25		35.50	46.50	
	6080	2-1/2″		24.20	.661		32.50	23.50		56	72.50	
	6090	3″		20.80	.769		38	27.50		65.50	84.50	
	9900	Minimum labor/equipment charge	1 Plum	4	2	Job		79		79	123	

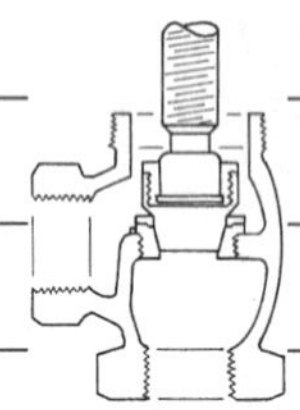

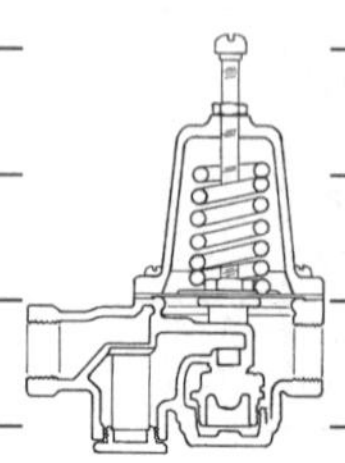

	15110 | Valves		CREW	DAILY OUTPUT	LABOR-HOURS	UNIT	MAT.	LABOR	EQUIP.	TOTAL	TOTAL INCL O&P	
160	0010	**VALVES, BRONZE**										160
	1020	Angle, 150 lb., rising stem, threaded										
	1070	3/4″	1 Plum	20	.400	Ea.	64	15.85		79.85	94.50	
	1080	1″		19	.421		91	16.65		107.65	126	
	1100	1-1/2″		13	.615		152	24.50		176.50	205	
	1110	2″		11	.727		247	29		276	315	
	1380	Ball, 150 psi, threaded										
	1460	3/4″	1 Plum	20	.400	Ea.	11.10	15.85		26.95	36.50	
	1750	Check, swing, class 150, regrinding disc, threaded										
	1850	1/2″	1 Plum	24	.333	Ea.	24.50	13.20		37.70	47.50	
	1860	3/4″		20	.400		33	15.85		48.85	61	
	1870	1″		19	.421		49.50	16.65		66.15	80.50	
	1890	1-1/2″		13	.615		81	24.50		105.50	127	
	1900	2″		11	.727		119	29		148	176	
	2850	Gate, N.R.S., soldered, 125 psi										
	2920	1/2″	1 Plum	24	.333	Ea.	16	13.20		29.20	38	
	2940	3/4″		20	.400		18.30	15.85		34.15	44.50	
	2950	1″		19	.421		26	16.65		42.65	54.50	
	2970	1-1/2″		13	.615		45	24.50		69.50	87.50	
	2980	2″		11	.727		62.50	29		91.50	114	
	4850	Globe, class 150, rising stem, threaded										
	4950	1/2″	1 Plum	24	.333	Ea.	34.50	13.20		47.70	58.50	
	4960	3/4″	″	20	.400	″	47.50	15.85		63.35	76.50	
	5600	Relief, pressure & temperature, self-closing, ASME, threaded										
	5640	3/4″	1 Plum	28	.286	Ea.	74	11.30		85.30	98.50	
	5650	1″		24	.333		107	13.20		120.20	139	
	5660	1-1/4″		20	.400		215	15.85		230.85	261	
	5670	1-1/2″		18	.444		410	17.60		427.60	480	
	5680	2″		16	.500		450	19.80		469.80	525	
	6400	Pressure, water, ASME, threaded										
	6440	3/4″	1 Plum	28	.286	Ea.	44.50	11.30		55.80	66.50	
	6450	1″		24	.333		93.50	13.20		106.70	124	
	6470	1-1/2″		18	.444		211	17.60		228.60	260	
	6900	Reducing, water pressure										
	6940	1/2″	1 Plum	24	.333	Ea.	125	13.20		138.20	159	
	6960	1″	″	19	.421	″	194	16.65		210.65	239	

15110 | Valves

			Crew	Daily Output	Labor-Hours	Unit	2004 Bare Costs Mat.	Labor	Equip.	Total	Total Incl O&P	
160	8350	Tempering, water, sweat connections										160
	8400	1/2″	1 Plum	24	.333	Ea.	45.50	13.20		58.70	70.50	
	8440	3/4″	″	20	.400	″	55.50	15.85		71.35	86	
	8650	Threaded connections										
	8700	1/2″	1 Plum	24	.333	Ea.	55.50	13.20		68.70	82	
	8740	3/4″		20	.400	″	213	15.85		228.85	259	
	9000	Minimum labor/equipment charge		4	2	Job		79		79	123	
200	0010	**VALVES, IRON BODY**										200
	1020	Butterfly, wafer type, gear actuator, 200 lb.										
	1030	2″	1 Plum	14	.571	Ea.	109	22.50		131.50	154	
	1060	4″	Q-1	5	3.200	″	145	114		259	335	
	1650	Gate, 125 lb., N.R.S.										
	2150	Flanged										
	2240	2-1/2″	Q-1	5	3.200	Ea.	305	114		419	510	
	2260	3″		4.50	3.556		340	127		467	570	
	2280	4″		3	5.333		485	190		675	830	
	3550	OS&Y, 125 lb., flanged										
	3680	4″	Q-1	3	5.333	Ea.	375	190		565	705	
	3690	5″	Q-2	3.40	7.059		615	261		876	1,075	
	3700	6″	″	3	8		615	296		911	1,150	
	9000	Minimum labor/equipment charge	1 Plum	3	2.667	Job		106		106	164	
500	0010	**VALVES, PLASTIC**										500
	0020											
	1150	Ball, PVC, socket or threaded, single union										
	1280	2″	1 Plum	17	.471	Ea.	114	18.65		132.65	154	
	1650	CPVC, socket or threaded, single union										
	1770	2″	1 Plum	17	.471	Ea.	122	18.65		140.65	163	
	3150	Ball check, PVC, socket or threaded										
	3290	2″	1 Plum	17	.471	Ea.	170	18.65		188.65	216	
	9000	Minimum labor/equipment charge	″	3.50	2.286	Job		90.50		90.50	140	

15120 | Piping Specialties

			Crew	Daily Output	Labor-Hours	Unit	Mat.	Labor	Equip.	Total	Total Incl O&P	
250	0010	**DIELECTRIC UNIONS** Standard gaskets for water and air										250
	0020	250 psi maximum pressure										
	0280	Female IPT to sweat, straight										
	0340	3/4″ pipe size	1 Plum	20	.400	Ea.	3.15	15.85		19	28	
	0780	Female IPT to female IPT, straight										
	0800	1/2″ pipe size	1 Plum	24	.333	Ea.	4.75	13.20		17.95	26	
	0840	3/4″ pipe size		20	.400	″	5.30	15.85		21.15	30.50	
	9000	Minimum labor/equipment charge		4	2	Job		79		79	123	
320	0010	**EXPANSION TANKS**										320
	2000	Steel, liquid expansion, ASME, painted, 15 gallon capacity	Q-5	17	.941	Ea.	365	33.50		398.50	450	
	2040	30 gallon capacity		12	1.333		405	47.50		452.50	520	
	2080	60 gallon capacity		8	2		555	71.50		626.50	720	
	2120	100 gallon capacity		6	2.667		735	95.50		830.50	960	
	3000	Steel ASME expansion, rubber diaphragm, 19 gal. cap. accept.		12	1.333		1,425	47.50		1,472.50	1,650	
	3020	31 gallon capacity		8	2		1,600	71.50		1,671.50	1,850	
	3040	61 gallon capacity		6	2.667		2,250	95.50		2,345.50	2,625	
	3080	119 gallon capacity		4	4		2,425	143		2,568	2,875	
	9000	Minimum labor/equipment charge		4	4	Job		143		143	222	
350	0010	**FLEXIBLE CONNECTORS**, Corrugated, 7/8″ O.D., 1/2″ I.D.										350
	0050	Gas, seamless brass, steel fittings										
	0200	12″ long	1 Plum	36	.222	Ea.	11.95	8.80		20.75	27	
	0220	18″ long		36	.222		14.85	8.80		23.65	30	

For expanded coverage of these items see *Means Mechanical or Plumbing Cost Data 2004*

15120 | Piping Specialties

		15120 Piping Specialties	CREW	DAILY OUTPUT	LABOR-HOURS	UNIT	2004 BARE COSTS MAT.	LABOR	EQUIP.	TOTAL	TOTAL INCL O&P	
350	0240	24" long	1 Plum	34	.235	Ea.	17.55	9.30		26.85	34	350
	0260	30" long		34	.235		18.95	9.30		28.25	35.50	
	9000	Minimum labor/equipment charge		4	2	Job		79		79	123	
580	0010	**MIXING VALVE** Automatic water tempering										580
	0040	1/2" size	1 Stpi	19	.421	Ea.	395	16.75		411.75	460	
	0050	3/4" size		18	.444	"	395	17.65		412.65	465	
	9000	Minimum labor/equipment charge		5	1.600	Job		63.50		63.50	98.50	
760	0010	**STEAM TRAP**										760
	0020											
	0030	Cast iron body, threaded										
	0040	Inverted bucket										
	0050	1/2" pipe size	1 Stpi	12	.667	Ea.	87	26.50		113.50	137	
	0100	1" pipe size		9	.889		231	35.50		266.50	310	
	0120	1-1/4" pipe size		8	1		350	40		390	445	
	1000	Float & thermostatic, 15 psi										
	1010	3/4" pipe size	1 Stpi	16	.500	Ea.	85	19.90		104.90	125	
	1020	1" pipe size		15	.533		101	21		122	144	
	1040	1-1/2" pipe size		9	.889		180	35.50		215.50	253	
	1060	2" pipe size		6	1.333		330	53		383	445	
	9000	Minimum labor/equipment charge		4	2	Job		79.50		79.50	123	
820	0010	**STRAINERS, Y TYPE** Bronze body										820
	0020											
	0050	Screwed, 150 lb., 1/4" pipe size	1 Stpi	24	.333	Ea.	11.40	13.25		24.65	33	
	0100	1/2" pipe size		20	.400		15.90	15.90		31.80	42	
	0120	3/4" pipe size		19	.421		17.40	16.75		34.15	45	
	0140	1" pipe size		17	.471		18.70	18.70		37.40	49.50	
	0160	1-1/2" pipe size		14	.571		40.50	22.50		63	79.50	
	0180	2" pipe size		13	.615		53.50	24.50		78	97	
	0182	3" pipe size		12	.667		310	26.50		336.50	380	
	0220	3" pipe size	Q-5	16	1		565	36		601	680	
	0240	4" pipe size	"	15	1.067		1,300	38		1,338	1,475	
	1000	Flanged, 150 lb., 1-1/2" pipe size	1 Stpi	11	.727		305	29		334	380	
	1020	2" pipe size	"	8	1		390	40		430	490	
	1030	2-1/2" pipe size	Q-5	5	3.200		505	115		620	735	
	1040	3" pipe size		4.50	3.556		620	127		747	880	
	1060	4" pipe size		3	5.333		945	191		1,136	1,325	
	1100	6" pipe size	Q-6	3	8		2,500	297		2,797	3,200	
	1106	8" pipe size	"	2.60	9.231		3,900	340		4,240	4,825	
	1500	For 300 lb rating, add					40%					
	9000	Minimum labor/equipment charge	1 Stpi	3.75	2.133	Job		85		85	132	
940	0010	**WATER SUPPLY METERS**										940
	2000	Domestic/commercial, bronze										
	2020	Threaded										
	2060	5/8" diameter, to 20 GPM	1 Plum	16	.500	Ea.	40	19.80		59.80	74.50	
	2080	3/4" diameter, to 30 GPM		14	.571		67.50	22.50		90	110	
	2100	1" diameter, to 50 GPM		12	.667		94	26.50		120.50	144	
	2300	Threaded/flanged										
	2340	1-1/2" diameter, to 100 GPM	1 Plum	8	1	Ea.	330	39.50		369.50	425	
	9000	Minimum labor/equipment charge	"	3.25	2.462	Job		97.50		97.50	151	

15140 | Domestic Water Piping

		15140 Domestic Water Piping	CREW	DAILY OUTPUT	LABOR-HOURS	UNIT	MAT.	LABOR	EQUIP.	TOTAL	TOTAL INCL O&P	
100	0010	**BACKFLOW PREVENTER** Includes valves										100
	0020	and four test cocks, corrosion resistant, automatic operation										

15140 | Domestic Water Piping

		15140 Domestic Water Piping	CREW	DAILY OUTPUT	LABOR-HOURS	UNIT	2004 BARE COSTS MAT.	LABOR	EQUIP.	TOTAL	TOTAL INCL O&P	
100	1000	Double check principle										100
	1010	Threaded, with ball valves										
	1020	3/4" pipe size	1 Plum	16	.500	Ea.	191	19.80		210.80	241	
	1030	1" pipe size		14	.571		194	22.50		216.50	248	
	1040	1-1/2" pipe size		10	.800		340	31.50		371.50	420	
	1050	2" pipe size		7	1.143		405	45.50		450.50	515	
	1080	Threaded, with gate valves										
	1100	3/4" pipe size	1 Plum	16	.500	Ea.	620	19.80		639.80	710	
	1120	1" pipe size		14	.571		645	22.50		667.50	740	
	1140	1-1/2" pipe size		10	.800		735	31.50		766.50	860	
	1160	2" pipe size		7	1.143		905	45.50		950.50	1,075	
	4000	Reduced pressure principle										
	4100	Threaded, valves are ball										
	4120	3/4" pipe size	1 Plum	16	.500	Ea.	715	19.80		734.80	815	
	4140	1" pipe size		14	.571		760	22.50		782.50	870	
	4160	1-1/2" pipe size		10	.800		1,025	31.50		1,056.50	1,175	
	4180	2" pipe size		7	1.143		1,100	45.50		1,145.50	1,275	
	4500	Minimum labor/equipment charge		4	2	Job		79		79	123	
	5000	Flanged, bronze, valves are OS&Y										
	5060	2-1/2" pipe size	Q-1	5	3.200	Ea.	2,750	114		2,864	3,200	
	5080	3" pipe size 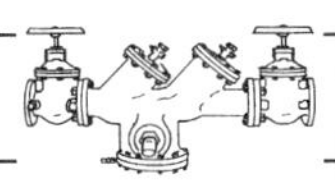		4.50	3.556		3,425	127		3,552	3,975	
	5100	4" pipe size		3	5.333		4,850	190		5,040	5,650	
	5120	6" pipe size	Q-2	3	8		10,300	296		10,596	11,800	
	5600	Flanged, iron, valves are OS&Y										
	5660	2-1/2" pipe size	Q-1	5	3.200	Ea.	1,900	114		2,014	2,275	
	5680	3" pipe size		4.50	3.556		2,025	127		2,152	2,425	
	5700	4" pipe size		3	5.333		2,675	190		2,865	3,250	
	5720	6" pipe size	Q-2	3	8		4,200	296		4,496	5,075	
	5800	Rebuild 4" diameter reduced pressure BFP	1 Plum	4	2		214	79		293	360	
	5810	6" diameter		2.66	3.008		227	119		346	435	
	5820	8" diameter		2	4		281	158		439	555	
	5830	10" diameter		1.60	5		330	198		528	665	
	9010	Minimum labor/equipment charge		2	4	Job		158		158	246	
600	0010	**VACUUM BREAKERS** Hot or cold water										600
	1030	Anti-siphon, brass										
	1060	1/2" size	1 Plum	24	.333	Ea.	16.35	13.20		29.55	38.50	
	1080	3/4" size		20	.400		28	15.85		43.85	55	
	1100	1" size		19	.421		43.50	16.65		60.15	74	
	1120	1-1/4" size		15	.533		76.50	21		97.50	117	
	1140	1-1/2" size		13	.615		89.50	24.50		114	137	
	1160	2" size		11	.727		140	29		169	199	
	9000	Minimum labor/equipment charge		4	2	Job		79		79	123	
800	0010	**WATER HAMMER ARRESTORS / SHOCK ABSORBERS**										800
	0490	Copper										
	0500	3/4" male I.P.S. For 1 to 11 fixtures	1 Plum	12	.667	Ea.	14.50	26.50		41	57	
	0600	1" male I.P.S., For 12 to 32 fixtures		8	1		37	39.50		76.50	103	
	0700	1-1/4" male I.P.S. For 33 to 60 fixtures		8	1		43	39.50		82.50	109	
	0800	1-1/2" male I.P.S. For 61 to 113 fixtures		8	1		58.50	39.50		98	126	
	0900	2" male I.P.S.For 114 to 154 fixtures		8	1		94	39.50		133.50	165	
	1000	2-1/2" male I.P.S. For 155 to 330 fixtures		4	2		315	79		394	470	
	9000	Minimum labor/equipment charge		3.50	2.286	Job		90.50		90.50	140	

For expanded coverage of these items see *Means Mechanical or Plumbing Cost Data 2004*

15150 | Sanitary Waste and Vent Piping

		15150 | Sanitary Waste and Vent Piping	CREW	DAILY OUTPUT	LABOR-HOURS	UNIT	2004 BARE COSTS MAT.	LABOR	EQUIP.	TOTAL	TOTAL INCL O&P	
200	0010	**CLEANOUTS**										200
	0060	Floor type										
	0080	Round or square, scoriated nickel bronze top										
	0100	2″ pipe size	1 Plum	10	.800	Ea.	75.50	31.50		107	132	
	0140	4″ pipe size	″	6	1.333	″	112	53		165	206	
	0980	Round top, recessed for terrazzo 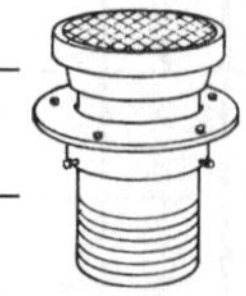										
	1000	2″ pipe size	1 Plum	9	.889	Ea.	75.50	35		110.50	138	
	1100	4″ pipe size	″	4	2		112	79		191	247	
	1120	5″ pipe size	Q-1	6	2.667	↓	143	95		238	305	
	9000	Minimum labor/equipment charge	1 Plum	3	2.667	Job		106		106	164	
250	0010	**CLEANOUT TEE**										250
	0100	Cast iron, B&S, with countersunk plug 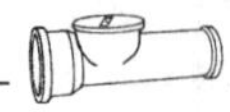										
	0220	3″ pipe size	1 Plum	3.60	2.222	Ea.	112	88		200	260	
	0240	4″ pipe size	″	3.30	2.424		140	96		236	305	
	0500	For round smooth access cover, same price				↓						
	4000	Plastic, tees and adapters. Add plugs										
	4010	ABS, DWV										
	4020	Cleanout tee, 1-1/2″ pipe size	1 Plum	15	.533	Ea.	2.94	21		23.94	36	
	9000	Minimum labor/equipment charge	″	2.75	2.909	Job		115		115	179	
300	0010	**FLOOR AND AREA DRAINS**										300
	0400	Deck, auto park, C.I., 13″ top 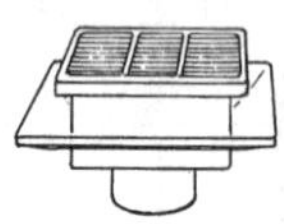										
	0440	3″, 4″, 5″, and 6″ pipe size	Q-1	8	2	Ea.	550	71.50		621.50	715	
	0480	For galvanized body, add				″	262			262	288	
	0500											
	2000	Floor, medium duty, C.I., deep flange, 7″ dia top										
	2040	2″ and 3″ pipe size	Q-1	12	1.333	Ea.	77.50	47.50		125	159	
	2080	For galvanized body, add					32.50			32.50	35.50	
	2120	For polished bronze top, add				↓	39			39	43	
	2500	Heavy duty, cleanout & trap w/bucket, C.I., 15″ top										
	2540	2″, 3″, and 4″ pipe size	Q-1	6	2.667	Ea.	2,025	95		2,120	2,375	
	2560	For galvanized body, add					570			570	625	
	2580	For polished bronze top, add				↓	630			630	695	
350	0010	**FLOOR RECEPTORS** For connection to 2″, 3″ & 4″ diameter pipe										350
	0200	12-1/2″ square top, 25 sq in open area	Q-1	10	1.600	Ea.	370	57		427	500	
	0300	For grate with 4″ diam. x 3-3/4″ high funnel, add					52.50			52.50	58	
	0400	For grate with 6″ diameter x 6″ high funnel, add					68			68	75	
	0700	For acid-resisting bucket, add					83			83	91.50	
	0900	For stainless steel mesh bucket liner, add					65.50			65.50	72	
	2000	12-5/8″ diameter top, 40 sq. in. open area	Q-1	10	1.600	↓	234	57		291	345	
	2100	For options, add same prices as square top										
	3000	8″ x 4″ rectangular top, 7.5 sq. in. open area	Q-1	14	1.143	Ea.	246	40.50		286.50	335	
	3100	For trap primer connections, add				″	26.50			26.50	29	
	9000	Minimum labor/equipment charge	Q-1	3	5.333	Job		190		190	295	
400	0010	**INTERCEPTORS**										400
	0150	Grease, cast iron, 4 GPM, 8 lb. fat capacity	1 Plum	4	2	Ea.	445	79		524	615	
	0200	7 GPM, 14 lb. fat capacity		4	2		615	79		694	805	
	1000	10 GPM, 20 lb. fat capacity	↓	4	2		725	79		804	925	
	1160	100 GPM, 200 lb. fat capacity	Q-1	2	8		5,475	285		5,760	6,475	
	1240	300 GPM, 600 lb. fat capacity	″	1	16	↓	12,100	570		12,670	14,200	
	9000	Minimum labor/equipment charge	1 Plum	3	2.667	Job		106		106	164	
800	0010	**TRAPS**										800
	0030	Cast iron, service weight										

15150 | Sanitary Waste and Vent Piping

		15150 | Sanitary Waste and Vent Piping	CREW	DAILY OUTPUT	LABOR-HOURS	UNIT	2004 BARE COSTS MAT.	LABOR	EQUIP.	TOTAL	TOTAL INCL O&P	
800	0050	Running P trap, without vent										800
	1100	2″	Q-1	16	1	Ea.	19.70	35.50		55.20	77	
	1150	4″	″	13	1.231		57	44		101	131	
	1160	6″	Q-2	17	1.412		264	52		316	370	
	3000	P trap, B&S, 2″ pipe size	Q-1	16	1		14.30	35.50		49.80	71.50	
	3040	3″ pipe size	″	14	1.143		21.50	40.50		62	86.50	
	3900											
	4700	Copper, drainage, drum trap										
	4840	3″ x 6″ swivel, 1-1/2″ pipe size	1 Plum	16	.500	Ea.	36	19.80		55.80	70	
	5100	P trap, standard pattern										
	5200	1-1/4″ pipe size	1 Plum	18	.444	Ea.	17.05	17.60		34.65	46.50	
	5240	1-1/2″ pipe size		17	.471		17.05	18.65		35.70	48	
	5260	2″ pipe size		15	.533		27	21		48	62.50	
	5280	3″ pipe size		11	.727		64.50	29		93.50	115	
	9000	Minimum labor/equipment charge		3	2.667	Job		106		106	164	
900	0010	**VENT FLASHING, CAPS**										900
	0120	Vent caps										
	0140	Cast iron										
	0180	2-1/2″ - 3-5/8″ pipe	1 Plum	21	.381	Ea.	33	15.10		48.10	60	
	0190	4″ - 4-1/8″ pipe	″	19	.421	″	39.50	16.65		56.15	69.50	
	0900	Vent flashing										
	1000	Aluminum with lead ring										
	1050	3″ pipe	1 Plum	17	.471	Ea.	7.90	18.65		26.55	37.50	
	1060	4″ pipe	″	16	.500	″	9.55	19.80		29.35	41	
	1350	Copper with neoprene ring										
	1440	2″ pipe	1 Plum	18	.444	Ea.	14.85	17.60		32.45	44	
	1450	3″ pipe		17	.471		17.40	18.65		36.05	48	
	1460	4″ pipe		16	.500		19.20	19.80		39	51.50	
	9000	Minimum labor/equipment charge		4	2	Job		79		79	123	

15160 | Storm Drainage Piping

		15160 | Storm Drainage Piping	CREW	DAILY OUTPUT	LABOR-HOURS	UNIT	MAT.	LABOR	EQUIP.	TOTAL	TOTAL INCL O&P	
500	0010	**STORM AREA DRAINS**										500
	0140	Cornice, C.I., 45° or 90° outlet										
	0200	3″ and 4″ pipe size	Q-1	12	1.333	Ea.	127	47.50		174.50	213	
	0260	For galvanized body, add					25			25	27.50	
	0280	For polished bronze dome, add					21.50			21.50	23.50	
	3860	Roof, flat metal deck, C.I. body, 12″ C.I. dome										
	3890	3″ pipe size	Q-1	14	1.143	Ea.	151	40.50		191.50	229	
	3900	4″ pipe size	″	13	1.231	″	151	44		195	234	
	4620	Main, all aluminum, 12″ low profile dome										
	4640	2″, 3″ and 4″ pipe size	Q-1	14	1.143	Ea.	168	40.50		208.50	247	
	9000	Minimum labor/equipment charge	1 Plum	4	2	Job		79		79	123	

15180 | Heating and Cooling Piping

		15180 | Heating and Cooling Piping	CREW	DAILY OUTPUT	LABOR-HOURS	UNIT	MAT.	LABOR	EQUIP.	TOTAL	TOTAL INCL O&P	
200	0010	**PUMPS, CIRCULATING** Heated or chilled water application										200
	0100											
	0600	Bronze, sweat connections, 1/40 HP, in line										
	0640	3/4″ size	Q-1	16	1	Ea.	120	35.50		155.50	188	
	1000	Flange connection, 3/4″ to 1-1/2″ size										
	1040	1/12 HP	Q-1	6	2.667	Ea.	325	95		420	505	
	1060	1/8 HP		6	2.667		560	95		655	760	
	1100	1/3 HP		6	2.667		630	95		725	835	
	1140	2″ size, 1/6 HP		5	3.200		805	114		919	1,050	
	1180	2-1/2″ size, 1/4 HP		5	3.200		1,050	114		1,164	1,325	

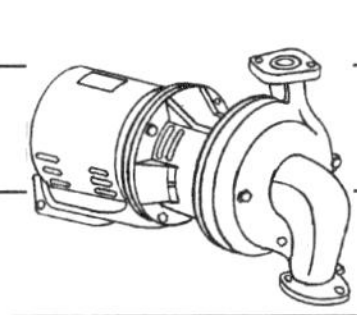

MECHANICAL 15

15100 | Building Services Piping

	15180 | Heating and Cooling Piping		CREW	DAILY OUTPUT	LABOR-HOURS	UNIT	2004 BARE COSTS MAT.	LABOR	EQUIP.	TOTAL	TOTAL INCL O&P	
200	1220	3″ size, 1/4 HP	Q-1	4	4	Ea.	1,100	143		1,243	1,425	200
	9000	Minimum labor/equipment charge	↓	3.25	4.923	Job		176		176	272	

15200 | Process Piping

	15210 | Process Air/Gas Piping		CREW	DAILY OUTPUT	LABOR-HOURS	UNIT	2004 BARE COSTS MAT.	LABOR	EQUIP.	TOTAL	TOTAL INCL O&P	
100	0010	AIR COMPRESSORS										100
	5250	Air, reciprocating air cooled, splash lubricated, tank mounted										
	5300	Single stage, 1 phase, 140 psi										
	5303	1/2 HP, 30 gal tank	1 Stpi	3	2.667	Ea.	925	106		1,031	1,200	
	5305	3/4 HP, 30 gal tank		2.60	3.077		955	122		1,077	1,250	
	5307	1 HP, 30 gal tank	↓	2.20	3.636	↓	1,175	145		1,320	1,525	

15400 | Plumbing Fixtures & Equipment

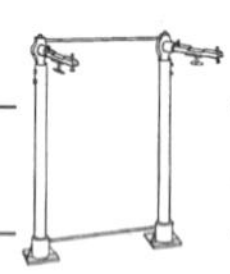

	15410 | Plumbing Fixtures		CREW	DAILY OUTPUT	LABOR-HOURS	UNIT	2004 BARE COSTS MAT.	LABOR	EQUIP.	TOTAL	TOTAL INCL O&P	
200	0010	**CARRIERS/SUPPORTS** For plumbing fixtures										200
	0020											
	3000	Lavatory, concealed arm										
	3050	Floor mounted, single										
	3100	High back fixture	1 Plum	6	1.333	Ea.	169	53		222	268	
	3200	Flat slab fixture		6	1.333		197	53		250	299	
	3220	Paraplegic	↓	6	1.333	↓	140	53		193	236	
	6980	Water closet, siphon jet										
	7000	Horizontal, adjustable, caulk										
	7040	Single, 4″ pipe size	1 Plum	6	1.333	Ea.	210	53		263	315	
	7060	5″ pipe size		6	1.333		278	53		331	385	
	7100	Double, 4″ pipe size		5	1.600		390	63.50		453.50	530	
	7120	5″ pipe size	↓	5	1.600	↓	278	63.50		341.50	405	
	8200	Water closet, residential										
	8220	Vertical centerline, floor mount										
	8240	Single, 3″ caulk, 2″ or 3″ vent	1 Plum	6	1.333	Ea.	195	53		248	296	
	8260	4″ caulk, 2″ or 4″ vent		6	1.333	″	251	53		304	360	
	9990	Minimum labor/equipment charge	↓	3.50	2.286	Job		90.50		90.50	140	
300	0010	**FAUCETS/FITTINGS**										300
	0020											
	0150	Bath, faucets, diverter spout combination, sweat	1 Plum	8	1	Ea.	69.50	39.50		109	138	
	0200	For integral stops, IPS unions, add					73			73	80.50	
	0420	Bath, press-bal mix valve w/diverter, spout, shower hd, arm/flange	1 Plum	8	1		109	39.50		148.50	182	
	0500	Drain, central lift, 1-1/2″ IPS male		20	.400		38	15.85		53.85	66.50	
	0600	Trip lever, 1-1/2″ IPS male	↓	20	.400	↓	38.50	15.85		54.35	67	
	0810	Bidet										
	0812	Fitting, over the rim, swivel spray/pop-up drain	1 Plum	8	1	Ea.	136	39.50		175.50	211	
	0840	Flush valves, with vacuum breaker										

	15410 | Plumbing Fixtures		CREW	DAILY OUTPUT	LABOR-HOURS	UNIT	2004 BARE COSTS MAT.	LABOR	EQUIP.	TOTAL	TOTAL INCL O&P	
300	0850	Water closet										300
	0860	Exposed, rear spud	1 Plum	8	1	Ea.	109	39.50		148.50	182	
	0870	Top spud		8	1		100	39.50		139.50	172	
	0880	Concealed, rear spud		8	1		143	39.50		182.50	219	
	0890	Top spud		8	1		164	39.50		203.50	242	
	0900	Wall hung		8	1		126	39.50		165.50	200	
	0920	Urinal										
	0930	Exposed, stall	1 Plum	8	1	Ea.	100	39.50		139.50	172	
	0940	Wall, (washout)		8	1		100	39.50		139.50	172	
	0950	Pedestal, top spud		8	1		102	39.50		141.50	174	
	0960	Concealed, stall		8	1		115	39.50		154.50	188	
	0970	Wall (washout)		8	1		118	39.50		157.50	192	
	0971	Automatic flush sensor and operator for										
	0972	urinals or water closets	1 Plum	5.33	1.501	Ea.	330	59.50		389.50	455	
	1000	Kitchen sink faucets, top mount, cast spout		10	.800		51.50	31.50		83	106	
	1100	For spray, add		24	.333		9.50	13.20		22.70	31	
	1200	Wall type, swing tube spout		10	.800		64	31.50		95.50	120	
	2000	Laundry faucets, shelf type, IPS or copper unions		12	.667		42.50	26.50		69	87.50	
	2100	Lavatory faucet, centerset, without drain		10	.800		37.50	31.50		69	90	
	2200	With pop-up drain		16	.500		52	19.80		71.80	88	
	2800	Self-closing, center set		10	.800		112	31.50		143.50	172	
	2810	Automatic sensor and operator, with faucet head		6.15	1.301		293	51.50		344.50	400	
	2850	Medical, bedpan cleanser, with pedal valve,		12	.667		460	26.50		486.50	545	
	2860	With screwdriver stop valve		12	.667		239	26.50		265.50	305	
	2870	With self-closing spray valve		12	.667		179	26.50		205.50	238	
	2900	Faucet, gooseneck spout, wrist handles, grid drain		10	.800		107	31.50		138.50	167	
	2940	Mixing valve, knee action, screwdriver stops		4	2		530	79		609	710	
	3000	Service sink faucet, cast spout, pail hook, hose end		14	.571		72	22.50		94.50	114	
	4000	Shower by-pass valve with union		18	.444		50.50	17.60		68.10	83	
	4100	Shower arm with flange and head		22	.364		58	14.40		72.40	86	
	4140	Shower, hand held, pin mount, massage action, chrome		22	.364		63.50	14.40		77.90	92.50	
	4142	Polished brass		22	.364		121	14.40		135.40	156	
	4144	Shower, hand held, wall mtd, adj. spray, 2 wall mounts, chrome		20	.400		128	15.85		143.85	166	
	4146	Polished brass		20	.400		235	15.85		250.85	284	
	4148	Shower, hand held head, bar mounted 24", adj. spray, chrome		20	.400		159	15.85		174.85	200	
	4150	Polished brass		20	.400		350	15.85		365.85	410	
	4200	Shower thermostatic mixing valve, concealed		8	1		233	39.50		272.50	320	
	4300	For inlet strainer, check, and stops, add					31			31	34	
	5000	Sillcock, compact, brass, IPS or copper to hose	1 Plum	24	.333		4.74	13.20		17.94	25.50	
	6000	Stop and waste valves, bronze										
	6100	Angle, solder end 1/2"	1 Plum	24	.333	Ea.	4.38	13.20		17.58	25.50	
	6110	3/4"		20	.400		4.49	15.85		20.34	29.50	
	6300	Straightway, solder end 3/8"		24	.333		3.59	13.20		16.79	24.50	
	6310	1/2"		24	.333		2.32	13.20		15.52	23	
	6330	1"		19	.421		4	16.65		20.65	30.50	
	6400	Straightway, threaded 3/8"		24	.333		4.07	13.20		17.27	25	
	6410	1/2"		24	.333		3.13	13.20		16.33	24	
	6420	3/4"		20	.400		3.73	15.85		19.58	28.50	
	6430	1"		19	.421		10.50	16.65		27.15	37.50	
	7800	Water closet, wax gasket		96	.083		.99	3.30		4.29	6.20	
	7820	Gasket toilet tank to bowl		32	.250		1.24	9.90		11.14	16.70	
	8000	Water supply stops, polished chrome plate										
	8200	Angle, 3/8"	1 Plum	24	.333	Ea.	4.73	13.20		17.93	25.50	
	8300	1/2"		22	.364		4.73	14.40		19.13	27.50	
	8400	Straight, 3/8"		26	.308		7.05	12.20		19.25	26.50	
300	8500	1/2"		24	.333		7.05	13.20		20.25	28.50	300

MECHANICAL 15

		15410 Plumbing Fixtures	Crew	Daily Output	Labor-Hours	Unit	2004 Bare Costs Mat.	Labor	Equip.	Total	Total Incl O&P	
300	8600	Water closet, angle, w/flex riser, 3/8"	1 Plum	24	.333	Ea.	11.05	13.20		24.25	32.50	300
	9000	Minimum labor/equipment charge		4	2	Job		79		79	123	
		15411 Commercial/Indust Fixtures										
500	0010	**HYDRANTS**										500
	0050	Wall type, moderate climate, bronze, encased										
	0200	3/4" IPS connection	1 Plum	16	.500	Ea.	232	19.80		251.80	286	
	1000	Non-freeze, bronze, exposed										
	1100	3/4" IPS connection, 4" to 9" thick wall	1 Plum	14	.571	Ea.	157	22.50		179.50	208	
	1120	10" to 14" thick wall	"	12	.667		171	26.50		197.50	229	
	1280	For anti-siphon type					41.50			41.50	45.50	
	9000	Minimum labor/equipment charge	1 Plum	3	2.667	Job		106		106	164	
700	0010	**URINALS**										700
	3000	Wall hung, vitreous china, with hanger & self-closing valve										
	3100	Siphon jet type	Q-1	3	5.333	Ea.	291	190		481	615	
	3120	Blowout type		3	5.333		310	190		500	640	
	3300	Rough-in, supply, waste & vent		2.83	5.654		92	202		294	415	
	5000	Stall type, vitreous china, includes valve		2.50	6.400		460	228		688	860	
	6980	Rough-in, supply, waste and vent		1.99	8.040		119	287		406	575	
	9000	Minimum labor/equipment charge		4	4	Job		143		143	221	
840	0010	**WASH FOUNTAINS**										840
	1900	Group, foot control										
	2000	Precast terrazzo, circular, 36" diam., 5 or 6 persons	Q-2	3	8	Ea.	2,650	296		2,946	3,350	
	2100	54" diameter for 8 or 10 persons		2.50	9.600		3,200	355		3,555	4,075	
	2400	Semi-circular, 36" diam. for 3 persons		3	8		2,350	296		2,646	3,050	
	2500	54" diam. for 4 or 5 persons		2.50	9.600		2,925	355		3,280	3,750	
	5610	Group, infrared control, barrier free										
	5614	Precast terrazzo										
	5620	Semi-circular 36" diam. for 3 persons	Q-2	3	8	Ea.	3,475	296		3,771	4,275	
	5630	46" diam. for 4 persons		2.80	8.571		3,900	315		4,215	4,775	
	5640	Circular, 54" diam. for 8 persons, button control		2.50	9.600		5,675	355		6,030	6,800	
	5700	Rough-in, supply, waste and vent for above wash fountains	Q-1	1.82	8.791		90.50	315		405.50	585	
	9000	Minimum labor/equipment charge	Q-2	3	8	Job		296		296	460	
		15412 Drinking Fountains										
200	0010	**DRINKING FOUNTAIN** For connection to cold water supply										200
	1000	Wall mounted, non-recessed										
	2700	Stainless steel, single bubbler, no back	1 Plum	4	2	Ea.	825	79		904	1,025	
	2740	With back		4	2		980	79		1,059	1,200	
	2780	Dual handle & wheelchair projection type		4	2		505	79		584	680	
	2820	Dual level for handicapped type		3.20	2.500		1,025	99		1,124	1,275	
	3980	For rough-in, supply and waste, add		2.21	3.620		42	143		185	269	
	4000	Wall mounted, semi-recessed										
	4200	Poly-marble, single bubbler	1 Plum	4	2	Ea.	680	79		759	870	
	4600	Stainless steel, satin finish, single bubbler	"	4	2	"	785	79		864	985	
	6000	Wall mounted, fully recessed										
	6400	Poly-marble, single bubbler	1 Plum	4	2	Ea.	765	79		844	965	
	6800	Stainless steel, single bubbler		4	2		660	79		739	850	
	7580	For rough-in, supply and waste, add		1.83	4.372		42	173		215	315	
	7590											
	7600	Floor mounted, pedestal type										
	8600	Enameled iron, heavy duty service, 2 bubblers	1 Plum	2	4	Ea.	1,100	158		1,258	1,450	
	8880	For freeze-proof valve system, add		2	4		335	158		493	615	
	8900	For rough-in, supply and waste, add		1.83	4.372		42	173		215	315	
	9000	Minimum labor/equipment charge		2	4	Job		158		158	246	

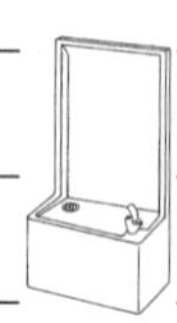

15400 | Plumbing Fixtures & Equipment

	15413 | Electric Water Coolers		CREW	DAILY OUTPUT	LABOR-HOURS	UNIT	2004 BARE COSTS MAT.	LABOR	EQUIP.	TOTAL	TOTAL INCL O&P	
900	0010	WATER COOLER										900
	0100	Wall mounted, non-recessed										
	0140	4 GPH	Q-1	4	4	Ea.	500	143		643	770	
	0160	8 GPH, barrier free, sensor operated		4	4		620	143		763	905	
	1000	Dual height, 8.2 GPH		3.80	4.211		805	150		955	1,125	
	1040	14.3 GPH		3.80	4.211		855	150		1,005	1,175	
	3300	Semi-recessed, 8.1 GPH		4	4		840	143		983	1,150	
	4600	Floor mounted, flush-to-wall										
	4640	4 GPH	1 Plum	3	2.667	Ea.	510	106		616	725	
	4980	For stainless steel cabinet, add					96			96	106	
	5000	Dual height, 8.2 GPH	1 Plum	2	4		870	158		1,028	1,200	
	9000	Minimum labor/equipment charge	"	2	4	Job		158		158	246	
	15414 | Emergency Fixtures											
200	0010	**INDUSTRIAL SAFETY FIXTURES** Rough-in not included										200
	0020											
	1000	Eye wash fountain										
	1400	Plastic bowl, pedestal mounted	Q-1	4	4	Ea.	182	143		325	420	
	5000	Shower, single head, drench, ball valve, pull, freestanding		4	4		255	143		398	500	
	5200	Horizontal or vertical supply		4	4		237	143		380	480	
	6000	Multi-nozzle, eye/face wash combination		4	4		480	143		623	750	
	6400	Multi-nozzle, 12 spray, shower only		4	4		1,150	143		1,293	1,475	
	6600	For freeze-proof, add		6	2.667		268	95		363	440	
	9000	Minimum labor/equipment charge		3	5.333	Job		190		190	295	
	15418 | Resi/Comm/Industrial Fixtures											
100	0010	**BATHS**										100
	0100	Tubs, recessed porcelain enamel on cast iron, with trim (R15100-420)										
	0180	48" x 42"	Q-1	4	4	Ea.	1,450	143		1,593	1,825	
	0220	72" x 36"		3	5.333		1,350	190		1,540	1,775	
	0300	Mat bottom, 4' long		5.50	2.909		990	104		1,094	1,250	
	0380	5' long		4.40	3.636		375	130		505	610	
	0560	Corner 48" x 44"		4.40	3.636		1,400	130		1,530	1,725	
	2000	Enameled formed steel, 4'-6" long		5.80	2.759		295	98.50		393.50	480	
	2200	5' long		5.50	2.909		286	104		390	475	
	4600	Module tub & showerwall surround, molded fiberglass										
	4610	5' long x 34" wide x 76" high	Q-1	4	4	Ea.	490	143		633	760	
	4750	Handicap with 1-1/2" OD grab bar, antiskid bottom										
	4760	60" x 32-3/4" x 72" high	Q-1	4	4	Ea.	665	143		808	950	
	4770	60" x 30" x 71" high with molded seat	"	3.50	4.571	"	825	163		988	1,175	
	6000	Whirlpool, bath with vented overflow, molded fiberglass										
	6100	66" x 48" x 24"	Q-1	1	16	Ea.	2,225	570		2,795	3,325	
	6400	72" x 36" x 24"		1	16		2,175	570		2,745	3,275	
	6500	60" x 30" x 21"		1	16		1,900	570		2,470	2,950	
	6600	72" x 42" x 22"		1	16		3,075	570		3,645	4,275	
	6700	83" x 65"		.30	53.333		4,150	1,900		6,050	7,525	
	6710	For color add					10%					
	6711	For designer colors and trim add					25%					
	7000	Redwood hot tub system										
	7050	4' diameter x 4' deep	Q-1	1	16	Ea.	1,300	570		1,870	2,300	
	7150	6' diameter x 4' deep		.80	20		2,000	715		2,715	3,300	
	7200	8' diameter x 4' deep		.80	20		3,025	715		3,740	4,425	
	9600	Rough-in, supply, waste and vent, for all above tubs, add		2.07	7.729		115	276		391	550	
	9900	Minimum labor/equipment charge		3	5.333	Job		190		190	295	
400	0010	**LAUNDRY SINKS** With trim										400
	0020	Porcelain enamel on cast iron, black iron frame										

	15418 | Resi/Comm/Industrial Fixtures		CREW	DAILY OUTPUT	LABOR-HOURS	UNIT	2004 BARE COSTS MAT.	LABOR	EQUIP.	TOTAL	TOTAL INCL O&P	
400	0050	24″ x 20″, single compartment	Q-1	6	2.667	Ea.	247	95		342	420	400
	0100	24″ x 23″, single compartment	″	6	2.667	″	269	95		364	445	
	2000	Molded stone, on wall hanger or legs										
	2020	22″ x 23″, single compartment	Q-1	6	2.667	Ea.	120	95		215	280	
	2100	45″ x 21″, double compartment	″	5	3.200	″	248	114		362	450	
	3000	Plastic, on wall hanger or legs										
	3020	18″ x 23″, single compartment	Q-1	6.50	2.462	Ea.	87.50	88		175.50	233	
	3100	20″ x 24″, single compartment		6.50	2.462		114	88		202	261	
	3200	36″ x 23″, double compartment		5.50	2.909		137	104		241	310	
	3300	40″ x 24″, double compartment		5.50	2.909		198	104		302	380	
	5000	Stainless steel, counter top, 22″ x 17″ single compartment		6	2.667		325	95		420	500	
	5100	22″ x 22″, single compartment		6	2.667		410	95		505	595	
	5200	33″ x 22″, double compartment		5	3.200		405	114		519	620	
	9600	Rough-in, supply, waste and vent, for all laundry sinks		2.14	7.477		71	267		338	495	
	9810	Minimum labor/equipment charge	1 Plum	3	2.667	Job		106		106	164	
450	0010	**LAVATORIES** With trim, white unless noted otherwise (R15100-420)										450
	0020											
	0500	Vanity top, porcelain enamel on cast iron										
	0600	20″ x 18″	Q-1	6.40	2.500	Ea.	206	89		295	365	
	0640	33″ x 19″ oval	″	6.40	2.500	″	405	89		494	585	
	0860	For color, add					25%					
	1000	Cultured marble, 19″ x 17″, single bowl	Q-1	6.40	2.500	Ea.	160	89		249	315	
	1120	25″ x 22″, single bowl		6.40	2.500		160	89		249	315	
	1900	Stainless steel, self-rimming, 25″ x 22″, single bowl, ledge		6.40	2.500		224	89		313	385	
	1960	17″ x 22″, single bowl		6.40	2.500		222	89		311	385	
	2600	Steel, enameled, 20″ x 17″, single bowl		5.80	2.759		129	98.50		227.50	295	
	2900	Vitreous china, 20″ x 16″, single bowl		5.40	2.963		222	106		328	410	
	2960	20″ x 17″, single bowl		5.40	2.963		146	106		252	325	
	3580	Rough-in, supply, waste and vent for all above lavatories		2.30	6.957		63	248		311	455	
	4000	Wall hung										
	4040	Porcelain enamel on cast iron, 16″ x 14″, single bowl	Q-1	8	2	Ea.	315	71.50		386.50	455	
	4180	20″ x 18″, single bowl	″	8	2	″	237	71.50		308.50	370	
	4580	For color, add					30%					
	6000	Vitreous china, 18″ x 15″, single bowl with backsplash	Q-1	7	2.286	Ea.	184	81.50		265.50	330	
	6960	Rough-in, supply, waste and vent for above lavatories	″	1.66	9.639	″	220	345		565	775	
	9000	Minimum labor/equipment charge	1 Plum	3	2.667	Job		106		106	164	
500	0011	**SHOWERS,** Stall, with drain only										500
	0020											
	1520	32″ square	Q-1	2	8	Ea.	325	285		610	800	
	1530	36″ square		2	8		410	285		695	890	
	1540	Terrazzo receptor, 32″ square		2	8		690	285		975	1,200	
	1580	36″ corner angle		1.80	8.889		755	315		1,070	1,325	
	3000	Fiberglass, one piece, with 3 walls, 32″ x 32″ square		2.40	6.667		350	238		588	760	
	3100	36″ x 36″ square		2.40	6.667		395	238		633	805	
	4200	Rough-in, supply, waste and vent for above showers		2.05	7.805		62.50	278		340.50	500	
	5500	Head, water economizer, 3.0 GPM	1 Plum	24	.333		65	13.20		78.20	92	
	9000	Minimum labor/equipment charge	Q-1	4	4	Job		143		143	221	
600	0010	**SINKS** With faucets and drain										600
	2000	Kitchen, counter top style, P.E. on C.I., 24″ x 21″ single bowl	Q-1	5.60	2.857	Ea.	201	102		303	380	
	2100	31″ x 22″ single bowl		5.60	2.857		249	102		351	430	
	2200	32″ x 21″ double bowl		4.80	3.333		288	119		407	500	
	3000	Stainless steel, self rimming, 19″ x 18″ single bowl		5.60	2.857		335	102		437	525	
	3100	25″ x 22″ single bowl		5.60	2.857		370	102		472	565	

Important: See the Reference Section for critical supporting data - Reference Nos., Crews, & City Cost Indexes

		15418 | Resi/Comm/Industrial Fixtures	CREW	DAILY OUTPUT	LABOR-HOURS	UNIT	2004 BARE COSTS MAT.	LABOR	EQUIP.	TOTAL	TOTAL INCL O&P	
600	3200	33" x 22" double bowl	Q-1	4.80	3.333	Ea.	530	119		649	770	600
	3300	43" x 22" double bowl		4.80	3.333		615	119		734	860	
	4000	Steel, enameled, with ledge, 24" x 21" single bowl		5.60	2.857		107	102		209	275	
	4100	32" x 21" double bowl		4.80	3.333		146	119		265	345	
	4960	For color sinks except stainless steel, add					10%					
	4980	For rough-in, supply, waste and vent, counter top sinks	Q-1	2.14	7.477		71	267		338	495	
	5790	For rough-in, supply, waste & vent, sinks		1.85	8.649		71	310		381	560	
	6650	Service, floor, corner, P.E. on C.I., 28" x 28"		4.40	3.636		490	130		620	740	
	6790	For rough-in, supply, waste & vent, floor service sinks		1.64	9.756		161	350		511	715	
	9000	Minimum labor/equipment charge		4	4	Job		143		143	221	
900	0010	**WATER CLOSETS** R15100-420										900
	0030	For automatic flush, see 15410-300-0972										
	0100											
	0150	Tank type, vitreous china, incl. seat, supply pipe w/stop										
	0200	Wall hung, one piece	Q-1	5.30	3.019	Ea.	390	108		498	595	
	0400	Two piece, close coupled		5.30	3.019		470	108		578	680	
	0960	For rough-in, supply, waste, vent and carrier		2.73	5.861		264	209		473	615	
	1000	Floor mounted, one piece		5.30	3.019		485	108		593	700	
	1100	Two piece, close coupled, water saver		5.30	3.019		160	108		268	345	
	1960	For color, add					30%					
	1980	For rough-in, supply, waste and vent	Q-1	3.05	5.246	Ea.	130	187		317	435	
	3000	Bowl only, with flush valve, seat										
	3100	Wall hung	Q-1	5.80	2.759	Ea.	340	98.50		438.50	530	
	3200	For rough-in, supply, waste and vent, single WC		2.56	6.250	"	274	223		497	645	
	9000	Minimum labor/equipment charge		4	4	Job		143		143	221	

		15440 | Plumbing Pumps	CREW	DAILY OUTPUT	LABOR-HOURS	UNIT	MAT.	LABOR	EQUIP.	TOTAL	TOTAL INCL O&P	
240	0010	**PUMPS, PRESSURE BOOSTER SYSTEM**										240
	0200	Pump system, with diaphragm tank, control, press. switch										
	0300	1 HP pump	Q-1	1.30	12.308	Ea.	3,650	440		4,090	4,675	
	0400	1-1/2 HP pump		1.25	12.800		3,675	455		4,130	4,750	
	0420	2 HP pump		1.20	13.333		3,775	475		4,250	4,875	
	0440	3 HP pump		1.10	14.545		3,825	520		4,345	5,000	
	0460	5 HP pump	Q-2	1.50	16		4,225	590		4,815	5,575	
	0480	7-1/2 HP pump		1.42	16.901		4,725	625		5,350	6,150	
	0500	10 HP pump		1.34	17.910		4,925	660		5,585	6,450	
	1000	Pump/ energy storage system, diaphragm tank, 3 HP pump										
	1100	motor, PRV, switch, gauge, control center, flow switch										
	1200	125 lb. working pressure	Q-2	.70	34.286	Ea.	10,600	1,275		11,875	13,700	
	1300	250 lb. working pressure		.64	37.500	"	11,600	1,375		12,975	15,000	
	9000	Minimum labor/equipment charge		.30	80	Job		2,950		2,950	4,575	
800	0010	**PUMPS, SEWAGE EJECTOR** With operating and level controls										800
	0100	Simplex system incl. tank, cover, pump 15' head										
	0500	37 gal PE tank, 12 GPM, 1/2 HP, 2" discharge	Q-1	3.20	5	Ea.	375	178		553	690	
	0510	3" discharge		3.10	5.161		405	184		589	730	
	0530	87 GPM, .7 HP, 2" discharge		3.20	5		570	178		748	905	
	0540	3" discharge		3.10	5.161		620	184		804	965	
	0600	45 gal. coated stl tank, 12 GPM, 1/2 HP, 2" discharge		3	5.333		670	190		860	1,025	
	0610	3" discharge		2.90	5.517		695	197		892	1,075	
	0630	87 GPM, .7 HP, 2" discharge		3	5.333		855	190		1,045	1,225	
	0640	3" discharge		2.90	5.517		905	197		1,102	1,300	
	0660	134 GPM, 1 HP, 2" discharge		2.80	5.714		925	204		1,129	1,350	
	0680	3" discharge		2.70	5.926		975	211		1,186	1,400	
	0700	70 gal. PE tank, 12 GPM, 1/2 HP, 2" discharge		2.60	6.154		735	219		954	1,150	
	0710	3" discharge		2.40	6.667		785	238		1,023	1,225	

	15440 | Plumbing Pumps		CREW	DAILY OUTPUT	LABOR-HOURS	UNIT	2004 BARE COSTS MAT.	LABOR	EQUIP.	TOTAL	TOTAL INCL O&P	
800	0730	87 GPM, 0.7 HP, 2" discharge	Q-1	2.50	6.400	Ea.	940	228		1,168	1,375	800
	0740	3" discharge		2.30	6.957		1,000	248		1,248	1,475	
	0760	134 GPM, 1 HP, 2" discharge		2.20	7.273		1,025	259		1,284	1,525	
	0770	3" discharge		2	8		1,100	285		1,385	1,650	
	9000	Minimum labor/equipment charge		2.50	6.400	Job		228		228	355	
940	0010	**PUMPS, SUBMERSIBLE** Sump										940
	7000	Sump pump, automatic										
	7100	Plastic, 1-1/4" discharge, 1/4 HP	1 Plum	6	1.333	Ea.	105	53		158	198	
	7140	1/3 HP		5	1.600		127	63.50		190.50	238	
	7180	1-1/2" discharge, 1/2 HP		4	2		171	79		250	310	
	7500	Cast iron, 1-1/4" discharge, 1/4 HP		6	1.333		125	53		178	219	
	7540	1/3 HP		6	1.333		140	53		193	236	
	7560	1/2 HP		5	1.600		176	63.50		239.50	291	
	9000	Minimum labor/equipment charge		4	2	Job		79		79	123	
	15480 | Domestic Water Heaters											
200	0010	**WATER HEATERS**										200
	0050											
	1000	Residential, electric, glass lined tank, 5 yr, 10 gal., single element	1 Plum	2.30	3.478	Ea.	210	138		348	445	
	1060	30 gallon, double element		2.20	3.636		280	144		424	535	
	1080	40 gallon, double element		2	4		300	158		458	575	
	1100	52 gallon, double element		2	4		345	158		503	625	
	1120	66 gallon, double element		1.80	4.444		475	176		651	800	
	1140	80 gallon, double element		1.60	5		535	198		733	895	
	2000	Gas fired, foam lined tank, 10 yr, vent not incl.,										
	2040	30 gallon	1 Plum	2	4	Ea.	385	158		543	670	
	2060	40 gallon		1.90	4.211		405	167		572	705	
	2100	75 gallon		1.50	5.333		745	211		956	1,150	
	3000	Oil fired, glass lined tank, 5 yr, vent not included, 30 gallon		2	4		785	158		943	1,100	
	3040	50 gallon		1.80	4.444		1,375	176		1,551	1,775	
	4000	Commercial, 100° rise. NOTE: for each size tank, a range of										
	4010	heaters between the ones shown are available										
	4020	Electric										
	4100	5 gal., 3 kW, 12 GPH, 208V	1 Plum	2	4	Ea.	1,450	158		1,608	1,825	
	4160	50 gal., 36 kW, 148 GPH, 208V	"	1.80	4.444		3,350	176		3,526	3,950	
	4480	400 gal., 210 kW, 860 GPH, 480V	Q-1	1	16		20,200	570		20,770	23,100	
	6000	Gas fired, flush jacket, std. controls, vent not incl.										
	6040	75 MBH input, 73 GPH	1 Plum	1.40	5.714	Ea.	1,375	226		1,601	1,875	
	6060	98 MBH input, 95 GPH		1.40	5.714		2,425	226		2,651	3,025	
	6180	200 MBH input, 192 GPH		.60	13.333		4,300	530		4,830	5,575	
	8000	Oil fired, flush jacket, std. controls, vent not incl.										
	8060	103 MBH gross output, 116 GPH	1 Plum	1.10	7.273	Ea.	1,200	288		1,488	1,775	
	8080	122 MBH gross output, 141 GPH	"	1	8		1,525	315		1,840	2,175	
	8160	225 MBH gross output, 256 GPH	Q-1	.80	20		2,300	715		3,015	3,650	
	8280	735 MBH gross output, 880 GPH	"	.40	40		7,550	1,425		8,975	10,500	
	9000	Minimum labor/equipment charge	1 Plum	1.75	4.571	Job		181		181	281	

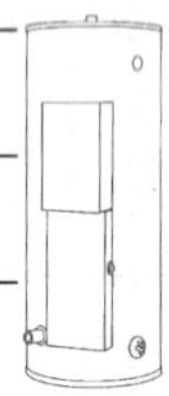

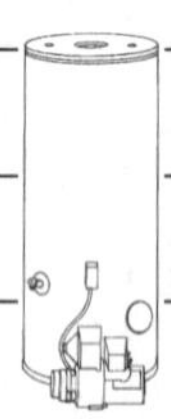

15 MECHANICAL

15510 | Heating Boilers and Accessories

			CREW	DAILY OUTPUT	LABOR-HOURS	UNIT	2004 BARE COSTS MAT.	LABOR	EQUIP.	TOTAL	TOTAL INCL O&P
120	0010	**BURNERS**									
	0990	Residential, conversion, gas fired, LP or natural									
	1000	Gun type, atmospheric input 72 to 200 MBH	Q-1	2.50	6.400	Ea.	655	228		883	1,075
	1020	120 to 360 MBH		2	8		725	285		1,010	1,250
	1040	280 to 800 MBH		1.70	9.412		1,400	335		1,735	2,075
	3000	Flame retention oil fired assembly, input									
	3040	2.0 to 5.0 GPH	Q-1	2	8	Ea.	525	285		810	1,025
300	0010	**BOILERS, ELECTRIC, ASME** Standard controls and trim									
	1000	Steam, 6 KW, 20.5 MBH	Q-19	1.20	20	Ea.	2,625	740		3,365	4,025
	1160	60 KW, 205 MBH		1	24		4,575	890		5,465	6,425
	1300	296 KW, 1010 MBH		.45	53.333		10,200	1,975		12,175	14,300
	1400	592 KW, 2020 MBH	Q-21	.34	94.118		19,300	3,550		22,850	26,700
	1540	1110 KW, 3788 MBH	″	.19	168		27,000	6,350		33,350	39,500
	2000	Hot water, 6 KW, 20.5 MBH	Q-19	1.30	18.462		2,450	685		3,135	3,750
	2040	24 KW, 82 MBH		1.20	20		2,925	740		3,665	4,375
	2060	30 KW, 103 MBH		1.20	20		3,200	740		3,940	4,650
	2070	36 KW, 123 MBH		1.20	20		3,425	740		4,165	4,900
	2080	48 KW, 164 MBH		1.10	21.818		4,775	805		5,580	6,500
	2100	74 KW, 253 MBH		1.10	21.818		4,950	805		5,755	6,700
	9000	Minimum labor/equipment charge	Q-20	1	20	Job		715		715	1,125
400	0010	**BOILERS, GAS FIRED** Natural or propane, standard controls									
	1000	Cast iron, with insulated jacket									
	2000	Steam, gross output, 81 MBH	Q-7	1.40	22.857	Ea.	1,275	865		2,140	2,750
	2060	163 MBH		.90	35.556		1,900	1,350		3,250	4,200
	2200	440 MBH		.51	62.500		3,500	2,375		5,875	7,525
	2320	1,875 MBH		.30	106		11,400	4,050		15,450	18,900
	2400	3570 MBH		.18	181		18,000	6,900		24,900	30,600
	2540	6,970 MBH		.10	320		59,500	12,100		71,600	84,500
	2800										
	3000	Hot water, gross output, 80 MBH	Q-7	1.46	21.918	Ea.	1,125	830		1,955	2,550
	3020	100 MBH		1.35	23.704		1,325	900		2,225	2,850
	3040	122 MBH		1.10	29.091		1,450	1,100		2,550	3,275
	3060	163 MBH		1	32		1,775	1,200		2,975	3,825
	3200	440 MBH		.58	54.983		3,325	2,075		5,400	6,875
	3320	2,000 MBH		.26	125		11,900	4,725		16,625	20,500
	3540	6,970 MBH		.09	359		60,500	13,600		74,100	87,500
	4000	Steel, insulating jacket									
	6000	Hot water, including burner & one zone valve, gross output									
	6010	51.2 MBH	Q-6	2	12	Ea.	1,650	445		2,095	2,525
	6020	72 MBH		2	12		1,850	445		2,295	2,725
	6040	89 MBH		1.90	12.632		1,900	470		2,370	2,800
	6060	105 MBH		1.80	13.333		2,125	495		2,620	3,100
	6080	132 MBH		1.70	14.118		2,425	525		2,950	3,475
	6100	155 MBH		1.50	16		2,800	595		3,395	4,000
	6110	186 MBH		1.40	17.143		3,375	635		4,010	4,700
	6120	227 MBH		1.30	18.462		4,300	685		4,985	5,775
	6140	292 MBH		1.20	20		4,850	740		5,590	6,475
	6180	480 MBH		.70	34.286		7,000	1,275		8,275	9,675
	6200	640 MBH		.60	40		8,325	1,475		9,800	11,500
	6220	800 MBH		.50	48		9,700	1,775		11,475	13,500
	6240	960 MBH		.45	53.333		12,000	1,975		13,975	16,400
	7000	For tankless water heater on smaller gas units, add					10%				
	7050	For additional zone valves up to 312 MBH add				Ea.	107			107	117
	7060										

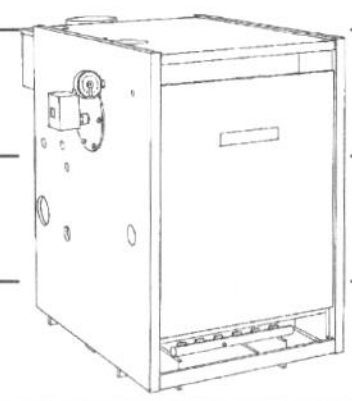

	15510 Heating Boilers and Accessories		CREW	DAILY OUTPUT	LABOR-HOURS	UNIT	2004 BARE COSTS MAT.	LABOR	EQUIP.	TOTAL	TOTAL INCL O&P	
400	9900	Minimum labor/equipment charge	Q-6	1	24	Job		890		890	1,375	400
460	0010	**BOILERS, GAS/OIL** Combination with burners and controls										460
	1000	Cast iron with insulated jacket										
	2000	Steam, gross output, 720 MBH	Q-7	.43	74.074	Ea.	6,725	2,800		9,525	11,800	
	2140	2,700 MBH		.19	165		15,700	6,275		21,975	27,100	
	2380	6,970 MBH		.09	372		63,500	14,100		77,600	92,000	
	2900	Hot water, gross output										
	2910	200 MBH	Q-6	.61	39.024	Ea.	5,400	1,450		6,850	8,200	
	2920	300 MBH		.49	49.080		5,400	1,825		7,225	8,775	
	2930	400 MBH		.41	57.971		6,325	2,150		8,475	10,300	
	2940	500 MBH		.36	67.039		6,800	2,475		9,275	11,300	
	3000	584 MBH	Q-7	.44	72.072		6,000	2,725		8,725	10,800	
	3060	1,460 MBH		.28	113		12,100	4,275		16,375	20,000	
	3300	13,500 MBH, 403.3 BHP		.04	727		97,500	27,600		125,100	150,000	
	4000	Steel, insulated jacket, skid base, tubeless										
	4500	Steam, 150 psi gross output, 335 MBH, 10 BHP	Q-6	.54	44.037	Ea.	10,900	1,625		12,525	14,500	
	4640	1,339 MBH, 40 BHP		.29	81.911		21,400	3,050		24,450	28,300	
	4720	2,511 MBH, 75 BHP		.17	141		29,200	5,250		34,450	40,200	
500	0010	**BOILERS, OIL FIRED** Standard controls, flame retention burner										500
	1000	Cast iron, with insulated flush jacket										
	2000	Steam, gross output, 109 MBH	Q-7	1.20	26.667	Ea.	1,500	1,000		2,500	3,200	
	2060	207 MBH		.90	35.556		2,025	1,350		3,375	4,350	
	2180	1,084 MBH		.38	85.106		7,050	3,225		10,275	12,800	
	2200	1,360 MBH		.33	98.160		8,225	3,725		11,950	14,800	
	2240	2,175 MBH		.24	133		11,200	5,075		16,275	20,200	
	2340	4,360 MBH		.15	214		35,300	8,125		43,425	51,500	
	2460	6,970 MBH		.09	363		60,500	13,800		74,300	88,000	
	3000	Hot water, same price as steam										
	5000	Steel, insulated jacket, burner										
	7000	Hot water, gross output, 103 MBH	Q-6	1.60	15	Ea.	1,200	555		1,755	2,175	
	7020	122 MBH		1.45	16.506		2,275	610		2,885	3,450	
	7040	137 MBH		1.36	17.595		2,575	655		3,230	3,850	
	7060	168 MBH		1.30	18.405		2,850	685		3,535	4,175	
	7080	225 MBH		1.22	19.704		2,925	730		3,655	4,350	
	7100	315 MBH		.96	25.105		3,450	930		4,380	5,250	
	7120	420 MBH		.70	34.483		4,500	1,275		5,775	6,925	
	9000	Minimum labor/equipment charge		1.75	13.714	Job		510		510	790	

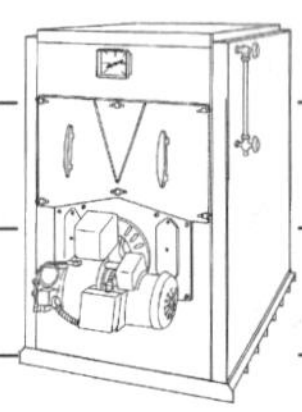

	15530 Furnaces		CREW	DAILY OUTPUT	LABOR-HOURS	UNIT	MAT.	LABOR	EQUIP.	TOTAL	TOTAL INCL O&P	
200	0010	**FURNACE COMPONENTS AND COMBINATIONS**										200
	0080	Coils, A/C evaporator, for gas or oil furnaces										
	0090	Add-on, with holding charge										
	0100	Upflow										
	0120	1-1/2 ton cooling	Q-5	4	4	Ea.	127	143		270	360	
	0130	2 ton cooling		3.70	4.324		153	155		308	410	
	0140	3 ton cooling		3.30	4.848		191	173		364	480	
	0150	4 ton cooling		3	5.333		270	191		461	595	
	0160	5 ton cooling		2.70	5.926		345	212		557	710	
	0300	Downflow										
	0330	2-1/2 ton cooling	Q-5	3	5.333	Ea.	178	191		369	490	
	0340	3-1/2 ton cooling		2.60	6.154		239	220		459	605	
	0350	5 ton cooling		2.20	7.273		345	260		605	785	
	0600	Horizontal										

Important: See the Reference Section for critical supporting data - Reference Nos., Crews, & City Cost Indexes

	15530 | Furnaces		Crew	Daily Output	Labor-Hours	Unit	2004 Bare Costs Mat.	Labor	Equip.	Total	Total Incl O&P	
200	0630	2 ton cooling	Q-5	3.90	4.103	Ea.	180	147		327	425	200
	0640	3 ton cooling		3.50	4.571		206	164		370	480	
	0650	4 ton cooling		3.20	5		259	179		438	560	
	0660	5 ton cooling	↓	2.90	5.517	↓	345	197		542	685	
	2000	Cased evaporator coils for air handlers										
	2100	1-1/2 ton cooling	Q-5	4.40	3.636	Ea.	191	130		321	410	
	2110	2 ton cooling		4.10	3.902		194	140		334	430	
	2120	2-1/2 ton cooling		3.90	4.103		218	147		365	470	
	2130	3 ton cooling		3.70	4.324		256	155		411	520	
	2140	3-1/2 ton cooling		3.50	4.571		240	164		404	520	
	2150	4 ton cooling		3.20	5		288	179		467	590	
	2160	5 ton cooling	↓	2.90	5.517	↓	330	197		527	670	
	3010	Air handler, modular										
	3100	With cased evaporator cooling coil										
	3120	1-1/2 ton cooling	Q-5	3.80	4.211	Ea.	545	151		696	835	
	3130	2 ton cooling		3.50	4.571		570	164		734	880	
	3140	2-1/2 ton cooling		3.30	4.848		620	173		793	950	
	3150	3 ton cooling		3.10	5.161		675	185		860	1,025	
	3160	3-1/2 ton cooling		2.90	5.517		810	197		1,007	1,200	
	3170	4 ton cooling		2.50	6.400		920	229		1,149	1,375	
	3180	5 ton cooling	↓	2.10	7.619	↓	975	273		1,248	1,500	
	3500	With no cooling coil										
	3520	1-1/2 ton coil size	Q-5	12	1.333	Ea.	350	47.50		397.50	455	
	3530	2 ton coil size		10	1.600		375	57.50		432.50	500	
	3540	2-1/2 ton coil size		10	1.600		415	57.50		472.50	545	
	3554	3 ton coil size		9	1.778		460	63.50		523.50	605	
	3560	3-1/2 ton coil size		9	1.778		515	63.50		578.50	665	
	3570	4 ton coil size		8.50	1.882		660	67.50		727.50	830	
	3580	5 ton coil size	↓	8	2	↓	710	71.50		781.50	890	
	4000	With heater										
	4120	5 kW, 17.1 MBH	Q-5	16	1	Ea.	315	36		351	400	
	4130	7.5 kW, 25.6 MBH		15.60	1.026		335	36.50		371.50	425	
	4140	10 kW, 34.2 MBH		15.20	1.053		390	37.50		427.50	490	
	4150	12.5 KW, 42.7 MBH	↓	14.80	1.081	↓	440	38.50		478.50	540	
400	0010	**FURNACES** Hot air heating, blowers, standard controls										400
	0020	not including gas, oil or flue piping										
	1000	Electric, UL listed										
	1020	10.2 MBH	Q-20	5	4	Ea.	315	143		458	570	
	1100	34.1 MBH	"	4.40	4.545	"	405	163		568	700	
	3000	Gas, AGA certified, upflow, direct drive models										
	3020	45 MBH input	Q-9	4	4	Ea.	435	140		575	700	
	3040	60 MBH input		3.80	4.211		580	147		727	875	
	3060	75 MBH input		3.60	4.444		615	155		770	925	
	3100	100 MBH input		3.20	5		655	175		830	995	
	3120	125 MBH input		3	5.333		755	186		941	1,125	
	3130	150 MBH input		2.80	5.714		875	200		1,075	1,275	
	3140	200 MBH input	↓	2.60	6.154	↓	1,675	215		1,890	2,175	
	6000	Oil, UL listed, atomizing gun type burner										
	6020	56 MBH output	Q-9	3.60	4.444	Ea.	745	155		900	1,075	
	6040	95 MBH output		3.40	4.706		790	164		954	1,125	
	6060	134 MBH output		3.20	5		1,100	175		1,275	1,475	
	6080	151 MBH output		3	5.333		1,225	186		1,411	1,625	
	6100	200 MBH input		2.60	6.154	↓	1,725	215		1,940	2,250	
	9000	Minimum labor/equipment charge	↓	2.75	5.818	Job		203		203	320	

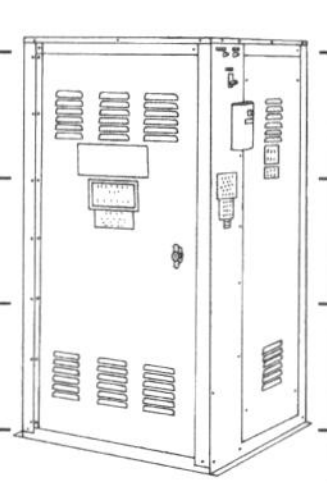

15540 | Fuel-Fired Heaters

			CREW	DAILY OUTPUT	LABOR-HOURS	UNIT	2004 BARE COSTS MAT.	LABOR	EQUIP.	TOTAL	TOTAL INCL O&P	
300	0010	**DUCT FURNACES** Includes burner, controls, stainless steel										300
	0020	heat exchanger. Gas fired, electric ignition										
	1000	Outdoor installation, with vent cap										
	1120	225 MBH output	Q-5	2.50	6.400	Ea.	3,250	229		3,479	3,925	
	1160	375 MBH output		1.60	10		4,325	360		4,685	5,300	
	1180	450 MBH output		1.40	11.429		4,525	410		4,935	5,600	
900	0010	**SPACE HEATERS** Cabinet, grilles, fan, controls, burner,										900
	0020	thermostat, no piping. For flue see 15550										
	1000	Gas fired, floor mounted										
	1100	60 MBH output	Q-5	10	1.600	Ea.	545	57.50		602.50	690	
	1180	180 MBH output		6	2.667		815	95.50		910.50	1,050	
	2000	Suspension mounted, propeller fan, 20 MBH output		8.50	1.882		425	67.50		492.50	570	
	2040	60 MBH output		7	2.286		525	82		607	700	
	2100	130 MBH output		5	3.200		795	115		910	1,050	
	2240	320 MBH output		2	8		1,625	286		1,911	2,250	
	2500	For powered venter and adapter, add					271			271	298	
	5000	Wall furnace, 17.5 MBH output	Q-5	6	2.667		590	95.50		685.50	795	
	5020	24 MBH output		5	3.200		585	115		700	825	
	5040	35 MBH output		4	4		790	143		933	1,075	
	9000	Minimum labor/equipment charge		3.50	4.571	Job		164		164	254	

15550 | Breechings, Chimneys & Stacks

			CREW	DAILY OUTPUT	LABOR-HOURS	UNIT	MAT.	LABOR	EQUIP.	TOTAL	TOTAL INCL O&P	
200	0010	**DRAFT CONTROL DEVICES**										200
	1000	Barometric,gas fired system only, 6" size for 5" and 6" pipes	1 Shee	20	.400	Ea.	41	15.50		56.50	69.50	
	1040	8" size, for 7" and 8" pipes	"	18	.444	"	56.50	17.25		73.75	89.50	
	2000	All fuel, oil, oil/gas, coal										
	2020	10" for 9" and 10" pipes	1 Shee	15	.533	Ea.	96	20.50		116.50	139	
	3260	For thermal switch for above, add	"	24	.333		39.50	12.95		52.45	63.50	
	5000	Vent damper, bi-metal, gas, 3" diameter	Q-9	24	.667		23.50	23.50		47	62.50	
	5010	4" diameter		24	.667		23.50	23.50		47	62.50	
	5020	5" diameter		23	.696		23.50	24.50		48	64	
	5030	6" diameter		22	.727		23.50	25.50		49	65.50	
	5040	7" diameter		21	.762		25	26.50		51.50	69.50	
	5050	8" diameter		20	.800		27	28		55	74.50	
	5101	Electric, automatic, gas, 4" diameter		24	.667		145	23.50		168.50	196	
	5110	5" diameter		23	.696		148	24.50		172.50	202	
	5121	6" diameter		22	.727		150	25.50		175.50	205	
	5130	7" diameter		21	.762		153	26.50		179.50	210	
	5140	8" diameter		20	.800		156	28		184	216	
	5150	9" diameter		20	.800		159	28		187	220	
	5160	10" diameter		19	.842		164	29.50		193.50	227	
	5170	12" diameter		19	.842		167	29.50		196.50	230	
	5250	Automatic, oil, 4" diameter		24	.667		162	23.50		185.50	215	
	5260	5" diameter		23	.696		165	24.50		189.50	220	
	5270	6" diameter		22	.727		169	25.50		194.50	226	
	5280	7" diameter		21	.762		172	26.50		198.50	231	
	5290	8" diameter		20	.800		179	28		207	242	
	5300	9" diameter		20	.800		182	28		210	245	
	5310	10" diameter		19	.842		182	29.50		211.50	248	
	5320	12" diameter		19	.842		185	29.50		214.50	251	
	5330	14" diameter		18	.889		248	31		279	320	
	5340	16" diameter		17	.941		288	33		321	365	
	5350	18" diameter		16	1		320	35		355	410	
	9000	Minimum labor/equipment charge	1 Shee	4	2	Job		77.50		77.50	123	
440	0010	**VENT CHIMNEY** Prefab metal, U.L. listed										440
	0020	Gas, double wall, galvanized steel										

15 MECHANICAL

15550 | Breechings, Chimneys & Stacks

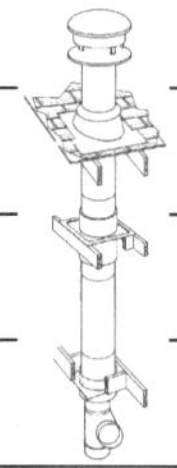

		15550 Breechings, Chimneys & Stacks	CREW	DAILY OUTPUT	LABOR-HOURS	UNIT	2004 BARE COSTS MAT.	LABOR	EQUIP.	TOTAL	TOTAL INCL O&P	
440	0080	3″ diameter	Q-9	72	.222	V.L.F.	3.68	7.75		11.43	16.35	440
	0100	4″ diameter		68	.235		4.57	8.20		12.77	18.05	
	0120	5″ diameter		64	.250		5.35	8.75		14.10	19.70	
	0140	6″ diameter		60	.267		6.20	9.30		15.50	21.50	
	0160	7″ diameter		56	.286		9.10	10		19.10	26	
	0180	8″ diameter		52	.308		10.15	10.75		20.90	28	
	0200	10″ diameter		48	.333		21.50	11.65		33.15	42	
	0220	12″ diameter		44	.364		28.50	12.70		41.20	51.50	
	0260	16″ diameter		40	.400		65	13.95		78.95	93.50	
	0300	20″ diameter	Q-10	36	.667		99.50	24		123.50	148	
	0340	24″ diameter	″	32	.750		155	27		182	213	
	5000	Vent damper bi-metal 6″ flue	Q-9	16	1	Ea.	105	35		140	172	
	5100	Gas, auto., electric		8	2	″	169	70		239	297	
	7800	All fuel, double wall, stainless steel, 6″ diameter		60	.267	V.L.F.	30	9.30		39.30	48	
	7802	7″ diameter		56	.286		38.50	10		48.50	58.50	
	7804	8″ diameter		52	.308		45.50	10.75		56.25	67	
	7806	10″ diameter		48	.333		66	11.65		77.65	91	
	7808	12″ diameter		44	.364		88.50	12.70		101.20	118	
	7810	14″ diameter		42	.381		116	13.30		129.30	149	
	8000	All fuel, double wall, stainless steel fittings										
	8010	Roof support 6″ diameter	Q-9	30	.533	Ea.	78	18.65		96.65	116	
	8020	7″ diameter		28	.571		88.50	19.95		108.45	129	
	8030	8″ diameter		26	.615		98.50	21.50		120	142	
	8040	10″ diameter		24	.667		116	23.50		139.50	164	
	8050	12″ diameter		22	.727		139	25.50		164.50	193	
	8060	14″ diameter		21	.762		177	26.50		203.50	236	
	8100	Elbow 15°, 6″ diameter		30	.533		62.50	18.65		81.15	98.50	
	8120	7″ diameter		28	.571		70.50	19.95		90.45	109	
	8140	8″ diameter		26	.615		80	21.50		101.50	123	
	8160	10″ diameter		24	.667		108	23.50		131.50	155	
	8180	12″ diameter		22	.727		127	25.50		152.50	180	
	8200	14″ diameter		21	.762		152	26.50		178.50	209	
	8300	Insulated tee with insulated tee cap, 6″ diameter		30	.533		118	18.65		136.65	160	
	8340	7″ diameter		28	.571		155	19.95		174.95	202	
	8360	8″ diameter		26	.615		175	21.50		196.50	226	
	8380	10″ diameter		24	.667		246	23.50		269.50	305	
	8400	12″ diameter		22	.727		345	25.50		370.50	420	
	8420	14″ diameter		21	.762		455	26.50		481.50	540	
	8500	Joist shield, 6″ diameter		30	.533		36	18.65		54.65	69	
	8510	7″ diameter		28	.571		39	19.95		58.95	74.50	
	8520	8″ diameter		26	.615		48	21.50		69.50	87	
	8530	10″ diameter		24	.667		65	23.50		88.50	109	
	8540	12″ diameter		22	.727		81	25.50		106.50	129	
	8550	14″ diameter		21	.762		101	26.50		127.50	153	
	8600	Round top, 6″ diameter		30	.533		40	18.65		58.65	73.50	
	8620	7″ diameter		28	.571		54.50	19.95		74.45	91.50	
	8640	8″ diameter		26	.615		73.50	21.50		95	115	
	8660	10″ diameter		24	.667		132	23.50		155.50	182	
	8680	12″ diameter		22	.727		188	25.50		213.50	247	
	8700	14″ diameter		21	.762		251	26.50		277.50	320	
	8800	Adjustable roof flashing, 6″ diameter		30	.533		47.50	18.65		66.15	82	
	8820	7″ diameter		28	.571		54.50	19.95		74.45	91.50	
	8840	8″ diameter		26	.615		59	21.50		80.50	99	
	8860	10″ diameter		24	.667		76	23.50		99.50	121	
	8880	12″ diameter		22	.727		98	25.50		123.50	148	
	8900	14″ diameter		21	.762		122	26.50		148.50	177	

MECHANICAL 15

15500 | Heat Generation Equipment

	15550 | Breechings, Chimneys & Stacks	CREW	DAILY OUTPUT	LABOR-HOURS	UNIT	2004 BARE COSTS MAT.	LABOR	EQUIP.	TOTAL	TOTAL INCL O&P	
440	9990 Minimum labor/equipment charge	Q-9	3	5.333	Job		186		186	295	440

15600 | Refrigeration Equipment

	15610 | Refrigeration Compressors	CREW	DAILY OUTPUT	LABOR-HOURS	UNIT	2004 BARE COSTS MAT.	LABOR	EQUIP.	TOTAL	TOTAL INCL O&P	
400	0010 **RECIPROCATING REFRIGERANT COMPRESSORS**										400
	1000 (Ratings are ARI standard 515 group IV using R-22) 10 ton	Q-5	1	16	Ea.	6,225	570		6,795	7,725	
	1100 20 ton	Q-6	.72	33.333		9,525	1,225		10,750	12,400	
	1400 50 ton	"	.20	120		13,100	4,450		17,550	21,300	
	1600 130 ton	Q-7	.21	152	↓	16,500	5,775		22,275	27,200	

	15620 | Packaged Water Chillers	CREW	DAILY OUTPUT	LABOR-HOURS	UNIT	MAT.	LABOR	EQUIP.	TOTAL	TOTAL INCL O&P	
100	0010 **ABSORPTION WATER CHILLERS**										100
	3000 Gas fired, air cooled										
	3180 3 ton	Q-5	1.30	12.308	Ea.	6,025	440		6,465	7,325	
	9000 Minimum labor/equipment charge	"	1	16	Job		570		570	890	
600	0010 **CENTRIFUGAL/SCREW/RECIP. WATER CHILLERS,** W/ standard controls										600
	0110 Screw, liquid chiller, air cooled, insulated evaporator										
	0120 130 ton	Q-7	.14	228	Ea.	59,500	8,650		68,150	79,000	
	0124 160 ton		.13	246		73,500	9,325		82,825	95,500	
	0490 Reciprocating, packaged w/integral air cooled condenser, 15 ton cool		.37	86.486		13,300	3,275		16,575	19,700	
	0500 20 ton cooling		.34	94.955		17,600	3,600		21,200	25,000	
	0520 40 ton cooling	↓	.30	108	↓	23,900	4,100		28,000	32,700	
	0680 Water cooled, single compressor, semi-hermetic, tower not incl.										
	0760 10 ton cooling	Q-6	.36	67.039	Ea.	10,000	2,475		12,475	14,900	
	0840 35 ton cooling	Q-7	.33	97.859	"	13,600	3,700		17,300	20,800	
	0980 Water cooled, multiple compress., semi-hermetic, tower not incl.										
	1100 50 ton cooling	Q-7	.28	113	Ea.	25,400	4,325		29,725	34,700	
	1160 100 ton cooling		.18	179		45,300	6,800		52,100	60,500	
	1200 140 ton cooling	↓	.16	202	↓	60,500	7,675		68,175	78,500	
	9000 Minimum labor/equipment charge	Q-6	1	24	Job		890		890	1,375	

	15640 | Packaged Cooling Towers	CREW	DAILY OUTPUT	LABOR-HOURS	UNIT	MAT.	LABOR	EQUIP.	TOTAL	TOTAL INCL O&P	
400	0010 **COOLING TOWERS** Packaged units										400
	0070 Galvanized steel										
	0080 Induced draft, crossflow										
	0100 Vertical, belt drive, 61 tons	Q-6	90	.267	TonAC	79	9.90		88.90	102	
	0150 100 ton		100	.240		62.50	8.90		71.40	83	
	0200 115 ton		109	.220		61	8.15		69.15	79.50	
	0250 131 ton		120	.200		54	7.40		61.40	70.50	
	0260 162 ton	↓	132	.182	↓	45.50	6.75		52.25	60.50	
	1000 For higher capacities, use multiples										
	1500 Induced air, double flow										
	1900 Vertical, gear drive, 167 ton	Q-6	126	.190	TonAC	84	7.05		91.05	103	
	2000 297 ton		129	.186		60	6.90		66.90	76.50	
	2100 582 ton		132	.182		49	6.75		55.75	64	
	2200 1016 ton	↓	150	.160	↓	47	5.95		52.95	60.50	

15600 | Refrigeration Equipment

15640 | Packaged Cooling Towers

		15640 Packaged Cooling Towers	CREW	DAILY OUTPUT	LABOR-HOURS	UNIT	2004 BARE COSTS MAT.	LABOR	EQUIP.	TOTAL	TOTAL INCL O&P	
400	3500	For pumps and piping, add	Q-6	38	.632	TonAC	39.50	23.50		63	80	400
	4000	For absorption systems, add				↓	75%	75%				
	5000	Fiberglass										
	5010	Draw thru										
	5100	60 ton	Q-6	1.50	16	Ea.	3,175	595		3,770	4,425	
	5120	125 ton		.99	24.242		6,500	900		7,400	8,550	
	5140	300 ton		.43	55.814		15,200	2,075		17,275	20,000	
	5160	600 ton		.22	109		27,400	4,050		31,450	36,400	
	5180	1000 ton	↓	.15	160	↓	47,000	5,925		52,925	60,500	
	6000	Stainless steel										
	6010	Induced draft, crossflow, horizontal, belt drive										
	6100	57 ton	Q-6	1.50	16	Ea.	7,975	595		8,570	9,675	
	6120	91 ton		.99	24.242		11,500	900		12,400	14,000	
	6140	111 ton		.43	55.814		13,500	2,075		15,575	18,000	
	6160	126 ton	↓	.22	109	↓	14,600	4,050		18,650	22,400	
	6170	Induced draft, crossflow, vertical, gear drive										
	6172	167 ton	Q-6	.75	32	Ea.	23,500	1,175		24,675	27,700	
	6174	297 ton		.43	55.814		30,200	2,075		32,275	36,400	
	6176	582 ton		.23	104		47,500	3,875		51,375	58,500	
	6180	1016 ton		.15	160	↓	83,000	5,925		88,925	100,500	
	9000	Minimum labor/equipment charge	↓	1	24	Job		890		890	1,375	

15670 | Refrigerant Condensing Units

		15670 Refrigerant Condensing Units	CREW	DAILY OUTPUT	LABOR-HOURS	UNIT	MAT.	LABOR	EQUIP.	TOTAL	TOTAL INCL O&P	
300	0010	**CONDENSING UNITS**										300
	0030	Air cooled, compressor, standard controls										
	0050	1.5 ton	Q-5	2.50	6.400	Ea.	555	229		784	965	
	0200	2.5 ton		1.70	9.412		695	335		1,030	1,275	
	0300	3 ton		1.30	12.308		825	440		1,265	1,600	
	0400	4 ton		.90	17.778		1,075	635		1,710	2,150	
	0500	5 ton		.60	26.667	↓	1,300	955		2,255	2,900	
	9000	Minimum labor/equipment charge	↓	1.50	10.667	Job		380		380	590	

15700 | Heating/Ventilating/Air Conditioning Equipment

15705 | Curbs/Pads/Stands Prefab

		15705 Curbs/Pads/Stands Prefab	CREW	DAILY OUTPUT	LABOR-HOURS	UNIT	2004 BARE COSTS MAT.	LABOR	EQUIP.	TOTAL	TOTAL INCL O&P	
600	0010	**CURBS/PADS PREFABRICATED**										600
	6000	Pad, fiberglass reinforced concrete with polystyrene foam core										
	6050	Condenser, 2″ thick, 20″ x 38″	1 Shee	8	1	Ea.	8.55	39		47.55	71	
	6090	24″ x 36″	″	12	.667		11.80	26		37.80	54	
	6280	36″ x 36″	Q-9	8	2		17	70		87	130	
	6300	36″ x 40″		7	2.286		19.30	80		99.30	147	
	6320	36″ x 48″	↓	7	2.286	↓	22.50	80		102.50	151	

15720 | Air Handling Units

		15720 Air Handling Units	CREW	DAILY OUTPUT	LABOR-HOURS	UNIT	MAT.	LABOR	EQUIP.	TOTAL	TOTAL INCL O&P	
500	0010	**MAKE-UP AIR UNIT**										500
	0020	Indoor suspension, natural/LP gas, direct fired,										
	0030	standard control. For flue see division 15550-440										
	0040	70°F temperature rise, MBH is input										

		15720 Air Handling Units	CREW	DAILY OUTPUT	LABOR-HOURS	UNIT	2004 BARE COSTS MAT.	LABOR	EQUIP.	TOTAL	TOTAL INCL O&P	
500	0100	2000 CFM, 168 MBH	Q-6	3	8	Ea.	5,950	297		6,247	7,000	500
	0220	12,000 CFM, 1005 MBH	"	1	24		9,325	890		10,215	11,700	
	0300	24,000 CFM, 2007 MBH	Q-7	1	32		12,200	1,200		13,400	15,300	
	0400	50,000 CFM, 4180 MBH	"	.80	40	↓	16,700	1,525		18,225	20,800	
	9000	Minimum labor/equipment charge	Q-6	2.75	8.727	Job		325		325	500	

		15730 Unitary Air Conditioning Equip	CREW	DAILY OUTPUT	LABOR-HOURS	UNIT	MAT.	LABOR	EQUIP.	TOTAL	TOTAL INCL O&P	
200	0010	**COMPUTER ROOM UNITS**										200
	1000	Air cooled, includes remote condenser but not										
	1020	interconnecting tubing or refrigerant										
	1160	6 ton	Q-5	.30	53.333	Ea.	20,500	1,900		22,400	25,600	
	1240	10 ton	"	.25	64		24,100	2,300		26,400	30,200	
	1320	20 ton	Q-6	.29	82.759	↓	32,200	3,075		35,275	40,200	
600	0010	**ROOF TOP AIR CONDITIONERS** Standard controls, curb, economizer										600
	0020											
	1000	Single zone, electric cool, gas heat										
	1090	2 ton cooling, 55 MBH heating	Q-5	.93	17.204	Ea.	2,325	615		2,940	3,500	
	1100	3 ton cooling, 60 MBH heating		.70	22.857		2,825	820		3,645	4,375	
	1120	4 ton cooling, 95 MBH heating	↓	.61	26.403		3,375	945		4,320	5,175	
	1160	10 ton cooling, 200 MBH heating	Q-6	.67	35.982		6,900	1,325		8,225	9,675	
	1220	30 ton cooling, 540 MBH heating	Q-7	.47	68.376		23,900	2,600		26,500	30,300	
	1240	40 ton cooling, 675 MBH heating	"	.35	91.168	↓	30,700	3,450		34,150	39,100	
	1700	Gas cool, gas heat										
	1720	3 ton cooling, 90 MBH heating	Q-5	.93	17.204	Ea.	7,400	615		8,015	9,100	
	2000	Multizone, electric cool, gas heat, economizer										
	2120	20 ton cooling, 360 MBH heating	Q-7	.53	60.038	Ea.	81,000	2,275		83,275	93,000	
	2180	30 ton cooling, 540 MBH heating		.37	85.562		105,000	3,250		108,250	120,500	
	2210	50 ton cooling, 540 MBH heating		.22	142		110,000	5,400		115,400	129,500	
	2220	70 ton cooling, 1500 MBH heating		.16	198		119,000	7,525		126,525	142,500	
	2240	80 ton cooling, 1500 MBH heating		.14	228		136,000	8,650		144,650	163,000	
	2280	105 ton cooling, 1500 MBH heating	↓	.11	290		157,500	11,000		168,500	190,500	
	2400	For hot water heat coil, deduct					5%					
	2500	For steam heat coil, deduct					2%					
	2600	For electric heat, deduct				↓	3%	5%				
	9000	Minimum labor/equipment charge	Q-5	1.50	10.667	Job		380		380	590	
800	0010	**WINDOW UNIT AIR CONDITIONERS**										800
	4500	10,000 BTUH	1 Carp	6	1.333	Ea.	485	44		529	605	
	4520	12,000 BTUH	L-2	8	2	"	525	57.50		582.50	675	
	9000	Minimum labor/equipment charge	1 Carp	2	4	Job		132		132	219	
840	0010	**SELF-CONTAINED SINGLE PACKAGE**										840
	0100	Air cooled, for free blow or duct, not incl. remote condenser										
	0200	3 ton cooling	Q-5	1	16	Ea.	2,425	570		2,995	3,550	
	0210	4 ton cooling	"	.80	20		2,625	715		3,340	4,000	
	0240	10 ton cooling	Q-7	1	32		5,100	1,200		6,300	7,475	
	0280	30 ton cooling	"	.80	40	↓	12,300	1,525		13,825	15,900	
	1000	Water cooled for free blow or duct, not including tower										
	1010	Constant volume										
	1100	3 ton cooling	Q-6	1	24	Ea.	2,350	890		3,240	3,950	
	1120	5 ton cooling	"	1	24		3,000	890		3,890	4,675	
	1140	10 ton cooling	Q-7	.90	35.556		6,100	1,350		7,450	8,800	
	1180	30 ton cooling	"	.70	45.714		26,900	1,725		28,625	32,300	
	1240	60 ton cooling	Q-8	.30	106	↓	49,500	4,050	259	53,809	61,000	
	9000	Minimum labor/equipment charge	Q-5	1	16	Job		570		570	890	
900	0010	**SPLIT DUCTLESS SYSTEM**										900
	0100	Cooling only, single zone										

15 MECHANICAL

Important: See the Reference Section for critical supporting data - Reference Nos., Crews, & City Cost Indexes

15730 | Unitary Air Conditioning Equip

			CREW	DAILY OUTPUT	LABOR-HOURS	UNIT	2004 BARE COSTS MAT.	LABOR	EQUIP.	TOTAL	TOTAL INCL O&P	
900	0110	Wall mount										900
	0120	3/4 ton cooling	Q-5	2	8	Ea.	1,125	286		1,411	1,675	
	0130	1 ton cooling		1.80	8.889		1,225	320		1,545	1,850	
	0140	1-1/2 ton cooling		1.60	10		1,575	360		1,935	2,300	
	0150	2 ton cooling	↓	1.40	11.429	↓	2,275	410		2,685	3,125	
	1000	Ceiling mount										
	1020	2 ton cooling	Q-5	1.40	11.429	Ea.	1,100	410		1,510	1,850	
	1030	3 ton cooling	"	1.20	13.333	"	3,225	475		3,700	4,300	
	2000	T-Bar mount										
	2010	2 ton cooling	Q-5	1.40	11.429	Ea.	2,400	410		2,810	3,275	
	2020	3 ton cooling		1.20	13.333		2,875	475		3,350	3,925	
	2030	3-1/2 ton cooling	↓	1.10	14.545	↓	3,475	520		3,995	4,625	
	3000	Multizone										
	3010	Wall mount										
	3020	2 @ 3/4 ton cooling	Q-5	1.80	8.889	Ea.	1,075	320		1,395	1,700	
	5000	Cooling / Heating										
	5010	Wall mount										
	5110	1 ton cooling	Q-5	1.70	9.412	Ea.	795	335		1,130	1,400	
	5120	1-1/2 ton cooling	"	1.50	10.667	"	1,275	380		1,655	2,000	
	5300	Ceiling mount										
	5310	3 ton cooling	Q-5	1	16	Ea.	3,750	570		4,320	5,000	
	7000	Accessories for all split ductless systems										
	7010	Add for ambient frost control	Q-5	8	2	Ea.	118	71.50		189.50	241	
	7020	Add for tube / wiring kit										
	7030	15' kit	Q-5	32	.500	Ea.	28.50	17.90		46.40	58.50	
	7040	35' kit	"	24	.667	"	91.50	24		115.50	137	

15740 | Heat Pumps

			CREW	DAILY OUTPUT	LABOR-HOURS	UNIT	MAT.	LABOR	EQUIP.	TOTAL	TOTAL INCL O&P	
100	0010	**AIR-SOURCE HEAT PUMPS** (Not including interconnecting tubing)										100
	1000	Air to air, split system, not including curbs, pads, or ductwork										
	1020	2 ton cooling, 8.5 MBH heat @ 0°F	Q-5	1.20	13.333	Ea.	985	475		1,460	1,825	
	1040	3 ton cooling, 13 MBH heat @ 0°F		.80	20		1,200	715		1,915	2,425	
	1080	7.5 ton cooling, 33 MBH heat @ 0°F	↓	.30	53.333		3,675	1,900		5,575	6,975	
	1120	15 ton cooling, 64 MBH heat @ 0°F	Q-6	.26	92.308		7,325	3,425		10,750	13,400	
	1130	20 ton cooling, 85 MBH heat @ 0°F		.20	120		9,850	4,450		14,300	17,700	
	1140	25 ton cooling, 119 MBH heat @ 0°F	↓	.20	120	↓	11,700	4,450		16,150	19,800	
	1500	Single package, not including curbs, pads, or plenums										
	1520	2 ton cooling, 6.5 MBH heat @ 0°F	Q-5	1.50	10.667	Ea.	1,950	380		2,330	2,725	
	1560	3 ton cooling, 10 MBH heat @ 0°F	"	1.20	13.333	"	2,275	475		2,750	3,250	
800	0010	**WATER-SOURCE HEAT PUMPS** (Not including interconnecting tubing)										800
	2000	Water source to air, single package										
	2100	1 ton cooling, 13 MBH heat @ 75°F	Q-5	2	8	Ea.	895	286		1,181	1,425	
	2140	2 ton cooling, 19 MBH heat @ 75°F		1.70	9.412		1,050	335		1,385	1,675	
	2220	5 ton cooling, 29 MBH heat @ 75°F		.90	17.778	↓	1,675	635		2,310	2,800	
	9000	Minimum labor/equipment charge	↓	1.75	9.143	Job		325		325	505	

15750 | Humidity Control Equipment

			CREW	DAILY OUTPUT	LABOR-HOURS	UNIT	MAT.	LABOR	EQUIP.	TOTAL	TOTAL INCL O&P	
500	0010	**HUMIDIFIERS**										500
	0520	Steam, room or duct, filter, regulators, auto. controls, 220 V										
	0560	22 lb. per hour	Q-5	5	3.200	Ea.	2,250	115		2,365	2,650	
	0580	33 lb. per hour		4	4		2,325	143		2,468	2,775	
	0620	100 lb. per hour		3	5.333	↓	3,400	191		3,591	4,050	
	9000	Minimum labor/equipment charge	↓	3.50	4.571	Job		164		164	254	

For expanded coverage of these items see *Means Mechanical or Plumbing Cost Data 2004*

15760 | Terminal Heating & Cooling Units

		15760 Terminal Heating & Cooling Units	CREW	DAILY OUTPUT	LABOR-HOURS	UNIT	2004 BARE COSTS MAT.	LABOR	EQUIP.	TOTAL	TOTAL INCL O&P	
100	0010	**COILS, FLANGED**										100
	1500	Hot water heating, 1 row, 24" x 48"	Q-5	4	4	Ea.	1,075	143		1,218	1,425	
	9000	Minimum labor/equipment charge	1 Plum	1	8	Job		315		315	490	
200	0010	**DUCT HEATERS** Electric, 480 V, 3 Ph										200
	0020	Finned tubular insert, 500°F										
	0100	8" wide x 6" high, 4.0 kW	Q-20	16	1.250	Ea.	540	45		585	665	
	0120	12" high, 8.0 kW		15	1.333		895	48		943	1,050	
	0140	18" high, 12.0 kW		14	1.429		1,250	51		1,301	1,450	
	0160	24" high, 16.0 kW		13	1.538		1,625	55		1,680	1,850	
	0180	30" high, 20.0 kW		12	1.667		1,975	59.50		2,034.50	2,275	
	0300	12" wide x 6" high, 6.7 kW		15	1.333		575	48		623	705	
	0320	12" high, 13.3 kW		14	1.429		925	51		976	1,100	
	0340	18" high, 20.0 kW		13	1.538		1,300	55		1,355	1,500	
	8000	To obtain BTU multiply kW by 3413										
	9900	Minimum labor/equipment charge	1 Shee	4	2	Job		77.50		77.50	123	
250	0010	**ELECTRIC HEATING**, not incl. conduit or feed wiring										250
	1300	Baseboard heaters, 2' long, 375 watt	1 Elec	8	1	Ea.	29	39.50		68.50	92.50	
	1400	3' long, 500 watt		8	1		34.50	39.50		74	98.50	
	1600	4' long, 750 watt		6.70	1.194		41	47		88	118	
	1800	5' long, 935 watt 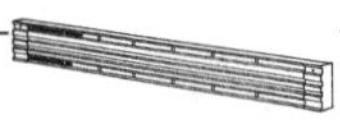		5.70	1.404		48.50	55.50		104	139	
	2000	6' long, 1125 watt		5	1.600		54	63		117	157	
	2400	8' long, 1500 watt		4	2		68	79		147	196	
	2600	9' long, 1680 watt		3.60	2.222		76	87.50		163.50	219	
	2800	10' long, 1875 watt		3.30	2.424		84.50	95.50		180	240	
	2950	Wall heaters with fan, 120 to 277 volt										
	3190	1500 watt	1 Elec	4	2	Ea.	95.50	79		174.50	226	
	3230	2500 watt		3.50	2.286		180	90		270	335	
	3250	4000 watt		2.70	2.963		330	117		447	540	
	3600	Thermostats, integral		16	.500		17.80	19.70		37.50	50	
	3800	Line voltage, 1 pole		8	1		21	39.50		60.50	84	
	3810	2 pole		8	1		27	39.50		66.50	90	
	9990	Minimum labor/equipment charge		4	2	Job		79		79	121	
300	0010	**FAN COIL AIR CONDITIONING** Cabinet mounted, filters, controls										300
	0020											
	0100	Chilled water, 1/2 ton cooling	Q-5	8	2	Ea.	520	71.50		591.50	685	
	0120	1 ton cooling		6	2.667		600	95.50		695.50	810	
	0160	2.5 ton cooling		5	3.200		1,125	115		1,240	1,400	
	0180	3 ton cooling		4	4		1,225	143		1,368	1,575	
	0262	For hot water coil, add					40%	10%				
	0940	Direct expansion, for use w/air cooled condensing, 1.5 ton cooling	Q-5	5	3.200	Ea.	400	115		515	620	
	1000	5 ton cooling	"	3	5.333		710	191		901	1,075	
	1040	10 ton cooling	Q-6	2.60	9.231		1,950	340		2,290	2,650	
	1060	20 ton cooling	"	.70	34.286		3,575	1,275		4,850	5,900	
	1510	For condensing unit add see division 15670.										
	3000	Chilled water, horizontal unit, housing, 2 pipe, controls										
	3100	1/2 ton cooling	Q-5	8	2	Ea.	875	71.50		946.50	1,075	
	3110	1 ton cooling		6	2.667		1,050	95.50		1,145.50	1,325	
	3120	1.5 ton cooling		5.50	2.909		1,275	104		1,379	1,550	
	3130	2 ton cooling		5.25	3.048		1,525	109		1,634	1,850	
	3140	3 ton cooling		4	4		1,750	143		1,893	2,150	
	3150	3.5 ton cooling		3.80	4.211		1,950	151		2,101	2,350	
	3160	4 ton cooling		3.80	4.211		1,950	151		2,101	2,350	
	4000	With electric heat, 2 pipe										
	4100	1/2 ton cooling	Q-5	8	2	Ea.	1,075	71.50		1,146.50	1,275	

15 MECHANICAL

Important: See the Reference Section for critical supporting data - Reference Nos., Crews, & City Cost Indexes

	15760 | Terminal Heating & Cooling Units		CREW	DAILY OUTPUT	LABOR-HOURS	UNIT	2004 BARE COSTS MAT.	LABOR	EQUIP.	TOTAL	TOTAL INCL O&P
300	4105	3/4 ton cooling	Q-5	7	2.286	Ea.	1,175	82		1,257	1,425
	4110	1 ton cooling		6	2.667		1,300	95.50		1,395.50	1,575
	4120	1.5 ton cooling		5.50	2.909		1,450	104		1,554	1,725
	4130	2 ton cooling		5.25	3.048		1,825	109		1,934	2,175
	4135	2.5 ton cooling		4.80	3.333		2,500	119		2,619	2,925
	4140	3 ton cooling		4	4		3,350	143		3,493	3,925
	4150	3.5 ton cooling		3.80	4.211		4,025	151		4,176	4,650
	4160	4 ton cooling		3.60	4.444		4,650	159		4,809	5,350
	9000	Minimum labor/equipment charge		3.75	4.267	Job		153		153	237
600	0010	**HYDRONIC HEATING** Terminal units, not incl. main supply pipe									
	1000	Radiation									
	1150	Fin tube, wall hung, 14" slope top cover, with damper									
	1200	1-1/4" copper tube, 4-1/4" alum. fin	Q-5	38	.421	L.F.	28	15.05		43.05	54
	1250	1-1/4" steel tube, 4-1/4" steel fin		36	.444		25.50	15.90		41.40	52.50
	1255	2" steel tube, 4-1/4" steel fin		32	.500		27.50	17.90		45.40	58
	1310	Baseboard, pkgd, 1/2" copper tube, alum. fin, 7" high		60	.267		5.35	9.55		14.90	20.50
	1320	3/4" copper tube, alum. fin, 7" high		58	.276		5.65	9.85		15.50	21.50
	1340	1" copper tube, alum. fin, 8-7/8" high		56	.286		11.70	10.20		21.90	29
	1360	1-1/4" copper tube, alum. fin, 8-7/8" high		54	.296		17.30	10.60		27.90	35.50
	1500	Note: fin tube may also require corners, caps, etc.									
	3000	Radiators, cast iron									
	3100	Free standing or wall hung, 6 tube, 25" high	Q-5	96	.167	Section	21.50	5.95		27.45	33
	9000	Minimum labor/equipment charge	"	3.20	5	Job		179		179	277
800	0010	**UNIT HEATERS**									
	3950	Unit heaters, propeller, 115 V 2 psi steam, 60°F entering air									
	4000	Horizontal, 12 MBH	Q-5	12	1.333	Ea.	252	47.50		299.50	350
	4060	43.9 MBH		8	2		415	71.50		486.50	565
	4240	286.9 MBH		2	8		1,275	286		1,561	1,850
	4250	326.0 MBH		1.90	8.421		1,375	300		1,675	2,000
	4260	364 MBH		1.80	8.889		1,500	320		1,820	2,150
	4270	404 MBH		1.60	10		2,025	360		2,385	2,800
	4300	For vertical diffuser, add					143			143	157
	4310	Vertical flow, 40 MBH	Q-5	11	1.455		400	52		452	520
	4326	131.0 MBH	"	4	4		635	143		778	920
	4354	420 MBH,(460 V)	Q-6	1.80	13.333		1,525	495		2,020	2,450
	4358	500 MBH, (460 V)		1.71	14.035		2,150	520		2,670	3,175
	4362	570 MBH, (460 V)		1.40	17.143		3,025	635		3,660	4,300
	4366	620 MBH, (460 V)		1.30	18.462		3,475	685		4,160	4,875
	4370	960 MBH, (460 V)		1.10	21.818		6,100	810		6,910	7,950
	15780 | Energy Recovery Equipment										
100	0010	**HEAT RECOVERY PACKAGES**									
	0100	Air to air									
	9000	Minimum labor/equipment charge	1 Shee	4	2	Job		77.50		77.50	123

MECHANICAL 15

15810 | Ducts

		15810 Ducts	CREW	DAILY OUTPUT	LABOR-HOURS	UNIT	2004 BARE COSTS MAT.	LABOR	EQUIP.	TOTAL	TOTAL INCL O&P	
100	0010	**METAL DUCTWORK** R15810-050										100
	0020	Fabricated rectangular, includes fittings, joints, supports,										
	0100	Aluminum, alloy 3003-H14, under 100 lb.	Q-10	75	.320	Lb.	2.45	11.60		14.05	21	
	0110	100 to 500 lb.		80	.300		1.95	10.85		12.80	19.35	
	0120	500 to 1,000 lb.		95	.253		1.65	9.15		10.80	16.30	
	0500	Galvanized steel, under 200 lb.		235	.102		.82	3.70		4.52	6.75	
	0520	200 to 500 lb.		245	.098		.62	3.55		4.17	6.30	
	0540	500 to 1,000 lb.		255	.094		.52	3.41		3.93	5.95	
	0560	1,000 to 2,000 lb.		265	.091		.40	3.28		3.68	5.65	
	0580	Over 5,000 lb.	↓	285	.084	↓	.32	3.05		3.37	5.20	
	0590											
	1000	Stainless steel, type 304, under 100 lb.	Q-10	165	.145	Lb.	2.03	5.25		7.28	10.60	
	1020	100 to 500 lb.		175	.137		1.53	4.97		6.50	9.55	
	1030	500 to 1,000 lb.	↓	190	.126	↓	1.23	4.58		5.81	8.60	
	9990	Minimum labor/equipment charge	1 Shee	3	2.667	Job		103		103	164	
300	0010	**FIBEROUS GLASS DUCTWORK** R15810-050										300
	9990	Minimum labor/equipment charge	1 Shee	3	2.667	Job		103		103	164	
500	0010	**FLEXIBLE DUCTS** R15810-050										500
	1300	Flexible, coated fiberglass fabric on corr. resist. metal helix										
	1400	pressure to 12" (WG) UL-181										
	1500	Non-insulated, 3" diameter	Q-9	400	.040	L.F.	1.03	1.40		2.43	3.34	
	1540	5" diameter		320	.050		1.29	1.75		3.04	4.19	
	1560	6" diameter		280	.057		1.52	2		3.52	4.83	
	1580	7" diameter		240	.067		1.81	2.33		4.14	5.70	
	1600	8" diameter		200	.080		2.09	2.79		4.88	6.75	
	1640	10" diameter		160	.100		2.63	3.49		6.12	8.45	
	1660	12" diameter		120	.133		3.18	4.66		7.84	10.90	
	1900	Insulated, 1" thick, PE jacket, 3" diameter		380	.042		1.11	1.47		2.58	3.55	
	1910	4" diameter		340	.047		1.19	1.64		2.83	3.91	
	1920	5" diameter		300	.053		1.31	1.86		3.17	4.39	
	1940	6" diameter		260	.062		1.51	2.15		3.66	5.05	
	1960	7" diameter		220	.073		1.71	2.54		4.25	5.90	
	1980	8" diameter		180	.089		1.90	3.10		5	7	
	2020	10" diameter		140	.114		2.33	3.99		6.32	8.85	
	2040	12" diameter		100	.160		2.86	5.60		8.46	12	
	2060	14" diameter	↓	80	.200	↓	3.25	7		10.25	14.65	
	9990	Minimum labor/equipment charge	1 Shee	3	2.667	Job		103		103	164	
700	0010	**GLASS FIBER-REINFORCED PLASTIC DUCTS** R15810-050										700
	9990	Minimum labor/equipment charge	1 Shee	3	2.667	Job		103		103	164	

15820 | Duct Accessories

		15820 Duct Accessories	CREW	DAILY OUTPUT	LABOR-HOURS	UNIT	MAT.	LABOR	EQUIP.	TOTAL	TOTAL INCL O&P	
300	0010	**DUCT ACCESSORIES**										300
	0050	Air extractors, 12" x 4"	1 Shee	24	.333	Ea.	15.55	12.95		28.50	37.50	
	0100	8" x 6"		22	.364		15.55	14.10		29.65	39.50	
	0200	20" x 8"		16	.500		35	19.40		54.40	69.50	
	0240	18" x 10"		14	.571		34.50	22		56.50	72.50	
	0280	24" x 12"		10	.800		48	31		79	102	
	3000	Fire damper, curtain type, 1-1/2 hr rated, vertical, 6" x 6"		24	.333		13.40	12.95		26.35	35	
	3020	8" x 6"		22	.364		21	14.10		35.10	45.50	
	3240	16" x 14"		18	.444		32.50	17.25		49.75	63.50	
	3400	24" x 20"		8	1		41.50	39		80.50	108	
	5990	Multi-blade dampers, opposed blade, 8" x 6"		24	.333		19.20	12.95		32.15	41.50	
	5994	8" x 8"		22	.364		20	14.10		34.10	44.50	
	5996	10" x 10"		21	.381		22.50	14.80		37.30	48	
	6000	12" x 12"	↓	21	.381	↓	26.50	14.80		41.30	52.50	

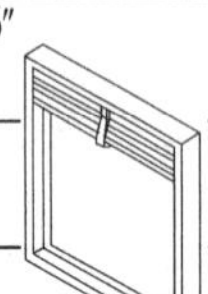

15820 | Duct Accessories

		15820 Duct Accessories	CREW	DAILY OUTPUT	LABOR-HOURS	UNIT	2004 BARE COSTS MAT.	LABOR	EQUIP.	TOTAL	TOTAL INCL O&P	
300	6020	12" x 18"	1 Shee	18	.444	Ea.	35	17.25		52.25	66	300
	6030	14" x 10"		20	.400		25.50	15.50		41	52.50	
	6031	14" x 14"		17	.471		31	18.25		49.25	63.50	
	6033	16" x 12"		17	.471		31	18.25		49.25	63.50	
	6035	16" x 16" 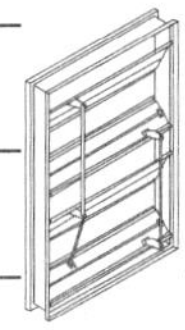		16	.500		38.50	19.40		57.90	72.50	
	6037	18" x 16"		15	.533		43	20.50		63.50	80.50	
	6038	18" x 18"		15	.533		46.50	20.50		67	84	
	6070	20" x 16"		14	.571		46.50	22		68.50	86	
	6072	20" x 20"		13	.615		55	24		79	98.50	
	6074	22" x 18"		14	.571		55	22		77	95.50	
	6076	24" x 16"		11	.727		54.50	28		82.50	105	
	6078	24" x 20"		8	1		65	39		104	133	
	6080	24" x 24"		8	1		76	39		115	145	
	6110	26" x 26"	↓	6	1.333		88	51.50		139.50	179	
	6133	30" x 30"	Q-9	6.60	2.424		118	84.50		202.50	263	
	6135	32" x 32"		6.40	2.500		134	87.50		221.50	286	
	6180	48" x 36"	↓	5.60	2.857	↓	225	100		325	405	
	7500	Variable volume modulating motorized damper, incl. elect. mtr.										
	7504	8" x 6"	1 Shee	15	.533	Ea.	106	20.50		126.50	149	
	7506	10" x 6"		14	.571		106	22		128	151	
	7510	10" x 10"		13	.615		110	24		134	159	
	7520	12" x 12"		12	.667		112	26		138	164	
	7522	12" x 16"		11	.727		106	28		134	161	
	7524	16" x 10"		12	.667		113	26		139	165	
	7526	16" x 14"		10	.800		118	31		149	178	
	7528	16" x 18"		9	.889		125	34.50		159.50	192	
	7542	18" x 18"		8	1		130	39		169	205	
	7544	20" x 14"		8	1		132	39		171	207	
	7546	20" x 18"		7	1.143		140	44.50		184.50	224	
	7560	24" x 12"		8	1		134	39		173	209	
	7562	24" x 18"		7	1.143		149	44.50		193.50	234	
	7568	28" x 10"		7	1.143		222	44.50		266.50	315	
	7590	30" x 14"		5	1.600		235	62		297	360	
	7600	30" x 18"		4	2		250	77.50		327.50	400	
	7610	30" x 24"		3.80	2.105		286	81.50		367.50	445	
	7700	For thermostat, add	↓	8	1	↓	27	39		66	91	
	8000	Multi-blade dampers, parallel blade										
	8100	8" x 8"	1 Shee	24	.333	Ea.	53.50	12.95		66.45	79.50	
	8140	16" x 10"		20	.400		71	15.50		86.50	103	
	8160	18" x 12"		18	.444		77	17.25		94.25	112	
	8220	28" x 16"	↓	10	.800	↓	105	31		136	164	
	9000	Silencers, noise control for air flow, duct				MCFM	48			48	53	
	9900	Minimum labor/equipment charge	1 Shee	4	2	Job		77.50		77.50	123	

15830 | Fans

		15830 Fans	CREW	DAILY OUTPUT	LABOR-HOURS	UNIT	MAT.	LABOR	EQUIP.	TOTAL	TOTAL INCL O&P	
100	0010	**FANS**										100
	0020	Air conditioning and process air handling										
	0030	Axial flow, compact, low sound, 2.5" S.P.										
	0050	3,800 CFM, 5 HP	Q-20	3.40	5.882	Ea.	3,700	211		3,911	4,400	
	0120	15,600 CFM, 10 HP	"	1.60	12.500	"	6,475	450		6,925	7,825	
	0200	In-line centrifugal, supply/exhaust booster										
	0220	aluminum wheel/hub, disconnect switch, 1/4" S.P.										
	0240	500 CFM, 10" diameter connection 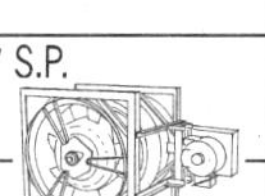	Q-20	3	6.667	Ea.	820	239		1,059	1,275	
	0280	1,520 CFM, 16" diameter connection		2	10		1,225	360		1,585	1,925	
	0320	3,480 CFM, 20" diameter connection	↓	.80	25	↓	1,225	895		2,120	2,750	

	15830 | Fans	CREW	DAILY OUTPUT	LABOR-HOURS	UNIT	2004 BARE COSTS MAT.	LABOR	EQUIP.	TOTAL	TOTAL INCL O&P	
100	0326 5,080 CFM, 20" diameter connection	Q-20	.75	26.667	Ea.	1,325	955		2,280	2,975	100
	2500 Ceiling fan, right angle, extra quiet, 0.10" S.P.										
	2540 210 CFM	Q-20	19	1.053	Ea.	153	37.50		190.50	229	
	2580 885 CFM		16	1.250		385	45		430	490	
	2620 2,960 CFM	↓	11	1.818		710	65		775	880	
	2640 For wall or roof cap, add	1 Shee	16	.500	↓	130	19.40		149.40	174	
	4500 Corrosive fume resistant, plastic										
	4600 roof ventilators, centrifugal, V belt drive, motor										
	4620 1/4" S.P., 250 CFM, 1/4 HP	Q-20	6	3.333	Ea.	2,425	119		2,544	2,850	
	4640 895 CFM, 1/3 HP		5	4		2,625	143		2,768	3,100	
	4660 1630 CFM, 1/2 HP		4	5		3,100	179		3,279	3,675	
	4680 2240 CFM, 1 HP	↓	3	6.667	↓	3,225	239		3,464	3,925	
	5000 Utility set, centrifugal, V belt drive, motor										
	5020 1/4" S.P., 1200 CFM, 1/4 HP	Q-20	6	3.333	Ea.	2,550	119		2,669	3,000	
	5040 1520 CFM, 1/3 HP		5	4		2,550	143		2,693	3,025	
	5060 1850 CFM, 1/2 HP		4	5		2,550	179		2,729	3,100	
	5080 2180 CFM, 3/4 HP	↓	3	6.667	↓	2,600	239		2,839	3,225	
	6650 Residential, bath exhaust, grille, back draft damper										
	6660 50 CFM	Q-20	24	.833	Ea.	30	30		60	80	
	6670 110 CFM		22	.909		50	32.50		82.50	107	
	6680 Light combination, squirrel cage, 100 watt, 70 CFM	↓	24	.833	↓	61.50	30		91.50	115	
	6700 Light/heater combination, ceiling mounted										
	6710 70 CFM, 1450 watt	Q-20	24	.833	Ea.	74.50	30		104.50	129	
	6800 Heater combination, recessed, 70 CFM		24	.833		35.50	30		65.50	86	
	6820 With 2 infrared bulbs		23	.870		53.50	31		84.50	108	
	6900 Kitchen exhaust, grille, complete, 160 CFM		22	.909		62.50	32.50		95	120	
	6910 180 CFM		20	1		53	36		89	115	
	6920 270 CFM	↓	18	1.111	↓	95.50	40		135.50	168	
	6940 Residential roof jacks and wall caps										
	6944 Wall cap with back draft damper										
	6946 3" & 4" dia. round duct	1 Shee	11	.727	Ea.	12.80	28		40.80	58.50	
	6948 6" dia. round duct	"	11	.727	"	31	28		59	78.50	
	6958 Roof jack with bird screen and back draft damper										
	6960 3" & 4" dia. round duct	1 Shee	11	.727	Ea.	12.30	28		40.30	58	
	6962 3-1/4" x 10" rectangular duct	"	10	.800	"	22.50	31		53.50	74	
	6980 Transition										
	6982 3-1/4" x 10" to 6" dia. round	1 Shee	20	.400	Ea.	13.85	15.50		29.35	40	
	7100 Direct drive, 320 CFM, 11" sq. damper	Q-20	7	2.857		320	102		422	510	
	7120 600 CFM, 11" sq. damper	"	6	3.333	↓	325	119		444	545	
	8000 Ventilation, residential										
	8020 Attic, roof type										
	8030 Aluminum dome, damper & curb										
	8040 6" diameter, 300 CFM	1 Elec	16	.500	Ea.	238	19.70		257.70	292	
	8050 7" diameter, 450 CFM		15	.533		259	21		280	320	
	8060 9" diameter, 900 CFM		14	.571		415	22.50		437.50	495	
	8080 12" diameter, 1000 CFM (gravity)		10	.800		268	31.50		299.50	345	
	8090 16" diameter, 1500 CFM (gravity)		9	.889		325	35		360	410	
	8100 20" diameter, 2500 CFM (gravity)		8	1		395	39.50		434.50	495	
	8110 26" diameter, 4000 CFM (gravity)		7	1.143		480	45		525	600	
	8120 32" diameter, 6500 CFM (gravity)		6	1.333		660	52.50		712.50	805	
	8130 38" diameter, 8000 CFM (gravity)		5	1.600		975	63		1,038	1,175	
	8140 50" diameter, 13,000 CFM (gravity)	↓	4	2	↓	1,425	79		1,504	1,675	
	8160 Plastic, ABS dome										
	8180 1050 CFM	1 Elec	14	.571	Ea.	79	22.50		101.50	121	
	8200 1600 CFM	"	12	.667	"	118	26.50		144.50	171	
	8240 Attic, wall type, with shutter, one speed										

15 MECHANICAL

15830 | Fans

	Line	Description	Crew	Daily Output	Labor-Hours	Unit	2004 Bare Costs Mat.	Labor	Equip.	Total	Total Incl O&P
100	8250	12" diameter, 1000 CFM	1 Elec	14	.571	Ea.	187	22.50		209.50	241
	8260	14" diameter, 1500 CFM		12	.667		203	26.50		229.50	264
	8270	16" diameter, 2000 CFM		9	.889		229	35		264	305
	8290	Whole house, wall type, with shutter, one speed									
	8300	30" diameter, 4800 CFM	1 Elec	7	1.143	Ea.	490	45		535	610
	8310	36" diameter, 7000 CFM		6	1.333		535	52.50		587.50	670
	8320	42" diameter, 10,000 CFM		5	1.600		600	63		663	755
	8330	48" diameter, 16,000 CFM		4	2		745	79		824	940
	8340	For two speed, add					44.50			44.50	49
	8350	Whole house, lay-down type, with shutter, one speed									
	8360	30" diameter, 4500 CFM	1 Elec	8	1	Ea.	525	39.50		564.50	635
	8370	36" diameter, 6500 CFM		7	1.143		560	45		605	690
	8380	42" diameter, 9000 CFM		6	1.333		620	52.50		672.50	760
	8390	48" diameter, 12,000 CFM		5	1.600		700	63		763	865
	8440	For two speed, add					33.50			33.50	36.50
	8450	For 12 hour timer switch, add	1 Elec	32	.250		33.50	9.85		43.35	51.50
	9900	Minimum labor/equipment charge	"	4	2	Job		79		79	121

15840 | Air Terminal Units

	Line	Description	Crew	Daily Output	Labor-Hours	Unit	2004 Bare Costs Mat.	Labor	Equip.	Total	Total Incl O&P
500	0010	**CONSTANT VOLUME MIXING BOXES**									
	5180	Mixing box, includes electric or pneumatic motor									
	5190	Recommend use with attenuator, see 15820-300-9000									
	5200	Constant volume, 150 to 270 CFM	Q-9	12	1.333	Ea.	500	46.50		546.50	630
	5210	270 to 600 CFM	"	11	1.455	"	515	51		566	650
700	0010	**VARIABLE VOLUME MIXING BOXES**									
	5500	VAV Cool only, pneum. press indep. 300 to 600 CFM	Q-9	11	1.455	Ea.	281	51		332	390
	5510	500 to 1000 CFM		9	1.778		293	62		355	425
	5520	800 to 1600 CFM		9	1.778		305	62		367	435
	5530	1100 to 2000 CFM		8	2		320	70		390	460
	5540	1500 to 3000 CFM		7	2.286		335	80		415	495
	5550	2000 to 4000 CFM		6	2.667		355	93		448	540
	5560	For electric, w/thermostat, press. dependent, add					20			20	22

15850 | Air Outlets & Inlets

	Line	Description	Crew	Daily Output	Labor-Hours	Unit	2004 Bare Costs Mat.	Labor	Equip.	Total	Total Incl O&P
300	0010	**DIFFUSERS** Aluminum, opposed blade damper unless noted									
	0100	Ceiling, linear, also for sidewall									
	0120	2" wide	1 Shee	32	.250	L.F.	27.50	9.70		37.20	45.50
	0160	4" wide		26	.308		36	11.95		47.95	58.50
	0180	6" wide		24	.333		43	12.95		55.95	68
	0200	8" wide		22	.364		49	14.10		63.10	76.50
	0500	Perforated, 24" x 24" lay-in panel size, 6" x 6"		16	.500	Ea.	74	19.40		93.40	112
	0520	8" x 8"		15	.533		76	20.50		96.50	117
	0530	9" x 9"		14	.571		78	22		100	121
	0540	10" x 10"		14	.571		78	22		100	121
	0560	12" x 12"		12	.667		80	26		106	129
	0590	16" x 16"		11	.727		94.50	28		122.50	149
	0600	18" x 18"		10	.800		99.50	31		130.50	159
	0610	20" x 20"		10	.800		114	31		145	174
	0620	24" x 24"		9	.889		132	34.50		166.50	200
	1000	Rectangular, 1 to 4 way blow, 6" x 6"		16	.500		39.50	19.40		58.90	74
	1010	8" x 8"		15	.533		47	20.50		67.50	85
	1014	9" x 9"		15	.533		50.50	20.50		71	88.50
	1016	10" x 10"		15	.533		59.50	20.50		80	98.50
	1020	12" x 6"		15	.533		67	20.50		87.50	107

15850 | Air Outlets & Inlets

			Crew	Daily Output	Labor-Hours	Unit	2004 Bare Costs Mat.	Labor	Equip.	Total	Total Incl O&P	
300	1040	12″ x 9″	1 Shee	14	.571	Ea.	73	22		95	115	300
	1060	12″ x 12″		12	.667		69.50	26		95.50	118	
	1070	14″ x 6″		13	.615		72	24		96	118	
	1074	14″ x 14″		12	.667		101	26		127	152	
	1150	18″ x 18″		9	.889		134	34.50		168.50	203	
	1170	24″ x 12″		10	.800		121	31		152	182	
	2000	T bar mounting, 24″ x 24″ lay-in frame, 6″ x 6″		16	.500		82	19.40		101.40	121	
	2020	9″ x 9″		14	.571		90.50	22		112.50	135	
	2040	12″ x 12″		12	.667		117	26		143	170	
	2060	15″ x 15″		11	.727		150	28		178	210	
	2080	18″ x 18″		10	.800		157	31		188	221	
	6000	For steel diffusers instead of aluminum, deduct					10%					
	9000	Minimum labor/equipment charge	1 Shee	4	2	Job		77.50		77.50	123	
500	0010	**GRILLES**										500
	0020	Aluminum										
	1000	Air return, 6″ x 6″	1 Shee	26	.308	Ea.	13	11.95		24.95	33	
	1020	10″ x 6″		24	.333		15.60	12.95		28.55	37.50	
	1080	16″ x 8″		22	.364		23.50	14.10		37.60	48	
	1100	12″ x 12″		22	.364		23.50	14.10		37.60	48	
	1120	24″ x 12″		18	.444		41.50	17.25		58.75	73.50	
	1300	48″ x 24″		12	.667		161	26		187	218	
	9000	Minimum labor/equipment charge		4	2	Job		77.50		77.50	123	
700	0010	**REGISTERS**										700
	0020											
	0980	Air supply										
	1000	Ceiling/wall, O.B. damper, anodized aluminum										
	1010	One or two way deflection, adj. curved face bars										
	1140	14″ x 8″	1 Shee	17	.471	Ea.	32	18.25		50.25	64	
	3000	Baseboard, hand adj. damper, enameled steel										
	3012	8″ x 6″	1 Shee	26	.308	Ea.	12.05	11.95		24	32	
	3020	10″ x 6″		24	.333		13.60	12.95		26.55	35.50	
	3040	12″ x 5″		23	.348		15.55	13.50		29.05	38.50	
	3060	12″ x 6″		23	.348		14.35	13.50		27.85	37.50	
	3080	12″ x 8″		22	.364		21	14.10		35.10	45.50	
	3100	14″ x 6″		20	.400		15.55	15.50		31.05	41.50	
	9000	Minimum labor/equipment charge		4	2	Job		77.50		77.50	123	
800	0010	**VENTILATORS** Base & damper										800
	9000	Minimum labor/equipment charge	1 Plum	2	4	Job		158		158	246	

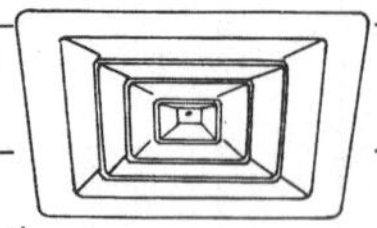

15860 | Air Cleaning Devices

			Crew	Daily Output	Labor-Hours	Unit	2004 Bare Costs Mat.	Labor	Equip.	Total	Total Incl O&P	
100	0010	**AIR FILTERS**										100
	0020											
	0050	Activated charcoal type, full flow				MCFM	600			600	660	
	0060	Activated charcoal type, full flow, impregnated media 12″ deep					175			175	193	
	0070	Activated charcoal type, HEPA filter & frame for field erection					175			175	193	
	0080	Activated charcoal type, HEPA filter-diffuser, ceiling install.					165			165	182	
	2000	Electronic air cleaner, duct mounted										
	2150	400 - 1000 CFM	1 Shee	2.30	3.478	Ea.	675	135		810	955	
	2200	1000 - 1400 CFM		2.20	3.636		705	141		846	1,000	
	2250	1400 - 2000 CFM		2.10	3.810		780	148		928	1,100	
	2950	Mechanical media filtration units										
	3000	High efficiency type, with frame, non-supported				MCFM	45			45	49.50	
	3100	Supported type					55			55	60.50	
	4000	Medium efficiency, extended surface					5			5	5.50	

Important: See the Reference Section for critical supporting data - Reference Nos., Crews, & City Cost Indexes

		15860 | Air Cleaning Devices	CREW	DAILY OUTPUT	LABOR-HOURS	UNIT	2004 BARE COSTS MAT.	LABOR	EQUIP.	TOTAL	TOTAL INCL O&P	
100	4500	Permanent washable				MCFM	20			20	22	100
	5000	Renewable disposable roll				↓	120			120	132	
	5500	Throwaway glass or paper media type				Ea.	4.60			4.60	5.05	
	9000	Minimum labor/equipment charge	1 Shee	2.25	3.556	Job		138		138	218	

For information about Means Estimating Seminars, see yellow pages 12 and 13 in back of book

Division Notes

		CREW	DAILY OUTPUT	LABOR-HOURS	UNIT	2004 BARE COSTS MAT.	LABOR	EQUIP.	TOTAL	TOTAL INCL O&P

Division 16 Electrical

Estimating Tips

16060 Grounding & Bonding

- When taking off grounding system, identify separately the type and size of wire and list each unique type of ground connection.

16100 Wiring Methods

- Conduit should be taken off in three main categories: power distribution, branch power, and branch lighting, so the estimator can concentrate on systems and components, therefore making it easier to ensure all items have been accounted for.
- For cost modifications for elevated conduit installation, add the percentages to labor according to the height of installation and only the quantities exceeding the different height levels, not to the total conduit quantities.
- Remember that aluminum wiring of equal ampacity is larger in diameter than copper and may require larger conduit.
- If more than three wires at a time are being pulled, deduct percentages from the labor hours of that grouping of wires.
- The estimator should take the weights of materials into consideration when completing a takeoff. Topics to consider include: How will the materials be supported? What methods of support are available? How high will the support structure have to reach? Will the final support structure be able to withstand the total burden? Is the support material included or separate from the fixture, equipment and material specified?

16200 Electrical Power

- Do not overlook the costs for equipment used in the installation. If scaffolding or highlifts are available in the field, contractors may use them in lieu of the proposed ladders and rolling staging.

16400 Low-Voltage Distribution

- Supports and concrete pads may be shown on drawings for the larger equipment, or the support system may be just a piece of plywood for the back of a panelboard. In either case, it must be included in the costs.

16500 Lighting

- Fixtures should be taken off room by room, using the fixture schedule, specifications, and the ceiling plan. For large concentrations of lighting fixtures in the same area deduct the percentages from labor hours.

16700 Communications
16800 Sound & Video

- When estimating material costs for special systems, it is always prudent to obtain manufacturers' quotations for equipment prices and special installation requirements which will affect the total costs.

Reference Numbers

Reference numbers are shown in bold squares at the beginning of some major classifications. These numbers refer to related items in the Reference Section. The reference information may be an estimating procedure, an alternate pricing method or technical information.

Note: Not all subdivisions listed here necessarily appear in this publication.

16055 | Selective Demolition

		Description	Crew	Daily Output	Labor-Hours	Unit	2004 Bare Costs Mat.	Labor	Equip.	Total	Total Incl O&P	
300	0010	**ELECTRICAL DEMOLITION** R16055-300										300
	0020	Conduit to 15′ high, including fittings & hangers										
	0100	Rigid galvanized steel, 1/2″ to 1″ diameter	1 Elec	242	.033	L.F.		1.30		1.30	2	
	0120	1-1/4″ to 2″		200	.040			1.58		1.58	2.42	
	0140	2″ to 4″		151	.053			2.09		2.09	3.21	
	0200	Electric metallic tubing (EMT) 1/2″ to 1″		394	.020			.80		.80	1.23	
	0220	1-1/4″ to 1-1/2″		326	.025			.97		.97	1.49	
	0240	2″ to 3″		236	.034			1.34		1.34	2.05	
	0400	Wiremold raceway, including fittings & hangers										
	0420	No. 3000	1 Elec	250	.032	L.F.		1.26		1.26	1.94	
	0440	No. 4000	″	217	.037	″		1.45		1.45	2.23	
	0500	Channels, steel, including fittings & hangers										
	0520	3/4″ x 1-1/2″	1 Elec	308	.026	L.F.		1.02		1.02	1.57	
	0540	1-1/2″ x 1-1/2″	″	269	.030	″		1.17		1.17	1.80	
	0600	Copper bus duct, indoor, 3 phase										
	0610	Including hangers & supports										
	0620	225 amp	2 Elec	135	.119	L.F.		4.67		4.67	7.20	
	0640	400 amp		106	.151			5.95		5.95	9.15	
	0660	600 amp		86	.186			7.35		7.35	11.25	
	0680	1000 amp		60	.267			10.50		10.50	16.15	
	0700	1600 amp		40	.400			15.75		15.75	24	
	0720	3000 amp		10	1.600			63		63	97	
	0800	Plug-in switches, 600V 3 ph, incl. disconnecting										
	0820	wire, conduit terminations, 30 amp	1 Elec	15.50	.516	Ea.		20.50		20.50	31.50	
	0840	60 amp		13.90	.576			22.50		22.50	35	
	0850	100 amp		10.40	.769			30.50		30.50	46.50	
	0860	200 amp		6.20	1.290			51		51	78	
	0940	1200 amp	2 Elec	2	8			315		315	485	
	0960	1600 amp	″	1.70	9.412			370		370	570	
	1010	Safety switches, 250 or 600V, incl. disconnection										
	1050	of wire & conduit terminations										
	1100	30 amp	1 Elec	12.30	.650	Ea.		25.50		25.50	39.50	
	1120	60 amp		8.80	.909			36		36	55	
	1140	100 amp		7.30	1.096			43		43	66.50	
	1160	200 amp		5	1.600			63		63	97	
	1210	Panel boards, incl. removal of all breakers,										
	1220	conduit terminations & wire connections										
	1230	3 wire, 120/240V, 100A, to 20 circuits	1 Elec	2.60	3.077	Ea.		121		121	186	
	1240	200 amps, to 42 circuits		1.30	6.154			242		242	375	
	1260	4 wire, 120/208V, 125A, to 20 circuits		2.40	3.333			131		131	202	
	1270	200 amps, to 42 circuits		1.20	6.667			263		263	405	
	1300	Transformer, dry type, 1 ph, incl. removal of										
	1320	supports, wire & conduit terminations										
	1340	1 kVA	1 Elec	7.70	1.039	Ea.		41		41	63	
	1360	5 kVA		4.70	1.702			67		67	103	
	1420	75 kVA		1.25	6.400			252		252	390	
	1440	3 Phase to 600V, primary										
	1460	3 kVA	1 Elec	3.85	2.078	Ea.		82		82	126	
	1480	15 kVA	2 Elec	4.20	3.810			150		150	231	
	1500	30 kVA	″	3.50	4.571			180		180	277	
	1530	112.5 kVA	R-3	2.90	6.897			267	55	322	470	
	1560	500 kVA		1.40	14.286			555	114	669	980	
	1570	750 kVA		1.10	18.182			705	145	850	1,225	
	1600	Pull boxes & cabinets, sheet metal, incl. removal										
	1620	of supports and conduit terminations										
	1640	6″ x 6″ x 4″	1 Elec	31.10	.257	Ea.		10.15		10.15	15.60	

16055 | Selective Demolition

300

	16055 Selective Demolition	CREW	DAILY OUTPUT	LABOR-HOURS	UNIT	2004 BARE COSTS MAT.	LABOR	EQUIP.	TOTAL	TOTAL INCL O&P
1660	12″ x 12″ x 4″ (R16055 -300)	1 Elec	23.30	.343	Ea.		13.55		13.55	21
1720	Junction boxes, 4″ sq. & oct.		80	.100			3.94		3.94	6.05
1740	Handy box		107	.075			2.95		2.95	4.53
1760	Switch box		107	.075			2.95		2.95	4.53
1780	Receptacle & switch plates		257	.031			1.23		1.23	1.88
1800	Wire, THW-THWN-THHN, removed from									
1810	in place conduit, to 15′ high									
1830	#14	1 Elec	65	.123	C.L.F.		4.85		4.85	7.45
1840	#12		55	.145			5.75		5.75	8.80
1850	#10		45.50	.176			6.95		6.95	10.65
1880	#4	2 Elec	53	.302			11.90		11.90	18.30
1890	#3		50	.320			12.60		12.60	19.40
1910	1/0		33.20	.482			19		19	29
1920	2/0		29.20	.548			21.50		21.50	33
1930	3/0		25	.640			25		25	39
1980	400 kcmil		17	.941			37		37	57
1990	500 kcmil		16.20	.988			39		39	60
2000	Interior fluorescent fixtures, incl. supports									
2010	& whips, to 15′ high									
2100	Recessed drop-in 2′ x 2′, 2 lamp	2 Elec	35	.457	Ea.		18		18	27.50
2120	2′ x 4′, 2 lamp		33	.485			19.10		19.10	29.50
2140	2′ x 4′, 4 lamp		30	.533			21		21	32.50
2160	4′ x 4′, 4 lamp		20	.800			31.50		31.50	48.50
2180	Surface mount, acrylic lens & hinged frame									
2200	1′ x 4′, 2 lamp	2 Elec	44	.364	Ea.		14.35		14.35	22
2220	2′ x 2′, 2 lamp		44	.364			14.35		14.35	22
2260	2′ x 4′, 4 lamp		33	.485			19.10		19.10	29.50
2280	4′ x 4′, 4 lamp		40	.400			15.75		15.75	24
2300	Strip fixtures, surface mount									
2320	4′ long, 1 lamp	2 Elec	53	.302	Ea.		11.90		11.90	18.30
2340	4′ long, 2 lamp		50	.320			12.60		12.60	19.40
2360	8′ long, 1 lamp		42	.381			15		15	23
2380	8′ long, 2 lamp		40	.400			15.75		15.75	24
2400	Pendant mount, industrial, incl. removal									
2410	of chain or rod hangers, to 15′ high									
2420	4′ long, 2 lamp	2 Elec	35	.457	Ea.		18		18	27.50
2440	8′ long, 2 lamp	"	27	.593	"		23.50		23.50	36
2460	Interior incandescent, surface, ceiling									
2470	or wall mount, to 12′ high									
2480	Metal cylinder type, 75 Watt	2 Elec	62	.258	Ea.		10.15		10.15	15.65
2500	150 Watt	"	62	.258	"		10.15		10.15	15.65
2520	Metal halide, high bay									
2540	400 Watt	2 Elec	15	1.067	Ea.		42		42	64.50
2560	1000 Watt		12	1.333			52.50		52.50	80.50
2580	150 Watt, low bay		20	.800			31.50		31.50	48.50
2600	Exterior fixtures, incandescent, wall mount									
2620	100 Watt	2 Elec	50	.320	Ea.		12.60		12.60	19.40
2640	Quartz, 500 Watt		33	.485			19.10		19.10	29.50
2660	1500 Watt		27	.593			23.50		23.50	36
2680	Wall pack, mercury vapor									
2700	175 Watt	2 Elec	25	.640	Ea.		25		25	39
2720	250 Watt	"	25	.640	"		25		25	39
9000	Minimum labor/equipment charge	1 Elec	4	2	Job		79		79	121

16050 | Basic Electrical Materials & Methods

16060 | Grounding & Bonding

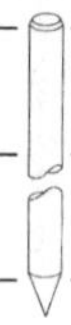

		16060 Grounding & Bonding	CREW	DAILY OUTPUT	LABOR-HOURS	UNIT	2004 BARE COSTS MAT.	LABOR	EQUIP.	TOTAL	TOTAL INCL O&P	
800	0010	**GROUNDING**										800
	0030	Rod, copper clad, 8′ long, 1/2″ diameter	1 Elec	5.50	1.455	Ea.	13.10	57.50		70.60	102	
	0040	5/8″ diameter		5.50	1.455		14.15	57.50		71.65	104	
	0050	3/4″ diameter		5.30	1.509		27	59.50		86.50	121	
	0080	10′ long, 1/2″ diameter		4.80	1.667		16.90	65.50		82.40	120	
	0090	5/8″ diameter		4.60	1.739		20.50	68.50		89	128	
	0100	3/4″ diameter		4.40	1.818		30	71.50		101.50	143	
	0130	15′ long, 3/4″ diameter		4	2		79	79		158	208	
	0260	Wire ground bare armored, #8-1 conductor		2	4	C.L.F.	72	158		230	320	
	0280	#4-1 conductor		1.60	5		133	197		330	450	
	0390	Bare copper wire, #8 stranded		11	.727		9.80	28.50		38.30	55	
	0400	#6		10	.800		16.60	31.50		48.10	67	
	0600	#2	2 Elec	10	1.600		38.50	63		101.50	140	
	0800	3/0		6.60	2.424		91.50	95.50		187	248	
	1000	4/0		5.70	2.807		114	111		225	295	
	1200	250 kcmil	3 Elec	7.20	3.333		135	131		266	350	
	1800	Water pipe ground clamps, heavy duty										
	2000	Bronze, 1/2″ to 1″ diameter	1 Elec	8	1	Ea.	13.05	39.50		52.55	75	
	2100	1-1/4″ to 2″ diameter		8	1		17.20	39.50		56.70	79.50	
	2200	2-1/2″ to 3″ diameter		6	1.333		35.50	52.50		88	120	
	2800	Brazed connections, #6 wire		12	.667		10.75	26.50		37.25	52.50	
	3000	#2 wire		10	.800		14.40	31.50		45.90	64.50	
	3100	3/0 wire		8	1		21.50	39.50		61	84.50	
	3200	4/0 wire		7	1.143		24.50	45		69.50	96	
	3400	250 kcmil wire		5	1.600		29	63		92	129	
	3600	500 kcmil wire		4	2		35.50	79		114.50	160	
	9000	Minimum labor/equipment charge		4	2	Job		79		79	121	

16100 | Wiring Methods

16120 | Conductors & Cables

		16120 Conductors & Cables	CREW	DAILY OUTPUT	LABOR-HOURS	UNIT	2004 BARE COSTS MAT.	LABOR	EQUIP.	TOTAL	TOTAL INCL O&P	
120	0010	**ARMORED CABLE**										120
	0020											
	0050	600 volt, copper (BX), #14, 2 conductor, solid	1 Elec	2.40	3.333	C.L.F.	52	131		183	260	
	0100	3 conductor, solid		2.20	3.636		69.50	143		212.50	297	
	0152	#12, 2 conductor, solid		2.10	3.810		53	150		203	290	
	0202	3 conductor, solid		1.80	4.444		83.50	175		258.50	360	
	0252	#10, 2 conductor, solid		1.80	4.444		96.50	175		271.50	375	
	0302	3 conductor, solid		1.50	5.333		130	210		340	470	
	0352	#8, 3 conductor, solid		1.20	6.667		203	263		466	630	
	9010	600 volt, copper (MC) steel clad, #14, 2 wire		2.40	3.333		50.50	131		181.50	258	
	9020	3 wire		2.20	3.636		67	143		210	294	
	9040	#12, 2 wire		2.30	3.478		51	137		188	267	
	9050	3 wire		2	4		79	158		237	330	
	9070	#10, 2 wire		2	4		90	158		248	340	
	9080	3 wire		1.60	5		140	197		337	460	
	9100	#8, 2 wire, stranded		1.80	4.444		172	175		347	460	
	9110	3 wire, stranded		1.30	6.154		251	242		493	650	
	9900	Minimum labor/equipment charge		4	2	Job		79		79	121	

16120 | Conductors & Cables

		16120 Conductors & Cables	CREW	DAILY OUTPUT	LABOR-HOURS	UNIT	2004 BARE COSTS MAT.	LABOR	EQUIP.	TOTAL	TOTAL INCL O&P	
500	0010	**MINERAL INSULATED CABLE** 600 volt										500
	0100	1 conductor, #12	1 Elec	1.60	5	C.L.F.	208	197		405	535	
	1500	2 conductor, #12		1.40	5.714		430	225		655	820	
	1600	#10		1.20	6.667		520	263		783	975	
	1800	#8		1.10	7.273		650	287		937	1,150	
	2000	#6		1.05	7.619		825	300		1,125	1,375	
	2100	#4	2 Elec	2	8		1,075	315		1,390	1,675	
	2200	3 conductor, #12	1 Elec	1.20	6.667		540	263		803	1,000	
	2400	#10		1.10	7.273		625	287		912	1,125	
	2600	#8		1.05	7.619		755	300		1,055	1,300	
	2800	#6		1	8		980	315		1,295	1,550	
	3000	#4	2 Elec	1.80	8.889		1,300	350		1,650	1,975	
	3100	4 conductor, #12	1 Elec	1.20	6.667		575	263		838	1,050	
	3200	#10		1.10	7.273		680	287		967	1,200	
	3400	#8		1	8		890	315		1,205	1,475	
	3600	#6		.90	8.889		1,125	350		1,475	1,775	
	3620	7 conductor, #12		1.10	7.273		725	287		1,012	1,250	
	3640	#10		1	8		910	315		1,225	1,475	
	3800	M.I. terminations, 600 volt, 1 conductor, #12		8	1	Ea.	8.45	39.50		47.95	70	
	5500	2 conductor, #12		6.70	1.194		8.45	47		55.45	82	
	5600	#10		6.40	1.250		12.70	49.50		62.20	89.50	
	5800	#8		6.20	1.290		12.70	51		63.70	92	
	6000	#6		5.70	1.404		12.70	55.50		68.20	99	
	6200	#4		5.30	1.509		28	59.50		87.50	123	
	6400	3 conductor, #12		5.70	1.404		12.70	55.50		68.20	99	
	6500	#10		5.50	1.455		12.70	57.50		70.20	102	
	6600	#8		5.20	1.538		12.70	60.50		73.20	107	
	6800	#6		4.80	1.667		12.70	65.50		78.20	115	
	7200	#4		4.60	1.739		29	68.50		97.50	137	
	7400	4 conductor, #12		4.60	1.739		14.10	68.50		82.60	121	
	7500	#10		4.40	1.818		14.10	71.50		85.60	126	
	7600	#8		4.20	1.905		14.10	75		89.10	131	
	8400	#6		4	2		29.50	79		108.50	154	
	8500	7 conductor, #12		3.50	2.286		14.10	90		104.10	154	
	8600	#10		3	2.667		29.50	105		134.50	194	
	8800	Crimping tool, plier type					40			40	44	
	9000	Stripping tool					100			100	110	
	9200	Hand vise					45			45	49.50	
	9500	Minimum labor/equipment charge	1 Elec	4	2	Job		79		79	121	
550	0010	**NON-METALLIC SHEATHED CABLE** 600 volt										550
	0100	Copper with ground wire, (Romex)										
	0152	#14, 2 wire	1 Elec	2.50	3.200	C.L.F.	12.25	126		138.25	207	
	0202	3 wire 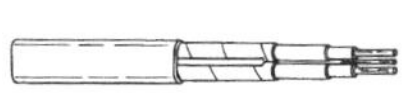		2.30	3.478		21.50	137		158.50	235	
	0252	#12, 2 wire		2.20	3.636		17.25	143		160.25	239	
	0302	3 wire		2	4		31	158		189	276	
	0352	#10, 2 wire		2	4		34	158		192	279	
	0402	3 wire		1.40	5.714		47.50	225		272.50	395	
	0452	#8, 3 conductor		1.30	6.154		91	242		333	475	
	0502	#6, 3 wire		1.20	6.667		145	263		408	565	
	0550	SE type SER aluminum cable, 3 RHW and										
	0602	1 bare neutral, 3 #8 & 1 #8	1 Elec	1.50	5.333	C.L.F.	92.50	210		302.50	425	
	0652	3 #6 & 1 #6	"	1.30	6.154		105	242		347	490	
	0702	3 #4 & 1 #6	2 Elec	2.20	7.273		117	287		404	570	
	0752	3 #2 & 1 #4		2	8		173	315		488	675	
	0802	3 #1/0 & 1 #2		1.80	8.889		261	350		611	825	

ELECTRICAL 16

		16120 | Conductors & Cables	CREW	DAILY OUTPUT	LABOR-HOURS	UNIT	2004 BARE COSTS MAT.	LABOR	EQUIP.	TOTAL	TOTAL INCL O&P	
550	0852	3 #2/0 & 1 #1	2 Elec	1.60	10	C.L.F.	310	395		705	945	550
	0902	3 #4/0 & 1 #2/0		1.40	11.429		440	450		890	1,175	
	6500	Service entrance cap for copper SEU										
	6700	150 amp	1 Elec	10	.800	Ea.	12.10	31.50		43.60	62	
	6800	200 amp		8	1	"	18.30	39.50		57.80	80.50	
	9000	Minimum labor/equipment charge		4	2	Job		79		79	121	
900	0010	**WIRE**										900
	0020	600 volt type THW, copper, solid, #14	1 Elec	13	.615	C.L.F.	2.95	24.50		27.45	41	
	0030	#12		11	.727		4.15	28.50		32.65	48.50	
	0040	#10 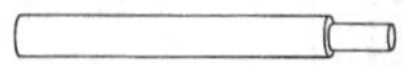		10	.800		6.50	31.50		38	55.50	
	0051	Wire, 600 volt, stranded R16132-210										
	0140	#8	1 Elec	8	1	C.L.F.	12.65	39.50		52.15	74.50	
	0160	#6	"	6.50	1.231		20	48.50		68.50	96.50	
	0180	#4	2 Elec	10.60	1.509		30.50	59.50		90	125	
	0200	#3		10	1.600		37.50	63		100.50	139	
	0220	#2		9	1.778		47	70		117	160	
	0240	#1		8	2		59.50	79		138.50	187	
	0260	1/0		6.60	2.424		72	95.50		167.50	226	
	0280	2/0		5.80	2.759		88.50	109		197.50	264	
	0300	3/0		5	3.200		109	126		235	315	
	0350	4/0		4.40	3.636		138	143		281	370	
	0400	250 kcmil	3 Elec	6	4		168	158		326	425	
	0420	300 kcmil		5.70	4.211		199	166		365	475	
	0450	350 kcmil		5.40	4.444		230	175		405	520	
	0480	400 kcmil		5.10	4.706		273	185		458	585	
	0490	500 kcmil		4.80	5		320	197		517	655	
	0510	750 kcmil		3.30	7.273		520	287		807	1,000	
	0530	Aluminum, stranded, #8	1 Elec	9	.889		12.45	35		47.45	67.50	
	0540	#6	"	8	1		17.05	39.50		56.55	79.50	
	0560	#4	2 Elec	13	1.231		21	48.50		69.50	98	
	0580	#2		10.60	1.509		29	59.50		88.50	123	
	0600	#1 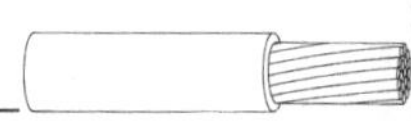		9	1.778		42	70		112	154	
	0620	1/0		8	2		50.50	79		129.50	177	
	0640	2/0		7.20	2.222		59.50	87.50		147	201	
	0680	3/0		6.60	2.424		74	95.50		169.50	229	
	0700	4/0		6.20	2.581		82.50	102		184.50	247	
	0720	250 kcmil	3 Elec	8.70	2.759		100	109		209	277	
	0740	300 kcmil		8.10	2.963		139	117		256	330	
	0760	350 kcmil		7.50	3.200		141	126		267	350	
	0780	400 kcmil		6.90	3.478		165	137		302	395	
	0800	500 kcmil		6	4		182	158		340	440	
	0850	600 kcmil		5.70	4.211		230	166		396	510	
	0880	700 kcmil		5.10	4.706		262	185		447	575	
	9000	Minimum labor/equipment charge	1 Elec	4	2	Job		79		79	121	
		16132 | Conduit & Tubing										
205	0010	**CONDUIT** To 15′ high, includes 2 terminations, 2 elbows and R16132-210										205
	0020	11 beam clamps per 100 L.F.										
	1161	Field bends, 45° to 90°, 1/2″ diameter	1 Elec	53	.151	Ea.		5.95		5.95	9.15	
	1162	3/4″ diameter		47	.170			6.70		6.70	10.30	
	1163	1″ diameter		44	.182			7.15		7.15	11	
	1164	1-1/4″ diameter		23	.348			13.70		13.70	21	
	1165	1-1/2″ diameter		21	.381			15		15	23	
	1166	2″ diameter		16	.500			19.70		19.70	30.50	
	1991	Field bends, 45° to 90°, 1/2″ diameter		44	.182			7.15		7.15	11	
	1992	3/4″ diameter		40	.200			7.90		7.90	12.10	

 Important: See the Reference Section for critical supporting data - Reference Nos., Crews, & City Cost Indexes

16132 | Conduit & Tubing

		16132 Conduit & Tubing	CREW	DAILY OUTPUT	LABOR-HOURS	UNIT	2004 BARE COSTS MAT.	LABOR	EQUIP.	TOTAL	TOTAL INCL O&P	
205	1993	1" diameter R16132-210	1 Elec	36	.222	Ea.		8.75		8.75	13.45	205
	1994	1-1/4" diameter		19	.421			16.60		16.60	25.50	
	1995	1-1/2" diameter		18	.444			17.50		17.50	27	
	1996	2" diameter		13	.615	↓		24.50		24.50	37.50	
	2500	Steel, intermediate conduit (IMC), 1/2" diameter		100	.080	L.F.	1.37	3.15		4.52	6.35	
	2530	3/4" diameter		90	.089		1.62	3.50		5.12	7.20	
	2550	1" diameter		70	.114		2.26	4.50		6.76	9.40	
	2570	1-1/4" diameter		65	.123		2.98	4.85		7.83	10.75	
	2600	1-1/2" diameter		60	.133		3.44	5.25		8.69	11.85	
	2630	2" diameter		50	.160		4.44	6.30		10.74	14.60	
	2650	2-1/2" diameter	↓	40	.200		8.95	7.90		16.85	22	
	2670	3" diameter	2 Elec	60	.267		11.50	10.50		22	29	
	2700	3-1/2" diameter		54	.296		13.15	11.65		24.80	32.50	
	2730	4" diameter	↓	50	.320	↓	16.35	12.60		28.95	37.50	
	2731	Field bends, 45° to 90°, 1/2" diameter	1 Elec	44	.182	Ea.		7.15		7.15	11	
	2732	3/4" diameter		40	.200			7.90		7.90	12.10	
	2733	1" diameter		36	.222			8.75		8.75	13.45	
	2734	1-1/4" diameter		19	.421			16.60		16.60	25.50	
	2735	1-1/2" diameter		18	.444			17.50		17.50	27	
	2736	2" diameter		13	.615	↓		24.50		24.50	37.50	
	5000	Electric metallic tubing (EMT), 1/2" diameter		170	.047	L.F.	.40	1.85		2.25	3.28	
	5020	3/4" diameter		130	.062		.60	2.42		3.02	4.39	
	5040	1" diameter		115	.070		1.02	2.74		3.76	5.35	
	5060	1-1/4" diameter		100	.080		1.53	3.15		4.68	6.55	
	5080	1-1/2" diameter		90	.089		1.95	3.50		5.45	7.55	
	5100	2" diameter 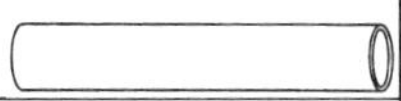		80	.100		2.50	3.94		6.44	8.80	
	5120	2-1/2" diameter	↓	60	.133		5.80	5.25		11.05	14.40	
	5140	3" diameter	2 Elec	100	.160		6.35	6.30		12.65	16.65	
	5160	3-1/2" diameter		90	.178		8.10	7		15.10	19.70	
	5180	4" diameter	↓	80	.200	↓	9	7.90		16.90	22	
	5200	Field bends, 45° to 90°, 1/2" diameter	1 Elec	89	.090	Ea.		3.54		3.54	5.45	
	5220	3/4" diameter		80	.100			3.94		3.94	6.05	
	5240	1" diameter		73	.110			4.32		4.32	6.65	
	5260	1-1/4" diameter		38	.211			8.30		8.30	12.75	
	5280	1-1/2" diameter		36	.222			8.75		8.75	13.45	
	5300	2" diameter		26	.308			12.10		12.10	18.65	
	5320	Offsets, 1/2" diameter		65	.123			4.85		4.85	7.45	
	5340	3/4" diameter		62	.129			5.10		5.10	7.80	
	5360	1" diameter		53	.151			5.95		5.95	9.15	
	5380	1-1/4" diameter		30	.267			10.50		10.50	16.15	
	5400	1-1/2" diameter		28	.286			11.25		11.25	17.30	
	7600	EMT, "T" fittings with covers, 1/2" diameter, set screw		16	.500	↓	12.40	19.70		32.10	44	
	9990	Minimum labor/equipment charge	↓	4	2	Job		79		79	121	
230	0010	**CONDUIT IN CONCRETE SLAB** Including terminations,										230
	0020	fittings and supports										
	3230	PVC, schedule 40, 1/2" diameter	1 Elec	270	.030	L.F.	.43	1.17		1.60	2.27	
	3250	3/4" diameter 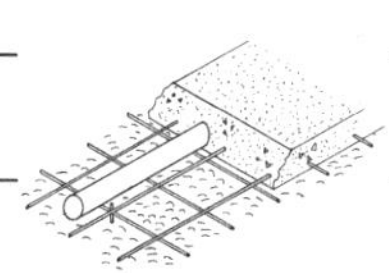		230	.035		.52	1.37		1.89	2.68	
	3270	1" diameter		200	.040		.69	1.58		2.27	3.18	
	3300	1-1/4" diameter		170	.047		.97	1.85		2.82	3.92	
	3330	1-1/2" diameter		140	.057		1.20	2.25		3.45	4.77	
	3350	2" diameter		120	.067		1.51	2.63		4.14	5.70	
	4350	Rigid galvanized steel, 1/2" diameter		200	.040		1.43	1.58		3.01	4	
	4400	3/4" diameter		170	.047		1.72	1.85		3.57	4.74	
	4450	1" diameter		130	.062		2.53	2.42		4.95	6.50	
	4500	1-1/4" diameter	↓	110	.073	↓	3.33	2.87		6.20	8.05	

	16132 Conduit & Tubing		CREW	DAILY OUTPUT	LABOR-HOURS	UNIT	2004 BARE COSTS MAT.	LABOR	EQUIP.	TOTAL	TOTAL INCL O&P	
230	4600	1-1/2" diameter	1 Elec	100	.080	L.F.	3.93	3.15		7.08	9.15	230
	4800	2" diameter		90	.089		5.25	3.50		8.75	11.15	
	9000	Minimum labor/equipment charge		4	2	Job		79		79	121	
240	0010	**CONDUIT IN TRENCH** Includes terminations and fittings										240
	0020	Does not include excavation or backfill, see div. 02315										
	0200	Rigid galvanized steel, 2" diameter	1 Elec	150	.053	L.F.	5.05	2.10		7.15	8.80	
	0400	2-1/2" diameter	"	100	.080		9.55	3.15		12.70	15.35	
	0600	3" diameter	2 Elec	160	.100		11.95	3.94		15.89	19.20	
	0800	3-1/2" diameter		140	.114		14.90	4.50		19.40	23.50	
	1000	4" diameter		100	.160		16.95	6.30		23.25	28.50	
	1200	5" diameter		80	.200		34.50	7.90		42.40	50	
	1400	6" diameter		60	.267		51	10.50		61.50	72	
	9000	Minimum labor/equipment charge	1 Elec	4	2	Job		79		79	121	
260	0010	**CUTTING AND DRILLING**										260
	0100	Hole drilling to 10' high, concrete wall										
	0110	8" thick, 1/2" pipe size	R-31	12	.667	Ea.	3.26	26.50	3.95	33.71	48.50	
	0120	3/4" pipe size		12	.667		3.26	26.50	3.95	33.71	48.50	
	0130	1" pipe size		9.50	.842		6.70	33	4.99	44.69	64	
	0140	1-1/4" pipe size		9.50	.842		6.70	33	4.99	44.69	64	
	0150	1-1/2" pipe size		9.50	.842		6.70	33	4.99	44.69	64	
	0160	2" pipe size		4.40	1.818		7.25	71.50	10.80	89.55	130	
	0170	2-1/2" pipe size		4.40	1.818		7.25	71.50	10.80	89.55	130	
	0180	3" pipe size		4.40	1.818		7.25	71.50	10.80	89.55	130	
	0190	3-1/2" pipe size		3.30	2.424		7.75	95.50	14.40	117.65	171	
	0200	4" pipe size		3.30	2.424		7.75	95.50	14.40	117.65	171	
	0500	12" thick, 1/2" pipe size		9.40	.851		4.98	33.50	5.05	43.53	62.50	
	0520	3/4" pipe size		9.40	.851		4.98	33.50	5.05	43.53	62.50	
	0540	1" pipe size		7.30	1.096		9.45	43	6.50	58.95	84	
	0560	1-1/4" pipe size		7.30	1.096		9.45	43	6.50	58.95	84	
	0570	1-1/2" pipe size		7.30	1.096		9.45	43	6.50	58.95	84	
	0580	2" pipe size		3.60	2.222		11.10	87.50	13.20	111.80	162	
	0590	2-1/2" pipe size		3.60	2.222		11.10	87.50	13.20	111.80	162	
	0600	3" pipe size		3.60	2.222		11.10	87.50	13.20	111.80	162	
	0610	3-1/2" pipe size		2.80	2.857		12.60	113	16.95	142.55	206	
	0630	4" pipe size		2.50	3.200		12.60	126	19	157.60	229	
	0650	16" thick, 1/2" pipe size		7.60	1.053		6.70	41.50	6.25	54.45	77.50	
	0670	3/4" pipe size		7	1.143		6.70	45	6.80	58.50	84	
	0690	1" pipe size		6	1.333		12.25	52.50	7.90	72.65	103	
	0710	1-1/4" pipe size		5.50	1.455		12.25	57.50	8.65	78.40	111	
	0730	1-1/2" pipe size		5.50	1.455		12.25	57.50	8.65	78.40	111	
	0750	2" pipe size		3	2.667		14.95	105	15.80	135.75	195	
	0770	2-1/2" pipe size		2.70	2.963		14.95	117	17.55	149.50	215	
	0790	3" pipe size		2.50	3.200		14.95	126	19	159.95	231	
	0810	3-1/2" pipe size		2.30	3.478		17.45	137	20.50	174.95	253	
	0830	4" pipe size		2	4		17.45	158	23.50	198.95	287	
	0850	20" thick, 1/2" pipe size		6.40	1.250		8.40	49.50	7.40	65.30	93	
	0870	3/4" pipe size		6	1.333		8.40	52.50	7.90	68.80	98.50	
	0890	1" pipe size		5	1.600		15	63	9.50	87.50	124	
	0910	1-1/4" pipe size		4.80	1.667		15	65.50	9.90	90.40	128	
	0930	1-1/2" pipe size		4.60	1.739		15	68.50	10.30	93.80	133	
	0950	2" pipe size		2.70	2.963		18.75	117	17.55	153.30	219	
	0970	2-1/2" pipe size		2.40	3.333		18.75	131	19.75	169.50	244	
	0990	3" pipe size		2.20	3.636		18.75	143	21.50	183.25	264	
	1010	3-1/2" pipe size		2	4		22.50	158	23.50	204	293	
	1030	4" pipe size		1.70	4.706		22.50	185	28	235.50	340	

16132 | Conduit & Tubing

260

		CREW	DAILY OUTPUT	LABOR-HOURS	UNIT	2004 BARE COSTS				TOTAL INCL O&P
						MAT.	LABOR	EQUIP.	TOTAL	
1050	24" thick, 1/2" pipe size	R-31	5.50	1.455	Ea.	10.15	57.50	8.65	76.30	109
1070	3/4" pipe size		5.10	1.569		10.15	62	9.30	81.45	116
1090	1" pipe size		4.30	1.860		17.75	73.50	11.05	102.30	145
1110	1-1/4" pipe size		4	2		17.75	79	11.85	108.60	154
1130	1-1/2" pipe size		4	2		17.75	79	11.85	108.60	154
1150	2" pipe size		2.40	3.333		22.50	131	19.75	173.25	249
1170	2-1/2" pipe size		2.20	3.636		22.50	143	21.50	187	269
1190	3" pipe size		2	4		22.50	158	23.50	204	293
1210	3-1/2" pipe size		1.80	4.444		27	175	26.50	228.50	330
1230	4" pipe size		1.50	5.333		27	210	31.50	268.50	390
1500	Brick wall, 8" thick, 1/2" pipe size		18	.444		3.26	17.50	2.64	23.40	33.50
1520	3/4" pipe size		18	.444		3.26	17.50	2.64	23.40	33.50
1540	1" pipe size		13.30	.601		6.70	23.50	3.57	33.77	48
1560	1-1/4" pipe size		13.30	.601		6.70	23.50	3.57	33.77	48
1580	1-1/2" pipe size		13.30	.601		6.70	23.50	3.57	33.77	48
1600	2" pipe size		5.70	1.404		7.25	55.50	8.30	71.05	102
1620	2-1/2" pipe size		5.70	1.404		7.25	55.50	8.30	71.05	102
1640	3" pipe size		5.70	1.404		7.25	55.50	8.30	71.05	102
1660	3-1/2" pipe size		4.40	1.818		7.75	71.50	10.80	90.05	130
1680	4" pipe size		4	2		7.75	79	11.85	98.60	143
1700	12" thick, 1/2" pipe size		14.50	.552		4.98	21.50	3.27	29.75	42.50
1720	3/4" pipe size		14.50	.552		4.98	21.50	3.27	29.75	42.50
1740	1" pipe size		11	.727		9.45	28.50	4.31	42.26	59
1760	1-1/4" pipe size		11	.727		9.45	28.50	4.31	42.26	59
1780	1-1/2" pipe size		11	.727		9.45	28.50	4.31	42.26	59
1800	2" pipe size		5	1.600		11.10	63	9.50	83.60	120
1820	2-1/2" pipe size		5	1.600		11.10	63	9.50	83.60	120
1840	3" pipe size		5	1.600		11.10	63	9.50	83.60	120
1860	3-1/2" pipe size		3.80	2.105		12.60	83	12.50	108.10	155
1880	4" pipe size		3.30	2.424		12.60	95.50	14.40	122.50	177
1900	16" thick, 1/2" pipe size		12.30	.650		6.70	25.50	3.86	36.06	51
1920	3/4" pipe size		12.30	.650		6.70	25.50	3.86	36.06	51
1940	1" pipe size		9.30	.860		12.25	34	5.10	51.35	71
1960	1-1/4" pipe size		9.30	.860		12.25	34	5.10	51.35	71
1980	1-1/2" pipe size		9.30	.860		12.25	34	5.10	51.35	71
2000	2" pipe size		4.40	1.818		14.95	71.50	10.80	97.25	138
2010	2-1/2" pipe size		4.40	1.818		14.95	71.50	10.80	97.25	138
2030	3" pipe size		4.40	1.818		14.95	71.50	10.80	97.25	138
2050	3-1/2" pipe size		3.30	2.424		17.45	95.50	14.40	127.35	182
2070	4" pipe size		3	2.667		17.45	105	15.80	138.25	198
2090	20" thick, 1/2" pipe size		10.70	.748		8.40	29.50	4.43	42.33	59.50
2110	3/4" pipe size		10.70	.748		8.40	29.50	4.43	42.33	59.50
2130	1" pipe size		8	1		15	39.50	5.95	60.45	83.50
2150	1-1/4" pipe size		8	1		15	39.50	5.95	60.45	83.50
2170	1-1/2" pipe size		8	1		15	39.50	5.95	60.45	83.50
2190	2" pipe size		4	2		18.75	79	11.85	109.60	155
2210	2-1/2" pipe size		4	2		18.75	79	11.85	109.60	155
2230	3" pipe size		4	2		18.75	79	11.85	109.60	155
2250	3-1/2" pipe size		3	2.667		22.50	105	15.80	143.30	203
2270	4" pipe size		2.70	2.963		22.50	117	17.55	157.05	223
2290	24" thick, 1/2" pipe size		9.40	.851		10.15	33.50	5.05	48.70	68
2310	3/4" pipe size		9.40	.851		10.15	33.50	5.05	48.70	68
2330	1" pipe size		7.10	1.127		17.75	44.50	6.70	68.95	95
2350	1-1/4" pipe size		7.10	1.127		17.75	44.50	6.70	68.95	95
2370	1-1/2" pipe size		7.10	1.127		17.75	44.50	6.70	68.95	95
2390	2" pipe size		3.60	2.222		22.50	87.50	13.20	123.20	175

	16132 | Conduit & Tubing		Crew	Daily Output	Labor-Hours	Unit	2004 Bare Costs Mat.	Labor	Equip.	Total	Total Incl O&P
260	2410	2-1/2" pipe size	R-31	3.60	2.222	Ea.	22.50	87.50	13.20	123.20	175
	2430	3" pipe size		3.60	2.222		22.50	87.50	13.20	123.20	175
	2450	3-1/2" pipe size		2.80	2.857		27	113	16.95	156.95	222
	2470	4" pipe size	↓	2.50	3.200	↓	27	126	19	172	245
	2480										
	3000	Knockouts to 8' high, metal boxes & enclosures									
	3020	With hole saw, 1/2" pipe size	1 Elec	53	.151	Ea.		5.95		5.95	9.15
	3040	3/4" pipe size		47	.170			6.70		6.70	10.30
	3050	1" pipe size		40	.200			7.90		7.90	12.10
	3060	1-1/4" pipe size		36	.222			8.75		8.75	13.45
	3070	1-1/2" pipe size		32	.250			9.85		9.85	15.15
	3080	2" pipe size		27	.296			11.65		11.65	17.95
	3090	2-1/2" pipe size		20	.400			15.75		15.75	24
	4010	3" pipe size		16	.500			19.70		19.70	30.50
	4030	3-1/2" pipe size		13	.615			24.50		24.50	37.50
	4050	4" pipe size		11	.727			28.50		28.50	44
	4070	With hand punch set, 1/2" pipe size		40	.200			7.90		7.90	12.10
	4090	3/4" pipe size		32	.250			9.85		9.85	15.15
	4110	1" pipe size		30	.267			10.50		10.50	16.15
	4130	1-1/4" pipe size		28	.286			11.25		11.25	17.30
	4150	1-1/2" pipe size		26	.308			12.10		12.10	18.65
	4170	2" pipe size		20	.400			15.75		15.75	24
	4190	2-1/2" pipe size		17	.471			18.55		18.55	28.50
	4200	3" pipe size		15	.533			21		21	32.50
	4220	3-1/2" pipe size		12	.667			26.50		26.50	40.50
	4240	4" pipe size		10	.800			31.50		31.50	48.50
	4260	With hydraulic punch, 1/2" pipe size		44	.182			7.15		7.15	11
	4280	3/4" pipe size		38	.211			8.30		8.30	12.75
	4300	1" pipe size		38	.211			8.30		8.30	12.75
	4320	1-1/4" pipe size		38	.211			8.30		8.30	12.75
	4340	1-1/2" pipe size		38	.211			8.30		8.30	12.75
	4360	2" pipe size		32	.250			9.85		9.85	15.15
	4380	2-1/2" pipe size		27	.296			11.65		11.65	17.95
	4400	3" pipe size		23	.348			13.70		13.70	21
	4420	3-1/2" pipe size	↓	20	.400	↓		15.75		15.75	24
	16133 | Multi-outlet Assemblies										
800	0010	SURFACE RACEWAY									
	0090	Metal, straight section									
	0100	No. 500	1 Elec	100	.080	L.F.	.73	3.15		3.88	5.65
	0110	No. 700		100	.080		.82	3.15		3.97	5.75
	0400	No. 1500, small pancake		90	.089		1.50	3.50		5	7.05
	0600	No. 2000, base & cover, blank		90	.089		1.47	3.50		4.97	7
	0800	No. 3000, base & cover, blank		75	.107		2.95	4.20		7.15	9.70
	1000	No. 4000, base & cover, blank		65	.123		4.80	4.85		9.65	12.75
	1200	No. 6000, base & cover, blank		50	.160	↓	8.05	6.30		14.35	18.55
	2400	Fittings, elbows, No. 500		40	.200	Ea.	1.33	7.90		9.23	13.55
	2800	Elbow cover, No. 2000		40	.200		2.55	7.90		10.45	14.90
	2880	Tee, No. 500		42	.190		2.56	7.50		10.06	14.35
	2900	No. 2000		27	.296		8.05	11.65		19.70	27
	3000	Switch box, No. 500		16	.500		8.40	19.70		28.10	40
	3400	Telephone outlet, No. 1500		16	.500		9.65	19.70		29.35	41
	3600	Junction box, No. 1500	↓	16	.500	↓	6.70	19.70		26.40	38
	3800	Plugmold wired sections, No. 2000									
	4000	1 circuit, 6 outlets, 3 ft. long	1 Elec	8	1	Ea.	24.50	39.50		64	87.50

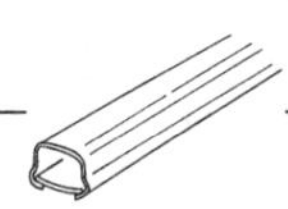

16 ELECTRICAL

16133 | Multi-outlet Assemblies

		16133 Multi-outlet Assemblies	CREW	DAILY OUTPUT	LABOR-HOURS	UNIT	2004 BARE COSTS MAT.	LABOR	EQUIP.	TOTAL	TOTAL INCL O&P	
800	4100	2 circuits, 8 outlets, 6 ft. long	1 Elec	5.30	1.509	Ea.	41	59.50		100.50	137	800
	9300	Non-metallic, straight section										
	9310	7/16" x 7/8", base & cover, blank	1 Elec	160	.050	L.F.	1.14	1.97		3.11	4.28	
	9320	Base & cover w/ adhesive		160	.050		1.33	1.97		3.30	4.49	
	9340	7/16" x 1-5/16", base & cover, blank		145	.055		1.33	2.17		3.50	4.80	
	9350	Base & cover w/ adhesive		145	.055		1.55	2.17		3.72	5.05	
	9370	11/16" x 2-1/4", base & cover, blank		130	.062		1.87	2.42		4.29	5.80	
	9380	Base & cover w/ adhesive		130	.062		2.15	2.42		4.57	6.10	
	9400	Fittings, elbows, 7/16" x 7/8"		50	.160	Ea.	1.33	6.30		7.63	11.15	
	9410	7/16" x 1-5/16"		45	.178		1.38	7		8.38	12.25	
	9420	11/16" x 2-1/4"		40	.200		1.50	7.90		9.40	13.75	
	9430	Tees, 7/16" x 7/8"		35	.229		1.75	9		10.75	15.80	
	9440	7/16" x 1-5/16"		32	.250		1.80	9.85		11.65	17.15	
	9450	11/16" x 2-1/4"		30	.267		1.82	10.50		12.32	18.15	
	9460	Cover clip, 7/16" x 7/8"		80	.100		.35	3.94		4.29	6.45	
	9470	7/16" x 1-5/16"		72	.111		.32	4.38		4.70	7.10	
	9480	11/16" x 2-1/4"		64	.125		.53	4.93		5.46	8.15	
	9490	Blank end, 7/16" x 7/8"		50	.160		.50	6.30		6.80	10.25	
	9510	11/16" x 2-1/4"		40	.200		.85	7.90		8.75	13.05	
	9520	Round fixture box 5.5" dia x 1"		25	.320		8.40	12.60		21	28.50	
	9530	Device box, 1 gang		30	.267		3.71	10.50		14.21	20	
	9540	2 gang		25	.320		5.50	12.60		18.10	25.50	
	9990	Minimum labor/equipment charge		5	1.600	Job		63		63	97	

16134 | Wireway & Aux Gutters

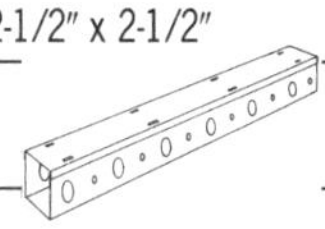

		16134 Wireway & Aux Gutters	CREW	DAILY OUTPUT	LABOR-HOURS	UNIT	MAT.	LABOR	EQUIP.	TOTAL	TOTAL INCL O&P	
150	0010	**WIREWAY** to 15' high										150
	0100	Screw cover, NEMA 1 w/ fittings and supports, 2-1/2" x 2-1/2"	1 Elec	45	.178	L.F.	9.90	7		16.90	21.50	
	0200	4" x 4"	"	40	.200		10.50	7.90		18.40	23.50	
	0400	6" x 6"	2 Elec	60	.267		17.85	10.50		28.35	36	
	0600	8" x 8"	"	40	.400		30	15.75		45.75	57	
	4475	Screw cover, NEMA 3R w/ fittings and supports, 4" x 4"	1 Elec	36	.222		22.50	8.75		31.25	38	
	4480	6" x 6"	2 Elec	55	.291		29.50	11.45		40.95	50	
	4485	8" x 8"		36	.444		47.50	17.50		65	79.50	
	4490	12" x 12"		18	.889		66	35		101	127	

16136 | Boxes

		16136 Boxes	CREW	DAILY OUTPUT	LABOR-HOURS	UNIT	MAT.	LABOR	EQUIP.	TOTAL	TOTAL INCL O&P	
600	0010	**OUTLET BOXES**										600
	0020	Pressed steel, octagon, 4"	1 Elec	20	.400	Ea.	1.48	15.75		17.23	25.50	
	0060	Covers, blank		64	.125		.62	4.93		5.55	8.25	
	0100	Extension rings		40	.200		2.39	7.90		10.29	14.75	
	0152	Square 4"		18	.444		2.10	17.50		19.60	29.50	
	0200	Extension rings		40	.200		2.48	7.90		10.38	14.85	
	0250	Covers, blank		64	.125		.70	4.93		5.63	8.30	
	0300	Plaster rings		64	.125		1.14	4.93		6.07	8.80	
	0652	Switchbox		24	.333		2.31	13.15		15.46	22.50	
	1102	Concrete, floor, 1 gang		4.80	1.667		61.50	65.50		127	169	
	9000	Minimum labor/equipment charge		4	2	Job		79		79	121	
620	0010	**OUTLET BOXES, PLASTIC**										620
	0050	4" diameter, round with 2 mounting nails	1 Elec	25	.320	Ea.	1.86	12.60		14.46	21.50	
	0102	Bar hanger mounted		23	.348		3.28	13.70		16.98	24.50	
	0202	Square with 2 mounting nails		23	.348		3	13.70		16.70	24.50	
	0300	Plaster ring		64	.125		1.03	4.93		5.96	8.70	
	0402	Switch box with 2 mounting nails, 1 gang		27	.296		1.36	11.65		13.01	19.45	

	16136 | Boxes		CREW	DAILY OUTPUT	LABOR-HOURS	UNIT	2004 BARE COSTS MAT.	LABOR	EQUIP.	TOTAL	TOTAL INCL O&P	
620	0502	2 gang	1 Elec	23	.348	Ea.	2.28	13.70		15.98	23.50	620
	0602	3 gang		18	.444		3.60	17.50		21.10	31	
	0702	Old work box		27	.296		2.40	11.65		14.05	20.50	
	9000	Minimum labor/equipment charge		4	2	Job		79		79	121	
700	0010	**PULL BOXES & CABINETS**										700
	0100	Sheet metal, pull box, NEMA 1, type SC, 6" W x 6" H x 4" D	1 Elec	8	1	Ea.	9.65	39.50		49.15	71	
	0200	8" W x 8" H x 4" D		8	1		13.20	39.50		52.70	75	
	0300	10" W x 12" H x 6" D		5.30	1.509		23.50	59.50		83	117	
	0400	16" W x 20" H x 8" D		4	2		88	79		167	218	
	0500	20" W x 24" H x 8" D		3.20	2.500		103	98.50		201.50	264	
	0600	24" W x 36" H x 8" D		2.70	2.963		145	117		262	340	
	0650	Hinged cabinets, NEMA 1, 6" W x 6" H x 4" D		8	1		9.80	39.50		49.30	71.50	
	0802	12" W x 16" H x 6" D		4	2		32	79		111	157	
	1000	20" W x 20" H x 6" D		3.60	2.222		65	87.50		152.50	207	
	1200	20" W x 20" H x 8" D		3.20	2.500		130	98.50		228.50	294	
	1400	24" W x 36" H x 8" D		2.70	2.963		208	117		325	410	
	1600	24" W x 42" H x 8" D		2	4		315	158		473	585	
	7000	Cabinets, current transformer										
	7050	Single door, 24" H x 24" W x 10" D	1 Elec	1.60	5	Ea.	121	197		318	440	
	7100	30" H x 24" W x 10" D		1.30	6.154		127	242		369	515	
	7150	36" H x 24" W x 10" D		1.10	7.273		142	287		429	595	
	7200	30" H x 30" W x 10" D		1	8		148	315		463	650	
	7250	36" H x 30" W x 10" D		.90	8.889		200	350		550	760	
	7300	36" H x 36" W x 10" D		.80	10		218	395		613	845	
	7500	Double door, 48" H x 36" W x 10" D		.60	13.333		410	525		935	1,250	
	7550	24" H x 24" W x 12" D		1	8		205	315		520	710	
	9990	Minimum labor/equipment charge		2	4	Job		158		158	242	

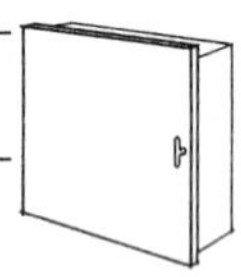

	16139 | Residential Wiring		CREW	DAILY OUTPUT	LABOR-HOURS	UNIT	MAT.	LABOR	EQUIP.	TOTAL	TOTAL INCL O&P	
700	0010	**RESIDENTIAL WIRING**										700
	0020	20' avg. runs and #14/2 wiring incl. unless otherwise noted										
	1000	Service & panel, includes 24' SE-AL cable, service eye, meter,										
	1010	Socket, panel board, main bkr., ground rod, 15 or 20 amp										
	1020	1-pole circuit breakers, and misc. hardware										
	1100	100 amp, with 10 branch breakers	1 Elec	1.19	6.723	Ea.	410	265		675	860	
	1110	With PVC conduit and wire		.92	8.696		445	345		790	1,025	
	1120	With RGS conduit and wire		.73	10.959		575	430		1,005	1,300	
	1150	150 amp, with 14 branch breakers		1.03	7.767		645	305		950	1,175	
	1170	With PVC conduit and wire		.82	9.756		715	385		1,100	1,375	
	1180	With RGS conduit and wire		.67	11.940		955	470		1,425	1,775	
	1200	200 amp, with 18 branch breakers	2 Elec	1.80	8.889		840	350		1,190	1,475	
	1220	With PVC conduit and wire		1.46	10.959		915	430		1,345	1,675	
	1230	With RGS conduit and wire		1.24	12.903		1,225	510		1,735	2,125	
	1800	Lightning surge suppressor for above services, add	1 Elec	32	.250		40	9.85		49.85	59	
	2000	Switch devices										
	2100	Single pole, 15 amp, Ivory, with a 1-gang box, cover plate,										
	2110	Type NM (Romex) cable	1 Elec	17.10	.468	Ea.	6.75	18.45		25.20	36	
	2120	Type MC (BX) cable		14.30	.559		18.05	22		40.05	54	
	2130	EMT & wire		5.71	1.401		17.85	55		72.85	105	
	2150	3-way, #14/3, type NM cable		14.55	.550		10.10	21.50		31.60	44.50	
	2170	Type MC cable		12.31	.650		23	25.50		48.50	65	
	2180	EMT & wire		5	1.600		19.95	63		82.95	119	
	2200	4-way, #14/3, type NM cable		14.55	.550		23	21.50		44.50	59	
	2220	Type MC cable		12.31	.650		36	25.50		61.50	79	
	2230	EMT & wire		5	1.600		33	63		96	133	

Important: See the Reference Section for critical supporting data - Reference Nos., Crews, & City Cost Indexes

16139 | Residential Wiring

		Description	CREW	DAILY OUTPUT	LABOR-HOURS	UNIT	2004 BARE COSTS MAT.	LABOR	EQUIP.	TOTAL	TOTAL INCL O&P	
700	2250	S.P., 20 amp, #12/2, type NM cable	1 Elec	13.33	.600	Ea.	11.90	23.50		35.40	49.50	700
	2270	Type MC cable		11.43	.700		22.50	27.50		50	67	
	2280	EMT & wire		4.85	1.649		23.50	65		88.50	126	
	2290	S.P. rotary dimmer, 600W, no wiring		17	.471		16	18.55		34.55	46	
	2300	S.P. rotary dimmer, 600W, type NM cable		14.55	.550		18.45	21.50		39.95	54	
	2320	Type MC cable		12.31	.650		29.50	25.50		55	72	
	2330	EMT & wire		5	1.600		30	63		93	130	
	2350	3-way rotary dimmer, type NM cable		13.33	.600		16.15	23.50		39.65	54.50	
	2370	Type MC cable		11.43	.700		27.50	27.50		55	72.50	
	2380	EMT & wire		4.85	1.649		28	65		93	131	
	2400	Interval timer wall switch, 20 amp, 1-30 min., #12/2										
	2410	Type NM cable	1 Elec	14.55	.550	Ea.	29	21.50		50.50	65.50	
	2420	Type MC cable		12.31	.650		37.50	25.50		63	80.50	
	2430	EMT & wire		5	1.600		40.50	63		103.50	142	
	2500	Decorator style										
	2510	S.P., 15 amp, type NM cable	1 Elec	17.10	.468	Ea.	10.30	18.45		28.75	40	
	2520	Type MC cable		14.30	.559		21.50	22		43.50	58	
	2530	EMT & wire		5.71	1.401		21.50	55		76.50	109	
	2550	3-way, #14/3, type NM cable		14.55	.550		13.65	21.50		35.15	48.50	
	2570	Type MC cable		12.31	.650		26.50	25.50		52	68.50	
	2580	EMT & wire		5	1.600		23.50	63		86.50	123	
	2600	4-way, #14/3, type NM cable		14.55	.550		26.50	21.50		48	62.50	
	2620	Type MC cable		12.31	.650		39.50	25.50		65	83	
	2630	EMT & wire		5	1.600		36.50	63		99.50	137	
	2650	S.P., 20 amp, #12/2, type NM cable		13.33	.600		15.45	23.50		38.95	53.50	
	2670	Type MC cable		11.43	.700		26	27.50		53.50	71	
	2680	EMT & wire		4.85	1.649		27	65		92	130	
	2700	S.P., slide dimmer, type NM cable		17.10	.468		25.50	18.45		43.95	56.50	
	2720	Type MC cable		14.30	.559		37	22		59	74.50	
	2730	EMT & wire		5.71	1.401		37	55		92	126	
	2770	Type MC cable		14.30	.559		33	22		55	70.50	
	2780	EMT & wire		5.71	1.401		33.50	55		88.50	122	
	2800	3-way touch dimmer, type NM cable		13.33	.600		40	23.50		63.50	80.50	
	2820	Type MC cable		11.43	.700		51	27.50		78.50	99	
	2830	EMT & wire		4.85	1.649		51.50	65		116.50	157	
	3100	S.P. switch/15 amp recpt., Ivory, 1-gang box, plate										
	3110	Type NM cable	1 Elec	11.43	.700	Ea.	14.80	27.50		42.30	59	
	3120	Type MC cable		10	.800		26	31.50		57.50	77	
	3130	EMT & wire		4.40	1.818		26.50	71.50		98	139	
	3150	S.P. switch/pilot light, type NM cable		11.43	.700		15.45	27.50		42.95	59.50	
	3170	Type MC cable		10	.800		26.50	31.50		58	78	
	3180	EMT & wire		4.43	1.806		27	71		98	139	
	3190	2-S.P. switches, 2-#14/2, no wiring		14	.571		5.70	22.50		28.20	41	
	3200	2-S.P. switches, 2-#14/2, type NM cables		10	.800		16.55	31.50		48.05	66.50	
	3220	Type MC cable		8.89	.900		36	35.50		71.50	94	
	3230	EMT & wire		4.10	1.951		27.50	77		104.50	149	
	3250	3-way switch/15 amp recpt., #14/3, type NM cable		10	.800		21	31.50		52.50	71.50	
	3270	Type MC cable		8.89	.900		34	35.50		69.50	92	
	3280	EMT & wire		4.10	1.951		31	77		108	152	
	3300	2-3 way switches, 2-#14/3, type NM cables		8.89	.900		27.50	35.50		63	84.50	
	3320	Type MC cable		8	1		50	39.50		89.50	116	
	3330	EMT & wire		4	2		34.50	79		113.50	159	
	3350	S.P. switch/20 amp recpt., #12/2, type NM cable		10	.800		25	31.50		56.50	76	
	3370	Type MC cable		8.89	.900		33.50	35.50		69	91	
	3380	EMT & wire		4.10	1.951		36.50	77		113.50	159	
	3400	Decorator style										

16139 | Residential Wiring		CREW	DAILY OUTPUT	LABOR-HOURS	UNIT	2004 BARE COSTS MAT.	LABOR	EQUIP.	TOTAL	TOTAL INCL O&P
700 3410	S.P. switch/15 amp recpt., type NM cable	1 Elec	11.43	.700	Ea.	18.35	27.50		45.85	62.50
3420	Type MC cable		10	.800		29.50	31.50		61	81
3430	EMT & wire		4.40	1.818		30	71.50		101.50	143
3450	S.P. switch/pilot light, type NM cable		11.43	.700		19	27.50		46.50	63.50
3470	Type MC cable		10	.800		30.50	31.50		62	82
3480	EMT & wire		4.40	1.818		30.50	71.50		102	144
3500	2-S.P. switches, 2-#14/2, type NM cables		10	.800		20	31.50		51.50	70.50
3520	Type MC cable		8.89	.900		39.50	35.50		75	98
3530	EMT & wire		4.10	1.951		31	77		108	152
3550	3-way/15 amp recpt., #14/3, type NM cable		10	.800		24.50	31.50		56	75.50
3580	EMT & wire		4.10	1.951		34.50	77		111.50	156
3650	2-3 way switches, 2-#14/3, type NM cables		8.89	.900		31	35.50		66.50	88.50
3670	Type MC cable		8	1		53.50	39.50		93	120
3680	EMT & wire		4	2		38.50	79		117.50	163
3700	S.P. switch/20 amp recpt., #12/2, type NM cable		10	.800		29	31.50		60.50	80
3720	Type MC cable		8.89	.900		37	35.50		72.50	95
3730	EMT & wire		4.10	1.951		40	77		117	162
4000	Receptacle devices									
4010	Duplex outlet, 15 amp recpt., Ivory, 1-gang box, plate									
4015	Type NM cable	1 Elec	14.55	.550	Ea.	5.25	21.50		26.75	39.50
4020	Type MC cable		12.31	.650		16.55	25.50		42.05	57.50
4030	EMT & wire		5.33	1.501		16.35	59		75.35	109
4050	With #12/2, type NM cable		12.31	.650		6.25	25.50		31.75	46.50
4070	Type MC cable		10.67	.750		16.75	29.50		46.25	64
4080	EMT & wire		4.71	1.699		17.65	67		84.65	122
4100	20 amp recpt., #12/2, type NM cable		12.31	.650		12.15	25.50		37.65	53
4120	Type MC cable		10.67	.750		22.50	29.50		52	70.50
4130	EMT & wire		4.71	1.699		23.50	67		90.50	129
4140	For GFI see line 4300 below									
4150	Decorator style, 15 amp recpt., type NM cable	1 Elec	14.55	.550	Ea.	8.80	21.50		30.30	43
4170	Type MC cable		12.31	.650		20	25.50		45.50	61.50
4180	EMT & wire		5.33	1.501		19.90	59		78.90	113
4200	With #12/2, type NM cable		12.31	.650		9.80	25.50		35.30	50.50
4220	Type MC cable		10.67	.750		20.50	29.50		50	68
4230	EMT & wire		4.71	1.699		21	67		88	127
4250	20 amp recpt. #12/2, type NM cable		12.31	.650		15.70	25.50		41.20	57
4270	Type MC cable		10.67	.750		26	29.50		55.50	74.50
4280	EMT & wire		4.71	1.699		27	67		94	133
4300	GFI, 15 amp recpt., type NM cable		12.31	.650		33	25.50		58.50	75.50
4320	Type MC cable		10.67	.750		44	29.50		73.50	94
4330	EMT & wire		4.71	1.699		44	67		111	152
4350	GFI with #12/2, type NM cable		10.67	.750		34	29.50		63.50	83
4370	Type MC cable		9.20	.870		44.50	34.50		79	102
4380	EMT & wire		4.21	1.900		45.50	75		120.50	165
4400	20 amp recpt., #12/2 type NM cable		10.67	.750		35.50	29.50		65	84.50
4420	Type MC cable		9.20	.870		46	34.50		80.50	103
4430	EMT & wire		4.21	1.900		47	75		122	167
4500	Weather-proof cover for above receptacles, add		32	.250		4.40	9.85		14.25	20
4550	Air conditioner outlet, 20 amp-240 volt recpt.									
4560	30′ of #12/2, 2 pole circuit breaker									
4570	Type NM cable	1 Elec	10	.800	Ea.	40.50	31.50		72	93
4580	Type MC cable		9	.889		54.50	35		89.50	114
4590	EMT & wire		4	2		51.50	79		130.50	178
4600	Decorator style, type NM cable		10	.800		44.50	31.50		76	97.50
4620	Type MC cable		9	.889		58.50	35		93.50	119
4630	EMT & wire		4	2		55.50	79		134.50	182 700

16139 | Residential Wiring		CREW	DAILY OUTPUT	LABOR-HOURS	UNIT	2004 BARE COSTS MAT.	LABOR	EQUIP.	TOTAL	TOTAL INCL O&P
700 4650	Dryer outlet, 30 amp-240 volt recpt., 20′ of #10/3									
4660	2 pole circuit breaker									
4670	Type NM cable	1 Elec	6.41	1.248	Ea.	47	49		96	128
4680	Type MC cable		5.71	1.401		60.50	55		115.50	152
4690	EMT & wire		3.48	2.299		52.50	90.50		143	197
4700	Range outlet, 50 amp-240 volt recpt., 30′ of #8/3									
4710	Type NM cable	1 Elec	4.21	1.900	Ea.	68.50	75		143.50	191
4720	Type MC cable		4	2		106	79		185	238
4730	EMT & wire		2.96	2.703		67	106		173	238
4750	Central vacuum outlet, Type NM cable		6.40	1.250		42	49.50		91.50	122
4770	Type MC cable		5.71	1.401		64	55		119	156
4780	EMT & wire		3.48	2.299		53.50	90.50		144	198
4800	30 amp-110 volt locking recpt., #10/2 circ. bkr.									
4810	Type NM cable	1 Elec	6.20	1.290	Ea.	50	51		101	133
4820	Type MC cable		5.40	1.481		76	58.50		134.50	173
4830	EMT & wire		3.20	2.500		61	98.50		159.50	218
4900	Low voltage outlets									
4910	Telephone recpt., 20′ of 4/C phone wire	1 Elec	26	.308	Ea.	6.70	12.10		18.80	26
4920	TV recpt., 20′ of RG59U coax wire, F type connector	"	16	.500	"	11.15	19.70		30.85	43
4950	Door bell chime, transformer, 2 buttons, 60′ of bellwire									
4970	Economy model	1 Elec	11.50	.696	Ea.	51.50	27.50		79	99
4980	Custom model		11.50	.696		82.50	27.50		110	133
4990	Luxury model, 3 buttons		9.50	.842		224	33		257	297
6000	Lighting outlets									
6050	Wire only (for fixture), type NM cable	1 Elec	32	.250	Ea.	3.31	9.85		13.16	18.80
6070	Type MC cable		24	.333		12.65	13.15		25.80	34
6080	EMT & wire		10	.800		11.45	31.50		42.95	61
6100	Box (4″), and wire (for fixture), type NM cable		25	.320		7.35	12.60		19.95	27.50
6120	Type MC cable		20	.400		16.70	15.75		32.45	42.50
6130	EMT & wire		11	.727		15.45	28.50		43.95	61
6200	Fixtures (use with lines 6050 or 6100 above)									
6210	Canopy style, economy grade	1 Elec	40	.200	Ea.	25.50	7.90		33.40	40
6220	Custom grade		40	.200		46	7.90		53.90	62.50
6250	Dining room chandelier, economy grade		19	.421		76	16.60		92.60	109
6270	Luxury grade		15	.533		495	21		516	580
6310	Kitchen fixture (fluorescent), economy grade		30	.267		51.50	10.50		62	72.50
6320	Custom grade		25	.320		161	12.60		173.60	196
6350	Outdoor, wall mounted, economy grade		30	.267		27	10.50		37.50	45.50
6360	Custom grade		30	.267		101	10.50		111.50	127
6370	Luxury grade		25	.320		227	12.60		239.60	269
6410	Outdoor PAR floodlights, 1 lamp, 150 watt		20	.400		21	15.75		36.75	47
6420	2 lamp, 150 watt each		20	.400		35	15.75		50.75	62.50
6430	For infrared security sensor, add		32	.250		87	9.85		96.85	111
6450	Outdoor, quartz-halogen, 300 watt flood		20	.400		38	15.75		53.75	66
6600	Recessed downlight, round, pre-wired, 50 or 75 watt trim		30	.267		35	10.50		45.50	54.50
6610	With shower light trim		30	.267		43	10.50		53.50	63.50
6620	With wall washer trim		28	.286		52.50	11.25		63.75	75.50
6630	With eye-ball trim		28	.286		52.50	11.25		63.75	75.50
6640	For direct contact with insulation, add					1.60			1.60	1.76
6700	Porcelain lamp holder	1 Elec	40	.200		3.50	7.90		11.40	15.95
6710	With pull switch		40	.200		3.84	7.90		11.74	16.30
6750	Fluorescent strip, 1-20 watt tube, wrap around diffuser, 24″		24	.333		51.50	13.15		64.65	76.50
6760	1-40 watt tube, 48″		24	.333		65	13.15		78.15	91.50
6770	2-40 watt tubes, 48″		20	.400		79	15.75		94.75	111
6780	With residential ballast		20	.400		89.50	15.75		105.25	123
700 6800	Bathroom heat lamp, 1-250 watt		28	.286		34.50	11.25		45.75	55.50

	16139 | Residential Wiring	CREW	DAILY OUTPUT	LABOR-HOURS	UNIT	2004 BARE COSTS MAT.	LABOR	EQUIP.	TOTAL	TOTAL INCL O&P	
700	6810 2-250 watt lamps	1 Elec	28	.286	Ea.	55	11.25		66.25	78	700
	6820 For timer switch, see line 2400										
	6900 Outdoor post lamp, incl. post, fixture, 35′ of #14/2										
	6910 Type NMC cable	1 Elec	3.50	2.286	Ea.	181	90		271	335	
	6920 Photo-eye, add	↓	27	.296	↓	29	11.65		40.65	50	
	6950 Clock dial time switch, 24 hr., w/enclosure, type NM cable	↓	11.43	.700	↓	52.50	27.50		80	100	
	6970 Type MC cable	↓	11	.727	↓	63.50	28.50		92	114	
	6980 EMT & wire	↓	4.85	1.649	↓	63.50	65		128.50	170	
	7000 Alarm systems										
	7050 Smoke detectors, box, #14/3, type NM cable	1 Elec	14.55	.550	Ea.	28	21.50		49.50	64	
	7070 Type MC cable	↓	12.31	.650	↓	38.50	25.50		64	81.50	
	7080 EMT & wire	↓	5	1.600	↓	35.50	63		98.50	136	
	7090 For relay output to security system, add				↓	11.75			11.75	12.95	
	8000 Residential equipment										
	8050 Disposal hook-up, incl. switch, outlet box, 3′ of flex										
	8060 20 amp-1 pole circ. bkr., and 25′ of #12/2										
	8070 Type NM cable	1 Elec	10	.800	Ea.	19.60	31.50		51.10	70	
	8080 Type MC cable	↓	8	1	↓	32	39.50		71.50	95.50	
	8090 EMT & wire	↓	5	1.600	↓	32.50	63		95.50	133	
	8100 Trash compactor or dishwasher hook-up, incl. outlet box,										
	8110 3′ of flex, 15 amp-1 pole circ. bkr., and 25′ of #14/2										
	8120 Type NM cable	1 Elec	10	.800	Ea.	13.95	31.50		45.45	64	
	8130 Type MC cable	↓	8	1	↓	27.50	39.50		67	90.50	
	8140 EMT & wire	↓	5	1.600	↓	27	63		90	127	
	8150 Hot water sink dispensor hook-up, use line 8100										
	8200 Vent/exhaust fan hook-up, type NM cable	1 Elec	32	.250	Ea.	3.31	9.85		13.16	18.80	
	8220 Type MC cable	↓	24	.333	↓	12.65	13.15		25.80	34	
	8230 EMT & wire	↓	10	.800	↓	11.45	31.50		42.95	61	
	8250 Bathroom vent fan, 50 CFM (use with above hook-up)										
	8260 Economy model	1 Elec	15	.533	Ea.	21.50	21		42.50	56	
	8270 Low noise model	↓	15	.533	↓	30	21		51	65.50	
	8280 Custom model	↓	12	.667	↓	111	26.50		137.50	163	
	8300 Bathroom or kitchen vent fan, 110 CFM										
	8310 Economy model	1 Elec	15	.533	Ea.	56.50	21		77.50	94.50	
	8320 Low noise model	"	15	.533	"	75	21		96	115	
	8350 Paddle fan, variable speed (w/o lights)										
	8360 Economy model (AC motor)	1 Elec	10	.800	Ea.	100	31.50		131.50	159	
	8370 Custom model (AC motor)	↓	10	.800	↓	173	31.50		204.50	239	
	8380 Luxury model (DC motor)	↓	8	1	↓	340	39.50		379.50	435	
	8390 Remote speed switch for above, add	↓	12	.667	↓	24.50	26.50		51	67.50	
	8500 Whole house exhaust fan, ceiling mount, 36″, variable speed										
	8510 Remote switch, incl. shutters, 20 amp-1 pole circ. bkr.										
	8520 30′ of #12/2, type NM cable	1 Elec	4	2	Ea.	610	79		689	790	
	8530 Type MC cable	↓	3.50	2.286	↓	625	90		715	825	
	8540 EMT & wire	↓	3	2.667	↓	625	105		730	850	
	8600 Whirlpool tub hook-up, incl. timer switch, outlet box										
	8610 3′ of flex, 20 amp-1 pole GFI circ. bkr.										
	8620 30′ of #12/2, type NM cable	1 Elec	5	1.600	Ea.	79.50	63		142.50	185	
	8630 Type MC cable	↓	4.20	1.905	↓	88.50	75		163.50	212	
	8640 EMT & wire	↓	3.40	2.353	↓	89.50	92.50		182	240	
	8650 Hot water heater hook-up, incl. 1-2 pole circ. bkr., box;										
	8660 3′ of flex, 20′ of #10/2, type NM cable	1 Elec	5	1.600	Ea.	20.50	63		83.50	120	
	8670 Type MC cable	↓	4.20	1.905	↓	36.50	75		111.50	155	
	8680 EMT & wire	↓	3.40	2.353	↓	28.50	92.50		121	174	
	9000 Heating/air conditioning										
	9050 Furnace/boiler hook-up, incl. firestat, local on-off switch										

Important: See the Reference Section for critical supporting data - Reference Nos., Crews, & City Cost Indexes

16139 | Residential Wiring

			CREW	DAILY OUTPUT	LABOR-HOURS	UNIT	2004 BARE COSTS MAT.	LABOR	EQUIP.	TOTAL	TOTAL INCL O&P	
700	9060	Emergency switch, and 40′ of type NM cable	1 Elec	4	2	Ea.	42.50	79		121.50	168	700
	9070	Type MC cable		3.50	2.286		62	90		152	206	
	9080	EMT & wire		1.50	5.333		61	210		271	395	
	9100	Air conditioner hook-up, incl. local 60 amp disc. switch										
	9110	3′ sealtite, 40 amp, 2 pole circuit breaker										
	9130	40′ of #8/2, type NM cable	1 Elec	3.50	2.286	Ea.	142	90		232	294	
	9140	Type MC cable		3	2.667		199	105		304	380	
	9150	EMT & wire		1.30	6.154		152	242		394	540	
	9200	Heat pump hook-up, 1-40 & 1-100 amp 2 pole circ. bkr.										
	9210	Local disconnect switch, 3′ sealtite										
	9220	40′ of #8/2 & 30′ of #3/2										
	9230	Type NM cable	1 Elec	1.30	6.154	Ea.	325	242		567	730	
	9240	Type MC cable		1.08	7.407		500	292		792	1,000	
	9250	EMT & wire		.94	8.511		335	335		670	885	
	9500	Thermostat hook-up, using low voltage wire										
	9520	Heating only	1 Elec	24	.333	Ea.	4.74	13.15		17.89	25	
	9530	Heating/cooling	″	20	.400	″	5.70	15.75		21.45	30.50	

16140 | Wiring Devices

			CREW	DAILY OUTPUT	LABOR-HOURS	UNIT	MAT.	LABOR	EQUIP.	TOTAL	TOTAL INCL O&P	
910	0010	WIRING DEVICES										910
	0200	Toggle switch, quiet type, single pole, 15 amp	1 Elec	40	.200	Ea.	4.69	7.90		12.59	17.25	
	0500	20 amp		27	.296		6.80	11.65		18.45	25.50	
	0550	Rocker, 15 amp		40	.200		5.05	7.90		12.95	17.65	
	0560	20 amp		27	.296		11.70	11.65		23.35	31	
	0600	3 way, 15 amp		23	.348		6.70	13.70		20.40	28.50	
	0850	Rocker, 15 amp		23	.348		7.10	13.70		20.80	29	
	0860	20 amp		18	.444		17	17.50		34.50	45.50	
	0900	4 way, 15 amp		15	.533		20	21		41	54.50	
	1030	Rocker, 15 amp		15	.533		27.50	21		48.50	62.50	
	1040	20 amp		11	.727		42	28.50		70.50	90	
	1650	Dimmer switch, 120 volt, incandescent, 600 watt, 1 pole		16	.500		10.80	19.70		30.50	42.50	
	2460	Receptacle, duplex, 120 volt, grounded, 15 amp		40	.200		1.14	7.90		9.04	13.35	
	2470	20 amp		27	.296		7.05	11.65		18.70	25.50	
	2490	Dryer, 30 amp		15	.533		10.35	21		31.35	44	
	2500	Range, 50 amp		11	.727		10.75	28.50		39.25	56	
	2600	Wall plates, stainless steel, 1 gang		80	.100		1.80	3.94		5.74	8.05	
	2800	2 gang		53	.151		4.10	5.95		10.05	13.65	
	3200	Lampholder, keyless		26	.308		9.20	12.10		21.30	29	
	3400	Pullchain with receptacle		22	.364		8.90	14.35		23.25	32	
	9000	Minimum labor/equipment charge		4	2	Job		79		79	121	

16150 | Wiring Connections

			CREW	DAILY OUTPUT	LABOR-HOURS	UNIT	MAT.	LABOR	EQUIP.	TOTAL	TOTAL INCL O&P	
275	0010	MOTOR CONNECTIONS										275
	0020	Flexible conduit and fittings, 115 volt, 1 phase, up to 1 HP motor	1 Elec	8	1	Ea.	4.34	39.50		43.84	65.50	
	9000	Minimum labor/equipment charge	″	4	2	Job		79		79	121	

16200 | Electrical Power

16210 | Electrical Utility Services

			CREW	DAILY OUTPUT	LABOR-HOURS	UNIT	2004 BARE COSTS MAT.	LABOR	EQUIP.	TOTAL	TOTAL INCL O&P	
600	0010	**METER CENTERS AND SOCKETS**										600
	0100	Sockets, single position, 4 terminal, 100 amp	1 Elec	3.20	2.500	Ea.	31.50	98.50		130	186	
	0200	150 amp		2.30	3.478		35.50	137		172.50	250	
	0300	200 amp		1.90	4.211		47.50	166		213.50	305	
	0400	20 amp		3.20	2.500		75.50	98.50		174	234	
	0500	Double position, 4 terminal, 100 amp		2.80	2.857		124	113		237	310	
	0600	150 amp		2.10	3.810		141	150		291	385	
	0700	200 amp		1.70	4.706	↓	300	185		485	620	
	9000	Minimum labor/equipment charge	↓	3	2.667	Job		105		105	161	

16220 | Motors & Generators

			CREW	DAILY OUTPUT	LABOR-HOURS	UNIT	MAT.	LABOR	EQUIP.	TOTAL	TOTAL INCL O&P	
900	0010	**VARIABLE FREQUENCY DRIVES/ADJUSTABLE FREQUENCY DRIVES**										900
	0100	Enclosed (NEMA 1), 460 volt, for 3 HP motor size	1 Elec	.80	10	Ea.	1,475	395		1,870	2,200	
	0110	5 HP motor size		.80	10		1,600	395		1,995	2,350	
	0120	7.5 HP motor size		.67	11.940		1,875	470		2,345	2,800	
	0130	10 HP motor size	↓	.67	11.940		1,875	470		2,345	2,800	
	0140	15 HP motor size	2 Elec	.89	17.978		2,175	710		2,885	3,500	
	0150	20 HP motor size		.89	17.978		3,225	710		3,935	4,650	
	0160	25 HP motor size		.67	23.881		3,750	940		4,690	5,575	
	0170	30 HP motor size		.67	23.881		4,575	940		5,515	6,475	
	0180	40 HP motor size		.67	23.881		6,700	940		7,640	8,825	
	0190	50 HP motor size	↓	.53	30.189		7,300	1,200		8,500	9,850	
	0200	60 HP motor size	R-3	.56	35.714		8,000	1,375	284	9,659	11,200	
	0210	75 HP motor size		.56	35.714		11,000	1,375	284	12,659	14,500	
	0220	100 HP motor size		.50	40		11,200	1,550	320	13,070	15,100	
	0230	125 HP motor size		.50	40		12,400	1,550	320	14,270	16,400	
	0240	150 HP motor size		.50	40		15,800	1,550	320	17,670	20,200	
	0250	200 HP motor size	↓	.42	47.619		19,600	1,850	380	21,830	24,800	
	1100	Custom-engineered, 460 volt, for 3 HP motor size	1 Elec	.56	14.286		2,425	565		2,990	3,525	
	1110	5 HP motor size		.56	14.286		2,550	565		3,115	3,675	
	1120	7.5 HP motor size		.47	17.021		3,050	670		3,720	4,375	
	1130	10 HP motor size	↓	.47	17.021		3,050	670		3,720	4,375	
	1140	15 HP motor size	2 Elec	.62	25.806		3,350	1,025		4,375	5,250	
	1150	20 HP motor size		.62	25.806		4,575	1,025		5,600	6,600	
	1160	25 HP motor size		.47	34.043		4,950	1,350		6,300	7,500	
	1170	30 HP motor size		.47	34.043		6,750	1,350		8,100	9,475	
	1180	40 HP motor size		.47	34.043		7,550	1,350		8,900	10,400	
	1190	50 HP motor size	↓	.37	43.243		7,550	1,700		9,250	10,900	
	1200	60 HP motor size	R-3	.39	51.282		11,300	1,975	410	13,685	15,900	
	1210	75 HP motor size		.39	51.282		13,000	1,975	410	15,385	17,800	
	1220	100 HP motor size		.35	57.143		13,600	2,200	455	16,255	18,800	
	1230	125 HP motor size		.35	57.143		14,300	2,200	455	16,955	19,600	
	1240	150 HP motor size		.35	57.143		16,100	2,200	455	18,755	21,600	
	1250	200 HP motor size	↓	.29	68.966	↓	22,200	2,675	550	25,425	29,100	
	2000	For complex & special design systems to meet specific										
	2010	requirements, obtain quote from vendor.										

16230 | Generator Assemblies

			CREW	DAILY OUTPUT	LABOR-HOURS	UNIT	MAT.	LABOR	EQUIP.	TOTAL	TOTAL INCL O&P	
450	0010	**GENERATOR SET**										450
	0020	Gas or gasoline operated, includes battery,										
	0050	charger, muffler & transfer switch										
	0200	3 phase 4 wire, 277/480 volt, 7.5 kW	R-3	.83	24.096	Ea.	6,000	930	192	7,122	8,250	
	0300	11.5 kW		.71	28.169		8,500	1,100	224	9,824	11,300	
	0400	20 kW	↓	.63	31.746	↓	10,000	1,225	252	11,477	13,200	

16200 | Electrical Power

16230 | Generator Assemblies

			Crew	Daily Output	Labor-Hours	Unit	2004 Bare Costs Mat.	Labor	Equip.	Total	Total Incl O&P	
450	0500	35 kW	R-3	.55	36.364	Ea.	12,000	1,400	289	13,689	15,700	450

16270 | Transformers

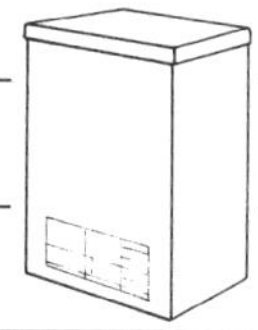

			Crew	Daily Output	Labor-Hours	Unit	2004 Bare Costs Mat.	Labor	Equip.	Total	Total Incl O&P	
200	0010	**DRY TYPE TRANSFORMER**										200
	0050	Single phase, 240/480 volt primary, 120/240 volt secondary										
	0100	1 kVA	1 Elec	2	4	Ea.	163	158		321	420	
	0300	2 kVA		1.60	5		244	197		441	575	
	0500	3 kVA		1.40	5.714		305	225		530	680	
	0700	5 kVA		1.20	6.667		415	263		678	865	
	0900	7.5 kVA	2 Elec	2.20	7.273		580	287		867	1,075	
	1100	10 kVA		1.60	10		740	395		1,135	1,425	
	1300	15 kVA		1.20	13.333		1,000	525		1,525	1,900	
	2300	3 phase, 480 volt primary 120/208 volt secondary										
	2310	Ventilated, 3 kVA	1 Elec	1	8	Ea.	525	315		840	1,050	
	2700	6 kVA		.80	10		720	395		1,115	1,400	
	2900	9 kVA		.70	11.429		820	450		1,270	1,600	
	3100	15 kVA	2 Elec	1.10	14.545		1,100	575		1,675	2,075	
	3300	30 kVA		.90	17.778		1,275	700		1,975	2,475	
	3500	45 kVA		.80	20		1,525	790		2,315	2,900	
	9000	Minimum labor/equipment charge	1 Elec	1	8	Job		315		315	485	

16400 | Low-Voltage Distribution

16410 | Encl Switches & Circuit Breakers

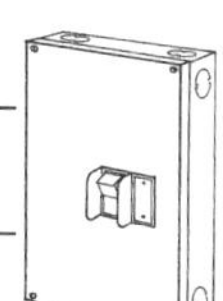

			Crew	Daily Output	Labor-Hours	Unit	2004 Bare Costs Mat.	Labor	Equip.	Total	Total Incl O&P	
200	0010	**CIRCUIT BREAKERS** (in enclosure)										200
	0100	Enclosed (NEMA 1), 600 volt, 3 pole, 30 amp	1 Elec	3.20	2.500	Ea.	410	98.50		508.50	605	
	0200	60 amp		2.80	2.857		410	113		523	630	
	0400	100 amp		2.30	3.478		470	137		607	730	
	0600	225 amp		1.50	5.333		1,100	210		1,310	1,525	
	0700	400 amp	2 Elec	1.60	10		1,850	395		2,245	2,650	
	9000	Minimum labor/equipment charge	1 Elec	4	2	Job		79		79	121	
800	0010	**SAFETY SWITCHES**										800
	0100	General duty 240 volt, 3 pole NEMA 1, fusible, 30 amp	1 Elec	3.20	2.500	Ea.	74.50	98.50		173	233	
	0200	60 amp		2.30	3.478		126	137		263	350	
	0300	100 amp		1.90	4.211		217	166		383	495	
	0400	200 amp		1.30	6.154		465	242		707	890	
	0500	400 amp	2 Elec	1.80	8.889		1,175	350		1,525	1,850	
	9990	Minimum labor/equipment charge	1 Elec	3	2.667	Job		105		105	161	
840	0010	**TIME SWITCHES**										840
	0100	Single pole, single throw, 24 hour dial	1 Elec	4	2	Ea.	82	79		161	211	
	0200	24 hour dial with reserve power		3.60	2.222		350	87.50		437.50	520	
	0300	Astronomic dial		3.60	2.222		141	87.50		228.50	290	
	0400	Astronomic dial with reserve power		3.30	2.424		455	95.50		550.50	645	
	0500	7 day calendar dial		3.30	2.424		126	95.50		221.50	286	
	0600	7 day calendar dial with reserve power		3.20	2.500		385	98.50		483.50	570	
	0700	Photo cell 2000 watt		8	1		15.15	39.50		54.65	77	

16400 | Low-Voltage Distribution

16410 | Encl Switches & Circuit Breakers

			CREW	DAILY OUTPUT	LABOR-HOURS	UNIT	2004 BARE COSTS MAT.	LABOR	EQUIP.	TOTAL	TOTAL INCL O&P	
840	1080	Load management device, 4 loads	1 Elec	2	4	Ea.	800	158		958	1,125	840
	1100	Load management device, 8 loads		1	8		1,300	315		1,615	1,900	
	9000	Minimum labor/equipment charge		3.50	2.286	Job		90		90	138	

16420 | Enclosed Controllers

			CREW	DAILY OUTPUT	LABOR-HOURS	UNIT	MAT.	LABOR	EQUIP.	TOTAL	TOTAL INCL O&P	
800	0010	**RELAYS** Enclosed (NEMA 1)										800
	0100	2 pole, 12 amp	1 Elec	5	1.600	Ea.	81	63		144	186	
	0200	4 pole, 10 amp	"	4.50	1.778	"	108	70		178	227	

16440 | Swbds, Panels & Control Centers

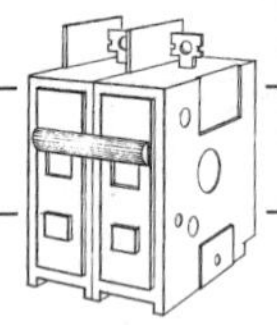

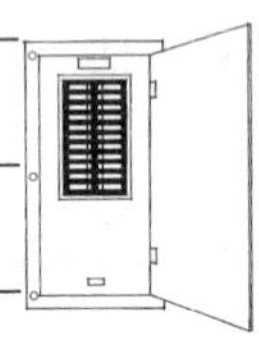

			CREW	DAILY OUTPUT	LABOR-HOURS	UNIT	MAT.	LABOR	EQUIP.	TOTAL	TOTAL INCL O&P	
700	0010	**PANELBOARD & LOAD CENTER CIRCUIT BREAKERS**										700
	0050	Bolt-on, 10,000 amp IC, 120 volt, 1 pole										
	0100	15 to 50 amp	1 Elec	10	.800	Ea.	10.60	31.50		42.10	60	
	0200	60 amp		8	1		10.60	39.50		50.10	72	
	0300	70 amp		8	1		20	39.50		59.50	82.50	
	0350	240 volt, 2 pole										
	0400	15 to 50 amp	1 Elec	8	1	Ea.	23.50	39.50		63	86	
	0500	60 amp		7.50	1.067		23.50	42		65.50	90	
	0600	80 to 100 amp		5	1.600		60	63		123	163	
	0700	3 pole, 15 to 60 amp		6.20	1.290		73.50	51		124.50	159	
	0800	70 amp		5	1.600		93	63		156	199	
	0900	80 to 100 amp		3.60	2.222		105	87.50		192.50	251	
	1000	22,000 amp I.C., 240 volt, 2 pole, 70 - 225 amp		2.70	2.963		455	117		572	680	
	1100	3 pole, 70 - 225 amp		2.30	3.478		505	137		642	765	
	1200	14,000 amp I.C., 277 volts, 1 pole, 15 - 30 amp		8	1		28	39.50		67.50	91.50	
	1300	22,000 amp I.C., 480 volts, 2 pole, 70 - 225 amp		2.70	2.963		455	117		572	680	
	1400	3 pole, 70 - 225 amp		2.30	3.478		560	137		697	825	
	2060	Plug-in tandem, 120/240 V, 2-15 A, 1 pole		11	.727		21	28.50		49.50	67	
	2070	1-15 A & 1-20 A		11	.727		21	28.50		49.50	67	
	2080	2-20 A		11	.727		21	28.50		49.50	67	
	9000	Minimum labor/equipment charge		3	2.667	Job		105		105	161	
720	0010	**PANELBOARDS** (Commercial use)										720
	0050	NQOD, w/20 amp 1 pole bolt-on circuit breakers										
	0100	3 wire, 120/240 volts, 100 amp main lugs										
	0150	10 circuits	1 Elec	1	8	Ea.	400	315		715	925	
	0200	14 circuits		.88	9.091		470	360		830	1,075	
	0250	18 circuits		.75	10.667		510	420		930	1,200	
	0300	20 circuits		.65	12.308		575	485		1,060	1,375	
	0600	4 wire, 120/208 volts, 100 amp main lugs, 12 circuits		1	8		450	315		765	980	
	0650	16 circuits		.75	10.667		520	420		940	1,225	
	0700	20 circuits		.65	12.308		605	485		1,090	1,425	
	0750	24 circuits		.60	13.333		660	525		1,185	1,525	
	0800	30 circuits		.53	15.094		760	595		1,355	1,750	
	0850	225 amp main lugs, 32 circuits	2 Elec	.90	17.778		855	700		1,555	2,025	
	0900	34 circuits		.84	19.048		875	750		1,625	2,125	
	0950	36 circuits		.80	20		895	790		1,685	2,175	
	1000	42 circuits		.68	23.529		1,000	925		1,925	2,525	
	1200	NEHB,w/20 amp, 1 pole bolt-on circuit breakers										
	1250	4 wire, 277/480 volts, 100 amp main lugs, 12 circuits	1 Elec	.88	9.091	Ea.	865	360		1,225	1,500	
	1300	20 circuits	"	.60	13.333		1,275	525		1,800	2,200	
	1350	225 amp main lugs, 24 circuits	2 Elec	.90	17.778		1,475	700		2,175	2,700	
	1400	30 circuits		.80	20		1,775	790		2,565	3,150	
	1450	36 circuits		.72	22.222		2,075	875		2,950	3,625	

16440 | Swbds, Panels & Control Centers

	Line	Description	Crew	Daily Output	Labor-Hours	Unit	2004 Bare Costs Mat.	2004 Bare Costs Labor	2004 Bare Costs Equip.	2004 Bare Costs Total	Total Incl O&P	
720	1600	NQOD panel, w/20 amp, 1 pole, circuit breakers										720
	2000	4 wire, 120/208 volts with main circuit breaker										
	2050	100 amp main, 24 circuits	1 Elec	.47	17.021	Ea.	830	670		1,500	1,925	
	2100	30 circuits	"	.40	20		935	790		1,725	2,225	
	2200	225 amp main, 32 circuits	2 Elec	.72	22.222		1,575	875		2,450	3,100	
	2250	42 circuits		.56	28.571		1,725	1,125		2,850	3,625	
	2300	400 amp main, 42 circuits		.48	33.333		2,350	1,325		3,675	4,600	
	2350	600 amp main, 42 circuits	↓	.40	40	↓	3,475	1,575		5,050	6,250	
	2400	NEHB, with 20 amp, 1 pole circuit breaker										
	2450	4 wire, 277/480 volts with main circuit breaker										
	2500	100 amp main, 24 circuits	1 Elec	.42	19.048	Ea.	1,700	750		2,450	3,025	
	2550	30 circuits	"	.38	21.053		2,000	830		2,830	3,475	
	2600	225 amp main, 30 circuits	2 Elec	.72	22.222		2,525	875		3,400	4,125	
	2650	42 circuits	"	.56	28.571	↓	3,100	1,125		4,225	5,150	
	9000	Minimum labor/equipment charge	1 Elec	1	8	Job		315		315	485	

16500 | Lighting

16510 | Interior Luminaires

	Line	Description	Crew	Daily Output	Labor-Hours	Unit	2004 Bare Costs Mat.	2004 Bare Costs Labor	2004 Bare Costs Equip.	2004 Bare Costs Total	Total Incl O&P	
440	0010	**INTERIOR LIGHTING FIXTURES** Including lamps, mounting										440
	0030	hardware and connections										
	0100	Fluorescent, C.W. lamps, troffer, recess mounted in grid, RS										
	0200	Acrylic lens, 1'W x 4'L, two 40 watt	1 Elec	5.70	1.404	Ea.	45	55.50		100.50	135	
	0300	2'W x 2'L, two U40 watt		5.70	1.404		48	55.50		103.50	138	
	0600	2'W x 4'L, four 40 watt 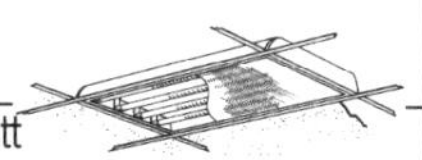		4.70	1.702		54.50	67		121.50	163	
	0910	Acrylic lens, 1'W x 4'L, two 32 watt		5.70	1.404		53.50	55.50		109	144	
	0930	2'W x 2'L, two U32 watt		5.70	1.404		76	55.50		131.50	169	
	0940	2'W x 4'L, two 32 watt		5.30	1.509		68	59.50		127.50	167	
	0950	2'W x 4'L, three 32 watt		5	1.600		72	63		135	176	
	0960	2'W x 4'L, four 32 watt	↓	4.70	1.702	↓	74	67		141	185	
	1000	Surface mounted, RS										
	1030	Acrylic lens with hinged & latched door frame										
	1100	1'W x 4'L, two 40 watt	1 Elec	7	1.143	Ea.	70	45		115	146	
	1200	2'W x 2'L, two U40 watt		7	1.143		75	45		120	152	
	1500	2'W x 4'L, four 40 watt	↓	5.30	1.509	↓	89	59.50		148.50	190	
	2100	Strip fixture										
	2130	Surface mounted										
	2200	4' long, one 40 watt RS	1 Elec	8.50	.941	Ea.	26.50	37		63.50	86	
	2300	4' long, two 40 watt RS	"	8	1		28.50	39.50		68	91.50	
	2600	8' long, one 75 watt, SL 	2 Elec	13.40	1.194		39.50	47		86.50	116	
	2700	8' long, two 75 watt, SL	"	12.40	1.290	↓	47.50	51		98.50	131	
	3000	Pendent mounted, industrial, white porcelain enamel										
	3030	C.W. lamps										
	3100	4' long, two 40 watt, RS	1 Elec	5.70	1.404	Ea.	43.50	55.50		99	133	
	3200	4' long, two 60 watt, HO	"	5	1.600		69	63		132	173	
	3300	8' long, two 75 watt, SL	2 Elec	8.80	1.818	↓	82	71.50		153.50	200	
	4220	Metal halide, integral ballast, ceiling, recess mounted 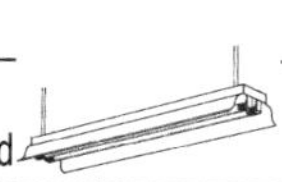										
	4230	prismatic glass lens, floating door										
	4240	2'W x 2'L, 250 watt	1 Elec	3.20	2.500	Ea.	267	98.50		365.50	445	

	16510 | Interior Luminaires		CREW	DAILY OUTPUT	LABOR-HOURS	UNIT	2004 BARE COSTS MAT.	LABOR	EQUIP.	TOTAL	TOTAL INCL O&P	
440	4250	2′W x 2′L, 400 watt	2 Elec	5.80	2.759	Ea.	305	109		414	500	440
	4260	Surface mounted, 2′W x 2′L, 250 watt	1 Elec	2.70	2.963		267	117		384	475	
	4270	400 watt	2 Elec	4.80	3.333		315	131		446	545	
	4280	High bay, aluminum reflector,										
	4290	Single unit, 400 watt	2 Elec	4.60	3.478	Ea.	320	137		457	560	
	4300	Single unit, 1000 watt		4	4		460	158		618	745	
	4310	Twin unit, 400 watt		3.20	5		640	197		837	1,000	
	4320	Low bay, aluminum reflector, 250W DX lamp	1 Elec	3.20	2.500		310	98.50		408.50	490	
	4340	High pressure sodium integral ballast ceiling, recess mounted										
	4350	prismatic glass lens, floating door										
	4360	2′W x 2′L, 150 watt lamp	1 Elec	3.20	2.500	Ea.	325	98.50		423.50	510	
	4370	2′W x 2′L, 400 watt lamp	2 Elec	5.80	2.759		390	109		499	595	
	4380	Surface mounted, 2′W x 2′L, 150 watt lamp	1 Elec	2.70	2.963		365	117		482	585	
	4390	400 watt lamp	2 Elec	4.80	3.333		410	131		541	650	
	4400	High bay, aluminum reflector,										
	4410	Single unit, 400 watt lamp	2 Elec	4.60	3.478	Ea.	295	137		432	535	
	4430	Single unit, 1000 watt lamp	″	4	4		425	158		583	710	
	4440	Low bay, aluminum reflector, 150 watt lamp	1 Elec	3.20	2.500		255	98.50		353.50	430	
	4450	Incandescent, high hat can, round alzak reflector, prewired										
	4470	100 watt	1 Elec	8	1	Ea.	56.50	39.50		96	123	
	4480	150 watt		8	1		81	39.50		120.50	150	
	4500	300 watt		6.70	1.194		184	47		231	275	
	4600	Square glass lens with metal trim, prewired										
	4630	100 watt	1 Elec	6.70	1.194	Ea.	45	47		92	122	
	4700	200 watt		6.70	1.194		74	47		121	154	
	6010	Vapor tight, incandescent, ceiling mounted, 200 watt		6.20	1.290		48	51		99	131	
	6100	Fluorescent, surface mounted, 2 lamps, 4′L, RS, 40 watt		3.20	2.500		90	98.50		188.50	250	
	6850	Vandalproof, surface mounted, fluorescent, two 40 watt		3.20	2.500		190	98.50		288.50	360	
	6860	Incandescent, one 150 watt		8	1		53	39.50		92.50	119	
	6900	Mirror light, fluorescent, RS, acrylic enclosure, two 40 watt		8	1		84.50	39.50		124	154	
	6910	One 40 watt		8	1		66	39.50		105.50	133	
	6920	One 20 watt		12	.667		52	26.50		78.50	98	
	7500	Ballast replacement, by weight of ballast, to 15′ high										
	7520	Indoor fluorescent, less than 2 lbs.	1 Elec	10	.800	Ea.		31.50		31.50	48.50	
	7540	Two 40W, watt reducer, 2 to 5 lbs.		9.40	.851		22	33.50		55.50	75.50	
	7560	Two F96 slimline, over 5 lbs.		8	1		32.50	39.50		72	96.50	
	7580	Vaportite ballast, less than 2 lbs.		9.40	.851			33.50		33.50	51.50	
	7600	2 lbs. to 5 lbs.		8.90	.899			35.50		35.50	54.50	
	7620	Over 5 lbs.		7.60	1.053			41.50		41.50	63.50	
	7630	Electronic ballast for two tubes		8	1		32.50	39.50		72	96.50	
	7640	Dimmable ballast one lamp		8	1		45	39.50		84.50	110	
	7650	Dimmable ballast two-lamp		7.60	1.053		73	41.50		114.50	144	
	9000	Minimum labor/equipment charge		3	2.667	Job		105		105	161	
800	0010	RESIDENTIAL FIXTURES										800
	0400	Fluorescent, interior, surface, circline, 32 watt & 40 watt	1 Elec	20	.400	Ea.	76.50	15.75		92.25	108	
	0500	2′ x 2′, two U 40 watt		8	1		93	39.50		132.50	163	
	0700	Shallow under cabinet, two 20 watt		16	.500		40.50	19.70		60.20	75.50	
	0900	Wall mounted, 4′L, one 40 watt, with baffle		10	.800		99	31.50		130.50	158	
	2000	Incandescent, exterior lantern, wall mounted, 60 watt		16	.500		31	19.70		50.70	64.50	
	2100	Post light, 150W, with 7′ post		4	2		110	79		189	242	
	2500	Lamp holder, weatherproof with 150W PAR		16	.500		16.90	19.70		36.60	49	
	2550	With reflector and guard		12	.667		52	26.50		78.50	97.50	
	2600	Interior pendent, globe with shade, 150 watt		20	.400		117	15.75		132.75	153	
	9000	Minimum labor/equipment charge		4	2	Job		79		79	121	

Important: See the Reference Section for critical supporting data - Reference Nos., Crews, & City Cost Indexes

16520 | Exterior Luminaires

			CREW	DAILY OUTPUT	LABOR-HOURS	UNIT	2004 BARE COSTS MAT.	LABOR	EQUIP.	TOTAL	TOTAL INCL O&P	
300	0010	**EXTERIOR FIXTURES** With lamps										300
	0400	Quartz, 500 watt	1 Elec	5.30	1.509	Ea.	53.50	59.50		113	151	
	1100	Wall pack, low pressure sodium, 35 watt		4	2		214	79		293	355	
	1150	55 watt		4	2		255	79		334	400	
	1160	High pressure sodium, 70 watt		4	2		185	79		264	325	
	1170	150 watt		4	2		228	79		307	370	
	1180	Metal Halide, 175 watt		4	2		255	79		334	400	
	1190	250 watt		4	2		260	79		339	405	
	1200	Floodlights with ballast and lamp,										
	1400	pole mounted, pole not included										
	2250	Low pressure sodium, 55 watt	1 Elec	2.70	2.963	Ea.	485	117		602	715	
	2270	90 watt		2	4	"	535	158		693	830	
	9000	Minimum labor/equipment charge		3.75	2.133	Job		84		84	129	

16530 | Emergency Lighting

			CREW	DAILY OUTPUT	LABOR-HOURS	UNIT	MAT.	LABOR	EQUIP.	TOTAL	TOTAL INCL O&P	
320	0010	**EXIT AND EMERGENCY LIGHTING**										320
	0080	Exit light ceiling or wall mount, incandescent, single face	1 Elec	8	1	Ea.	38.50	39.50		78	103	
	0100	Double face	"	6.70	1.194	"	44	47		91	121	
	0300	Emergency light units, battery operated										
	0350	Twin sealed beam light, 25 watt, 6 volt each										
	0500	Lead battery operated	1 Elec	4	2	Ea.	110	79		189	242	
	0700	Nickel cadmium battery operated		4	2	"	505	79		584	675	
	9000	Minimum labor/equipment charge		4	2	Job		79		79	121	

16550 | Special Purpose Lighting

			CREW	DAILY OUTPUT	LABOR-HOURS	UNIT	MAT.	LABOR	EQUIP.	TOTAL	TOTAL INCL O&P	
820	0010	**TRACK LIGHTING**										820
	0080	Track, 1 circuit, 4′ section	1 Elec	6.70	1.194	Ea.	36.50	47		83.50	113	
	0100	8′ section		5.30	1.509		62	59.50		121.50	160	
	0300	3 circuits, 4′ section		6.70	1.194		48	47		95	125	
	0400	8′ section		5.30	1.509		74	59.50		133.50	173	
	9000	Minimum labor/equipment charge		3	2.667	Job		105		105	161	

16585 | Lamps

			CREW	DAILY OUTPUT	LABOR-HOURS	UNIT	MAT.	LABOR	EQUIP.	TOTAL	TOTAL INCL O&P	
600	0010	**LAMPS**										600
	0080	Fluorescent, rapid start, cool white, 2′ long, 20 watt	1 Elec	1	8	C	300	315		615	815	
	0100	4′ long, 40 watt		.90	8.889		273	350		623	840	
	0170	4′ long, 34 watt energy saver		.90	8.889		273	350		623	840	
	0176	2′ long, T8, 17 W engergy saver		1	8		380	315		695	905	
	0178	3′ long, T8, 25 W energy saver		.90	8.889		380	350		730	960	
	0180	4′ long, T8, 32 watt energy saver		.90	8.889		238	350		588	800	
	0560	Twin tube compact lamp		.90	8.889		495	350		845	1,075	
	0570	Double twin tube compact lamp		.80	10		1,200	395		1,595	1,925	
	0600	Mercury vapor, mogul base, deluxe white, 100 watt		.30	26.667		3,200	1,050		4,250	5,150	
	0800	400 watt		.30	26.667		3,400	1,050		4,450	5,350	
	1000	Metal halide, mogul base, 175 watt		.30	26.667		4,000	1,050		5,050	6,000	
	1100	250 watt		.30	26.667		4,500	1,050		5,550	6,575	
	1350	High pressure sodium, 70 watt		.30	26.667		4,625	1,050		5,675	6,725	
	1370	150 watt		.30	26.667		4,975	1,050		6,025	7,100	
	1500	Low pressure sodium, 35 watt		.30	26.667		7,250	1,050		8,300	9,600	
	1600	90 watt		.30	26.667		8,925	1,050		9,975	11,500	
	1800	Incandescent, interior, A21, 100 watt		1.60	5		183	197		380	505	
	1900	A21, 150 watt		1.60	5		188	197		385	510	
	2300	R30, 75 watt		1.30	6.154		760	242		1,002	1,200	
	2500	Exterior, PAR 38, 75 watt		1.30	6.154		1,350	242		1,592	1,850	
	2600	PAR 38, 150 watt		1.30	6.154		1,500	242		1,742	2,025	

ELECTRICAL 16

16500 | Lighting

	16585 | Lamps		CREW	DAILY OUTPUT	LABOR-HOURS	UNIT	2004 BARE COSTS MAT.	LABOR	EQUIP.	TOTAL	TOTAL INCL O&P	
600	9000	Minimum labor/equipment charge	1 Elec	4	2	Job		79		79	121	600

16800 | Sound & Video

	16810 | Sound and Video Circuits		CREW	DAILY OUTPUT	LABOR-HOURS	UNIT	2004 BARE COSTS MAT.	LABOR	EQUIP.	TOTAL	TOTAL INCL O&P	
750	0010	**SOUND AND VIDEO CABLES & FITTINGS**										750
	1250	Nonshielded, #22-2 conductor	1 Elec	10	.800	C.L.F.	8.60	31.50		40.10	58	

	16820 | Sound Reinforcement		CREW	DAILY OUTPUT	LABOR-HOURS	UNIT	MAT.	LABOR	EQUIP.	TOTAL	TOTAL INCL O&P	
300	0010	**DOORBELL SYSTEM** Incl. transformer, button & signal										300
	0020											
	1000	Door chimes, 2 notes, minimum	1 Elec	16	.500	Ea.	22	19.70		41.70	54.50	
	1020	Maximum		12	.667		115	26.50		141.50	168	
	1100	Tube type, 3 tube system		12	.667		163	26.50		189.50	220	
	1180	4 tube system		10	.800		261	31.50		292.50	335	
	1900	For transformer & button, minimum add		5	1.600		12.40	63		75.40	111	
	1960	Maximum, add		4.50	1.778		37	70		107	149	
	3000	For push button only, minimum		24	.333		2.42	13.15		15.57	22.50	
	3100	Maximum		20	.400		19.45	15.75		35.20	45.50	
	9000	Minimum labor/equipment charge		4	2	Job		79		79	121	
800	0010	**PUBLIC ADDRESS SYSTEM**										800
	0100	Conventional, office	1 Elec	5.33	1.501	Speaker	97.50	59		156.50	198	
	0200	Industrial		2.70	2.963	"	188	117		305	385	
	9000	Minimum labor/equipment charge		3.50	2.286	Job		90		90	138	
840	0010	**SOUND SYSTEM** not including rough-in wires, cables & conduits										840
	2020	11 station capacity	2 Elec	4	4	Ea.	690	158		848	1,000	
	3600	House telephone, talking station	1 Elec	1.60	5		335	197		532	675	
	3800	Press to talk, release to listen	"	5.30	1.509		78.50	59.50		138	178	
	4000	System-on button					47			47	51.50	
	4200	Door release	1 Elec	4	2		84	79		163	214	
	4400	Combination speaker and microphone		8	1		143	39.50		182.50	218	
	4600	Termination box		3.20	2.500		45	98.50		143.50	201	
	4800	Amplifier or power supply		5.30	1.509		515	59.50		574.50	660	
	5000	Vestibule door unit		16	.500	Name	95	19.70		114.70	136	
	5200	Strip cabinet		27	.296	Ea.	180	11.65		191.65	215	
	5400	Directory		16	.500	"	84.50	19.70		104.20	124	
	9000	Minimum labor/equipment charge		3.50	2.286	Job		90		90	138	

	16850 | Television Equipment		CREW	DAILY OUTPUT	LABOR-HOURS	UNIT	MAT.	LABOR	EQUIP.	TOTAL	TOTAL INCL O&P	
600	0010	**T.V. SYSTEMS** not including rough-in wires, cables & conduits										600
	5000	T.V. Antenna only, minimum	1 Elec	6	1.333	Ea.	34	52.50		86.50	118	
	5100	Maximum	"	4	2	"	143	79		222	278	

Assemblies Section

Table of Contents

How to Use the Assemblies Cost Tables

The following is a detailed explanation of a sample Assemblies Cost Table. Most Assembly Tables are separated into three parts: 1) an illustration of the system to be estimated; 2) the components and related costs of a typical system; and 3) the costs for similar systems with dimensional and/or size variations. For costs of the components that comprise these systems or "assemblies," refer to the Unit Price Section. Next to each bold number below is the item being described with the appropriate component of the sample entry following in parenthesis. In most cases, if the work is to be subcontracted, the general contractor will need to add an additional markup (RSMeans suggests using 10%) to the "Total" figures.

1 System/Line Numbers (B1010 263 1700)

Each Assemblies Cost Line has been assigned a unique identification number based on the UNIFORMAT II classification system.

UNIFORMAT II Major Group

B1010 263 1700

UNIFORMAT II Level 3

Means Major Classification

Means Individual Line Number

B10 Superstructure

B1010 Floor Construction

This page illustrates and describes a wood joist floor system including wood joist, oak floor, sub-floor, bridging, sand and finish floor, furring, plasterboard, taped, finished and painted ceiling. Lines within System Components give the unit price and total price per square foot for this system. Prices for alternate wood joist floor systems are on Line Items B1010 263 1700 thru 2300. Both material quantities and labor costs have been adjusted for the system listed.

Factors: To adjust for job conditions other than normal working situations use Lines B1010 263 2700 thru 4000.

Example: You are to install the system during off peak hours, 6 P.M. to 2 A.M. Go to Line B1010 263 3800 and apply this percentage to the appropriate INST. costs.

System Components	QUANTITY	UNIT	COST PER S.F. MAT.	INST.	TOTAL
Wood joists, 2" x 8", 16" O.C.,oak floor (sanded & finished),1/2" sub Floor, 1"x 3" bridging, furring, 5/8" drywall, taped, finished and painted.					
Wood joists, 2" x 8", 16" O.C.	1.000	L.F.	.88	.80	1.68
Subfloor plywood CDX 1/2"	1.000	SF Flr.	.43	.58	1.01
Bridging 1" x 3"	.150	Pr.	.05	.51	.56
Oak flooring, No. 1 common	1.000	S.F.	4.09	2.57	6.66
Sand and finish floor	1.000	S.F.	.76	1.17	1.93
Wood furring, 1" x 3", 16" O.C.	1.000	L.F.	.23	1.25	1.48
Gypsum drywall, 5/8" thick	1.000	S.F.	[illegible]	[illegible]	[illegible]
Tape and finishing	1.000	S.F.	.04	[illegible]	[illegible]
Paint ceiling	1.000	S.F.	.15	[illegible]	[illegible]
TOTAL			6.87	8.31	15.18

B1010 263	Floor-Ceiling, Wood Joists	COST PER S.F. MAT.	INST.	TOTAL
1600	For alternate wood joist systems:			
1700	16" on center, 2" x 6" joists	6.55	8.20	14.75
1800	2" x 10"	7.20	8.50	15.70
1900	2" x 12"	7.60	8.50	16.10
2000	2" x 14"	7.95	8.65	16.60
2100	12" on center, 2" x 10" joists	7.50	8.70	16.20
2200	2" x 12"	8	8.75	16.75
2300	2" x 14"	8.45	8.95	17.40

2

Illustration

At the top of most assembly tables is an illustration, a brief description, and the design criteria used to develop the cost.

3

System Components

The components of a typical system are listed separately to show what has been included in the development of the total system price. The table below contains prices for other similar systems with dimensional and/or size variations.

4

Quantity

This is the number of line item units required for one system unit. For example, we assume that it will take .15 pair of 1″ × 3″ bridging on a square foot basis.

5

Unit of Measure for Each Item

The abbreviated designation indicates the unit of measure, as defined by industry standards, upon which the price of the component is based. For example, wood joists are priced by the L.F. (linear foot) while plywood is priced by S.F. (square foot).

6

Unit of Measure for Each System (Cost per S.F.)

Costs shown in the three right hand columns have been adjusted by the component quantity and unit of measure for the entire system. In this example, "Cost per S.F." is the unit of measure for this system or "assembly."

7

Materials (6.55)

This column contains the Materials Cost of each component. These cost figures are bare costs plus 10% for profit.

8

Installation (8.20)

Installation includes labor and equipment plus the installing contractor's overhead and profit. Equipment costs are the bare rental costs plus 10% for profit. The labor overhead and profit is defined on the inside back cover of this book.

9

Total (14.75)

The figure in this column is the sum of the material and installation costs.

Material Cost	+	Installation Cost	=	Total
$6.55	+	$8.20	=	$14.75

A SUBSTRUCTURE

A1010 Standard Foundations

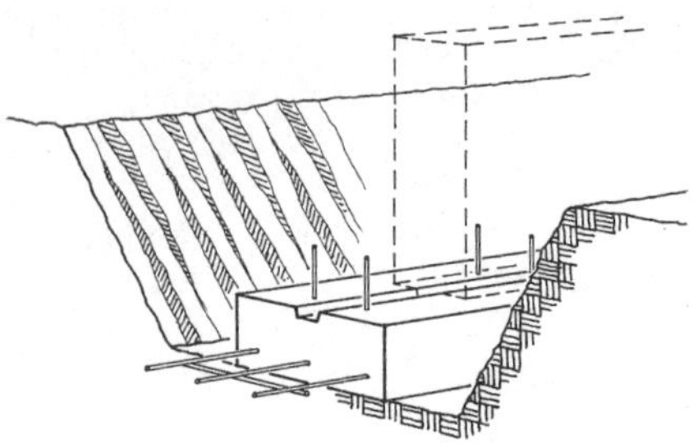

This page illustrates and describes a strip footing system including concrete, forms, reinforcing, keyway and dowels. Lines within System Components give the unit price and total price per linear foot for this system. Prices for alternate strip footing systems are on Line Items A1010 120 1500 thru 2500. Both material quantities and labor costs have been adjusted for the system listed.

Factors: To adjust for job conditions other than normal working situations use Lines A1010 120 2900 thru 4000.

Example: You are to install this footing, and due to a lack of accessibility, only hand tools can be used. Material handling is also a problem. Go to Lines A1010 120 3400 and 3600 and apply these percentages to the appropriate MAT. and INST. costs.

System Components	QUANTITY	UNIT	COST PER L.F. MAT.	INST.	TOTAL
Strip footing, 2′-0″ wide x 1′-0″ thick, 2000 psi concrete including forms Reinforcing, keyway, and dowels.					
Concrete, 2000 psi	.074	C.Y.	5.66		5.66
Placing concrete	.074	C.Y.		1.36	1.36
Forms, footing, 4 uses	2.000	S.F.	1.58	6.84	8.42
Reinforcing	3.170	Lb.	1.05	1.55	2.60
Keyway, 2″ x 4″, 4 uses	1.000	L.F.	.18	.83	1.01
Dowels, #4 bars, 2′ long, 24″ O.C.	.500	Ea.	.23	1.07	1.30
TOTAL			8.70	11.65	20.35

A1010 120	Strip Footing	COST PER L.F. MAT.	INST.	TOTAL
1400	Above system with the following:			
1500	2′-0″ wide x 1′ thick, 3000 psi concrete	8.95	11.65	20.60
1600	4000 psi concrete	9.25	11.65	20.90
1800	For alternate footing systems:			
1900	2′-6″ wide x 1′ thick, 2000 psi concrete	10.40	12.40	22.80
2000	3000 psi concrete	10.75	12.40	23.15
2100	4000 psi concrete	11.10	12.40	23.50
2300	3′-0″ wide x 1′ thick, 2000 psi concrete	12	13.10	25.10
2400	3000 psi concrete	12.35	13.10	25.45
2500	4000 psi concrete	12.80	13.10	25.90
2700				
2800				
2900	Cut & patch to match existing construction, add, minimum	2%	3%	
3000	Maximum	5%	9%	
3100	Dust protection, add, minimum	1%	2%	
3200	Maximum	4%	11%	
3300	Equipment usage curtailment, add, minimum	1%	1%	
3400	Maximum	3%	10%	
3500	Material handling & storage limitation, add, minimum	1%	1%	
3600	Maximum	6%	7%	
3700	Protection of existing work, add, minimum	2%	2%	
3800	Maximum	5%	7%	
3900	Shift work requirements, add, minimum		5%	
4000	Maximum		30%	

A1010 Standard Foundations

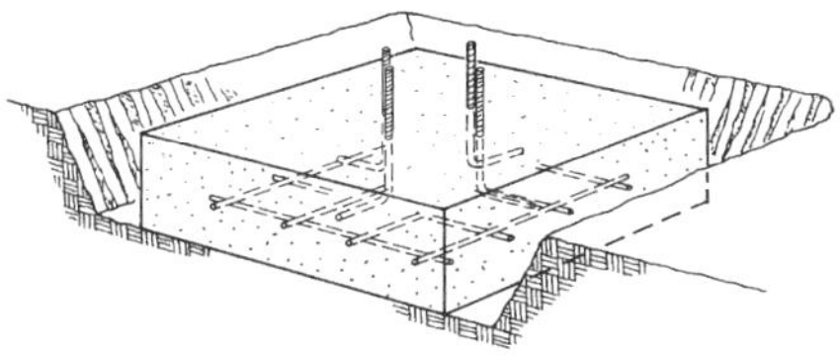

This page illustrates and describes a spread footing system including concrete, forms, reinforcing and anchor bolts. Lines within System Components give the unit price and total price on a cost each basis for this system. Prices for alternate spread footing systems are on Line Items A1010 230 1300 thru 2300. Both material quantities and labor costs have been adjusted for the system listed.

Factors: To adjust for job conditions other than normal working situations use Lines A1010 230 2900 thru 4000.

Example: You are to install the system in an existing occupied building. Access to the site and protection to the building are mandatory. Go to Lines A1010 230 3600 and 3800 and apply these percentages to the appropriate MAT. and INST. costs.

System Components	QUANTITY	UNIT	COST EACH MAT.	COST EACH INST.	COST EACH TOTAL
Interior column footing, 3′ square 1′ thick, 2000 psi concrete, Including forms, reinforcing, and anchor bolts.					
Concrete, 2000 psi	.330	C.Y.	25.25		25.25
Placing concrete	.330	C.Y.		13.21	13.21
Forms, footing, 4 uses	12.000	SFCA	6.84	48	54.84
Reinforcing	11.000	Lb.	3.63	5.39	9.02
Anchor bolts, 3/4″ diameter	2.000	Ea.	1.14	4.44	5.58
TOTAL			36.86	71.04	107.90

A1010 230	Spread Footing	COST EACH MAT.	COST EACH INST.	COST EACH TOTAL
1200	Above system with the following:			
1300	3′ square x 1′ thick, 3000 psi concrete	38	71	109
1400	4000 psi concrete	39.50	71	110.50
1600	For alternate footing systems:			
1700	4′ square x 1′ thick, 2000 psi concrete	62	102	164
1800	3000 psi concrete	64	102	166
1900	4000 psi concrete	66.50	102	168.50
2100	5′ square x 1′-3″ thick, 2000 psi concrete	117	169	286
2200	3000 psi concrete	121	169	290
2300	4000 psi concrete	125	169	294
2600				
2700				
2900	Cut & patch to match existing construction, add, minimum	2%	3%	
3000	Maximum	5%	9%	
3100	Dust protection, add, minimum	1%	2%	
3200	Maximum	4%	11%	
3300	Equipment usage curtailment, add, minimum	1%	1%	
3400	Maximum	3%	10%	
3500	Material handling & storage limitation, add, minimum	1%	1%	
3600	Maximum	6%	7%	
3700	Protection of existing work, add, minimum	2%	2%	
3800	Maximum	5%	7%	
3900	Shift work requirements, add, minimum		5%	
4000	Maximum		30%	

A1010 Standard Foundations

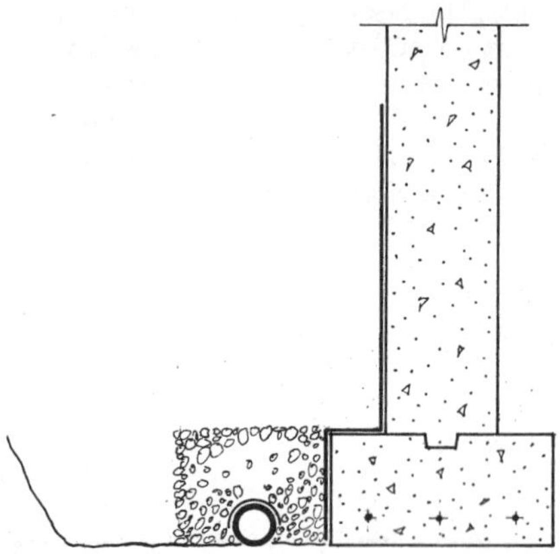

General: Footing drains can be placed either inside or outside of foundation walls depending upon the source of water to be intercepted. If the source of subsurface water is principally from grade or a subsurface stream above the bottom of the footing, outside drains should be used. For high water tables, use inside drains or both inside and outside.

The effectiveness of underdrains depends on good waterproofing. This must be carefully installed and protected during construction.

Costs below include the labor and materials for the pipe and 6″ only of gravel or crushed stone around pipe. Excavation and backfill are not included.

System Components	QUANTITY	UNIT	COST PER L.F. MAT.	INST.	TOTAL
Foundation underdrain, outside, PVC 4″ diameter.					
PVC pipe 4″ diam. S.D.R. 35	1.000	L.F.	1.93	3.08	5.01
Pipe bedding, graded gravel 3/4″ to 1/2″	.070	C.Y.	1.89	.62	2.51
TOTAL			3.82	3.70	7.52

A1010 310	Foundation Underdrain	COST PER L.F. MAT.	INST.	TOTAL
0900	For alternate drain systems:			
1000	Foundation underdrain, outside only, PVC, 4″ diameter	3.82	3.70	7.52
1100	6″ diameter	5.85	4.10	9.95
1200	Bituminous fiber, 4″ diameter	3.82	3.70	7.52
1300	6″ diameter	5.85	4.10	9.95
1400	Porous concrete, 6″ diameter	5.10	4.47	9.57
1450	8″ diameter	6.25	6	12.25
1500	12″ diameter	11.25	6.85	18.10
1600	Corrugated metal, 16 ga. asphalt coated, 6″ diameter	5.35	7.05	12.40
1650	8″ diameter	7.25	7.45	14.70
1700	10″ diameter	8.85	7.80	16.65
2000	Vitrified clay, C-211, 4″ diameter	4.09	6.60	10.69
2050	6″ diameter	6.05	8.35	14.40
2100	12″ diameter	15.20	10.05	25.25
3000	Outside and inside, PVC, 4″ diameter	7.65	7.40	15.05
3100	6″ diameter	11.75	8.15	19.90
3200	Bituminous fiber, 4″ diameter	7.65	7.40	15.05
3300	6″ diameter	11.75	8.15	19.90
3400	Porous concrete, 6″ diameter	10.20	8.90	19.10
3450	8″ diameter	12.50	11.95	24.45
3500	12″ diameter	22.50	13.70	36.20
3600	Corrugated metal, 16 ga., asphalt coated, 6″ diameter	10.75	14.05	24.80
3650	8″ diameter	14.50	14.85	29.35
3700	10″ diameter	17.70	15.55	33.25
4000	Vitrified clay, C-211, 4″ diameter	8.20	13.15	21.35
4050	6″ diameter	12.15	16.75	28.90
4100	12″ diameter	30.50	20	50.50
4700	Cut & patch to match existing construction, add, minimum	2%	3%	
4800	Maximum	5%	9%	
4900	Dust protection, add, minimum	1%	2%	
5000	Maximum	4%	11%	
5100	Equipment usage curtailment, add, minimum	1%	1%	

Important: See the Reference Section for critical supporting data - Reference Numbers and City Cost Indexes

A1010 Standard Foundations

A1010 310	Foundation Underdrain	COST PER L.F.		
		MAT.	INST.	TOTAL
5200	Maximum	3%	10%	
5300	Material handling & storage limitation, add, minimum	1%	1%	
5400	Maximum	6%	7%	
5500	Protection of existing work, add, minimum	2%	2%	
5600	Maximum	5%	7%	
5700	Shift work requirements, add, minimum		5%	
5800	Maximum		30%	
5900	Temporary shoring and bracing, add, minimum	2%	5%	
6000	Maximum	5%	12%	

A1030 Slab on Grade

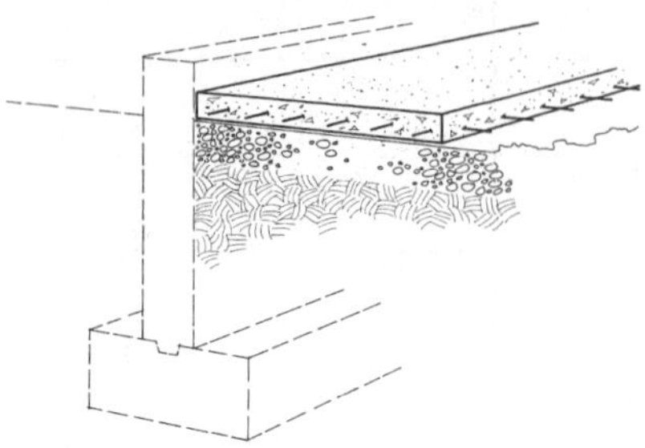

This page illustrates and describes a slab on grade system including slab, bank run gravel, bulkhead forms, placing concrete, welded wire fabric, vapor barrier, steel trowel finish and curing paper. Lines within System Components give the unit price and total price per square foot for this system. Prices for alternate slab on grade systems are on Line Items A1030 110 1500 thru 2600. Both material quantities and labor costs have been adjusted for the system listed.

Factors: To adjust for job conditions other than normal working situations use Lines A1030 110 2900 thru 4000.

Example: You are to install the system at a site where protection of the existing building is required. Go to Line A1030 110 3800 and apply these percentages to the appropriate MAT. and INST. costs.

System Components	QUANTITY	UNIT	COST PER S.F. MAT.	COST PER S.F. INST.	COST PER S.F. TOTAL
Ground slab, 4″ thick, 3000 psi concrete, 4″ granular base, vapor barrier Welded wire fabric, screed and steel trowel finish.					
Concrete, 4″ thick, 3000 psi concrete	.012	C.Y.	.96		.96
Bank run gravel, 4″ deep	.074	C.Y.	.20	.07	.27
Polyethylene vapor barrier, 10 mil.	.011	C.S.F.	.06	.13	.19
Bulkhead forms, expansion material	.100	L.F.	.04	.28	.32
Welded wire fabric, 6 x 6 - #10/10	.011	C.S.F.	.09	.32	.41
Place concrete	.012	C.Y.		.24	.24
Screed & steel trowel finish	1.000	S.F.		.72	.72
TOTAL			1.35	1.76	3.11

A1030 110	Interior Slab on Grade	COST PER S.F. MAT.	COST PER S.F. INST.	COST PER S.F. TOTAL
1400	Above system with the following:			
1500	4″ thick slab, 3000 psi concrete, 6″ deep bank run gravel	1.55	1.75	3.30
1600	12″ deep bank run gravel	1.95	1.77	3.72
1700				
1800				
1900				
2000	For alternate slab systems:			
2100	5″ thick slab, 3000 psi concrete, 6″ deep bank run gravel	1.79	1.81	3.60
2200	12″ deep bank run gravel	2.19	1.83	4.02
2300				
2400				
2500	6″ thick slab, 3000 psi concrete, 6″ deep bank run gravel	2.11	1.89	4
2600	12″ deep bank run gravel	2.51	1.91	4.42
2700				
2900	Cut & patch to match existing construction, add, minimum	2%	3%	
3000	Maximum	5%	9%	
3100	Dust protection, add, minimum	1%	2%	
3200	Maximum	4%	11%	
3300	Equipment usage curtailment, add, minimum	1%	1%	
3400	Maximum	3%	10%	
3500	Material handling & storage limitation, add, minimum	1%	1%	
3600	Maximum	6%	7%	
3700	Protection of existing work, add, minimum	2%	2%	
3800	Maximum	5%	7%	
3900	Shift work requirements, add, minimum		5%	
4000	Maximum		30%	

A2010 Basement Excavation

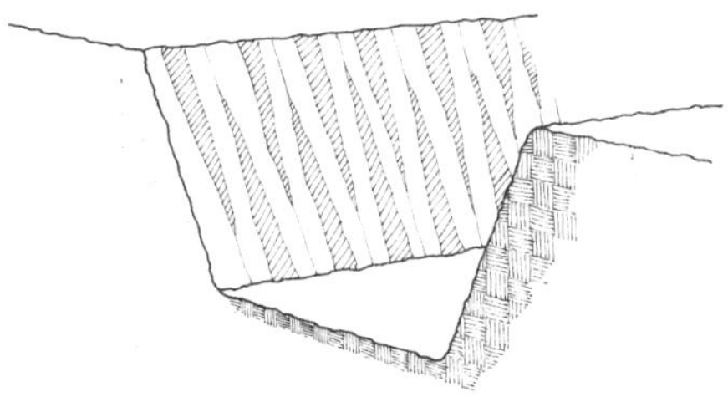

This page illustrates and describes continuous footing and trench excavation systems including a wheel mounted backhoe, operator, equipment rental, fuel, oil, mobilization, no hauling or backfill. Lines within System Components give the unit price and total price per linear foot for this system. Prices for alternate continuous footing and trench excavation systems are on Line Items A2010 130 1300 thru 2700. Both material quantities and labor costs have been adjusted for the system listed.

Factors: To adjust for job conditions other than normal working situations use Lines A2010 130 2900 thru 4000.

Example: You are to install the system and supply dust protection. Go to Line A2010 130 3000 and apply this percentage to the appropriate TOTAL costs.

System Components	QUANTITY	UNIT	COST PER L.F. EQUIP.	LABOR	TOTAL
Continuous footing or trench excav. w/ 3/4 C.Y. wheel mntd backhoe Including operator, equipment rental, fuel, oil and mobilization. Trench Is 4′ wide at bottom, 3′ deep w/sides sloped 1 to 2 and 200′ long. Costs Are based on a production rate of 240 C.Y./day. No hauling/backfill incl.					
Equipment operator	4.000	Hr.		212.50	212.50
3/4 C.Y. backhoe, wheel mounted	.500	Day	74.25		74.25
Operating expense (fuel, oil)	4.000	Hr.	40.04		40.04
Mobilization	1.000	Ea.	154	95.50	249.50
TOTAL			268.29	308	576.29
COST PER L.F.		L.F.	1.34	1.54	2.88

A2010 130	Excavation, Footings or Trench	COST PER L.F. EQUIP.	LABOR	TOTAL
1200	For alternate trench sizes, 4′ bottom with sloped sides:			
1300	2′ deep, 50′ long	3.08	6.15	9.23
1400	100′ long	1.54	3.09	4.63
1500	300′ long	.51	1.03	1.54
1600	4′ deep, 50′ long	3.08	6.15	9.23
1700	100′ long	1.54	3.09	4.63
1800	300′ long	.51	1.83	2.34
1900	6′ deep, 50′ long	3.08	6.15	9.23
2000	100′ long	1.54	3.88	5.42
2100	300′ long	.51	3.15	3.66
2200	8′ deep, 50′ long	3.08	6.15	9.23
2300	100′ long	1.54	5.20	6.74
2400	300′ long	.51	3.15	3.66
2500	10′ deep, 50′ long	3.08	8.30	11.38
2600	100′ long	1.54	7.05	8.59
2700	300′ long	.51	6.25	6.76
2800				
2900	Dust protection, add, minimum		2%	2%
3000	Maximum		11%	11%
3100	Equipment usage curtailment, add, minimum		1%	1%
3200	Maximum		10%	10%
3300	Material handling & storage limitation, add, minimum		1%	1%
3400	Maximum		7%	7%
3500	Protection of existing work, add, minimum		2%	2%
3600	Maximum		7%	7%
3700	Shift work requirements, add, minimum		5%	5%
3800	Maximum		30%	30%
3900	Temporary shoring and bracing, add, minimum		5%	5%
4000	Maximum		12%	12%

A2010 Basement Excavation

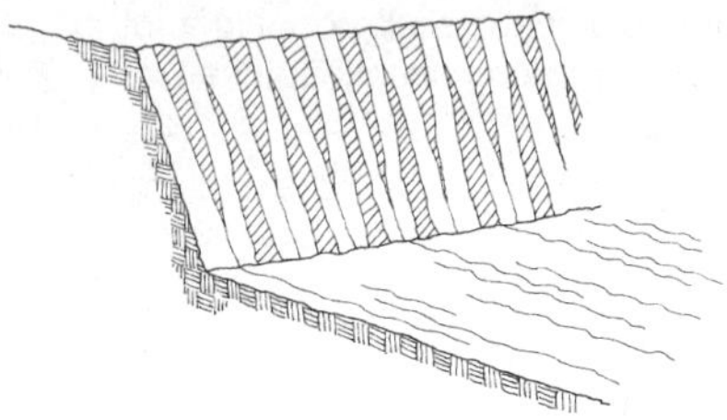

This page illustrates and describes foundation excavation systems including a backhoe-loader, operator, equipment rental, fuel, oil, mobilization, hauling material, no backfilling. Lines within System Components give the unit price and total price per cubic yard for this system. Prices for alternate foundation excavation systems are on Line Items A2010 140 1600 thru 2200. Both material quantities and labor costs have been adjusted for the system listed.

Factors: To adjust for job conditions other than normal working situations use Lines A2010 140 2900 thru 4000.

Example: You are to install the system with the use of temporary shoring and bracing. Go to Line A2010 140 3900 and apply these percentages to the appropriate TOTAL costs.

System Components	QUANTITY	UNIT	COST PER C.Y. EQUIP.	LABOR	TOTAL
Foundation excav w/ 3/4 C.Y. backhoe-loader, incl. operator, equip Rental, fuel, oil, and mobilization. Hauling of excavated material is Included. Prices based on one day production of 360 C.Y. in medium soil Without backfilling.					
Equipment operator	8.000	Hr.		425	425
Backhoe-loader, 3/4 C.Y.	1.000	Day	148.50		148.50
Operating expense (fuel, oil)	8.000	Hr.	80.08		80.08
Hauling, 12 C.Y. trucks, 1 mile round trip	360.000	C.Y.	727.20	475.20	1,202.40
Mobilization	1.000	Ea.	154	95.50	249.50
TOTAL			1,109.78	995.70	2,105.48
COST PER C.Y.		C.Y.	3.07	2.76	5.84

A2010 140	Excavation, Foundation	COST PER C.Y. EQUIP.	LABOR	TOTAL
1400	For alternate size excavations:			
1500				
1600	100 C.Y.	3.56	4.67	8.23
1700	200 C.Y.	2.79	3	5.79
1800	300 C.Y.	2.53	2.83	5.36
1900	400 C.Y.	2.41	2.73	5.14
2000	500 C.Y.	2.33	2.70	5.03
2100	600 C.Y.	2.28	2.66	4.94
2200	700 C.Y.	2.24	2.61	4.85
2300				
2400				
2500				
2900	Dust protection, add, minimum		2%	2%
3000	Maximum		11%	11%
3100	Equipment usage curtailment, add, minimum		1%	1%
3200	Maximum		10%	10%
3300	Material handling & storage limitation, add, minimum		1%	1%
3400	Maximum		7%	7%
3500	Protection of existing work, add, minimum		2%	2%
3600	Maximum		7%	7%
3700	Shift work requirements, add, minimum		5%	5%
3800	Maximum		30%	30%
3900	Temporary shoring and bracing, add, minimum		5%	5%
4000	Maximum		12%	12%

SUBSTRUCTURE A

A2020 Basement Walls

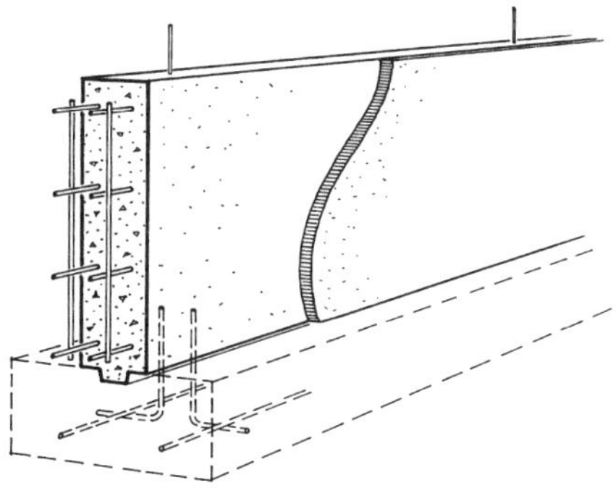

This page illustrates and describes a concrete wall system including concrete, placing concrete, forms, reinforcing, insulation, waterproofing and anchor bolts. Lines within System Components give the unit price and total price per linear foot for this system. Prices for alternate concrete wall systems are on Line Items A2020 120 1500 thru 2600. Both material quantities and labor costs have been adjusted for the system listed.

Factors: To adjust for job conditions other than normal working situations use Lines A2020 120 2900 thru 4000.

Example: You are to install this wall system where delivery of material is difficult. Go to Line A2020 120 3600 and apply these percentages to the appropriate MAT. and INST. costs.

System Components	QUANTITY	UNIT	COST PER L.F.		
			MAT.	INST.	TOTAL
Cast in place concrete foundation wall, 8″ thick, 3′ high, 2500 psi					
Concrete including forms, reinforcing, waterproofing, and anchor bolts.					
Concrete, 2500 psi, 8″ thick, 3′ high	.070	C.Y.	5.46		5.46
Forms, wall, 4 uses	6.000	S.F.	4.44	35.40	39.84
Reinforcing	6.000	Lb.	1.98	2.04	4.02
Placing concrete	.070	C.Y.		2.63	2.63
Waterproofing	3.000	S.F.	.33	2.46	2.79
Rigid insulaton, 1″ polystyrene	3.000	S.F.	.63	1.92	2.55
Anchor bolts, 1/2″ diameter, 4′ O.C.	.250	Ea.	.27	.59	.86
TOTAL			13.11	45.04	58.15

A2020 120	Concrete Wall	COST PER L.F.		
		MAT.	INST.	TOTAL
1400	For alternate wall systems:			
1500	8″ thick, 2500 psi concrete, 4′ high	17.90	60.50	78.40
1600	6′ high	26.50	90	116.50
1700	8′ high	35.50	120	155.50
1800	3500 psi concrete, 4′ high	18.25	60.50	78.75
1900	6′ high	27.50	90	117.50
2000	8′ high	36.50	120	156.50
2100	12″ thick, 2500 psi concrete, 4′ high	23	63.50	86.50
2200	6′ high	33.50	94	127.50
2300	8′ high	44.50	125	169.50
2400	3500 psi concrete, 4′ high	23.50	63.50	87
2500	8′ high	45.50	125	170.50
2600	10′ high	56.50	155	211.50
2700				
2900	Cut & patch to match existing construction, add, minimum	2%	3%	
3000	Maximum	5%	9%	
3100	Dust protection, add, minimum	1%	2%	
3200	Maximum	4%	11%	
3300	Equipment usage curtailment, add, minimum	1%	1%	
3400	Maximum	3%	10%	
3500	Material handling & storage limitation, add, minimum	1%	1%	
3600	Maximum	6%	7%	
3700	Protection of existing work, add, minimum	2%	2%	
3800	Maximum	5%	7%	
3900	Shift work requirements, add, minimum		5%	
4000	Maximum		30%	
4100				

A2020 Basement Walls

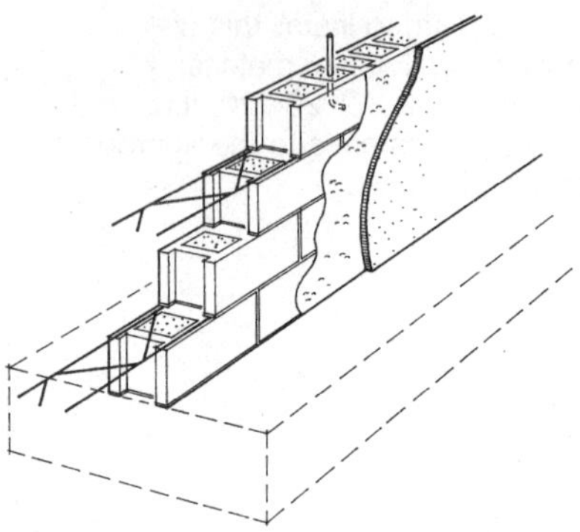

This page illustrates and describes a concrete block wall system including concrete block, masonry reinforcing, parging, waterproofing insulation and anchor bolts. Lines within System Components give the unit price and total price per linear foot for this system. Prices for alternate concrete block wall systems are on Line Items A2020 140 1300 thru 2600. Both material quantities and labor costs have been adjusted for the system listed.

Factors: To adjust for job conditions other than normal working situations use Lines A2020 140 2900 thru 4000.

Example: You are to install the system to match an existing foundation wall. Go to Line A2020 140 3000 and apply these percentages to the appropriate MAT. and INST. costs.

System Components	QUANTITY	UNIT	COST PER L.F.		
			MAT.	INST.	TOTAL
Concrete block, 8″ thick, masonry reinforcing, parged and Waterproofed, insulation and anchor bolts, wall 2′-8″ high.					
Concrete block, 8″ x 8″ x 16″	2.670	S.F.	5.15	12.55	17.70
Masonry reinforcing	2.000	L.F.	.24	.30	.54
Parging	.193	S.Y.	.69	1.87	2.56
Waterproofing	2.670	S.F.	.29	2.19	2.48
Insulation, 1″ rigid polystyrene	2.670	S.F.	.56	1.71	2.27
Anchor bolts, 1/2″ diameter, 4′ O.C.	.250	Ea.	.27	.59	.86
TOTAL			7.20	19.21	26.41

A2020 140	Concrete Block Wall	COST PER L.F.		
		MAT.	INST.	TOTAL
1200	For alternate wall systems:			
1300	8″ thick block, 4′ high	10.70	28.50	39.20
1400	6′ high	15.80	42	57.80
1500	8′ high	21	56.50	77.50
1600	Grouted solid, 4′ high	14.45	38.50	52.95
1700	6′ high	21.50	57	78.50
1800	8′ high	28.50	76	104.50
2100	12″ thick block, 4′ high	13.80	41	54.80
2200	6′ high	20.50	61	81.50
2300	8′ high	27.50	81.50	109
2400	Grouted solid, 4′ high	19.95	51.50	71.45
2500	6′ high	29.50	76.50	106
2600	8′ high	39.50	102	141.50
2700				
2900	Cut & patch to match existing construction, add, minimum	2%	3%	
3000	Maximum	5%	9%	
3100	Dust protection, add, minimum	1%	2%	
3200	Maximum	4%	11%	
3300	Equipment usage curtailment, add, minimum	1%	1%	
3400	Maximum	3%	10%	
3500	Material handling & storage limitation, add, minimum	1%	1%	
3600	Maximum	6%	7%	
3700	Protection of existing work, add, minimum	2%	2%	
3800	Maximum	5%	7%	
3900	Shift work requirements, add, minimum		5%	
4000	Maximum		30%	

B SHELL

B1010 Floor Construction

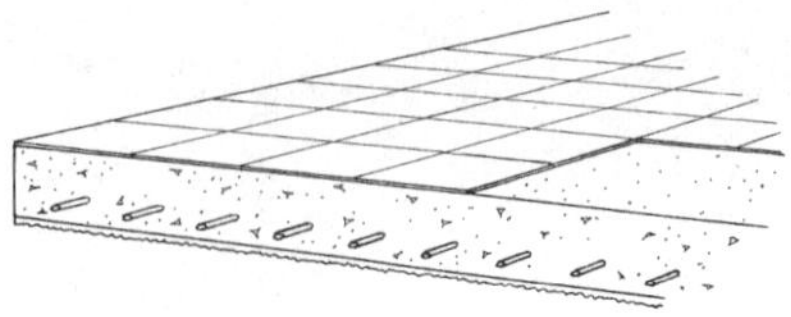

This page illustrates and describes a reinforced concrete slab system including concrete, placing concrete, formwork, reinforcing, steel trowel finish, V.A. floor tile and acoustical spray ceiling finish. Lines within System Components give the unit price per square foot for this system. Prices for alternate reinforced concrete slab systems are on Line Items B1010 225 1500 thru 1800. Both material quantities and labor costs have been adjusted for the system listed.

Factors: To adjust for job conditions other than normal working situations use Lines B1010 225 2700 thru 4000.

Example: You are to install the system to match an existing floor system. Go to Line B1010 225 2800 and apply these percentages to the appropriate MAT. and INST. costs.

System Components	QUANTITY	UNIT	COST PER S.F.		
			MAT.	INST.	TOTAL
Flat slab system, reinforced concrete with vinyl tile floor, sprayed Acoustical ceiling finish, not including columns.					
Concrete, 4000 psi, 6″ thick	.020	C.Y.	1.68		1.68
Placing concrete	.020	C.Y.		.54	.54
Formwork, 4 use	1.000	S.F.	.89	4.57	5.46
Edge form	.120	S.F.	.09	.88	.97
Reinforcing steel	3.000	Lb.	.99	1.47	2.46
Steel trowel finish	1.000	S.F.		.66	.66
Vinyl asbestos floor tile	1.000	S.F.	2.33	.79	3.12
Acoustical spray ceiling finish	1.000	S.F.	.23	.13	.36
TOTAL			6.21	9.04	15.25

B1010 225	Floor - Ceiling, Concrete Slab	COST PER S.F.		
		MAT.	INST.	TOTAL
1400	For alternate slab systems:			
1500	Concrete, 4000 psi, 7″ thick	6.50	9.45	15.95
1600	8″ thick	7	9.95	16.95
1700	9″ thick	7.40	10.35	17.75
1800	10″ thick	7.85	10.90	18.75
1900				
2000				
2100				
2200				
2300				
2400				
2500				
2700	Cut & patch to match existing construction, add, minimum	2%	3%	
2800	Maximum	5%	9%	
2900	Dust protection, add, minimum	1%	2%	
3000	Maximum	4%	11%	
3100	Equipment usage curtailment, add, minimum	1%	1%	
3200	Maximum	3%	10%	
3300	Material handling & storage limitation, add, minimum	1%	1%	
3400	Maximum	6%	7%	
3500	Protection of existing work, add, minimum	2%	2%	
3600	Maximum	5%	7%	
3700	Shift work requirements, add, minimum		5%	
3800	Maximum		30%	
3900	Temporary shoring and bracing, add, minimum	2%	5%	
4000	Maximum	5%	12%	

B1010 Floor Construction

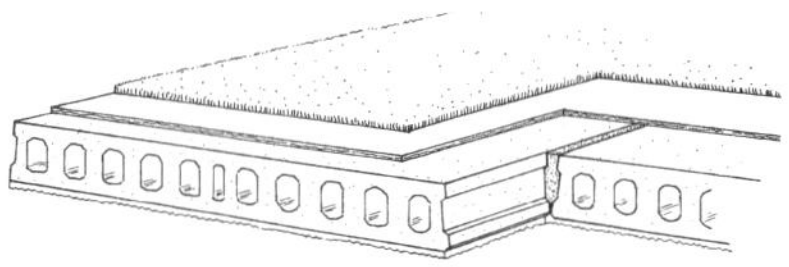

This page illustrates and describes a hollow core prestressed concrete panel system including hollow core slab, grout, carpet, carpet padding, and sprayed ceiling. Lines within System Components give the unit price and total price per square foot for this system. Prices for alternate hollow core prestressed concrete panel systems are on Line Items B1010 232 1300 thru 1600. Both material and labor costs have been adjusted for the system listed.

Factors: To adjust for job conditions other than normal working situations use Lines B1010 232 2700 thru 3800.

Example: You are to install the system where dust control is a major concern. Go to Line B1010 232 2800 and apply these percentages to the appropriate MAT. and INST. costs.

System Components	QUANTITY	UNIT	COST PER S.F. MAT.	INST.	TOTAL
Precast hollow core plank with carpeted floors, padding And sprayed textured ceiling.					
Hollow core plank, 4″ thick with grout topping	1.000	S.F.	4.20	2.70	6.90
Nylon carpet, 26 oz. medium traffic	.110	S.Y.	2.59	.58	3.17
Carpet padding, minimum quality	.110	S.Y.	.39	.29	.68
Sprayed texture ceiling	1.000	S.F.	.04	.48	.52
TOTAL			7.22	4.05	11.27

B1010 232	Floor - Ceiling, Conc. Panel	COST PER S.F. MAT.	INST.	TOTAL
1200	For alternate floor systems:			
1300	Hollow core concrete plank, with grout, 6″ thick	8.15	3.66	11.81
1400	8″ thick	8.60	3.37	11.97
1500	10″ thick	8	3.15	11.15
1600	12″ thick	9.20	2.97	12.17
1700				
1800				
1900				
2000				
2100				
2200				
2300				
2400				
2500				
2700	Dust protection, add, minimum	1%	2%	
2800	Maximum	4%	11%	
2900	Equipment usage curtailment, add, minimum	1%	1%	
3000	Maximum	3%	10%	
3100	Material handling & storage limitation, add, minimum	1%	1%	
3200	Maximum	6%	7%	
3300	Protection of existing work, add, minimum	2%	2%	
3400	Maximum	5%	7%	
3500	Shift work requirements, add, minimum		5%	
3600	Maximum		30%	
3700	Temporary shoring and bracing, add, minimum	2%	5%	
3800	Maximum	5%	12%	

B1010 Floor Construction

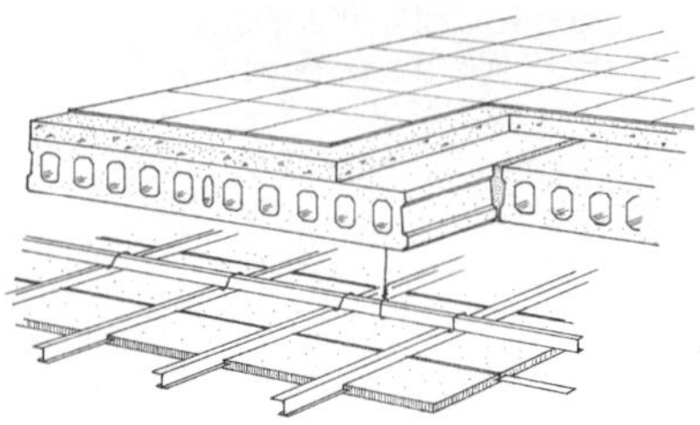

This page illustrates and describes a hollow core plank system including precast hollow core plank, concrete topping, V.A. tile, concealed suspension system and ceiling tile. Lines within System Components give the unit price and total price per square foot for this system. Prices for alternate hollow core plank systems are on Line Items B1010 233 1500 thru 1800. Both material quantities and labor costs have been adjusted for the system listed.

Factors: To adjust for job conditions other than normal working situations use Lines B1010 233 2900 thru 4000.

Example: You are to install the system where equipment usage is a problem. Go to Line B1010 233 3200 and apply these percentages to the appropriate MAT. and INST. costs.

System Components	QUANTITY	UNIT	COST PER S.F.		
			MAT.	INST.	TOTAL
Precast hollow core plank with 2″ topping, vinyl composition floor tile And suspended acoustical tile ceiling.					
Hollow core concrete plank, 4″ thick	1.000	S.F.	4.20	2.70	6.90
Concrete topping, 2″ thick	1.000	S.F.	.78	4.01	4.79
Vinyl composition tile, .08″ thick	1.000	S.F.	1.06	.79	1.85
Concealed Z bar suspension system, 12″ module	1.000	S.F.	.41	.84	1.25
Ceiling tile, mineral fiber, 3/4″ thick	1.000	S.F.	1.43	.44	1.87
TOTAL			7.88	8.78	16.66

B1010 233	Floor - Ceiling, Conc. Plank	COST PER S.F.		
		MAT.	INST.	TOTAL
1400	For alternate floor systems:			
1500	Hollow core concrete plank, 6″ thick	8.85	8.40	17.25
1600	8″ thick	9.30	8.10	17.40
1700	10″ thick	8.70	7.90	16.60
1800	12″ thick	9.90	7.70	17.60
1900				
2000				
2100				
2200				
2300				
2400				
2500				
2600				
2700				
2900	Dust protection, add, minimum	1%	2%	
3000	Maximum	4%	11%	
3100	Equipment usage curtailment, add, minimum	1%	1%	
3200	Maximum	3%	10%	
3300	Material handling & storage limitation, add, minimum	1%	1%	
3400	Maximum	6%	7%	
3500	Protection of existing work, add, minimum	2%	2%	
3600	Maximum	5%	7%	
3700	Shift work requirements, add, minimum		5%	
3800	Maximum		30%	
3900	Temporary shoring and bracing, add, minimum	2%	5%	
4000	Maximum	5%	12%	

B1010 Floor Construction

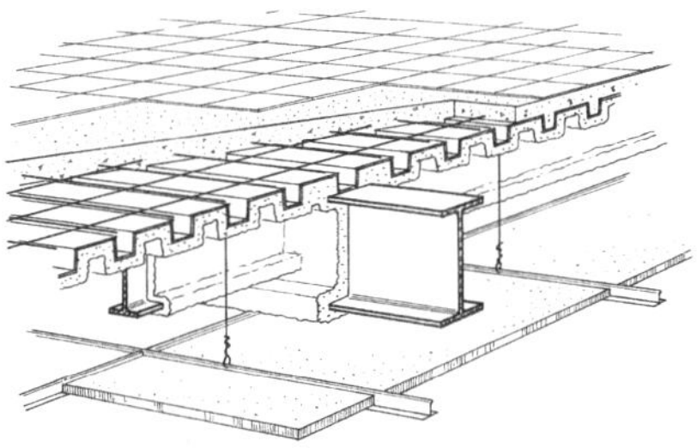

This page illustrates and describes a structural steel w/metal decking and concrete system including steel beams, steel decking, shear studs, concrete, placing concrete, edge form, steel trowel finish, curing, wire fabric, fireproofing, beams and decking, tile floor, and suspended ceiling. Lines within System Components give the unit price and total price per square foot for this system. Prices for alternate structural steel w/metal decking and concrete systems are on Line Items B1010 243 1900 thru 2400. Both material quantities and labor costs have been adjusted for the system listed.

Factors: To adjust for job conditions other than normal working situations use Lines B1010 243 2900 thru 4000.

Example: You are to install the system where material handling and storage are a problem. Go to Line B1010 243 3400 and apply these percentages to the appropriate MAT. and INST. costs.

System Components	QUANTITY	UNIT	COST PER S.F. MAT.	INST.	TOTAL
Composite structural beams, 20 ga. 3″ deep steel decking, shear studs, 3000 psi, concrete, placing concrete, edge forms, steel trowel finish, Welded wire fabric fireproofing, tile floor and suspended ceiling.					
Steel deck, 20 gage, 3″ deep, galvanized	1.000	S.F.	1.06	.78	1.84
Structural steel framing	.004	Ton	5.60	2.15	7.75
Shear studs, 3/4″	.250	Ea.	.08	.34	.42
Concrete, 3000 psi, 5″ thick	.015	C.Y.	1.20		1.20
Place concrete	.015	C.Y.		.41	.41
Edge form	.140	L.F.	.05	.46	.51
Steel trowel finish	1.000	S.F.		.66	.66
Welded wire fabric 6 x 6 - #10/10	.011	C.S.F.	.09	.32	.41
Fireproofing, sprayed	1.880	S.F.	.88	1.48	2.36
Vinyl composition floor tile	1.000	S.F.	.83	.79	1.62
Suspended acoustical ceiling	1.000	S.F.	1.70	1.15	2.85
TOTAL			11.49	8.54	20.03

B1010 243	Floor - Ceiling, Struc. Steel	COST PER S.F. MAT.	INST.	TOTAL
1800	For alternate floor systems:			
1900	Composite deck, galvanized, 3″ deep, 22 ga.	11.40	8.50	19.90
2000	18 ga.	11.75	8.60	20.35
2100	16 ga.	12.20	8.65	20.85
2200	Composite deck, galvanized, 2″ deep, 22 ga.	11.30	8.40	19.70
2300	20 ga.	11.40	8.40	19.80
2400	16 ga.	12	8.50	20.50
2500				
2900	Dust protection, add, minimum	1%	2%	
3000	Maximum	4%	11%	
3100	Equipment usage curtailment, add, minimum	1%	1%	
3200	Maximum	3%	10%	
3300	Material handling & storage limitation, add, minimum	1%	1%	
3400	Maximum	6%	7%	
3500	Protection of existing work, add, minimum	2%	2%	
3600	Maximum	5%	7%	
3700	Shift work requirements, add, minimum		5%	
3800	Maximum		30%	
3900	Temporary shoring and bracing, add, minimum	2%	5%	
4000	Maximum	5%	12%	

B1010 Floor Construction

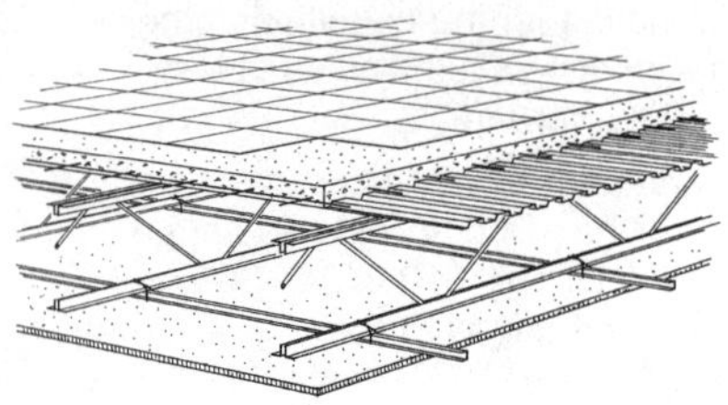

This page illustrates and describes an open web joist and steel slab-form system including open web steel joist, slab form, concrete, placing concrete, wire fabric, steel trowel finish, tile floor and plasterboard. Lines within System Components give the unit price and total price per square foot for this system. Prices for alternate open web joists and steel slab-form systems are on Line Items B1010 245 1900 thru 2200. Both material quantities and labor costs have been adjusted for the systems listed.

Factors: To adjust for job conditions other than normal working situations use Lines B1010 245 2900 thru 4000.

Example: You are to install the system in a congested commercial area and most work will be done at night. Go to Line B1010 245 3800 and apply this percentage to the appropriate INST. cost.

System Components	QUANTITY	UNIT	COST PER S.F.		
			MAT.	INST.	TOTAL
Open web steel joists, 24″ O.C., slab form, 3000 psi concrete, welded Wire fabric, steel trowel finish, floor tile, 5/8″ drywall ceiling.					
Open web steel joists, 12″ deep, 5.2#/L.F., 24″ O.C.	2.630	Lb.	1.32	.92	2.24
Slab form, 28 gage, 9/16″ deep, galvanized	1.000	S.F.	.54	.58	1.12
Concrete, 3000 psi, 2-1/2″ thick	.008	C.Y.	.64		.64
Placing concrete	.008	C.Y.		.22	.22
Welded wire fabric 6 x 6 - #10/10	.011	C.S.F.	.09	.32	.41
Steel trowel finish	1.000	S.F.		.66	.66
Vinyl composition floor tile	1.000	S.F.	.83	.79	1.62
Ceiling furring, 3/4″ channels, 24″ O.C.	1.000	S.F.	.11	.92	1.03
Gypsum drywall, 5/8″ thick, finished	1.000	S.F.	.28	1.14	1.42
Paint ceiling	1.000	S.F.	.15	.50	.65
TOTAL			3.96	6.05	10.01

B1010 245	Floor - Ceiling, Steel Joists	COST PER S.F.		
		MAT.	INST.	TOTAL
1800	For alternate floor systems:			
1900	Open web joists, 16″ deep, 6.6#/L.F.	4.29	6.25	10.54
2000	20″ deep, 8.4# L.F.	4.74	6.60	11.34
2100	24″ deep, 11.5# L.F.	5.50	7.15	12.65
2200	26″ deep, 12.8# L.F.	5.85	7.35	13.20
2300				
2400				
2500				
2600				
2700				
2900	Dust protection, add, minimum	1%	2%	
3000	Maximum	4%	11%	
3100	Equipment usage curtailment, add, minimum	1%	1%	
3200	Maximum	3%	10%	
3300	Material handling & storage limitation, add, minimum	1%	1%	
3400	Maximum	6%	7%	
3500	Protection of existing work, add, minimum	2%	2%	
3600	Maximum	5%	7%	
3700	Shift work requirements, add, minimum		5%	
3800	Maximum		30%	
3900	Temporary shoring and bracing, add, minimum	2%	5%	
4000	Maximum	5%	12%	

B1010 Floor Construction

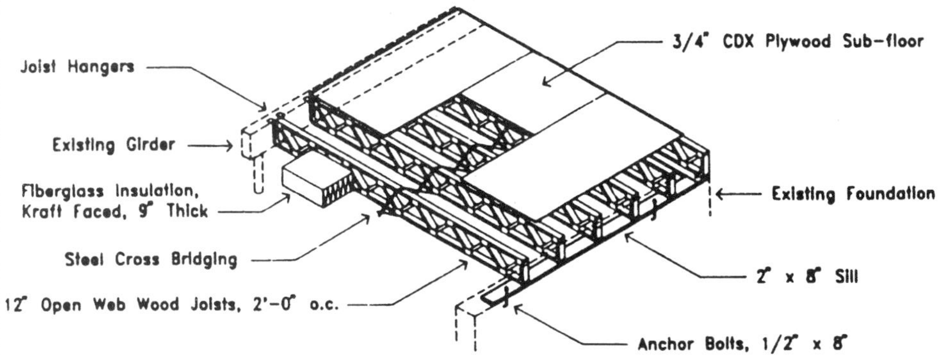

Open Web Wood Joists

This page illustrates and describes floor framing systems including sills, joist hangers, joists, bridging, subfloor and insulation. Lines within systems components give the unit price and total price per square foot for this system. Prices for alternate joist types are on line items B1010 262 0100 through 0700. Both material quantities and labor costs have been adjusted for the system listed.

Factors: To adjust for job conditions other than normal working situations use lines B1010 262 2700 through 4000.

Example: You are to install the system where delivery of material is difficult. Go to line B1010 262 3400 and apply these percentages to the appropriate MAT. and INST. costs.

System Components	QUANTITY	UNIT	COST PER S.F. MAT.	COST PER S.F. INST.	COST PER S.F. TOTAL
Open web wood joists, sill, butt to girder, 5/8" subfloor, insul.					
Anchor bolt, hook type, with nut and washer, 1/2" x 8" long	.250	Ea.	.01	.06	.07
Joist & beam hanger, 18 ga. galvanized	.750	Ea.	.04	.20	.24
Bridging, steel, compression, for 2" x 12" joists	.001	C.Pr.	.12	.22	.34
Framing, open web wood joists, 12" deep	.001	M.L.F.	1.85	1	2.85
Framing, sills, 2" x 8"	.001	M.B.F.	.66	1.30	1.96
Sub-floor, plywood, CDX, 3/4" thick	1.000	S.F.	.50	.65	1.15
Insulation, fiberglass, kraft face, 9" thick x 23" wide	1.000	S.F.	.66	.32	.98
TOTAL			3.84	3.75	7.59

B1010 262	Open Web Wood Joists	COST PER S.F. MAT.	COST PER S.F. INST.	COST PER S.F. TOTAL
0080	For alternate floor systems:			
0100	12" open web joists, sill, butt to girder, 5/8"subflr	3.84	3.75	7.59
0200	14" joists	4.14	3.83	7.97
0300	16" joists	4.22	3.88	8.10
0400	18" joists	4.24	3.93	8.17
0450	Wood struc."I"joists,24"O.C.,to 24'span,50 PSF LL, butt to girder	3.98	3.49	7.47
0500	55 PSF LL	4.12	3.54	7.66
0600	to 30'span, 45 PSF LL	4.33	3.43	7.76
0700	55 PSF LL	5.20	3.49	8.69
2700	Cut & patch to match existing construction, add, minimum	2%	3%	
2800	Maximum	5%	9%	
2900	Dust protection, add, minimum	1%	2%	
3000	Maximum	4%	11%	
3100	Equipment usage curtailment, add, minimum	1%	1%	
3200	Maximum	3%	10%	
3300	Material handling & storage limitation, add, minimum	1%	1%	
3400	Maximum	6%	7%	
3500	Protection of existing work, add, minimum	2%	2%	
3600	Maximum	5%	7%	
3700	Shift work requirements, add, minimum		5%	
3800	Maximum		30%	
3900	Temporary shoring and bracing, add, minimum	2%	5%	
4000	Maximum	5%	12%	

B1010 Floor Construction

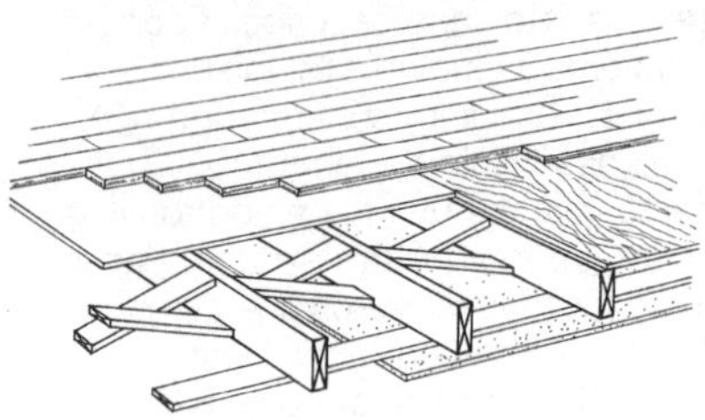

This page illustrates and describes a wood joist floor system including wood joist, oak floor, sub-floor, bridging, sand and finish floor, furring, plasterboard, taped, finished and painted ceiling. Lines within System Components give the unit price and total price per square foot for this system. Prices for alternate wood joist floor systems are on Line Items B1010 263 1700 thru 2300. Both material quantities and labor costs have been adjusted for the system listed.

Factors: To adjust for job conditions other than normal working situations use Lines B1010 263 2700 thru 4000.

Example: You are to install the system during off peak hours, 6 P.M. to 2 A.M. Go to Line B1010 263 3800 and apply this percentage to the appropriate INST. costs.

System Components	QUANTITY	UNIT	COST PER S.F. MAT.	INST.	TOTAL
Wood joists, 2″ x 8″, 16″ O.C.,oak floor (sanded & finished),1/2″ sub Floor, 1″x 3″ bridging, furring, 5/8″ drywall, taped, finished and painted.					
Wood joists, 2″ x 8″, 16″ O.C.	1.000	L.F.	.88	.80	1.68
Subfloor plywood CDX 1/2″	1.000	SF Flr.	.43	.58	1.01
Bridging 1″ x 3″	.150	Pr.	.05	.51	.56
Oak flooring, No. 1 common	1.000	S.F.	4.09	2.57	6.66
Sand and finish floor	1.000	S.F.	.76	1.17	1.93
Wood furring, 1″ x 3″, 16″ O.C.	1.000	L.F.	.23	1.25	1.48
Gypsum drywall, 5/8″ thick	1.000	S.F.	.24	.49	.73
Tape and finishing	1.000	S.F.	.04	.44	.48
Paint ceiling	1.000	S.F.	.15	.50	.65
TOTAL			6.87	8.31	15.18

B1010 263	Floor-Ceiling, Wood Joists	COST PER S.F. MAT.	INST.	TOTAL
1600	For alternate wood joist systems:			
1700	16″ on center, 2″ x 6″ joists	6.55	8.20	14.75
1800	2″ x 10″	7.20	8.50	15.70
1900	2″ x 12″	7.60	8.50	16.10
2000	2″ x 14″	7.95	8.65	16.60
2100	12″ on center, 2″ x 10″ joists	7.50	8.70	16.20
2200	2″ x 12″	8	8.75	16.75
2300	2″ x 14″	8.45	8.95	17.40
2400				
2500				
2700	Cut & patch to match existing construction, add, minimum	2%	3%	
2800	Maximum	5%	9%	
2900	Dust protection, add, minimum	1%	2%	
3000	Maximum	4%	11%	
3100	Equipment usage curtailment, add, minimum	1%	1%	
3200	Maximum	3%	10%	
3300	Material handling & storage limitation, add, minimum	1%	1%	
3400	Maximum	6%	7%	
3500	Protection of existing work, add, minimum	2%	2%	
3600	Maximum	5%	7%	
3700	Shift work requirements, add, minimum		5%	
3800	Maximum		30%	
3900	Temporary shoring and bracing, add, minimum	2%	5%	
4000	Maximum	5%	12%	

B1020 Roof Coverings

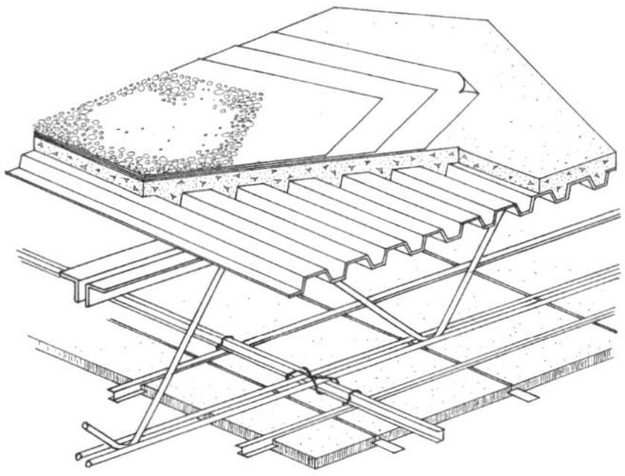

This page illustrates and describes a flat roof bar joist system including asphalt and gravel roof, lightweight concrete, metal decking, open web joists, and suspended acoustic ceiling. Lines within System Components give the unit price and total price per square foot for this system. Prices for alternate flat roof bar joist systems are on Line Items B1020 106 1300 thru 1600. Both material quantities and labor costs have been adjusted for the system listed.

Factors: To adjust for job conditions other than normal working situations use Lines B1020 106 2700 thru 4000.

Example: You are to install the system where material handling and storage present some problem. Go to Line B1020 106 3300 and apply these percentages to the appropriate MAT. and INST. costs.

System Components	QUANTITY	UNIT	COST PER S.F.		
			MAT.	INST.	TOTAL
Three ply asphalt and gravel roof on lightweight concrete, metal decking on Web joist 4′ O.C., with suspended acoustic ceiling.					
Open web joist, 10″ deep, 4′ O.C.	1.250	Lb.	.63	.44	1.07
Metal decking, 1-1/2″ deep, 22 Ga., galvanized	1.000	S.F.	.92	.52	1.44
Gravel roofing, 3 ply asphalt	.010	C.S.F.	.68	1.36	2.04
Perlite roof fill, 3″ thick	1.000	S.F.	.83	.44	1.27
Suspended ceiling, 3/4″ mineral fiber on "Z" bar suspension	1.000	S.F.	1.82	2.92	4.74
TOTAL			4.88	5.68	10.56

B1020 106	Steel Joist Roof & Ceiling	COST PER S.F.		
		MAT.	INST.	TOTAL
1200	For alternate roof systems:			
1300	Open web bar joist, 16″ deep, 5#/L.F.	5.10	5.80	10.90
1400	18″ deep, 6.6#/L.F.	5.25	5.95	11.20
1500	20″ deep, 9.6#/L.F.	5.45	6.10	11.55
1600	24″ deep, 11.52#/L.F.	5.70	6.25	11.95
1700				
1800				
1900				
2000				
2100				
2200				
2300				
2400				
2500				
2700	Cut & patch to match existing construction, add, minimum	2%	3%	
2800	Maximum	5%	9%	
2900	Dust protection, add, minimum	1%	2%	
3000	Maximum	4%	11%	
3100	Equipment usage curtailment, add, minimum	1%	1%	
3200	Maximum	3%	10%	
3300	Material handling & storage limitation, add, minimum	1%	1%	
3400	Maximum	6%	7%	
3500	Protection of existing work, add, minimum	2%	2%	
3600	Maximum	5%	7%	
3700	Shift work requirements, add, minimum		5%	
3800	Maximum		30%	
3900	Temporary shoring and bracing, add, minimum	2%	5%	
4000	Maximum	5%	12%	

B1020 Roof Coverings

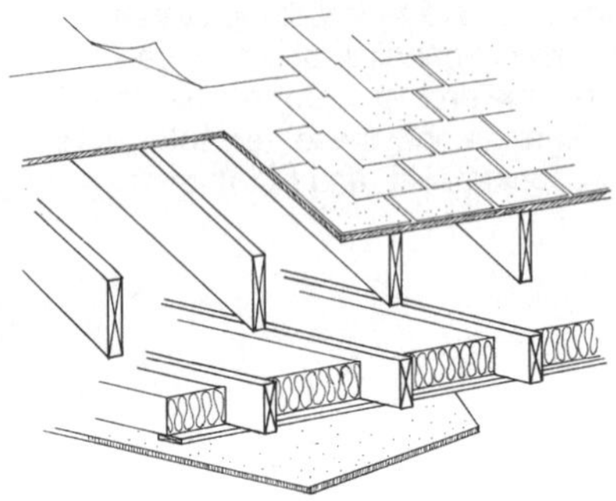

This page illustrates and describes a wood frame roof system including rafters, ceiling joists, sheathing, building paper, asphalt shingles, roof trim, furring, insulation, plaster and paint. Lines within System Components give the unit price and total price per square foot for this system. Prices for alternate wood frame roof systems are on Line Items B1020 202 1900 thru 2700. Both material quantities and labor costs have been adjusted for the system listed.

Factors: To adjust for job conditions other than normal working situations use Lines B1020 202 3300 thru 4000.

Example: You are to install the system while protecting existing work. Go to Line B1020 202 3800 and apply these percentages to the appropriate MAT. and INST. costs.

System Components	QUANTITY	UNIT	COST PER S.F.		
			MAT.	INST.	TOTAL
Wood frame roof system, 4 in 12 pitch, including rafters, sheathing, Shingles, insulation, drywall, thin coat plaster, and painting.					
Rafters, 2" x 6", 16" O.C., 4 in 12 pitch	1.080	L.F.	.62	.95	1.57
Ceiling joists, 2" x 6", 16 O.C.	1.000	L.F.	.57	.70	1.27
Sheathing, 1/2" CDX	1.080	S.F.	.46	.68	1.14
Building paper, 15# felt	.011	C.S.F.	.03	.13	.16
Asphalt shingles, 240#	.011	C.S.F.	.46	.90	1.36
Roof trim	.100	L.F.	.17	.20	.37
Furring, 1" x 3", 16" O.C.	1.000	L.F.	.23	1.25	1.48
Fiberglass insulation, 6" batts	1.000	S.F.	.45	.32	.77
Gypsum board, 1/2" thick	1.000	S.F.	.21	.49	.70
Thin coat plaster	1.000	S.F.	.07	.54	.61
Paint, roller, 2 coats	1.000	S.F.	.15	.50	.65
TOTAL			3.42	6.66	10.08

B1020 202	Wood Frame Roof & Ceiling	COST PER S.F.		
		MAT.	INST.	TOTAL
1800	For alternate roof systems:			
1900	Rafters 16" O.C., 2" x 8"	3.75	6.70	10.45
2000	2" x 10"	4.09	7.25	11.34
2100	2" x 12"	4.55	7.35	11.90
2200	Rafters 24" O.C., 2" x 6"	3.12	6.25	9.37
2300	2" x 8"	3.36	6.30	9.66
2400	2" x 10"	3.61	6.65	10.26
2500	2" x 12"	3.96	6.80	10.76
2600	Roof pitch, 6 in 12, add	3%	10%	
2700	8 in 12, add	5%	12%	
2800				
2900				
3000				
3100				
3300	Cut & patch to match existing construction, add, minimum	2%	3%	
3400	Maximum	5%	9%	
3500	Material handling & storage limitation, add, minimum	1%	1%	
3600	Maximum	6%	7%	
3700	Protection of existing work, add, minimum	2%	2%	
3800	Maximum	5%	7%	
3900	Shift work requirements, add, minimum		5%	
4000	Maximum		30%	

Important: See the Reference Section for critical supporting data - Reference Numbers and City Cost Indexes

B1020 Roof Construction

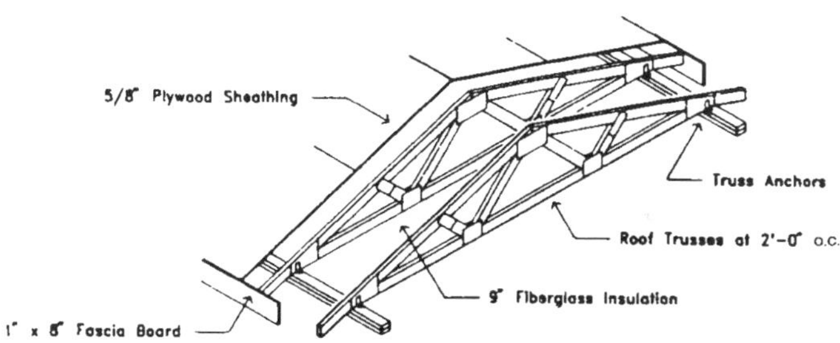

Wood Fabricated Roof Truss System

This page illustrates and describes a sloped roof truss framing system including trusses, truss anchors, sheathing, fascia and insulation. Lines B1020 210 1500 through 1800 are for a flat roof framing system including flat roof trusses, truss anchors, sheathing, fascia and insulation. Lines within system components give the unit price and total price per square foot for this system. Prices for alternate sizes and systems are given in line items B1020 210 0100 through 1800. Both material quantities and labor costs have been adjusted for the system listed.

Factors: To adjust for job conditions other than normal working situations use lines B1020 210 2700 through 4000.

Example: You are to install the system where crane placement and movement will be difficult. Go to line B1020 210 3200 and apply these percentages to the appropriate MAT. and INST. costs.

System Components	QUANTITY	UNIT	COST PER S.F. MAT.	INST.	TOTAL
Roof truss, 2′ O.C., 4/12 slope, 12′ span, 5/8 sheath., fascia,insul.					
Timber connectors, rafter anchors, galv., 1 1/2″ x 5 1/4″	.084	Ea.	.04	.25	.29
Plywood sheathing, CDX, 5/8″ thick	1.054	S.F.	.53	.71	1.24
Truss, 2′ O.C., metal plate connectors, 12′ span	.042	Ea.	1.11	1.37	2.48
Molding, fascia trim, 1″ x 8″	.167	L.F.	.28	.33	.61
Insulation, fiberglass, kraft face, 9″ thick x 23″ wide	1.000	S.F.	.66	.32	.98
TOTAL			2.62	2.98	5.60

B1020 210	Roof Truss, Wood	COST PER S.F. MAT.	INST.	TOTAL
0090	For alternate roof systems:			
0100	Roof truss, 2′O.C., 4/12p, 1′ovhg, 12′span, 5/8″sheath, fascia, insul.	2.62	2.98	5.60
0200	20′ span	2.48	2.56	5.04
0300	24′ span	2.44	2.38	4.82
0400	26′ span	4.08	3.21	7.29
0500	28′ span	2.47	2.33	4.80
0600	30′ span	2.76	2.30	5.06
0700	32′ span	2.72	2.26	4.98
0800	34′ span	2.94	2.23	5.17
0900	8/12 pitch, 20′ span	3.12	2.76	5.88
1000	24′ span	3.08	2.57	5.65
1100	26′ span	4.64	3.48	8.12
1200	28′ span	3.06	2.51	5.57
1300	32′ span	3.11	2.45	5.56
1400	36′ span	3.17	2.41	5.58
1500	Flat roof frame, fab.strl.joists, 2′ O.C., 15′-24′ span, 50 PSF LL	3.31	2	5.31
1600	55 PSF LL	3.45	2.05	5.50
1700	24′-30′span, 45 PSF LL	3.63	1.91	5.54
1800	55 PSF LL	4.48	1.97	6.45
2700	Cut & patch to match existing construction, add, minimum	2%	3%	
2800	Maximum	5%	9%	
2900	Dust protection, add, minimum	1%	2%	
3000	Maximum	4%	11%	
3100	Equipment usage curtailment, add, minimum	1%	1%	
3200	Maximum	3%	10%	
3300	Material handling & storage limitation, add, minimum	1%	1%	
3400	Maximum	6%	7%	
3500	Protection of existing work, add, minimum	2%	2%	
3600	Maximum	5%	7%	
3700	Shift work requirements, add, minimum		5%	

B SHELL

B1020 Roof Construction

B1020 210	Roof Truss, Wood	COST PER S.F.		
		MAT.	INST.	TOTAL
3800	Maximum		30%	
3900	Temporary shoring and bracing, add, minimum	2%	5%	
4000	Maximum	5%	12%	

SHELL B

B2010 Exterior Walls

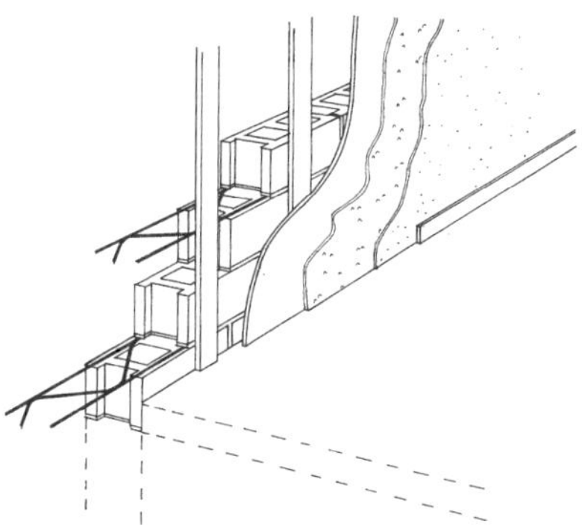

This page illustrates and describes a masonry concrete block system including concrete block wall, pointed, reinforcing, waterproofing, gypsum plaster on gypsum lath, and furring. Lines within System Components give the unit price and total price per square foot for this system. Prices for alternate masonry concrete block wall systems are on Line Items B2010 108 1500 thru 2200. Both material quantities and labor costs have been adjusted for the system listed.

Factors: To adjust for job conditions other than normal working situations use Lines B2010 108 2700 thru 4000.

Example: You are to install the system and match existing construction at several locations. Go to Line B2010 108 2800 and apply these percentages to the appropriate MAT. and INST. costs.

System Components	QUANTITY	UNIT	COST PER S.F.		
			MAT.	INST.	TOTAL
Concrete block, 8″ thick, reinforced every 2 courses, waterproofing gypsum Plaster over gypsum lath on 1″ x 3″ furring, interior painting & baseboard.					
Concrete block, 8″ x 8″ x 16″, reinforced	1.000	S.F.	2.93	4.82	7.75
Silicone waterproofing, 2 coats	1.000	S.F.	.60	.14	.74
Bituminous coating, 1/16″ thick	1.000	S.F.	.18	.82	1
Furring, 1″ x 3″, 16″ O.C.	1.000	L.F.	.23	1.68	1.91
Gypsum lath, 3/8″ thick	.110	S.Y.	.41	.50	.91
Gypsum plaster, 2 coats	.110	S.Y.	.38	2.02	2.40
Painting, 2 coats	1.000	S.F.	.10	.39	.49
Baseboard wood, 9/16″ x 2-5/8″	.100	L.F.	.13	.18	.31
Paint baseboard, primer + 1 coat enamel	.100	L.F.	.01	.05	.06
TOTAL			4.97	10.60	15.57

B2010 108	Masonry Wall, Concrete Block	COST PER S.F.		
		MAT.	INST.	TOTAL
1400	For alternate exterior wall systems:			
1500	8″ thick block, fluted 2 sides	6.80	12.45	19.25
1600	Deep grooved	5.30	12.45	17.75
1700	Slump block	6.95	10.80	17.75
1800	Split rib	5.10	12.45	17.55
1900				
2000	12″ thick block, regular	4.44	12.65	17.09
2100	Slump block	11.45	11.80	23.25
2200	Split rib	5.65	14.30	19.95
2300				
2700	Cut & patch to match existing construction, add, minimum	2%	3%	
2800	Maximum	5%	9%	
2900	Dust protection, add, minimum	1%	2%	
3000	Maximum	4%	11%	
3100	Equipment usage curtailment, add, minimum	1%	1%	
3200	Maximum	3%	10%	
3300	Material handling & storage limitation, add, minimum	1%	1%	
3400	Maximum	6%	7%	
3500	Protection of existing work, add, minimum	2%	2%	
3600	Maximum	5%	7%	
3700	Shift work requirements, add, minimum		5%	
3800	Maximum		30%	
3900	Temporary shoring and bracing, add, minimum	2%	5%	
4000	Maximum	5%	12%	

B SHELL

B2010 Exterior Walls

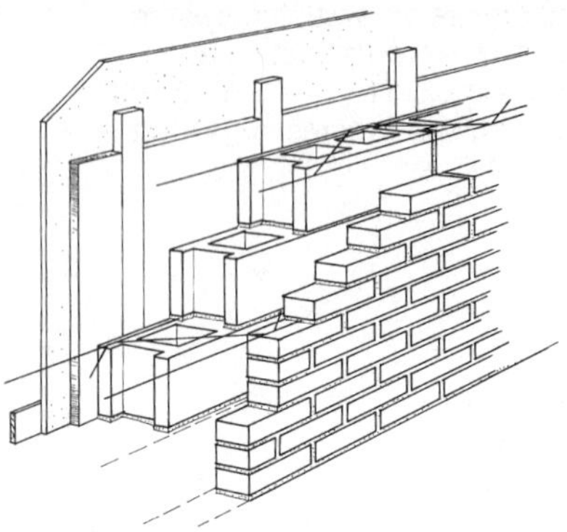

This page illustrates and describes a masonry wall, brick-stone system including brick, concrete block, durawall, insulation, plasterboard, taped and finished, furring, baseboard and painting interior. Lines within System Components give the unit price and total price per square foot for this system. Prices for alternate masonry wall, brick-stone systems are on Line Item B2010 131 1500 thru 2500. Both material quantities and labor costs have been adjusted for the system listed.

Factors: To adjust for job conditions other than normal working situations use Lines B2010 131 3100 thru 4200.

Example: You are to install the system without damaging the existing work. Go to Line B2010 131 3900 and apply these percentages to the appropriate MAT. and INST. costs.

System Components	QUANTITY	UNIT	COST PER S.F.		
			MAT.	INST.	TOTAL
Face brick, 4"thick, concrete block back-up, reinforce every second course, 3/4"insulation, furring, 1/2"drywall, taped, finish, and painted, baseboard					
Face brick, 4" brick	1.000	S.F.	3.03	9.10	12.13
Concrete back-up block, reinforced 8" thick	1.000	S.F.	1.78	5.05	6.83
3/4" rigid polystyrene insulation	1.000	S.F.	.40	.55	.95
Furring, 1" x 3", wood, 16" O.C.	1.000	L.F.	.23	.88	1.11
Drywall, 1/2" thick	1.000	S.F.	.21	.44	.65
Taping & finishing	1.000	S.F.	.04	.44	.48
Painting, 2 coats	1.000	S.F.	.15	.50	.65
Baseboard, wood, 9/16" x 2-5/8"	.100	L.F.	.13	.18	.31
Paint baseboard, primer + 1 coat enamel	.100	L.F.	.01	.05	.06
TOTAL			5.98	17.19	23.17

B2010 131	Masonry Wall, Brick - Stone	COST PER S.F.		
		MAT.	INST.	TOTAL
1400	For alternate exterior wall systems:			
1500	Face brick, Norman, 4" x 2-2/3" x 12" (4.5 per S.F.)	7.15	14.35	21.50
1600	Roman, 4" x 2" x 12" (6.0 per S.F.)	8.05	16.10	24.15
1700	Engineer, 4" x 3-1/5" x 8" (5.63 per S.F.)	6.10	15.80	21.90
1800	S.C.R., 6" x 2-2/3" x 12" (4.5 per S.F.)	7.60	14.55	22.15
1900	Jumbo, 6" x 4" x 12" (3.0 per S.F.)	6.90	12.70	19.60
2000	Norwegian, 6" x 3-1/5" x 12" (3.75 per S.F.)	6.20	13.45	19.65
2100				
2200				
2300	Stone, veneer, fieldstone, 6" thick	64.50	14.55	79.05
2400	Marble, 2" thick	39.50	25.50	65
2500	Limestone, 2" thick	12.10	17.40	29.50
3100	Cut & patch to match existing construction, add, minimum	2%	3%	
3200	Maximum	5%	9%	
3300	Dust protection, add, minimum	1%	2%	
3400	Maximum	4%	11%	
3500	Equipment usage curtailment, add, minimum	1%	1%	
3600	Maximum	3%	10%	
3700	Material handling & storage limitation, add, minimum	1%	1%	
3800	Maximum	6%	7%	
3900	Protection of existing work, add, minimum	2%	2%	
4000	Maximum	5%	7%	
4100	Shift work requirements, add, minimum		5%	
4200	Maximum		30%	

B2010 Exterior Walls

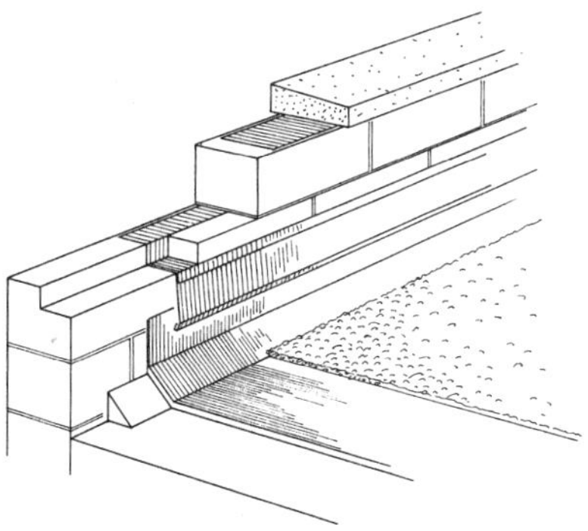

This page illustrates and describes a parapet wall system including wall, coping, flashing and cant strip. Lines within System Components give the unit price and cost per lineal foot for this system. Prices for alternate parapet wall systems are on Lines B2010 142 1300 thru 2400. Both material quantities and labor costs have been adjusted for the system listed.

Factors: To adjust for job conditions other than normal working situations, use Lines B2010 142 2900 thru 4000.

Example: You are to install the system without damaging the adjacent property. Go to Line B2010 142 3600 and apply these percentages to the appropriate MAT. and INST. costs.

System Components	QUANTITY	UNIT	COST PER L.F.		
			MAT.	INST.	TOTAL
Concrete block parapet, incl. reinf., coping, flashing & cant strip, 2′high					
8″ concrete block	2.000	S.F.	5.86	9.64	15.50
Masonry reinforcing	1.000	L.F.	.12	.15	.27
Coping, precast	1.000	L.F.	11.55	11.10	22.65
Roof cant	1.000	L.F.	2.37	1.25	3.62
Through wall flashing	1.000	L.F.	1.78	1.95	3.73
Cap flashing	1.000	L.F.	2.45	3.54	5.99
TOTAL			24.13	27.63	51.76

B2010 142	Parapet Wall	COST PER L.F.		
		MAT.	INST.	TOTAL
1200	For alternate systems:			
1300	Concrete block			
1400	12″ thick	31.50	36	67.50
1500	Split rib block, 8″ thick	24.50	31.50	56
1600	12″ thick	30	37	67
1700	Brick, common brick, 8″ wall	27.50	47.50	75
1800	12″ wall	32.50	60	92.50
1900	4″ brick, 4″ backup block	26.50	45.50	72
2000	8″ backup block	28	46.50	74.50
2100	Stucco on masonry	24.50	30.50	55
2200	On wood frame	9.50	24.50	34
2300	Wood, T 1-11 siding	20.50	30.50	51
2400	Boards, 1″ x 6″ cedar	21.50	34.50	56
2500				
2600				
2700				
2900	Dust protection, add, minimum	1%	2%	
3000	Maximum	4%	11%	
3100	Equipment usage curtailment, add, minimum	1%	1%	
3200	Maximum	3%	10%	
3300	Material handling & storage limitation, add, minimum	1%	1%	
3400	Maximum	6%	7%	
3500	Protection of existing work, add, minimum	2%	2%	
3600	Maximum	5%	7%	
3700	Shift work requirements, add, minimum		5%	
3800	Maximum		30%	
3900	Temporary shoring and bracing, add, minimum	2%	5%	
4000	Maximum	5%	12%	

B2010 Exterior Walls

This page illustrates and describes a masonry cleaning and restoration system, including staging, cleaning, repointing. Lines within System Components give the unit price and cost per square foot for this system. Prices for alternate systems are on Lines B2010 143 1300 thru 2500. Both material quantities and labor costs have been adjusted for the system listed.

Factors: To adjust for conditions other than normal working situations, use Lines B2010 143 2900 thru 4200.

Example: You are to clean a wall and be concerned about dust control. Go to line B2010 143 3100 and apply these percentages to the appropriate MAT. and INST. costs.

System Components	QUANTITY	UNIT	COST PER S.F. MAT.	INST.	TOTAL
Repoint existing building, brick, running bond, high pressure cleaning, Water only, soft old mortar.					
Scaffold, building exterior	.010	C.S.F.	.27	.55	.82
Cleaning, high pressure water only	1.000	S.F.		1.42	1.42
Repoint, brick running bond	1.000	S.F.	.30	4.45	4.75
TOTAL			.57	6.42	6.99

B2010 143	Masonry Restoration - Cleaning	COST PER S.F. MAT.	INST.	TOTAL
1200	For alternate masonry surfaces:			
1300	Brick, common bond	.56	6.60	7.16
1400	Flemish bond	.58	6.95	7.53
1500	English bond	.58	7.40	7.98
1600	Add for wire cut face brick	.15		.15
1700	Stone, 2′ x 2′ blocks	.66	4.75	5.41
1800	2′ x 4′ blocks	.56	4.06	4.62
2000	Add to above prices for alternate cleaning systems:			
2100	Chemical brush and wash	.03	.49	.52
2200	High pressure chemical and water	.04	.49	.53
2300	Sandblasting, wet system	.18	1.31	1.49
2400	Dry system	.17	.77	.94
2500	Steam cleaning		.39	.39
2600				
2900	Cut & patch to match existing construction, add, minimum	2%	3%	
3000	Maximum	5%	9%	
3100	Dust protection, add, minimum	1%	2%	
3200	Maximum	4%	11%	
3300	Equipment usage curtailment, add, minimum	1%	1%	
3400	Maximum	3%	10%	
3500	Material handling & storage limitation, add, minimum	1%	1%	
3600	Maximum	6%	7%	
3800				
3900	Protection of existing work, add, minimum	2%	2%	
4000	Maximum	5%	7%	
4100	Shift work requirements, add, minimum		5%	
4200	Maximum		30%	

B2010 Exterior Walls

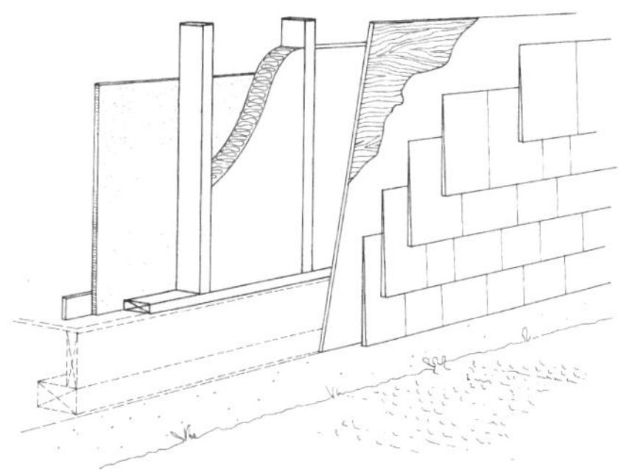

This page illustrates and describes a wood frame exterior wall system including wood studs, sheathing, felt, insulation, plasterboard, taped and finished, baseboard and painted interior. Lines within System Components give the unit price and total price per square foot for this system. Prices for alternate wood frame exterior wall systems are on Line Items B2010 150 1700 thru 2700. Both material quantities and labor costs have been adjusted for the system listed.

Factors: To adjust for job conditions other than normal working situations use Lines B2010 150 3100 thru 4200.

Example: You are to install the system with need for complete temporary bracing. Go to Line B2010 150 4200 and apply these percentages to the appropriate MAT. and INST. costs.

System Components	QUANTITY	UNIT	COST PER S.F.		
			MAT.	INST.	TOTAL
Wood stud wall, cedar shingle siding, building paper, plywood sheathing, Insulation, 5/8″ drywall, taped, finished and painted, baseboard.					
2″ x 4″ wood studs, 16″ O.C.	.100	L.F.	.39	.88	1.27
1/2″ CDX sheathing	1.000	S.F.	.43	.63	1.06
18″ No. 1 red cedar shingles, 7-1/2″ exposure	.008	C.S.F.	1.03	1.55	2.58
15# felt paper	.010	C.S.F.	.03	.12	.15
3-1/2″ fiberglass insulation	1.000	S.F.	.37	.27	.64
5/8″ drywall	1.000	S.F.	.24	.44	.68
Drywall finishing adder	1.000	S.F.	.04	.44	.48
Baseboard trim, stock pine, 9/16″ x 3-1/2″, painted	.100	L.F.	.13	.18	.31
Paint baseboard, primer + 1 coat enamel	.100	L.F.	.01	.05	.06
Paint, 2 coats, interior	1.000	S.F.	.15	.50	.65
TOTAL			2.82	5.06	7.88

B2010 150	Wood Frame Exterior Wall	COST PER S.F.		
		MAT.	INST.	TOTAL
1600	For alternate exterior wall systems:			
1700	Aluminum siding, horizontal clapboard	3.14	5.20	8.34
1800	Cedar bevel siding, 1/2″ x 6″, vertical , painted	3.89	5.25	9.14
1900	Redwood siding 1″ x 4″ to 1″ x 6″ vertical, T & G	5.20	5.70	10.90
2000	Board and batten	4.01	5.20	9.21
2100	Ship lap siding	3.95	5.25	9.20
2200	Plywood, grooved (T1-11) fir	2.81	4.81	7.62
2300	Redwood	4.34	4.81	9.15
2400	Southern yellow pine	2.68	4.81	7.49
2500	Masonry on stud wall, stucco, wire and plaster	2.27	7.50	9.77
2600	Stone veneer	63.50	10	73.50
2700	Brick veneer, brick $275 per M	4.82	12.60	17.42
3100	Cut & patch to match existing construction, add, minimum	2%	3%	
3200	Maximum	5%	9%	
3300	Dust protection, add, minimum	1%	2%	
3400	Maximum	4%	11%	
3500	Material handling & storage limitation, add, minimum	1%	1%	
3600	Maximum	6%	7%	
3700	Protection of existing work, add, minimum	2%	2%	
3800	Maximum	5%	7%	
3900	Shift work requirements, add, minimum		5%	
4000	Maximum		30%	
4100	Temporary shoring and bracing, add, minimum	2%	5%	
4200	Maximum	5%	12%	

B2010 Exterior Walls

B2010 190	Selective Price Sheet	COST PER S.F.		
		MAT.	INST.	TOTAL
0100	Exterior surface, masonry, concrete block, standard 4" thick	1.02	4.65	5.67
0200	6" thick	1.52	5	6.52
0300	8" thick	1.65	5.35	7
0400	12" thick	2.40	6.85	9.25
0500	Split rib, 4" thick	2.30	5.80	8.10
0600	8" thick	3.05	6.65	9.70
0700	Brick running bond, standard size, 6.75/S.F.	3.03	9.10	12.13
0800	Buff, 6.75/S.F.	3.22	9.10	12.32
0900	Stucco, on frame	.48	4.06	4.54
1000	On masonry	.25	3.21	3.46
1100	Metal, aluminum, horizontal, plain	1.35	1.70	3.05
1200	Insulated	1.53	1.70	3.23
1300	Vertical, plain	1.19	1.70	2.89
1400	Insulated	1.24	1.77	3.01
1500	Wood, beveled siding, "A" grade cedar, 1/2" x 6"	2.10	1.75	3.85
1600	1/2" x 8"	2.71	1.59	4.30
1700	Shingles, 16" #1 red, 7-1/2" exposure	1.19	2.13	3.32
1800	18" perfections, 7-1/2" exposure	1.61	1.79	3.40
1900	Handsplit, 10" exposure	1.52	1.75	3.27
2000	White cedar, 7-1/2" exposure	.93	2.19	3.12
2100	Vertical, board & batten, redwood	3.43	2.19	5.62
2200	White pine	.74	1.59	2.33
2300	T. & G. boards, redwood, 1" x 4"	2.93	2.92	5.85
2400	1' x 8"	2.52	2.33	4.85
2500				
2600	Interior surface, drywall, taped & finished, standard, 1/2"	.25	.91	1.16
2700	5/8" thick	.28	.91	1.19
2800	Fire resistant, 1/2" thick	.26	.91	1.17
2900	5/8" thick	.29	.91	1.20
3000	Moisture resistant, 1/2" thick	.31	.91	1.22
3100	5/8" thick	.34	.91	1.25
3200	Core board, 1" thick	.50	1.82	2.32
3300	Plaster, gypsum, 2 coats	.38	2.04	2.42
3400	3 coats	.54	2.46	3
3500	Perlite or vermiculite, 2 coats	.40	2.33	2.73
3600	3 coats	.66	2.93	3.59
3700	Gypsum lath, standard, 3/8" thick	.42	.50	.92
3800	1/2" thick	.43	.54	.97
3900	Fire resistant, 3/8" thick	.43	.61	1.04
4000	1/2" thick	.48	.66	1.14
4100	Metal lath, diamond, 2.5 lb.	.18	.50	.68
4200	Rib, 3.4 lb.	.40	.61	1.01
4300	Framing metal studs including top and bottom			
4400	Runners, walls 10' high			
4500	24" O.C., non load bearing 20 gauge, 2-1/2" wide	.25	.73	.98
4600	3-5/8" wide	.30	.74	1.04
4700	4" wide	.31	.74	1.05
4800	6" wide	.41	.75	1.16
4900	Load bearing 18 gauge, 2-1/2" wide	.41	.82	1.23
5000	3-5/8" wide	.49	.84	1.33
5100	4" wide	4.43	8.50	12.93
5200	6" wide	.66	.87	1.53
5300	16" O.C., non load bearing 20 gauge, 2-1/2" wide	.33	1.12	1.45
5400	3-5/8" wide	.40	1.14	1.54
5500	4" wide	.42	1.15	1.57
5600	6" wide	.55	1.17	1.72
5700	Load bearing 18 gauge, 2-1/2" wide	.56	1.14	1.70
5800	3-5/8" wide	.67	1.15	1.82

B2010 Exterior Walls

B2010 190	Selective Price Sheet	COST PER S.F.		
		MAT.	INST.	TOTAL
5900	4" wide	.70	1.19	1.89
6000	6" wide	.89	1.20	2.09
6100	Framing wood studs incl. double top plate and			
6200	Single bottom plate, walls 10' high			
6300	24" O.C., 2" x 4"	.30	.70	1
6400	2" x 6"	.51	.76	1.27
6500	16" O.C., 2" x 4"	.39	.88	1.27
6600	2" x 6"	.66	.97	1.63
6700	Sheathing, boards, 1" x 6"	.90	1.35	2.25
6800	1"x 8"	.88	1.14	2.02
6900	Plywood, 3/8" thick	.46	.73	1.19
7000	1/2" thick	.43	.78	1.21
7100	5/8" thick	.50	.83	1.33
7200	3/4" thick	.63	.90	1.53
7300	Wood fiber, 5/8" thick	.78	.73	1.51
7400	Gypsum weatherproof, 1/2" thick	.31	.83	1.14
7500	Insulation, fiberglass batts, 3-1/2" thick, R11	.31	.63	.94
7600	6" thick, R19	.40	.73	1.13
7700	Poured 4" thick, fiberglass wool, R4/inch	.41	2.19	2.60
7800	Mineral wool, R3/inch	.36	2.19	2.55
7900	Polystyrene, R4/inch	2.45	2.19	4.64
8000	Perlite or vermiculite, R2.7/inch	1.82	2.19	4.01
8100	Rigid, fiberglass, R4.3/inch, 1" thick	.40	.44	.84
8200	R8.7/inch, 2" thick	.91	.49	1.40

B2020 Exterior Windows

This page illustrates and describes an aluminum window system including double hung aluminum window, exterior and interior trim, hardware and insulating glass. Lines within System Components give the unit price and total price on a cost each basis for this system. Prices for alternate aluminum window systems are on Line Items B2020 110 1100 thru 2400. Both material quantities and labor costs have been adjusted for the system listed.

Factors: To adjust for job conditions other than normal working situations use Lines B2020 110 3100 thru 4000.

Example: You are to install the system and cut and patch to match existing construction. Go to Line B2020 110 3200 and apply these percentages to the appropriate MAT. and INST. costs.

System Components	QUANTITY	UNIT	COST EACH MAT.	COST EACH INST.	COST EACH TOTAL
Double hung aluminum window, 2′-4″ x 2′-6″ exterior and interior trim, Hardware glazed with insulating glass.					
Double hung aluminum window, 2′-4″ x 2′ 6″	1.000	Ea.	171.99	32.36	204.35
Exterior and interior trim	1.000	Set	22	33.50	55.50
Hardware	1.000	Set	2.02	18.25	20.27
Insulating glass	5.830	S.F.	60.05	24.08	84.13
TOTAL			256.06	108.19	364.25

B2020 110	Windows - Aluminum	COST EACH MAT.	COST EACH INST.	COST EACH TOTAL
1000	For alternate window systems:			
1100	Aluminum, double hung, 2′-8″ x 4′-6″	505	178	683
1200	3′- 4″ x 5′-6″	790	269	1,059
1300	Casement, 3′-6″ x 2′-4″	345	131	476
1400	4′-6″ x 2′-4″	440	164	604
1500	5′-6″ x 2′-4″	565	215	780
1600	Projected window, 2′-1″ x 3′-0″	257	112	369
1700	3′-5″ x 3′-0″	410	161	571
1800	4′-0″ x 4′-0″	660	246	906
1900	Horizontal sliding 3′-0″ x 3′-0″	279	139	418
2000	3′-6″ x 4′-0″	425	198	623
2100	5′-0″ x 6′-0″	910	380	1,290
2200	Picture window, 3′-8″ x 3′-1″	282	161	443
2300	4′-4″ x 4′-5″	465	247	712
2400	5′-8″ x 4′-9″	675	350	1,025
2500				
2600				
2700				
2800				
2900				
3100	Cut & patch to match existing construction, add, minimum	2%	3%	
3200	Maximum	5%	9%	
3300	Dust protection, add, minimum	1%	2%	
3400	Maximum	4%	11%	
3500	Material handling & storage limitation, add, minimum	1%	1%	
3600	Maximum	6%	7%	
3700	Protection of existing work, add, minimum	2%	2%	
3800	Maximum	5%	7%	
3900	Shift work requirements, add, minimum		5%	
4000	Maximum		30%	

SHELL B

Important: See the Reference Section for critical supporting data - Reference Numbers and City Cost Indexes

B2020 Exterior Windows

This page illustrates and describes a wood window system including wood picture window, exterior and interior trim, hardware and insulating glass. Lines within System Components give the unit price and total price on a cost each basis for this system. Prices for alternate wood window systems are on Line Items B2020 112 1300 thru 1800. Both material quantities and labor costs have been adjusted for the system listed.

Factors: To adjust for job conditions other than normal working situations use Lines B2020 112 3100 thru 4000.

Example: You are to install the above system where dust control is vital. Go to Line B2020 112 3400 and apply these percentages to the appropriate MAT. and INST. costs.

System Components	QUANTITY	UNIT	COST EACH MAT.	COST EACH INST.	COST EACH TOTAL
Wood picture window 4′-0″ x 4′-6″, exterior and interior trim, hardware, Glazed with insulating glass.					
4′-0″ x 4′-6″ wood picture window with insulating glass	1.000	Ea.	325	79.50	404.50
Exterior and interior trim	1.000	Set	60.50	73	133.50
Hardware	1.000	Set	2.02	18.25	20.27
TOTAL			387.52	170.75	558.27

B2020 112	Windows - Wood	COST EACH MAT.	COST EACH INST.	COST EACH TOTAL
1200	For alternate window systems:			
1300	Picture window 5′-0″ x 4′-0″	475	171	646
1400	6′-0″ x 4′-6″	585	179	764
1500	Bow, bay window, 8′-0″ x 5′-0″, standard	1,950	179	2,129
1600	Deluxe	1,450	179	1,629
1700	Bow, bay window, 12′-0″ x 6′-0″ standard	2,500	237	2,737
1800	Deluxe	2,025	237	2,262
1900	Double hung, 3′-0″ x 4′-0″	305	111	416
2000	4′-0″ x 4′-6″	395	146	541
2100	Casement 2′-0″ x 3′-0″	281	96	377
2200	2 leaf, 4′-0″ x 4′-0″	730	159	889
2300	3 leaf, 6′-0″ x 6′-0″	1,150	255	1,405
2400	Awning, 2′-10″ x 1′-10″	267	96	363
2500	3′-6″ x 2′-4″	340	111	451
2600	4′-0″ x 3′-0″	510	146	656
2700	Horizontal sliding 3′-0″ x 2′-0″	218	96	314
2800	4′-0″ x 3′-6″	257	111	368
2900	6′-0″ x 5′-0″	465	146	611
3100	Cut & patch to match existing construction, add, minimum	2%	3%	
3200	Maximum	5%	9%	
3300	Dust protection, add, minimum	1%	2%	
3400	Maximum	4%	11%	
3500	Material handling & storage limitation, add, minimum	1%	1%	
3600	Maximum	6%	7%	
3700	Protection of existing work, add, minimum	2%	2%	
3800	Maximum	5%	7%	
3900	Shift work requirements, add, minimum		5%	
4000	Maximum		30%	

B2020 Exterior Windows

This page illustrates and describes storm window and door systems based on a cost each price. Prices for alternate storm window and door systems are on Line Items B2020 116 0400 thru 2400. Both material quantities and labor costs have been adjusted for the system listed.

Factors: To adjust for job conditions other than normal working situations use Lines B2020 116 3100 thru 4000 and apply these percentages to the appropriate MAT and INST. costs.

Example: You are to install the system and protect all existing construction. Go to Line B2020 116 3800 and apply these percentages to the appropriate MAT. and INST. costs.

B2020 116	Storm Windows & Doors	COST EACH		
		MAT.	INST.	TOTAL
0350	Storm Window and door systems, single glazed			
0400	Window, custom aluminum anodized, 2′-0″ x 3′-5″	80	29	109
0500	2′-6″ x 5′-0″	107	31.50	138.50
0600	4′-0″ x 6′-0″	227	35	262
0700	White painted aluminum, 2′-0″ x 3′-5″	95.50	29	124.50
0800	2′-6″ x 5′-0″	153	31.50	184.50
0900	4′-0″ x 6′-0″	275	35	310
1000	Average quality aluminum, anodized, 2′-0″ x 3′-5″	81.50	29	110.50
1100	2′-6″ x 5′-0″	98	31.50	129.50
1200	4′-0″ x 6′-0″	121	35	156
1300	White painted aluminum, 2′-0″ x 3′-5″	80	29	109
1400	2′-6″ x 5′-0″	88.50	31.50	120
1500	4′-0″ x 6′-0″	97	35	132
1600	Mill finish, 2′-0″ x 3′-5″	73	29	102
1700	2′-6″ x 5′-0″	81.50	31.50	113
1800	4′-0″ x 6′-0″	91	35	126
1900				
2000	Door, aluminum anodized 3′-0″ x 6′-8″	191	62.50	253.50
2100	White painted aluminum, 3′-0″ x 6′-8″	235	62.50	297.50
2200	Mill finish, 3′-0″ x 6′-8″	240	62.50	302.50
2300				
2400	Wood, storm and screen, painted	287	97	384
2500				
2600				
2700				
2800				
3100	Cut & patch to match existing construction, add, minimum	2%	3%	
3200	Maximum	5%	9%	
3300	Dust protection, add, minimum	1%	2%	
3400	Maximum	4%	11%	
3500	Material handling & storage limitation, add, minimum	1%	1%	
3600	Maximum	6%	7%	
3700	Protection of existing work, add, minimum	2%	2%	
3800	Maximum	5%	7%	
3900	Shift work requirements, add, minimum		5%	
4000	Maximum		30%	

B2020 Exterior Windows

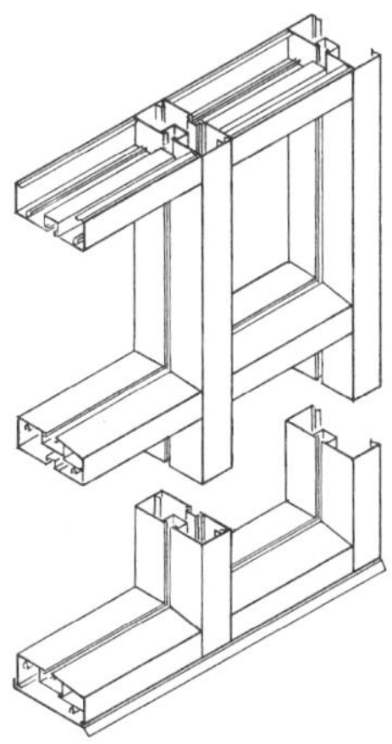

This page illustrates and describes a window wall system including aluminum tube framing, caulking, and glass. Lines within System Components give the unit price and total price per square foot for this system. Prices for alternate window wall systems are on Line Items B2020 124 1400 thru 2600. Both material quantities and labor costs have been adjusted for the system listed.

Factors: To adjust for job conditions other than normal working situations use Lines B2020 124 3100 thru 4000.

Example: You are to install the system with need for complete dust protection. Go to Line B2020 124 3400 and apply these percentages to the appropriate MAT. and INST. costs.

System Components	QUANTITY	UNIT	COST PER S.F. MAT.	COST PER S.F. INST.	COST PER S.F. TOTAL
Window wall, including aluminum header, sill, mullions, caulking					
And glass					
Header, mill finish, 1-3/4" x 4-1/2" deep	.167	L.F.	2.26	1.66	3.92
Sill, mill finish, 1-3/4" x 4-1/2" deep	.167	L.F.	2.15	1.62	3.77
Vertical mullion, 1-3/4" x 4-1/2" deep, 6' O.C.	.191	L.F.	5.46	4.05	9.51
Caulking	.381	L.F.	.06	.68	.74
Glass, 1/4" plate	1.000	S.F.	5.30	6.90	12.20
TOTAL			15.23	14.91	30.14

B2020 124	Aluminum Frame, Window Wall	COST PER S.F. MAT.	COST PER S.F. INST.	COST PER S.F. TOTAL
1200	For alternate systems:			
1300				
1400	Mill finish, 2" x 4-1/2" deep, insulating glass	17.70	17.25	34.95
1500	Thermo-break, 2-1/4" x 4-1/2" deep, insulating glass	18.05	17.50	35.55
1600	Bronze finish, 1-3/4" x 4-1/2" deep, 1/4" plate glass	17	16.20	33.20
1700	2" x 4-1/2" deep, insulating glass	19.55	18.65	38.20
1800	Thermo-break, 2-1/4" x 4-1/2" deep, insulating glass	19.95	18.95	38.90
1900				
2000	Black finish, 1-3/4" x 4-1/2" deep, 1/4" plate glass	17.90	16.90	34.80
2100	2" x 4-1/2" deep, insulating glass	20.50	19.35	39.85
2200	Thermo-break, 2-1/4" x 4-1/2" deep, insulating glass	21	19.70	40.70
2300				
2400	Stainless steel, 1-3/4" x 4-1/2" deep, 1/4" plate glass	22.50	20.50	43
2500	2" x 4-1/2" deep, insulating glass	25.50	23	48.50
2600	Thermo-break, 2-1/4" x 4-1/2" deep, insulating glass	26	23.50	49.50
2700				
3100	Cut & patch to match existing construction, add, minimum	2%	3%	
3200	Maximum	5%	9%	
3300	Dust protection, add, minimum	1%	2%	
3400	Maximum	4%	11%	
3500	Material handling, & storage limitation, add, minimum	1%	1%	
3600	Maximum	6%	7%	
3700	Protection of existing work, add, minimum	2%	2%	
3800	Maximum	5%	7%	
3900	Shift work requirements, add, minimum		5%	
4000	Maximum		30%	

B2030 Exterior Doors

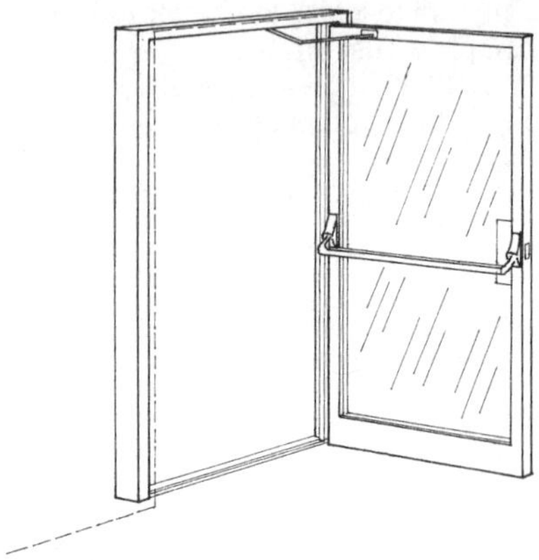

This page illustrates and describes a commercial metal door system, including a single aluminum and glass door, narrow stiles, jamb, hardware weatherstripping, panic hardware and closer. Lines within System Components give the unit price and total price on a cost each basis for this system. Prices for alternate commercial metal door systems are on Line Items B2030 125 1300 thru 2500. Both material quantities and labor costs have been adjusted for the system listed.

Factors: To adjust for job conditions other than normal working situations, use Lines B2030 125 3100 thru 4000.

Example: You are to install the system and cut and patch to match existing construction. Go to Line B2030 125 3200 and apply these percentages to the appropriate MAT. and INST. costs.

System Components	QUANTITY	UNIT	COST EACH		
			MAT.	INST.	TOTAL
Single aluminum and glass door, 3'-0"x7'-0", with narrow stiles, ext. jamb. Weatherstripping, 1/2" tempered insul. glass, panic hardware, and closer.					
Aluminum door, 3'-0" x 7'-0" x 1-3/4", narrow stiles	1.000	Ea.	565	555	1,120
Tempered insulating glass, 1/2" thick	20.000	S.F.	394	301	695
Panic hardware	1.000	Set	375	73	448
Automatic closer	1.000	Ea.	93	67.50	160.50
TOTAL			1,427	996.50	2,423.50

B2030 125	Doors, Metal - Commercial	COST EACH		
		MAT.	INST.	TOTAL
1200	For alternate door systems:			
1300	Single aluminum and glass with transom, 3'-0" x 10'-0"	1,800	1,175	2,975
1400	Anodized aluminum and glass, 3'-0" x 7'-0"	1,625	1,200	2,825
1500	With transom, 3'-0" x 10'-0"	2,075	1,400	3,475
1600	Steel, deluxe, hollow metal 3'-0" x 7'-0"	920	380	1,300
1700	With transom 3'-0" x 10'-0"	1,175	430	1,605
1800	Fire door, "A" label, 3'-0" x 7'-0"	975	380	1,355
1900	Double, aluminum and glass, 6'-0" x 7'-0"	2,325	1,700	4,025
2000	With transom, 6'-0" x 10'-0"	2,475	2,075	4,550
2100	Anodized aluminum and glass 6'-0" x 7'-0"	2,625	2,000	4,625
2200	With transom, 6'-0" x 10'-0"	2,725	2,425	5,150
2300	Steel, deluxe, hollow metal, 6'-0" x 7'-0"	1,500	660	2,160
2400	With transom, 6'-0" x 10'-0"	2,050	760	2,810
2500	Fire door, "A" label, 6'-0" x 7'-0"	1,575	655	2,230
2800				
2900				
3100	Cut & patch to match existing construction, add, minimum	2%	3%	
3200	Maximum	5%	9%	
3300	Dust protection, add, minimum	1%	2%	
3400	Maximum	4%	11%	
3500	Material handling & storage limitation, add, minimum	1%	1%	
3600	Maximum	6%	7%	
3700	Protection of existing work, add, minimum	2%	2%	
3800	Maximum	5%	7%	
3900	Shift work requirements, add, minimum		5%	
4000	Maximum		30%	

SHELL B

B2030 Exterior Doors

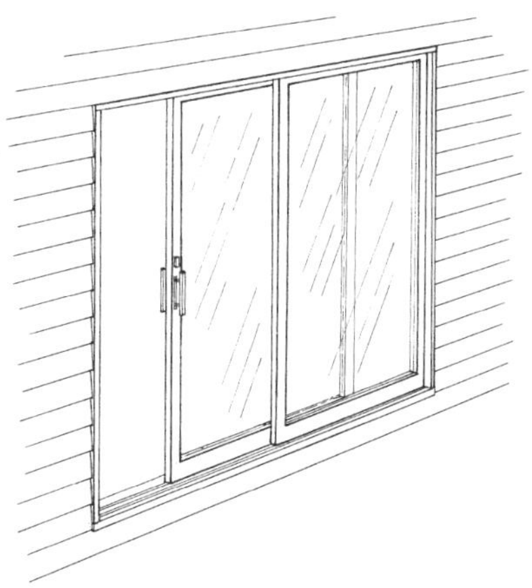

This page illustrates and describes sliding door systems including a sliding door, frame, interior and exterior trim with exterior staining. Lines within System Components give the unit price and total price on a cost each basis for this system. Prices for alternate sliding door systems are on Line Items B2030 150 1100 thru 2400. Both material quantities and labor costs have been adjusted for the system listed.

Factors: To adjust for job conditions other than normal working situations use Lines B2030 150 2700 thru 4000.

Example: You are to install the system with temporary shoring and bracing. Go to Line B2030 150 3900 and apply these percentages to the appropriate MAT. and INST. costs.

System Components	QUANTITY	UNIT	COST EACH		
			MAT.	INST.	TOTAL
Sliding wood door, 6′-0″ x 6′-8″, with wood frame, interior and exterior Trim and exterior staining.					
Sliding wood door, standard, 6′-0″ x 6′-8″, insulated glass	1.000	Ea.	765	219	984
Interior & exterior trim	1.000	Set	18.80	36.40	55.20
Stain door	1.000	Ea.	1.85	22.50	24.35
Stain trim	1.000	Ea.	2	9.40	11.40
TOTAL			787.65	287.30	1,074.95

B2030 150	Doors, Sliding - Patio	COST EACH		
		MAT.	INST.	TOTAL
1000	For alternate sliding door systems:			
1100	Wood, standard, 8′-0″ x 6′-8″, insulated glass	910	365	1,275
1200	12′-0″ x 6′-8″	3,000	445	3,445
1300	Vinyl coated, 6′-0″ x 6′-8″	1,100	287	1,387
1400	8′-0″ x 6′-8″	1,275	365	1,640
1500	12′-0″ x 6′-8″	2,050	445	2,495
1700	Aluminum, standard, 6′-0″ x 6′-8″, insulated glass	820	310	1,130
1800	8′-0″ x 6′-8″	1,250	390	1,640
1900	12′-0″ x 6′-8″	1,550	465	2,015
2000	Anodized, 6′-0″ x 6′-8″	1,550	310	1,860
2100	8′-0″ x 6′-8″	1,475	390	1,865
2200	12′-0″ x 6′-8″	2,425	465	2,890
2300				
2400	Deduct for single glazing	71.50		71.50
2500				
2700	Cut & patch to match existing construction, add, minimum	2%	3%	
2800	Maximum	5%	9%	
2900	Dust protection, add, minimum	1%	2%	
3000	Maximum	4%	11%	
3100	Equipment usage curtailment, add, minimum	1%	1%	
3200	Maximum	3%	10%	
3300	Material handling & storage limitation, add, minimum	1%	1%	
3400	Maximum	6%	7%	
3500	Protection of existing, work, add, minimum	2%	2%	
3600	Maximum	5%	7%	
3700	Shift work requirements, add, minimum		5%	
3800	Maximum		30%	
3900	Temporary shoring and bracing, add, minimum	2%	5%	
4000	Maximum	5%	12%	

B SHELL

B2030 Exterior Doors

This page illustrates and describes residential door systems including a door, frame, trim, hardware, weatherstripping, stained and finished. Lines within System Components give the unit price and total price on a cost each basis for this system. Prices for alternate residential door systems are on Line Items B2030 240 1500 thru 2200. Both material quantities and labor costs have been adjusted for the system listed.

Factors: To adjust for job conditions other than normal working situations use Lines B2030 240 3100 thru 4000.

Example: You are to install the system with a material handling and storage limitation. Go to Line B2030 240 3600 and apply these percentages to the appropriate MAT. and INST. costs.

System Components	QUANTITY	UNIT	COST EACH MAT.	COST EACH INST.	COST EACH TOTAL
Single solid wood colonial door, 3′ x 6′-8″ with wood frame and trim, Stained and finished, including hardware and weatherstripping.					
Solid wood colonial door, fir 3′ x 6′-8″ x 1-3/4″, hinges	1.000	Ea.	390	58.50	448.50
Hinges, ball bearing	1.000	Ea.	29		29
Exterior frame, with trim	1.000	Set	127.50	39.61	167.11
Interior trim	1.000	Set	15.90	74	89.90
Lockset	1.000	Set	33	31.50	64.50
Sill, oak 8″ deep	3.500	Ea.	51.28	30.63	81.91
Weatherstripping	1.000	Set	17.35	57.50	74.85
Stained and finished	1.000	Ea.	3.96	47.50	51.46
TOTAL			667.99	339.24	1,007.23

B2030 240	Doors, Residential - Exterior	COST EACH MAT.	COST EACH INST.	COST EACH TOTAL
1400	For alternate exterior door system:			
1500	Single doors			
1600	Hollow metal exterior door, plain	580	325	905
1700	Solid core wood door, plain	380	325	705
1800				
1900	Double doors, 6′ x 6′-8″			
2000	Solid colonial double doors, fir	1,125	445	1,570
2100	Hollow metal exterior doors, plain	945	435	1,380
2200	Hollow core wood doors, plain	550	435	985
2300				
2800				
2900				
3100	Cut & patch to match existing construction, add, minimum	2%	3%	
3200	Maximum	5%	9%	
3300	Dust protection, add, minimum	1%	2%	
3400	Maximum	4%	11%	
3500	Material handling & storage limitation, add, minimum	1%	1%	
3600	Maximum	6%	7%	
3700	Protection of existing work, add, minimum	2%	2%	
3800	Maximum	5%	7%	
3900	Shift work requirements, add, minimum		5%	
4000	Maximum		30%	

B2030 Exterior Doors

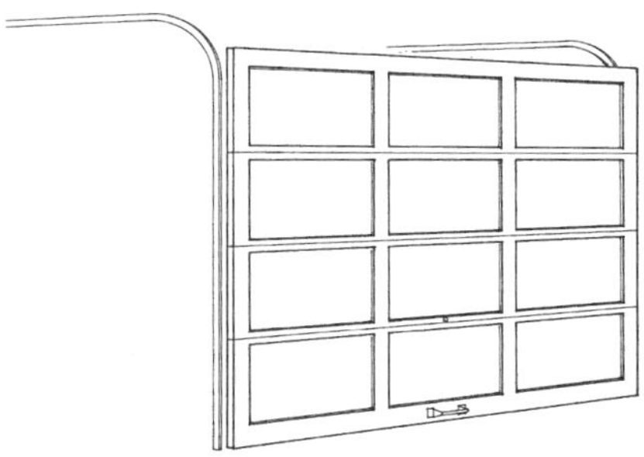

This page illustrates and describes overhead door systems including an overhead door, track, hardware, trim, and electric door opener. Lines within System Components give the unit price and total price on a cost each basis for this system. Prices for alternate overhead door systems are on Line Items B2030 410 1300 thru 2300. Both material quantities and labor costs have been adjusted for the system listed.

Factors: To adjust for job conditions other than normal working situations use Lines B2030 410 3100 thru 4000.

Example: You are to install the system and match existing construction. Go to Line B2030 410 3200 and apply these percentages to the appropriate MAT. and INST. costs.

System Components	QUANTITY	UNIT	COST EACH MAT.	COST EACH INST.	COST EACH TOTAL
Wood overhead door, commercial sectional, including track, door, hardware And trim, electrically operated.					
Commercial, heavy duty wood door, 8′ x 8′ x 1-3/4″ thick	1.000	Ea.	610	440	1,050
Frame, 2 x 8, pressure treated	1.000	Set	25.20	55.20	80.40
Wood trim	1.000	Set	32.16	43.68	75.84
Painting, two coats	1.000	Ea.	17.28	151.20	168.48
Electric trolley operator	1.000	Ea.	715	219	934
TOTAL			1,399.64	909.08	2,308.72

B2030 410	Doors, Overhead	COST EACH MAT.	COST EACH INST.	COST EACH TOTAL
1200	For alternate sectional overhead door systems:			
1300	Commercial, wood, 1-3/4″ thick, 12′ x 12′	2,175	1,275	3,450
1400	14′ x 14′	2,750	1,500	4,250
1500	Fiberglass & aluminum 12′ x 12′	2,775	1,300	4,075
1600	20′ x 20′	5,725	3,325	9,050
1700	Residential, wood, 9′ x 7′	550	345	895
1800	16′ x 7′	1,075	515	1,590
1900	Hardboard faced, 9′ x 7′	475	345	820
2000	16′ x 7′	865	515	1,380
2100	Fiberglass & aluminum, 9′ x 7′	675	405	1,080
2200	16′ x 7′	1,150	515	1,665
2300	For residential electric opener, add	410	54.50	464.50
2400				
2500				
2600				
2700				
2800				
2900				
3100	Cut & patch to match existing construction, add, minumum	2%	3%	
3200	Maximum	5%	9%	
3300	Dust protection, add, minimum	1%	2%	
3400	Maximum	4%	11%	
3500	Material handling & storage limitation, add, minimum	1%	1%	
3600	Maximum	6%	7%	
3700	Protection of existing work, add, minimum	2%	2%	
3800	Maximum	5%	7%	
3900	Shift work requirements, add, minimum		5%	
4000	Maximum		30%	

B2030 Exterior Doors

B2030 810	Selective Price Sheet	COST EACH		
		MAT.	INST.	TOTAL
0100	Door closer, rack and pinion	135	67.50	202.50
0200	Backcheck and adjustable power	137	73	210
0300	Regular, hinge face mount, all sizes, regular arm	170	67.50	237.50
0400	Hold open arm	183	67.50	250.50
0500	Top jamb mount, all sizes, regular arm	170	73	243
0600	Hold open arm	183	73	256
0700	Stop face mount, all sizes, regular arm	170	67.50	237.50
0800	Hold open arm	183	67.50	250.50
0900	Fusible link, hinge face mount, all sizes, regular arm	182	67.50	249.50
1000	Hold open arm	195	67.50	262.50
1100	Top jamb mount, all sizes, regular arm	182	73	255
1200	Hold open arm	195	73	268
1300	Stop face mount, all sizes, regular arm	182	67.50	249.50
1400	Hold open arm	195	67.50	262.50
1500				
1600				
1700	Door stops			
1800				
1900	Holder & bumper, floor or wall	31	13.70	44.70
2000	Wall bumper	9.35	13.70	23.05
2100	Floor bumper	4.88	13.70	18.58
2200	Plunger type, door mounted	26	13.70	39.70
2300	Hinges, full mortise, material only, per pair			
2400	Low frequency, 4-1/2" x 4-1/2", steel base, USP	9.70		9.70
2500	Brass base, US10	36.50		36.50
2600	Stainless steel base, US32	63		63
2700	Average frequency, 4-1/2" x 4-1/2", steel base, USP	21.50		21.50
2800	Brass base, US10	44		44
2900	Stainless steel base, US32	67.50		67.50
3000	High frequency, 4-1/2" x 4-1/2", steel base, USP	44.50		44.50
3100	Brass base, US10	69.50		69.50
3200	Stainless steel base, US32	110		110
3300	Kick plate			
3400				
3500	6" high, for 3'-0" door, aluminum	28	29	57
3600	Bronze	34.50	29	63.50
3700	Panic device			
3800				
3900	For rim locks, single door, exit	375	73	448
4000	Outside key and pull	425	87.50	512.50
4100	Bar and vertical rod, exit only	540	87.50	627.50
4200	Outside key and pull	645	109	754
4300	Lockset			
4400				
4500	Heavy duty, cylindrical, passage doors	122	36.50	158.50
4600	Classroom	272	54.50	326.50
4700	Bedroom, bathroom, and inner office doors	154	36.50	190.50
4800	Apartment, office, and corridor doors	216	44	260
4900	Standard duty, cylindrical, exit doors	75.50	44	119.50
5100	Passage doors	53	36.50	89.50
5200	Public restroom, classroom, & office doors	105	54.50	159.50
5300	Deadlock, mortise, heavy duty	129	48.50	177.50
5400	Double cylinder	143	48.50	191.50
5500	Entrance lock, cylinder, deadlocking latch	119	48.50	167.50
5600	Deadbolt	144	54.50	198.50
5700	Commercial, mortise, wrought knob, keyed, minimum	165	54.50	219.50
5800	Maximum	305	62.50	367.50
5900	Cast knob, keyed, minimum	223	48.50	271.50

Important: See the Reference Section for critical supporting data - Reference Numbers and City Cost Indexes

B2030 Exterior Doors

B2030 810	Selective Price Sheet	COST EACH		
		MAT.	INST.	TOTAL
6000	Maximum	435	48.50	483.50
6100	Push-pull			
6200				
6300	Aluminum	6.80	36.50	43.30
6400	Bronze	17.25	36.50	53.75
6500	Door pull, designer style, minimum	64.50	36.50	101
6600	Maximum	325	54.50	379.50
6700	Threshold			
6800				
6900	3′-0″ long door saddles, aluminum, minimum	3.77	9.10	12.87
7000	Maximum	32	36.50	68.50
7100	Bronze, minimum	30	7.30	37.30
7200	Maximum	62.50	36.50	99
7300	Rubber, 1/2″ thick, 5-1/2″ wide	33	22	55
7400	2-3/4″ wide	15.05	22	37.05
7500	Weatherstripping, per set			
7600				
7700	Doors, wood frame, interlocking for 3′ x 7′ door, zinc	13.60	146	159.60
7800	Bronze	21.50	146	167.50
7900	Wood frame, spring type for 3′ x 7′ door, bronze	17.35	57.50	74.85
8000	Metal frame, spring type for 3′ x 7′ door, bronze	29	146	175
8100	For stainless steel, spring type, add	133%		
8200				
8300	Metal frame, extruded sections, 3′ x 7′ door, aluminum	39.50	219	258.50
8400	Bronze	100	219	319

B3010 Roof Coverings

B3010 160	Selective Price Sheet	COST PER S.F.		
		MAT.	INST.	TOTAL
0100	Roofing, built-up, asphalt roll roof, 3 ply organic/mineral surface	.43	1.11	1.54
0200	3 plies glass fiber felt type iv, 1 ply mineral surface	.67	1.20	1.87
0300	Cold applied, 3 ply		.40	.40
0400	Coal tar pitch, 4 ply tarred felt	1.11	1.43	2.54
0500	Mopped, 3 ply glass fiber	.92	1.57	2.49
0600	4 ply organic felt	1.11	1.43	2.54
0700	Elastomeric, hypalon, neoprene unreinforced	2.26	2.64	4.90
0800	Polyester reinforced	2.28	3.12	5.40
0900	Neoprene, 5 coats 60 mils	5	9.20	14.20
1000	Over 10,000 S.F.	4.68	4.80	9.48
1100	PVC, traffic deck sprayed	1.46	4.80	6.26
1200	With neoprene	1.55	1.93	3.48
1300	Shingles, fiber cement, strip, 14" x 30", 325#/sq.	2.68	1.85	4.53
1500	Shake, 9.35" x 16" 500#/sq.	2.43	1.85	4.28
1600				
1700	Asphalt, strip, 210-235#/sq.	.34	.74	1.08
1800	235-240#/sq.	.42	.82	1.24
1900	Class A laminated	.47	.91	1.38
2000	Class C laminated	.54	1.02	1.56
2100	Slate, buckingham, 3/16" thick	5.50	2.35	7.85
2200	Black, 1/4" thick	6.25	2.35	8.60
2300	Wood, shingles, 16" no. 1, 5" exp.	1.79	1.75	3.54
2400	Red cedar, 18" perfections	1.76	1.59	3.35
2500	Shakes, 24", 10" exposure	1.52	1.75	3.27
2600	18", 8-1/2" exposure	1.07	2.19	3.26
2700	Insulation, ceiling batts, fiberglass, 3-1/2" thick, R11	.25	.27	.52
2800	6" thick, R19	.34	.32	.66
2900	9" thick, R30	.66	.38	1.04
3000	12" thick, R38	.84	.44	1.28
3100	Mineral, 3-1/2" thick, R13	.29	.27	.56
3200	Fiber, 6" thick, R19	.37	.27	.64
3300	Roof deck, fiberboard, 1" thick, R2.78	.37	.51	.88
3400	Mineral, 2" thick, R5.26	.75	.51	1.26
3500	Perlite boards, 3/4" thick, R2.08	.32	.51	.83
3600	2" thick, R5.26	.67	.58	1.25
3700	Polystyrene extruded, R5.26, 1" thick,	.25	.27	.52
3800	2" thick R10	.39	.33	.72
3900	40 PSI compressive strength, 1" thick R5	.39	.27	.66
4000	Tapered for drainage	.55	.29	.84
4100	Foamglass, 1 1/2" thick R4.55	1.39	.51	1.90
4200	3" thick R9.00	2.83	.58	3.41
4300	Ceiling, plaster, gypsum, 2 coats	.38	2.33	2.71
4400	3 coats	.54	2.75	3.29
4500	Perlite or vermiculite, 2 coats	.40	2.69	3.09
4600	3 coats	.66	3.39	4.05
4700	Gypsum lath, plain 3/8" thick	.42	.50	.92
4800	1/2" thick	.43	.54	.97
4900	Firestop, 3/8" thick	.43	.61	1.04
5000	1/2" thick	.48	.66	1.14
5100	Metal lath, rib, 2.75 lb.	.28	.57	.85
5200	3.40 lb.	.40	.61	1.01
5300	Diamond, 2.50 lb.	.18	.57	.75
5400	3.40 lb.	.30	.72	1.02
5500	Drywall, taped and finished standard, 1/2" thick	.25	1.14	1.39
5600	5/8" thick	.28	1.14	1.42
5700	Fire resistant, 1/2" thick	.26	1.14	1.40
5800	5/8" thick	.29	1.14	1.43
5900	Water resist., 1/2" thick	.31	1.14	1.45

B3010 Roof Coverings

B3010 160	Selective Price Sheet	COST PER S.F.		
		MAT.	INST.	TOTAL
6000	5/8" thick	.34	1.14	1.48
6100	Finish, instead of taping			
6200	For thin coat plaster, add	.07	.54	.61
6300	Finish textured spray, add	.04	.48	.52
6400	Drywall, no finish included, see system A6.9-700			
6500	Tile, stapled glued, mineral fiber plastic coated, 5/8" thick	.98	.44	1.42
6600	3/4" thick	1.43	.44	1.87
6700	Wood fiber, 1/2" thick	.84	1.09	1.93
6800	3/4" thick	1.13	1.09	2.22
6900	Suspended, fiberglass film faced, 5/8" thick	.56	.70	1.26
7000	3" thick	1.39	.97	2.36
7100	Mineral fiber 5/8" thick, standard face	.68	.65	1.33
7200	Aluminum faced	1.62	.73	2.35
7300	Wood fiber reveal edge, 1" thick			
7400	3" thick			
7500	Ceiling suspension systems, for tile, "T" bar, class "A", 2' x 4' grid	.45	.55	1
7600	2' x 2' grid	.56	.67	1.23
7700	Concealed "Z" bar, 12" module	.41	.84	1.25
7800				
7900	Plaster/drywall, 3/4" channels, steel furring, 16" O.C.	.16	1.33	1.49
8000	24" O.C.	.11	.92	1.03
8100	1-1/2" channels, 16" O.C.	.23	1.48	1.71
8200	24" O.C.	.16	.99	1.15
8300	Ceiling framing, 2" x 4" studs, 16" O.C.	.26	.73	.99
8400	24" O.C.	.17	.49	.66

B3020 Roof Openings

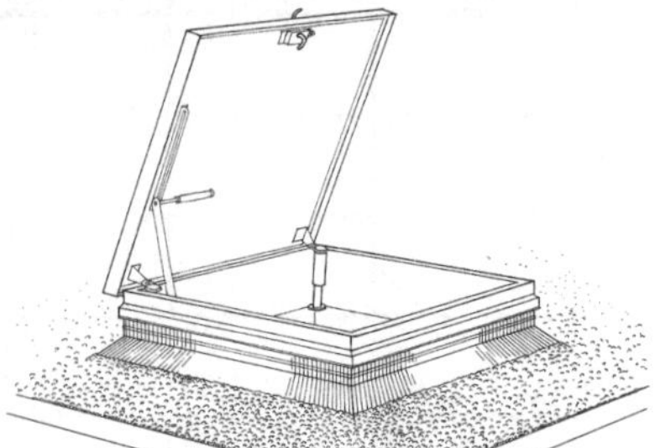

This page illustrates and describes a roof hatch system. Lines within System Components give the unit price and total cost each for this system. Prices for alternate systems are on Lines B3020 230 1500 thru 2400. Both material quantities and labor costs have been adjusted for the system listed.

Factors: To adjust for job conditions other than normal working situtations use Lines B3020 230 2900 thru 4000.

Example: You are to install the system and cut and patch to match existing construction. Use line B3020 230 3000 and apply these percentages to the appropriate MAT. and INST. costs.

System Components	QUANTITY	UNIT	COST EACH MAT.	INST.	TOTAL
Roof hatch, 2′-6″ x 3′-0″, aluminum, curb and cover included.					
Through steel construction.					
Roof hatch, 2′-6″ x 3′-0″, aluminum	1.000	Ea.	470	167	637
Cutout decking	11.000	L.F.		23.21	23.21
Frame opening	44.000	L.F.	18.48	295.24	313.72
Flashing, 16 oz. copper	16.000	S.F.	39.20	56.64	95.84
Cant strip, 4 x 4, foamglass	12.000	L.F.	28.44	15	43.44
TOTAL			556.12	557.09	1,113.21

B3020 230	Roof Hatches, Skylights	COST EACH MAT.	INST.	TOTAL
1400	For alternate systems:			
1500	Roof hatch, aluminum, curb and cover, 2′-6″ x 4′-6″	755	680	1,435
1600	2′-6″ x 8′-0″	1,450	980	2,430
1700	Skylight, plexiglass dome, with curb mounting, 30″ x 32″	490	525	1,015
1800	30″ x 45″	585	660	1,245
1900	40″ x 45″	670	895	1,565
2000	Smoke hatch, unlabeled, 2′-6″ x 3′-0″	675	595	1,270
2100	2′-6″ x 4′-6″	915	725	1,640
2200	2′-6″ x 8′-0″	1,575	1,000	2,575
2300				
2400	For steel ladder, add per vertical linear foot	29	27.50	56.50
2500				
2600				
2700				
2900	Cut & patch to match existing construction, add, minimum	2%	3%	
3000	Maximum	5%	9%	
3100	Dust protection, add, minimum	1%	2%	
3200	Maximum	4%	11%	
3300	Equipment usage curtailment, add, minimum	1%	1%	
3400	Maximum	3%	10%	
3500	Material handling & storage limitation, add, minimum	1%	1%	
3600	Maximum	6%	7%	
3700	Protection of existing work, add, minimum	2%	2%	
3800	Maximum	5%	7%	
3900	Shift work requirements, add, minimum		5%	
4000	Maximum		30%	

Important: See the Reference Section for critical supporting data - Reference Numbers and City Cost Indexes

B3020 Roof Openings

B3020 240	Selective Price Sheet	COST PER L.F.		
		MAT.	INST.	TOTAL
0100	Downspouts per L.F., aluminum, enameled .024" thick, 2" x 3"	1.23	2.73	3.96
0200	3" x 4"	1.75	3.51	5.26
0300	Round .025" thick, 3" diam.	1.08	2.59	3.67
0400	4" diam.	1.76	3.51	5.27
0500	Copper, round 16 oz. stock, 2" diam.	4.98	2.59	7.57
0600	3" diam.	4.96	2.59	7.55
0700	4" diam.	5.10	3.39	8.49
0800	5" diam.	7.05	3.78	10.83
0900	Rectangular, 2" x 3"	5.25	2.59	7.84
1000	3" x 4"	6.90	3.39	10.29
1100	Lead coated copper, round, 2" diam.	5.90	2.59	8.49
1200	3" diam.	6.30	2.59	8.89
1300	4" diam.	8.40	3.39	11.79
1400	5" diam.	11.55	3.78	15.33
1500	Rectangular, 2" x 3"	8.10	2.59	10.69
1600	3" x 4"	8.75	3.39	12.14
1700	Steel galvanized, round 28 gauge, 3" diam.	.81	2.59	3.40
1800	4" diam.	1.11	3.39	4.50
1900	5" diam.	1.50	3.78	5.28
2000	6" diam.	2.17	4.68	6.85
2100	Rectangular, 2" x 3"	.72	2.59	3.31
2200	3" x 4"	1.35	3.39	4.74
2300	Elbows, aluminum, round, 3" diam.	2.18	4.92	7.10
2400	4" diam.	3.29	4.92	8.21
2500	Rectangular, 2" x 3"	1	4.92	5.92
2600	3" x 4"	3.52	4.92	8.44
2700	Copper, round 16 oz., 2" diam.	12.30	4.92	17.22
2800	3" diam.	5.25	4.92	10.17
2900	4" diam.	8.60	4.92	13.52
3100	Rectangular, 2" x 3"	5.65	4.92	10.57
3200	3" x 4"	7.05	4.92	11.97
3300	Drip edge per L.F., aluminum, 5" wide	.28	1.09	1.37
3400	8" wide	.33	1.09	1.42
3500	28" wide	3.76	4.38	8.14
3600				
3700	Steel galvanized, 5" wide	.24	1.09	1.33
3800	8" wide	.36	1.09	1.45
3900				
4000				
4100				
4200				
4300	Flashing 12" wide per S.F., aluminum, mill finish, .013" thick	.39	2.81	3.20
4400	.019" thick	.73	2.81	3.54
4500	.040" thick	1.53	2.81	4.34
4600	.050" thick	2.02	2.81	4.83
4700	Copper, mill finish, 16 oz.	2.45	3.54	5.99
4800	20 oz.	4.29	3.71	8
4900	24 oz.	5.15	3.88	9.03
5000	32 oz.	6.85	4.08	10.93
5100	Lead, 2.5 lb./S.F., 12" wide	3.08	3.02	6.10
5200	Over 12" wide	3.58	3.02	6.60
5300	Lead-coated copper, fabric backed, 2 oz.	1.55	1.24	2.79
5400	5 oz.	1.78	1.24	3.02
5500	Mastic backed, 2 oz.	1.21	1.24	2.45
5600	5 oz.	1.50	1.24	2.74
5700	Paper backed, 2 oz.	1.05	1.24	2.29
5800	3 oz.	1.23	1.24	2.47
5900	Polyvinyl chloride, black, .010" thick	.15	1.43	1.58

B3020 Roof Openings

B3020 240	Selective Price Sheet	COST PER L.F.		
		MAT.	INST.	TOTAL
6000	.020" thick	.24	1.43	1.67
6100	.030" thick	.31	1.43	1.74
6200	.056" thick	.76	1.43	2.19
6300	Steel, galvanized, 20 gauge	.80	3.14	3.94
6400	30 gauge	.34	2.55	2.89
6500	Stainless, 32 gauge, .010" thick	2.38	2.63	5.01
6600	28 gauge, .015" thick	2.95	2.63	5.58
6700	26 gauge, .018" thick	3.58	2.63	6.21
6800	24 gauge, .025" thick	4.64	2.63	7.27
6900	Gutters per L.F., aluminum, 5" box, .027" thick	1.34	4.10	5.44
7000	.032" thick	1.34	4.10	5.44
7100	Copper, half round, 4" wide	4.41	4.10	8.51
7200	6" wide	6.80	4.28	11.08
7300	Steel, 26 gauge galvanized, 5" wide	1.05	4.10	5.15
7400	6" wide	1.57	4.10	5.67
7500	Wood, treated hem-fir, 3" x 4"	6.90	4.38	11.28
7600	4" x 5"	8	4.38	12.38
7700	Reglet per L.F., aluminum, .025" thick	1.03	1.95	2.98
7800	Copper, 10 oz.	1.78	1.95	3.73
7900	Steel, galvanized, 24 gauge	.76	1.95	2.71
8000	Stainless, .020" thick	1.76	1.95	3.71
8100	Counter flashing 12" wide per L.F., aluminum, .032" thick	1.33	3.28	4.61
8200	Copper, 10 oz.	3.73	3.28	7.01
8300	Steel, galvanized, 24 gauge	.69	3.28	3.97
8400	Stainless, .020" thick	3.10	3.28	6.38

C INTERIORS

C1010 Partitions

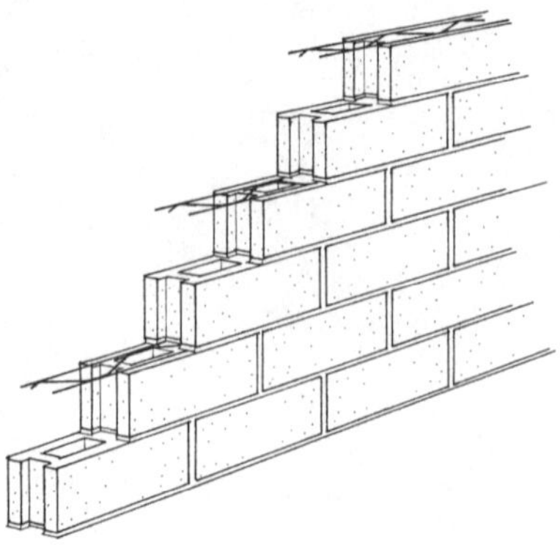

This page illustrates and describes a concrete block wall system including concrete block, horizontal reinforcing alternate courses, mortar, and tooled joints both sides. Lines within System Components give the unit price and total price per square foot for this system. Prices for alternate concrete block wall systems are on Line Items C1010 106 0900 thru 2100. Both material quantities and labor costs have been adjusted for the system listed.

Factors: To adjust for job conditions other than normal working situations use Lines C1010 106 2900 thru 4000.

Example: You are to install the system and protect all existing work. Go to Line C1010 106 3600 and apply these percentages to the appropriate MAT. and INST. costs.

System Components	QUANTITY	UNIT	COST PER S.F. MAT.	COST PER S.F. INST.	COST PER S.F. TOTAL
Concrete block partition, including horizontal reinforcing every second Course, mortar, tooled joints both sides.					
8″ x 16″ concrete block, normal weight, 4″ thick	1.000	S.F.	1.02	4.65	5.67
Horizontal reinforcing every second course	.750	L.F.	.09	.11	.20
TOTAL			1.11	4.76	5.87

C1010 106	Partitions, Concrete Block	COST PER S.F. MAT.	COST PER S.F. INST.	COST PER S.F. TOTAL
0800	For alternate block partition systems:			
0900	8″ x 16″ concrete block, normal weight, 6″ thick	1.61	5.10	6.71
1000	8″ thick	1.74	5.45	7.19
1100	10″ thick	2.42	5.65	8.07
1200	12″ thick	2.49	6.95	9.44
1300	8″ x 16″ concrete block, lightweight, 4″ thick	1.31	4.65	5.96
1400	6″ thick	1.75	4.99	6.74
1500	8″ thick	2.13	5.30	7.43
1600	10″ thick	2.78	5.50	8.28
1700	12″ thick	3.25	6.75	10
1800	8″ x 16″ glazed concrete block, 4″ thick	6.60	5.90	12.50
1900	8″ thick	7.95	6.55	14.50
2000	Structural facing tile, 6T series, glazed 2 sides, 4″ thick	12.75	10.35	23.10
2100	6″ thick	14.60	10.90	25.50
2200				
2300				
2400				
2500				
2600				
2900	Cut & patch to match existing construction, add, minimum	2%	3%	
3000	Maximum	5%	9%	
3100	Dust protection, add, minimum	1%	2%	
3200	Maximum	4%	11%	
3300	Material handling & storage limitation, add, minimum	1%	1%	
3400	Maximum	6%	7%	
3500	Protection of existing work, add, minimum	2%	2%	
3600	Maximum	5%	7%	
3700	Shift work requirements, add, minimum		5%	
3800	Maximum		30%	
3900	Temporary shoring and bracing, add, minimum	2%	5%	
4000	Maximum	5%	12%	

Important: See the Reference Section for critical supporting data - Reference Numbers and City Cost Indexes

C1010 Partitions

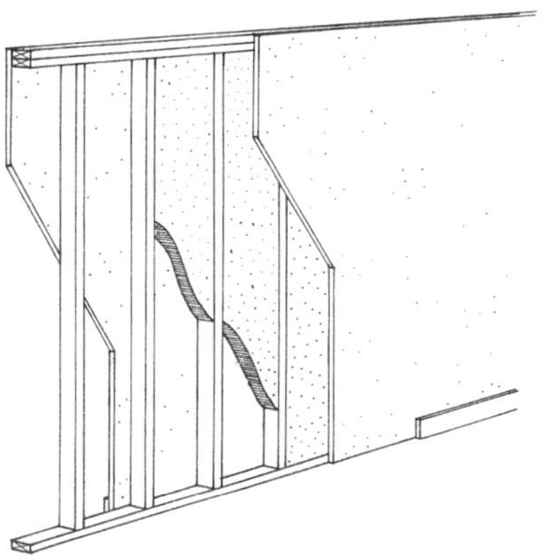

This page illustrates and describes a wood stud partition system including wood studs with plates, gypsum plasterboard – taped and finished, insulation, baseboard and painting. Lines within System Components give the unit price and total price per square foot for this system. Prices for alternate wood stud partition systems are on Line Items C1010 132 1300 thru 2700. Both material quantities and labor costs have been adjusted for the system listed.

Factors: To adjust for job conditions other than normal working situations use Lines C1010 132 2900 thru 4000.

Example: You are to install the system where material handling and storage present a serious problem. Go to Line C1010 132 3400 and apply these percentages to the appropriate MAT. and INST. costs.

System Components	QUANTITY	UNIT	COST PER S.F. MAT.	COST PER S.F. INST.	COST PER S.F. TOTAL
Wood stud wall, 2″x4″, 16″ O.C., dbl. top plate, sngl bot. plate, 5/8″ dwl.					
Taped, finished and painted on 2 faces, insulation, baseboard, wall 8′ high					
Wood studs, 2″ x 4″, 16″ O.C., 8′ high	1.000	S.F.	.42	1.09	1.51
Gypsum drywall, 5/8″ thick	2.000	S.F.	.48	.88	1.36
Taping and finishing	2.000	S.F.	.08	.88	.96
Insulation, 3-1/2″ fiberglass batts	1.000	S.F.	.37	.27	.64
Baseboard	.200	L.F.	.34	.46	.80
Paint baseboard, primer + 2 coats	.200	L.F.	.04	.28	.32
Painting, roller, 2 coats	2.000	S.F.	.30	1	1.30
TOTAL			2.03	4.86	6.89

C1010 132	Partitions, Wood Stud	COST PER S.F. MAT.	COST PER S.F. INST.	COST PER S.F. TOTAL
1200	For alternate wood stud systems:			
1300	2″ x 3″ studs, 8′ high, 16″ O.C.	2.01	4.80	6.81
1400	24″ O.C.	1.92	4.60	6.52
1500	10′ high, 16″ O.C.	1.98	4.60	6.58
1600	24″ O.C.	1.89	4.44	6.33
1700	2″ x 4″ studs, 8′ high, 24″ O.C.	1.94	4.65	6.59
1800	10′ high, 16″ O.C.	2	4.65	6.65
1900	24″ O.C.	1.91	4.47	6.38
2000	12′ high, 16″ O.C.	1.97	4.65	6.62
2100	24″ O.C.	1.85	4.47	6.32
2200	2″ x 6″ studs, 8′ high, 16″ O.C.	2.32	4.98	7.30
2300	24″ O.C.	2.16	4.72	6.88
2400	10′ high, 16″ O.C.	2.27	4.74	7.01
2500	24″ O.C.	2.12	4.53	6.65
2600	12′ high, 16″ O.C.	2.21	4.77	6.98
2700	24″ O.C.	2.07	4.55	6.62
2900	Cut & patch to match existing construction, add, minimum	2%	3%	
3000	Maximum	5%	9%	
3100	Dust protection, add, minimum	1%	2%	
3200	Maximum	4%	11%	
3300	Material handling & storage limitation, add, minimum	1%	1%	
3400	Maximum	6%	7%	
3500	Protection of existing work, add, minimum	2%	2%	
3600	Maximum	5%	7%	
3700	Shift work requirements, add, minimum		5%	
3800	Maximum		30%	
3900	Temporary shoring and bracing, add, minimum	2%	5%	
4000	Maximum	5%	12%	

C1010 Partitions

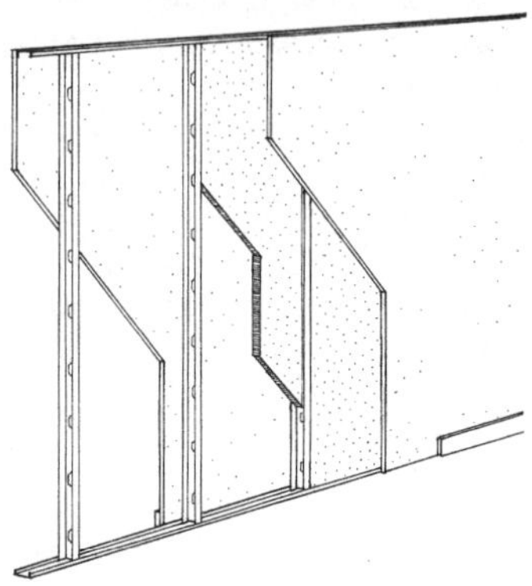

This page illustrates and describes a non-load bearing metal stud partition system including metal studs with runners, gypsum plasterboard, taped and finished, insulation, baseboard and painting. Lines within System Components give the unit price and total price per square foot for this system. Prices for alternate non-load bearing metal stud partition systems are on Line Items C1010 134 1300 thru 2300. Both material quantities and labor costs have been adjusted for the system listed.

Factors: To adjust for job conditions other than normal working situations use Lines C1010 134 2900 thru 4000.

Example: You are to install the system and cut and patch to match existing construction. Go to Line C1010 134 3000 and apply these percentages to the appropriate MAT. and INST. costs.

System Components	QUANTITY	UNIT	COST PER S.F. MAT.	COST PER S.F. INST.	COST PER S.F. TOTAL
Non-load bearing metal studs, including top & bottom runners, 5/8″ drywall, Taped, finished and painted 2 faces, insulation, painted baseboard.					
Metal studs, 25 ga., 3-5/8″ wide, 24″ O.C.	1.000	S.F.	.17	.59	.76
Gypsum drywall, 5/8″ thick	2.000	S.F.	.48	.88	1.36
Taping & finishing	2.000	S.F.	.08	.88	.96
Insulation, 3-1/2″ fiberglass batts	1.000	S.F.	.37	.27	.64
Baseboard	.200	L.F.	.27	.36	.63
Paint baseboard, primer + 2 coats	.200	L.F.	.03	.22	.25
Painting, roller work, 2 coats	2.000	S.F.	.30	1	1.30
TOTAL			1.70	4.20	5.90

C1010 134	Partitions, Metal Stud, NLB	COST PER S.F. MAT.	COST PER S.F. INST.	COST PER S.F. TOTAL
1200	For alternate metal stud systems:			
1300	Non-load bearing, 25 ga., 24″ O.C., 2-1/2″ wide	1.69	4.19	5.88
1400	6″ wide	1.80	4.21	6.01
1500	16″ O.C., 2-1/2″ wide	1.73	4.34	6.07
1600	3-5/8″ wide	1.74	4.35	6.09
1700	6″ wide	1.87	4.36	6.23
1800	20 ga., 24″ O.C., 2-1/2″ wide	1.78	4.34	6.12
1900	3-5/8″ wide	1.84	4.35	6.19
2000	6″ wide	1.94	4.36	6.30
2100	16″ O.C., 2-1/2″ wide	1.84	4.52	6.36
2200	3-5/8″ wide	1.92	4.54	6.46
2300	6″ wide	2.04	4.55	6.59
2400				
2500				
2600				
2700				
2900	Cut & patch to match existing construction, add, minimum	2%	3%	
3000	Maximum	5%	9%	
3100	Dust protection, add, minimum	1%	2%	
3200	Maximum	4%	11%	
3300	Material handling & storage limitation, add, minimum	1%	1%	
3400	Maximum	6%	7%	
3500	Protection of existing work, add, minimum	2%	2%	
3600	Maximum	5%	7%	
3700	Shift work requirements, add, minimum		5%	
3800	Maximum		30%	
3900	Temporary shoring and bracing, add, minimum	2%	5%	
4000	Maximum	5%	12%	

C1010 Partitions

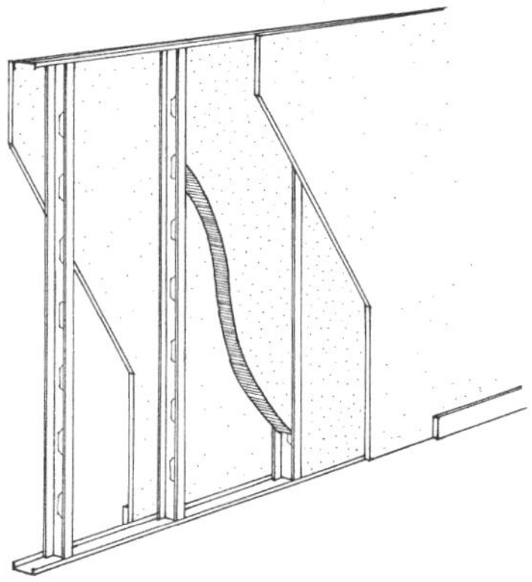

This page illustrates and describes a drywall system including gypsum plasterboard, taped and finished, metal studs with runners, insulation, baseboard and painting. Lines within System Components give the unit price and total price per square foot for this system. Prices for alternate drywall systems are on Line Items C1010 136 1300 thru 1900. Both material quantities and labor costs have been adjusted for the system listed.

Factors: To adjust for job conditions other than normal working situations use Lines C1010 136 2900 thru 4000.

Example: You are to install the system and control dust in the work area. Go to Line C1010 136 3100 and apply these percentages to the appropriate MAT. and INST. costs.

System Components	QUANTITY	UNIT	COST PER S.F. MAT.	COST PER S.F. INST.	COST PER S.F. TOTAL
Gypsum drywall, taped, finished and painted 2 faces, galvanized metal studs					
Including top & bottom runners, insulation, painted baseboard, wall 10′high					
Gypsum drywall, 5/8″ thick, standard	2.000	S.F.	.48	.88	1.36
Taping and finishing	2.000	S.F.	.08	.88	.96
Metal studs, 20 ga., 3-5/8″ wide, 24″ O.C.	1.000	S.F.	.17	.59	.76
Insulation, 3-1/2″ fiberglass batts	1.000	S.F.	.37	.27	.64
Baseboard	.200	L.F.	.27	.36	.63
Paint baseboard, primer + 2 coats	.200	L.F.	.02	.09	.11
Painting, roller 2 coats	2.000	S.F.	.20	.94	1.14
TOTAL			1.59	4.01	5.60

C1010 136	Partitions, Drywall	COST PER S.F. MAT.	COST PER S.F. INST.	COST PER S.F. TOTAL
1200	For alternate drywall systems:			
1300	Gypsum drywall, 5/8″ thick, fire resistant	1.61	4.01	5.62
1400	Water resistant	1.71	4.01	5.72
1500	1/2″ thick, standard	1.53	4.01	5.54
1600	Fire resistant	1.55	4.01	5.56
1700	Water resistant	1.63	4.01	5.64
1800	3/8″ thick, vinyl faced, standard	2.35	5.05	7.40
1900	5/8″ thick, vinyl faced, fire resistant	2.55	5.05	7.60
2000				
2100				
2200				
2300				
2400				
2500				
2600				
2700				
2900	Cut & patch to match existing construction, add, minimum	2%	3%	
3000	Maximum	5%	9%	
3100	Dust protection, add, minimum	1%	2%	
3200	Maximum	4%	11%	
3300	Material handling & storage limitation, add, minimum	1%	1%	
3400	Maximum	6%	7%	
3500	Protection of existing work, add, minimum	2%	2%	
3600	Maximum	5%	7%	
3700	Shift work requirements, add, minimum		5%	
3800	Maximum		30%	
3900	Temporary shoring and bracing, add, minimum	2%	5%	
4000	Maximum	5%	12%	

C1010 Partitions

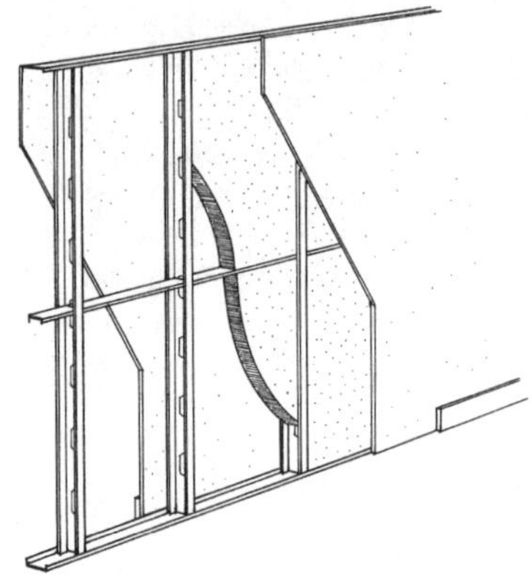

This page illustrates and describes a load bearing metal stud wall system including metal studs, sheetrock–taped and finished, insulation, baseboard and painting. Lines within System Components give the unit price and total price per square foot for this system. Prices for alternate load bearing metal stud wall systems are on Line Items C1010 138 1500 thru 2500. Both material quantities and labor costs have been adjusted for the system listed.

Factors: To adjust for job conditions other than normal working situations use Lines C1010 138 2900 thru 4000.

Example: You are to install the system using temporary shoring and bracing. Go to Line C1010 138 3900 and apply these percentages to the appropriate MAT. and INST. costs.

System Components	QUANTITY	UNIT	COST PER S.F.		
			MAT.	INST.	TOTAL
Load bearing, 18 ga., 3-5/8″, galvanized metal studs, 24″ O.C.,including Top and bottom runners, 1/2″ drywall, taped, finished and painted 2 Faces, 3″ insulation, and painted baseboard, wall 10′ high.					
Metal studs, 24″ O.C., 18 ga., 3-5/8″ wide, galvanized	1.000	S.F.	.49	.84	1.33
Gypsum drywall 1/2″ thick	2.000	S.F.	.42	.88	1.30
Taping and finishing	2.000	S.F.	.08	.88	.96
Insulation, 3-1/2″ fiberglass batts	1.000	S.F.	.37	.27	.64
Baseboard	.200	L.F.	.27	.36	.63
Paint baseboard, primer + 2 coats	.200	L.F.	.03	.22	.25
Painting, roller 2 coats	2.000	S.F.	.30	1	1.30
TOTAL			1.96	4.45	6.41

C1010 138	Partitions, Metal Stud, LB	COST PER S.F.		
		MAT.	INST.	TOTAL
1400	For alternate metal stud systems:			
1500	Load bearing, 18 ga., 24″ O.C., 2-1/2″ wide	1.88	4.43	6.31
1600	6″ wide	2.13	4.48	6.61
1700	16″ O.C. 2-1/2″ wide	1.99	4.64	6.63
1800	3-5/8″ wide	2.09	4.65	6.74
1900	6″ wide	2.29	4.69	6.98
2000	16 ga., 24″ O.C., 2-1/2″ wide	1.95	4.54	6.49
2100	3-5/8″ wide	2.05	4.56	6.61
2200	6″ wide	2.23	4.61	6.84
2300	16″ O.C., 2-1/2″ wide	2.07	4.77	6.84
2400	3-5/8″ wide	2.19	4.80	6.99
2500	6″ wide	2.42	4.85	7.27
2600				
2700				
2900	Cut & patch to match existing construction, add, minimum	2%	3%	
3000	Maximum	5%	9%	
3100	Dust protection, add, minumum	1%	2%	
3200	Maximum	4%	11%	
3300	Material handling & storage limitation, add, minimum	1%	1%	
3400	Maximum	6%	7%	
3500	Protection of existing work, add, minimum	2%	2%	
3600	Maximum	5%	7%	
3700	Shift work requirements, add, minimum		5%	
3800	Maximum		30%	
3900	Temporary shoring and bracing, add, minimum	2%	5%	
4000	Maximum	5%	12%	

C1010 Partitions

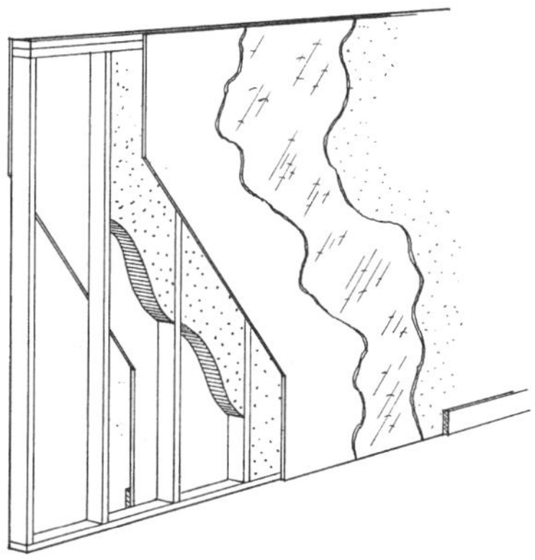

This page illustrates and describes a plaster and lath system including gypsum plaster, gypsum lath, wood studs with plates, insulation, baseboard and painting. Lines within System Components give the unit price and price per square foot for this system. Prices for alternate plaster and lath systems are on Line Items C1010 148 1500 thru 2500. Both material quantities and labor costs have been adjusted for the system listed.

Factors: To adjust for job conditions other than normal working situations use Lines C1010 148 2900 thru 4000.

Example: You are to install the system during evening hours only. Go to Line C1010 148 3800 and apply this percentage to the appropriate INST. costs.

System Components	QUANTITY	UNIT	COST PER S.F. MAT.	COST PER S.F. INST.	COST PER S.F. TOTAL
Gypsum plaster, 2 coats, over 3/8" lath, 2 faces, 2" x 4" wood Stud partition, 24"O.C. incl. double top plate,single bottom plate,3-1/2" Insulation, baseboard and painting.					
Gypsum plaster, 2 coats	.220	S.Y.	.76	4.05	4.81
Lath, gypsum, 3/8" thick	.220	S.Y.	.83	1	1.83
Wood studs, 2" x 4", 24" O.C.	1.000	S.F.	.30	.70	1
Insulation, 3-1/2" fiberglass batts	1.000	S.F.	.37	.27	.64
Baseboard, 9/16" x 3-1/2"	.200	L.F.	.27	.36	.63
Paint baseboard, primer + 2 coats	.200	L.F.	.02	.09	.11
Paint, 2 coats	2.000	S.F.	.30	1	1.30
TOTAL			2.85	7.47	10.32

C1010 148	Partitions, Plaster & Lath	COST PER S.F. MAT.	COST PER S.F. INST.	COST PER S.F. TOTAL
1400	For alternate plaster systems:			
1500	Gypsum plaster, 3 coats	3.17	8.30	11.47
1600	Perlite plaster, 2 coats	2.87	8.05	10.92
1700	3 coats	3.40	9.20	12.60
1800				
1900				
2000	For alternate lath systems:			
2100	Gypsum lath, 1/2" thick	2.89	8.10	10.99
2200	Foil back, 3/8" thick	2.91	8.20	11.11
2300	1/2" thick	3	8.25	11.25
2400	Metal lath, 2.5 Lb. diamond	2.40	8.05	10.45
2500	3.4 Lb. diamond	2.63	8.20	10.83
2600				
2700				
2900	Cut & patch to match existing construction, add, minimum	2%	3%	
3000	Maximum	5%	9%	
3100	Dust protection, add, minimum	1%	2%	
3200	Maximum	4%	11%	
3300	Material handling & storage limitation, add, minimum	1%	1%	
3400	Maximum	6%	7%	
3500	Protection of existing work, add, minimum	2%	2%	
3600	Maximum	5%	7%	
3700	Shift work requirements, add, minimum		5%	
3800	Maximum		30%	
3900	Temporary shoring and bracing, add, minimum	2%	5%	
4000	Maximum	5%	12%	

C1010 Partitions

C1010 170	Selective Price Sheet	COST PER S.F.		
		MAT.	INST.	TOTAL
0100	Studs			
0200				
0300	24" O.C. metal, 10' high wall, including			
0400	Top and bottom runners			
0500	Non load bearing, galvanized 25 Ga., 1-5/8" wide	.15	.58	.73
0600	2-1/2" wide	.16	.58	.74
0700	3-5/8" wide	.17	.59	.76
0800	4" wide	.19	.59	.78
0900	6" wide	.27	.60	.87
1000				
1100	Galvanized 20 Ga., 2-1/2" wide	.25	.73	.98
1200	3-5/8" wide	.30	.74	1.04
1300	4" wide	.31	.74	1.05
1400	6" wide	.41	.75	1.16
1500	Load bearing, painted 18 Ga., 2-1/2" wide	.41	.82	1.23
1600	3-5/8" wide	.49	.84	1.33
1700	4" wide	.44	.85	1.29
1800	6" wide	.66	.87	1.53
1900	Galvanized 18 Ga., 2-1/2" wide	.41	.82	1.23
2000	3-5/8" wide	.49	.84	1.33
2100	4" wide	.44	.85	1.29
2200	6" wide	.66	.87	1.53
2300	Galvanized 16 Ga., 2-1/2" wide	.48	.93	1.41
2400	3-5/8" wide	.58	.95	1.53
2500	4" wide	.61	.97	1.58
2600	6" wide	.76	1	1.76
2700				
2800				
2900	24" O.C. wood, 10' high wall, including			
3000	Double top plate and shoe			
3100	2" x 4"	.35	.83	1.18
3200	2" x 6"	.59	.88	1.47
3300				
3400				
3500	Furring, 24" O.C. 10' high wall, metal, 3/4" channels	.11	.92	1.03
3600	1-1/2" channels	.16	.99	1.15
3700	Wood, on wood, 1" x 2" strips	.11	.40	.51
3800	1" x 3" strips	.12	.40	.52
3900	On masonry, 1" x 2" strips	.11	.44	.55
4000	1" x 3" strips	.12	.44	.56
4100	On concrete, 1" x 2" strips	.11	.84	.95
4200	1" x 3" strips	.12	.84	.96
4300	Studs			
4400				
4500	16" O.C., metal, 10' high wall, including			
4600	Top and bottom runners			
4700	Non load bearing, galvanized 25 Ga., 1-5/8" wide	.20	.88	1.08
4800	2-1/2" wide	.22	.89	1.11
4900	3-5/8" wide	.23	.91	1.14
5000	4" wide	.26	.92	1.18
5100	6" wide	.37	.93	1.30
5200				
5300	Galvanized 20 Ga., 2-1/2" wide	.33	1.12	1.45
5400	3-5/8" wide	.40	1.14	1.54
5500	4" wide	.42	1.15	1.57
5600	6" wide	.55	1.17	1.72

Important: See the Reference Section for critical supporting data - Reference Numbers and City Cost Indexes

C1010 Partitions

C1010 170	Selective Price Sheet	COST PER S.F. MAT.	COST PER S.F. INST.	COST PER S.F. TOTAL
5700	Load bearing, painted 18 Ga., 2-1/2" wide	.56	1.14	1.70
5800	3-5/8" wide	.67	1.15	1.82
5900	4" wide	.70	1.19	1.89
6000	6" wide	.89	1.20	2.09
6100	Galvanized 18 Ga., 2-1/2" wide	.56	1.14	1.70
6200	3-5/8" wide	.67	1.15	1.82
6300	4" wide	.70	1.19	1.89
6400	6" wide	.89	1.20	2.09
6500	Galvanized 16 Ga., 2-1/2" wide	.66	1.29	1.95
6600	3-5/8" wide	.79	1.33	2.12
6700	4" wide	.84	1.35	2.19
6800	6" wide	1.05	1.37	2.42
6900				
7000				
7100	16" O.C., wood, 10' high wall, including			
7200	Double top plate and shoe			
7300	2" x 4"	.43	.95	1.38
7400	2" x 6"	.58	.85	1.43
7500				
7600				
7700	Furring, 16" O.C. 10' high wall, metal, 3/4" channels	.16	1.33	1.49
7800	1-1/2" channels	.23	1.48	1.71
7900	Wood, on wood, 1" x 2" strips	.16	.60	.76
8000	1" x 3" strips	.17	.60	.77
8100	On masonry, 1" x 2" strips	.16	.66	.82
8200	1" x 3" strips	.17	.66	.83
8300	On concrete, 1" x 2" strips	.16	1.26	1.42
8400	1" x 3" strips	.17	1.26	1.43

C1010 Wall Finishes

C1010 180	Selective Price Sheet	COST PER S.F.		
		MAT.	INST.	TOTAL
0100	Lath, gypsum perforated			
0200				
0300	Lath, gypsum, perforated, regular, 3/8″ thick	.42	.50	.92
0400	1/2″ thick	.43	.61	1.04
0500	Fire resistant, 3/8″ thick	.43	.61	1.04
0600	1/2″ thick	.48	.66	1.14
0700	Foil back, 3/8″ thick	.44	.57	1.01
0800	1/2″ thick	.48	.61	1.09
0900				
1000	Metal lath			
1100	Metal lath, diamond, painted, 2.5 lb.	.18	.50	.68
1200	3.4 lb.	.30	.57	.87
1300	Rib painted, 2.75 lb	.28	.57	.85
1400	3.40 lb	.40	.61	1.01
1500				
1600				
1700	Plaster, gypsum, 2 coats	.38	2.04	2.42
1800	3 coats	.54	2.46	3
1900	Perlite/vermiculite, 2 coats	.40	2.33	2.73
2000	3 coats	.66	2.93	3.59
2100	Bondcrete, 1 coat	.40	1.08	1.48
2200				
2500				
2600				
2700	Drywall, standard, 3/8″ thick, no finish included	.22	.44	.66
2800	1/2″ thick, no finish included	.21	.44	.65
2900	Taped and finished	.25	.91	1.16
3000	5/8″ thick, no finish included	.24	.44	.68
3100	Taped and finished	.28	.91	1.19
3200	Fire resistant, 1/2″ thick, no finish included	.22	.44	.66
3300	Taped and finished	.26	.91	1.17
3400	5/8″ thick, no finish included	.25	.44	.69
3500	Taped and finished	.29	.91	1.20
3600	Water resistant, 1/2″ thick, no finish included	.26	.44	.70
3700	Taped and finished	.31	.91	1.22
3800	5/8″ thick, no finish included	.30	.44	.74
3900	Taped and finished	.34	.91	1.25
4000	Finish, instead of taping			
4100	For thin coat plaster, add	.07	.54	.61
4200	Finish, textured spray, add	.04	.48	.52

C1010 Partitions

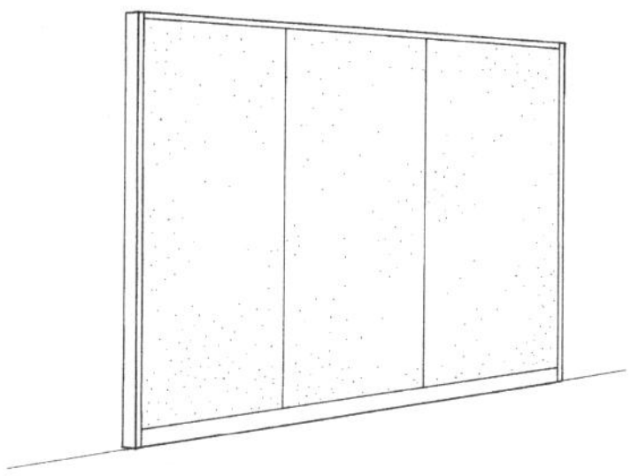

This page illustrates and describes a movable office partition system including demountable office partitions of various styles and sizes based on a cost per square foot basis. Prices for alternate movable office partition systems are on Line Items C1010 210 0300 thru 3600.

C1010 210	Partitions, Movable Office	COST PER S.F.		
		MAT.	INST.	TOTAL
0250	Office partition, demountable, no deduction for door opening, add for doors			
0300	Air wall, cork finish, semi acoustic, 1-5/8″ thick, minimum	31	2.69	33.69
0400	Maximum	29	4.61	33.61
0500	Acoustic, 2″ thick, minimum	25	2.87	27.87
0600	Maximum	43	3.89	46.89
0700				
0800				
0900	In-plant modular office system, w/prehung steel door			
1000	3″ thick honeycomb core panels			
1100	12′ x 12′, 2 wall	6.35	.36	6.71
1200	4 wall	13.20	1.10	14.30
1300				
1400				
1500	Gypsum, demountable, 3″ to 3-3/4″ thick x 9′ high, vinyl clad	4.83	2.03	6.86
1600	Fabric clad	12	2.21	14.21
1700	1.75 system, vinyl clad hardboard, paper honeycomb core panel			
1800	1-3/4″ to 2-1/2″ thick x 9′ high	8.10	2.03	10.13
1900	Unitized gypsum panel, 2″ to 2-1/2″ thick x 9′ high, vinyl clad	10.45	2.03	12.48
2000	Fabric clad	17.20	2.21	19.41
2100				
2200				
2300	Unitized mineral fiber panel system, 2-1/4″ thick x 9′ high			
2400	Vinyl clad mineral fiber	10.40	2.03	12.43
2500	Fabric clad mineral fiber	15.45	2.21	17.66
2600				
2700	Movable steel walls, modular system			
2800	Unitized panels, 48″ wide x 9′ high			
2900	Baked enamel, pre-finished	11.75	1.62	13.37
3000	Fabric clad	17	1.74	18.74
3100	For acoustical partitions, add, minimum	1.79		1.79
3200	Maximum	8.35		8.35
3300				
3400				
3500	Note: For door prices, see divisions 08100 & 08200			
3600	For door hardware prices, see division 08700			
3700				
3800				
3900				
4000				

C1020 Interior Doors

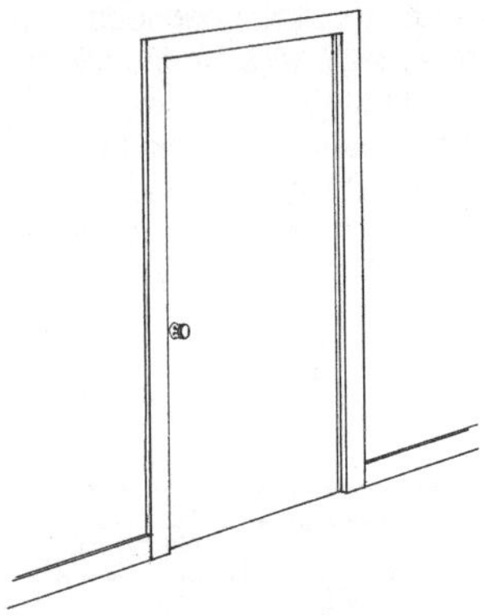

This page illustrates and describes flush interior door systems including hollow core door, jamb, header and trim with hardware. Lines within System Components give the unit price and total price on a cost each basis for this system. Prices for alternate flush interior door systems are on Line items C1020 106 1100 thru 2400. Both material quantities and labor costs have been adjusted for the system listed.

Factors: To adjust for job conditions other than normal working situations use Lines C1020 106 2700 thru 3800.

Example: You are to install the system in an area where dust protection is vital. Go to Line C1020 106 3000 and apply these percentages to the appropriate MAT. and INST. costs.

System Components	QUANTITY	UNIT	COST EACH		
			MAT.	INST.	TOTAL
Single hollow core door, include jamb, header, trim and hardware, painted.					
Hollow core Lauan, 1-3/8" thick, 2'-0" x 6'-8", painted	1.000	Ea.	28	48.50	76.50
Paint door and frame, 1 coat	1.000	Ea.	1.85	22.50	24.35
Wood jamb, 4-9/16" deep	1.000	Set	96	37.28	133.28
Trim, casing	1.000	Set	30.08	58.24	88.32
Hardware, hinges	1.000	Set	9.70		9.70
Hardware, lockset	1.000	Set	14.70	27.50	42.20
TOTAL			180.33	194.02	374.35

C1020 106	Doors, Interior Flush, Wood	COST EACH		
		MAT.	INST.	TOTAL
1000	For alternate door systems:			
1100	Lauan (Mahogany) hollow core, 1-3/8" x 2'-6" x 6'-8"	188	197	385
1200	2'-8" x 6'-8"	192	199	391
1300	3'-0" x 6'-8"	197	207	404
1500	Birch, hollow core, 1-3/8" x 2'-0" x 6'-8"	193	194	387
1600	2'-6" x 6'-8"	202	197	399
1700	2'-8" x 6'-8"	206	199	405
1800	3'-0" x 6'-8"	214	207	421
1900	Solid core, pre-hung, 1-3/8" x 2'-6" x 6'-8"	325	193	518
2000	2'-8" x 6'-8"	350	195	545
2100	3'-0" x 6'-8"	365	201	566
2200				
2400	For metal frame instead of wood, add	50%	20%	
2600				
2700	Cut & patch to match existing construction, add, minimum	2%	3%	
2800	Maximum	5%	9%	
2900	Dust protection, add, minimum	1%	2%	
3000	Maximum	4%	11%	
3100	Equipment usage curtailment, add, minimum	1%	1%	
3200	Maximum	3%	10%	
3300	Material handling & storage limitation, add, minimum	1%	1%	
3400	Maximum	6%	7%	
3500	Protection of existing work, add, minimum	2%	2%	
3600	Maximum	5%	7%	
3700	Shift work requirements, add, minimum		5%	
3800	Maximum		30%	
3900				

C1020 Interior Doors

This page illustrates and describes interior, solid and louvered door systems including a pine panel door, wood jambs, header, and trim with hardware. Lines within System Components give the unit price and total price on a cost each basis for this system. Prices for alternate interior, solid and louvered systems are on Line Items C1020 108 1200 thru 2400. Both material quantities and labor costs have been adjusted for the system listed.

Factors: To adjust for job conditions other than normal working situations use Lines C1020 108 2900 thru 4000.

Example: You are to install the system during night hours only. Go to Line C1020 108 4000 and apply these percentages to the appropriate INST. costs.

System Components	QUANTITY	UNIT	COST EACH MAT.	COST EACH INST.	COST EACH TOTAL
Single interior door, including jamb, header, trim and hardware, painted.					
Solid pine panel door, 1-3/8″ thick, 2′-0″ x 6′-8″	1.000	Ea.	128	48.50	176.50
Paint door and frame, 1 coat	1.000	Ea.	1.85	22.50	24.35
Wooden jamb, 4-5/8″ deep	1.000	Set	96	37.28	133.28
Trim, casing	1.000	Set	30.08	58.24	88.32
Hardware, hinges	1.000	Set	9.70		9.70
Hardware, lockset	1.000	Set	14.70	27.50	42.20
TOTAL			280.33	194.02	474.35

C1020 108	Doors, Interior Solid & Louvered	COST EACH MAT.	COST EACH INST.	COST EACH TOTAL
1000				
1100	For alternate door systems:			
1200	Solid pine, painted raised panel, 1-3/8″ x 2′-6″ x 6′-8″	300	197	497
1300	2′-8″ x 6′-8″	315	199	514
1400	3′-0″ x 6′-8″	325	207	532
1500				
1600	Louvered pine, painted 1′-6″ x 6′-8″	251	192	443
1700	2′-0″ x 6′-8″	280	197	477
1800	2′-6″ x 6′-8″	294	199	493
1900	3′-0″ x 6′-8″	315	207	522
2200	For prehung door, deduct	5%	30%	
2300				
2400	For metal frame instead of wood, add	50%	20%	
2500				
2900	Cut & patch to match existing construction, add, minimum	2%	3%	
3000	Maximum	5%	9%	
3100	Dust protection, add, minimum	1%	2%	
3200	Maximum	4%	11%	
3300	Equipment usage curtailment, add, minimum	1%	1%	
3400	Maximum	3%	10%	
3500	Material handling & storage limitation, add, minimum	1%	1%	
3600	Maximum	6%	7%	
3700	Protection of existing work, add, minimum	2%	2%	
3800	Maximum	5%	7%	
3900	Shift work requirements, add, minimum		5%	
4000	Maximum		30%	

C1020 Interior Doors

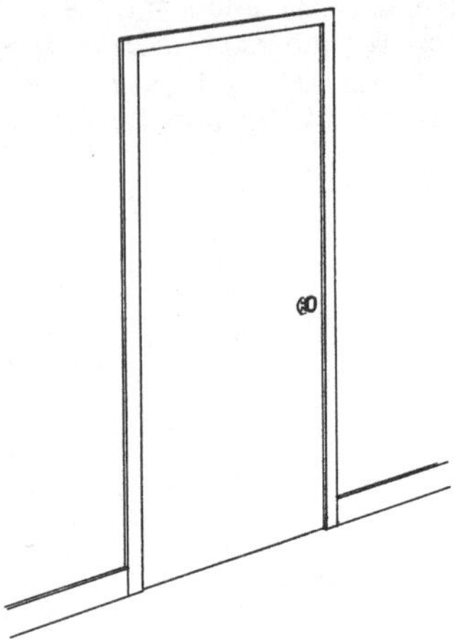

This page illustrates and describes interior metal door systems including a metal door, metal frame and hardware. Lines within System Components give the unit price and total price on a cost each for this system. Prices for alternate interior metal door systems are on Line Items C1020 110 1100 thru 2100. Both material quantities and labor costs have been adjusted for the system listed.

Factors: To adjust for job conditions other than normal working situations use Lines C1020 110 2900 thru 4000.

Example: You are to install the system while protecting existing construction. Go to Line C1020 110 3700 and apply these percentages to the appropriate MAT. and INST. costs.

System Components	QUANTITY	UNIT	COST EACH MAT.	COST EACH INST.	COST EACH TOTAL
Single metal door, including frame and hardware.					
Hollow metal door, 1-3/8″ thick, 2′-6″ x 6′-8″, painted	1.000	Ea.	217	58.50	275.50
Metal frame, 5-3/4″ deep	1.000	Set	84.50	54.50	139
Paint door and frame, 1 coat	1.000	Ea.	1.85	22.50	24.35
Hardware, hinges	1.000	Set	62.50		62.50
Hardware, passage lockset	1.000	Set	14.70	27.50	42.20
TOTAL			380.55	163	543.55

C1020 110	Doors, Interior Flush, Metal	COST EACH MAT.	COST EACH INST.	COST EACH TOTAL
1000	For alternate systems:			
1100	Hollow metal doors, 1-3/8″ thick, 2′-8″ x 6′-8″	385	163	548
1200	3′-0″ x 7′-0″	375	171	546
1300				
1400	Interior fire door, 1-3/8″ thick, 2′-6″ x 6′-8″	460	163	623
1500	2′-8″ x 6′-8″	465	163	628
1600	3′-0″ x 7′-0″	465	171	636
1700				
1800	Add to fire doors:			
1900	Baked enamel finish	30%	15%	
2000	Galvanizing	15%		
2200				
2300				
2400				
2900	Cut & patch to match existing construction, add, minimum	2%	3%	
3000	Maximum	5%	9%	
3100	Dust protection, add, minimum	1%	2%	
3200	Maximum	4%	11%	
3300	Equipment usage curtailment, add, minimum	1%	1%	
3400	Maximum	3%	10%	
3500	Material handling & storage limitation, add, minimum	1%	1%	
3600	Maximum	6%	7%	
3700	Protection of existing work, add, minimum	2%	2%	
3800	Maximum	5%	7%	
3900	Shift work requirements, add, minimum		5%	
4000	Maximum		30%	

C1020 Interior Doors

This page illustrates and describes an interior closet door system including an interior closet door, painted, with trim and hardware. Prices for alternate interior closet door systems are on Line Items C1020 112 0500 thru 2200. Both material quantities and labor costs have been adjusted for the system listed.

Factors: To adjust for job conditions other than normal working situations use Lines C1020 112 2900 thru 4000.

Example: You are to install the system and match the existing construction. Go to Line C1020 112 2900 and apply these percentages to the appropriate MAT. and INST. costs.

C1020 112	Doors, Closet	COST PER SET		
		MAT.	INST.	TOTAL
0350	Interior closet door painted, including frame, trim and hardware, prehung.			
0400	Bi-fold doors			
0500	Pine paneled, 3′-0″ x 6′-8″	279	161	440
0600	6′-0″ x 6′-8″	420	216	636
0700	Birch, hollow core, 3′-0″ x 6′-8″	158	174	332
0800	6′-0″ x 6′-8″	240	233	473
0900	Lauan, hollow core, 3′-0″ x 6′-8″	147	161	308
1000	6′-0″ x 6′-8″	218	216	434
1100	Louvered pine, 3′-0″ x 6′-8″	254	161	415
1200	6′-0″ x 6′-8″	335	216	551
1300				
1400	Sliding, bi-passing closet doors			
1500	Pine paneled, 4′-0″ x 6′-8″	490	170	660
1600	6′-0″ x 6′-8″	590	216	806
1700	Birch, hollow core, 4′-0″ x 6′-8″	305	170	475
1800	6′-0″ x 6′-8″	360	216	576
1900	Lauan, hollow core, 4′-0″ x 6′-8″	270	170	440
2000	6′-0″ x 6′-8″	310	216	526
2100	Louvered pine, 4′-0″ x 6′-8″	505	170	675
2200	6′-0″ x 6′-8″	610	216	826
2300				
2400				
2500				
2600				
2700				
2800				
2900	Cut & patch to match existing construction, add, minimum	2%	3%	
3000	Maximum	5%	9%	
3100	Dust protection, add, minimum	1%	2%	
3200	Maximum	4%	11%	
3300	Equipment usage curtailment, add, minimum	1%	1%	
3400	Maximum	3%	10%	
3500	Material handling & storage limitation, add, minimum	1%	1%	
3600	Maximum	6%	7%	
3700	Protection of existing work, add, minimum	2%	2%	
3800	Maximum	5%	7%	
3900	Shift work requirements, add, minimum		5%	
4000	Maximum		30%	

C2010 Stair Construction

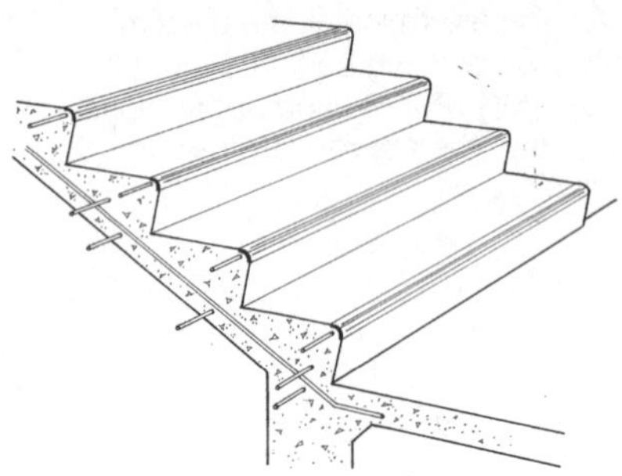

This page illustrates and describes a stair system based on a cost per flight price. Prices for various stair systems are on Line Items C2010 130 0700 thru 3200. Both material quantities and labor costs have been adjusted for the system listed.

Factors: To adjust for job conditions other than normal working situations use Lines C2010 130 3500 thru 4200.

Example: You are to install the system during evenings only. Go to Line C2010 130 4200 and apply this percentage to the appropriate MAT. and INST. costs.

System Components	QUANTITY	UNIT	COST PER FLIGHT		
			MAT.	INST.	TOTAL
Below are various stair systems based on cost per flight of stairs, no side Walls. Stairs are 4′-0″ wide, railings are included unless otherwise noted					

C2010 130	Stairs	COST PER FLIGHT		
		MAT.	INST.	TOTAL
0700	Concrete, cast in place, no nosings, no railings, 12 risers	365	1,525	1,890
0800	24 risers	730	3,000	3,730
0900	Add for 1 intermediate landing	93.50	420	513.50
1000	Concrete, cast in place, with nosings, no railings, 12 risers	760	1,775	2,535
1100	24 risers	1,525	3,525	5,050
1200	Add for 1 intermediate landing	126	440	566
1300	Steel, grating tread, safety nosing, 12 risers	1,625	875	2,500
1400	24 risers	3,250	1,775	5,025
1500	Add for intermediate landing	390	232	622
1600	Steel, cement fill pan tread, 12 risers	2,425	875	3,300
1700	24 risers	4,850	1,775	6,625
1800	Add for intermediate landing	390	232	622
1900	Spiral, industrial, 4′ - 6″ diameter, 12 risers	2,375	620	2,995
2000	24 risers	4,750	1,250	6,000
2100	Wood, box stairs, oak treads, 12 risers	1,700	495	2,195
2200	24 risers	3,375	995	4,370
2300	Add for 1 intermediate landing	163	112	275
2400	Wood, basement stairs, no risers, 12 steps	55.50	18.55	74.05
2500	24 steps	111	37	148
2600	Add for 1 intermediate landing	24	21	45
2700	Wood, open, rough sawn cedar, 12 steps	1,925	400	2,325
2800	24 steps	3,850	805	4,655
2900	Add for 1 intermediate landing	19.20	22	41.20
3000	Wood, residential, oak treads, 12 risers	2,025	2,200	4,225
3100	24 risers	4,075	4,400	8,475
3200	Add for 1 intermediate landing	151	101	252
3500	Dust protection, add, minimum	1%	2%	
3600	Maximum	4%	11%	
3700	Material handling & storage limitation, add, minimum	1%	1%	
3800	Maximum	6%	7%	
3900	Protection of existing work, add, minimum	2%	2%	
4000	Maximum	5%	7%	
4100	Shift work requirements, add, minimum		5%	
4200	Maximum		30%	

INTERIORS C

Important: See the Reference Section for critical supporting data - Reference Numbers and City Cost Indexes

C3010 Wall Finishes

C3010 210	Selective Price Sheet	COST PER S.F.		
		MAT.	INST.	TOTAL
0100	Painting, on plaster or drywall, brushwork, primer and 1 ct.	.10	.56	.66
0200	Primer and 2 ct.	.15	.89	1.04
0300	Rollerwork, primer and 1 ct.	.10	.47	.57
0400	Primer and 2 ct.	.15	.66	.81
0500	Woodwork incl. puttying, brushwork, primer and 1 ct.	.10	.84	.94
0600	Primer and 2 ct.	.15	1.12	1.27
0700	Wood trim to 6″ wide, enamel, primer and 1 ct.	.10	.47	.57
0800	Primer and 2 ct.	.15	.60	.75
0900	Cabinets and casework, enamel, primer and 1 ct.	.10	.95	1.05
1000	Primer and 2 ct.	.15	1.17	1.32
1100	On masonry or concrete, latex, brushwork, primer and 1 ct.	.19	.79	.98
1200	Primer and 2 ct.	.25	1.13	1.38
1300	For block filler, add	.11	.97	1.08
1400				
1500	Varnish, wood trim, sealer 1 ct., sanding, puttying, quality work	.13	1.71	1.84
1600	Meduim work	.10	1.33	1.43
1700	Without sanding	.13	.20	.33
1800				
1900	Wall coverings, wall paper, at $9.70 per double roll, average workmanship	.32	.59	.91
2000	At $20.00 per double roll, average workmanship	.70	.70	1.40
2100	At $44.00 per double roll, quality workmanship	1.67	.86	2.53
2200				
2300	Grass cloths with lining paper, minimum	.69	.94	1.63
2400	Maximum	2.23	1.07	3.30
2500	Vinyl, fabric backed, light weight	.64	.59	1.23
2600	Medium weight	.79	.78	1.57
2700	Heavy weight	1.28	.86	2.14
2800				
2900	Cork tiles, 12″ x 12″, 3/16″ thick	2.86	1.57	4.43
3000	5/16″ thick	3.25	1.60	4.85
3100	Granular surface, 12″ x 36″, 1/2″ thick	1.08	.98	2.06
3200	1″ thick	1.40	1.02	2.42
3300	Aluminum foil	.91	1.37	2.28
3400				
3500	Tile, ceramic, adhesive set, 4-1/4″ x 4-1/4″	2.30	3.70	6
3600	6″ x 6″	2.89	3.51	6.40
3700	Decorated, 4-1/4″ x 4-1/4″, minimum	3.85	2.60	6.45
3800	Maximum	42.50	3.90	46.40
3900	For epoxy grout, add	.36	.88	1.24
4000	Pregrouted sheets	4.57	2.93	7.50
4100	Glass mosaics, 3/4″ tile on 12″ sheets, minimum	16.80	9.60	26.40
4200	Color group 8	55	10.95	65.95
4300	Metal, tile pattern, 4′ x 4′ sheet, 24 ga., nailed			
4400	Stainless steel	24.50	1.71	26.21
4500	Aluminized steel	13.10	1.71	14.81
4600				
4700	Brick, interior veneer, 4″ face brick, running bond, minimum	3	9.30	12.30
4800	Maximum	3	9.30	12.30
4900	Simulated, urethane pieces, set in mastic	5.25	2.92	8.17
5000	Fiberglass panels	3.03	2.19	5.22
5100	Wall coating, on drywall, thin coat, plain	.07	.54	.61
5200	Stipple	.07	.54	.61
5300	Textured spray	.04	.48	.52
5400				
5500	Paneling not incl. furring or trim, hardboard, tempered, 1/8″ thick	.33	1.75	2.08
5600	1/4″ thick	.42	1.75	2.17
5700	Plastic faced, 1/8″ thick	.59	1.75	2.34
5800	1/4″ thick	.79	1.75	2.54

C3010 Wall Finishes

C3010 210	Selective Price Sheet	COST PER S.F.		
		MAT.	INST.	TOTAL
5900	Woodgrained, 1/4" thick, minimum	.53	1.75	2.28
6000	Maximum	1.11	2.06	3.17
6100	Plywood, 4' x 8' shts. 1/4" thick, prefin., birch faced, min.	.87	1.75	2.62
6200	Maximum	1.94	2.50	4.44
6300	Walnut, minimum	2.72	1.75	4.47
6400	Maximum	5.15	2.19	7.34
6500	Mahogany, african	2.48	2.19	4.67
6600	Philippine	1.07	1.75	2.82
6700	Chestnut	4.71	2.33	7.04
6800	Pecan	2.04	2.19	4.23
6900	Rosewood	4.52	2.74	7.26
7000	Teak	3.19	2.19	5.38
7100	Aromatic cedar, plywood	2.06	2.19	4.25
7200	Particle board	1	2.19	3.19
7300	Wood board, 3/4" thick, knotty pine	1.39	2.92	4.31
7400	Rough sawn cedar	1.77	2.92	4.69
7500	Redwood, clear	4.14	2.92	7.06
7600	Aromatic cedar	3.19	3.18	6.37

C3020 Floor Finishes

C3020 430	Selective Price Sheet	COST PER S.F.		
		MAT.	INST.	TOTAL
0100	Flooring, carpet, acrylic, 26 oz. light traffic	1.83	.58	2.41
0200	35 oz. heavy traffic	3.83	.58	4.41
0300	Nylon anti-static, 15 oz. light traffic	1.21	.77	1.98
0400	22 oz. medium traffic	2.39	.58	2.97
0500	26 oz. heavy traffic	3.33	.63	3.96
0600	28 oz. heavy traffic	3.72	.63	4.35
0700	Tile, foamed back, needle punch	2.84	.69	3.53
0800	Tufted loop	1.20	.69	1.89
0900	Wool, 36 oz. medium traffic	8.40	.63	9.03
1000	42 oz. heavy traffic	8.50	.63	9.13
1100	Composition, epoxy, with colored chips, minimum	2.41	3.26	5.67
1200	Maximum	2.93	4.50	7.43
1300	Trowelled, minimum	3.11	3.93	7.04
1400	Maximum	4.53	4.59	9.12
1500	Terrazzo, 1/4" thick, chemical resistant, minimum	5.70	4.57	10.27
1600	Maximum	8.75	6.10	14.85
1700	Resilient, asphalt tile, 1/8" thick	1.08	.99	2.07
1800	Conductive flooring, rubber, 1/8" thick	3.01	1.25	4.26
1900	Cork tile 1/8" thick, standard finish	3.93	1.25	5.18
2000	Urethane finish	4.95	1.25	6.20
2100	PVC sheet goods for gyms, 1/4" thick	4.24	4.94	9.18
2200	3/8" thick	4.77	6.60	11.37
2300	Vinyl composition 12" x 12" tile, plain, 1/16" thick	.83	.79	1.62
2400	1/8" thick	2.10	.79	2.89
2500	Vinyl tile, 12" x 12" x 1/8" thick, minimum	2.33	.79	3.12
2600	Maximum	8.80	.79	9.59
2700	Vinyl sheet goods, backed, .093" thick	2.33	1.72	4.05
2900	Slate, random rectangular, 1/4" thick	13.25	5.20	18.45
3000	1/2" thick	12.30	7.40	19.70
3100	Natural cleft, irregular, 3/4" thick	5.70	8.45	14.15
3200	For sand rubbed finish, add	6.15		6.15
3300	Terrazzo, cast in place, bonded 1-3/4" thick, gray cement	2.68	7.05	9.73
3400	White cement	3.06	7.05	10.11
3500	Not bonded 3" thick, gray cement	3.36	7.95	11.31
3600	White cement	3.66	7.95	11.61
3700	Precast, 12" x 12" x 1" thick	16.95	27	43.95
3800	1-1/4" thick	18.80	27	45.80
3900	16" x 16" x 1-1/4" thick	20.50	34	54.50
4000	1-1/2" thick	18.65	37	55.65
4100	Marble travertine, standard, 12" x 12" x 3/4" thick	10.60	11.70	22.30
4200				
4300	Tile, ceramic, natural clay, thin set	3.98	3.84	7.82
4400	Porcelain, thin set	4.41	3.70	8.11
4500	Specialty, decorator finish	9.85	3.84	13.69
4600				
4700	Quarry, red, mud set, 4" x 4" x 1/2" thick	4.19	5.85	10.04
4800	6" x 6" x 1/2" thick	3.26	5	8.26
4900	Brown, imported, 6" x 6" x 3/4" thick	4.98	5.85	10.83
5000	8" x 8" x 1" thick	5.65	6.40	12.05
5100	Slate, vermont, thin set, 6" x 6" x 1/4" thick	4.55	3.90	8.45
5200				
5300	Wood, maple strip, 25/32" x 2-1/4", finished, select	5.25	2.77	8.02
5400	2nd and better	3.22	2.77	5.99
5500	Oak, 25/32" x 2-1/4" finished, clear	3.28	2.77	6.05
5600	No. 1 common	4.22	2.77	6.99
5700	Parquet, standard, 5/16", finished, minimum	3.27	2.94	6.21
5800	Maximum	5.90	4.58	10.48
5900	Custom, finished, minimum	15.60	4.38	19.98

C3020 Floor Finishes

C3020 430	Selective Price Sheet	COST PER S.F.		
		MAT.	INST.	TOTAL
6000	Maximum	20.50	8.75	29.25
6100	Prefinished, oak, 2-1/4" wide	6.80	2.57	9.37
6200	Ranch plank	8.45	3.02	11.47
6300	Sleepers on concrete, treated, 24" O.C., 1" x 2"	.08	.29	.37
6400	1" x 3"	.16	.37	.53
6500	2" x 4"	.41	.51	.92
6600	2" x 6"	.74	.59	1.33
6700	Refinish old floors, minimum	.76	.86	1.62
6800	Maximum	1.14	2.65	3.79
6900	Subflooring, plywood, CDX, 1/2" thick	.43	.58	1.01
7000	5/8" thick	.50	.65	1.15
7100	3/4" thick	.63	.70	1.33
7200				
7300	1" x 10" boards, S4S, laid regular	1.16	.80	1.96
7400	Laid diagonal	1.16	.97	2.13
7500	1" x 8" boards, S4S, laid regular	.88	.88	1.76
7600	Laid diagonal	.88	1.03	1.91
7700	Underlayment, plywood, underlayment grade, 3/8" thick	.58	.58	1.16
7800	1/2" thick	.68	.60	1.28
7900	5/8" thick	.79	.63	1.42
8000	3/4" thick	.91	.67	1.58
8100	Particle board, 3/8" thick	.41	.58	.99
8200	1/2" thick	.43	.60	1.03
8300	5/8" thick	.54	.63	1.17
8400	3/4" thick	.59	.67	1.26

C3030 Ceiling Finishes

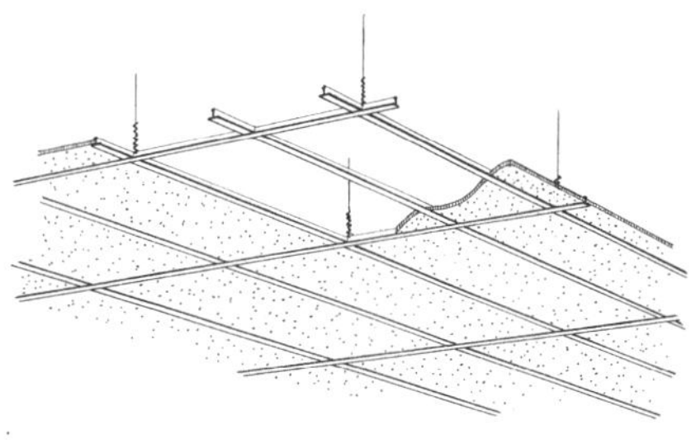

This page illustrates suspended acoustical board systems including acoustic ceiling board, hangers, and T bar suspension. Lines within System Components give the unit price and total price per square foot for this system. Prices for alternate suspended acoustical board systems are on Line Items C3030 210 1100 thru 2400. Both material quantities and labor costs have been adjusted for the system listed.

Factors: To adjust for job conditions other than normal working situations use Lines C3030 210 2900 thru 4000.

Example: You are to install the system and protect existing construction. Go to Line C3030 210 3800 and apply these percentages to the appropriate MAT. and INST. costs.

System Components	QUANTITY	UNIT	COST PER S.F. MAT.	COST PER S.F. INST.	COST PER S.F. TOTAL
Suspended acoustical ceiling board installed on exposed grid system.					
Fiberglass boards, film faced, 2′ x 4′, 5/8″ thick	1.000	S.F.	.56	.70	1.26
Hangers, #12 wire	1.000	S.F.	.05	.07	.12
T bar suspension system, 2′ x 4′ grid	1.000	S.F.	.45	.55	1
TOTAL			1.06	1.32	2.38

C3030 210	Ceiling, Suspended Acoustical	COST PER S.F. MAT.	COST PER S.F. INST.	COST PER S.F. TOTAL
1000	For alternate suspended ceiling systems:			
1100	2′ x 4′ grid, mineral fiber board, aluminum faced, 5/8″ thick	1.59	1.29	2.88
1200	Standard faced	1.18	1.27	2.45
1300	Plastic faced	1.56	1.71	3.27
1400	Fiberglass, film faced, 3″ thick, R11	1.89	1.59	3.48
1500	Grass cloth faced, 3/4″ thick	2.28	1.50	3.78
1600	1″ thick	2.47	1.52	3.99
1700	1-1/2″ thick, nubby face	2.95	1.54	4.49
2200				
2300				
2400	Add for 2′ x 2′ grid system	.11	.13	.24
2500				
2600				
2700				
2900	Cut & patch to match existing construction, add, minimum	2%	3%	
3000	Maximum	5%	9%	
3100	Dust protection, add, minimum	1%	2%	
3200	Maximum	4%	11%	
3300	Equipment usage curtailment, add, minimum	1%	1%	
3400	Maximum	3%	10%	
3500	Material handling & storage limitation, add, minimum	1%	1%	
3600	Maximum	6%	7%	
3700	Protection of existing work, add, minimum	2%	2%	
3800	Maximum	5%	7%	
3900	Shift work requirements, add, minimum		5%	
4000	Maximum		30%	

C3030 Ceiling Finishes

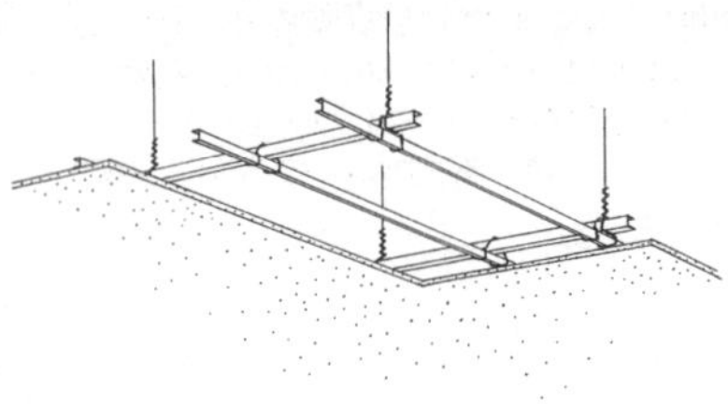

This page illustrates and describes suspended gypsum board systems including gypsum board, metal furring, taping, finished and painted. Lines within System Components give the unit price and total price per square foot for this system. Prices for alternate suspended gypsum board systems are on Line Items C3030 220 1500 thru 1700. Both material quantities and labor costs have been adjusted for the system listed.

Factors: To adjust for job conditions other than normal working situations use Lines C3030 220 2900 thru 4000.

Example: You are to install the system and control dust in the working area. Go to Line C3030 220 3200 and apply these percentages to the appropriate MAT. and INST. costs.

System Components	QUANTITY	UNIT	COST PER S.F.		
			MAT.	INST.	TOTAL
Suspended ceiling gypsum board,4′ x 8′ x 5/8″ thick,					
On metal furring, taped, finished and painted.					
Gypsum drywall, 4′ x 8′, 5/8″ thick, screwed	1.000	S.F.	.24	.49	.73
Main runners, 1-1/2″ C.R.C., 4′ O.C.	.500	S.F.	.08	.50	.58
25 ga., channels, 2′ O.C.	1.000	S.F.	.11	.92	1.03
Taped and finished	1.000	S.F.	.04	.44	.48
Paint, 2 coats, roller work	1.000	S.F.	.15	.50	.65
TOTAL			.62	2.85	3.47

C3030 220	Ceilings, Suspended Gypsum Board	COST PER S.F.		
		MAT.	INST.	TOTAL
1400	For alternate drywall ceiling systems:			
1500	Thin coat plaster, 2 coats paint	.50	2.45	2.95
1600	Spray-on sand finish, no paint	.47	2.39	2.86
1700	12″ x 12″ x 3/4″ acoustical wood fiber tile	1.56	3	4.56
1800				
1900				
2000				
2100				
2200				
2300				
2400				
2500				
2600				
2700				
2900	Cut & patch to match existing construction, add, minimum	2%	3%	
3000	Maximum	5%	9%	
3100	Dust protection, add, minimum	1%	2%	
3200	Maximum	4%	11%	
3300	Equipment usage curtailment, add, minimum	1%	1%	
3400	Maximum	3%	10%	
3500	Material handling & storage limitation, add, minimum	1%	1%	
3600	Maximum	6%	7%	
3700	Protection of existing work, add, minimum	2%	2%	
3800	Maximum	5%	7%	
3900	Shift work requirements, add, minimum		5%	
4000	Maximum		30%	

C3030 Ceiling Finishes

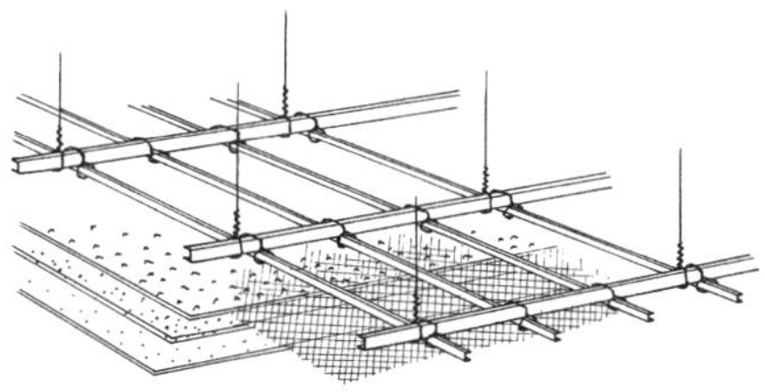

This page illustrates and describes suspended plaster and lath systems including gypsum plaster, lath, furring and runners with ceiling painted. Lines within System Components give the unit price and total price per square foot for this system. Prices for alternate suspended plaster and lath systems are on Line Items C3030 230 1300 thru 2300. Both material quantities and labor costs have been adjusted for the system listed.

Factors: To adjust for job conditions other than normal working situations use Lines C3030 230 2900 thru 4000.

Example: You are to install the system to match existing construction. Go to Line C3030 230 2900 and apply these percentages to the appropriate MAT. and INST. costs.

System Components	QUANTITY	UNIT	COST PER S.F.		
			MAT.	INST.	TOTAL
Gypsum plaster, 3 coats, on 3.4# rib lath, on 3/4″ C.R.C. furring on 1-1/2″ main runners, ceiling painted.					
Gypsum plaster, 3 coats	.110	S.Y.	.54	2.73	3.27
3.4# rib lath	.110	S.Y.	.39	.61	1
Main runners, 1-1/2″ C.R.C. 24″ O.C.	.333	S.F.	.16	.99	1.15
Furring 3/4″ C.R.C. 16″ O.C.	1.000	S.F.	.16	1.33	1.49
Painting, 2 coats, roller work	1.000	S.F.	.15	.50	.65
TOTAL			1.40	6.16	7.56

C3030 230 Ceiling, Suspended Plaster

C3030 230	Ceiling, Suspended Plaster	COST PER S.F.		
		MAT.	INST.	TOTAL
1200	For alternate plaster ceiling systems:			
1300	Gypsum plaster, 3 coats, on 2.5# diamond lath	1.19	6.15	7.34
1400	On 3/8″ gypsum lath	1.42	6.15	7.57
1500	2 coats, on 3.4# rib lath	1.43	6.15	7.58
1600	On 2.5# diamond lath	1.03	5.70	6.73
1700	On 3/8″ gypsum lath	1.26	5.70	6.96
1800	Perlite plaster, 3 coats, on 3.4# rib lath	1.51	6.75	8.26
1900	On 2.5# diamond lath	1.30	6.75	8.05
2000	On 3/8″ gypsum lath	1.53	6.75	8.28
2100	2 coats, on 3.4# rib lath	1.25	6.10	7.35
2200	On 2.5# diamond lath	1.04	6.05	7.09
2300	On 3/8″ gypsum lath	1.27	6.05	7.32
2400				
2500				
2600				
2700				
2900	Cut & patch to match existing construction, add, minimum	2%	3%	
3000	Maximum	5%	9%	
3100	Dust protection, add, minimum	1%	2%	
3200	Maximum	4%	11%	
3300	Equipment usage curtailment, add, minimum	1%	1%	
3400	Maximum	3%	10%	
3500	Material handling & storage limitation, add, minimum	1%	1%	
3600	Maximum	6%	7%	
3700	Protection of existing work, add, minimum	2%	2%	
3800	Maximum	5%	7%	
3900	Shift work requirements, add, minimum		5%	
4000	Maximum		30%	

C3030 Ceiling Finishes

C3030 250	Selective Price Sheet	COST PER S.F.		
		MAT.	INST.	TOTAL
0100	Ceiling, plaster, gypsum, 2 coats	.38	2.33	2.71
0200	3 coats	.54	2.75	3.29
0300	Perlite or vermiculite, 2 coats	.40	2.69	3.09
0400	3 coats	.66	3.39	4.05
0500	Gypsum lath, plain, 3/8" thick	.42	.50	.92
0600	1/2" thick	.43	.54	.97
0700	Firestop, 3/8" thick	.43	.61	1.04
0800	1/2" thick	.48	.66	1.14
0900	Metal lath, rib, 2.75 lb.	.28	.57	.85
1000	3.40 lb.	.40	.61	1.01
1100	Diamond, 2.50 lb.	.18	.57	.75
1200	3.40 lb.	.30	.72	1.02
1300				
1400				
1500	Drywall, standard, 1/2" thick, no finish included	.21	.49	.70
1600	Taped and finished	.25	1.14	1.39
1700	5/8" thick, no finish included	.24	.49	.73
1800	Taped and finished	.28	1.14	1.42
1900	Fire resistant, 1/2" thick, no finish included	.22	.49	.71
2000	Taped and finished	.26	1.14	1.40
2100	5/8" thick, no finish included	.25	.49	.74
2200	Taped and finished	.29	1.14	1.43
2300	Water resistant, 1/2" thick, no finish included	.26	.49	.75
2400	Taped and finished	.31	1.14	1.45
2500	5/8" thick, no finish included	.30	.49	.79
2600	Taped and finished	.34	1.14	1.48
2700	Finish, instead of taping			
2800	For thin coat plaster, add	.07	.54	.61
2900	Finish, textured spray, add	.04	.48	.52
3000				
3100				
3200				
3300	Tile, stapled or glued, mineral fiber plastic coated, 5/8" thick	.98	.44	1.42
3400	3/4" thick	1.43	.44	1.87
3500	Wood fiber, 1/2" thick	.84	1.09	1.93
3600	3/4" thick	1.13	1.09	2.22
3700	Suspended, fiberglass, film faced, 5/8" thick	.56	.70	1.26
3800	3" thick	1.39	.97	2.36
3900	Mineral fiber, 5/8" thick, standard	.68	.65	1.33
4000	Aluminum	1.62	.73	2.35
4100	Wood fiber, reveal edge, 1" thick			
4200	3" thick			
4300	Framing, metal furring, 3/4" channels, 12" O.C.	.18	1.84	2.02
4400	16" O.C.	.16	1.33	1.49
4500	24" O.C.	.11	.92	1.03
4600				
4700	1-1/2" channels, 12" O.C.	.26	2.03	2.29
4800	16" O.C.	.23	1.48	1.71
4900	24" O.C.	.16	.99	1.15
5000				
5100				
5200				
5300	Ceiling suspension systems, for tile,			
5400	Concealed "Z" bar, 12" module	.41	.84	1.25
5500	Class A, "T" bar 2'-0" x 4'-0" grid	.45	.55	1
5600	2'-0" x 2'-0" grid	.56	.67	1.23
5700	Carrier channels for lighting fixtures, add	.15	.95	1.10
5800				

C3030 Ceiling Finishes

C3030 250	Selective Price Sheet	COST PER S.F.		
		MAT.	INST.	TOTAL
5900				
6000				
6100	Wood, furring 1" x 3", on wood, 12" O.C.	.23	1.25	1.48
6200	16" O.C..	.17	.94	1.11
6300	24" O.C.	.12	.63	.75
6400				
6500	On concrete, 12" O.C.	.23	2.08	2.31
6600	16" O.C.	.17	1.56	1.73
6700	24" O.C.	.12	1.04	1.16
6800				
6900	Joists, 2" x 4", 12" O.C.	.44	.90	1.34
7000	16" O.C.	.37	.76	1.13
7100	24" O.C.	.30	.61	.91
7200	32" O.C.	.23	.47	.70
7300	2" x 6", 12" O.C.	.74	.91	1.65
7400	16" O.C.	.62	.76	1.38
7500	24" O.C.	.49	.60	1.09
7600	32" O.C.	.36	.45	.81

D SERVICES

D1010 Elevators and Lifts

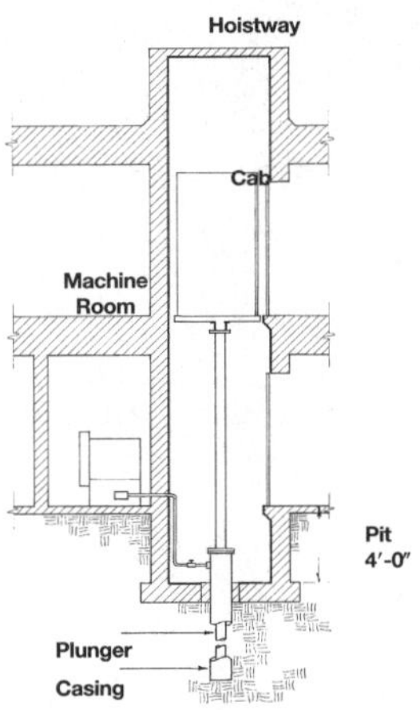

This page illustrates and describes oil hydraulic elevator systems. Prices for various oil hydraulic elevator systems are on Line Items D1010 130 0400 thru 2000. Both material quantities and labor costs have been adjusted for the system listed.

Factors: To adjust for job conditions other than normal working situations use Lines D1010 130 3100 thru 4000.

Example: You are to install the system with limited equipment usage. Go to Line D1010 130 3400 and apply this percentage to the appropriate TOTAL costs.

D1010 130	**Oil Hydraulic Elevators**	COST EACH		
		MAT.	INST.	TOTAL
0310	Oil-hydraulic elevator systems			
0320	Including piston and piston shaft			
0330				
0400	1500 Lb. passenger, 2 floors	31,500	10,300	41,800
0500	3 floors	36,700	15,900	52,600
0600	4 floors	51,000	22,400	73,400
0700	5 floors	55,500	27,800	83,300
0800	6 floors	60,000	33,200	93,200
0900				
1000	2500 Lb. passenger, 2 floors	33,600	10,300	43,900
1100	3 floors	38,700	15,900	54,600
1200	4 floors	49,700	22,400	72,100
1300	5 floors	54,500	27,800	82,300
1400	6 floors	62,000	33,200	95,200
1500				
1600	4000 Lb. passenger, 2 floors	38,200	10,300	48,500
1700	3 floors	44,600	15,900	60,500
1800	4 floors	57,500	22,400	79,900
1900	5 floors	62,500	27,800	90,300
2000	6 floors	67,000	33,200	100,200
2100				
2200				
2300				
2400				
2500				
2600				
2700				
2800				
2900				
3000				
3100	Dust protection, add, minimum	1%	2%	
3200	Maximum	4%	11%	
3300	Equipment usage curtailment, add, minimum	1%	1%	
3400	Maximum	3%	10%	
3500	Material handling & storage limitation, add, minimum	1%	1%	
3600	Maximum	6%	7%	
3700	Protection of existing work, add, minimum	2%	2%	
3800	Maximum	5%	7%	
3900	Shift work requirements, add, minimum		5%	
4000	Maximum		30%	

Important: See the Reference Section for critical supporting data - Reference Numbers and City Cost Indexes

D2010 Plumbing Fixtures

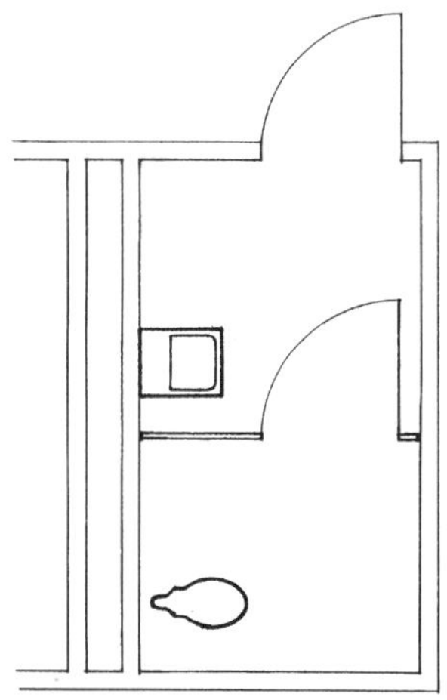

This page illustrates and describes a women's public restroom system including a water closet, lavatory, accessories, and service piping. Lines within System Components give the unit price and total price on a cost each basis for this system. Prices for alternate women's public restroom systems are on Line Items D2010 956 1400 thru 1700. Both material quantities and labor costs have been adjusted for the system listed.

Factors: To adjust for job conditions other than normal working situations use Lines D2010 956 2900 thru 4000.

Example: You are to install the system and protect surrounding area from dust. Go to Line D2010 956 3100 and apply these percentages to the MAT. and INST. costs.

System Components	QUANTITY	UNIT	COST EACH		
			MAT.	INST.	TOTAL
Public women's restroom incl. water closet, lavatory, accessories and Necessary service piping to install this system in one wall.					
Water closet, wall mounted, one piece	1.000	Ea.	375	153	528
Rough-in waste and vent for water closet	1.000	Set	300	345	645
Lavatory, 20" x 18" P.E. cast iron with accessories	1.000	Ea.	261	111	372
Rough-in waste and vent for lavatory	1.000	Set	242	535	777
Toilet partition, painted metal between walls, floor mounted	1.000	Ea.	430	146	576
For handicap unit, add	1.000	Ea.	300		300
Grab bar, 36" long	1.000	Ea.	58	44	102
Mirror, 18" x 24", with stainless steel shelf	1.000	Ea.	99.50	22	121.50
Napkin/tampon dispenser, recessed	1.000	Ea.	345	29	374
Soap Dispenser, chrome, surface mounted, liquid	1.000	Ea.	45.50	22	67.50
Toilet tissue dispenser, surface mounted, stainless steel	1.000	Ea.	12.30	14.60	26.90
Towel dispenser, surface mounted, stainless steel	1.000	Ea.	40	27.50	67.50
TOTAL			2,508.30	1,449.10	3,957.40

D2010 956	Plumbing - Public Restroom	COST EACH		
		MAT.	INST.	TOTAL
1200				
1300	For alternate size restrooms:			
1400	Two water closets, two lavatories	4,275	2,750	7,025
1500				
1600	For each additional water closet over 2, add	1,125	640	1,765
1700	For each additional lavatory over 2, add	650	690	1,340
1800				
1900				
2400	NOTE: PLUMBING APPROXIMATIONS			
2500	WATER CONTROL: water meter, backflow preventer,			
2600	Shock absorbers, vacuum breakers, mixer....10 to 15% of fixtures			
2700	PIPE AND FITTINGS: .... 30 to 60% of fixtures			
2800				
2900	Cut & patch to match existing construction, add, minimum	2%	3%	
3000	Maximum	5%	9%	
3100	Dust protection, add, minimum	1%	2%	
3200	Maximum	4%	11%	
3300	Equipment usage curtailment, add, minimum	1%	1%	
3400	Maximum	3%	10%	
3500	Material handling & storage limitation, add, minimum	1%	1%	

D2010 Plumbing Fixtures

D2010 956	Plumbing - Public Restroom	COST EACH		
		MAT.	INST.	TOTAL
3600	Maximum	6%	7%	
3700	Protection of existing work, add, minimum	2%	2%	
3800	Maximum	5%	7%	
3900	Shift work requirements, add, minimum		5%	
4000	Maximum		30%	

D2010 Plumbing Fixtures

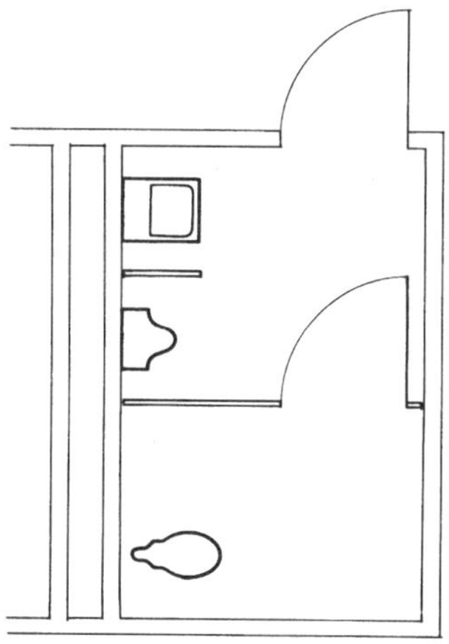

This page illustrates and describes a men's public restroom system including a water closet, urinal, lavatory, accessories and service piping. Lines within System Components give the unit price and total price on a cost each basis for this system. Prices for alternate men's public restroom systems are on Line Items D2010 957 1800 thru 2200. Both material quantities and labor costs have been adjusted for the system listed.

Factors: To adjust for job conditions other than normal working situations use Lines D2010 957 2900 thru 4000.

Example: You are to install the system and match existing construction. Go to Line D2010 957 3000 and apply these percentages to the appropriate MAT. and INST. costs.

System Components	QUANTITY	UNIT	COST EACH		
			MAT.	INST.	TOTAL
Public men's restroom incl. water closet, urinal, lavatory, accessories, And necessary service piping to install this system in one wall.					
Water closet, wall mounted, one piece	1.000	Ea.	375	153	528
Rough-in waste & vent for water closet	1.000	Set	300	345	645
Urinal, wall hung	1.000	Ea.	320	295	615
Rough-in waste & vent for urinal	1.000	Set	101	315	416
Lavatory, 20″ x 18″, P.E. cast iron with accessories	1.000	Ea.	261	111	372
Rough-in waste & vent for lavatory	1.000	Set	242	535	777
Partition, painted mtl., between walls, floor mntd.	1.000	Ea.	430	146	576
For handicap unit, add	1.000	Ea.	300		300
Grab bars	1.000	Ea.	58	44	102
Urinal screen, painted metal, wall mounted	1.000	Ea.	272	87.50	359.50
Mirror, 18″ x 24″ with stainless steel shelf	1.000	Ea.	99.50	22	121.50
Soap dispenser, surface mounted, liquid	1.000	Ea.	45.50	22	67.50
Toilet tissue dispenser, surface mounted, stainless steel	1.000	Ea.	12.30	14.60	26.90
Towel dispenser, surface mounted, stainless steel	1.000	Ea.	40	27.50	67.50
TOTAL			2,856.30	2,117.60	4,973.90

D2010 957	Plumbing - Public Restroom	COST EACH		
		MAT.	INST.	TOTAL
1600				
1700	For alternate size restrooms:			
1800	Two water closets, two urinals, two lavatories	5,325	4,125	9,450
1900				
2000	For each additional water closet over 2, add	1,125	640	1,765
2100	For each additional urinal over 2, add	695	700	1,395
2200	For each additional lavatory over 2, add	650	690	1,340
2300				
2400	NOTE: PLUMBING APPROXIMATIONS			
2500	WATER CONTROL: water meter, backflow preventer,			
2600	Shock absorbers, vacuum breakers, mixer....10 to 15% of fixtures			
2700	PIPE AND FITTINGS: .... 30 to 60% of fixtures			
2800				
2900	Cut & patch to match existing construction, add, minimum	2%	3%	
3000	Maximum	5%	9%	
3100	Dust protection, add, minimum	1%	2%	
3200	Maximum	4%	11%	
3300	Equipment usage curtailment, add, minimum	1%	1%	

D2010 Plumbing Fixtures

D2010 957	Plumbing - Public Restroom	COST EACH		
		MAT.	INST.	TOTAL
3400	Maximum	3%	10%	
3500	Material handling, & storage limitation, add, minimum	1%	1%	
3600	Maximum	6%	7%	
3700	Protection of existing work, add, minimum	2%	2%	
3800	Maximum	5%	7%	
3900	Shift work requirements, add, minimum		5%	
4000	Maximum		30%	

Important: See the Reference Section for critical supporting data - Reference Numbers and City Cost Indexes

D2010 Plumbing Fixtures

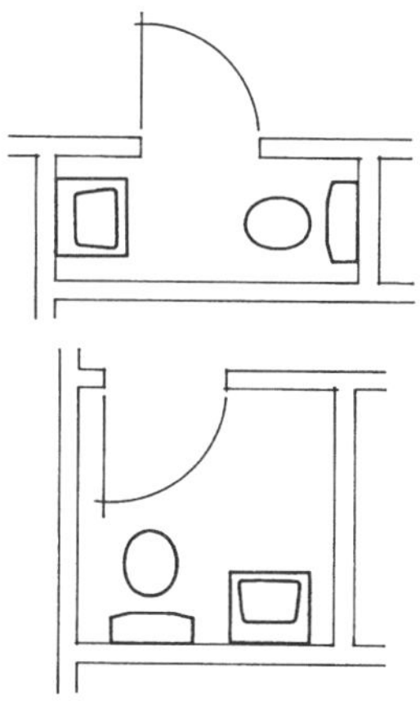

This page illustrates and describes a two fixture lavatory system including a water closet, lavatory, accessories and all service piping. Lines within System Components give the unit price and total price on a cost each basis for this system. Prices for an alternate two fixture lavatory system are on Line Item D2010 958 1900. Both material quantities and labor costs have been adjusted for the system listed.

Factors: To adjust for job conditions other than normal working situations use Lines D2010 958 2900 thru 4000.

Example: You are to install the system while controlling dust in the work area. Go to Line D2010 958 3200 and apply these percentages to the appropriate MAT. and INST. costs.

System Components	QUANTITY	UNIT	COST EACH MAT.	COST EACH INST.	COST EACH TOTAL
Two fixture bathroom incl. water closet, lavatory, accessories and Necessary service piping to install this system in 2 walls.					
Water closet, floor mounted, 2 piece, close coupled	1.000	Ea.	176	167	343
Rough in waste & vent for water closet	1.000	Set	143	290	433
Lavatory, 20″ x 18″, P.E. cast iron with accessories	1.000	Ea.	261	111	372
Rough in waste & vent for lavatory	1.000	Set	242	535	777
Additional service piping					
1/2″ copper pipe with sweat solder joints	10.000	L.F.	11.10	60.50	71.60
2″ black schedule 40 steel pipe with threaded couplings	12.000	L.F.	48.12	165.60	213.72
4″ cast iron soil pipe with lead and oakum joints	7.000	L.F.	63.70	112.70	176.40
Accessories					
Toilet tissue dispenser, chrome, single roll	1.000	Ea.	12.30	14.60	26.90
18″ long stainless steel towel bar	1.000	Ea.	33	19.05	52.05
Medicine cabinet with mirror, 20″ x 16″, unlighted	1.000	Ea.	76.50	31.50	108
TOTAL			1,066.72	1,506.95	2,573.67

D2010 958	Plumbing - Two Fixture Bathroom	COST EACH MAT.	COST EACH INST.	COST EACH TOTAL
1800				
1900	Above system installed in 1 wall with all necessary service piping	1,025	1,300	2,325
2000				
2400	NOTE: PLUMBING APPROXIMATIONS			
2500	WATER CONTROL: water meter, backflow preventer,			
2600	Shock absorbers, vacuum breakers, mixer....10 to 15% of fixtures			
2700	PIPE AND FITTINGS: .... 30 to 60% of fixtures			
2800				
2900	Cut & patch to match existing construction, add, minimum	2%	3%	
3000	Maximum	5%	9%	
3100	Dust protection, add, minimum	1%	2%	
3200	Maximum	4%	11%	
3300	Equipment usage curtailment, add, minimum	1%	1%	
3400	Maximum	3%	10%	
3500	Material handling & storage limitation, add, minimum	1%	1%	
3600	Maximum	6%	7%	
3700	Protection of existing work, add, minimum	2%	2%	
3800	Maximum	5%	7%	
3900	Shift work requirements, add, minimum		5%	
4000	Maximum		30%	

D2010 Plumbing Fixtures

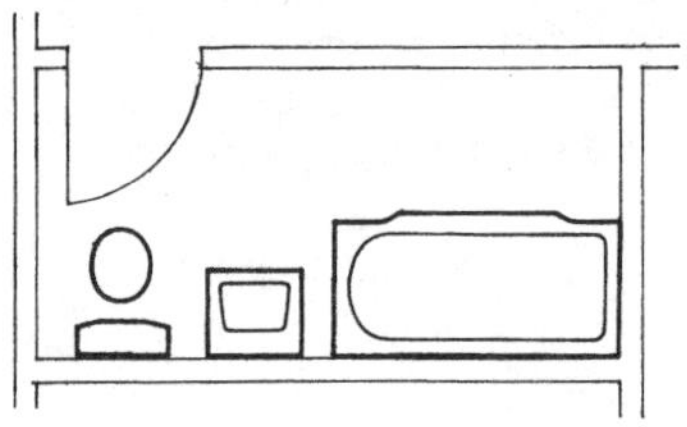

This page illustrates and describes a three fixture bathroom system including a water closet, tub, lavatory, accessories and service piping. Lines within System Components give the unit price and total price on a cost each basis for this system. Prices for an alternate three fixture bathroom system are on Line Item D2010 959 1700. Both material quantities and labor costs have been adjusted for the system listed.

Factors: To adjust for job conditions other than normal working situations use Lines D2010 959 2900 thru 4000.

Example: You are to install the system and protect all existing work. Go to Line D2010 959 3800 and apply these percentages to the appropriate MAT. and INST. costs.

System Components	QUANTITY	UNIT	COST EACH		
			MAT.	INST.	TOTAL
Three fixture bathroom incl. water closet, bathtub, lavatory, accessories,					
And necessary service piping to install this system in 1 wall.					
Water closet, floor mounted, 2 piece, close coupled	1.000	Ea.	176	167	343
Rough-in waste & vent for water closet	1.000	Set	132.99	269.70	402.69
Bathtub, P.E. cast iron 5′ long with accessories	1.000	Ea.	410	201	611
Rough-in waste & vent for bathtub	1.000	Set	120.65	403.75	524.40
Lavatory, 20″ x 18″ P.E. cast iron with accessories	1.000	Ea.	261	111	372
Rough-in waste & vent for lavatory	1.000	Set	242	535	777
Accessories					
Toilet tissue dispenser, chrome, single roll	1.000	Ea.	12.30	14.60	26.90
18″ long stainless steel towel bar	2.000	Ea.	66	38.10	104.10
Medicine cabinet with mirror, 20″ x 16″, unlighted	1.000	Ea.	76.50	31.50	108
TOTAL			1,497.44	1,771.65	3,269.09

D2010 959	Plumbing - Three Fixture Bathroom	COST EACH		
		MAT.	INST.	TOTAL
1600				
1700	Above system installed in one wall with all necessary service piping	1,475	1,725	3,200
2400	NOTE: PLUMBING APPROXIMATIONS			
2500	WATER CONTROL: water meter, backflow preventer,			
2600	Shock absorbers, vacuum breakers, mixer....10 to 15% of fixtures			
2700	PIPE AND FITTINGS: .... 30 to 60% of fixtures			
2800				
2900	Cut & patch to match existing construction, add, minimum	2%	3%	
3000	Maximum	5%	9%	
3100	Dust protection, add, minimum	1%	2%	
3200	Maximum	4%	11%	
3300	Equipment usage curtailment, add, minimum	1%	1%	
3400	Maximum	3%	10%	
3500	Material handling & storage limitation, add, minimum	1%	1%	
3600	Maximum	6%	7%	
3700	Protection of existing work, add, minimum	2%	2%	
3800	Maximum	5%	7%	
3900	Shift work requirements, add, minimum		5%	
4000	Maximum		30%	

Important: See the Reference Section for critical supporting data - Reference Numbers and City Cost Indexes

D2010 Plumbing Fixtures

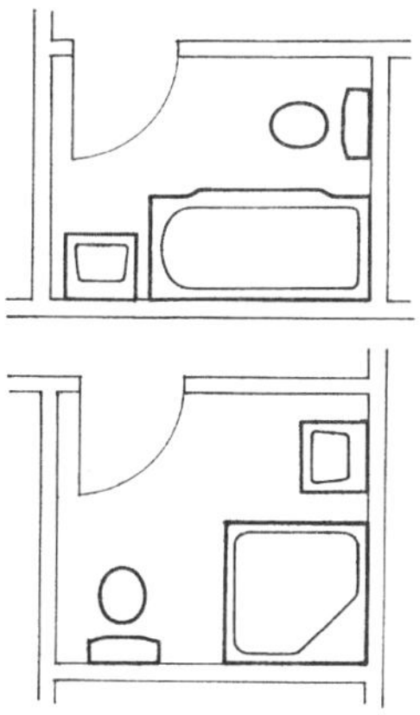

This page illustrates and describes a three fixture bathroom system including a water closet, tub, lavatory, accessories, and service piping. Lines within System Components give the unit price and total price on a cost each basis for this system. Prices for an alternate three fixture bathroom system are on Line Item D2010 960 2000. Both material quantities and labor costs have been adjusted for the system listed.

Factors: To adjust for job conditions other than normal working situations use Lines D2010 960 2900 thru 4000.

Example: You are to install the system and protect the surrounding area from dust. Go to Line D2010 960 3100 and apply these percentages to the appropriate MAT. and INST. costs.

System Components	QUANTITY	UNIT	COST EACH MAT.	COST EACH INST.	COST EACH TOTAL
Three fixture bathroom incl. water closet, bathtub, lavatory, accessories,					
And necessary service piping to install this system in 2 walls.					
Water closet, floor mounted, 2 piece, close coupled	1.000	Ea.	176	167	343
Rough-in waste & vent for water closet	1.000	Set	143	290	433
Bathtub, P.E. cast iron, 5′ long with accessories	1.000	Ea.	410	201	611
Rough-in waste & vent for bathtub	1.000	Set	127	425	552
Lavatory, 20″ x 18″ P.E. cast iron with accessories	1.000	Ea.	261	111	372
Rough-in waste & vent for lavatory	1.000	Set	242	535	777
Additional service piping					
1-1/4″ copper DWV type tubing, sweat solder joints	6.000	L.F.	13.74	49.20	62.94
2″ black schedule 40 steel pipe with threaded couplings	12.000	L.F.	48.12	165.60	213.72
Accessories					
Toilet tissue dispenser, chrome, single roll	1.000	Ea.	12.30	14.60	26.90
18″ long stainless steel towel bar	2.000	Ea.	66	38.10	104.10
Medicine cabinet with mirror, 20″ x 16″, unlighted	1.000	Ea.	76.50	31.50	108
TOTAL			1,575.66	2,028	3,603.66

D2010 960	Plumbing - Three Fixture Bathroom	COST EACH MAT.	COST EACH INST.	COST EACH TOTAL
2000	Above system with corner contour tub, P.E. cast iron	2,700	2,025	4,725
2300				
2400	NOTE: PLUMBING APPROXIMATIONS			
2500	WATER CONTROL: water meter, backflow preventer,			
2600	Shock absorbers, vacuum breakers, mixer....10 to 15% of fixtures			
2700	PIPE AND FITTINGS: .... 30 to 60% of fixtures			
2800				
2900	Cut & patch to match existing construction, add, minimum	2%	3%	
3000	Maximum	5%	9%	
3100	Dust protection, add, minimum	1%	2%	
3200	Maximum	4%	11%	
3300	Equipment usage curtailment, add, minimum	1%	1%	
3400	Maximum	3%	10%	
3500	Material handling & storage limitation, add, minimum	1%	1%	
3600	Maximum	6%	7%	
3700	Protection of existing work, add, minimum	2%	2%	
3800	Maximum	5%	7%	
3900	Shift work requirements, add, minimum		5%	
4000	Maximum		30%	

D2010 Plumbing Fixtures

This page illustrates and describes a three fixture bathroom system including a water closet, shower, lavatory, accessories, and service piping, Lines within System Components give the unit price and total price on a cost each basis for this system. Prices for an alternate three fixture bathroom system are on Line Item D2010 961 2200. Both material quantities and labor costs have been adjusted for the system listed.

Factors: To adjust for job conditions other than normal working situations use Lines D2010 961 2900 thru 4000.

Example: You are to install the system and protect existing construction. Go to Line D2010 961 3700 and apply these percentages to the appropriate MAT. and INST. costs.

System Components	QUANTITY	UNIT	COST EACH MAT.	COST EACH INST.	COST EACH TOTAL
Three fixture bathroom incl. water closet, shower, lavatory, accessories,					
And necessary service piping to install this system in 2 walls.					
Water closet, floor mounted, 2 piece, close coupled	1.000	Ea.	176	167	343
Rough-in waste & vent for water closet	1.000	Set	143	290	433
32" shower, enameled stall, molded stone receptor	1.000	Ea.	360	440	800
Rough-in waste & vent for shower	1.000	Set	68.50	430	498.50
Lavatory, 20" x 18" P.E. cast iron with accessories	1.000	Ea.	261	111	372
Rough-in waste & vent for lavatory	1.000	Set	242	535	777
Additional service piping					
1-1/4" Copper DWV type tubing, sweat solder joints	4.000	L.F.	9.16	32.80	41.96
2" black schedule 40 steel pipe with threaded couplings	6.000	L.F.	24.06	82.80	106.86
4" cast iron soil with lead and oakum joints	7.000	L.F.	63.70	112.70	176.40
Accessories					
Toilet tissue dispenser, chrome, single roll	1.000	Ea.	12.30	14.60	26.90
18" long stainless steel towel bar	2.000	Ea.	66	38.10	104.10
Medicine cabinet with mirror, 20" x 16", unlighted	1.000	Ea.	76.50	31.50	108
TOTAL			1,502.22	2,285.50	3,787.72

D2010 961	Plumbing - Three Fixture Bathroom	COST EACH MAT.	COST EACH INST.	COST EACH TOTAL
2100	Above system installed with 36" corner angle shower, enameled steel,			
2200	Molded stone receptor	1,975	2,325	4,300
2300				
2400	NOTE: PLUMBING APPROXIMATIONS			
2500	WATER CONTROL: water meter, backflow preventer,			
2600	Shock absorbers, vacuum breakers, mixer....10 to 15% of fixtures			
2700	PIPE AND FITTINGS: .... 30 to 60% of fixtures			
2800				
2900	Cut & patch to match existing construction, add, minimum	2%	3%	
3000	Maximum	5%	9%	
3100	Dust protection, add, minimum	1%	2%	
3200	Maximum	4%	11%	
3300	Equipment usage curtailment, add, minimum	1%	1%	
3400	Maximum	3%	10%	
3500	Material handling & storage limitation, add, minimum	1%	1%	
3600	Maximum	6%	7%	
3700	Protection of existing work, add, minimum	2%	2%	
3800	Maximum	5%	7%	
3900	Shift work requirements, add, minimum		5%	
4000	Maximum		30%	

Important: See the Reference Section for critical supporting data - Reference Numbers and City Cost Indexes

D2010 Plumbing Fixtures

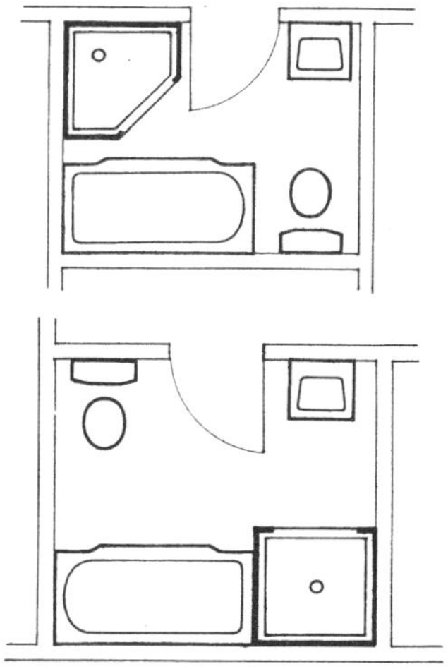

This page illustrates and describes a four fixture bathroom system including a water closet, shower, bathtub, lavatory, accessories, and service piping. Lines within System Components give the unit price and total price on a cost each basis for this system. Prices for an alternate four fixture bathroom system are on Line Item D2010 962 1900. Both material quantities and labor costs have been adjusted for the system listed.

Factors: To adjust for job conditions other than normal working situations use Lines D2010 962 2900 thru 4000.

Example: You are to install the system during weekends and evenings. Go to Line D2010 962 4000 and apply these percentages to the appropriate MAT. and INST. costs.

System Components	QUANTITY	UNIT	COST EACH		
			MAT.	INST.	TOTAL
Four fixture bathroom incl. water closet, shower, bathtub, lavatory					
Accessories and necessary service piping to install this system in 2 walls.					
Water closet, floor mounted, 2 piece, close coupled	1.000	Ea.	176	167	343
Rough-in waste & vent for water closet	1.000	Set	143	290	433
32″ shower, enameled steel stall, molded stone receptor	1.000	Ea.	360	440	800
Rough-in waste & vent for shower	1.000	Set	68.50	430	498.50
Bathtub, P.E. cast iron, 5′ long with accessories	1.000	Ea.	410	201	611
Rough-in waste & vent for bathtub	1.000	Set	127	425	552
Lavatory, 20″ x 18″ P.E. cast iron with accessories	1.000	Ea.	261	111	372
Rough-in waste & vent for lavatory	1.000	Set	242	535	777
Accessories					
Toilet tissue dispenser, chrome, single roll	1.000	Ea.	12.30	14.60	26.90
18″ long stainless steel towel bar	2.000	Ea.	66	38.10	104.10
Medicine cabinet with mirror, 20″ x 16″ unlighted	1.000	Ea.	76.50	31.50	108
TOTAL			1,942.30	2,683.20	4,625.50

D2010 962	Plumbing - Four Fixture Bathroom	COST EACH		
		MAT.	INST.	TOTAL
1900	Above system with 36″ corner angle shower, plumbing in 3 walls	2,475	2,950	5,425
2000				
2300				
2400	NOTE: PLUMBING APPROXIMATIONS			
2500	WATER CONTROL: water meter, backflow preventer,			
2600	Shock absorbers, vacuum breakers, mixer....10 to 15% of fixtures			
2700	PIPE AND FITTINGS: .... 30 to 60% of fixtures			
2800				
2900	Cut & patch to match existing construction, add, minimum	2%	3%	
3000	Maximum	5%	9%	
3100	Dust protection, add, minimum	1%	2%	
3200	Maximum	4%	11%	
3300	Equipment usage curtailment, add, minimum	1%	1%	
3400	Maximum	3%	10%	
3500	Material handling & storage limitation, add, minimum	1%	1%	
3600	Maximum	6%	7%	
3700	Protection of existing work, add, minimum	2%	2%	
3800	Maximum	5%	7%	
3900	Shift work requirements, add, minimum		5%	
4000	Maximum		30%	

D2010 Plumbing Fixtures

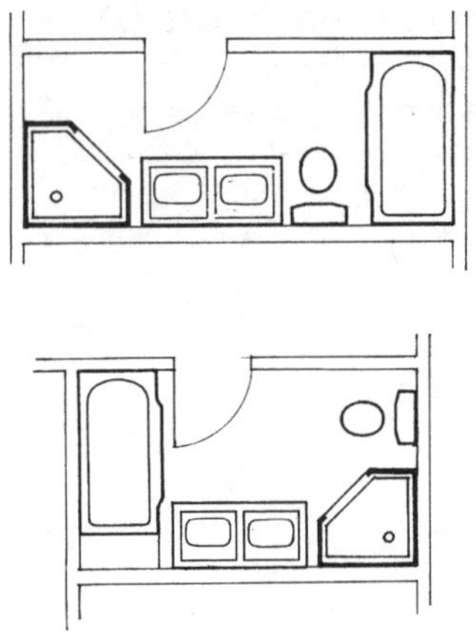

This page illustrates and describes a five fixture bathroom system including a water closet, shower, bathtub, lavatories, accessories, and service piping. Lines within System Components give the unit price and total price on a cost each basis for this system. Prices for an alternate five fixture bathroom system are on Line Item D2010 963 1900. Both material quantities and labor costs have been adjusted for the system listed.

Factors: To adjust for job conditions other than normal working situations use Lines D2010 963 2900 thru 4000.

Example: You are to install and match any existing construction. Go to Line D2010 963 2900 and apply these percentages to the appropriate MAT. and INST. costs.

System Components	QUANTITY	UNIT	COST EACH		
			MAT.	INST.	TOTAL
Five fixture bathroom incl. water closet, shower, bathtub, 2 lavatories, Accessories and necessary service piping to install this system in 1 wall.					
Water closet, floor mounted, 2 piece, close coupled	1.000	Ea.	176	167	343
Rough-in waste & vent for water closet	1.000	Set	143	290	433
36″ corner angle shower, enameled steel stall, molded stone receptor	1.000	Ea.	830	490	1,320
Rough-in waste & vent for shower	1.000	Set	68.50	430	498.50
Bathtub, P.E. cast iron, 5′ long with accessories	1.000	Ea.	410	201	611
Rough-in waste & vent for bathtub	1.000	Set	127	425	552
Countertop, plastic laminate, with backsplash	2.000	Ea.	122	58.40	180.40
Vanity base, 2 doors, 24″ wide	2.000	Ea.	180	79.50	259.50
Lavatories, 20″ x 18″ cabinet mntd, PECI	2.000	Ea.	454	276	730
Rough-in waste & vent for lavatories	1.600	Set	110.40	616	726.40
Accessories					
Toilet tissue dispenser, chrome, single roll	1.000	Ea.	12.30	14.60	26.90
18″ long stainless steel towel bars	2.000	Ea.	66	38.10	104.10
Medicine cabinet with mirror, 20″ x 16″, unlighted	2.000	Ea.	153	63	216
TOTAL			2,852.20	3,148.60	6,000.80

D2010 963	Plumbing - Five Fixture Bathroom	COST EACH		
		MAT.	INST.	TOTAL
1900	Above system installed in 2 walls with all necessary service piping	2,875	3,250	6,125
2000				
2100				
2200				
2300				
2400	NOTE: PLUMBING APPROXIMATIONS			
2500	WATER CONTROL: water meter, backflow preventer,			
2600	Shock absorbers, vacuum breakers, mixer....10 to 15% of fixtures			
2700	PIPE AND FITTINGS: .... 30 to 60% of fixtures			
2800				
2900	Cut & patch to match existing construction, add, minimum	2%	3%	
3000	Maximum	5%	9%	
3100	Dust protection, add, minimum	1%	2%	
3200	Maximum	4%	11%	
3300	Equipment usage curtailment, add, minimum	1%	1%	
3400	Maximum	3%	10%	
3500	Material handling & storage limitation, add, minimum	1%	1%	
3600	Maximum	6%	7%	

SERVICES D

D2010 Plumbing Fixtures

D2010 963	Plumbing - Five Fixture Bathroom	COST EACH		
		MAT.	INST.	TOTAL
3700	Protection of existing work, add, minimum	2%	2%	
3800	Maximum	5%	7%	
3900	Shift work requirements, add, minimum		5%	
4000	Maximum		30%	

D SERVICES

D3010 Energy Supply

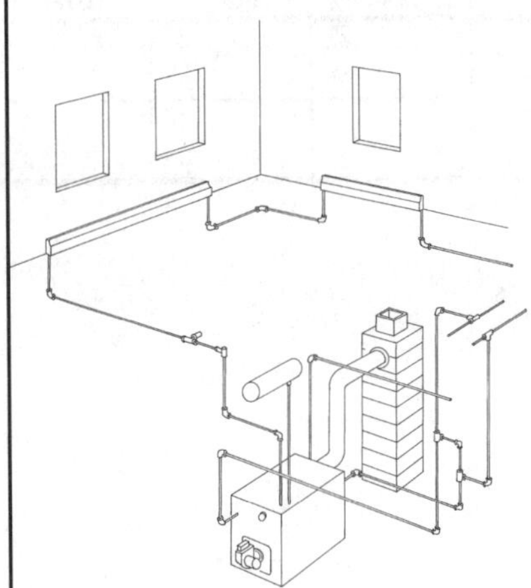

This page illustrates and describes an oil fired hot water baseboard system including an oil fired boiler, fin tube radiation and all fittings and piping. Lines within System Components give the unit price and total price per square foot for this system. Prices for alternate oil fired hot water baseboard systems are on Line Items D3010 540 1700 thru 2100. Both material quantities and labor costs have been adjusted for the system listed.

Factors: To adjust for job conditions other than normal working situations use Lines D3010 540 2900 thru 4000.

Example: You are to install the system while protecting all existing work. Go to Line D3010 540 3700 and apply these percentages to the appropriate MAT. and INST. costs.

System Components	QUANTITY	UNIT	COST PER S.F. MAT.	COST PER S.F. INST.	COST PER S.F. TOTAL
Oil fired hot water baseboard system including boiler, fin tube radiation And all necessary fittings and piping.					
Area to 800 S.F.					
Boiler, cast iron, w/oil piping, 97 MBH	1.000	Ea.	1,625	1,081.25	2,706.25
Copper piping	130.000	L.F.	349.70	1,098.50	1,448.20
Fin tube radiation	68.000	L.F.	425	1,040.40	1,465.40
Circulator	1.000	Ea.	360	147	507
Oil tank	1.000	Ea.	276	178	454
Expansion tank, ASME	1.000	Ea.	405	63.50	468.50
TOTAL			3,440.70	3,608.65	7,049.35
COST PER S.F.		S.F.	4.30	4.51	8.81

D3010 540	Heating - Oil Fired Hot Water	COST PER S.F. MAT.	COST PER S.F. INST.	COST PER S.F. TOTAL
1600	For alternate hot water systems:			
1700	Cast iron boiler, area to 1000 S.F.	3.72	3.89	7.61
1800	To 1200 S.F.	3.21	3.52	6.73
1900	To 1600 S.F.	3.72	3.51	7.23
2000	To 2000 S.F.	3.10	3.13	6.23
2100	To 3000 S.F.	2.27	2.63	4.90
2200				
2300				
2700				
2800				
3000	Maximum	5%	9%	
3100	Dust protection, add, minimum	1%	2%	
3200	Maximum	4%	11%	
3300	Equipment usage curtailment, add, minimum	1%	1%	
3400	Maximum	3%	10%	
3500	Material handling & storage limitation, add, minimum	1%	1%	
3600	Maximum	6%	7%	
3700	Protection of existing work, add, minimum	2%	2%	
3800	Maximum	5%	7%	
3900	Shift work requirements, add, minimum		5%	
4000	Maximum		30%	

D3010 Energy Supply

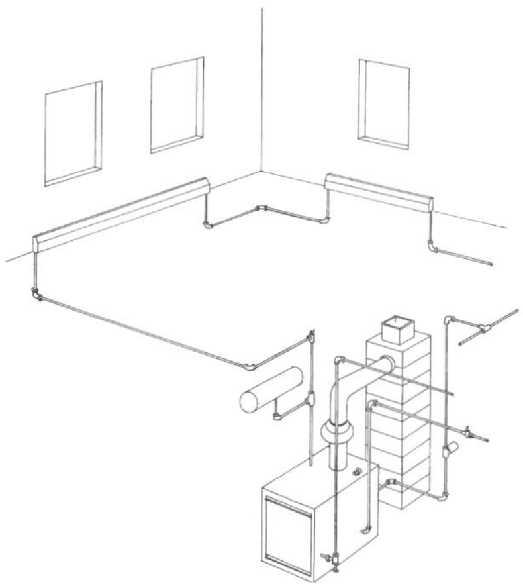

This page illustrates and describes a gas fired hot water baseboard system including a gas fired boiler, fin tube radiation, fittings and piping. Lines within System Components give the unit price and total price per square foot for this system. Prices for alternate gas fired hot water baseboard systems are on Line Items D3010 550 1500 thru 1900. Both material quantities and labor costs have been adjusted for the system listed.

Factors: To adjust for job conditions other than normal working situations use Lines D3010 550 2900 thru 4000.

Example: You are to install the system with minimal equipment usage. Go to Line D3010 550 3400 and apply these percentages to the appropriate MAT. and INST. costs.

System Components	QUANTITY	UNIT	COST PER S.F. MAT.	COST PER S.F. INST.	COST PER S.F. TOTAL
Gas fired hot water baseboard system including boiler, fin tube radiation, All necessary fittings and piping.					
Area to 800 S.F.					
Cast iron boiler, insulating jacket, gas piping, 80 MBH	1.000	Ea.	1,562.50	1,625	3,187.50
Copper piping	130.000	L.F.	349.70	1,098.50	1,448.20
Fin tube radiation	68.000	L.F.	425	1,040.40	1,465.40
Circulator, flange connection	1.000	Ea.	360	147	507
Expansion tank, ASME	1.000	Ea.	405	63.50	468.50
TOTAL			3,102.20	3,974.40	7,076.60
COST PER S.F.		S.F.	3.88	4.97	8.85

D3010 550	Heating - Gas Fired Hot Water	COST PER S.F. MAT.	COST PER S.F. INST.	COST PER S.F. TOTAL
1400	For alternate hot water systems:			
1500	Cast iron boiler, area to 1000 S.F.	3.25	4.37	7.62
1600	To 1200 S.F.	2.82	3.91	6.73
1700	To 1600 S.F.	3.43	3.81	7.24
1800	To 2000 S.F.	2.98	3.42	6.40
1900	To 3000 S.F.	2.34	2.99	5.33
2000				
2100				
2900	Cut & patch to match existing construction, add, minimum	2%	3%	
3000	Maximum	5%	9%	
3100	Dust protection, add, minimum	1%	2%	
3200	Maximum	4%	11%	
3300	Equipment usage curtailment, add, minimum	1%	1%	
3400	Maximum	3%	10%	
3500	Material handling & storage limitation, add, minimum	1%	1%	
3600	Maximum	6%	7%	
3700	Protection of existing work, add, minimum	2%	2%	
3800	Maximum	5%	7%	
3900	Shift work requirements, add, minimum		5%	
4000	Maximum		30%	

D3010 Energy Supply

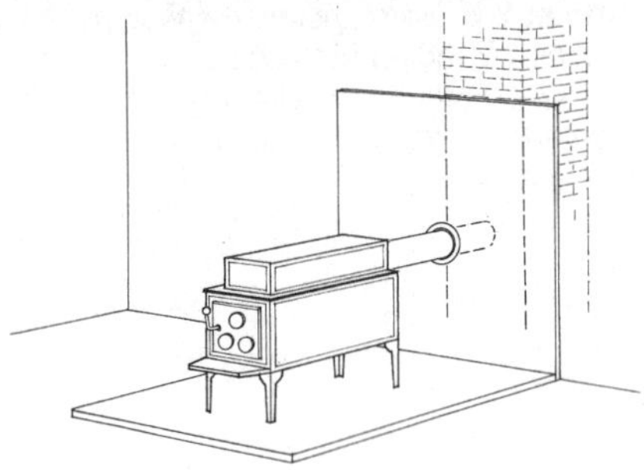

This page illustrates and describes a wood burning stove system including a free standing stove, preformed hearth, masonry chimney, and necessary piping and fittings. Lines within System Components give the unit price and total price on a cost each basis for this system. Prices for alternate wood burning stove systems are on Line Items D3010 991 1600 thru 2000. Both material quantities and labor costs have been adjusted for the system listed.

Factors: To adjust for job conditions other than normal working situations use Lines D3010 991 2900 thru 4000.

Example: You are to install the system and protect existing construction. Go to Line D3010 991 3500 and apply these percentages to the appropriate MAT., and INST. costs.

System Components	QUANTITY	UNIT	COST EACH		
			MAT.	INST.	TOTAL
Cast iron, free standing,wood burning stove with preformed hearth, masonry Chimney, and all necessary piping and fittings to install in chimney.					
Cast iron wood burning stove, stove pipe	1.000	Ea.	850	675	1,525
Wall panel or hearth, non-combustible	1.000	Ea.	408.50	123.50	532
16″ x 16″ brick chimney, 8″ x 8″ flue	20.000	V.L.F.	408.75	1,062.50	1,471.25
Foundation	.500	C.Y.	79.50	91.95	171.45
TOTAL			1,746.75	1,952.95	3,699.70

D3010 991	Wood Burning Stoves	COST EACH		
		MAT.	INST.	TOTAL
1400	For alternate wood burning systems:			
1500				
1600	Installed in existing fireplace	1,075	740	1,815
1700				
1800	Installed with insulated metal chimney			
1900	System including ceiling package,			
2000	Metal chimney to 10 L.F.	1,800	1,050	2,850
2100				
2200				
2300				
2400				
2500				
2600				
2700				
2900	Dust protection, add, minimum	1%	2%	
3000	Maximum	4%	11%	
3100	Equipment usage curtailment, add, minimum	1%	1%	
3200	Maximum	3%	10%	
3300	Material handling & storage limitation, add, minimum	1%	1%	
3400	Maximum	6%	7%	
3500	Protection of existing work, add, minimum	2%	2%	
3600	Maximum	5%	7%	
3700	Shift work requirements, add, minimum		5%	
3800	Maximum		30%	
3900	Temporary shoring and bracing, add, minimum	2%	5%	
4000	Maximum	5%	12%	

D3010 Energy Supply

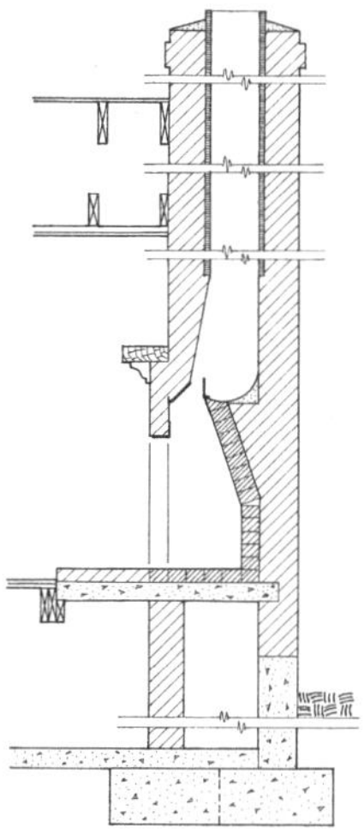

This page illustrates and describes masonry fireplace systems including a brick fireplace, footing, foundation, hearth, firebox, chimney and flue.
Lines within System Components give the unit price and total price on a cost each basis for this system. Prices for alternate masonry fireplace systems are on Line Items D3010 992 1300 thru 1500. Both material quantities and labor costs have been adjusted for the system listed.

Factors: To adjust for job conditions other than normal working situations use Lines D3010 992 2700 thru 4000.

Example: You are to install the system with some temporary shoring and bracing. Go to Line D3010 992 3900 and apply these percentages to the appropriate MAT. and INST. costs.

System Components	QUANTITY	UNIT	COST EACH		
			MAT.	INST.	TOTAL
Brick masonry fireplace, including footing, foundation, hearth, firebox, Chimney and flue, chimney 12′ above firebox.					
Footing, 4′ x 7′ x 12″ thick, 3000 psi concrete	1.040	C.Y.	116.48	159.21	275.69
Foundation, 12″ concrete block	180.000	S.F.	486	1,395	1,881
Fireplace, brick faced, 6′-0″ wide x 5′-0″ high	1.000	Ea.	420	1,950	2,370
Hearth	1.000	Ea.	154	390	544
Chimney, 20″ x 20″, one 12″ x 12″ flue, 12 V.L.F.	12.000	V.L.F.	342	678	1,020
Mantle, wood	7.000	L.F.	38.50	85.05	123.55
TOTAL			1,556.98	4,657.26	6,214.24

D3010 992	Masonry Fireplace	COST EACH		
		MAT.	INST.	TOTAL
1200	Above system with the following:			
1300	Chimney, 20″ x 20″, one 12″ x 12″ flue, 18 V.L.F.	1,725	5,000	6,725
1400	24 V.L.F.	1,900	5,325	7,225
1500	Fieldstone face instead of brick	1,875	4,650	6,525
1600				
1700				
1800				
1900				
2000				
2100				
2200				
2300				
2700	Cut & patch to match existing construction, add, minimum	2%	3%	
2800	Maximum	5%	9%	
2900	Dust protection, add, minimum	1%	2%	
3000	Maximum	4%	11%	
3100	Equipment usage curtailment, add, minimum	1%	1%	
3200	Maximum	3%	10%	
3300	Material handling & storage limitation, add, minimum	1%	1%	
3400	Maximum	6%	7%	
3500	Protection of existing work, add, minimum	2%	2%	
3600	Maximum	5%	7%	
3700	Shift work requirements, add, minimum		5%	
3800	Maximum		30%	
3900	Temporary shoring and bracing, add, minimum	2%	5%	
4000	Maximum	5%	12%	

D3020 Heat Generating Systems

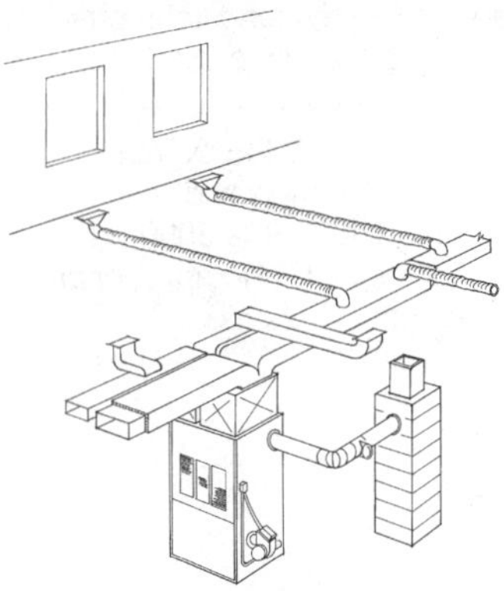

This page illustrates and describes an oil fired forced air system including an oil fired furnace, ductwork, registers and hookups. Lines within System Components give the unit price and total price per square foot for this system. Prices for alternate oil fired forced air systems are on Line Items D3020 122 1700 thru 2800. Both material quantities and labor costs have been adjusted for the system listed.

Factors: To adjust for job conditions other than normal working situations use Lines D3020 122 3100 thru 4200.

Example: You are to install the system during evenings and weekends. Go to Line D3020 122 4200 and apply this percentage to the appropriate INST. cost.

System Components	QUANTITY	UNIT	COST PER S.F. MAT.	INST.	TOTAL
Oil fired hot air heating system including furnace, ductwork, registers And all necessary hookups.					
Area to 800 S.F., heat only					
Furnace, oil, atomizing gun type burner, w/oil piping	1.000	Ea.	1,025	307.50	1,332.50
Oil tank, steel, 275 gallon	1.000	Ea.	276	178	454
Duct, galvanized, steel	312.000	Lb.	280.80	1,825.20	2,106
Insulation, blanket type, duct work	270.000	S.F.	143.10	650.70	793.80
Flexible duct, 6″ diameter, insulated	100.000	L.F.	166	341	507
Registers, baseboard, gravity, 12″ x 6″	8.000	Ea.	126	172	298
Return, damper, 36″ x 18″	1.000	Ea.	149	49	198
TOTAL			2,165.90	3,523.40	5,689.30
COST PER S.F.		S.F.	2.71	4.40	7.11

D3020 122	Heating-Cooling, Oil, Forced Air	COST PER S.F. MAT.	INST.	TOTAL
1600	For alternate heating systems:			
1700	Oil fired, area to 1000 S.F.	2.21	3.58	5.79
1800	To 1200 S.F.	1.92	3.16	5.08
1900	To 1600 S.F.	1.59	2.63	4.22
2000	To 2000 S.F.	1.66	3.38	5.04
2100	To 3000 S.F.	1.41	2.69	4.10
2200	For combined heating and cooling systems:			
2300	Oil fired, heating and cooling, area to 800 S.F.	4.52	5.25	9.77
2400	To 1000 S.F.	4.13	4.28	8.41
2500	To 1200 S.F.	3.52	3.74	7.26
2600	To 1600 S.F.	2.87	3.09	5.96
2700	To 2000 S.F.	2.74	3.80	6.54
2800	To 3000 S.F.	2.24	2.99	5.23
3100	Cut & patch to match existing construction, add, minimum	2%	3%	
3200	Maximum	5%	9%	
3300	Dust protection, add, minimum	1%	2%	
3400	Maximum	4%	11%	
3500	Equipment usage curtailment, add, minimum	1%	1%	
3600	Maximum	3%	10%	
3700	Material handling & storage limitation, add, minimum	1%	1%	
3800	Maximum	6%	7%	
3900	Protection of existing work, add, minimum	2%	2%	
4000	Maximum	5%	7%	
4100	Shift work requirements, add, minimum		5%	
4200	Maximum		30%	

SERVICES D

D3020 Heat Generating Systems

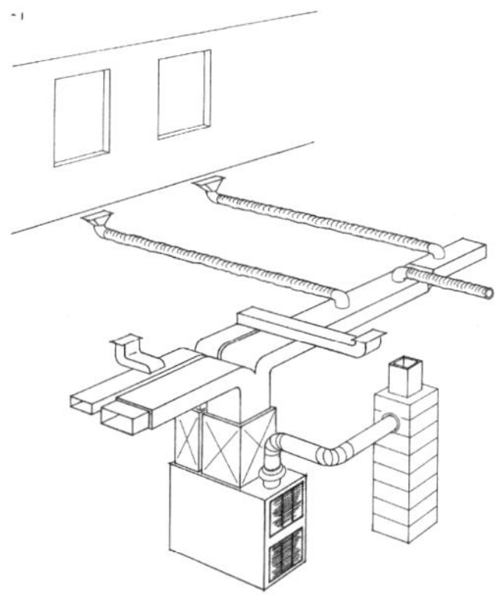

This page illustrates and describes a gas fired forced air system including a gas fired furnace, ductwork, registers and hookups. Lines within System Components give the unit price and total price per square foot for this system. Prices for alternate gas fired forced air systems are on Line Items D3020 124 1500 thru 2600. Both material quantities and labor costs have been adjusted for the system listed.

Factors: To adjust for job conditions other than normal working situations use Lines D3020 124 2900 thru 4000.

Example: You are to install the system with material handling and storage limitations. Go to Line D3020 124 3500 and apply these percentages to the appropriate MAT. and INST. costs.

System Components	QUANTITY	UNIT	COST PER S.F. MAT.	COST PER S.F. INST.	COST PER S.F. TOTAL
Gas fired hot air heating system including furnace, ductwork, registers And all necessary hookups.					
Area to 800 S.F., heat only					
Furnace, gas, AGA certified, direct drive, w/gas piping, 44 MBH	1.000	Ea.	600	276.25	876.25
Duct, galvanized steel	312.000	Lb.	280.80	1,825.20	2,106
Insulation, blanket type, ductwork	270.000	S.F.	143.10	650.70	793.80
Flexible duct, 6″ diameter, insulated	100.000	L.F.	166	341	507
Registers, baseboard, gravity, 12″ x 6″	8.000	Ea.	126	172	298
Return, damper, 36″ x 18″	1.000	Ea.	149	49	198
TOTAL			1,464.90	3,314.15	4,779.05
COST PER S.F.		S.F.	1.83	4.14	5.97

D3020 124	Heating-Cooling, Gas, Forced Air	COST PER S.F. MAT.	COST PER S.F. INST.	COST PER S.F. TOTAL
1400	For alternate heating systems:			
1500	Gas fired, area to 1000 S.F.	1.66	3.38	5.04
1600	To 1200 S.F.	1.47	2.99	4.46
1700	To 1600 S.F.	1.42	2.79	4.21
1800	To 2000 S.F.	1.40	3.30	4.70
1900	To 3000 S.F.	1.13	2.63	3.76
2000	For combined heating and cooling systems:			
2100	Gas fired, heating and cooling, area to 800 S.F.	3.64	4.97	8.61
2200	To 1000 S.F.	3.58	4.08	7.66
2300	To 1200 S.F.	3.07	3.58	6.65
2400	To 1600 S.F.	2.69	3.26	5.95
2500	To 2000 S.F.	2.48	3.71	6.19
2600	To 3000 S.F.	1.96	2.92	4.88
2900	Cut & patch to match existing construction, add, minimum	2%	3%	
3000	Maximum	5%	9%	
3100	Dust protection, add, minimum	1%	2%	
3200	Maximum	4%	11%	
3300	Equipment usage curtailment, add, minimum	1%	1%	
3400	Maximum	3%	10%	
3500	Material handling & storage limitation, add, minimum	1%	1%	
3600	Maximum	6%	7%	
3700	Protection of existing work, add, minimum	2%	2%	
3800	Maximum	5%	7%	
3900	Shift work requirements, add, minimum		5%	
4000	Maximum		30%	

D3020 Heat Generating Systems

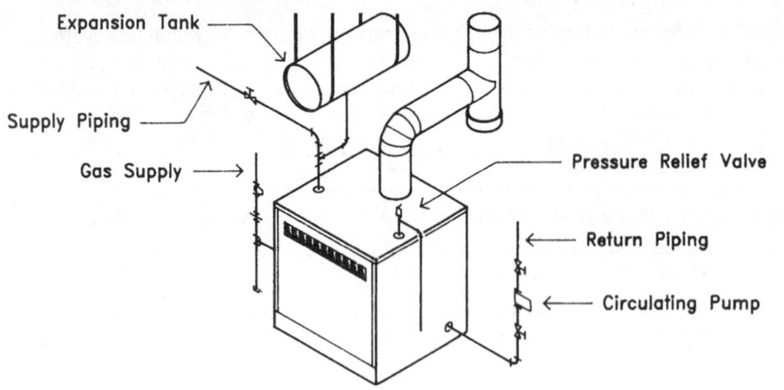

Cast Iron Boiler, Hot Water, Gas Fired

This page illustrates and describes boilers including expansion tank, circulating pump and all service piping. Lines within Systems Components give the material and installation price on a cost each basis for the components. Prices for alternate boiler systems are on Line Items D3020 130 1010 thru D3020 138 1040. Material quantities and labor costs have been adjusted for the system listed.

Factors: To adjust for job conditions other than normal working situations use Line D3020 138 2700 thru 4000.

System Components	QUANTITY	UNIT	COST EACH MAT.	COST EACH INST.	COST EACH TOTAL
Cast iron boiler, including piping, chimney and accessories					
Boilers, gas fired, std controls, CI, insulated, HW, gross output 100 MBH	1.000	Ea.	1,450	1,400	2,850
Pipe, black steel, Sch 40, threaded, W/coupling & hangers, 10′ OC, 3/4″ dia	20.000	L.F.	35.33	181.13	216.46
Pipe, black steel, Sch 40, threaded, W/coupling & hangers, 10′ OC, 1″ dia.	20.000	L.F.	48.16	198.88	247.04
Elbow, 90°, black, straight, 3/4″ dia.	9.000	Ea.	15.57	315	330.57
Elbow, 90°, black, straight, 1″ dia.	6.000	Ea.	18	228	246
Tee, black, straight, 3/4″ dia.	2.000	Ea.	5.50	109	114.50
Tee, black, straight, 1″ dia.	2.000	Ea.	8.40	123	131.40
Tee, black, reducing, 1″ dia.	2.000	Ea.	10.70	123	133.70
Pipe cap, black, 3/4″ dia.	1.000	Ea.	1.96	15.35	17.31
Union, black with brass seat, 3/4″ dia.	2.000	Ea.	13.80	76	89.80
Union, black with brass seat, 1″ dia.	2.000	Ea.	19.80	82	101.80
Pipe nipples, black, 3/4″ dia.	5.000	Ea.	3.93	20.13	24.06
Pipe nipples, black, 1″ dia.	3.000	Ea.	3.36	13.88	17.24
Valves, bronze, gate, N.R.S., threaded, class 150, 3/4″ size	2.000	Ea.	61	49	110
Valves, bronze, gate, N.R.S., threaded, class 150, 1″ size	2.000	Ea.	79	52	131
Gas cock, brass, 3/4″ size	1.000	Ea.	11.90	22.50	34.40
Thermometer, stem type, 9″ case, 8″ stem, 3/4″ NPT	2.000	Ea.	260	35.20	295.20
Tank, steel, liquid expansion, ASME, painted, 15 gallon capacity	1.000	Ea.	400	52	452
Pump, circulating, bronze, flange connection, 3/4″ to 1-1/2″ size, 1/8 HP	1.000	Ea.	615	147	762
Vent chimney, all fuel, pressure tight, double wall, SS, 6″ dia.	20.000	L.F.	800	295	1,095
Vent chimney, elbow, 90° fixed, 6″ dia.	2.000	Ea.	454	59	513
Vent chimney, Tee, 6″ dia.	2.000	Ea.	300	74	374
Vent chimney, ventilated roof thimble, 6″ dia.	1.000	Ea.	201	34	235
Vent chimney, adjustable roof flashing, 6″ dia.	1.000	Ea.	52.50	29.50	82
Vent chimney, stack cap, 6″ diameter	1.000	Ea.	158	19.25	177.25
Insulation, fiberglass pipe covering, 1″ wall, 1″ IPS	20.000	L.F.	17.60	76.60	94.20
TOTAL			5,044.51	3,830.42	8,874.93

D3020 130	Boiler, Cast Iron, Hot Water, Gas	COST EACH MAT.	COST EACH INST.	COST EACH TOTAL
1000	For alternate boilers:			
1010	Boiler, cast iron, gas, hot water, 100 MBH	5,050	3,825	8,875
1020	200 MBH	5,850	4,375	10,225
1030	320 MBH	6,175	4,750	10,925
1040	440 MBH	8,225	6,125	14,350
1050	544 MBH	12,300	9,700	22,000

SERVICES D

Important: See the Reference Section for critical supporting data - Reference Numbers and City Cost Indexes

D3020 Heat Generating Systems

D3020 130	Boiler, Cast Iron, Hot Water, Gas	COST EACH		
		MAT.	INST.	TOTAL
1060	765 MBH	15,700	10,500	26,200
1070	1088 MBH	17,500	11,100	28,600
1080	1530 MBH	21,600	14,800	36,400
1090	2312 MBH	26,100	17,300	43,400

D3020 134	Boiler, Cast Iron, Steam, Gas	COST EACH		
		MAT.	INST.	TOTAL
1010	Boiler, cast iron, gas, steam, 100 MBH	4,250	3,700	7,950
1020	200 MBH	10,200	6,975	17,175
1030	320 MBH	11,400	7,800	19,200
1040	544 MBH	14,900	12,400	27,300
1050	765 MBH	17,000	12,900	29,900
1060	1275 MBH	20,200	14,100	34,300
1070	2675 MBH	29,100	20,600	49,700

D3020 136	Boiler, Cast Iron, Hot Water, Gas/Oil	COST EACH		
		MAT.	INST.	TOTAL
1010	Boiler, cast iron, gas & oil, hot water, 584 MBH	14,400	10,800	25,200
1020	876 MBH	20,900	11,600	32,500
1030	1168 MBH	21,400	13,000	34,400
1040	1460 MBH	24,300	15,800	40,100
1050	2044 MBH	27,700	16,300	44,000
1060	2628 MBH	37,200	19,700	56,900

D3020 138	Boiler, Cast Iron, Steam, Gas/Oil	COST EACH		
		MAT.	INST.	TOTAL
1010	Boiler, cast iron, gas & oil, steam, 810 MBH	18,900	14,000	32,900
1020	1360 MBH	21,900	14,900	36,800
1030	2040 MBH	27,100	19,200	46,300
1040	2700 MBH	30,100	21,300	51,400
2700	Cut & patch to match existing construction, add, minimum	2%	3%	
2800	Maximum	5%	9%	
2900	Dust protection, add, minimum	1%	2%	
3000	Maximum	4%	11%	
3100	Equipment usage curtailment, add, minimum	1%	1%	
3200	Maximum	3%	10%	
3300	Material handling & storage limitation, add, minimum	1%	1%	
3400	Maximum	6%	7%	
3500	Protection of existing work, add, minimum	2%	2%	
3600	Maximum	5%	7%	
3700	Shift work requirements, add, minimum		5%	
3800	Maximum		30%	
3900	Temporary shoring and bracing, add, minimum	2%	5%	
4000	Maximum	5%	12%	

D3020 Heat Generating Systems

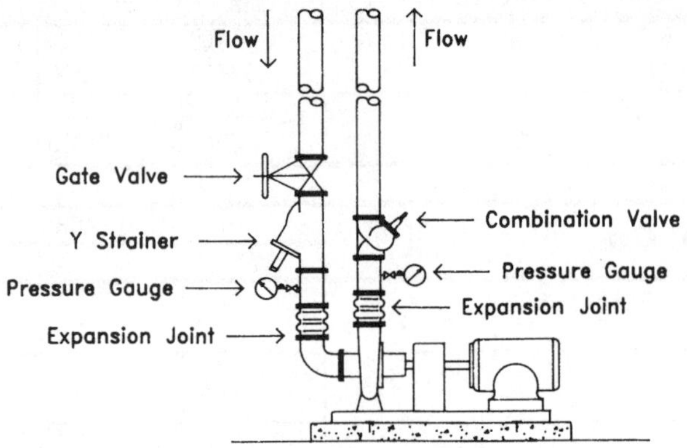

Base Mounted End–Suction Pump

This page illustrates and describes end suction base mounted circulating pumps, including strainer and piping. Lines within system components give the material and installation price on a cost each basis for the components. Prices for alternate circulating pumps are on Line items D3020 330 1010 thru 1050. Material quantities and labor costs have been adjusting for the system listed.

Factors: To adjust for job conditions other than normal working situations uses Lines D3020 330 2700 thru 4000.

System Components	QUANTITY	UNIT	COST EACH MAT.	COST EACH INST.	COST EACH TOTAL
Circulator pump, end suction, including pipe, Valves, fittings and guages.					
Pump, circulating, CI, base mounted, 2-1/2" size, 3 HP, to 150 GPM	1.000	Ea.	2,125	490	2,615
Pipe, black steel, Sch. 40, on yoke & roll hangers, 10' O.C., 2-1/2" dia	12.000	L.F.	56.04	247.44	303.48
Elbow, 90°, weld joint, steel, 2-1/2" pipe size	1.000	Ea.	10.20	121.70	131.90
Flange, weld neck, 150 LB, 2-1/2" pipe size	8.000	Ea.	176	486.80	662.80
Valve, iron body, gate, 125 lb., N.R.S., flanged, 2-1/2" size	1.000	Ea.	335	177	512
Strainer, Y type, iron body, flanged, 125 lb., 2-1/2" pipe size	1.000	Ea.	80	178	258
Multipurpose valve, CI body, 2" size	1.000	Ea.	390	61.50	451.50
Expansion joint, flanged spool, 6" F to F, 2-1/2" dia	2.000	Ea.	374	143	517
T-O-L, weld joint, socket, 1/4" pipe size, nozzle	2.000	Ea.	8.54	84.44	92.98
T-O-L, weld joint, socket, 1/2" pipe size, nozzle	2.000	Ea.	8.54	87.78	96.32
Control gauges, pressure or vacuum, 3-1/2" diameter dial	2.000	Ea.	44	30.80	74.80
Pressure/temperature relief plug, 316 SS, 3/4" OD, 7-1/2" insertion	2.000	Ea.	126	30.80	156.80
Insulation, fiberglass pipe covering, 1-1/2" wall, 2-1/2" IPS	12.000	L.F.	27.72	56.16	83.88
Pump control system	1.000	Ea.	965	455	1,420
Pump balancing	1.000	Ea.		217	217
TOTAL			4,726.04	2,867.42	7,593.46

D3020 330	Circulating Pump Systems, End Suction	COST EACH MAT.	COST EACH INST.	COST EACH TOTAL
1000	For alternate pump sizes:			
1010	Pump, base mtd with motor, end-suction, 2-1/2" size, 3 HP, to 150 GPM	4,725	2,875	7,600
1020	3" size, 5 HP, to 225 GPM	5,225	3,275	8,500
1030	4" size, 7-1/2 HP, to 350 GPM	6,475	3,950	10,425
1040	5" size, 15 HP, to 1000 GPM	8,975	6,000	14,975
1050	6" size, 25 HP, to 1550 GPM	12,000	7,275	19,275
2700	Cut & patch to match existing construction, add, minimum	2%	3%	
2800	Maximum	5%	9%	
2900	Dust protection, add, minimum	1%	2%	
3000	Maximum	4%	11%	
3100	Equipment usage curtailment, add, minimum	1%	1%	
3200	Maximum	3%	10%	
3300	Material handling & storage limitation, add, minimum	1%	1%	
3400	Maximum	6%	7%	
3500	Protection of existing work, add, minimum	2%	2%	
3600	Maximum	5%	7%	

Important: See the Reference Section for critical supporting data - Reference Numbers and City Cost Indexes

D3020 Heat Generating Systems

D3020 330	Circulating Pump Systems, End Suction	COST EACH		
		MAT.	INST.	TOTAL
3700	Shift work requirements, add, minimum		5%	
3800	Maximum		30%	
3900	Temporary shoring and bracing, add, minimum	2%	5%	
4000	Maximum	5%	12%	

D3040 Distribution Systems

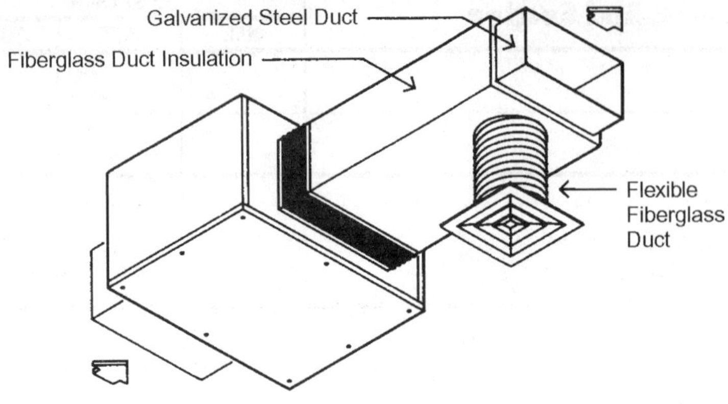

Horizontal Fan Coil Air Conditioning System

This page illustrates and describes fan coil air conditioners including duct work, duct installation, piping and diffusers. Lines within Systems Components give the material and installation price on a cost each basis for the components. Prices for alternate fan coil A/C unit systems are on Line Items D3040 118 1010 thru D3040 126 1120. Material quantities and labor costs have been adjusted for the system listed.

Factors: To adjust for job conditions other than normal working situations use Lines D3040 126 2700 thru 4000.

System Components	QUANTITY	UNIT	COST EACH MAT.	COST EACH INST.	COST EACH TOTAL
Fan coil A/C system, horizontal, with housing, controls, 2 pipe, 1/2 ton					
Fan coil A/C, horizontal housing, filters, chilled water, 1/2 ton cooling	1.000	Ea.	965	111	1,076
Pipe, black steel, Sch 40, threaded, W/cplgs & hangers, 10′ OC, 3/4″ dia.	20.000	L.F.	33.76	173.08	206.84
Elbow, 90°, black, straight, 3/4″ dia.	6.000	Ea.	10.38	210	220.38
Tee, black, straight, 3/4″ dia.	2.000	Ea.	5.50	109	114.50
Union, black with brass seat, 3/4″ dia.	2.000	Ea.	13.80	76	89.80
Pipe nipples, 3/4″ diam	3.000	Ea.	2.36	12.08	14.44
Valves, bronze, gate, N.R.S., threaded, class 150, 3/4″ size	2.000	Ea.	61	49	110
Circuit setter, balance valve, bronze body, threaded, 3/4″ pipe size	1.000	Ea.	53	24.50	77.50
Insulation, fiberglass pipe covering, 1″ wall, 3/4″ IPS	20.000	L.F.	17.40	73.20	90.60
Ductwork, 12″ x 8″ fabricated, galvanized steel, 12 LF	55.000	Lb.	49.50	321.75	371.25
Insulation, ductwork, blanket type, fiberglass, 1″ thk, 1-1/2 LB density	40.000	S.F.	21.20	96.40	117.60
Diffusers, aluminum, OB damper, ceiling, perf, 24″x24″ panel size, 6″x6″	2.000	Ea.	163	61	224
Ductwork, flexible, fiberglass fabric, insulated, 1″thk, PE jacket, 6″ dia	16.000	L.F.	26.56	54.56	81.12
Round volume control damper 6″ dia.	2.000	Ea.	49	45	94
Fan coil unit balancing	1.000	Ea.		64	64
Re-heat coil balancing	1.000	Ea.		91.50	91.50
Diffuser/register, high, balancing	2.000	Ea.		177	177
TOTAL			1,471.46	1,749.07	3,220.53

D3040 118	Fan Coil A/C Unit, Two Pipe	COST EACH MAT.	COST EACH INST.	COST EACH TOTAL
1000	For alternate A/C units:			
1010	Fan coil A/C system, cabinet mounted, controls, 2 pipe, 1/2 Ton	770	925	1,695
1020	1 Ton	925	1,050	1,975
1030	1-1/2 Ton	995	1,050	2,045
1040	2 Ton	1,525	1,275	2,800
1050	3 Ton	2,000	1,325	3,325

D3040 120	Fan Coil A/C Unit, Two Pipe, Electric Heat	COST EACH MAT.	COST EACH INST.	COST EACH TOTAL
1010	Fan coil A/C system, cabinet mntd, elect. ht, controls, 2 pipe, 1/2 Ton	1,225	925	2,150
1020	1 Ton	1,500	1,050	2,550
1030	1-1/2 Ton	1,775	1,050	2,825
1040	2 Ton	2,650	1,300	3,950
1050	3 Ton	4,500	1,325	5,825

D3040 Distribution Systems

D3040 122	Fan Coil A/C Unit, Four Pipe	COST EACH		
		MAT.	INST.	TOTAL
1010	Fan coil A/C system, cabinet mounted, controls, 4 pipe, 1/2 Ton	1,200	1,800	3,000
1020	1 Ton	1,375	1,925	3,300
1030	1-1/2 Ton	1,475	1,925	3,400
1040	2 Ton	1,800	2,000	3,800
1050	3 Ton	2,525	2,150	4,675

D3040 124	Fan Coil A/C, Horizontal, Duct Mount, 2 Pipe	COST EACH		
		MAT.	INST.	TOTAL
1010	Fan coil A/C system, horizontal w/housing, controls, 2 pipe, 1/2 Ton	1,475	1,750	3,225
1020	1 Ton	1,925	2,550	4,475
1030	1-1/2 Ton	2,425	3,325	5,750
1040	2 Ton	3,100	3,975	7,075
1050	3 Ton	3,750	5,275	9,025
1060	3-1/2 Ton	3,950	5,475	9,425
1070	4 Ton	4,075	6,200	10,275
1080	5 Ton	4,800	7,900	12,700
1090	6 Ton	4,875	8,475	13,350
1100	7 Ton	4,900	8,800	13,700
1110	8 Ton	4,900	8,975	13,875
1120	10 Ton	5,275	9,325	14,600

D3040 126	Fan Coil A/C, Horiz., Duct Mount, 2 Pipe, Elec. Ht.	COST EACH		
		MAT.	INST.	TOTAL
1010	Fan coil A/C system, horiz. hsng, elect. ht, ctrls, 2 pipe, 1/2 Ton	1,675	1,750	3,425
1020	1 Ton	2,175	2,550	4,725
1030	1-1/2 Ton	2,600	3,325	5,925
1040	2 Ton	3,425	3,975	7,400
1050	3 Ton	5,525	5,450	10,975
1060	3-1/2 Ton	6,250	5,475	11,725
1070	4 Ton	7,050	6,225	13,275
1080	5 Ton	8,025	7,900	15,925
1090	6 Ton	8,100	8,500	16,600
1100	7 Ton	8,225	8,450	16,675
1110	8 Ton	9,000	8,550	17,550
1120	10 Ton	9,575	8,800	18,375
2700	Cut & patch to match existing construction, add, minimum	2%	3%	
2800	Maximum	5%	9%	
2900	Dust protection, add, minimum	1%	2%	
3000	Maximum	4%	11%	
3100	Equipment usage curtailment, add, minimum	1%	1%	
3200	Maximum	3%	10%	
3300	Material handling & storage limitation, add, minimum	1%	1%	
3400	Maximum	6%	7%	
3500	Protection of existing work, add, minimum	2%	2%	
3600	Maximum	5%	7%	
3700	Shift work requirements, add, minimum		5%	
3800	Maximum		30%	
3900	Temporary shoring and bracing, add, minimum	2%	5%	
4000	Maximum	5%	12%	

D3040 Distribution Systems

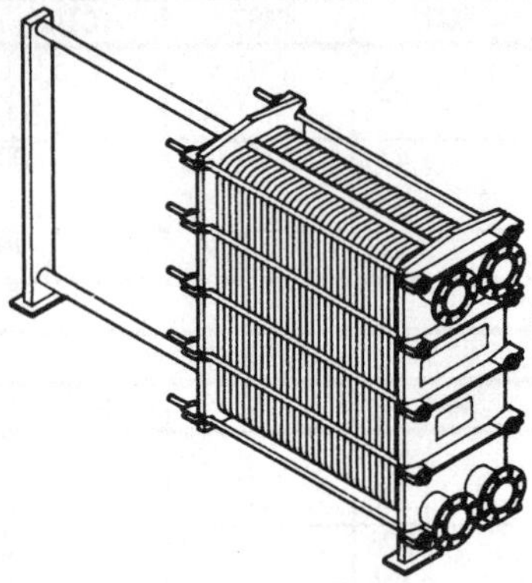

Plate Heat Exchanger

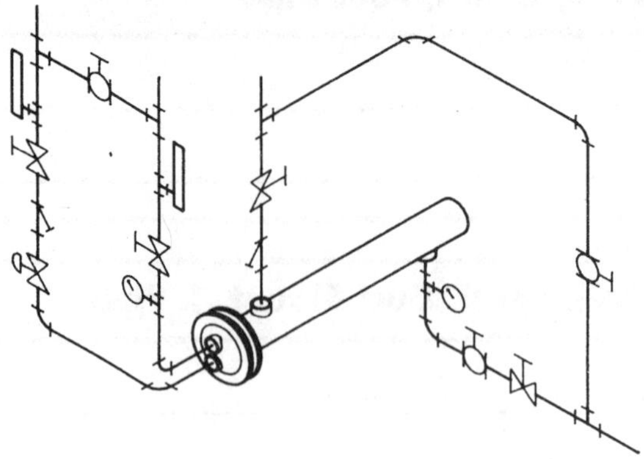

Shell and Tube Heat Exchanger

This page illustrates and describes plate and shell and tube type heat exchangers including pressure gauge and exchanger control system. Lines within Systems Components give the material and installation price on a cost each basis for the components. Prices for alternate heat exchanger systems are on Line Items D3040 610 1010 thru D3040 620 1040. Material quantities and labor costs have been adjusted for the system listed.

Factors To adjust for job conditions other then normal working situations use Lines D3040 620 2700 thru 4000.

System Components	QUANTITY	UNIT	COST EACH MAT.	COST EACH INST.	COST EACH TOTAL
Shell and tube type heat exchanger with related valves and piping.					
Heat exchanger, 4 pass, 3/4" O.D. copper tubes, by steam at 10 psi, 40 GPM	1.000	Ea.	2,475	222	2,697
Pipe, black steel, Sch 40, threaded, W/couplings & hangers, 10′ OC, 2" dia.	40.000	L.F.	202.51	696.90	899.41
Elbow, 90°, straight, 2" dia.	16.000	Ea.	179.20	784	963.20
Reducer, black steel, concentric, 2" dia.	2.000	Ea.	19.60	84	103.60
Valves, bronze, gate, rising stem, threaded, class 150, 2" size	6.000	Ea.	522	267	789
Valves, bronze, globe, class 150, rising stem, threaded, 2" size	2.000	Ea.	466	89	555
Strainers, Y type, bronze body, screwed, 150 lb., 2" pipe size	2.000	Ea.	118	76	194
Valve, electric motor actuated, brass, 2 way, screwed, 1-1/2" pipe size	1.000	Ea.	410	38.50	448.50
Union, black with brass seat, 2" dia.	8.000	Ea.	164	416	580
Tee, black, straight, 2" dia.	9.000	Ea.	118.80	724.50	843.30
Tee, black, reducing run and outlet, 2" dia.	4.000	Ea.	96	322	418
Pipe nipple, black, 2" dia.	21.000	Ea.	2.01	6.90	8.91
Thermometers, stem type, 9" case, 8" stem, 3/4" NPT	2.000	Ea.	260	35.20	295.20
Gauges, pressure or vacuum, 3-1/2" diameter dial	2.000	Ea.	44	30.80	74.80
Insulation, fiberglass pipe covering, 1" wall, 2" IPS	40.000	L.F.	47.60	168.40	216
Heat exchanger control system	1.000	Ea.	2,025	1,600	3,625
Coil balancing	1.000	Ea.		91	91
TOTAL			7,149.72	5,652.20	12,801.92

D3040 610	Heat Exchanger, Plate Type	COST EACH MAT.	COST EACH INST.	COST EACH TOTAL
1000	For alternate heat exchangers:			
1010	Plate heat exchanger, 400 GPM	26,200	11,600	37,800
1020	800 GPM	41,400	14,700	56,100
1030	1200 GPM	53,500	19,900	73,400
1040	1800 GPM	78,500	24,700	103,200

D3040 620	Heat Exchanger, Shell & Tube	COST EACH MAT.	COST EACH INST.	COST EACH TOTAL
1010	Shell & tube heat exchanger, 40 GPM	7,150	5,650	12,800
1020	96 GPM	12,400	10,200	22,600
1030	240 GPM	22,400	13,900	36,300
1040	600 GPM	42,300	19,400	61,700
2700	Cut & patch to match existing construction, add, minimum	2%	3%	
2800	Maximum	5%	9%	
2900	Dust protection, add, minimum	1%	2%	
3000	Maximum	4%	11%	

D3040 Distribution Systems

D3040 620	Heat Exchanger, Shell & Tube	COST EACH		
		MAT.	INST.	TOTAL
3100	Equipment usage curtailment, add, minimum	1%	1%	
3200	Maximum	3%	10%	
3300	Material handling & storage limitation, add, minimum	1%	1%	
3400	Maximum	6%	7%	
3500	Protection of existing work, add, minimum	2%	2%	
3600	Maximum	5%	7%	
3700	Shift work requirements, add, minimum		5%	
3800	Maximum		30%	
3900	Temporary shoring and bracing, add, minimum	2%	5%	
4000	Maximum	5%	12%	

D3050 Terminal & Package Units

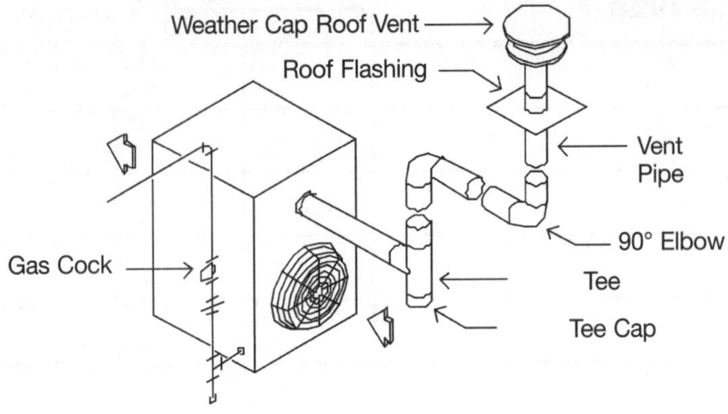

Gas Unit Heater

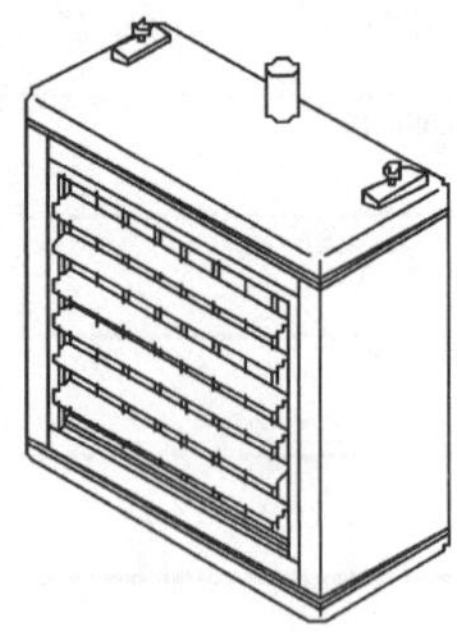

Hydronic Unit Heater

This page illustrates and describes unit heaters including piping and vents. Lines within Systems Components give the material and installation price on a cost each basis for this system. Prices for alternate unit heater systems are on Line items D3050 120 1010 thru D3050 140 1040. Material quantities and labor costs have been adjusted for the system listed.

Factors: To adjust for job conditions other than normal working situations use Lines D3050 140 2700 thru 4000.

System Components	QUANTITY	UNIT	COST EACH MAT.	COST EACH INST.	COST EACH TOTAL
Unit heaters complete with piping, thermostats and chimney as required					
Space heater, propeller fan, 20 MBH output	1.000	Ea.	465	104	569
Pipe, black steel, Sch 40, threaded, W/coupling & hangers, 10′ OC, 3/4″ dia	20.000	L.F.	32.97	169.05	202.02
Elbow, 90°, black steel, straight, 1/2″ dia.	3.000	Ea.	4.29	99	103.29
Elbow, 90°, black steel, straight, 3/4″ dia.	3.000	Ea.	5.19	105	110.19
Tee, black steel, reducing, 3/4″ dia.	1.000	Ea.	4.73	54.50	59.23
Union, black with brass seat, 3/4″ dia.	1.000	Ea.	6.90	38	44.90
Pipe nipples, black, 1/2″ dia	2.000	Ea.	2.68	15.60	18.28
Pipe nipples, black, 3/4″ dia	2.000	Ea.	1.57	8.05	9.62
Pipe cap, black, 3/4″ dia.	1.000	Ea.	1.96	15.35	17.31
Gas cock, brass, 3/4″ size	1.000	Ea.	11.90	22.50	34.40
Thermostat, 1 set back, electric, timed	1.000	Ea.	89	61.50	150.50
Wiring, thermostat hook-up	1.000	Ea.	5.20	20	25.20
Vent chimney, prefab metal, U.L. listed, gas, double wall, galv. st, 4″ dia	12.000	L.F.	60.60	156	216.60
Vent chimney, gas, double wall, galv. steel, elbow 90°, 4″ dia.	2.000	Ea.	40	52	92
Vent chimney, gas, double wall, galv. steel, Tee, 4″ dia.	1.000	Ea.	25.50	34	59.50
Vent chimney, gas, double wall, galv. steel, T cap, 4″ dia.	1.000	Ea.	1.97	21	22.97
Vent chimney, gas, double wall, galv. steel, roof flashing, 4″ dia.	1.000	Ea.	6.45	26	32.45
Vent chimney, gas, double wall, galv. steel, top, 4″ dia.	1.000	Ea.	9.70	20	29.70
TOTAL			775.61	1,021.55	1,797.16

D3050 120	Unit Heaters, Gas	COST EACH MAT.	COST EACH INST.	COST EACH TOTAL
1000	For alternate unit heaters:			
1010	Space heater, suspended, gas fired, propeller fan, 20 MBH	775	1,025	1,800
1020	60 MBH	915	1,075	1,990
1030	100 MBH	1,150	1,125	2,275
1040	160 MBH	1,375	1,225	2,600
1050	200 MBH	1,650	1,375	3,025
1060	280 MBH	2,150	1,450	3,600
1070	320 MBH	2,850	1,650	4,500

D3050 130	Unit Heaters, Hydronic	COST EACH MAT.	COST EACH INST.	COST EACH TOTAL
1010	Space heater, suspended, horiz. mount, HW, prop. fan, 20 MBH	1,350	1,075	2,425
1020	60 MBH	1,750	1,225	2,975
1030	100 MBH	2,075	1,325	3,400
1040	150 MBH	2,600	1,475	4,075
1050	200 MBH	2,750	1,650	4,400
1060	300 MBH	3,025	1,950	4,975

Important: See the Reference Section for critical supporting data - Reference Numbers and City Cost Indexes

D3050 Terminal & Package Units

D3050 140	Cabinet Unit Heaters, Hydronic	COST EACH		
		MAT.	INST.	TOTAL
1010	Unit heater, cabinet type, horizontal blower, hot water, 20 MBH	1,650	1,075	2,725
1020	60 MBH	2,375	1,175	3,550
1030	100 MBH	2,600	1,300	3,900
1040	120 MBH	2,675	1,325	4,000
2700	Cut & patch to match existing construction, add, minimum	2%	3%	
2800	Maximum	5%	9%	
2900	Dust protection, add, minimum	1%	2%	
3000	Maximum	4%	11%	
3100	Equipment usage curtailment, add, minimum	1%	1%	
3200	Maximum	3%	10%	
3300	Material handling & storage limitation, add, minimum	1%	1%	
3400	Maximum	6%	7%	
3500	Protection of existing work, add, minimum	2%	2%	
3600	Maximum	5%	7%	
3700	Shift work requirements, add, minimum		5%	
3800	Maximum		30%	
3900	Temporary shoring and bracing, add, minimum	2%	5%	
4000	Maximum	5%	12%	

D3050 Cooling Generating Systems

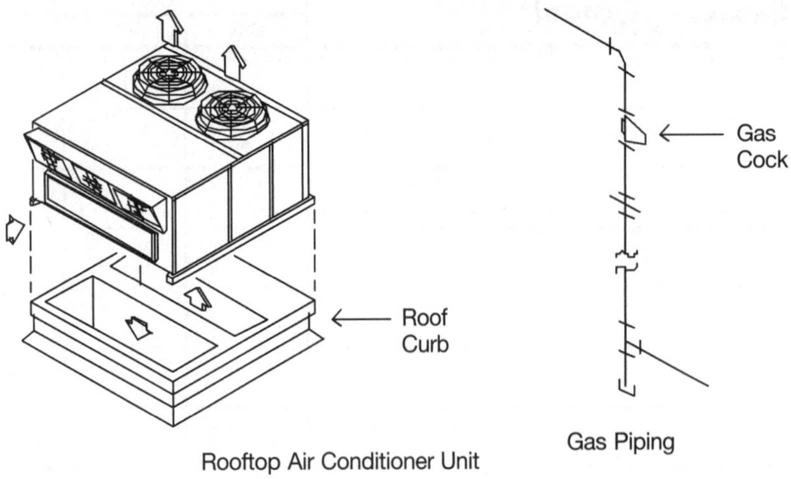

Rooftop Air Conditioner Unit

Gas Piping

This page illustrates and describes packaged air conditioners including rooftop DX unit, pipes and fittings. Lines with Systems Components and give the material and installation price on a cost each basis for the components. Prices for alternate A/C systems are on Line Items D3050 175 1010 thru 1060. Material quantities and labor costs have been adjusted for the system listed.

Factors: To adjust for job conditions other than normal working situations use Lines D3050 175 2700 thru 4000.

System Components	QUANTITY	UNIT	COST EACH MAT.	COST EACH INST.	COST EACH TOTAL
Packaged rooftop DX unit, gas, with related piping and valves.					
Roof top A/C, curb, economizer, sgl zone, elec cool, gas ht, 5 ton, 112 MBH	1.000	Ea.	4,050	1,575	5,625
Pipe, black steel, Sch 40, threaded, W/cplgs & hangers, 10′ OC, 1″ dia	20.000	L.F.	47.04	194.25	241.29
Elbow, 90°, black, straight, 3/4″ dia.	3.000	Ea.	5.19	105	110.19
Elbow, 90°, black, straight, 1″ dia.	3.000	Ea.	9	114	123
Tee, black, reducing, 1″ dia.	1.000	Ea.	5.35	61.50	66.85
Union, black with brass seat, 1″ dia.	1.000	Ea.	9.90	41	50.90
Pipe nipple, black, 3/4″ dia	2.000	Ea.	3.14	16.10	19.24
Pipe nipple, black, 1″ dia	2.000	Ea.	2.24	9.25	11.49
Cap, black, 1″ dia.	1.000	Ea.	2.37	16.35	18.72
Gas cock, brass, 1″ size	1.000	Ea.	12	26	38
Control system, pneumatic, Rooftop A/C unit	1.000	Ea.	2,100	1,750	3,850
Rooftop unit heat/cool balancing	1.000	Ea.		345	345
TOTAL			6,246.23	4,253.45	10,499.68

D3050 175	Rooftop Air Conditioner, Const. Volume	COST EACH MAT.	COST EACH INST.	COST EACH TOTAL
1000	For alternate A/C systems:			
1010	A/C, Rooftop, DX cool, gas heat, curb, economizer, fltrs, 5 Ton	6,250	4,250	10,500
1020	7-1/2 Ton	8,100	4,450	12,550
1030	12-1/2 Ton	10,900	4,850	15,750
1040	18 Ton	16,300	5,325	21,625
1050	25 Ton	23,300	6,050	29,350
1060	40 Ton	36,000	8,075	44,075
1010	A/C, Rooftop, DX cool, gas heat, curb, ecmizr, fltrs, VAV, 12-1/2 Ton	12,900	5,200	18,100
1020	18 Ton	18,700	5,700	24,400
1030	25 Ton	26,800	6,550	33,350
1040	40 Ton	43,400	8,900	52,300
1050	60 Ton	60,500	12,100	72,600
1060	80 Ton	159,000	15,200	174,200
2700	Cut & patch to match existing construction, add, minimum	2%	3%	
2800	Maximum	5%	9%	
2900	Dust protection, add, minimum	1%	2%	
3000	Maximum	4%	11%	
3100	Equipment usage curtailment, add, minimum	1%	1%	
3200	Maximum	3%	10%	
3300	Material handling & storage limitation, add, minimum	1%	1%	
3400	Maximum	6%	7%	
3500	Protection of existing work, add, minimum	2%	2%	

D3050 Cooling Generating Systems

D3050 175	Rooftop Air Conditioner, Const. Volume	COST EACH		
		MAT.	INST.	TOTAL
3600	Maximum	5%	7%	
3700	Shift work requirements, add, minimum		5%	
3800	Maximum		30%	
3900	Temporary shoring and bracing, add, minimum	2%	5%	
4000	Maximum	5%	12%	

D3050 Terminal & Package Units

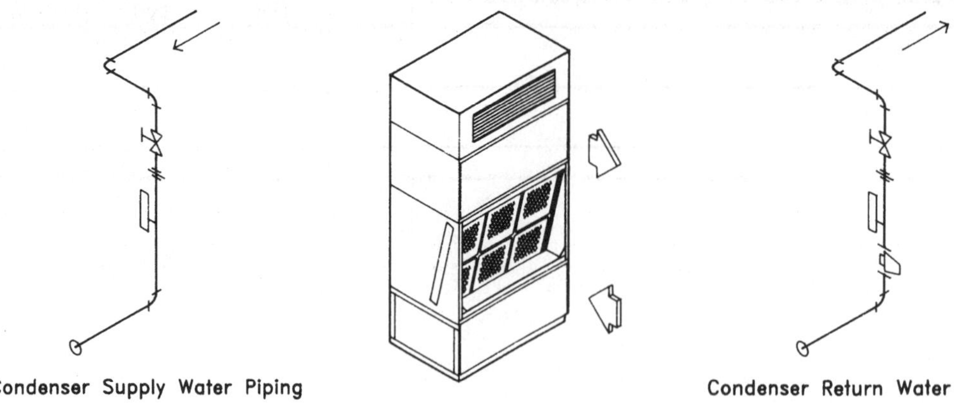

Self–Contained Air Conditioner Unit, Water Cooled

This page illustrates and describes packaged electrical air conditioning units including multizone control and related piping. Lines within Systems Components give the material and installation price on a cost each basis for the components. Prices for alternate packaged electrical A/C unit systems are on Line Items D3050 210 1010 thru D3050 225 1060. Material quantities and labor costs have been adjusted for the system listed.

Factors: To adjust for job conditions other than normal working situations use Lines D3050 225 2700 thru 4000.

System Components	QUANTITY	UNIT	COST EACH MAT.	COST EACH INST.	COST EACH TOTAL
Packaged water cooled electric air conditioning unit with related piping and valves.					
Self-contained, water cooled, elect. heat, not inc tower, 5 ton, const vol	1.000	Ea.	6,025	1,150	7,175
Pipe, black steel, Sch 40, threaded, W/cplg & hangers, 10′ OC, 1-1/4″ dia.	20.000	L.F.	58.70	213.93	272.63
Elbow, 90°, black, straight, 1-1/4″ dia.	6.000	Ea.	29.70	240	269.70
Tee, black, straight, 1-1/4″ dia.	2.000	Ea.	15.20	126	141.20
Tee, black, reducing, 1-1/4″ dia.	2.000	Ea.	18.60	126	144.60
Thermometers, stem type, 9″ case, 8″ stem, 3/4″ NPT	2.000	Ea.	260	35.20	295.20
Union, black with brass seat, 1-1/4″ dia.	2.000	Ea.	27.80	84	111.80
Pipe nipple, black, 1-1/4″ dia	3.000	Ea.	4.10	14.93	19.03
Valves, bronze, gate, N.R.S., threaded, class 150, 1-1/4″ size	2.000	Ea.	106	66	172
Circuit setter, bal valve, bronze body, threaded, 1-1/4″ pipe size	1.000	Ea.	101	33	134
Insulation, fiberglass pipe covering, 1″ wall, 1-1/4″ IPS	20.000	L.F.	20.20	80.20	100.40
Control system, pneumatic, A/C Unit with heat	1.000	Ea.	2,100	1,750	3,850
Re-heat coil balancing	1.000	Ea.		91.50	91.50
Rooftop unit heat/cool balancing	1.000	Ea.		345	345
TOTAL			8,766.30	4,355.76	13,122.06

D3050 210	AC Unit, Package, Elec. Ht., Water Cooled	COST EACH MAT.	COST EACH INST.	COST EACH TOTAL
1000	For alternate A/C systems:			
1010	A/C, Self contained, single pkg., water cooled, elect. heat, 5 Ton	8,775	4,350	13,125
1020	10 Ton	13,900	5,350	19,250
1030	20 Ton	18,600	6,900	25,500
1040	30 Ton	26,900	7,500	34,400
1050	40 Ton	32,900	8,275	41,175
1060	50 Ton	41,700	9,925	51,625

D3050 215	AC Unit, Package, Elec. Ht., Water Cooled, VAV	COST EACH MAT.	COST EACH INST.	COST EACH TOTAL
1010	A/C, Self contained, single pkg., water cooled, elect. ht, VAV, 10 Ton	15,900	5,350	21,250
1020	20 Ton	22,100	6,900	29,000
1030	30 Ton	31,900	7,500	39,400
1040	40 Ton	40,300	8,275	48,575
1050	50 Ton	49,800	9,925	59,725
1060	60 Ton	57,000	12,100	69,100

D3050 Terminal & Package Units

D3050 220	AC Unit, Package, Hot Water Coil, Water Cooled	COST EACH		
		MAT.	INST.	TOTAL
1010	A/C, Self contn'd, single pkg., water cool, H/W ht, const. vol, 5 Ton	7,325	6,025	13,350
1020	10 Ton	11,100	7,200	18,300
1030	20 Ton	29,300	9,075	38,375
1040	30 Ton	38,300	11,600	49,900
1050	40 Ton	47,400	14,100	61,500
1060	50 Ton	58,500	17,300	75,800

D3050 225	AC Unit, Package, HW Coil, Water Cooled, VAV	COST EACH		
		MAT.	INST.	TOTAL
1010	A/C, Self contn'd, single pkg., water cool, H/W ht, VAV, 10 Ton	13,100	6,775	19,875
1020	20 Ton	18,600	8,850	27,450
1030	30 Ton	25,700	11,200	36,900
1040	40 Ton	34,100	12,200	46,300
1050	50 Ton	40,500	14,300	54,800
1060	60 Ton	48,800	15,600	64,400
2700	Cut & patch to match existing construction, add, minimum	2%	3%	
2800	Maximum	5%	9%	
2900	Dust protection, add, minimum	1%	2%	
3000	Maximum	4%	11%	
3100	Equipment usage curtailment, add, minimum	1%	1%	
3200	Maximum	3%	10%	
3300	Material handling & storage limitation, add, minimum	1%	1%	
3400	Maximum	6%	7%	
3500	Protection of existing work, add, minimum	2%	2%	
3600	Maximum	5%	7%	
3700	Shift work requirements, add, minimum		5%	
3800	Maximum		30%	
3900	Temporary shoring and bracing, add, minimum	2%	5%	
4000	Maximum	5%	12%	

D3050 Terminal & Package Units

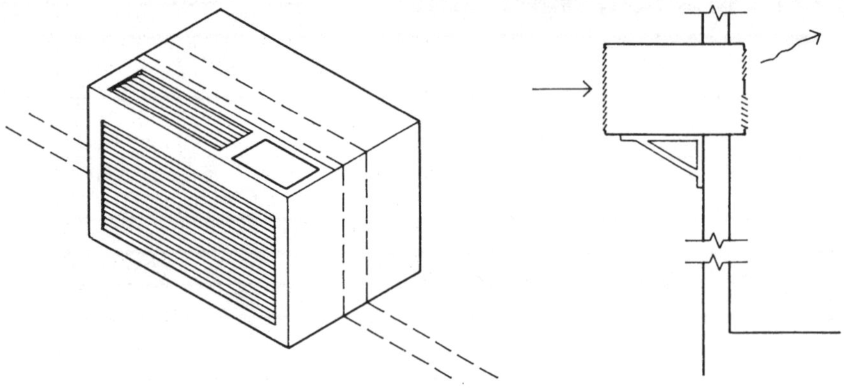

Thru–Wall Heat Pump   Thru–Wall Air Conditioner

This page illustrates and describes thru-wall air conditioning and heat pump units. Lines within Systems Components give the material and installation price on a cost each basis for the components. Prices for alternate thru-wall unit systems are on Line Items D3050 255 1010 thru 1050. Material quantities and labor costs have been adjusted for the system listed.

Factors: To adjust for job conditions other than normal working conditions use Lines D3050 255 2700 thru 4000.

System Components	QUANTITY	UNIT	COST EACH MAT.	COST EACH INST.	COST EACH TOTAL
Electric thru-wall air conditioning unit.					
A/C unit, thru-wall, electric heat., cabinet, louver, 1/2 ton	1.000	Ea.	1,050	148	1,198
TOTAL			1,050	148	1,198

D3050 255	Thru-Wall A/C Unit	COST EACH MAT.	COST EACH INST.	COST EACH TOTAL
1000	For alternate A/C units:			
1010	A/C Unit, thru-the-wall, sup. elect. heat, cabinet, louver, 1/2 Ton	1,050	148	1,198
1020	3/4 Ton	1,100	178	1,278
1030	1 Ton	1,225	222	1,447
1040	1-1/2 Ton	1,500	370	1,870
1050	2 Ton	1,500	465	1,965
1010	Heat pump, thru-the-wall, cabinet, louver, 1/2 Ton	1,400	111	1,511
1020	3/4 Ton	1,450	148	1,598
1030	1 Ton	1,550	222	1,772
1040	Sup. elect. heat, 1-1/2 Ton	2,025	575	2,600
1050	2 Ton	2,125	590	2,715
2700	Cut & patch to match existing construction, add, minimum	2%	3%	
2800	Maximum	5%	9%	
2900	Dust protection, add, minimum	1%	2%	
3000	Maximum	4%	11%	
3100	Equipment usage curtailment, add, minimum	1%	1%	
3200	Maximum	3%	10%	
3300	Material handling & storage limitation, add, minimum	1%	1%	
3400	Maximum	6%	7%	
3500	Protection of existing work, add, minimum	2%	2%	
3600	Maximum	5%	7%	
3700	Shift work requirements, add, minimum		5%	
3800	Maximum		30%	
3900	Temporary shoring and bracing, add, minimum	2%	5%	
4000	Maximum	5%	12%	

D4010 Sprinklers

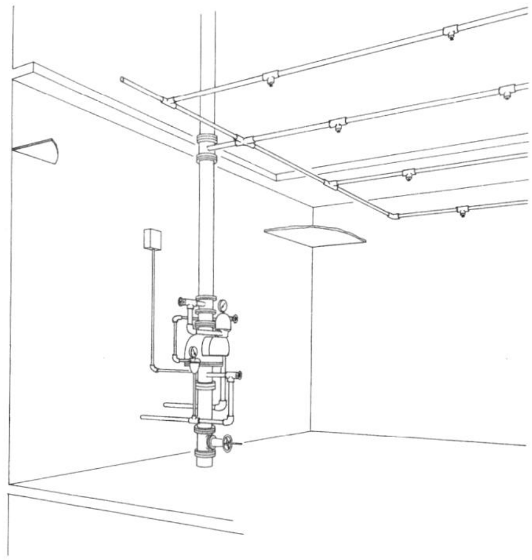

This page illustrates and describes a dry type fire sprinkler system. Lines within System Components give the unit cost of a system for a 2,000 square foot building. Lines D4010 320 1700 thru 2400 give the square foot costs for alternate systems. Both material quantities and labor costs have been adjusted for the system listed.

Factors: To adjust for conditions other than normal working conditions, use Lines D4010 320 2900 thru 4000.

System Components	QUANTITY	UNIT	COST EACH MAT.	COST EACH INST.	COST EACH TOTAL
Pre-action fire sprinkler system, ordinary hazard, open area to 2000 S.F. On one floor.					
Air compressor	1.000	Ea.	790	380	1,170
4" OS & Y valve	1.000	Ea.	410	295	705
Pipe riser 4" diameter	10.000	L.F.	113.50	263.10	376.60
Sprinkler head supply piping, 1"	163.000	L.F.	168	693.75	861.75
Sprinkler head supply piping, 1-1/4"	163.000	L.F.	150.15	547.25	697.40
Sprinkler head supply piping, 2-1/2"	163.000	L.F.	260.30	672.60	932.90
Pipe fittings, 1"	8.000	Ea.	33.60	492	525.60
Pipe fittings, 1-1/4"	8.000	Ea.	60.80	504	564.80
Pipe fittings, 2-1/2"	4.000	Ea.	140	394	534
Pipe fittings, 4"	2.000	Ea.	248	442	690
Detectors	2.000	Ea.	700	61	761
Sprinkler heads	16.000	Ea.	88	488	576
Fire department connection	1.000	Ea.	320	177	497
TOTAL			3,482.35	5,409.70	8,892.05

D4010 320	Fire Sprinkler Systems, Dry	COST PER S.F. MAT.	COST PER S.F. INST.	COST PER S.F. TOTAL
1700	Ordinary hazard, one floor, area to 2000 S.F./floor	1.75	2.71	4.46
1800	For each additional floor, add per floor	.63	2.25	2.88
1900	Area to 3200 S.F./ floor	1.35	2.71	4.06
2000	For each additional floor, add per floor	.65	2.42	3.07
2100	Area to 5000 S.F./ floor	1.42	2.73	4.15
2200	For each additional floor, add per floor	.84	2.53	3.37
2300	Area to 8000 S.F./ floor	1.06	2.30	3.36
2400	For each additional floor, add per floor	.65	2.17	2.82
2500				
2600				
2700				
2800				
2900	Cut & patch to match existing construction, add, minimum	2%	3%	
3000	Maximum	5%	9%	
3100	Dust protection, add, minimum	1%	2%	
3200	Maximum	4%	11%	
3300	Equipment usage curtailment, add, minimum	1%	1%	
3400	Maximum	3%	10%	

D4010 Sprinklers

D4010 320	Fire Sprinkler Systems, Dry	COST PER S.F.		
		MAT.	INST.	TOTAL
3500	Material handling & storage limitation, add, minimum	1%	1%	
3600	Maximum	6%	7%	
3700	Protection of existing work, add, minimum	2%	2%	
3800	Maximum	5%	7%	
3900	Shift work requirements, add, minimum		5%	
4000	Maximum		30%	

Important: See the Reference Section for critical supporting data - Reference Numbers and City Cost Indexes

D4010 Sprinklers

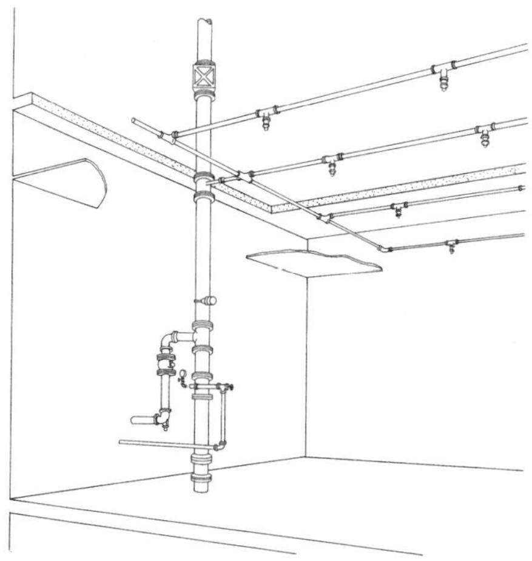

This page illustrates and describes a wet type fire sprinkler system. Lines within System Components give the unit cost of a system for a 2,000 square foot building. Lines D4010 420 1900 thru 2600 give the square foot costs for alternate systems. Both material quantities and labor costs have been adjusted for the system listed.

Factors: To adjust for conditions other than normal working conditions, use Lines D4010 420 2900 thru 4000.

System Components	QUANTITY	UNIT	COST EACH		
			MAT.	INST.	TOTAL
Wet pipe fire sprinkler system, ordinary hazard, open area to 2000 S.F.					
On one floor.					
4″ OS & Y valve	1.000	Ea.	410	295	705
Wet pipe alarm valve	1.000	Ea.	930	440	1,370
Water motor alarm	1.000	Ea.	203	123	326
3″ check valve	1.000	Ea.	176	295	471
Pipe riser, 4″ diameter	10.000	L.F.	113.50	263.10	376.60
Water gauges and trim	1.000	Set	1,400	885	2,285
Electric fire horn	1.000	Ea.	40	72.50	112.50
Sprinkler head supply piping, 1″ diameter	168.000	L.F.	168	693.75	861.75
Sprinkler head supply piping, 1-1/2″ diameter	168.000	L.F.	171.05	607.75	778.80
Sprinkler head supply piping, 2-1/2″ diameter	168.000	L.F.	260.30	672.60	932.90
Pipe fittings, 1″	8.000	Ea.	33.60	492	525.60
Pipe fittings, 1-1/2″	8.000	Ea.	75.60	544	619.60
Pipe fittings, 2-1/2″	4.000	Ea.	140	394	534
Pipe fittings, 4″	2.000	Ea.	248	442	690
Sprinkler heads	16.000	Ea.	88	488	576
Fire department connection	1.000	Ea.	320	177	497
TOTAL			4,777.05	6,884.70	11,661.75

D4010 420	Fire Sprinkler Systems, Wet	COST PER S.F.		
		MAT.	INST.	TOTAL
2100	Ordinary hazard, one floor, area to 2000 S.F./floor	2.40	3.45	5.85
2000	For each additional floor, add per floor	.65	2.30	2.95
2100	Area to 3200 S.F./floor	1.74	3.14	4.88
2200	For each additional floor, add per floor	.65	2.42	3.07
2300	Area to 5000 S.F./ floor	1.54	2.99	4.53
2400	For each additional floor, add per floor	.84	2.53	3.37
2500	Area to 8000 S.F./ floor	1.10	2.47	3.57
2600	For each additional floor, add per floor	.70	2.30	3.00
2900	Cut & patch, to match existing construction, add, minimum	2%	3.%	
3000	Maximin	5%	9%	
3100	Dust protection, add, minimum	1%	2%	
3200	Maximum	4%	11%	
3300	Equipment usage curtailment, add, minimum	1%	1%	
3400	Maximum	3%	10%	
3500	Material handling & storage limitation, add, minimum	1%	1%	

D5010 Electrical Service/Distribution

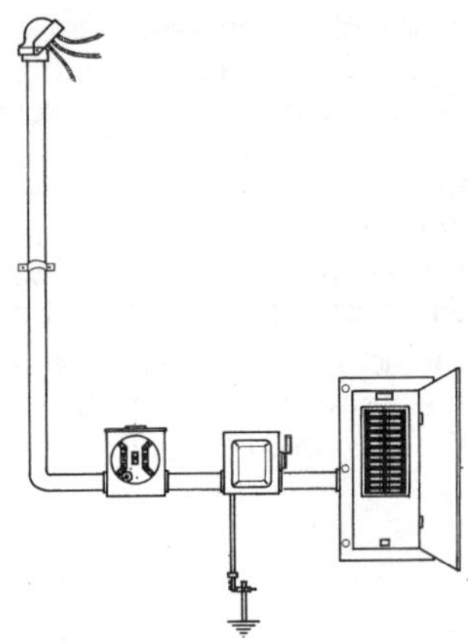

This page illustrates and describes commercial service systems including a meter socket, service head and cable, entrance switch, steel conduit, copper wire, panel board, ground rod, wire, and conduit. Lines within System Components give the unit price and total price on a cost each basis for this system. Prices for alternate commercial service systems are on Line Items D5010 210 1500 thru 1700. Both material quantities and labor costs have been adjusted for the system listed.

Factors: To adjust for job conditions other than normal working situations use Lines D5010 210 2900 thru 4000.

Example: You are to install the system with maximum equipment usage curtailment. Go to Line D5010 210 3400 and apply these percentages to the appropriate MAT. and INST. costs.

System Components	QUANTITY	UNIT	COST EACH		
			MAT.	INST.	TOTAL
Commercial electric service including service breakers, metering 120/208 Volt, 3 phase, 4 wire, feeder, and panel board.					
100 Amp Service					
Meter socket	1.000	Ea.	35	151	186
Service head	.200	C.L.F.	38	97	135
Service entrance switch	1.000	Ea.	238	255	493
Rigid steel conduit	20.000	L.F.	65.60	149	214.60
600 volt copper wire #3	1.000	C.L.F.	41.50	97	138.50
Panel board, 24 circuits, 20 Amp breakers	1.000	Ea.	725	805	1,530
Ground rod	1.000	Ea.	14.40	88	102.40
TOTAL			1,157.50	1,642	2,799.50

D5010 210	Commercial Service - 3 Phase	COST EACH		
		MAT.	INST.	TOTAL
1400	For alternate size services:			
1500	120/208 Volt, 3 phase, 4 wire service, 60 Amp	790	1,275	2,065
1600	200 Amp	2,200	2,900	5,100
1700	400 Amp	4,475	3,625	8,100
1800				
1900				
2000				
2100				
2200				
2300				
2400				
2500				
2900	Cut & patch to match existing construction, add, minimum	2%	3%	
3000	Maximum	5%	9%	
3100	Dust protection, add, minimum	1%	2%	
3200	Maximum	4%	11%	
3300	Equipment usage curtailment, add, minimum	1%	1%	
3400	Maximum	3%	10%	
3500	Material handling & storage limitation, add, minimum	1%	1%	
3600	Maximum	6%	7%	
3700	Protection of existing work, add, minimum	2%	2%	
3800	Maximum	5%	7%	
3900	Shift work requirements, add, minimum		5%	
4000	Maximum		30%	

D5010 Electrical Service/Distribution

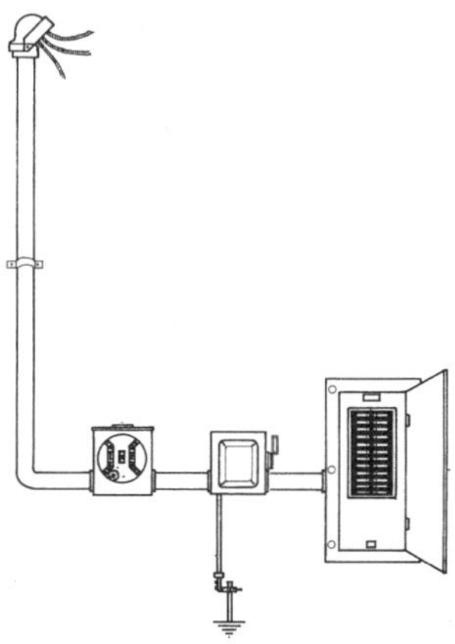

This page illustrates and describes a residential, single phase system including a weather cap, service entrance cable, meter socket, entrance switch, ground rod, ground cable, EMT, and panelboard. Lines with System Components give the unit price and total price on a cost each basis for this system. Prices for an alternate residential, single phase system are also given. Both material quantities and labor costs have been adjusted for the system listed.

Factors: To adjust for job conditions other than normal working situations use Lines D5010 220 2900 thru 4000.

Example: You are to install the system with a minimum equipment usage curtailment. Go to Line D5010 220 3300 and apply these percentages to the appropriate MAT. and INST. costs.

System Components	QUANTITY	UNIT	COST EACH		
			MAT.	INST.	TOTAL
100 Amp Service, single phase					
Weathercap	1.000	Ea.	9.15	40.50	49.65
Service entrance cable	.200	C.L.F.	38	97	135
Meter socket	1.000	Ea.	35	151	186
Entrance disconnect switch	1.000	Ea.	238	255	493
Ground rod, with clamp	1.000	Ea.	18.60	101	119.60
Ground cable	.100	C.L.F.	14.60	30.50	45.10
Panelboard, 12 circuit	1.000	Ea.	182	405	587
TOTAL			535.35	1,080	1,615.35
200 Amp Service , single phase					
Weathercap	1.000	Ea.	20	60.50	80.50
Service entrance cable	.200	C.L.F.	97	138	235
Meter socket	1.000	Ea.	52	255	307
Entrance disconnect switch	1.000	Ea.	515	375	890
Ground rod, with clamp	1.000	Ea.	33	110	143
Ground cable	.100	C.L.F.	9.70	16.70	26.40
3/4" EMT	10.000	L.F.	6.60	37.30	43.90
Panelboard, 24 circuit	1.000	Ea.	490	620	1,110
TOTAL			1,223.30	1,612.50	2,835.80

D5010 220	Residential Service - Single Phase	COST EACH		
		MAT.	INST.	TOTAL
2800				
2900	Cut & patch to match existing construction, add, minimum	2%	3%	
3000	Maximum	5%	9%	
3100	Dust protection, add, minimum	1%	2%	
3200	Maximum	4%	11%	
3300	Equipment usage curtailment, add, minimum	1%	1%	
3400	Maximum	3%	10%	
3500	Material handling & storage limitation, add, minimum	1%	1%	
3600	Maximum	6%	7%	
3700	Protection of existing work, add, minimum	2%	2%	
3800	Maximum	5%	7%	
3900	Shift work requirements, add, minimum		5%	
4000	Maximum		30%	

D5020 Lighting and Branch Wiring

D5020 180	Selective Price Sheet	COST EACH		
		MAT.	INST.	TOTAL
0100	Using non-metallic sheathed, cable, air conditioning receptacle	13.40	48.50	61.90
0200	Disposal wiring	10.55	54	64.55
0300	Dryer circuit	31.50	88	119.50
0400	Duplex receptacle	13.40	37.50	50.90
0500	Fire alarm or smoke detector	66	48.50	114.50
0600	Furnace circuit & switch	19.80	80.50	100.30
0700	Ground fault receptacle	42	60.50	102.50
0800	Heater circuit	12.05	60.50	72.55
0900	Lighting wiring	12.75	30.50	43.25
1000	Range circuit	65.50	121	186.50
1100	Switches single pole	13.10	30.50	43.60
1200	3-way	17.15	40.50	57.65
1300	Water heater circuit	19.40	97	116.40
1400	Weatherproof receptacle	104	80.50	184.50
1500	Using BX cable, air conditioning receptacle	25	58.50	83.50
1600	Disposal wiring	21.50	64.50	86
1700	Dryer circuit	46	105	151
1800	Duplex receptacle	25	45	70
1900	Fire alarm or smoke detector	66	48.50	114.50
2000	Furnace circuit & switch	32	97	129
2100	Ground fault receptacle	54.50	73.50	128
2200	Heater circuit	21	73.50	94.50
2300	Lighting wiring	21.50	36.50	58
2400	Range circuit	94.50	147	241.50
2500	Switches, single pole	24.50	36.50	61
2600	3-way	28.50	48.50	77
2700	Water heater circuit	37	115	152
2800	Weatherproof receptacle	113	97	210
2900	Using EMT conduit, air conditioning receptacle	26	72.50	98.50
3000	Disposal wiring	23.50	80.50	104
3100	Dryer circuit	37	131	168
3200	Duplex receptacle	26	55.50	81.50
3300	Fire alarm or smoke detector	75.50	72.50	148
3400	Furnace circuit & switch	32.50	121	153.50
3500	Ground fault receptacle	66	89.50	155.50
3600	Heater circuit	22.50	89.50	112
3700	Lighting wiring	21.50	45.50	67
3800	Range circuit	64.50	179	243.50
3900	Switches, single pole	25.50	45.50	71
4000	3-way	24	60.50	84.50
4100	Water heater circuit	28.50	142	170.50
4200	Weatherproof receptacle	114	121	235
4300	Using aluminum conduit, air conditioning receptacle	29.50	97	126.50
4400	Disposal wiring	27	108	135
4500	Dryer circuit	48	173	221
4600	Duplex receptacle	28.50	74.50	103
4700	Fire alarm or smoke detector	83.50	97	180.50
4800	Furnace circuit & switch	39.50	161	200.50
4900	Ground fault receptacle	63.50	121	184.50
5000	Heater circuit	28	121	149
5100	Lighting wiring	30.50	60.50	91
5200	Range circuit	79	242	321
5300	Switches, single pole	37.50	60.50	98
5400	3-way	36	80.50	116.50
5500	Water heater circuit	39	194	233
5600	Weatherproof receptacle	129	161	290
5700	Using galvanized steel conduit	31.50	103	134.50
5800	Disposal wiring	26.50	115	141.50

D5020 Lighting and Branch Wiring

D5020 180	Selective Price Sheet	COST EACH		
		MAT.	INST.	TOTAL
5900	Dryer circuit	54.50	186	240.50
6000	Duplex receptacle	31.50	79.50	111
6100	Fire alarm or smoke detector	83.50	103	186.50
6200	Furnace circuit & switch	39.50	173	212.50
6300	Ground fault receptacle	64.50	127	191.50
6400	Heater circuit	29.50	127	156.50
6500	Lighting wiring	31.50	64.50	96
6600	Range circuit	80	255	335
6700	Switches, single pole	38.50	64.50	103
6800	3-way	41	83.50	124.50
6900	Water heater circuit	40	202	242
7000	Weatherproof receptacle	127	173	300
7100				
7200				
7300				
7400				
7500				
7600				
7700				
7800				
7900				
8000				
8100				
8200				
8300				
8400				

D5020 Lighting and Branch Wiring

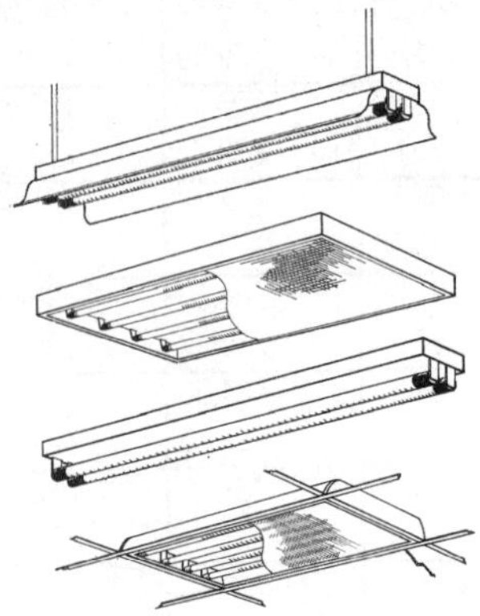

This page illustrates and describes fluorescent lighting systems including a fixture, lamp, outlet box and wiring. Lines within System Components give the unit price and total price on a cost each basis for this system. Prices for alternate fluorescent lighting systems are on Line Items D5020 248 1300 thru 1500. Both material quantities and labor costs have been adjusted for the system listed.

Factors: To adjust for job conditions other than normal working situations use Lines D5020 248 2900 thru 4000.

Example: You are to install the system during evening hours. Go to Line D5020 248 3900 and apply this percentage to the appropriate INST. cost.

System Components	QUANTITY	UNIT	COST EACH MAT.	COST EACH INST.	COST EACH TOTAL
Fluorescent lighting, including fixture, lamp, outlet box and wiring.					
Recessed lighting fixture, on suspended system	1.000	Ea.	60	103	163
Outlet box	1.000	Ea.	2.31	27	29.31
#12 wire	.660	C.L.F.	3.02	29.04	32.06
Conduit, EMT, 1/2″ conduit	20.000	L.F.	8.60	57	65.60
TOTAL			73.93	216.04	289.97

D5020 248	Lighting, Fluorescent	COST EACH MAT.	COST EACH INST.	COST EACH TOTAL
1200	For alternate lighting fixtures:			
1300	Surface mounted, 2′ x 4′, acrylic prismatic diffuser	113	227	340
1400	Strip fixture, 8′ long, two 8′ lamps	68	214	282
1500	Pendant mounted, industrial, 8′ long, with reflectors	105	246	351
1600				
1700				
1800				
1900				
2000				
2100				
2200				
2300				
2900	Cut & patch to match existing construction, add, minimum	2%	3%	
3000	Maximum	5%	9%	
3100	Dust protection, add, minimum	1%	2%	
3200	Maximum	4%	11%	
3300	Equipment usage curtailment, add, minimum	1%	1%	
3400	Maximum	3%	10%	
3500	Material handling & storage limitation, add, minimum	1%	1%	
3600	Maximum	6%	7%	
3700	Protection of existing work, add, minimum	2%	2%	
3800	Maximum	5%	7%	
3900	Shift work requirements, add, minimum		5%	
4000	Maximum		30%	

Important: See the Reference Section for critical supporting data - Reference Numbers and City Cost Indexes

D5020 Lighting and Branch Wiring

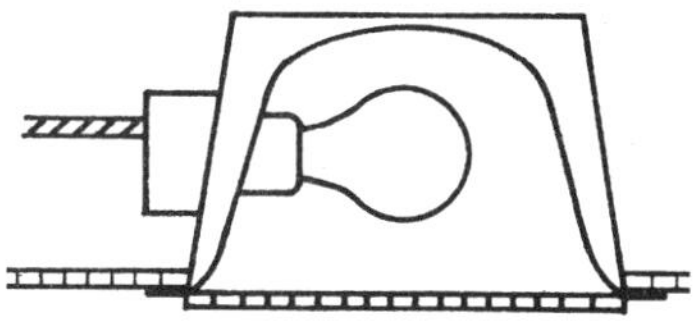

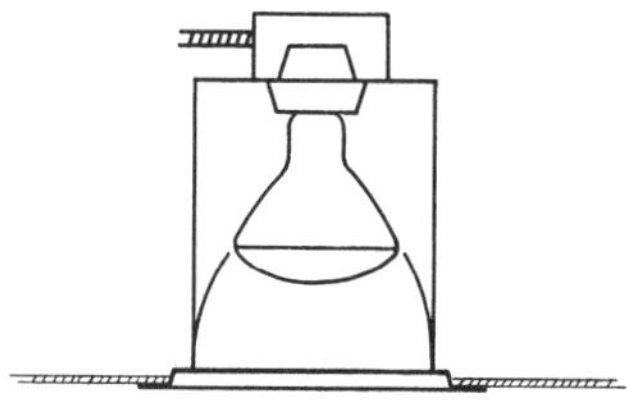

This page illustrates and describes incandescent lighting systems including a fixture, lamp, outlet box, conduit and wiring. Lines within System Components give the unit price and total price on a cost each basis for this system. Prices for an alternate incandescent lighting system are also given. Both material quantities and labor costs have been adjusted for the system listed.

Factors: To adjust for conditions other than normal working situations use Lines D5020 250 2900 thru 4000.

Example: You are to install the system and cut and match existing construction. Go to Line D5020 250 3000 and apply this percentage to the appropriate INST. costs.

System Components	QUANTITY	UNIT	COST EACH		
			MAT.	INST.	TOTAL
Incandescent light fixture, including lamp, outlet box, conduit and wiring.					
Recessed wide reflector with flat glass lens	1.000	Ea.	81.50	72.50	154
Outlet box	1.000	Ea.	2.31	27	29.31
Armored cable, 3 wire	.200	C.L.F.	15.30	44	59.30
TOTAL			99.11	143.50	242.61
Recessed flood light fixture, including lamp, outlet box, conduit & wire					
Recessed, R-40 flood lamp with refractor	1.000	Ea.	89	60.50	149.50
150 watt R-40 flood lamp	.010	Ea.	8.35	3.75	12.10
Outlet box	1.000	Ea.	2.31	27	29.31
Outlet box cover	1.000	Ea.	.77	7.55	8.32
Romex, 12-2 with ground	.200	C.L.F.	3.80	44	47.80
Conduit, 1/2" EMT	20.000	L.F.	8.60	57	65.60
TOTAL			112.83	199.80	312.63

D5020 250	Lighting, Incandescent	COST EACH		
		MAT.	INST.	TOTAL
2900	Cut & patch to match existing construction, add, minimum	2%	3%	
3000	Maximum	5%	9%	
3100	Dust protection, add, minimum	1%	2%	
3200	Maximum	4%	11%	
3300	Equipment usage curtailment, add, minimum	1%	1%	
3400	Maximum	3%	10%	
3500	Material handling & storage limitation, add, minimum	1%	1%	
3600	Maximum	6%	7%	
3700	Protection of existing work, add, minimum	2%	2%	
3800	Maximum	5%	7%	
3900	Shift work requirements, add, minimum		5%	
4000	Maximum		30%	

D5020 Lighting and Branch Wiring

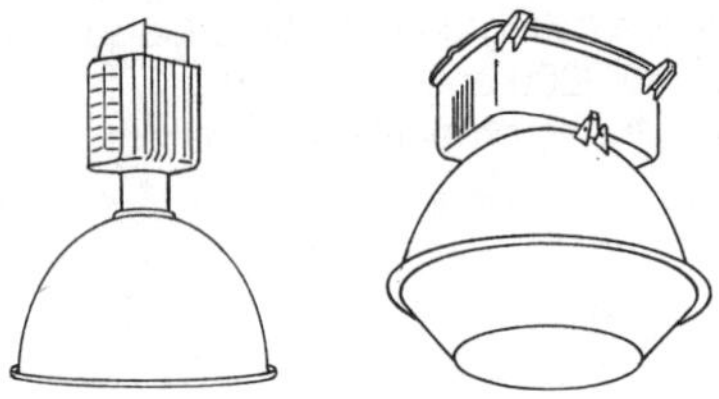

This page illustrates and describes high intensity lighting systems including a lamp, EMT conduit, EMT "T" fitting with cover, and wire. Lines within System Components give the unit price and total price on a cost each basis for this system. Prices for alternate high intensity lighting systems are on Line Items D5020 252 1100 thru 1800. Both material quantities and labor costs have been adjusted for the system listed.

Factors: To adjust for job conditions other than normal working situations use Lines D5020 252 2900 thru 4000.

Example: You are to install the system and protect existing construction. Go to Line D5020 252 3700 and apply these percentages to the appropriate MAT. and INST. costs.

System Components	QUANTITY	UNIT	COST EACH		
			MAT.	INST.	TOTAL
High intensity lighting system consisting of 400 watt metal halide fixture					
And lamp with 1/2″ EMT conduit and fittings using #12 wire, high bay.		Ea.			
Single unit, 400 watt	1.000	Ea.	350	211	561
Electric metallic tubing (EMT), 1/2″ diameter	30.000	Ea.	12.90	85.50	98.40
1/2″ EMT "T" fitting with cover	1.000	Ea.	13.65	30.50	44.15
Wire, 600 volt, type THWN-THHN, copper, solid, #12	.600	Ea.	2.74	26.40	29.14
TOTAL			379.29	353.40	732.69

D5020 252	Lighting, High Intensity	COST EACH		
		MAT.	INST.	TOTAL
1000	For alternate high intensity systems:			
1200	High bay: 400 watt, high pressure sodium and lamp	355	355	710
1400	1000 watt, metal halide	535	385	920
1500	High pressure sodium	500	385	885
1700	Low bay: 250 watt, metal halide	370	293	663
1800	150 watt, high pressure sodium	310	293	603
1900				
2000				
2100				
2200				
2300				
2400				
2500				
2600				
2700				
2900	Cut & patch to match existing construction, add, minimum	2%	3%	
3000	Maximum	5%	9%	
3100	Dust protection, add, minimum	1%	2%	
3200	Maximum	4%	11%	
3300	Equipment usage curtailment, add, minimum	1%	1%	
3400	Maximum	3%	10%	
3500	Material handling & storage limitation, add, minimum	1%	1%	
3600	Maximum	6%	7%	
3700	Protection of existing work, add, minimum	2%	2%	
3800	Maximum	5%	7%	
3900	Shift work requirements, add, minimum		5%	
4000	Maximum		30%	

D5090 Other Electrical Systems

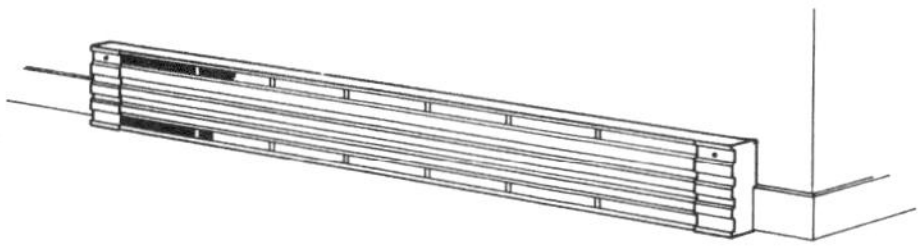

This page illustrates and describes baseboard heat systems including a thermostat, outlet box, breaker, and feed. Lines within System Components give the unit price and total price on a cost each basis for this system. Prices for alternate baseboard heat systems are on Line Items D5090 530 1500 thru 1800. Both material quantities and labor costs have been adjusted for the system listed.

Factors: To adjust for job conditions other than normal working situations use Lines D5090 530 2900 thru 4000.

Example: You are to install the system during evenings and weekends only. Go to Line D5090 530 4000 and apply this percentage to the appropriate INST. cost.

System Components	QUANTITY	UNIT	COST EACH MAT.	COST EACH INST.	COST EACH TOTAL
Baseboard heat including thermostat, outlet box, breaker and feed.					
Electric baseboard heater, 4′ long	1.000	Ea.	45	72.50	117.50
Thermostat, integral	1.000	Ea.	19.60	30.50	50.10
Romex, 12-3 with ground	.400	C.L.F.	13.60	96.80	110.40
Panel board breaker, 20 Amp	1.000	Ea.	25.50	60.50	86
Outlet box	1.000	Ea.	2.31	27	29.31
TOTAL			106.01	287.30	393.31

D5090 530	Heat, Baseboard	COST EACH MAT.	COST EACH INST.	COST EACH TOTAL
1400	For alternate baseboard heating systems:			
1500	Electric baseboard, 2′ long	93	275	368
1600	6′ long	121	310	431
1700	8′ long	136	335	471
1800	10′ long	154	360	514
1900				
2000				
2100				
2200				
2300				
2400				
2500				
2600				
2700				
2900	Cut & patch to match existing construction, add, minimum	2%	3%	
3000	Maximum	5%	9%	
3100	Dust protection, add, minimum	1%	2%	
3200	Maximum	4%	11%	
3300	Equipment usage curtailment, add, minimum	1%	1%	
3400	Maximum	3%	10%	
3500	Material handling & storage limitation, add, minimum	1%	1%	
3600	Maximum	6%	7%	
3700	Protection of existing work, add, minimum	2%	2%	
3800	Maximum	5%	7%	
3900	Shift work requirements, add, minimum		5%	
4000	Maximum		30%	

E EQUIPMENT & FURNISHINGS

E1090 Other Equipment

This page illustrates and describes kitchen systems including top and bottom cabinets, custom laminated plastic top, single bowl sink, and appliances. Lines within System Components give the unit price and total price on a cost each basis for this system. Prices for alternate kitchen systems are on Line Items E1090 310 1500 and 1600. Both material quantities and labor costs have been adjusted for the system listed.

Factors: To adjust for job conditions other than normal working situations use Lines E1090 310 2900 thru 4000.

Example: You are to install the system and protect the work area from dust. Go to Line E1090 310 3200 and apply these percentages to the appropriate MAT. and INST. costs.

System Components	QUANTITY	UNIT	COST EACH MAT.	COST EACH INST.	COST EACH TOTAL
Kitchen cabinets including wall and base cabinets, custom laminated Plastic top,sink & appliances,no plumbing or electrical rough-in included.					
Prefinished wood cabinets, average quality, wall and base	20.000	L.F.	1,940	580	2,520
Custom laminated plastic counter top	20.000	L.F.	183	292	475
Stainless steel sink, 22" x 25"	1.000	Ea.	405	158	563
Faucet, top mount	1.000	Ea.	56.50	49	105.50
Dishwasher, built-in	1.000	Ea.	270	244	514
Compactor, built-in	1.000	Ea.	455	87.50	542.50
Range hood, vented, 30" wide	1.000	Ea.	41	185	226
TOTAL			3,350.50	1,595.50	4,946

E1090 310	Kitchens	COST EACH MAT.	COST EACH INST.	COST EACH TOTAL
1400	For alternate kitchen systems:			
1500	Prefinished wood cabinets, high quality	6,675	1,825	8,500
1600	Custom cabinets, built in place, high quality	8,600	2,125	10,725
1700				
1800				
1900				
2000				
2100				
2200	NOTE: No plumbing or electric rough-ins are included in the above			
2300	Prices, for plumbing see Division 15, for electric see Division 16.			
2400				
2500				
2600				
2700				
2900	Cut & patch to match existing construction, add, minimum	2%	3%	
3000	Maximum	5%	9%	
3100	Dust protection, add, minimum	1%	2%	
3200	Maximum	4%	11%	
3300	Equipment usage curtailment, add, minimum	1%	1%	
3400	Maximum	3%	10%	
3500	Material handling & storage limitation, add, minimum	1%	1%	
3600	Maximum	6%	7%	
3700	Protection of existing work, add, minimum	2%	2%	
3800	Maximum	5%	7%	
3900	Shift work requirements, add, minimum		5%	
4000	Maximum		30%	

Important: See the Reference Section for critical supporting data - Reference Numbers and City Cost Indexes

E1090 Other Equipment

E1090 315 Selective Price Sheet

		COST EACH		
		MAT.	INST.	TOTAL
0100	Cabinets standard wood, base, one drawer one door, 12″ wide	139	35.50	174.50
0200	15″ wide	190	36.50	226.50
0300	18″ wide	207	37.50	244.50
0400	21″ wide	216	38.50	254.50
0500	24″ wide	248	39.50	287.50
0600				
0700	Two drawers two doors, 27″ wide	275	40	315
0800	30″ wide	295	41	336
0900	33″ wide	305	42	347
1000	36″ wide	315	43	358
1100	42″ wide	340	44	384
1200	48″ wide	360	46.50	406.50
1300	Drawer base (4 drawers), 12″ wide	330	35.50	365.50
1400	15″ wide	254	36.50	290.50
1500	18″ wide	283	37.50	320.50
1600	24″ wide	310	39.50	349.50
1700	Sink or range base, 30″ wide	243	41	284
1800	33″ wide	261	42	303
1900	36″ wide	273	43	316
2000	42″ wide	292	44	336
2100	Corner base, 36″ wide	400	48.50	448.50
2200	Lazy susan with revolving door	390	53	443
2300	Cabinets standard wood, wall two doors, 12″ high, 30″ wide	153	35.50	188.50
2400	36″ wide	178	36.50	214.50
2500	15″ high, 30″ high	160	36.50	196.50
2600	36″ wide	181	38.50	219.50
2700	24″ high, 30″ wide	199	37.50	236.50
2800	36″ wide	220	38.50	258.50
2900	30″ high, 30″ wide	224	45.50	269.50
3000	36″ wide	256	46.50	302.50
3100	42″ wide	279	47.50	326.50
3200	48″ wide	315	47.50	362.50
3300	One door, 30″ high, 12″ wide	134	40	174
3400	15″ wide	152	41	193
3500	18″ wide	166	42	208
3600	24″ wide	187	43	230
3700	Corner, 30″ high, 24″ wide	153	48.50	201.50
3800	36″ wide	198	53	251
3900	Broom, 84″ high, 24″ deep, 18″ wide	410	87.50	497.50
4000	Oven, 84″ high, 24″ deep, 27″ wide	595	109	704
4100	Valance board, 4′ long	34	8.85	42.85
4200	6″ long	51.50	13.25	64.75
4300	Counter tops, laminated plastic, stock 25″ wide w/backsplash, min.	9.15	14.60	23.75
4400	Maximum	17.45	17.50	34.95
4500	Custom, 7/8″ thick, no splasn	18.70	14.60	33.30
4600	Cove splash	24.50	14.60	39.10
4700	1-1/4″ thick, no splash	22	15.65	37.65
4800	Square splash	27.50	15.65	43.15
4900	Post formed	9.70	14.60	24.30
5000				
5100	Maple laminated 1-1/2″ thick, no splash	57.50	15.65	73.15
5200	Square splash	68.50	15.65	84.15
5300				
5400				
5500	Appliances, range, free standing, minimum	265	69	334
5600	Maximum	1,575	172	1,747
5700	Built-in, minimum	490	80.50	570.50
5800	Maximum	1,450	440	1,890

E10 Equipment

E1090 Other Equipment

E1090 315	Selective Price Sheet	MAT.	INST.	TOTAL
		COST EACH		
5900	Counter top range 4 burner, maximum	193	80.50	273.50
6000	Maximum	485	161	646
6100	Compactor, built-in, minimum	455	87.50	542.50
6200	Maximum	515	146	661
6300	Dishwasher, built-in, minimum	270	244	514
6400	Maximum	320	490	810
6500	Garbage disposer, sink-pipe, minimum	46.50	97.50	144
6600	Maximum	160	97.50	257.50
6700	Range hood, 30″ wide, 2 speed, minimum	41	185	226
6800	Maximum	655	310	965
6900	Refrigerator, no frost, 12 cu. ft.	485	69	554
7000	20 cu. ft.	750	115	865
7100	Plumb. not incl. rough-ins, sinks porc. C.I., single bowl, 21″ x 24″	221	158	379
7200	21″ x 30″	273	158	431
7300	Double bowl, 20″ x 32″	315	184	499
7400				
7500	Stainless steel, single bowl, 19″ x 18″	365	158	523
7600	22″ x 25″	405	158	563
7700				
7800				
7900				
8000				
8100				
8200				
8300				
8400				

G BUILDING SITEWORK

G1030 Site Earthwork

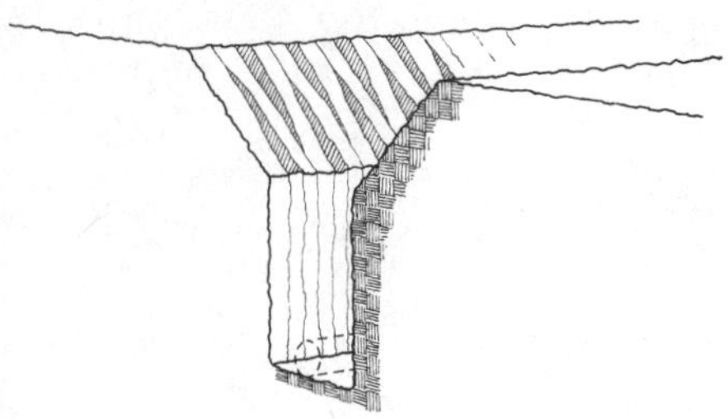

This page illustrates and describes utility trench excavation with backfill systems including a trench, backfill, excavated material, utility pipe or ductwork not included. Lines within System Components give the unit price and total price on a linear foot basis for this system. Prices for alternate utility trench excavation with backfill systems are on Line Items G1030 810 1100 thru 2500. Both material quantities and labor costs have been adjusted for the system listed.

Factors: To adjust for job conditions other than normal working situations use Lines G1030 810 3100 thru 4000.

Example: You are to install the system and protect existing construction. Go to Line G1030 810 3600 and apply these percentages to the appropriate TOTAL cost.

System Components	QUANTITY	UNIT	COST PER L.F. EQUIP.	LABOR	TOTAL
Cont. utility trench excav. with 3/8 C.Y. wheel mtd. backhoe, Backfilling with excav. mat. and comp. in 12″ lifts. Trench is 2′ wide by 4′ dp. Cost of utility piping or ductwork is not incl. Cost based on an Excavation production rate of 150 C.Y. daily no hauling included.					
Machine excavate trench, 2′ wide by 4′ deep	.296	C.Y.	.45	1.52	1.97
Dozer backfill with excavated material	.296	C.Y.	.23	.41	.64
Compact in 12″ lifts, vibrating plate	.296	C.Y.	.07	1.75	1.82
TOTAL			.75	3.68	4.43

G1030 810	Excavation, Utility Trench	COST PER L.F. EQUIP.	LABOR	TOTAL
1000	Alternate size trenches:			
1100	2′ wide x 2′ deep	.38	1.84	2.22
1200	3′ deep	.57	2.76	3.33
1300	With sloping sides, 2′ wide x 5′ deep	2.13	10.35	12.50
1400	6′ deep	2.84	13.85	16.70
1500	7′ deep	3.64	17.75	21.50
1600	8′ deep	4.53	22	26.50
1700	9′ deep	5.50	27	32.50
1800	10′ deep	6.60	32.50	39

G1030 810	Excavation, Utility Trench	COST PER L.F. MAT.	INST.	TOTAL
2000	For hauling excavated material up to 2 miles & backfilling w/ gravel			
2100	Gravel, 2′ wide by 2′ deep, add	3.11	2.78	5.89
2200	4′ deep, add	6.20	5.55	11.75
2300	6′ deep, add	23.50	21	44.50
2400	8′ deep, add	37.50	33.50	71
2500	10′ deep, add	54.50	48.50	103
2600				
2700				
2800	For shallow, hand excavated trenches, no backfill, to 6′ dp, light soil		43	43
2900	Heavy soil		86	86
3000				
3100	Equipment usage curtailment, add, minimum	1%	1%	
3200	Maximum	3%	10%	
3300	Material handling & storage limitation, add, minimum	1%	1%	
3400	Maximum	6%	7%	
3500	Protection of existing work, add, minimum	2%	2%	
3600	Maximum	5%	7%	
3700	Shift work requirements, add, minimum		5%	
3800	Maximum		30%	
3900	Temporary shoring and bracing, add, minimum	2%	5%	
4000	Maximum	5%	12%	

Important: See the Reference Section for critical supporting data - Reference Numbers and City Cost Indexes

G2010 Roadways

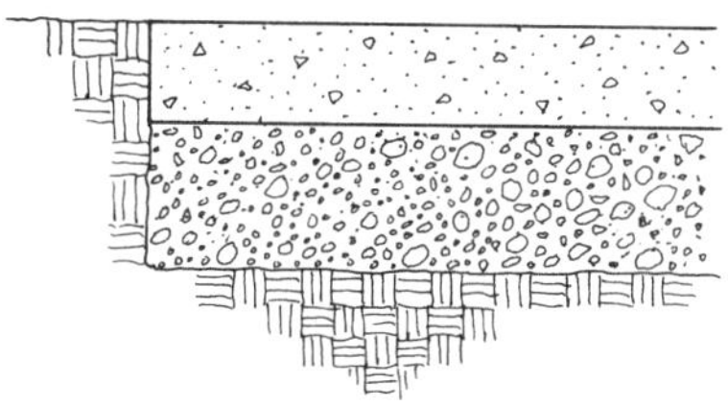

This page illustrates and describes driveway systems including concrete slab, gravel base, broom finish, compaction, and joists. Lines within System Components give the unit price and total price on a cost each basis for this system. Prices for alternate driveway systems are on Line Items G2010 240 1300 thru 2700. Both material quantities and labor costs have been adjusted for the system listed.

Factors: To adjust for job conditions other than normal working situations use Lines G2010 240 2900 thru 4000.

Example: You are to install the system and match existing construction. Go to Line G2010 240 3000 and apply these percentages to the appropriate MAT. and INST. costs.

System Components	QUANTITY	UNIT	COST EACH		
			MAT.	INST.	TOTAL
Complete driveway, 10′ x 30′, 4″ concrete slab, 3000 psi, on 6″ compacted Gravel base, concrete broom finished and cured with 10′ x 10′ joints.					
Grade and compact subgrade	33.000	S.Y.		47.52	47.52
Place and compact 6″ crushed stone base	33.000	S.Y.	189.75	30.69	220.44
Place and remove edgeforms (4 use)	80.000	L.F.	35.20	220.80	256
Concrete for slab, 3000 psi	4.000	C.Y.	320		320
Place concrete, 4″ thick slab, direct chute	4.000	C.Y.		80.08	80.08
Screed, float and broom finish slab	4.000	C.Y.		189	189
Spray on membrane curing compound	3.000	S.F.	17.10	21.75	38.85
Saw cut 3″ deep joints	60.000	L.F.	23.40	63	86.40
TOTAL			585.45	652.84	1,238.29

G2010 240	Driveways	COST EACH		
		MAT.	INST.	TOTAL
1200	For alternate driveway systems:			
1300	Concrete: 10′ x 30′ with 6″ concrete on 6″ crushed stone	745	695	1,440
1400	20′ x 30′ with 4″ concrete	1,175	1,250	2,425
1500	With 6″ concrete	1,500	1,325	2,825
1600	10′ wide, for each additional 10′ length over 30′, 4″ thick, add	190	209	399
1700	6″ thick, add	250	223	473
1800	20′ wide, for each additional 10′ length over 30′, 4″ thick, add	385	395	780
1900	6″ thick, add	485	420	905
2000	Asphalt: 10′ x 30′ with 2″ binder, 1″ topping on 6″ cr. st., sealed	365	420	785
2100	3″ binder, 1″ topping	415	430	845
2200	20′ x 30′ with 2″ binder, 1″ topping	725	590	1,315
2300	3″ binding, 1″ topping	830	610	1,440
2400	10′ wide for each add'l. 10′ length over 30′, 2″ b + 1″ t, add	121	57	178
2500	3″ binder, 1″ topping	138	60	198
2600	20′ wide for each add'l. 10′ length over 30′, 2″ b + 1″ t, add	242	114	356
2700	3″ binder, 1″ topping	276	120	396
2800				
2900	Cut & patch to match existing construction, add, minimum	2%	3%	
3000	Maximum	5%	9%	
3100	Dust protection, add, minimum	1%	2%	
3200	Maximum	4%	11%	
3300	Equipment usage curtailment, add, minimum	1%	1%	
3400	Maximum	3%	10%	
3500	Material handling & storage limitation, add, minimum	1%	1%	
3600	Maximum	6%	7%	
3700	Protection of existing work, add, minimum	2%	2%	
3800	Maximum	5%	7%	
3900	Shift work requirements, add, minimum		5%	
4000	Maximum		30%	

G2020 Parking Lots

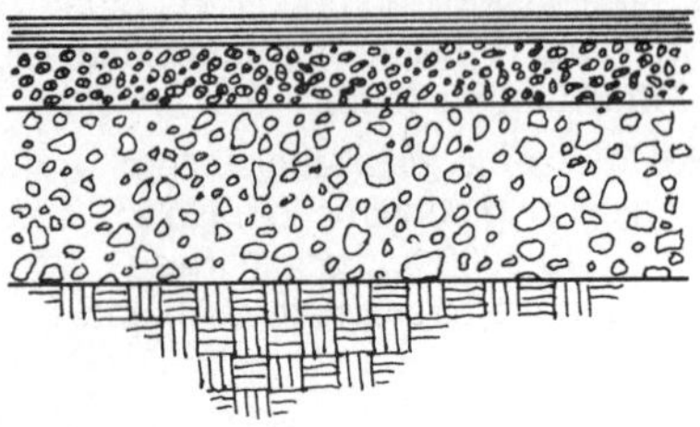

This page illustrates and describes asphalt parking lot systems including asphalt binder, topping, crushed stone base, painted parking stripes and concrete parking blocks. Lines within System Components give the unit price and total price per square yard for this system. Prices for alternate asphalt parking lot systems are on Line Items G2020 230 1500 thru 2500. Both material quantities and labor costs have been adjusted for the system listed.

Factors: To adjust for job conditions other than normal working situations use Lines G2020 230 2900 thru 4000.

Example: You are to install the system and match existing construction. Go to Line G2020 230 2900 and apply these percentages to the appropriate MAT. and INST. costs.

System Components	QUANTITY	UNIT	COST PER S.Y. MAT.	INST.	TOTAL
Parking lot consisting of 2″ asphalt binder and 1″ topping on 6″ Crushed stone base with painted parking stripes and concrete parking blocks					
Fine grade and compact subgrade	1.000	S.Y.		1.44	1.44
6″ crushed stone base, stone	.320	Ton	5.75	.93	6.68
2″ asphalt binder	1.000	S.Y.	3.16	.97	4.13
1″ asphalt topping	1.000	S.Y.	1.75	.64	2.39
Paint parking stripes	.500	L.F.	.09	.06	.15
6″ x 10″ x 6′ precast concrete parking blocks	.020	Ea.	.71	.29	1
Mobilization of equipment	.005	Ea.		1.25	1.25
TOTAL			11.46	5.58	17.04

G2020 230	Parking Lots, Asphalt	COST PER S.Y. MAT.	INST.	TOTAL
1400	For alternate parking lot systems:			
1500	Above system on 9″ crushed stone	14.30	5.65	19.95
1600	12″ crushed stone	17.15	5.75	22.90
1700	On bank run gravel, 6″ deep	9.35	5.20	14.55
1800	9″ deep	11.15	5.30	16.45
1900	12″ deep	12.95	5.45	18.40
2000	3″ binder plus 1″ topping on 6″ crushed stone	13	5.85	18.85
2100	9″ deep crushed stone	15.85	5.95	21.80
2200	12″ deep crushed stone	18.70	6.05	24.75
2300	On bank run gravel, 6″ deep	10.85	5.45	16.30
2400	9″ deep	12.70	5.60	18.30
2500	12″ deep	14.50	5.70	20.20
2600				
2700				
2900	Cut & patch to match existing construction, add, minimum	2%	3%	
3000	Maximum	5%	9%	
3100	Dust protection, add, minimum	1%	2%	
3200	Maximum	4%	11%	
3300	Equipment usage curtailment, add, minimum	1%	1%	
3400	Maximum	3%	10%	
3500	Material handling & storage limitation, add, minimum	1%	1%	
3600	Maximum	6%	7%	
3700	Protection of existing work, add, minimum	2%	2%	
3800	Maximum	5%	7%	
3900	Shift work requirements, add, minimum		5%	
4000	Maximum		30%	

G2020 Parking Lots

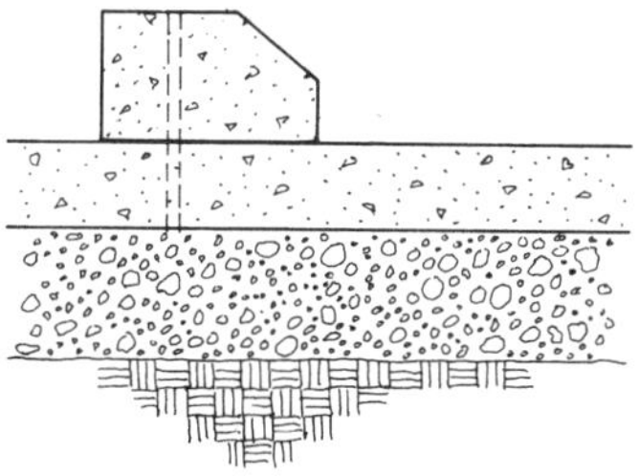

This page illustrates and describes concrete parking lot systems including a concrete slab, unreinforced, compacted gravel base, broom finished, cured, joints, painted parking stripes, and parking blocks. Lines within System Components give the unit price and total price per square yard for this system. Prices for alternate concrete parking lot systems are on Line Items G2020 240 1500 thru 2500. All materials have been adjusted according to the system listed.

Factors: To adjust for job conditions other than normal working situations use Lines G2020 240 2900 thru 4000.

Example: You are to install the system and protect existing construction. Go to Line G2020 240 3800 and apply these percentages to the appropriate MAT. and INST. costs.

System Components	QUANTITY	UNIT	COST PER S.Y. MAT.	INST.	TOTAL
Parking lot, 4″ slab, 3000 psi concrete on 6″ compacted gravel base					
With 12′ x 12′ joints, painted parking strips, conc. parking blocks.					
Grade and compact subgrade	1.000	S.Y.		1.44	1.44
6″ crushed stone base, stone	.320	Ton	5.75	.93	6.68
Place and remove edge forms, 4 use	.250	L.F.	.11	.69	.80
Concrete for slab, 3000 psi	1.000	S.F.	8.80		8.80
Place concrete, 4″ thick slab, direct chute	1.000	S.F.		2.21	2.21
Spray on membrane curing compound	1.000	S.F.	.51	.46	.97
Saw cut 3″ deep joints	2.250	L.F.	.88	2.37	3.25
Paint parking stripes	.500	L.F.	.09	.06	.15
6″ x 10″ x 6′ precast concrete parking blocks	.020	Ea.	.71	.29	1
TOTAL			16.85	8.45	25.30

G2020 240	Parking Lots, Concrete	COST PER S.Y. MAT.	INST.	TOTAL
1400	For alternate parking lot systems:			
1500	Above system on 3″ crushed stone	16.85	8.45	25.30
1600	9″ crushed stone	19.70	8.50	28.20
1700	On bank run gravel, 6″ deep	14.75	8.05	22.80
1800	9″ deep	16.55	8.15	24.70
1900	On compacted subgrade only	11.10	7.55	18.65
2000	6″ concrete slab, 3000 psi on 3″ crushed stone	21	9.50	30.50
2100	6″ crushed stone	21	9.50	30.50
2200	9″ crushed stone	24	9.60	33.60
2300	On bank run gravel, 6″ deep	18.95	9.15	28.10
2400	9″ deep	21	9.25	30.25
2500	On compacted subgrade only	21	9.50	30.50
2600				
2700				
2900	Cut & patch to match existing construction, add, minimum	2%	3%	
3000	Maximum	5%	9%	
3100	Dust protection, add, minimum	1%	2%	
3200	Maximum	4%	11%	
3300	Equipment usage curtailment, add, minimum	1%	1%	
3400	Maximum	3%	10%	
3500	Material handling & storage limitation, add, minimum	1%	1%	
3600	Maximum	6%	7%	
3700	Protection of existing work, add, minimum	2%	2%	
3800	Maximum	5%	7%	
3900	Shift work requirements, add, minimum		5%	
4000	Maximum		30%	

G2030 Pedestrian Paving

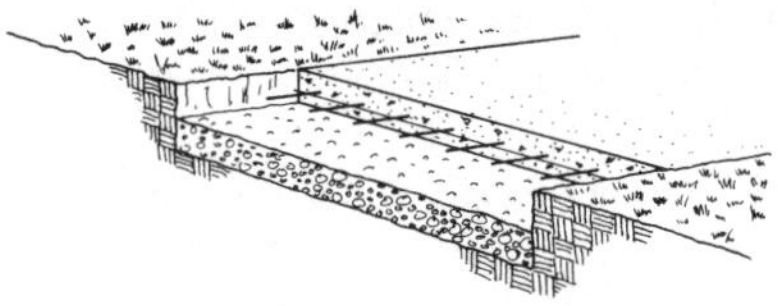

This page illustrates and describes sidewalk systems including concrete, welded wire and broom finish. Lines within System Components give unit price and total price per square foot for this system. Prices for alternate sidewalk systems are on Line Items G2030 105 1700 thru 1900. Both material quantities and labor costs have been adjusted for the system listed.

Factors: To adjust for job conditions other than normal working situations use Lines G2030 105 2900 thru 4000.

Example: You are to install the system and match existing construction. Go to Line G2030 105 2900 and apply these percentages to the appropriate MAT. and INST. costs.

System Components	QUANTITY	UNIT	COST PER S.F.		
			MAT.	INST.	TOTAL
4″ thick concrete sidewalk with welded wire fabric					
3000 psi air entrained concrete, broom finish.					
Gravel fill, 4″ deep	.012	C.Y.	.20	.07	.27
Compact fill	.012	C.Y.		.02	.02
Hand grade	1.000	S.F.		1.59	1.59
Edge form	.250	L.F.	.11	.69	.80
Welded wire fabric	.011	S.F.	.09	.32	.41
Concrete, 3000 psi air entrained	.012	C.Y.	.96		.96
Place concrete	.012	C.Y.		.24	.24
Broom finish	1.000	S.F.		.63	.63
TOTAL			1.36	3.56	4.92

G2030 105	Sidewalks	COST PER S.F.		
		MAT.	INST.	TOTAL
1600	For alternate sidewalk systems:			
1700	Asphalt (bituminous), 2″ thick	.60	2.03	2.63
1800	Brick, on sand, bed, 4.5 brick per S.F.	2.62	7.75	10.37
1900	Flagstone, slate, 1″ thick, rectangular	7.70	8.45	16.15
2000				
2100				
2200				
2300				
2400				
2500				
2600				
2700				
2900	Cut & patch to match existing construction, add, minimum	2%	3%	
3000	Maximum	5%	9%	
3100	Dust protection, add, minimum	1%	2%	
3200	Maximum	4%	11%	
3300	Equipment usage curtailment, add, minimum	1%	1%	
3400	Maximum	3%	10%	
3500	Material handling & storage limitation, add, minimum	1%	1%	
3600	Maximum	6%	7%	
3700	Protection of existing work, add, minimum	2%	2%	
3800	Maximum	5%	7%	
3900	Shift work requirements, add, minimum		5%	
4000	Maximum		30%	

Important: See the Reference Section for critical supporting data - Reference Numbers and City Cost Indexes

G2050 Landscaping

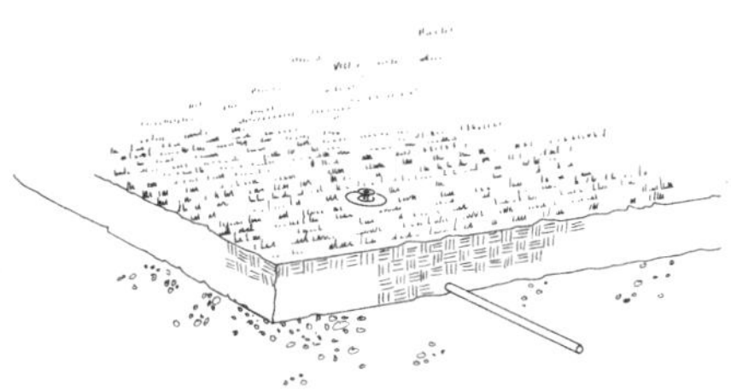

This page describes landscaping—lawn establishment systems including loam, lime, fertilizer, side and top mulching. Lines within system components give the unit price and total price per square yard for this system. Prices for alternate landscaping—lawn establishment systems are on Line Items G2050 420 1100 and 1200. Both material quantities and labor costs have been adjusted for the system listed.

Factors: To adjust for job conditions other than normal working situations use Lines G2050 420 2900 thru 4000.

Example: You are to install the system and provide dust protection. Go to Line G2050 420 3200 and apply these percentages to the appropriate MAT. and INST. costs.

System Components	QUANTITY	UNIT	COST PER S.Y.		
			MAT.	INST.	TOTAL
Establishing lawns with loam, lime, fertilizer, seed and top mulching					
On rough graded areas.					
Furnish and place loam 4″ deep	.110	C.Y.	2.78	.66	3.44
Fine grade, lime, fertilize and seed	1.000	S.Y.	.37	2.39	2.76
Hay mulch, 1 bale/M.S.F.	1.000	S.Y.	.08	.42	.50
Rolling with hand roller	1.000	S.Y.		.74	.74
TOTAL			3.23	4.21	7.44

G2050 420	Landscaping - Lawn Establishment	COST PER S.Y.		
		MAT.	INST.	TOTAL
1000	For alternate lawn systems:			
1100	Above system with jute mesh in place of hay mulch	3.88	4.29	8.17
1200	Above system with sod in place of seed	5	1.49	6.49
1300				
1400				
1500				
1600				
1700				
1800				
1900				
2000				
2100				
2200				
2300				
2400				
2500				
2600				
2700				
2900	Cut & patch to match existing construction, add, minimum	2%	3%	
3000	Maximum	5%	9%	
3100	Dust protection, add, minimum	1%	2%	
3200	Maximum	4%	11%	
3300	Equipment usage curtailment, add, minimum	1%	1%	
3400	Maximum	3%	10%	
3500	Material handling & storage limitation, add, minimum	1%	1%	
3600	Maximum	6%	7%	
3700	Protection of existing work, add, minimum	2%	2%	
3800	Maximum	5%	7%	
3900	Shift work requirements, add, minimum		5%	
4000	Maximum		30%	

G2050 Landscaping

G2050 510	Selective Price Sheet	MAT.	INST.	TOTAL
		COST PER UNIT		
0100	Shrubs and trees, Evergreen, in prepared beds			
0200				
0300	Arborvitae pyramidal, 4'-5'	47	67.50	114.50
0400	Globe, 12"-15"	11.80	11.05	22.85
0500	Cedar, blue, 8'-10'	129	113	242
0600	Hemlock, Canadian, 2-1/2'-3'	19.15	29.50	48.65
0700	Juniper, andora, 18"-24"	17.15	13.25	30.40
0800	Wiltoni, 15"-18"	13.20	13.25	26.45
0900	Skyrocket, 4-1/2'-5'	51.50	36.50	88
1000	Blue pfitzer, 2'-2-1/2'	23	24	47
1100	Ketleerie, 2-1/2'-3'	34.50	21	55.50
1200	Pine, black, 2-1/2'-3'	36	21	57
1300	Mugo, 18"-24"	37	17.70	54.70
1400	White, 4'-5'	52.50	27	79.50
1500	Spruce, blue, 18"-24"	35.50	17.70	53.20
1600	Norway, 4'-5'	67	27	94
1700	Yew, denisforma, 12"-15"	24	17.70	41.70
1800	Capitata, 18"-24"	21	35.50	56.50
1900	Hicksi, 2'-2-1/2'	31	35.50	66.50
2000				
2100	Trees, Deciduous, in prepared beds			
2200				
2300	Beech, 5'-6'	237	40.50	277.50
2400	Dogwood, 4'-5'	72.50	50.50	123
2500	Elm, 8'-10'	124	102	226
2600	Magnolia, 4'-5'	67	102	169
2700	Maple, red, 8'-10', 1-1/2" caliper	167	203	370
2800	Oak, 2-1/2"-3" caliper	252	340	592
2900	Willow, 6'-8', 1" caliper	65	102	167
3000				
3100	Shrubs, Broadleaf Evergreen, in prepared beds			
3200				
3300	Andromeda, 15"-18", container	18.90	11.05	29.95
3400	Azalea, 15"-18", container	24.50	11.05	35.55
3500	Barberry, 9"-12", container	10.70	8.15	18.85
3600	Boxwood, 15"-18", B & B	23	11.05	34.05
3700	Euonymus, emerald gaiety, 12"-15", container	15.75	9.20	24.95
3800	Holly, 15"-18", B & B	16.90	11.05	27.95
3900	Mount laurel, 18"-24", B & B	53.50	13.25	66.75
4000	Privet, 18"-24", B & B	17.35	8.15	25.50
4100	Rhodendron, 18"-24", container	29.50	22	51.50
4200	Rosemary, 1 gal. container	63.50	1.77	65.27
4300	Deciduous, amalanchier, 2'-3', B & B	86	18.60	104.60
4400	Azalea, 15"-18", B & B	22.50	11.05	33.55
4500	Bayberry, 2'-3', B & B	26.50	18.60	45.10
4600	Cotoneaster, 15"-18", B & B	15.70	13.25	28.95
4700	Dogwood, 3'-4', B & B	25	50.50	75.50
4800	Euonymus, alatus compacta, 15"-18", container	20.50	13.25	33.75
4900	Forsythia, 2'-3', container	18.15	17.70	35.85
5000	Honeysuckle, 3'-4', B & B	18.80	17.70	36.50
5100	Hydrangea, 2'-3', B & B	22.50	18.60	41.10
5200	Lilac, 3'-4', B & B	27.50	50.50	78
5300	Quince, 2'-3', B & B	21	18.60	39.60
5400				
5500	Ground Cover			
5600				
5700	Plants, Pachysandra, prepared beds, per hundred	27.50	70.50	98
5800	Vinca minor or English ivy, per hundred	28.50	88.50	117

G2050 Landscaping

G2050 510	Selective Price Sheet	COST PER UNIT		
		MAT.	INST.	TOTAL
5900	Plant bed prep. 18″ dp., mach., per square foot	2.17	.31	2.48
6000	By hand, per square foot	2.17	2.22	4.39
6100	Stone chips, Georgia marble, 50# bags, per bag	2.59	2.04	4.63
6200	Onyx gemstone, per bag	18.70	4.08	22.78
6300	Quartz, per bag	7	4.08	11.08
6400	Pea gravel, truck load lots, per cubic yard	25.50	38	63.50
6500	Mulch, polyethylene mulch, per square yard	.17	.34	.51
6600	Wood chips, 2″ deep, per square yard	1.87	1.57	3.44
6700	Peat moss, 1″ deep, per square yard	1.84	.38	2.22
6800	Erosion control per square yard, Jute mesh stapled	.73	.50	1.23
6900	Plastic netting, stapled	.66	.42	1.08
7000	Polypropylene mesh, stapled	1.36	.42	1.78
7100	Tobacco netting, stapled	.08	.42	.50
7200	Lawns per square yard, seeding incl. fine grade, limestone, fert. and seed	.37	2.39	2.76
7300	Sodding, incl. fine grade, level ground	239	88	327
7400	On slope			
7500	Edging, per linear foot, Redwood, untreated, 1″ x 4″	1.54	1.75	3.29
7600	2″ x 4″	3.07	2.65	5.72
7700	Stl. edge strips, 1/4″ x 5″ inc. stakes	3.44	2.65	6.09
7800	3/16″ x 4″	2.75	2.65	5.40
7900	Brick edging, set on edge	2.05	5.75	7.80
8000	Set flat	1.02	2.10	3.12
8100				
8200				
8300				
8400				

G3020 Sanitary Sewer

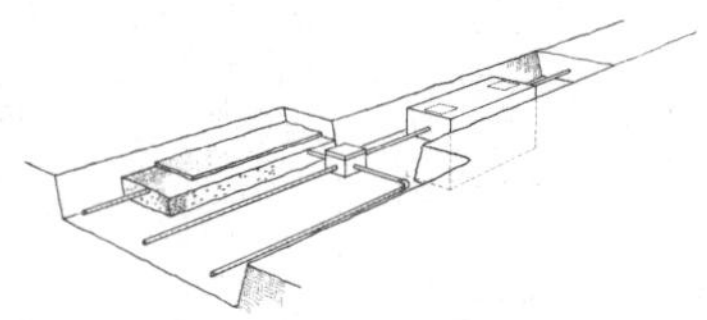

This page illustrates and describes drainage and utilities – septic system including a tank, distribution box, excavation, piping, crushed stone and backfill. Lines within System Components give the unit price and total price on a cost each basis for this system. Prices for alternate drainage and utilities – septic systems are on Line Items G3020 610 1700 thru 2000. Both material quantities and labor costs have been adjusted for the system listed.

Factors: To adjust for job conditions other than normal working situations use Lines G3020 610 3100 thru 4000.

Example: You are to install the system with a material handling limitation. Go to Line G3020 610 3500 and apply these percentages to the appropriate MAT. and INST. costs.

System Components	QUANTITY	UNIT	COST EACH		
			MAT.	INST.	TOTAL
Septic system including tank, distribution box, excavation, piping, crushed Stone and backfill for a 1000 S.F. leaching field.					
Precast concrete septic tank, 1000 gal. capacity	1.000	Ea.	585	194	779
Concrete distribution box	1.000	Ea.	109	43	152
Sch. 40 sewer pipe, PVC, 4″ diameter	25.000	L.F.	328.10	523.60	851.70
Sch. 40 sewer pipe fittings, PVC, 4″ diameter	8.000	Ea.	51.20	772	823.20
Excavation for septic tank	120.000	C.Y.		162.54	162.54
Excavation for disposal field	120.000	C.Y.		395.92	395.92
Crushed stone backfill	76.000	C.Y.	1,288.20	670.32	1,958.52
Backfill with excavated material	26.000	C.Y.		33.54	33.54
Building paper	6.000	C.S.F.	13.44	71.10	84.54
TOTAL			2,374.94	2,866.02	5,240.96

G3020 610	Septic Systems	COST EACH		
		MAT.	INST.	TOTAL
1600	For alternate septic systems:			
1700	1000 gal. tank with 2000 S.F. field	3,525	4,075	7,600
1800	With leaching pits	2,125	875	3,000
1900	2000 gal. tank with 2000 S.F. field	4,075	4,200	8,275
2000	With leaching pits	2,800	1,300	4,100
2100				
2200				
2300				
2400				
2500				
2600				
2700				
2800				
2900				
3100	Dust protection, add, minimum	1%	2%	
3200	Maximum	4%	11%	
3300	Equipment usage curtailment, add, minimum	1%	1%	
3400	Maximum	3%	10%	
3500	Material handling & storage limitation, add, minimum	1%	1%	
3600	Maximum	6%	7%	
3700	Protection of existing work, add, minimum	2%	2%	
3800	Maximum	5%	7%	
3900	Shift work requirements, add, minimum		5%	
4000	Maximum		30%	

G3060 Fuel Distribution

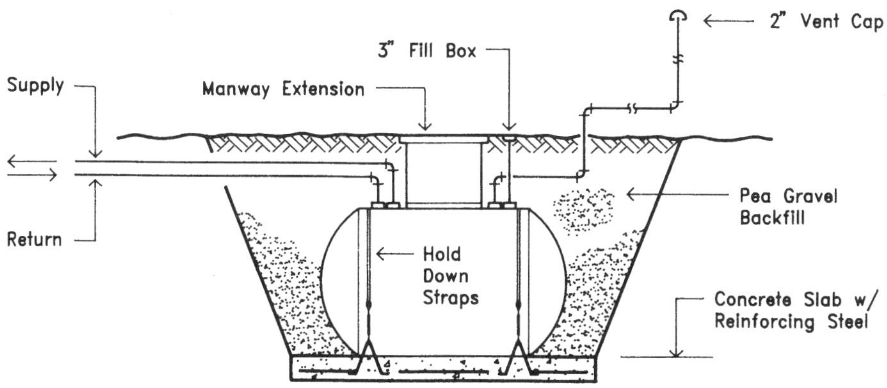

Fiberglass Underground Fuel Storage Tank, Single Wall

This page illustrates and describes fiberglass tanks including hold down slab, reinforcing steel and peastone gravel. Lines within Systems Components give the material and installation price on a cost each basis for the components. Prices for alternate fiberglass tank systems are on Line Items G3060 305 1010 thru G3060 310 1100. Material quantities and labor costs have been adjusted for the system listed.

Factors: To adjust for job conditions other than normal working situations use Lines G3060 310 2700 thru 4000.

System Components	QUANTITY	UNIT	COST EACH MAT.	COST EACH INST.	COST EACH TOTAL
Single wall underground fiberglass storage tanks including, excavation, hold down slab, backfill, manway extension and related piping.					
Tank, fiberglass, underground, single wall, U.L. listed, 550 gal	1.000	Ea.	1,625	330	1,955
Tank, fiberglass, manways, add	1.000	Ea.	935		935
Tank, for hold-downs 500-4000 gal, add	1.000	Ea.	213	118	331
Foot valve, single poppet, 3/4" dia	1.000	Ea.	41.50	27.50	69
Tubing, copper, Type L, 3/4" dia	120.000	L.F.	187.20	774	961.20
Elbow 90°, copper, wrought, cu x cu, 3/4" dia.	8.000	Ea.	5.92	208	213.92
Fuel oil specialties, valve, ball chk, globe type, fusible, 3/4" dia	2.000	Ea.	78	49	127
Elbow 90°, black steel, straight, 2" dia.	3.000	Ea.	33.60	147	180.60
Union, black steel, with brass seat, 2" dia.	1.000	Ea.	20.50	52	72.50
Pipe and nipples, steel, Sch. 40, threaded, black, 2" dia	2.000	Ea.	152.38	524.40	676.78
Vent protector / breather, 2" dia.	1.000	Ea.	14.10	15.40	29.50
Pipe, black steel, welded, Sch. 40, 3" dia.	30.000	L.F.	165	674.70	839.70
Elbow 90°, steel, weld joint, butt, 3" dia.	3.000	Ea.	35.85	414.75	450.60
Flange, weld neck, 150 lb., 3" pipe size	1.000	Ea.	24	69.10	93.10
Fuel fill box, locking inner cover, 3" dia.	1.000	Ea.	112	98.50	210.50
Remote tank gauging system, 30', 5" pointer travel	1.000	Ea.	4,075	247	4,322
Tank leak detector system, 8 channel, external monitoring	1.000	Ea.	1,750		1,750
Tank leak detection system, probes, well monitoring, liquid phase detection	2.000	Ea.	1,570		1,570
Excavating, trench, NO sheeting or dewatering, 6'-10' D,3/4 CY hyd backhoe	18.000	C.Y.		104.94	104.94
Concrete hold down pad, 6' x 5' x 8" thick	30.000	C.Y.	63.30	34.50	97.80
Reinforcing in hold down pad, #3 to #7	120.000	Lb.	37.20	58.80	96
Stone back-fill, hand spread, pea gravel	9.000	C.Y.	553.50	342	895.50
Corrosion resistance, wrap & coat, small diam pipe, 1" diam, add	120.000	L.F.	139.20		139.20
Corrosion resistance, wrap & coat, small diameter pipe, 2" diam, add	36.000	L.F.	45.72		45.72
Corrosion resistance wrap & coat, 4" diam, add	30.000	L.F.	45.90		45.90
TOTAL			11,922.87	4,289.59	16,212.46

G3060 305	Fiberglass Fuel Tank, Single Wall	COST EACH MAT.	COST EACH INST.	COST EACH TOTAL
1000	For alternate storage tanks:			
1010	Storage tank, fuel, underground, single wall fiberglass, 550 Gal.	11,900	4,300	16,200
1020	2000 Gal.	14,900	5,875	20,775
1030	4000 Gal.	19,900	8,050	27,950
1040	6000 Gal.	21,300	8,975	30,275
1050	8000 Gal.	25,900	11,000	36,900
1060	10,000 Gal.	29,000	11,900	40,900
1070	15,000 Gal.	33,600	14,900	48,500

G3060 Fuel Distribution

G3060 305	Fiberglass Fuel Tank, Single Wall	COST EACH		
		MAT.	INST.	TOTAL
1080	20,000 Gal.	42,600	18,000	60,600
1090	25,000 Gal.	54,000	21,000	75,000
1100	30,000 Gal.	61,500	22,700	84,200
1110	40,000 Gal.	77,500	28,700	106,200
1120	48,000 Gal.	105,000	33,000	138,000

G3060 310	Fiberglass Fuel Tank, Double Wall	COST EACH		
		MAT.	INST.	TOTAL
1010	Storage tank, fuel, underground, double wall fiberglass, 550 Gal.	14,300	4,900	19,200
1020	2500 Gal.	20,200	6,850	27,050
1030	4000 Gal.	25,900	8,525	34,425
1040	6000 Gal.	27,700	9,600	37,300
1050	8000 Gal.	32,700	11,600	44,300
1060	10,000 Gal.	36,600	13,500	50,100
1070	15,000 Gal.	47,500	15,500	63,000
1080	20,000 Gal.	58,000	18,800	76,800
1090	25,000 Gal.	69,500	21,900	91,400
1100	30,000 Gal.	78,500	23,100	101,600
2700	Cut & patch to match existing construction, add, minimum	2%	3%	
2800	Maximum	5%	9%	
2900	Dust protection, add, minimum	1%	2%	
3000	Maximum	4%	11%	
3100	Equipment usage curtailment, add, minimum	1%	1%	
3200	Maximum	3%	10%	
3300	Material handling & storage limitation, add, minimum	1%	1%	
3400	Maximum	6%	7%	
3500	Protection of existing work, add, minimum	2%	2%	
3600	Maximum	5%	7%	
3700	Shift work requirements, add, minimum		5%	
3800	Maximum		30%	
3900	Temporary shoring and bracing, add, minimum	2%	5%	
4000	Maximum	5%	12%	

G3060 Fuel Distribution

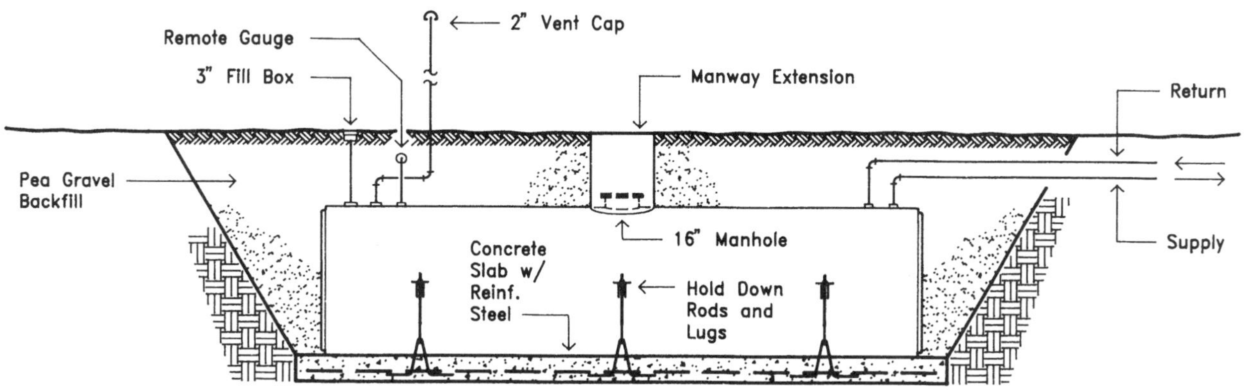

Steel Underground Fuel Storage Tank, Single Wall

System Components	QUANTITY	UNIT	COST EACH MAT.	COST EACH INST.	COST EACH TOTAL
Single wall underground steel storage tank including, excavation, hold down slab, backfill, manway extension and related piping.					
Tank, steel, underground, sti-P3, set in place, 500 gal, 7 Ga. shell	1.000	Ea.	905	330	1,235
Tank, for manways, add	1.000	Ea.	580		580
Tanks, for hold-downs, 500-2000 gal, add	1.000	Ea.	335	118	453
Foot valve, single poppet, 3/4" dia	1.000	Ea.	41.50	27.50	69
Tubing, copper, Type L, 3/4" dia.	120.000	L.F.	187.20	774	961.20
Elbow 90°, copper, wrought, cu x cu, 3/4" dia.	8.000	Ea.	5.92	208	213.92
Fuel oil valve, ball chk, globe type, fusible, 3/4" dia.	2.000	Ea.	78	49	127
Elbow 90°, black steel, straight, 2" dia.	3.000	Ea.	33.60	147	180.60
Union, black steel, with brass seat, 3/4" dia.	1.000	Ea.	6.90	38	44.90
Pipe and nipples, steel, Sch. 40, threaded, black, 2" dia	2.000	Ea.	152.38	524.40	676.78
Vent protector / breather, 2" dia.	1.000	Ea.	14.10	15.40	29.50
Pipe, black steel, welded, Sch. 40, 3" dia.	30.000	L.F.	165	674.70	839.70
Elbow 90°, steel, weld joint, butt, 90° elbow, 3" dia.	3.000	Ea.	35.85	414.75	450.60
Flange, steel, weld neck, 150 lb., 3" pipe size	1.000	Ea.	24	69.10	93.10
Fuel fill box, locking inner cover, 3" dia.	1.000	Ea.	112	98.50	210.50
Remote tank gauging system, 30', 5" pointer travel	1.000	Ea.	4,075	247	4,322
Tank leak detector system, 8 channel, external monitoring	1.000	Ea.	1,750		1,750
Tank leak detection system, probes, well monitoring, liquid phase detection	2.000	Ea.	1,570		1,570
Excavation, NO sheeting or dewatering,6'-10' D,3/4 CY hyd backhoe	18.000	C.Y.		104.94	104.94
Concrete hold down pad, 6' x 5' x 8" thick	30.000	C.Y.	63.30	34.50	97.80
Reinforcing in hold down pad, #3 to #7	120.000	Lb.	37.20	58.80	96
Stone back-fill, hand spread, pea gravel	9.000	C.Y.	553.50	342	895.50
Corrosion resistance, wrap & coat, small diam pipe, 1" dia., add	120.000	L.F.	139.20		139.20
Corrosion resistance, wrap & coat, small diameter pipe, 2" dia., add	36.000	L.F.	45.72		45.72
Corrosion resistance wrap & coat, 4" dia., add	30.000	L.F.	45.90		45.90
TOTAL			10,956.27	4,275.59	15,231.86

G3060 315	Steel Fuel Tank, Single Wall	COST EACH MAT.	COST EACH INST.	COST EACH TOTAL
1000	For alternate storage tanks:			
1010	Storage tank, fuel, underground, single wall steel, 550 Gal.	11,000	4,275	15,275
1020	2000 Gal.	14,600	5,875	20,475
1030	5000 Gal.	21,500	8,350	29,850
1040	10,000 Gal.	32,200	12,800	45,000
1050	15,000 Gal.	37,600	14,800	52,400

G3060 Fuel Distribution

G3060 315	Steel Fuel Tank, Single Wall	COST EACH		
		MAT.	INST.	TOTAL
1060	20,000 Gal.	46,900	17,800	64,700
1070	25,000 Gal.	54,000	20,900	74,900
1080	30,000 Gal.	62,000	22,400	84,400
1090	40,000 Gal.	77,500	28,300	105,800

G3060 320	Steel Fuel Tank, Double Wall	COST EACH		
		MAT.	INST.	TOTAL
1010	Storage tank, fuel, underground, double wall steel, 500 Gal.	13,800	5,025	18,825
1020	2000 Gal.	18,200	6,600	24,800
1030	4000 Gal.	24,500	8,625	33,125
1040	6000 Gal.	29,300	9,525	38,825
1050	8000 Gal.	34,600	11,800	46,400
1060	10,000 Gal.	38,600	13,600	52,200
1070	15,000 Gal.	45,400	15,900	61,300
1080	20,000 Gal.	55,000	19,000	74,000
1090	25,000 Gal.	72,000	21,800	93,800
1100	30,000 Gal.	82,000	23,300	105,300
1110	40,000 Gal.	101,000	29,300	130,300
1120	50,000 Gal.	125,000	35,200	160,200

G3060 325	Steel Tank, Above Ground, Single Wall	COST EACH		
		MAT.	INST.	TOTAL
1010	Storage tank, fuel, above ground, single wall steel, 550 Gal.	8,125	3,550	11,675
1020	2000 Gal.	10,600	3,650	14,250
1030	5000 Gal.	12,500	5,600	18,100
1040	10,000 Gal.	20,600	5,950	26,550
1050	15,000 Gal.	22,800	6,100	28,900
1060	20,000 Gal.	26,700	6,300	33,000
1070	25,000 Gal.	28,800	6,450	35,250
1080	30,000 Gal.	34,000	6,700	40,700

G3060 330	Steel Tank, Above Ground, Double Wall	COST EACH		
		MAT.	INST.	TOTAL
1010	Storage tank, fuel, above ground, double wall steel, 500 Gal.	11,100	3,600	14,700
1020	2000 Gal.	14,000	3,675	17,675
1030	4000 Gal.	18,200	3,750	21,950
1040	6000 Gal.	20,900	5,775	26,675
1050	8000 Gal.	24,100	5,950	30,050
1060	10,000 Gal.	25,500	6,050	31,550
1070	15,000 Gal.	33,300	6,250	39,550
1080	20,000 Gal.	36,400	6,450	42,850
1090	25,000 Gal.	42,000	6,625	48,625
1100	30,000 Gal.	45,000	6,875	51,875
2700	Cut & patch to match existing construction, add, minimum	2%	3%	
2800	Maximum	5%	9%	
2900	Dust protection, add, minimum	1%	2%	
3000	Maximum	4%	11%	
3100	Equipment usage curtailment, add, minimum	1%	1%	
3200	Maximum	3%	10%	
3300	Material handling & storage limitation, add, minimum	1%	1%	
3400	Maximum	6%	7%	
3500	Protection of existing work, add, minimum	2%	2%	
3600	Maximum	5%	7%	
3700	Shift work requirements, add, minimum		5%	
3800	Maximum		30%	
3900	Temporary shoring and bracing, add, minimum	2%	5%	
4000	Maximum	5%	12%	

Important: See the Reference Section for critical supporting data - Reference Numbers and City Cost Indexes

G4020 Lighting and Branch Wiring

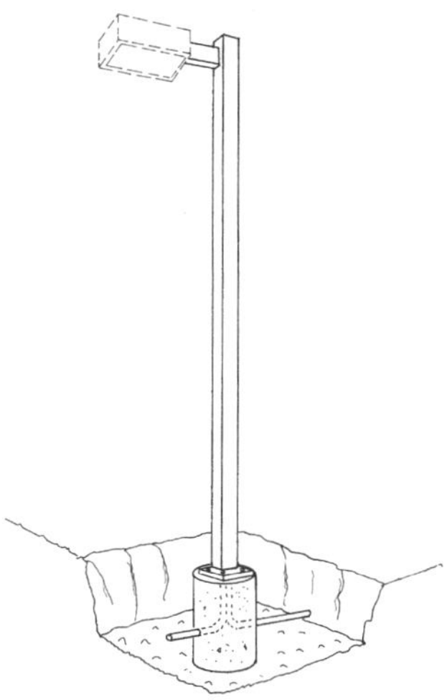

This page illustrates and describes light poles for parking or walkway area lighting. Included are aluminum or steel light poles, single or multiple fixture bracket arms, excavation, concrete footing, backfill and compaction. Lines within system components give the unit price and total price on a cost per each basis. Prices for alternate systems are shown on lines G4020 210 0200 through 1440. Both materialquantities and labor costs have been adjusted for the system listed.

Factors: To adjust for job conditions other than normal working situations use lines G4020 210 2700 through 3600.

Example: You are to install the system where you must be careful not to damage existing walks or parking areas. Go to line G4020 210 3300 and apply these percentages to the appropriate MAT. and INST. costs.

System Components	QUANTITY	UNIT	COST EACH		
			MAT.	INST.	TOTAL
Light poles, aluminum, 20′ high, 1 arm bracket					
Aluminum light pole, 20′, no concrete base	1.000	Ea.	675	470.50	1,145.50
Bracket arm for Aluminum light pole	1.000	Ea.	88.50	60.50	149
Excavation by hand, pits to 6′ deep, heavy soil or clay	2.368	C.Y.		203.65	203.65
Footing, concrete incl forms, reinforcing, spread, under 1 C.Y.	.465	C.Y.	52.08	71.19	123.27
Backfill by hand	1.903	C.Y.		59.94	59.94
Compaction vibrating plate	1.903	C.Y.		8.26	8.26
TOTAL			815.58	874.04	1,689.62

G4020 210	Light Pole (Installed)	COST EACH		
		MAT.	INST.	TOTAL
0195	For alternate light pole systems:			
0200	Light pole, aluminum, 20′ high, 1 arm bracket	815	870	1,685
0240	2 arm brackets	905	870	1,775
0280	3 arm brackets	995	905	1,900
0320	4 arm brackets	1,075	905	1,980
0360	30′ high, 1 arm bracket	1,475	1,100	2,575
0400	2 arm brackets	1,575	1,100	2,675
0440	3 arm brackets	1,650	1,150	2,800
0480	4 arm brackets	1,750	1,150	2,900
0680	40′ high, 1 arm bracket	1,800	1,475	3,275
0720	2 arm brackets	1,875	1,475	3,350
0760	3 arm brackets	1,975	1,525	3,500
0800	4 arm brackets	2,050	1,525	3,575
0840	Steel, 20′ high, 1 arm bracket	1,000	930	1,930
0880	2 arm brackets	1,075	930	2,005
0920	3 arm brackets	1,100	965	2,065
0960	4 arm brackets	1,175	965	2,140
1000	30′ high, 1 arm bracket	1,175	1,175	2,350
1040	2 arm brackets	1,250	1,175	2,425
1080	3 arm brackets	1,250	1,200	2,450
1120	4 arm brackets	1,350	1,200	2,550
1320	40′ high, 1 arm bracket	1,500	1,600	3,100
1360	2 arm brackets	1,575	1,600	3,175
1400	3 arm brackets	1,600	1,625	3,225
1440	4 arm brackets	1,700	1,625	3,325
2700	Cut & patch to match existing construction, add, minimum	2%	3%	

G4020 Lighting and Branch Wiring

G4020 210	Light Pole (Installed)	COST EACH		
		MAT.	INST.	TOTAL
2800	Maximum	5%	9%	
2900	Equipment usage curtailment, add, minimum	1%	1%	
3000	Maximum	3%	10%	
3100	Material handling & storage limitation, add, minimum	1%	1%	
3200	Maximum	6%	7%	
3300	Protection of existing work, add, minimum	2%	2%	
3400	Maximum	5%	7%	
3500	Shift work requirements, add, minimum		5%	
3600	Maximum		30%	

For information about Means Estimating Seminars, see yellow pages 12 and 13 in back of book

Reference Section

We've put all the reference information into one section, making it easy to find what you need to know . . . and easy to use the book on a daily basis. This section is visually identified by a vertical gray bar on the edge of pages.

In the reference number information that follows, you'll see the background that relates to the "reference numbers" that appeared in the Unit Price Section. You'll find reference tables, explanations and estimating information that support how we arrived at the unit price data. Also included are alternate pricing methods, technical data and estimating procedures along with information on design and economy in construction.

Also in this Reference Section, we've included Change Orders, information on pricing changes to contract documents; Crew Listings, a full listing of all the crews, equipment and their costs; Historical Cost Indexes for cost comparisons over time; City Cost Indexes and Location Factors for adjusting costs to the region you are in; and an explanation of all abbreviations used in the book.

Table of Contents

R01100-005 Tips for Accurate Estimating

1. Use pre-printed or columnar forms for orderly sequence of dimensions and locations and for recording telephone quotations.
2. Use only the front side of each paper or form except for certain pre-printed summary forms.
3. Be consistent in listing dimensions: For example, length x width x height. This helps in rechecking to ensure that, the total length of partitions is appropriate for the building area.
4. Use printed (rather than measured) dimensions where given.
5. Add up multiple printed dimensions for a single entry where possible.
6. Measure all other dimensions carefully.
7. Use each set of dimensions to calculate multiple related quantities.
8. Convert foot and inch measurements to decimal feet when listing. Memorize decimal equivalents to .01 parts of a foot (1/8″ equals approximately .01′).
9. Do not "round off" quantities until the final summary.
10. Mark drawings with different colors as items are taken off.
11. Keep similar items together, different items separate.
12. Identify location and drawing numbers to aid in future checking for completeness.
13. Measure or list everything on the drawings or mentioned in the specifications.
14. It may be necessary to list items not called for to make the job complete.
15. Be alert for: Notes on plans such as N.T.S. (not to scale); changes in scale throughout the drawings; reduced size drawings; discrepancies between the specifications and the drawings.
16. Develop a consistent pattern of performing an estimate. For example:
    a. Start the quantity takeoff at the lower floor and move to the next higher floor.
    b. Proceed from the main section of the building to the wings.
    c. Proceed from south to north or vice versa, clockwise or counterclockwise.
    d. Take off floor plan quantities first, elevations next, then detail drawings.
17. List all gross dimensions that can be either used again for different quantities, or used as a rough check of other quantities for verification (exterior perimeter, gross floor area, individual floor areas, etc.).
18. Utilize design symmetry or repetition (repetitive floors, repetitive wings, symmetrical design around a center line, similar room layouts, etc.). Note: Extreme caution is needed here so as not to omit or duplicate an area.
19. Do not convert units until the final total is obtained. For instance, when estimating concrete work, keep all units to the nearest cubic foot, then summarize and convert to cubic yards.
20. When figuring alternatives, it is best to total all items involved in the basic system, then total all items involved in the alternates. Therefore you work with positive numbers in all cases. When adds and deducts are used, it is often confusing whether to add or subtract a portion of an item; especially on a complicated or involved alternate.

R01100-040 Builder's Risk Insurance

Builder's Risk Insurance is insurance on a building during construction. Premiums are paid by the owner or the contractor. Blasting, collapse and underground insurance would raise total insurance costs above those listed. Floater policy for materials delivered to the job runs $.75 to $1.25 per $100 value. Contractor equipment insurance runs $.50 to $1.50 per $100 value. Insurance for miscellaneous tools to $1,500 value runs from $3.00 to $7.50 per $100 value.

Tabulated below are New England Builder's Risk insurance rates in dollars per $100 value for $1,000 deductible. For $25,000 deductible, rates can be reduced 13% to 34%. On contracts over $1,000,000, rates may be lower than those tabulated. Policies are written annually for the total completed value in place. For "all risk" insurance (excluding flood, earthquake and certain other perils) add $.025 to total rates below.

Coverage	Frame Construction (Class 1)		Brick Construction (Class 4)		Fire Resistive (Class 6)	
	Range	Average	Range	Average	Range	Average
Fire Insurance	$.350 to $ .850	$.600	$.158 to $.189	$.174	$.052 to $.080	$.070
Extended Coverage	.115 to .200	.158	.080 to .105	.101	.081 to .105	.100
Vandalism	.012 to .016	.014	.008 to .011	.011	.008 to .011	.010
Total Annual Rate	$.477 to $1.066	$.772	$.246 to $.305	$.286	$.141 to $.196	$.180

R01100-050 General Contractor's Overhead

There are two distinct types of overhead on a construction project: Project Overhead and Main Office Overhead. Project Overhead includes those costs at a construction site not directly associated with the installation of construction materials. Examples of Project Overhead costs include the following:

1. Superintendent
2. Construction office and storage trailers
3. Temporary sanitary facilities
4. Temporary utilities
5. Security fencing
6. Photographs
7. Clean up
8. Performance and payment bonds

The above Project Overhead items are also referred to as General Requirements and therefore are estimated in Division 1. Division 1 is the first division listed in the CSI MasterFormat but it is usually the last division estimated. The sum of the costs in Divisions 1 through 16 is referred to as the sum of the direct costs.

All construction projects also include indirect costs. The primary components of indirect costs are the contractor's Main Office Overhead and profit. The amount of the Main Office Overhead expense varies depending on the the following:

1. Owner's compensation
2. Project managers and estimator's wages
3. Clerical support wages
4. Office rent and utilities
5. Corporate legal and accounting costs
6. Advertising
7. Automobile expenses
8. Association dues
9. Travel and entertainment expenses

These costs are usually calculated as a percentage of annual sales volume. This percentage can range from 35% for a small contractor doing less than $500,000 to 5% for a large contractor with sales in excess of $100 million.

R01100-060 Workers' Compensation Insurance Rates by Trade

The table below tabulates the national averages for Workers' Compensation insurance rates by trade and type of building. The average "Insurance Rate" is multiplied by the "% of Building Cost" for each trade. This produces the "Workers' Compensation Cost" by % of total labor cost, to be added for each trade by building type to determine the weighted average Workers' Compensation rate for the building types analyzed.

Trade	Insurance Rate (% Labor Cost) Range	Insurance Rate (% Labor Cost) Average	% of Building Cost Office Bldgs.	% of Building Cost Schools & Apts.	% of Building Cost Mfg.	Workers' Compensation Office Bldgs.	Workers' Compensation Schools & Apts.	Workers' Compensation Mfg.
Excavation, Grading, etc.	3.5 % to 18.5%	10.3%	4.8%	4.9%	4.5%	.49%	.50%	.46%
Piles & Foundations	7.3 to 76.7	22.9	7.1	5.2	8.7	1.63	1.19	1.99
Concrete	5.7 to 35.2	15.8	5.0	14.8	3.7	.79	2.34	.58
Masonry	5.1 to 31.0	15.0	6.9	7.5	1.9	1.04	1.13	.29
Structural Steel	7.2 to 112.0	38.9	10.7	3.9	17.6	4.16	1.52	6.85
Miscellaneous & Ornamental Metals	5.2 to 25.4	12.6	2.8	4.0	3.6	.35	.50	.45
Carpentry & Millwork	6.7 to 53.2	18.5	3.7	4.0	0.5	.68	.74	.09
Metal or Composition Siding	5.3 to 35.5	16.1	2.3	0.3	4.3	.37	.05	.69
Roofing	7.3 to 77.1	31.8	2.3	2.6	3.1	.73	.83	.99
Doors & Hardware	4.5 to 25.3	10.9	0.9	1.4	0.4	.10	.15	.04
Sash & Glazing	4.9 to 38.0	13.8	3.5	4.0	1.0	.48	.55	.14
Lath & Plaster	4.0 to 45.5	14.6	3.3	6.9	0.8	.48	1.01	.12
Tile, Marble & Floors	3.7 to 23.3	9.6	2.6	3.0	0.5	.25	.29	.05
Acoustical Ceilings	2.5 to 24.5	10.7	2.4	0.2	0.3	.26	.02	.03
Painting	4.3 to 29.6	12.9	1.5	1.6	1.6	.19	.21	.21
Interior Partitions	6.7 to 53.2	18.5	3.9	4.3	4.4	.72	.80	.81
Miscellaneous Items	2.5 to 110.1	17.3	5.2	3.7	9.7	.90	.64	1.68
Elevators	2.5 to 15.3	7.2	2.1	1.1	2.2	.15	.08	.16
Sprinklers	2.8 to 23.1	9.0	0.5	—	2.0	.05	—	.18
Plumbing	3.0 to 12.5	7.8	4.9	7.2	5.2	.38	.56	.41
Heat., Vent., Air Conditioning	4.0 to 28.1	11.1	13.5	11.0	12.9	1.50	1.22	1.43
Electrical	2.7 to 12.5	6.4	10.1	8.4	11.1	.65	.54	.71
Total	2.5 % to 110.1%	—	100.0%	100.0%	100.0%	16.35%	14.87%	18.36%
		Overall Weighted Average	16.53%					

Workers' Compensation Insurance Rates by States

The table below lists the weighted average Workers' Compensation base rate for each state with a factor comparing this with the national average of 16.2%.

State	Weighted Average	Factor	State	Weighted Average	Factor	State	Weighted Average	Factor
Alabama	28.0%	173	Kentucky	16.6%	102	North Dakota	12.9%	80
Alaska	17.9	110	Louisiana	28.2	174	Ohio	12.7	78
Arizona	7.1	44	Maine	20.8	128	Oklahoma	22.2	137
Arkansas	14.7	91	Maryland	12.0	74	Oregon	14.3	88
California	18.8	116	Massachusetts	14.6	90	Pennsylvania	16.0	99
Colorado	14.0	86	Michigan	19.0	117	Rhode Island	21.2	131
Connecticut	24.9	154	Minnesota	27.7	171	South Carolina	13.9	86
Delaware	11.9	73	Mississippi	17.0	105	South Dakota	14.5	90
District of Columbia	19.1	118	Missouri	20.7	128	Tennessee	16.4	101
Florida	31.4	194	Montana	20.6	127	Texas	14.6	90
Georgia	23.0	142	Nebraska	20.1	124	Utah	12.3	76
Hawaii	16.9	104	Nevada	16.2	100	Vermont	20.0	123
Idaho	10.4	64	New Hampshire	22.6	140	Virginia	12.5	77
Illinois	18.2	112	New Jersey	10.6	65	Washington	10.6	65
Indiana	6.2	38	New Mexico	15.2	94	West Virginia	12.8	79
Iowa	12.0	74	New York	13.5	83	Wisconsin	15.9	98
Kansas	8.8	54	North Carolina	13.7	85	Wyoming	8.0	49
			Weighted Average for U.S. is	16.5% of payroll = 100%				

Rates in the following table are the base or manual costs per $100 of payroll for Workers' Compensation in each state. Rates are usually applied to straight time wages only and not to premium time wages and bonuses.

The weighted average skilled worker rate for 35 trades is 16.2%. For bidding purposes, apply the full value of Workers' Compensation directly to total labor costs, or if labor is 38%, materials 42% and overhead and profit 20% of total cost, carry 38/80 x 16.2% =7.7% of cost (before overhead and profit) into overhead. Rates vary not only from state to state but also with the experience rating of the contractor.

Rates are the most current available at the time of publication.

R01100-060 Workers' Compensation Insurance Rates by Trade and State (cont.)

State	Carpentry — 3 stories or less 5651	Carpentry — interior cab. work 5437	Carpentry — general 5403	Concrete Work — NOC 5213	Concrete Work — flat (flr., sdwk.) 5221	Electrical Wiring — inside 5190	Excavation — earth NOC 6217	Excavation — rock 6217	Glaziers 5462	Insulation Work 5479	Lathing 5443	Masonry 5022	Painting & Decorating 5474	Pile Driving 6003	Plastering 5480	Plumbing 5183	Roofing 5551	Sheet Metal Work (HVAC) 5538	Steel Erection — door & sash 5102	Steel Erection — inter., ornam. 5102	Steel Erection — structure 5040	Steel Erection — NOC 5057	Tile Work — (interior ceramic) 5348	Waterproofing 9014	Wrecking 5701
AL	23.03	13.25	33.49	13.23	10.32	9.17	15.14	15.14	38.02	19.34	13.00	30.97	28.83	29.60	45.45	11.88	68.79	28.14	25.40	25.40	44.99	44.16	15.68	7.34	44.99
AK	12.28	13.50	11.97	11.89	9.38	9.65	18.53	18.53	16.81	21.09	8.51	16.75	13.92	51.57	13.48	8.75	36.97	9.15	11.46	11.46	29.37	18.17	8.49	8.15	29.37
AZ	6.40	4.59	13.14	6.28	3.94	3.56	5.17	5.17	6.37	9.71	4.46	6.18	4.25	8.37	5.21	3.64	11.20	4.75	7.99	7.99	12.75	6.86	3.76	2.71	50.31
AR	12.83	10.62	14.66	14.60	6.75	6.24	9.07	9.07	15.62	26.86	10.49	10.75	9.85	13.56	13.96	6.36	22.13	10.19	8.32	8.32	37.69	30.03	7.13	4.00	37.69
CA	28.28	9.07	28.28	13.18	13.18	9.93	7.88	7.88	16.37	23.34	10.90	14.55	19.45	21.18	18.12	11.59	39.79	15.33	13.55	13.55	23.93	21.91	7.74	19.45	21.91
CO	17.74	9.16	12.15	13.17	8.11	5.71	11.90	11.90	10.18	15.27	5.87	14.10	11.24	19.75	11.07	8.33	25.02	13.10	8.13	8.13	34.92	15.33	8.02	6.34	34.92
CT	21.93	17.69	29.90	28.54	17.17	9.55	12.88	12.88	18.03	32.51	15.16	28.12	18.82	29.36	25.85	11.47	52.53	14.89	16.46	16.46	70.71	21.51	13.29	7.07	50.86
DE	11.48	11.48	9.88	9.99	7.61	5.06	7.82	7.82	9.69	9.88	10.52	9.25	12.70	16.84	10.52	6.05	21.69	8.41	10.11	10.11	23.77	10.11	8.23	9.25	23.37
DC	12.49	11.98	16.79	21.64	21.97	7.25	8.71	8.71	31.70	12.33	8.71	19.06	9.35	23.54	15.51	11.65	25.17	8.78	17.99	17.99	54.08	22.86	23.27	3.99	54.08
FL	35.52	25.27	36.16	35.18	17.49	12.49	16.03	16.03	27.03	25.63	15.21	27.00	24.86	68.52	42.18	12.45	53.69	21.04	17.19	17.19	67.47	46.35	12.40	10.76	67.47
GA	32.80	16.96	24.66	16.82	12.15	9.13	14.76	14.76	18.12	19.52	19.99	21.28	18.47	30.93	20.75	11.27	40.25	15.09	14.49	14.49	38.29	55.91	10.34	9.92	38.29
HI	17.38	11.62	28.20	13.74	12.27	7.10	7.73	7.73	20.55	21.19	10.94	17.87	11.33	20.15	15.61	6.06	33.24	8.02	11.11	11.11	31.71	21.70	9.78	11.54	31.71
ID	9.86	5.74	13.99	9.13	6.98	4.97	5.36	5.36	9.28	7.65	4.97	8.01	7.10	11.84	8.68	3.92	27.36	8.02	9.87	9.87	26.44	12.60	4.78	4.98	26.44
IL	15.59	13.15	18.76	27.06	10.81	8.23	9.43	9.43	15.78	13.70	9.70	16.83	10.09	25.35	13.11	9.87	29.38	14.82	15.55	15.55	50.68	23.78	15.66	4.69	50.68
IN	5.29	4.52	6.96	5.65	3.48	2.67	3.50	3.50	4.94	3.74	2.67	5.12	4.84	7.38	4.04	2.97	11.69	4.41	5.18	5.18	21.38	10.19	3.69	2.46	21.38
IA	10.40	5.30	11.50	12.21	6.38	3.96	6.30	6.30	13.38	8.84	5.29	7.81	8.26	9.58	8.44	4.85	17.87	6.26	10.24	10.24	48.08	29.00	6.20	4.52	29.58
KS	10.94	7.55	10.15	7.52	5.93	3.67	5.22	5.22	6.91	8.78	4.57	7.23	6.46	8.35	9.08	4.91	18.79	6.05	7.00	7.00	20.07	12.71	4.25	3.02	20.07
KY	17.33	12.77	19.21	16.77	6.05	6.85	17.98	17.98	17.33	19.31	11.00	10.81	10.91	19.93	17.05	5.48	17.33	13.46	13.29	13.29	36.48	24.51	12.79	3.95	36.48
LA	23.14	24.93	53.17	26.43	15.61	9.88	17.49	17.49	20.14	21.43	24.51	28.33	29.58	31.25	22.37	8.64	77.12	21.20	18.98	18.98	51.76	24.21	13.77	13.78	66.41
ME	13.43	10.42	43.59	24.29	11.12	5.03	12.19	12.19	13.35	15.39	14.03	17.26	16.07	31.44	17.95	7.32	32.07	8.82	15.08	15.08	37.84	61.81	11.66	6.06	37.84
MD	10.55	5.95	10.55	11.35	5.05	5.15	9.25	9.25	13.20	13.05	6.45	11.35	6.75	27.45	6.65	5.55	22.80	7.00	9.15	9.15	26.80	18.50	6.35	3.50	26.80
MA	10.62	6.96	16.60	17.84	9.34	3.69	6.41	6.41	8.63	14.02	6.80	14.44	8.26	14.80	5.69	5.09	33.29	8.79	14.31	14.31	40.78	35.81	11.47	3.86	38.14
MI	21.68	13.59	19.80	22.12	9.35	5.71	11.69	11.69	13.87	12.38	13.59	19.33	14.93	39.06	15.59	8.24	41.33	10.41	11.36	11.36	39.06	29.15	11.31	6.40	39.06
MN	20.92	21.58	36.74	16.39	13.64	6.79	15.72	15.72	18.94	22.75	23.83	25.13	18.89	29.83	23.83	10.56	65.54	10.92	16.66	16.66	112.03	36.34	17.83	6.43	132.92
MS	15.88	14.44	19.71	11.45	9.13	7.19	10.97	10.97	11.49	12.48	7.98	13.65	12.76	23.19	17.21	7.92	31.71	19.53	10.85	10.85	42.19	33.21	9.10	7.01	42.19
MO	18.67	12.07	17.22	17.39	12.13	7.88	11.53	11.53	13.12	21.70	14.67	18.39	15.05	25.80	16.18	9.66	36.85	13.49	14.36	14.36	69.65	43.83	6.53	8.15	69.65
MT	22.03	10.06	20.00	12.80	10.83	5.54	17.20	17.20	11.51	16.19	19.02	14.58	12.03	76.71	13.82	8.93	59.20	9.44	9.73	9.73	41.23	17.93	6.54	5.33	41.23
NE	21.22	10.97	20.95	28.60	9.65	8.32	12.20	12.20	15.50	24.47	11.30	18.50	12.32	23.35	14.75	11.17	36.72	13.05	13.15	13.15	52.27	38.35	9.97	6.15	50.40
NV	17.85	9.06	13.89	10.87	10.44	7.67	11.47	11.47	14.46	17.06	7.20	12.16	11.24	13.14	12.28	10.71	23.68	20.06	16.04	16.04	36.06	33.10	9.57	7.25	44.29
NH	16.93	11.48	21.49	32.83	11.47	6.81	17.55	17.55	12.44	28.41	8.80	21.86	15.11	17.52	21.89	11.37	61.69	12.49	14.85	14.85	59.93	32.72	14.21	6.54	59.93
NJ	11.19	7.77	11.19	9.15	6.78	3.85	6.97	6.97	7.19	10.59	9.00	12.02	8.97	13.42	9.00	5.75	28.36	6.67	9.38	9.38	17.22	10.28	4.46	4.54	23.36
NM	22.18	6.84	14.66	12.96	8.24	6.41	8.27	8.27	12.43	11.53	6.24	12.92	10.22	16.44	9.38	7.70	28.79	10.65	19.01	19.01	43.96	21.54	5.89	6.30	43.96
NY	13.46	6.24	14.87	17.22	11.81	5.86	8.25	8.25	10.46	8.73	15.59	17.70	12.38	15.54	8.98	7.46	28.94	14.33	8.80	8.80	14.52	20.59	8.83	6.21	29.99
NC	13.68	10.74	18.04	13.14	6.63	7.71	8.44	8.44	10.19	11.90	7.68	9.44	9.58	15.60	15.16	8.08	25.69	10.77	8.00	8.00	41.99	18.79	7.12	4.14	41.99
ND	9.60	9.60	19.60	5.81	5.81	4.00	5.39	5.39	9.60	9.60	9.04	8.48	6.30	20.12	9.04	5.35	21.53	5.35	20.12	20.12	20.12	20.12	9.60	21.53	14.48
OH	6.07	12.00	9.68	12.78	9.99	5.55	8.44	8.44	9.99	12.41	2.48	11.45	15.04	25.09	4.63	6.40	21.67	8.11	10.30	10.30	26.21	16.39	8.70	15.16	26.21
OK	26.16	11.79	19.33	15.51	11.50	7.41	18.45	18.45	13.55	18.00	12.05	16.80	17.10	36.26	18.29	8.72	42.26	12.45	11.17	11.17	69.45	46.76	11.42	8.20	69.45
OR	17.06	9.31	16.98	13.61	8.81	5.00	10.55	10.55	16.83	8.87	7.87	15.13	14.15	16.07	11.45	6.07	22.99	11.17	10.25	10.25	35.39	16.67	10.80	5.00	35.39
PA	8.69	8.69	12.27	13.48	8.94	6.46	7.91	7.91	15.69	12.63	18.94	11.65	13.04	16.76	10.78	7.18	25.87	12.58	13.96	13.96	59.05	13.96	8.22	18.96	81.34
RI	19.53	11.65	18.07	18.23	16.24	4.43	10.38	10.38	12.85	22.78	11.97	25.11	24.13	37.66	17.25	8.52	33.92	10.29	14.07	14.07	59.49	37.50	14.36	7.70	78.79
SC	19.10	13.03	19.24	12.88	6.19	7.02	8.32	8.32	12.22	10.69	7.27	9.55	11.19	16.64	18.89	6.88	28.98	11.42	9.49	9.49	21.80	23.66	6.23	4.65	21.80
SD	15.40	7.40	15.04	18.64	5.48	5.64	11.42	11.42	9.88	13.70	7.72	8.28	13.08	25.78	11.65	8.56	20.66	8.33	11.93	11.93	45.92	17.38	6.33	4.17	45.92
TN	26.71	12.31	20.38	16.92	8.67	7.33	12.26	12.26	10.17	12.94	10.97	14.30	12.72	15.20	15.23	9.52	33.46	12.19	9.28	9.28	39.46	21.96	7.22	5.73	39.46
TX	16.89	11.33	13.25	12.10	8.85	7.36	10.49	10.49	11.61	15.68	9.53	13.68	9.89	13.95	20.99	8.06	22.80	14.58	11.23	11.23	33.39	14.78	7.29	8.09	17.86
UT	9.45	6.23	10.52	7.68	7.99	7.05	6.38	6.38	9.55	10.95	12.32	13.75	15.29	17.24	10.60	6.41	26.27	6.30	8.99	8.99	23.51	23.51	6.06	5.83	26.53
VT	19.90	11.34	19.46	34.13	12.01	6.10	10.61	10.61	21.25	25.54	10.85	19.53	10.41	23.91	18.23	10.10	28.73	12.62	13.94	13.94	47.05	37.23	8.52	9.59	47.05
VA	11.10	8.05	11.35	12.89	6.35	4.15	7.42	7.42	7.80	7.18	12.99	8.00	10.15	16.73	7.63	6.04	22.40	7.96	13.93	13.93	35.15	20.77	9.28	3.45	35.15
WA	7.57	7.57	8.66	6.22	7.51	2.85	8.31	8.31	13.01	7.88	8.66	12.82	9.07	18.93	10.38	4.68	19.63	4.02	12.96	12.96	9.25	9.25	9.89	10.23	9.25
WV	12.00	12.00	12.00	23.80	23.80	5.13	7.74	7.74	6.08	6.08	16.88	11.57	12.20	12.89	12.20	4.77	13.44	6.08	14.98	14.98	9.84	13.15	16.02	3.88	13.27
WI	11.29	9.54	18.46	11.96	10.19	4.78	7.16	7.16	14.43	14.33	10.61	17.44	11.44	19.25	12.41	6.40	40.73	7.27	16.38	16.38	39.18	19.60	13.98	5.38	39.18
WY	7.29	7.29	7.29	7.29	7.29	7.29	7.29	7.29	7.29	7.29	7.29	7.29	7.29	7.29	7.29	7.29	7.29	7.29	7.29	7.29	7.29	7.29	7.29	7.29	7.29
AVG.	16.06	10.91	18.51	15.79	9.94	6.40	10.34	10.34	13.82	15.24	10.71	14.97	12.89	22.94	14.62	7.78	31.75	11.09	12.61	12.61	38.86	24.78	9.63	7.27	40.51

R01100-060 Workers' Compensation (cont.) (Canada in Canadian dollars)

Province		Alberta	British Columbia	Manitoba	Ontario	New Brunswick	Newfndld. & Labrador	Northwest Territories	Nova Scotia	Prince Edward Island	Quebec	Saskat-chewan	Yukon
Carpentry—3 stories or less	Rate	2.72	4.63	3.80	5.00	4.40	7.20	4.62	8.06	7.25	12.69	5.01	3.25
	Code	25401	721028	40102	723	422	403	4-41	4226	401	80110	B12-02	202
Carpentry—interior cab. work	Rate	1.17	4.85	3.80	5.00	3.48	7.20	4.62	5.69	3.49	12.69	3.25	3.25
	Code	42133	721021	40102	723	427	403	4-41	4274	402	80110	B11-27	202
CARPENTRY—general	Rate	2.72	4.63	3.80	5.00	4.40	7.20	4.62	8.06	7.25	12.69	5.01	3.25
	Code	25401	721028	40102	723	422	403	4-41	4226	401	80110	B12-02	202
CONCRETE WORK—NOC	Rate	7.65	6.48	5.76	17.18	4.40	7.20	4.62	4.97	7.25	13.89	6.50	3.25
	Code	42104	721010	40110	748	422	403	4-41	4224	401	80100	B13-14	203
CONCRETE WORK—flat (flr. sidewalk)	Rate	7.65	6.48	5.76	17.18	4.40	7.20	4.62	4.97	7.25	13.89	6.50	3.25
	Code	42104	721010	40110	748	422	403	4-41	4224	401	80100	B13-14	203
ELECTRICAL Wiring—inside	Rate	2.98	2.90	2.03	3.03	1.74	4.83	3.46	2.61	3.49	5.69	3.25	2.35
	Code	42124	721019	40203	704	426	400	4-46	4261	402	80170	B11-05	206
EXCAVATION—earth NOC	Rate	2.57	3.40	3.80	4.21	2.93	7.20	3.46	4.25	3.61	7.71	3.70	3.25
	Code	40604	721031	40706	711	421	403	4-43	4214	404	80030	R11-06	207
EXCAVATION—rock	Rate	2.57	3.40	3.80	4.21	2.93	7.20	3.46	4.25	3.61	7.71	3.70	3.25
	Code	40604	721031	40706	711	421	403	4-43	4214	404	80030	R11-06	207
GLAZIERS	Rate	2.79	3.21	3.77	8.42	6.01	5.12	4.62	8.06	3.49	14.40	6.50	2.35
	Code	42121	715020	40109	751	423	402	4-41	4233	402	80150	B13-04	212
INSULATION WORK	Rate	2.89	5.80	3.80	8.42	6.01	5.12	4.62	8.06	7.25	14.15	5.01	3.25
	Code	42184	721029	40102	751	423	402	4-41	4234	401	80120	B12-07	202
LATHING	Rate	8.79	8.29	3.80	5.00	3.48	5.12	4.62	5.69	3.49	14.15	6.50	3.25
	Code	42135	721033	40102	723	427	402	4-41	4271	402	80120	B13-16	202
MASONRY	Rate	7.65	8.29	3.80	12.36	6.01	7.20	4.62	8.06	7.25	13.89	6.50	3.25
	Code	42102	721037	40102	741	423	403	4-41	4231	401	80100	B13-18	202
PAINTING & DECORATING	Rate	4.07	5.80	4.20	7.09	3.28	5.12	4.62	5.69	3.49	12.69	5.01	3.25
	Code	42111	721041	40105	719	427	402	4-41	4275	402	80110	B12-01	202
PILE DRIVING	Rate	7.65	13.55	3.80	5.84	4.40	10.89	3.46	4.97	7.25	7.71	6.50	3.25
	Code	42159	722004	40706	732	422	404	4-43	4221	401	80030	B13-10	202
PLASTERING	Rate	8.79	8.29	4.46	7.09	3.28	5.12	4.62	5.69	3.49	12.69	5.01	3.25
	Code	42135	721042	40108	719	427	402	4-41	4271	402	80110	B12-21	202
PLUMBING	Rate	1.99	4.30	2.37	3.96	2.30	4.60	3.46	2.94	3.49	7.13	3.25	1.40
	Code	42122	721043	40204	707	424	401	4-46	4241	402	80160	B11-01	214
ROOFING	Rate	9.70	8.74	5.42	12.36	8.05	7.20	4.62	8.06	7.25	22.17	6.50	3.25
	Code	42118	721036	40403	728	430	403	4-41	4236	401	80130	B13-20	202
SHEET METAL WORK (HVAC)	Rate	1.99	4.30	5.14	3.96	2.30	4.60	3.46	2.94	3.49	7.13	3.25	2.35
	Code	42117	721043	40402	707	424	401	4-46	4244	402	80160	B11-07	208
STEEL ERECTION—door & sash	Rate	3.84	13.55	6.56	17.18	4.40	10.89	4.62	8.06	7.25	32.47	6.50	3.25
	Code	42106	722005	40502	748	422	404	4-41	4227	401	80080	B13-22	202
STEEL ERECTION—inter., ornam.	Rate	3.84	13.55	6.56	17.18	4.40	7.20	4.62	8.06	7.25	32.47	6.50	3.25
	Code	42106	722005	40502	748	422	403	4-41	4227	401	80080	B13-22	202
STEEL ERECTION—structure	Rate	3.84	13.55	6.56	17.18	4.40	10.89	4.62	8.06	7.25	32.47	6.50	3.25
	Code	42106	722005	40502	748	422	404	4-41	4227	401	80080	B13-22	202
STEEL ERECTION—NOC	Rate	3.84	13.55	6.56	17.18	4.40	10.89	4.62	8.06	7.25	32.47	6.50	3.25
	Code	42106	722005	40502	748	422	404	4-41	4227	401	80080	B13-22	202
TILE WORK—inter. (ceramic)	Rate	2.94	3.27	1.88	7.09	3.28	5.12	4.62	5.69	3.49	12.69	6.50	3.25
	Code	42113	721054	40103	719	427	402	4-41	4276	402	80110	B13-01	202
WATERPROOFING	Rate	4.07	5.80	3.80	5.00	6.01	7.20	4.62	8.06	3.49	22.17	5.01	3.25
	Code	42139	721016	40102	723	423	403	4-41	4239	402	80130	B12-17	202
WRECKING	Rate	2.57	6.08	5.15	17.18	2.93	7.20	3.46	4.25	7.25	12.69	6.50	3.25
	Code	40604	721005	40106	748	421	403	4-43	4211	401	80110	B13-09	202

R01100-070 Contractor's Overhead & Profit

Below are the **average** installing contractor's percentage mark-ups applied to base labor rates to arrive at typical billing rates.

Column A: Labor rates are based on union wages averaged for 30 major U.S. cities. Base rates including fringe benefits are listed hourly and daily. These figures are the sum of the wage rate and employer-paid fringe benefits such as vacation pay, employer-paid health and welfare costs, pension costs, plus appropriate training and industry advancement funds costs.

Column B: Workers' Compensation rates are the national average of state rates established for each trade.

Column C: Column C lists average fixed overhead figures for all trades. Included are Federal and State Unemployment costs set at 6.2%; Social Security Taxes (FICA) set at 7.65%; Builder's Risk Insurance costs set at 0.44%; and Public Liability costs set at 2.02%. All the percentages except those for Social Security Taxes vary from state to state as well as from company to company.

Columns D and E: Percentages in Columns D and E are based on the presumption that the installing contractor has annual billing of $4,000,000 and up. Overhead percentages may increase with smaller annual billing. The overhead percentages for any give such as the contractor's annual volume, engineering and logistical support costs, and staff requirements. The figures for overhead and profit will also vary depending on the type of job, the job location, and the prevailing economic conditions. All factors should be examined very carefully for each job.

Column F: Column F lists the total of Columns B, C, D, and E.

Column G: Column G is Column A (hourly base labor rate) multiplied by the percentage in Column F (O&P percentage).

Column H: Column H is the total of Column A (hourly base labor rate) plus Column G (Total O&P).

Column I: Column I is Column H multiplied by eight hours.

		A		B	C	D	E	F	G	H	I
		Base Rate Incl. Fringes		Work-ers' Comp. Ins.	Average Fixed Over-head	Over-head	Profit	Total Overhead & Profit		Rate with O & P	
Abbr.	Trade	Hourly	Daily					%	Amount	Hourly	Daily
Skwk	Skilled Workers Average (35 trades)	$33.65	$269.20	16.2%	16.3%	16.0%	15.0%	63.5%	$21.35	$55.00	$440.00
	Helpers Average (5 trades)	24.55	196.40	17.9				65.2	16.00	40.55	324.40
	Foreman Average, Inside ($.50 over trade)	34.15	273.20	16.2				63.5	21.70	55.85	446.80
	Foreman Average, Outside ($2.00 over trade)	35.65	285.20	16.2				63.5	22.65	58.30	466.40
Clab	Common Building Laborers	26.00	208.00	18.5				65.8	17.10	43.10	344.80
Asbe	Asbestos Workers	36.00	288.00	15.2				62.5	22.50	58.50	468.00
Boil	Boilermakers	41.25	330.00	13.1				60.4	24.90	66.15	529.20
Bric	Bricklayers	34.25	274.00	15.0				62.3	21.35	55.60	444.80
Brhe	Bricklayer Helpers	25.60	204.80	15.0				62.3	15.95	41.55	332.40
Carp	Carpenters	33.00	264.00	18.5				65.8	21.70	54.70	437.60
Cefi	Cement Finishers	31.55	252.40	9.9				57.2	18.05	49.60	396.80
Elec	Electricians	39.40	315.20	6.4				53.7	21.15	60.55	484.40
Elev	Elevator Constructors	41.60	332.80	7.2				54.5	22.65	64.25	514.00
Eqhv	Equipment Operators, Crane or Shovel	34.80	278.40	10.3				57.6	20.05	54.85	438.80
Eqmd	Equipment Operators, Medium Equipment	33.65	269.20	10.3				57.6	19.40	53.05	424.40
Eqlt	Equipment Operators, Light Equipment	32.15	257.20	10.3				57.6	18.50	50.65	405.20
Eqol	Equipment Operators, Oilers	29.20	233.60	10.3				57.6	16.80	46.00	368.00
Eqmm	Equipment Operators, Master Mechanics	35.20	281.60	10.3				57.6	20.30	55.50	444.00
Glaz	Glaziers	32.05	256.40	13.8				61.1	19.60	51.65	413.20
Lath	Lathers	30.55	244.40	10.7				58.0	17.70	48.25	386.00
Marb	Marble Setters	31.95	255.60	15.0				62.3	19.90	51.85	414.80
Mill	Millwrights	34.35	274.80	10.3				57.6	19.80	54.15	433.20
Mstz	Mosaic and Terrazzo Workers	31.60	252.80	9.6				56.9	18.00	49.60	396.80
Pord	Painters, Ordinary	29.60	236.80	12.9				60.2	17.80	47.40	379.20
Psst	Painters, Structural Steel	30.05	240.40	50.0				97.3	29.25	59.30	474.40
Pape	Paper Hangers	29.35	234.80	12.9				60.2	17.65	47.00	376.00
Pile	Pile Drivers	32.05	256.40	22.9				70.2	22.50	54.55	436.40
Plas	Plasterers	29.95	239.60	14.6				61.9	18.55	48.50	388.00
Plah	Plasterer Helpers	25.80	206.40	14.6				61.9	15.95	41.75	334.00
Plum	Plumbers	39.60	316.80	7.8				55.1	21.80	61.40	491.20
Rodm	Rodmen (Reinforcing)	37.10	296.80	24.8				72.1	26.75	63.85	510.80
Rofc	Roofers, Composition	28.45	227.60	31.8				79.1	22.50	50.95	407.60
Rots	Roofers, Tile and Slate	28.65	229.20	31.8				79.1	22.65	51.30	410.40
Rohe	Roofer Helpers (Composition)	20.95	167.60	31.8				79.1	16.55	37.50	300.00
Shee	Sheet Metal Workers	38.80	310.40	11.1				58.4	22.65	61.45	491.60
Spri	Sprinkler Installers	39.25	314.00	9.0				56.3	22.10	61.35	490.80
Stpi	Steamfitters or Pipefitters	39.75	318.00	7.8				55.1	21.90	61.65	493.20
Ston	Stone Masons	33.85	270.80	15.0				62.3	21.10	54.95	439.60
Sswk	Structural Steel Workers	37.15	297.20	38.9				86.2	32.00	69.15	553.20
Tilf	Tile Layers (Floor)	31.50	252.00	9.6				56.9	17.90	49.40	395.20
Tilh	Tile Layer Helpers	24.45	195.60	9.6				56.9	13.90	38.35	306.80
Trlt	Truck Drivers, Light	25.70	205.60	15.1				62.4	16.05	41.75	334.00
Trhv	Truck Drivers, Heavy	26.45	211.60	15.1				62.4	16.50	42.95	343.60
Sswl	Welders, Structural Steel	37.15	297.20	38.9				86.2	32.00	69.15	553.20
Wrck	*Wrecking	26.00	208.00	40.5				87.8	22.85	48.85	390.80

*Not included in Averages.

R01100-080 Performance Bond

This table shows the cost of a Performance Bond for a construction job scheduled to be completed in 12 months. Add 1% of the premium cost per month for jobs requiring more than 12 months to complete. The rates are "standard" rates offered to contractors that the bonding company considers financially sound and capable of doing the work. Preferred rates are offered by some bonding companies based upon financial strength of the contractor. Actual rates vary from contractor to contractor and from bonding company to bonding company. Contractors should prequalify through a bonding agency before submitting a bid on a contract that requires a bond.

Contract Amount			Building Construction				Highways & Bridges							
			Class B Projects				Class A New Construction				Class A-1 Highway Resurfacing			
First	$100,000	bid	$25.00 per M				$15.00 per M				$9.40 per M			
Next	400,000	bid	$2,500	plus	$15.00	per M	$1,500	plus	$10.00	per M	$940	plus	$7.20	per M
Next	2,000,000	bid	8,500	plus	10.00	per M	5,500	plus	7.00	per M	3,820	plus	5.00	per M
Next	2,500,000	bid	28,500	plus	7.50	per M	19,500	plus	5.50	per M	15,820	plus	4.50	per M
Next	2,500,000	bid	47,250	plus	7.00	per M	33,250	plus	5.00	per M	28,320	plus	4.50	per M
Over	7,500,000	bid	64,750	plus	6.00	per M	45,750	plus	4.50	per M	39,570	plus	4.00	per M

R01100-090 Sales Tax by State

State sales tax on materials is tabulated below (5 states have no sales tax). Many states allow local jurisdictions, such as a county or city, to levy additional sales tax.

Some projects may be sales tax exempt, particularly those constructed with public funds.

State	Tax (%)	State	Tax (%)	State	Tax (%)	State	Tax (%)
Alabama	4	Illinois	6.25	Montana	0	Rhode Island	7
Alaska	0	Indiana	5	Nebraska	5	South Carolina	5
Arizona	5	Iowa	5	Nevada	6.5	South Dakota	4
Arkansas	4.625	Kansas	4.9	New Hampshire	0	Tennessee	6
California	6	Kentucky	6	New Jersey	6	Texas	6.25
Colorado	3	Louisiana	4	New Mexico	5	Utah	4.75
Connecticut	6	Maine	5	New York	4	Vermont	5
Delaware	0	Maryland	5	North Carolina	4	Virginia	3.5
District of Columbia	5.75	Massachusetts	5	North Dakota	5	Washington	6.5
Florida	6	Michigan	6	Ohio	5	West Virginia	6
Georgia	4	Minnesota	6.5	Oklahoma	4.5	Wisconsin	5
Hawaii	4	Mississippi	7	Oregon	0	Wyoming	4
Idaho	5	Missouri	4.225	Pennsylvania	6	Average	4.65 %

Sales Tax by Province (Canada)

GST - a value-added tax, which the government imposes on most goods and services provided in or imported into Canada.
PST - a retail sales tax, which five of the provinces impose on the price of most goods and some services.
QST - a value-added tax, similar to the federal GST, which Quebec imposes.
HST - Three provinces have combined their retail sales tax with the federal GST into one harmonized tax.

Province	PST (%)	QST (%)	GST(%)	HST(%)
Alberta	0	0	7	0
British Columbia	7.5	0	7	0
Manitoba	7	0	7	0
New Brunswick	0	0	0	15
Newfoundland	0	0	0	15
Northwest Territories	0	0	7	0
Nova Scotia	0	0	0	15
Ontario	8	0	7	0
Prince Edward Island	10	0	7	0
Quebec	0	7.5	7	0
Saskatchewan	6	0	7	0
Yukon	0	0	7	0

R01100-100 Unemployment Taxes and Social Security Taxes

Mass. State Unemployment tax ranges from 1.325% to 7.225% plus an experience rating assessment the following year, on the first $10,800 of wages. Federal Unemployment tax is 6.2% of the first $7,000 of wages. This is reduced by a credit for payment to the state. The minimum Federal Unemployment tax is 0.8% after all credits.

Combined rates in Mass. thus vary from 2.125% to 8.025% of the first $10,800 of wages. Combined average U.S. rate is about 6.2% of the first $7,000. Contractors with permanent workers will pay less since the average annual wages for skilled workers is $33.65 x 2,000 hours or about $67,300 per year. The average combined rate for U.S. would thus be 6.2% x $7,000 ÷ $67,300 = 0.6% of total wages for permanent employees.

Rates vary not only from state to state but also with the experience rating of the contractor.

Social Security (FICA) for 2004 is estimated at time of publication to be 7.65% of wages up to $87,000.

R01107-010 Architectural Fees

Tabulated below are typical percentage fees by project size, for good professional architectural service. Fees may vary from those listed depending upon degree of design difficulty and economic conditions in any particular area.

Rates can be interpolated horizontally and vertically. Various portions of the same project requiring different rates should be adjusted proportionately. For alterations, add 50% to the fee for the first $500,000 of project cost and add 25% to the fee for project cost over $500,000.

Architectural fees tabulated below include Structural, Mechanical and Electrical Engineering Fees. They do not include the fees for special consultants such as kitchen planning, security, acoustical, interior design, etc.

Civil Engineering fees are included in the Architectural fee for project sites requiring minimal design such as city sites. However, separate Civil Engineering fees must be added when utility connections require design, drainage calculations are needed, stepped foundations are required, or provisions are required to protect adjacent wetlands.

Building Types	Total Project Size in Thousands of Dollars						
	100	250	500	1,000	5,000	10,000	50,000
Factories, garages, warehouses, repetitive housing	9.0%	8.0%	7.0%	6.2%	5.3%	4.9%	4.5%
Apartments, banks, schools, libraries, offices, municipal buildings	12.2	12.3	9.2	8.0	7.0	6.6	6.2
Churches, hospitals, homes, laboratories, museums, research	15.0	13.6	12.7	11.9	9.5	8.8	8.0
Memorials, monumental work, decorative furnishings	—	16.0	14.5	13.1	10.0	9.0	8.3

R01107-030 Engineering Fees

Typical **Structural Engineering Fees** based on type of construction and total project size. These fees are included in Architectural Fees.

Type of Construction	Total Project Size (in thousands of dollars)			
	$500	$500-$1,000	$1,000-$5,000	Over $5000
Industrial buildings, factories & warehouses	Technical payroll times 2.0 to 2.5	1.60%	1.25%	1.00%
Hotels, apartments, offices, dormitories, hospitals, public buildings, food stores	↓	2.00%	1.70%	1.20%
Museums, banks, churches and cathedrals	↓	2.00%	1.75%	1.25%
Thin shells, prestressed concrete, earthquake resistive	↓	2.00%	1.75%	1.50%
Parking ramps, auditoriums, stadiums, convention halls, hangars & boiler houses	↓	2.50%	2.00%	1.75%
Special buildings, major alterations, underpinning & future expansion	↓	Add to above 0.5%	Add to above 0.5%	Add to above 0.5%

For complex reinforced concrete or unusually complicated structures, add 20% to 50%.

Typical **Mechanical and Electrical Engineering Fees** are based on the size of the subcontract. The fee structure for Mechanical Engineering is shown below; Electrical Engineering fees range from 2 to 5%. These fees are included in Architectural Fees.

Type of Construction	Subcontract Size							
	$25,000	$50,000	$100,000	$225,000	$350,000	$500,000	$750,000	$1,000,000
Simple structures	6.4%	5.7%	4.8%	4.5%	4.4%	4.3%	4.2%	4.1%
Intermediate structures	8.0	7.3	6.5	5.6	5.1	5.0	4.9	4.8
Complex structures	12.0	9.0	9.0	8.0	7.5	7.5	7.0	7.0

For renovations, add 15% to 25% to applicable fee.

R01250-010 Repair and Remodeling

Cost figures are based on new construction utilizing the most cost-effective combination of labor, equipment and material with the work scheduled in proper sequence to allow the various trades to accomplish their work in an efficient manner.

The costs for repair and remodeling work must be modified due to the following factors that may be present in any given repair and remodeling project.

1. Equipment usage curtailment due to the physical limitations of the project, with only hand-operated equipment being used.
2. Increased requirement for shoring and bracing to hold up the building while structural changes are being made and to allow for temporary storage of construction materials on above-grade floors.
3. Material handling becomes more costly due to having to move within the confines of an enclosed building. For multi-story construction, low capacity elevators and stairwells may be the only access to the upper floors.
4. Large amount of cutting and patching and attempting to match the existing construction is required. It is often more economical to remove entire walls rather than create many new door and window openings. This sort of trade-off has to be carefully analyzed.
5. Cost of protection of completed work is increased since the usual sequence of construction usually cannot be accomplished.
6. Economies of scale usually associated with new construction may not be present. If small quantities of components must be custom fabricated due to job requirements, unit costs will naturally increase. Also, if only small work areas are available at a given time, job scheduling between trades becomes difficult and subcontractor quotations may reflect the excessive start-up and shut-down phases of the job.
7. Work may have to be done on other than normal shifts and may have to be done around an existing production facility which has to stay in production during the course of the repair and remodeling.
8. Dust and noise protection of adjoining non-construction areas can involve substantial special protection and alter usual construction methods.
9. Job may be delayed due to unexpected conditions discovered during demolition or removal. These delays ultimately increase construction costs.
10. Piping and ductwork runs may not be as simple as for new construction. Wiring may have to be snaked through walls and floors.
11. Matching "existing construction" may be impossible because materials may no longer be manufactured. Substitutions may be expensive.
12. Weather protection of existing structure requires additional temporary structures to protect building at openings.
13. On small projects, because of local conditions, it may be necessary to pay a tradesman for a minimum of four hours for a task that is completed in one hour.

All of the above areas can contribute to increased costs for a repair and remodeling project. Each of the above factors should be considered in the planning, bidding and construction stage in order to minimize the increased costs associated with repair and remodeling jobs.

R01540-100 Steel Tubular Scaffolding

On new construction, tubular scaffolding is efficient up to 60′ high or five stories. Above this it is usually better to use a hung scaffolding if construction permits. Swing scaffolding operations may interfere with tenants. In this case, the tubular is more practical at all heights.

In repairing or cleaning the front of an existing building the cost of tubular scaffolding per S.F. of building front increases as the height increases above the first tier. The first tier cost is relatively high due to leveling and alignment.

The minimum efficient crew for erection is three workers. For heights over 50′, a crew of four is more efficient. Use two or more on top and two at the bottom for handing up or hoisting. Four workers can erect and dismantle about nine frames per hour up to five stories. From five to eight stories they will average six frames per hour. With 7′ horizontal spacing this will run about 400 S.F. and 265 S.F. of wall surface, respectively. Time for placing planks must be added to the above. On heights above 50′, five planks can be placed per labor-hour.

The table below shows the number of pieces required to erect tubular steel scaffolding for 1000 S.F. of building frontage. This area is made up of a scaffolding system that is 12 frames (11 bays) long by 2 frames high.

For jobs under twenty-five frames, add 50% to rental cost. Rental rates will be lower for jobs over three months duration. Large quantities for long periods can reduce rental rates by 20%.

Description of Component	CSI Line Item	Number of Pieces for 1000 S.F. of Building Front	Unit
5′ Wide Standard Frame, 6′-4″ High	01540-750-2200	24	Ea.
Leveling Jack & Plate	01540-750-2650	24	↓
Cross Brace	01540-750-2500	44	
Side Arm Bracket, 21″	01540-750-2700	12	
Guardrail Post	01540-750-2550	12	
Guardrail, 7′ section	01540-750-2600	22	
Stairway Section	01540-750-2900	2	
Stairway Starter Bar	01540-750-2910	1	
Stairway Inside Handrail	01540-750-2920	2	
Stairway Outside Handrail	01540-750-2930	2	
Walk-Thru Frame Guardrail	01540-750-2940	2	↓

Scaffolding is often used as falsework over 15′ high during construction of cast-in-place concrete beams and slabs. Two foot wide scaffolding is generally used for heavy beam construction. The span between frames depends upon the load to be carried with a maximum span of 5′.

Heavy duty shoring frames with a capacity of 10,000#/leg can be spaced up to 10′ O.C. depending upon form support design and loading.

Scaffolding used as horizontal shoring requires less than half the material required with conventional shoring.

On new construction, erection is done by carpenters.

Rolling towers supporting horizontal shores can reduce labor and speed the job. For maintenance work, catwalks with spans up to 70′ can be supported by the rolling towers.

R01540-200 Pump Staging

Pump staging is generally not available for rent. The table below shows the number of pieces required to erect pump staging for 2400 S.F. of building frontage. This area is made up of a pump jack system that is 3 poles (2 bays) wide by 2 poles high.

Item	CSI Line Item	Number of Pieces for 2400 S.F. of Building Front	Unit
Aluminum pole section, 24′ long	01540-550-0200	6	Ea.
Aluminum splice joint, 6′ long	01540-550-0600	3	↓
Aluminum foldable brace	01540-550-0900	3	
Aluminum pump jack	01540-550-0700	3	
Aluminum support for workbench/back safety rail	01540-550-1000	3	
Aluminum scaffold plank/workbench, 14″ wide x 24′ long	01540-550-1100	4	
Safety net, 22′ long	01540-550-1250	2	
Aluminum plank end safety rail	01540-550-1200	2	↓

The cost in place for this 2400 S.F. will depend on how many uses are realized during the life of the equipment. Several options are given in Division 01540-550.

R01590-100 Contractor Equipment

Rental Rates shown in Division 01590 pertain to late model high quality machines in excellent working condition, rented from equipment dealers. Rental rates from contractors may be substantially lower than the rental rates from equipment dealers depending upon economic conditions; for older, less productive machines, reduce rates by a maximum of 15%. Any overtime must be added to the base rates. For shift work, rates are lower. Usual rule of thumb is 150% of one shift rate for two shifts; 200% for three shifts.

For periods of less than one week, operated equipment is usually more economical to rent than renting bare equipment and hiring an operator.

Costs to move equipment to a job site (mobilization) or from a job site (demobilization) are not included in rental rates in Division 01590, nor in any Equipment costs on any Unit Price line items or crew listings. These costs can be found in section 02305-250. If a piece of equipment is already at a job site, it is not appropriate to utilize mob/demob costs in an estimate again.

Rental rates vary throughout the country with larger cities generally having lower rates. Lease plans for new equipment are available for periods in excess of six months with a percentage of payments applying toward purchase.

Monthly rental rates vary from 2% to 5% of the cost of the equipment depending on the anticipated life of the equipment and its wearing parts. Weekly rates are about 1/3 the monthly rates and daily rental rates about 1/3 the weekly rate.

The hourly operating costs for each piece of equipment include costs to the user such as fuel, oil, lubrication, normal expendables for the equipment, and a percentage of mechanic's wages chargeable to maintenance. The hourly operating costs listed do not include the operator's wages.

The daily cost for equipment used in the standard crews is figured by dividing the weekly rate by five, then adding eight times the hourly operating cost to give the total daily equipment cost, not including the operator. This figure is in the right hand column of Division 01590 under Crew Equipment Cost/Day.

Pile Driving rates shown for pile hammer and extractor do not include leads, crane, boiler or compressor. Vibratory pile driving requires an added field specialist during set-up and pile driving operation for the electric model. The hydraulic model requires a field specialist for set-up only. Up to 125 reuses of sheet piling are possible using vibratory drivers. For normal conditions, crane capacity for hammer type and size are as follows.

Crane Capacity	Hammer Type and Size		
	Air or Steam	Diesel	Vibratory
25 ton	to 8,750 ft.-lb.		70 H.P.
40 ton	15,000 ft.-lb.	to 32,000 ft.-lb.	170 H.P.
60 ton	25,000 ft.-lb.		300 H.P.
100 ton		112,000 ft.-lb.	

Cranes should be specified for the job by size, building and site characteristics, availability, performance characteristics, and duration of time required.

Backhoes & Shovels rent for about the same as equivalent size cranes but maintenance and operating expense is higher. Crane operators rate must be adjusted for high boom heights. Average adjustments: for 150′ boom add 2% per hour; over 185′, add 4% per hour; over 210′, add 6% per hour; over 250′, add 8% per hour and over 295′, add 12% per hour.

Tower Cranes of the climbing or static type have jibs from 50′ to 200′ and capacities at maximum reach range from 4,000 to 14,000 pounds. Lifting capacities increase up to maximum load as the hook radius decreases.

Typical rental rates, based on purchase price are about 2% to 3% per month.

Erection and dismantling runs between 500 and 2000 labor hours. Climbing operation takes 10 labor hours per 20′ climb. Crane dead time is about 5 hours per 40′ climb. If crane is bolted to side of the building add cost of ties and extra mast sections. Climbing cranes have from 80′ to 180′ of mast while static cranes have 80′ to 800′ of mast.

Truck Cranes can be converted to tower cranes by using tower attachments. Mast heights over 400′ have been used. See Division 01590-600 for rental rates of high boom cranes.

A single 100′ high material **Hoist and Tower** can be erected and dismantled in about 400 labor hours; a double 100′ high hoist and tower in about 600 labor hours. Erection times for additional heights are 3 and 4 labor hours per vertical foot respectively up to 150′, and 4 to 5 labor hours per vertical foot over 150′ high. A 40′ high portable Buck hoist takes about 160 labor hours to erect and dismantle. Additional heights take 2 labor hours per vertical foot to 80′ and 3 labor hours per vertical foot for the next 100′. Most material hoists do not meet local code requirements for carrying personnel.

A 150′ high **Personnel Hoist** requires about 500 to 800 labor hours to erect and dismantle. Budget erection time at 5 labor hours per vertical foot for all trades. Local code requirements or labor scarcity requiring overtime can add up to 50% to any of the above erection costs.

Earthmoving Equipment: The selection of earthmoving equipment depends upon the type and quantity of material, moisture content, haul distance, haul road, time available, and equipment available. Short haul cut and fill operations may require dozers only, while another operation may require excavators, a fleet of trucks, and spreading and compaction equipment. Stockpiled material and granular material are easily excavated with front end loaders. Scrapers are most economically used with hauls between 300′ and 1-1/2 miles if adequate haul roads can be maintained. Shovels are often used for blasted rock and any material where a vertical face of 8′ or more can be excavated. Special conditions may dictate the use of draglines, clamshells, or backhoes. Spreading and compaction equipment must be matched to the soil characteristics, the compaction required and the rate the fill is being supplied.

R02065-300 Bituminous Paving

City	Sidewalk Mix Bituminous Asphalt per Ton*	Sidewalks (2″) 9.2 S.Y./ton				Pavement (3″) 6.13 S.Y./ton			
		Cost per S.Y.			Per Ton	Cost per S.Y.			Per Ton
		Material*	Installation	Total	Total	Material*	Installation	Total	Total
Atlanta	$30.00	$3.26	$1.91	$5.17	$47.56	$4.89	$ .95	$5.84	$35.80
Baltimore	34.73	3.78	1.84	5.62	51.70	5.67	.92	6.59	40.40
Boston	39.50	4.29	1.96	6.25	57.50	6.44	.98	7.42	45.48
Buffalo	34.60	3.76	1.86	5.62	51.70	5.64	.93	6.57	40.27
Chicago	31.75	3.45	2.03	5.48	50.42	5.18	1.01	6.19	37.94
Cincinnati	41.00	4.46	2.16	6.62	60.90	6.69	1.08	7.77	47.63
Cleveland	29.50	3.21	2.12	5.33	49.04	4.81	1.06	5.87	35.98
Columbus	30.50	3.32	2.03	5.35	49.22	4.98	1.01	5.99	36.72
Dallas	29.25	3.18	1.71	4.89	44.99	4.77	.85	5.62	34.45
Denver	30.00	3.26	2.12	5.38	49.50	4.89	1.06	5.95	36.47
Detroit	32.00	3.48	1.93	5.41	49.77	5.22	.96	6.18	37.88
Houston	34.50	3.75	1.76	5.51	50.69	5.63	.88	6.51	39.91
Indianapolis	29.75	3.23	1.99	5.22	48.02	4.85	.99	5.84	35.80
Kansas City	29.50	3.21	1.81	5.02	46.18	4.81	.90	5.71	35.00
Los Angeles	36.00	3.91	2.13	6.04	55.57	5.87	1.06	6.93	42.48
Memphis	38.75	4.21	1.82	6.03	55.48	6.32	.91	7.23	44.32
Milwaukee	32.50	3.53	1.85	5.38	49.50	5.30	.92	6.22	38.13
Minneapolis	31.38	3.41	2.16	5.57	51.24	5.12	1.08	6.20	38.01
Nashville	27.75	3.02	1.97	4.99	45.91	4.53	.98	5.51	33.78
New Orleans	33.75	3.67	1.75	5.42	49.86	5.51	.87	6.38	39.11
New York City	49.50	5.38	2.28	7.66	70.47	8.08	1.14	9.22	56.52
Philadelphia	32.50	3.53	1.87	5.40	49.68	5.30	.93	6.23	38.19
Phoenix	30.00	3.26	2.08	5.34	49.13	4.89	1.04	5.93	36.35
Pittsburgh	35.00	3.80	2.17	5.97	54.92	5.71	1.08	6.79	41.62
St. Louis	32.00	3.48	1.92	5.40	49.68	5.22	.96	6.18	37.88
San Antonio	33.25	3.61	1.77	5.38	49.50	5.42	.88	6.30	38.62
San Diego	37.00	4.02	2.00	6.02	55.38	6.04	1.00	7.04	43.16
San Francisco	40.25	4.38	2.21	6.59	60.63	6.57	1.10	7.67	47.02
Seattle	39.50	4.29	2.25	6.54	60.17	6.44	1.12	7.56	46.34
Washington, D.C.	35.41	3.85	1.78	5.63	51.80	5.78	.89	6.67	40.89
Average	$34.00	$3.70	$1.97	$5.67	$52.16	$5.55	$ .98	$6.53	$40.03

Assumed density is 145 lb. per C.F.
*Includes delivery within 20 miles

Table below shows quantities and bare costs for 1000 S.Y. of Bituminous Paving.

Item	Sidewalks, 2″ Thick (02775-275-0020)		Roads and Parking Areas, 3″ Thick (02740-300-0460)	
	Quantities	Cost	Quantities	Cost
Bituminous asphalt Installation using	108.7 tons @ $34.00 per ton Crew B-37 @ $1,421.40 /720 SY/day x 1000	$3,696.00 1,974.17	163.1 tons @ $34.00 per ton Crew B-25B @ $4,825.60 /4900SY/ day x 1000	$5,546.00 984.80
Total per 1000 S.Y.		$5,670.17		$6,530.80
Total per S.Y.		$ 5.67		$ 6.53
Total per Ton		$ 52.16		$ 40.03

R02115-200 Underground Storage Tank Removal

Underground Storage Tank Removal can be divided into two categories: Non-Leaking and Leaking. Prior to removing an underground storage tank, tests should be made, with the proper authorities present, to determine whether a tank has been leaking or the surrounding soil has been contaminated.

To safely remove Liquid Underground Storage Tanks:

1. Excavate to the top of the tank.
2. Disconnect all piping.
3. Open all tank vents and access ports.
4. Remove all liquids and/or sludge.
5. Purge the tank with an inert gas.
6. Provide access to the inside of the tank and clean out the interior using proper personal protective equipment (PPE).
7. Excavate soil surrounding the tank using proper PPE for on-site personnel.
8. Pull and properly dispose of the tank.
9. Clean up the site of all contaminated material.
10. Install new tanks or close the excavation.

R02220-510 General Demolition

When estimating demolition the authors recommend getting a bid from a local contractor, if possible.Variables including disposal sites, protection of adjacent structures, salvage, and economic conditions can affect costs by 100%.

In calculating the line item costs of this major classification, three preliminary qualifications are assumed:

1. The tools used for demolition are generally of the hand or pneumatic hand type (note crew size).
2. The cost of rubbish handling (removing rubbish to on-site containers or trucks) is not included.
3. The cost of hauling rubbish to an approved dumpsite is not included.

For total general demolition cost, add rubbish handling and hauling. (See Division 02220-350)

R02250-400 Wood Sheet Piling

Wood sheet piling may be used for depths to 20′ where there is no ground water. If moderate ground water is encountered Tongue & Groove sheeting will help to keep it out. When considerable ground water is present, steel sheeting must be used.

For estimating purposes on trench excavation, sizes are as follows:

Depth	Sheeting	Wales	Braces	B.F. per S.F.
To 8′	3 x 12′s	6 x 8′s, 2 line	6 x 8′s, @ 10′	4.0 @ 8′
8′ x 12′	3 x 12′s	10 x 10′s, 2 line	10 x 10′s, @ 9′	5.0 average
12′ to 20′	3 x 12′s	12 x 12′s, 3 line	12 x 12′s, @ 8′	7.0 average

Sheeting to be toed in at least 2′ depending upon soil conditions. A five person crew with an air compressor and sheeting driver can drive and brace 440 SF/day at 8′ deep, 360 SF/day at 12′ deep, and 320 SF/day at 16′ deep. For normal soils, piling can be pulled in 1/3 the time to install. Pulling difficulty increases with the time in the ground. Production can be increased by high pressure jetting. Figures below assume 50% of lumber is salvaged and includes pulling costs. Some jurisdictions require an equipment operator in addition to Crew B-31.

Sheeting Pulled

Daily Cost Crew B-31	L.H./ Day	Hourly Cost	Daily Cost	8′ Depth, 440 S.F./Day			16′ Depth, 320 S.F./Day		
				To Drive (1 Day)		To Pull (1/3 Day)	To Drive (1 Day)		To Pull (1/3 Day)
1 Foreman	8	$28.00	$ 224.00		$ 224.00	$ 74.59		$ 224.00	$ 74.59
3 Laborers	24	26.00	624.00		624.00	207.79		624.00	207.79
1 Carpenter	8	33.00	264.00		264.00	87.91		264.00	87.91
1 Air Compressor			127.60		127.60	42.49		127.60	42.49
1 Sheeting Driver			7.20		7.20	2.40		7.20	2.40
2 -50 Ft. Air Hoses, 1-1/2″ Diam.			20.00		20.00	6.66		20.00	6.66
Lumber (50% salvage)				1.76 MBF	563.20		2.24 MBF	716.80	
Bare Total			$1,266.80		$1,830.00	$421.84		$1,983.60	$421.84
Bare Total/S.F.					$ 4.16	$ .96		$ 4.51	$ 1.32
Bare Total (Drive and Pull)/S.F.					$ 5.12			$ 5.83	

Sheeting Left in Place

Daily Cost	8′ Depth 440 S.F./Day		10′ Depth 400 S.F./Day		12′ Depth 360 S.F./Day		16′ Depth 320 S.F./Day		18′ Depth 305 S.F./Day		20′ Depth 280 S.F./Day	
Crew B-31		$1,256.80		$1,256.80		$1,256.80		$1,256.80		$1,256.80		$1,256.80
Lumber	1.76 MBF	1,126.40	1.8 MBF	1,152.00	1.8 MBF	1,152.00	2.24 MBF	1,433.60	2.1 MBF	1,344.00	1.9 MBF	1,216.00
Bare Total in Place		$2,383.20		$2,408.80		$2,408.80		$2,690.40		$2,600.80		$2,472.80
Bare Total/S.F.		$ 5.42		$ 6.02		$ 6.69		$ 8.41		$ 8.53		$ 8.83
Bare Total/M.B.F.		$1,354.09		$1,338.22		$1,338.22		$1,201.07		$1,238.48		$1,301.47

R02315-300 Compacting Backfill

Compaction of fill in embankments, around structures, in trenches, and under slabs is important to control settlement. Factors affecting compaction are:

1. Soil gradation
2. Moisture content
3. Equipment used
4. Depth of fill per lift
5. Density required

Production Rate:

$$\frac{1.75' \text{ plate width x 50 F.P.M. x 50 min./hr. x .67' lift}}{27 \text{ C.F. per C.Y.}} = 108.5 \text{ C.Y./hr.}$$

Production Rate for 4 Passes:

$$\frac{108.5 \text{ C.Y.}}{4 \text{ passes}} = 27.125 \text{ C.Y./hr. x 8 hrs.} = 217 \text{ C.Y./day}$$

The costs for testing and soil analyses are listed in Division 01450-500. Also, see Division 02315 for further backfill, borrow, and compaction costs.

Example:

Compact granular fill around a building foundation using a 21″ wide x 24″ vibratory plate in 8″ lifts. Operator moves at 50 FPM working a 50 minute hour to develop 95% Modified Proctor Density with 4 passes.

Compacting 217 C.Y. with 21″ Wide Vibratory Plate	L.H./ Day	Hourly Cost	Daily Cost	C.Y. Cost
1 Laborer	8	$26.00	$208.00	$ .96
1 Vibratory Plate Compactor			41.40	.19
Bare Total for 217 C.Y./day			$249.40	$1.15

R02315-400 Excavating

The selection of equipment used for structural excavation and bulk excavation or for grading is determined by the following factors.

1. Quantity of material.
2. Type of material.
3. Depth or height of cut.
4. Length of haul.
5. Condition of haul road.
6. Accessibility of site.
7. Moisture content and dewatering requirements.
8. Availability of excavating and hauling equipment.

Some additional costs must be allowed for hand trimming the sides and bottom of concrete pours and other excavation below the general excavation.

When planning excavation and fill, the following should also be considered.

1. Swell factor.
2. Compaction factor.
3. Moisture content.
4. Density requirements.

A typical example for scheduling and estimating the cost of excavation of a 15′ deep basement on a dry site when the material must be hauled off the site, is outlined below.

Assumptions:

1. Swell factor, 18%.
2. No mobilization or demobilization.
3. Allowance included for idle time and moving on job.
4. No dewatering, sheeting, or bracing.
5. No truck spotter or hand trimming.

Number of B.C.Y. per truck = 1.5 C.Y. bucket x 8 passes = 12 loose C.Y.

$$= 12 \text{ x } \frac{100}{118} = 10.2 \text{ B.C.Y. per truck}$$

Truck Haul Cycle:

Load truck 8 passes	=	4 minutes
Haul distance 1 mile	=	9 minutes
Dump time	=	2 minutes
Return 1 mile	=	7 minutes
Spot under machine	=	1 minute
		23 minute cycle

Fleet Haul Production per day in B.C.Y.

$$4 \text{ trucks x } \frac{50 \text{ min. hour}}{23 \text{ min. haul cycle}} \text{ x 8 hrs. x 10.2 B.C.Y.}$$

$$= 4 \text{ x } 2.2 \text{ x } 8 \text{ x } 10.2 = 718 \text{ B.C.Y./day}$$

Excavating Cost with a 1-1/2 C.Y. Hydraulic Excavator 15′ Deep, 2 Mile Round Trip Haul		L.H./Day	Hourly Cost	Daily Cost	Subtotal	Unit Price
1	Equipment Operator	8	$34.80	$ 278.40		
1	Oiler	8	29.20	233.60		
4	Truck Drivers	32	26.45	846.40	$1,358.40	$1.89
1	Hydraulic Excavator			683.40		
4	Dump Trucks			1,906.40	2,589.80	$3.60
Bare Total for 720 B.C.Y.				$3,948.20	$3,948.20	$5.49

Description	1-1/2 C.Y. Hyd. Backhoe 15′ Deep		1-1/2 C.Y. Power Shovel 7′ Bank		1-1/2 C.Y. Dragline 7′ Deep		2-1/2 C.Y. Trackloader Stockpile	
Operator (and Oiler, if required)		$ 512.00		$ 512.00		$ 512.00		$ 512.00
Truck Drivers	3 Ea.	634.80	4 Ea.	846.40	3 Ea.	634.80	4 Ea.	846.40
Equipment Rental		683.40		980.80		1,290.20		761.00
20 C.Y. Trailer Dump Trucks	3 Ea.	1,429.80	4 Ea.	1,906.40	3 Ea.	1,429.80	4 Ea.	1,906.40
Bare Total Cost per Day		$3,260.00		$4,245.60		$3,866.80		$4,025.80
Daily Production, C.Y. Bank Measure		720.00		960.00		640.00		1,000.00
Bare Cost per C.Y.		$ 4.53		$ 4.42		$ 6.04		$ 4.03

Add the mobilization and demobilization costs to the total excavation costs. When equipment is rented for more than three days, there is often no mobilization charge by the equipment dealer. On larger jobs outside of urban areas, scrapers can move earth economically provided a dump site or fill area and adequate haul roads are available. Excavation within sheeting bracing or cofferdam bracing is usually done with a clamshell and production is low, since the clamshell may have to be guided by hand between the bracing. When excavating or filling an area enclosed with a wellpoint system, add 10% to 15% to the cost to allow for restricted access. When estimating earth excavation quantities for structures, allow work space outside the building footprint for construction of the foundation, and a slope of 1:1 unless sheeting is used.

R02315-450 Excavating Equipment

The table below lists THEORETICAL hourly production in C.Y./hr. bank measure for some typical excavation equipment. Figures assume 50 minute hours, 83% job efficiency, 100% operator efficiency, 90° swing and properly sized hauling units, which must be modified for adverse digging and loading conditions. Actual production costs in the front of the book average about 50% of the theoretical values listed here.

Equipment	Soil Type	B.C.Y. Weight	% Swell	1 C.Y.	1-1/2 C.Y.	2 C.Y.	2-1/2 C.Y.	3 C.Y.	3-1/2 C.Y.	4 C.Y.
Hydraulic Excavator "Backhoe" 15′ Deep Cut	Moist loam, sandy clay	3400 lb.	40%	85	125	175	220	275	330	380
	Sand and gravel	3100	18	80	120	160	205	260	310	365
	Common earth	2800	30	70	105	150	190	240	280	330
	Clay, hard, dense	3000	33	65	100	130	170	210	255	300
Power Shovel Optimum Cut (Ft.)	Moist loam, sandy clay	3400	40	170 (6.0)	245 (7.0)	295 (7.8)	335 (8.4)	385 (8.8)	435 (9.1)	475 (9.4)
	Sand and gravel	3100	18	165 (6.0)	225 (7.0)	275 (7.8)	325 (8.4)	375 (8.8)	420 (9.1)	460 (9.4)
	Common earth	2800	30	145 (7.8)	200 (9.2)	250 (10.2)	295 (11.2)	335 (12.1)	375 (13.0)	425 (13.8)
	Clay, hard, dense	3000	33	120 (9.0)	175 (10.7)	220 (12.2)	255 (13.3)	300 (14.2)	335 (15.1)	375 (16.0)
Drag Line Optimum Cut (Ft.)	Moist loam, sandy clay	3400	40	130 (6.6)	180 (7.4)	220 (8.0)	250 (8.5)	290 (9.0)	325 (9.5)	385 (10.0)
	Sand and gravel	3100	18	130 (6.6)	175 (7.4)	210 (8.0)	245 (8.5)	280 (9.0)	315 (9.5)	375 (10.0)
	Common earth	2800	30	110 (8.0)	160 (9.0)	190 (9.9)	220 (10.5)	250 (11.0)	280 (11.5)	310 (12.0)
	Clay, hard, dense	3000	33	90 (9.3)	130 (10.7)	160 (11.8)	190 (12.3)	225 (12.8)	250 (13.3)	280 (12.0)
				Wheel Loaders				Track Loaders		
				3 C.Y.	4 C.Y.	6 C.Y.	8 C.Y.	2-1/4 C.Y.	3 C.Y.	4 C.Y.
Loading Tractors	Moist loam, sandy clay	3400	40	260	340	510	690	135	180	250
	Sand and gravel	3100	18	245	320	480	650	130	170	235
	Common earth	2800	30	230	300	460	620	120	155	220
	Clay, hard, dense	3000	33	200	270	415	560	110	145	200
	Rock, well-blasted	4000	50	180	245	380	520	100	130	180

R02455-900 Wood Bearing Piles

Untreated Southern Yellow Pine is most generally used for pile foundations cut off below the low water line. These are driven with the bark on. All piles cut off above the low water line should be treated with a preservative or encased in concrete for the section above water.

Item	Unit Cost	Units	Quantity	Total Cost	Cost/Pile	Cost/L.F.
50′ Treated Piles, 13″ Butt, 7″ Tip	$ 8.75	L.F.	200	$ 87,500.00	$437.50	$ 8.75
Installation, Crew B-19	3,657.60	Day	13	47,548.80	237.74	4.75
Mobilization & Demobilization, Crew B-19	3,657.60	Day	3	10,972.80	54.86	1.10
Transportation of Equip. One Way	715.00	Day	1	715.00	3.58	.07
Bare Totals				$146,736.60	$733.68	$14.67

The above figures are based on driving 800 L.F. daily which can be considered average. Time is included for moving rig, cutoff & ordinary delays. A general observation is that the cost of a pile in place complete is about two times the cost of pile only. See also equipment rental division 01590 and R01590-100 for equipment capacities.

R02465-600 Caissons

The three principal types of cassions are:

(1) **Belled Caissons,** which except for shallow depths and poor soil conditions, are generally recommended. They provide more bearing than shaft area. Because of its conical shape, no horizontal reinforcement of the bell is required.

(2) **Straight Shaft Caissons** are used where relatively light loads are to be supported by caissons that rest on high value bearing strata. While the shaft is larger in diameter than for belled types this is more than offset by the saving in time and labor.

(3) **Keyed Caissons** are used when extremely heavy loads are to be carried. A keyed or socketed caisson transfers its load into rock by a combination of end-bearing and shear reinforcing of the shaft. The most economical shaft often consists of a steel casing, a steel wide flange core and concrete. Allowable compressive stresses of .225 f′c for concrete, 16,000 psi for the wide flange core, and 9,000 psi for the steel casing are commonly used. The usual range of shaft diameter is 18″ to 84″. The number of sizes specified for any one project should be limited due to the problems of casing and auger storage. When hand work is to be performed, shaft diameters should not be less than 32″. When inspection of borings is required a minimum shaft diameter of 30″ is recommended. Concrete caissons are intended to be poured against earth excavation so permanent forms which add to cost should not be used if the excavation is clean and the earth sufficiently impervious to prevent excessive loss of concrete.

Soil Conditions for Belling		
Good	**Requires Handwork**	**Not Recommended**
Clay	Hard Shale	Silt
Sandy Clay	Limestone	Sand
Silty Clay	Sandstone	Gravel
Clayey Silt	Weathered Mica	Igneous Rock
Hard-pan		
Soft Shale		
Decomposed Rock		

R02510-800 Piping Designations

There are several systems currently in use to describe pipe and fittings. The following paragraphs will help to identify and clarify classifications of piping systems used for water distribution.

Piping may be classified by schedule. Piping schedules include 5S, 10S, 10, 20, 30, Standard, 40, 60, Extra Strong, 80, 100, 120, 140, 160 and Double Extra Strong. These schedules are dependent upon the pipe wall thickness. The wall thickness of a particular schedule may vary with pipe size.

Ductile iron pipe for water distribution is classified by Pressure Classes such as Class 150, 200, 250, 300 and 350. These classes are actually the rated water working pressure of the pipe in pounds per square inch (psi). The pipe in these pressure classes is designed to withstand the rated water working pressure plus a surge allowance of 100 psi.

The American Water Works Association (AWWA) provides standards for various types of **plastic pipe.** C-900 is the specification for polyvinyl chloride (PVC) piping used for water distribution in sizes ranging from 4″ through 12″. C-901 is the specification for polyethylene (PE) pressure pipe, tubing and fittings used for water distribution in sizes ranging from 1/2″ through 3″. C-905 is the specification for PVC piping sizes 14″ and greater.

PVC pressure-rated pipe is identified using the standard dimensional ratio (SDR) method. This method is defined by the American Society for Testing and Materials (ASTM) Standard D 2241. This pipe is available in SDR numbers 64, 41, 32.5, 26, 21, 17, and 13.5. Pipe with an SDR of 64 will have the thinnest wall while pipe with an SDR of 13.5 will have the thickest wall. When the pressure rating (PR) of a pipe is given in psi, it is based on a line supplying water at 73 degrees F.

The National Sanitation Foundation (NSF) seal of approval is applied to products that can be used with potable water. These products have been tested to ANSI/NSF Standard 14.

Valves and strainers are classified by American National Standards Institute (ANSI) Classes. These Classes are 125, 150, 200, 250, 300, 400, 600, 900, 1500 and 2500. Within each class there is an operating pressure range dependent upon temperature. Design parameters should be compared to the appropriate material dependent, pressure-temperature rating chart for accurate valve selection.

R02510-810 Concrete Pipe

Prices given are for inside 20 mile delivery zone. Add $2.09 per ton of pipe for each additional 10 miles. Minimum truckload is 10 tons. The non-reinforced pipe listed in the front of the book is designation ASTM C14-59 extra strength. The reinforced pipe listed is ASTM C76-65T class 3, no gaskets. The installation cost given includes shaping bottom of the trench, placing the pipe, and backfilling and tamping to the top of the pipe only.

Site Construction | R029 | Landscaping

R02920-500 Seeding

The type of grass is determined by light, shade and moisture content of soil plus intended use. Fertilizer should be disked 4″ before seeding. For steep slopes disk five tons of mulch and lay two tons of hay or straw on surface per acre after seeding. Surface mulch can be staked, lightly disked or tar emulsion sprayed. Material for mulch can be wood chips, peat moss, partially rotted hay or straw, wood fibers and sprayed emulsions. Hemp seed blankets with fertilizer are also available. For spring seeding, watering is necessary. Late fall seeding may have to be reseeded in the spring. Hydraulic seeding, power mulching, and aerial seeding can be used on large areas.

R02930-900 Cost of Trees: Based on Pin Oak (Quercus palustris)

Tree Diameter	Normal Height	Catalog List Price of Tree	Guying Material	Bare Equipment Charge	Bare Installation Labor	Bare Total
2 to 3 inch	14 feet	$ 146	$15.35	$ 53.28	$ 88.48	$ 303.11
3 to 4 inch	16 feet	323	18.35	88.80	147.47	577.62
4 to 5 inch	18 feet	373	75.50	106.56	176.96	732.02
6 to 7 inch	22 feet	813	80.00	133.20	221.20	1,247.40
8 to 9 inch	26 feet	1,023	90.00	177.60	294.93	1,585.53

Installation Time & Cost for Planting Trees, Bare Costs														
Ball Size Diam. X Depth	Soil in Ball	Weight of Ball	Hole Diam. Req'd	Hole Excavation	Amount of Soil Displ.	Topsoil Handled	Time Required in Labor-Hours						Cost	
							Dig & Lace	Handle Ball	Dig Hole	Plant & Prune	Water & Guy	Total L.H.		Bare Total per Tree
Inches	C.F.	Lbs.	Feet	C.F.	C.F.	C.F.							Crew	
12 x 12	.70	56.00	2.00	4.00	3.00	11.00	.25	.17	.33	.25	.07	1.10	1 Clab	$ 28.60
18 x 16	2.00	160.00	2.50	8.00	6.00	21.00	.50	.33	.47	.35	.08	1.70	2 Clab	44.20
24 x 18	4.00	320.00	3.00	13.00	9.00	38.00	1.00	.67	1.08	.82	.20	3.80	3 Clab	98.80
30 x 21	7.50	600.00	4.00	27.00	19.50	76.00	.82	.71	.79	1.22	.26	3.80	B-6	139.50
36 x 24	12.50	980.00	4.50	38.00	25.50	114.00	1.08	.95	1.11	1.32	.30	4.76		174.74
42 x 27	19.00	1,520.00	5.50	64.00	45.00	185.00	1.90	1.27	1.87	1.43	.34	6.80		249.63
48 x 30	28.00	2,040.00	6.00	85.00	57.00	254.00	2.41	1.60	2.06	1.55	.39	8.00		293.68
54 x 33	38.50	3,060.00	7.00	127.00	88.50	370.00	2.86	1.90	2.39	1.76	.45	9.40		345.07
60 x 36	52.00	4,160.00	7.50	159.00	107.00	474.00	3.26	2.17	2.73	2.00	.51	10.70		392.80
66 x 39	68.00	5,440.00	8.00	196.00	128.00	596.00	3.61	2.41	3.07	2.26	.58	11.90		436.85
72 x 42	87.00	7,160.00	9.00	267.00	180.00	785.00	3.90	2.60	3.71	2.78	.70	13.70		502.93

R03110-010 Wall Form Materials

Aluminum Forms

Approximate weight is 3 lbs. per S.F.C.A. Standard widths are available from 4″ to 36″ with 36″ most common. Standard lengths of 2′, 4′, 6′ to 8′ are available. Forms are lightweight and fewer ties are needed with the wider widths. The form face is either smooth or textured.

Cost of bare forms per S.F.C.A. with different facing surface is listed below. Forms may also be rented.

Finish	Purchase Cost Per S.F.							
	3′ x 8′	2′ x 8′	12″ x 8′	6″ x 8′	3′ x 4′	2′ x 4′	12″ x 4′	6″ x 4′
Smooth Aluminum (.096″ Face Sheet)	$11.30	$13.95	$18.83	$31.39	$10.87	$15.10	$19.93	$30.96
Textured Brick Aluminum	$12.56	$19.50	$27.19	$34.19	$14.55	$19.15	$24.51	$39.26

Metal Framed Plywood Forms

Manufacturers claim over 75 reuses of plywood and over 300 reuses of steel frames. Many specials such as corners, fillers, pilasters, etc. are available. Monthly rental is generally about 15% of purchase price for first month and 9% per month thereafter with 90% of rental applied to purchase for the first month and decreasing percentages thereafter. Aluminum framed forms cost 25% to 30% more than steel framed.

After the first month, extra days may be prorated from the monthly charge. Rental rates do not include ties, accessories, cleaning, loss of hardware or freight in and out. Approximate weight is 5 lbs. per S.F. for steel; 3 lbs. per S.F. for aluminum.

Forms can be rented with option to buy.

Plywood Forms, Job Fabricated

There are two types of plywood used for concrete forms.

1. Exterior plyform which is completely waterproof. This is face oiled to facilitate stripping. Ten reuses can be expected with this type with 25 reuses possible.
2. An overlaid type consists of a resin fiber fused to exterior plyform. No oiling is required except to facilitate cleaning. This is available in both high density (HDO) and medium density overlaid (MDO). Using HDO, 50 reuses can be expected with 200 possible.

Plyform is available in 5/8″ and 3/4″ thickness. High density overlaid is available in 3/8″, 1/2″, 5/8″ and 3/4″ thickness.

5/8″ thick is sufficient for most building forms, while 3/4″ is best on heavy construction.

For prices on plywood and framing lumber see R06100-010 and R06160-020.

Plywood Forms, Modular, Prefabricated

There are many plywood forming systems without frames. Most of these are manufactured from 1-1/8″ (HDO) plywood and have some hardware attached. These are used principally for foundation walls 8′ or less high. With care and maintenance, 100 reuses can be attained with decreasing quality of surface finish.

Steel Forms

Approximate weight is 6-1/2 lbs. per S.F.C.A. including accessories. Standard widths are available from 2″ to 24″, with 24″ most common. Standard lengths are from 2′ to 8′, with 4′ the most common. Forms are easily ganged into modular units.

Forms are usually leased for 15% of the purchase price per month prorated daily over 30 days.

Rental may be applied to sale price and usually rental forms are bought. With careful handling and cleaning 200 to 400 reuses are possible.

Straight wall gang forms up to 12′ x 20′ or 8′ x 30′ can be fabricated. These crane handled forms usually lease for approx. 9% per month.

Individual job analysis is available from the manufacturer at no charge.

R03110-020 Floor Pans and Domes

For 8′ to 15′ Ceiling Heights Using Crew C-2 at Bare Cost per Labor-Hour	20″ Pans (03110-420)				19″ Domes (03110-420)			
	1 Use		4 Use		1 Use		4 Use	
Purchase 20″ pans or rent 19″ domes		$2,250.00		$ 562.50		$222.00		$177.60
Adjustable shores and accessories		75.00		60.00		75.00		60.00
110 S.F. 3/4″ plyform		102.74		33.39		102.74		33.39
210 B.F. supporting joists, girts and braces		97.44		31.67		97.44		31.67
Labor handle, place, strip, oil and clean pans and domes	1.5 L.H.	49.50	1.0 L.H.	33.00	1.8 L.H.	59.40	1.2 L.H.	39.60
Make up, erect, remove wood decking	10 L.H.	330.00	8.5 L.H.	280.50	10.0 L.H.	330.00	8.5 L.H.	280.50
Bare Total per 100 S.F. of floor area		$2,904.68		$1,001.06		$886.58		$622.76

The figures above are for closed deck forming. For open deck forming deduct $23 per 100 S.F. for four uses.

For slab height from 15′ to 20′ add $56 per 100 S.F. and for 20′ to 35′ add $75 per 100 S.F., both for four uses.

R03110-040 Forms for Reinforced Concrete

Design Economy

Avoid many sizes in proportioning beams and columns.

From story to story avoid changing column dimensions. Gain strength by adding steel or using a richer mix. If a change in size of column is necessary, vary one dimension only to minimize form alterations. Keep beams and columns the same width.

From floor to floor in a multi-story building vary beam depth, not width, as that will leave slab panel form unchanged. It is cheaper to vary the strength of a beam from floor to floor by means of steel area than by 2″ changes in either width or depth.

Cost Factors

Material includes the cost of lumber, cost of rent for metal pans or forms if used, nails, form ties, form oil, bolts and accessories.

Labor includes the cost of carpenters to make up, erect, remove and repair, plus common labor to clean and move. Having carpenters remove forms minimizes repairs.

Improper alignment and condition of forms will increase finishing cost. When forms are heavily oiled, concrete surfaces must be neutralized before finishing. Special curing compounds will cause spillages to spall off in first frost. Gang forming methods will reduce costs on large projects.

Materials Used

Boards are seldom used unless their architectural finish is required. Generally, steel, fiberglass and plywood are used for contact surfaces. Labor on plywood is 10% less than with boards. The plywood is backed up with 2 x 4′s at 12″ to 32″ O.C. Walers are generally 2 - 2 x 4′s. Column forms are held together with steel yokes or bands. Shoring is with adjustable shoring or scaffolding for high ceilings.

Reuse

Floor and column forms can be reused four or possibly five times without excessive repair. Remember to allow for 10% waste on each reuse.

When modular sized wall forms are made, up to twenty uses can be expected with exterior plyform.

When forms are reused, the cost to erect, strip, clean and move will not be affected. 10% replacement of lumber should be included and about one hour of carpenter time for repairs on each reuse per 100 S.F.

The reuse cost for certain accessory items normally rented on a monthly basis will be lower than the cost for the first use.

After fifth use, new material required plus time needed for repair prevent form cost from dropping further and it may go up. Much depends on care in stripping, the number of special bays, changes in beam or column sizes and other factors.

Costs for multiple use of formwork may be developed as follows:

2 Uses

$$\frac{(\text{1st Use + Reuse})}{2} = \text{avg. cost/2 uses}$$

3 Uses

$$\frac{(\text{1st Use + 2 Reuse})}{3} = \text{avg. cost/3 uses}$$

4 Uses

$$\frac{(\text{1st use + 3 Reuse})}{4} = \text{avg. cost/4 uses}$$

R03110-050 Forms In Place

This section assumes that all cuts are made with power saws, that adjustable shores are employed and that maximum use is made of commercial form ties and accessories. Bare costs are used in the table below.

BEAM AND GIRDER, INTERIOR, 12″ Wide (Line 03110-405-2000)			First Use			Reuse		
Item	Cost	Unit	Quantity	Material	Installation	Quantity	Material	Installation
5/8″ exterior plyform	$880.00	M.S.F.	115 S.F.	$101.20		11.5 S.F.	$10.12	
Lumber	464.00	M.B.F.	200 B.F.	92.80		20.0 B.F.	9.28	
Accessories, incl. adjustable shores			Allow	25.00		Allow	25.00	
Make up, crew C-2	32.17	L.H.	6.4 L.H.		$205.89	1.0 L.H.		$ 32.17
Erect and strip			8.3 L.H.		267.01	8.3 L.H.		267.01
Clean and move			1.3 L.H.		41.82	1.3 L.H.		41.82
Bare Total per 100 S.F.C.A.			16.0 L.H.	$219.00	$514.72	10.6 L.H.	$44.40	$341.00

For structural steel frame with beams encased, subtract 1.2 labor-hours, and 25% of the lumber per 100 S.F.C.A.

BOX CULVERT, 5′ to 8′ Square or Rectangular (Line 03110-415-0010)			First Use			Reuse		
Item	Cost	Unit	Quantity	Material	Installation	Quantity	Material	Installation
3/4″ exterior plyform	$934.00	M.S.F.	110 S.F.	$102.74		11.0 S.F.	$10.27	
Lumber	464.00	M.B.F.	170 B.F.	78.88		17.0 B.F.	7.89	
Accessories			Allow	20.00		Allow	20.00	
Build in place, crew C-1	31.25	L.H.	14.5 L.H.		$453.13	1.0 L.H.		$ 31.25
Strip and salvage			4.3 L.H.		134.38	4.3 L.H.		134.38
Bare Total per 100 S.F.C.A.			18.8 L.H.	$201.62	$587.51	18.8 L.H.	$38.16	$165.63

COLUMNS, 24″ x 24″ (Line 03110-410-6500)			First Use			Reuse		
Item	Cost	Unit	Quantity	Material	Installation	Quantity	Material	Installation
5/8″ exterior plyform	$880.00	M.S.F.	120 S.F.	$105.60		12.0 S.F.	$10.56	
Lumber	464.00	M.B.F.	125 B.F.	58.00		12.5 B.F.	5.80	
Clamps, chamfer strips and accessories			Allow	25.00		Allow	25.00	
Make up, crew C-1	31.25	L.H.	5.8 L.H.		$181.25	1.0 L.H.		$ 31.25
Erect and strip			9.8 L.H.		306.25	9.8 L.H.		306.25
Clean and move			1.2 L.H.		37.50	1.2 L.H.		37.50
Bare Total per 100 S.F.C.A.			16.8 L.H.	$188.60	$525.00	12.0 L.H.	$41.36	$375.00

FLAT SLAB WITH DROP PANELS (Line 03110-420-2000)			First Use			Reuse		
Item	Cost	Unit	Quantity	Material	Installation	Quantity	Material	Installation
5/8″ exterior plyform	$880.00	M.S.F.	115 S.F.	$101.20		11.5 S.F.	$10.12	
Lumber	464.00	M.B.F.	210 B.F.	97.44		21.0 B.F.	9.74	
Accessories, incl. adjustable shores			Allow	25.00		Allow	25.00	
Make up, crew C-2	32.17	L.H.	3.5 L.H.		$112.60	1.0 L.H.		$ 32.17
Erect and strip			6.0 L.H.		193.02	6.0 L.H.		193.02
Clean and move			1.2 L.H.		38.60	1.2 L.H.		38.60
Bare Total per 100 S.F.C.A.			10.7 L.H.	$223.64	$344.22	8.2 L.H.	$44.86	$263.79

Drop panels included but column caps figure with columns.

FOOTINGS, SPREAD (Line 03110-430-5000)			First Use			Reuse		
Item	Cost	Unit	Quantity	Materials	Installation	Quantity	Material	Installation
Lumber	$464.00	M.B.F.	260 B.F.	$120.64		26 B.F.	$12.06	
Accessories			Allow	7.50		Allow	7.50	
Make up, crew C-1	31.25	L.H.	4.7 L.H.		$146.88	1.0 L.H.		$ 31.25
Erect and strip			4.2 L.H.		131.25	4.2 L.H.		131.25
Clean and move			1.6 L.H.		50.00	1.6 L.H.		50.00
Bare Total per 100 S.F.C.A.			10.5 L.H.	$128.14	$328.13	6.8 L.H.	$19.56	$212.50

R03110-050 Forms In Place (cont.)

FOUNDATION WALL, 8′ High (Line 03110-455-2000)			First Use			Reuse		
Item	Cost	Unit	Quantity	Material	Installation	Quantity	Material	Installation
5/8″ exterior plyform	$880.00	M.S.F.	110 S.F.	$ 96.80		11.0 S.F.	$ 9.68	
Lumber	464.00	M.B.F.	140 B.F.	64.96		14.0 B.F.	6.50	
Accessories			Allow	25.00		Allow	25.00	
Make up, crew C-2	32.17	L.H.	5.0 L.H.		$160.85	1.0 L.H.		$ 32.17
Erect and strip			6.5 L.H.		209.11	6.5 L.H.		209.11
Clean and move			1.5 L.H.		48.26	1.5 L.H.		48.26
Bare Total per 100 S.F.C.A.			13.0 L.H.	$186.76	$418.22	9.0 L.H.	$41.18	$289.54

PILE CAPS, Square or Rectangular (Line 03110-430-3000)			First Use			Reuse		
Item	Cost	Unit	Quantity	Material	Installation	Quantity	Material	Installation
5/8″ exterior plyform	$880.00	M.S.F.	110 S.F.	$ 96.80		11.0 S.F.	$ 9.68	
Lumber	464.00	M.B.F.	160 B.F.	74.24		16.0 B.F.	7.42	
Accessories			Allow	7.50		Allow	7.50	
Make up, crew C-1	31.25	L.H.	4.5 L.H.		$140.63	1.0 L.H.		$ 31.25
Erect and strip			5.0 L.H.		156.25	5.0 L.H.		156.25
Clean and move			1.5 L.H.		46.88	1.5 L.H.		46.88
Bare Total per 100 S.F.C.A.			11.0 L.H.	$178.54	$343.76	7.5 L.H.	$24.60	$234.38

STAIRS, Average Run (Inclined Length x Width) (Line 03110-450-0010)			First Use			Reuse		
Item	Cost	Unit	Quantity	Materials	Installation	Quantity	Material	Installation
5/8″ exterior plyform	$880.00	M.S.F.	110 S.F.	$ 96.80		11.0 S.F.	$ 9.68	
Lumber	464.00	M.B.F.	425 B.F.	197.20		42.5 B.F.	19.72	
Accessories			Allow	20.00		Allow	20.00	
Build in place, crew C-2	32.17	L.H.	25.0 L.H.		$804.25	1.0 L.H.		$ 32.17
Strip and salvage			4.0 L.H.		128.68	4.0 L.H.		128.68
Bare Total per 100 S.F.C.A.			29.0 L.H.	$314.00	$932.93	29.0 L.H.	$49.40	$160.85

R03110-060 Formwork Labor Hours

Item	Hours Required				Total Hours	Multiple Use		
	Unit	Fabricate	Erect & Strip	Clean & Move	1 Use	2 Use	3 Use	4 Use
Beam and Girder, interior beams, 12″ wide	100 S.F.	6.4	8.3	1.3	16.0	13.3	12.4	12.0
Hung from steel beams		5.8	7.7	1.3	14.8	12.4	11.6	11.2
Beam sides only, 36″ high		5.8	7.2	1.3	14.3	11.9	11.1	10.7
Beam bottoms only, 24″ wide		6.6	13.0	1.3	20.9	18.1	17.2	16.7
Box out for openings		9.9	10.0	1.1	21.0	16.6	15.1	14.3
Buttress forms, to 8′ high		6.0	6.5	1.2	13.7	11.2	10.4	10.0
Centering, steel, 3/4″ rib lath			1.0		1.0			
3/8″ rib lath or slab form			0.9		0.9			
Chamfer strip or keyway	100 L.F.		1.5		1.5	1.5	1.5	1.5
Columns, fiber tube 8″ diameter			20.6		20.6			
12″			21.3		21.3			
16″			22.9		22.9			
20″			23.7		23.7			
24″			24.6		24.6			
30″			25.6		25.6			
Round Steel, 12″ diameter			22.0		22.0	22.0	22.0	22.0
16″			25.6		25.6	25.6	25.6	25.6
20″			30.5		30.5	30.5	30.5	30.5
24″			37.7		37.7	37.7	37.7	37.7
Plywood 8″ x 8″	100 S.F.	7.0	11.0	1.2	19.2	16.2	15.2	14.7
12″ x 12″		6.0	10.5	1.2	17.7	15.2	14.4	14.0
16″ x 16″		5.9	10.0	1.2	17.1	14.7	13.8	13.4
24″ x 24″		5.8	9.8	1.2	16.8	14.4	13.6	13.2
Steel framed plywood 8″ x 8″			10.0	1.0	11.0	11.0	11.0	11.0
12″ x 12″			9.3	1.0	10.3	10.3	10.3	10.3
16″ x 16″			8.5	1.0	9.5	9.5	9.5	9.5
24″ x 24″			7.8	1.0	8.8	8.8	8.8	8.8
Drop head forms, plywood		9.0	12.5	1.5	23.0	19.0	17.7	17.0
Coping forms		8.5	15.0	1.5	25.0	21.3	20.0	19.4
Culvert, box			14.5	4.3	18.8	18.8	18.8	18.8
Curb forms, 6″ to 12″ high, on grade		5.0	8.5	1.2	14.7	12.7	12.1	11.7
On elevated slabs		6.0	10.8	1.2	18.0	15.5	14.7	14.3
Edge forms to 6″ high, on grade	100 L.F.	2.0	3.5	0.6	6.1	5.6	5.4	5.3
7″ to 12″ high	100 S.F.	2.5	5.0	1.0	8.5	7.8	7.5	7.4
Equipment foundations		10.0	18.0	2.0	30.0	25.5	24.0	23.3
Flat slabs, including drops		3.5	6.0	1.2	10.7	9.5	9.0	8.8
Hung from steel		3.0	5.5	1.2	9.7	8.7	8.4	8.2
Closed deck for domes		3.0	5.8	1.2	10.0	9.0	8.7	8.5
Open deck for pans		2.2	5.3	1.0	8.5	7.9	7.7	7.6
Footings, continuous, 12″ high		3.5	3.5	1.5	8.5	7.3	6.8	6.6
Spread, 12″ high		4.7	4.2	1.6	10.5	8.7	8.0	7.7
Pile caps, square or rectangular		4.5	5.0	1.5	11.0	9.3	8.7	8.4
Grade beams, 24″ deep		2.5	5.3	1.2	9.0	8.3	8.0	7.9
Lintel or Sill forms		8.0	17.0	2.0	27.0	23.5	22.3	21.8
Spandrel beams, 12″ wide		9.0	11.2	1.3	21.5	17.5	16.2	15.5
Stairs			25.0	4.0	29.0	29.0	29.0	29.0
Trench forms in floor		4.5	14.0	1.5	20.0	18.3	17.7	17.4
Walls, Plywood, at grade, to 8′ high		5.0	6.5	1.5	13.0	11.0	9.7	9.5
8′ to 16′		7.5	8.0	1.5	17.0	13.8	12.7	12.1
16′ to 20′		9.0	10.0	1.5	20.5	16.5	15.2	14.5
Foundation walls, to 8′ high		4.5	6.5	1.0	12.0	10.3	9.7	9.4
8′ to 16′ high		5.5	7.5	1.0	14.0	11.8	11.0	10.6
Retaining wall to 12′ high, battered		6.0	8.5	1.5	16.0	13.5	12.7	12.3
Radial walls to 12′ high, smooth		8.0	9.5	2.0	19.5	16.0	14.8	14.3
2′ chords		7.0	8.0	1.5	16.5	13.5	12.5	12.0
Prefabricated modular, to 8′ high		—	4.3	1.0	5.3	5.3	5.3	5.3
Steel, to 8′ high		—	6.8	1.2	8.0	8.0	8.0	8.0
8′ to 16′ high		—	9.1	1.5	10.6	10.3	10.2	10.2
Steel framed plywood to 8′ high		—	6.8	1.2	8.0	7.5	7.3	7.2
8′ to 16′ high		—	9.3	1.2	10.5	9.5	9.2	9.0

R03210-010 Reinforcing Steel Weights and Measures

Bar Designation No.**	Nominal Weight Lb./Ft.	U.S. Customary Units			SI Units			
		Nominal Dimensions*			Nominal Dimensions*			
		Diameter in.	Cross Sectional Area, in.^{2}	Perimeter in.	Nominal Weight kg/m	Diameter mm	Cross Sectional Area, cm^2	Perimeter mm
3	.376	.375	.11	1.178	.560	9.52	.71	29.9
4	.668	.500	.20	1.571	.994	12.70	1.29	39.9
5	1.043	.625	.31	1.963	1.552	15.88	2.00	49.9
6	1.502	.750	.44	2.356	2.235	19.05	2.84	59.8
7	2.044	.875	.60	2.749	3.042	22.22	3.87	69.8
8	2.670	1.000	.79	3.142	3.973	25.40	5.10	79.8
9	3.400	1.128	1.00	3.544	5.059	28.65	6.45	90.0
10	4.303	1.270	1.27	3.990	6.403	32.26	8.19	101.4
11	5.313	1.410	1.56	4.430	7.906	35.81	10.06	112.5
14	7.650	1.693	2.25	5.320	11.384	43.00	14.52	135.1
18	13.600	2.257	4.00	7.090	20.238	57.33	25.81	180.1

* The nominal dimensions of a deformed bar are equivalent to those of a plain round bar having the same weight per foot as the deformed bar.
** Bar numbers are based on the number of eighths of an inch included in the nominal diameter of the bars.

R03310-010 Proportionate Quantities

The tables below show both quantities per S.F. of floor areas as well as form and reinforcing quantities per C.Y. Unusual structural requirements would increase the ratios below. High strength reinforcing would reduce the steel weights. Figures are for 3000 psi concrete and 60,000 psi reinforcing unless specified otherwise.

Type of Construction	Live Load	Span	Per S.F. of Floor Area				Per C.Y. of Concrete		
			Concrete	Forms	Reinf.	Pans	Forms	Reinf.	Pans
Flat Plate	50 psf	15 Ft.	.46 C.F.	1.06 S.F.	1.71 lb.		62 S.F.	101 lb.	
		20	.63	1.02	2.40		44	104	
		25	.79	1.02	3.03		35	104	
	100	15	.46	1.04	2.14		61	126	
		20	.71	1.02	2.72		39	104	
		25	.83	1.01	3.47		33	113	
Flat Plate (waffle construction) 20″ domes	50	20	.43	1.00	2.10	.84 S.F.	63	135	53 S.F.
		25	.52	1.00	2.90	.89	52	150	46
		30	.64	1.00	3.70	.87	42	155	37
	100	20	.51	1.00	2.30	.84	53	125	45
		25	.64	1.00	3.20	.83	42	135	35
		30	.76	1.00	4.40	.81	36	160	29
Waffle Construction 30″ domes	50	25	.69	1.06	1.83	.68	42	72	40
		30	.74	1.06	2.39	.69	39	87	39
		35	.86	1.05	2.71	.69	33	85	39
		40	.78	1.00	4.80	.68	35	165	40
Flat Slab (two way with drop panels)	50	20	.62	1.03	2.34		45	102	
		25	.77	1.03	2.99		36	105	
		30	.95	1.03	4.09		29	116	
	100	20	.64	1.03	2.83		43	119	
		25	.79	1.03	3.88		35	133	
		30	.96	1.03	4.66		29	131	
	200	20	.73	1.03	3.03		38	112	
		25	.86	1.03	4.23		32	133	
		30	1.06	1.03	5.30		26	135	
One Way Joists 20″ Pans	50	15	.36	1.04	1.40	.93	78	105	70
		20	.42	1.05	1.80	.94	67	120	60
		25	.47	1.05	2.60	.94	60	150	54
	100	15	.38	1.07	1.90	.93	77	140	66
		20	.44	1.08	2.40	.94	67	150	58
		25	.52	1.07	3.50	.94	55	185	49
One Way Joists 8″ x 16″ filler blocks	50	15	.34	1.06	1.80	.81 Ea.	84	145	64 Ea.
		20	.40	1.08	2.20	.82	73	145	55
		25	.46	1.07	3.20	.83	63	190	49
	100	15	.39	1.07	1.90	.81	74	130	56
		20	.46	1.09	2.80	.82	64	160	48
		25	.53	1.10	3.60	.83	56	190	42
One Way Beam & Slab	50	15	.42	1.30	1.73		84	111	
		20	.51	1.28	2.61		68	138	
		25	.64	1.25	2.78		53	117	
	100	15	.42	1.30	1.90		84	122	
		20	.54	1.35	2.69		68	154	
		25	.69	1.37	3.93		54	145	
	200	15	.44	1.31	2.24		80	137	
		20	.58	1.40	3.30		65	163	
		25	.69	1.42	4.89		53	183	
Two Way Beam & Slab	100	15	.47	1.20	2.26		69	130	
		20	.63	1.29	3.06		55	131	
		25	.83	1.33	3.79		43	123	
	200	15	.49	1.25	2.70		41	149	
		20	.66	1.32	4.04		54	165	
		25	.88	1.32	6.08		41	187	

R03310-010 Proportionate Quantities (cont.)

4000 psi Concrete and 60,000 psi Reinforcing—Form and Reinforcing Quantities per C.Y.

Item	Size	Forms	Reinforcing	Minimum	Maximum
Columns (square tied)	10″ x 10″	130 S.F.C.A.	#5 to #11	220 lbs.	875 lbs.
	12″ x 12″	108	#6 to #14	200	955
	14″ x 14″	92	#7 to #14	190	900
	16″ x 16″	81	#6 to #14	187	1082
	18″ x 18″	72	#6 to #14	170	906
	20″ x 20″	65	#7 to #18	150	1080
	22″ x 22″	59	#8 to #18	153	902
	24″ x 24″	54	#8 to #18	164	884
	26″ x 26″	50	#9 to #18	169	994
	28″ x 28″	46	#9 to #18	147	864
	30″ x 30″	43	#10 to #18	146	983
	32″ x 32″	40	#10 to #18	175	866
	34″ x 34″	38	#10 to #18	157	772
	36″ x 36″	36	#10 to #18	175	852
	38″ x 38″	34	#10 to #18	158	765
	40″ x 40″	32 ▼	#10 to #18	143	692

Item	Size	Form	Spiral	Reinforcing	Minimum	Maximum
Columns (spirally reinforced)	12″ diameter	34.5 L.F.	190 lbs.	#4 to #11	165 lbs.	1505 lb.
		34.5	190	#14 & #18	—	1100
	14″	25	170	#4 to #11	150	970
		25	170	#14 & #18	800	1000
	16″	19	160	#4 to #11	160	950
		19	160	#14 & #18	605	1080
	18″	15	150	#4 to #11	160	915
		15	150	#14 & #18	480	1075
	20″	12	130	#4 to #11	155	865
		12	130	#14 & #18	385	1020
	22″	10	125	#4 to #11	165	775
		10	125	#14 & #18	320	995
	24″	9	120	#4 to #11	195	800
		9	120	#14 & #18	290	1150
	26″	7.3	100	#4 to #11	200	729
		7.3	100	#14 & #18	235	1035
	28″	6.3	95	#4 to #11	175	700
		6.3	95	#14 & #18	200	1075
	30″	5.5	90	#4 to #11	180	670
		5.5	90	#14 & #18	175	1015
	32″	4.8	85	#4 to #11	185	615
		4.8	85	#14 & #18	155	955
	34″	4.3	80	#4 to #11	180	600
		4.3	80	#14 & #18	170	855
	36″	3.8	75	#4 to #11	165	570
		3.8	75	#14 & #18	155	865
	40″	3.0	70	#4 to #11	165	500
		3.0	70	#14 & #18	145	765

R03310-010 Proportionate Quantities (cont.)

3000 psi Concrete and 60,000 psi Reinforcing—Form and Reinforcing Quantities per C.Y.						
Item	**Type**	**Loading**	**Height**	**C.Y./L.F.**	**Forms/C.Y.**	**Reinf./C.Y.**
Retaining Walls	Cantilever	Level Backfill	4 Ft.	0.2 C.Y.	49 S.F.	35 lbs.
			8	0.5	42	45
			12	0.8	35	70
			16	1.1	32	85
			20	1.6	28	105
		Highway Surcharge	4	0.3	41	35
			8	0.5	36	55
			12	0.8	33	90
			16	1.2	30	120
			20	1.7	27	155
		Railroad Surcharge	4	0.4	28	45
			8	0.8	25	65
			12	1.3	22	90
			16	1.9	20	100
			20	2.6	18	120
	Gravity, with Vertical Face	Level Backfill	4	0.4	37	None
			7	0.6	27	↓
			10	1.2	20	↓
		Sloping Backfill	4	0.3	31	↓
			7	0.8	21	↓
			10	1.6	15	↓

Item	Span	Live Load in Kips per Linear Foot							
		Under 1 Kip		**2 to 3 Kips**		**4 to 5 Kips**		**6 to 7 Kips**	
		Forms	**Reinf.**	**Forms**	**Reinf.**	**Forms**	**Reinf.**	**Forms**	**Reinf.**
Beams	10 Ft.	—	—	90 S.F.	170 #	85 S.F.	175 #	75 S.F.	185 #
	16	130 S.F.	165 #	85	180	75	180	65	225
	20	110	170	75	185	62	200	51	200
	26	90	170	65	215	62	215	—	—
	30	85	175	60	200	—	—	—	—

Item	Size	Type	Forms per C.Y.	Reinforcing per C.Y.
Spread Footings	Under 1 C.Y.	1,000 psf soil	24 S.F.	44 lbs.
		5,000	24	42
		10,000	24	52
	1 C.Y. to 5 C.Y.	1,000	14	49
		5,000	14	50
		10,000	14	50
	Over 5 C.Y.	1,000	9	54
		5,000	9	52
		10,000	9	56
Pile Caps (30 Ton Concrete Piles)	Under 5 C.Y.	shallow caps	20	65
		medium	20	50
		deep	20	40
	5 C.Y. to 10 C.Y.	shallow	14	55
		medium	15	45
		deep	15	40
	10 C.Y. to 20 C.Y.	shallow	11	60
		medium	11	45
		deep	12	35
	Over 20 C.Y.	shallow	9	60
		medium	9	45
		deep	10	40

R03310-010 Proportionate Quantities (cont.)

3000 psi Concrete and 60,000 psi Reinforcing — Form and Reinforcing Quantities per C.Y.						
Item	Size	Pile Spacing	50 T Pile	100 T Pile	50 T Pile	100 T Pile
Pile Caps (Steel H Piles)	Under 5 C.Y.	24″ O.C.	24 S.F.	24 S.F.	75 lbs.	90 lbs.
		30″	25	25	80	100
		36″	24	24	80	110
	5 C.Y. to 10 C.Y.	24″	15	15	80	110
		30″	15	15	85	110
		36″	15	15	75	90
	Over 10 C.Y.	24″	13	13	85	90
		30″	11	11	85	95
		36″	10	10	85	90

Item		8″ Thick		10″ Thick		12″ Thick		15″ Thick	
	Height	Forms	Reinf.	Forms	Reinf.	Forms	Reinf.	Forms	Reinf.
Basement Walls	7 Ft.	81 S.F.	44 lbs.	65 S.F.	45 lbs.	54 S.F.	44 lbs.	41 S.F.	43 lbs.
	8	↓	44	↓	45	↓	44	↓	43
	9	↓	46	↓	45	↓	44	↓	43
	10	↓	57	↓	45	↓	44	↓	43
	12	↓	83	↓	50	↓	52	↓	43
	14	↓	116	↓	65	↓	64	↓	51
	16	↓		↓	86	↓	90	↓	65
	18	↓		↓		↓	106	↓	70

R03310-020 Materials for One C.Y. of Concrete

This is an approximate method of figuring quantities of cement, sand and coarse aggregate for a field mix with waste allowance included.

With crushed gravel as coarse aggregate, to determine barrels of cement required, divide 10 by total mix; that is, for 1:2:4 mix, 10 divided by 7 = 1-3/7 barrels. If the coarse aggregate is crushed stone, use 10-1/2 instead of 10 as given for gravel.

To determine tons of sand required, multiply barrels of cement by parts of sand and then by 0.2; that is, for the 1:2:4 mix, as above, 1-3/7 x 2 x .2 = .57 tons.

Tons of crushed gravel are in the same ratio to tons of sand as parts in the mix, or 4/2 x .57 = 1.14 tons.

1 bag cement = 94#
4 bags = 1 barrel

1 C.Y. sand or crushed gravel = 2700#
1 ton sand or crushed gravel = 20 C.F.

1 C.Y. crushed stone = 2575#
1 ton crushed stone = 21 C.F.

Average carload of cement is 692 bags; of sand or gravel is 56 tons.

Do not stack stored cement over 10 bags high.

R03310-060 Concrete Material Net Prices

Costs below are C.Y. of concrete delivered; per ton of bulk cement; per bag cement delivered T.L.L.; per ton for stone and sand aggregates loaded at plant (no trucking included) and per 4 C.F. bag for perlite or vermiculite aggregate delivered T.L.L.

City	Ready Mix Concrete Regular Weight per C.Y.		Cement T.L. Lots		Aggregates per Ton			Vermiculite or Perlite 4 C.F. Bag
			Bulk	Bags	Crushed Stone			
	3000 psi	5000 psi	per Ton	per Bag	1-1/2"	3/4"	Sand	
Atlanta	$72.00	$ 80.00	$ 87.20	$7.45	$17.95	$18.50	$21.50	$6.15
Baltimore	69.00	77.00	80.15	6.85	11.95	12.30	23.50	5.65
Boston	78.00	88.00	103.60	8.85	14.45	14.90	13.60	7.30
Buffalo	84.50	93.00	85.45	7.30	13.35	13.75	19.65	6.05
Chicago	80.00	88.00	81.95	7.00	18.10	18.65	12.80	5.80
Cincinnati	61.50	71.00	99.50	8.50	8.60	8.85	13.60	7.05
Cleveland	71.00	80.00	95.95	8.20	15.55	16.05	16.90	6.80
Columbus	64.50	74.00	105.90	9.05	10.95	11.30	14.75	7.50
Dallas	71.00	81.00	106.50	9.10	19.90	20.50	19.25	7.55
Denver	69.00	80.00	112.95	9.65	13.80	14.25	16.50	8.00
Detroit	73.00	81.00	86.60	7.40	11.10	11.45	17.54	6.10
Houston	64.50	73.00	87.20	7.45	16.85	17.35	15.50	6.15
Indianapolis	71.00	81.00	105.35	9.00	11.40	11.75	18.20	7.45
Kansas City	69.00	78.00	91.30	7.80	10.55	10.90	18.25	6.45
Los Angeles	62.50	73.00	108.25	9.25	12.90	13.30	25.90	7.65
Memphis	65.50	74.00	94.80	8.10	16.50	17.00	17.45	6.70
Milwaukee	75.00	85.00	103.60	8.85	12.75	13.15	14.30	7.30
Minneapolis	77.00	87.00	105.35	9.00	14.25	14.70	15.35	7.45
Nashville	67.50	76.00	87.80	7.50	10.75	11.10	21.80	6.20
New Orleans	66.50	76.00	97.15	8.30	19.30	19.90	13.45	6.85
New York City	90.50	99.00	88.35	7.55	21.85	22.50	15.25	6.25
Philadelphia	71.00	79.00	87.20	7.45	14.50	14.95	18.85	6.15
Phoenix	74.00	83.00	91.90	7.85	11.85	12.20	12.50	6.50
Pittsburgh	70.00	79.00	96.55	8.25	15.25	15.70	19.50	6.80
St. Louis	62.50	71.00	93.05	7.95	10.80	11.15	14.35	6.60
San Antonio	57.50	65.00	84.25	7.20	11.20	11.55	18.50	5.95
San Diego	82.50	92.00	101.25	8.65	15.55	16.05	21.00	7.15
San Francisco	96.00	107.00	114.70	9.80	19.90	20.50	26.50	8.10
Seattle	75.00	81.00	67.30	5.75	17.00	17.55	25.50	4.75
Washington, D.C.	82.50	91.00	91.90	7.85	14.05	14.50	22.50	6.50
Average	$72.00	$ 81.00	$ 94.75	$8.10	$14.45	$14.90	$18.15	$6.70

R03310-090 Placing Ready Mixed Concrete

For ground pours allow for 5% waste when figuring quantities.

Prices in the front of the book assume normal deliveries. If deliveries are made before 8 A.M. or after 5 P.M. or on Saturday afternoons add $26 per C.Y. Large volume discounts are not included in prices in front of book.

For the lower floors without truck access, concrete may be wheeled in rubber tired buggies, conveyer handled, crane handled or pumped. Pumping is economical if there is top steel. Conveyers are more efficient for thick slabs. Concrete pump with an operator can be rented from $800 per day for up to 25 C.Y. to $1750 per day for a 400 C.Y. pour. Figures include travel time if done at straight time. Pumping lightweight concrete costs an extra $2.70 per C.Y.

At higher floors the rubber tired buggies may be hoisted by a hoisting tower then wheeled to location. Placement by a conveyer is limited to three floors and is best for high volume pours. Pumped concrete is best when building has no crane access. Concrete may be pumped directly as high as thirty-six stories using special pumping techniques. Normal maximum height is about fifteen stories.

Best pumping aggregate is screened and graded bank gravel rather than crushed stone.

Pumping downward is more difficult than pumping upwards. Horizontal distance from pump to pour may increase preparation time prior to pour. Placing by cranes, either mobile, climbing or tower types continues as the most efficient method for high rise concrete buildings.

Cost per C.Y. for Wheeled Concrete, Dumped Only (Add to appropriate placing cost)								
	10 C.F. Walking Cart				18 C.F. Riding Cart			
Item	Hourly Cost	Wheeled up to 50 ft.	Wheeled up to 150 ft.	Wheeled up to 250 ft.	Hourly Cost	Wheeled up to 50 ft.	Wheeled up to 150 ft.	Wheeled up to 250 ft.
Laborer	$26.00	$6.50	$ 8.68	$11.57	$26.00	$2.60	$3.38	$4.42
.125 Labor foreman	3.50	.88	1.17	1.56	3.50	.35	.46	.60
Concrete cart	6.10	1.53	2.04	2.71	.96	.10	.13	.16
Total Cost/C.Y.		$8.91	$11.89	$15.84		$3.05	$3.97	$5.18
Hourly production		4 C.Y.				10 C.Y.		

R03310-100 Average C.Y. of Concrete

Rubbing and floor finish not included — 4 uses of forms assumed, 5 story building.

Item	Description	Strength in psi	Cost for 1 C.Y. of Concrete				Total
			Ready Mix	Place	Forms	Reinforcing	
Beams	10′ span	4000	$76.50	$63.92	$325.68	$116.62	$582.72
5 kip/L.F.	25′		76.50	42.61	297.36	143.28	559.75
Beam & Slab, 1 way	15′	4000	76.50	19.67	424.20	60.56	580.93
125 psf Sup. L.	25′		76.50	17.05	272.70	76.45	442.70
Beam & Slab, 2 way	15′	4000	76.50	19.67	357.42	64.53	518.12
125 psf Sup. L.	25′		76.50	17.05	222.74	61.06	377.35
Columns,	16″ x 16″	4000	76.50	42.61	397.71	359.10	875.92
square tied	24″ x 24″		76.50	27.79	265.14	296.79	666.22
Columns, Tied	16″ diameter	4000	76.50	42.61	201.78	343.24	664.13
reinforced	24″ diameter		76.50	27.79	130.50	339.84	574.63
Flat Plate,	15′ span	4000	76.50	23.24	212.89	62.55	375.18
125 psf Sup. L.	25′		76.50	19.67	115.17	56.09	267.43
Flat Slab with drops,	20′	4000	76.50	23.24	160.82	59.07	319.63
125 psf Sup. L.	30′		76.50	19.67	108.46	65.03	269.66
Grade Wall,	8″ thick	3000	72.50	15.12	324.00	21.84	433.46
8′ high	15″ thick		72.50	12.96	164.00	21.35	270.81
Fiberglass Pan Joists, 24″	15′ span	4000	76.50	19.67	461.56	23.33	581.06
125 psf Sup. L.	25′ span		76.50	17.05	383.36	40.21	517.12
Pile Caps	under 5 C.Y.	3000	72.50	15.12	65.00	23.73	176.35
″	over 10 C.Y.		72.50	6.33	32.50	21.91	133.24
Slab	4″ thick	3500	74.20	12.37	17.66	17.01	121.24
on Grade	6″ thick		74.20	8.24	11.52	11.34	105.30
Spread	under 1 C.Y.	3000	72.50	24.73	69.12	26.51	192.86
Footings	over 5 C.Y.		72.50	12.37	25.92	31.13	141.92
Strip	9″ x 18″ plain	3000	72.50	12.37	86.76	—	171.63
Footings	12″ x 36″ reinforced		72.50	8.78	43.38	23.06	147.72
Waffle, 24″ x 24″	20′ span	4000	76.50	19.67	262.08	69.55	427.80
125 psf Sup. L.	30′ span		76.50	17.05	208.32	66.30	368.17

*Placement by direct chute assumed. All others, placement by pump assumed.

When form and reinforcing quantity ratios are available compute the C.Y. costs directly from the unit prices in the front of the book. See also R03310-010 for relative quantities per C.Y. for additional items. Higher strength concrete would change the ratios below. The tables below show typical examples of how the total bare costs per C.Y. have been derived. Ready mix concrete prices are U.S. average and equipment handling is assumed.

Beams 5 kip/L.F.	10′ Span						25′ Span					
	Material			Installation			Material			Installation		
4000 psi concrete	1 C.Y.		$ 76.50	1 C.Y.		$ 63.92	1 C.Y.		$ 76.50	1 C.Y.		$ 42.61
Formwork, 4 uses	69 S.F. @	.88	60.72	69 S.F. @	3.84	264.96	63 S.F. @	.88	55.44	63 S.F.@	3.84	241.92
Reinforcing steel	175 # @	.30	51.87	175 # @	.37	64.75	215 # @	.30	63.73	215 # @	.37	79.55
Total per C.Y.			$189.09			$393.63			$195.67			$364.08

Beam & Slab, One Way	125 psf, 15′ Span						125 psf, 25′ Span					
	Material			Installation			Material			Installation		
4000 psi concrete	1 C.Y.		$ 76.50	1 C.Y.		$ 19.67	1 C.Y.		$ 76.50	1 C.Y.		$ 17.05
Formwork, 4 uses	84 S.F. @	.92	77.28	84 S.F. @	4.13	346.92	54 S.F. @	.92	49.68	54 S.F. @	4.13	223.02
Reinforcing steel	122 # @	.30	36.16	122 # @	.20	24.40	154 # @	.30	45.65	154 # @	.20	30.80
Total per C.Y.			$189.94			$390.99			$171.83			$270.87

Beam & Slab, Two Way	125 psf, 15′ Span						125 psf, 25′ Span					
	Material			Installation			Material			Installation		
4000 psi concrete	1 C.Y.		$ 76.50	1 C.Y.		$ 19.67	1 C.Y.		$ 76.50	1 C.Y.		$ 17.05
Formwork, 4 uses	69 S.F. @	.84	57.96	69 S.F. @	4.34	299.46	43 S.F. @	.84	36.12	43 S.F. @	4.34	186.62
Reinforcing steel	130 # @	.30	38.53	130 # @	.20	26.00	123 # @	.30	36.46	123 # @	.20	24.60
Total per C.Y.			$172.99			$345.13			$149.08			$228.27

R03310-100 Average C.Y. of Concrete (cont.)

Columns, Square Tied	16″ Square						24″ Square					
	Material			Installation			Material			Installation		
4000 psi concrete	1 C.Y.		$ 76.50	1 C.Y.		$ 42.61	1 C.Y.		$ 76.50	1 C.Y.		$ 27.79
Formwork, 4 uses	81 S.F. @	.78	63.18	81 S.F. @	4.13	334.53	54 S.F. @	.78	42.12	54 S.F. @	4.13	223.02
Reinforcing steel, avg.	634 # @	.30	187.92	634 # @	.27	171.18	524 # @	.30	155.31	524 # @	.27	141.48
Total per C.Y.			$327.60			$548.32			$273.93			$392.29

Note: Reinforcing of 16″ and 24″ square columns can vary from 144 lb. to 972 lb. per C.Y.

Columns, Round Tied Reinforced	16″ Diameter						24″ Diameter					
	Material			Installation			Material			Installation		
4000 psi concrete	1 C.Y.		$ 76.50	1 C.Y.		$ 42.61	1 C.Y.		$ 76.50	1 C.Y.		$ 27.79
Formwork, fiber forms	19 L.F. @	3.48	66.12	19 L.F. @	7.14	135.66	9 L.F. @	6.80	61.20	9 L.F. @	7.70	69.30
Reinforcing steel, avg.	606 # @	.30	179.62	606 # @	.27	163.62	600 # @	.30	177.84	600 # @	.27	162.00
Ties	70 # @	.59	41.30	70 # @	.40	28.00	18 # @	.59	10.62	18 # @	.40	7.20
Total per C.Y.			$363.54			$369.89			$326.16			$266.29

Note: Reinforcing of 16″ and 24″ diameter columns vary from 160 lb. to 1150 lb. per C.Y.

Flat Plate	125 psf, 15′ Span						125 psf, 25′ Span					
	Material			Installation			Material			Installation		
4000 psi concrete	1 C.Y.		$ 76.50	1 C.Y.		$ 23.24	1 C.Y.		$ 76.50	1 C.Y.		$ 19.67
Formwork, 4 uses	61 S.F. @	.73	44.53	61 S.F. @	2.76	168.36	33 S.F. @	.73	24.09	33 S.F. @	2.76	91.08
Reinforcing steel	126 # @	.30	37.35	126 # @	.20	25.20	113 # @	.30	33.49	113 # @	.20	22.60
Total per C.Y.			$158.38			$216.80			$134.08			$133.35

Flat Slab with Drops	125 psf, 20′ Span						125 psf, 30′ Span					
	Material			Installation			Material			Installation		
4000 psi concrete	1 C.Y.		$ 76.50	1 C.Y.		$ 23.24	1 C.Y.		$ 76.50	1 C.Y.		$ 19.67
Formwork, 4 uses	43 S.F. @	.90	38.70	43 S.F. @	2.84	122.12	29 S.F. @	.90	26.10	29 S.F. @	2.84	82.36
Reinforcing steel	119 # @	.30	35.27	119 # @	.20	23.80	131 # @	.30	38.83	131 # @	.20	26.20
Total per C.Y.			$150.47			$169.16			$141.43			$128.23

Grade Wall, 8′ High	8″ Thick						15″ Thick					
	Material			Installation			Material			Installation		
3000 psi concrete	1 C.Y.		$ 72.50	1 C.Y.		$ 15.12	1 C.Y.		$ 72.50	1 C.Y.		$ 12.96
Formwork, 4 uses	81 S.F. @	.78	63.18	81 S.F. @	3.22	260.82	41 S.F. @	.78	31.98	41 S.F. @	3.22	132.02
Reinforcing steel	44 # @	.30	13.04	44 # @	.20	8.80	43 # @	.30	12.75	43 # @	.20	8.60
Total per C.Y.			$148.72			$284.74			$117.23			$153.58

Fiberglass Pan Joists, 24″ One Way	125 psf, 15′ Span						125 psf, 25′ Span					
	Material			Installation			Material			Installation		
4000 psi concrete	1 C.Y.		$ 76.50	1 C.Y.		$ 19.67	1 C.Y.		$ 76.50	1 C.Y.		$ 17.05
Formwork, 4 uses	49 S.F. @	.75	36.75	49 S.F. @	3.21	157.29	40 S.F. @	.75	30.00	40 S.F. @	3.21	128.40
Reinforcing steel	47 # @	.30	13.93	47 # @	.20	9.40	81 # @	.30	24.01	81 # @	.20	16.20
19″ fiberglass pans	44 S.F. @	5.63	247.72	44 S.F. @	.45	19.80	37 S.F. @	5.63	208.31	37 # @	.45	16.65
Total per C.Y.			$374.90			$206.16			$338.82			$178.30

Pile Caps	Under 5 C.Y.						Over 10 C.Y.					
	Material			Installation			Material			Installation		
3000 psi concrete	1.05 C.Y.		$ 72.50	1 C.Y.		$15.12	1.05 C.Y.		$72.50	1 C.Y.		$ 6.33
Formwork, 4 uses	20 S.F. @	.63	12.60	20 S.F. @	2.62	52.40	10 S.F. @	.63	6.30	10 S.F. @	2.62	26.20
Reinforcing steel	52 # @	.30	15.41	52 # @	.16	8.32	48 # @	.30	14.23	48 # @	.16	7.68
Total per C.Y.			$100.51			$75.84			$93.03			$40.21

R03310-100 Average C.Y. of Concrete (cont.)

Slab on Grade	4″ Thick				6″ Thick			
	Material		Installation		Material		Installation	
3500 psi concrete	1.05 C.Y.	$74.20	1 C.Y.	$12.37	1.05 C.Y.	$74.20	1 C.Y.	$ 8.24
Formwork, 4 uses	9.2 L.F. @ .25	2.30	9.2 L.F. @ 1.67	15.36	6 L.F. @ .25	1.50	6 L.F. @ 1.67	10.02
W.W. Fabric	81 S.F. @ .07	5.67	81 S.F. @ .14	11.34	54 S.F. @ .07	3.78	54 S.F. @ .14	7.56
Total per C.Y.		$82.17		$39.07		$79.48		$25.82

Spread Footings	Under 1 C.Y.				Over 5 C.Y.			
	Material		Installation		Material		Installation	
3000 psi concrete	1.05 C.Y.	$72.50	1.05 C.Y.	$24.73	1.05 C.Y.	$72.50	1 C.Y.	$12.37
Formwork, 4 uses	24 S.F. @ .47	11.28	24 S.F. @ 2.41	57.84	9 S.F. @ .47	4.23	9 S.F. @ 2.41	21.69
Reinforcing steel	46 # @ .30	13.63	46 # @ .28	12.88	54 # @ .30	16.01	54 # @ .28	15.12
Total per C.Y.		$97.41		$95.45		$92.74		$49.18

Strip Footings	9′ x 18″ Plain				12″ x 36″ Reinforced			
	Material		Installation		Material		Installation	
3000 psi concrete	1.05 C.Y.	$72.50	1.05 C.Y.	$12.37	1.05 C.Y.	$72.50	1 C.Y.	$ 8.78
Formwork, 4 uses	36 S.F. @ .35	12.60	36 S.F. @ 2.06	74.16	18 S.F. @ .35	6.30	18 S.F. @ 2.06	37.08
Reinforcing steel		—		—	40 # @ .30	11.86	40 # @ .28	11.20
Total per C.Y.		$85.10		$86.53		$90.66		$57.06

Waffle, 24″ x 24″	125 psf, 20′ Span				125 psf, 30′ Span			
	Material		Installation		Material		Installation	
4000 psi concrete	1 C.Y.	$ 76.50	1 C.Y.	$ 19.67	1 C.Y.	$ 76.50	1 C.Y.	$ 17.05
Formwork, 4 uses	39 S.F. @ .75	29.25	39 S.F. @ 3.29	128.31	31 S.F. @ .75	23.25	31 S.F. @ 3.29	101.99
Reinforcing steel & mesh	107 # @ .37	39.59	107 # @ .28	29.96	102 # @ .37	37.74	102 # @ .28	28.56
19″ x 19″ Fiberglass Domes	39 S.F. @ 2.22	86.58	39 S.F. @ .46	17.94	31 S.F. @ 2.22	68.82	31 S.F. @ .46	14.26
Total per C.Y.		$231.92		$195.88		$206.31		$161.86

R03310-120 Lift Slabs

The cost advantage of the lift slab method is due to placing all concrete, reinforcing steel, inserts and electrical conduit at ground level and in reduction of formwork. Minimum economical project size is about 30,000 S.F. Slabs may be tilted for parking garage ramps.

It is now used in all types of buildings and has gone up to 22 stories high in apartment buildings. Current trend is to use post-tensioned flat plate slabs with spans from 22′ to 35′. Cylindrical void forms are used when deep slabs are required. One pound of prestressing steel is about equal to seven pounds of conventional reinforcing.

To be considered cured for stressing and lifting, a slab must have attained 75% of design strength. Seven days are usually sufficient with four to five days possible if high early strength cement is used. Slabs can be stacked using two coats of a non-bonding agent to insure that slabs do not stick to each other. Lifting is done by companies specializing in this work. Lift rate is 5′ to 15′ per hour with an average of 10′ per hour. Total areas up to 33,000 S.F. have been lifted at one time. 24 to 36 jacking columns are common. Most economical bay sizes are 24′ to 28′ with four to fourteen stories most efficient. Continuous design reduces reinforcing steel cost. Use of post-tensioned slabs allows larger bay sizes.

R03410-030 Prestressed Precast Concrete Structural Units

See also R03410-090 for post-tensioned prestressed concrete.

Type	Location	Depth	Span in Ft.	Live Load Lb. per S.F.
Double Tee 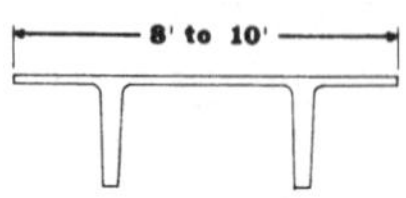	Floor	28″ to 34″	60 to 80	50 to 80
	Roof	12″ to 24″	30 to 50	40
	Wall	Width 8′	Up to 55′ high	Wind
Multiple Tee 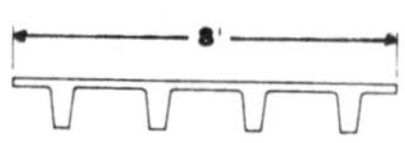	Roof	8″ to 12″	15 to 40	40
	Floor	8″ to 12″	15 to 30	100
Plank 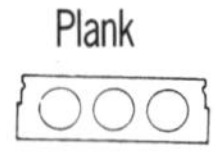 or 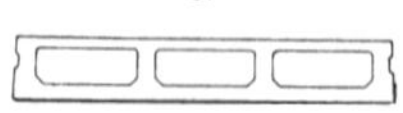	Roof or Floor	4″ 6″ 8″ 10″ 12″	Roof / Floor 13 / 12 22 / 18 26 / 25 33 / 29 42 / 32	40 for Roof 100 for Floor
Single Tee 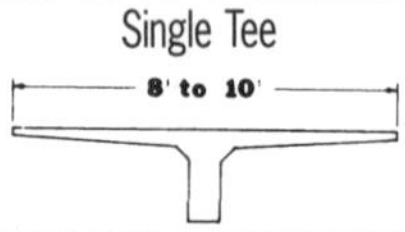	Roof	28″ 32″ 36″ 48″	40 80 100 120	40
AASHO Girder 	Bridges	Type 4 5 6	100 110 125	Highway
Box Beam 	Bridges	15″ 27″ 33″	40 to 100	Highway

The majority of precast projects today utilize double tees rather than single tees because of speed and ease of installation. As a result casting beds at manufacturing plants are normally formed for double tees. Single tee projects will therefore require an initial set up charge to be spread over the individual single tee costs.

For floors, a 2″ to 3″ topping is field cast over the shapes. For roofs, insulating concrete or rigid insulation is placed over the shapes.

Member lengths up to 40′ are standard haul, 40′ to 60′ require special permits and lengths over 60′ must be escorted. Over width and/or over length can add up to 100% on hauling costs.

Large heavy members may require two cranes for lifting which would increase erection costs by about 45%. An eight man crew can install 12 to 20 double tees, or 45 to 70 quad tees or planks per day.

Grouting of connections must also be included.

Several system buildings utilizing precast members are available. Heights can go up to 22 stories for apartment buildings. Optimum design ratio is 3 S.F. of surface to 1 S.F. of floor area.

R03520-010 Lightweight Concrete

Vermiculite or Perlite come in bags of 4 C.F. under various trade names. Weight is about 8 lbs. per C.F. For insulating roof fill use 1:6 mix. For structural deck use 1:4 mix over gypsum boards, steeltex, steel centering, etc. supported by closely spaced joists or bulb trees. For structural slabs use 1:3:2 vermiculite sand concrete over steeltex, metal lath, steel centering, etc. on joists spaced 2′-0″ O.C. for maximum L.L. of 80 P.S.F. Use same mix for slab base fill over steel flooring or regular reinforced concrete slab when tile, terrazzo or other finish is to be laid over.

For slabs on grade use 1:3:2 mix when tile, etc. finish is to be laid over. If radiant heating units are installed use a 1:6 mix for a base. After coils are in place, cover with a regular granolithic finish (mix 1:3:2) to a minimum depth of 1-1/2″ over top of units.

Reinforce all slabs with 6 x 6 or 10 x 10 welded wire mesh.

Vermiculite concrete can be purchased ready mixed but the following breakdown is included for field mix. Prices given below are for a one story building and assume 50 C.Y. or more. For less than 50 C.Y. add 10%. For over one story add 5% per C.Y. Screed finish cost is included below.

Quantities per C.Y., Field Mix	1:6 Mix Insulating Roof Fill		1:3:2 Mix Lightweight Structural Concrete	
Portland cement	5.0 bags	$ 40.50	6.2 bags	$ 50.22
Vermiculite or Perlite	7.5 bags	50.25		
treated type			4.7 bags	32.90
Sand			12.5 C.F.	12.20
Plant and Water		7.37		7.37
Labor, machine mix, hoist and place, Crew C-8	1.12 L.H.	46.82	.91 L.H.	38.04
Total bare cost in place, per C.Y.		$144.94		$140.73

R04210-120 Common and Face Brick Prices

Prices are based on truckload lot purchases for Common Building Brick and Facing Brick. Prices are per M, (thousand), brick.

	Material			Installation				Total			
	Brick per M Delivered		Mortar	Common in 8″ Wall		Face Brick, 4″ Veneer		Common in 8″ Wall		Face Brick, 4″ Veneer	
City	Common	Face	3/8″ Joint	Bare Costs	Incl. O & P	Bare Costs	Incl. O & P	Bare Costs	Incl. O & P	Bare Costs	Incl. O & P
Atlanta	$185	$259	$44.00	$ 472	$ 765	$ 566	$ 918	$ 707	$1,000	$ 869	$1,221
Baltimore	315	443	for 8″ Wall	551	894	661	1,073	919	1,262	1,153	1,565
Boston	410	575	and	986	1,600	1,183	1,920	1,452	2,066	1,811	2,548
Buffalo	380	532	$36.00	773	1,254	927	1,505	1,208	1,689	1,511	2,089
Chicago	245	343	for 4″ Wall	892	1,447	1,070	1,736	1,188	1,743	1,459	2,125
Cincinnati	180	254		668	1,085	802	1,301	897	1,314	1,100	1,599
Cleveland	225	319		748	1,214	898	1,457	1,024	1,490	1,263	1,822
Columbus	180	255		653	1,061	784	1,273	882	1,290	1,083	1,572
Dallas	290	408		452	733	542	880	795	1,076	998	1,336
Denver	240	339		546	886	655	1,063	837	1,177	1,040	1,448
Detroit	315	443		829	1,346	995	1,615	1,197	1,714	1,487	2,107
Houston	340	480		463	752	556	902	857	1,146	1,086	1,432
Indianapolis	215	300		652	1,058	782	1,269	917	1,323	1,127	1,614
Kansas City	355	503		712	1,155	854	1,386	1,122	1,565	1,408	1,940
Los Angeles	230	325		782	1,269	939	1,523	1,063	1,550	1,310	1,894
Memphis	180	251		564	915	677	1,098	793	1,144	972	1,393
Milwaukee	300	425		786	1,276	943	1,531	1,139	1,629	1,417	2,005
Minneapolis	340	475		867	1,408	1,041	1,689	1,261	1,802	1,566	2,214
Nashville	190	265		471	764	565	917	711	1,004	874	1,226
New Orleans	230	325		410	666	492	799	691	947	863	1,170
New York City	265	375		1,078	1,750	1,294	2,100	1,395	2,067	1,716	2,522
Philadelphia	290	405		872	1,416	1,047	1,699	1,215	1,759	1,500	2,152
Phoenix	230	325		492	798	590	958	773	1,079	961	1,329
Pittsburgh	295	418		681	1,105	817	1,326	1,029	1,453	1,284	1,793
St. Louis	245	346		420	681	504	817	716	977	896	1,209
San Antonio	225	315		701	1,137	841	1,364	977	1,413	1,201	1,724
San Diego	275	390		881	1,430	1,057	1,716	1,208	1,757	1,495	2,154
San Francisco	270	380		771	1,251	925	1,501	1,093	1,573	1,352	1,928
Seattle	405	567		783	1,271	939	1,525	1,244	1,732	1,559	2,145
Washington, D.C.	265	375		569	923	683	1,108	886	1,240	1,105	1,530
Average	$270	$380	▼	$ 685	$1,110	$ 820	$1,330	$1,005	$1,435	$1,250	$1,760

Common building brick manufactured according to ASTM C62 and facing brick manufactured according to ASTM C216 are the two standard bricks available for general building use.

Building brick is made in three grades; SW, where high resistance to damage caused by cyclic freezing is required; MW, where moderate resistance to cyclic freezing is needed; and NW, where little resistance to cyclic freezing is needed. Facing brick is made in only the two grades SW and MW. Additionally, facing brick is available in three types; FBS, for general use; FBX, for general use where a higher degree of precision and lower permissible variation in size than FBS is needed; and FBA, for general use to produce characteristic architectural effects resulting from non-uniformity in size and texture of the units.

In figuring above installation costs, a D-8 Crew (with a daily output of 1.5 M) was used for the 4″ veneer. A D-8 Crew (with a daily output of 1.8 M) was used for the 8″ solid wall.

In figuring the total cost including overhead and profit, an allowance of 10% was added to the sum of the cost of the brick and mortar. Also, 3% breakage was included for both the bare costs and the costs with overhead and profit. If bricks are delivered palletized with 280 to 300 per pallet, or packaged, allow only 1-1/2% for breakage. Then add $10 per M to the cost of brick and deduct two hours helper time. The net result is a savings of $30 to $40 per M in place. Packaged or palletized delivery is practical when a job is big enough to have a crane or other equipment available to handle a package of brick. This is so on all industrial work but not always true on small commercial buildings.

The prices above are for bricks used in commercial, apartment house or industial construction. If it is possible to obtain the price of the actual brick to be used, it should be done and substituted in the table. The use of buff and gray face is increasing, and there is a continuing trend to the Norman, Roman, Jumbo and SCR brick.

See R04210-500 for brick quantities per S.F. and mortar quantities per M brick. (Average prices for the various sizes are listed in Division 4.)

Common red clay brick for backup is not used that often. Concrete block is the most usual backup material with occasional use of sand lime or cement brick. Sand lime cost about $15 per M less than red clay and cement brick are about $5 per M less than red clay. These figures may be substituted in the common brick breakdown for the cost of these items in place, as labor is about the same. Building brick is commonly used in solid walls for strength and as a fire stop.

Brick panels built on the ground and then crane erected to the upper floors have proven to be economical. This allows the work to be done under cover and without scaffolding.

R04210-500 Brick, Block & Mortar Quantities

Running Bond						For Other Bonds Standard Size Add to S.F. Quantities in Table to Left		
Number of Brick per S.F. of Wall - Single Wythe with 3/8″ Joints				C.F. of Mortar per M Bricks, Waste Included				
Type Brick	Nominal Size (incl. mortar) L H W	Modular Coursing	Number of Brick per S.F.	3/8″ Joint	1/2″ Joint	Bond Type	Description	Factor
Standard	8 x 2-2/3 x 4	3C=8″	6.75	10.3	12.9	Common	full header every fifth course	+20%
Economy	8 x 4 x 4	1C=4″	4.50	11.4	14.6		full header every sixth course	+16.7%
Engineer	8 x 3-1/5 x 4	5C=16″	5.63	10.6	13.6	English	full header every second course	+50%
Fire	9 x 2-1/2 x 4-1/2	2C=5″	6.40	550 # Fireclay	—	Flemish	alternate headers every course	+33.3%
Jumbo	12 x 4 x 6 or 8	1C=4″	3.00	23.8	30.8		every sixth course	+5.6%
Norman	12 x 2-2/3 x 4	3C=8″	4.50	14.0	17.9	Header = W x H exposed		+100%
Norwegian	12 x 3-1/5 x 4	5C=16″	3.75	14.6	18.6	Rowlock = H x W exposed		+100%
Roman	12 x 2 x 4	2C=4″	6.00	13.4	17.0	Rowlock stretcher = L x W exposed		+33.3%
SCR	12 x 2-2/3 x 6	3C=8″	4.50	21.8	28.0	Soldier = H x L exposed		—
Utility	12 x 4 x 4	1C=4″	3.00	15.4	19.6	Sailor = W x L exposed		-33.3%

Concrete Blocks Nominal Size	Approximate Weight per S.F.		Blocks per 100 S.F.	Mortar per M block, waste included	
	Standard	Lightweight		Partitions	Back up
2″ x 8″ x 16″	20 PSF	15 PSF	113	27 C.F.	36 C.F.
4″	30	20	↓	41	51
6″	42	30	↓	56	66
8″	55	38	↓	72	82
10″	70	47	↓	87	97
12″	85	55	↓	102	112

Metals | R050 | Materials, Coatings, & Fastenings

R05090-520 Welded Structural Steel

Usual weight reductions with welded design run 10% to 20% compared with bolted or riveted connections. This amounts to about the same total cost compared with bolted structures since field welding is more expensive than bolts. For normal spans of 18′ to 24′ figure 6 to 7 connections per ton.

Trusses — For welded trusses add 4% to weight of main members for connections. Up to 15% less steel can be expected in a welded truss compared to one that is shop bolted. Cost of erection is the same whether shop bolted or welded.

General — Typical electrodes for structural steel welding are E6010, E6011, E60T and E70T. Typical buildings vary between 2# to 8# of weld rod per ton of steel. Buildings utilizing continuous design require about three times as much welding as conventional welded structures. In estimating field erection by welding, it is best to use the average linear feet of weld per ton to arrive at the welding cost per ton. The type, size and position of the weld will have a direct bearing on the cost per linear foot. A typical field welder will deposit 1.8# to 2# of weld rod per hour manually. Using semiautomatic methods can increase production by as much as 50% to 75%.

R05120-210 Structural Steel

1 to 2 Story Building	Bare Costs for 100 Ton Job with High Strength Bolts					Including Subs O&P		
City	Fabricated and Delivered	Unloading and Sorting	Erection Equipment	Field Erection Labor	Total per Ton in Place	Material	Installation	Total
Atlanta	$1,150	$29	$118	$197	$1,494	$1,265	$ 551	$1,816
Baltimore	1,125	35	144	237	1,541	1,238	665	1,903
Boston	1,300	46	182	306	1,834	1,430	856	2,286
Buffalo	1,375	38	137	256	1,806	1,513	698	2,211
Chicago	1,200	49	179	327	1,755	1,320	897	2,217
Cincinnati	1,175	35	129	237	1,576	1,293	648	1,941
Cleveland	1,175	40	124	269	1,608	1,293	712	2,005
Columbus	1,175	36	120	238	1,569	1,293	642	1,935
Dallas	1,225	23	124	154	1,526	1,348	466	1,814
Denver	1,300	28	130	187	1,645	1,430	543	1,973
Detroit	1,300	46	151	309	1,806	1,430	827	2,257
Houston	1,225	25	134	170	1,554	1,348	510	1,858
Indianapolis	1,375	35	115	233	1,758	1,513	626	2,139
Kansas City	1,375	38	161	256	1,830	1,513	725	2,238
Los Angeles	1,475	44	143	292	1,954	1,623	783	2,406
Memphis	1,225	26	138	173	1,562	1,348	522	1,870
Milwaukee	1,275	39	133	259	1,706	1,403	701	2,104
Minneapolis	1,150	46	202	305	1,703	1,265	876	2,141
Nashville	1,300	25	134	171	1,630	1,430	512	1,942
New Orleans	1,325	23	109	151	1,608	1,458	444	1,902
New York City	1,475	71	206	477	2,229	1,623	1,247	2,870
Philadelphia	1,300	50	181	332	1,863	1,430	910	2,340
Phoenix	1,200	32	105	213	1,550	1,320	572	1,892
Pittsburgh	1,250	40	181	267	1,738	1,375	771	2,146
St. Louis	1,350	40	183	266	1,839	1,485	771	2,256
San Antonio	1,200	21	101	139	1,461	1,320	409	1,729
San Diego	1,425	44	144	292	1,905	1,568	784	2,352
San Francisco	1,425	44	156	292	1,917	1,568	797	2,365
Seattle	1,225	37	129	249	1,640	1,348	674	2,022
Washington, DC	1,175	32	158	213	1,578	1,293	630	1,923
U.S. Average	$1,275	$37	$145	$247	$1,704	$1,403	$ 672	$2,075

Adjustments to the Base Price Above	Bare Costs	Incl. Subs O&P
For 3 to 6 story building, add per ton	$ 32	$ 75
For 7 to 15 stories, add per ton	64	100
For over 15 stories, add per ton	121	175
For field welded connections, add per ton	187	275
For multi-story masonry wall bearing construction, add to erection costs	30%	30%

General Average per Ton in Place for A36 Steel (High Strength Steel Base Price May Be Substituted)				
		Bare Costs		Incl. Subs O&P
Item	Description	Itemized	Summary	
Material	Base price	$ 360	$1,275	$1,403
	Extras and delivery to shop	155		
	Drafting	105		
	Shop fabrication & warehouse rehandling	565		
	Shop coat paint	40		
	Trucking to job site	50		
Installation	Unload and shake out, 1.0 hours @ $37.15	37	$ 429	$ 672
	Erect and plumb, 5.0 hours @ $37.15	184		
	Field bolt, 1.7 hours @ $37.15	63		
	Crane & minor erection equipment	145		
	Total per ton in place	$1,704	$1,704	$2,075

R05120-215 Structural Steel Estimating for Repair and Remodeling Projects

The correct approach to estimating structural steel is dependent upon the amount of steel required for the particular project. If the project is a sizable addition to a building, the data can be used directly from the cost data book. This is not the case however if the project requires a small amount of steel, for instance to reinforce existing roof or floor structural systems. To better understand this, please refer to the unit price line for a W16x31 beam with bolted connections in Section 05120-720 or 05120-640. Assume your project requires the reinforcement of the structural members of the roof system of a 3 story building required for the installation a new piece of HVAC equipment. The project will need 4 pieces of W16x31 that are 30′ long. After pricing this 120 L.F. job using the unit prices for a W16x31, an analysis will reveal that the price is wholly inadequate.

The first problem is apparent if you examine the amount of steel that can be installed per day. The unit price line indicates 900 linear feet per day can be installed by a 5 man crew with an 90 ton crane. This productivity is correct for new construction but certainly is not for repair and remodeling work. Installation of new structural steel considers that each member is installed from the foundation to the roof of the structure in a planned and systematic manner with a crane having unrestricted access to all parts of the project. Additionally each connection is planned and detailed with full field access for fit-up and final bolting. The erection is planned and progresses such that interferences and conflicts with other structural members are minimized, if not completely eliminated. All of these assumptions are clearly not the case with a repair and remodeling job, and a significant decrease in the stated productivity will be observed.

A crane will certainly be needed to lift the members into the general area of the project but in most cases will not be able to place the beams into their final position. An opening in the existing roof may not be large enough to permit the beams to pass through and it may be necessary to bring them into the building through existing windows or doors. Moving the beams to the actual area where they will be installed may involve hand labor and the use of dollies. Finally, hoists and/or jacks may be needed for final positioning.

The connection of new members to existing can often be accomplished by field bolting with accurate field measurements and good planning but in many cases access to both sides of existing members is not possible and field welding becomes the only alternative. In addition to the cost of the actual welding, protection of existing finishes, systems and structure and fire protection must be considered.

New beams can never be installed tight to the existing decks or floors which they must support and the use of shims and tack welding becomes necessary. Additionally, further planning and cost is involved in assuring that existing loads are minimized during the installation and shimming process.

It is apparent that installation of structural steel as part of a repair and remodeling project involves more than simply installing the members and estimating the cost in the same manner as new construction. The best procedure for estimating the total cost is adequate planning and coordination of each process and activity that will be needed. Unit costs for the materials, labor and equipment can then be attached to each needed activity and a final, complete price can be determined.

R06100-010 Thirty City Lumber Prices

Prices for boards are for #2 or better or sterling, whichever is in best supply. Dimension lumber is "Standard or Better" either Southern Yellow Pine (S.Y.P.), Spruce-Pine-Fir (S.P.F.), Hem-Fir (H.F.) or Douglas Fir (D.F.). The species of lumber used in a geographic area is listed by city. Plyform is 3/4" BB oil sealed fir or S.Y.P. whichever prevails locally, 3/4" CDX is S.Y.P. or Fir.

These are prices at the time of publication and should be checked against the current market price. Relative differences between cities will stay approximately constant.

City	Species	Contractor Purchases per M.B.F. S4S								Contractor Purchases per M.S.F.	
		Dimensions						Boards		3/4" Ext. Plyform	3/4" Thick CDX T&G
		2"x4"	2"x6"	2"x8"	2"x10"	2"x12"	4"x4"	1"x6"	1"x12"		
Atlanta	S.Y.P.	$461	$488	$562	$644	$732	$851	$1,065	$1,445	$ 928	$809
Baltimore	S.P.F.	432	457	527	603	686	798	998	1354	869	758
Boston	S.P.F.	417	441	508	582	662	770	963	1307	839	731
Buffalo	S.P.F.	411	434	501	573	652	758	949	1287	826	720
Chicago	H.F.	495	523	603	691	785	913	1143	1551	995	868
Cincinnati	S.P.F.	443	468	540	618	703	818	1023	1388	891	777
Cleveland	S.P.F.	429	454	523	599	681	792	991	1345	863	752
Columbus	S.P.F.	438	464	534	612	696	809	1012	1374	882	768
Dallas	S.Y.P.	455	481	555	635	722	840	1051	1426	916	798
Denver	H.F.	493	522	601	689	783	911	1139	1546	992	865
Detroit	H.F.	513	543	626	716	815	947	1185	1608	1032	900
Houston	S.Y.P.	450	476	549	629	715	832	1040	1412	906	790
Indianapolis	S.P.F.	499	529	609	698	793	923	1154	1566	1005	876
Kansas City	D.F.	493	522	601	689	783	911	1139	1546	992	865
Los Angeles	D.F.	485	514	592	678	771	896	1121	1522	977	851
Memphis	S.Y.P.	461	488	562	644	732	851	1065	1445	928	809
Milwaukee	H.F.	495	523	603	691	785	913	1143	1551	995	868
Minneapolis	H.F.	493	522	601	689	783	911	1139	1546	992	865
Nashville	S.Y.P.	461	488	562	644	732	851	1065	1445	928	809
New Orleans	S.Y.P.	464	491	566	648	737	857	1072	1455	934	814
New York City	S.P.F.	429	454	523	599	681	792	991	1345	863	752
Philadelphia	S.P.F.	432	457	527	603	686	798	998	1354	869	758
Phoenix	D.F.	508	538	620	710	807	939	1174	1594	1023	892
Pittsburgh	S.P.F.	438	464	534	612	696	809	1012	1374	882	768
St. Louis	H.F.	499	529	609	698	793	923	1154	1566	1005	876
San Antonio	S.Y.P.	459	486	560	642	730	848	1061	1441	925	806
San Diego	D.F.	490	519	598	685	779	905	1133	1537	987	860
San Francisco	D.F.	464	491	566	648	737	857	1072	1455	934	814
Seattle	D.F.	459	486	560	642	730	848	1061	1440	925	806
Washington, DC	S.P.F.	436	462	532	609	693	806	1008	1368	878	765
Average		$464	$491	$566	$648	$737	$857	$1,072	$1,455	$ 934	$814

To convert square feet of surface to board feet, 4% waste included.

S4S Size	Multiply S.F. by	T & G Size	Multiply S.F. by	Flooring Size	Multiply S.F. by
				25/32" x 2-1/4"	1.37
1 x 4	1.18	1 x 4	1.27	25/32" x 3-1/4"	1.29
1 x 6	1.13	1 x 6	1.18	15/32" x 1-1/2"	1.54
1 x 8	1.11	1 x 8	1.14	1" x 3"	1.28
1 x 10	1.09	2 x 6	2.36	1" x 4"	1.24

R06110-030 Lumber Product Material Prices

The price of forest products fluctuates widely from location to location and from season to season depending upon economic conditions. The table below indicates National Average material prices in effect Jan. 1 of this book year. The table shows relative differences between various sizes, grades and species. These percentage differentials remain fairly constant even though lumber prices in general may change significantly during the year.

Availability of certain items depends upon geographic location and must be checked prior to firm price bidding.

National Average Contractor Price, Quantity Purchase

	Dimension Lumber, S4S, #2 & Better, KD					
	Species	**2″x4″**	**2″x6″**	**2″x8″**	**2″x10″**	**2″x12″**
Framing Lumber per MBF	Douglas Fir	$ 392	$ 385	$ 412	$ 446	$ 432
	Spruce	311	293	302	359	367
	Southern Yellow Pine	383	362	369	452	531
	Hem-Fir	324	301	359	373	351

Heavy Timbers, Fir	
3″ x 4″ thru 3″ x 12″	$527
4″ x 4″ thru 4″ x 12″	527
6″ x 6″ thru 6″ x 12″	770
8″ x 8″ thru 8″ x 12″	1,060
10″ x 10″ and 10″ x 12″	1,094

	S4S "D" Quality or Clear, KD					
	Species	**1″x4″**	**1″x6″**	**1″x8″**	**1″x10″**	**1″x12″**
Boards per MBF *See also Cedar Siding	Sugar Pine	$1,337	$1,553	$1,553	$2,261	$3,038
	Idaho Pine	803	770	857	878	878
	Engleman Spruce	844	1,080	1,073	1,080	1,600
	So. Yellow Pine	830	959	938	857	1,087
	Ponderosa Pine	770	1,094	945	1,195	1,890

S4S # 2 & Better or Sterling, KD					
Species	**1″x4″**	**1″x6″**	**1″x8″**	**1″x10″**	**1″x12″**
Sugar Pine	$560	$581	$601	$608	$864
Idaho Pine	803	770	857	878	837
Engleman Spruce	601	628	614	540	810
So. Yellow Pine	533	513	533	513	668
Ponderosa Pine	594	601	628	493	736

R06160-020 Plywood

There are two types of plywood used in construction: interior, which is moisture resistant but not waterproofed, and exterior, which is waterproofed.

The grade of the exterior surface of the plywood sheets is designated by the first letter: A, for smooth surface with patches allowed; B, for solid surface with patches and plugs allowed; C, which may be surface plugged or may have knot holes up to 1″ wide; and D, which is used only for interior type plywood and may have knot holes up to 2-1/2″ wide. "Structural Grade" is specifically designed for engineered applications such as box beams. All CC & DD grades have roof and floor spans marked on them.

Underlayment grade plywood runs from 1/4″ to 1-1/4″ thick. Thicknesses 5/8″ and over have optional tongue and groove joints which eliminates the need for blocking the edges. Underlayment 19/32″ and over may be referred to as Sturd-i-Floor.

The price of plywood can fluctuate widely due to geographic and economic conditions. When one or two local prices are known, the relative prices for other types and sizes may be found by direct factoring of the prices in the table below.

Typical uses for various plywood grades are as follows:

AA-AD Interior — cupboards, shelving, paneling, furniture
BB Plyform — concrete form plywood
CDX — wall and roof sheathing
Structural — box beams, girders, stressed skin panels
AA-AC Exterior — fences, signs, siding, soffits, etc.
Underlayment — base for resilient floor coverings
Overlaid HDO — high density for concrete forms & highway signs
Overlaid MDO — medium density for painting, siding, soffits & signs
303 Siding — exterior siding, textured, striated, embossed, etc.

National Average Price in Lots of 10 MSF, per MSF-January

Grade	Type	4′x8′	Type	4′x8′	4′x10′
Sanded Grade	1/4″ Interior AD	$ 448	1/4″ Exterior AC	$ 450	$ 448
	3/8″	530	3/8″	530	530
	1/2″	619	1/2″	620	619
	5/8″	724	5/8″	724	724
	3/4″	800	3/4″	800	800
	1″	1,221	1″	1,342	1,342
	1-1/4″	1,464	Exterior AA, add	150	155
	Interior AA, add	150	Exterior AB, add	125	130
			CD Structural 1	**Underlayment**	
Unsanded Grade 4′ x 8′ Sheets	5/16″ CDX	$ 420			
	3/8″	435	3/8″ 4′x8′ sheets $340	3/8″ 4′x8′ sheets	$ 530
	1/2″	392	1/2″ 455	1/2″	619
	5/8″	450	3/4″ 655	5/8″T&G	540
	3/4″	574		3/4″T&G	839
	3/4″ T&G	839		1-1/8″ 2-4-1T&G	1,026
Form Plywood	5/8″ Exterior, oiled BB, plyform	$ 880	5/8″ HDO (overlay 2 sides)		$2,148
	3/4″ Exterior, oiled BB, plyform	934	3/4″ HDO (overlay 2 sides)		2,214
Overlaid 4′x8′ Sheets	Overlay 2 Sides MDO		Overlay 1 Side MDO		
	3/8″ thick	$1,345	3/8″ thick		$1,142
	1/2″	1,576	1/2″		1,316
	5/8″	1,776	5/8″		1,478
	3/4″	2,034	3/4″		1,734
303 Siding	Fir, rough sawn, natural finish, 3/8″ thick	$ 518	Texture 1-11	5/8″ thick, Fir	$ 945
	Redwood	1,796		Redwood	1,890
	Cedar	1,701		Cedar	1,537
	Southern Yellow Pine	418		Southern Yellow	763
Waferboard/O.S.B.	1/4″ sheathing	$ 250	19/32″ T&G		$ 424
	7/16″ sheathing	281	23/32″ T&G		510

For 2 MSF to 10 MSF, add 10%. For less than 2 MSF, add 15%.

Wood & Plastics | R061 | Rough Carpentry

R06170-100 Wood Roof Trusses

Loading figures represent live load. An additional load of 10 psf on the top chord and 10 psf on the bottom chord is included in the truss design. Spacing is 24″ O.C.

Span in Feet	Cost per Truss for Different Live Loads and Roof Pitches					
	Flat	4 in 12 Pitch		5 in 12 Pitch		8 in 12 Pitch
	40 psf	30 psf	40 psf	30 psf	40 psf	30 psf
20	$ 64	$ 49	$ 51	$ 59	$ 54	$ 60
22	70	53	55	61	57	60
24	77	45	47	52	50	55
26	80	46	48	52	51	65
28	88	58	60	61	63	66
30	111	59	61	63	64	78
32	119	71	74	70	130	115
34	126	104	106	113	113	124
36	133	110	112	115	115	130
38	141	118	120	121	122	140
40	148	120	122	121	122	146

Thermal & Moist. Protection | R075 | Membrane Roofing

R07510-020 Built-Up Roofing

Asphalt is available in kegs of 100 lbs. each; coal tar pitch in 560 lb. kegs. Prepared roofing felts are available in a wide range of sizes, weights & characteristics. However, the most commonly used are #15 (432 S.F. per roll, 13 lbs. per square) and #30 (216 S.F. per roll, 27 lbs. per square).

Inter-ply bitumen varies from 24 lbs. per sq. (asphalt) to 30 lbs. per sq. (coal tar) per ply, ± 25%. Flood coat bitumen also varies from 60 lbs. per sq. (asphalt) to 75 lbs. per sq. (coal tar), ± 25%. Expendable equipment (mops, brooms, screeds, etc.) runs about 16% of the bitumen cost. For new, inexperienced crews this factor may be much higher.

Rigid insulation board is typically applied in two layers. The first is mechanically attached to nailable decks or spot or solid mopped to non-nailable decks; the second layer is then spot or solid mopped to the first layer. Membrane application follows the insulation, except in protected membrane roofs, where the membrane goes down first and the insulation on top, followed with ballast (stone or concrete pavers). Insulation and related labor costs are NOT included in Div. 07510-300; they can be found in Div. 07220-700.

Prices shown are bare costs only.

4-Ply Felt on Insulated Deck	Glass Fiber Felt w/ Hot Asphalt		Organic Felt w/ Hot Coal Tar	
4 ply #15 felt (1/4 roll per ply)	$3.99 per ply	$ 15.96	$4.24 per ply	$ 16.96
156 Lbs. hot asphalt/195 lbs. hot coal tar	$420.00 per ton	32.76	$605.00 per ton	58.99
500 Lbs. roofing stone (3/8″ to 1/2″)	$14.75 per ton	3.69	$14.75 per ton	3.69
Installation, crew G-1, @ $1,789.20 per day	.050 days	89.46	.048 days	85.88
Total Bare Cost per Square in-place		$141.87		$165.52

Doors & Windows | R085 | Windows

R08510-100 Steel Sash

Ironworker crew will erect 25 S.F. or 1.3 sash unit per hour, whichever is less.

Mechanic will point 30 L.F. per hour.

Painter will paint 90 S.F. per coat per hour.

Glazier production depends on light size.

Allow 1 lb. special steel sash putty per 16″ x 20″ light.

Doors & Windows | R085 | Windows

R08550-200 Replacement Windows

Replacement windows are typically measured per United Inch.

United Inches are calculated by rounding the width and height of the window opening up to the nearest inch, then adding the two figures.

The labor cost for replacement windows includes removal of sash, existing sash balance or weights, parting bead where necessary and installation of new window.

Debris hauling and dump fees are not included.

Doors & Windows | R087 | Hardware

R08700-100 Hinges

All closer equipped doors should have ball bearing hinges. Lead lined or extremely heavy doors require special strength hinges. Usually 1-1/2 pair of hinges are used per door up to 7′-6″ high openings. Table below shows typical hinge requirements.

Use Frequency	Type Hinge Required	Type of Opening	Type of Structure
High	Heavy weight ball bearing	Entrances Toilet Rooms	Banks, Office buildings, Schools, Stores & Theaters Office buildings and Schools
Average	Standard weight ball bearing	Entrances Corridors Toilet Rooms	Dwellings Office buildings and Schools Stores
Low	Plain bearing	Interior	Dwellings

Door Thickness	Weight of Doors in Pounds per Square Foot				
	White Pine	Oak	Hollow Core	Solid Core	Hollow Metal
1-3/8″	3 psf	6 psf	1-1/2 psf	3-1/2 — 4 psf	6-1/2 psf
1-3/4″	3-1/2	7	2-1/2	4-1/2 — 5-1/4	6-1/2
2-1/4″	4-1/2	9	—	5-1/2 — 6-3/4	6-1/2

Doors & Windows | R088 | Glazing

R08810-010 Glazing Labor

Glass sizes are estimated by the "united inch" (height + width). Table below shows the number of lights glazed in an eight hour period by the crew size indicated, for glass up to 1/4″ thick. Square or nearly square lights are more economical on a S.F. basis. Long slender lights will have a high S.F. installation cost. For insulated glass reduce production by 33%. For 1/2″ float glass reduce production by 50%. Production time for glazing with two glaziers per day averages: 1/4″ float glass 120 S.F.; 1/2″ float glass 55 S.F.; 1/2″ insulated glass 95 S.F.; 3/4″ insulated glass 75 S.F.

Glazing Method	United Inches per Light							
	40″	60″	80″	100″	135″	165″	200″	240″
Number of Men in Crew	1	1	1	1	2	3	3	4
Industrial sash, putty	60	45	24	15	18	—	—	—
With stops, putty bed	50	36	21	12	16	8	4	3
Wood stops, rubber	40	27	15	9	11	6	3	2
Metal stops, rubber	30	24	14	9	9	6	3	2
Structural glass	10	7	4	3	—	—	—	—
Corrugated glass	12	9	7	4	4	4	3	—
Storefronts	16	15	13	11	7	6	4	4
Skylights, putty glass	60	36	21	12	16	—	—	—
Thiokol set	15	15	11	9	9	6	3	2
Vinyl set, snap on	18	18	13	12	12	7	5	4
Maximum area per light	2.8 S.F.	6.3 S.F.	11.1 S.F.	17.4 S.F.	31.6 S.F.	47 S.F.	69 S.F.	100 S.F.
Daily Bare Crew Cost	$264.00	$264.00	$264.00	$264.00	$528.00	$792.00	$792.00	$1,056.00

R09250-050 Lath, Plaster and Gypsum Board

Gypsum board lath is available in 3/8″ thick x 16″ wide x 4′ long sheets as a base material for multi-layer plaster applications. It is also available as a base for either multi-layer or veneer plaster applications in 1/2″ and 5/8″ thick–4′ wide x 8′, 10′ or 12′ long sheets. Fasteners are screws or blued ring shank nails for wood framing and screws for metal framing.

Metal lath is available in diamond mesh pattern with flat or self-furring profiles. Paper backing is available for applications where excessive plaster waste needs to be avoided. A slotted mesh ribbed lath should be used in areas where the span between structural supports is greater than normal. Most metal lath comes in 27″ x 96″ sheets. Diamond mesh weighs 1.75, 2.5 or 3.4 pounds per square yard, slotted mesh lath weighs 2.75 or 3.4 pounds per square yard. Metal lath can be nailed, screwed or tied in place.

Many **accessories** are available. Corner beads, flat reinforcing strips, casing beads, control and expansion joints, furring brackets and channels are some examples. Note that accessories are not included in plaster or stucco line items.

Plaster is defined as a material or combination of materials that when mixed with a suitable amount of water, forms a plastic mass or paste. When applied to a surface, the paste adheres to it and subsequently hardens, preserving in a rigid state the form or texture imposed during the period of elasticity.

Gypsum plaster is made from ground calcined gypsum. It is mixed with aggregates and water for use as a base coat plaster.

Vermiculite plaster is a fire-retardant plaster covering used on steel beams, concrete slabs and other heavy construction materials. Vermiculite is a group name for certain clay minerals, hydrous silicates or aluminum, magnesium and iron that have been expanded by heat.

Perlite plaster is a plaster using perlite as an aggregate instead of sand. Perlite is a volcanic glass that has been expanded by heat.

Gauging plaster is a mix of gypsum plaster and lime putty that when applied produces a quick drying finish coat.

Veneer plaster is a one or two component gypsum plaster used as a thin finish coat over special gypsum board.

Keenes cement is a white cementitious material manufactured from gypsum that has been burned at a high temperature and ground to a fine powder. Alum is added to accelerate the set. The resulting plaster is hard and strong and accepts and maintains a high polish, hence it is used as a finishing plaster.

Stucco is a Portland cement based plaster used primarily as an exterior finish.

Plaster is used on both interior and exterior surfaces. Generally it is applied in multiple-coat systems. A three-coat system uses the terms scratch, brown and finish to identify each coat. A two-coat system uses base and finish to describe each coat. Each type of plaster and application system has attributes that are chosen by the designer to best fit the intended use.

Gypsum Plaster Quantities for 100 S.Y.	2 Coat, 5/8″ Thick		3 Coat, 3/4″ Thick		
	Base	Finish	Scratch	Brown	Finish
	1:3 Mix	2:1 Mix	1:2 Mix	1:3 Mix	2:1 Mix
Gypsum plaster	1,300 lb.		1,350 lb.	650 lb.	
Sand	1.75 C.Y.		1.85 C.Y.	1.35 C.Y.	
Finish hydrated lime		340 lb.			340 lb.
Gauging plaster		170 lb.			170 lb.

Vermiculite or Perlite Plaster Quantities for 100 S.Y.	2 Coat, 5/8″ Thick		3 Coat, 3/4″ Thick		
	Base	Finish	Scratch	Brown	Finish
Gypsum plaster	1,250 lb.		1,450 lb.	800 lb.	
Vermiculite or perlite	7.8 bags		8.0 bags	3.3 bags	
Finish hydrated lime		340 lb.			340 lb.
Gauging plaster		170 lb.			170 lb.

Stucco–Three-Coat System Quantities for 100 S.Y.	On Wood Frame	On Masonry
Portland cement	29 bags	21 bags
Sand	2.6 C.Y.	2.0 C.Y.
Hydrated lime	180 lb.	120 lb.

R09250-100 Levels of Gypsum Drywall Finish

In the past, contract documents often used phrases such as "industry standard" and "workmanlike finish" to specify the expected quality of gypsum board wall and ceiling installations. The vagueness of these descriptions led to unacceptable work and disputes.

In order to resolve this problem, four major trade associations concerned with the manufacture, erection, finish and decoration of gypsum board wall and ceiling systems have developed an industry-wide *Recommended Levels of Gypsum Board Finish*.

The finish of gypsum board walls and ceilings for specific final decoration is dependent on a number of factors. A primary consideration is the location of the surface and the degree of decorative treatment desired. Painted and unpainted surfaces in warehouses and other areas where appearance is normally not critical may simply require the taping of wallboard joints and 'spotting' of fastener heads. Blemish-free, smooth, monolithic surfaces often intended for painted and decorated walls and ceilings in habitated structures, ranging from single-family dwellings through monumental buildings, require additional finishing prior to the application of the final decoration.

Other factors to be considered in determining the level of finish of the gypsum board surface are (1) the type of angle of surface illumination (both natural and artificial lighting), and (2) the paint and method of application or the type and finish of wallcovering specified as the final decoration. Critical lighting conditions, gloss paints, and thin wallcoverings require a higher level of gypsum board finish than do heavily textured surfaces which are subsequently painted or surfaces which are to be decorated with heavy grade wallcoverings.

The following descriptions were developed jointly by the Association of the Wall and Ceiling Industries-International (AWCI), Ceiling & Interior Systems Construction Association (CISCA), Gypsum Association (GA), and Painting and Decorating Contractors of America (PDCA) as a guide.

Level 0: No taping, finishing, or accessories required. This level of finish may be useful in temporary construction or whenever the final decoration has not been determined.

Level 1: All joints and interior angles shall have tape set in joint compound. Surface shall be free of excess joint compound. Tool marks and ridges are acceptable. Frequently specified in plenum areas above ceilings, in attics, in areas where the assembly would generally be concealed or in building service corridors, and other areas not normally open to public view.

Level 2: All joints and interior angles shall have tape embedded in joint compound and wiped with a joint knife leaving a thin coating of joint compound over all joints and interior angles. Fastener heads and accessories shall be covered with a coat of joint compound. Surface shall be free of excess joint compound. Tool marks and ridges are acceptable. Joint compound applied over the body of the tape at the time of tape embedment shall be considered a separate coat of joint compound and shall satisfy the conditions of this level. Specified where water-resistant gypsum backing board is used as a substrate for tile; may be specified in garages, warehouse storage, or other similar areas where surface appearance is not of primary concern.

Level 3: All joints and interior angles shall have tape embedded in joint compound and one additional coat of joint compound applied over all joints and interior angles. Fastener heads and accessories shall be covered with two separate coats of joint compound. All joint compound shall be smooth and free of tool marks and ridges. Typically specified in appearance areas which are to receive heavy- or medium-texture (spray or hand applied) finishes before final painting, or where heavy-grade wallcoverings are to be applied as the final decoration. This level of finish is not recommended where smooth painted surfaces or light to medium wallcoverings are specified.

Level 4: All joints and interior angles shall have tape embedded in joint compound and two separate coats of joint compound applied over all flat joints and one separate coat of joint compound applied over interior angles. Fastener heads and accessories shall be covered with three separate coats of joint compound. All joint compound shall be smooth and free of tool marks and ridges. This level should be specified where flat paints, light textures, or wallcoverings are to be applied. In critical lighting areas, flat paints applied over light textures tend to reduce joint photographing. Gloss, semi-gloss, and enamel paints are not recommended over this level of finish. The weight, texture, and sheen level of wallcoverings applied over this level of finish should be carefully evaluated. Joints and fasteners must be adequately concealed if the wallcovering material is lightweight, contains limited pattern, has a gloss finish, or any combination of these finishes is present. Unbacked vinyl wallcoverings are not recommended over this level of finish.

Level 5: All joints and interior angles shall have tape embedded in joint compound and two separate coats of joint compound applied over all flat joints and one separate coat of joint compound applied over interior angles. Fastener heads and accessories shall be covered with three separate coats of joint compound. A thin skim coat of joint compound or a material manufactured especially for this purpose, shall be applied to the entire surface. The surface shall be smooth and free of tool marks and ridges. This level of finish is highly recommended where gloss, semi-gloss, enamel, or nontextured flat paints are specified or where severe lighting conditions occur. This highest quality finish is the most effective method to provide a uniform surface and minimize the possibility of joint photographing and of fasteners showing through the final decoration.

R13281-120 Asbestos Removal Process

Asbestos removal is accomplished by a specialty contractor who understands the federal and state regulations regarding the handling and disposal of the material. The process of asbestos removal is divided into many individual steps. An accurate estimate can be calculated only after all the steps have been priced.

The steps are generally as follows:

1. Obtain an asbestos abatement plan from an industrial hygienist.
2. Monitor the air quality in and around the removal area and along the path of travel between the removal area and transport area. This establishes the background contamination.
3. Construct a two part decontamination chamber at entrance to removal area.
4. Install a HEPA filter to create a negative pressure in the removal area.
5. Install wall, floor and ceiling protection as required by the plan, usually 2 layers of fireproof 6 mil polyethylene.
6. Industrial hygienist visually inspects work area to verify compliance with plan.
7. Provide temporary supports for conduit and piping affected by the removal process.
8. Proceed with asbestos removal and bagging process. Monitor air quality as described in Step #2. Discontinue operations when contaminate levels exceed applicable standards.
9. Document the legal disposal of materials in accordance with EPA standards.
10. Thoroughly clean removal area including all ledges, crevices and surfaces.
11. Post abatement inspection by industrial hygienist to verify plan compliance.
12. Provide a certificate from a licensed industrial hygienist attesting that contaminate levels are within acceptable standards before returning area to regular use.

R13281-460 Lead Paint Remediation Methods

Lead paint remediation can be accomplished by the following methods.

1. Abrasive blast
2. Chemical stripping
3. Power tool cleaning with vacuum collection system
4. Encapsulation
5. Remove and replace
6. Enclosure

Each of these methods has strengths and weakness depending on the specific circumstances of the project. The following is an overview of each method.

1. **Abrasive blasting** is usually accomplished with sand or recyclable metallic blast. Before work can begin, the area must be contained to ensure the blast material with lead does not escape to the atmosphere. The use of vacuum blast greatly reduces the containment requirements. Lead abatement equipment that may be associated with this work includes a negative air machine. In addition, it is necessary to have an industrial hygienist monitor the project on a continual basis. When the work is complete, the spent blast sand with lead must be disposed of as a hazardous material. If metallic shot was used, the lead is separated from the shot and disposed of as hazardous material. Worker protection includes disposable clothing and respiratory protection.
2. **Chemical stripping** requires strong chemicals be applied to the surface to remove the lead paint. Before the work can begin, the area under/adjacent to the work area must be covered to catch the chemical and removed lead. After the chemical is applied to the painted surface it is usually covered with paper. The chemical is left in place for the specified period, then the paper with lead paint is pulled or scraped off. The process may require several chemical applications. The paper with chemicals and lead paint adhered to it, plus the containment and loose scrapings collected by a HEPA (High Efficiency Particulate Air Filter) vac, must be disposed of as a hazardous material. The chemical stripping process usually requires a neutralizing agent and several wash downs after the paint is removed. Worker protection includes a neoprene or other compatible protective clothing and respiratory protection with face shield. An industrial hygienist is required intermittently during the process.
3. **Power tool cleaning** is accomplished using shrouded needle blasting guns. The shrouding with different end configurations is held up against the surface to be cleaned. The area is blasted with hardened needles and the shroud captures the lead with a HEPA vac and deposits it in a holding tank. An industrial hygienist monitors the project, protective clothing and a respirator is required until air samples prove otherwise. When the work is complete the lead must be disposed of as a hazardous material.
4. **Encapsulation** is a method that leaves the well bonded lead paint in place after the peeling paint has been removed. Before the work can begin, the area under/adjacent to the work must be covered to catch the scrapings. The scraped surface is then washed with a detergent and rinsed. The prepared surface is covered with approximately 10 mils of paint. A reinforcing fabric can also be embedded in the paint covering. The scraped paint and containment must be disposed of as a hazardous material. Workers must wear protective clothing and respirators.
5. **Remove and replace** is an effective way to remove lead paint from windows, gypsum walls and concrete masonry surfaces. The painted materials are removed and new materials are installed. Workers should wear a respirator and tyvek suit. The demolished materials must be disposed of as hazardous waste if it fails the TCLP (Toxicity Characteristic Leachate Process) test.
6. **Enclosure** is the process that permanently seals lead painted materials in place. This process has many applications such as covering lead painted drywall with new drywall, covering exterior construction with tyvek paper then residing, or covering lead painted structural members with aluminum or plastic. The seams on all enclosing materials must be securely sealed. An industrial hygienist monitors the project, and protective clothing and a respirator is required until air samples prove otherwise.

All the processes require clearance monitoring and wipe testing as required by the hygienist.

Conveying Systems | R142 | Elevators

R14200-100 Freight Elevators

Capacities run from 2,000 lbs. to over 100,000 lbs. with 3,000 lbs. to 10,000 lbs. most common. Travel speeds are generally lower and control less intricate than on passenger elevators. Unit prices in division 14210 are for hydraulic and geared elevators.

R14200-200 Elevator Selective Costs

See R14200-400 for cost development.

	Passenger		Freight		Hospital	
A. Base Unit	Hydraulic	Electric	Hydraulic	Electric	Hydraulic	Electric
Capacity	1,500 lb.	2,000 lb.	2,000 lb.	4,000 lb.	4,000 lb.	4,000 lb.
Speed	100 F.P.M.	200 F.P.M.	100 F.P.M.	200 F.P.M.	100 F.P.M.	200 F.P.M.
#Stops/Travel Ft.	2/12	4/40	2/20	4/40	2/20	4/40
Push Button Oper.	Yes	Yes	Yes	Yes	Yes	Yes
Telephone Box & Wire	"	"	"	"	"	"
Emergency Lighting	"	"	No	No	"	"
Cab	Plastic Lam. Walls	Plastic Lam. Walls	Painted Steel	Painted Steel	Plastic Lam. Walls	Plastic Lam. Walls
Cove Lighting	Yes	Yes	No	No	Yes	Yes
Floor	V.C.T.	V.C.T.	Wood w/Safety Treads	Wood w/Safety Treads	V.C.T.	V.C.T.
Doors, & Speedside Slide	Yes	Yes	Yes	Yes	Yes	Yes
Gates, Manual	No	No	No	No	No	No
Signals, Lighted Buttons	Car and Hall	Car and Hall	Car and Hall	Car and Hall	Car and Hall	Car and Hall
O.H. Geared Machine	N.A.	Yes	N.A.	Yes	N.A.	Yes
Variable Voltage Contr.	"	"	N.A.	"	"	"
Emergency Alarm	Yes	"	Yes	"	Yes	"
Class "A" Loading	N.A.	N.A.	"	"	N.A.	N.A.

R14200-300 Passenger Elevators

Electric elevators are used generally but hydraulic elevators can be used for lifts up to 70′ and where large capacities are required. Hydraulic speeds are limited to 200 F.P.M. but cars are self leveling at the stops. On low rises, hydraulic installation runs about 15% less than standard electric types but on higher rises this installation cost advantage is reduced. Maintenance of hydraulic elevators is about the same as electric type but underground portion is not included in the maintenance contract.

In electric elevators there are several control systems available, the choice of which will be based upon elevator use, size, speed and cost criteria. The two types of drives are geared for low speeds and gearless for 450 F.P.M. and over.

The tables on the preceding pages illustrate typical installed costs of the various types of elevators available.

R14200-400 Elevator Cost Development

Requirement: One passenger elevator, five story hydraulic, 2,500 lb. capacity, 12′ floor to floor, speed 150 F.P.M., emergency power switching and maintenance contract.

Description	Adjustment		Unit Cost	Total Cost
A. Base Elevator, Hyd Pass, Base Unit, 1500 lb, 100 fpm, 2 Stops, Std. Finish				$41,500
B. Capacity Adjustment (2,500 lb.)	1	Ea.	$2,050	2,050
C. Excess Travel Over Base (4 x 12′) = 48′ — (12′ for Base Unit) =	36	V.L.F.	495	17,820
D. Stops Over Base 5 — (2 for Base Unit) =	3	Stops	4,200	12,600
E. Speed Adjustment (150 F.P.M.)	1	Ea.	2,375	2,375
F. Options:				
1. Intercom Service	1	Ea.	665	665
2. Emergency Power Switching, Automatic	1	Ea.	2,400	2,400
3. Stainless Steel Entrance Doors	5	Ea.	1675	8,375
4. Maintenance Contract (12 Months)	1	Ea.	2,650	2,650
5. Position Indicators	2	Ea.	350	700
Total Cost				$91,135

R15050-720 Demolition (Selective vs. Removal for Replacement)

Demolition can be divided into two basic categories.

One type of demolition involves the removal of material with no concern for its replacement. The labor-hours to estimate this work are found in Div. 15055 under "Selective Demolition". It is selective in that individual items or all the material installed as a system or trade grouping such as plumbing or heating systems are removed. This may be accomplished by the easiest way possible, such as sawing, torch cutting, or sledge hammer as well as simple unbolting.

The second type of demolition is the removal of some item for repair or replacement. This removal may involve careful draining, opening of unions, disconnecting and tagging of electrical connections, capping of pipes/ducts to prevent entry of debris or leakage of the material contained as well as transport of the item away from its in-place location to a truck/dumpster. An approximation of the time required to accomplish this type of demolition is to use half of the time indicated as necessary to install a new unit. For example; installation of a new pump might be listed as requiring 6 labor-hours so if we had to estimate the removal of the old pump we would allow an additional 3 hours for a total of 9 hours. That is, the complete replacement of a defective pump with a new pump would be estimated to take 9 labor-hours.

R15100-050 Pipe Material Considerations

1. Malleable fittings should be used for gas service.
2. Malleable fittings are used where there are stresses/strains due to expansion and vibration.
3. Cast fittings may be broken as an aid to disassembling of heating lines frozen by long use, temperature and minerals.
4. Cast iron pipe is extensively used for underground and submerged service.
5. Type M (light wall) copper tubing is available in hard temper only and is used for nonpressure and less severe applications than K and L.
6. Type L (medium wall) copper tubing, available hard or soft for interior service.
7. Type K (heavy wall) copper tubing, available in hard or soft temper for use where conditions are severe. For underground and interior service.
8. Hard drawn tubing requires fewer hangers or supports but should not be bent. Silver brazed fittings are recommended, however soft solder is normally used.
9. Type DMV (very light wall) copper tubing designed for drainage, waste and vent plus other non-critical pressure services.

Domestic/Imported Pipe and Fittings Cost

The prices shown in this publication for steel/cast iron pipe and steel, cast iron, malleable iron fittings are based on domestic production sold at the normal trade discounts. The above listed items of foreign manufacture may be available at prices of 1/3 to 1/2 those shown. Some imported items after minor machining or finishing operations are being sold as domestic to further complicate the system.

Caution: Most pipe prices in this book also include a coupling and pipe hangers which for the larger sizes can add significantly to the per foot cost and should be taken into account when comparing "book cost" with quoted supplier's cost.

R15100-420 Plumbing Fixture Installation Time

Item	Rough-In	Set	Total Hours	Item	Rough-In	Set	Total Hours
Bathtub	5	5	10	Shower head only	2	1	3
Bathtub and shower, cast iron	6	6	12	Shower drain	3	1	4
Fire hose reel and cabinet	4	2	6	Shower stall, slate		15	15
Floor drain to 4 inch diameter	3	1	4	Slop sink	5	3	8
Grease trap, single, cast iron	5	3	8	Test 6 fixtures			14
Kitchen gas range		4	4	Urinal, wall	6	2	8
Kitchen sink, single	4	4	8	Urinal, pedestal or floor	6	4	10
Kitchen sink, double	6	6	12	Water closet and tank	4	3	7
Laundry tubs	4	2	6	Water closet and tank, wall hung	5	3	8
Lavatory wall hung	5	3	8	Water heater, 45 gals. gas, automatic	5	2	7
Lavatory pedestal	5	3	8	Water heaters, 65 gals. gas, automatic	5	2	7
Shower and stall	6	4	10	Water heaters, electric, plumbing only	4	2	6

Fixture prices in front of book are based on the cost per fixture set in place. The rough-in cost, which must be added for each fixture, includes carrier, if required, some supply, waste and vent pipe connecting fittings and stops. The lengths of rough-in pipe are nominal runs which would connect to the larger runs and stacks. The supply runs and DWV runs and stacks must be accounted for in separate entries. In the eastern half of the United States it is common for the plumber to carry these to a point 5′ outside the building.

R15810-050 Ductwork

Duct Weight in Pounds per L.F., Straight Runs

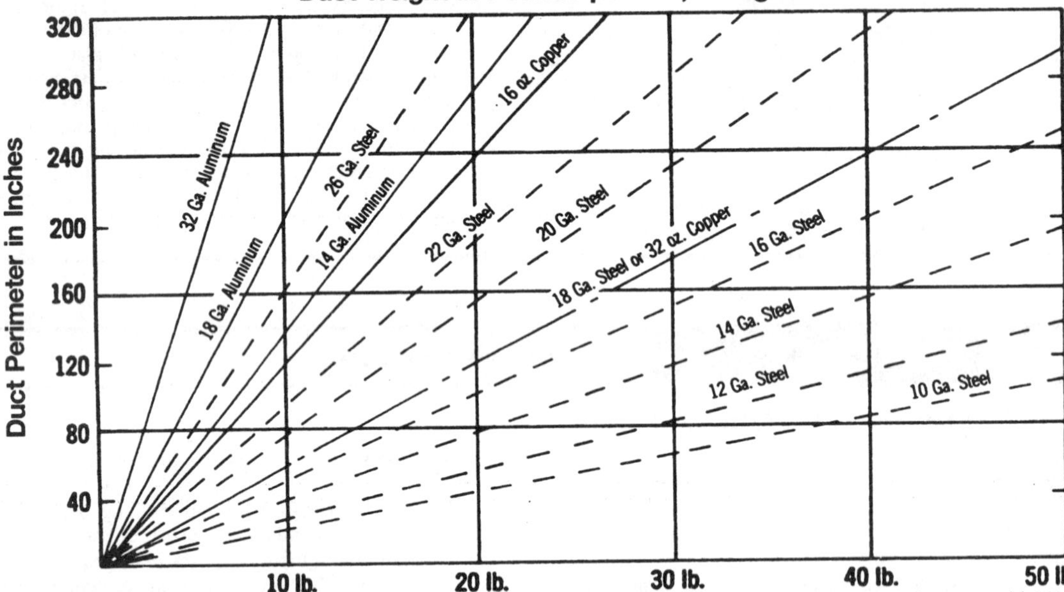

Add to the above for fittings; 90° elbow is 3 L.F.; 45° elbow is 2.5 L.F.; offset is 4 L.F.; transition offset is 6 L.F.; square-to-round transition is 4 L.F.; 90° reducing elbow is 5 L.F. For bracing and waste, add 20% to aluminum and copper, 15% to steel.

R16055-300 Electrical Demolition (Removal for Replacement)

The purpose of this reference number is to provide a guide to users for electrical "removal for replacement" by applying the rule of thumb: 1/3 of new installation time (typical range from 20% to 50%) for removal. Remember to use reasonable judgment when applying the suggested percentage factor. For example:

Contractors have been requested to remove an existing fluorescent lighting fixture and replace with a new fixture utilizing energy saver lamps and electronic ballast:

In order to fully understand the extent of the project, contractors should visit the job site and estimate the time to perform the renovation work in accordance with applicable national, state and local regulation and codes.

The contractor may need to add extra labor hours to his estimate if he discovers unknown concealed conditions such as: contaminated asbestos ceiling, broken acoustical ceiling tile and need to repair, patch and touch-up paint all the damaged or disturbed areas, tasks normally assigned to general contractors. In addition, the owner could request that the contractors salvage the materials removed and turn over the materials to the owner or dispose of the materials to a reclamation station. The normal removal item is 0.5 labor hour for a lighting fixture and 1.5 labor-hours for new installation time. Revise the estimate times from 2 labor-hours work up to a minimum 4 labor-hours work for just fluorescent lighting fixture.

For removal of large concentrations of lighting fixtures in the same area, apply an "economy of scale" to reduce estimating labor hours.

R16132-210 Conductors in Conduit

Table below lists maximum number of conductors for various sized conduit using THW, TW or THWN insulations.

Copper Wire Size	Conduit Size																										
	1/2"			3/4"			1"			1-1/4"			1-1/2"			2"			2-1/2"			3"		3-1/2"		4"	
	TW	THW	THWN	TW	THW	THWN	TW	THW	THWN	TW	THW	THWN	TW	THW	THWN	TW	THW	THWN	TW	THW	THWN	THW	THWN	THW	THWN	THW	THWN
#14	9	6	13	15	10	24	25	16	39	44	29	69	60	40	94	99	65	154	142	93		143		192			
#12	7	4	10	12	8	18	19	13	29	35	24	51	47	32	70	78	53	114	111	76	164	117		157			
#10	5	4	6	9	6	11	15	11	18	26	19	32	36	26	44	60	43	73	85	61	104	95	160	127		163	
#8	2	1	3	4	3	5	7	5	9	12	10	16	17	13	22	28	22	36	40	32	51	49	79	66	106	85	136
#6		1	1		2	4		4	6		7	11		10	15		16	26		23	37	36	57	48	76	62	98
#4		1	1		1	2		3	4		5	7		7	9		12	16		17	22	27	35	36	47	47	60
#3		1	1		1	1		2	3		4	6		6	8		10	13		15	19	23	29	31	39	40	51
#2		1	1		1	1		2	3		4	5		5	7		9	11		13	16	20	25	27	33	34	43
#1					1	1		1	1		3	3		4	5		6	8		9	12	14	18	19	25	25	32
1/0					1	1		1	1		2	3		3	4		5	7		8	10	12	15	16	21	21	27
2/0					1	1		1	1		1	2		3	3		5	6		7	8	10	13	14	17	18	22
3/0					1	1		1	1		1	1		2	3		4	5		6	7	9	11	12	14	15	18
4/0						1		1	1		1	1		1	2		3	4		5	6	7	9	10	12	13	15
250 kcmil								1	1		1	1		1	1		2	3		4	4	6	7	8	10	10	12
300								1	1		1	1		1	1		2	3		3	4	5	6	7	8	9	11
350									1		1	1		1	1		1	2		3	3	4	5	6	7	8	9
400											1	1		1	1		1	1		2	3	4	5	5	6	7	8
500											1	1		1	1		1	1		1	2	3	4	4	5	6	7
600												1		1	1		1	1		1	1	3	3	4	4	5	5
700														1	1		1	1		1	1	2	3	3	4	4	5
750 ↓														1	1		1	1		1	1	2	2	3	3	4	4

For information about Means Estimating Seminars, see yellow pages 12 and 13 in back of book

Change Orders

Change Order Considerations

A Change Order is a written document, usually prepared by the design professional, and signed by the owner, the architect/engineer and the contractor. A change order states the agreement of the parties to: an addition, deletion, or revision in the work; an adjustment in the contract sum, if any; or an adjustment in the contract time, if any. Change orders, or "extras" in the construction process occur after execution of the construction contract and impact architects/engineers, contractors and owners.

Change orders that are properly recognized and managed can ensure orderly, professional and profitable progress for all who are involved in the project. There are many causes for change orders and change order requests. In all cases, change orders or change order requests should be addressed promptly and in a precise and prescribed manner. The following paragraphs include information regarding change order pricing and procedures.

The Causes of Change Orders

Reasons for issuing change orders include:

- Unforeseen field conditions that require a change in the work
- Correction of design discrepancies, errors or omissions in the contract documents
- Owner-requested changes, either by design criteria, scope of work, or project objectives
- Completion date changes for reasons unrelated to the construction process
- Changes in building code interpretations, or other public authority requirements that require a change in the work
- Changes in availability of existing or new materials and products

Procedures

Properly written contract documents must include the correct change order procedures for all parties—owners, design professionals and contractors—to follow in order to avoid costly delays and litigation.

Being "in the right" is not always a sufficient or acceptable defense. The contract provisions requiring notification and documentation must be adhered to within a defined or reasonable time frame.

The appropriate method of handling change orders is by a written proposal and acceptance by all parties involved. Prior to starting work on a project, all parties should identify their authorized agents who may sign and accept change orders, as well as any limits placed on their authority.

Time may be a critical factor when the need for a change arises. For such cases, the contractor might be directed to proceed on a "time and materials" basis, rather than wait for all paperwork to be processed—a delay that could impede progress. In this situation, the contractor must still follow the prescribed change order procedures, including but not limited to, notification and documentation.

All forms used for change orders should be dated and signed by the proper authority. Lack of documentation can be very costly, especially if legal judgments are to be made and if certain field personnel are no longer available. For time and material change orders, the contractor should keep accurate daily records of all labor and material allocated to the change. Forms that can be used to document change order work are available in *Means Forms for Building Construction Professionals.*

Owners or awarding authorities who do considerable and continual building construction (such as the federal government) realize the inevitability of change orders for numerous reasons, both predictable and unpredictable. As a result, the federal government, the American Institute of Architects (AIA), the Engineers Joint Contract Documents Committee (EJCDC) and other contractor, legal and technical organizations have developed standards and procedures to be followed by all parties to achieve contract continuance and timely completion, while being financially fair to all concerned.

In addition to the change order standards put forth by industry associations, there are also many books available on the subject.

Pricing Change Orders

When pricing change orders, regardless of their cause, the most significant factor is *when* the change occurs. The need for a change may be perceived in the field or requested by the architect/engineer *before* any of the actual installation has begun, or may evolve or appear *during* construction when the item of work in question is partially installed. In the latter cases, the original sequence of construction is disrupted, along with all contiguous and supporting systems. Change orders cause the greatest impact when they occur *after* the installation has been completed and must be uncovered, or even replaced. Post-completion changes may be caused by necessary design changes, product failure, or changes in the owner's requirements that are not discovered until the building or the systems begin to function.

Specified procedures of notification and record keeping must be adhered to and enforced regardless of the stage of construction: *before, during,* or *after* installation. Some bidding documents anticipate change orders by requiring that unit prices including overhead and profit percentages—for additional as well as deductible changes—be listed. Generally these unit prices do not fully take into account the ripple effect, or impact on other trades, and should be used for general guidance only.

When pricing change orders, it is important to classify the time frame in which the change occurs. There are two basic time frames for change orders: *pre-installation change orders*, which occur before the start of construction, and *post-installation change orders*, which involve reworking after the original installation. Change orders that occur between these stages may be priced according to the extent of work completed using a combination of techniques developed for pricing *pre-* and *post-installation* changes.

The following factors are the basis for a check list to use when preparing a change order estimate.

Factors To Consider When Pricing Change Orders

As an estimator begins to prepare a change order, the following questions should be reviewed to determine their impact on the final price.

General

- Is the change order work *pre-installation* or *post-installation*?

  Change order work costs vary according to how much of the installation has been completed. Once workers have the project scoped in their mind, even though they have not started, it can be difficult to refocus. Consequently they may spend more than the normal amount of time understanding the change. Also, modifications to work in place such as trimming or refitting usually take more time than was initially estimated. The greater the amount of work in place, the more reluctant workers are to change it. Psychologically they may resent the change and as a result the rework takes longer than normal. Post-installation change order estimates must include demolition of existing work as required to accomplish the change. If the work is performed at a later time, additional obstacles such as building finishes may be present which must be protected. Regardless of whether the change occurs pre-installation or post-installation, attempt to isolate the identifiable factors and price them separately. For example, add shipping costs that may be required pre-installation or any demolition required post-installation. Then analyze the potential impact on productivity of psychological and/or learning curve factors and adjust the output rates accordingly. One approach is to break down the typical workday into segments and quantify the impact on each segment. The following chart may be useful as a guide:

Activities (Productivity) Expressed as Percentages of a Workday			
Task	Means Mechanical Cost Data (for New Construction)	Pre-Installation Change Orders	Post-Installation Change Orders
1. Study plans	3%	6%	6%
2. Material procurement	3%	3%	3%
3. Receiving and storing	3%	3%	3%
4. Mobilization	5%	5%	5%
5. Site movement	5%	5%	8%
6. Layout and marking	8%	10%	12%
7. Actual installation	64%	59%	54%
8. Clean-up	3%	3%	3%
9. Breaks—non-productive	6%	6%	6%
Total	100%	100%	100%

Change Order Installation Efficiency

The labor-hours expressed (for new construction) are based on average installation time, using an efficiency level of approximately 60-65%. For change order situations, adjustments to this efficiency level should reflect the daily labor-hour allocation for that particular occurrence.

If any of the specific percentages expressed in the above chart do not apply to a particular project situation, then those percentage points should be reallocated to the appropriate task(s). Example: Using data for new construction, assume there is no new material being utilized. The percentages for Tasks 2 and 3 would therefore be reallocated to other tasks. If the time required for Tasks 2 and 3 can now be applied to installation, we can add the time allocated for *Material Procurement* and *Receiving and Storing* to the *Actual Installation* time for new construction, thereby increasing the Actual Installation percentage.

This chart shows that, due to reduced productivity, labor costs will be higher than those for new construction by 5% to 15% for pre-installation change orders and by 15% to 25% for post-installation change orders. Each job and change order is unique and must be examined individually. Many factors, covered elsewhere in this section, can each have a significant impact on productivity and change order costs. All such factors should be considered in every case.

- Will the change substantially delay the original completion date?

  A significant change in the project may cause the original completion date to be extended. The extended schedule may subject the contractor to new wage rates dictated by relevant labor contracts. Project supervision and other project overhead must also be extended beyond the original completion date. The schedule extension may also put installation into a new weather season. For example, underground piping scheduled for October installation was delayed until January. As a result, frost penetrated the trench area, thereby changing the degree of difficulty of the task. Changes and delays may have a ripple effect throughout the project. This effect must be analyzed and negotiated with the owner.

- What is the net effect of a deduct change order?

  In most cases, change orders resulting in a deduction or credit reflect only bare costs. The contractor may retain the overhead and profit based on the original bid.

Materials

- Will you have to pay more or less for the new material, required by the change order, than you paid for the original purchase?

  The same material prices or discounts will usually apply to materials purchased for change orders as new construction. In some instances, however, the contractor may forfeit the advantages of competitive pricing for change orders. Consider the following example:

  A contractor purchased over $20,000 worth of fan coil units for an installation, and obtained the maximum discount. Some time later it was determined the project required an additional matching unit. The contractor has to purchase this unit from the original supplier to ensure a match. The supplier at this time may not discount the unit because of the small quantity, and the fact that he is no longer in a competitive situation. The impact of quantity on purchase can add between 0% and 25% to material prices and/or subcontractor quotes.

- If materials have been ordered or delivered to the job site, will they be subject to a cancellation charge or restocking fee?

  Check with the supplier to determine if ordered materials are subject to a cancellation charge. Delivered materials not used as result of a change order may be subject to a restocking fee if returned to the supplier. Common restocking charges run between 20% and 40%. Also, delivery charges to return the goods to the supplier must be added.

Labor

- How efficient is the existing crew at the actual installation?

  Is the same crew that performed the initial work going to do the change order? Possibly the change consists of the installation of a unit identical to one already installed; therefore the change should take less time. Be sure to consider this potential productivity increase and modify the productivity rates accordingly.

- If the crew size is increased, what impact will that have on supervision requirements?

  Under most bargaining agreements or management practices, there is a point at which a working foreman is replaced by a nonworking foreman. This replacement increases project overhead by adding a nonproductive worker. If additional workers are added to accelerate the project or to perform changes while maintaining the schedule, be sure to add additional supervision time if warranted. Calculate the hours involved and the additional cost directly if possible.

- What are the other impacts of increased crew size?

  The larger the crew, the greater the potential for productivity to decrease. Some of the factors that cause this productivity loss are: overcrowding (producing restrictive conditions in the working space), and possibly a shortage of any special tools and equipment required. Such factors affect not only the crew working on the elements directly involved in the change order, but other crews whose movement may also be hampered.

  As the crew increases, check its basic composition for changes by the addition or deletion of apprentices or nonworking foreman and quantify the potential effects of equipment shortages or other logistical factors.

- As new crews, unfamiliar with the project, are brought onto the site, how long will it take them to become oriented to the project requirements?

  The orientation time for a new crew to become 100% effective varies with the site and type of project. Orientation is easiest at a new construction site, and most difficult at existing, very restrictive renovation sites. The type of work also affects orientation time. When all elements of the work are exposed, such as concrete or masonry work, orientation is decreased. When the work is concealed or less visible, such as existing electrical systems, orientation takes longer. Usually orientation can be accomplished in one day or less. Costs for added orientation should be itemized and added to the total estimated cost.

- How much actual production can be gained by working overtime?

  Short term overtime can be used effectively to accomplish more work in a day. However, as overtime is scheduled to run beyond several weeks, studies have shown marked decreases in output. The following chart shows the effect of long term overtime on worker efficiency. If the anticipated change requires extended overtime to keep the job on schedule, these factors can be used as a guide to predict the impact on time and cost. Add project overhead, particularly supervision, that may also be incurred.

Days per Week	Hours per Day	Production Efficiency					Payroll Cost Factors	
		1 Week	2 Weeks	3 Weeks	4 Weeks	Average 4 Weeks	@ 1-1/2 Times	@ 2 Times
5	8	100%	100%	100%	100%	100%	100%	100%
	9	100	100	95	90	96.25	105.6	111.1
	10	100	95	90	85	91.25	110.0	120.0
	11	95	90	75	65	81.25	113.6	127.3
	12	90	85	70	60	76.25	116.7	133.3
6	8	100	100	95	90	96.25	108.3	116.7
	9	100	95	90	85	92.50	113.0	125.9
	10	95	90	85	80	87.50	116.7	133.3
	11	95	85	70	65	78.75	119.7	139.4
	12	90	80	65	60	73.75	122.2	144.4
7	8	100	95	85	75	88.75	114.3	128.6
	9	95	90	80	70	83.75	118.3	136.5
	10	90	85	75	65	78.75	121.4	142.9
	11	85	80	65	60	72.50	124.0	148.1
	12	85	75	60	55	68.75	126.2	152.4

Effects of Overtime

Caution: Under many labor agreements, Sundays and holidays are paid at a higher premium than the normal overtime rate.

The use of long-term overtime is counterproductive on almost any construction job; that is, the longer the period of overtime, the lower the actual production rate. Numerous studies have been conducted, and while they have resulted in slightly different numbers, all reach the same conclusion. The figure above tabulates the effects of overtime work on efficiency.

As illustrated, there can be a difference between the *actual* payroll cost per hour and the *effective* cost per hour for overtime work. This is due to the reduced production efficiency with the increase in weekly hours beyond 40. This difference between actual and effective cost results from overtime work over a prolonged period. Short-term overtime work does not result in as great a reduction in efficiency, and in such cases, effective cost may not vary significantly from the actual payroll cost. As the total hours per week are increased on a regular basis, more time is lost because of fatigue, lowered morale, and an increased accident rate.

As an example, assume a project where workers are working 6 days a week, 10 hours per day. From the figure above (based on productivity studies), the average effective productive hours over a four-week period are:

$$0.875 \text{ x } 60 = 52.5$$

Depending upon the locale and day of week, overtime hours may be paid at time and a half or double time. For time and a half, the overall (average) *actual* payroll cost (including regular and overtime hours) is determined as follows:

$$\frac{\text{40 reg. hrs.} + (\text{20 overtime hrs. x } 1.5)}{\text{60 hrs.}} = 1.167$$

Based on 60 hours, the payroll cost per hour will be 116.7% of the normal rate at 40 hours per week. However, because the effective production (efficiency) for 60 hours is reduced to the equivalent of 52.5 hours, the effective cost of overtime is calculated as follows:

For time and a half:

$$\frac{\text{40 reg. hrs.} + (\text{20 overtime hrs. x } 1.5)}{\text{52.5 hrs.}} = 1.33$$

Installed cost will be 133% of the normal rate (for labor).

Thus, when figuring overtime, the actual cost per unit of work will be higher than the apparent overtime payroll dollar increase, due to the reduced productivity of the longer workweek. These efficiency calculations are true only for those cost factors determined by hours worked. Costs that are applied weekly or monthly, such as equipment rentals, will not be similarly affected.

Equipment

- What equipment is required to complete the change order?

  Change orders may require extending the rental period of equipment already on the job site, or the addition of special equipment brought in to accomplish the change work. In either case, the additional rental charges and operator labor charges must be added.

Summary

The preceding considerations and others you deem appropriate should be analyzed and applied to a change order estimate. The impact of each should be quantified and listed on the estimate to form an audit trail.

Change orders that are properly identified, documented, and managed help to ensure the orderly, professional and profitable progress of the work. They also minimize potential claims or disputes at the end of the project.

Crew No.	Bare Costs		Incl. Subs O & P		Cost Per Labor-Hour	
Crew A-1	**Hr.**	**Daily**	**Hr.**	**Daily**	**Bare Costs**	**Incl. O&P**
1 Building Laborer	$26.00	$208.00	$43.10	$344.80	$26.00	$43.10
1 Concrete saw, gas manual		42.80		47.10	5.35	5.89
8 L.H., Daily Totals		$250.80		$391.90	$31.35	$48.99

Crew A-1A	Hr.	Daily	Hr.	Daily	Bare Costs	Incl. O&P
1 Skilled Worker	$33.65	$269.20	$55.00	$440.00	$33.65	$55.00
1 Shot Blaster, 20"		274.30		301.75	34.29	37.72
8 L.H., Daily Totals		$543.50		$741.75	$67.94	$92.72

Crew A-1B	Hr.	Daily	Hr.	Daily	Bare Costs	Incl. O&P
1 Laborers, (Semi-Skilled)	$26.00	$208.00	$43.10	$344.80	$26.00	$43.10
1 Concr. saw, gas, self-prop.		92.20		101.40	11.53	12.68
8 L.H., Daily Totals		$300.20		$446.20	$37.53	$55.78

Crew A-1C	Hr.	Daily	Hr.	Daily	Bare Costs	Incl. O&P
1 Building Laborer	$26.00	$208.00	$43.10	$344.80	$26.00	$43.10
1 Brush saw		18.80		20.70	2.35	2.59
8 L.H., Daily Totals		$226.80		$365.50	$28.35	$45.69

Crew A-1D	Hr.	Daily	Hr.	Daily	Bare Costs	Incl. O&P
1 Building Laborer	$26.00	$208.00	$43.10	$344.80	$26.00	$43.10
1 Vibrating plate, gas, 18"		28.60		31.45	3.58	3.93
8 L.H., Daily Totals		$236.60		$376.25	$29.58	$47.03

Crew A-1E	Hr.	Daily	Hr.	Daily	Bare Costs	Incl. O&P
1 Building Laborer	$26.00	$208.00	$43.10	$344.80	$26.00	$43.10
1 Vibrating plate, gas, 21"		41.40		45.55	5.18	5.69
8 L.H., Daily Totals		$249.40		$390.35	$31.18	$48.79

Crew A-1F	Hr.	Daily	Hr.	Daily	Bare Costs	Incl. O&P
1 Building Laborer	$26.00	$208.00	$43.10	$344.80	$26.00	$43.10
1 Rammer/tamper, gas, 6" - 11"		35.40		38.95	4.43	4.87
8 L.H., Daily Totals		$243.40		$383.75	$30.43	$47.97

Crew A-1G	Hr.	Daily	Hr.	Daily	Bare Costs	Incl. O&P
1 Building Laborer	$26.00	$208.00	$43.10	$344.80	$26.00	$43.10
1 Rammer/tamper, gas, 13" - 18"		35.40		38.95	4.43	4.87
8 L.H., Daily Totals		$243.40		$383.75	$30.43	$47.97

Crew A-1H	Hr.	Daily	Hr.	Daily	Bare Costs	Incl. O&P
1 Building Laborer	$26.00	$208.00	$43.10	$344.80	$26.00	$43.10
1 Pressure washer		55.60		61.15	6.95	7.65
8 L.H., Daily Totals		$263.60		$405.95	$32.95	$50.75

Crew A-1J	Hr.	Daily	Hr.	Daily	Bare Costs	Incl. O&P
1 Building Laborer	$26.00	$208.00	$43.10	$344.80	$26.00	$43.10
1 Rototiller		70.90		78.00	8.86	9.75
8 L.H., Daily Totals		$278.90		$422.80	$34.86	$52.85

Crew A-1K	Hr.	Daily	Hr.	Daily	Bare Costs	Incl. O&P
1 Building Laborer	$26.00	$208.00	$43.10	$344.80	$26.00	$43.10
1 Lawn aerator		24.45		26.90	3.06	3.36
8 L.H., Daily Totals		$232.45		$371.70	$29.06	$46.46

Crew A-1L	Hr.	Daily	Hr.	Daily	Bare Costs	Incl. O&P
1 Building Laborer	$26.00	$208.00	$43.10	$344.80	$26.00	$43.10
1 Powwr blower/vacuum		24.45		26.90	3.06	3.36
8 L.H., Daily Totals		$232.45		$371.70	$29.06	$46.46

Crew No.	Bare Costs		Incl. Subs O & P		Cost Per Labor-Hour	
Crew A-1M	**Hr.**	**Daily**	**Hr.**	**Daily**	**Bare Costs**	**Incl. O&P**
1 Building Laborer	$26.00	$208.00	$43.10	$344.80	$26.00	$43.10
1 Snow blower		70.90		78.00	8.86	9.75
8 L.H., Daily Totals		$278.90		$422.80	$34.86	$52.85

Crew A-2	Hr.	Daily	Hr.	Daily	Bare Costs	Incl. O&P
2 Laborers	$26.00	$416.00	$43.10	$689.60	$25.90	$42.65
1 Truck Driver (light)	25.70	205.60	41.75	334.00		
1 Light Truck, 1.5 Ton		122.00		134.20	5.08	5.59
24 L.H., Daily Totals		$743.60		$1157.80	$30.98	$48.24

Crew A-2A	Hr.	Daily	Hr.	Daily	Bare Costs	Incl. O&P
2 Laborers	$26.00	$416.00	$43.10	$689.60	$25.90	$42.65
1 Truck Driver (light)	25.70	205.60	41.75	334.00		
1 Light Truck, 1.5 Ton		122.00		134.20		
1 Concrete Saw		92.20		101.40	8.93	9.82
24 L.H., Daily Totals		$835.80		$1259.20	$34.83	$52.47

Crew A-3	Hr.	Daily	Hr.	Daily	Bare Costs	Incl. O&P
1 Truck Driver (heavy)	$26.45	$211.60	$42.95	$343.60	$26.45	$42.95
1 Dump Truck, 12 Ton		325.00		357.50	40.63	44.69
8 L.H., Daily Totals		$536.60		$701.10	$67.08	$87.64

Crew A-3A	Hr.	Daily	Hr.	Daily	Bare Costs	Incl. O&P
1 Truck Driver (light)	$25.70	$205.60	$41.75	$334.00	$25.70	$41.75
1 Pickup Truck (4x4)		82.00		90.20	10.25	11.28
8 L.H., Daily Totals		$287.60		$424.20	$35.95	$53.03

Crew A-3B	Hr.	Daily	Hr.	Daily	Bare Costs	Incl. O&P
1 Equip. Oper. (medium)	$33.65	$269.20	$53.05	$424.40	$30.05	$48.00
1 Truck Driver (heavy)	26.45	211.60	42.95	343.60		
1 Dump Truck, 16 Ton		476.60		524.25		
1 F.E. Loader, 3 C.Y.		295.60		325.15	48.26	53.09
16 L.H., Daily Totals		$1253.00		$1617.40	$78.31	$101.09

Crew A-3C	Hr.	Daily	Hr.	Daily	Bare Costs	Incl. O&P
1 Equip. Oper. (light)	$32.15	$257.20	$50.65	$405.20	$32.15	$50.65
1 Wheeled Skid Steer Loader		183.80		202.20	22.98	25.27
8 L.H., Daily Totals		$441.00		$607.40	$55.13	$75.92

Crew A-3D	Hr.	Daily	Hr.	Daily	Bare Costs	Incl. O&P
1 Truck Driver, Light	$25.70	$205.60	$41.75	$334.00	$25.70	$41.75
1 Pickup Truck (4x4)		82.00		90.20		
1 Flatbed Trailer, 25 Ton		88.60		97.45	21.33	23.46
8 L.H., Daily Totals		$376.20		$521.65	$47.03	$65.21

Crew A-3E	Hr.	Daily	Hr.	Daily	Bare Costs	Incl. O&P
1 Equip. Oper. (crane)	$34.80	$278.40	$54.85	$438.80	$30.63	$48.90
1 Truck Driver (heavy)	26.45	211.60	42.95	343.60		
1 Pickup Truck (4x4)		82.00		90.20	5.13	5.64
16 L.H., Daily Totals		$572.00		$872.60	$35.76	$54.54

Crew A-3F	Hr.	Daily	Hr.	Daily	Bare Costs	Incl. O&P
1 Equip. Oper. (crane)	$34.80	$278.40	$54.85	$438.80	$30.63	$48.90
1 Truck Driver (heavy)	26.45	211.60	42.95	343.60		
1 Pickup Truck (4x4)		82.00		90.20		
1 Tractor, 6x2, 40 Ton Cap.		303.00		333.30		
1 Lowbed Trailer, 75 Ton		171.80		189.00	34.80	38.28
16 L.H., Daily Totals		$1046.80		$1394.90	$65.43	$87.18

Crew No.	Bare Costs		Incl. Subs O & P		Cost Per Labor-Hour	
Crew A-3G	**Hr.**	**Daily**	**Hr.**	**Daily**	**Bare Costs**	**Incl. O&P**
1 Equip. Oper. (crane)	$34.80	$278.40	$54.85	$438.80	$30.63	$48.90
1 Truck Driver (heavy)	26.45	211.60	42.95	343.60		
1 Pickup Truck (4x4)		82.00		90.20		
1 Tractor, 6x4, 45 Ton Cap.		331.60		364.75		
1 Lowbed Trailer, 75 Ton		171.80		189.00	36.59	40.25
16 L.H., Daily Totals		$1075.40		$1426.35	$67.22	$89.15
Crew A-4	**Hr.**	**Daily**	**Hr.**	**Daily**	**Bare Costs**	**Incl. O&P**
2 Carpenters	$33.00	$528.00	$54.70	$875.20	$31.87	$52.27
1 Painter, Ordinary	29.60	236.80	47.40	379.20		
24 L.H., Daily Totals		$764.80		$1254.40	$31.87	$52.27
Crew A-5	**Hr.**	**Daily**	**Hr.**	**Daily**	**Bare Costs**	**Incl. O&P**
2 Laborers	$26.00	$416.00	$43.10	$689.60	$25.97	$42.95
.25 Truck Driver (light)	25.70	51.40	41.75	83.50		
.25 Light Truck, 1.5 Ton		30.50		33.55	1.69	1.86
18 L.H., Daily Totals		$497.90		$806.65	$27.66	$44.81
Crew A-6	**Hr.**	**Daily**	**Hr.**	**Daily**	**Bare Costs**	**Incl. O&P**
1 Instrument Man	$33.65	$269.20	$55.00	$440.00	$32.85	$53.33
1 Rodman/Chainman	32.05	256.40	51.65	413.20		
1 Laser Transit/Level		59.50		65.45	3.72	4.09
16 L.H., Daily Totals		$585.10		$918.65	$36.57	$57.42
Crew A-7	**Hr.**	**Daily**	**Hr.**	**Daily**	**Bare Costs**	**Incl. O&P**
1 Chief Of Party	$41.25	$330.00	$66.15	$529.20	$35.65	$57.60
1 Instrument Man	33.65	269.20	55.00	440.00		
1 Rodman/Chainman	32.05	256.40	51.65	413.20		
1 Laser Transit/Level		59.50		65.45	2.48	2.73
24 L.H., Daily Totals		$915.10		$1447.85	$38.13	$60.33
Crew A-8	**Hr.**	**Daily**	**Hr.**	**Daily**	**Bare Costs**	**Incl. O&P**
1 Chief Of Party	$41.25	$330.00	$66.15	$529.20	$34.75	$56.11
1 Instrument Man	33.65	269.20	55.00	440.00		
2 Rodmen/Chainmen	32.05	512.80	51.65	826.40		
1 Laser Transit/Level		59.50		65.45	1.86	2.05
32 L.H., Daily Totals		$1171.50		$1861.05	$36.61	$58.16
Crew A-9	**Hr.**	**Daily**	**Hr.**	**Daily**	**Bare Costs**	**Incl. O&P**
1 Asbestos Foreman	$36.50	$292.00	$59.30	$474.40	$36.06	$58.60
7 Asbestos Workers	36.00	2016.00	58.50	3276.00		
64 L.H., Daily Totals		$2308.00		$3750.40	$36.06	$58.60
Crew A-10	**Hr.**	**Daily**	**Hr.**	**Daily**	**Bare Costs**	**Incl. O&P**
1 Asbestos Foreman	$36.50	$292.00	$59.30	$474.40	$36.06	$58.60
7 Asbestos Workers	36.00	2016.00	58.50	3276.00		
64 L.H., Daily Totals		$2308.00		$3750.40	$36.06	$58.60
Crew A-10A	**Hr.**	**Daily**	**Hr.**	**Daily**	**Bare Costs**	**Incl. O&P**
1 Asbestos Foreman	$36.50	$292.00	$59.30	$474.40	$36.17	$58.77
2 Asbestos Workers	36.00	576.00	58.50	936.00		
24 L.H., Daily Totals		$868.00		$1410.40	$36.17	$58.77
Crew A-10B	**Hr.**	**Daily**	**Hr.**	**Daily**	**Bare Costs**	**Incl. O&P**
1 Asbestos Foreman	$36.50	$292.00	$59.30	$474.40	$36.13	$58.70
3 Asbestos Workers	36.00	864.00	58.50	1404.00		
32 L.H., Daily Totals		$1156.00		$1878.40	$36.13	$58.70

Crew No.	Bare Costs		Incl. Subs O & P		Cost Per Labor-Hour	
Crew A-10C	**Hr.**	**Daily**	**Hr.**	**Daily**	**Bare Costs**	**Incl. O&P**
3 Asbestos Workers	$36.00	$864.00	$58.50	$1404.00	$36.00	$58.50
1 Flatbed Truck		122.00		134.20	5.08	5.59
24 L.H., Daily Totals		$986.00		$1538.20	$41.08	$64.09
Crew A-10D	**Hr.**	**Daily**	**Hr.**	**Daily**	**Bare Costs**	**Incl. O&P**
2 Asbestos Workers	$36.00	$576.00	$58.50	$936.00	$34.00	$54.46
1 Equip. Oper. (crane)	34.80	278.40	54.85	438.80		
1 Equip. Oper. Oiler	29.20	233.60	46.00	368.00		
1 Hydraulic Crane, 33 Ton		649.00		713.90	20.28	22.31
32 L.H., Daily Totals		$1737.00		$2456.70	$54.28	$76.77
Crew A-11	**Hr.**	**Daily**	**Hr.**	**Daily**	**Bare Costs**	**Incl. O&P**
1 Asbestos Foreman	$36.50	$292.00	$59.30	$474.40	$36.06	$58.60
7 Asbestos Workers	36.00	2016.00	58.50	3276.00		
2 Chipping Hammers		32.00		35.20	.50	.55
64 L.H., Daily Totals		$2340.00		$3785.60	$36.56	$59.15
Crew A-12	**Hr.**	**Daily**	**Hr.**	**Daily**	**Bare Costs**	**Incl. O&P**
1 Asbestos Foreman	$36.50	$292.00	$59.30	$474.40	$36.06	$58.60
7 Asbestos Workers	36.00	2016.00	58.50	3276.00		
1 Large Prod. Vac. Loader		503.05		553.35	7.86	8.65
64 L.H., Daily Totals		$2811.05		$4303.75	$43.92	$67.25
Crew A-13	**Hr.**	**Daily**	**Hr.**	**Daily**	**Bare Costs**	**Incl. O&P**
1 Equip. Oper. (light)	$32.15	$257.20	$50.65	$405.20	$32.15	$50.65
1 Large Prod. Vac. Loader		503.05		553.35	62.88	69.17
8 L.H., Daily Totals		$760.25		$958.55	$95.03	$119.82
Crew B-1	**Hr.**	**Daily**	**Hr.**	**Daily**	**Bare Costs**	**Incl. O&P**
1 Labor Foreman (outside)	$28.00	$224.00	$46.40	$371.20	$26.67	$44.20
2 Laborers	26.00	416.00	43.10	689.60		
24 L.H., Daily Totals		$640.00		$1060.80	$26.67	$44.20
Crew B-2	**Hr.**	**Daily**	**Hr.**	**Daily**	**Bare Costs**	**Incl. O&P**
1 Labor Foreman (outside)	$28.00	$224.00	$46.40	$371.20	$26.40	$43.76
4 Laborers	26.00	832.00	43.10	1379.20		
40 L.H., Daily Totals		$1056.00		$1750.40	$26.40	$43.76
Crew B-3	**Hr.**	**Daily**	**Hr.**	**Daily**	**Bare Costs**	**Incl. O&P**
1 Labor Foreman (outside)	$28.00	$224.00	$46.40	$371.20	$27.76	$45.26
2 Laborers	26.00	416.00	43.10	689.60		
1 Equip. Oper. (med.)	33.65	269.20	53.05	424.40		
2 Truck Drivers (heavy)	26.45	423.20	42.95	687.20		
1 F.E. Loader, T.M., 2.5 C.Y.		761.00		837.10		
2 Dump Trucks, 16 Ton		953.20		1048.50	35.71	39.28
48 L.H., Daily Totals		$3046.60		$4058.00	$63.47	$84.54
Crew B-3A	**Hr.**	**Daily**	**Hr.**	**Daily**	**Bare Costs**	**Incl. O&P**
4 Laborers	$26.00	$832.00	$43.10	$1379.20	$27.53	$45.09
1 Equip. Oper. (med.)	33.65	269.20	53.05	424.40		
1 Hyd. Excavator, 1.5 C.Y.		683.40		751.75	17.09	18.79
40 L.H., Daily Totals		$1784.60		$2555.35	$44.62	$63.88
Crew B-3B	**Hr.**	**Daily**	**Hr.**	**Daily**	**Bare Costs**	**Incl. O&P**
2 Laborers	$26.00	$416.00	$43.10	$689.60	$28.03	$45.55
1 Equip. Oper. (med.)	33.65	269.20	53.05	424.40		
1 Truck Driver (heavy)	26.45	211.60	42.95	343.60		
1 Backhoe Loader, 80 H.P.		230.80		253.90		
1 Dump Truck, 16 Ton		476.60		524.25	22.11	24.32
32 L.H., Daily Totals		$1604.20		$2235.75	$50.14	$69.87

Crew No.	Bare Costs		Incl. Subs O & P		Cost Per Labor-Hour	
Crew B-3C	**Hr.**	**Daily**	**Hr.**	**Daily**	**Bare Costs**	**Incl. O&P**
3 Laborers	$26.00	$624.00	$43.10	$1034.40	$27.91	$45.59
1 Equip. Oper. (med.)	33.65	269.20	53.05	424.40		
1 F.E. Crawler Ldr, 4 C.Y.		1061.00		1167.10	33.16	36.47
32 L.H., Daily Totals		$1954.20		$2625.90	$61.07	$82.06

Crew B-4	Hr.	Daily	Hr.	Daily	Bare Costs	Incl. O&P
1 Labor Foreman (outside)	$28.00	$224.00	$46.40	$371.20	$26.41	$43.63
4 Laborers	26.00	832.00	43.10	1379.20		
1 Truck Driver (heavy)	26.45	211.60	42.95	343.60		
1 Tractor, 4 x 2, 195 H.P.		215.20		236.70		
1 Platform Trailer		120.60		132.65	7.00	7.70
48 L.H., Daily Totals		$1603.40		$2463.35	$33.41	$51.33

Crew B-5	Hr.	Daily	Hr.	Daily	Bare Costs	Incl. O&P
1 Labor Foreman (outside)	$28.00	$224.00	$46.40	$371.20	$28.47	$46.41
4 Laborers	26.00	832.00	43.10	1379.20		
2 Equip. Oper. (med.)	33.65	538.40	53.05	848.80		
1 Air Compr., 250 C.F.M.		127.60		140.35		
2 Air Tools & Accessories		22.40		24.65		
2-50 Ft. Air Hoses, 1.5" Dia.		10.00		11.00		
1 F.E. Loader, T.M., 2.5 C.Y.		761.00		837.10	16.45	18.09
56 L.H., Daily Totals		$2515.40		$3612.30	$44.92	$64.50

Crew B-6	Hr.	Daily	Hr.	Daily	Bare Costs	Incl. O&P
2 Laborers	$26.00	$416.00	$43.10	$689.60	$28.05	$45.62
1 Equip. Oper. (light)	32.15	257.20	50.65	405.20		
1 Backhoe Loader, 48 H.P.		207.80		228.60	8.66	9.52
24 L.H., Daily Totals		$881.00		$1323.40	$36.71	$55.14

Crew B-6B	Hr.	Daily	Hr.	Daily	Bare Costs	Incl. O&P
2 Labor Foremen (out)	$28.00	$448.00	$46.40	$742.40	$26.67	$44.20
4 Laborers	26.00	832.00	43.10	1379.20		
1 Winch Truck		317.80		349.60		
1 Flatbed Truck		122.00		134.20		
1 Butt Fusion Machine		433.60		476.95	18.20	20.02
48 L.H., Daily Totals		$2153.40		$3082.35	$44.87	$64.22

Crew B-7	Hr.	Daily	Hr.	Daily	Bare Costs	Incl. O&P
1 Labor Foreman (outside)	$28.00	$224.00	$46.40	$371.20	$27.61	$45.31
4 Laborers	26.00	832.00	43.10	1379.20		
1 Equip. Oper. (med.)	33.65	269.20	53.05	424.40		
1 Chipping Machine		164.80		181.30		
1 F.E. Loader, T.M., 2.5 C.Y.		761.00		837.10		
2 Chain Saws, 36"		66.80		73.50	20.68	22.75
48 L.H., Daily Totals		$2317.80		$3266.70	$48.29	$68.06

Crew B-7A	Hr.	Daily	Hr.	Daily	Bare Costs	Incl. O&P
2 Laborers	$26.00	$416.00	$43.10	$689.60	$28.05	$45.62
1 Equip. Oper. (light)	32.15	257.20	50.65	405.20		
1 Rake w/Tractor		184.90		203.40		
2 Chain Saws, 18"		37.60		41.35	9.27	10.20
24 L.H., Daily Totals		$895.70		$1339.55	$37.32	$55.82

Crew No.	Bare Costs		Incl. Subs O & P		Cost Per Labor-Hour	
Crew B-8	**Hr.**	**Daily**	**Hr.**	**Daily**	**Bare Costs**	**Incl. O&P**
1 Labor Foreman (outside)	$28.00	$224.00	$46.40	$371.20	$28.68	$46.33
2 Laborers	26.00	416.00	43.10	689.60		
2 Equip. Oper. (med.)	33.65	538.40	53.05	848.80		
1 Equip. Oper. Oiler	29.20	233.60	46.00	368.00		
2 Truck Drivers (heavy)	26.45	423.20	42.95	687.20		
1 Hyd. Crane, 25 Ton		624.20		686.60		
1 F.E. Loader, T.M., 2.5 C.Y.		761.00		837.10		
2 Dump Trucks, 16 Ton		953.20		1048.50	36.54	40.19
64 L.H., Daily Totals		$4173.60		$5537.00	$65.22	$86.52

Crew B-9	Hr.	Daily	Hr.	Daily	Bare Costs	Incl. O&P
1 Labor Foreman (outside)	$28.00	$224.00	$46.40	$371.20	$26.40	$43.76
4 Laborers	26.00	832.00	43.10	1379.20		
1 Air Compr., 250 C.F.M.		127.60		140.35		
2 Air Tools & Accessories		22.40		24.65		
2-50 Ft. Air Hoses, 1.5" Dia.		10.00		11.00	4.00	4.40
40 L.H., Daily Totals		$1216.00		$1926.40	$30.40	$48.16

Crew B-9A	Hr.	Daily	Hr.	Daily	Bare Costs	Incl. O&P
2 Laborers	$26.00	$416.00	$43.10	$689.60	$26.15	$43.05
1 Truck Driver (heavy)	26.45	211.60	42.95	343.60		
1 Water Tanker		118.00		129.80		
1 Tractor		215.20		236.70		
2-50 Ft. Disch. Hoses		7.50		8.25	14.20	15.62
24 L.H., Daily Totals		$968.30		$1407.95	$40.35	$58.67

Crew B-9B	Hr.	Daily	Hr.	Daily	Bare Costs	Incl. O&P
2 Laborers	$26.00	$416.00	$43.10	$689.60	$26.15	$43.05
1 Truck Driver (heavy)	26.45	211.60	42.95	343.60		
2-50 Ft. Disch. Hoses		7.50		8.25		
1 Water Tanker		118.00		129.80		
1 Tractor		215.20		236.70		
1 Pressure Washer		46.40		51.05	16.13	17.74
24 L.H., Daily Totals		$1014.70		$1459.00	$42.28	$60.79

Crew B-9C	Hr.	Daily	Hr.	Daily	Bare Costs	Incl. O&P
1 Labor Foreman (outside)	$28.00	$224.00	$46.40	$371.20	$26.40	$43.76
4 Laborers	26.00	832.00	43.10	1379.20		
1 Air Compr., 250 C.F.M.		127.60		140.35		
2-50 Ft. Air Hoses, 1.5" Dia.		10.00		11.00		
2 Breaker, Pavement, 60 lb.		22.40		24.65	4.00	4.40
40 L.H., Daily Totals		$1216.00		$1926.40	$30.40	$48.16

Crew B-9D	Hr.	Daily	Hr.	Daily	Bare Costs	Incl. O&P
1 Labor Foreman (Outside)	$28.00	$224.00	$46.40	$371.20	$26.40	$43.76
4 Common Laborers	26.00	832.00	43.10	1379.20		
1 Air Compressor, 250 CFM		127.60		140.35		
2 Air hoses, 1.5" x 50'		10.00		11.00		
2 Air tamper		47.00		51.70	4.62	5.08
40 L.H., Daily Totals		$1240.60		$1953.45	$31.02	$48.84

Crew B-10	Hr.	Daily	Hr.	Daily	Bare Costs	Incl. O&P
1 Equip. Oper. (med.)	$33.65	$269.20	$53.05	$424.40	$31.10	$49.73
.5 Laborer	26.00	104.00	43.10	172.40		
12 L.H., Daily Totals		$373.20		$596.80	$31.10	$49.73

Crew B-10A	Hr.	Daily	Hr.	Daily	Bare Costs	Incl. O&P
1 Equip. Oper. (med.)	$33.65	$269.20	$53.05	$424.40	$31.10	$49.73
.5 Laborer	26.00	104.00	43.10	172.40		
1 Roll. Compact., 2K Lbs.		130.80		143.90	10.90	11.99
12 L.H., Daily Totals		$504.00		$740.70	$42.00	$61.72

Crews

CREWS

Crew No.	Bare Costs		Incl. Subs O & P		Cost Per Labor-Hour	

Crew B-10B	Hr.	Daily	Hr.	Daily	Bare Costs	Incl. O&P
1 Equip. Oper. (med.)	$33.65	$269.20	$53.05	$424.40	$31.10	$49.73
.5 Laborer	26.00	104.00	43.10	172.40		
1 Dozer, 200 H.P.		863.40		949.75	71.95	79.15
12 L.H., Daily Totals		$1236.60		$1546.55	$103.05	$128.88

Crew B-10C	Hr.	Daily	Hr.	Daily	Bare Costs	Incl. O&P
1 Equip. Oper. (med.)	$33.65	$269.20	$53.05	$424.40	$31.10	$49.73
.5 Laborer	26.00	104.00	43.10	172.40		
1 Dozer, 200 H.P.		863.40		949.75		
1 Vibratory Roller, Towed		610.20		671.20	122.80	135.08
12 L.H., Daily Totals		$1846.80		$2217.75	$153.90	$184.81

Crew B-10D	Hr.	Daily	Hr.	Daily	Bare Costs	Incl. O&P
1 Equip. Oper. (med.)	$33.65	$269.20	$53.05	$424.40	$31.10	$49.73
.5 Laborer	26.00	104.00	43.10	172.40		
1 Dozer, 200 H.P.		863.40		949.75		
1 Sheepsft. Roller, Towed		77.80		85.60	78.43	86.28
12 L.H., Daily Totals		$1314.40		$1632.15	$109.53	$136.01

Crew B-10E	Hr.	Daily	Hr.	Daily	Bare Costs	Incl. O&P
1 Equip. Oper. (med.)	$33.65	$269.20	$53.05	$424.40	$31.10	$49.73
.5 Laborer	26.00	104.00	43.10	172.40		
1 Tandem Roller, 5 Ton		108.20		119.00	9.02	9.92
12 L.H., Daily Totals		$481.40		$715.80	$40.12	$59.65

Crew B-10F	Hr.	Daily	Hr.	Daily	Bare Costs	Incl. O&P
1 Equip. Oper. (med.)	$33.65	$269.20	$53.05	$424.40	$31.10	$49.73
.5 Laborer	26.00	104.00	43.10	172.40		
1 Tandem Roller, 10 Ton		183.60		201.95	15.30	16.83
12 L.H., Daily Totals		$556.80		$798.75	$46.40	$66.56

Crew B-10G	Hr.	Daily	Hr.	Daily	Bare Costs	Incl. O&P
1 Equip. Oper. (med.)	$33.65	$269.20	$53.05	$424.40	$31.10	$49.73
.5 Laborer	26.00	104.00	43.10	172.40		
1 Sheepsft. Roll., 130 H.P.		762.60		838.85	63.55	69.91
12 L.H., Daily Totals		$1135.80		$1435.65	$94.65	$119.64

Crew B-10H	Hr.	Daily	Hr.	Daily	Bare Costs	Incl. O&P
1 Equip. Oper. (med.)	$33.65	$269.20	$53.05	$424.40	$31.10	$49.73
.5 Laborer	26.00	104.00	43.10	172.40		
1 Diaphr. Water Pump, 2″		46.20		50.80		
1-20 Ft. Suction Hose, 2″		3.55		3.90		
2-50 Ft. Disch. Hoses, 2″		6.30		6.95	4.67	5.14
12 L.H., Daily Totals		$429.25		$658.45	$35.77	$54.87

Crew B-10I	Hr.	Daily	Hr.	Daily	Bare Costs	Incl. O&P
1 Equip. Oper. (med.)	$33.65	$269.20	$53.05	$424.40	$31.10	$49.73
.5 Laborer	26.00	104.00	43.10	172.40		
1 Diaphr. Water Pump, 4″		87.40		96.15		
1-20 Ft. Suction Hose, 4″		7.30		8.05		
2-50 Ft. Disch. Hoses, 4″		10.50		11.55	8.77	9.64
12 L.H., Daily Totals		$478.40		$712.55	$39.87	$59.37

Crew B-10J	Hr.	Daily	Hr.	Daily	Bare Costs	Incl. O&P
1 Equip. Oper. (med.)	$33.65	$269.20	$53.05	$424.40	$31.10	$49.73
.5 Laborer	26.00	104.00	43.10	172.40		
1 Centr. Water Pump, 3″		53.80		59.20		
1-20 Ft. Suction Hose, 3″		5.45		6.00		
2-50 Ft. Disch. Hoses, 3″		7.50		8.25	5.56	6.12
12 L.H., Daily Totals		$439.95		$670.25	$36.66	$55.85

Crew B-10K	Hr.	Daily	Hr.	Daily	Bare Costs	Incl. O&P
1 Equip. Oper. (med.)	$33.65	$269.20	$53.05	$424.40	$31.10	$49.73
.5 Laborer	26.00	104.00	43.10	172.40		
1 Centr. Water Pump, 6″		229.80		252.80		
1-20 Ft. Suction Hose, 6″		14.50		15.95		
2-50 Ft. Disch. Hoses, 6″		23.80		26.20	22.34	24.58
12 L.H., Daily Totals		$641.30		$891.75	$53.44	$74.31

Crew B-10L	Hr.	Daily	Hr.	Daily	Bare Costs	Incl. O&P
1 Equip. Oper. (med.)	$33.65	$269.20	$53.05	$424.40	$31.10	$49.73
.5 Laborer	26.00	104.00	43.10	172.40		
1 Dozer, 80 H.P.		300.20		330.20	25.02	27.52
12 L.H., Daily Totals		$673.40		$927.00	$56.12	$77.25

Crew B-10M	Hr.	Daily	Hr.	Daily	Bare Costs	Incl. O&P
1 Equip. Oper. (med.)	$33.65	$269.20	$53.05	$424.40	$31.10	$49.73
.5 Laborer	26.00	104.00	43.10	172.40		
1 Dozer, 300 H.P.		1099.00		1208.90	91.58	100.74
12 L.H., Daily Totals		$1472.20		$1805.70	$122.68	$150.47

Crew B-10N	Hr.	Daily	Hr.	Daily	Bare Costs	Incl. O&P
1 Equip. Oper. (med.)	$33.65	$269.20	$53.05	$424.40	$31.10	$49.73
.5 Laborer	26.00	104.00	43.10	172.40		
1 F.E. Loader, T.M., 1.5 C.Y		300.60		330.65	25.05	27.56
12 L.H., Daily Totals		$673.80		$927.45	$56.15	$77.29

Crew B-10O	Hr.	Daily	Hr.	Daily	Bare Costs	Incl. O&P
1 Equip. Oper. (med.)	$33.65	$269.20	$53.05	$424.40	$31.10	$49.73
.5 Laborer	26.00	104.00	43.10	172.40		
1 F.E. Loader, T.M., 2.25 C.Y.		538.00		591.80	44.83	49.32
12 L.H., Daily Totals		$911.20		$1188.60	$75.93	$99.05

Crew B-10P	Hr.	Daily	Hr.	Daily	Bare Costs	Incl. O&P
1 Equip. Oper. (med.)	$33.65	$269.20	$53.05	$424.40	$31.10	$49.73
.5 Laborer	26.00	104.00	43.10	172.40		
1 F.E. Loader, T.M., 2.5 C.Y.		761.00		837.10	63.42	69.76
12 L.H., Daily Totals		$1134.20		$1433.90	$94.52	$119.49

Crew B-10Q	Hr.	Daily	Hr.	Daily	Bare Costs	Incl. O&P
1 Equip. Oper. (med.)	$33.65	$269.20	$53.05	$424.40	$31.10	$49.73
.5 Laborer	26.00	104.00	43.10	172.40		
1 F.E. Loader, T.M., 5 C.Y.		1061.00		1167.10	88.42	97.26
12 L.H., Daily Totals		$1434.20		$1763.90	$119.52	$146.99

Crew B-10R	Hr.	Daily	Hr.	Daily	Bare Costs	Incl. O&P
1 Equip. Oper. (med.)	$33.65	$269.20	$53.05	$424.40	$31.10	$49.73
.5 Laborer	26.00	104.00	43.10	172.40		
1 F.E. Loader, W.M., 1 C.Y.		192.00		211.20	16.00	17.60
12 L.H., Daily Totals		$565.20		$808.00	$47.10	$67.33

Crew B-10S	Hr.	Daily	Hr.	Daily	Bare Costs	Incl. O&P
1 Equip. Oper. (med.)	$33.65	$269.20	$53.05	$424.40	$31.10	$49.73
.5 Laborer	26.00	104.00	43.10	172.40		
1 F.E. Loader, W.M., 1.5 C.Y.		236.20		259.80	19.68	21.65
12 L.H., Daily Totals		$609.40		$856.60	$50.78	$71.38

Crew B-10T	Hr.	Daily	Hr.	Daily	Bare Costs	Incl. O&P
1 Equip. Oper. (med.)	$33.65	$269.20	$53.05	$424.40	$31.10	$49.73
.5 Laborer	26.00	104.00	43.10	172.40		
1 F.E. Loader, W.M.,2.5 C.Y.		295.60		325.15	24.63	27.10
12 L.H., Daily Totals		$668.80		$921.95	$55.73	$76.83

Crews

Crew No.	Bare Costs		Incl. Subs O & P		Cost Per Labor-Hour	
Crew B-10U	**Hr.**	**Daily**	**Hr.**	**Daily**	**Bare Costs**	**Incl. O&P**
1 Equip. Oper. (med.)	$33.65	$269.20	$53.05	$424.40	$31.10	$49.73
.5 Laborer	26.00	104.00	43.10	172.40		
1 F.E. Loader, W.M., 5.5 C.Y.		674.80		742.30	56.23	61.86
12 L.H., Daily Totals		$1048.00		$1339.10	$87.33	$111.59
Crew B-10V	**Hr.**	**Daily**	**Hr.**	**Daily**	**Bare Costs**	**Incl. O&P**
1 Equip. Oper. (med.)	$33.65	$269.20	$53.05	$424.40	$31.10	$49.73
.5 Laborer	26.00	104.00	43.10	172.40		
1 Dozer, 700 H.P.		3047.00		3351.70	253.92	279.31
12 L.H., Daily Totals		$3420.20		$3948.50	$285.02	$329.04
Crew B-10W	**Hr.**	**Daily**	**Hr.**	**Daily**	**Bare Costs**	**Incl. O&P**
1 Equip. Oper. (med.)	$33.65	$269.20	$53.05	$424.40	$31.10	$49.73
.5 Laborer	26.00	104.00	43.10	172.40		
1 Dozer, 105 H.P.		434.00		477.40	36.17	39.78
12 L.H., Daily Totals		$807.20		$1074.20	$67.27	$89.51
Crew B-10X	**Hr.**	**Daily**	**Hr.**	**Daily**	**Bare Costs**	**Incl. O&P**
1 Equip. Oper. (med.)	$33.65	$269.20	$53.05	$424.40	$31.10	$49.73
.5 Laborer	26.00	104.00	43.10	172.40		
1 Dozer, 410 H.P.		1473.00		1620.30	122.75	135.03
12 L.H., Daily Totals		$1846.20		$2217.10	$153.85	$184.76
Crew B-10Y	**Hr.**	**Daily**	**Hr.**	**Daily**	**Bare Costs**	**Incl. O&P**
1 Equip. Oper. (med.)	$33.65	$269.20	$53.05	$424.40	$31.10	$49.73
.5 Laborer	26.00	104.00	43.10	172.40		
1 Vibratory Drum Roller		323.60		355.95	26.97	29.66
12 L.H., Daily Totals		$696.80		$952.75	$58.07	$79.39
Crew B-11A	**Hr.**	**Daily**	**Hr.**	**Daily**	**Bare Costs**	**Incl. O&P**
1 Equipment Oper. (med.)	$33.65	$269.20	$53.05	$424.40	$29.83	$48.08
1 Laborer	26.00	208.00	43.10	344.80		
1 Dozer, 200 H.P.		863.40		949.75	53.96	59.36
16 L.H., Daily Totals		$1340.60		$1718.95	$83.79	$107.44
Crew B-11B	**Hr.**	**Daily**	**Hr.**	**Daily**	**Bare Costs**	**Incl. O&P**
1 Equipment Oper. (light)	$32.15	$257.20	$50.65	$405.20	$29.08	$46.88
1 Laborer	26.00	208.00	43.10	344.80		
1 Air Powered Tamper		23.50		25.85		
1 Air Compr. 365 C.F.M.		157.60		173.35		
2-50 Ft. Air Hoses, 1.5" Dia.		10.00		11.00	11.94	13.14
16 L.H., Daily Totals		$656.30		$960.20	$41.02	$60.02
Crew B-11C	**Hr.**	**Daily**	**Hr.**	**Daily**	**Bare Costs**	**Incl. O&P**
1 Equipment Oper. (med.)	$33.65	$269.20	$53.05	$424.40	$29.83	$48.08
1 Laborer	26.00	208.00	43.10	344.80		
1 Backhoe Loader, 48 H.P.		207.80		228.60	12.99	14.29
16 L.H., Daily Totals		$685.00		$997.80	$42.82	$62.37
Crew B-11K	**Hr.**	**Daily**	**Hr.**	**Daily**	**Bare Costs**	**Incl. O&P**
1 Equipment Oper. (med.)	$33.65	$269.20	$53.05	$424.40	$29.83	$48.08
1 Laborer	26.00	208.00	43.10	344.80		
1 Trencher, 8' D., 16" W.		1310.00		1441.00	81.88	90.06
16 L.H., Daily Totals		$1787.20		$2210.20	$111.71	$138.14
Crew B-11L	**Hr.**	**Daily**	**Hr.**	**Daily**	**Bare Costs**	**Incl. O&P**
1 Equipment Oper. (med.)	$33.65	$269.20	$53.05	$424.40	$29.83	$48.08
1 Laborer	26.00	208.00	43.10	344.80		
1 Grader, 30,000 Lbs.		432.00		475.20	27.00	29.70
16 L.H., Daily Totals		$909.20		$1244.40	$56.83	$77.78

Crew No.	Bare Costs		Incl. Subs O & P		Cost Per Labor-Hour	
Crew B-11M	**Hr.**	**Daily**	**Hr.**	**Daily**	**Bare Costs**	**Incl. O&P**
1 Equipment Oper. (med.)	$33.65	$269.20	$53.05	$424.40	$29.83	$48.08
1 Laborer	26.00	208.00	43.10	344.80		
1 Backhoe Loader, 80 H.P.		230.80		253.90	14.43	15.87
16 L.H., Daily Totals		$708.00		$1023.10	$44.26	$63.95
Crew B-11W	**Hr.**	**Daily**	**Hr.**	**Daily**	**Bare Costs**	**Incl. O&P**
1 Equipment Operator (med.)	$33.65	$269.20	$53.05	$424.40	$27.01	$43.80
1 Common Laborer	26.00	208.00	43.10	344.80		
10 Truck Drivers, Heavy	26.45	2116.00	42.95	3436.00		
1 Dozer, 200 H.P.		863.40		949.75		
1 Vib. roller, smth, towed, 23 Ton		610.20		671.20		
10 Dump Truck, 10 Ton		3250.00		3575.00	49.20	54.12
96 L.H., Daily Totals		$7316.80		$9401.15	$76.21	$97.92
Crew B-11Y	**Hr.**	**Daily**	**Hr.**	**Daily**	**Bare Costs**	**Incl. O&P**
1 Labor Foreman (Outside)	$28.00	$224.00	$46.40	$371.20	$28.77	$46.78
5 Common Laborers	26.00	1040.00	43.10	1724.00		
3 Equipment Operator (med.)	33.65	807.60	53.05	1273.20		
1 Dozer, 80 H.P.		300.20		330.20		
2 Wlk-Beh. Comp., 2-Drum, 1 Ton		261.60		287.75		
4 Vibratory plate, gas, 21"		165.60		182.15	10.10	11.11
72 L.H., Daily Totals		$2799.00		$4168.50	$38.87	$57.89
Crew B-12A	**Hr.**	**Daily**	**Hr.**	**Daily**	**Bare Costs**	**Incl. O&P**
1 Equip. Oper. (crane)	$34.80	$278.40	$54.85	$438.80	$32.00	$50.42
1 Equip. Oper. Oiler	29.20	233.60	46.00	368.00		
1 Hyd. Excavator, 1 C.Y.		487.40		536.15	30.46	33.51
16 L.H., Daily Totals		$999.40		$1342.95	$62.46	$83.93
Crew B-12B	**Hr.**	**Daily**	**Hr.**	**Daily**	**Bare Costs**	**Incl. O&P**
1 Equip. Oper. (crane)	$34.80	$278.40	$54.85	$438.80	$32.00	$50.42
1 Equip. Oper. Oiler	29.20	233.60	46.00	368.00		
1 Hyd. Excavator, 1.5 C.Y.		683.40		751.75	42.71	46.98
16 L.H., Daily Totals		$1195.40		$1558.55	$74.71	$97.40
Crew B-12C	**Hr.**	**Daily**	**Hr.**	**Daily**	**Bare Costs**	**Incl. O&P**
1 Equip. Oper. (crane)	$34.80	$278.40	$54.85	$438.80	$32.00	$50.42
1 Equip. Oper. Oiler	29.20	233.60	46.00	368.00		
1 Hyd. Excavator, 2 C.Y.		887.20		975.90	55.45	61.00
16 L.H., Daily Totals		$1399.20		$1782.70	$87.45	$111.42
Crew B-12D	**Hr.**	**Daily**	**Hr.**	**Daily**	**Bare Costs**	**Incl. O&P**
1 Equip. Oper. (crane)	$34.80	$278.40	$54.85	$438.80	$32.00	$50.42
1 Equip. Oper. Oiler	29.20	233.60	46.00	368.00		
1 Hyd. Excavator, 3.5 C.Y.		1965.00		2161.50	122.81	135.09
16 L.H., Daily Totals		$2477.00		$2968.30	$154.81	$185.51
Crew B-12E	**Hr.**	**Daily**	**Hr.**	**Daily**	**Bare Costs**	**Incl. O&P**
1 Equip. Oper. (crane)	$34.80	$278.40	$54.85	$438.80	$32.00	$50.42
1 Equip. Oper. Oiler	29.20	233.60	46.00	368.00		
1 Hyd. Excavator, .5 C.Y.		323.00		355.30	20.19	22.21
16 L.H., Daily Totals		$835.00		$1162.10	$52.19	$72.63
Crew B-12F	**Hr.**	**Daily**	**Hr.**	**Daily**	**Bare Costs**	**Incl. O&P**
1 Equip. Oper. (crane)	$34.80	$278.40	$54.85	$438.80	$32.00	$50.42
1 Equip. Oper. Oiler	29.20	233.60	46.00	368.00		
1 Hyd. Excavator, .75 C.Y.		458.80		504.70	28.68	31.54
16 L.H., Daily Totals		$970.80		$1311.50	$60.68	$81.96

Crews

Crew No.	Bare Costs		Incl. Subs O & P		Cost Per Labor-Hour	
Crew B-12G	**Hr.**	**Daily**	**Hr.**	**Daily**	**Bare Costs**	**Incl. O&P**
1 Equip. Oper. (crane)	$34.80	$278.40	$54.85	$438.80	$32.00	$50.42
1 Equip. Oper. Oiler	29.20	233.60	46.00	368.00		
1 Power Shovel, .5 C.Y.		450.90		496.00		
1 Clamshell Bucket, .5 C.Y.		33.40		36.75	30.27	33.30
16 L.H., Daily Totals		$996.30		$1339.55	$62.27	$83.72

Crew B-12H	Hr.	Daily	Hr.	Daily	Bare Costs	Incl. O&P
1 Equip. Oper. (crane)	$34.80	$278.40	$54.85	$438.80	$32.00	$50.42
1 Equip. Oper. Oiler	29.20	233.60	46.00	368.00		
1 Power Shovel, 1 C.Y.		792.00		871.20		
1 Clamshell Bucket, 1 C.Y.		44.00		48.40	52.25	57.48
16 L.H., Daily Totals		$1348.00		$1726.40	$84.25	$107.90

Crew B-12I	Hr.	Daily	Hr.	Daily	Bare Costs	Incl. O&P
1 Equip. Oper. (crane)	$34.80	$278.40	$54.85	$438.80	$32.00	$50.42
1 Equip. Oper. Oiler	29.20	233.60	46.00	368.00		
1 Power Shovel, .75 C.Y.		604.50		664.95		
1 Dragline Bucket, .75 C.Y.		18.60		20.45	38.94	42.84
16 L.H., Daily Totals		$1135.10		$1492.20	$70.94	$93.26

Crew B-12J	Hr.	Daily	Hr.	Daily	Bare Costs	Incl. O&P
1 Equip. Oper. (crane)	$34.80	$278.40	$54.85	$438.80	$32.00	$50.42
1 Equip. Oper. Oiler	29.20	233.60	46.00	368.00		
1 Gradall, 3 Ton, .5 C.Y.		823.40		905.75	51.46	56.61
16 L.H., Daily Totals		$1335.40		$1712.55	$83.46	$107.03

Crew B-12K	Hr.	Daily	Hr.	Daily	Bare Costs	Incl. O&P
1 Equip. Oper. (crane)	$34.80	$278.40	$54.85	$438.80	$32.00	$50.42
1 Equip. Oper. Oiler	29.20	233.60	46.00	368.00		
1 Gradall, 3 Ton, 1 C.Y.		958.80		1054.70	59.93	65.92
16 L.H., Daily Totals		$1470.80		$1861.50	$91.93	$116.34

Crew B-12L	Hr.	Daily	Hr.	Daily	Bare Costs	Incl. O&P
1 Equip. Oper. (crane)	$34.80	$278.40	$54.85	$438.80	$32.00	$50.42
1 Equip. Oper. Oiler	29.20	233.60	46.00	368.00		
1 Power Shovel, .5 C.Y.		450.90		496.00		
1 F.E. Attachment, .5 C.Y.		45.20		49.70	31.01	34.11
16 L.H., Daily Totals		$1008.10		$1352.50	$63.01	$84.53

Crew B-12M	Hr.	Daily	Hr.	Daily	Bare Costs	Incl. O&P
1 Equip. Oper. (crane)	$34.80	$278.40	$54.85	$438.80	$32.00	$50.42
1 Equip. Oper. Oiler	29.20	233.60	46.00	368.00		
1 Power Shovel, .75 C.Y.		604.50		664.95		
1 F.E. Attachment, .75 C.Y.		50.00		55.00	40.91	45.00
16 L.H., Daily Totals		$1166.50		$1526.75	$72.91	$95.42

Crew B-12N	Hr.	Daily	Hr.	Daily	Bare Costs	Incl. O&P
1 Equip. Oper. (crane)	$34.80	$278.40	$54.85	$438.80	$32.00	$50.42
1 Equip. Oper. Oiler	29.20	233.60	46.00	368.00		
1 Power Shovel, 1 C.Y.		792.00		871.20		
1 F.E. Attachment, 1 C.Y.		56.80		62.50	53.05	58.36
16 L.H., Daily Totals		$1360.80		$1740.50	$85.05	$108.78

Crew B-12O	Hr.	Daily	Hr.	Daily	Bare Costs	Incl. O&P
1 Equip. Oper. (crane)	$34.80	$278.40	$54.85	$438.80	$32.00	$50.42
1 Equip. Oper. Oiler	29.20	233.60	46.00	368.00		
1 Power Shovel, 1.5 C.Y.		915.20		1006.70		
1 F.E. Attachment, 1.5 C.Y.		65.60		72.15	61.30	67.43
16 L.H., Daily Totals		$1492.80		$1885.65	$93.30	$117.85

Crew B-12P	Hr.	Daily	Hr.	Daily	Bare Costs	Incl. O&P
1 Equip. Oper. (crane)	$34.80	$278.40	$54.85	$438.80	$32.00	$50.42
1 Equip. Oper. Oiler	29.20	233.60	46.00	368.00		
1 Crawler Crane, 40 Ton		915.20		1006.70		
1 Dragline Bucket, 1.5 C.Y.		30.20		33.20	59.09	65.00
16 L.H., Daily Totals		$1457.40		$1846.70	$91.09	$115.42

Crew B-12Q	Hr.	Daily	Hr.	Daily	Bare Costs	Incl. O&P
1 Equip. Oper. (crane)	$34.80	$278.40	$54.85	$438.80	$32.00	$50.42
1 Equip. Oper. Oiler	29.20	233.60	46.00	368.00		
1 Hyd. Excavator, 5/8 C.Y.		418.00		459.80	26.13	28.74
16 L.H., Daily Totals		$930.00		$1266.60	$58.13	$79.16

Crew B-12R	Hr.	Daily	Hr.	Daily	Bare Costs	Incl. O&P
1 Equip. Oper. (crane)	$34.80	$278.40	$54.85	$438.80	$32.00	$50.42
1 Equip. Oper. Oiler	29.20	233.60	46.00	368.00		
1 Hyd. Excavator, 1.5 C.Y.		683.40		751.75	42.71	46.98
16 L.H., Daily Totals		$1195.40		$1558.55	$74.71	$97.40

Crew B-12S	Hr.	Daily	Hr.	Daily	Bare Costs	Incl. O&P
1 Equip. Oper. (crane)	$34.80	$278.40	$54.85	$438.80	$32.00	$50.42
1 Equip. Oper. Oiler	29.20	233.60	46.00	368.00		
1 Hyd. Excavator, 2.5 C.Y.		1181.00		1299.10	73.81	81.19
16 L.H., Daily Totals		$1693.00		$2105.90	$105.81	$131.61

Crew B-12T	Hr.	Daily	Hr.	Daily	Bare Costs	Incl. O&P
1 Equip. Oper. (crane)	$34.80	$278.40	$54.85	$438.80	$32.00	$50.42
1 Equip. Oper. Oiler	29.20	233.60	46.00	368.00		
1 Crawler Crane, 75 Ton		1217.00		1338.70		
1 F.E. Attachment, 3 C.Y.		89.40		98.35	81.65	89.82
16 L.H., Daily Totals		$1818.40		$2243.85	$113.65	$140.24

Crew B-12V	Hr.	Daily	Hr.	Daily	Bare Costs	Incl. O&P
1 Equip. Oper. (crane)	$34.80	$278.40	$54.85	$438.80	$32.00	$50.42
1 Equip. Oper. Oiler	29.20	233.60	46.00	368.00		
1 Crawler Crane, 75 Ton		1217.00		1338.70		
1 Dragline Bucket, 3 C.Y.		47.60		52.35	79.04	86.94
16 L.H., Daily Totals		$1776.60		$2197.85	$111.04	$137.36

Crew B-13	Hr.	Daily	Hr.	Daily	Bare Costs	Incl. O&P
1 Labor Foreman (outside)	$28.00	$224.00	$46.40	$371.20	$28.00	$45.66
4 Laborers	26.00	832.00	43.10	1379.20		
1 Equip. Oper. (crane)	34.80	278.40	54.85	438.80		
1 Equip. Oper. Oiler	29.20	233.60	46.00	368.00		
1 Hyd. Crane, 25 Ton		624.20		686.60	11.15	12.26
56 L.H., Daily Totals		$2192.20		$3243.80	$39.15	$57.92

Crew B-13A	Hr.	Daily	Hr.	Daily	Bare Costs	Incl. O&P
1 Foreman	$28.00	$224.00	$46.40	$371.20	$28.60	$46.37
2 Laborers	26.00	416.00	43.10	689.60		
2 Equipment Operators	33.65	538.40	53.05	848.80		
2 Truck Drivers (heavy)	26.45	423.20	42.95	687.20		
1 Crane, 75 Ton		1217.00		1338.70		
1 F.E. Lder, 3.75 C.Y.		1061.00		1167.10		
2 Dump Trucks, 12 Ton		650.00		715.00	52.29	57.51
56 L.H., Daily Totals		$4529.60		$5817.60	$80.89	$103.88

Crew No.	Bare Costs	Incl. Subs O & P	Cost Per Labor-Hour

Crew B-13B	Hr.	Daily	Hr.	Daily	Bare Costs	Incl. O&P
1 Labor Foreman (outside)	$28.00	$224.00	$46.40	$371.20	$28.00	$45.66
4 Laborers	26.00	832.00	43.10	1379.20		
1 Equip. Oper. (crane)	34.80	278.40	54.85	438.80		
1 Equip. Oper. Oiler	29.20	233.60	46.00	368.00		
1 Hyd. Crane, 55 Ton		911.00		1002.10	16.27	17.89
56 L.H., Daily Totals		$2479.00		$3559.30	$44.27	$63.55

Crew B-13C	Hr.	Daily	Hr.	Daily	Bare Costs	Incl. O&P
1 Labor Foreman (outside)	$28.00	$224.00	$46.40	$371.20	$28.00	$45.66
4 Laborers	26.00	832.00	43.10	1379.20		
1 Equip. Oper. (crane)	34.80	278.40	54.85	438.80		
1 Equip. Oper. Oiler	29.20	233.60	46.00	368.00		
1 Crawler Crane, 100 Ton		1602.00		1762.20	28.61	31.47
56 L.H., Daily Totals		$3170.00		$4319.40	$56.61	$77.13

Crew B-14	Hr.	Daily	Hr.	Daily	Bare Costs	Incl. O&P
1 Labor Foreman (outside)	$28.00	$224.00	$46.40	$371.20	$27.36	$44.91
4 Laborers	26.00	832.00	43.10	1379.20		
1 Equip. Oper. (light)	32.15	257.20	50.65	405.20		
1 Backhoe Loader, 48 H.P.		207.80		228.60	4.33	4.76
48 L.H., Daily Totals		$1521.00		$2384.20	$31.69	$49.67

Crew B-15	Hr.	Daily	Hr.	Daily	Bare Costs	Incl. O&P
1 Equipment Oper. (med)	$33.65	$269.20	$53.05	$424.40	$28.44	$45.86
.5 Laborer	26.00	104.00	43.10	172.40		
2 Truck Drivers (heavy)	26.45	423.20	42.95	687.20		
2 Dump Trucks, 16 Ton		953.20		1048.50		
1 Dozer, 200 H.P.		863.40		949.75	64.88	71.37
28 L.H., Daily Totals		$2613.00		$3282.25	$93.32	$117.23

Crew B-16	Hr.	Daily	Hr.	Daily	Bare Costs	Incl. O&P
1 Labor Foreman (outside)	$28.00	$224.00	$46.40	$371.20	$26.61	$43.89
2 Laborers	26.00	416.00	43.10	689.60		
1 Truck Driver (heavy)	26.45	211.60	42.95	343.60		
1 Dump Truck, 16 Ton		476.60		524.25	14.89	16.38
32 L.H., Daily Totals		$1328.20		$1928.65	$41.50	$60.27

Crew B-17	Hr.	Daily	Hr.	Daily	Bare Costs	Incl. O&P
2 Laborers	$26.00	$416.00	$43.10	$689.60	$27.65	$44.95
1 Equip. Oper. (light)	32.15	257.20	50.65	405.20		
1 Truck Driver (heavy)	26.45	211.60	42.95	343.60		
1 Backhoe Loader, 48 H.P.		207.80		228.60		
1 Dump Truck, 12 Ton		325.00		357.50	16.65	18.32
32 L.H., Daily Totals		$1417.60		$2024.50	$44.30	$63.27

Crew B-18	Hr.	Daily	Hr.	Daily	Bare Costs	Incl. O&P
1 Labor Foreman (outside)	$28.00	$224.00	$46.40	$371.20	$26.67	$44.20
2 Laborers	26.00	416.00	43.10	689.60		
1 Vibrating Compactor		41.40		45.55	1.73	1.90
24 L.H., Daily Totals		$681.40		$1106.35	$28.40	$46.10

Crew B-19	Hr.	Daily	Hr.	Daily	Bare Costs	Incl. O&P
1 Pile Driver Foreman	$34.05	$272.40	$57.95	$463.60	$32.63	$53.98
4 Pile Drivers	32.05	1025.60	54.55	1745.60		
2 Equip. Oper. (crane)	34.80	556.80	54.85	877.60		
1 Equip. Oper. Oiler	29.20	233.60	46.00	368.00		
1 Crane, 40 Ton & Access.		915.20		1006.70		
60 L.F. Pile Leads		90.00		99.00		
1 Hammer, Diesel, 22k Ft-Lb		564.00		620.40	24.52	26.97
64 L.H., Daily Totals		$3657.60		$5180.90	$57.15	$80.95

Crew B-19A	Hr.	Daily	Hr.	Daily	Bare Costs	Incl. O&P
1 Pile Driver Foreman	$34.05	$272.40	$57.95	$463.60	$32.63	$53.98
4 Pile Drivers	32.05	1025.60	54.55	1745.60		
2 Equip. Oper. (crane)	34.80	556.80	54.85	877.60		
1 Equip. Oper. Oiler	29.20	233.60	46.00	368.00		
1 Crawler Crane, 75 Ton		1217.00		1338.70		
60 L.F. Leads, 25K Ft. Lbs.		132.00		145.20		
1 Hammer, Diesel, 41k Ft-Lb		662.60		728.85	31.43	34.57
64 L.H., Daily Totals		$4100.00		$5667.55	$64.06	$88.55

Crew B-20	Hr.	Daily	Hr.	Daily	Bare Costs	Incl. O&P
1 Labor Foreman (out)	$28.00	$224.00	$46.40	$371.20	$29.22	$48.17
1 Skilled Worker	33.65	269.20	55.00	440.00		
1 Laborer	26.00	208.00	43.10	344.80		
24 L.H., Daily Totals		$701.20		$1156.00	$29.22	$48.17

Crew B-20A	Hr.	Daily	Hr.	Daily	Bare Costs	Incl. O&P
1 Labor Foreman	$28.00	$224.00	$46.40	$371.20	$31.33	$50.01
1 Laborer	26.00	208.00	43.10	344.80		
1 Plumber	39.60	316.80	61.40	491.20		
1 Plumber Apprentice	31.70	253.60	49.15	393.20		
32 L.H., Daily Totals		$1002.40		$1600.40	$31.33	$50.01

Crew B-21	Hr.	Daily	Hr.	Daily	Bare Costs	Incl. O&P
1 Labor Foreman (out)	$28.00	$224.00	$46.40	$371.20	$30.01	$49.12
1 Skilled Worker	33.65	269.20	55.00	440.00		
1 Laborer	26.00	208.00	43.10	344.80		
.5 Equip. Oper. (crane)	34.80	139.20	54.85	219.40		
.5 S.P. Crane, 5 Ton		158.90		174.80	5.68	6.24
28 L.H., Daily Totals		$999.30		$1550.20	$35.69	$55.36

Crew B-21A	Hr.	Daily	Hr.	Daily	Bare Costs	Incl. O&P
1 Labor Foreman	$28.00	$224.00	$46.40	$371.20	$32.02	$50.98
1 Laborer	26.00	208.00	43.10	344.80		
1 Plumber	39.60	316.80	61.40	491.20		
1 Plumber Apprentice	31.70	253.60	49.15	393.20		
1 Equip. Oper. (crane)	34.80	278.40	54.85	438.80		
1 S.P. Crane, 12 Ton		482.20		530.40	12.06	13.26
40 L.H., Daily Totals		$1763.00		$2569.60	$44.08	$64.24

Crew B-22	Hr.	Daily	Hr.	Daily	Bare Costs	Incl. O&P
1 Labor Foreman (out)	$28.00	$224.00	$46.40	$371.20	$30.33	$49.50
1 Skilled Worker	33.65	269.20	55.00	440.00		
1 Laborer	26.00	208.00	43.10	344.80		
.75 Equip. Oper. (crane)	34.80	208.80	54.85	329.10		
.75 S.P. Crane, 5 Ton		238.35		262.20	7.95	8.74
30 L.H., Daily Totals		$1148.35		$1747.30	$38.28	$58.24

Crew B-22A	Hr.	Daily	Hr.	Daily	Bare Costs	Incl. O&P
1 Labor Foreman (out)	$28.00	$224.00	$46.40	$371.20	$29.42	$48.16
1 Skilled Worker	33.65	269.20	55.00	440.00		
2 Laborers	26.00	416.00	43.10	689.60		
.75 Equipment Oper. (crane)	34.80	208.80	54.85	329.10		
.75 Crane, 5 Ton		238.35		262.20		
1 Generator, 5 KW		37.00		40.70		
1 Butt Fusion Machine		433.60		476.95	18.66	20.52
38 L.H., Daily Totals		$1826.95		$2609.75	$48.08	$68.68

Crew No.	Bare Costs		Incl. Subs O & P		Cost Per Labor-Hour	
Crew B-22B	**Hr.**	**Daily**	**Hr.**	**Daily**	**Bare Costs**	**Incl. O&P**
1 Skilled Worker	$33.65	$269.20	$55.00	$440.00	$29.83	$49.05
1 Laborer	26.00	208.00	43.10	344.80		
1 Electro Fusion Machine		172.80		190.10	10.80	11.88
16 L.H., Daily Totals		$650.00		$974.90	$40.63	$60.93
Crew B-23	**Hr.**	**Daily**	**Hr.**	**Daily**	**Bare Costs**	**Incl. O&P**
1 Labor Foreman (outside)	$28.00	$224.00	$46.40	$371.20	$26.40	$43.76
4 Laborers	26.00	832.00	43.10	1379.20		
1 Drill Rig, Wells		2877.00		3164.70		
1 Light Truck, 3 Ton		160.40		176.45	75.94	83.53
40 L.H., Daily Totals		$4093.40		$5091.55	$102.34	$127.29
Crew B-23A	**Hr.**	**Daily**	**Hr.**	**Daily**	**Bare Costs**	**Incl. O&P**
1 Labor Foreman (outside)	$28.00	$224.00	$46.40	$371.20	$29.22	$47.52
1 Laborer	26.00	208.00	43.10	344.80		
1 Equip. Operator (medium)	33.65	269.20	53.05	424.40		
1 Drill Rig, Wells		2877.00		3164.70		
1 Pickup Truck, 3/4 Ton		75.80		83.40	123.03	135.34
24 L.H., Daily Totals		$3654.00		$4388.50	$152.25	$182.86
Crew B-23B	**Hr.**	**Daily**	**Hr.**	**Daily**	**Bare Costs**	**Incl. O&P**
1 Labor Foreman (outside)	$28.00	$224.00	$46.40	$371.20	$29.22	$47.52
1 Laborer	26.00	208.00	43.10	344.80		
1 Equip. Operator (medium)	33.65	269.20	53.05	424.40		
1 Drill Rig, Wells		2877.00		3164.70		
1 Pickup Truck, 3/4 Ton		75.80		83.40		
1 Pump, Cntfgl, 6"		229.80		252.80	132.61	145.87
24 L.H., Daily Totals		$3883.80		$4641.30	$161.83	$193.39
Crew B-24	**Hr.**	**Daily**	**Hr.**	**Daily**	**Bare Costs**	**Incl. O&P**
1 Cement Finisher	$31.55	$252.40	$49.60	$396.80	$30.18	$49.13
1 Laborer	26.00	208.00	43.10	344.80		
1 Carpenter	33.00	264.00	54.70	437.60		
24 L.H., Daily Totals		$724.40		$1179.20	$30.18	$49.13
Crew B-25	**Hr.**	**Daily**	**Hr.**	**Daily**	**Bare Costs**	**Incl. O&P**
1 Labor Foreman	$28.00	$224.00	$46.40	$371.20	$28.27	$46.11
7 Laborers	26.00	1456.00	43.10	2413.60		
3 Equip. Oper. (med.)	33.65	807.60	53.05	1273.20		
1 Asphalt Paver, 130 H.P		1457.00		1602.70		
1 Tandem Roller, 10 Ton		183.60		201.95		
1 Roller, Pneumatic Wheel		244.60		269.05	21.42	23.57
88 L.H., Daily Totals		$4372.80		$6131.70	$49.69	$69.68
Crew B-25B	**Hr.**	**Daily**	**Hr.**	**Daily**	**Bare Costs**	**Incl. O&P**
1 Labor Foreman	$28.00	$224.00	$46.40	$371.20	$28.72	$46.69
7 Laborers	26.00	1456.00	43.10	2413.60		
4 Equip. Oper. (medium)	33.65	1076.80	53.05	1697.60		
1 Asphalt Paver, 130 H.P.		1457.00		1602.70		
2 Rollers, Steel Wheel		367.20		403.90		
1 Roller, Pneumatic Wheel		244.60		269.05	21.55	23.71
96 L.H., Daily Totals		$4825.60		$6758.05	$50.27	$70.40
Crew B-25C	**Hr.**	**Daily**	**Hr.**	**Daily**	**Bare Costs**	**Incl. O&P**
1 Labor Foreman	$28.00	$224.00	$46.40	$371.20	$28.88	$46.97
3 Laborers	26.00	624.00	43.10	1034.40		
2 Equip. Oper. (medium)	33.65	538.40	53.05	848.80		
1 Asphalt Paver, 130 H.P.		1457.00		1602.70		
1 Rollers, Steel Wheel		183.60		201.95	34.18	37.60
48 L.H., Daily Totals		$3027.00		$4059.05	$63.06	$84.57

Crew No.	Bare Costs		Incl. Subs O & P		Cost Per Labor-Hour	
Crew B-26	**Hr.**	**Daily**	**Hr.**	**Daily**	**Bare Costs**	**Incl. O&P**
1 Labor Foreman (outside)	$28.00	$224.00	$46.40	$371.20	$29.09	$47.69
6 Laborers	26.00	1248.00	43.10	2068.80		
2 Equip. Oper. (med.)	33.65	538.40	53.05	848.80		
1 Rodman (reinf.)	37.10	296.80	63.85	510.80		
1 Cement Finisher	31.55	252.40	49.60	396.80		
1 Grader, 30,000 Lbs.		432.00		475.20		
1 Paving Mach. & Equip.		1580.00		1738.00	22.86	25.15
88 L.H., Daily Totals		$4571.60		$6409.60	$51.95	$72.84
Crew B-27	**Hr.**	**Daily**	**Hr.**	**Daily**	**Bare Costs**	**Incl. O&P**
1 Labor Foreman (outside)	$28.00	$224.00	$46.40	$371.20	$26.50	$43.93
3 Laborers	26.00	624.00	43.10	1034.40		
1 Berm Machine		188.40		207.25	5.89	6.48
32 L.H., Daily Totals		$1036.40		$1612.85	$32.39	$50.41
Crew B-28	**Hr.**	**Daily**	**Hr.**	**Daily**	**Bare Costs**	**Incl. O&P**
2 Carpenters	$33.00	$528.00	$54.70	$875.20	$30.67	$50.83
1 Laborer	26.00	208.00	43.10	344.80		
24 L.H., Daily Totals		$736.00		$1220.00	$30.67	$50.83
Crew B-29	**Hr.**	**Daily**	**Hr.**	**Daily**	**Bare Costs**	**Incl. O&P**
1 Labor Foreman (outside)	$28.00	$224.00	$46.40	$371.20	$28.00	$45.66
4 Laborers	26.00	832.00	43.10	1379.20		
1 Equip. Oper. (crane)	34.80	278.40	54.85	438.80		
1 Equip. Oper. Oiler	29.20	233.60	46.00	368.00		
1 Gradall, 3 Ton, 1/2 C.Y.		823.40		905.75	14.70	16.17
56 L.H., Daily Totals		$2391.40		$3462.95	$42.70	$61.83
Crew B-30	**Hr.**	**Daily**	**Hr.**	**Daily**	**Bare Costs**	**Incl. O&P**
1 Equip. Oper. (med.)	$33.65	$269.20	$53.05	$424.40	$28.85	$46.32
2 Truck Drivers (heavy)	26.45	423.20	42.95	687.20		
1 Hyd. Excavator, 1.5 C.Y.		683.40		751.75		
2 Dump Trucks, 16 Ton		953.20		1048.50	68.19	75.01
24 L.H., Daily Totals		$2329.00		$2911.85	$97.04	$121.33
Crew B-31	**Hr.**	**Daily**	**Hr.**	**Daily**	**Bare Costs**	**Incl. O&P**
1 Labor Foreman (outside)	$28.00	$224.00	$46.40	$371.20	$27.80	$46.08
3 Laborers	26.00	624.00	43.10	1034.40		
1 Carpenter	33.00	264.00	54.70	437.60		
1 Air Compr., 250 C.F.M.		127.60		140.35		
1 Sheeting Driver		7.20		7.90		
2-50 Ft. Air Hoses, 1.5" Dia.		10.00		11.00	3.62	3.98
40 L.H., Daily Totals		$1256.80		$2002.45	$31.42	$50.06
Crew B-32	**Hr.**	**Daily**	**Hr.**	**Daily**	**Bare Costs**	**Incl. O&P**
1 Laborer	$26.00	$208.00	$43.10	$344.80	$31.74	$50.56
3 Equip. Oper. (med.)	33.65	807.60	53.05	1273.20		
1 Grader, 30,000 Lbs.		432.00		475.20		
1 Tandem Roller, 10 Ton		183.60		201.95		
1 Dozer, 200 H.P.		863.40		949.75	46.22	50.84
32 L.H., Daily Totals		$2494.60		$3244.90	$77.96	$101.40
Crew B-32A	**Hr.**	**Daily**	**Hr.**	**Daily**	**Bare Costs**	**Incl. O&P**
1 Laborer	$26.00	$208.00	$43.10	$344.80	$31.10	$49.73
2 Equip. Oper. (medium)	33.65	538.40	53.05	848.80		
1 Grader, 30,000 Lbs.		432.00		475.20		
1 Roller, Vibratory, 29,000 Lbs.		436.80		480.50	36.20	39.82
24 L.H., Daily Totals		$1615.20		$2149.30	$67.30	$89.55

Crew B-32B	Bare Costs Hr.	Bare Costs Daily	Incl. Subs O & P Hr.	Incl. Subs O & P Daily	Bare Costs	Incl. O&P
1 Laborer	$26.00	$208.00	$43.10	$344.80	$31.10	$49.73
2 Equip. Oper. (medium)	33.65	538.40	53.05	848.80		
1 Dozer, 200 H.P.		863.40		949.75		
1 Roller, Vibratory, 29,000 Lbs.		436.80		480.50	54.18	59.59
24 L.H., Daily Totals		$2046.60		$2623.85	$85.28	$109.32

Crew B-32C	Hr.	Daily	Hr.	Daily	Bare Costs	Incl. O&P
1 Labor Foreman	$28.00	$224.00	$46.40	$371.20	$30.16	$48.63
2 Laborers	26.00	416.00	43.10	689.60		
3 Equip. Oper. (medium)	33.65	807.60	53.05	1273.20		
1 Grader, 30,000 Lbs.		432.00		475.20		
1 Roller, Steel Wheel		183.60		201.95		
1 Dozer, 200 H.P.		863.40		949.75	30.81	33.89
48 L.H., Daily Totals		$2926.60		$3960.90	$60.97	$82.52

Crew B-33A	Hr.	Daily	Hr.	Daily	Bare Costs	Incl. O&P
1 Equip. Oper. (med.)	$33.65	$269.20	$53.05	$424.40	$31.46	$50.21
.5 Laborer	26.00	104.00	43.10	172.40		
.25 Equip. Oper. (med.)	33.65	67.30	53.05	106.10		
1 Scraper, Towed, 7 C.Y.		170.00		187.00		
1.25 Dozer, 300 H.P.		1373.75		1511.15	110.27	121.29
14 L.H., Daily Totals		$1984.25		$2401.05	$141.73	$171.50

Crew B-33B	Hr.	Daily	Hr.	Daily	Bare Costs	Incl. O&P
1 Equip. Oper. (med.)	$33.65	$269.20	$53.05	$424.40	$31.46	$50.21
.5 Laborer	26.00	104.00	43.10	172.40		
.25 Equip. Oper. (med.)	33.65	67.30	53.05	106.10		
1 Scraper, Towed, 10 C.Y.		189.10		208.00		
1.25 Dozer, 300 H.P.		1373.75		1511.15	111.63	122.80
14 L.H., Daily Totals		$2003.35		$2422.05	$143.09	$173.01

Crew B-33C	Hr.	Daily	Hr.	Daily	Bare Costs	Incl. O&P
1 Equip. Oper. (med.)	$33.65	$269.20	$53.05	$424.40	$31.46	$50.21
.5 Laborer	26.00	104.00	43.10	172.40		
.25 Equip. Oper. (med.)	33.65	67.30	53.05	106.10		
1 Scraper, Towed, 12 C.Y.		189.10		208.00		
1.25 Dozer, 300 H.P.		1373.75		1511.15	111.63	122.80
14 L.H., Daily Totals		$2003.35		$2422.05	$143.09	$173.01

Crew B-33D	Hr.	Daily	Hr.	Daily	Bare Costs	Incl. O&P
1 Equip. Oper. (med.)	$33.65	$269.20	$53.05	$424.40	$31.46	$50.21
.5 Laborer	26.00	104.00	43.10	172.40		
.25 Equip. Oper. (med.)	33.65	67.30	53.05	106.10		
1 S.P. Scraper, 14 C.Y.		1468.00		1614.80		
.25 Dozer, 300 H.P.		274.75		302.25	124.48	136.93
14 L.H., Daily Totals		$2183.25		$2619.95	$155.94	$187.14

Crew B-33E	Hr.	Daily	Hr.	Daily	Bare Costs	Incl. O&P
1 Equip. Oper. (med.)	$33.65	$269.20	$53.05	$424.40	$31.46	$50.21
.5 Laborer	26.00	104.00	43.10	172.40		
.25 Equip. Oper. (med.)	33.65	67.30	53.05	106.10		
1 S.P. Scraper, 24 C.Y.		2297.00		2526.70		
.25 Dozer, 300 H.P.		274.75		302.25	183.70	202.07
14 L.H., Daily Totals		$3012.25		$3531.85	$215.16	$252.28

Crew B-33F	Hr.	Daily	Hr.	Daily	Bare Costs	Incl. O&P
1 Equip. Oper. (med.)	$33.65	$269.20	$53.05	$424.40	$31.46	$50.21
.5 Laborer	26.00	104.00	43.10	172.40		
.25 Equip. Oper. (med.)	33.65	67.30	53.05	106.10		
1 Elev. Scraper, 11 C.Y.		805.40		885.95		
.25 Dozer, 300 H.P.		274.75		302.25	77.15	84.87
14 L.H., Daily Totals		$1520.65		$1891.10	$108.61	$135.08

Crew B-33G	Hr.	Daily	Hr.	Daily	Bare Costs	Incl. O&P
1 Equip. Oper. (med.)	$33.65	$269.20	$53.05	$424.40	$31.46	$50.21
.5 Laborer	26.00	104.00	43.10	172.40		
.25 Equip. Oper. (med.)	33.65	67.30	53.05	106.10		
1 Elev. Scraper, 20 C.Y.		1597.00		1756.70		
.25 Dozer, 300 H.P.		274.75		302.25	133.70	147.07
14 L.H., Daily Totals		$2312.25		$2761.85	$165.16	$197.28

Crew B-34A	Hr.	Daily	Hr.	Daily	Bare Costs	Incl. O&P
1 Truck Driver (heavy)	$26.45	$211.60	$42.95	$343.60	$26.45	$42.95
1 Dump Truck, 12 Ton		325.00		357.50	40.63	44.69
8 L.H., Daily Totals		$536.60		$701.10	$67.08	$87.64

Crew B-34B	Hr.	Daily	Hr.	Daily	Bare Costs	Incl. O&P
1 Truck Driver (heavy)	$26.45	$211.60	$42.95	$343.60	$26.45	$42.95
1 Dump Truck, 16 Ton		476.60		524.25	59.58	65.53
8 L.H., Daily Totals		$688.20		$867.85	$86.03	$108.48

Crew B-34C	Hr.	Daily	Hr.	Daily	Bare Costs	Incl. O&P
1 Truck Driver (heavy)	$26.45	$211.60	$42.95	$343.60	$26.45	$42.95
1 Truck Tractor, 40 Ton		303.00		333.30		
1 Dump Trailer, 16.5 C.Y.		103.20		113.50	50.78	55.85
8 L.H., Daily Totals		$617.80		$790.40	$77.23	$98.80

Crew B-34D	Hr.	Daily	Hr.	Daily	Bare Costs	Incl. O&P
1 Truck Driver (heavy)	$26.45	$211.60	$42.95	$343.60	$26.45	$42.95
1 Truck Tractor, 40 Ton		303.00		333.30		
1 Dump Trailer, 20 C.Y.		116.40		128.05	52.43	57.67
8 L.H., Daily Totals		$631.00		$804.95	$78.88	$100.62

Crew B-34E	Hr.	Daily	Hr.	Daily	Bare Costs	Incl. O&P
1 Truck Driver (heavy)	$26.45	$211.60	$42.95	$343.60	$26.45	$42.95
1 Truck, Off Hwy., 25 Ton		920.40		1012.45	115.05	126.56
8 L.H., Daily Totals		$1132.00		$1356.05	$141.50	$169.51

Crew B-34F	Hr.	Daily	Hr.	Daily	Bare Costs	Incl. O&P
1 Truck Driver (heavy)	$26.45	$211.60	$42.95	$343.60	$26.45	$42.95
1 Truck, Off Hwy., 22 C.Y.		946.60		1041.25	118.33	130.16
8 L.H., Daily Totals		$1158.20		$1384.85	$144.78	$173.11

Crew B-34G	Hr.	Daily	Hr.	Daily	Bare Costs	Incl. O&P
1 Truck Driver (heavy)	$26.45	$211.60	$42.95	$343.60	$26.45	$42.95
1 Truck, Off Hwy., 34 C.Y.		1226.00		1348.60	153.25	168.58
8 L.H., Daily Totals		$1437.60		$1692.20	$179.70	$211.53

Crew B-34H	Hr.	Daily	Hr.	Daily	Bare Costs	Incl. O&P
1 Truck Driver (heavy)	$26.45	$211.60	$42.95	$343.60	$26.45	$42.95
1 Truck, Off Hwy., 42 C.Y.		1310.00		1441.00	163.75	180.13
8 L.H., Daily Totals		$1521.60		$1784.60	$190.20	$223.08

Crew No.	Bare Costs Hr.	Bare Costs Daily	Incl. Subs O & P Hr.	Incl. Subs O & P Daily	Cost Per Labor-Hour Bare Costs	Cost Per Labor-Hour Incl. O&P
Crew B-34J	**Hr.**	**Daily**	**Hr.**	**Daily**	**Bare Costs**	**Incl. O&P**
1 Truck Driver (heavy)	$26.45	$211.60	$42.95	$343.60	$26.45	$42.95
1 Truck, Off Hwy., 60 C.Y.		1663.00		1829.30	207.88	228.66
8 L.H., Daily Totals		$1874.60		$2172.90	$234.33	$271.61
Crew B-34K	**Hr.**	**Daily**	**Hr.**	**Daily**	**Bare Costs**	**Incl. O&P**
1 Truck Driver (heavy)	$26.45	$211.60	$42.95	$343.60	$26.45	$42.95
1 Truck Tractor, 240 H.P.		331.60		364.75		
1 Low Bed Trailer		171.80		189.00	62.93	69.22
8 L.H., Daily Totals		$715.00		$897.35	$89.38	$112.17
Crew B-34N	**Hr.**	**Daily**	**Hr.**	**Daily**	**Bare Costs**	**Incl. O&P**
1 Truck Driver (heavy)	$26.45	$211.60	$42.95	$343.60	$26.45	$42.95
1 Dump Truck, 12 Ton		325.00		357.50		
1 Flatbed Trailer, 40 Ton		120.60		132.65	55.70	61.27
8 L.H., Daily Totals		$657.20		$833.75	$82.15	$104.22
Crew B-35	**Hr.**	**Daily**	**Hr.**	**Daily**	**Bare Costs**	**Incl. O&P**
1 Laborer Foreman (out)	$28.00	$224.00	$46.40	$371.20	$31.88	$51.13
1 Skilled Worker	33.65	269.20	55.00	440.00		
1 Welder (plumber)	39.60	316.80	61.40	491.20		
1 Laborer	26.00	208.00	43.10	344.80		
1 Equip. Oper. (crane)	34.80	278.40	54.85	438.80		
1 Equip. Oper. Oiler	29.20	233.60	46.00	368.00		
1 Electric Welding Mach.		77.75		85.55		
1 Hyd. Excavator, .75 C.Y.		458.80		504.70	11.18	12.30
48 L.H., Daily Totals		$2066.55		$3044.25	$43.06	$63.43
Crew B-35A	**Hr.**	**Daily**	**Hr.**	**Daily**	**Bare Costs**	**Incl. O&P**
1 Laborer Foreman (out)	$28.00	$224.00	$46.40	$371.20	$31.04	$49.98
2 Laborers	26.00	416.00	43.10	689.60		
1 Skilled Worker	33.65	269.20	55.00	440.00		
1 Welder (plumber)	39.60	316.80	61.40	491.20		
1 Equip. Oper. (crane)	34.80	278.40	54.85	438.80		
1 Equip. Oper. Oiler	29.20	233.60	46.00	368.00		
1 Welder, 300 amp		75.20		82.70		
1 Crane, 75 Ton		1217.00		1338.70	23.08	25.38
56 L.H., Daily Totals		$3030.20		$4220.20	$54.12	$75.36
Crew B-36	**Hr.**	**Daily**	**Hr.**	**Daily**	**Bare Costs**	**Incl. O&P**
1 Labor Foreman (outside)	$28.00	$224.00	$46.40	$371.20	$29.46	$47.74
2 Laborers	26.00	416.00	43.10	689.60		
2 Equip. Oper. (med.)	33.65	538.40	53.05	848.80		
1 Dozer, 200 H.P.		863.40		949.75		
1 Aggregate Spreader		41.40		45.55		
1 Tandem Roller, 10 Ton		183.60		201.95	27.21	29.93
40 L.H., Daily Totals		$2266.80		$3106.85	$56.67	$77.67
Crew B-36A	**Hr.**	**Daily**	**Hr.**	**Daily**	**Bare Costs**	**Incl. O&P**
1 Labor Foreman (outside)	$28.00	$224.00	$46.40	$371.20	$30.66	$49.26
2 Laborers	26.00	416.00	43.10	689.60		
4 Equip. Oper. (med.)	33.65	1076.80	53.05	1697.60		
1 Dozer, 200 H.P.		863.40		949.75		
1 Aggregate Spreader		41.40		45.55		
1 Roller, Steel Wheel		183.60		201.95		
1 Roller, Pneumatic Wheel		244.60		269.05	23.80	26.18
56 L.H., Daily Totals		$3049.80		$4224.70	$54.46	$75.44

Crew No.	Bare Costs Hr.	Bare Costs Daily	Incl. Subs O & P Hr.	Incl. Subs O & P Daily	Cost Per Labor-Hour Bare Costs	Cost Per Labor-Hour Incl. O&P
Crew B-36B	**Hr.**	**Daily**	**Hr.**	**Daily**	**Bare Costs**	**Incl. O&P**
1 Labor Foreman (outside)	$28.00	$224.00	$46.40	$371.20	$30.13	$48.47
2 Laborers	26.00	416.00	43.10	689.60		
4 Equip. Oper. (medium)	33.65	1076.80	53.05	1697.60		
1 Truck Driver, Heavy	26.45	211.60	42.95	343.60		
1 Grader, 30,000 Lbs.		432.00		475.20		
1 F.E. Loader, crl, 1.5 C.Y.		354.40		389.85		
1 Dozer, 300 H.P.		1099.00		1208.90		
1 Roller, Vibratory		436.80		480.50		
1 Truck, Tractor, 240 H.P.		331.60		364.75		
1 Water Tanker, 5000 Gal.		118.00		129.80	43.31	47.64
64 L.H., Daily Totals		$4700.20		$6151.00	$73.44	$96.11
Crew B-36C	**Hr.**	**Daily**	**Hr.**	**Daily**	**Bare Costs**	**Incl. O&P**
1 Labor Foreman (outside)	$28.00	$224.00	$46.40	$371.20	$31.08	$49.70
3 Equip. Oper. (medium)	33.65	807.60	53.05	1273.20		
1 Truck Driver, Heavy	26.45	211.60	42.95	343.60		
1 Grader, 30,000 Lbs.		432.00		475.20		
1 Dozer, 300 H.P.		1099.00		1208.90		
1 Roller, Vibratory		436.80		480.50		
1 Truck, Tractor, 240 H.P.		331.60		364.75		
1 Water Tanker, 5000 Gal.		118.00		129.80	60.44	66.48
40 L.H., Daily Totals		$3660.60		$4647.15	$91.52	$116.18
Crew B-37	**Hr.**	**Daily**	**Hr.**	**Daily**	**Bare Costs**	**Incl. O&P**
1 Labor Foreman (outside)	$28.00	$224.00	$46.40	$371.20	$27.36	$44.91
4 Laborers	26.00	832.00	43.10	1379.20		
1 Equip. Oper. (light)	32.15	257.20	50.65	405.20		
1 Tandem Roller, 5 Ton		108.20		119.00	2.25	2.48
48 L.H., Daily Totals		$1421.40		$2274.60	$29.61	$47.39
Crew B-38	**Hr.**	**Daily**	**Hr.**	**Daily**	**Bare Costs**	**Incl. O&P**
1 Labor Foreman (outside)	$28.00	$224.00	$46.40	$371.20	$29.16	$47.26
2 Laborers	26.00	416.00	43.10	689.60		
1 Equip. Oper. (light)	32.15	257.20	50.65	405.20		
1 Equip. Oper. (medium)	33.65	269.20	53.05	424.40		
1 Backhoe Loader, 48 H.P.		207.80		228.60		
1 Hyd.Hammer, (1200 lb)		112.00		123.20		
1 F.E. Loader (170 H.P.)		438.40		482.25		
1 Pavt. Rem. Bucket		50.00		55.00	20.21	22.23
40 L.H., Daily Totals		$1974.60		$2779.45	$49.37	$69.49
Crew B-39	**Hr.**	**Daily**	**Hr.**	**Daily**	**Bare Costs**	**Incl. O&P**
1 Labor Foreman (outside)	$28.00	$224.00	$46.40	$371.20	$27.36	$44.91
4 Laborers	26.00	832.00	43.10	1379.20		
1 Equip. Oper. (light)	32.15	257.20	50.65	405.20		
1 Air Compr., 250 C.F.M.		127.60		140.35		
2 Air Tools & Accessories		22.40		24.65		
2-50 Ft. Air Hoses, 1.5" Dia.		10.00		11.00	3.33	3.67
48 L.H., Daily Totals		$1473.20		$2331.60	$30.69	$48.58
Crew B-40	**Hr.**	**Daily**	**Hr.**	**Daily**	**Bare Costs**	**Incl. O&P**
1 Pile Driver Foreman (out)	$34.05	$272.40	$57.95	$463.60	$32.63	$53.98
4 Pile Drivers	32.05	1025.60	54.55	1745.60		
2 Equip. Oper. (crane)	34.80	556.80	54.85	877.60		
1 Equip. Oper. Oiler	29.20	233.60	46.00	368.00		
1 Crane, 40 Ton		915.20		1006.70		
1 Vibratory Hammer & Gen.		1403.00		1543.30	36.22	39.84
64 L.H., Daily Totals		$4406.60		$6004.80	$68.85	$93.82

Crew No.	Bare Costs		Incl. Subs O & P		Cost Per Labor-Hour	
Crew B-41	**Hr.**	**Daily**	**Hr.**	**Daily**	**Bare Costs**	**Incl. O&P**
1 Labor Foreman (outside)	$28.00	$224.00	$46.40	$371.20	$26.91	$44.37
4 Laborers	26.00	832.00	43.10	1379.20		
.25 Equip. Oper. (crane)	34.80	69.60	54.85	109.70		
.25 Equip. Oper. Oiler	29.20	58.40	46.00	92.00		
.25 Crawler Crane, 40 Ton		228.80		251.70	5.20	5.72
44 L.H., Daily Totals		$1412.80		$2203.80	$32.11	$50.09

Crew B-42	**Hr.**	**Daily**	**Hr.**	**Daily**	**Bare Costs**	**Incl. O&P**
1 Labor Foreman (outside)	$28.00	$224.00	$46.40	$371.20	$29.14	$48.60
4 Laborers	26.00	832.00	43.10	1379.20		
1 Equip. Oper. (crane)	34.80	278.40	54.85	438.80		
1 Equip. Oper. Oiler	29.20	233.60	46.00	368.00		
1 Welder	37.15	297.20	69.15	553.20		
1 Hyd. Crane, 25 Ton		624.20		686.60		
1 Gas Welding Machine		75.20		82.70		
1 Horz. Boring Csg. Mch.		381.60		419.75	16.89	18.58
64 L.H., Daily Totals		$2946.20		$4299.45	$46.03	$67.18

Crew B-43	**Hr.**	**Daily**	**Hr.**	**Daily**	**Bare Costs**	**Incl. O&P**
1 Labor Foreman (outside)	$28.00	$224.00	$46.40	$371.20	$28.33	$46.09
3 Laborers	26.00	624.00	43.10	1034.40		
1 Equip. Oper. (crane)	34.80	278.40	54.85	438.80		
1 Equip. Oper. Oiler	29.20	233.60	46.00	368.00		
1 Drill Rig & Augers		2877.00		3164.70	59.94	65.93
48 L.H., Daily Totals		$4237.00		$5377.10	$88.27	$112.02

Crew B-44	**Hr.**	**Daily**	**Hr.**	**Daily**	**Bare Costs**	**Incl. O&P**
1 Pile Driver Foreman	$34.05	$272.40	$57.95	$463.60	$32.23	$53.62
4 Pile Drivers	32.05	1025.60	54.55	1745.60		
2 Equip. Oper. (crane)	34.80	556.80	54.85	877.60		
1 Laborer	26.00	208.00	43.10	344.80		
1 Crane, 40 Ton, & Access.		915.20		1006.70		
45 L.F. Leads, 15K Ft. Lbs.		67.50		74.25	15.35	16.89
64 L.H., Daily Totals		$3045.50		$4512.55	$47.58	$70.51

Crew B-45	**Hr.**	**Daily**	**Hr.**	**Daily**	**Bare Costs**	**Incl. O&P**
1 Equip. Oper. (med.)	$33.65	$269.20	$53.05	$424.40	$30.05	$48.00
1 Truck Driver (heavy)	26.45	211.60	42.95	343.60		
1 Dist. Tank Truck, 3K Gal.		240.40		264.45		
1 Tractor, 4 x 2, 250 H.P.		314.20		345.60	34.66	38.13
16 L.H., Daily Totals		$1035.40		$1378.05	$64.71	$86.13

Crew B-46	**Hr.**	**Daily**	**Hr.**	**Daily**	**Bare Costs**	**Incl. O&P**
1 Pile Driver Foreman	$34.05	$272.40	$57.95	$463.60	$29.36	$49.39
2 Pile Drivers	32.05	512.80	54.55	872.80		
3 Laborers	26.00	624.00	43.10	1034.40		
1 Chain Saw, 36" Long		33.40		36.75	.70	.77
48 L.H., Daily Totals		$1442.60		$2407.55	$30.06	$50.16

Crew B-47	**Hr.**	**Daily**	**Hr.**	**Daily**	**Bare Costs**	**Incl. O&P**
1 Blast Foreman	$28.00	$224.00	$46.40	$371.20	$28.72	$46.72
1 Driller	26.00	208.00	43.10	344.80		
1 Equip. Oper. (light)	32.15	257.20	50.65	405.20		
1 Crawler Type Drill, 4"		634.20		697.60		
1 Air Compr., 600 C.F.M.		286.00		314.60		
2-50 Ft. Air Hoses, 3" Dia.		35.40		38.95	39.82	43.80
24 L.H., Daily Totals		$1644.80		$2172.35	$68.54	$90.52

Crew No.	Bare Costs		Incl. Subs O & P		Cost Per Labor-Hour	
Crew B-47A	**Hr.**	**Daily**	**Hr.**	**Daily**	**Bare Costs**	**Incl. O&P**
1 Drilling Foreman	$28.00	$224.00	$46.40	$371.20	$30.67	$49.08
1 Equip. Oper. (heavy)	34.80	278.40	54.85	438.80		
1 Oiler	29.20	233.60	46.00	368.00		
1 Quarry Drill		827.40		910.15	34.48	37.92
24 L.H., Daily Totals		$1563.40		$2088.15	$65.15	$87.00

Crew B-47C	**Hr.**	**Daily**	**Hr.**	**Daily**	**Bare Costs**	**Incl. O&P**
1 Laborer	$26.00	$208.00	$43.10	$344.80	$29.08	$46.88
1 Equip. Oper. (light)	32.15	257.20	50.65	405.20		
1 Air Compressor, 750 CFM		305.00		335.50		
2-50' Air Hoses, 3"		35.40		38.95		
1 Air Track Drill, 4"		634.20		697.60	60.91	67.00
16 L.H., Daily Totals		$1439.80		$1822.05	$89.99	$113.88

Crew B-47E	**Hr.**	**Daily**	**Hr.**	**Daily**	**Bare Costs**	**Incl. O&P**
1 Laborer Foreman	$28.00	$224.00	$46.40	$371.20	$26.50	$43.93
3 Laborers	26.00	624.00	43.10	1034.40		
1 Truck, Flatbed, 3 Ton		160.40		176.45	5.01	5.51
32 L.H., Daily Totals		$1008.40		$1582.05	$31.51	$49.44

Crew B-48	**Hr.**	**Daily**	**Hr.**	**Daily**	**Bare Costs**	**Incl. O&P**
1 Labor Foreman (outside)	$28.00	$224.00	$46.40	$371.20	$28.88	$46.74
3 Laborers	26.00	624.00	43.10	1034.40		
1 Equip. Oper. (crane)	34.80	278.40	54.85	438.80		
1 Equip. Oper. Oiler	29.20	233.60	46.00	368.00		
1 Equip. Oper. (light)	32.15	257.20	50.65	405.20		
1 Centr. Water Pump, 6"		229.80		252.80		
1-20 Ft. Suction Hose, 6"		14.50		15.95		
1-50 Ft. Disch. Hose, 6"		11.90		13.10		
1 Drill Rig & Augers		2877.00		3164.70	55.95	61.55
56 L.H., Daily Totals		$4750.40		$6064.15	$84.83	$108.29

Crew B-49	**Hr.**	**Daily**	**Hr.**	**Daily**	**Bare Costs**	**Incl. O&P**
1 Labor Foreman (outside)	$28.00	$224.00	$46.40	$371.20	$30.02	$48.83
3 Laborers	26.00	624.00	43.10	1034.40		
2 Equip. Oper. (crane)	34.80	556.80	54.85	877.60		
2 Equip. Oper. Oilers	29.20	467.20	46.00	736.00		
1 Equip. Oper. (light)	32.15	257.20	50.65	405.20		
2 Pile Drivers	32.05	512.80	54.55	872.80		
1 Hyd. Crane, 25 Ton		624.20		686.60		
1 Centr. Water Pump, 6"		229.80		252.80		
1-20 Ft. Suction Hose, 6"		14.50		15.95		
1-50 Ft. Disch. Hose, 6"		11.90		13.10		
1 Drill Rig & Augers		2877.00		3164.70	42.70	46.97
88 L.H., Daily Totals		$6399.40		$8430.35	$72.72	$95.80

Crew B-50	**Hr.**	**Daily**	**Hr.**	**Daily**	**Bare Costs**	**Incl. O&P**
2 Pile Driver Foremen	$34.05	$544.80	$57.95	$927.20	$31.23	$52.01
6 Pile Drivers	32.05	1538.40	54.55	2618.40		
2 Equip. Oper. (crane)	34.80	556.80	54.85	877.60		
1 Equip. Oper. Oiler	29.20	233.60	46.00	368.00		
3 Laborers	26.00	624.00	43.10	1034.40		
1 Crane, 40 Ton		915.20		1006.70		
60 L.F. Leads, 15K Ft. Lbs.		90.00		99.00		
1 Hammer, 15K Ft. Lbs.		359.80		395.80		
1 Air Compr., 600 C.F.M.		286.00		314.60		
2-50 Ft. Air Hoses, 3" Dia.		35.40		38.95		
1 Chain Saw, 36" Long		33.40		36.75	15.36	16.89
112 L.H., Daily Totals		$5217.40		$7717.40	$46.59	$68.90

Crew B-51	Hr.	Daily	Hr.	Daily	Bare Costs	Incl. O&P
1 Labor Foreman (outside)	$28.00	$224.00	$46.40	$371.20	$26.28	$43.42
4 Laborers	26.00	832.00	43.10	1379.20		
1 Truck Driver (light)	25.70	205.60	41.75	334.00		
1 Light Truck, 1.5 Ton		122.00		134.20	2.54	2.80
48 L.H., Daily Totals		$1383.60		$2218.60	$28.82	$46.22

Crew B-52	Hr.	Daily	Hr.	Daily	Bare Costs	Incl. O&P
1 Carpenter Foreman	$35.00	$280.00	$58.05	$464.40	$30.42	$50.01
1 Carpenter	33.00	264.00	54.70	437.60		
3 Laborers	26.00	624.00	43.10	1034.40		
1 Cement Finisher	31.55	252.40	49.60	396.80		
.5 Rodman (reinf.)	37.10	148.40	63.85	255.40		
.5 Equip. Oper. (med.)	33.65	134.60	53.05	212.20		
.5 F.E. Ldr., T.M., 2.5 C.Y.		380.50		418.55	6.79	7.47
56 L.H., Daily Totals		$2083.90		$3219.35	$37.21	$57.48

Crew B-53	Hr.	Daily	Hr.	Daily	Bare Costs	Incl. O&P
1 Equip. Oper. (light)	$32.15	$257.20	$50.65	$405.20	$32.15	$50.65
1 Trencher, Chain, 12 H.P.		59.40		65.35	7.43	8.17
8 L.H., Daily Totals		$316.60		$470.55	$39.58	$58.82

Crew B-54	Hr.	Daily	Hr.	Daily	Bare Costs	Incl. O&P
1 Equip. Oper. (light)	$32.15	$257.20	$50.65	$405.20	$32.15	$50.65
1 Trencher, Chain, 40 H.P.		208.60		229.45	26.08	28.68
8 L.H., Daily Totals		$465.80		$634.65	$58.23	$79.33

Crew B-54A	Hr.	Daily	Hr.	Daily	Bare Costs	Incl. O&P
.17 Labor Foreman (outside)	$28.00	$38.08	$46.40	$63.10	$32.83	$52.08
1 Equipment Operator (med.)	33.65	269.20	53.05	424.40		
1 Wheel Trencher, 67 H.P.		767.40		844.15	81.99	90.19
9.36 L.H., Daily Totals		$1074.68		$1331.65	$114.82	$142.27

Crew B-54B	Hr.	Daily	Hr.	Daily	Bare Costs	Incl. O&P
.25 Labor Foreman (outside)	$28.00	$56.00	$46.40	$92.80	$32.52	$51.72
1 Equipment Operator (med.)	33.65	269.20	53.05	424.40		
1 Wheel Trencher, 150 H.P.		1326.00		1458.60	132.60	145.86
10 L.H., Daily Totals		$1651.20		$1975.80	$165.12	$197.58

Crew B-55	Hr.	Daily	Hr.	Daily	Bare Costs	Incl. O&P
2 Laborers	$26.00	$416.00	$43.10	$689.60	$25.90	$42.65
1 Truck Driver (light)	25.70	205.60	41.75	334.00		
1 Auger, 4" to 36" Dia		614.80		676.30		
1 Flatbed 3 Ton Truck		160.40		176.45	32.30	35.53
24 L.H., Daily Totals		$1396.80		$1876.35	$58.20	$78.18

Crew B-56	Hr.	Daily	Hr.	Daily	Bare Costs	Incl. O&P
1 Laborer	$26.00	$208.00	$43.10	$344.80	$29.08	$46.88
1 Equip. Oper. (light)	32.15	257.20	50.65	405.20		
1 Crawler Type Drill, 4"		634.20		697.60		
1 Air Compr., 600 C.F.M.		286.00		314.60		
1-50 Ft. Air Hose, 3" Dia.		17.70		19.45	58.62	64.48
16 L.H., Daily Totals		$1403.10		$1781.65	$87.70	$111.36

Crew B-57	Hr.	Daily	Hr.	Daily	Bare Costs	Incl. O&P
1 Labor Foreman (outside)	$28.00	$224.00	$46.40	$371.20	$29.36	$47.35
2 Laborers	26.00	416.00	43.10	689.60		
1 Equip. Oper. (crane)	34.80	278.40	54.85	438.80		
1 Equip. Oper. (light)	32.15	257.20	50.65	405.20		
1 Equip. Oper. Oiler	29.20	233.60	46.00	368.00		
1 Power Shovel, 1 C.Y.		792.00		871.20		
1 Clamshell Bucket, 1 C.Y.		44.00		48.40		
1 Centr. Water Pump, 6"		229.80		252.80		
1-20 Ft. Suction Hose, 6"		14.50		15.95		
20-50 Ft. Disch. Hoses, 6"		238.00		261.80	27.46	30.21
48 L.H., Daily Totals		$2727.50		$3722.95	$56.82	$77.56

Crew B-58	Hr.	Daily	Hr.	Daily	Bare Costs	Incl. O&P
2 Laborers	$26.00	$416.00	$43.10	$689.60	$28.05	$45.62
1 Equip. Oper. (light)	32.15	257.20	50.65	405.20		
1 Backhoe Loader, 48 H.P.		207.80		228.60		
1 Small Helicopter, w/pilot		2108.00		2318.80	96.49	106.14
24 L.H., Daily Totals		$2989.00		$3642.20	$124.54	$151.76

Crew B-59	Hr.	Daily	Hr.	Daily	Bare Costs	Incl. O&P
1 Truck Driver (heavy)	$26.45	$211.60	$42.95	$343.60	$26.45	$42.95
1 Truck, 30 Ton		215.20		236.70		
1 Water tank, 6000 Gal.		118.00		129.80	41.65	45.82
8 L.H., Daily Totals		$544.80		$710.10	$68.10	$88.77

Crew B-60	Hr.	Daily	Hr.	Daily	Bare Costs	Incl. O&P
1 Labor Foreman (outside)	$28.00	$224.00	$46.40	$371.20	$29.76	$47.82
2 Laborers	26.00	416.00	43.10	689.60		
1 Equip. Oper. (crane)	34.80	278.40	54.85	438.80		
2 Equip. Oper. (light)	32.15	514.40	50.65	810.40		
1 Equip. Oper. Oiler	29.20	233.60	46.00	368.00		
1 Crawler Crane, 40 Ton		915.20		1006.70		
45 L.F. Leads, 15K Ft. Lbs.		67.50		74.25		
1 Backhoe Loader, 48 H.P.		207.80		228.60	21.26	23.38
56 L.H., Daily Totals		$2856.90		$3987.55	$51.02	$71.20

Crew B-61	Hr.	Daily	Hr.	Daily	Bare Costs	Incl. O&P
1 Labor Foreman (outside)	$28.00	$224.00	$46.40	$371.20	$27.63	$45.27
3 Laborers	26.00	624.00	43.10	1034.40		
1 Equip. Oper. (light)	32.15	257.20	50.65	405.20		
1 Cement Mixer, 2 C.Y.		161.00		177.10		
1 Air Compr., 160 C.F.M.		82.00		90.20	6.08	6.68
40 L.H., Daily Totals		$1348.20		$2078.10	$33.71	$51.95

Crew B-62	Hr.	Daily	Hr.	Daily	Bare Costs	Incl. O&P
2 Laborers	$26.00	$416.00	$43.10	$689.60	$28.05	$45.62
1 Equip. Oper. (light)	32.15	257.20	50.65	405.20		
1 Loader, Skid Steer		143.20		157.50	5.97	6.56
24 L.H., Daily Totals		$816.40		$1252.30	$34.02	$52.18

Crew B-63	Hr.	Daily	Hr.	Daily	Bare Costs	Incl. O&P
4 Laborers	$26.00	$832.00	$43.10	$1379.20	$27.23	$44.61
1 Equip. Oper. (light)	32.15	257.20	50.65	405.20		
1 Loader, Skid Steer		143.20		157.50	3.58	3.94
40 L.H., Daily Totals		$1232.40		$1941.90	$30.81	$48.55

Crew No.	Bare Costs		Incl. Subs O & P		Cost Per Labor-Hour	
Crew B-64	**Hr.**	**Daily**	**Hr.**	**Daily**	**Bare Costs**	**Incl. O&P**
1 Laborer	$26.00	$208.00	$43.10	$344.80	$25.85	$42.42
1 Truck Driver (light)	25.70	205.60	41.75	334.00		
1 Power Mulcher (small)		103.20		113.50		
1 Light Truck, 1.5 Ton		122.00		134.20	14.08	15.48
16 L.H., Daily Totals		$638.80		$926.50	$39.93	$57.90
Crew B-65	**Hr.**	**Daily**	**Hr.**	**Daily**	**Bare Costs**	**Incl. O&P**
1 Laborer	$26.00	$208.00	$43.10	$344.80	$25.85	$42.42
1 Truck Driver (light)	25.70	205.60	41.75	334.00		
1 Power Mulcher (large)		202.20		222.40		
1 Light Truck, 1.5 Ton		122.00		134.20	20.26	22.29
16 L.H., Daily Totals		$737.80		$1035.40	$46.11	$64.71
Crew B-66	**Hr.**	**Daily**	**Hr.**	**Daily**	**Bare Costs**	**Incl. O&P**
1 Equip. Oper. (light)	$32.15	$257.20	$50.65	$405.20	$32.15	$50.65
1 Backhoe Ldr. w/Attchmt.		168.40		185.25	21.05	23.16
8 L.H., Daily Totals		$425.60		$590.45	$53.20	$73.81
Crew B-67	**Hr.**	**Daily**	**Hr.**	**Daily**	**Bare Costs**	**Incl. O&P**
1 Millwright	$34.35	$274.80	$54.15	$433.20	$33.25	$52.40
1 Equip. Oper. (light)	32.15	257.20	50.65	405.20		
1 Forklift		239.80		263.80	14.99	16.49
16 L.H., Daily Totals		$771.80		$1102.20	$48.24	$68.89
Crew B-68	**Hr.**	**Daily**	**Hr.**	**Daily**	**Bare Costs**	**Incl. O&P**
2 Millwrights	$34.35	$549.60	$54.15	$866.40	$33.62	$52.98
1 Equip. Oper. (light)	32.15	257.20	50.65	405.20		
1 Forklift		239.80		263.80	9.99	10.99
24 L.H., Daily Totals		$1046.60		$1535.40	$43.61	$63.97
Crew B-69	**Hr.**	**Daily**	**Hr.**	**Daily**	**Bare Costs**	**Incl. O&P**
1 Labor Foreman (outside)	$28.00	$224.00	$46.40	$371.20	$28.33	$46.09
3 Laborers	26.00	624.00	43.10	1034.40		
1 Equip Oper. (crane)	34.80	278.40	54.85	438.80		
1 Equip Oper. Oiler	29.20	233.60	46.00	368.00		
1 Truck Crane, 80 Ton		1020.00		1122.00	21.25	23.38
48 L.H., Daily Totals		$2380.00		$3334.40	$49.58	$69.47
Crew B-69A	**Hr.**	**Daily**	**Hr.**	**Daily**	**Bare Costs**	**Incl. O&P**
1 Labor Foreman	$28.00	$224.00	$46.40	$371.20	$28.53	$46.39
3 Laborers	26.00	624.00	43.10	1034.40		
1 Equip. Oper. (medium)	33.65	269.20	53.05	424.40		
1 Concrete Finisher	31.55	252.40	49.60	396.80		
1 Curb Paver		573.60		630.95	11.95	13.15
48 L.H., Daily Totals		$1943.20		$2857.75	$40.48	$59.54
Crew B-69B	**Hr.**	**Daily**	**Hr.**	**Daily**	**Bare Costs**	**Incl. O&P**
1 Labor Foreman	$28.00	$224.00	$46.40	$371.20	$28.53	$46.39
3 Laborers	26.00	624.00	43.10	1034.40		
1 Equip. Oper. (medium)	33.65	269.20	53.05	424.40		
1 Cement Finisher	31.55	252.40	49.60	396.80		
1 Curb/Gutter Paver		687.40		756.15	14.32	15.75
48 L.H., Daily Totals		$2057.00		$2982.95	$42.85	$62.14

Crew No.	Bare Costs		Incl. Subs O & P		Cost Per Labor-Hour	
Crew B-70	**Hr.**	**Daily**	**Hr.**	**Daily**	**Bare Costs**	**Incl. O&P**
1 Labor Foreman (outside)	$28.00	$224.00	$46.40	$371.20	$29.56	$47.84
3 Laborers	26.00	624.00	43.10	1034.40		
3 Equip. Oper. (med.)	33.65	807.60	53.05	1273.20		
1 Motor Grader, 30,000 Lb.		432.00		475.20		
1 Grader Attach., Ripper		70.60		77.65		
1 Road Sweeper, S.P.		420.00		462.00		
1 F.E. Loader, 1-3/4 C.Y.		236.20		259.80	20.69	22.76
56 L.H., Daily Totals		$2814.40		$3953.45	$50.25	$70.60
Crew B-71	**Hr.**	**Daily**	**Hr.**	**Daily**	**Bare Costs**	**Incl. O&P**
1 Labor Foreman (outside)	$28.00	$224.00	$46.40	$371.20	$29.56	$47.84
3 Laborers	26.00	624.00	43.10	1034.40		
3 Equip. Oper. (med.)	33.65	807.60	53.05	1273.20		
1 Pvmt. Profiler, 750 H.P.		4351.00		4786.10		
1 Road Sweeper, S.P.		420.00		462.00		
1 F.E. Loader, 1-3/4 C.Y.		236.20		259.80	89.41	98.36
56 L.H., Daily Totals		$6662.80		$8186.70	$118.97	$146.20
Crew B-72	**Hr.**	**Daily**	**Hr.**	**Daily**	**Bare Costs**	**Incl. O&P**
1 Labor Foreman (outside)	$28.00	$224.00	$46.40	$371.20	$30.08	$48.49
3 Laborers	26.00	624.00	43.10	1034.40		
4 Equip. Oper. (med.)	33.65	1076.80	53.05	1697.60		
1 Pvmt. Profiler, 750 H.P.		4351.00		4786.10		
1 Hammermill, 250 H.P.		1315.00		1446.50		
1 Windrow Loader		805.60		886.15		
1 Mix Paver 165 H.P.		1674.00		1841.40		
1 Roller, Pneu. Tire, 12 T.		244.60		269.05	131.10	144.21
64 L.H., Daily Totals		$10315.00		$12332.40	$161.18	$192.70
Crew B-73	**Hr.**	**Daily**	**Hr.**	**Daily**	**Bare Costs**	**Incl. O&P**
1 Labor Foreman (outside)	$28.00	$224.00	$46.40	$371.20	$31.03	$49.73
2 Laborers	26.00	416.00	43.10	689.60		
5 Equip. Oper. (med.)	33.65	1346.00	53.05	2122.00		
1 Road Mixer, 310 H.P.		1610.00		1771.00		
1 Roller, Tandem, 12 Ton		183.60		201.95		
1 Hammermill, 250 H.P.		1315.00		1446.50		
1 Motor Grader, 30,000 Lb.		432.00		475.20		
.5 F.E. Loader, 1-3/4 C.Y.		118.10		129.90		
.5 Truck, 30 Ton		107.60		118.35		
.5 Water Tank 5000 Gal.		59.00		64.90	59.77	65.75
64 L.H., Daily Totals		$5811.30		$7390.60	$90.80	$115.48
Crew B-74	**Hr.**	**Daily**	**Hr.**	**Daily**	**Bare Costs**	**Incl. O&P**
1 Labor Foreman (outside)	$28.00	$224.00	$46.40	$371.20	$30.19	$48.45
1 Laborer	26.00	208.00	43.10	344.80		
4 Equip. Oper. (med.)	33.65	1076.80	53.05	1697.60		
2 Truck Drivers (heavy)	26.45	423.20	42.95	687.20		
1 Motor Grader, 30,000 Lb.		432.00		475.20		
1 Grader Attach., Ripper		70.60		77.65		
2 Stabilizers, 310 H.P.		2122.00		2334.20		
1 Flatbed Truck, 3 Ton		160.40		176.45		
1 Chem. Spreader, Towed		66.00		72.60		
1 Vibr. Roller, 29,000 Lb.		436.80		480.50		
1 Water Tank 5000 Gal.		118.00		129.80		
1 Truck, 30 Ton		215.20		236.70	56.58	62.24
64 L.H., Daily Totals		$5553.00		$7083.90	$86.77	$110.69

Crew No.	Bare Costs		Incl. Subs O & P		Cost Per Labor-Hour	
Crew B-75	**Hr.**	**Daily**	**Hr.**	**Daily**	**Bare Costs**	**Incl. O&P**
1 Labor Foreman (outside)	$28.00	$224.00	$46.40	$371.20	$30.72	$49.24
1 Laborer	26.00	208.00	43.10	344.80		
4 Equip. Oper. (med.)	33.65	1076.80	53.05	1697.60		
1 Truck Driver (heavy)	26.45	211.60	42.95	343.60		
1 Motor Grader, 30,000 Lb.		432.00		475.20		
1 Grader Attach., Ripper		70.60		77.65		
2 Stabilizers, 310 H.P.		2122.00		2334.20		
1 Dist. Truck, 3000 Gal.		240.40		264.45		
1 Vibr. Roller, 29,000 Lb.		436.80		480.50	58.96	64.86
56 L.H., Daily Totals		$5022.20		$6389.20	$89.68	$114.10

Crew No.	Bare Costs		Incl. Subs O & P		Cost Per Labor-Hour	
Crew B-76	**Hr.**	**Daily**	**Hr.**	**Daily**	**Bare Costs**	**Incl. O&P**
1 Dock Builder Foreman	$34.05	$272.40	$57.95	$463.60	$32.57	$54.04
5 Dock Builders	32.05	1282.00	54.55	2182.00		
2 Equip. Oper. (crane)	34.80	556.80	54.85	877.60		
1 Equip. Oper. Oiler	29.20	233.60	46.00	368.00		
1 Crawler Crane, 50 Ton		1260.00		1386.00		
1 Barge, 400 Ton		273.20		300.50		
1 Hammer, 15K Ft. Lbs.		359.80		395.80		
60 L.F. Leads, 15K Ft. Lbs.		90.00		99.00		
1 Air Compr., 600 C.F.M.		286.00		314.60		
2-50 Ft. Air Hoses, 3" Dia.		35.40		38.95	32.01	35.21
72 L.H., Daily Totals		$4649.20		$6426.05	$64.58	$89.25

Crew No.	Bare Costs		Incl. Subs O & P		Cost Per Labor-Hour	
Crew B-77	**Hr.**	**Daily**	**Hr.**	**Daily**	**Bare Costs**	**Incl. O&P**
1 Labor Foreman	$28.00	$224.00	$46.40	$371.20	$26.34	$43.49
3 Laborers	26.00	624.00	43.10	1034.40		
1 Truck Driver (light)	25.70	205.60	41.75	334.00		
1 Crack Cleaner, 25 H.P.		44.40		48.85		
1 Crack Filler, Trailer Mtd.		157.00		172.70		
1 Flatbed Truck, 3 Ton		160.40		176.45	9.05	9.95
40 L.H., Daily Totals		$1415.40		$2137.60	$35.39	$53.44

Crew No.	Bare Costs		Incl. Subs O & P		Cost Per Labor-Hour	
Crew B-78	**Hr.**	**Daily**	**Hr.**	**Daily**	**Bare Costs**	**Incl. O&P**
1 Labor Foreman	$28.00	$224.00	$46.40	$371.20	$26.28	$43.42
4 Laborers	26.00	832.00	43.10	1379.20		
1 Truck Driver (light)	25.70	205.60	41.75	334.00		
1 Paint Striper, S.P.		134.20		147.60		
1 Flatbed Truck, 3 Ton		160.40		176.45		
1 Pickup Truck, 3/4 Ton		75.80		83.40	7.72	8.49
48 L.H., Daily Totals		$1632.00		$2491.85	$34.00	$51.91

Crew No.	Bare Costs		Incl. Subs O & P		Cost Per Labor-Hour	
Crew B-79	**Hr.**	**Daily**	**Hr.**	**Daily**	**Bare Costs**	**Incl. O&P**
1 Labor Foreman	$28.00	$224.00	$46.40	$371.20	$26.34	$43.49
3 Laborers	26.00	624.00	43.10	1034.40		
1 Truck Driver (light)	25.70	205.60	41.75	334.00		
1 Thermo. Striper, T.M.		549.20		604.10		
1 Flatbed Truck, 3 Ton		160.40		176.45		
2 Pickup Trucks, 3/4 Ton		151.60		166.75	21.53	23.68
40 L.H., Daily Totals		$1914.80		$2686.90	$47.87	$67.17

Crew No.	Bare Costs		Incl. Subs O & P		Cost Per Labor-Hour	
Crew B-80	**Hr.**	**Daily**	**Hr.**	**Daily**	**Bare Costs**	**Incl. O&P**
1 Labor Foreman	$28.00	$224.00	$46.40	$371.20	$27.96	$45.48
1 Laborer	26.00	208.00	43.10	344.80		
1 Truck Driver (light)	25.70	205.60	41.75	334.00		
1 Equip. Oper. (light)	32.15	257.20	50.65	405.20		
1 Flatbed Truck, 3 Ton		160.40		176.45		
1 Fence Post Auger, T.M.		357.40		393.15	16.18	17.80
32 L.H., Daily Totals		$1412.60		$2024.80	$44.14	$63.28

Crew No.	Bare Costs		Incl. Subs O & P		Cost Per Labor-Hour	
Crew B-80A	**Hr.**	**Daily**	**Hr.**	**Daily**	**Bare Costs**	**Incl. O&P**
3 Laborers	$26.00	$624.00	$43.10	$1034.40	$26.00	$43.10
1 Flatbed Truck, 3 Ton		160.40		176.45	6.68	7.35
24 L.H., Daily Totals		$784.40		$1210.85	$32.68	$50.45

Crew No.	Bare Costs		Incl. Subs O & P		Cost Per Labor-Hour	
Crew B-80B	**Hr.**	**Daily**	**Hr.**	**Daily**	**Bare Costs**	**Incl. O&P**
3 Laborers	$26.00	$624.00	$43.10	$1034.40	$27.54	$44.99
1 Equip. Oper. (light)	32.15	257.20	50.65	405.20		
1 Crane, Flatbed Mnt.		201.80		222.00	6.31	6.94
32 L.H., Daily Totals		$1083.00		$1661.60	$33.85	$51.93

Crew No.	Bare Costs		Incl. Subs O & P		Cost Per Labor-Hour	
Crew B-81	**Hr.**	**Daily**	**Hr.**	**Daily**	**Bare Costs**	**Incl. O&P**
1 Laborer	$26.00	$208.00	$43.10	$344.80	$28.70	$46.37
1 Equip. Oper. (med.)	33.65	269.20	53.05	424.40		
1 Truck Driver (heavy)	26.45	211.60	42.95	343.60		
1 Hydromulcher, T.M.		196.00		215.60		
1 Tractor Truck, 4x2		215.20		236.70	17.13	18.85
24 L.H., Daily Totals		$1100.00		$1565.10	$45.83	$65.22

Crew No.	Bare Costs		Incl. Subs O & P		Cost Per Labor-Hour	
Crew B-82	**Hr.**	**Daily**	**Hr.**	**Daily**	**Bare Costs**	**Incl. O&P**
1 Laborer	$26.00	$208.00	$43.10	$344.80	$29.08	$46.88
1 Equip. Oper. (light)	32.15	257.20	50.65	405.20		
1 Horiz. Borer, 6 H.P.		62.80		69.10	3.93	4.32
16 L.H., Daily Totals		$528.00		$819.10	$33.01	$51.20

Crew No.	Bare Costs		Incl. Subs O & P		Cost Per Labor-Hour	
Crew B-83	**Hr.**	**Daily**	**Hr.**	**Daily**	**Bare Costs**	**Incl. O&P**
1 Tugboat Captain	$33.65	$269.20	$53.05	$424.40	$29.83	$48.08
1 Tugboat Hand	26.00	208.00	43.10	344.80		
1 Tugboat, 250 H.P.		440.60		484.65	27.54	30.29
16 L.H., Daily Totals		$917.80		$1253.85	$57.37	$78.37

Crew No.	Bare Costs		Incl. Subs O & P		Cost Per Labor-Hour	
Crew B-84	**Hr.**	**Daily**	**Hr.**	**Daily**	**Bare Costs**	**Incl. O&P**
1 Equip. Oper. (med.)	$33.65	$269.20	$53.05	$424.40	$33.65	$53.05
1 Rotary Mower/Tractor		222.20		244.40	27.78	30.55
8 L.H., Daily Totals		$491.40		$668.80	$61.43	$83.60

Crew No.	Bare Costs		Incl. Subs O & P		Cost Per Labor-Hour	
Crew B-85	**Hr.**	**Daily**	**Hr.**	**Daily**	**Bare Costs**	**Incl. O&P**
3 Laborers	$26.00	$624.00	$43.10	$1034.40	$27.62	$45.06
1 Equip. Oper. (med.)	33.65	269.20	53.05	424.40		
1 Truck Driver (heavy)	26.45	211.60	42.95	343.60		
1 Aerial Lift Truck, 80'		501.80		552.00		
1 Brush Chipper, 130 H.P.		164.80		181.30		
1 Pruning Saw, Rotary		10.55		11.60	16.93	18.62
40 L.H., Daily Totals		$1781.95		$2547.30	$44.55	$63.68

Crew No.	Bare Costs		Incl. Subs O & P		Cost Per Labor-Hour	
Crew B-86	**Hr.**	**Daily**	**Hr.**	**Daily**	**Bare Costs**	**Incl. O&P**
1 Equip. Oper. (med.)	$33.65	$269.20	$53.05	$424.40	$33.65	$53.05
1 Stump Chipper, S.P.		87.20		95.90	10.90	11.99
8 L.H., Daily Totals		$356.40		$520.30	$44.55	$65.04

Crew No.	Bare Costs		Incl. Subs O & P		Cost Per Labor-Hour	
Crew B-86A	**Hr.**	**Daily**	**Hr.**	**Daily**	**Bare Costs**	**Incl. O&P**
1 Equip. Oper. (medium)	$33.65	$269.20	$53.05	$424.40	$33.65	$53.05
1 Grader, 30,000 Lbs.		432.00		475.20	54.00	59.40
8 L.H., Daily Totals		$701.20		$899.60	$87.65	$112.45

Crew No.	Bare Costs		Incl. Subs O & P		Cost Per Labor-Hour	
Crew B-86B	**Hr.**	**Daily**	**Hr.**	**Daily**	**Bare Costs**	**Incl. O&P**
1 Equip. Oper. (medium)	$33.65	$269.20	$53.05	$424.40	$33.65	$53.05
1 Dozer, 200 H.P.		863.40		949.75	107.93	118.72
8 L.H., Daily Totals		$1132.60		$1374.15	$141.58	$171.77

Crews

Crew No.	Bare Costs		Incl. Subs O & P		Cost Per Labor-Hour	

Crew B-87	Hr.	Daily	Hr.	Daily	Bare Costs	Incl. O&P
1 Laborer	$26.00	$208.00	$43.10	$344.80	$32.12	$51.06
4 Equip. Oper. (med.)	33.65	1076.80	53.05	1697.60		
2 Feller Bunchers, 50 H.P.		864.80		951.30		
1 Log Chipper, 22″ Tree		1207.00		1327.70		
1 Dozer, 105 H.P.		434.00		477.40		
1 Chainsaw, Gas, 36″ Long		33.40		36.75	63.48	69.83
40 L.H., Daily Totals		$3824.00		$4835.55	$95.60	$120.89

Crew B-88	Hr.	Daily	Hr.	Daily	Bare Costs	Incl. O&P
1 Laborer	$26.00	$208.00	$43.10	$344.80	$32.56	$51.63
6 Equip. Oper. (med.)	33.65	1615.20	53.05	2546.40		
2 Feller Bunchers, 50 H.P.		864.80		951.30		
1 Log Chipper, 22″ Tree		1207.00		1327.70		
2 Log Skidders, 50 H.P.		1490.80		1639.90		
1 Dozer, 105 H.P.		434.00		477.40		
1 Chainsaw, Gas, 36″ Long		33.40		36.75	71.96	79.16
56 L.H., Daily Totals		$5853.20		$7324.25	$104.52	$130.79

Crew B-89	Hr.	Daily	Hr.	Daily	Bare Costs	Incl. O&P
1 Equip. Oper. (light)	$32.15	$257.20	$50.65	$405.20	$28.93	$46.20
1 Truck Driver (light)	25.70	205.60	41.75	334.00		
1 Truck, Stake Body, 3 Ton		160.40		176.45		
1 Concrete Saw		92.20		101.40		
1 Water Tank, 65 Gal.		13.60		14.95	16.64	18.30
16 L.H., Daily Totals		$729.00		$1032.00	$45.57	$64.50

Crew B-89A	Hr.	Daily	Hr.	Daily	Bare Costs	Incl. O&P
1 Skilled Worker	$33.65	$269.20	$55.00	$440.00	$29.83	$49.05
1 Laborer	26.00	208.00	43.10	344.80		
1 Core Drill (large)		106.00		116.60	6.63	7.29
16 L.H., Daily Totals		$583.20		$901.40	$36.46	$56.34

Crew B-89B	Hr.	Daily	Hr.	Daily	Bare Costs	Incl. O&P
1 Equip. Oper. (light)	$32.15	$257.20	$50.65	$405.20	$28.93	$46.20
1 Truck Driver, Light	25.70	205.60	41.75	334.00		
1 Wall Saw, Hydraulic, 10 H.P.		75.10		82.60		
1 Generator, Diesel, 100 KW		179.60		197.55		
1 Water Tank, 65 Gal.		13.60		14.95		
1 Flatbed Truck, 3 Ton		160.40		176.45	26.79	29.47
16 L.H., Daily Totals		$891.50		$1210.75	$55.72	$75.67

Crew B-90	Hr.	Daily	Hr.	Daily	Bare Costs	Incl. O&P
1 Labor Foreman (outside)	$28.00	$224.00	$46.40	$371.20	$27.90	$45.36
3 Laborers	26.00	624.00	43.10	1034.40		
2 Equip. Oper. (light)	32.15	514.40	50.65	810.40		
2 Truck Drivers (heavy)	26.45	423.20	42.95	687.20		
1 Road Mixer, 310 H.P.		1610.00		1771.00		
1 Dist. Truck, 2000 Gal.		208.00		228.80	28.41	31.25
64 L.H., Daily Totals		$3603.60		$4903.00	$56.31	$76.61

Crew B-90A	Hr.	Daily	Hr.	Daily	Bare Costs	Incl. O&P
1 Labor Foreman	$28.00	$224.00	$46.40	$371.20	$30.66	$49.26
2 Laborers	26.00	416.00	43.10	689.60		
4 Equip. Oper. (medium)	33.65	1076.80	53.05	1697.60		
2 Graders, 30,000 Lbs.		864.00		950.40		
1 Roller, Steel Wheel		183.60		201.95		
1 Roller, Pneumatic Wheel		244.60		269.05	23.08	25.38
56 L.H., Daily Totals		$3009.00		$4179.80	$53.74	$74.64

Crew B-90B	Hr.	Daily	Hr.	Daily	Bare Costs	Incl. O&P
1 Labor Foreman	$28.00	$224.00	$46.40	$371.20	$30.16	$48.63
2 Laborers	26.00	416.00	43.10	689.60		
3 Equip. Oper. (medium)	33.65	807.60	53.05	1273.20		
1 Roller, Steel Wheel		183.60		201.95		
1 Roller, Pneumatic Wheel		244.60		269.05		
1 Road Mixer, 310 H.P.		1610.00		1771.00	42.46	46.71
48 L.H., Daily Totals		$3485.80		$4576.00	$72.62	$95.34

Crew B-91	Hr.	Daily	Hr.	Daily	Bare Costs	Incl. O&P
1 Labor Foreman (outside)	$28.00	$224.00	$46.40	$371.20	$30.13	$48.47
2 Laborers	26.00	416.00	43.10	689.60		
4 Equip. Oper. (med.)	33.65	1076.80	53.05	1697.60		
1 Truck Driver (heavy)	26.45	211.60	42.95	343.60		
1 Dist. Truck, 3000 Gal.		240.40		264.45		
1 Aggreg. Spreader, S.P.		771.20		848.30		
1 Roller, Pneu. Tire, 12 Ton		244.60		269.05		
1 Roller, Steel, 10 Ton		183.60		201.95	22.50	24.75
64 L.H., Daily Totals		$3368.20		$4685.75	$52.63	$73.22

Crew B-92	Hr.	Daily	Hr.	Daily	Bare Costs	Incl. O&P
1 Labor Foreman (outside)	$28.00	$224.00	$46.40	$371.20	$26.50	$43.93
3 Laborers	26.00	624.00	43.10	1034.40		
1 Crack Cleaner, 25 H.P.		44.40		48.85		
1 Air Compressor		52.40		57.65		
1 Tar Kettle, T.M.		40.95		45.05		
1 Flatbed Truck, 3 Ton		160.40		176.45	9.32	10.25
32 L.H., Daily Totals		$1146.15		$1733.60	$35.82	$54.18

Crew B-93	Hr.	Daily	Hr.	Daily	Bare Costs	Incl. O&P
1 Equip. Oper. (med.)	$33.65	$269.20	$53.05	$424.40	$33.65	$53.05
1 Feller Buncher, 50 H.P.		432.40		475.65	54.05	59.46
8 L.H., Daily Totals		$701.60		$900.05	$87.70	$112.51

Crew B-94A	Hr.	Daily	Hr.	Daily	Bare Costs	Incl. O&P
1 Laborer	$26.00	$208.00	$43.10	$344.80	$26.00	$43.10
1 Diaph. Water Pump, 2″		46.20		50.80		
1-20 Ft. Suction Hose, 2″		3.55		3.90		
2-50 Ft. Disch. Hoses, 2″		6.30		6.95	7.01	7.71
8 L.H., Daily Totals		$264.05		$406.45	$33.01	$50.81

Crew B-94B	Hr.	Daily	Hr.	Daily	Bare Costs	Incl. O&P
1 Laborer	$26.00	$208.00	$43.10	$344.80	$26.00	$43.10
1 Diaph. Water Pump, 4″		87.40		96.15		
1-20 Ft. Suction Hose, 4″		7.30		8.05		
2-50 Ft. Disch. Hoses, 4″		10.50		11.55	13.15	14.47
8 L.H., Daily Totals		$313.20		$460.55	$39.15	$57.57

Crew B-94C	Hr.	Daily	Hr.	Daily	Bare Costs	Incl. O&P
1 Laborer	$26.00	$208.00	$43.10	$344.80	$26.00	$43.10
1 Centr. Water Pump, 3″		53.80		59.20		
1-20 Ft. Suction Hose, 3″		5.45		6.00		
2-50 Ft. Disch. Hoses, 3″		7.50		8.25	8.34	9.18
8 L.H., Daily Totals		$274.75		$418.25	$34.34	$52.28

Crew B-94D	Hr.	Daily	Hr.	Daily	Bare Costs	Incl. O&P
1 Laborer	$26.00	$208.00	$43.10	$344.80	$26.00	$43.10
1 Centr. Water Pump, 6″		229.80		252.80		
1-20 Ft. Suction Hose, 6″		14.50		15.95		
2-50 Ft. Disch. Hoses, 6″		23.80		26.20	33.51	36.86
8 L.H., Daily Totals		$476.10		$639.75	$59.51	$79.96

Crew No.	Bare Costs		Incl. Subs O & P		Cost Per Labor-Hour	
Crew B-95A	**Hr.**	**Daily**	**Hr.**	**Daily**	**Bare Costs**	**Incl. O&P**
1 Equip. Oper. (crane)	$34.80	$278.40	$54.85	$438.80	$30.40	$48.97
1 Laborer	26.00	208.00	43.10	344.80		
1 Hyd. Excavator, 5/8 C.Y.		418.00		459.80	26.13	28.74
16 L.H., Daily Totals		$904.40		$1243.40	$56.53	$77.71

Crew B-95B	Hr.	Daily	Hr.	Daily	Bare Costs	Incl. O&P
1 Equip. Oper. (crane)	$34.80	$278.40	$54.85	$438.80	$30.40	$48.97
1 Laborer	26.00	208.00	43.10	344.80		
1 Hyd. Excavator, 1.5 C.Y.		683.40		751.75	42.71	46.98
16 L.H., Daily Totals		$1169.80		$1535.35	$73.11	$95.95

Crew B-95C	Hr.	Daily	Hr.	Daily	Bare Costs	Incl. O&P
1 Equip. Oper. (crane)	$34.80	$278.40	$54.85	$438.80	$30.40	$48.97
1 Laborer	26.00	208.00	43.10	344.80		
1 Hyd. Excavator, 2.5 C.Y.		1181.00		1299.10	73.81	81.19
16 L.H., Daily Totals		$1667.40		$2082.70	$104.21	$130.16

Crew C-1	Hr.	Daily	Hr.	Daily	Bare Costs	Incl. O&P
3 Carpenters	$33.00	$792.00	$54.70	$1312.80	$31.25	$51.80
1 Laborer	26.00	208.00	43.10	344.80		
32 L.H., Daily Totals		$1000.00		$1657.60	$31.25	$51.80

Crew C-2	Hr.	Daily	Hr.	Daily	Bare Costs	Incl. O&P
1 Carpenter Foreman (out)	$35.00	$280.00	$58.05	$464.40	$32.17	$53.33
4 Carpenters	33.00	1056.00	54.70	1750.40		
1 Laborer	26.00	208.00	43.10	344.80		
48 L.H., Daily Totals		$1544.00		$2559.60	$32.17	$53.33

Crew C-2A	Hr.	Daily	Hr.	Daily	Bare Costs	Incl. O&P
1 Carpenter Foreman (out)	$35.00	$280.00	$58.05	$464.40	$31.93	$52.48
3 Carpenters	33.00	792.00	54.70	1312.80		
1 Cement Finisher	31.55	252.40	49.60	396.80		
1 Laborer	26.00	208.00	43.10	344.80		
48 L.H., Daily Totals		$1532.40		$2518.80	$31.93	$52.48

Crew C-3	Hr.	Daily	Hr.	Daily	Bare Costs	Incl. O&P
1 Rodman Foreman	$39.10	$312.80	$67.30	$538.40	$33.96	$57.44
4 Rodmen (reinf.)	37.10	1187.20	63.85	2043.20		
1 Equip. Oper. (light)	32.15	257.20	50.65	405.20		
2 Laborers	26.00	416.00	43.10	689.60		
3 Stressing Equipment		48.60		53.45		
.5 Grouting Equipment		78.47		86.30	1.99	2.18
64 L.H., Daily Totals		$2300.27		$3816.15	$35.95	$59.62

Crew C-4	Hr.	Daily	Hr.	Daily	Bare Costs	Incl. O&P
1 Rodman Foreman	$39.10	$312.80	$67.30	$538.40	$37.60	$64.71
3 Rodmen (reinf.)	37.10	890.40	63.85	1532.40		
3 Stressing Equipment		48.60		53.45	1.52	1.67
32 L.H., Daily Totals		$1251.80		$2124.25	$39.12	$66.38

Crew C-5	Hr.	Daily	Hr.	Daily	Bare Costs	Incl. O&P
1 Rodman Foreman	$39.10	$312.80	$67.30	$538.40	$35.93	$60.51
4 Rodmen (reinf.)	37.10	1187.20	63.85	2043.20		
1 Equip. Oper. (crane)	34.80	278.40	54.85	438.80		
1 Equip. Oper. Oiler	29.20	233.60	46.00	368.00		
1 Hyd. Crane, 25 Ton		624.20		686.60	11.15	12.26
56 L.H., Daily Totals		$2636.20		$4075.00	$47.08	$72.77

Crew C-6	Hr.	Daily	Hr.	Daily	Bare Costs	Incl. O&P
1 Labor Foreman (outside)	$28.00	$224.00	$46.40	$371.20	$27.26	$44.73
4 Laborers	26.00	832.00	43.10	1379.20		
1 Cement Finisher	31.55	252.40	49.60	396.80		
2 Gas Engine Vibrators		52.00		57.20	1.08	1.19
48 L.H., Daily Totals		$1360.40		$2204.40	$28.34	$45.92

Crew C-7	Hr.	Daily	Hr.	Daily	Bare Costs	Incl. O&P
1 Labor Foreman (outside)	$28.00	$224.00	$46.40	$371.20	$28.04	$45.62
5 Laborers	26.00	1040.00	43.10	1724.00		
1 Cement Finisher	31.55	252.40	49.60	396.80		
1 Equip. Oper. (med.)	33.65	269.20	53.05	424.40		
1 Equip. Oper. (oiler)	29.20	233.60	46.00	368.00		
2 Gas Engine Vibrators		52.00		57.20		
1 Concrete Bucket, 1 C.Y.		15.60		17.15		
1 Hyd. Crane, 55 Ton		911.00		1002.10	13.59	14.95
72 L.H., Daily Totals		$2997.80		$4360.85	$41.63	$60.57

Crew C-8	Hr.	Daily	Hr.	Daily	Bare Costs	Incl. O&P
1 Labor Foreman (outside)	$28.00	$224.00	$46.40	$371.20	$28.96	$46.85
3 Laborers	26.00	624.00	43.10	1034.40		
2 Cement Finishers	31.55	504.80	49.60	793.60		
1 Equip. Oper. (med.)	33.65	269.20	53.05	424.40		
1 Concrete Pump (small)		719.20		791.10	12.84	14.13
56 L.H., Daily Totals		$2341.20		$3414.70	$41.80	$60.98

Crew C-8A	Hr.	Daily	Hr.	Daily	Bare Costs	Incl. O&P
1 Labor Foreman (outside)	$28.00	$224.00	$46.40	$371.20	$28.18	$45.82
3 Laborers	26.00	624.00	43.10	1034.40		
2 Cement Finishers	31.55	504.80	49.60	793.60		
48 L.H., Daily Totals		$1352.80		$2199.20	$28.18	$45.82

Crew C-8B	Hr.	Daily	Hr.	Daily	Bare Costs	Incl. O&P
1 Labor Foreman (outside)	$28.00	$224.00	$46.40	$371.20	$27.93	$45.75
3 Laborers	26.00	624.00	43.10	1034.40		
1 Equipment Operator	33.65	269.20	53.05	424.40		
1 Vibrating Screed		37.65		41.40		
1 Vibratory Roller		436.80		480.50		
1 Dozer, 200 H.P.		863.40		949.75	33.45	36.79
40 L.H., Daily Totals		$2455.05		$3301.65	$61.38	$82.54

Crew C-8C	Hr.	Daily	Hr.	Daily	Bare Costs	Incl. O&P
1 Labor Foreman (outside)	$28.00	$224.00	$46.40	$371.20	$28.53	$46.39
3 Laborers	26.00	624.00	43.10	1034.40		
1 Cement Finisher	31.55	252.40	49.60	396.80		
1 Equipment Operator (med.)	33.65	269.20	53.05	424.40		
1 Shotcrete Rig, 12 CY/hr		227.40		250.15	4.74	5.21
48 L.H., Daily Totals		$1597.00		$2476.95	$33.27	$51.60

Crew C-8D	Hr.	Daily	Hr.	Daily	Bare Costs	Incl. O&P
1 Labor Foreman (outside)	$28.00	$224.00	$46.40	$371.20	$29.43	$47.44
1 Laborer	26.00	208.00	43.10	344.80		
1 Cement Finisher	31.55	252.40	49.60	396.80		
1 Equipment Operator (light)	32.15	257.20	50.65	405.20		
1 Compressor, 250 CFM		127.60		140.35		
2 Hoses, 1", 50'		6.90		7.60	4.20	4.62
32 L.H., Daily Totals		$1076.10		$1665.95	$33.63	$52.06

Crew No.	Bare Costs		Incl. Subs O & P		Cost Per Labor-Hour	
Crew C-8E	**Hr.**	**Daily**	**Hr.**	**Daily**	**Bare Costs**	**Incl. O&P**
1 Labor Foreman (outside)	$28.00	$224.00	$46.40	$371.20	$29.43	$47.44
1 Laborer	26.00	208.00	43.10	344.80		
1 Cement Finisher	31.55	252.40	49.60	396.80		
1 Equipment Operator (light)	32.15	257.20	50.65	405.20		
1 Compressor, 250 CFM		127.60		140.35		
2 Hoses, 1", 50'		6.90		7.60		
1 Concrete Pump (small)		719.20		791.10	26.68	29.35
32 L.H., Daily Totals		$1795.30		$2457.05	$56.11	$76.79
Crew C-10	**Hr.**	**Daily**	**Hr.**	**Daily**	**Bare Costs**	**Incl. O&P**
1 Laborer	$26.00	$208.00	$43.10	$344.80	$29.70	$47.43
2 Cement Finishers	31.55	504.80	49.60	793.60		
24 L.H., Daily Totals		$712.80		$1138.40	$29.70	$47.43
Crew C-10B	**Hr.**	**Daily**	**Hr.**	**Daily**	**Bare Costs**	**Incl. O&P**
3 Laborers	$26.00	$624.00	$43.10	$1034.40	$28.22	$45.70
2 Cement Finishers	31.55	504.80	49.60	793.60		
1 Concrete mixer, 10 CF		105.20		115.70		
2 Concrete finisher, 48" dia		54.40		59.85	3.99	4.39
40 L.H., Daily Totals		$1288.40		$2003.55	$32.21	$50.09
Crew C-11	**Hr.**	**Daily**	**Hr.**	**Daily**	**Bare Costs**	**Incl. O&P**
1 Struc. Steel Foreman	$39.15	$313.20	$72.90	$583.20	$36.23	$65.41
6 Struc. Steel Workers	37.15	1783.20	69.15	3319.20		
1 Equip. Oper. (crane)	34.80	278.40	54.85	438.80		
1 Equip. Oper. Oiler	29.20	233.60	46.00	368.00		
1 Truck Crane, 150 Ton		1610.00		1771.00	22.36	24.60
72 L.H., Daily Totals		$4218.40		$6480.20	$58.59	$90.01
Crew C-12	**Hr.**	**Daily**	**Hr.**	**Daily**	**Bare Costs**	**Incl. O&P**
1 Carpenter Foreman (out)	$35.00	$280.00	$58.05	$464.40	$32.47	$53.35
3 Carpenters	33.00	792.00	54.70	1312.80		
1 Laborer	26.00	208.00	43.10	344.80		
1 Equip. Oper. (crane)	34.80	278.40	54.85	438.80		
1 Hyd. Crane, 12 Ton		612.00		673.20	12.75	14.03
48 L.H., Daily Totals		$2170.40		$3234.00	$45.22	$67.38
Crew C-13	**Hr.**	**Daily**	**Hr.**	**Daily**	**Bare Costs**	**Incl. O&P**
1 Struc. Steel Worker	$37.15	$297.20	$69.15	$553.20	$35.77	$64.33
1 Welder	37.15	297.20	69.15	553.20		
1 Carpenter	33.00	264.00	54.70	437.60		
1 Gas Welding Machine		75.20		82.70	3.13	3.45
24 L.H., Daily Totals		$933.60		$1626.70	$38.90	$67.78
Crew C-14	**Hr.**	**Daily**	**Hr.**	**Daily**	**Bare Costs**	**Incl. O&P**
1 Carpenter Foreman (out)	$35.00	$280.00	$58.05	$464.40	$32.19	$53.30
5 Carpenters	33.00	1320.00	54.70	2188.00		
4 Laborers	26.00	832.00	43.10	1379.20		
4 Rodmen (reinf.)	37.10	1187.20	63.85	2043.20		
2 Cement Finishers	31.55	504.80	49.60	793.60		
1 Equip. Oper. (crane)	34.80	278.40	54.85	438.80		
1 Equip. Oper. Oiler	29.20	233.60	46.00	368.00		
1 Crane, 80 Ton, & Tools		1020.00		1122.00	7.08	7.79
144 L.H., Daily Totals		$5656.00		$8797.20	$39.27	$61.09

Crew No.	Bare Costs		Incl. Subs O & P		Cost Per Labor-Hour	
Crew C-14A	**Hr.**	**Daily**	**Hr.**	**Daily**	**Bare Costs**	**Incl. O&P**
1 Carpenter Foreman (out)	$35.00	$280.00	$58.05	$464.40	$33.14	$55.10
16 Carpenters	33.00	4224.00	54.70	7001.60		
4 Rodmen (reinf.)	37.10	1187.20	63.85	2043.20		
2 Laborers	26.00	416.00	43.10	689.60		
1 Cement Finisher	31.55	252.40	49.60	396.80		
1 Equip. Oper. (med.)	33.65	269.20	53.05	424.40		
1 Gas Engine Vibrator		26.00		28.60		
1 Concrete Pump (small)		719.20		791.10	3.73	4.10
200 L.H., Daily Totals		$7374.00		$11839.70	$36.87	$59.20
Crew C-14B	**Hr.**	**Daily**	**Hr.**	**Daily**	**Bare Costs**	**Incl. O&P**
1 Carpenter Foreman (out)	$35.00	$280.00	$58.05	$464.40	$33.08	$54.89
16 Carpenters	33.00	4224.00	54.70	7001.60		
4 Rodmen (reinf.)	37.10	1187.20	63.85	2043.20		
2 Laborers	26.00	416.00	43.10	689.60		
2 Cement Finishers	31.55	504.80	49.60	793.60		
1 Equip. Oper. (med.)	33.65	269.20	53.05	424.40		
1 Gas Engine Vibrator		26.00		28.60		
1 Concrete Pump (small)		719.20		791.10	3.58	3.94
208 L.H., Daily Totals		$7626.40		$12236.50	$36.66	$58.83
Crew C-14C	**Hr.**	**Daily**	**Hr.**	**Daily**	**Bare Costs**	**Incl. O&P**
1 Carpenter Foreman (out)	$35.00	$280.00	$58.05	$464.40	$31.63	$52.57
6 Carpenters	33.00	1584.00	54.70	2625.60		
2 Rodmen (reinf.)	37.10	593.60	63.85	1021.60		
4 Laborers	26.00	832.00	43.10	1379.20		
1 Cement Finisher	31.55	252.40	49.60	396.80		
1 Gas Engine Vibrator		26.00		28.60	.23	.26
112 L.H., Daily Totals		$3568.00		$5916.20	$31.86	$52.83
Crew C-14D	**Hr.**	**Daily**	**Hr.**	**Daily**	**Bare Costs**	**Incl. O&P**
1 Carpenter Foreman (out)	$35.00	$280.00	$58.05	$464.40	$32.82	$54.37
18 Carpenters	33.00	4752.00	54.70	7876.80		
2 Rodmen (reinf.)	37.10	593.60	63.85	1021.60		
2 Laborers	26.00	416.00	43.10	689.60		
1 Cement Finisher	31.55	252.40	49.60	396.80		
1 Equip. Oper. (med.)	33.65	269.20	53.05	424.40		
1 Gas Engine Vibrator		26.00		28.60		
1 Concrete Pump (small)		719.20		791.10	3.73	4.10
200 L.H., Daily Totals		$7308.40		$11693.30	$36.55	$58.47
Crew C-14E	**Hr.**	**Daily**	**Hr.**	**Daily**	**Bare Costs**	**Incl. O&P**
1 Carpenter Foreman (out)	$35.00	$280.00	$58.05	$464.40	$32.63	$54.70
2 Carpenters	33.00	528.00	54.70	875.20		
4 Rodmen (reinf.)	37.10	1187.20	63.85	2043.20		
3 Laborers	26.00	624.00	43.10	1034.40		
1 Cement Finisher	31.55	252.40	49.60	396.80		
1 Gas Engine Vibrator		26.00		28.60	.30	.33
88 L.H., Daily Totals		$2897.60		$4842.60	$32.93	$55.03
Crew C-14F	**Hr.**	**Daily**	**Hr.**	**Daily**	**Bare Costs**	**Incl. O&P**
1 Laborer Foreman (out)	$28.00	$224.00	$46.40	$371.20	$29.92	$47.80
2 Laborers	26.00	416.00	43.10	689.60		
6 Cement Finishers	31.55	1514.40	49.60	2380.80		
1 Gas Engine Vibrator		26.00		28.60	.36	.40
72 L.H., Daily Totals		$2180.40		$3470.20	$30.28	$48.20

Crews

Crew No.	Bare Costs		Incl. Subs O & P		Cost Per Labor-Hour	
Crew C-14G	**Hr.**	**Daily**	**Hr.**	**Daily**	**Bare Costs**	**Incl. O&P**
1 Laborer Foreman (out)	$28.00	$224.00	$46.40	$371.20	$29.46	$47.29
2 Laborers	26.00	416.00	43.10	689.60		
4 Cement Finishers	31.55	1009.60	49.60	1587.20		
1 Gas Engine Vibrator		26.00		28.60	.46	.51
56 L.H., Daily Totals		$1675.60		$2676.60	$29.92	$47.80

Crew C-14H	**Hr.**	**Daily**	**Hr.**	**Daily**	**Bare Costs**	**Incl. O&P**
1 Carpenter Foreman (out)	$35.00	$280.00	$58.05	$464.40	$32.61	$54.00
2 Carpenters	33.00	528.00	54.70	875.20		
1 Rodman (reinf.)	37.10	296.80	63.85	510.80		
1 Laborer	26.00	208.00	43.10	344.80		
1 Cement Finisher	31.55	252.40	49.60	396.80		
1 Gas Engine Vibrator		26.00		28.60	.54	.60
48 L.H., Daily Totals		$1591.20		$2620.60	$33.15	$54.60

Crew C-15	**Hr.**	**Daily**	**Hr.**	**Daily**	**Bare Costs**	**Incl. O&P**
1 Carpenter Foreman (out)	$35.00	$280.00	$58.05	$464.40	$31.02	$51.09
2 Carpenters	33.00	528.00	54.70	875.20		
3 Laborers	26.00	624.00	43.10	1034.40		
2 Cement Finishers	31.55	504.80	49.60	793.60		
1 Rodman (reinf.)	37.10	296.80	63.85	510.80		
72 L.H., Daily Totals		$2233.60		$3678.40	$31.02	$51.09

Crew C-16	**Hr.**	**Daily**	**Hr.**	**Daily**	**Bare Costs**	**Incl. O&P**
1 Labor Foreman (outside)	$28.00	$224.00	$46.40	$371.20	$30.77	$50.63
3 Laborers	26.00	624.00	43.10	1034.40		
2 Cement Finishers	31.55	504.80	49.60	793.60		
1 Equip. Oper. (med.)	33.65	269.20	53.05	424.40		
2 Rodmen (reinf.)	37.10	593.60	63.85	1021.60		
1 Concrete Pump (small)		719.20		791.10	9.99	10.99
72 L.H., Daily Totals		$2934.80		$4436.30	$40.76	$61.62

Crew C-17	**Hr.**	**Daily**	**Hr.**	**Daily**	**Bare Costs**	**Incl. O&P**
2 Skilled Worker Foremen	$35.65	$570.40	$58.30	$932.80	$34.05	$55.66
8 Skilled Workers	33.65	2153.60	55.00	3520.00		
80 L.H., Daily Totals		$2724.00		$4452.80	$34.05	$55.66

Crew C-17A	**Hr.**	**Daily**	**Hr.**	**Daily**	**Bare Costs**	**Incl. O&P**
2 Skilled Worker Foremen	$35.65	$570.40	$58.30	$932.80	$34.06	$55.65
8 Skilled Workers	33.65	2153.60	55.00	3520.00		
.125 Equip. Oper. (crane)	34.80	34.80	54.85	54.85		
.125 Crane, 80 Ton, & Tools		127.50		140.25	1.57	1.73
81 L.H., Daily Totals		$2886.30		$4647.90	$35.63	$57.38

Crew C-17B	**Hr.**	**Daily**	**Hr.**	**Daily**	**Bare Costs**	**Incl. O&P**
2 Skilled Worker Foremen	$35.65	$570.40	$58.30	$932.80	$34.07	$55.64
8 Skilled Workers	33.65	2153.60	55.00	3520.00		
.25 Equip. Oper. (crane)	34.80	69.60	54.85	109.70		
.25 Crane, 80 Ton, & Tools		255.00		280.50		
.25 Walk Behind Power Tools		6.80		7.50	3.19	3.51
82 L.H., Daily Totals		$3055.40		$4850.50	$37.26	$59.15

Crew C-17C	**Hr.**	**Daily**	**Hr.**	**Daily**	**Bare Costs**	**Incl. O&P**
2 Skilled Worker Foremen	$35.65	$570.40	$58.30	$932.80	$34.08	$55.63
8 Skilled Workers	33.65	2153.60	55.00	3520.00		
.375 Equip. Oper. (crane)	34.80	104.40	54.85	164.55		
.375 Crane, 80 Ton & Tools		382.50		420.75	4.61	5.07
83 L.H., Daily Totals		$3210.90		$5038.10	$38.69	$60.70

Crew C-17D	**Hr.**	**Daily**	**Hr.**	**Daily**	**Bare Costs**	**Incl. O&P**
2 Skilled Worker Foremen	$35.65	$570.40	$58.30	$932.80	$34.09	$55.62
8 Skilled Workers	33.65	2153.60	55.00	3520.00		
.5 Equip. Oper. (crane)	34.80	139.20	54.85	219.40		
.5 Crane, 80 Ton & Tools		510.00		561.00	6.07	6.68
84 L.H., Daily Totals		$3373.20		$5233.20	$40.16	$62.30

Crew C-17E	**Hr.**	**Daily**	**Hr.**	**Daily**	**Bare Costs**	**Incl. O&P**
2 Skilled Worker Foremen	$35.65	$570.40	$58.30	$932.80	$34.05	$55.66
8 Skilled Workers	33.65	2153.60	55.00	3520.00		
1 Hyd. Jack with Rods		77.55		85.30	.97	1.07
80 L.H., Daily Totals		$2801.55		$4538.10	$35.02	$56.73

Crew C-18	**Hr.**	**Daily**	**Hr.**	**Daily**	**Bare Costs**	**Incl. O&P**
.125 Labor Foreman (out)	$28.00	$28.00	$46.40	$46.40	$26.22	$43.47
1 Laborer	26.00	208.00	43.10	344.80		
1 Concrete Cart, 10 C.F.		48.80		53.70	5.42	5.96
9 L.H., Daily Totals		$284.80		$444.90	$31.64	$49.43

Crew C-19	**Hr.**	**Daily**	**Hr.**	**Daily**	**Bare Costs**	**Incl. O&P**
.125 Labor Foreman (out)	$28.00	$28.00	$46.40	$46.40	$26.22	$43.47
1 Laborer	26.00	208.00	43.10	344.80		
1 Concrete Cart, 18 C.F.		74.80		82.30	8.31	9.14
9 L.H., Daily Totals		$310.80		$473.50	$34.53	$52.61

Crew C-20	**Hr.**	**Daily**	**Hr.**	**Daily**	**Bare Costs**	**Incl. O&P**
1 Labor Foreman (outside)	$28.00	$224.00	$46.40	$371.20	$27.90	$45.57
5 Laborers	26.00	1040.00	43.10	1724.00		
1 Cement Finisher	31.55	252.40	49.60	396.80		
1 Equip. Oper. (med.)	33.65	269.20	53.05	424.40		
2 Gas Engine Vibrators		52.00		57.20		
1 Concrete Pump (small)		719.20		791.10	12.05	13.26
64 L.H., Daily Totals		$2556.80		$3764.70	$39.95	$58.83

Crew C-21	**Hr.**	**Daily**	**Hr.**	**Daily**	**Bare Costs**	**Incl. O&P**
1 Labor Foreman (outside)	$28.00	$224.00	$46.40	$371.20	$27.90	$45.57
5 Laborers	26.00	1040.00	43.10	1724.00		
1 Cement Finisher	31.55	252.40	49.60	396.80		
1 Equip. Oper. (med.)	33.65	269.20	53.05	424.40		
2 Gas Engine Vibrators		52.00		57.20		
1 Concrete Conveyer		145.20		159.70	3.08	3.39
64 L.H., Daily Totals		$1982.80		$3133.30	$30.98	$48.96

Crew C-22	**Hr.**	**Daily**	**Hr.**	**Daily**	**Bare Costs**	**Incl. O&P**
1 Rodman Foreman	$39.10	$312.80	$67.30	$538.40	$37.24	$63.87
4 Rodmen (reinf.)	37.10	1187.20	63.85	2043.20		
.125 Equip. Oper. (crane)	34.80	34.80	54.85	54.85		
.125 Equip. Oper. Oiler	29.20	29.20	46.00	46.00		
.125 Hyd. Crane, 25 Ton		78.03		85.85	1.86	2.04
42 L.H., Daily Totals		$1642.03		$2768.30	$39.10	$65.91

Crew C-23	**Hr.**	**Daily**	**Hr.**	**Daily**	**Bare Costs**	**Incl. O&P**
2 Skilled Worker Foremen	$35.65	$570.40	$58.30	$932.80	$33.72	$54.75
6 Skilled Workers	33.65	1615.20	55.00	2640.00		
1 Equip. Oper. (crane)	34.80	278.40	54.85	438.80		
1 Equip. Oper. Oiler	29.20	233.60	46.00	368.00		
1 Crane, 90 Ton		1416.00		1557.60	17.70	19.47
80 L.H., Daily Totals		$4113.60		$5937.20	$51.42	$74.22

Crews

Crew C-24	Bare Costs Hr.	Bare Costs Daily	Incl. Subs O & P Hr.	Incl. Subs O & P Daily	Bare Costs	Incl. O&P
2 Skilled Worker Foremen	$35.65	$570.40	$58.30	$932.80	$33.72	$54.75
6 Skilled Workers	33.65	1615.20	55.00	2640.00		
1 Equip. Oper. (crane)	34.80	278.40	54.85	438.80		
1 Equip. Oper. Oiler	29.20	233.60	46.00	368.00		
1 Truck Crane, 150 Ton		1610.00		1771.00	20.13	22.14
80 L.H., Daily Totals		$4307.60		$6150.60	$53.85	$76.89

Crew C-25	Hr.	Daily	Hr.	Daily	Bare Costs	Incl. O&P
2 Rodmen (reinf.)	$37.10	$593.60	$63.85	$1021.60	$29.02	$50.67
2 Rodmen Helpers	20.95	335.20	37.50	600.00		
32 L.H., Daily Totals		$928.80		$1621.60	$29.02	$50.67

Crew C-27	Hr.	Daily	Hr.	Daily	Bare Costs	Incl. O&P
2 Cement Finishers	$31.55	$504.80	$49.60	$793.60	$31.55	$49.60
1 Concrete Saw		92.20		101.40	5.76	6.34
16 L.H., Daily Totals		$597.00		$895.00	$37.31	$55.94

Crew C-28	Hr.	Daily	Hr.	Daily	Bare Costs	Incl. O&P
1 Cement Finisher	$31.55	$252.40	$49.60	$396.80	$31.55	$49.60
1 Portable Air Compressor		18.60		20.45	2.33	2.56
8 L.H., Daily Totals		$271.00		$417.25	$33.88	$52.16

Crew D-1	Hr.	Daily	Hr.	Daily	Bare Costs	Incl. O&P
1 Bricklayer	$34.25	$274.00	$55.60	$444.80	$29.93	$48.58
1 Bricklayer Helper	25.60	204.80	41.55	332.40		
16 L.H., Daily Totals		$478.80		$777.20	$29.93	$48.58

Crew D-2	Hr.	Daily	Hr.	Daily	Bare Costs	Incl. O&P
3 Bricklayers	$34.25	$822.00	$55.60	$1334.40	$30.99	$50.41
2 Bricklayer Helpers	25.60	409.60	41.55	664.80		
.5 Carpenter	33.00	132.00	54.70	218.80		
44 L.H., Daily Totals		$1363.60		$2218.00	$30.99	$50.41

Crew D-3	Hr.	Daily	Hr.	Daily	Bare Costs	Incl. O&P
3 Bricklayers	$34.25	$822.00	$55.60	$1334.40	$30.90	$50.20
2 Bricklayer Helpers	25.60	409.60	41.55	664.80		
.25 Carpenter	33.00	66.00	54.70	109.40		
42 L.H., Daily Totals		$1297.60		$2108.60	$30.90	$50.20

Crew D-4	Hr.	Daily	Hr.	Daily	Bare Costs	Incl. O&P
1 Bricklayer	$34.25	$274.00	$55.60	$444.80	$29.40	$47.34
2 Bricklayer Helpers	25.60	409.60	41.55	664.80		
1 Equip. Oper. (light)	32.15	257.20	50.65	405.20		
1 Grout Pump, 50 C.F./hr		110.90		122.00		
1 Hoses & Hopper		15.60		17.15		
1 Accessories		11.70		12.85	4.32	4.75
32 L.H., Daily Totals		$1079.00		$1666.80	$33.72	$52.09

Crew D-5	Hr.	Daily	Hr.	Daily	Bare Costs	Incl. O&P
1 Bricklayer	$34.25	$274.00	$55.60	$444.80	$34.25	$55.60
8 L.H., Daily Totals		$274.00		$444.80	$34.25	$55.60

Crew D-6	Hr.	Daily	Hr.	Daily	Bare Costs	Incl. O&P
3 Bricklayers	$34.25	$822.00	$55.60	$1334.40	$30.05	$48.82
3 Bricklayer Helpers	25.60	614.40	41.55	997.20		
.25 Carpenter	33.00	66.00	54.70	109.40		
50 L.H., Daily Totals		$1502.40		$2441.00	$30.05	$48.82

Crew D-7	Hr.	Daily	Hr.	Daily	Bare Costs	Incl. O&P
1 Tile Layer	$31.50	$252.00	$49.40	$395.20	$27.98	$43.88
1 Tile Layer Helper	24.45	195.60	38.35	306.80		
16 L.H., Daily Totals		$447.60		$702.00	$27.98	$43.88

Crew D-8	Hr.	Daily	Hr.	Daily	Bare Costs	Incl. O&P
3 Bricklayers	$34.25	$822.00	$55.60	$1334.40	$30.79	$49.98
2 Bricklayer Helpers	25.60	409.60	41.55	664.80		
40 L.H., Daily Totals		$1231.60		$1999.20	$30.79	$49.98

Crew D-9	Hr.	Daily	Hr.	Daily	Bare Costs	Incl. O&P
3 Bricklayers	$34.25	$822.00	$55.60	$1334.40	$29.93	$48.58
3 Bricklayer Helpers	25.60	614.40	41.55	997.20		
48 L.H., Daily Totals		$1436.40		$2331.60	$29.93	$48.58

Crew D-10	Hr.	Daily	Hr.	Daily	Bare Costs	Incl. O&P
1 Bricklayer Foreman	$36.25	$290.00	$58.85	$470.80	$31.30	$50.48
1 Bricklayer	34.25	274.00	55.60	444.80		
2 Bricklayer Helpers	25.60	409.60	41.55	664.80		
1 Equip. Oper. (crane)	34.80	278.40	54.85	438.80		
1 Truck Crane, 12.5 Ton		482.20		530.40	12.06	13.26
40 L.H., Daily Totals		$1734.20		$2549.60	$43.36	$63.74

Crew D-11	Hr.	Daily	Hr.	Daily	Bare Costs	Incl. O&P
1 Bricklayer Foreman	$36.25	$290.00	$58.85	$470.80	$32.03	$52.00
1 Bricklayer	34.25	274.00	55.60	444.80		
1 Bricklayer Helper	25.60	204.80	41.55	332.40		
24 L.H., Daily Totals		$768.80		$1248.00	$32.03	$52.00

Crew D-12	Hr.	Daily	Hr.	Daily	Bare Costs	Incl. O&P
1 Bricklayer Foreman	$36.25	$290.00	$58.85	$470.80	$30.43	$49.39
1 Bricklayer	34.25	274.00	55.60	444.80		
2 Bricklayer Helpers	25.60	409.60	41.55	664.80		
32 L.H., Daily Totals		$973.60		$1580.40	$30.43	$49.39

Crew D-13	Hr.	Daily	Hr.	Daily	Bare Costs	Incl. O&P
1 Bricklayer Foreman	$36.25	$290.00	$58.85	$470.80	$31.58	$51.18
1 Bricklayer	34.25	274.00	55.60	444.80		
2 Bricklayer Helpers	25.60	409.60	41.55	664.80		
1 Carpenter	33.00	264.00	54.70	437.60		
1 Equip. Oper. (crane)	34.80	278.40	54.85	438.80		
1 Truck Crane, 12.5 Ton		482.20		530.40	10.05	11.05
48 L.H., Daily Totals		$1998.20		$2987.20	$41.63	$62.23

Crew E-1	Hr.	Daily	Hr.	Daily	Bare Costs	Incl. O&P
1 Welder Foreman	$39.15	$313.20	$72.90	$583.20	$36.15	$64.23
1 Welder	37.15	297.20	69.15	553.20		
1 Equip. Oper. (light)	32.15	257.20	50.65	405.20		
1 Gas Welding Machine		75.20		82.70	3.13	3.45
24 L.H., Daily Totals		$942.80		$1624.30	$39.28	$67.68

Crew E-2	Hr.	Daily	Hr.	Daily	Bare Costs	Incl. O&P
1 Struc. Steel Foreman	$39.15	$313.20	$72.90	$583.20	$35.96	$64.34
4 Struc. Steel Workers	37.15	1188.80	69.15	2212.80		
1 Equip. Oper. (crane)	34.80	278.40	54.85	438.80		
1 Equip. Oper. Oiler	29.20	233.60	46.00	368.00		
1 Crane, 90 Ton		1416.00		1557.60	25.29	27.81
56 L.H., Daily Totals		$3430.00		$5160.40	$61.25	$92.15

Crew No.	Bare Costs		Incl. Subs O & P		Cost Per Labor-Hour	
Crew E-3	**Hr.**	**Daily**	**Hr.**	**Daily**	**Bare Costs**	**Incl. O&P**
1 Struc. Steel Foreman	$39.15	$313.20	$72.90	$583.20	$37.82	$70.40
1 Struc. Steel Worker	37.15	297.20	69.15	553.20		
1 Welder	37.15	297.20	69.15	553.20		
1 Gas Welding Machine		75.20		82.70	3.13	3.45
24 L.H., Daily Totals		$982.80		$1772.30	$40.95	$73.85
Crew E-4	**Hr.**	**Daily**	**Hr.**	**Daily**	**Bare Costs**	**Incl. O&P**
1 Struc. Steel Foreman	$39.15	$313.20	$72.90	$583.20	$37.65	$70.09
3 Struc. Steel Workers	37.15	891.60	69.15	1659.60		
1 Gas Welding Machine		75.20		82.70	2.35	2.59
32 L.H., Daily Totals		$1280.00		$2325.50	$40.00	$72.68
Crew E-5	**Hr.**	**Daily**	**Hr.**	**Daily**	**Bare Costs**	**Incl. O&P**
2 Struc. Steel Foremen	$39.15	$626.40	$72.90	$1166.40	$36.52	$66.15
5 Struc. Steel Workers	37.15	1486.00	69.15	2766.00		
1 Equip. Oper. (crane)	34.80	278.40	54.85	438.80		
1 Welder	37.15	297.20	69.15	553.20		
1 Equip. Oper. Oiler	29.20	233.60	46.00	368.00		
1 Crane, 90 Ton		1416.00		1557.60		
1 Gas Welding Machine		75.20		82.70	18.64	20.50
80 L.H., Daily Totals		$4412.80		$6932.70	$55.16	$86.65
Crew E-6	**Hr.**	**Daily**	**Hr.**	**Daily**	**Bare Costs**	**Incl. O&P**
3 Struc. Steel Foremen	$39.15	$939.60	$72.90	$1749.60	$36.57	$66.36
9 Struc. Steel Workers	37.15	2674.80	69.15	4978.80		
1 Equip. Oper. (crane)	34.80	278.40	54.85	438.80		
1 Welder	37.15	297.20	69.15	553.20		
1 Equip. Oper. Oiler	29.20	233.60	46.00	368.00		
1 Equip. Oper. (light)	32.15	257.20	50.65	405.20		
1 Crane, 90 Ton		1416.00		1557.60		
1 Gas Welding Machine		75.20		82.70		
1 Air Compr., 160 C.F.M.		82.00		90.20		
2 Impact Wrenches		28.80		31.70	12.52	13.77
128 L.H., Daily Totals		$6282.80		$10255.80	$49.09	$80.13
Crew E-7	**Hr.**	**Daily**	**Hr.**	**Daily**	**Bare Costs**	**Incl. O&P**
1 Struc. Steel Foreman	$39.15	$313.20	$72.90	$583.20	$36.52	$66.15
4 Struc. Steel Workers	37.15	1188.80	69.15	2212.80		
1 Equip. Oper. (crane)	34.80	278.40	54.85	438.80		
1 Equip. Oper. Oiler	29.20	233.60	46.00	368.00		
1 Welder Foreman	39.15	313.20	72.90	583.20		
2 Welders	37.15	594.40	69.15	1106.40		
1 Crane, 90 Ton		1416.00		1557.60		
2 Gas Welding Machines		150.40		165.45	19.58	21.54
80 L.H., Daily Totals		$4488.00		$7015.45	$56.10	$87.69
Crew E-8	**Hr.**	**Daily**	**Hr.**	**Daily**	**Bare Costs**	**Incl. O&P**
1 Struc. Steel Foreman	$39.15	$313.20	$72.90	$583.20	$36.28	$65.42
4 Struc. Steel Workers	37.15	1188.80	69.15	2212.80		
1 Welder Foreman	39.15	313.20	72.90	583.20		
4 Welders	37.15	1188.80	69.15	2212.80		
1 Equip. Oper. (crane)	34.80	278.40	54.85	438.80		
1 Equip. Oper. Oiler	29.20	233.60	46.00	368.00		
1 Equip. Oper. (light)	32.15	257.20	50.65	405.20		
1 Crane, 90 Ton		1416.00		1557.60		
4 Gas Welding Machines		300.80		330.90	16.51	18.16
104 L.H., Daily Totals		$5490.00		$8692.50	$52.79	$83.58

Crew No.	Bare Costs		Incl. Subs O & P		Cost Per Labor-Hour	
Crew E-9	**Hr.**	**Daily**	**Hr.**	**Daily**	**Bare Costs**	**Incl. O&P**
2 Struc. Steel Foremen	$39.15	$626.40	$72.90	$1166.40	$36.57	$66.36
5 Struc. Steel Workers	37.15	1486.00	69.15	2766.00		
1 Welder Foreman	39.15	313.20	72.90	583.20		
5 Welders	37.15	1486.00	69.15	2766.00		
1 Equip. Oper. (crane)	34.80	278.40	54.85	438.80		
1 Equip. Oper. Oiler	29.20	233.60	46.00	368.00		
1 Equip. Oper. (light)	32.15	257.20	50.65	405.20		
1 Crane, 90 Ton		1416.00		1557.60		
5 Gas Welding Machines		376.00		413.60	14.00	15.40
128 L.H., Daily Totals		$6472.80		$10464.80	$50.57	$81.76
Crew E-10	**Hr.**	**Daily**	**Hr.**	**Daily**	**Bare Costs**	**Incl. O&P**
1 Welder Foreman	$39.15	$313.20	$72.90	$583.20	$38.15	$71.03
1 Welder	37.15	297.20	69.15	553.20		
1 Gas Welding Machines		75.20		82.70		
1 Truck, 3 Ton		160.40		176.45	14.73	16.20
16 L.H., Daily Totals		$846.00		$1395.55	$52.88	$87.23
Crew E-11	**Hr.**	**Daily**	**Hr.**	**Daily**	**Bare Costs**	**Incl. O&P**
2 Painters, Struc. Steel	$30.05	$480.80	$59.30	$948.80	$29.56	$53.09
1 Building Laborer	26.00	208.00	43.10	344.80		
1 Equip. Oper. (light)	32.15	257.20	50.65	405.20		
1 Air Compressor 250 C.F.M.		127.60		140.35		
1 Sand Blaster		15.60		17.15		
1 Sand Blasting Accessories		11.70		12.85	4.84	5.32
32 L.H., Daily Totals		$1100.90		$1869.15	$34.40	$58.41
Crew E-12	**Hr.**	**Daily**	**Hr.**	**Daily**	**Bare Costs**	**Incl. O&P**
1 Welder Foreman	$39.15	$313.20	$72.90	$583.20	$35.65	$61.78
1 Equip. Oper. (light)	32.15	257.20	50.65	405.20		
1 Gas Welding Machine		75.20		82.70	4.70	5.17
16 L.H., Daily Totals		$645.60		$1071.10	$40.35	$66.95
Crew E-13	**Hr.**	**Daily**	**Hr.**	**Daily**	**Bare Costs**	**Incl. O&P**
1 Welder Foreman	$39.15	$313.20	$72.90	$583.20	$36.82	$65.48
.5 Equip. Oper. (light)	32.15	128.60	50.65	202.60		
1 Gas Welding Machine		75.20		82.70	6.27	6.89
12 L.H., Daily Totals		$517.00		$868.50	$43.09	$72.37
Crew E-14	**Hr.**	**Daily**	**Hr.**	**Daily**	**Bare Costs**	**Incl. O&P**
1 Welder Foreman	$39.15	$313.20	$72.90	$583.20	$39.15	$72.90
1 Gas Welding Machine		75.20		82.70	9.40	10.34
8 L.H., Daily Totals		$388.40		$665.90	$48.55	$83.24
Crew E-16	**Hr.**	**Daily**	**Hr.**	**Daily**	**Bare Costs**	**Incl. O&P**
1 Welder Foreman	$39.15	$313.20	$72.90	$583.20	$38.15	$71.03
1 Welder	37.15	297.20	69.15	553.20		
1 Gas Welding Machine		75.20		82.70	4.70	5.17
16 L.H., Daily Totals		$685.60		$1219.10	$42.85	$76.20
Crew E-17	**Hr.**	**Daily**	**Hr.**	**Daily**	**Bare Costs**	**Incl. O&P**
1 Structural Steel Foreman	$39.15	$313.20	$72.90	$583.20	$38.15	$71.03
1 Structural Steel Worker	37.15	297.20	69.15	553.20		
1 Power Tool		4.40		4.85	.28	.30
16 L.H., Daily Totals		$614.80		$1141.25	$38.43	$71.33

Crew No.	Bare Costs		Incl. Subs O & P		Cost Per Labor-Hour	
Crew E-18	**Hr.**	**Daily**	**Hr.**	**Daily**	**Bare Costs**	**Incl. O&P**
1 Structural Steel Foreman	$39.15	$313.20	$72.90	$583.20	$36.85	$66.68
3 Structural Steel Workers	37.15	891.60	69.15	1659.60		
1 Equipment Operator (med.)	33.65	269.20	53.05	424.40		
1 Crane, 20 Ton		647.45		712.20	16.19	17.80
40 L.H., Daily Totals		$2121.45		$3379.40	$53.04	$84.48
Crew E-19	**Hr.**	**Daily**	**Hr.**	**Daily**	**Bare Costs**	**Incl. O&P**
1 Structural Steel Worker	$37.15	$297.20	$69.15	$553.20	$36.15	$64.23
1 Structural Steel Foreman	39.15	313.20	72.90	583.20		
1 Equip. Oper. (light)	32.15	257.20	50.65	405.20		
1 Power Tool		4.40		4.85		
1 Crane, 20 Ton		647.45		712.20	27.16	29.88
24 L.H., Daily Totals		$1519.45		$2258.65	$63.31	$94.11
Crew E-20	**Hr.**	**Daily**	**Hr.**	**Daily**	**Bare Costs**	**Incl. O&P**
1 Structural Steel Foreman	$39.15	$313.20	$72.90	$583.20	$36.11	$64.94
5 Structural Steel Workers	37.15	1486.00	69.15	2766.00		
1 Equip. Oper. (crane)	34.80	278.40	54.85	438.80		
1 Oiler	29.20	233.60	46.00	368.00		
1 Power Tool		4.40		4.85		
1 Crane, 40 Ton		772.50		849.75	12.14	13.35
64 L.H., Daily Totals		$3088.10		$5010.60	$48.25	$78.29
Crew E-22	**Hr.**	**Daily**	**Hr.**	**Daily**	**Bare Costs**	**Incl. O&P**
1 Skilled Worker Foreman	$35.65	$285.20	$58.30	$466.40	$34.32	$56.10
2 Skilled Workers	33.65	538.40	55.00	880.00		
24 L.H., Daily Totals		$823.60		$1346.40	$34.32	$56.10
Crew E-24	**Hr.**	**Daily**	**Hr.**	**Daily**	**Bare Costs**	**Incl. O&P**
3 Structural Steel Workers	$37.15	$891.60	$69.15	$1659.60	$36.28	$65.13
1 Equipment Operator (medium)	33.65	269.20	53.05	424.40		
1-25 Ton Crane		624.20		686.60	19.51	21.46
32 L.H., Daily Totals		$1785.00		$2770.60	$55.79	$86.59
Crew E-25	**Hr.**	**Daily**	**Hr.**	**Daily**	**Bare Costs**	**Incl. O&P**
1 Welder Foreman	$39.15	$313.20	$72.90	$583.20	$39.15	$72.90
1 Cutting Torch		20.00		22.00		
1 Gases		64.80		71.30	10.60	11.66
8 L.H., Daily Totals		$398.00		$676.50	$49.75	$84.56
Crew F-3	**Hr.**	**Daily**	**Hr.**	**Daily**	**Bare Costs**	**Incl. O&P**
4 Carpenters	$33.00	$1056.00	$54.70	$1750.40	$33.36	$54.73
1 Equip. Oper. (crane)	34.80	278.40	54.85	438.80		
1 Hyd. Crane, 12 Ton		612.00		673.20	15.30	16.83
40 L.H., Daily Totals		$1946.40		$2862.40	$48.66	$71.56
Crew F-4	**Hr.**	**Daily**	**Hr.**	**Daily**	**Bare Costs**	**Incl. O&P**
4 Carpenters	$33.00	$1056.00	$54.70	$1750.40	$32.67	$53.27
1 Equip. Oper. (crane)	34.80	278.40	54.85	438.80		
1 Equip. Oper. Oiler	29.20	233.60	46.00	368.00		
1 Hyd. Crane, 55 Ton		911.00		1002.10	18.98	20.88
48 L.H., Daily Totals		$2479.00		$3559.30	$51.65	$74.15
Crew F-5	**Hr.**	**Daily**	**Hr.**	**Daily**	**Bare Costs**	**Incl. O&P**
1 Carpenter Foreman	$35.00	$280.00	$58.05	$464.40	$33.50	$55.54
3 Carpenters	33.00	792.00	54.70	1312.80		
32 L.H., Daily Totals		$1072.00		$1777.20	$33.50	$55.54

Crew No.	Bare Costs		Incl. Subs O & P		Cost Per Labor-Hour	
Crew F-6	**Hr.**	**Daily**	**Hr.**	**Daily**	**Bare Costs**	**Incl. O&P**
2 Carpenters	$33.00	$528.00	$54.70	$875.20	$30.56	$50.09
2 Building Laborers	26.00	416.00	43.10	689.60		
1 Equip. Oper. (crane)	34.80	278.40	54.85	438.80		
1 Hyd. Crane, 12 Ton		612.00		673.20	15.30	16.83
40 L.H., Daily Totals		$1834.40		$2676.80	$45.86	$66.92
Crew F-7	**Hr.**	**Daily**	**Hr.**	**Daily**	**Bare Costs**	**Incl. O&P**
2 Carpenters	$33.00	$528.00	$54.70	$875.20	$29.50	$48.90
2 Building Laborers	26.00	416.00	43.10	689.60		
32 L.H., Daily Totals		$944.00		$1564.80	$29.50	$48.90
Crew G-1	**Hr.**	**Daily**	**Hr.**	**Daily**	**Bare Costs**	**Incl. O&P**
1 Roofer Foreman	$30.45	$243.60	$54.55	$436.40	$26.59	$47.62
4 Roofers, Composition	28.45	910.40	50.95	1630.40		
2 Roofer Helpers	20.95	335.20	37.50	600.00		
1 Application Equipment		138.40		152.25		
1 Tar Kettle/Pot		52.50		57.75		
1 Crew Truck		109.10		120.00	5.36	5.89
56 L.H., Daily Totals		$1789.20		$2996.80	$31.95	$53.51
Crew G-2	**Hr.**	**Daily**	**Hr.**	**Daily**	**Bare Costs**	**Incl. O&P**
1 Plasterer	$29.95	$239.60	$48.50	$388.00	$27.25	$44.45
1 Plasterer Helper	25.80	206.40	41.75	334.00		
1 Building Laborer	26.00	208.00	43.10	344.80		
1 Grouting Equipment		110.90		122.00	4.62	5.08
24 L.H., Daily Totals		$764.90		$1188.80	$31.87	$49.53
Crew G-3	**Hr.**	**Daily**	**Hr.**	**Daily**	**Bare Costs**	**Incl. O&P**
2 Sheet Metal Workers	$38.80	$620.80	$61.45	$983.20	$32.40	$52.28
2 Building Laborers	26.00	416.00	43.10	689.60		
32 L.H., Daily Totals		$1036.80		$1672.80	$32.40	$52.28
Crew G-4	**Hr.**	**Daily**	**Hr.**	**Daily**	**Bare Costs**	**Incl. O&P**
1 Labor Foreman (outside)	$28.00	$224.00	$46.40	$371.20	$26.67	$44.20
2 Building Laborers	26.00	416.00	43.10	689.60		
1 Light Truck, 1.5 Ton		122.00		134.20		
1 Air Compr., 160 C.F.M.		82.00		90.20	8.50	9.35
24 L.H., Daily Totals		$844.00		$1285.20	$35.17	$53.55
Crew G-5	**Hr.**	**Daily**	**Hr.**	**Daily**	**Bare Costs**	**Incl. O&P**
1 Roofer Foreman	$30.45	$243.60	$54.55	$436.40	$25.85	$46.29
2 Roofers, Composition	28.45	455.20	50.95	815.20		
2 Roofer Helpers	20.95	335.20	37.50	600.00		
1 Application Equipment		138.40		152.25	3.46	3.81
40 L.H., Daily Totals		$1172.40		$2003.85	$29.31	$50.10
Crew G-6A	**Hr.**	**Daily**	**Hr.**	**Daily**	**Bare Costs**	**Incl. O&P**
2 Roofers Composition	$28.45	$455.20	$50.95	$815.20	$28.45	$50.95
1 Small Compressor		13.10		14.40		
2 Pneumatic Nailers		39.00		42.90	3.26	3.58
16 L.H., Daily Totals		$507.30		$872.50	$31.71	$54.53
Crew G-7	**Hr.**	**Daily**	**Hr.**	**Daily**	**Bare Costs**	**Incl. O&P**
1 Carpenter	$33.00	$264.00	$54.70	$437.60	$33.00	$54.70
1 Small Compressor		13.10		14.40		
1 Pneumatic Nailer		19.50		21.45	4.08	4.48
8 L.H., Daily Totals		$296.60		$473.45	$37.08	$59.18

Crew No.	Bare Costs		Incl. Subs O & P		Cost Per Labor-Hour	
Crew H-1	**Hr.**	**Daily**	**Hr.**	**Daily**	**Bare Costs**	**Incl. O&P**
2 Glaziers	$32.05	$512.80	$51.65	$826.40	$34.60	$60.40
2 Struc. Steel Workers	37.15	594.40	69.15	1106.40		
32 L.H., Daily Totals		$1107.20		$1932.80	$34.60	$60.40
Crew H-2	**Hr.**	**Daily**	**Hr.**	**Daily**	**Bare Costs**	**Incl. O&P**
2 Glaziers	$32.05	$512.80	$51.65	$826.40	$30.03	$48.80
1 Building Laborer	26.00	208.00	43.10	344.80		
24 L.H., Daily Totals		$720.80		$1171.20	$30.03	$48.80
Crew H-3	**Hr.**	**Daily**	**Hr.**	**Daily**	**Bare Costs**	**Incl. O&P**
1 Glazier	$32.05	$256.40	$51.65	$413.20	$28.30	$46.10
1 Helper	24.55	196.40	40.55	324.40		
16 L.H., Daily Totals		$452.80		$737.60	$28.30	$46.10
Crew J-1	**Hr.**	**Daily**	**Hr.**	**Daily**	**Bare Costs**	**Incl. O&P**
3 Plasterers	$29.95	$718.80	$48.50	$1164.00	$28.29	$45.80
2 Plasterer Helpers	25.80	412.80	41.75	668.00		
1 Mixing Machine, 6 C.F.		90.00		99.00	2.25	2.48
40 L.H., Daily Totals		$1221.60		$1931.00	$30.54	$48.28
Crew J-2	**Hr.**	**Daily**	**Hr.**	**Daily**	**Bare Costs**	**Incl. O&P**
3 Plasterers	$29.95	$718.80	$48.50	$1164.00	$28.67	$46.21
2 Plasterer Helpers	25.80	412.80	41.75	668.00		
1 Lather	30.55	244.40	48.25	386.00		
1 Mixing Machine, 6 C.F.		90.00		99.00	1.88	2.06
48 L.H., Daily Totals		$1466.00		$2317.00	$30.55	$48.27
Crew J-3	**Hr.**	**Daily**	**Hr.**	**Daily**	**Bare Costs**	**Incl. O&P**
1 Terrazzo Worker	$31.60	$252.80	$49.60	$396.80	$28.45	$44.65
1 Terrazzo Helper	25.30	202.40	39.70	317.60		
1 Terrazzo Grinder, Electric		59.55		65.50		
1 Terrazzo Mixer		121.60		133.75	11.32	12.45
16 L.H., Daily Totals		$636.35		$913.65	$39.77	$57.10
Crew J-4	**Hr.**	**Daily**	**Hr.**	**Daily**	**Bare Costs**	**Incl. O&P**
1 Tile Layer	$31.50	$252.00	$49.40	$395.20	$27.98	$43.88
1 Tile Layer Helper	24.45	195.60	38.35	306.80		
16 L.H., Daily Totals		$447.60		$702.00	$27.98	$43.88
Crew K-1	**Hr.**	**Daily**	**Hr.**	**Daily**	**Bare Costs**	**Incl. O&P**
1 Carpenter	$33.00	$264.00	$54.70	$437.60	$29.35	$48.23
1 Truck Driver (light)	25.70	205.60	41.75	334.00		
1 Truck w/Power Equip.		160.40		176.45	10.03	11.03
16 L.H., Daily Totals		$630.00		$948.05	$39.38	$59.26
Crew K-2	**Hr.**	**Daily**	**Hr.**	**Daily**	**Bare Costs**	**Incl. O&P**
1 Struc. Steel Foreman	$39.15	$313.20	$72.90	$583.20	$34.00	$61.27
1 Struc. Steel Worker	37.15	297.20	69.15	553.20		
1 Truck Driver (light)	25.70	205.60	41.75	334.00		
1 Truck w/Power Equip.		160.40		176.45	6.68	7.35
24 L.H., Daily Totals		$976.40		$1646.85	$40.68	$68.62
Crew L-1	**Hr.**	**Daily**	**Hr.**	**Daily**	**Bare Costs**	**Incl. O&P**
1 Electrician	$39.40	$315.20	$60.55	$484.40	$39.50	$60.98
1 Plumber	39.60	316.80	61.40	491.20		
16 L.H., Daily Totals		$632.00		$975.60	$39.50	$60.98

Crew No.	Bare Costs		Incl. Subs O & P		Cost Per Labor-Hour	
Crew L-2	**Hr.**	**Daily**	**Hr.**	**Daily**	**Bare Costs**	**Incl. O&P**
1 Carpenter	$33.00	$264.00	$54.70	$437.60	$28.77	$47.63
1 Carpenter Helper	24.55	196.40	40.55	324.40		
16 L.H., Daily Totals		$460.40		$762.00	$28.77	$47.63
Crew L-3	**Hr.**	**Daily**	**Hr.**	**Daily**	**Bare Costs**	**Incl. O&P**
1 Carpenter	$33.00	$264.00	$54.70	$437.60	$36.05	$57.85
.5 Electrician	39.40	157.60	60.55	242.20		
.5 Sheet Metal Worker	38.80	155.20	61.45	245.80		
16 L.H., Daily Totals		$576.80		$925.60	$36.05	$57.85
Crew L-3A	**Hr.**	**Daily**	**Hr.**	**Daily**	**Bare Costs**	**Incl. O&P**
1 Carpenter Foreman (outside)	$35.00	$280.00	$58.05	$464.40	$36.27	$59.18
.5 Sheet Metal Worker	38.80	155.20	61.45	245.80		
12 L.H., Daily Totals		$435.20		$710.20	$36.27	$59.18
Crew L-4	**Hr.**	**Daily**	**Hr.**	**Daily**	**Bare Costs**	**Incl. O&P**
2 Skilled Workers	$33.65	$538.40	$55.00	$880.00	$30.62	$50.18
1 Helper	24.55	196.40	40.55	324.40		
24 L.H., Daily Totals		$734.80		$1204.40	$30.62	$50.18
Crew L-5	**Hr.**	**Daily**	**Hr.**	**Daily**	**Bare Costs**	**Incl. O&P**
1 Struc. Steel Foreman	$39.15	$313.20	$72.90	$583.20	$37.10	$67.64
5 Struc. Steel Workers	37.15	1486.00	69.15	2766.00		
1 Equip. Oper. (crane)	34.80	278.40	54.85	438.80		
1 Hyd. Crane, 25 Ton		624.20		686.60	11.15	12.26
56 L.H., Daily Totals		$2701.80		$4474.60	$48.25	$79.90
Crew L-5A	**Hr.**	**Daily**	**Hr.**	**Daily**	**Bare Costs**	**Incl. O&P**
1 Structural Steel Foreman	$39.15	$313.20	$72.90	$583.20	$37.06	$66.51
2 Structural Steel Workers	37.15	594.40	69.15	1106.40		
1 Equip. Oper. (crane)	34.80	278.40	54.85	438.80		
1 Crane,SP, 25 Ton		575.20		632.70	17.98	19.77
32 L.H., Daily Totals		$1761.20		$2761.10	$55.04	$86.28
Crew L-6	**Hr.**	**Daily**	**Hr.**	**Daily**	**Bare Costs**	**Incl. O&P**
1 Plumber	$39.60	$316.80	$61.40	$491.20	$39.53	$61.12
.5 Electrician	39.40	157.60	60.55	242.20		
12 L.H., Daily Totals		$474.40		$733.40	$39.53	$61.12
Crew L-7	**Hr.**	**Daily**	**Hr.**	**Daily**	**Bare Costs**	**Incl. O&P**
2 Carpenters	$33.00	$528.00	$54.70	$875.20	$31.91	$52.22
1 Building Laborer	26.00	208.00	43.10	344.80		
.5 Electrician	39.40	157.60	60.55	242.20		
28 L.H., Daily Totals		$893.60		$1462.20	$31.91	$52.22
Crew L-8	**Hr.**	**Daily**	**Hr.**	**Daily**	**Bare Costs**	**Incl. O&P**
2 Carpenters	$33.00	$528.00	$54.70	$875.20	$34.32	$56.04
.5 Plumber	39.60	158.40	61.40	245.60		
20 L.H., Daily Totals		$686.40		$1120.80	$34.32	$56.04
Crew L-9	**Hr.**	**Daily**	**Hr.**	**Daily**	**Bare Costs**	**Incl. O&P**
1 Labor Foreman (inside)	$26.50	$212.00	$43.95	$351.60	$30.08	$51.02
2 Building Laborers	26.00	416.00	43.10	689.60		
1 Struc. Steel Worker	37.15	297.20	69.15	553.20		
.5 Electrician	39.40	157.60	60.55	242.20		
36 L.H., Daily Totals		$1082.80		$1836.60	$30.08	$51.02

Crew No.	Bare Costs		Incl. Subs O & P		Cost Per Labor-Hour	
Crew L-10	**Hr.**	**Daily**	**Hr.**	**Daily**	**Bare Costs**	**Incl. O&P**
1 Structural Steel Foreman	$39.15	$313.20	$72.90	$583.20	$37.03	$65.63
1 Structural Steel Worker	37.15	297.20	69.15	553.20		
1 Equip. Oper. (crane)	34.80	278.40	54.85	438.80		
1 Hyd. Crane, 12 Ton		612.00		673.20	25.50	28.05
24 L.H., Daily Totals		$1500.80		$2248.40	$62.53	$93.68

Crew M-1	Hr.	Daily	Hr.	Daily	Bare Costs	Incl. O&P
3 Elevator Constructors	$41.60	$998.40	$64.25	$1542.00	$39.52	$61.05
1 Elevator Apprentice	33.30	266.40	51.45	411.60		
5 Hand Tools		70.00		77.00	2.19	2.41
32 L.H., Daily Totals		$1334.80		$2030.60	$41.71	$63.46

Crew M-3	Hr.	Daily	Hr.	Daily	Bare Costs	Incl. O&P
1 Electrician Foreman (out)	$41.40	$331.20	$63.65	$509.20	$35.46	$55.46
1 Common Laborer	26.00	208.00	43.10	344.80		
.25 Equipment Operator, Medium	33.65	67.30	53.05	106.10		
1 Elevator Constructor	41.60	332.80	64.25	514.00		
1 Elevator Apprentice	33.30	266.40	51.45	411.60		
.25 Crane, SP, 4 x 4, 20 ton		143.60		157.95	4.22	4.65
34 L.H., Daily Totals		$1349.30		$2043.65	$39.68	$60.11

Crew M-4	Hr.	Daily	Hr.	Daily	Bare Costs	Incl. O&P
1 Electrician Foreman (out)	$41.40	$331.20	$63.65	$509.20	$35.18	$55.04
1 Common Laborer	26.00	208.00	43.10	344.80		
.25 Equipment Operator, Crane	34.80	69.60	54.85	109.70		
.25 Equipment Operator, Oiler	29.20	58.40	46.00	92.00		
1 Elevator Constructor	41.60	332.80	64.25	514.00		
1 Elevator Apprentice	33.30	266.40	51.45	411.60		
.25 Crane, Hyd, SP, 4WD, 40 Ton		215.30		236.85	5.98	6.58
36 L.H., Daily Totals		$1481.70		$2218.15	$41.16	$61.62

Crew Q-1	Hr.	Daily	Hr.	Daily	Bare Costs	Incl. O&P
1 Plumber	$39.60	$316.80	$61.40	$491.20	$35.65	$55.28
1 Plumber Apprentice	31.70	253.60	49.15	393.20		
16 L.H., Daily Totals		$570.40		$884.40	$35.65	$55.28

Crew Q-1C	Hr.	Daily	Hr.	Daily	Bare Costs	Incl. O&P
1 Plumber	$39.60	$316.80	$61.40	$491.20	$34.98	$54.53
1 Plumber Apprentice	31.70	253.60	49.15	393.20		
1 Equip. Oper. (medium)	33.65	269.20	53.05	424.40		
1 Trencher, Chain		1310.00		1441.00	54.58	60.04
24 L.H., Daily Totals		$2149.60		$2749.80	$89.56	$114.57

Crew Q-2	Hr.	Daily	Hr.	Daily	Bare Costs	Incl. O&P
2 Plumbers	$39.60	$633.60	$61.40	$982.40	$36.97	$57.32
1 Plumber Apprentice	31.70	253.60	49.15	393.20		
24 L.H., Daily Totals		$887.20		$1375.60	$36.97	$57.32

Crew Q-3	Hr.	Daily	Hr.	Daily	Bare Costs	Incl. O&P
1 Plumber Foreman (inside)	$40.10	$320.80	$62.20	$497.60	$37.75	$58.54
2 Plumbers	39.60	633.60	61.40	982.40		
1 Plumber Apprentice	31.70	253.60	49.15	393.20		
32 L.H., Daily Totals		$1208.00		$1873.20	$37.75	$58.54

Crew Q-4	Hr.	Daily	Hr.	Daily	Bare Costs	Incl. O&P
1 Plumber Foreman (inside)	$40.10	$320.80	$62.20	$497.60	$37.75	$58.54
1 Plumber	39.60	316.80	61.40	491.20		
1 Welder (plumber)	39.60	316.80	61.40	491.20		
1 Plumber Apprentice	31.70	253.60	49.15	393.20		
1 Electric Welding Mach.		77.75		85.55	2.43	2.67
32 L.H., Daily Totals		$1285.75		$1958.75	$40.18	$61.21

Crew Q-5	Hr.	Daily	Hr.	Daily	Bare Costs	Incl. O&P
1 Steamfitter	$39.75	$318.00	$61.65	$493.20	$35.78	$55.48
1 Steamfitter Apprentice	31.80	254.40	49.30	394.40		
16 L.H., Daily Totals		$572.40		$887.60	$35.78	$55.48

Crew Q-6	Hr.	Daily	Hr.	Daily	Bare Costs	Incl. O&P
2 Steamfitters	$39.75	$636.00	$61.65	$986.40	$37.10	$57.53
1 Steamfitter Apprentice	31.80	254.40	49.30	394.40		
24 L.H., Daily Totals		$890.40		$1380.80	$37.10	$57.53

Crew Q-7	Hr.	Daily	Hr.	Daily	Bare Costs	Incl. O&P
1 Steamfitter Foreman (inside)	$40.25	$322.00	$62.45	$499.60	$37.89	$58.76
2 Steamfitters	39.75	636.00	61.65	986.40		
1 Steamfitter Apprentice	31.80	254.40	49.30	394.40		
32 L.H., Daily Totals		$1212.40		$1880.40	$37.89	$58.76

Crew Q-8	Hr.	Daily	Hr.	Daily	Bare Costs	Incl. O&P
1 Steamfitter Foreman (inside)	$40.25	$322.00	$62.45	$499.60	$37.89	$58.76
1 Steamfitter	39.75	318.00	61.65	493.20		
1 Welder (steamfitter)	39.75	318.00	61.65	493.20		
1 Steamfitter Apprentice	31.80	254.40	49.30	394.40		
1 Electric Welding Mach.		77.75		85.55	2.43	2.67
32 L.H., Daily Totals		$1290.15		$1965.95	$40.32	$61.43

Crew Q-9	Hr.	Daily	Hr.	Daily	Bare Costs	Incl. O&P
1 Sheet Metal Worker	$38.80	$310.40	$61.45	$491.60	$34.92	$55.33
1 Sheet Metal Apprentice	31.05	248.40	49.20	393.60		
16 L.H., Daily Totals		$558.80		$885.20	$34.92	$55.33

Crew Q-10	Hr.	Daily	Hr.	Daily	Bare Costs	Incl. O&P
2 Sheet Metal Workers	$38.80	$620.80	$61.45	$983.20	$36.22	$57.37
1 Sheet Metal Apprentice	31.05	248.40	49.20	393.60		
24 L.H., Daily Totals		$869.20		$1376.80	$36.22	$57.37

Crew Q-11	Hr.	Daily	Hr.	Daily	Bare Costs	Incl. O&P
1 Sheet Metal Foreman (inside)	$39.30	$314.40	$62.25	$498.00	$36.99	$58.59
2 Sheet Metal Workers	38.80	620.80	61.45	983.20		
1 Sheet Metal Apprentice	31.05	248.40	49.20	393.60		
32 L.H., Daily Totals		$1183.60		$1874.80	$36.99	$58.59

Crew Q-12	Hr.	Daily	Hr.	Daily	Bare Costs	Incl. O&P
1 Sprinkler Installer	$39.25	$314.00	$61.35	$490.80	$35.33	$55.22
1 Sprinkler Apprentice	31.40	251.20	49.10	392.80		
16 L.H., Daily Totals		$565.20		$883.60	$35.33	$55.22

Crew Q-13	Hr.	Daily	Hr.	Daily	Bare Costs	Incl. O&P
1 Sprinkler Foreman (inside)	$39.75	$318.00	$62.15	$497.20	$37.41	$58.49
2 Sprinkler Installers	39.25	628.00	61.35	981.60		
1 Sprinkler Apprentice	31.40	251.20	49.10	392.80		
32 L.H., Daily Totals		$1197.20		$1871.60	$37.41	$58.49

Crew No.	Bare Costs		Incl. Subs O & P		Cost Per Labor-Hour	

Crew Q-14	Hr.	Daily	Hr.	Daily	Bare Costs	Incl. O&P
1 Asbestos Worker	$36.00	$288.00	$58.50	$468.00	$32.40	$52.65
1 Asbestos Apprentice	28.80	230.40	46.80	374.40		
16 L.H., Daily Totals		$518.40		$842.40	$32.40	$52.65

Crew Q-15	Hr.	Daily	Hr.	Daily	Bare Costs	Incl. O&P
1 Plumber	$39.60	$316.80	$61.40	$491.20	$35.65	$55.28
1 Plumber Apprentice	31.70	253.60	49.15	393.20		
1 Electric Welding Mach.		77.75		85.55	4.86	5.35
16 L.H., Daily Totals		$648.15		$969.95	$40.51	$60.63

Crew Q-16	Hr.	Daily	Hr.	Daily	Bare Costs	Incl. O&P
2 Plumbers	$39.60	$633.60	$61.40	$982.40	$36.97	$57.32
1 Plumber Apprentice	31.70	253.60	49.15	393.20		
1 Electric Welding Mach.		77.75		85.55	3.24	3.56
24 L.H., Daily Totals		$964.95		$1461.15	$40.21	$60.88

Crew Q-17	Hr.	Daily	Hr.	Daily	Bare Costs	Incl. O&P
1 Steamfitter	$39.75	$318.00	$61.65	$493.20	$35.78	$55.48
1 Steamfitter Apprentice	31.80	254.40	49.30	394.40		
1 Electric Welding Mach.		77.75		85.55	4.86	5.35
16 L.H., Daily Totals		$650.15		$973.15	$40.64	$60.83

Crew Q-17A	Hr.	Daily	Hr.	Daily	Bare Costs	Incl. O&P
1 Steamfitter	$39.75	$318.00	$61.65	$493.20	$35.45	$55.27
1 Steamfitter Apprentice	31.80	254.40	49.30	394.40		
1 Equip. Oper. (crane)	34.80	278.40	54.85	438.80		
1 Truck Crane, 12 Ton		612.00		673.20		
1 Electric Welding Mach.		77.75		85.55	28.74	31.61
24 L.H., Daily Totals		$1540.55		$2085.15	$64.19	$86.88

Crew Q-18	Hr.	Daily	Hr.	Daily	Bare Costs	Incl. O&P
2 Steamfitters	$39.75	$636.00	$61.65	$986.40	$37.10	$57.53
1 Steamfitter Apprentice	31.80	254.40	49.30	394.40		
1 Electric Welding Mach.		77.75		85.55	3.24	3.56
24 L.H., Daily Totals		$968.15		$1466.35	$40.34	$61.09

Crew Q-19	Hr.	Daily	Hr.	Daily	Bare Costs	Incl. O&P
1 Steamfitter	$39.75	$318.00	$61.65	$493.20	$36.98	$57.17
1 Steamfitter Apprentice	31.80	254.40	49.30	394.40		
1 Electrician	39.40	315.20	60.55	484.40		
24 L.H., Daily Totals		$887.60		$1372.00	$36.98	$57.17

Crew Q-20	Hr.	Daily	Hr.	Daily	Bare Costs	Incl. O&P
1 Sheet Metal Worker	$38.80	$310.40	$61.45	$491.60	$35.82	$56.37
1 Sheet Metal Apprentice	31.05	248.40	49.20	393.60		
.5 Electrician	39.40	157.60	60.55	242.20		
20 L.H., Daily Totals		$716.40		$1127.40	$35.82	$56.37

Crew Q-21	Hr.	Daily	Hr.	Daily	Bare Costs	Incl. O&P
2 Steamfitters	$39.75	$636.00	$61.65	$986.40	$37.67	$58.29
1 Steamfitter Apprentice	31.80	254.40	49.30	394.40		
1 Electrician	39.40	315.20	60.55	484.40		
32 L.H., Daily Totals		$1205.60		$1865.20	$37.67	$58.29

Crew Q-22	Hr.	Daily	Hr.	Daily	Bare Costs	Incl. O&P
1 Plumber	$39.60	$316.80	$61.40	$491.20	$35.65	$55.28
1 Plumber Apprentice	31.70	253.60	49.15	393.20		
1 Truck Crane, 12 Ton		612.00		673.20	38.25	42.08
16 L.H., Daily Totals		$1182.40		$1557.60	$73.90	$97.36

Crew Q-22A	Hr.	Daily	Hr.	Daily	Bare Costs	Incl. O&P
1 Plumber	$39.60	$316.80	$61.40	$491.20	$33.03	$52.13
1 Plumber Apprentice	31.70	253.60	49.15	393.20		
1 Laborer	26.00	208.00	43.10	344.80		
1 Equip. Oper. (crane)	34.80	278.40	54.85	438.80		
1 Truck Crane, 12 Ton		612.00		673.20	19.13	21.04
32 L.H., Daily Totals		$1668.80		$2341.20	$52.16	$73.17

Crew Q-23	Hr.	Daily	Hr.	Daily	Bare Costs	Incl. O&P
1 Plumber Foreman	$41.60	$332.80	$64.50	$516.00	$38.28	$59.65
1 Plumber	39.60	316.80	61.40	491.20		
1 Equip. Oper. (medium)	33.65	269.20	53.05	424.40		
1 Power Tool		4.40		4.85		
1 Crane, 20 Ton		647.45		712.20	27.16	29.88
24 L.H., Daily Totals		$1570.65		$2148.65	$65.44	$89.53

Crew R-1	Hr.	Daily	Hr.	Daily	Bare Costs	Incl. O&P
1 Electrician Foreman	$39.90	$319.20	$61.35	$490.80	$34.53	$54.02
3 Electricians	39.40	945.60	60.55	1453.20		
2 Helpers	24.55	392.80	40.55	648.80		
48 L.H., Daily Totals		$1657.60		$2592.80	$34.53	$54.02

Crew R-1A	Hr.	Daily	Hr.	Daily	Bare Costs	Incl. O&P
1 Electrician	$39.40	$315.20	$60.55	$484.40	$31.98	$50.55
1 Helper	24.55	196.40	40.55	324.40		
16 L.H., Daily Totals		$511.60		$808.80	$31.98	$50.55

Crew R-2	Hr.	Daily	Hr.	Daily	Bare Costs	Incl. O&P
1 Electrician Foreman	$39.90	$319.20	$61.35	$490.80	$34.57	$54.14
3 Electricians	39.40	945.60	60.55	1453.20		
2 Helpers	24.55	392.80	40.55	648.80		
1 Equip. Oper. (crane)	34.80	278.40	54.85	438.80		
1 S.P. Crane, 5 Ton		317.80		349.60	5.68	6.24
56 L.H., Daily Totals		$2253.80		$3381.20	$40.25	$60.38

Crew R-3	Hr.	Daily	Hr.	Daily	Bare Costs	Incl. O&P
1 Electrician Foreman	$39.90	$319.20	$61.35	$490.80	$38.68	$59.73
1 Electrician	39.40	315.20	60.55	484.40		
.5 Equip. Oper. (crane)	34.80	139.20	54.85	219.40		
.5 S.P. Crane, 5 Ton		158.90		174.80	7.95	8.74
20 L.H., Daily Totals		$932.50		$1369.40	$46.63	$68.47

Crew R-4	Hr.	Daily	Hr.	Daily	Bare Costs	Incl. O&P
1 Struc. Steel Foreman	$39.15	$313.20	$72.90	$583.20	$38.00	$68.18
3 Struc. Steel Workers	37.15	891.60	69.15	1659.60		
1 Electrician	39.40	315.20	60.55	484.40		
1 Gas Welding Machine		75.20		82.70	1.88	2.07
40 L.H., Daily Totals		$1595.20		$2809.90	$39.88	$70.25

Crew R-5	Hr.	Daily	Hr.	Daily	Bare Costs	Incl. O&P
1 Electrician Foreman	$39.90	$319.20	$61.35	$490.80	$34.05	$53.35
4 Electrician Linemen	39.40	1260.80	60.55	1937.60		
2 Electrician Operators	39.40	630.40	60.55	968.80		
4 Electrician Groundmen	24.55	785.60	40.55	1297.60		
1 Crew Truck		109.10		120.00		
1 Tool Van		129.35		142.30		
1 Pickup Truck, 3/4 Ton		75.80		83.40		
.2 Crane, 55 Ton		182.20		200.40		
.2 Crane, 12 Ton		122.40		134.65		
.2 Auger, Truck Mtd.		575.40		632.95		
1 Tractor w/Winch		254.40		279.85	16.46	18.11
88 L.H., Daily Totals		$4444.65		$6288.35	$50.51	$71.46

Crew R-6	Hr.	Daily	Hr.	Daily	Bare Costs	Incl. O&P
1 Electrician Foreman	$39.90	$319.20	$61.35	$490.80	$34.05	$53.35
4 Electrician Linemen	39.40	1260.80	60.55	1937.60		
2 Electrician Operators	39.40	630.40	60.55	968.80		
4 Electrician Groundmen	24.55	785.60	40.55	1297.60		
1 Crew Truck		109.10		120.00		
1 Tool Van		129.35		142.30		
1 Pickup Truck, 3/4 Ton		75.80		83.40		
.2 Crane, 55 Ton		182.20		200.40		
.2 Crane, 12 Ton		122.40		134.65		
.2 Auger, Truck Mtd.		575.40		632.95		
1 Tractor w/Winch		254.40		279.85		
3 Cable Trailers		480.75		528.85		
.5 Tensioning Rig		157.45		173.20		
.5 Cable Pulling Rig		916.00		1007.60	34.12	37.54
88 L.H., Daily Totals		$5998.85		$7998.00	$68.17	$90.89

Crew R-7	Hr.	Daily	Hr.	Daily	Bare Costs	Incl. O&P
1 Electrician Foreman	$39.90	$319.20	$61.35	$490.80	$27.11	$44.02
5 Electrician Groundmen	24.55	982.00	40.55	1622.00		
1 Crew Truck		109.10		120.00	2.27	2.50
48 L.H., Daily Totals		$1410.30		$2232.80	$29.38	$46.52

Crew R-8	Hr.	Daily	Hr.	Daily	Bare Costs	Incl. O&P
1 Electrician Foreman	$39.90	$319.20	$61.35	$490.80	$34.53	$54.02
3 Electrician Linemen	39.40	945.60	60.55	1453.20		
2 Electrician Groundmen	24.55	392.80	40.55	648.80		
1 Pickup Truck, 3/4 Ton		75.80		83.40		
1 Crew Truck		109.10		120.00	3.85	4.24
48 L.H., Daily Totals		$1842.50		$2796.20	$38.38	$58.26

Crew R-9	Hr.	Daily	Hr.	Daily	Bare Costs	Incl. O&P
1 Electrician Foreman	$39.90	$319.20	$61.35	$490.80	$32.04	$50.65
1 Electrician Lineman	39.40	315.20	60.55	484.40		
2 Electrician Operators	39.40	630.40	60.55	968.80		
4 Electrician Groundmen	24.55	785.60	40.55	1297.60		
1 Pickup Truck, 3/4 Ton		75.80		83.40		
1 Crew Truck		109.10		120.00	2.89	3.18
64 L.H., Daily Totals		$2235.30		$3445.00	$34.93	$53.83

Crew R-10	Hr.	Daily	Hr.	Daily	Bare Costs	Incl. O&P
1 Electrician Foreman	$39.90	$319.20	$61.35	$490.80	$37.01	$57.35
4 Electrician Linemen	39.40	1260.80	60.55	1937.60		
1 Electrician Groundman	24.55	196.40	40.55	324.40		
1 Crew Truck		109.10		120.00		
3 Tram Cars		340.35		374.40	9.36	10.30
48 L.H., Daily Totals		$2225.85		$3247.20	$46.37	$67.65

Crew R-11	Hr.	Daily	Hr.	Daily	Bare Costs	Incl. O&P
1 Electrician Foreman	$39.90	$319.20	$61.35	$490.80	$35.36	$55.26
4 Electricians	39.40	1260.80	60.55	1937.60		
1 Equip. Oper. (crane)	34.80	278.40	54.85	438.80		
1 Helper	24.55	196.40	40.55	324.40		
1 Common Laborer	26.00	208.00	43.10	344.80		
1 Crew Truck		109.10		120.00		
1 Crane, 12 Ton		612.00		673.20	11.27	12.39
64 L.H., Daily Totals		$2983.90		$4329.60	$46.63	$67.65

Crew R-12	Hr.	Daily	Hr.	Daily	Bare Costs	Incl. O&P
1 Carpenter Foreman	$33.50	$268.00	$55.55	$444.40	$30.94	$51.72
4 Carpenters	33.00	1056.00	54.70	1750.40		
4 Common Laborers	26.00	832.00	43.10	1379.20		
1 Equip. Oper. (med.)	33.65	269.20	53.05	424.40		
1 Steel Worker	37.15	297.20	69.15	553.20		
1 Dozer, 200 H.P.		863.40		949.75		
1 Pickup Truck, 3/4 Ton		75.80		83.40	10.67	11.74
88 L.H., Daily Totals		$3661.60		$5584.75	$41.61	$63.46

Crew R-13	Hr.	Daily	Hr.	Daily	Bare Costs	Incl. O&P
1 Electrician Foreman	$39.90	$319.20	$61.35	$490.80	$37.33	$57.66
3 Electricians	39.40	945.60	60.55	1453.20		
.25 Equip. Oper. (crane)	34.80	69.60	54.85	109.70		
1 Equipment Oiler	29.20	233.60	46.00	368.00		
.25-1 Hyd. Crane, 33 Ton		162.25		178.50	3.86	4.25
42 L.H., Daily Totals		$1730.25		$2600.20	$41.19	$61.91

Crew R-15	Hr.	Daily	Hr.	Daily	Bare Costs	Incl. O&P
1 Electrician Foreman	$39.90	$319.20	$61.35	$490.80	$38.28	$59.03
4 Electricians	39.40	1260.80	60.55	1937.60		
1 Equipment Operator	32.15	257.20	50.65	405.20		
1 Aerial Lift Truck		250.20		275.20	5.21	5.73
48 L.H., Daily Totals		$2087.40		$3108.80	$43.49	$64.76

Crew R-18	Hr.	Daily	Hr.	Daily	Bare Costs	Incl. O&P
.25 Electrician Foreman	$39.90	$79.80	$61.35	$122.70	$30.30	$48.30
1 Electrician	39.40	315.20	60.55	484.40		
2 Helpers	24.55	392.80	40.55	648.80		
26 L.H., Daily Totals		$787.80		$1255.90	$30.30	$48.30

Crew R-19	Hr.	Daily	Hr.	Daily	Bare Costs	Incl. O&P
.5 Electrician Foreman	$39.90	$159.60	$61.35	$245.40	$39.50	$60.71
2 Electricians	39.40	630.40	60.55	968.80		
20 L.H., Daily Totals		$790.00		$1214.20	$39.50	$60.71

Crew R-21	Hr.	Daily	Hr.	Daily	Bare Costs	Incl. O&P
1 Electrician Foreman	$39.90	$319.20	$61.35	$490.80	$39.38	$60.56
3 Electricians	39.40	945.60	60.55	1453.20		
.1 Equip. Oper. (med.)	33.65	26.92	53.05	42.44		
.1 Hyd. Crane 25 Ton		57.52		63.25	1.75	1.93
32. L.H., Daily Totals		$1349.24		$2049.69	$41.13	$62.49

Crew R-22	Hr.	Daily	Hr.	Daily	Bare Costs	Incl. O&P
.66 Electrician Foreman	$39.90	$210.67	$61.35	$323.93	$33.10	$52.08
2 Helpers	24.55	392.80	40.55	648.80		
2 Electricians	39.40	630.40	60.55	968.80		
37.28 L.H., Daily Totals		$1233.87		$1941.53	$33.10	$52.08

Crew R-30	Hr.	Daily	Hr.	Daily	Bare Costs	Incl. O&P
.25 Electrician	$41.40	$82.80	$63.65	$127.30	$31.31	$50.05
1 Electrician	39.40	315.20	60.55	484.40		
2 Laborers, (Semi-Skilled)	26.00	416.00	43.10	689.60		
26 L.H., Daily Totals		$814.00		$1301.30	$31.31	$50.05

Crew R-31	Hr.	Daily	Hr.	Daily	Bare Costs	Incl. O&P
1 Electrician	$39.40	$315.20	$60.55	$484.40	$39.40	$60.55
1 Core Drill, Elec, 2.5 HP		47.45		52.20	5.93	6.52
8 L.H., Daily Totals		$362.65		$536.60	$45.33	$67.07

Historical Cost Indexes

The table below lists both the Means Historical Cost Index based on Jan. 1, 1993 = 100 as well as the computed value of an index based on Jan. 1, 2004 costs. Since the Jan. 1, 2004 figure is estimated, space is left to write in the actual index figures as they become available through either the quarterly "Means Construction Cost Indexes" or as printed in the "Engineering News-Record." To compute the actual index based on Jan. 1, 2004 = 100, divide the Historical Cost Index for a particular year by the actual Jan. 1, 2004 Construction Cost Index. Space has been left to advance the index figures as the year progresses.

Year	Historical Cost Index Jan. 1, 1993 = 100		Current Index Based on Jan. 1, 2004 = 100	
	Est.	Actual	Est.	Actual
Oct 2004				
July 2004				
April 2004				
Jan 2004	133.0		100.0	100.0
July 2003		132.0	99.2	
2002		128.7	96.8	
2001		125.1	94.1	
2000		120.9	90.9	
1999		117.6	88.4	
1998		115.1	86.5	
1997		112.8	84.8	
1996		110.2	82.9	
1995		107.6	80.9	
1994		104.4	78.5	
1993		101.7	76.5	
1992		99.4	74.8	
1991		96.8	72.8	
1990		94.3	70.9	

Year	Historical Cost Index Jan. 1, 1993 = 100	Current Index Based on Jan. 1, 2004 = 100	
	Actual	Est.	Actual
July 1989	92.1	69.3	
1988	89.9	67.6	
1987	87.7	65.9	
1986	84.2	63.3	
1985	82.6	62.1	
1984	82.0	61.6	
1983	80.2	60.3	
1982	76.1	57.3	
1981	70.0	52.6	
1980	62.9	47.3	
1979	57.8	43.5	
1978	53.5	40.2	
1977	49.5	37.2	
1976	46.9	35.3	
1975	44.8	33.7	
1974	41.4	31.1	
1973	37.7	28.3	
1972	34.8	26.2	

Year	Historical Cost Index Jan. 1, 1993 = 100	Current Index Based on Jan. 1, 2004 = 100	
	Actual	Est.	Actual
July 1971	32.1	24.1	
1970	28.7	21.6	
1969	26.9	20.2	
1968	24.9	18.7	
1967	23.5	17.7	
1966	22.7	17.1	
1965	21.7	16.3	
1964	21.2	15.9	
1963	20.7	15.6	
1962	20.2	15.2	
1961	19.8	14.9	
1960	19.7	14.8	
1959	19.3	14.5	
1958	18.8	14.1	
1957	18.4	13.8	
1956	17.6	13.2	
1955	16.6	12.5	
1954	16.0	12.0	

Adjustments to Costs

The Historical Cost Index can be used to convert National Average building costs at a particular time to the approximate building costs for some other time.

Time Adjustment using the Historical Cost Indexes:

$$\frac{\text{Index for Year A}}{\text{Index for Year B}} \times \text{Cost in Year B} = \text{Cost in Year A}$$

Example:

Estimate and compare construction costs for different years in the same city.

To estimate the National Average construction cost of a building in 1970, knowing that it cost $900,000 in 2004:

INDEX in 1970 = 28.7

INDEX in 2004 = 133.0

$$\frac{\text{INDEX 1970}}{\text{INDEX 2004}} \times \text{Cost 2004} = \text{Cost 1970}$$

$$\frac{28.7}{133.0} \times \$900{,}000 = .216 \times \$900{,}000 = \$194{,}400$$

The construction cost of the building in 1970 is $194,400.

Note: The City Cost Indexes for Canada can be used to convert U.S. National averages to local costs in Canadian dollars.

How to Use the City Cost Indexes

What you should know before you begin

Means City Cost Indexes (CCI) are an extremely useful tool to use when you want to compare costs from city to city and region to region.

This publication contains average construction cost indexes for 316 major U.S. and Canadian cities and Location Factors covering over 930 three-digit zip code locations.

Keep in mind that a City Cost Index number is a *percentage ratio* of a specific city's cost to the national average cost of the same item at a stated time period.

In other words, these index figures represent relative construction *factors* (or, if you prefer, multipliers) for Material and Installation costs, as well as the weighted average for Total In Place costs for each CSI MasterFormat division. Installation costs include both labor and equipment rental costs. When estimating equipment rental rates only, for a specific location, use 01590 EQUIPMENT RENTAL index.

The 30 City Average Index is the average of 30 major U.S. cities and serves as a National Average.

Index figures for both material and installation are based on the 30 major city average of 100 and represent the cost relationship as of July 1, 2003. The index for each division is computed from representative material and labor quantities for that division. The weighted average for each city is a weighted total of the components listed above it, but does not include relative productivity between trades or cities.

As changes occur in local material prices, labor rates and equipment rental rates, the impact of these changes should be accurately measured by the change in the City Cost Index for each particular city (as compared to the 30 City Average).

Therefore, if you know (or have estimated) building costs in one city today, you can easily convert those costs to expected building costs in another city.

In addition, by using the Historical Cost Index, you can easily convert National Average building costs at a particular time to the approximate building costs for some other time. The City Cost Indexes can then be applied to calculate the costs for a particular city.

Quick Calculations

Location Adjustment Using the City Cost Indexes:

$$\frac{\text{Index for City A}}{\text{Index for City B}} \times \text{Cost in City B} = \text{Cost in City A}$$

Time Adjustment for the National Average Using the Historical Cost Index:

$$\frac{\text{Index for Year A}}{\text{Index for Year B}} \times \text{Cost in Year B} = \text{Cost in Year A}$$

Adjustment from the National Average:

$$\frac{\text{Index for City A}}{100} \times \text{National Average Cost} = \text{Cost in City A}$$

Since each of the other RSMeans publications contains many different items, any *one* item multiplied by the particular city index may give incorrect results. However, the larger the number of items compiled, the closer the results should be to actual costs for that particular city.

The City Cost Indexes for Canadian cities are calculated using Canadian material and equipment prices and labor rates, in Canadian dollars. Therefore, indexes for Canadian cities can be used to convert U.S. National Average prices to local costs in Canadian dollars.

How to use this section

1. Compare costs from city to city.

In using the Means Indexes, remember that an index number is not a fixed number but a *ratio:* It's a percentage ratio of a building component's cost at any stated time to the National Average cost of that same component at the same time period. Put in the form of an equation:

$$\frac{\text{Specific City Cost}}{\text{National Average Cost}} \times 100 = \text{City Index Number}$$

Therefore, when making cost comparisons between cities, do not subtract one city's index number from the index number of another city and read the result as a percentage difference. Instead, divide one city's index number by that of the other city. The resulting number may then be used as a multiplier to calculate cost differences from city to city.

The formula used to find cost differences between cities for the purpose of comparison is as follows:

$$\frac{\text{City A Index}}{\text{City B Index}} \times \text{City B Cost (Known)} = \text{City A Cost (Unknown)}$$

In addition, you can use *Means CCI* to calculate and compare costs division by division between cities using the same basic formula. (Just be sure that you're comparing similar divisions.)

2. Compare a specific city's construction costs with the National Average.

When you're studying construction location feasibility, it's advisable to compare a prospective project's cost index with an index of the National Average cost.

For example, divide the weighted average index of construction costs of a specific city by that of the 30 City Average, which = 100.

$$\frac{\text{City Index}}{100} = \text{\% of National Average}$$

As a result, you get a ratio that indicates the relative cost of construction in that city in comparison with the National Average.

3. Convert U.S. National Average to actual costs in Canadian City.

$$\frac{\text{Index for Canadian City}}{100} \times \text{National Average Cost} = \text{Cost in Canadian City in \\$ CAN}$$

4. Adjust construction cost data based on a National Average.

When you use a source of construction cost data which is based on a National Average (such as *Means cost data publications*), it is necessary to adjust those costs to a specific location.

$$\frac{\text{City Index}}{100} \times \frac{\text{"Book" Cost Based on}}{\text{National Average Costs}} = \frac{\text{City Cost}}{\text{(Unknown)}}$$

5. When applying the City Cost Indexes to demolition projects, use the appropriate division index. For example, for removal of existing doors and windows, use the Division 8 index.

What you might like to know about how we developed the Indexes

To create a reliable index, RSMeans researched the building type most often constructed in the United States and Canada. Because it was concluded that no one type of building completely represented the building construction industry, nine different types of buildings were combined to create a composite model.

The exact material, labor and equipment quantities are based on detailed analysis of these nine building types, then each quantity is weighted in proportion to expected usage. These various material items, labor hours, and equipment rental rates are thus combined to form a composite building representing as closely as possible the actual usage of materials, labor and equipment used in the North American Building Construction Industry.

The following structures were chosen to make up that composite model:

1. Factory, 1 story
2. Office, 2–4 story
3. Store, Retail
4. Town Hall, 2–3 story
5. High School, 2–3 story
6. Hospital, 4–8 story
7. Garage, Parking
8. Apartment, 1–3 story
9. Hotel/Motel, 2–3 story

For the purposes of ensuring the timeliness of the data, the components of the index for the composite model have been streamlined. They currently consist of:

- specific quantities of 66 commonly used construction materials;
- specific labor-hours for 21 building construction trades; and
- specific days of equipment rental for 6 types of construction equipment (normally used to install the 66 material items by the 21 trades.)

A sophisticated computer program handles the updating of all costs for each city on a quarterly basis. Material and equipment price quotations are gathered quarterly from over 316 cities in the United States and Canada. These prices and the latest negotiated labor wage rates for 21 different building trades are used to compile the quarterly update of the City Cost Index.

The 30 major U.S. cities used to calculate the National Average are:

Atlanta, GA	Memphis, TN
Baltimore, MD	Milwaukee, WI
Boston, MA	Minneapolis, MN
Buffalo, NY	Nashville, TN
Chicago, IL	New Orleans, LA
Cincinnati, OH	New York, NY
Cleveland, OH	Philadelphia, PA
Columbus, OH	Phoenix, AZ
Dallas, TX	Pittsburgh, PA
Denver, CO	St. Louis, MO
Detroit, MI	San Antonio, TX
Houston, TX	San Diego, CA
Indianapolis, IN	San Francisco, CA
Kansas City, MO	Seattle, WA
Los Angeles, CA	Washington, DC

F.Y.I.: The CSI MasterFormat Divisions

1. General Requirements
2. Site Construction
3. Concrete
4. Masonry
5. Metals
6. Wood & Plastics
7. Thermal & Moisture Protection
8. Doors & Windows
9. Finishes
10. Specialties
11. Equipment
12. Furnishings
13. Special Construction
14. Conveying Systems
15. Mechanical
16. Electrical

The information presented in the CCI is organized according to the Construction Specifications Institute (CSI) MasterFormat.

What the CCI does not indicate

The weighted average for each city is a total of the components listed above weighted to reflect typical usage, but it does *not* include the productivity variations between trades or cities.

In addition, the CCI does not take into consideration factors such as the following:

- managerial efficiency
- competitive conditions
- automation
- restrictive union practices
- unique local requirements
- regional variations due to specific building codes

	DIVISION	UNITED STATES 30 CITY AVERAGE			ALABAMA BIRMINGHAM			ALABAMA HUNTSVILLE			ALABAMA MOBILE			ALABAMA MONTGOMERY			ALABAMA TUSCALOOSA		
		MAT.	INST.	TOTAL	MAT.	INST.	TOTAL	MAT.	INST.	TOTAL	MAT.	INST.	TOTAL	MAT.	INST.	TOTAL	MAT.	INST.	TOTAL
01590	EQUIPMENT RENTAL	.0	100.0	100.0	.0	101.4	101.4	.0	101.3	101.3	.0	97.8	97.8	.0	97.8	97.8	.0	101.3	101.3
02	SITE CONSTRUCTION	100.0	100.0	100.0	85.9	93.3	91.5	83.9	92.8	90.6	95.3	86.3	88.5	95.7	86.8	89.0	84.4	92.1	90.2
03100	CONCRETE FORMS & ACCESSORIES	100.0	100.0	100.0	91.9	76.1	78.1	93.7	70.5	73.3	93.7	54.8	59.6	92.2	49.6	54.9	93.6	40.9	47.4
03200	CONCRETE REINFORCEMENT	100.0	100.0	100.0	92.4	87.1	89.2	92.4	79.2	84.4	95.4	55.2	71.1	95.4	85.6	89.5	92.4	85.9	88.5
03300	CAST-IN-PLACE CONCRETE	100.0	100.0	100.0	93.5	69.8	83.6	88.3	67.4	79.6	93.3	56.3	77.8	95.0	50.7	76.5	92.0	48.4	73.8
03	CONCRETE	100.0	100.0	100.0	91.0	77.1	83.9	88.6	72.3	80.2	91.4	57.0	73.8	92.1	58.6	74.9	90.4	54.0	71.7
04	MASONRY	100.0	100.0	100.0	85.7	78.1	80.9	85.6	66.6	73.7	86.2	54.0	66.0	86.8	37.5	55.9	85.9	39.8	57.1
05	METALS	100.0	100.0	100.0	97.7	95.8	97.0	97.3	91.4	95.1	95.9	80.9	90.3	95.4	92.5	94.3	96.4	93.1	95.1
06	WOOD & PLASTICS	100.0	100.0	100.0	93.1	76.5	84.2	92.4	71.2	81.0	92.4	54.9	72.2	90.7	50.3	68.9	92.4	39.9	64.1
07	THERMAL & MOISTURE PROTECTION	100.0	100.0	100.0	95.7	81.5	88.8	95.4	75.7	85.8	95.4	70.2	83.1	95.1	63.4	79.7	95.3	62.2	79.2
08	DOORS & WINDOWS	100.0	100.0	100.0	98.5	79.1	93.6	98.5	67.8	90.7	98.5	55.1	87.5	98.5	59.4	88.6	98.5	59.3	88.5
09200	PLASTER & GYPSUM BOARD	100.0	100.0	100.0	102.1	76.4	84.6	98.7	71.0	79.8	98.7	54.2	68.4	98.7	49.4	65.2	98.7	38.7	57.9
095,098	CEILINGS & ACOUSTICAL TREATMENT	100.0	100.0	100.0	95.0	76.4	82.4	95.0	71.0	78.7	95.0	54.2	67.4	95.0	49.4	64.2	95.0	38.7	56.9
09600	FLOORING	100.0	100.0	100.0	104.3	52.5	91.0	104.3	45.4	89.1	114.7	56.9	99.8	112.9	29.4	91.4	104.3	43.2	88.6
097,099	WALL FINISHES, PAINTS & COATINGS	100.0	100.0	100.0	95.3	72.0	81.4	95.3	66.2	77.9	100.2	58.7	75.3	95.3	57.7	72.8	95.3	49.7	68.0
09	FINISHES	100.0	100.0	100.0	99.0	71.1	83.9	98.4	65.1	80.4	103.5	55.2	77.4	102.6	45.6	71.8	98.4	40.7	67.2
10 - 14	TOTAL DIV. 10000 - 14000	100.0	100.0	100.0	100.0	80.8	95.9	100.0	78.9	95.5	100.0	73.3	94.3	100.0	71.2	93.8	100.0	68.5	93.2
15	MECHANICAL	100.0	100.0	100.0	99.8	67.8	85.1	99.8	64.9	83.8	99.8	63.2	83.0	99.8	43.1	73.7	99.8	37.9	71.3
16	ELECTRICAL	100.0	100.0	100.0	97.6	65.6	78.9	97.8	69.2	81.1	97.8	53.3	71.8	97.6	66.8	79.6	97.8	65.6	79.0
01 - 16	WEIGHTED AVERAGE	100.0	100.0	100.0	96.6	76.5	86.9	96.2	72.6	84.8	97.2	62.8	80.6	97.1	59.0	78.7	96.3	57.1	77.4

	DIVISION	ALASKA ANCHORAGE			ALASKA FAIRBANKS			ALASKA JUNEAU			ARIZONA FLAGSTAFF			ARIZONA MESA/TEMPE			ARIZONA PHOENIX		
		MAT.	INST.	TOTAL	MAT.	INST.	TOTAL	MAT.	INST.	TOTAL	MAT.	INST.	TOTAL	MAT.	INST.	TOTAL	MAT.	INST.	TOTAL
01590	EQUIPMENT RENTAL	.0	118.4	118.4	.0	118.4	118.4	.0	118.4	118.4	.0	94.7	94.7	.0	97.5	97.5	.0	98.1	98.1
02	SITE CONSTRUCTION	143.9	134.0	136.5	127.3	134.0	132.3	139.5	134.0	135.4	82.2	101.2	96.4	85.5	104.0	99.3	85.9	104.9	100.1
03100	CONCRETE FORMS & ACCESSORIES	133.3	118.0	119.9	135.2	124.0	125.4	134.9	118.0	120.1	103.1	63.4	68.3	100.7	65.3	69.6	101.9	70.6	74.4
03200	CONCRETE REINFORCEMENT	137.9	108.1	119.9	115.9	108.1	111.2	102.6	108.1	105.9	101.6	75.2	85.7	100.0	73.4	83.9	98.1	75.9	84.7
03300	CAST-IN-PLACE CONCRETE	193.1	117.7	161.6	160.9	118.2	143.1	193.9	117.7	162.0	94.7	79.8	88.5	101.4	70.6	88.5	101.5	79.9	92.5
03	CONCRETE	151.3	115.4	132.9	127.9	118.3	123.0	147.7	115.4	131.2	119.0	71.2	94.5	100.2	68.5	84.0	99.8	74.6	86.9
04	MASONRY	220.7	124.7	160.6	214.1	124.7	158.2	223.4	124.7	161.6	102.5	57.0	74.0	107.2	54.4	74.1	95.1	68.1	78.2
05	METALS	129.4	103.2	119.5	129.5	103.5	119.7	129.7	103.2	119.7	97.8	69.1	86.9	98.6	69.1	87.5	100.0	71.8	89.4
06	WOOD & PLASTICS	115.0	116.0	115.5	115.2	123.8	119.8	115.0	116.0	115.5	108.6	62.5	83.8	103.7	71.4	86.3	104.8	71.5	86.9
07	THERMAL & MOISTURE PROTECTION	199.5	119.0	160.3	195.5	121.7	159.5	196.2	119.0	158.6	107.4	67.8	88.1	105.8	65.5	86.2	105.7	71.0	88.8
08	DOORS & WINDOWS	127.0	111.8	123.1	124.1	116.0	122.0	124.1	111.8	121.0	102.2	66.4	93.1	99.1	67.1	91.0	100.2	71.3	92.9
09200	PLASTER & GYPSUM BOARD	136.6	116.3	122.8	136.6	124.3	128.2	136.6	116.3	122.8	95.2	61.5	72.2	99.6	70.5	79.8	100.1	70.7	80.1
095,098	CEILINGS & ACOUSTICAL TREATMENT	126.4	116.3	119.5	126.4	124.3	124.9	126.4	116.3	119.5	107.2	61.5	76.2	108.0	70.5	82.6	108.0	70.7	82.7
09600	FLOORING	164.6	132.1	156.2	164.6	132.1	156.2	164.6	132.1	156.2	95.7	53.9	84.9	97.8	67.2	89.9	98.1	70.2	90.9
097,099	WALL FINISHES, PAINTS & COATINGS	167.8	110.4	133.4	167.8	124.7	141.9	167.8	110.4	133.4	94.8	50.4	68.2	105.2	54.8	75.0	105.2	62.3	79.5
09	FINISHES	157.0	119.9	137.0	154.5	126.1	139.1	155.4	119.9	136.2	98.1	59.5	77.2	100.4	64.3	80.9	100.6	69.2	83.7
10 - 14	TOTAL DIV. 10000 - 14000	100.0	119.7	104.2	100.0	120.8	104.5	100.0	119.7	104.2	100.0	76.5	95.0	100.0	73.0	94.2	100.0	78.1	95.3
15	MECHANICAL	100.4	108.8	104.3	100.4	114.7	107.0	100.4	105.9	102.9	100.2	77.2	89.6	100.2	69.1	85.9	100.2	77.3	89.6
16	ELECTRICAL	153.3	116.7	131.9	155.8	116.7	133.0	155.8	116.7	133.0	100.6	49.9	70.9	96.4	62.6	76.7	104.2	65.0	81.3
01 - 16	WEIGHTED AVERAGE	135.1	116.5	126.1	131.0	119.2	125.3	134.4	115.9	125.5	102.0	68.9	86.1	99.7	69.2	85.0	100.0	74.8	87.9

	DIVISION	ARIZONA PRESCOTT			ARIZONA TUCSON			ARKANSAS FORT SMITH			ARKANSAS JONESBORO			ARKANSAS LITTLE ROCK			ARKANSAS PINE BLUFF		
		MAT.	INST.	TOTAL	MAT.	INST.	TOTAL	MAT.	INST.	TOTAL	MAT.	INST.	TOTAL	MAT.	INST.	TOTAL	MAT.	INST.	TOTAL
01590	EQUIPMENT RENTAL	.0	94.7	94.7	.0	97.5	97.5	.0	85.9	85.9	.0	107.7	107.7	.0	85.9	85.9	.0	85.9	85.9
02	SITE CONSTRUCTION	70.0	100.6	92.9	82.3	104.6	99.0	77.4	84.0	82.3	101.0	99.5	99.9	77.2	84.0	82.3	79.5	84.0	82.9
03100	CONCRETE FORMS & ACCESSORIES	98.4	58.0	63.0	101.4	70.1	73.9	99.8	43.5	50.4	85.7	49.6	54.0	93.6	61.4	65.3	77.8	61.2	63.2
03200	CONCRETE REINFORCEMENT	101.6	73.0	84.3	97.1	75.2	83.9	97.6	76.4	84.8	93.3	50.8	67.6	97.8	73.9	83.4	97.7	73.9	83.3
03300	CAST-IN-PLACE CONCRETE	94.6	65.8	82.6	104.3	79.8	94.1	89.9	68.7	81.0	85.6	58.7	74.3	89.9	68.8	81.1	82.4	68.7	76.7
03	CONCRETE	103.7	63.6	83.2	101.1	74.2	87.3	87.0	59.2	72.7	83.7	54.7	68.8	86.6	66.7	76.4	84.6	66.5	75.3
04	MASONRY	103.0	60.6	76.5	97.0	56.9	71.9	96.8	55.0	70.6	91.9	48.3	64.6	95.1	55.0	70.0	117.3	55.0	78.3
05	METALS	97.8	67.8	86.5	98.6	69.8	87.7	97.2	72.6	87.9	91.4	77.3	86.1	96.8	72.1	87.5	95.9	71.9	86.8
06	WOOD & PLASTICS	104.2	56.7	78.6	104.0	71.5	86.5	104.0	42.2	70.7	90.0	50.8	68.8	100.9	65.8	82.0	82.7	65.8	73.6
07	THERMAL & MOISTURE PROTECTION	105.6	63.6	85.2	107.0	64.6	86.4	99.3	50.0	75.3	109.1	53.4	82.0	98.0	52.5	75.8	98.0	52.5	75.8
08	DOORS & WINDOWS	102.2	60.3	91.6	96.1	71.3	89.8	97.1	47.4	84.5	98.7	49.6	86.2	97.1	60.4	87.8	92.5	60.4	84.3
09200	PLASTER & GYPSUM BOARD	93.1	55.5	67.6	100.4	70.7	80.2	86.1	41.3	55.6	93.5	49.7	63.7	86.1	65.6	72.1	80.1	65.6	70.2
095,098	CEILINGS & ACOUSTICAL TREATMENT	107.2	55.5	72.2	109.4	70.7	83.1	98.1	41.3	59.6	98.7	49.7	65.5	98.1	65.6	76.1	94.2	65.6	74.8
09600	FLOORING	94.0	53.7	83.6	97.2	53.9	86.1	116.5	72.9	105.2	77.8	47.4	70.0	117.7	72.9	106.2	105.4	72.9	97.0
097,099	WALL FINISHES, PAINTS & COATINGS	94.8	50.4	68.2	102.8	50.4	71.3	96.6	60.2	74.7	85.0	54.1	66.5	96.6	61.7	75.7	96.6	61.7	75.7
09	FINISHES	96.1	55.9	74.3	100.3	64.8	81.1	99.0	50.2	72.6	89.7	49.7	68.1	99.3	64.2	80.4	94.0	64.2	77.9
10 - 14	TOTAL DIV. 10000 - 14000	100.0	75.4	94.7	100.0	78.1	95.3	100.0	67.3	93.0	100.0	58.2	91.0	100.0	70.4	93.6	100.0	70.4	93.6
15	MECHANICAL	100.2	75.0	88.6	100.1	71.1	86.8	100.1	44.7	74.6	100.3	44.6	74.6	100.1	58.6	81.0	100.1	47.2	75.8
16	ELECTRICAL	100.2	48.0	69.7	99.2	63.3	78.2	95.9	67.0	79.0	101.8	49.6	71.3	96.5	72.8	82.7	95.0	72.8	82.0
01 - 16	WEIGHTED AVERAGE	99.6	66.4	83.6	99.0	71.2	85.6	96.4	58.6	78.2	95.6	56.7	76.9	96.3	65.9	81.6	95.8	63.6	80.3

DIVISION		ARKANSAS			CALIFORNIA																	
		TEXARKANA			ANAHEIM			BAKERSFIELD			FRESNO			LOS ANGELES			OAKLAND		
		MAT.	INST.	TOTAL	MAT.	INST.	TOTAL	MAT.	INST.	TOTAL	MAT.	INST.	TOTAL	MAT.	INST.	TOTAL	MAT.	INST.	TOTAL
01590	EQUIPMENT RENTAL	.0	86.6	86.6	.0	102.1	102.1	.0	99.5	99.5	.0	99.5	99.5	.0	97.6	97.6	.0	103.5	103.5
02	SITE CONSTRUCTION	96.1	84.8	87.7	102.6	109.4	107.7	107.6	106.5	106.7	109.1	106.1	106.9	96.4	108.0	105.1	154.4	104.7	117.2
03100	CONCRETE FORMS & ACCESSORIES	86.1	41.1	46.6	104.3	124.1	121.7	97.5	124.0	120.7	100.2	131.5	127.7	106.2	123.8	121.7	107.4	136.1	132.5
03200	CONCRETE REINFORCEMENT	97.2	49.2	68.2	109.0	113.7	111.8	104.0	113.7	109.8	107.6	113.7	111.3	106.0	113.9	110.8	97.6	114.6	107.9
03300	CAST-IN-PLACE CONCRETE	89.8	45.8	71.4	105.8	120.4	111.9	101.0	119.4	108.7	110.6	115.1	112.4	85.9	117.8	99.2	132.7	120.3	127.5
03	CONCRETE	83.6	45.2	64.0	109.4	119.9	114.8	106.0	119.4	112.9	111.4	121.3	116.5	101.2	118.8	110.2	123.8	125.4	124.6
04	MASONRY	96.8	33.9	57.4	89.9	119.9	108.7	108.8	119.7	115.6	112.5	109.7	110.8	99.7	123.9	114.9	149.0	124.9	133.9
05	METALS	88.3	60.5	77.7	110.0	100.5	106.4	104.4	100.0	102.7	106.3	100.3	104.0	111.3	98.8	106.6	101.5	105.9	103.2
06	WOOD & PLASTICS	91.7	44.0	66.0	91.8	121.4	107.8	84.1	121.5	104.3	97.2	135.1	117.6	89.8	121.0	106.6	100.9	138.0	120.9
07	THERMAL & MOISTURE PROTECTION	98.8	43.7	72.0	121.4	117.9	119.7	105.6	114.2	109.8	101.3	111.2	106.2	111.6	120.0	115.7	113.2	127.3	120.1
08	DOORS & WINDOWS	97.5	42.5	83.6	102.8	116.4	106.2	101.6	112.9	104.5	101.9	121.7	106.9	96.5	116.2	101.5	106.3	126.5	111.5
09200	PLASTER & GYPSUM BOARD	83.9	43.2	56.2	99.0	122.0	114.6	99.0	122.0	114.6	97.9	135.9	123.8	90.6	122.0	112.0	98.7	138.4	125.7
095,098	CEILINGS & ACOUSTICAL TREATMENT	99.4	43.2	61.3	117.3	122.0	120.4	117.3	122.0	120.4	117.3	135.9	129.9	111.4	122.0	118.6	113.8	138.4	130.5
09600	FLOORING	108.2	45.2	92.0	122.6	104.9	118.1	116.8	95.4	111.3	133.2	140.0	135.0	114.5	104.9	112.0	115.1	121.5	116.8
097,099	WALL FINISHES, PAINTS & COATINGS	96.6	37.9	61.4	111.1	108.3	109.4	110.6	95.4	101.5	134.3	98.2	112.6	99.8	109.8	105.8	115.8	140.6	130.7
09	FINISHES	96.9	41.6	67.0	113.9	118.9	116.6	114.2	116.0	115.2	121.5	130.9	126.6	107.6	118.7	113.6	116.7	134.8	126.5
10 - 14	TOTAL DIV. 10000 - 14000	100.0	41.6	87.4	100.0	116.2	103.5	100.0	128.0	106.0	100.0	128.4	106.1	100.0	115.0	103.2	100.0	131.5	106.8
15	MECHANICAL	100.1	39.4	72.2	100.1	112.1	105.7	100.1	105.7	102.7	100.1	111.2	105.2	100.0	112.0	105.6	100.2	128.0	113.0
16	ELECTRICAL	97.6	41.3	64.7	95.3	107.1	102.2	95.2	97.2	96.4	94.6	95.6	95.2	107.7	112.3	110.4	109.7	139.6	127.2
01 - 16	WEIGHTED AVERAGE	95.2	46.7	71.8	103.8	113.3	108.4	103.2	109.9	106.5	105.1	112.4	108.6	102.9	114.0	108.3	111.1	125.7	118.2

DIVISION		CALIFORNIA																	
		OXNARD			REDDING			RIVERSIDE			SACRAMENTO			SAN DIEGO			SAN FRANCISCO		
		MAT.	INST.	TOTAL	MAT.	INST.	TOTAL	MAT.	INST.	TOTAL	MAT.	INST.	TOTAL	MAT.	INST.	TOTAL	MAT.	INST.	TOTAL
01590	EQUIPMENT RENTAL	.0	98.1	98.1	.0	99.1	99.1	.0	100.6	100.6	.0	103.1	103.1	.0	97.4	97.4	.0	108.9	108.9
02	SITE CONSTRUCTION	109.1	104.2	105.4	114.2	105.4	107.6	100.2	107.1	105.4	117.4	111.1	112.7	102.9	100.8	101.3	156.7	111.3	122.7
03100	CONCRETE FORMS & ACCESSORIES	103.0	124.1	121.5	102.4	131.6	128.0	104.7	124.0	121.6	105.7	132.1	128.9	106.3	106.0	106.1	107.7	137.3	133.7
03200	CONCRETE REINFORCEMENT	104.0	113.4	109.7	100.8	113.8	108.7	103.1	113.5	109.4	98.7	113.9	107.9	106.2	113.4	110.6	111.2	115.0	113.5
03300	CAST-IN-PLACE CONCRETE	107.2	119.5	112.4	121.4	115.1	118.8	104.7	120.4	111.3	114.4	116.2	115.1	112.9	107.5	110.6	132.6	122.0	128.1
03	CONCRETE	109.5	119.4	114.6	119.9	121.3	120.6	108.3	119.8	114.2	114.6	121.9	118.3	114.8	107.3	111.0	125.3	126.6	126.0
04	MASONRY	113.7	114.2	114.0	116.4	111.6	113.4	86.4	117.4	105.8	119.4	111.6	114.5	100.5	106.2	104.1	149.2	128.1	136.0
05	METALS	103.9	99.6	102.3	108.3	100.6	105.4	109.9	100.2	106.2	97.6	100.0	98.5	109.8	99.4	105.9	107.0	107.6	107.2
06	WOOD & PLASTICS	91.6	121.5	107.7	92.2	135.1	115.3	91.8	121.4	107.8	95.1	135.4	116.8	97.8	102.1	100.1	100.9	138.2	121.0
07	THERMAL & MOISTURE PROTECTION	110.6	113.7	112.1	111.4	112.0	111.7	119.2	114.5	116.9	124.3	113.7	119.1	113.9	103.9	109.0	113.2	131.8	122.3
08	DOORS & WINDOWS	100.4	116.5	104.5	103.2	123.0	108.3	102.8	116.4	106.2	117.8	123.9	119.4	104.3	104.9	104.5	110.8	130.3	115.8
09200	PLASTER & GYPSUM BOARD	99.0	122.0	114.6	98.1	135.9	123.8	98.4	122.0	114.4	96.8	135.9	123.4	100.9	101.9	101.6	100.5	138.4	126.3
095,098	CEILINGS & ACOUSTICAL TREATMENT	117.3	122.0	120.4	122.5	135.9	131.6	114.7	122.0	119.6	120.3	135.9	130.9	112.6	101.9	105.3	118.7	138.4	132.1
09600	FLOORING	116.8	104.9	113.8	116.1	111.0	114.8	121.3	104.9	117.1	117.9	114.8	117.1	110.8	107.6	110.0	115.1	127.7	118.4
097,099	WALL FINISHES, PAINTS & COATINGS	110.0	101.6	104.9	110.0	115.8	113.5	107.4	108.3	108.0	113.1	115.8	114.7	109.6	108.3	108.8	115.8	153.4	138.3
09	FINISHES	114.0	118.2	116.2	115.1	127.8	122.0	112.3	118.9	115.8	116.6	128.8	123.2	110.7	106.1	108.2	118.0	137.9	128.8
10 - 14	TOTAL DIV. 10000 - 14000	100.0	116.4	103.5	100.0	128.5	106.1	100.0	116.1	103.5	100.0	129.4	106.3	100.0	110.5	102.3	100.0	132.2	106.9
15	MECHANICAL	100.1	112.2	105.7	100.1	109.4	104.4	100.0	112.1	105.6	100.1	113.0	106.0	100.2	107.3	103.5	100.2	156.1	125.9
16	ELECTRICAL	99.1	99.8	99.5	103.4	106.7	105.3	95.4	103.5	100.1	101.7	106.7	104.6	98.9	98.4	98.6	106.3	154.6	134.5
01 - 16	WEIGHTED AVERAGE	104.3	110.9	107.5	107.3	113.4	110.3	103.2	112.2	107.6	107.5	115.0	111.1	105.1	104.3	104.7	112.5	135.5	123.6

DIVISION		CALIFORNIA												COLORADO					
		SAN JOSE			SANTA BARBARA			STOCKTON			VALLEJO			COLORADO SPRINGS			DENVER		
		MAT.	INST.	TOTAL	MAT.	INST.	TOTAL	MAT.	INST.	TOTAL	MAT.	INST.	TOTAL	MAT.	INST.	TOTAL	MAT.	INST.	TOTAL
01590	EQUIPMENT RENTAL	.0	99.9	99.9	.0	99.5	99.5	.0	99.1	99.1	.0	103.7	103.7	.0	95.5	95.5	.0	100.6	100.6
02	SITE CONSTRUCTION	148.7	99.7	112.0	109.1	106.5	107.1	107.0	105.4	105.8	114.0	111.2	111.9	95.9	97.3	97.0	94.4	106.6	103.5
03100	CONCRETE FORMS & ACCESSORIES	105.2	136.2	132.4	103.8	124.1	121.6	103.2	131.4	128.0	106.8	134.5	131.1	90.4	85.4	86.0	99.5	85.5	87.2
03200	CONCRETE REINFORCEMENT	101.1	114.6	109.3	104.0	113.5	109.8	104.5	113.8	110.1	97.4	114.7	107.8	98.7	86.5	91.3	98.1	86.7	91.2
03300	CAST-IN-PLACE CONCRETE	126.8	119.6	123.8	106.8	119.4	112.1	106.7	115.0	110.2	119.0	117.3	118.3	101.1	89.4	96.2	94.6	88.9	92.2
03	CONCRETE	119.1	125.3	122.3	109.3	119.4	114.5	109.3	121.2	115.4	116.9	123.5	120.2	108.4	87.1	97.5	101.8	87.0	94.2
04	MASONRY	149.9	124.4	134.0	109.2	117.2	114.2	113.3	111.5	112.2	84.7	125.3	110.1	108.8	84.5	93.6	108.5	87.1	95.1
05	METALS	109.8	107.2	108.8	103.7	99.9	102.3	105.9	100.4	103.8	100.8	101.9	101.3	100.8	89.2	96.4	104.3	89.4	98.7
06	WOOD & PLASTICS	99.5	137.7	120.1	91.6	121.5	107.7	93.5	135.1	115.9	92.9	137.7	117.1	93.1	87.2	89.9	100.7	87.1	93.3
07	THERMAL & MOISTURE PROTECTION	107.2	130.0	118.3	106.9	111.7	109.2	110.9	110.3	110.6	126.6	125.5	126.1	103.8	86.5	95.4	103.1	84.5	94.0
08	DOORS & WINDOWS	92.7	128.4	101.8	101.6	116.5	105.4	100.9	123.7	106.7	119.2	130.0	122.0	98.4	90.6	96.4	99.8	90.5	97.4
09200	PLASTER & GYPSUM BOARD	99.2	138.4	125.9	99.0	122.0	114.6	100.0	135.9	124.4	97.3	138.4	125.3	92.5	86.9	88.7	102.3	86.9	91.8
095,098	CEILINGS & ACOUSTICAL TREATMENT	108.1	138.4	128.6	117.3	122.0	120.4	117.3	135.9	129.9	120.3	138.4	132.6	102.8	86.9	92.1	101.5	86.9	91.6
09600	FLOORING	113.8	127.7	117.4	116.8	104.5	113.6	116.8	110.7	115.2	122.5	127.7	123.9	108.3	92.7	104.3	109.3	97.9	106.3
097,099	WALL FINISHES, PAINTS & COATINGS	111.6	138.7	127.9	110.0	101.6	104.9	110.0	104.5	106.7	112.4	140.3	129.1	106.9	55.3	76.0	106.9	76.6	88.7
09	FINISHES	113.1	135.7	125.3	114.2	118.1	116.3	114.1	126.3	120.7	115.5	135.4	126.2	100.1	83.4	91.1	101.3	86.8	93.5
10 - 14	TOTAL DIV. 10000 - 14000	100.0	130.6	106.6	100.0	116.5	103.5	100.0	128.5	106.1	100.0	129.3	106.3	100.0	91.4	98.1	100.0	90.9	98.0
15	MECHANICAL	100.1	143.6	120.1	100.1	112.2	105.7	100.1	104.8	102.3	100.1	119.1	108.8	100.2	83.1	92.3	100.1	84.9	93.1
16	ELECTRICAL	107.9	146.6	130.5	92.3	108.6	101.8	102.7	113.9	109.2	98.1	102.7	100.8	98.4	91.5	94.4	100.0	95.2	97.2
01 - 16	WEIGHTED AVERAGE	109.3	129.7	119.2	103.4	112.6	107.9	104.9	113.3	108.9	106.0	118.9	112.2	101.3	87.6	94.7	101.3	89.9	95.8

	DIVISION	COLORADO												CONNECTICUT					
		FORT COLLINS			GRAND JUNCTION			GREELEY			PUEBLO			BRIDGEPORT			BRISTOL		
		MAT.	INST.	TOTAL	MAT.	INST.	TOTAL	MAT.	INST.	TOTAL	MAT.	INST.	TOTAL	MAT.	INST.	TOTAL	MAT.	INST.	TOTAL
01590	**EQUIPMENT RENTAL**	.0	97.0	97.0	.0	99.7	99.7	.0	97.0	97.0	.0	96.6	96.6	.0	101.8	101.8	.0	101.8	101.8
02	**SITE CONSTRUCTION**	105.8	99.9	101.4	124.1	102.3	107.8	92.5	98.6	97.0	115.9	96.6	101.5	101.9	104.8	104.0	101.0	104.7	103.8
03100	CONCRETE FORMS & ACCESSORIES	98.9	79.9	82.2	109.5	79.9	83.5	96.4	51.1	56.6	104.3	85.5	87.8	104.0	115.0	113.6	104.0	114.6	113.3
03200	CONCRETE REINFORCEMENT	102.9	84.7	91.9	106.6	84.6	93.3	102.4	80.0	88.9	99.5	86.5	91.7	109.1	125.7	119.1	109.1	125.6	119.1
03300	CAST-IN-PLACE CONCRETE	109.8	77.8	96.4	116.8	83.4	102.8	91.8	62.7	79.6	104.7	90.1	98.6	107.1	121.4	113.1	100.4	121.3	109.1
03	**CONCRETE**	115.7	80.2	97.5	116.7	82.1	99.0	99.4	61.4	79.9	105.6	87.4	96.3	110.2	119.1	114.8	106.8	118.9	113.0
04	**MASONRY**	123.4	62.8	85.5	148.1	63.9	95.4	116.9	41.7	69.8	106.8	84.5	92.8	106.7	121.0	115.7	98.3	121.0	112.5
05	**METALS**	100.1	84.6	94.2	101.1	83.8	94.6	100.1	81.0	92.8	102.7	90.1	97.9	98.4	122.6	107.6	98.4	122.3	107.4
06	**WOOD & PLASTICS**	100.5	82.6	90.8	108.7	82.7	94.7	98.0	50.3	72.3	104.5	87.6	95.4	101.0	112.4	107.1	101.0	112.4	107.1
07	**THERMAL & MOISTURE PROTECTION**	103.7	73.5	89.0	105.1	68.6	87.3	102.8	61.2	82.5	104.5	85.9	95.4	102.1	121.5	111.6	102.3	117.4	109.7
08	**DOORS & WINDOWS**	95.8	87.9	93.8	100.8	87.3	97.3	95.8	69.2	89.0	95.2	90.8	94.1	107.2	125.3	111.8	107.2	117.0	109.7
09200	PLASTER & GYPSUM BOARD	101.1	82.3	88.3	112.8	82.3	92.0	100.0	49.1	65.4	89.7	86.9	87.8	108.5	111.8	110.7	108.5	111.8	110.7
095,098	CEILINGS & ACOUSTICAL TREATMENT	96.3	82.3	86.8	102.0	82.3	88.7	96.3	49.1	64.3	112.4	86.9	95.1	105.9	111.8	109.9	105.9	111.8	109.9
09600	FLOORING	109.1	74.4	100.2	116.7	74.4	105.8	107.7	74.4	99.1	113.0	97.9	109.1	100.3	119.8	105.3	100.3	119.8	105.3
097,099	WALL FINISHES, PAINTS & COATINGS	106.9	54.8	75.7	117.3	73.0	90.7	106.9	34.2	63.3	117.3	51.2	77.7	93.0	113.0	105.0	93.0	113.0	105.0
09	**FINISHES**	100.7	76.2	87.5	108.9	78.7	92.6	99.4	52.1	73.9	106.4	84.3	94.4	104.3	114.9	110.1	104.4	114.9	110.1
10 - 14	**TOTAL DIV. 10000 - 14000**	100.0	89.1	97.7	100.0	89.8	97.8	100.0	80.3	95.8	100.0	92.2	98.3	100.0	117.0	103.7	100.0	117.0	103.7
15	**MECHANICAL**	100.1	83.0	92.2	100.1	73.1	87.7	100.1	74.9	88.5	100.1	72.4	87.3	100.2	107.7	103.6	100.2	107.6	103.6
16	**ELECTRICAL**	93.7	87.2	89.9	94.8	68.0	79.1	93.7	87.2	89.9	95.9	78.1	85.5	102.8	109.9	107.0	102.8	112.7	108.6
01 - 16	**WEIGHTED AVERAGE**	102.5	82.1	92.7	106.3	78.1	92.7	99.6	70.5	85.6	101.8	83.7	93.1	103.1	114.3	108.5	102.2	114.2	108.0

	DIVISION	CONNECTICUT																	
		HARTFORD			NEW BRITAIN			NEW HAVEN			NORWALK			STAMFORD			WATERBURY		
		MAT.	INST.	TOTAL	MAT.	INST.	TOTAL	MAT.	INST.	TOTAL	MAT.	INST.	TOTAL	MAT.	INST.	TOTAL	MAT.	INST.	TOTAL
01590	**EQUIPMENT RENTAL**	.0	101.8	101.8	.0	101.8	101.8	.0	102.4	102.4	.0	101.8	101.8	.0	101.8	101.8	.0	101.8	101.8
02	**SITE CONSTRUCTION**	101.4	104.7	103.9	101.2	104.7	103.9	101.1	105.6	104.5	101.7	104.8	104.0	102.3	104.8	104.2	101.4	104.8	103.9
03100	CONCRETE FORMS & ACCESSORIES	102.7	114.6	113.1	104.3	114.7	113.4	103.8	114.9	113.5	104.0	115.0	113.7	103.9	115.2	113.9	104.0	114.9	113.5
03200	CONCRETE REINFORCEMENT	109.1	125.6	119.1	109.1	125.6	119.1	109.1	125.6	119.1	109.1	125.8	119.2	109.1	125.8	119.2	109.1	125.6	119.1
03300	CAST-IN-PLACE CONCRETE	101.5	121.3	109.8	102.0	121.3	110.1	103.7	121.4	111.1	105.4	123.0	112.8	107.1	123.1	113.8	107.1	121.4	113.1
03	**CONCRETE**	107.3	118.8	113.2	107.7	118.9	113.4	124.2	119.0	121.6	109.4	119.7	114.6	110.2	119.8	115.1	110.2	119.0	114.7
04	**MASONRY**	98.4	121.0	112.5	98.4	121.0	112.5	98.5	121.0	112.6	98.6	122.8	113.7	98.8	122.8	113.8	98.8	121.0	112.7
05	**METALS**	99.0	122.2	107.8	95.1	122.3	105.4	95.3	122.5	105.6	98.4	122.8	107.7	98.4	123.1	107.7	98.4	122.5	107.5
06	**WOOD & PLASTICS**	101.0	112.4	107.1	101.0	112.4	107.1	101.0	112.4	107.1	101.0	112.4	107.1	101.0	112.4	107.1	101.0	112.4	107.1
07	**THERMAL & MOISTURE PROTECTION**	100.9	117.4	108.9	102.3	119.1	110.5	102.4	119.1	110.6	102.3	122.2	112.0	102.3	122.2	112.0	102.3	119.1	110.5
08	**DOORS & WINDOWS**	107.2	117.0	109.7	107.2	117.0	109.7	107.2	125.3	111.8	107.2	125.3	111.8	107.2	125.3	111.8	107.2	125.3	111.8
09200	PLASTER & GYPSUM BOARD	108.5	111.8	110.7	108.5	111.8	110.7	108.5	111.8	110.7	108.5	111.8	110.7	108.5	111.8	110.7	108.5	111.8	110.7
095,098	CEILINGS & ACOUSTICAL TREATMENT	105.9	111.8	109.9	105.9	111.8	109.9	105.9	111.8	109.9	105.9	111.8	109.9	105.9	111.8	109.9	105.9	111.8	109.9
09600	FLOORING	100.3	119.8	105.3	100.3	119.8	105.3	100.3	119.8	105.3	100.3	119.8	105.3	100.3	119.8	105.3	100.3	119.8	105.3
097,099	WALL FINISHES, PAINTS & COATINGS	93.0	113.0	105.0	93.0	113.0	105.0	93.0	113.0	105.0	93.0	113.0	105.0	93.0	113.0	105.0	93.0	113.0	105.0
09	**FINISHES**	104.3	114.9	110.1	104.4	114.9	110.1	104.4	114.9	110.1	104.4	114.9	110.1	104.4	114.9	110.1	104.2	114.9	110.0
10 - 14	**TOTAL DIV. 10000 - 14000**	100.0	117.0	103.7	100.0	117.0	103.7	100.0	117.0	103.7	100.0	117.0	103.7	100.0	117.2	103.7	100.0	117.0	103.7
15	**MECHANICAL**	100.2	107.6	103.6	100.2	107.6	103.6	100.2	107.6	103.6	100.2	107.7	103.6	100.2	107.7	103.6	100.2	107.6	103.6
16	**ELECTRICAL**	102.2	107.1	105.1	102.9	112.7	108.6	102.7	112.7	108.5	102.8	112.7	108.6	102.8	143.1	126.3	102.1	109.9	106.7
01 - 16	**WEIGHTED AVERAGE**	102.2	113.4	107.6	101.8	114.2	107.8	103.8	114.7	109.1	102.5	115.0	108.5	102.6	119.3	110.7	102.5	114.2	108.2

	DIVISION	D.C.			DELAWARE			FLORIDA											
		WASHINGTON			WILMINGTON			DAYTONA BEACH			FORT LAUDERDALE			JACKSONVILLE			MELBOURNE		
		MAT.	INST.	TOTAL	MAT.	INST.	TOTAL	MAT.	INST.	TOTAL	MAT.	INST.	TOTAL	MAT.	INST.	TOTAL	MAT.	INST.	TOTAL
01590	**EQUIPMENT RENTAL**	.0	102.4	102.4	.0	118.2	118.2	.0	97.8	97.8	.0	89.4	89.4	.0	97.8	97.8	.0	97.8	97.8
02	**SITE CONSTRUCTION**	101.5	89.4	92.4	85.7	111.5	105.0	118.1	86.8	94.7	103.4	74.0	81.4	118.2	87.4	95.2	126.2	87.3	97.1
03100	CONCRETE FORMS & ACCESSORIES	100.1	80.2	82.6	98.3	102.5	102.0	94.2	69.7	72.7	92.2	70.8	73.4	93.9	55.3	60.0	88.7	73.0	75.0
03200	CONCRETE REINFORCEMENT	99.9	88.5	93.0	96.9	98.8	98.1	95.4	83.3	88.1	95.4	74.6	82.8	95.4	52.4	69.4	96.4	83.4	88.6
03300	CAST-IN-PLACE CONCRETE	112.7	84.7	101.0	79.0	97.0	86.5	90.5	70.2	82.0	95.0	71.4	85.1	91.4	60.5	78.5	105.2	74.6	92.4
03	**CONCRETE**	110.2	84.7	97.1	97.5	100.9	99.2	90.2	73.7	81.7	92.3	73.0	82.4	90.6	58.2	74.0	100.0	76.8	88.1
04	**MASONRY**	91.4	81.5	85.2	108.7	86.5	94.8	86.5	64.1	72.5	86.8	69.7	76.1	86.1	53.0	65.4	84.0	71.6	76.2
05	**METALS**	97.0	107.8	101.1	98.2	115.8	104.9	98.2	95.5	97.2	97.6	91.4	95.2	97.2	81.5	91.3	106.1	96.1	102.3
06	**WOOD & PLASTICS**	97.4	80.6	88.3	96.5	104.0	100.5	92.9	72.1	81.7	89.0	68.6	78.0	92.9	54.2	72.1	87.7	72.1	79.3
07	**THERMAL & MOISTURE PROTECTION**	97.0	82.3	89.8	101.2	105.6	103.4	95.4	70.2	83.1	95.4	70.1	83.1	95.7	59.4	78.0	95.6	76.0	86.1
08	**DOORS & WINDOWS**	102.0	90.0	99.0	94.2	107.5	97.6	100.9	69.5	92.9	98.5	65.6	90.1	100.9	51.4	88.3	100.1	74.1	93.5
09200	PLASTER & GYPSUM BOARD	106.1	80.0	88.4	109.0	103.9	105.5	98.7	71.9	80.4	98.2	68.2	77.8	98.7	53.5	67.9	95.2	71.9	79.3
095,098	CEILINGS & ACOUSTICAL TREATMENT	101.9	80.0	87.1	101.9	103.9	103.2	95.0	71.9	79.4	95.0	68.2	76.9	95.0	53.5	66.9	91.1	71.9	78.1
09600	FLOORING	112.0	97.1	108.1	83.5	93.1	86.0	118.6	72.1	106.6	118.6	63.4	104.4	118.6	52.1	101.4	115.2	72.1	104.1
097,099	WALL FINISHES, PAINTS & COATINGS	113.2	89.7	99.1	90.3	100.0	96.1	111.3	74.4	89.1	107.5	51.5	73.9	111.3	49.6	74.3	111.3	95.6	101.9
09	**FINISHES**	102.1	84.1	92.4	100.9	100.2	100.5	107.5	70.6	87.6	105.3	66.9	84.6	107.5	53.5	78.4	105.7	75.0	89.1
10 - 14	**TOTAL DIV. 10000 - 14000**	100.0	91.6	98.2	100.0	105.8	101.2	100.0	82.0	96.1	100.0	89.9	97.8	100.0	77.2	95.1	100.0	85.1	96.8
15	**MECHANICAL**	100.0	90.6	95.6	100.4	120.7	109.7	99.8	75.1	88.4	99.8	72.2	87.1	99.8	53.2	78.3	99.8	79.5	90.5
16	**ELECTRICAL**	95.4	95.7	95.6	98.6	104.7	102.1	97.8	66.3	79.4	97.8	72.8	83.2	97.4	66.9	79.6	97.6	74.1	83.9
01 - 16	**WEIGHTED AVERAGE**	100.2	89.9	95.2	98.9	106.8	102.7	98.7	74.7	87.2	97.9	73.5	86.2	98.6	62.4	81.1	100.8	79.0	90.3

	DIVISION	FLORIDA																	
		MIAMI			ORLANDO			PANAMA CITY			PENSACOLA			ST. PETERSBURG			TALLAHASSEE		
		MAT.	INST.	TOTAL	MAT.	INST.	TOTAL	MAT.	INST.	TOTAL	MAT.	INST.	TOTAL	MAT.	INST.	TOTAL	MAT.	INST.	TOTAL
01590	**EQUIPMENT RENTAL**	.0	89.4	89.4	.0	97.8	97.8	.0	97.8	97.8	.0	97.8	97.8	.0	97.8	97.8	.0	97.8	97.8
02	**SITE CONSTRUCTION**	102.6	73.8	81.1	118.7	86.5	94.6	133.4	84.1	96.5	130.9	86.5	97.7	119.1	86.0	94.3	119.6	86.0	94.4
03100	CONCRETE FORMS & ACCESSORIES	92.5	70.8	73.5	93.9	55.3	60.0	93.0	28.8	36.6	83.9	53.0	56.8	91.6	49.1	54.3	93.8	40.5	47.0
03200	CONCRETE REINFORCEMENT	95.4	74.6	82.8	95.4	80.6	86.5	99.6	51.3	70.4	102.1	51.6	71.6	98.8	59.0	74.8	95.4	51.9	69.1
03300	CAST-IN-PLACE CONCRETE	92.5	71.6	83.8	98.4	70.0	86.5	96.1	35.6	70.8	96.1	56.5	79.5	102.4	57.0	83.4	94.7	49.8	76.0
03	**CONCRETE**	91.0	73.1	81.8	91.8	66.8	79.0	99.2	37.2	67.5	98.1	55.7	76.4	96.4	55.5	75.4	92.2	47.9	69.6
04	**MASONRY**	84.6	70.1	75.5	90.7	64.1	74.1	90.9	28.4	51.8	88.0	51.4	65.1	134.4	50.8	82.1	89.0	40.0	58.3
05	**METALS**	101.4	90.9	97.5	107.3	93.9	102.2	96.4	66.9	85.3	96.4	80.9	90.5	100.2	82.2	93.4	98.1	79.8	91.2
06	**WOOD & PLASTICS**	89.0	68.6	78.0	92.9	53.5	71.6	91.8	28.9	57.9	83.1	53.5	67.2	90.5	49.1	68.2	91.1	38.9	63.0
07	**THERMAL & MOISTURE PROTECTION**	99.4	71.5	85.8	95.7	68.3	82.4	96.0	32.4	65.0	95.7	54.5	75.6	95.3	51.2	73.8	95.7	48.0	72.5
08	**DOORS & WINDOWS**	98.5	66.2	90.3	100.9	58.0	90.0	98.5	27.6	80.5	98.4	51.9	86.6	99.5	48.1	86.4	99.5	42.5	85.0
09200	PLASTER & GYPSUM BOARD	98.2	68.2	77.8	102.1	52.7	68.5	96.9	27.5	49.7	92.8	52.8	65.6	96.4	48.2	63.6	98.7	37.7	57.2
095,098	CEILINGS & ACOUSTICAL TREATMENT	95.0	68.2	76.9	95.0	52.7	66.4	89.8	27.5	47.6	89.8	52.8	64.7	89.8	48.2	61.6	95.0	37.7	56.2
09600	FLOORING	126.0	64.2	110.0	118.6	72.1	106.6	118.0	20.0	92.7	112.2	55.3	97.5	116.9	55.2	101.0	118.6	40.5	98.5
097,099	WALL FINISHES, PAINTS & COATINGS	107.5	51.5	73.9	111.3	57.0	78.7	111.3	25.4	59.8	111.3	57.6	79.1	111.3	48.0	73.3	111.3	40.8	69.1
09	**FINISHES**	107.8	67.1	85.8	108.0	57.7	80.8	107.3	26.3	63.6	104.7	53.8	77.2	105.7	49.9	75.5	107.7	39.7	70.9
10 - 14	**TOTAL DIV. 10000 - 14000**	100.0	89.9	97.8	100.0	79.5	95.6	100.0	48.9	89.0	100.0	54.7	90.3	100.0	57.3	90.8	100.0	64.0	92.2
15	**MECHANICAL**	99.8	76.0	88.9	99.8	56.4	79.8	99.8	26.1	65.9	99.8	51.8	77.7	99.8	51.7	77.6	99.8	41.3	72.9
16	**ELECTRICAL**	98.2	76.2	85.4	98.4	46.3	67.9	96.2	34.7	60.2	100.8	56.0	74.6	98.2	50.0	70.0	98.4	42.9	65.9
01 - 16	**WEIGHTED AVERAGE**	98.6	74.8	87.1	100.6	64.6	83.2	99.8	39.1	70.6	99.5	59.2	80.1	102.2	57.6	80.7	99.1	50.6	75.7

	DIVISION	FLORIDA			GEORGIA														
		TAMPA			ALBANY			ATLANTA			AUGUSTA			COLUMBUS			MACON		
		MAT.	INST.	TOTAL	MAT.	INST.	TOTAL	MAT.	INST.	TOTAL	MAT.	INST.	TOTAL	MAT.	INST.	TOTAL	MAT.	INST.	TOTAL
01590	**EQUIPMENT RENTAL**	.0	97.8	97.8	.0	90.2	90.2	.0	92.9	92.9	.0	92.2	92.2	.0	90.2	90.2	.0	102.6	102.6
02	**SITE CONSTRUCTION**	119.3	86.6	94.8	103.6	75.9	82.9	104.6	95.9	98.1	100.9	92.9	94.9	103.5	76.3	83.2	104.3	94.1	96.6
03100	CONCRETE FORMS & ACCESSORIES	95.3	77.7	79.9	93.6	44.4	50.5	91.4	81.4	82.6	88.6	53.9	58.1	94.8	56.9	61.5	92.5	57.5	61.8
03200	CONCRETE REINFORCEMENT	95.4	92.8	93.8	94.8	90.8	92.4	99.1	92.6	95.2	104.8	79.2	89.3	95.4	91.0	92.7	97.8	91.0	93.7
03300	CAST-IN-PLACE CONCRETE	100.1	63.5	84.8	96.7	42.9	74.2	99.9	76.6	90.2	94.4	50.4	76.0	96.3	43.8	74.4	95.0	47.1	75.0
03	**CONCRETE**	95.0	76.7	85.6	93.0	54.3	73.2	98.5	81.7	89.9	94.4	57.9	75.7	93.0	60.2	76.2	92.5	61.7	76.7
04	**MASONRY**	87.6	78.1	81.6	88.7	32.9	53.8	89.8	72.1	78.8	90.0	41.5	59.7	88.8	36.3	55.9	102.6	39.6	63.2
05	**METALS**	101.1	97.4	99.7	95.5	89.4	93.2	91.4	80.7	87.4	89.8	71.0	82.7	95.9	91.1	94.1	90.8	91.5	91.1
06	**WOOD & PLASTICS**	94.2	80.6	86.9	92.4	44.3	66.4	90.5	84.2	87.1	88.1	55.5	70.5	93.8	61.0	76.1	101.1	59.9	78.9
07	**THERMAL & MOISTURE PROTECTION**	95.6	64.3	80.4	95.5	52.2	74.4	91.7	77.5	84.8	91.1	51.4	71.8	95.1	54.8	75.5	93.9	58.6	76.7
08	**DOORS & WINDOWS**	100.9	73.7	94.0	98.5	51.2	86.5	100.0	79.6	94.8	94.1	56.6	84.6	98.5	60.5	88.8	96.9	60.8	87.7
09200	PLASTER & GYPSUM BOARD	98.7	80.6	86.4	98.4	43.3	60.9	117.1	84.2	94.7	116.1	54.7	74.3	98.4	60.5	72.6	107.6	59.3	74.7
095,098	CEILINGS & ACOUSTICAL TREATMENT	95.0	80.6	85.3	93.7	43.3	59.6	96.0	84.2	88.0	97.5	54.7	68.5	93.7	60.5	71.2	87.8	59.3	68.5
09600	FLOORING	118.6	55.2	102.2	118.6	33.8	96.7	84.4	81.1	83.5	83.2	43.1	72.9	118.6	39.4	98.2	92.5	39.7	78.9
097,099	WALL FINISHES, PAINTS & COATINGS	111.3	48.0	73.3	107.5	42.7	68.6	90.4	80.6	84.5	90.4	40.6	60.5	107.5	40.9	67.6	109.2	50.0	73.7
09	**FINISHES**	107.6	71.0	87.8	105.1	41.1	70.5	93.1	81.5	86.8	92.6	50.1	69.6	105.0	51.8	76.2	91.7	53.0	70.8
10 - 14	**TOTAL DIV. 10000 - 14000**	100.0	80.8	95.9	100.0	74.0	94.4	100.0	84.6	96.7	100.0	71.8	93.9	100.0	76.1	94.9	100.0	77.4	95.1
15	**MECHANICAL**	99.8	76.5	89.1	99.8	46.9	75.5	99.9	80.1	90.8	99.9	68.6	85.5	99.8	38.4	71.6	99.8	43.2	73.8
16	**ELECTRICAL**	97.2	50.0	69.6	92.3	62.8	75.1	96.6	85.3	90.0	98.4	51.3	70.8	93.3	40.7	62.6	91.9	62.0	74.4
01 - 16	**WEIGHTED AVERAGE**	99.8	74.8	87.8	97.4	55.6	77.2	96.8	82.0	89.7	95.4	61.1	78.9	97.5	54.2	76.6	95.9	60.7	78.9

	DIVISION	GEORGIA						HAWAII			IDAHO								
		SAVANNAH			VALDOSTA			HONOLULU			BOISE			LEWISTON			POCATELLO		
		MAT.	INST.	TOTAL	MAT.	INST.	TOTAL	MAT.	INST.	TOTAL	MAT.	INST.	TOTAL	MAT.	INST.	TOTAL	MAT.	INST.	TOTAL
01590	**EQUIPMENT RENTAL**	.0	91.3	91.3	.0	90.2	90.2	.0	99.3	99.3	.0	101.4	101.4	.0	94.3	94.3	.0	101.4	101.4
02	**SITE CONSTRUCTION**	104.1	77.3	84.0	114.4	76.5	86.1	138.9	106.5	114.7	78.8	103.0	96.9	83.9	96.9	93.7	80.0	103.0	97.2
03100	CONCRETE FORMS & ACCESSORIES	93.3	54.6	59.3	81.0	47.2	51.4	106.3	145.7	140.8	98.4	82.2	84.2	113.7	69.3	74.8	98.3	81.7	83.7
03200	CONCRETE REINFORCEMENT	101.1	79.4	88.0	101.2	45.0	67.3	104.0	120.7	114.1	98.2	77.9	85.9	111.5	96.4	102.4	98.6	77.4	85.8
03300	CAST-IN-PLACE CONCRETE	93.3	52.8	76.4	94.6	53.4	77.4	197.0	130.3	169.1	99.0	93.4	96.6	107.3	88.1	99.3	98.2	93.2	96.1
03	**CONCRETE**	92.0	60.1	75.7	97.8	50.7	73.7	154.7	134.2	144.2	104.7	85.1	94.6	113.1	81.2	96.8	100.9	84.7	92.6
04	**MASONRY**	91.9	54.9	68.7	94.9	49.9	66.7	134.4	131.7	132.7	136.9	76.9	99.3	138.0	87.8	106.5	132.6	68.5	92.5
05	**METALS**	96.2	85.9	92.3	95.7	74.0	87.5	115.7	108.4	112.9	112.8	78.2	99.7	96.5	88.2	93.3	113.1	77.0	99.4
06	**WOOD & PLASTICS**	106.8	52.3	77.4	80.1	43.2	60.2	95.0	149.6	124.4	96.8	80.8	88.2	106.6	62.9	83.0	96.8	80.8	88.2
07	**THERMAL & MOISTURE PROTECTION**	95.5	54.5	75.5	95.3	59.0	77.6	113.2	131.8	122.2	97.1	82.1	89.8	167.6	81.7	125.7	97.3	72.3	85.1
08	**DOORS & WINDOWS**	99.6	52.7	87.7	93.8	40.0	80.1	105.9	137.7	113.9	94.7	77.6	90.4	115.0	68.9	103.3	94.7	70.7	88.6
09200	PLASTER & GYPSUM BOARD	98.7	51.5	66.6	91.3	42.1	57.8	119.3	150.9	140.8	88.2	80.1	82.7	155.2	61.8	91.6	88.2	80.1	82.7
095,098	CEILINGS & ACOUSTICAL TREATMENT	95.0	51.5	65.5	89.8	42.1	57.5	117.3	150.9	140.0	112.4	80.1	90.5	159.6	61.8	93.4	112.4	80.1	90.5
09600	FLOORING	118.6	50.8	101.1	110.4	40.6	92.4	167.8	136.4	159.7	100.3	58.5	89.5	138.4	97.3	127.8	100.6	58.5	89.8
097,099	WALL FINISHES, PAINTS & COATINGS	108.2	50.7	73.7	107.5	37.0	65.2	104.5	149.9	131.8	103.9	53.8	73.9	129.9	76.6	97.9	103.9	56.0	75.2
09	**FINISHES**	105.6	53.2	77.3	101.4	44.2	70.5	135.5	145.9	141.1	99.7	74.8	86.2	168.8	74.0	117.6	99.8	75.0	86.4
10 - 14	**TOTAL DIV. 10000 - 14000**	100.0	74.8	94.6	100.0	73.5	94.3	100.0	122.9	104.9	100.0	91.1	98.1	100.0	93.1	98.5	100.0	91.0	98.1
15	**MECHANICAL**	99.8	49.2	76.5	99.8	43.3	73.8	100.1	120.6	109.5	100.1	79.5	90.6	101.4	86.8	94.7	100.1	79.5	90.6
16	**ELECTRICAL**	94.7	54.6	71.3	90.8	31.4	56.1	113.2	122.9	118.9	84.0	79.7	81.5	85.3	87.7	86.7	86.6	76.8	80.9
01 - 16	**WEIGHTED AVERAGE**	98.1	59.5	79.5	97.5	50.4	74.8	117.6	126.1	121.7	101.6	81.8	92.1	112.2	84.7	99.0	101.3	79.8	90.9

DIVISION		IDAHO			ILLINOIS														
		TWIN FALLS			CHICAGO			DECATUR			EAST ST. LOUIS			JOLIET			PEORIA		
		MAT.	INST.	TOTAL	MAT.	INST.	TOTAL	MAT.	INST.	TOTAL	MAT.	INST.	TOTAL	MAT.	INST.	TOTAL	MAT.	INST.	TOTAL
01590	**EQUIPMENT RENTAL**	.0	101.4	101.4	.0	91.5	91.5	.0	102.2	102.2	.0	108.8	108.8	.0	89.3	89.3	.0	101.4	101.4
02	**SITE CONSTRUCTION**	86.8	100.9	97.3	94.7	90.5	91.6	85.9	95.9	93.4	103.0	96.4	98.1	94.7	89.6	90.9	94.8	95.0	95.0
03100	CONCRETE FORMS & ACCESSORIES	101.2	36.5	44.5	103.4	138.7	134.3	98.1	106.9	105.8	91.9	106.4	104.6	104.3	136.6	132.6	96.2	110.6	108.9
03200	CONCRETE REINFORCEMENT	107.1	50.3	72.8	100.0	142.6	125.7	98.4	98.4	98.4	104.1	105.0	104.6	100.0	131.1	118.8	98.4	101.1	100.0
03300	CAST-IN-PLACE CONCRETE	100.8	46.2	78.0	111.1	134.4	120.9	98.7	106.1	101.8	92.4	111.4	100.3	111.0	124.3	116.6	96.4	107.7	101.1
03	**CONCRETE**	109.9	43.6	75.9	109.6	137.0	123.6	97.4	105.3	101.5	88.1	108.8	98.7	109.6	130.3	120.2	96.2	108.0	102.3
04	**MASONRY**	136.0	40.2	76.0	92.4	135.5	119.4	68.1	94.6	84.7	72.7	111.2	96.8	93.9	125.1	113.5	111.4	108.1	109.4
05	**METALS**	113.0	63.3	94.2	96.2	122.9	106.3	95.6	105.9	99.5	93.1	120.9	103.6	94.3	115.1	102.2	95.6	108.9	100.7
06	**WOOD & PLASTICS**	99.5	36.5	65.5	104.5	137.7	122.4	100.3	106.3	103.5	97.0	104.0	100.8	106.4	137.7	123.3	100.3	108.6	104.7
07	**THERMAL & MOISTURE PROTECTION**	98.4	48.1	73.9	100.4	131.2	115.4	97.3	96.1	96.7	92.3	104.7	98.3	100.2	125.5	112.5	97.2	106.8	101.9
08	**DOORS & WINDOWS**	98.0	37.3	82.6	103.7	138.7	112.6	96.2	104.0	98.2	86.0	111.5	92.5	101.6	135.6	110.2	96.2	105.9	98.7
09200	PLASTER & GYPSUM BOARD	87.7	34.5	51.5	101.4	138.6	126.7	105.0	106.3	105.9	101.8	104.0	103.3	98.8	138.7	125.9	105.0	108.6	107.5
095,098	CEILINGS & ACOUSTICAL TREATMENT	105.9	34.5	57.6	100.6	138.6	126.3	96.2	106.3	103.0	91.0	104.0	99.8	100.6	138.7	126.4	96.2	108.6	104.6
09600	FLOORING	101.9	58.5	90.7	83.6	132.8	96.2	96.8	88.0	94.5	105.5	108.7	106.4	83.1	120.9	92.9	96.8	103.8	98.6
097,099	WALL FINISHES, PAINTS & COATINGS	103.9	33.0	61.4	80.6	133.6	112.4	91.3	100.6	96.9	99.4	93.2	95.7	78.4	119.1	102.8	91.3	99.4	96.2
09	**FINISHES**	99.5	39.9	67.3	90.6	137.5	115.9	95.4	103.2	99.6	95.2	104.0	99.9	90.0	132.8	113.1	95.4	108.4	102.4
10 - 14	**TOTAL DIV. 10000 - 14000**	100.0	51.4	89.5	100.0	127.6	105.9	100.0	99.2	99.8	100.0	100.7	100.1	100.0	125.8	105.5	100.0	99.5	99.9
15	**MECHANICAL**	100.1	37.6	71.4	100.0	123.5	110.8	100.0	101.4	100.7	100.0	99.2	99.6	100.2	121.8	110.1	100.0	103.6	101.7
16	**ELECTRICAL**	88.7	32.7	56.0	100.0	124.6	114.4	102.6	84.7	92.1	99.3	95.3	97.0	99.5	110.5	105.9	101.6	92.2	96.1
01 - 16	**WEIGHTED AVERAGE**	103.3	47.2	76.3	99.6	126.4	112.5	96.2	99.0	97.5	93.8	104.3	98.8	99.1	120.4	109.4	98.6	103.5	101.0

DIVISION		ILLINOIS						INDIANA											
		ROCKFORD			SPRINGFIELD			ANDERSON			BLOOMINGTON			EVANSVILLE			FORT WAYNE		
		MAT.	INST.	TOTAL	MAT.	INST.	TOTAL	MAT.	INST.	TOTAL	MAT.	INST.	TOTAL	MAT.	INST.	TOTAL	MAT.	INST.	TOTAL
01590	**EQUIPMENT RENTAL**	.0	101.4	101.4	.0	102.2	102.2	.0	96.9	96.9	.0	86.1	86.1	.0	121.5	121.5	.0	96.9	96.9
02	**SITE CONSTRUCTION**	93.9	95.3	94.9	92.2	95.9	95.0	87.5	96.8	94.5	77.4	95.9	91.2	82.5	130.4	118.4	88.4	96.8	94.7
03100	CONCRETE FORMS & ACCESSORIES	101.1	117.1	115.2	100.4	107.6	106.7	92.6	78.4	80.1	103.3	79.0	82.0	92.3	78.2	80.0	92.0	77.9	79.6
03200	CONCRETE REINFORCEMENT	98.4	133.3	119.5	98.4	98.4	98.4	96.4	83.1	88.3	88.4	83.0	85.1	96.9	83.2	88.6	96.4	79.9	86.4
03300	CAST-IN-PLACE CONCRETE	98.7	100.0	99.2	91.7	102.7	96.3	104.2	83.1	95.4	102.2	79.8	92.9	97.6	92.5	95.5	110.7	78.6	97.3
03	**CONCRETE**	97.6	114.4	106.2	94.1	104.5	99.4	97.8	81.5	89.4	105.5	79.7	92.3	105.2	84.4	94.5	101.0	79.1	89.8
04	**MASONRY**	82.7	112.2	101.2	69.3	95.6	85.8	91.7	79.9	84.3	90.3	78.3	82.8	86.5	82.4	83.9	91.8	80.2	84.5
05	**METALS**	95.6	124.1	106.4	95.6	106.2	99.6	102.6	90.2	97.9	97.7	76.6	89.7	91.1	88.0	89.9	102.6	88.6	97.3
06	**WOOD & PLASTICS**	100.3	116.5	109.0	103.3	106.3	104.9	111.2	77.1	92.8	123.2	77.8	98.7	97.8	75.2	85.6	112.1	76.6	93.0
07	**THERMAL & MOISTURE PROTECTION**	97.2	112.5	104.6	96.8	101.5	99.1	100.6	78.7	89.9	91.9	83.0	87.5	95.1	86.5	90.9	100.3	85.2	93.0
08	**DOORS & WINDOWS**	96.2	118.9	102.0	96.2	104.0	98.2	99.3	80.8	94.6	103.2	81.2	97.6	95.6	78.5	91.3	99.3	77.4	93.8
09200	PLASTER & GYPSUM BOARD	105.0	116.8	113.1	105.0	106.3	105.9	102.4	77.1	85.2	99.4	77.9	84.8	95.3	73.6	80.5	101.3	76.6	84.5
095,098	CEILINGS & ACOUSTICAL TREATMENT	96.2	116.8	110.2	96.2	106.3	103.0	97.8	77.1	83.8	84.2	77.9	79.9	95.0	73.6	80.5	97.8	76.6	83.4
09600	FLOORING	96.8	97.1	96.9	97.0	88.2	94.7	86.9	92.2	88.3	103.5	92.2	100.6	96.7	87.0	94.2	86.9	83.1	85.9
097,099	WALL FINISHES, PAINTS & COATINGS	91.3	112.2	103.8	91.3	93.0	92.4	91.1	69.7	78.3	89.4	89.0	89.1	96.1	90.9	93.0	91.1	77.1	82.7
09	**FINISHES**	95.5	113.0	104.9	95.5	102.6	99.3	90.7	79.7	84.8	96.2	82.3	88.7	95.4	80.6	87.4	90.5	78.5	84.0
10 - 14	**TOTAL DIV. 10000 - 14000**	100.0	109.6	102.1	100.0	99.7	99.9	100.0	86.8	97.2	100.0	86.4	97.1	100.0	89.1	97.7	100.0	88.9	97.6
15	**MECHANICAL**	100.0	108.4	103.9	100.0	104.0	101.9	99.8	83.7	92.4	99.7	85.9	93.3	99.9	87.3	94.1	99.8	81.4	91.3
16	**ELECTRICAL**	101.6	114.3	109.0	102.5	92.5	96.7	85.6	90.1	88.2	100.3	83.7	90.6	96.0	86.9	90.7	86.2	84.0	84.9
01 - 16	**WEIGHTED AVERAGE**	97.2	111.9	104.3	96.0	100.7	98.3	97.0	85.0	91.2	99.1	83.2	91.4	96.6	89.1	93.0	97.5	83.1	90.6

DIVISION		INDIANA															IOWA		
		GARY			INDIANAPOLIS			MUNCIE			SOUTH BEND			TERRE HAUTE			CEDAR RAPIDS		
		MAT.	INST.	TOTAL	MAT.	INST.	TOTAL	MAT.	INST.	TOTAL	MAT.	INST.	TOTAL	MAT.	INST.	TOTAL	MAT.	INST.	TOTAL
01590	**EQUIPMENT RENTAL**	.0	96.9	96.9	.0	91.7	91.7	.0	96.3	96.3	.0	107.0	107.0	.0	121.5	121.5	.0	95.7	95.7
02	**SITE CONSTRUCTION**	88.1	98.7	96.0	87.5	100.5	97.2	77.5	96.0	91.4	87.7	96.5	94.3	83.9	130.4	118.7	86.3	94.8	92.7
03100	CONCRETE FORMS & ACCESSORIES	93.0	104.5	103.1	93.2	87.3	88.0	90.1	78.1	79.6	97.8	80.6	82.7	93.7	80.6	82.2	103.8	81.2	84.0
03200	CONCRETE REINFORCEMENT	96.4	102.0	99.8	95.7	85.0	89.2	97.8	83.1	88.9	96.4	78.3	85.4	96.9	82.7	88.3	98.3	84.6	90.0
03300	CAST-IN-PLACE CONCRETE	108.8	104.8	107.1	100.2	87.8	95.0	107.4	82.7	97.1	101.3	85.1	94.5	94.5	93.3	94.0	106.2	82.6	96.3
03	**CONCRETE**	100.1	104.2	102.2	100.2	86.6	93.2	103.3	81.3	92.0	92.1	83.2	87.6	108.3	85.6	96.7	99.8	83.0	91.2
04	**MASONRY**	93.1	96.0	94.9	98.7	85.9	90.7	92.5	79.9	84.6	90.1	79.9	83.7	93.4	80.7	85.4	106.9	82.1	91.4
05	**METALS**	102.6	103.1	102.8	104.7	79.6	95.2	99.3	90.5	96.0	101.2	102.9	101.9	91.9	88.1	90.4	88.4	94.1	90.5
06	**WOOD & PLASTICS**	110.2	105.0	107.4	107.7	87.3	96.7	113.0	77.0	93.6	111.4	80.2	94.6	99.9	79.3	88.8	110.4	80.1	94.1
07	**THERMAL & MOISTURE PROTECTION**	99.6	100.5	100.1	99.4	87.0	93.4	93.5	79.0	86.4	99.4	85.1	92.4	95.2	84.3	89.9	99.8	83.0	91.6
08	**DOORS & WINDOWS**	99.3	106.2	101.1	107.1	86.8	101.9	98.2	80.7	93.7	92.7	79.7	89.4	96.2	81.8	92.5	99.4	83.1	95.3
09200	PLASTER & GYPSUM BOARD	98.3	105.8	103.4	98.2	87.4	90.8	94.1	77.1	82.5	101.9	80.2	87.1	95.3	77.9	83.4	103.3	79.8	87.3
095,098	CEILINGS & ACOUSTICAL TREATMENT	97.8	105.8	103.2	96.5	87.4	90.3	85.5	77.1	79.8	97.8	80.2	85.9	95.0	77.9	83.4	110.3	79.8	89.7
09600	FLOORING	86.9	103.4	91.1	86.7	92.2	88.1	95.8	92.2	94.9	86.9	80.5	85.2	96.7	90.8	95.2	123.6	60.5	107.3
097,099	WALL FINISHES, PAINTS & COATINGS	91.1	105.0	99.4	91.1	92.2	91.8	89.4	69.7	77.6	91.1	83.5	86.6	96.1	90.5	92.7	108.0	79.9	91.1
09	**FINISHES**	90.1	104.4	97.8	90.2	88.9	89.5	92.9	79.6	85.7	90.6	80.7	85.3	95.4	83.3	88.9	111.6	76.3	92.5
10 - 14	**TOTAL DIV. 10000 - 14000**	100.0	104.7	101.0	100.0	89.7	97.8	100.0	86.3	97.1	100.0	82.5	96.2	100.0	90.7	98.0	100.0	86.9	97.2
15	**MECHANICAL**	99.8	98.0	99.0	99.8	86.2	93.5	99.6	83.7	92.3	99.8	79.6	90.5	99.9	81.2	91.3	100.4	86.4	94.0
16	**ELECTRICAL**	91.6	106.6	100.4	101.3	90.0	94.7	90.7	79.4	84.1	96.9	86.7	90.9	93.8	83.7	87.9	97.0	85.3	90.2
01 - 16	**WEIGHTED AVERAGE**	97.9	101.9	99.8	100.2	88.0	94.3	97.3	83.4	90.6	96.3	85.2	91.0	97.4	87.9	92.8	99.3	85.3	92.6

DIVISION		IOWA																	
		COUNCIL BLUFFS			DAVENPORT			DES MOINES			DUBUQUE			SIOUX CITY			WATERLOO		
		MAT.	INST.	TOTAL	MAT.	INST.	TOTAL	MAT.	INST.	TOTAL	MAT.	INST.	TOTAL	MAT.	INST.	TOTAL	MAT.	INST.	TOTAL
01590	EQUIPMENT RENTAL	.0	95.4	95.4	.0	99.8	99.8	.0	101.7	101.7	.0	94.3	94.3	.0	99.8	99.8	.0	99.8	99.8
02	SITE CONSTRUCTION	91.3	91.5	91.5	84.6	98.7	95.2	78.1	100.3	94.7	84.5	91.8	90.0	93.6	95.7	95.2	85.1	94.9	92.4
03100	CONCRETE FORMS & ACCESSORIES	80.6	60.6	63.1	103.4	89.9	91.5	105.5	81.1	84.1	82.5	69.9	71.5	103.8	64.6	69.4	104.4	51.2	57.7
03200	CONCRETE REINFORCEMENT	100.3	80.0	88.1	98.3	98.0	98.1	98.3	81.3	88.0	96.9	84.3	89.3	98.3	69.0	80.6	98.3	84.1	89.7
03300	CAST-IN-PLACE CONCRETE	110.5	76.6	96.3	102.3	103.9	103.0	106.5	81.3	96.0	104.0	90.6	98.4	105.3	59.0	85.9	106.2	50.0	82.7
03	CONCRETE	101.7	71.0	86.0	97.8	96.8	97.3	98.8	82.0	90.2	96.1	80.7	88.2	99.3	64.6	81.5	99.8	58.5	78.6
04	MASONRY	106.5	69.4	83.3	104.4	90.4	95.6	100.6	82.8	89.4	108.0	74.2	86.8	99.7	61.4	75.7	101.2	61.4	76.3
05	METALS	93.7	90.7	92.5	88.4	103.2	94.0	88.3	94.1	90.5	86.9	93.2	89.3	88.4	85.3	87.2	88.4	91.9	89.7
06	WOOD & PLASTICS	87.2	56.7	70.8	110.4	86.9	97.8	111.9	80.2	94.8	89.3	67.4	77.5	110.4	63.2	84.9	111.1	51.3	78.9
07	THERMAL & MOISTURE PROTECTION	99.0	67.3	83.5	99.2	90.4	94.9	100.1	81.2	90.9	99.3	71.6	85.8	99.2	59.4	79.8	98.8	55.5	77.7
08	DOORS & WINDOWS	98.5	63.8	89.7	99.4	92.0	97.5	99.4	83.2	95.3	98.5	78.1	93.3	99.4	64.5	90.6	95.0	63.7	87.1
09200	PLASTER & GYPSUM BOARD	94.1	55.8	68.0	103.3	86.6	91.9	99.6	79.6	86.0	94.6	66.8	75.7	103.3	62.2	75.3	103.3	50.0	67.0
095,098	CEILINGS & ACOUSTICAL TREATMENT	106.4	55.8	72.1	110.3	86.6	94.2	109.0	79.6	89.1	106.4	66.8	79.6	110.3	62.2	77.7	110.3	50.0	69.5
09600	FLOORING	98.5	49.4	85.9	109.9	82.8	102.9	109.7	45.1	93.0	112.6	42.1	94.4	110.4	63.8	98.4	111.6	64.1	99.3
097,099	WALL FINISHES, PAINTS & COATINGS	99.5	74.4	84.4	103.7	97.9	100.2	103.7	81.0	90.1	106.7	79.9	90.6	104.9	65.0	81.0	104.9	38.0	64.8
09	FINISHES	100.7	58.6	77.9	106.7	88.8	97.0	105.1	73.8	88.2	105.9	64.6	83.6	108.1	64.0	84.3	107.4	51.4	77.2
10 - 14	TOTAL DIV. 10000 - 14000	100.0	78.9	95.5	100.0	91.0	98.1	100.0	87.7	97.4	100.0	83.7	96.5	100.0	82.5	96.2	100.0	76.7	95.0
15	MECHANICAL	100.4	79.6	90.8	100.4	97.3	99.0	100.4	83.5	92.6	100.4	81.7	91.8	100.4	84.8	93.2	100.4	48.5	76.5
16	ELECTRICAL	103.9	84.5	92.5	93.4	92.6	92.9	97.7	86.4	91.1	102.0	73.1	85.1	97.0	80.7	87.5	97.0	66.3	79.1
01 - 16	WEIGHTED AVERAGE	99.6	76.3	88.4	98.1	94.7	96.5	98.1	85.0	91.7	98.3	78.9	88.9	98.7	75.7	87.6	98.0	64.0	81.6

DIVISION		KANSAS															KENTUCKY		
		DODGE CITY			KANSAS CITY			SALINA			TOPEKA			WICHITA			BOWLING GREEN		
		MAT.	INST.	TOTAL	MAT.	INST.	TOTAL	MAT.	INST.	TOTAL	MAT.	INST.	TOTAL	MAT.	INST.	TOTAL	MAT.	INST.	TOTAL
01590	EQUIPMENT RENTAL	.0	103.3	103.3	.0	99.9	99.9	.0	103.3	103.3	.0	101.4	101.4	.0	103.3	103.3	.0	96.2	96.2
02	SITE CONSTRUCTION	109.3	92.8	96.9	89.8	91.0	90.7	99.0	93.2	94.7	91.7	90.6	90.9	93.0	93.5	93.4	69.4	100.4	92.6
03100	CONCRETE FORMS & ACCESSORIES	94.7	45.0	51.1	100.2	95.1	95.7	90.3	47.7	52.9	100.0	46.7	53.2	96.7	60.3	64.7	82.7	83.0	82.9
03200	CONCRETE REINFORCEMENT	106.6	59.9	78.4	101.3	84.1	90.9	106.0	81.9	91.4	98.4	92.0	94.5	98.4	83.1	89.2	87.8	70.2	77.2
03300	CAST-IN-PLACE CONCRETE	114.6	58.9	91.3	89.8	94.1	91.6	99.4	50.4	78.9	90.8	54.9	75.8	86.6	59.2	75.1	89.2	82.8	86.5
03	CONCRETE	115.1	54.2	83.9	95.7	93.1	94.4	101.6	56.7	78.6	93.6	59.7	76.3	91.3	65.5	78.1	96.6	80.6	88.4
04	MASONRY	103.1	43.5	65.8	102.4	84.6	91.2	117.6	41.9	70.2	97.2	52.4	69.2	90.9	67.8	76.5	94.0	75.5	82.4
05	METALS	92.9	79.6	87.9	97.8	95.6	97.0	92.7	89.8	91.6	95.7	95.9	95.8	95.7	92.0	94.3	95.1	78.3	88.7
06	WOOD & PLASTICS	93.4	45.1	67.4	99.8	96.0	97.8	89.7	48.6	67.5	97.2	43.0	67.9	95.2	59.1	75.7	94.1	82.5	87.8
07	THERMAL & MOISTURE PROTECTION	98.2	50.7	75.1	96.2	91.5	93.9	97.6	56.4	77.5	97.5	70.0	84.1	97.1	67.9	82.9	85.0	80.2	82.6
08	DOORS & WINDOWS	96.1	48.7	84.1	95.0	92.5	94.4	96.1	54.3	85.5	96.2	61.2	87.4	96.2	65.7	88.5	95.3	75.5	90.3
09200	PLASTER & GYPSUM BOARD	102.5	43.4	62.3	98.3	95.8	96.6	101.5	47.0	64.4	105.0	41.2	61.6	105.0	57.8	72.9	90.8	82.3	85.0
095,098	CEILINGS & ACOUSTICAL TREATMENT	89.7	43.4	58.4	96.2	95.8	95.9	89.7	47.0	60.8	96.2	41.2	58.9	96.2	57.8	70.2	91.1	82.3	85.2
09600	FLOORING	96.1	53.2	85.1	87.1	56.8	79.3	94.1	38.9	79.9	97.7	49.4	85.2	96.8	82.5	93.1	92.5	65.5	85.6
097,099	WALL FINISHES, PAINTS & COATINGS	91.3	51.5	67.4	99.4	92.3	95.2	91.3	40.6	60.9	91.3	69.3	78.1	91.3	65.4	75.8	96.1	77.1	84.7
09	FINISHES	94.7	46.8	68.8	93.9	87.8	90.6	93.3	44.5	66.9	95.7	47.5	69.6	95.5	64.6	78.8	92.8	79.2	85.4
10 - 14	TOTAL DIV. 10000 - 14000	100.0	53.3	89.9	100.0	89.5	97.7	100.0	70.1	93.6	100.0	74.3	94.5	100.0	75.0	94.6	100.0	67.3	93.0
15	MECHANICAL	100.0	49.6	76.8	99.9	91.2	95.9	100.0	37.5	71.3	100.0	72.9	87.6	100.0	68.8	85.7	99.9	84.4	92.8
16	ELECTRICAL	98.1	52.3	71.3	106.0	106.1	106.0	97.7	75.4	84.7	103.2	78.0	88.5	100.4	75.4	85.8	94.1	83.8	88.1
01 - 16	WEIGHTED AVERAGE	100.0	56.5	79.0	98.3	92.9	95.7	98.6	59.0	79.5	97.6	69.4	84.0	96.7	73.0	85.3	95.4	82.2	89.0

DIVISION		KENTUCKY									LOUISIANA								
		LEXINGTON			LOUISVILLE			OWENSBORO			ALEXANDRIA			BATON ROUGE			LAKE CHARLES		
		MAT.	INST.	TOTAL	MAT.	INST.	TOTAL	MAT.	INST.	TOTAL	MAT.	INST.	TOTAL	MAT.	INST.	TOTAL	MAT.	INST.	TOTAL
01590	EQUIPMENT RENTAL	.0	103.7	103.7	.0	96.2	96.2	.0	121.5	121.5	.0	86.6	86.6	.0	86.4	86.4	.0	86.4	86.4
02	SITE CONSTRUCTION	77.4	102.5	96.2	66.7	100.3	91.8	82.0	130.0	117.9	103.2	85.6	90.0	111.7	85.6	92.2	113.3	85.3	92.4
03100	CONCRETE FORMS & ACCESSORIES	97.1	70.6	73.9	91.5	82.6	83.7	88.3	67.2	69.7	80.7	45.8	50.1	100.7	50.1	56.3	101.1	51.0	57.1
03200	CONCRETE REINFORCEMENT	96.9	92.8	94.4	96.9	93.8	95.0	87.9	96.2	92.9	99.1	64.0	77.9	98.3	52.5	70.6	98.3	52.4	70.6
03300	CAST-IN-PLACE CONCRETE	96.3	82.5	90.6	93.2	81.4	88.3	92.0	79.2	86.6	93.9	45.0	73.5	87.3	54.0	73.4	92.5	55.5	77.0
03	CONCRETE	98.5	79.3	88.7	96.6	84.4	90.3	106.7	77.2	91.6	88.6	50.1	68.9	95.1	52.7	73.4	97.7	53.6	75.1
04	MASONRY	90.5	49.6	64.9	92.1	79.3	84.1	90.9	53.6	67.6	114.7	55.6	77.7	95.5	57.3	71.6	93.3	58.4	71.5
05	METALS	96.6	88.1	93.4	96.6	88.8	93.7	86.5	89.0	87.5	87.1	73.8	82.1	100.6	66.7	87.7	100.1	66.9	87.5
06	WOOD & PLASTICS	103.9	73.3	87.4	102.6	82.5	91.7	94.1	66.2	79.1	86.4	45.6	64.4	110.0	50.3	77.8	108.0	51.0	77.3
07	THERMAL & MOISTURE PROTECTION	95.7	79.8	87.9	85.2	81.8	83.5	95.2	69.9	82.9	99.4	56.0	78.2	97.4	57.1	77.8	100.6	57.3	79.5
08	DOORS & WINDOWS	96.2	79.7	92.0	96.2	86.2	93.7	93.6	77.9	89.6	99.2	51.4	87.1	104.8	50.4	91.0	104.8	52.4	91.5
09200	PLASTER & GYPSUM BOARD	96.8	71.7	79.7	98.7	82.3	87.5	90.2	64.4	72.7	81.2	44.8	56.4	102.5	49.5	66.4	102.5	50.2	66.9
095,098	CEILINGS & ACOUSTICAL TREATMENT	95.0	71.7	79.2	95.0	82.3	86.4	82.0	64.4	70.1	96.8	44.8	61.6	101.9	49.5	66.4	101.9	50.2	66.9
09600	FLOORING	97.5	45.5	84.1	96.0	73.7	90.3	94.9	65.5	87.4	105.6	71.7	96.9	110.0	70.3	99.7	110.2	55.0	95.9
097,099	WALL FINISHES, PAINTS & COATINGS	96.1	61.6	75.4	96.1	80.7	86.8	96.1	102.8	100.1	96.6	43.5	64.7	98.9	49.8	69.5	98.9	45.9	67.1
09	FINISHES	96.0	64.4	78.9	95.4	80.5	87.3	91.8	70.0	80.0	95.7	49.9	70.9	104.3	53.7	77.0	104.4	50.9	75.5
10 - 14	TOTAL DIV. 10000 - 14000	100.0	86.7	97.1	100.0	89.8	97.8	100.0	89.5	97.7	100.0	75.9	94.8	100.0	69.8	93.5	100.0	70.1	93.6
15	MECHANICAL	99.9	50.4	77.1	99.9	83.1	92.1	99.9	55.3	79.4	100.1	39.8	72.3	100.1	45.2	74.9	100.1	58.8	81.1
16	ELECTRICAL	94.7	52.7	70.2	94.8	83.8	88.4	93.4	77.5	84.1	95.8	58.5	74.0	100.2	52.3	72.2	100.2	65.6	80.0
01 - 16	WEIGHTED AVERAGE	96.7	67.9	82.9	95.9	85.0	90.7	95.5	75.5	85.9	96.7	56.0	77.1	100.6	56.5	79.4	100.9	61.1	81.7

DIVISION		LOUISIANA									MAINE								
		MONROE			NEW ORLEANS			SHREVEPORT			AUGUSTA			BANGOR			LEWISTON		
		MAT.	INST.	TOTAL	MAT.	INST.	TOTAL	MAT.	INST.	TOTAL	MAT.	INST.	TOTAL	MAT.	INST.	TOTAL	MAT.	INST.	TOTAL
01590	**EQUIPMENT RENTAL**	.0	86.6	86.6	.0	87.6	87.6	.0	86.6	86.6	.0	101.8	101.8	.0	101.8	101.8	.0	101.8	101.8
02	**SITE CONSTRUCTION**	103.2	85.4	89.9	115.8	87.8	94.8	101.9	85.3	89.5	82.8	102.7	97.7	82.7	101.1	96.5	81.3	101.1	96.1
03100	CONCRETE FORMS & ACCESSORIES	80.0	46.5	50.6	100.6	70.0	73.8	100.9	49.4	55.7	102.5	67.5	71.7	96.4	83.3	84.9	102.6	83.3	85.7
03200	CONCRETE REINFORCEMENT	98.0	64.2	77.6	98.3	66.3	79.0	97.6	65.2	78.0	88.8	115.5	104.9	88.8	115.8	105.1	109.1	115.8	113.1
03300	CAST-IN-PLACE CONCRETE	93.9	49.4	75.3	91.6	61.4	78.9	92.6	53.2	76.1	83.0	65.2	75.6	83.0	67.2	76.4	94.5	67.2	83.1
03	**CONCRETE**	88.5	51.9	69.7	97.2	66.7	81.6	88.4	54.7	71.1	100.6	75.7	87.9	99.7	83.3	91.3	103.8	83.3	93.3
04	**MASONRY**	109.5	59.9	78.4	95.3	62.0	74.4	101.7	52.7	71.0	94.6	54.0	69.1	112.6	62.0	80.9	96.2	62.0	74.8
05	**METALS**	87.1	73.5	81.9	103.0	74.6	92.2	87.9	73.5	82.5	94.4	86.5	91.4	94.2	84.7	90.6	97.5	84.7	92.7
06	**WOOD & PLASTICS**	85.6	46.1	64.3	104.3	73.4	87.6	105.9	50.1	75.8	100.0	66.0	81.6	94.1	86.0	89.7	100.0	86.0	92.4
07	**THERMAL & MOISTURE PROTECTION**	99.4	58.9	79.7	101.4	65.7	84.0	98.3	57.5	78.4	102.7	54.4	79.1	102.5	61.6	82.6	102.3	61.6	82.5
08	**DOORS & WINDOWS**	99.2	55.9	88.2	105.4	70.6	96.6	97.1	54.3	86.2	104.0	64.3	93.9	103.9	77.0	97.1	107.2	77.0	99.5
09200	PLASTER & GYPSUM BOARD	80.7	45.3	56.6	101.4	73.2	82.2	86.1	49.4	61.2	107.2	64.0	77.8	104.3	84.6	90.9	109.1	84.6	92.4
095,098	CEILINGS & ACOUSTICAL TREATMENT	96.8	45.3	61.9	100.4	73.2	82.0	98.1	49.4	65.1	98.1	64.0	75.0	96.8	84.6	88.5	105.9	84.6	91.5
09600	FLOORING	105.2	45.4	89.8	110.6	53.9	96.0	116.3	67.2	103.6	100.3	45.6	86.2	97.8	56.9	87.3	100.3	56.9	89.1
097,099	WALL FINISHES, PAINTS & COATINGS	96.6	51.9	69.8	100.7	65.9	79.8	96.6	43.5	64.7	93.0	37.6	59.7	93.0	34.6	58.0	93.0	34.6	58.0
09	**FINISHES**	95.5	45.9	68.7	104.6	67.0	84.3	100.1	51.6	73.9	101.4	59.3	78.7	100.0	73.6	85.8	102.8	73.6	87.0
10 - 14	**TOTAL DIV. 10000 - 14000**	100.0	67.4	93.0	100.0	74.9	94.6	100.0	69.7	93.5	100.0	67.9	93.1	100.0	88.2	97.5	100.0	88.2	97.5
15	**MECHANICAL**	100.1	53.8	78.8	100.1	64.7	83.8	100.1	56.6	80.1	100.2	74.9	88.5	100.2	73.9	88.1	100.2	73.9	88.1
16	**ELECTRICAL**	98.3	58.0	74.8	100.5	69.6	82.5	96.6	69.6	80.8	102.6	84.1	91.8	100.1	84.0	90.7	102.7	84.0	91.8
01 - 16	**WEIGHTED AVERAGE**	96.6	59.0	78.5	101.5	69.4	86.0	96.5	61.6	79.6	99.4	74.6	87.4	99.9	79.2	89.9	100.8	79.2	90.4

DIVISION		MAINE			MARYLAND						MASSACHUSETTS								
		PORTLAND			BALTIMORE			HAGERSTOWN			BOSTON			BROCKTON			FALL RIVER		
		MAT.	INST.	TOTAL	MAT.	INST.	TOTAL	MAT.	INST.	TOTAL	MAT.	INST.	TOTAL	MAT.	INST.	TOTAL	MAT.	INST.	TOTAL
01590	**EQUIPMENT RENTAL**	.0	101.8	101.8	.0	102.9	102.9	.0	98.8	98.8	.0	108.7	108.7	.0	104.0	104.0	.0	105.2	105.2
02	**SITE CONSTRUCTION**	80.0	101.1	95.8	97.8	93.1	94.3	87.5	88.7	88.4	87.8	109.0	103.7	85.5	104.9	100.0	84.6	105.0	99.9
03100	CONCRETE FORMS & ACCESSORIES	101.6	83.3	85.6	103.3	75.0	78.4	91.2	76.2	78.1	108.4	132.5	129.5	108.1	120.4	118.9	108.1	120.5	119.0
03200	CONCRETE REINFORCEMENT	109.1	115.8	113.1	99.9	89.8	93.8	88.3	75.9	80.8	107.9	141.0	127.9	109.1	140.6	128.1	109.1	125.8	119.2
03300	CAST-IN-PLACE CONCRETE	86.0	67.2	78.2	95.8	78.9	88.7	81.5	62.5	73.5	107.1	140.1	120.9	102.0	132.4	114.7	98.6	135.9	114.2
03	**CONCRETE**	99.5	83.3	91.2	99.1	80.3	89.5	84.7	72.5	78.5	113.5	135.9	125.0	110.5	127.7	119.3	108.8	126.1	117.7
04	**MASONRY**	93.7	62.0	73.9	93.1	74.9	81.7	97.4	74.5	83.1	115.5	141.6	131.8	111.3	130.3	123.2	111.1	133.4	125.1
05	**METALS**	97.5	84.7	92.7	95.3	100.4	97.2	94.3	91.7	93.3	100.4	124.5	109.5	97.4	120.9	106.3	97.4	115.4	104.2
06	**WOOD & PLASTICS**	100.0	86.0	92.4	101.6	76.0	87.8	89.8	75.8	82.3	103.8	132.4	119.2	102.7	119.5	111.8	102.7	119.9	112.0
07	**THERMAL & MOISTURE PROTECTION**	102.1	61.6	82.4	93.7	83.4	88.7	92.9	72.4	82.9	102.9	138.0	120.0	102.7	127.2	114.6	102.6	122.0	112.1
08	**DOORS & WINDOWS**	107.2	77.0	99.5	93.8	84.7	91.5	91.0	77.4	87.6	103.1	133.5	110.9	101.1	122.9	106.6	101.1	119.7	105.8
09200	PLASTER & GYPSUM BOARD	109.1	84.6	92.4	110.6	75.7	86.8	105.5	75.7	85.2	112.5	132.4	126.0	105.7	119.1	114.8	105.7	119.1	114.8
095,098	CEILINGS & ACOUSTICAL TREATMENT	105.9	84.6	91.5	98.0	75.7	82.9	99.4	75.7	83.3	104.5	132.4	123.4	107.4	119.1	115.3	107.4	119.1	115.3
09600	FLOORING	100.3	56.9	89.1	93.2	83.2	90.6	88.3	77.5	85.5	100.9	145.3	112.3	101.4	145.3	112.7	101.1	145.3	112.5
097,099	WALL FINISHES, PAINTS & COATINGS	93.0	34.6	58.0	96.2	73.3	82.5	96.2	42.4	63.9	96.1	147.2	126.8	95.5	123.0	112.0	95.5	123.0	112.0
09	**FINISHES**	102.9	73.6	87.1	95.4	75.7	84.8	92.7	72.8	81.9	103.2	136.6	121.3	103.0	125.2	115.0	103.0	125.5	115.1
10 - 14	**TOTAL DIV. 10000 - 14000**	100.0	88.2	97.5	100.0	86.9	97.2	100.0	89.4	97.7	100.0	124.1	105.2	100.0	120.7	104.4	100.0	121.6	104.7
15	**MECHANICAL**	100.2	73.9	88.1	99.9	84.5	92.8	99.9	88.0	94.4	100.2	123.8	111.1	100.2	108.6	104.0	100.2	108.5	104.0
16	**ELECTRICAL**	103.0	84.0	91.9	100.3	90.5	94.6	97.1	81.8	88.2	98.5	121.3	111.9	98.2	100.1	99.3	97.7	100.0	99.1
01 - 16	**WEIGHTED AVERAGE**	100.1	79.2	90.0	97.5	84.9	91.4	94.5	81.1	88.1	102.9	128.1	115.0	101.5	116.6	108.8	101.2	116.0	108.3

DIVISION		MASSACHUSETTS																	
		HYANNIS			LAWRENCE			LOWELL			NEW BEDFORD			PITTSFIELD			SPRINGFIELD		
		MAT.	INST.	TOTAL	MAT.	INST.	TOTAL	MAT.	INST.	TOTAL	MAT.	INST.	TOTAL	MAT.	INST.	TOTAL	MAT.	INST.	TOTAL
01590	**EQUIPMENT RENTAL**	.0	104.0	104.0	.0	104.0	104.0	.0	101.8	101.8	.0	105.2	105.2	.0	101.8	101.8	.0	101.8	101.8
02	**SITE CONSTRUCTION**	82.3	104.9	99.2	86.1	104.9	100.2	85.2	104.9	99.9	84.2	105.0	99.8	86.3	103.6	99.3	85.7	103.8	99.2
03100	CONCRETE FORMS & ACCESSORIES	98.1	120.1	117.4	107.9	120.6	119.0	104.3	120.7	118.7	108.1	120.5	119.0	104.2	94.5	95.7	104.4	112.6	111.6
03200	CONCRETE REINFORCEMENT	87.5	125.8	110.6	108.2	127.4	119.8	109.1	127.4	120.2	109.1	125.8	119.2	90.7	107.9	101.1	109.1	123.9	118.0
03300	CAST-IN-PLACE CONCRETE	93.0	135.2	110.7	102.8	136.1	116.7	93.5	136.2	111.3	95.7	135.9	112.5	102.0	109.9	105.3	97.3	111.4	103.2
03	**CONCRETE**	100.7	125.7	113.5	110.8	126.5	118.9	100.8	126.4	113.9	107.4	126.1	117.0	102.9	102.0	102.5	102.7	113.6	108.3
04	**MASONRY**	110.3	133.5	124.8	110.5	130.3	122.9	96.1	130.1	117.4	110.8	133.4	125.0	96.7	104.5	101.6	96.4	106.8	102.9
05	**METALS**	93.9	114.8	101.8	94.9	115.7	102.8	94.9	112.8	101.7	97.4	115.4	104.2	94.7	99.6	96.6	97.3	106.5	100.8
06	**WOOD & PLASTICS**	93.5	119.5	107.5	102.7	119.5	111.8	101.9	119.5	111.4	102.7	119.9	112.0	101.9	92.2	96.7	101.9	115.2	109.0
07	**THERMAL & MOISTURE PROTECTION**	102.1	123.0	112.3	102.6	127.7	114.9	102.4	127.4	114.5	102.6	122.0	112.1	102.4	105.4	103.9	102.4	108.7	105.5
08	**DOORS & WINDOWS**	97.2	119.5	102.9	101.1	119.4	105.7	107.2	119.4	110.3	101.1	119.7	105.8	107.2	100.2	105.4	107.2	116.9	109.7
09200	PLASTER & GYPSUM BOARD	99.9	119.1	113.0	109.1	119.1	115.9	109.1	119.1	115.9	105.7	119.1	114.8	109.1	91.0	96.8	109.1	114.6	112.8
095,098	CEILINGS & ACOUSTICAL TREATMENT	98.3	119.1	112.4	105.9	119.1	114.8	105.9	119.1	114.8	107.4	119.1	115.3	105.9	91.0	95.8	105.9	114.6	111.8
09600	FLOORING	97.3	145.3	109.7	100.3	145.3	111.9	100.3	145.3	111.9	101.1	145.3	112.5	100.5	116.4	104.6	100.2	116.4	104.4
097,099	WALL FINISHES, PAINTS & COATINGS	95.5	123.0	112.0	93.1	123.0	111.0	93.0	123.0	111.0	95.5	123.0	112.0	93.0	98.9	96.5	94.8	98.9	97.2
09	**FINISHES**	99.0	125.2	113.2	102.6	125.2	114.8	102.6	125.2	114.8	102.9	125.5	115.1	102.7	98.4	100.3	102.7	112.4	107.9
10 - 14	**TOTAL DIV. 10000 - 14000**	100.0	120.7	104.4	100.0	120.8	104.5	100.0	120.8	104.5	100.0	121.6	104.7	100.0	101.8	100.4	100.0	105.8	101.2
15	**MECHANICAL**	100.2	108.4	104.0	100.2	109.8	104.6	100.2	120.8	109.7	100.2	108.5	104.0	100.2	97.3	98.9	100.2	98.8	99.5
16	**ELECTRICAL**	94.4	100.0	97.7	102.1	124.4	115.1	102.7	124.4	115.4	98.5	100.0	99.4	102.7	87.4	93.8	102.7	91.4	96.1
01 - 16	**WEIGHTED AVERAGE**	98.4	115.8	106.8	101.5	119.5	110.2	100.2	121.4	110.4	101.1	116.0	108.3	100.5	98.6	99.6	100.8	104.9	102.8

	DIVISION	MASSACHUSETTS			MICHIGAN														
		WORCESTER			ANN ARBOR			DEARBORN			DETROIT			FLINT			GRAND RAPIDS		
		MAT.	INST.	TOTAL	MAT.	INST.	TOTAL	MAT.	INST.	TOTAL	MAT.	INST.	TOTAL	MAT.	INST.	TOTAL	MAT.	INST.	TOTAL
01590	EQUIPMENT RENTAL	.0	101.8	101.8	.0	112.4	112.4	.0	112.4	112.4	.0	97.8	97.8	.0	112.4	112.4	.0	105.3	105.3
02	SITE CONSTRUCTION	85.2	104.8	99.9	79.6	95.3	91.4	79.4	95.5	91.5	94.1	97.3	96.5	69.7	94.3	88.2	82.6	88.1	86.7
03100	CONCRETE FORMS & ACCESSORIES	104.9	120.1	118.2	97.5	114.5	112.4	97.4	123.7	120.5	98.8	123.8	120.7	101.9	96.9	97.5	97.3	82.1	84.0
03200	CONCRETE REINFORCEMENT	109.1	139.6	127.5	95.7	122.3	111.8	95.7	122.8	112.1	95.1	122.8	111.8	95.7	121.9	111.5	96.9	88.8	92.0
03300	CAST-IN-PLACE CONCRETE	93.5	131.6	109.4	94.6	114.7	103.0	92.5	120.5	104.2	101.5	120.5	109.5	95.3	96.5	95.8	96.5	103.1	99.2
03	CONCRETE	100.9	126.9	114.2	92.7	116.6	104.9	91.6	122.7	107.6	96.2	121.3	109.0	93.3	102.4	98.0	98.6	90.0	94.2
04	MASONRY	96.1	130.1	117.4	99.6	109.6	105.8	99.4	120.5	112.6	98.4	120.5	112.2	99.6	95.5	97.0	95.0	57.8	71.7
05	METALS	97.3	117.1	104.8	100.8	126.8	110.6	100.9	128.2	111.2	101.4	105.6	103.0	100.9	123.8	109.6	97.6	82.0	91.7
06	WOOD & PLASTICS	102.4	119.5	111.6	102.0	114.1	108.5	102.0	123.9	113.8	102.5	123.9	114.1	106.1	98.0	101.7	98.2	82.9	90.0
07	THERMAL & MOISTURE PROTECTION	102.4	121.9	111.9	97.9	112.0	104.7	96.8	124.1	110.1	95.3	124.1	109.3	96.1	95.6	95.9	92.0	64.6	78.7
08	DOORS & WINDOWS	107.2	126.3	112.0	96.4	114.1	100.9	96.4	119.4	102.2	98.0	120.4	103.7	96.4	101.6	97.7	94.2	73.6	89.0
09200	PLASTER & GYPSUM BOARD	109.1	119.1	115.9	104.7	113.2	110.5	104.7	123.2	117.3	104.7	123.2	117.3	106.3	96.5	99.6	95.3	77.3	83.1
095,098	CEILINGS & ACOUSTICAL TREATMENT	105.9	119.1	114.8	102.0	113.2	109.6	102.0	123.2	116.4	103.5	123.2	116.9	102.0	96.5	98.3	95.0	77.3	83.0
09600	FLOORING	100.3	142.3	111.1	87.6	114.2	94.5	87.2	125.0	96.9	87.3	125.0	97.0	87.4	77.9	85.0	96.2	46.0	83.3
097,099	WALL FINISHES, PAINTS & COATINGS	93.0	123.0	111.0	86.5	106.5	98.5	86.5	113.2	102.5	88.4	113.2	103.3	86.5	83.3	84.6	96.1	44.0	64.9
09	FINISHES	102.6	124.6	114.5	95.1	113.7	105.1	94.9	123.2	110.2	96.1	123.2	110.7	94.6	91.6	93.0	95.3	71.2	82.3
10 - 14	TOTAL DIV. 10000 - 14000	100.0	109.1	102.0	100.0	113.3	102.9	100.0	116.0	103.4	100.0	116.0	103.4	100.0	98.2	99.6	100.0	107.7	101.7
15	MECHANICAL	100.2	107.9	103.7	100.0	106.9	103.1	100.0	117.5	108.0	100.0	119.2	108.8	100.0	96.8	98.5	99.9	59.0	81.1
16	ELECTRICAL	102.7	102.5	102.6	94.2	91.3	92.5	94.2	118.1	108.2	95.3	118.0	108.6	94.2	92.4	93.1	94.3	63.7	76.4
01 - 16	WEIGHTED AVERAGE	100.5	116.0	108.0	97.1	108.6	102.7	97.0	118.5	107.4	98.3	116.8	107.2	96.9	98.6	97.7	96.8	72.4	85.0

	DIVISION	MICHIGAN												MINNESOTA					
		KALAMAZOO			LANSING			MUSKEGON			SAGINAW			DULUTH			MINNEAPOLIS		
		MAT.	INST.	TOTAL	MAT.	INST.	TOTAL	MAT.	INST.	TOTAL	MAT.	INST.	TOTAL	MAT.	INST.	TOTAL	MAT.	INST.	TOTAL
01590	EQUIPMENT RENTAL	.0	105.3	105.3	.0	112.4	112.4	.0	105.3	105.3	.0	112.4	112.4	.0	102.0	102.0	.0	106.4	106.4
02	SITE CONSTRUCTION	83.2	88.3	87.0	86.1	94.1	92.1	81.2	88.2	86.4	72.0	94.1	88.5	82.6	102.9	97.8	82.4	109.2	102.5
03100	CONCRETE FORMS & ACCESSORIES	96.5	91.9	92.4	101.1	94.4	95.2	97.4	88.6	89.7	97.5	95.6	95.9	99.4	123.9	120.9	100.1	138.1	133.4
03200	CONCRETE REINFORCEMENT	96.9	90.4	93.0	95.7	121.7	111.4	97.5	90.2	93.1	95.7	121.3	111.2	98.1	107.2	103.6	98.3	133.1	119.3
03300	CAST-IN-PLACE CONCRETE	98.2	103.6	100.4	94.7	96.8	95.6	95.9	100.6	97.9	93.5	96.5	94.7	107.3	107.4	107.4	105.7	130.8	116.2
03	CONCRETE	102.0	94.8	98.3	93.0	101.4	97.3	96.5	92.3	94.3	92.1	101.7	97.0	99.3	115.6	107.6	100.3	135.0	118.0
04	MASONRY	97.7	86.9	91.0	92.9	95.9	94.8	94.8	81.3	86.4	101.1	88.8	93.4	108.5	117.3	114.0	108.5	140.8	128.7
05	METALS	97.9	86.0	93.4	99.6	123.6	108.7	95.8	85.6	92.0	100.9	122.5	109.1	95.2	125.4	106.6	97.1	142.3	114.2
06	WOOD & PLASTICS	99.9	91.9	95.6	104.7	93.2	98.5	96.5	88.7	92.3	97.6	96.2	96.9	111.9	126.3	119.7	112.3	136.1	125.2
07	THERMAL & MOISTURE PROTECTION	91.6	87.3	89.5	96.8	94.9	95.9	90.7	80.8	85.9	96.7	91.9	94.4	100.6	117.0	108.6	100.5	137.5	118.5
08	DOORS & WINDOWS	90.9	86.0	89.7	96.4	100.2	97.4	90.2	84.9	88.8	95.9	100.7	97.1	97.2	123.6	103.9	100.3	145.6	111.8
09200	PLASTER & GYPSUM BOARD	95.3	86.5	89.3	108.1	91.7	96.9	88.2	83.2	84.8	104.7	94.7	97.9	100.0	127.6	118.8	100.0	137.5	125.6
095,098	CEILINGS & ACOUSTICAL TREATMENT	95.0	86.5	89.3	102.0	91.7	95.0	100.2	83.2	88.7	102.0	94.7	97.1	85.5	127.6	114.0	85.5	137.5	120.7
09600	FLOORING	96.2	73.0	90.2	95.8	86.3	93.3	95.4	84.7	92.6	87.6	61.9	81.0	106.3	126.6	111.5	103.7	124.9	109.1
097,099	WALL FINISHES, PAINTS & COATINGS	96.1	79.9	86.4	100.5	97.4	98.6	95.3	63.8	76.4	86.5	83.6	84.8	93.5	106.2	101.1	99.8	131.1	118.6
09	FINISHES	95.3	86.9	90.8	99.9	92.7	96.0	94.6	84.9	89.4	94.8	87.6	90.9	99.6	123.6	112.6	99.1	135.3	118.7
10 - 14	TOTAL DIV. 10000 - 14000	100.0	110.3	102.2	100.0	102.6	100.6	100.0	109.0	101.9	100.0	98.7	99.7	100.0	104.7	101.0	100.0	112.0	102.6
15	MECHANICAL	99.9	89.3	95.0	100.0	95.6	98.0	99.7	92.2	96.3	100.0	94.2	97.3	100.1	98.1	99.2	100.0	121.9	110.1
16	ELECTRICAL	94.0	74.6	82.6	93.1	91.6	92.2	94.4	70.0	80.1	95.1	86.4	90.1	101.6	98.9	100.0	102.6	111.9	108.0
01 - 16	WEIGHTED AVERAGE	97.0	87.4	92.4	97.2	98.3	97.7	95.6	85.9	90.9	96.9	95.7	96.3	99.1	110.8	104.8	99.9	127.9	113.4

	DIVISION	MINNESOTA									MISSISSIPPI								
		ROCHESTER			SAINT PAUL			ST. CLOUD			BILOXI			GREENVILLE			JACKSON		
		MAT.	INST.	TOTAL	MAT.	INST.	TOTAL	MAT.	INST.	TOTAL	MAT.	INST.	TOTAL	MAT.	INST.	TOTAL	MAT.	INST.	TOTAL
01590	EQUIPMENT RENTAL	.0	102.0	102.0	.0	102.0	102.0	.0	101.7	101.7	.0	98.3	98.3	.0	98.3	98.3	.0	98.3	98.3
02	SITE CONSTRUCTION	81.7	102.1	97.0	84.6	103.7	98.9	77.5	105.6	98.5	105.4	86.4	91.2	109.8	86.2	92.1	101.8	86.2	90.1
03100	CONCRETE FORMS & ACCESSORIES	100.0	108.4	107.3	90.7	135.7	130.2	83.8	131.6	125.7	96.0	44.0	50.4	77.3	36.4	41.4	91.3	41.6	47.7
03200	CONCRETE REINFORCEMENT	98.1	132.4	118.8	94.7	133.1	117.9	101.5	132.4	120.1	95.4	63.9	76.4	103.4	46.0	68.7	95.4	49.9	67.9
03300	CAST-IN-PLACE CONCRETE	103.6	101.2	102.6	106.6	129.9	116.4	96.4	127.8	109.5	105.2	47.7	81.1	105.6	41.8	78.9	103.2	44.3	78.6
03	CONCRETE	97.5	111.5	104.7	102.1	133.6	118.2	89.8	130.8	110.8	97.5	51.0	73.7	100.4	42.2	70.6	96.2	46.1	70.5
04	MASONRY	107.7	112.9	111.0	118.5	140.8	132.4	110.1	126.1	120.1	89.2	39.3	58.0	132.0	40.8	74.9	91.3	40.8	59.7
05	METALS	95.1	139.6	111.9	93.9	141.8	112.0	95.2	138.8	111.7	97.2	84.8	92.5	95.5	76.4	88.3	97.1	78.3	90.0
06	WOOD & PLASTICS	112.3	108.0	110.0	103.1	133.4	119.4	94.3	129.6	113.4	96.4	44.1	68.2	77.3	35.8	54.9	91.6	42.5	65.1
07	THERMAL & MOISTURE PROTECTION	100.4	107.2	103.7	100.4	135.6	117.5	98.3	121.3	109.5	95.3	48.8	72.6	95.4	43.7	70.2	94.9	45.0	70.6
08	DOORS & WINDOWS	97.2	130.3	105.6	94.4	144.1	107.0	91.7	142.0	104.5	98.5	50.2	86.2	97.9	40.6	83.3	98.9	45.3	85.3
09200	PLASTER & GYPSUM BOARD	100.0	108.7	106.0	95.3	134.8	122.2	88.2	131.1	117.4	100.7	43.1	61.5	91.3	34.5	52.7	100.7	41.4	60.4
095,098	CEILINGS & ACOUSTICAL TREATMENT	85.5	108.7	101.2	82.9	134.8	118.1	70.1	131.1	111.4	95.0	43.1	59.9	89.8	34.5	52.4	95.0	41.4	58.7
09600	FLOORING	106.1	85.4	100.8	99.1	124.9	105.7	97.9	124.9	104.8	118.6	43.0	99.1	109.0	36.8	90.4	118.6	42.6	99.0
097,099	WALL FINISHES, PAINTS & COATINGS	95.9	102.3	99.8	99.8	124.3	114.5	103.6	131.1	120.1	107.5	37.6	65.6	107.5	37.0	65.2	107.5	37.0	65.2
09	FINISHES	99.6	103.9	102.0	96.8	132.9	116.3	92.6	130.7	113.2	105.6	42.7	71.6	100.6	36.3	65.8	105.6	41.4	70.9
10 - 14	TOTAL DIV. 10000 - 14000	100.0	100.0	100.0	100.0	111.2	102.4	100.0	109.2	102.0	100.0	58.6	91.1	100.0	56.8	90.7	100.0	57.7	90.9
15	MECHANICAL	100.0	102.6	101.2	100.0	117.1	107.9	99.7	120.2	109.1	99.8	57.9	80.5	99.8	36.9	70.8	99.8	36.5	70.7
16	ELECTRICAL	101.5	87.7	93.5	100.8	110.3	106.4	99.4	110.3	105.8	97.8	47.5	68.4	96.9	39.5	63.4	98.3	39.5	63.9
01 - 16	WEIGHTED AVERAGE	98.8	107.6	103.0	99.2	125.5	111.8	96.1	123.3	109.2	98.8	56.1	78.2	100.6	47.4	75.0	98.7	49.0	74.7

DIVISION	MISSISSIPPI			MISSOURI															
	MERIDIAN			CAPE GIRARDEAU			COLUMBIA			JOPLIN			KANSAS CITY			SPRINGFIELD		
	MAT.	INST.	TOTAL	MAT.	INST.	TOTAL	MAT.	INST.	TOTAL	MAT.	INST.	TOTAL	MAT.	INST.	TOTAL	MAT.	INST.	TOTAL
01590 EQUIPMENT RENTAL	.0	98.3	98.3	.0	108.6	108.6	.0	108.8	108.8	.0	106.0	106.0	.0	103.9	103.9	.0	102.5	102.5
02 SITE CONSTRUCTION	100.7	86.2	89.8	93.3	92.9	93.0	101.0	94.6	96.2	100.4	98.5	99.0	93.8	97.1	96.3	94.2	93.3	93.5
03100 CONCRETE FORMS & ACCESSORIES	75.8	35.4	40.4	84.1	77.8	78.6	85.6	75.0	76.3	103.5	67.8	72.1	102.8	103.9	103.8	101.5	68.0	72.1
03200 CONCRETE REINFORCEMENT	102.1	49.3	70.2	102.9	86.7	93.1	102.4	110.3	107.2	107.9	75.3	88.2	102.5	110.9	107.6	98.4	109.5	105.1
03300 CAST-IN-PLACE CONCRETE	99.7	44.3	76.5	85.9	87.2	86.4	88.3	70.7	81.0	101.4	74.5	90.2	94.2	105.3	98.9	101.2	65.3	86.2
03 CONCRETE	92.7	43.3	67.4	86.7	84.5	85.6	84.7	82.0	83.3	99.0	72.8	85.6	95.2	106.1	100.8	98.9	76.1	87.2
04 MASONRY	88.9	30.3	52.2	113.2	78.2	91.3	130.4	86.0	102.6	97.1	62.2	75.3	101.7	100.8	101.1	86.4	67.4	74.5
05 METALS	95.5	77.3	88.6	92.8	111.5	99.9	93.1	121.3	103.8	97.4	91.4	95.2	100.9	113.1	105.5	96.5	105.6	99.9
06 WOOD & PLASTICS	76.2	34.4	53.6	80.7	75.3	77.8	91.4	71.1	80.4	102.8	66.0	83.0	102.7	103.7	103.2	100.3	69.3	83.6
07 THERMAL & MOISTURE PROTECTION	94.8	41.5	68.8	95.1	82.5	89.0	92.4	83.9	88.3	99.0	65.5	82.7	98.2	104.4	101.2	96.5	72.3	84.7
08 DOORS & WINDOWS	97.8	39.3	83.0	93.8	77.7	89.7	90.8	93.6	91.5	91.6	73.5	87.0	97.8	106.7	100.0	96.2	79.6	92.0
09200 PLASTER & GYPSUM BOARD	91.3	33.0	51.6	92.1	74.4	80.0	100.2	70.1	79.7	109.5	64.9	79.1	103.9	103.6	103.7	105.0	68.2	80.0
095,098 CEILINGS & ACOUSTICAL TREATMENT	89.8	33.0	51.4	88.2	74.4	78.9	91.0	70.1	76.9	94.0	64.9	74.3	99.3	103.6	102.2	96.2	68.2	77.3
09600 FLOORING	107.9	29.9	87.8	92.9	81.3	89.9	102.4	94.1	100.3	112.2	54.7	97.4	92.1	94.1	92.6	105.4	54.7	92.4
097,099 WALL FINISHES, PAINTS & COATINGS	107.5	36.6	64.9	101.0	81.0	89.0	99.4	77.2	86.1	92.4	47.8	65.7	97.3	104.5	101.7	93.7	68.5	78.6
09 FINISHES	99.6	33.8	64.0	90.2	77.8	83.5	93.8	76.1	84.2	101.8	62.8	80.7	96.9	101.9	99.6	98.6	65.0	80.4
10 - 14 TOTAL DIV. 10000 - 14000	100.0	56.8	90.7	100.0	74.7	94.5	100.0	92.1	98.3	100.0	84.4	96.6	100.0	97.7	99.5	100.0	85.5	96.9
15 MECHANICAL	99.8	36.6	70.7	100.0	97.2	98.7	100.0	95.4	97.9	100.2	59.9	81.6	100.0	105.6	102.6	100.0	70.0	86.2
16 ELECTRICAL	96.5	62.6	76.7	100.5	113.1	107.9	96.8	85.6	90.2	93.6	66.1	77.5	105.5	108.7	107.4	102.3	71.4	84.2
01 - 16 WEIGHTED AVERAGE	96.8	49.3	73.9	96.0	92.2	94.1	96.7	90.4	93.6	98.0	71.1	85.0	99.4	104.8	102.0	98.0	76.4	87.6

DIVISION	MISSOURI						MONTANA											
	ST. JOSEPH			ST. LOUIS			BILLINGS			BUTTE			GREAT FALLS			HELENA		
	MAT.	INST.	TOTAL	MAT.	INST.	TOTAL	MAT.	INST.	TOTAL	MAT.	INST.	TOTAL	MAT.	INST.	TOTAL	MAT.	INST.	TOTAL
01590 EQUIPMENT RENTAL	.0	102.4	102.4	.0	109.8	109.8	.0	98.7	98.7	.0	98.4	98.4	.0	98.4	98.4	.0	98.4	98.4
02 SITE CONSTRUCTION	95.3	93.5	93.9	92.9	96.7	95.8	84.2	96.4	93.3	90.9	95.1	94.0	94.4	95.7	95.4	96.3	95.6	95.8
03100 CONCRETE FORMS & ACCESSORIES	102.7	89.1	90.7	99.4	107.2	106.2	97.8	67.7	71.4	84.3	65.4	67.7	103.7	66.2	70.8	103.8	66.8	71.4
03200 CONCRETE REINFORCEMENT	101.3	99.6	100.3	94.1	111.6	104.7	98.3	77.7	85.8	106.6	77.8	89.2	98.3	77.7	85.9	101.7	67.4	81.0
03300 CAST-IN-PLACE CONCRETE	94.2	100.0	96.6	85.9	111.3	96.5	117.2	70.8	97.8	120.7	70.9	99.9	127.5	58.6	98.7	129.9	69.2	104.5
03 CONCRETE	95.0	95.6	95.3	86.7	110.4	98.9	104.9	71.6	87.9	104.4	70.6	87.1	110.4	66.7	88.1	112.1	68.7	89.9
04 MASONRY	101.2	83.5	90.1	94.6	110.9	104.8	121.1	68.5	88.1	121.0	69.5	88.8	124.7	70.4	90.7	120.1	67.7	87.3
05 METALS	99.5	105.7	101.8	95.2	125.0	106.5	97.1	88.4	93.8	96.4	88.2	93.3	97.5	88.2	94.0	96.7	82.6	91.4
06 WOOD & PLASTICS	103.4	89.3	95.8	95.0	105.0	100.4	102.7	67.8	83.9	90.9	66.2	77.6	110.3	66.0	86.4	110.4	66.0	86.5
07 THERMAL & MOISTURE PROTECTION	98.8	92.0	95.5	95.0	106.5	100.6	99.8	69.5	85.0	99.0	69.7	84.7	100.0	67.9	84.3	100.0	69.1	84.9
08 DOORS & WINDOWS	96.5	96.1	96.4	92.7	112.0	97.6	99.2	66.4	90.9	96.0	65.5	88.2	99.4	65.5	90.8	98.9	62.7	89.7
09200 PLASTER & GYPSUM BOARD	110.4	88.8	95.7	99.2	104.9	103.1	103.3	67.2	78.7	95.0	65.6	75.0	103.3	65.3	77.5	102.7	65.3	77.3
095,098 CEILINGS & ACOUSTICAL TREATMENT	98.0	88.8	91.7	92.1	104.9	100.8	110.3	67.2	81.1	107.7	65.6	79.2	110.3	65.3	79.9	107.7	65.3	79.0
09600 FLOORING	94.4	87.6	92.7	100.2	99.4	100.0	108.3	53.3	94.1	99.6	42.4	84.9	108.3	57.6	95.3	108.3	55.7	94.8
097,099 WALL FINISHES, PAINTS & COATINGS	93.0	82.7	86.8	101.0	108.3	105.4	99.5	63.8	78.1	99.5	43.5	65.9	99.5	48.9	69.1	99.5	54.3	72.4
09 FINISHES	98.0	87.7	92.4	94.3	105.4	100.3	105.8	64.3	83.4	101.2	58.2	77.9	105.9	62.3	82.4	105.4	63.2	82.6
10 - 14 TOTAL DIV. 10000 - 14000	100.0	92.4	98.4	100.0	102.5	100.5	100.0	76.5	94.9	100.0	75.3	94.7	100.0	76.3	94.9	100.0	62.7	92.0
15 MECHANICAL	100.2	93.2	97.0	100.0	104.8	102.3	100.4	78.3	90.2	100.4	73.0	87.8	100.4	76.4	89.4	100.4	75.6	89.0
16 ELECTRICAL	104.9	86.6	94.2	103.9	113.1	109.3	96.0	77.6	85.2	106.1	72.3	86.4	96.7	72.5	82.6	96.7	72.5	82.5
01 - 16 WEIGHTED AVERAGE	99.2	92.1	95.8	96.0	108.8	102.2	101.1	76.2	89.1	101.1	73.4	87.8	102.5	74.2	88.9	102.3	73.2	88.3

DIVISION	MONTANA			NEBRASKA												NEVADA		
	MISSOULA			GRAND ISLAND			LINCOLN			NORTH PLATTE			OMAHA			CARSON CITY		
	MAT.	INST.	TOTAL	MAT.	INST.	TOTAL	MAT.	INST.	TOTAL	MAT.	INST.	TOTAL	MAT.	INST.	TOTAL	MAT.	INST.	TOTAL
01590 EQUIPMENT RENTAL	.0	98.4	98.4	.0	101.4	101.4	.0	101.4	101.4	.0	101.4	101.4	.0	90.6	90.6	.0	101.4	101.4
02 SITE CONSTRUCTION	74.3	94.9	89.7	97.0	91.2	92.6	88.0	91.2	90.4	97.8	90.0	91.9	77.7	89.9	86.8	61.9	104.4	93.7
03100 CONCRETE FORMS & ACCESSORIES	88.5	61.2	64.6	97.8	52.7	58.2	102.8	48.5	55.2	97.7	49.0	55.0	96.7	75.1	77.8	98.2	98.6	98.6
03200 CONCRETE REINFORCEMENT	108.5	78.6	90.4	107.2	76.3	88.5	98.4	77.3	85.7	108.6	76.1	89.0	103.1	77.8	87.8	106.2	112.2	109.8
03300 CAST-IN-PLACE CONCRETE	88.9	67.2	79.8	115.3	60.2	92.3	104.1	62.8	86.8	115.3	53.9	89.6	108.1	78.0	95.5	115.2	87.9	103.8
03 CONCRETE	82.2	67.6	74.8	108.3	61.1	84.1	100.5	60.3	79.9	108.5	57.3	82.2	101.6	76.8	88.9	110.3	97.4	103.7
04 MASONRY	146.6	62.4	93.9	104.5	51.5	71.3	94.8	68.7	78.5	89.9	41.8	59.8	100.0	77.7	86.1	134.6	83.1	102.3
05 METALS	96.5	87.5	93.1	93.0	84.9	89.9	96.0	86.4	92.4	93.2	84.8	90.0	98.6	77.5	90.6	103.6	102.4	103.1
06 WOOD & PLASTICS	95.3	62.3	77.5	97.1	47.3	70.2	101.4	41.3	69.0	96.9	47.3	70.2	94.8	76.3	84.8	96.1	100.0	98.2
07 THERMAL & MOISTURE PROTECTION	98.1	72.0	85.4	96.9	59.3	78.6	97.4	62.2	80.2	96.9	48.9	73.5	92.5	72.6	82.8	103.0	90.2	96.8
08 DOORS & WINDOWS	96.0	64.3	87.9	90.0	51.8	80.3	95.4	51.4	84.2	89.3	55.0	80.6	98.7	69.4	91.3	94.4	108.3	97.9
09200 PLASTER & GYPSUM BOARD	97.0	61.5	72.9	101.5	45.6	63.5	105.0	39.5	60.4	101.8	45.6	63.6	112.4	76.1	87.7	86.6	99.8	95.6
095,098 CEILINGS & ACOUSTICAL TREATMENT	107.7	61.5	76.4	89.7	45.6	59.8	96.2	39.5	57.8	91.0	45.6	60.3	134.0	76.1	94.8	105.9	99.8	101.8
09600 FLOORING	101.7	66.3	92.6	95.0	37.8	80.3	96.8	45.5	83.6	94.9	39.3	80.6	123.3	50.3	104.5	101.7	69.4	93.4
097,099 WALL FINISHES, PAINTS & COATINGS	99.5	48.9	69.1	91.3	44.2	63.1	91.3	44.6	63.3	91.3	53.4	68.6	148.6	75.9	105.0	103.9	84.2	92.1
09 FINISHES	101.3	60.4	79.2	93.8	47.4	68.7	95.9	45.6	68.7	94.0	47.0	68.6	121.3	70.5	93.8	98.0	91.3	94.4
10 - 14 TOTAL DIV. 10000 - 14000	100.0	58.7	91.1	100.0	78.4	95.4	100.0	77.6	95.2	100.0	58.5	91.1	100.0	80.5	95.8	100.0	120.5	104.4
15 MECHANICAL	100.4	71.0	86.9	100.0	80.3	90.9	100.0	80.3	90.9	100.0	76.0	89.0	99.7	81.1	91.2	100.1	90.1	95.5
16 ELECTRICAL	103.4	75.2	86.9	94.6	69.2	79.7	102.4	69.2	83.0	97.3	45.8	67.2	91.8	85.3	88.0	93.8	91.4	92.4
01 - 16 WEIGHTED AVERAGE	99.2	72.0	86.1	97.8	68.3	83.6	98.1	69.8	84.5	97.2	61.8	80.1	100.2	79.1	90.0	101.2	94.8	98.2

DIVISION		NEVADA						NEW HAMPSHIRE									NEW JERSEY		
		LAS VEGAS			RENO			MANCHESTER			NASHUA			PORTSMOUTH			CAMDEN		
		MAT.	INST.	TOTAL	MAT.	INST.	TOTAL	MAT.	INST.	TOTAL	MAT.	INST.	TOTAL	MAT.	INST.	TOTAL	MAT.	INST.	TOTAL
01590	**EQUIPMENT RENTAL**	.0	101.4	101.4	.0	101.4	101.4	.0	101.8	101.8	.0	101.8	101.8	.0	101.8	101.8	.0	99.7	99.7
02	**SITE CONSTRUCTION**	61.3	106.1	94.8	61.6	104.4	93.6	86.1	100.6	96.9	87.4	100.6	97.3	81.8	99.6	95.1	90.7	105.2	101.6
03100	CONCRETE FORMS & ACCESSORIES	93.7	108.9	107.0	98.6	98.7	98.7	103.8	66.4	70.9	104.4	66.4	71.0	91.0	61.5	65.2	104.4	127.6	124.7
03200	CONCRETE REINFORCEMENT	98.1	119.9	111.2	98.1	119.8	111.2	109.1	94.2	100.1	109.1	94.2	100.1	87.2	94.0	91.3	109.1	107.1	107.9
03300	CAST-IN-PLACE CONCRETE	108.7	111.3	109.8	117.3	87.9	105.1	102.9	100.5	101.9	94.5	100.5	97.0	89.6	93.8	91.3	82.2	127.5	101.1
03	**CONCRETE**	105.8	111.4	108.7	110.5	98.9	104.5	108.1	83.6	95.6	103.9	83.6	93.5	95.4	79.1	87.0	97.8	122.4	110.4
04	**MASONRY**	125.4	92.5	104.8	134.0	83.1	102.1	96.2	95.7	95.9	96.4	95.7	96.0	91.9	84.8	87.4	89.7	123.7	111.0
05	**METALS**	104.3	106.8	105.2	104.2	105.6	104.7	97.3	88.8	94.1	97.3	88.8	94.1	93.9	86.6	91.1	97.2	98.5	97.7
06	**WOOD & PLASTICS**	92.1	106.9	100.1	96.4	100.0	98.4	101.9	58.2	78.4	101.9	58.2	78.3	88.6	58.2	72.2	101.9	127.9	115.9
07	**THERMAL & MOISTURE PROTECTION**	105.5	100.1	102.9	103.0	90.2	96.8	102.2	92.0	97.2	102.6	92.0	97.4	102.1	90.0	96.2	102.0	123.1	112.3
08	**DOORS & WINDOWS**	94.7	114.1	99.7	94.7	108.7	98.3	107.2	67.7	97.2	107.2	67.7	97.2	108.1	61.5	96.3	107.2	117.9	109.9
09200	PLASTER & GYPSUM BOARD	87.1	107.0	100.6	88.2	99.8	96.1	109.1	56.0	73.0	109.1	56.0	73.0	100.7	56.0	70.3	109.1	127.7	121.8
095,098	CEILINGS & ACOUSTICAL TREATMENT	112.4	107.0	108.7	112.4	99.8	103.9	105.9	56.0	72.1	105.9	56.0	72.1	96.8	56.0	69.2	105.9	127.7	120.7
09600	FLOORING	101.7	82.6	96.8	101.7	69.4	93.4	100.5	107.9	102.4	100.3	107.9	102.2	94.8	107.9	98.2	100.3	128.8	107.6
097,099	WALL FINISHES, PAINTS & COATINGS	103.9	108.4	106.6	103.9	84.2	92.1	93.0	93.4	93.2	93.0	93.4	93.2	93.0	41.6	62.2	93.0	123.4	111.2
09	**FINISHES**	99.2	103.5	101.5	99.4	91.3	95.0	102.8	75.3	88.0	103.0	75.3	88.1	98.0	66.7	81.1	103.3	128.0	116.6
10 - 14	**TOTAL DIV. 10000 - 14000**	100.0	110.4	102.2	100.0	120.5	104.4	100.0	85.0	96.8	100.0	85.0	96.8	100.0	80.6	95.8	100.0	116.3	103.5
15	**MECHANICAL**	100.1	111.9	105.5	100.1	90.2	95.5	100.2	89.0	95.0	100.2	89.0	95.0	100.2	83.1	92.3	100.2	113.4	106.3
16	**ELECTRICAL**	95.7	106.3	101.9	93.8	91.4	92.4	102.9	77.4	88.0	102.7	77.4	87.9	100.1	77.4	86.9	102.7	122.2	114.1
01 - 16	**WEIGHTED AVERAGE**	100.6	106.7	103.5	101.5	95.4	98.5	101.5	85.4	93.7	101.0	85.4	93.5	98.3	80.7	89.8	100.0	117.3	108.4

DIVISION		NEW JERSEY															NEW MEXICO		
		ELIZABETH			JERSEY CITY			NEWARK			PATERSON			TRENTON			ALBUQUERQUE		
		MAT.	INST.	TOTAL	MAT.	INST.	TOTAL	MAT.	INST.	TOTAL	MAT.	INST.	TOTAL	MAT.	INST.	TOTAL	MAT.	INST.	TOTAL
01590	**EQUIPMENT RENTAL**	.0	101.8	101.8	.0	99.7	99.7	.0	101.8	101.8	.0	101.8	101.8	.0	99.2	99.2	.0	116.0	116.0
02	**SITE CONSTRUCTION**	105.1	105.6	105.5	90.7	105.1	101.5	109.4	105.6	106.5	103.3	105.0	104.6	92.5	105.6	102.3	79.1	111.6	103.4
03100	CONCRETE FORMS & ACCESSORIES	115.0	126.4	125.0	104.4	127.3	124.5	102.5	126.5	123.5	103.1	126.4	123.5	103.2	127.7	124.7	96.4	70.1	73.3
03200	CONCRETE REINFORCEMENT	84.0	125.2	108.9	109.1	125.1	118.8	109.1	125.2	118.8	109.1	125.2	118.8	109.1	117.5	114.2	98.8	66.5	79.3
03300	CAST-IN-PLACE CONCRETE	90.5	125.5	105.1	82.2	127.0	100.9	95.9	125.6	108.3	104.7	125.5	113.4	97.3	123.4	108.2	107.8	75.6	94.3
03	**CONCRETE**	101.6	124.8	113.5	97.8	125.5	112.0	104.5	124.8	114.9	109.0	124.8	117.1	105.2	123.0	114.3	105.7	72.4	88.6
04	**MASONRY**	111.3	119.9	116.7	89.7	122.6	110.3	98.5	119.9	111.9	94.1	119.9	110.3	89.3	119.6	108.3	122.0	65.8	86.8
05	**METALS**	93.7	108.7	99.4	97.2	105.6	100.4	97.2	108.8	101.6	97.2	108.7	101.6	98.0	102.4	99.7	105.3	88.1	98.8
06	**WOOD & PLASTICS**	116.1	127.9	122.5	101.9	127.9	115.9	103.8	127.9	116.8	103.8	127.9	116.8	101.9	127.8	115.9	96.8	72.2	83.6
07	**THERMAL & MOISTURE PROTECTION**	102.9	117.3	110.0	102.0	118.5	110.0	102.2	117.3	109.6	102.8	120.9	111.6	100.6	119.9	110.0	101.8	72.9	87.7
08	**DOORS & WINDOWS**	108.2	122.4	111.8	107.2	122.4	111.0	113.3	122.4	115.6	113.3	122.4	115.6	107.2	120.5	110.6	94.7	73.9	89.4
09200	PLASTER & GYPSUM BOARD	112.1	127.7	122.7	109.1	127.7	121.8	109.1	127.7	121.8	109.1	127.7	121.8	109.1	127.7	121.8	91.5	70.9	77.5
095,098	CEILINGS & ACOUSTICAL TREATMENT	96.8	127.7	117.7	105.9	127.7	120.7	105.9	127.7	120.7	105.9	127.7	120.7	105.9	127.7	120.7	112.4	70.9	84.3
09600	FLOORING	105.0	128.8	111.1	100.3	128.8	107.6	100.5	128.8	107.8	100.3	128.8	107.6	100.5	121.7	105.9	101.7	66.4	92.6
097,099	WALL FINISHES, PAINTS & COATINGS	92.9	123.5	111.2	93.0	123.5	111.3	92.9	123.5	111.2	92.9	123.5	111.2	93.0	123.5	111.3	103.9	56.8	75.7
09	**FINISHES**	104.3	127.0	116.6	103.3	127.7	116.5	103.9	127.0	116.4	103.5	127.0	116.2	103.4	126.5	115.9	100.3	67.9	82.8
10 - 14	**TOTAL DIV. 10000 - 14000**	100.0	91.8	98.2	100.0	92.9	98.5	100.0	91.8	98.2	100.0	91.8	98.2	100.0	116.2	103.5	100.0	80.1	95.7
15	**MECHANICAL**	100.2	121.6	110.1	100.2	122.1	110.3	100.2	122.7	110.6	100.2	120.7	109.6	100.2	120.2	109.4	100.1	72.5	87.4
16	**ELECTRICAL**	98.7	124.1	113.6	104.2	128.4	118.3	104.0	128.4	118.2	104.2	124.1	115.8	102.9	134.9	121.6	89.0	76.7	81.8
01 - 16	**WEIGHTED AVERAGE**	101.6	119.4	110.2	100.1	120.3	109.9	102.8	120.3	111.2	102.9	119.3	110.8	101.0	120.4	110.3	100.5	77.2	89.3

DIVISION		NEW MEXICO												NEW YORK					
		FARMINGTON			LAS CRUCES			ROSWELL			SANTA FE			ALBANY			BINGHAMTON		
		MAT.	INST.	TOTAL	MAT.	INST.	TOTAL	MAT.	INST.	TOTAL	MAT.	INST.	TOTAL	MAT.	INST.	TOTAL	MAT.	INST.	TOTAL
01590	**EQUIPMENT RENTAL**	.0	116.0	116.0	.0	86.6	86.6	.0	116.0	116.0	.0	116.0	116.0	.0	115.8	115.8	.0	115.9	115.9
02	**SITE CONSTRUCTION**	84.9	111.6	104.9	90.8	86.8	87.8	89.5	111.6	106.0	78.9	111.6	103.4	70.3	106.8	97.6	92.6	90.0	90.6
03100	CONCRETE FORMS & ACCESSORIES	96.3	70.1	73.3	93.6	68.4	71.5	96.3	70.0	73.2	96.3	70.1	73.3	98.0	92.6	93.3	103.0	79.4	82.3
03200	CONCRETE REINFORCEMENT	107.9	66.5	82.9	101.5	60.2	76.5	107.1	60.6	79.0	105.9	66.5	82.1	102.6	92.0	96.2	101.6	89.9	94.5
03300	CAST-IN-PLACE CONCRETE	108.4	75.6	94.7	94.3	65.8	82.4	99.8	75.6	89.7	101.9	75.6	90.9	88.9	103.6	95.1	106.2	89.3	99.2
03	**CONCRETE**	109.1	72.4	90.3	85.8	66.7	76.0	109.5	71.2	89.9	103.5	72.4	87.6	101.5	97.2	99.3	101.0	86.9	93.8
04	**MASONRY**	126.8	65.8	88.6	110.7	60.6	79.4	127.1	65.8	88.7	119.1	65.8	85.7	89.6	100.0	96.1	106.4	81.5	90.8
05	**METALS**	104.4	88.1	98.2	98.6	76.2	90.2	104.4	85.2	97.1	104.4	88.1	98.2	100.8	107.0	103.1	95.8	116.3	103.6
06	**WOOD & PLASTICS**	96.8	72.2	83.6	87.7	70.8	78.6	96.8	72.2	83.6	96.8	72.2	83.6	97.8	90.9	94.1	106.7	78.1	91.3
07	**THERMAL & MOISTURE PROTECTION**	102.7	72.9	88.2	88.5	65.9	77.5	103.0	72.9	88.4	102.1	72.9	87.9	90.6	95.5	92.9	101.4	83.0	92.4
08	**DOORS & WINDOWS**	97.6	73.9	91.6	87.1	71.3	83.1	93.6	72.1	88.1	93.8	73.9	88.7	95.6	85.8	93.1	90.7	76.0	87.0
09200	PLASTER & GYPSUM BOARD	86.0	70.9	75.8	88.5	70.9	76.5	86.0	70.9	75.8	86.0	70.9	75.8	101.7	90.4	94.0	108.4	76.9	86.9
095,098	CEILINGS & ACOUSTICAL TREATMENT	103.3	70.9	81.4	102.1	70.9	81.0	103.3	70.9	81.4	103.3	70.9	81.4	98.9	90.4	93.1	98.9	76.9	84.0
09600	FLOORING	101.7	66.4	92.6	134.0	67.9	117.0	101.7	66.4	92.6	101.7	66.4	92.6	86.5	97.9	89.4	97.4	83.4	93.8
097,099	WALL FINISHES, PAINTS & COATINGS	103.9	56.8	75.7	97.3	56.8	73.0	103.9	56.8	75.7	103.9	56.8	75.7	85.1	80.5	82.3	90.3	81.5	85.0
09	**FINISHES**	98.2	67.9	81.8	115.0	67.2	89.1	98.6	67.9	82.0	98.0	67.9	81.7	95.4	92.1	93.6	96.8	80.0	87.7
10 - 14	**TOTAL DIV. 10000 - 14000**	100.0	80.1	95.7	100.0	75.9	94.8	100.0	80.1	95.7	100.0	80.1	95.7	100.0	92.3	98.3	100.0	89.9	97.8
15	**MECHANICAL**	100.1	72.5	87.4	100.3	72.0	87.3	100.1	72.4	87.4	100.1	72.5	87.4	100.4	92.0	96.5	100.6	85.3	93.6
16	**ELECTRICAL**	86.7	76.7	80.9	87.5	62.7	73.0	87.7	76.7	81.3	89.0	76.7	81.8	100.9	92.9	96.2	101.4	83.3	90.8
01 - 16	**WEIGHTED AVERAGE**	101.2	77.2	89.6	96.9	69.9	83.9	101.0	76.7	89.3	99.6	77.2	88.8	97.8	96.2	97.0	98.7	87.0	93.1

	DIVISION	NEW YORK																	
		BUFFALO			HICKSVILLE			NEW YORK			RIVERHEAD			ROCHESTER			SCHENECTADY		
		MAT.	INST.	TOTAL	MAT.	INST.	TOTAL	MAT.	INST.	TOTAL	MAT.	INST.	TOTAL	MAT.	INST.	TOTAL	MAT.	INST.	TOTAL
01590	**EQUIPMENT RENTAL**	.0	93.4	93.4	.0	117.4	117.4	.0	117.8	117.8	.0	117.4	117.4	.0	116.8	116.8	.0	115.8	115.8
02	**SITE CONSTRUCTION**	94.6	93.5	93.8	111.5	130.4	125.7	132.4	128.4	129.4	111.7	130.4	125.7	72.5	107.5	98.7	70.0	106.6	97.4
03100	CONCRETE FORMS & ACCESSORIES	102.6	113.9	112.5	89.6	155.5	147.4	111.2	180.8	172.2	93.8	155.5	148.0	102.6	97.3	98.0	102.6	91.5	92.9
03200	CONCRETE REINFORCEMENT	101.0	102.8	102.1	103.1	188.7	154.8	109.4	189.2	157.6	105.1	188.8	155.6	101.9	88.6	93.9	101.3	92.0	95.7
03300	CAST-IN-PLACE CONCRETE	119.6	119.1	119.4	99.4	156.2	123.1	124.0	161.1	139.5	97.5	156.2	122.0	113.8	103.0	109.3	97.5	102.0	99.4
03	**CONCRETE**	108.6	112.8	110.8	105.0	160.4	133.4	121.7	173.3	148.1	103.9	160.5	132.9	114.6	98.7	106.4	106.0	96.2	101.0
04	**MASONRY**	102.6	116.2	111.1	109.2	157.2	139.2	106.4	163.0	141.8	115.8	157.2	141.7	100.9	99.7	100.1	91.0	97.1	94.8
05	**METALS**	103.2	93.8	99.6	110.4	138.7	121.1	110.2	141.6	122.1	110.4	140.2	121.7	106.7	109.4	107.7	100.8	107.0	103.1
06	**WOOD & PLASTICS**	108.7	113.6	111.3	88.9	157.5	125.9	108.4	187.7	151.2	92.7	157.5	127.6	105.1	97.6	101.1	103.1	90.9	96.5
07	**THERMAL & MOISTURE PROTECTION**	98.9	105.2	102.0	103.9	149.8	126.3	104.9	158.4	131.0	103.9	149.8	126.3	95.5	99.6	97.5	90.7	94.2	92.4
08	**DOORS & WINDOWS**	93.1	102.5	95.5	88.3	158.1	106.0	97.1	175.4	117.0	88.3	159.0	106.3	96.4	89.3	94.6	95.6	85.8	93.1
09200	PLASTER & GYPSUM BOARD	96.5	113.7	108.2	96.2	159.1	139.0	111.4	189.7	164.7	97.3	159.1	139.3	98.2	97.5	97.7	101.7	90.4	94.0
095,098	CEILINGS & ACOUSTICAL TREATMENT	93.7	113.7	107.2	83.3	159.1	134.6	113.0	189.7	165.0	83.3	159.1	134.6	97.8	97.5	97.6	98.9	90.4	93.1
09600	FLOORING	95.8	115.1	100.8	97.2	93.7	96.3	97.9	165.3	115.3	98.5	93.7	97.3	85.1	89.3	86.2	86.5	97.9	89.4
097,099	WALL FINISHES, PAINTS & COATINGS	94.9	114.3	106.5	112.8	151.3	135.9	97.4	153.2	130.9	112.8	151.3	135.9	92.6	99.0	96.4	85.1	80.5	82.3
09	**FINISHES**	94.0	115.0	105.3	106.1	142.8	125.9	108.8	176.5	145.4	106.6	142.8	126.2	94.7	96.1	95.4	95.1	91.4	93.1
10 - 14	**TOTAL DIV. 10000 - 14000**	100.0	109.8	102.1	100.0	127.9	106.0	100.0	142.5	109.2	100.0	127.9	106.0	100.0	96.8	99.3	100.0	91.1	98.1
15	**MECHANICAL**	99.9	97.3	98.7	99.8	150.0	122.9	100.3	163.9	129.6	99.8	150.1	123.0	99.9	91.1	95.9	100.4	90.4	95.8
16	**ELECTRICAL**	99.8	97.3	98.4	102.1	150.5	130.4	107.6	176.8	148.1	103.0	150.5	130.8	104.0	92.5	97.3	101.2	92.9	96.3
01 - 16	**WEIGHTED AVERAGE**	100.1	103.8	101.9	102.3	148.3	124.5	106.9	163.2	134.0	102.8	148.5	124.8	101.3	97.4	99.4	98.4	95.3	96.9

	DIVISION	NEW YORK															NORTH CAROLINA		
		SYRACUSE			UTICA			WATERTOWN			WHITE PLAINS			YONKERS			ASHEVILLE		
		MAT.	INST.	TOTAL	MAT.	INST.	TOTAL	MAT.	INST.	TOTAL	MAT.	INST.	TOTAL	MAT.	INST.	TOTAL	MAT.	INST.	TOTAL
01590	**EQUIPMENT RENTAL**	.0	115.8	115.8	.0	115.8	115.8	.0	115.8	115.8	.0	117.5	117.5	.0	117.5	117.5	.0	93.1	93.1
02	**SITE CONSTRUCTION**	91.3	106.8	102.9	68.0	105.7	96.2	76.1	107.8	99.8	122.4	125.4	124.7	131.8	125.2	126.9	102.6	72.4	80.0
03100	CONCRETE FORMS & ACCESSORIES	101.8	88.1	89.8	103.3	83.8	86.2	84.8	91.4	90.6	111.9	140.4	136.9	111.0	140.4	136.8	95.9	44.6	50.8
03200	CONCRETE REINFORCEMENT	102.6	90.5	95.3	102.6	82.6	90.5	103.2	90.6	95.6	102.1	188.0	154.0	106.1	188.0	155.5	94.3	40.4	61.8
03300	CAST-IN-PLACE CONCRETE	98.6	96.5	97.7	90.2	93.8	91.7	105.0	88.3	98.0	109.3	137.1	120.9	122.3	137.1	128.5	94.3	48.6	75.2
03	**CONCRETE**	104.6	92.5	98.4	102.5	88.2	95.2	116.5	91.2	103.5	109.1	147.0	128.5	120.6	147.0	134.1	98.9	47.2	72.4
04	**MASONRY**	96.5	95.2	95.7	89.6	91.9	91.0	90.7	96.9	94.6	101.2	136.1	123.1	106.0	136.1	124.9	78.9	38.2	53.4
05	**METALS**	100.6	105.9	102.6	98.8	102.6	100.2	98.9	105.6	101.4	99.7	136.9	113.8	106.9	137.0	118.3	93.6	75.8	86.9
06	**WOOD & PLASTICS**	103.1	86.8	94.3	103.1	82.6	92.1	85.1	92.1	88.9	110.5	141.3	127.1	109.2	141.3	126.5	96.6	46.4	69.5
07	**THERMAL & MOISTURE PROTECTION**	98.9	95.0	97.0	90.7	93.8	92.2	91.0	95.5	93.2	105.7	139.3	122.1	106.1	139.3	122.2	95.8	43.7	70.4
08	**DOORS & WINDOWS**	93.6	81.6	90.5	95.6	77.2	90.9	95.6	83.0	92.4	91.7	150.3	106.6	95.2	150.3	109.2	91.9	42.8	79.4
09200	PLASTER & GYPSUM BOARD	101.7	86.2	91.1	101.7	81.9	88.2	94.4	91.6	92.5	106.8	142.0	130.8	111.1	142.0	132.1	105.4	44.7	64.1
095,098	CEILINGS & ACOUSTICAL TREATMENT	98.9	86.2	90.3	98.9	81.9	87.4	98.9	91.6	94.0	90.9	142.0	125.5	111.7	142.0	132.2	89.5	44.7	59.1
09600	FLOORING	88.0	93.0	89.3	86.5	93.1	88.2	79.0	93.1	82.7	94.1	165.3	112.4	93.7	165.3	112.1	100.7	43.0	85.8
097,099	WALL FINISHES, PAINTS & COATINGS	90.4	87.6	88.7	85.1	87.3	86.4	85.1	75.7	79.5	95.0	151.3	128.7	95.0	151.3	128.7	108.7	40.3	67.7
09	**FINISHES**	96.6	88.7	92.3	95.2	85.3	89.8	92.3	89.6	90.9	101.9	146.5	126.0	106.9	146.5	128.3	96.5	43.9	68.1
10 - 14	**TOTAL DIV. 10000 - 14000**	100.0	102.6	100.6	100.0	88.9	97.6	100.0	88.5	97.5	100.0	124.0	105.2	100.0	132.0	106.9	100.0	66.2	92.7
15	**MECHANICAL**	100.4	87.3	94.4	100.4	83.6	92.7	100.4	79.7	90.9	100.6	130.2	114.2	100.6	130.3	114.2	100.0	44.6	74.5
16	**ELECTRICAL**	101.2	85.7	92.2	101.2	84.9	91.6	101.2	82.1	90.0	95.2	139.8	121.3	105.5	139.8	125.5	97.7	42.0	65.1
01 - 16	**WEIGHTED AVERAGE**	99.3	92.7	96.1	97.6	89.4	93.6	99.1	90.5	95.0	101.0	137.7	118.7	105.8	137.9	121.2	96.2	49.8	73.8

	DIVISION	NORTH CAROLINA																	
		CHARLOTTE			DURHAM			FAYETTEVILLE			GREENSBORO			RALEIGH			WILMINGTON		
		MAT.	INST.	TOTAL	MAT.	INST.	TOTAL	MAT.	INST.	TOTAL	MAT.	INST.	TOTAL	MAT.	INST.	TOTAL	MAT.	INST.	TOTAL
01590	**EQUIPMENT RENTAL**	.0	93.1	93.1	.0	99.5	99.5	.0	99.5	99.5	.0	99.5	99.5	.0	99.5	99.5	.0	93.1	93.1
02	**SITE CONSTRUCTION**	102.7	72.4	80.0	102.5	82.4	87.4	100.5	82.4	86.9	102.3	82.4	87.4	103.5	82.4	87.7	103.5	72.4	80.2
03100	CONCRETE FORMS & ACCESSORIES	103.7	44.6	51.9	98.0	44.9	51.4	93.4	44.9	50.8	98.1	44.9	51.4	100.8	44.9	51.8	97.5	44.9	51.3
03200	CONCRETE REINFORCEMENT	94.7	40.4	61.9	94.7	56.0	71.3	93.7	56.0	71.0	94.7	56.0	71.3	94.7	56.0	71.3	95.0	56.0	71.5
03300	CAST-IN-PLACE CONCRETE	96.7	48.6	76.6	94.5	48.7	75.4	91.2	48.7	73.5	93.7	48.7	74.9	100.0	48.7	78.6	93.8	48.7	75.0
03	**CONCRETE**	99.9	47.2	72.9	98.4	50.3	73.8	96.0	50.3	72.6	98.0	50.3	73.6	101.4	50.3	75.2	98.8	50.3	74.0
04	**MASONRY**	83.5	38.2	55.1	80.2	38.2	53.9	83.9	38.2	55.3	80.0	38.2	53.9	84.6	38.2	55.6	69.6	38.2	50.0
05	**METALS**	94.8	75.9	87.7	94.9	82.6	90.2	96.0	82.6	90.9	95.7	82.5	90.7	94.9	82.6	90.3	93.9	82.6	89.6
06	**WOOD & PLASTICS**	105.2	46.4	73.5	98.6	46.4	70.4	93.8	46.4	68.2	98.6	46.4	70.4	101.8	46.4	71.9	98.2	46.4	70.2
07	**THERMAL & MOISTURE PROTECTION**	95.8	44.7	70.9	96.5	44.3	71.0	96.1	44.3	70.8	96.5	44.3	71.0	96.3	44.3	70.9	95.8	44.3	70.7
08	**DOORS & WINDOWS**	96.0	42.8	82.5	96.0	48.8	84.0	92.0	48.8	81.0	96.0	48.8	84.0	92.8	48.8	81.6	92.1	48.8	81.1
09200	PLASTER & GYPSUM BOARD	111.2	44.7	65.9	111.2	44.7	65.9	105.7	44.7	64.2	111.2	44.7	65.9	111.2	44.7	65.9	106.2	44.7	64.3
095,098	CEILINGS & ACOUSTICAL TREATMENT	94.7	44.7	60.8	94.7	44.7	60.8	90.8	44.7	59.6	94.7	44.7	60.8	94.7	44.7	60.8	90.8	44.7	59.6
09600	FLOORING	104.5	43.0	88.7	104.7	43.0	88.8	100.8	43.0	85.9	104.7	43.0	88.8	104.7	43.0	88.8	101.7	43.0	86.6
097,099	WALL FINISHES, PAINTS & COATINGS	108.7	40.3	67.7	108.7	40.3	67.7	108.7	40.3	67.7	108.7	40.3	67.7	108.7	40.3	67.7	108.7	40.3	67.7
09	**FINISHES**	99.5	43.9	69.5	99.7	43.9	69.5	96.9	43.9	68.2	99.7	43.9	69.5	99.7	43.9	69.6	97.2	43.9	68.4
10 - 14	**TOTAL DIV. 10000 - 14000**	100.0	66.2	92.7	100.0	69.1	93.3	100.0	69.1	93.3	100.0	66.2	92.7	100.0	69.1	93.3	100.0	69.1	93.3
15	**MECHANICAL**	100.0	44.6	74.5	100.0	44.8	74.6	100.0	44.8	74.6	100.0	44.8	74.6	100.0	44.8	74.6	100.0	44.8	74.6
16	**ELECTRICAL**	99.8	37.7	63.5	96.0	40.2	63.4	92.3	40.2	61.8	96.9	40.2	63.7	96.7	40.2	63.6	97.5	40.2	64.0
01 - 16	**WEIGHTED AVERAGE**	97.8	49.2	74.4	97.1	51.9	75.3	96.0	51.9	74.7	97.2	51.8	75.3	97.4	51.9	75.5	95.8	51.0	74.2

DIVISION		NORTH CAROLINA			NORTH DAKOTA													OHIO		
		WINSTON-SALEM			BISMARCK			FARGO			GRAND FORKS			MINOT			AKRON		
		MAT.	INST.	TOTAL	MAT.	INST.	TOTAL	MAT.	INST.	TOTAL	MAT.	INST.	TOTAL	MAT.	INST.	TOTAL	MAT.	INST.	TOTAL
01590	EQUIPMENT RENTAL	.0	99.5	99.5	.0	98.4	98.4	.0	98.4	98.4	.0	98.4	98.4	.0	98.4	98.4	.0	97.6	97.6
02	SITE CONSTRUCTION	102.7	82.3	87.5	87.1	96.0	93.7	86.6	96.0	93.6	95.1	94.4	94.6	92.6	96.0	95.1	101.1	107.0	105.5
03100	CONCRETE FORMS & ACCESSORIES	99.6	44.7	51.4	96.4	52.8	58.1	97.2	53.1	58.5	93.0	46.2	51.9	87.7	59.9	63.3	100.0	99.5	99.5
03200	CONCRETE REINFORCEMENT	94.7	40.5	61.9	106.6	77.3	88.9	98.3	76.4	85.0	105.0	76.6	87.9	108.6	77.4	89.8	94.9	97.4	96.4
03300	CAST-IN-PLACE CONCRETE	96.7	48.6	76.6	103.3	57.9	84.3	108.3	60.2	88.2	105.8	56.0	85.0	105.8	56.4	85.1	99.4	106.2	102.2
03	CONCRETE	99.6	47.2	72.8	98.8	60.2	79.1	110.4	61.0	85.1	105.3	56.5	80.3	103.7	62.9	82.8	100.1	100.4	100.2
04	MASONRY	80.2	38.2	53.9	104.6	68.9	82.2	107.0	51.5	72.2	106.2	53.0	72.9	107.3	70.3	84.1	90.5	103.1	98.4
05	METALS	94.9	75.9	87.7	94.3	81.8	89.6	94.2	80.4	89.0	94.3	79.0	88.5	94.6	81.9	89.8	94.3	82.9	90.0
06	WOOD & PLASTICS	98.6	46.4	70.4	88.6	48.1	66.7	88.6	48.8	67.1	85.3	44.2	63.1	80.6	57.7	68.2	94.0	99.0	96.7
07	THERMAL & MOISTURE PROTECTION	96.5	44.3	71.0	100.3	58.4	79.9	101.0	55.1	78.7	101.1	53.2	77.8	100.8	59.6	80.7	104.5	100.4	102.5
08	DOORS & WINDOWS	96.0	44.5	82.9	99.6	52.3	87.6	99.6	52.7	87.7	99.6	47.3	86.3	99.7	57.6	89.0	104.2	99.4	103.0
09200	PLASTER & GYPSUM BOARD	111.2	44.7	65.9	120.3	46.9	70.3	120.3	47.6	70.8	118.7	42.9	67.1	117.2	56.8	76.1	100.3	98.5	99.1
095,098	CEILINGS & ACOUSTICAL TREATMENT	94.7	44.7	60.8	142.3	46.9	77.7	142.3	47.6	78.2	142.3	42.9	75.0	142.3	56.8	84.4	100.6	98.5	99.1
09600	FLOORING	104.7	43.0	88.8	108.9	81.6	101.8	108.7	48.7	93.2	107.1	48.7	92.0	104.3	84.3	99.2	103.4	95.4	101.4
097,099	WALL FINISHES, PAINTS & COATINGS	108.7	40.3	67.7	99.5	40.1	63.9	99.5	75.9	85.4	99.5	30.7	58.2	99.5	33.6	60.0	106.7	122.7	116.3
09	FINISHES	99.7	43.9	69.5	114.5	55.8	82.8	114.3	53.6	81.5	114.2	44.3	76.4	112.9	61.3	85.0	102.6	101.2	101.9
10 - 14	TOTAL DIV. 10000 - 14000	100.0	66.2	92.7	100.0	78.8	95.4	100.0	78.9	95.5	100.0	51.3	89.5	100.0	80.1	95.7	100.0	97.0	99.4
15	MECHANICAL	100.0	44.6	74.5	100.6	66.5	84.9	100.6	70.8	86.9	100.6	48.5	76.6	100.6	63.9	83.8	99.9	101.7	100.7
16	ELECTRICAL	96.9	40.2	63.7	97.7	72.8	83.2	97.5	72.5	82.9	101.6	74.2	85.6	105.7	70.6	85.2	98.0	94.0	95.7
01 - 16	WEIGHTED AVERAGE	97.3	50.6	74.8	100.2	68.9	85.1	101.6	67.6	85.2	101.6	60.3	81.7	101.6	69.6	86.2	99.3	99.0	99.2

DIVISION		OHIO																	
		CANTON			CINCINNATI			CLEVELAND			COLUMBUS			DAYTON			LORAIN		
		MAT.	INST.	TOTAL	MAT.	INST.	TOTAL	MAT.	INST.	TOTAL	MAT.	INST.	TOTAL	MAT.	INST.	TOTAL	MAT.	INST.	TOTAL
01590	EQUIPMENT RENTAL	.0	97.6	97.6	.0	102.6	102.6	.0	97.9	97.9	.0	96.6	96.6	.0	96.9	96.9	.0	97.6	97.6
02	SITE CONSTRUCTION	101.2	107.3	105.7	72.9	108.7	99.7	101.1	106.9	105.4	89.1	102.1	98.8	71.7	108.4	99.2	100.5	106.1	104.7
03100	CONCRETE FORMS & ACCESSORIES	100.0	88.8	90.2	99.7	91.6	92.6	100.2	106.4	105.7	94.9	87.9	88.8	99.6	83.2	85.2	100.1	107.8	106.9
03200	CONCRETE REINFORCEMENT	94.9	81.7	86.9	93.2	88.7	90.4	95.4	97.6	96.7	104.1	88.6	94.7	93.2	84.1	87.7	94.9	97.5	96.5
03300	CAST-IN-PLACE CONCRETE	100.3	104.2	101.9	84.7	93.8	88.5	97.5	114.6	104.6	90.4	96.0	92.7	77.2	87.6	81.5	94.6	109.0	100.6
03	CONCRETE	100.6	92.0	96.2	92.2	91.8	92.0	99.2	106.4	102.9	96.6	90.4	93.5	88.5	84.5	86.4	97.7	105.0	101.5
04	MASONRY	91.3	95.0	93.6	82.8	96.1	91.1	94.9	110.5	104.6	95.7	95.8	95.8	82.2	87.9	85.8	87.5	102.3	96.8
05	METALS	94.3	75.8	87.3	93.6	89.5	92.1	95.7	85.3	91.8	95.5	83.4	90.9	93.0	78.7	87.6	94.8	83.3	90.5
06	WOOD & PLASTICS	94.1	87.6	90.6	98.7	90.3	94.1	93.1	103.5	98.7	100.7	86.1	92.8	99.9	81.4	89.9	94.0	110.2	102.7
07	THERMAL & MOISTURE PROTECTION	105.2	97.1	101.2	89.6	97.6	93.5	103.3	114.5	108.8	99.8	97.2	98.5	95.3	89.6	92.5	105.1	109.2	107.1
08	DOORS & WINDOWS	98.6	79.6	93.8	95.9	87.2	93.7	95.0	101.8	96.8	99.5	85.1	95.9	96.2	80.1	92.1	98.6	105.5	100.4
09200	PLASTER & GYPSUM BOARD	100.8	86.8	91.3	97.6	90.3	92.6	99.5	103.1	102.0	101.3	85.6	90.6	97.6	81.2	86.4	100.3	110.1	106.9
095,098	CEILINGS & ACOUSTICAL TREATMENT	100.6	86.8	91.2	92.4	90.3	91.0	99.3	103.1	101.9	94.9	85.6	88.6	93.9	81.2	85.3	100.6	110.1	107.0
09600	FLOORING	103.7	86.9	99.3	105.1	104.3	104.9	103.2	108.5	104.6	98.1	94.7	97.2	107.8	87.4	102.5	103.7	113.9	106.3
097,099	WALL FINISHES, PAINTS & COATINGS	106.7	88.5	95.8	103.8	97.7	100.1	106.7	122.7	116.3	99.5	102.7	101.5	103.8	92.4	97.0	106.7	122.7	116.3
09	FINISHES	102.8	87.5	94.5	96.3	94.5	95.3	102.3	108.4	105.6	99.5	90.3	94.5	97.5	84.5	90.4	102.7	111.2	107.3
10 - 14	TOTAL DIV. 10000 - 14000	100.0	84.6	96.7	100.0	94.9	98.9	100.0	109.0	101.9	100.0	94.8	98.9	100.0	92.4	98.4	100.0	106.5	101.4
15	MECHANICAL	99.9	87.7	94.3	99.8	93.1	96.7	99.9	109.2	104.2	100.0	92.6	96.6	100.9	87.4	94.7	99.9	91.9	96.2
16	ELECTRICAL	96.9	93.7	95.0	97.2	83.1	88.9	97.4	109.4	104.4	100.3	90.5	94.6	95.9	87.2	90.8	97.0	109.4	104.2
01 - 16	WEIGHTED AVERAGE	98.7	90.4	94.7	95.0	93.0	94.0	98.5	106.2	102.2	98.3	91.9	95.2	94.8	87.6	91.3	98.2	101.8	99.9

DIVISION		OHIO									OKLAHOMA								
		SPRINGFIELD			TOLEDO			YOUNGSTOWN			ENID			LAWTON			MUSKOGEE		
		MAT.	INST.	TOTAL	MAT.	INST.	TOTAL	MAT.	INST.	TOTAL	MAT.	INST.	TOTAL	MAT.	INST.	TOTAL	MAT.	INST.	TOTAL
01590	EQUIPMENT RENTAL	.0	96.9	96.9	.0	99.7	99.7	.0	97.6	97.6	.0	77.0	77.0	.0	78.1	78.1	.0	86.6	86.6
02	SITE CONSTRUCTION	72.2	107.4	98.6	88.3	103.1	99.4	100.9	107.2	105.6	109.7	89.4	94.5	104.8	91.0	94.5	95.1	85.4	87.8
03100	CONCRETE FORMS & ACCESSORIES	99.6	86.3	87.9	94.9	101.2	100.4	100.0	93.0	93.8	95.6	41.9	48.5	99.7	57.4	62.6	101.3	37.9	45.7
03200	CONCRETE REINFORCEMENT	93.2	84.1	87.7	104.1	91.2	96.3	94.9	97.5	96.5	97.3	84.7	89.7	97.6	84.7	89.8	97.3	40.3	62.8
03300	CAST-IN-PLACE CONCRETE	80.9	87.6	83.7	90.4	107.7	97.6	98.4	105.3	101.3	92.6	53.0	76.1	89.6	53.1	74.3	83.3	40.8	65.6
03	CONCRETE	90.3	85.9	88.0	96.6	101.0	98.9	99.6	97.2	98.4	90.4	54.0	71.8	86.8	60.9	73.5	82.4	40.5	60.9
04	MASONRY	82.5	88.0	85.9	105.8	101.5	103.1	90.6	97.6	95.0	102.6	62.1	77.2	96.7	62.1	75.0	112.4	44.6	70.0
05	METALS	92.9	78.6	87.5	95.3	88.3	92.7	94.3	82.7	89.9	93.9	68.5	84.3	95.9	68.6	85.6	93.8	60.1	81.1
06	WOOD & PLASTICS	101.2	85.7	92.8	100.7	101.2	100.9	94.0	91.0	92.4	100.9	39.0	67.5	104.0	59.9	80.2	105.9	38.6	69.6
07	THERMAL & MOISTURE PROTECTION	95.2	90.1	92.7	101.5	106.8	104.1	105.3	99.7	102.6	99.7	62.3	81.5	99.4	64.5	82.4	99.2	47.1	73.8
08	DOORS & WINDOWS	94.2	83.1	91.3	97.5	95.3	96.9	98.6	97.6	98.4	95.6	53.3	84.8	97.1	64.7	88.9	95.5	37.9	80.9
09200	PLASTER & GYPSUM BOARD	97.6	85.6	89.4	101.3	101.1	101.2	100.3	90.3	93.5	84.3	38.2	52.9	86.1	59.7	68.1	85.3	37.6	52.8
095,098	CEILINGS & ACOUSTICAL TREATMENT	93.9	85.6	88.2	94.9	101.1	99.1	100.6	90.3	93.6	90.3	38.2	55.0	98.1	59.7	72.1	90.3	37.6	54.6
09600	FLOORING	107.8	87.4	102.5	97.2	99.0	97.7	103.7	101.9	103.2	113.5	55.9	98.7	116.5	55.9	100.8	117.7	45.0	99.0
097,099	WALL FINISHES, PAINTS & COATINGS	103.8	92.4	97.0	99.6	113.9	108.1	106.7	103.5	104.8	96.6	58.9	74.0	96.6	58.9	74.0	96.6	37.4	61.1
09	FINISHES	97.5	87.0	91.8	99.2	102.1	100.8	102.7	95.2	98.7	98.3	44.4	69.2	100.5	56.6	76.8	98.7	39.1	66.5
10 - 14	TOTAL DIV. 10000 - 14000	100.0	93.0	98.5	100.0	98.3	99.6	100.0	96.7	99.3	100.0	67.8	93.1	100.0	70.5	93.7	100.0	66.9	92.9
15	MECHANICAL	100.9	87.4	94.7	100.0	102.1	101.0	99.9	93.1	96.8	100.1	65.9	84.4	100.1	65.9	84.4	100.1	31.2	68.4
16	ELECTRICAL	95.9	87.3	90.9	101.0	103.3	102.3	97.0	92.4	94.3	95.3	71.2	81.2	97.6	66.5	79.4	94.8	36.5	60.7
01 - 16	WEIGHTED AVERAGE	94.8	88.3	91.7	98.7	100.6	99.6	98.6	95.1	96.9	97.3	63.5	81.0	97.3	66.3	82.4	96.5	45.0	71.7

DIVISION		OKLAHOMA						OREGON											
		OKLAHOMA CITY			TULSA			EUGENE			MEDFORD			PORTLAND			SALEM		
		MAT.	INST.	TOTAL	MAT.	INST.	TOTAL	MAT.	INST.	TOTAL	MAT.	INST.	TOTAL	MAT.	INST.	TOTAL	MAT.	INST.	TOTAL
01590	EQUIPMENT RENTAL	.0	78.3	78.3	.0	86.6	86.6	.0	99.5	99.5	.0	99.5	99.5	.0	99.5	99.5	.0	99.5	99.5
02	SITE CONSTRUCTION	106.1	91.6	95.3	101.6	86.4	90.2	108.3	104.8	105.7	118.2	104.8	108.1	109.4	104.8	106.0	108.4	104.8	105.7
03100	CONCRETE FORMS & ACCESSORIES	100.9	49.4	55.7	101.0	50.0	56.2	103.2	106.8	106.4	98.0	106.6	105.5	104.5	107.1	106.8	104.4	107.0	106.6
03200	CONCRETE REINFORCEMENT	97.6	84.7	89.8	97.6	84.6	89.8	106.8	98.3	101.7	98.6	98.2	98.4	102.1	98.5	99.9	107.9	98.4	102.2
03300	CAST-IN-PLACE CONCRETE	96.9	56.6	80.0	90.8	51.8	74.5	104.6	105.6	105.0	108.0	105.5	106.9	112.3	105.7	109.5	108.5	105.6	107.3
03	CONCRETE	90.6	58.6	74.2	87.5	58.2	72.5	108.5	104.4	106.4	115.4	104.2	109.7	111.9	104.6	108.2	110.7	104.5	107.5
04	MASONRY	98.6	64.3	77.1	96.0	64.4	76.2	113.5	103.6	107.3	110.3	103.6	106.1	114.7	106.3	109.4	118.3	106.3	110.8
05	METALS	96.1	68.6	85.7	96.6	83.4	91.6	94.8	96.8	95.6	94.3	96.5	95.2	93.5	97.3	94.9	95.4	97.1	96.1
06	WOOD & PLASTICS	105.9	48.0	74.7	104.7	49.5	74.9	91.6	106.8	99.8	86.4	106.8	97.4	93.1	106.8	100.5	93.1	106.8	100.5
07	THERMAL & MOISTURE PROTECTION	98.4	64.2	81.8	99.2	61.9	81.0	110.3	94.3	102.5	111.3	91.2	101.5	110.2	102.3	106.3	110.2	98.5	104.5
08	DOORS & WINDOWS	97.1	58.2	87.2	97.1	58.5	87.3	98.8	106.7	100.8	101.5	106.7	102.8	96.3	106.7	98.9	98.1	106.7	100.3
09200	PLASTER & GYPSUM BOARD	86.1	47.4	59.8	86.1	48.8	60.7	99.0	106.9	104.3	95.3	106.9	103.2	97.9	106.9	104.0	99.5	106.9	104.5
095,098	CEILINGS & ACOUSTICAL TREATMENT	98.1	47.4	63.8	98.1	48.8	64.7	117.3	106.9	110.2	114.7	106.9	109.4	117.3	106.9	110.2	123.8	106.9	112.3
09600	FLOORING	116.5	55.9	100.8	116.3	58.1	101.3	115.3	101.5	111.8	112.8	101.5	109.9	115.3	101.5	111.8	115.3	101.5	111.8
097,099	WALL FINISHES, PAINTS & COATINGS	96.6	58.9	74.0	96.6	55.3	71.8	115.4	74.6	91.0	115.4	64.7	85.0	115.4	74.6	91.0	115.4	74.6	91.0
09	FINISHES	100.6	50.2	73.4	99.9	51.1	73.6	113.9	102.5	107.7	112.8	101.3	106.6	113.8	102.5	107.7	115.1	102.5	108.3
10 - 14	TOTAL DIV. 10000 - 14000	100.0	69.9	93.5	100.0	70.6	93.7	100.0	109.8	102.1	100.0	109.7	102.1	100.0	109.8	102.1	100.0	109.8	102.1
15	MECHANICAL	100.1	67.1	84.9	100.1	64.8	83.8	100.1	107.0	103.3	100.1	106.9	103.3	100.1	107.0	103.3	100.1	107.0	103.3
16	ELECTRICAL	96.5	71.2	81.7	97.6	47.6	68.3	103.2	101.5	102.2	107.9	91.1	98.1	103.1	108.3	106.2	102.9	101.5	102.1
01 - 16	WEIGHTED AVERAGE	97.9	65.9	82.5	97.3	63.1	80.8	103.1	103.5	103.3	104.5	101.8	103.2	103.1	105.0	104.0	103.7	103.9	103.8

DIVISION		PENNSYLVANIA																	
		ALLENTOWN			ALTOONA			ERIE			HARRISBURG			PHILADELPHIA			PITTSBURGH		
		MAT.	INST.	TOTAL	MAT.	INST.	TOTAL	MAT.	INST.	TOTAL	MAT.	INST.	TOTAL	MAT.	INST.	TOTAL	MAT.	INST.	TOTAL
01590	EQUIPMENT RENTAL	.0	115.8	115.8	.0	115.8	115.8	.0	115.8	115.8	.0	115.0	115.0	.0	94.5	94.5	.0	115.8	115.8
02	SITE CONSTRUCTION	90.2	106.6	102.4	94.2	106.8	103.6	90.8	107.4	103.2	81.0	105.0	99.0	100.0	94.2	95.6	98.7	109.6	106.9
03100	CONCRETE FORMS & ACCESSORIES	101.7	109.1	108.2	82.8	89.4	88.6	100.4	98.2	98.4	93.5	85.0	86.1	101.5	131.3	127.7	96.9	99.8	99.4
03200	CONCRETE REINFORCEMENT	102.6	104.5	103.7	99.6	96.9	98.0	101.6	96.5	98.5	102.6	96.8	99.1	98.6	138.7	122.8	108.2	106.6	107.2
03300	CAST-IN-PLACE CONCRETE	89.3	97.7	92.8	99.5	89.5	95.3	97.9	87.4	93.5	98.6	88.5	94.4	97.4	125.7	109.2	97.3	98.7	97.9
03	CONCRETE	98.8	105.3	102.1	93.7	92.3	93.0	92.5	95.4	94.0	101.0	90.0	95.4	107.1	130.1	118.9	95.1	102.0	98.6
04	MASONRY	94.4	93.5	93.8	96.9	83.2	88.3	86.8	93.2	90.8	94.9	84.3	88.3	97.2	129.2	117.3	89.9	99.5	95.9
05	METALS	98.7	122.1	107.6	92.3	117.9	102.0	92.5	118.6	102.4	100.5	117.8	107.1	100.7	124.0	109.5	97.2	124.9	107.6
06	WOOD & PLASTICS	103.1	112.9	108.4	81.0	91.8	86.8	98.5	99.1	98.8	96.4	84.5	90.0	99.3	131.6	116.7	94.0	99.0	96.7
07	THERMAL & MOISTURE PROTECTION	98.9	110.4	104.5	97.5	92.7	95.1	97.6	98.4	98.0	102.8	103.1	102.9	101.8	131.0	116.0	96.9	101.9	99.4
08	DOORS & WINDOWS	93.6	109.0	97.5	88.1	97.8	90.5	88.2	96.0	90.2	93.6	93.9	93.6	96.6	135.7	106.5	93.9	109.2	97.8
09200	PLASTER & GYPSUM BOARD	101.7	113.0	109.4	93.0	91.2	91.8	101.7	98.8	99.7	101.7	83.7	89.4	100.5	132.3	122.2	90.6	98.7	96.1
095,098	CEILINGS & ACOUSTICAL TREATMENT	98.9	113.0	108.5	95.0	91.2	92.5	98.9	98.8	98.8	98.9	83.7	88.6	100.7	132.3	122.1	89.4	98.7	95.7
09600	FLOORING	88.0	86.7	87.7	82.1	61.9	76.9	90.0	92.1	90.5	88.2	81.6	86.5	85.6	130.5	97.1	96.8	102.2	98.2
097,099	WALL FINISHES, PAINTS & COATINGS	90.4	96.2	93.9	85.6	115.3	103.4	97.1	91.3	93.6	90.4	86.9	88.3	92.1	136.2	118.6	96.5	117.4	109.0
09	FINISHES	96.6	103.9	100.5	93.0	86.6	89.5	98.2	96.3	97.2	95.5	84.0	89.3	100.2	131.4	117.1	95.4	101.6	98.8
10 - 14	TOTAL DIV. 10000 - 14000	100.0	102.8	100.6	100.0	100.6	100.1	100.0	104.8	101.0	100.0	92.8	98.4	100.0	122.3	104.8	100.0	105.2	101.1
15	MECHANICAL	100.4	103.2	101.7	99.8	82.5	91.8	99.8	93.5	96.9	100.4	88.6	95.0	100.2	127.1	112.6	100.0	99.7	99.9
16	ELECTRICAL	101.2	85.9	92.3	88.9	102.6	96.9	90.8	86.8	88.4	100.4	79.7	88.3	96.4	124.8	113.0	95.8	102.6	99.8
01 - 16	WEIGHTED AVERAGE	98.2	102.8	100.4	94.4	94.3	94.3	94.5	97.3	95.9	98.3	91.4	95.0	100.2	125.0	112.1	96.7	104.5	100.5

DIVISION		PENNSYLVANIA									PUERTO RICO			RHODE ISLAND			SOUTH CAROLINA		
		READING			SCRANTON			YORK			SAN JUAN			PROVIDENCE			CHARLESTON		
		MAT.	INST.	TOTAL	MAT.	INST.	TOTAL	MAT.	INST.	TOTAL	MAT.	INST.	TOTAL	MAT.	INST.	TOTAL	MAT.	INST.	TOTAL
01590	EQUIPMENT RENTAL	.0	118.0	118.0	.0	115.8	115.8	.0	115.0	115.0	.0	89.3	89.3	.0	103.7	103.7	.0	99.1	99.1
02	SITE CONSTRUCTION	97.6	110.7	107.4	90.7	106.6	102.6	80.3	105.0	98.8	116.2	92.4	98.4	80.2	104.1	98.1	95.7	81.4	85.0
03100	CONCRETE FORMS & ACCESSORIES	101.0	86.4	88.2	101.8	87.7	89.4	81.1	82.5	82.3	103.2	20.9	31.0	106.7	114.0	113.1	98.0	39.8	46.9
03200	CONCRETE REINFORCEMENT	96.9	96.7	96.8	102.6	99.4	100.7	101.6	96.8	98.7	196.0	12.8	85.4	109.1	117.7	114.3	94.7	67.2	78.1
03300	CAST-IN-PLACE CONCRETE	71.2	89.9	79.0	93.3	91.4	92.5	87.9	86.3	87.2	107.1	32.5	75.9	87.7	115.0	99.1	83.4	50.6	69.7
03	CONCRETE	93.7	91.1	92.4	100.8	92.7	96.7	99.1	88.2	93.5	111.1	24.1	66.5	103.3	114.6	109.1	92.9	50.6	71.2
04	MASONRY	98.6	84.7	89.9	94.7	95.6	95.2	94.2	84.2	87.9	212.0	17.9	90.5	106.3	123.7	117.2	88.2	37.3	56.3
05	METALS	97.5	117.8	105.2	100.6	119.5	107.7	97.8	117.7	105.4	129.0	31.9	92.2	97.3	110.2	102.2	94.9	82.7	90.3
06	WOOD & PLASTICS	99.3	84.6	91.4	103.1	85.3	93.5	87.9	81.2	84.3	92.8	21.2	54.2	102.6	112.8	108.1	98.6	39.1	66.5
07	THERMAL & MOISTURE PROTECTION	101.8	103.2	102.5	98.8	101.5	100.1	97.1	101.1	99.1	169.2	24.8	98.9	101.1	113.7	107.2	96.0	45.4	71.3
08	DOORS & WINDOWS	94.2	91.4	93.5	93.6	95.0	93.9	90.4	92.1	90.8	154.6	17.7	119.8	101.1	114.7	104.5	96.0	44.1	82.8
09200	PLASTER & GYPSUM BOARD	109.0	84.0	92.0	101.7	84.6	90.0	96.5	80.4	85.5	278.6	18.7	101.8	105.7	112.2	110.1	111.2	37.2	60.9
095,098	CEILINGS & ACOUSTICAL TREATMENT	101.9	84.0	89.8	98.9	84.6	89.2	92.4	80.4	84.3	383.3	18.7	136.4	107.4	112.2	110.6	94.7	37.2	55.8
09600	FLOORING	83.7	87.9	84.8	88.0	85.4	87.3	84.0	87.2	84.8	201.7	17.8	154.4	101.1	116.3	105.0	104.7	42.7	88.7
097,099	WALL FINISHES, PAINTS & COATINGS	90.3	98.3	95.1	90.4	98.3	95.1	90.4	78.8	83.4	202.2	16.3	90.7	95.5	117.0	108.4	108.7	40.8	68.0
09	FINISHES	100.8	87.1	93.4	96.6	87.7	91.8	92.4	82.1	86.8	255.2	20.1	128.1	102.9	114.9	109.4	99.9	39.9	67.5
10 - 14	TOTAL DIV. 10000 - 14000	100.0	97.4	99.4	100.0	98.9	99.8	100.0	92.3	98.3	100.0	22.4	83.3	100.0	107.5	101.6	100.0	67.6	93.0
15	MECHANICAL	100.4	101.3	100.8	100.4	90.4	95.8	100.4	88.5	94.9	105.0	16.4	64.2	100.2	104.2	102.1	100.0	46.1	75.2
16	ELECTRICAL	98.8	83.5	89.8	101.2	87.8	93.4	93.0	74.5	82.2	138.8	16.3	67.2	98.2	97.8	98.0	96.8	34.5	60.4
01 - 16	WEIGHTED AVERAGE	98.2	95.7	97.0	98.7	95.4	97.1	96.1	90.0	93.1	140.4	27.1	85.8	100.1	109.5	104.6	96.7	50.4	74.4

DIVISION		SOUTH CAROLINA												SOUTH DAKOTA					
		COLUMBIA			FLORENCE			GREENVILLE			SPARTANBURG			ABERDEEN			PIERRE		
		MAT.	INST.	TOTAL	MAT.	INST.	TOTAL	MAT.	INST.	TOTAL	MAT.	INST.	TOTAL	MAT.	INST.	TOTAL	MAT.	INST.	TOTAL
01590	**EQUIPMENT RENTAL**	.0	99.1	99.1	.0	99.1	99.1	.0	99.1	99.1	.0	99.1	99.1	.0	98.4	98.4	.0	98.4	98.4
02	**SITE CONSTRUCTION**	95.0	81.4	84.8	106.8	81.4	87.8	101.6	81.0	86.2	101.4	81.0	86.1	83.1	93.9	91.2	81.4	93.9	90.8
03100	CONCRETE FORMS & ACCESSORIES	102.8	42.2	49.6	82.8	42.1	47.1	97.8	41.8	48.7	101.7	41.8	49.2	97.0	41.6	48.4	95.2	43.3	49.6
03200	CONCRETE REINFORCEMENT	94.7	67.2	78.1	94.3	67.2	77.9	94.3	40.9	62.0	94.3	40.9	62.0	105.2	50.5	72.1	104.7	62.7	79.3
03300	CAST-IN-PLACE CONCRETE	78.7	50.3	66.8	71.0	50.5	62.5	71.0	50.4	62.4	71.0	50.4	62.4	100.8	52.9	80.8	97.9	48.5	77.2
03	**CONCRETE**	90.8	51.5	70.7	94.0	51.6	72.3	92.4	46.5	68.9	92.7	46.5	69.0	98.2	48.7	72.8	95.8	50.2	72.4
04	**MASONRY**	86.9	38.2	56.4	73.0	37.3	50.6	70.9	37.3	49.9	73.0	37.3	50.6	108.4	59.8	78.0	104.8	57.9	75.5
05	**METALS**	94.8	82.6	90.2	93.8	82.5	89.5	93.6	72.9	85.8	93.7	72.9	85.8	106.0	72.2	93.2	105.9	77.4	95.1
06	**WOOD & PLASTICS**	104.2	42.6	71.0	83.2	42.6	61.3	98.4	42.6	68.3	102.6	42.6	70.2	102.6	39.7	68.7	100.8	42.0	69.1
07	**THERMAL & MOISTURE PROTECTION**	95.9	45.2	71.2	96.3	45.7	71.7	96.3	45.7	71.6	96.3	45.7	71.7	98.8	54.0	77.0	99.0	51.7	75.9
08	**DOORS & WINDOWS**	96.0	45.9	83.3	92.0	45.9	80.3	92.0	39.8	78.7	92.0	39.8	78.7	95.2	42.6	81.8	98.4	47.3	85.5
09200	PLASTER & GYPSUM BOARD	111.2	40.7	63.3	100.5	40.7	59.8	106.4	40.7	61.7	108.5	40.7	62.4	104.6	38.3	59.5	103.3	40.6	60.7
095,098	CEILINGS & ACOUSTICAL TREATMENT	94.7	40.7	58.2	90.8	40.7	56.9	89.5	40.7	56.5	89.5	40.7	56.5	107.2	38.3	60.6	105.9	40.6	61.7
09600	FLOORING	104.5	42.7	88.6	94.7	42.7	81.3	102.1	43.5	87.0	103.8	43.5	88.3	109.2	61.0	96.8	108.3	43.5	91.6
097,099	WALL FINISHES, PAINTS & COATINGS	108.7	40.8	68.0	108.7	40.8	68.0	108.7	40.8	68.0	108.7	40.8	68.0	99.5	36.7	61.8	99.5	42.0	65.0
09	**FINISHES**	99.9	41.9	68.5	95.6	41.9	66.5	98.2	42.0	67.8	99.0	42.0	68.2	105.5	44.2	72.4	104.7	42.6	71.1
10 - 14	**TOTAL DIV. 10000 - 14000**	100.0	68.1	93.1	100.0	68.0	93.1	100.0	68.1	93.1	100.0	68.1	93.1	100.0	61.1	91.6	100.0	72.7	94.1
15	**MECHANICAL**	100.0	38.9	71.9	100.0	38.9	71.9	100.0	38.8	71.8	100.0	38.8	71.8	100.3	41.5	73.2	100.3	39.7	72.4
16	**ELECTRICAL**	97.8	36.4	61.9	94.4	20.8	51.4	97.2	33.1	59.7	97.2	33.1	59.7	102.2	63.1	79.4	97.0	53.9	71.8
01 - 16	**WEIGHTED AVERAGE**	96.5	49.8	74.0	94.9	47.5	72.1	95.1	47.4	72.1	95.4	47.4	72.3	100.8	56.3	79.4	100.1	55.4	78.5

DIVISION		SOUTH DAKOTA						TENNESSEE											
		RAPID CITY			SIOUX FALLS			CHATTANOOGA			JACKSON			JOHNSON CITY			KNOXVILLE		
		MAT.	INST.	TOTAL	MAT.	INST.	TOTAL	MAT.	INST.	TOTAL	MAT.	INST.	TOTAL	MAT.	INST.	TOTAL	MAT.	INST.	TOTAL
01590	**EQUIPMENT RENTAL**	.0	98.4	98.4	.0	99.6	99.6	.0	104.5	104.5	.0	104.9	104.9	.0	97.4	97.4	.0	97.4	97.4
02	**SITE CONSTRUCTION**	81.5	93.7	90.6	82.5	95.8	92.5	104.6	96.6	98.6	102.7	95.9	97.6	115.0	85.4	92.8	90.9	85.4	86.8
03100	CONCRETE FORMS & ACCESSORIES	107.8	38.5	46.9	95.3	44.2	50.5	95.1	48.9	54.5	85.3	39.9	45.4	80.4	49.7	53.5	93.9	49.8	55.2
03200	CONCRETE REINFORCEMENT	98.3	62.8	76.8	98.3	62.8	76.8	95.2	45.7	65.3	99.3	45.4	66.8	99.3	45.9	67.0	95.2	45.9	65.4
03300	CAST-IN-PLACE CONCRETE	97.2	45.8	75.7	101.3	51.5	80.4	101.6	50.9	80.4	101.2	44.0	77.3	81.8	55.7	70.9	95.4	51.1	76.9
03	**CONCRETE**	95.5	47.2	70.7	95.5	51.6	73.0	93.8	50.9	71.8	95.4	44.5	69.3	100.0	52.9	75.9	90.9	51.4	70.7
04	**MASONRY**	104.8	53.8	72.9	102.1	57.9	74.4	97.7	45.8	65.2	111.7	35.6	64.1	109.8	45.6	69.6	75.3	45.6	56.7
05	**METALS**	108.0	77.8	96.5	108.5	77.8	96.9	96.2	79.2	89.8	94.2	77.4	87.9	93.9	78.7	88.1	96.4	79.1	89.9
06	**WOOD & PLASTICS**	110.3	37.5	71.1	100.8	42.6	69.5	98.4	50.1	72.3	83.8	39.8	60.1	73.8	51.2	61.6	86.6	51.2	67.5
07	**THERMAL & MOISTURE PROTECTION**	99.4	49.8	75.3	98.9	51.3	75.7	100.6	53.3	77.6	99.4	41.9	71.4	96.4	50.9	74.2	93.9	51.1	73.0
08	**DOORS & WINDOWS**	99.4	44.9	85.6	99.4	47.7	86.3	101.4	49.9	88.3	102.1	42.0	86.9	97.1	51.1	85.4	93.7	51.1	82.9
09200	PLASTER & GYPSUM BOARD	104.5	36.1	57.9	104.5	41.3	61.5	84.6	49.2	60.5	88.1	38.6	54.4	88.9	50.4	62.7	96.1	50.4	65.0
095,098	CEILINGS & ACOUSTICAL TREATMENT	111.1	36.1	60.3	111.1	41.3	63.9	102.3	49.2	66.4	95.7	38.6	57.0	92.9	50.4	64.1	99.4	50.4	66.2
09600	FLOORING	108.3	76.6	100.1	108.3	75.0	99.8	103.9	53.7	90.9	92.8	26.0	75.6	97.8	55.2	86.8	103.4	55.2	90.9
097,099	WALL FINISHES, PAINTS & COATINGS	99.5	42.0	65.0	99.5	42.0	65.0	109.5	48.9	73.2	96.6	33.2	58.5	107.3	57.4	77.4	107.3	57.4	77.4
09	**FINISHES**	105.9	45.4	73.2	105.8	49.3	75.3	100.0	49.8	72.8	95.9	36.6	63.9	99.6	51.5	73.6	94.6	51.5	71.3
10 - 14	**TOTAL DIV. 10000 - 14000**	100.0	70.1	93.6	100.0	72.9	94.2	100.0	58.9	91.2	100.0	55.5	90.4	100.0	67.7	93.0	100.0	67.7	93.0
15	**MECHANICAL**	100.3	37.1	71.2	100.3	38.7	71.9	99.9	48.2	76.1	100.0	59.7	81.4	99.8	60.2	81.6	99.8	61.9	82.3
16	**ELECTRICAL**	97.6	53.9	72.1	96.7	76.6	84.9	103.9	56.2	76.0	101.5	54.4	74.0	92.3	54.4	70.2	100.0	63.6	78.7
01 - 16	**WEIGHTED AVERAGE**	100.7	54.1	78.3	100.5	59.7	80.8	99.2	57.5	79.1	99.2	55.0	77.9	98.6	59.3	79.7	95.1	60.8	78.6

DIVISION		TENNESSEE						TEXAS											
		MEMPHIS			NASHVILLE			ABILENE			AMARILLO			AUSTIN			BEAUMONT		
		MAT.	INST.	TOTAL	MAT.	INST.	TOTAL	MAT.	INST.	TOTAL	MAT.	INST.	TOTAL	MAT.	INST.	TOTAL	MAT.	INST.	TOTAL
01590	**EQUIPMENT RENTAL**	.0	102.3	102.3	.0	105.5	105.5	.0	86.6	86.6	.0	86.6	86.6	.0	86.1	86.1	.0	88.9	88.9
02	**SITE CONSTRUCTION**	96.4	91.8	93.0	96.5	99.3	98.6	102.0	85.6	89.7	102.4	86.5	90.5	88.3	84.6	85.5	96.3	87.7	89.8
03100	CONCRETE FORMS & ACCESSORIES	95.4	64.4	68.2	94.1	67.6	70.9	97.9	45.5	51.9	101.6	55.2	60.9	99.5	58.6	63.6	103.5	56.5	62.2
03200	CONCRETE REINFORCEMENT	93.6	67.6	77.9	96.2	66.0	78.0	97.6	54.9	71.8	97.6	57.5	73.3	96.0	55.3	71.4	92.8	46.8	65.1
03300	CAST-IN-PLACE CONCRETE	90.3	66.6	80.4	94.6	71.8	85.0	94.2	44.7	73.5	99.0	51.2	79.0	88.1	52.3	73.2	86.2	58.1	74.4
03	**CONCRETE**	88.3	67.5	77.6	91.0	70.3	80.4	89.0	48.1	68.0	91.6	55.1	72.9	79.6	56.4	67.8	85.1	56.0	70.2
04	**MASONRY**	84.7	77.5	80.2	86.2	67.9	74.8	99.6	56.1	72.3	101.6	54.5	72.1	95.0	58.1	71.9	104.2	60.9	77.1
05	**METALS**	95.7	95.3	95.6	98.9	92.4	96.4	97.2	68.9	86.5	97.2	70.1	87.0	96.1	67.9	85.4	99.1	66.9	86.9
06	**WOOD & PLASTICS**	93.1	65.4	78.2	93.9	68.9	80.4	101.1	45.1	70.9	103.9	57.9	79.1	99.8	62.2	79.5	113.1	57.4	83.1
07	**THERMAL & MOISTURE PROTECTION**	96.8	73.8	85.6	97.3	68.7	83.4	99.4	52.8	76.7	102.0	51.8	77.5	96.5	57.3	77.4	103.0	61.5	82.8
08	**DOORS & WINDOWS**	100.8	68.1	92.5	98.7	69.1	91.1	92.8	48.0	81.4	92.8	53.2	82.8	93.9	61.3	85.6	97.4	52.8	86.1
09200	PLASTER & GYPSUM BOARD	94.3	64.8	74.2	85.8	68.5	74.0	86.1	44.3	57.6	86.1	57.4	66.6	89.8	61.8	70.8	88.2	56.8	66.9
095,098	CEILINGS & ACOUSTICAL TREATMENT	95.0	64.8	74.6	91.0	68.5	75.7	98.1	44.3	61.6	98.1	57.4	70.6	93.4	61.8	72.0	99.6	56.8	70.6
09600	FLOORING	98.7	49.0	85.9	106.0	80.7	99.5	116.5	71.9	105.0	116.3	69.4	104.2	101.4	53.2	89.0	110.6	75.5	101.6
097,099	WALL FINISHES, PAINTS & COATINGS	99.1	61.2	76.4	112.8	72.4	88.6	95.4	58.4	73.2	95.4	40.5	62.5	92.3	45.4	64.2	91.6	54.5	69.4
09	**FINISHES**	94.3	60.6	76.1	102.9	70.7	85.5	100.0	51.4	73.7	100.0	56.3	76.4	93.5	56.2	73.4	93.1	59.7	75.0
10 - 14	**TOTAL DIV. 10000 - 14000**	100.0	74.4	94.5	100.0	75.3	94.7	100.0	73.4	94.3	100.0	70.0	93.5	100.0	70.4	93.6	100.0	78.1	95.3
15	**MECHANICAL**	99.9	75.9	88.9	99.8	75.9	88.8	100.1	48.4	76.3	100.1	54.8	79.2	99.9	58.9	81.0	100.0	66.7	84.7
16	**ELECTRICAL**	100.6	79.5	88.3	100.7	68.1	81.6	97.6	51.9	70.9	98.4	63.9	78.2	95.2	72.1	81.7	94.7	67.9	79.0
01 - 16	**WEIGHTED AVERAGE**	96.4	76.2	86.7	97.9	75.7	87.2	97.3	56.1	77.4	97.9	60.9	80.1	94.5	63.6	79.6	97.0	65.4	81.7

DIVISION		TEXAS																	
		CORPUS CHRISTI			DALLAS			EL PASO			FORT WORTH			HOUSTON			LAREDO		
		MAT.	INST.	TOTAL	MAT.	INST.	TOTAL	MAT.	INST.	TOTAL	MAT.	INST.	TOTAL	MAT.	INST.	TOTAL	MAT.	INST.	TOTAL
01590	EQUIPMENT RENTAL	.0	94.9	94.9	.0	96.9	96.9	.0	86.6	86.6	.0	86.6	86.6	.0	99.2	99.2	.0	86.1	86.1
02	SITE CONSTRUCTION	123.3	79.9	90.8	129.6	86.2	97.1	102.5	85.3	89.6	101.9	85.9	89.9	125.1	85.9	95.8	88.6	84.4	85.5
03100	CONCRETE FORMS & ACCESSORIES	103.5	41.3	48.9	97.8	59.5	64.2	98.4	50.4	56.2	100.2	59.1	64.1	91.2	68.5	71.3	91.9	41.3	47.5
03200	CONCRETE REINFORCEMENT	94.1	53.2	69.4	102.4	59.0	76.2	97.6	51.9	70.0	97.6	58.9	74.2	95.5	64.5	76.8	96.0	53.3	70.2
03300	CAST-IN-PLACE CONCRETE	104.5	48.5	81.1	97.4	57.9	80.9	96.5	39.2	72.5	92.6	53.5	76.3	88.8	70.2	81.0	80.4	55.8	70.1
03	CONCRETE	89.8	48.0	68.4	88.7	60.4	74.2	90.2	47.6	68.4	88.4	57.8	72.7	84.4	69.9	77.0	79.4	49.6	64.1
04	MASONRY	86.2	51.0	64.1	101.7	62.1	76.9	98.6	50.9	68.8	95.0	62.0	74.3	99.8	65.8	78.5	95.3	52.6	68.5
05	METALS	95.6	81.2	90.1	95.5	85.1	91.6	97.0	64.8	84.8	97.0	70.8	87.1	103.6	91.8	99.2	97.0	66.1	85.3
06	WOOD & PLASTICS	113.9	40.9	74.6	102.7	60.4	79.9	102.3	54.2	76.3	109.9	60.4	83.2	98.0	69.1	82.4	90.3	40.8	63.6
07	THERMAL & MOISTURE PROTECTION	99.4	49.5	75.1	94.8	61.3	78.5	98.9	53.9	76.9	99.8	56.8	78.8	94.5	69.0	82.1	95.3	53.3	74.8
08	DOORS & WINDOWS	101.0	42.8	86.2	105.7	57.3	93.4	92.8	49.1	81.7	87.3	57.3	79.7	107.1	67.7	97.1	94.5	43.5	81.6
09200	PLASTER & GYPSUM BOARD	91.1	39.8	56.1	90.9	60.0	69.9	86.1	53.6	64.0	86.1	60.0	68.4	86.2	68.8	74.3	87.3	39.8	54.9
095,098	CEILINGS & ACOUSTICAL TREATMENT	98.6	39.8	58.8	96.5	60.0	71.8	98.1	53.6	68.0	98.1	60.0	72.3	99.3	68.8	78.6	93.4	39.8	57.1
09600	FLOORING	114.6	52.6	98.6	106.5	64.5	95.7	116.5	69.1	104.3	150.9	51.6	125.3	98.7	70.9	91.5	97.9	52.6	86.2
097,099	WALL FINISHES, PAINTS & COATINGS	105.8	44.5	69.0	104.9	57.8	76.7	95.4	38.8	61.5	96.6	57.7	73.3	100.0	63.9	78.4	92.3	49.4	66.6
09	FINISHES	101.0	43.0	69.7	101.5	60.1	79.2	100.0	53.1	74.7	111.8	57.5	82.4	98.2	68.4	82.1	92.1	43.5	65.9
10 - 14	TOTAL DIV. 10000 - 14000	100.0	74.4	94.5	100.0	77.6	95.2	100.0	68.2	93.1	100.0	77.6	95.2	100.0	82.8	96.3	100.0	66.4	92.8
15	MECHANICAL	99.9	49.5	76.7	99.8	67.5	85.0	100.1	36.9	71.0	100.1	62.8	82.9	100.0	73.3	87.7	99.8	44.5	74.4
16	ELECTRICAL	93.0	55.7	71.2	98.7	66.5	79.9	96.5	59.1	74.7	96.4	64.8	77.9	96.0	70.0	80.8	95.2	65.1	77.6
01 - 16	WEIGHTED AVERAGE	97.7	55.5	77.4	99.5	67.9	84.3	97.3	54.0	76.4	97.5	64.5	81.6	99.4	73.7	87.0	94.4	55.2	75.5

DIVISION		TEXAS															UTAH		
		LUBBOCK			ODESSA			SAN ANTONIO			WACO			WICHITA FALLS			LOGAN		
		MAT.	INST.	TOTAL	MAT.	INST.	TOTAL	MAT.	INST.	TOTAL	MAT.	INST.	TOTAL	MAT.	INST.	TOTAL	MAT.	INST.	TOTAL
01590	EQUIPMENT RENTAL	.0	96.8	96.8	.0	86.6	86.6	.0	89.0	89.0	.0	86.6	86.6	.0	86.6	86.6	.0	100.4	100.4
02	SITE CONSTRUCTION	133.0	83.8	96.1	102.0	86.0	90.0	88.3	88.8	88.7	100.9	85.9	89.6	101.6	85.6	89.6	90.9	101.1	98.5
03100	CONCRETE FORMS & ACCESSORIES	98.7	44.7	51.3	97.9	42.6	49.4	91.9	57.0	61.2	99.8	43.3	50.2	99.8	45.8	52.4	101.9	63.7	68.4
03200	CONCRETE REINFORCEMENT	97.0	54.8	71.5	99.7	54.7	72.5	96.0	53.5	70.3	97.6	55.1	71.9	97.6	53.4	70.9	98.6	77.5	85.9
03300	CAST-IN-PLACE CONCRETE	94.4	51.3	76.4	94.2	47.0	74.5	78.9	61.1	71.5	85.0	55.2	72.5	90.8	50.8	74.1	91.1	74.6	84.2
03	CONCRETE	87.8	51.0	68.9	89.3	47.5	67.9	78.6	58.5	68.3	84.5	50.7	67.2	87.4	50.1	68.3	109.1	70.4	89.3
04	MASONRY	98.9	52.8	70.1	99.6	53.2	70.5	95.2	59.3	72.7	96.4	61.2	74.4	96.8	59.8	73.6	121.1	64.2	85.5
05	METALS	101.3	84.0	94.7	97.4	68.1	86.3	97.6	68.9	86.7	97.1	68.7	86.4	97.1	69.5	86.7	105.0	76.9	94.3
06	WOOD & PLASTICS	102.7	45.2	71.7	101.1	42.8	69.7	90.3	57.0	72.4	108.8	39.8	71.6	108.8	45.0	74.4	88.4	62.1	74.2
07	THERMAL & MOISTURE PROTECTION	90.9	54.1	73.0	99.4	49.3	74.9	95.3	65.1	80.6	100.1	52.9	77.1	100.1	56.8	79.0	102.5	70.8	87.0
08	DOORS & WINDOWS	104.2	46.1	89.5	92.8	44.4	80.5	96.5	56.6	86.4	87.3	41.6	75.6	87.3	47.9	77.3	89.3	62.7	82.6
09200	PLASTER & GYPSUM BOARD	86.4	44.2	57.7	86.1	41.9	56.1	87.3	56.4	66.3	86.1	38.8	53.9	86.1	44.2	57.6	88.2	60.8	69.6
095,098	CEILINGS & ACOUSTICAL TREATMENT	99.4	44.2	62.0	98.1	41.9	60.1	93.4	56.4	68.3	98.1	38.8	58.0	98.1	44.2	61.6	112.4	60.8	77.5
09600	FLOORING	107.8	43.7	91.3	116.5	43.2	97.6	97.9	56.6	87.3	151.2	40.1	122.5	151.7	78.4	132.8	101.7	64.8	92.2
097,099	WALL FINISHES, PAINTS & COATINGS	107.0	36.3	64.6	95.4	36.3	60.0	92.3	49.4	66.6	96.6	37.4	61.1	98.4	59.3	75.0	103.9	51.2	72.3
09	FINISHES	101.7	42.8	69.9	100.0	41.4	68.3	92.2	55.6	72.4	111.8	40.8	73.4	112.1	52.8	80.1	101.1	61.8	79.8
10 - 14	TOTAL DIV. 10000 - 14000	100.0	73.1	94.2	100.0	67.3	93.0	100.0	71.4	93.8	100.0	74.9	94.6	100.0	68.3	93.2	100.0	80.4	95.8
15	MECHANICAL	99.7	51.8	77.7	100.1	40.5	72.7	99.8	69.9	86.0	100.1	58.3	80.9	100.1	52.8	78.3	100.1	73.7	88.0
16	ELECTRICAL	96.1	48.9	68.5	97.6	45.5	67.1	95.3	65.1	77.7	97.6	65.7	78.9	101.5	59.7	77.1	93.8	79.4	85.4
01 - 16	WEIGHTED AVERAGE	99.6	56.5	78.8	97.3	51.5	75.2	94.6	65.7	80.7	97.2	59.3	78.9	97.9	58.9	79.1	101.0	74.0	88.0

DIVISION		UTAH									VERMONT						VIRGINIA		
		OGDEN			PROVO			SALT LAKE CITY			BURLINGTON			RUTLAND			ALEXANDRIA		
		MAT.	INST.	TOTAL	MAT.	INST.	TOTAL	MAT.	INST.	TOTAL	MAT.	INST.	TOTAL	MAT.	INST.	TOTAL	MAT.	INST.	TOTAL
01590	EQUIPMENT RENTAL	.0	100.4	100.4	.0	99.2	99.2	.0	100.4	100.4	.0	101.8	101.8	.0	101.8	101.8	.0	101.0	101.0
02	SITE CONSTRUCTION	79.0	101.1	95.6	87.3	99.1	96.2	78.7	101.0	95.4	78.9	99.5	94.3	78.9	99.5	94.3	113.7	85.8	92.9
03100	CONCRETE FORMS & ACCESSORIES	101.9	63.7	68.4	103.4	63.8	68.6	100.0	63.8	68.2	93.8	53.9	58.8	104.8	54.0	60.2	92.8	74.5	76.8
03200	CONCRETE REINFORCEMENT	102.8	77.5	87.5	106.6	77.5	89.0	100.3	77.5	86.6	109.1	57.3	77.8	109.1	57.4	77.9	84.1	83.6	83.8
03300	CAST-IN-PLACE CONCRETE	92.4	74.6	85.0	91.2	74.6	84.2	100.3	74.6	89.6	99.0	67.0	85.6	94.3	67.0	82.9	97.6	79.7	90.1
03	CONCRETE	98.8	70.4	84.3	108.5	70.5	89.0	118.2	70.5	93.7	105.5	59.6	82.0	103.9	59.7	81.2	102.0	79.2	90.3
04	MASONRY	114.3	64.2	82.9	127.0	64.2	87.7	132.4	64.2	89.7	97.1	64.0	76.4	85.0	64.0	71.8	85.7	74.1	78.4
05	METALS	105.4	76.9	94.6	102.3	77.0	92.7	104.2	77.0	93.9	98.0	68.9	87.0	97.3	69.0	86.6	94.9	97.1	95.7
06	WOOD & PLASTICS	88.4	62.1	74.2	89.9	62.1	74.9	88.5	62.1	74.3	89.9	52.2	69.6	102.4	52.2	75.3	97.6	75.1	85.4
07	THERMAL & MOISTURE PROTECTION	100.9	70.8	86.2	104.9	70.8	88.3	103.8	70.8	87.7	102.0	58.9	81.0	101.7	60.1	81.4	95.0	79.6	87.5
08	DOORS & WINDOWS	89.3	62.7	82.6	93.7	62.7	85.8	89.3	62.7	82.6	107.2	48.5	92.3	107.2	48.5	92.3	96.0	77.6	91.3
09200	PLASTER & GYPSUM BOARD	88.2	60.8	69.6	86.8	60.8	69.1	88.2	60.8	69.6	109.1	49.8	68.7	109.1	49.8	68.7	111.2	74.2	86.0
095,098	CEILINGS & ACOUSTICAL TREATMENT	112.4	60.8	77.5	104.6	60.8	75.0	112.4	60.8	77.5	105.9	49.8	67.9	105.9	49.8	67.9	94.7	74.2	80.8
09600	FLOORING	101.7	64.8	92.2	102.5	64.8	92.8	101.2	64.8	91.8	100.3	74.9	93.7	100.3	74.9	93.7	104.7	86.1	99.9
097,099	WALL FINISHES, PAINTS & COATINGS	103.9	51.2	72.3	103.9	65.3	80.8	103.9	65.3	80.8	93.0	42.4	62.6	93.0	42.4	62.6	119.8	87.4	100.3
09	FINISHES	99.9	61.8	79.3	100.1	63.4	80.3	100.3	63.4	80.4	102.0	55.9	77.1	101.9	55.9	77.0	100.4	77.9	88.3
10 - 14	TOTAL DIV. 10000 - 14000	100.0	80.4	95.8	100.0	80.4	95.8	100.0	80.4	95.8	100.0	90.1	97.9	100.0	90.1	97.9	100.0	82.8	96.3
15	MECHANICAL	100.1	73.7	88.0	100.1	73.7	88.0	100.2	73.7	88.0	100.2	70.9	86.7	100.2	70.9	86.7	100.0	85.5	93.3
16	ELECTRICAL	93.8	79.4	85.4	94.5	79.4	85.6	97.2	79.4	86.8	103.7	70.1	84.1	102.7	70.1	83.6	97.0	92.6	94.4
01 - 16	WEIGHTED AVERAGE	98.9	74.0	86.9	101.3	74.0	88.1	102.5	74.2	88.9	101.0	68.2	85.1	100.0	68.2	84.7	98.2	83.9	91.3

DIVISION		VIRGINIA																	
		ARLINGTON			NEWPORT NEWS			NORFOLK			PORTSMOUTH			RICHMOND			ROANOKE		
		MAT.	INST.	TOTAL	MAT.	INST.	TOTAL	MAT.	INST.	TOTAL	MAT.	INST.	TOTAL	MAT.	INST.	TOTAL	MAT.	INST.	TOTAL
01590	EQUIPMENT RENTAL	.0	99.5	99.5	.0	104.6	104.6	.0	105.3	105.3	.0	104.5	104.5	.0	104.5	104.5	.0	99.5	99.5
02	SITE CONSTRUCTION	124.8	82.9	93.4	106.5	84.7	90.2	105.7	85.8	90.8	105.0	84.3	89.5	107.1	85.1	90.6	104.1	81.2	86.9
03100	CONCRETE FORMS & ACCESSORIES	91.6	71.9	74.3	98.0	49.3	55.2	102.5	49.5	56.0	85.3	49.3	53.7	98.8	59.3	64.2	97.8	38.3	45.6
03200	CONCRETE REINFORCEMENT	94.9	66.6	77.8	94.7	71.6	80.7	94.7	71.6	80.7	94.3	71.6	80.6	94.7	70.4	80.0	94.7	66.2	77.5
03300	CAST-IN-PLACE CONCRETE	94.9	76.3	87.1	95.5	57.6	79.7	98.3	57.6	81.3	94.6	57.3	79.0	101.6	57.4	83.1	107.3	49.0	82.9
03	CONCRETE	106.8	73.6	89.8	99.0	58.1	78.0	100.7	58.2	78.9	97.6	58.0	77.3	102.1	62.3	81.7	104.9	49.2	76.3
04	MASONRY	98.6	68.7	79.9	89.1	54.6	67.5	95.8	54.6	70.0	94.2	54.5	69.3	88.0	52.4	65.7	89.5	41.3	59.3
05	METALS	93.8	89.0	92.0	94.9	90.1	93.1	94.1	90.3	92.7	94.1	89.5	92.4	94.9	90.1	93.1	94.7	83.5	90.5
06	WOOD & PLASTICS	94.2	75.1	83.9	98.6	49.4	72.0	103.9	49.4	74.5	85.8	49.4	66.2	99.7	61.9	79.3	98.6	37.8	65.8
07	THERMAL & MOISTURE PROTECTION	97.0	74.2	85.9	96.0	52.0	74.5	95.7	52.0	74.4	96.0	52.0	74.5	95.5	56.1	76.3	95.9	47.5	72.3
08	DOORS & WINDOWS	93.9	72.8	88.6	96.0	52.4	84.9	96.0	55.2	85.6	96.0	55.2	85.7	96.0	57.9	86.3	96.0	45.2	83.1
09200	PLASTER & GYPSUM BOARD	107.7	74.2	84.9	111.2	46.9	67.5	111.2	46.9	67.5	103.0	46.9	64.8	111.2	59.8	76.2	111.2	35.9	60.0
095,098	CEILINGS & ACOUSTICAL TREATMENT	90.8	74.2	79.6	94.7	46.9	62.3	94.7	46.9	62.3	94.7	46.9	62.3	94.7	59.8	71.1	94.7	35.9	54.9
09600	FLOORING	102.9	73.4	95.3	104.7	42.8	88.8	104.5	42.8	88.6	95.9	42.8	82.2	104.5	68.0	95.1	104.7	38.4	87.6
097,099	WALL FINISHES, PAINTS & COATINGS	119.8	89.1	101.4	108.7	44.9	70.4	108.7	58.0	78.3	108.7	39.3	67.1	108.7	58.7	78.7	108.7	35.0	64.5
09	FINISHES	99.7	74.3	86.0	99.7	46.9	71.2	99.7	48.4	72.0	95.7	46.3	69.0	99.6	61.1	78.8	99.6	37.3	65.9
10 - 14	TOTAL DIV. 10000 - 14000	100.0	76.7	95.0	100.0	72.2	94.0	100.0	72.2	94.0	100.0	72.2	94.0	100.0	73.5	94.3	100.0	66.7	92.8
15	MECHANICAL	100.0	79.4	90.5	100.0	60.5	81.8	100.0	64.5	83.6	100.0	64.4	83.6	100.0	64.4	83.6	100.0	42.2	73.4
16	ELECTRICAL	94.2	89.6	91.5	96.9	58.8	74.6	97.0	61.0	75.9	94.7	61.0	75.0	97.8	72.4	82.9	96.9	36.4	61.5
01 - 16	WEIGHTED AVERAGE	99.2	79.0	89.4	97.8	62.2	80.6	98.3	63.7	81.7	97.1	63.2	80.8	98.2	67.6	83.5	98.4	49.9	75.0

DIVISION		WASHINGTON																	
		EVERETT			RICHLAND			SEATTLE			SPOKANE			TACOMA			VANCOUVER		
		MAT.	INST.	TOTAL	MAT.	INST.	TOTAL	MAT.	INST.	TOTAL	MAT.	INST.	TOTAL	MAT.	INST.	TOTAL	MAT.	INST.	TOTAL
01590	EQUIPMENT RENTAL	.0	104.2	104.2	.0	89.9	89.9	.0	104.0	104.0	.0	89.9	89.9	.0	104.2	104.2	.0	97.4	97.4
02	SITE CONSTRUCTION	95.0	115.4	110.3	104.6	89.4	93.3	99.3	113.2	109.7	105.4	89.5	93.5	97.9	115.4	111.0	111.5	100.9	103.6
03100	CONCRETE FORMS & ACCESSORIES	109.8	100.2	101.4	116.7	82.0	86.3	94.1	100.8	100.0	123.8	82.1	87.2	94.1	100.4	99.6	98.5	98.1	98.1
03200	CONCRETE REINFORCEMENT	105.4	91.4	96.9	98.9	86.9	91.7	99.4	91.5	94.6	99.5	87.0	91.9	99.4	91.5	94.6	102.3	91.3	95.6
03300	CAST-IN-PLACE CONCRETE	98.2	106.6	101.7	115.9	85.3	103.1	103.1	108.2	105.2	120.0	85.4	105.5	101.0	108.0	104.0	112.8	100.5	107.7
03	CONCRETE	98.0	100.2	99.1	107.2	83.9	95.3	100.3	101.1	100.7	109.9	84.0	96.6	99.2	100.8	100.0	110.5	97.3	103.8
04	MASONRY	133.2	102.2	113.8	114.8	85.3	96.4	127.7	102.2	111.7	116.4	86.0	97.4	127.5	102.2	111.7	128.2	100.5	110.8
05	METALS	98.5	87.2	94.2	92.8	82.6	89.0	100.0	89.5	96.0	92.6	82.7	88.9	99.5	87.4	94.9	98.0	89.3	94.7
06	WOOD & PLASTICS	99.4	99.1	99.2	97.5	81.6	88.9	84.4	99.1	92.3	106.2	81.6	92.9	83.2	99.1	91.8	80.9	97.8	90.0
07	THERMAL & MOISTURE PROTECTION	107.4	99.1	103.3	173.4	81.6	128.7	107.2	100.3	103.8	170.6	83.3	128.1	107.0	100.3	103.7	109.8	92.8	101.5
08	DOORS & WINDOWS	98.4	97.1	98.1	112.1	76.6	103.1	100.6	97.1	99.7	111.9	77.0	103.0	99.0	97.1	98.5	95.9	95.3	95.7
09200	PLASTER & GYPSUM BOARD	103.0	98.9	100.2	129.6	80.9	96.4	97.2	98.9	98.4	133.2	80.9	97.6	100.3	98.9	99.4	101.8	98.0	99.3
095,098	CEILINGS & ACOUSTICAL TREATMENT	112.5	98.9	103.3	118.6	80.9	93.0	113.5	98.9	103.6	118.6	80.9	93.0	115.1	98.9	104.2	123.1	98.0	106.1
09600	FLOORING	118.8	105.7	115.5	110.3	47.0	94.0	110.8	105.7	109.5	114.0	76.7	104.4	111.4	105.7	109.9	117.1	86.5	109.2
097,099	WALL FINISHES, PAINTS & COATINGS	111.8	83.9	95.1	113.2	76.4	91.1	111.8	95.0	101.7	113.2	76.4	91.1	111.8	95.0	101.7	119.3	70.6	90.1
09	FINISHES	114.2	99.4	106.2	127.7	74.2	98.8	111.2	100.6	105.5	129.4	80.2	102.8	112.0	100.6	105.9	113.1	92.8	102.1
10 - 14	TOTAL DIV. 10000 - 14000	100.0	100.9	100.2	100.0	93.5	98.6	100.0	106.1	101.3	100.0	93.5	98.6	100.0	106.1	101.3	100.0	90.2	97.9
15	MECHANICAL	99.9	99.0	99.5	100.5	97.1	98.9	99.9	113.0	105.9	100.5	85.5	93.6	100.0	97.1	98.6	100.1	103.2	101.5
16	ELECTRICAL	103.2	99.1	100.8	102.5	90.1	95.3	102.9	112.8	108.7	94.8	85.8	89.5	102.9	100.9	101.7	112.7	105.9	108.7
01 - 16	WEIGHTED AVERAGE	103.0	100.0	101.5	107.4	86.6	97.4	103.0	105.2	104.1	107.2	84.6	96.3	102.7	100.3	101.5	105.0	98.7	102.0

DIVISION		WASHINGTON			WEST VIRGINIA												WISCONSIN		
		YAKIMA			CHARLESTON			HUNTINGTON			PARKERSBURG			WHEELING			EAU CLAIRE		
		MAT.	INST.	TOTAL	MAT.	INST.	TOTAL	MAT.	INST.	TOTAL	MAT.	INST.	TOTAL	MAT.	INST.	TOTAL	MAT.	INST.	TOTAL
01590	EQUIPMENT RENTAL	.0	104.2	104.2	.0	99.5	99.5	.0	99.5	99.5	.0	99.5	99.5	.0	99.5	99.5	.0	100.7	100.7
02	SITE CONSTRUCTION	101.4	114.2	111.0	101.8	85.9	89.9	103.2	86.7	90.8	108.2	85.9	91.5	108.8	85.6	91.4	83.5	102.6	97.8
03100	CONCRETE FORMS & ACCESSORIES	96.4	94.8	95.0	105.8	91.3	93.1	99.0	91.1	92.1	88.0	90.8	90.5	90.0	87.7	88.0	99.5	98.3	98.5
03200	CONCRETE REINFORCEMENT	101.9	90.3	94.9	94.7	86.3	89.6	94.7	89.1	91.3	93.3	92.1	92.6	92.7	92.4	92.5	101.8	102.6	102.3
03300	CAST-IN-PLACE CONCRETE	107.9	84.2	98.0	93.5	108.2	99.6	100.8	108.9	104.2	94.2	97.3	95.5	94.2	98.4	96.0	98.6	97.7	98.2
03	CONCRETE	104.7	89.8	97.0	98.4	96.9	97.6	101.7	97.6	99.6	102.1	94.0	98.0	102.1	93.1	97.5	97.1	99.2	98.2
04	MASONRY	119.3	75.5	91.9	86.7	88.9	88.1	89.1	96.3	93.6	76.6	82.4	80.2	99.8	87.9	92.3	92.7	99.2	96.8
05	METALS	98.4	82.5	92.4	94.9	99.9	96.8	95.0	101.4	97.4	93.8	101.3	96.6	93.9	102.3	97.1	92.8	104.3	97.2
06	WOOD & PLASTICS	85.6	99.1	92.8	106.9	90.3	97.9	98.6	89.1	93.5	88.2	90.2	89.3	90.0	87.0	88.4	114.9	97.8	105.6
07	THERMAL & MOISTURE PROTECTION	107.2	79.9	93.9	95.9	90.0	93.0	96.2	92.5	94.4	96.4	85.1	90.9	96.5	85.7	91.2	98.5	91.2	94.9
08	DOORS & WINDOWS	98.6	83.9	94.8	97.2	83.8	93.8	96.0	83.7	92.8	96.6	83.1	93.2	97.5	86.4	94.7	100.9	96.8	99.9
09200	PLASTER & GYPSUM BOARD	100.4	98.9	99.4	111.2	89.9	96.7	111.2	88.6	95.9	104.6	89.8	94.5	105.2	86.4	92.4	114.4	98.0	103.3
095,098	CEILINGS & ACOUSTICAL TREATMENT	111.2	98.9	102.9	94.7	89.9	91.4	94.7	88.6	90.6	90.8	89.8	90.1	90.8	86.4	87.9	106.4	98.0	100.7
09600	FLOORING	112.7	70.7	101.9	104.5	101.2	103.7	104.5	113.4	106.8	99.1	87.8	96.2	100.2	106.0	101.7	91.4	109.5	96.0
097,099	WALL FINISHES, PAINTS & COATINGS	111.8	76.4	90.6	108.7	99.6	103.3	108.7	93.0	99.3	108.7	97.7	102.1	108.7	88.6	96.6	89.7	72.0	79.1
09	FINISHES	111.9	88.8	99.5	99.7	94.1	96.7	99.5	95.4	97.3	96.6	91.0	93.6	97.1	91.0	93.8	99.9	97.7	98.7
10 - 14	TOTAL DIV. 10000 - 14000	100.0	101.9	100.4	100.0	96.8	99.3	100.0	97.1	99.4	100.0	96.7	99.3	100.0	102.6	100.6	100.0	98.7	99.7
15	MECHANICAL	100.0	96.7	98.5	100.0	85.2	93.2	100.0	84.3	92.8	100.0	90.4	95.6	100.0	89.9	95.3	100.3	87.4	94.4
16	ELECTRICAL	106.8	90.1	97.0	97.0	86.6	90.9	96.9	85.1	90.0	97.5	93.6	95.2	93.8	91.0	92.2	103.5	88.5	94.7
01 - 16	WEIGHTED AVERAGE	103.2	91.2	97.4	97.7	90.3	94.2	98.0	91.2	94.8	97.1	90.9	94.1	98.3	91.2	94.9	98.3	95.6	97.0

DIVISION		WISCONSIN																	
		GREEN BAY			KENOSHA			LA CROSSE			MADISON			MILWAUKEE			RACINE		
		MAT.	INST.	TOTAL	MAT.	INST.	TOTAL	MAT.	INST.	TOTAL	MAT.	INST.	TOTAL	MAT.	INST.	TOTAL	MAT.	INST.	TOTAL
01590	EQUIPMENT RENTAL	.0	98.4	98.4	.0	97.9	97.9	.0	100.7	100.7	.0	100.2	100.2	.0	86.4	86.4	.0	100.2	100.2
02	SITE CONSTRUCTION	85.9	98.5	95.3	90.9	101.6	98.9	77.7	102.6	96.3	85.6	105.0	100.1	86.9	93.3	91.7	85.9	105.5	100.6
03100	CONCRETE FORMS & ACCESSORIES	116.8	97.6	99.9	113.1	105.0	106.0	83.1	97.9	96.1	102.7	98.5	99.0	105.3	109.3	108.8	102.2	104.9	104.6
03200	CONCRETE REINFORCEMENT	98.3	91.6	94.3	104.8	96.6	99.9	100.4	90.4	94.3	98.3	90.6	93.7	96.6	96.8	96.7	98.3	96.6	97.3
03300	CAST-IN-PLACE CONCRETE	102.0	99.8	101.1	114.1	101.1	108.7	88.7	97.1	92.2	102.9	100.3	101.8	102.9	107.4	104.7	102.9	100.7	102.0
03	CONCRETE	98.5	97.5	98.0	105.5	102.0	103.7	88.0	96.5	92.4	98.5	97.8	98.1	98.7	105.5	102.2	98.4	101.9	100.2
04	MASONRY	124.3	98.2	108.0	99.0	109.7	105.7	91.9	99.2	96.5	103.3	102.2	102.6	102.8	115.1	110.5	102.8	109.6	107.1
05	METALS	95.3	99.4	96.8	95.7	100.4	97.4	92.7	98.7	95.0	96.0	97.0	96.4	97.6	91.0	95.1	96.1	100.3	97.7
06	WOOD & PLASTICS	125.2	97.8	110.4	120.5	104.0	111.6	98.1	97.8	97.9	115.6	97.7	105.9	118.3	108.2	112.8	115.8	104.0	109.4
07	THERMAL & MOISTURE PROTECTION	100.8	89.3	95.2	97.2	104.7	100.9	97.7	90.3	94.1	96.8	98.4	97.6	96.1	107.1	101.4	97.5	102.5	99.9
08	DOORS & WINDOWS	99.2	90.1	96.9	96.8	103.1	98.4	100.9	86.7	97.3	101.7	97.0	100.5	104.0	105.3	104.3	101.7	103.1	102.0
09200	PLASTER & GYPSUM BOARD	98.6	98.0	98.2	99.5	104.6	103.0	106.7	98.0	100.8	112.2	98.0	102.6	111.3	108.6	109.5	111.3	104.6	106.7
095,098	CEILINGS & ACOUSTICAL TREATMENT	101.5	98.0	99.2	88.9	104.6	99.5	99.9	98.0	98.6	101.4	98.0	99.1	97.5	108.6	105.0	97.5	104.6	102.3
09600	FLOORING	109.3	109.5	109.4	110.5	115.0	111.6	83.9	109.5	90.5	94.0	110.3	98.2	96.7	114.7	101.4	94.0	115.0	99.4
097,099	WALL FINISHES, PAINTS & COATINGS	99.5	78.2	86.7	103.9	94.8	98.4	89.7	72.0	79.1	93.5	94.6	94.2	96.0	105.9	101.9	93.5	90.7	91.8
09	FINISHES	104.1	98.1	100.9	101.4	106.1	104.0	94.9	97.7	96.4	99.1	100.4	99.8	99.4	110.4	105.4	98.3	105.7	102.3
10 - 14	TOTAL DIV. 10000 - 14000	100.0	95.0	98.9	100.0	97.5	99.5	100.0	98.7	99.7	100.0	98.4	99.7	100.0	99.7	99.9	100.0	97.5	99.5
15	MECHANICAL	100.6	86.1	93.9	100.2	97.5	99.0	100.3	89.3	95.2	99.9	92.9	96.7	99.9	99.0	99.5	99.9	97.5	98.8
16	ELECTRICAL	97.0	88.1	91.8	95.3	89.3	91.8	103.8	88.5	94.9	96.5	88.6	91.9	94.4	99.1	97.2	92.7	97.5	95.5
01 - 16	WEIGHTED AVERAGE	100.6	93.7	97.2	99.2	100.4	99.8	96.4	94.6	95.5	98.8	96.9	97.9	99.2	102.3	100.7	98.4	101.8	100.1

DIVISION		WYOMING									CANADA								
		CASPER			CHEYENNE			ROCK SPRINGS			CALGARY, ALBERTA			EDMONTON, ALBERTA			HALIFAX, NOVA SCOTIA		
		MAT.	INST.	TOTAL	MAT.	INST.	TOTAL	MAT.	INST.	TOTAL	MAT.	INST.	TOTAL	MAT.	INST.	TOTAL	MAT.	INST.	TOTAL
01590	EQUIPMENT RENTAL	.0	101.4	101.4	.0	101.4	101.4	.0	101.4	101.4	.0	107.0	107.0	.0	107.0	107.0	.0	98.6	98.6
02	SITE CONSTRUCTION	79.9	100.6	95.3	79.7	100.5	95.3	83.0	100.1	95.8	113.1	105.4	107.3	116.9	105.4	108.3	96.1	96.0	96.1
03100	CONCRETE FORMS & ACCESSORIES	98.9	45.2	51.8	99.0	45.0	51.6	104.1	31.7	40.6	122.9	79.3	84.7	120.7	79.3	84.4	90.7	68.8	71.5
03200	CONCRETE REINFORCEMENT	105.1	53.3	73.8	98.6	53.4	71.3	107.1	53.0	74.5	161.6	68.4	105.3	161.6	68.4	105.3	148.5	50.9	89.5
03300	CAST-IN-PLACE CONCRETE	102.5	59.4	84.5	102.5	59.3	84.5	103.0	44.6	78.6	191.5	89.7	149.0	206.9	89.7	157.9	178.4	70.3	133.2
03	CONCRETE	103.9	52.4	77.5	103.2	52.3	77.1	104.7	41.3	72.3	160.8	81.5	120.1	168.3	81.5	123.8	152.1	66.7	108.4
04	MASONRY	109.4	40.5	66.2	106.6	42.7	66.6	174.5	33.0	86.0	180.3	76.4	115.3	181.4	76.4	115.7	160.2	71.8	104.9
05	METALS	104.7	64.2	89.4	106.0	64.2	90.2	104.7	62.7	88.8	124.2	87.3	110.2	125.9	87.3	111.3	107.9	73.9	95.0
06	WOOD & PLASTICS	97.4	43.4	68.3	96.7	43.4	68.0	103.4	30.8	64.2	113.3	79.1	94.9	110.1	79.1	93.4	85.1	68.2	76.0
07	THERMAL & MOISTURE PROTECTION	102.7	56.0	79.9	102.7	56.5	80.2	103.4	42.5	73.7	126.1	82.5	104.9	126.1	82.5	104.9	105.3	70.0	88.1
08	DOORS & WINDOWS	93.5	45.9	81.4	94.3	45.9	82.0	100.6	39.0	85.0	91.7	75.7	87.7	91.7	75.7	87.7	82.4	63.2	77.6
09200	PLASTER & GYPSUM BOARD	86.3	41.7	55.9	88.2	41.7	56.5	91.5	28.6	48.7	155.9	77.9	102.8	157.5	77.9	103.4	185.1	67.4	105.0
095,098	CEILINGS & ACOUSTICAL TREATMENT	104.6	41.7	62.0	112.4	41.7	64.5	104.6	28.6	53.2	121.4	77.9	92.0	122.7	77.9	92.4	99.3	67.4	77.7
09600	FLOORING	101.7	44.8	87.0	101.7	72.0	94.0	104.0	43.9	88.5	132.7	89.6	121.6	132.7	89.6	121.6	110.0	65.1	98.4
097,099	WALL FINISHES, PAINTS & COATINGS	103.9	58.7	76.8	103.8	58.7	76.8	103.9	35.7	63.0	109.3	78.9	91.1	109.5	78.9	91.1	109.4	62.3	81.1
09	FINISHES	98.5	45.5	69.9	100.1	51.0	73.6	99.8	33.5	64.0	124.5	81.0	101.0	125.1	81.0	101.3	115.4	67.5	89.5
10 - 14	TOTAL DIV. 10000 - 14000	100.0	80.3	95.8	100.0	80.3	95.8	100.0	53.6	90.0	140.0	90.9	129.4	140.0	90.9	129.4	140.0	66.1	124.1
15	MECHANICAL	100.1	70.8	86.6	100.1	46.3	75.4	100.1	32.7	69.1	101.5	80.4	91.8	101.6	80.4	91.8	101.3	71.4	87.6
16	ELECTRICAL	93.9	65.3	77.2	93.9	54.6	70.9	93.0	47.1	66.2	117.6	89.7	101.3	113.5	89.7	99.6	127.3	72.7	95.4
01 - 16	WEIGHTED AVERAGE	99.7	61.9	81.5	99.9	56.4	78.9	104.5	46.2	76.4	123.4	84.6	104.7	124.4	84.6	105.2	116.5	72.5	95.3

DIVISION		CANADA																	
		HAMILTON, ONTARIO			KITCHENER, ONTARIO			LAVAL, QUEBEC			LONDON, ONTARIO			MONTREAL, QUEBEC			OSHAWA, ONTARIO		
		MAT.	INST.	TOTAL	MAT.	INST.	TOTAL	MAT.	INST.	TOTAL	MAT.	INST.	TOTAL	MAT.	INST.	TOTAL	MAT.	INST.	TOTAL
01590	EQUIPMENT RENTAL	.0	103.1	103.1	.0	103.0	103.0	.0	99.2	99.2	.0	103.2	103.2	.0	101.0	101.0	.0	103.0	103.0
02	SITE CONSTRUCTION	109.3	103.8	105.2	98.3	103.2	102.0	91.1	97.1	95.5	109.3	103.5	105.0	90.8	97.4	95.7	110.9	103.4	105.3
03100	CONCRETE FORMS & ACCESSORIES	122.3	90.1	94.1	116.4	82.6	86.7	131.0	84.4	90.1	128.4	85.5	90.8	131.2	84.7	90.4	123.6	86.4	91.0
03200	CONCRETE REINFORCEMENT	174.8	91.3	124.4	110.4	91.1	98.8	157.2	85.8	114.0	123.8	89.7	103.2	159.8	85.8	115.1	174.8	91.9	124.7
03300	CAST-IN-PLACE CONCRETE	157.2	97.7	132.3	146.7	78.7	118.3	140.1	92.2	120.1	157.2	95.3	131.3	137.5	93.8	119.2	169.7	84.3	134.0
03	CONCRETE	145.1	93.0	118.4	125.0	83.1	103.6	135.0	87.5	110.7	139.5	89.8	114.1	134.0	88.2	110.6	151.4	86.8	118.3
04	MASONRY	167.1	94.0	121.3	162.9	89.8	117.2	160.4	84.0	112.6	167.1	91.2	119.6	163.7	84.1	113.9	165.6	93.3	120.3
05	METALS	115.6	91.0	106.3	114.5	90.4	105.4	106.9	87.1	99.4	113.5	90.5	104.8	116.5	87.6	105.6	106.9	90.7	100.8
06	WOOD & PLASTICS	114.1	89.1	100.6	108.5	81.0	93.7	126.5	84.5	103.9	114.1	83.9	97.8	126.5	84.8	104.0	115.1	84.8	98.8
07	THERMAL & MOISTURE PROTECTION	110.6	91.2	101.2	108.7	86.6	98.0	105.2	88.6	97.1	113.4	89.4	101.7	106.1	89.2	97.9	110.1	87.2	98.9
08	DOORS & WINDOWS	91.7	88.7	91.0	83.1	82.7	83.0	91.7	74.8	87.4	92.6	84.9	90.7	91.7	75.0	87.5	90.8	86.0	89.6
09200	PLASTER & GYPSUM BOARD	210.5	88.8	127.7	162.7	80.6	106.8	158.1	84.1	107.7	210.2	83.5	124.0	163.6	84.1	109.5	166.3	84.4	110.6
095,098	CEILINGS & ACOUSTICAL TREATMENT	124.0	88.8	100.2	99.3	80.6	86.6	99.3	84.1	89.0	122.7	83.5	96.2	122.7	84.1	96.5	99.3	84.4	89.2
09600	FLOORING	132.7	95.6	123.2	128.4	95.6	119.9	132.7	96.3	123.4	134.9	95.6	124.7	132.7	96.3	123.4	132.7	98.1	123.8
097,099	WALL FINISHES, PAINTS & COATINGS	109.4	102.4	105.2	109.4	93.8	100.0	109.4	89.1	97.2	109.4	102.4	105.2	109.4	89.1	97.2	109.4	107.6	108.3
09	FINISHES	131.5	92.6	110.5	118.6	86.2	101.1	119.5	87.5	102.2	132.0	89.2	108.8	124.7	87.7	104.7	121.2	91.0	104.9
10 - 14	TOTAL DIV. 10000 - 14000	140.0	97.1	130.8	140.0	94.8	130.3	140.0	84.1	128.0	140.0	95.8	130.5	140.0	84.9	128.1	140.0	96.3	130.6
15	MECHANICAL	101.7	88.6	95.7	101.3	84.8	93.7	101.3	77.2	90.2	101.6	85.0	94.0	101.8	77.3	90.5	101.3	86.5	94.5
16	ELECTRICAL	128.8	97.6	110.6	125.5	94.2	107.2	125.3	76.2	96.6	128.7	94.4	108.7	123.4	76.2	95.8	125.7	95.0	107.7
01 - 16	WEIGHTED AVERAGE	120.8	93.4	107.6	114.9	89.0	102.4	115.9	83.7	100.4	120.1	90.8	105.9	117.7	84.0	101.5	118.8	91.2	105.5

DIVISION		CANADA																	
		OTTAWA, ONTARIO			QUEBEC, QUEBEC			REGINA, SASKATCHEWAN			SASKATOON, SASKATCHEWAN			ST CATHARINES, ONTARIO			ST JOHNS, NEWFOUNDLAND		
		MAT.	INST.	TOTAL	MAT.	INST.	TOTAL	MAT.	INST.	TOTAL	MAT.	INST.	TOTAL	MAT.	INST.	TOTAL	MAT.	INST.	TOTAL
01590	**EQUIPMENT RENTAL**	.0	103.0	103.0	.0	101.4	101.4	.0	98.6	98.6	.0	98.6	98.6	.0	100.4	100.4	.0	98.6	98.6
02	**SITE CONSTRUCTION**	109.6	103.4	104.9	92.5	97.4	96.2	109.4	95.0	98.6	104.5	95.1	97.5	98.8	99.1	99.0	111.9	94.7	99.1
03100	CONCRETE FORMS & ACCESSORIES	121.1	86.4	90.7	131.1	85.0	90.6	102.8	58.1	63.6	102.8	58.0	63.5	113.9	84.4	88.0	100.6	59.1	64.2
03200	CONCRETE REINFORCEMENT	174.8	89.6	123.3	148.5	85.9	110.7	114.1	67.4	85.9	119.2	67.4	87.9	111.4	91.2	99.2	148.5	52.8	90.7
03300	CAST-IN-PLACE CONCRETE	159.5	95.2	132.6	151.5	94.3	127.6	162.8	67.6	123.0	148.0	67.6	114.4	140.1	83.3	116.3	179.6	66.6	132.4
03	**CONCRETE**	146.2	90.2	117.5	139.7	88.5	113.5	132.5	63.9	97.3	125.6	63.8	94.0	121.7	85.5	103.1	166.1	61.4	112.4
04	**MASONRY**	167.2	90.8	119.3	164.7	84.1	114.2	166.8	62.6	101.6	167.1	62.6	101.7	162.4	87.2	115.3	160.3	61.9	98.7
05	**METALS**	115.6	90.5	106.1	115.5	87.9	105.1	106.0	74.2	94.0	106.1	74.1	94.0	105.9	90.5	100.1	107.8	71.6	94.1
06	**WOOD & PLASTICS**	113.5	85.6	98.4	127.0	84.9	104.3	94.7	56.8	74.2	93.1	56.8	73.5	106.2	85.4	95.0	94.1	58.2	74.8
07	**THERMAL & MOISTURE PROTECTION**	110.6	87.9	99.5	105.5	89.4	97.7	105.3	63.8	85.1	104.2	62.9	84.1	108.7	87.9	98.6	108.2	61.4	85.4
08	**DOORS & WINDOWS**	91.7	85.7	90.2	91.7	82.4	89.3	86.6	55.6	78.8	85.6	55.6	78.0	82.6	85.4	83.3	98.0	55.7	87.2
09200	PLASTER & GYPSUM BOARD	256.2	85.2	139.9	215.5	84.1	126.1	146.0	55.6	84.5	146.0	55.6	84.5	147.5	85.0	105.0	191.4	57.1	100.0
095,098	CEILINGS & ACOUSTICAL TREATMENT	109.7	85.2	93.1	105.8	84.1	91.1	96.7	55.6	68.8	96.7	55.6	68.8	96.7	85.0	88.8	101.9	57.1	71.5
09600	FLOORING	132.7	94.3	122.8	132.7	96.3	123.4	120.1	61.2	104.9	120.1	61.2	104.9	126.6	95.6	118.6	114.9	55.4	99.6
097,099	WALL FINISHES, PAINTS & COATINGS	109.4	95.8	101.2	109.4	89.1	97.2	109.4	64.9	82.7	109.4	55.3	77.0	109.4	95.8	101.3	109.4	60.0	79.8
09	**FINISHES**	134.6	89.0	110.0	128.1	87.8	106.3	114.5	59.1	84.6	114.3	58.0	83.9	115.6	88.0	100.7	119.4	58.5	86.5
10 - 14	**TOTAL DIV. 10000 - 14000**	140.0	93.7	130.0	140.0	85.1	128.2	140.0	63.8	123.6	140.0	63.8	123.6	140.0	72.2	125.4	140.0	64.5	123.8
15	**MECHANICAL**	101.7	85.3	94.1	101.6	77.4	90.5	101.3	64.2	84.2	101.3	64.2	84.2	101.3	83.1	92.9	101.3	60.7	82.7
16	**ELECTRICAL**	124.4	95.4	107.5	127.2	76.2	97.4	128.4	65.9	91.9	128.6	65.9	92.0	127.5	95.5	108.8	124.2	61.2	87.4
01 - 16	**WEIGHTED AVERAGE**	120.8	90.9	106.4	119.0	84.4	102.3	115.2	66.9	91.9	114.1	66.8	91.3	113.1	88.3	101.2	120.7	64.8	93.7

DIVISION		CANADA																	
		THUNDER BAY, ONTARIO			TORONTO, ONTARIO			VANCOUVER, B C			WINDSOR, ONTARIO			WINNIPEG, MANITOBA					
		MAT.	INST.	TOTAL	MAT.	INST.	TOTAL	MAT.	INST.	TOTAL	MAT.	INST.	TOTAL	MAT.	INST.	TOTAL	MAT.	INST.	TOTAL
01590	**EQUIPMENT RENTAL**	.0	100.4	100.4	.0	103.2	103.2	.0	111.3	111.3	.0	100.4	100.4	.0	105.3	105.3	.0	.0	.0
02	**SITE CONSTRUCTION**	104.6	99.2	100.6	110.5	104.2	105.8	114.9	107.8	109.6	94.7	99.4	98.2	108.9	100.6	102.7	.0	.0	.0
03100	CONCRETE FORMS & ACCESSORIES	123.6	85.5	90.1	123.9	96.9	100.2	115.5	88.1	91.5	123.6	86.1	90.7	123.1	69.9	76.4	.0	.0	.0
03200	CONCRETE REINFORCEMENT	99.5	90.5	94.1	171.2	92.2	123.5	161.6	83.7	114.5	109.1	89.8	97.4	161.6	57.9	99.0	.0	.0	.0
03300	CAST-IN-PLACE CONCRETE	154.2	95.3	129.6	163.2	105.1	139.0	153.2	97.4	129.9	143.5	96.5	123.8	157.5	74.7	122.9	.0	.0	.0
03	**CONCRETE**	131.6	89.9	110.3	147.8	98.7	122.7	145.3	90.9	117.4	123.8	90.5	106.7	143.4	69.9	105.8	.0	.0	.0
04	**MASONRY**	163.3	88.0	116.2	186.2	102.2	133.6	175.1	91.5	122.8	162.7	94.6	120.1	169.0	65.4	104.1	.0	.0	.0
05	**METALS**	105.7	89.6	99.6	117.6	92.8	108.2	122.2	92.7	111.0	105.7	90.8	100.1	124.9	77.3	106.9	.0	.0	.0
06	**WOOD & PLASTICS**	115.1	85.8	99.3	115.1	95.5	104.5	113.3	86.1	98.6	115.1	84.3	98.5	113.2	70.8	90.4	.0	.0	.0
07	**THERMAL & MOISTURE PROTECTION**	109.0	87.1	98.3	110.6	98.2	104.6	113.9	92.5	103.5	108.8	91.0	100.1	105.2	72.5	89.3	.0	.0	.0
08	**DOORS & WINDOWS**	81.6	84.9	82.5	90.8	94.0	91.6	94.2	84.0	91.7	81.4	84.7	82.2	91.7	63.5	84.6	.0	.0	.0
09200	PLASTER & GYPSUM BOARD	175.6	85.4	114.2	166.9	95.4	118.2	152.4	85.2	106.6	168.8	83.9	111.0	157.6	69.4	97.6	.0	.0	.0
095,098	CEILINGS & ACOUSTICAL TREATMENT	95.4	85.4	88.6	101.9	95.4	97.5	101.9	85.2	90.5	95.4	83.9	87.6	101.9	69.4	79.9	.0	.0	.0
09600	FLOORING	132.7	55.1	112.7	132.7	101.4	124.7	132.7	92.8	122.5	132.7	96.4	123.4	132.7	68.0	116.1	.0	.0	.0
097,099	WALL FINISHES, PAINTS & COATINGS	109.4	97.9	102.5	112.4	107.6	109.5	109.3	101.8	104.8	109.4	97.5	102.2	109.5	57.0	78.0	.0	.0	.0
09	**FINISHES**	121.3	82.1	100.1	122.1	99.0	109.7	121.1	90.0	104.3	120.1	89.2	103.4	121.0	68.4	92.6	.0	.0	.0
10 - 14	**TOTAL DIV. 10000 - 14000**	140.0	73.3	125.6	140.0	99.6	131.3	140.0	94.8	130.3	140.0	74.6	125.9	140.0	67.8	124.5	.0	.0	.0
15	**MECHANICAL**	101.3	83.8	93.3	101.6	94.6	98.4	101.5	87.5	95.1	101.3	86.5	94.5	101.6	70.9	87.4	.0	.0	.0
16	**ELECTRICAL**	125.5	93.6	106.9	130.9	98.3	111.8	130.3	86.0	104.4	125.7	95.5	108.0	128.4	73.1	96.0	.0	.0	.0
01 - 16	**WEIGHTED AVERAGE**	114.9	88.1	102.0	121.7	98.0	110.2	121.8	91.0	107.0	113.5	90.8	102.5	120.7	73.2	97.8	.0	.0	.0

Location Factors

Costs shown in *Means cost data publications* are based on National Averages for materials and installation. To adjust these costs to a specific location, simply multiply the base cost by the factor and divide by 100 for that city. The data is arranged alphabetically by state and postal zip code numbers. For a city not listed, use the factor for a nearby city with similar economic characteristics.

STATE/ZIP	CITY	MAT.	INST.	TOTAL
ALABAMA				
350-352	Birmingham	96.6	76.5	86.9
354	Tuscaloosa	96.3	57.1	77.4
355	Jasper	97.4	54.2	76.6
356	Decatur	96.3	59.4	78.5
357-358	Huntsville	96.2	72.6	84.8
359	Gadsden	97.1	60.6	79.5
360-361	Montgomery	97.1	59.0	78.7
362	Anniston	96.1	48.9	73.4
363	Dothan	97.0	52.1	75.3
364	Evergreen	96.3	53.4	75.6
365-366	Mobile	97.2	62.8	80.6
367	Selma	96.6	54.3	76.2
368	Phenix City	97.4	57.8	78.3
369	Butler	96.8	52.1	75.2
ALASKA				
995-996	Anchorage	135.1	116.5	126.1
997	Fairbanks	131.0	119.2	125.3
998	Juneau	134.4	115.9	125.5
999	Ketchikan	145.5	115.9	131.2
ARIZONA				
850,853	Phoenix	100.0	74.8	87.9
852	Mesa/Tempe	99.7	69.2	85.0
855	Globe	100.6	64.5	83.2
856-857	Tucson	99.0	71.2	85.6
859	Show Low	100.7	65.2	83.5
860	Flagstaff	102.0	68.9	86.1
863	Prescott	99.6	66.4	83.6
864	Kingman	98.2	67.6	83.4
865	Chambers	98.2	64.4	81.9
ARKANSAS				
716	Pine Bluff	95.8	63.6	80.3
717	Camden	94.1	41.4	68.7
718	Texarkana	95.2	46.7	71.8
719	Hot Springs	93.2	40.8	68.0
720-722	Little Rock	96.3	65.9	81.6
723	West Memphis	95.6	56.7	76.9
724	Jonesboro	95.6	56.7	76.9
725	Batesville	94.2	51.4	73.6
726	Harrison	95.6	51.4	74.3
727	Fayetteville	92.6	36.1	65.4
728	Russellville	94.1	49.5	72.6
729	Fort Smith	96.4	58.6	78.2
CALIFORNIA				
900-902	Los Angeles	102.9	114.0	108.3
903-905	Inglewood	99.2	108.3	103.6
906-908	Long Beach	100.9	108.2	104.4
910-912	Pasadena	101.2	108.4	104.7
913-916	Van Nuys	105.0	108.5	106.7
917-918	Alhambra	103.8	108.5	106.1
919-921	San Diego	105.1	104.3	104.7
922	Palm Springs	101.2	105.1	103.1
923-924	San Bernardino	98.8	105.5	102.0
925	Riverside	103.2	112.2	107.6
926-927	Santa Ana	101.0	108.0	104.4
928	Anaheim	103.8	113.3	108.4
930	Oxnard	104.3	110.9	107.5
931	Santa Barbara	103.4	112.6	107.9
932-933	Bakersfield	103.2	109.9	106.5
934	San Luis Obispo	105.0	107.8	106.4
935	Mojave	101.6	105.1	103.3
936-938	Fresno	105.1	112.4	108.6
939	Salinas	105.9	117.5	111.5
940-941	San Francisco	112.5	135.5	123.6
942,956-958	Sacramento	107.5	115.0	111.1
943	Palo Alto	106.0	127.2	116.2
944	San Mateo	109.2	127.3	117.9
945	Vallejo	106.0	118.9	112.2
946	Oakland	111.1	125.7	118.2
947	Berkeley	110.7	124.6	117.4
948	Richmond	110.4	123.9	116.9
949	San Rafael	111.8	124.6	118.0
950	Santa Cruz	110.1	117.9	113.9

STATE/ZIP	CITY	MAT.	INST.	TOTAL
CALIFORNIA (CONT'D)				
951	San Jose	109.3	129.7	119.2
952	Stockton	104.9	113.3	108.9
953	Modesto	104.9	112.9	108.7
954	Santa Rosa	106.1	123.8	114.6
955	Eureka	107.3	106.6	107.0
959	Marysville	106.3	113.0	109.5
960	Redding	107.3	113.4	110.3
961	Susanville	107.3	112.9	110.0
COLORADO				
800-802	Denver	101.3	89.9	95.8
803	Boulder	99.2	85.8	92.8
804	Golden	101.8	85.4	93.9
805	Fort Collins	102.5	82.1	92.7
806	Greeley	99.6	70.5	85.6
807	Fort Morgan	100.1	85.5	93.1
808-809	Colorado Springs	101.3	87.6	94.7
810	Pueblo	101.8	83.7	93.1
811	Alamosa	104.3	80.7	92.9
812	Salida	104.3	80.6	92.9
813	Durango	104.8	79.7	92.7
814	Montrose	103.2	79.6	91.8
815	Grand Junction	106.3	78.1	92.7
816	Glenwood Springs	104.1	85.6	95.2
CONNECTICUT				
060	New Britain	101.8	114.2	107.8
061	Hartford	102.2	113.4	107.6
062	Willimantic	102.5	113.1	107.6
063	New London	99.2	114.2	106.5
064	Meriden	101.8	114.7	108.0
065	New Haven	103.8	114.7	109.1
066	Bridgeport	103.1	114.3	108.5
067	Waterbury	102.5	114.2	108.2
068	Norwalk	102.5	115.0	108.5
069	Stamford	102.6	119.3	110.7
D.C.				
200-205	Washington	100.2	89.9	95.2
DELAWARE				
197	Newark	99.5	106.8	103.0
198	Wilmington	98.9	106.8	102.7
199	Dover	99.6	106.8	103.1
FLORIDA				
320,322	Jacksonville	98.6	62.4	81.1
321	Daytona Beach	98.7	74.7	87.2
323	Tallahassee	99.1	50.6	75.7
324	Panama City	99.8	39.1	70.6
325	Pensacola	99.5	59.2	80.1
326,344	Gainesville	100.3	58.5	80.2
327-328,347	Orlando	100.6	64.6	83.2
329	Melbourne	100.8	79.0	90.3
330-332,340	Miami	98.6	74.8	87.1
333	Fort Lauderdale	97.9	73.5	86.2
334,349	West Palm Beach	96.8	70.6	84.2
335-336,346	Tampa	99.8	74.8	87.8
337	St. Petersburg	102.2	57.6	80.7
338	Lakeland	98.9	74.5	87.1
339,341	Fort Myers	98.2	66.2	82.8
342	Sarasota	100.0	68.7	84.9
GEORGIA				
300-303,399	Atlanta	96.8	82.0	89.7
304	Statesboro	97.0	46.4	72.6
305	Gainesville	95.8	59.1	78.1
306	Athens	95.0	64.0	80.0
307	Dalton	97.2	52.5	75.7
308-309	Augusta	95.4	61.1	78.9
310-312	Macon	95.9	60.7	78.9
313-314	Savannah	98.1	59.5	79.5
315	Waycross	97.9	53.2	76.4
316	Valdosta	97.5	50.4	74.8
317	Albany	97.4	55.6	77.2
318-319	Columbus	97.5	54.2	76.6

STATE/ZIP	CITY	MAT.	INST.	TOTAL
HAWAII				
967	Hilo	115.9	126.1	120.8
968	Honolulu	117.6	126.1	121.7
STATES & POSS.				
969	Guam	196.3	55.8	128.6
IDAHO				
832	Pocatello	101.3	79.8	90.9
833	Twin Falls	103.3	47.2	76.3
834	Idaho Falls	100.5	57.0	79.5
835	Lewiston	112.2	84.7	99.0
836-837	Boise	101.6	81.8	92.1
838	Coeur d'Alene	111.9	58.5	86.2
ILLINOIS				
600-603	North Suburban	99.1	119.6	109.0
604	Joliet	99.1	120.4	109.4
605	South Suburban	99.1	117.4	107.9
606	Chicago	99.6	126.4	112.5
609	Kankakee	95.6	108.6	101.9
610-611	Rockford	97.2	111.9	104.3
612	Rock Island	95.2	99.6	97.3
613	La Salle	96.7	102.0	99.2
614	Galesburg	96.4	102.2	99.2
615-616	Peoria	98.6	103.5	101.0
617	Bloomington	95.7	102.9	99.2
618-619	Champaign	99.6	101.3	100.4
620-622	East St. Louis	93.8	104.3	98.8
623	Quincy	95.7	97.1	96.4
624	Effingham	95.0	99.5	97.1
625	Decatur	96.2	99.0	97.5
626-627	Springfield	96.0	100.7	98.3
628	Centralia	93.1	100.4	96.6
629	Carbondale	92.7	98.1	95.3
INDIANA				
460	Anderson	97.0	85.0	91.2
461-462	Indianapolis	100.2	88.0	94.3
463-464	Gary	97.9	101.9	99.8
465-466	South Bend	96.3	85.2	91.0
467-468	Fort Wayne	97.5	83.1	90.6
469	Kokomo	95.6	84.8	90.4
470	Lawrenceburg	93.4	82.6	88.2
471	New Albany	95.0	77.1	86.3
472	Columbus	97.3	82.8	90.3
473	Muncie	97.3	83.4	90.6
474	Bloomington	99.1	83.2	91.4
475	Washington	95.6	87.9	91.9
476-477	Evansville	96.6	89.1	93.0
478	Terre Haute	97.4	87.9	92.8
479	Lafayette	97.0	81.8	89.7
IOWA				
500-503,509	Des Moines	98.1	85.0	91.7
504	Mason City	96.8	65.3	81.6
505	Fort Dodge	97.0	61.3	79.8
506-507	Waterloo	98.0	64.0	81.6
508	Creston	97.5	68.5	83.5
510-511	Sioux City	98.7	75.7	87.6
512	Sibley	97.9	54.6	77.0
513	Spencer	100.0	52.6	77.1
514	Carroll	96.6	58.0	78.0
515	Council Bluffs	99.6	76.3	88.4
516	Shenandoah	96.7	53.9	76.0
520	Dubuque	98.3	78.9	88.9
521	Decorah	97.7	56.2	77.7
522-524	Cedar Rapids	99.3	85.3	92.6
525	Ottumwa	97.8	72.7	85.7
526	Burlington	96.7	76.1	86.8
527-528	Davenport	98.1	94.7	96.5
KANSAS				
660-662	Kansas City	98.3	92.9	95.7
664-666	Topeka	97.6	69.4	84.0
667	Fort Scott	97.4	67.5	83.0
668	Emporia	97.2	61.3	79.9
669	Belleville	99.3	56.5	78.6
670-672	Wichita	96.7	73.0	85.3
673	Independence	98.9	53.5	77.0
674	Salina	98.6	59.0	79.5
675	Hutchinson	94.0	52.6	74.0
676	Hays	98.6	56.5	78.3
677	Colby	99.2	56.5	78.6

STATE/ZIP	CITY	MAT.	INST.	TOTAL
KANSAS (CONT'D)				
678	Dodge City	100.0	56.5	79.0
679	Liberal	98.4	47.0	73.6
KENTUCKY				
400-402	Louisville	95.9	85.0	90.7
403-405	Lexington	96.7	67.9	82.9
406	Frankfort	95.9	70.1	83.4
407-409	Corbin	94.6	46.0	71.2
410	Covington	95.6	94.8	95.2
411-412	Ashland	94.0	97.6	95.7
413-414	Campton	95.5	46.5	71.9
415-416	Pikeville	96.8	64.1	81.0
417-418	Hazard	94.8	46.2	71.4
420	Paducah	93.7	88.8	91.3
421-422	Bowling Green	95.4	82.2	89.0
423	Owensboro	95.5	75.5	85.9
424	Henderson	93.2	91.7	92.4
425-426	Somerset	92.5	48.0	71.0
427	Elizabethtown	92.0	85.0	88.6
LOUISIANA				
700-701	New Orleans	101.5	69.4	86.0
703	Thibodaux	101.7	66.1	84.5
704	Hammond	98.6	65.3	82.6
705	Lafayette	100.7	56.5	79.4
706	Lake Charles	100.9	61.1	81.7
707-708	Baton Rouge	100.6	56.5	79.4
710-711	Shreveport	96.5	61.6	79.6
712	Monroe	96.6	59.0	78.5
713-714	Alexandria	96.7	56.0	77.1
MAINE				
039	Kittery	97.6	74.8	86.6
040-041	Portland	100.1	79.2	90.0
042	Lewiston	100.8	79.2	90.4
043	Augusta	99.4	74.6	87.4
044	Bangor	99.9	79.2	89.9
045	Bath	98.9	74.8	87.3
046	Machias	98.4	74.1	86.7
047	Houlton	98.5	77.9	88.6
048	Rockland	97.4	72.8	85.6
049	Waterville	98.9	69.4	84.7
MARYLAND				
206	Waldorf	98.3	74.0	86.6
207-208	College Park	98.2	81.2	90.0
209	Silver Spring	97.5	78.6	88.4
210-212	Baltimore	97.5	84.9	91.4
214	Annapolis	97.0	80.9	89.2
215	Cumberland	94.3	79.2	87.0
216	Easton	95.9	46.8	72.2
217	Hagerstown	94.5	81.1	88.1
218	Salisbury	96.4	56.4	77.1
219	Elkton	93.4	69.2	81.7
MASSACHUSETTS				
010-011	Springfield	100.8	104.9	102.8
012	Pittsfield	100.5	98.6	99.6
013	Greenfield	98.8	101.2	99.9
014	Fitchburg	97.5	115.4	106.1
015-016	Worcester	100.5	116.0	108.0
017	Framingham	96.8	121.2	108.6
018	Lowell	100.2	121.4	110.4
019	Lawrence	101.5	119.5	110.2
020-022, 024	Boston	102.9	128.1	115.0
023	Brockton	101.5	116.6	108.8
025	Buzzards Bay	96.0	115.8	105.6
026	Hyannis	98.4	115.8	106.8
027	New Bedford	101.1	116.0	108.3
MICHIGAN				
480,483	Royal Oak	94.8	109.4	101.8
481	Ann Arbor	97.1	108.6	102.7
482	Detroit	98.3	116.8	107.2
484-485	Flint	96.9	98.6	97.7
486	Saginaw	96.9	95.7	96.3
487	Bay City	96.6	95.8	96.2
488-489	Lansing	97.2	98.3	97.7
490	Battle Creek	96.6	90.9	93.8
491	Kalamazoo	97.0	87.4	92.4
492	Jackson	95.3	94.2	94.8
493,495	Grand Rapids	96.8	72.4	85.0
494	Muskegon	95.6	85.9	90.9

Location Factors

STATE/ZIP	CITY	MAT.	INST.	TOTAL
MICHIGAN (CONT'D)				
496	Traverse City	94.7	76.3	85.8
497	Gaylord	96.1	80.4	88.5
498-499	Iron Mountain	98.2	89.5	94.0
MINNESOTA				
550-551	Saint Paul	99.2	125.5	111.8
553-555	Minneapolis	99.9	127.9	113.4
556-558	Duluth	99.1	110.8	104.8
559	Rochester	98.8	107.6	103.0
560	Mankato	97.3	104.6	100.8
561	Windom	96.0	83.2	89.8
562	Willmar	95.3	89.9	92.7
563	St. Cloud	96.1	123.3	109.2
564	Brainerd	97.1	103.5	100.2
565	Detroit Lakes	99.3	101.6	100.4
566	Bemidji	98.6	100.2	99.4
567	Thief River Falls	97.6	96.4	97.0
MISSISSIPPI				
386	Clarksdale	97.0	31.3	65.3
387	Greenville	100.6	47.4	75.0
388	Tupelo	98.6	39.6	70.2
389	Greenwood	98.4	33.8	67.3
390-392	Jackson	98.7	49.0	74.7
393	Meridian	96.8	49.3	73.9
394	Laurel	98.5	34.7	67.7
395	Biloxi	98.8	56.1	78.2
396	McComb	96.8	53.4	75.9
397	Columbus	98.5	39.0	69.8
MISSOURI				
630-631	St. Louis	96.0	108.8	102.2
633	Bowling Green	95.8	92.2	94.1
634	Hannibal	94.6	87.8	91.3
635	Kirksville	97.2	80.8	89.3
636	Flat River	96.9	96.6	96.8
637	Cape Girardeau	96.0	92.2	94.1
638	Sikeston	94.6	87.0	90.9
639	Poplar Bluff	94.1	87.7	91.0
640-641	Kansas City	99.4	104.8	102.0
644-645	St. Joseph	99.2	92.1	95.8
646	Chillicothe	96.7	72.8	85.2
647	Harrisonville	96.2	100.7	98.3
648	Joplin	98.0	71.1	85.0
650-651	Jefferson City	95.3	89.1	92.3
652	Columbia	96.7	90.4	93.6
653	Sedalia	96.3	86.9	91.8
654-655	Rolla	94.9	80.2	87.8
656-658	Springfield	98.0	76.4	87.6
MONTANA				
590-591	Billings	101.1	76.2	89.1
592	Wolf Point	102.1	71.8	87.5
593	Miles City	99.8	72.4	86.6
594	Great Falls	102.5	74.2	88.9
595	Havre	100.7	72.1	86.9
596	Helena	102.3	73.2	88.3
597	Butte	101.1	73.4	87.8
598	Missoula	99.2	72.0	86.1
599	Kalispell	98.6	71.3	85.4
NEBRASKA				
680-681	Omaha	100.2	79.1	90.0
683-685	Lincoln	98.1	69.8	84.5
686	Columbus	96.9	52.7	75.6
687	Norfolk	99.1	64.5	82.4
688	Grand Island	97.8	68.3	83.6
689	Hastings	97.5	60.3	79.6
690	Mccook	97.3	50.3	74.7
691	North Platte	97.2	61.8	80.1
692	Valentine	100.7	40.9	71.9
693	Alliance	100.4	38.1	70.4
NEVADA				
889-891	Las Vegas	100.6	106.7	103.5
893	Ely	101.7	83.7	93.0
894-895	Reno	101.5	95.4	98.5
897	Carson City	101.2	94.8	98.2
898	Elko	100.3	83.3	92.1
NEW HAMPSHIRE				
030	Nashua	101.0	85.4	93.5
031	Manchester	101.5	85.4	93.7

STATE/ZIP	CITY	MAT.	INST.	TOTAL
NEW HAMPSHIRE (CONT'D)				
032-033	Concord	99.0	85.4	92.4
034	Keene	98.3	54.2	77.0
035	Littleton	98.3	63.9	81.8
036	Charleston	97.8	50.9	75.2
037	Claremont	96.9	50.9	74.7
038	Portsmouth	98.3	80.7	89.8
NEW JERSEY				
070-071	Newark	102.8	120.3	111.2
072	Elizabeth	101.6	119.4	110.2
073	Jersey City	100.1	120.3	109.9
074-075	Paterson	102.9	119.3	110.8
076	Hackensack	100.1	119.9	109.6
077	Long Branch	99.7	119.8	109.4
078	Dover	100.4	119.2	109.5
079	Summit	100.5	119.4	109.6
080,083	Vineland	98.0	118.9	108.1
081	Camden	100.0	117.3	108.4
082,084	Atlantic City	98.9	117.7	108.0
085-086	Trenton	101.0	120.4	110.3
087	Point Pleasant	100.4	119.2	109.4
088-089	New Brunswick	100.8	118.4	109.3
NEW MEXICO				
870-872	Albuquerque	100.5	77.2	89.3
873	Gallup	101.1	77.2	89.6
874	Farmington	101.2	77.2	89.6
875	Santa Fe	99.6	77.2	88.8
877	Las Vegas	99.4	77.2	88.7
878	Socorro	98.9	77.2	88.4
879	Truth/Consequences	98.7	72.1	85.9
880	Las Cruces	96.9	69.9	83.9
881	Clovis	99.5	76.6	88.5
882	Roswell	101.0	76.7	89.3
883	Carrizozo	101.8	77.2	89.9
884	Tucumcari	100.5	76.6	89.0
NEW YORK				
100-102	New York	106.9	163.2	134.0
103	Staten Island	103.0	159.3	130.2
104	Bronx	100.4	159.3	128.8
105	Mount Vernon	101.3	138.4	119.2
106	White Plains	101.0	137.7	118.7
107	Yonkers	105.8	137.9	121.2
108	New Rochelle	101.9	138.4	119.5
109	Suffern	101.9	127.3	114.1
110	Queens	102.4	159.2	129.8
111	Long Island City	104.3	159.2	130.8
112	Brooklyn	104.5	160.3	131.4
113	Flushing	105.2	159.2	131.3
114	Jamaica	103.1	159.4	130.2
115,117,118	Hicksville	102.3	148.3	124.5
116	Far Rockaway	105.4	159.2	131.3
119	Riverhead	102.8	148.5	124.8
120-122	Albany	97.8	96.2	97.0
123	Schenectady	98.4	95.3	96.9
124	Kingston	102.8	116.5	109.4
125-126	Poughkeepsie	101.9	121.4	111.3
127	Monticello	101.1	116.0	108.3
128	Glens Falls	93.2	90.1	91.7
129	Plattsburgh	98.6	85.5	92.3
130-132	Syracuse	99.3	92.7	96.1
133-135	Utica	97.6	89.4	93.6
136	Watertown	99.1	90.5	95.0
137-139	Binghamton	98.7	87.0	93.1
140-142	Buffalo	100.1	103.8	101.9
143	Niagara Falls	98.4	106.0	102.1
144-146	Rochester	101.3	97.4	99.4
147	Jamestown	97.6	88.1	93.0
148-149	Elmira	97.0	87.4	92.4
NORTH CAROLINA				
270,272-274	Greensboro	97.2	51.8	75.3
271	Winston-Salem	97.3	50.6	74.8
275-276	Raleigh	97.4	51.9	75.5
277	Durham	97.1	51.9	75.3
278	Rocky Mount	95.8	37.3	67.6
279	Elizabeth City	96.6	39.8	69.2
280	Gastonia	97.3	48.9	74.0
281-282	Charlotte	97.8	49.2	74.4
283	Fayetteville	96.0	51.9	74.7
284	Wilmington	95.8	51.0	74.2
285	Kinston	94.4	36.8	66.7

Location Factors

STATE/ZIP	CITY	MAT.	INST.	TOTAL
NORTH CAROLINA (CONT'D)				
286	Hickory	94.6	36.0	66.3
287-288	Asheville	96.2	49.8	73.8
289	Murphy	95.9	35.1	66.6
NORTH DAKOTA				
580-581	Fargo	101.6	67.6	85.2
582	Grand Forks	101.6	60.3	81.7
583	Devils Lake	101.6	62.4	82.7
584	Jamestown	101.6	55.9	79.6
585	Bismarck	100.2	68.9	85.1
586	Dickinson	102.5	64.3	84.0
587	Minot	101.6	69.6	86.2
588	Williston	100.8	63.4	82.7
OHIO				
430-432	Columbus	98.3	91.9	95.2
433	Marion	95.5	90.8	93.2
434-436	Toledo	98.7	100.6	99.6
437-438	Zanesville	95.7	85.8	90.9
439	Steubenville	97.5	95.2	96.4
440	Lorain	98.2	101.8	99.9
441	Cleveland	98.5	106.2	102.2
442-443	Akron	99.3	99.0	99.2
444-445	Youngstown	98.6	95.1	96.9
446-447	Canton	98.7	90.4	94.7
448-449	Mansfield	96.3	94.2	95.3
450	Hamilton	94.7	91.6	93.2
451-452	Cincinnati	95.0	93.0	94.0
453-454	Dayton	94.8	87.6	91.3
455	Springfield	94.8	88.3	91.7
456	Chillicothe	94.3	94.8	94.5
457	Athens	97.8	79.5	88.9
458	Lima	98.4	88.2	93.5
OKLAHOMA				
730-731	Oklahoma City	97.9	65.9	82.5
734	Ardmore	95.7	65.3	81.0
735	Lawton	97.3	66.3	82.4
736	Clinton	97.3	63.5	81.0
737	Enid	97.3	63.5	81.0
738	Woodward	95.9	63.6	80.3
739	Guymon	97.1	34.9	67.1
740-741	Tulsa	97.3	63.1	80.8
743	Miami	94.6	67.4	81.5
744	Muskogee	96.5	45.0	71.7
745	Mcalester	94.2	56.1	75.8
746	Ponca City	94.7	63.4	79.6
747	Durant	94.8	63.7	79.8
748	Shawnee	96.5	62.0	79.9
749	Poteau	93.7	64.5	79.6
OREGON				
970-972	Portland	103.1	105.0	104.0
973	Salem	103.7	103.9	103.8
974	Eugene	103.1	103.5	103.3
975	Medford	104.5	101.8	103.2
976	Klamath Falls	105.5	101.8	103.7
977	Bend	104.3	103.6	103.9
978	Pendleton	98.3	102.1	100.1
979	Vale	95.8	95.0	95.4
PENNSYLVANIA				
150-152	Pittsburgh	96.7	104.5	100.5
153	Washington	94.2	103.5	98.7
154	Uniontown	94.4	100.5	97.4
155	Bedford	95.5	92.3	93.9
156	Greensburg	95.6	99.6	97.5
157	Indiana	94.3	98.6	96.4
158	Dubois	95.8	95.7	95.7
159	Johnstown	95.4	97.0	96.2
160	Butler	92.6	102.2	97.2
161	New Castle	92.6	102.2	97.3
162	Kittanning	93.2	103.8	98.3
163	Oil City	92.6	97.5	94.9
164-165	Erie	94.5	97.3	95.9
166	Altoona	94.4	94.3	94.3
167	Bradford	96.1	94.9	95.6
168	State College	95.8	94.7	95.3
169	Wellsboro	96.9	89.9	93.5
170-171	Harrisburg	98.3	91.4	95.0
172	Chambersburg	96.4	88.5	92.6
173-174	York	96.1	90.0	93.1
175-176	Lancaster	95.0	88.3	91.8

STATE/ZIP	CITY	MAT.	INST.	TOTAL
PENNSYLVANIA (CONT'D)				
177	Williamsport	93.4	83.9	88.8
178	Sunbury	95.8	90.6	93.3
179	Pottsville	94.8	92.1	93.5
180	Lehigh Valley	96.5	105.4	100.8
181	Allentown	98.2	102.8	100.4
182	Hazleton	95.8	94.4	95.1
183	Stroudsburg	95.7	98.8	97.2
184-185	Scranton	98.7	95.4	97.1
186-187	Wilkes-Barre	95.4	93.3	94.4
188	Montrose	95.0	94.6	94.8
189	Doylestown	95.3	116.1	105.3
190-191	Philadelphia	100.2	125.0	112.1
193	Westchester	97.2	114.5	105.6
194	Norristown	96.0	119.5	107.4
195-196	Reading	98.2	95.7	97.0
PUERTO RICO				
009	San Juan	140.4	27.1	85.8
RHODE ISLAND				
028	Newport	99.8	109.5	104.4
029	Providence	100.1	109.5	104.6
SOUTH CAROLINA				
290-292	Columbia	96.5	49.8	74.0
293	Spartanburg	95.4	47.4	72.3
294	Charleston	96.7	50.4	74.4
295	Florence	94.9	47.5	72.1
296	Greenville	95.1	47.4	72.1
297	Rock Hill	95.3	34.9	66.2
298	Aiken	96.4	72.5	84.9
299	Beaufort	97.2	38.5	68.9
SOUTH DAKOTA				
570-571	Sioux Falls	100.5	59.7	80.8
572	Watertown	99.7	53.6	77.5
573	Mitchell	98.6	53.2	76.7
574	Aberdeen	100.8	56.3	79.4
575	Pierre	100.1	55.4	78.5
576	Mobridge	99.3	53.6	77.2
577	Rapid City	100.7	54.1	78.3
TENNESSEE				
370-372	Nashville	97.9	75.7	87.2
373-374	Chattanooga	99.2	57.5	79.1
375,380-381	Memphis	96.4	76.2	86.7
376	Johnson City	98.6	59.3	79.7
377-379	Knoxville	95.1	60.8	78.6
382	Mckenzie	97.5	51.3	75.2
383	Jackson	99.2	55.0	77.9
384	Columbia	95.9	58.7	77.9
385	Cookeville	97.3	53.1	76.0
TEXAS				
750	Mckinney	99.6	58.7	79.9
751	Waxahackie	99.5	59.4	80.2
752-753	Dallas	99.5	67.9	84.3
754	Greenville	99.7	43.0	72.4
755	Texarkana	99.1	54.3	77.5
756	Longview	99.3	45.1	73.2
757	Tyler	100.0	57.6	79.5
758	Palestine	95.5	45.5	71.4
759	Lufkin	96.5	49.5	73.8
760-761	Fort Worth	97.5	64.5	81.6
762	Denton	97.9	54.7	77.1
763	Wichita Falls	97.9	58.9	79.1
764	Eastland	97.0	44.6	71.7
765	Temple	95.7	53.0	75.1
766-767	Waco	97.2	59.3	78.9
768	Brownwood	98.0	42.0	71.0
769	San Angelo	97.6	50.4	74.9
770-772	Houston	99.4	73.7	87.0
773	Huntsville	98.3	42.9	71.6
774	Wharton	99.8	48.0	74.8
775	Galveston	97.6	71.8	85.2
776-777	Beaumont	97.0	65.4	81.7
778	Bryan	94.9	66.2	81.1
779	Victoria	100.1	50.7	76.3
780	Laredo	94.4	55.2	75.5
781-782	San Antonio	94.6	65.7	80.7
783-784	Corpus Christi	97.7	55.5	77.4
785	Mc Allen	98.3	49.5	74.8
786-787	Austin	94.5	63.6	79.6

Location Factors

STATE/ZIP	CITY	MAT.	INST.	TOTAL
TEXAS (CONT'D)				
788	Del Rio	97.5	35.3	67.5
789	Giddings	94.6	44.8	70.6
790-791	Amarillo	97.9	60.9	80.1
792	Childress	97.6	55.1	77.1
793-794	Lubbock	99.6	56.5	78.8
795-796	Abilene	97.3	56.1	77.4
797	Midland	100.5	52.9	77.6
798-799,885	El Paso	97.3	54.0	76.4
UTAH				
840-841	Salt Lake City	102.5	74.2	88.9
842,844	Ogden	98.9	74.0	86.9
843	Logan	101.0	74.0	88.0
845	Price	101.7	52.6	78.1
846-847	Provo	101.3	74.0	88.1
VERMONT				
050	White River Jct.	99.9	49.4	75.6
051	Bellows Falls	98.2	51.9	75.9
052	Bennington	98.6	52.3	76.2
053	Brattleboro	99.1	52.4	76.6
054	Burlington	101.0	68.2	85.1
056	Montpelier	98.4	68.2	83.8
057	Rutland	100.0	68.2	84.7
058	St. Johnsbury	100.1	52.2	77.0
059	Guildhall	98.5	51.7	75.9
VIRGINIA				
220-221	Fairfax	98.5	80.3	89.7
222	Arlington	99.2	79.0	89.4
223	Alexandria	98.2	83.9	91.3
224-225	Fredericksburg	96.9	68.4	83.1
226	Winchester	97.7	56.9	78.0
227	Culpeper	97.5	57.8	78.3
228	Harrisonburg	97.9	49.6	74.6
229	Charlottesville	98.1	62.0	80.7
230-232	Richmond	98.2	67.6	83.5
233-235	Norfolk	98.3	63.7	81.7
236	Newport News	97.8	62.2	80.6
237	Portsmouth	97.1	63.2	80.8
238	Petersburg	97.9	67.6	83.3
239	Farmville	97.2	43.6	71.4
240-241	Roanoke	98.4	49.9	75.0
242	Bristol	96.9	50.2	74.4
243	Pulaski	96.6	44.4	71.5
244	Staunton	97.4	48.0	73.6
245	Lynchburg	97.5	52.6	75.9
246	Grundy	96.9	46.6	72.6
WASHINGTON				
980-981,987	Seattle	103.0	105.2	104.1
982	Everett	103.0	100.0	101.5
983-984	Tacoma	102.7	100.3	101.5
985	Olympia	102.5	100.2	101.4
986	Vancouver	105.0	98.7	102.0
988	Wenatchee	103.8	86.4	95.4
989	Yakima	103.2	91.2	97.4
990-992	Spokane	107.2	84.6	96.3
993	Richland	107.4	86.6	97.4
994	Clarkston	106.1	84.7	95.8
WEST VIRGINIA				
247-248	Bluefield	95.7	80.1	88.2
249	Lewisburg	97.3	83.3	90.5
250-253	Charleston	97.7	90.3	94.2
254	Martinsburg	97.2	78.0	87.9
255-257	Huntington	98.0	91.2	94.8
258-259	Beckley	95.4	88.9	92.3
260	Wheeling	98.3	91.2	94.9
261	Parkersburg	97.1	90.9	94.1
262	Buckhannon	97.1	90.7	94.0
263-264	Clarksburg	97.6	90.3	94.1
265	Morgantown	97.7	90.8	94.4
266	Gassaway	96.9	90.6	93.9
267	Romney	96.9	85.0	91.1
268	Petersburg	96.8	87.9	92.5
WISCONSIN				
530,532	Milwaukee	99.2	102.3	100.7
531	Kenosha	99.2	100.4	99.8
534	Racine	98.4	101.8	100.1
535	Beloit	98.3	98.8	98.6
537	Madison	98.8	96.9	97.9

STATE/ZIP	CITY	MAT.	INST.	TOTAL
WISCONSIN (CONT'D)				
538	Lancaster	96.7	94.2	95.5
539	Portage	95.1	95.7	95.4
540	New Richmond	96.8	97.2	97.0
541-543	Green Bay	100.6	93.7	97.2
544	Wausau	95.8	92.4	94.2
545	Rhinelander	99.4	92.7	96.1
546	La Crosse	96.4	94.6	95.5
547	Eau Claire	98.3	95.6	97.0
548	Superior	96.7	97.5	97.1
549	Oshkosh	96.4	91.7	94.1
WYOMING				
820	Cheyenne	99.9	56.4	78.9
821	Yellowstone Nat'l Park	99.3	50.6	75.8
822	Wheatland	101.0	49.5	76.2
823	Rawlins	102.4	45.5	75.0
824	Worland	100.1	46.2	74.1
825	Riverton	101.3	48.8	76.0
826	Casper	99.7	61.9	81.5
827	Newcastle	99.9	45.5	73.7
828	Sheridan	100.9	54.4	78.5
829-831	Rock Springs	104.5	46.2	76.4
CANADIAN FACTORS (reflect Canadian currency)				
ALBERTA				
	Calgary	123.4	84.6	104.7
	Edmonton	124.4	84.6	105.2
	Fort McMurray	117.1	84.1	101.2
	Lethbridge	118.0	83.6	101.4
	Lloydminster	117.1	84.1	101.2
	Medicine Hat	117.2	83.6	101.0
	Red Deer	117.6	83.6	101.2
BRITISH COLUMBIA				
	Kamloops	117.9	87.0	103.0
	Prince George	119.6	87.0	103.9
	Vancouver	121.8	91.0	107.0
	Victoria	119.8	87.5	104.2
MANITOBA				
	Brandon	117.1	72.9	95.8
	Portage la Prairie	117.1	72.9	95.8
	Winnipeg	120.7	73.2	97.8
NEW BRUNSWICK				
	Bathurst	115.2	64.8	90.9
	Dalhousie	115.2	64.8	90.9
	Fredericton	115.6	69.0	93.2
	Moncton	114.9	64.8	90.7
	Newcastle	115.3	64.8	90.9
	Saint John	117.0	69.0	93.9
NEWFOUNDLAND				
	Corner Brook	121.1	64.8	93.9
	St. John's	120.7	64.8	93.7
NORTHWEST TERRITORIES				
	Yellowknife	114.1	82.3	98.8
NOVA SCOTIA				
	Dartmouth	116.8	71.4	94.9
	Halifax	116.5	72.5	95.3
	New Glasgow	116.5	71.4	94.7
	Sydney	113.7	71.4	93.3
	Yarmouth	116.3	71.4	94.7
ONTARIO				
	Barrie	120.2	89.4	105.4
	Brantford	119.3	93.2	106.7
	Cornwall	119.3	89.7	105.0
	Hamilton	120.8	93.4	107.6
	Kingston	119.8	90.1	105.5
	Kitchener	114.9	89.0	102.4
	London	120.1	90.8	105.9
	North Bay	119.3	87.7	104.0
	Oshawa	118.8	91.2	105.5
	Ottawa	120.8	90.9	106.4
	Owen Sound	120.4	87.7	104.6
	Peterborough	119.3	89.5	104.9
	Sarnia	119.5	93.5	107.0
	St. Catharines	113.1	88.3	101.2
	Sudbury	113.4	87.9	101.1

STATE/ZIP	CITY	MAT.	INST.	TOTAL
ONTARIO (CONT'D)				
	Thunder Bay	114.9	88.1	102.0
	Toronto	121.7	98.0	110.2
	Windsor	113.5	90.8	102.5
PRINCE EDWARD ISLAND				
	Charlottetown	118.4	60.5	90.5
	Summerside	118.6	60.5	90.6
QUEBEC				
	Cap-de-la-Madeleine	116.6	83.9	100.8
	Charlesbourg	116.6	83.9	100.8
	Chicoutimi	115.7	83.6	100.2
	Gatineau	115.8	83.7	100.3
	Laval	115.9	83.7	100.4
	Montreal	117.7	84.0	101.5
	Quebec	119.0	84.4	102.3
	Sherbrooke	116.2	83.7	100.5
	Trois Rivieres	116.8	83.9	100.9
SASKATCHEWAN				
	Moose Jaw	114.5	66.9	91.5
	Prince Albert	113.8	66.7	91.1
	Regina	115.2	66.9	91.9
	Saskatoon	114.1	66.8	91.3
YUKON				
	Whitehorse	114.1	65.8	90.8

Abbreviations

Abbreviation	Meaning
A	Area Square Feet; Ampere
ABS	Acrylonitrile Butadiene Stryrene; Asbestos Bonded Steel
A.C.	Alternating Current; Air-Conditioning; Asbestos Cement; Plywood Grade A & C
A.C.I.	American Concrete Institute
AD	Plywood, Grade A & D
Addit.	Additional
Adj.	Adjustable
af	Audio-frequency
A.G.A.	American Gas Association
Agg.	Aggregate
A.H.	Ampere Hours
A hr.	Ampere-hour
A.H.U.	Air Handling Unit
A.I.A.	American Institute of Architects
AIC	Ampere Interrupting Capacity
Allow.	Allowance
alt.	Altitude
Alum.	Aluminum
a.m.	Ante Meridiem
Amp.	Ampere
Anod.	Anodized
Approx.	Approximate
Apt.	Apartment
Asb.	Asbestos
A.S.B.C.	American Standard Building Code
Asbe.	Asbestos Worker
A.S.H.R.A.E.	American Society of Heating, Refrig. & AC Engineers
A.S.M.E.	American Society of Mechanical Engineers
A.S.T.M.	American Society for Testing and Materials
Attchmt.	Attachment
Avg.	Average
A.W.G.	American Wire Gauge
AWWA	American Water Works Assoc.
Bbl.	Barrel
B. & B.	Grade B and Better; Balled & Burlapped
B. & S.	Bell and Spigot
B. & W.	Black and White
b.c.c.	Body-centered Cubic
B.C.Y.	Bank Cubic Yards
BE	Bevel End
B.F.	Board Feet
Bg. cem.	Bag of Cement
BHP	Boiler Horsepower; Brake Horsepower
B.I.	Black Iron
Bit.; Bitum.	Bituminous
Bk.	Backed
Bkrs.	Breakers
Bldg.	Building
Blk.	Block
Bm.	Beam
Boil.	Boilermaker
B.P.M.	Blows per Minute
BR	Bedroom
Brg.	Bearing
Brhe.	Bricklayer Helper
Bric.	Bricklayer
Brk.	Brick
Brng.	Bearing
Brs.	Brass
Brz.	Bronze
Bsn.	Basin
Btr.	Better
BTU	British Thermal Unit
BTUH	BTU per Hour
B.U.R.	Built-up Roofing
BX	Interlocked Armored Cable
c	Conductivity, Copper Sweat
C	Hundred; Centigrade
C/C	Center to Center, Cedar on Cedar
Cab.	Cabinet
Cair.	Air Tool Laborer
Calc	Calculated
Cap.	Capacity
Carp.	Carpenter
C.B.	Circuit Breaker
C.C.A.	Chromate Copper Arsenate
C.C.F.	Hundred Cubic Feet
cd	Candela
cd/sf	Candela per Square Foot
CD	Grade of Plywood Face & Back
CDX	Plywood, Grade C & D, exterior glue
Cefi.	Cement Finisher
Cem.	Cement
CF	Hundred Feet
C.F.	Cubic Feet
CFM	Cubic Feet per Minute
c.g.	Center of Gravity
CHW	Chilled Water; Commercial Hot Water
C.I.	Cast Iron
C.I.P.	Cast in Place
Circ.	Circuit
C.L.	Carload Lot
Clab.	Common Laborer
C.L.F.	Hundred Linear Feet
CLF	Current Limiting Fuse
CLP	Cross Linked Polyethylene
cm	Centimeter
CMP	Corr. Metal Pipe
C.M.U.	Concrete Masonry Unit
CN	Change Notice
Col.	Column
CO_2	Carbon Dioxide
Comb.	Combination
Compr.	Compressor
Conc.	Concrete
Cont.	Continuous; Continued
Corr.	Corrugated
Cos	Cosine
Cot	Cotangent
Cov.	Cover
C/P	Cedar on Paneling
CPA	Control Point Adjustment
Cplg.	Coupling
C.P.M.	Critical Path Method
CPVC	Chlorinated Polyvinyl Chloride
C.Pr.	Hundred Pair
CRC	Cold Rolled Channel
Creos.	Creosote
Crpt.	Carpet & Linoleum Layer
CRT	Cathode-ray Tube
CS	Carbon Steel, Constant Shear Bar Joist
Csc	Cosecant
C.S.F.	Hundred Square Feet
CSI	Construction Specifications Institute
C.T.	Current Transformer
CTS	Copper Tube Size
Cu	Copper, Cubic
Cu. Ft.	Cubic Foot
cw	Continuous Wave
C.W.	Cool White; Cold Water
Cwt.	100 Pounds
C.W.X.	Cool White Deluxe
C.Y.	Cubic Yard (27 cubic feet)
C.Y./Hr.	Cubic Yard per Hour
Cyl.	Cylinder
d	Penny (nail size)
D	Deep; Depth; Discharge
Dis.;Disch.	Discharge
Db.	Decibel
Dbl.	Double
DC	Direct Current
DDC	Direct Digital Control
Demob.	Demobilization
d.f.u.	Drainage Fixture Units
D.H.	Double Hung
DHW	Domestic Hot Water
Diag.	Diagonal
Diam.	Diameter
Distrib.	Distribution
Dk.	Deck
D.L.	Dead Load; Diesel
DLH	Deep Long Span Bar Joist
Do.	Ditto
Dp.	Depth
D.P.S.T.	Double Pole, Single Throw
Dr.	Driver
Drink.	Drinking
D.S.	Double Strength
D.S.A.	Double Strength A Grade
D.S.B.	Double Strength B Grade
Dty.	Duty
DWV	Drain Waste Vent
DX	Deluxe White, Direct Expansion
dyn	Dyne
e	Eccentricity
E	Equipment Only; East
Ea.	Each
E.B.	Encased Burial
Econ.	Economy
E.C.Y	Embankment Cubic Yards
EDP	Electronic Data Processing
EIFS	Exterior Insulation Finish System
E.D.R.	Equiv. Direct Radiation
Eq.	Equation
Elec.	Electrician; Electrical
Elev.	Elevator; Elevating
EMT	Electrical Metallic Conduit; Thin Wall Conduit
Eng.	Engine, Engineered
EPDM	Ethylene Propylene Diene Monomer
EPS	Expanded Polystyrene
Eqhv.	Equip. Oper., Heavy
Eqlt.	Equip. Oper., Light
Eqmd.	Equip. Oper., Medium
Eqmm.	Equip. Oper., Master Mechanic
Eqol.	Equip. Oper., Oilers
Equip.	Equipment
ERW	Electric Resistance Welded
E.S.	Energy Saver
Est.	Estimated
esu	Electrostatic Units
E.W.	Each Way
EWT	Entering Water Temperature
Excav.	Excavation
Exp.	Expansion, Exposure
Ext.	Exterior
Extru.	Extrusion
f.	Fiber stress
F	Fahrenheit; Female; Fill
Fab.	Fabricated
FBGS	Fiberglass
F.C.	Footcandles
f.c.c.	Face-centered Cubic
f'c.	Compressive Stress in Concrete; Extreme Compressive Stress
F.E.	Front End
FEP	Fluorinated Ethylene Propylene (Teflon)
F.G.	Flat Grain
F.H.A.	Federal Housing Administration
Fig.	Figure
Fin.	Finished
Fixt.	Fixture
Fl. Oz.	Fluid Ounces
Flr.	Floor
F.M.	Frequency Modulation; Factory Mutual
Fmg.	Framing
Fndtn.	Foundation
Fori.	Foreman, Inside

Abbreviations

Foro.	Foreman, Outside
Fount.	Fountain
FPM	Feet per Minute
FPT	Female Pipe Thread
Fr.	Frame
F.R.	Fire Rating
FRK	Foil Reinforced Kraft
FRP	Fiberglass Reinforced Plastic
FS	Forged Steel
FSC	Cast Body; Cast Switch Box
Ft.	Foot; Feet
Ftng.	Fitting
Ftg.	Footing
Ft. Lb.	Foot Pound
Furn.	Furniture
FVNR	Full Voltage Non-Reversing
FXM	Female by Male
Fy.	Minimum Yield Stress of Steel
g	Gram
G	Gauss
Ga.	Gauge
Gal.	Gallon
Gal./Min.	Gallon per Minute
Galv.	Galvanized
Gen.	General
G.F.I.	Ground Fault Interrupter
Glaz.	Glazier
GPD	Gallons per Day
GPH	Gallons per Hour
GPM	Gallons per Minute
GR	Grade
Gran.	Granular
Grnd.	Ground
H	High; High Strength Bar Joist; Henry
H.C.	High Capacity
H.D.	Heavy Duty; High Density
H.D.O.	High Density Overlaid
Hdr.	Header
Hdwe.	Hardware
Help.	Helpers Average
HEPA	High Efficiency Particulate Air Filter
Hg	Mercury
HIC	High Interrupting Capacity
HM	Hollow Metal
H.O.	High Output
Horiz.	Horizontal
H.P.	Horsepower; High Pressure
H.P.F.	High Power Factor
Hr.	Hour
Hrs./Day	Hours per Day
HSC	High Short Circuit
Ht.	Height
Htg.	Heating
Htrs.	Heaters
HVAC	Heating, Ventilation & Air-Conditioning
Hvy.	Heavy
HW	Hot Water
Hyd.;Hydr.	Hydraulic
Hz.	Hertz (cycles)
I.	Moment of Inertia
I.C.	Interrupting Capacity
ID	Inside Diameter
I.D.	Inside Dimension; Identification
I.F.	Inside Frosted
I.M.C.	Intermediate Metal Conduit
In.	Inch
Incan.	Incandescent
Incl.	Included; Including
Int.	Interior
Inst.	Installation
Insul.	Insulation/Insulated
I.P.	Iron Pipe
I.P.S.	Iron Pipe Size
I.P.T.	Iron Pipe Threaded
I.W.	Indirect Waste
J	Joule
J.I.C.	Joint Industrial Council
K	Thousand; Thousand Pounds; Heavy Wall Copper Tubing, Kelvin
K.A.H.	Thousand Amp. Hours
KCMIL	Thousand Circular Mils
KD	Knock Down
K.D.A.T.	Kiln Dried After Treatment
kg	Kilogram
kG	Kilogauss
kgf	Kilogram Force
kHz	Kilohertz
Kip.	1000 Pounds
KJ	Kiljoule
K.L.	Effective Length Factor
K.L.F.	Kips per Linear Foot
Km	Kilometer
K.S.F.	Kips per Square Foot
K.S.I.	Kips per Square Inch
kV	Kilovolt
kVA	Kilovolt Ampere
K.V.A.R.	Kilovar (Reactance)
KW	Kilowatt
KWh	Kilowatt-hour
L	Labor Only; Length; Long; Medium Wall Copper Tubing
Lab.	Labor
lat	Latitude
Lath.	Lather
Lav.	Lavatory
lb.; #	Pound
L.B.	Load Bearing; L Conduit Body
L. & E.	Labor & Equipment
lb./hr.	Pounds per Hour
lb./L.F.	Pounds per Linear Foot
lbf/sq.in.	Pound-force per Square Inch
L.C.L.	Less than Carload Lot
L.C.Y.	Loose Cubic Yards
Ld.	Load
LE	Lead Equivalent
LED	Light Emitting Diode
L.F.	Linear Foot
Lg.	Long; Length; Large
L & H	Light and Heat
LH	Long Span Bar Joist
L.H.	Labor Hours
L.L.	Live Load
L.L.D.	Lamp Lumen Depreciation
L-O-L	Lateralolet
lm	Lumen
lm/sf	Lumen per Square Foot
lm/W	Lumen per Watt
L.O.A.	Length Over All
log	Logarithm
L.P.	Liquefied Petroleum; Low Pressure
L.P.F.	Low Power Factor
LR	Long Radius
L.S.	Lump Sum
Lt.	Light
Lt. Ga.	Light Gauge
L.T.L.	Less than Truckload Lot
Lt. Wt.	Lightweight
L.V.	Low Voltage
M	Thousand; Material; Male; Light Wall Copper Tubing
M^2CA	Meters Squared Contact Area
m/hr; M.H.	Man-hour
mA	Milliampere
Mach.	Machine
Mag. Str.	Magnetic Starter
Maint.	Maintenance
Marb.	Marble Setter
Mat; Mat'l.	Material
Max.	Maximum
MBF	Thousand Board Feet
MBH	Thousand BTU's per hr.
MC	Metal Clad Cable
M.C.F.	Thousand Cubic Feet
M.C.F.M.	Thousand Cubic Feet per Minute
M.C.M.	Thousand Circular Mils
M.C.P.	Motor Circuit Protector
MD	Medium Duty
M.D.O.	Medium Density Overlaid
Med.	Medium
MF	Thousand Feet
M.F.B.M.	Thousand Feet Board Measure
Mfg.	Manufacturing
Mfrs.	Manufacturers
mg	Milligram
MGD	Million Gallons per Day
MGPH	Thousand Gallons per Hour
MH, M.H.	Manhole; Metal Halide; Man-Hour
MHz	Megahertz
Mi.	Mile
MI	Malleable Iron; Mineral Insulated
mm	Millimeter
Mill.	Millwright
Min., min.	Minimum, minute
Misc.	Miscellaneous
ml	Milliliter, Mainline
M.L.F.	Thousand Linear Feet
Mo.	Month
Mobil.	Mobilization
Mog.	Mogul Base
MPH	Miles per Hour
MPT	Male Pipe Thread
MRT	Mile Round Trip
ms	Millisecond
M.S.F.	Thousand Square Feet
Mstz.	Mosaic & Terrazzo Worker
M.S.Y.	Thousand Square Yards
Mtd.	Mounted
Mthe.	Mosaic & Terrazzo Helper
Mtng.	Mounting
Mult.	Multi; Multiply
M.V.A.	Million Volt Amperes
M.V.A.R.	Million Volt Amperes Reactance
MV	Megavolt
MW	Megawatt
MXM	Male by Male
MYD	Thousand Yards
N	Natural; North
nA	Nanoampere
NA	Not Available; Not Applicable
N.B.C.	National Building Code
NC	Normally Closed
N.E.M.A.	National Electrical Manufacturers Assoc.
NEHB	Bolted Circuit Breaker to 600V.
N.L.B.	Non-Load-Bearing
NM	Non-Metallic Cable
nm	Nanometer
No.	Number
NO	Normally Open
N.O.C.	Not Otherwise Classified
Nose.	Nosing
N.P.T.	National Pipe Thread
NQOD	Combination Plug-on/Bolt on Circuit Breaker to 240V.
N.R.C.	Noise Reduction Coefficient
N.R.S.	Non Rising Stem
ns	Nanosecond
nW	Nanowatt
OB	Opposing Blade
OC	On Center
OD	Outside Diameter
O.D.	Outside Dimension
ODS	Overhead Distribution System
O.G.	Ogee
O.H.	Overhead
O & P	Overhead and Profit
Oper.	Operator
Opng.	Opening
Orna.	Ornamental
OSB	Oriented Strand Board
O. S. & Y.	Outside Screw and Yoke

Abbreviations

Abbreviation	Meaning
Ovhd.	Overhead
OWG	Oil, Water or Gas
Oz.	Ounce
P.	Pole; Applied Load; Projection
p.	Page
Pape.	Paperhanger
P.A.P.R.	Powered Air Purifying Respirator
PAR	Parabolic Reflector
Pc., Pcs.	Piece, Pieces
P.C.	Portland Cement; Power Connector
P.C.F.	Pounds per Cubic Foot
P.C.M.	Phase Contract Microscopy
P.E.	Professional Engineer; Porcelain Enamel; Polyethylene; Plain End
Perf.	Perforated
Ph.	Phase
P.I.	Pressure Injected
Pile.	Pile Driver
Pkg.	Package
Pl.	Plate
Plah.	Plasterer Helper
Plas.	Plasterer
Pluh.	Plumbers Helper
Plum.	Plumber
Ply.	Plywood
p.m.	Post Meridiem
Pntd.	Painted
Pord.	Painter, Ordinary
pp	Pages
PP; PPL	Polypropylene
P.P.M.	Parts per Million
Pr.	Pair
P.E.S.B.	Pre-engineered Steel Building
Prefab.	Prefabricated
Prefin.	Prefinished
Prop.	Propelled
PSF; psf	Pounds per Square Foot
PSI; psi	Pounds per Square Inch
PSIG	Pounds per Square Inch Gauge
PSP	Plastic Sewer Pipe
Pspr.	Painter, Spray
Psst.	Painter, Structural Steel
P.T.	Potential Transformer
P. & T.	Pressure & Temperature
Ptd.	Painted
Ptns.	Partitions
Pu	Ultimate Load
PVC	Polyvinyl Chloride
Pvmt.	Pavement
Pwr.	Power
Q	Quantity Heat Flow
Quan.;Qty.	Quantity
Q.C.	Quick Coupling
r	Radius of Gyration
R	Resistance
R.C.P.	Reinforced Concrete Pipe
Rect.	Rectangle
Reg.	Regular
Reinf.	Reinforced
Req'd.	Required
Res.	Resistant
Resi.	Residential
Rgh.	Rough
RGS	Rigid Galvanized Steel
R.H.W.	Rubber, Heat & Water Resistant; Residential Hot Water
rms	Root Mean Square
Rnd.	Round
Rodm.	Rodman
Rofc.	Roofer, Composition
Rofp.	Roofer, Precast
Rohe.	Roofer Helpers (Composition)
Rots.	Roofer, Tile & Slate
R.O.W.	Right of Way
RPM	Revolutions per Minute
R.S.	Rapid Start
Rsr	Riser
RT	Round Trip
S.	Suction; Single Entrance; South
SCFM	Standard Cubic Feet per Minute
Scaf.	Scaffold
Sch.; Sched.	Schedule
S.C.R.	Modular Brick
S.D.	Sound Deadening
S.D.R.	Standard Dimension Ratio
S.E.	Surfaced Edge
Sel.	Select
S.E.R.;	Service Entrance Cable
S.E.U.	Service Entrance Cable
S.F.	Square Foot
S.F.C.A.	Square Foot Contact Area
S.F.G.	Square Foot of Ground
S.F. Hor.	Square Foot Horizontal
S.F.R.	Square Feet of Radiation
S.F. Shlf.	Square Foot of Shelf
S4S	Surface 4 Sides
Shee.	Sheet Metal Worker
Sin.	Sine
Skwk.	Skilled Worker
SL	Saran Lined
S.L.	Slimline
Sldr.	Solder
SLH	Super Long Span Bar Joist
S.N.	Solid Neutral
S-O-L	Socketolet
sp	Standpipe
S.P.	Static Pressure; Single Pole; Self-Propelled
Spri.	Sprinkler Installer
spwg	Static Pressure Water Gauge
Sq.	Square; 100 Square Feet
S.P.D.T.	Single Pole, Double Throw
SPF	Spruce Pine Fir
S.P.S.T.	Single Pole, Single Throw
SPT	Standard Pipe Thread
Sq. Hd.	Square Head
Sq. In.	Square Inch
S.S.	Single Strength; Stainless Steel
S.S.B.	Single Strength B Grade
sst	Stainless Steel
Sswk.	Structural Steel Worker
Sswl.	Structural Steel Welder
St.;Stl.	Steel
S.T.C.	Sound Transmission Coefficient
Std.	Standard
STK	Select Tight Knot
STP	Standard Temperature & Pressure
Stpi.	Steamfitter, Pipefitter
Str.	Strength; Starter; Straight
Strd.	Stranded
Struct.	Structural
Sty.	Story
Subj.	Subject
Subs.	Subcontractors
Surf.	Surface
Sw.	Switch
Swbd.	Switchboard
S.Y.	Square Yard
Syn.	Synthetic
S.Y.P.	Southern Yellow Pine
Sys.	System
t.	Thickness
T	Temperature; Ton
Tan	Tangent
T.C.	Terra Cotta
T & C	Threaded and Coupled
T.D.	Temperature Difference
T.E.M.	Transmission Electron Microscopy
TFE	Tetrafluoroethylene (Teflon)
T. & G.	Tongue & Groove; Tar & Gravel
Th.;Thk.	Thick
Thn.	Thin
Thrded	Threaded
Tilf.	Tile Layer, Floor
Tilh.	Tile Layer, Helper
THHN	Nylon Jacketed Wire
THW.	Insulated Strand Wire
THWN;	Nylon Jacketed Wire
T.L.	Truckload
T.M.	Track Mounted
Tot.	Total
T-O-L	Threadolet
T.S.	Trigger Start
Tr.	Trade
Transf.	Transformer
Trhv.	Truck Driver, Heavy
Trlr	Trailer
Trlt.	Truck Driver, Light
TV	Television
T.W.	Thermoplastic Water Resistant Wire
UCI	Uniform Construction Index
UF	Underground Feeder
UGND	Underground Feeder
U.H.F.	Ultra High Frequency
U.L.	Underwriters Laboratory
Unfin.	Unfinished
URD	Underground Residential Distribution
US	United States
USP	United States Primed
UTP	Unshielded Twisted Pair
V	Volt
V.A.	Volt Amperes
V.C.T.	Vinyl Composition Tile
VAV	Variable Air Volume
VC	Veneer Core
Vent.	Ventilation
Vert.	Vertical
V.F.	Vinyl Faced
V.G.	Vertical Grain
V.H.F.	Very High Frequency
VHO	Very High Output
Vib.	Vibrating
V.L.F.	Vertical Linear Foot
Vol.	Volume
VRP	Vinyl Reinforced Polyester
W	Wire; Watt; Wide; West
w/	With
W.C.	Water Column; Water Closet
W.F.	Wide Flange
W.G.	Water Gauge
Wldg.	Welding
W. Mile	Wire Mile
W-O-L	Weldolet
W.R.	Water Resistant
Wrck.	Wrecker
W.S.P.	Water, Steam, Petroleum
WT., Wt.	Weight
WWF	Welded Wire Fabric
XFER	Transfer
XFMR	Transformer
XHD	Extra Heavy Duty
XHHW; XLPE	Cross-Linked Polyethylene Wire Insulation
XLP	Cross-linked Polyethylene
Y	Wye
yd	Yard
yr	Year
Δ	Delta
%	Percent
~	Approximately
Ø	Phase
@	At
#	Pound; Number
<	Less Than
>	Greater Than

A

B

Index

INDEX

D

Index

F

Index

G

H

I

J

INDEX

N

O

P

T

U

V

X

Y

Z

Means Project Cost Report

By filling out and returning the Project Description, you can receive a discount of $20.00 off any one of the Means products advertised in the following pages. The cost information required includes all items marked (✔) except those where no costs occurred. The sum of all major items should equal the Total Project Cost.

$20.00 Discount per product for each report you submit.

DISCOUNT PRODUCTS AVAILABLE—FOR U.S. CUSTOMERS ONLY—STRICTLY CONFIDENTIAL

Project Description (No remodeling projects, please.)

✔ Type Building ______
✔ Location ______
Capacity ______
✔ Frame ______
✔ Exterior ______
✔ Basement: full ☐ partial ☐ none ☐ crawl ☐
✔ Height in Stories ______
✔ Total Floor Area ______
Ground Floor Area ______
✔ Volume in C.F. ______
% Air Conditioned ______ Tons ______
Comments ______

Owner ______
Architect ______
General Contractor ______
✔ Bid Date ______
Typical Bay Size ______
✔ Labor Force: ______ % Union ______ % Non-Union
✔ Project Description (Circle one number in each line)
1. Economy 2. Average 3. Custom 4. Luxury
1. Square 2. Rectangular 3. Irregular 4. Very Irregular

	✔ **Total Project Cost**			$
A	✔ **General Conditions**			$
B	✔ **Site Work**			$
BS	**Site Clearing & Improvement**			
BE	*Excavation*	(	C.Y.)	
BF	*Caissons & Piling*	(	L.F.)	
BU	*Site Utilities*			
BP	*Roads & Walks Exterior Paving*	(	S.Y.)	
C	✔ **Concrete**			$
C	*Cast in Place*	(	C.Y.)	
CP	*Precast*	(	S.F.)	
D	✔ **Masonry**			$
DB	*Brick*	(	M)	
DC	*Block*	(	M)	
DT	*Tile*	(	S.F.)	
DS	*Stone*	(	S.F.)	
E	✔ **Metals**			$
ES	*Structural Steel*	(	Tons)	
EM	*Misc. & Ornamental Metals*			
F	✔ **Wood & Plastics**			$
FR	*Rough Carpentry*	(	MBF)	
FF	*Finish Carpentry*			
FM	*Architectural Millwork*			
G	✔ **Thermal & Moisture Protection**			$
GW	*Waterproofing-Dampproofing*	(	S.F.)	
GN	*Insulation*	(	S.F.)	
GR	*Roofing & Flashing*	(	S.F.)	
GM	*Metal Siding/Curtain Wall*	(	S.F.)	
H	✔ **Doors and Windows**			$
HD	*Doors*	(	Ea.)	
HW	*Windows*	(	S.F.)	
HH	*Finish Hardware*			
HG	*Glass & Glazing*	(	S.F.)	
HS	*Storefronts*	(	S.F.)	

J	✔ **Finishes**			$
JL	*Lath & Plaster*	(	S.Y.)	
JD	*Drywall*	(	S.F.)	
JM	*Tile & Marble*	(	S.F.)	
JT	*Terrazzo*	(	S.F.)	
JA	*Acoustical Treatment*	(	S.F.)	
JC	*Carpet*	(	S.Y.)	
JF	*Hard Surface Flooring*	(	S.F.)	
JP	*Painting & Wall Covering*	(	S.F.)	
K	✔ **Specialties**			$
KB	*Bathroom Partitions & Access.*	(	S.F.)	
KF	*Other Partitions*	(	S.F.)	
KL	*Lockers*	(	Ea.)	
L	✔ **Equipment**			$
LK	*Kitchen*			
LS	*School*			
LO	*Other*			
M	✔ **Furnishings**			$
MW	*Window Treatment*			
MS	*Seating*	(	Ea.)	
N	✔ **Special Construction**			$
NA	*Acoustical*	(	S.F.)	
NB	*Prefab. Bldgs.*	(	S.F.)	
NO	*Other*			
P	✔ **Conveying Systems**			$
PE	*Elevators*	(	Ea.)	
PS	*Escalators*	(	Ea.)	
PM	*Material Handling*			
Q	✔ **Mechanical**			$
QP	*Plumbing*	*(No. of fixtures*	*)*	
QS	*Fire Protection (Sprinklers)*			
QF	*Fire Protection (Hose Standpipes)*			
QB	*Heating, Ventilating & A.C.*			
QH	*Heating & Ventilating*	*(BTU Output*	*)*	
QA	*Air Conditioning*	(	*Tons)*	
R	✔ **Electrical**			$
RL	*Lighting*	(	S.F.)	
RP	*Power Service*			
RD	*Power Distribution*			
RA	*Alarms*			
RG	*Special Systems*			
S	✔ **Mech./Elec. Combined**			$

Product Name ______
Product Number ______
Your Name ______
Title ______
Company ______
☐ Company
☐ Home Street Address ______
City, State, Zip ______
☐ Please send ______ forms.

Please specify the Means product you wish to receive. Complete the address information as requested and return this form with your check (product cost less $20.00) to address below.

RSMeans Company, Inc.,
Square Foot Costs Department
P.O. Box 800
Kingston, MA 02364-9988

For more information visit Means Web Site at www.rsmeans.com

Reed Construction Data/RSMeans . . . a tradition of excellence in Construction Cost Information and Services since 1942.

Table of Contents

Book Selection Guide

The following table provides definitive information on the content of each cost data publication. The number of lines of data provided in each unit price or assemblies division, as well as the number of reference tables and crews is listed for each book. The presence of other elements such as an historical cost index, city cost indexes, square foot models or cross-referenced index is also indicated. You can use the table to help select the Means' book that has the quantity and type of information you most need in your work.

Unit Cost Divisions	Building Construction Costs	Mechanical	Electrical	Repair & Remodel.	Square Foot	Site Work Landsc.	Assemblies	Interior	Concrete Masonry	Open Shop	Heavy Construc.	Light Commercial	Facil. Construc.	Plumbing	Western Construction Costs	Resi-dential
1	1102	821	906	1012		1044		846	1013	1101	1050	751	1527	873	1101	709
2	2536	1320	460	1455		7430		545	1383	2495	4645	679	4194	1574	2518	758
3	1404	89	69	725		1239		204	1781	1400	1401	207	1315	47	1404	238
4	809	22	0	617		650		564	1017	785	589	373	1032	0	795	303
5	1788	212	167	916		723		909	647	1756	991	754	1770	287	1770	726
6	1232	82	78	1215		448		1216	312	1223	598	1371	1343	47	1571	1473
7	1260	155	71	1232		464		486	404	1259	346	954	1314	165	1260	746
8	1785	28	0	1862		294		1621	620	1771	49	1187	1980	0	1787	1166
9	1575	47	0	1417		234		1671	359	1532	156	1318	1792	49	1566	1206
10	858	47	25	493		194		697	179	860	0	391	879	232	858	215
11	1018	322	169	500		134		812	44	925	107	217	1077	291	925	103
12	307	0	0	46		210		1416	27	298	0	62	1420	0	298	63
13	1084	950	359	446		367		846	75	1068	258	445	1698	889	1050	193
14	342	55	0	255		36		291	0	342	30	12	341	33	340	6
15	1981	12970	617	1769		1562		1174	59	1992	1762	1198	10672	9474	2011	826
16	1282	470	9614	1003		739		1108	55	1300	760	1081	9381	415	1231	552
17	427	354	427	0		0		0	0	427	0	0	427	356	427	0
Totals	**20790**	**17950**	**12967**	**14963**		**15768**		**14406**	**7975**	**20534**	**12642**	**11000**	**42162**	**14742**	**20912**	**9283**

Assembly Divisions	Building Construction Costs	Mechanical	Elec-trical	Repair & Remodel.	Square Foot	Site Work Landsc.	Assemblies	Interior	Concrete Masonry	Open Shop	Heavy Construc.	Light Commercial	Facil. Construc.	Plumbing	Western Construction Costs	Asm Div	Resi-dential
A		19	0	192	150	540	612	0	550		542	149	24	0		1	374
B		0	0	810	2470	0	5578	333	1914		0	2020	145	0		2	217
C		0	0	635	810	0	1204	1514	145		0	710	266	0		3	588
D		1031	782	685	1814	0	2430	752	0		0	1310	1031	881		4	871
E		0	0	85	255	0	292	5	0		0	255	5	0		5	393
F		0	0	0	123	0	126	0	0		0	123	3	0		6	350
G		465	172	332	111	1836	584	0	482		432	110	87	540		7	299
																8	760
																9	80
																10	0
																11	0
																12	0
Totals		**1515**	**954**	**2739**	**5733**	**2376**	**10826**	**2604**	**3091**		**974**	**4677**	**1561**	**1421**			**3932**

Reference Section	Building Construction Costs	Mechanical	Electrical	Repair & Remodel.	Square Foot	Site Work Landsc.	Assemblies	Interior	Concrete Masonry	Open Shop	Heavy Construc.	Light Commercial	Facil. Construc.	Plumbing	Western Construction Costs	Resi-dential
Tables	129	45	85	69	4	80	219	46	71	130	61	57	104	51	131	42
Models					102							43				32
Crews	410	410	410	391		410		410	410	393	410	393	391	410	410	393
City Cost Indexes	yes	yes	yes	yes	yes	yes	yes	yes	yes	yes	yes	yes	yes	yes	yes	yes
Historical Cost Indexes	yes	yes	yes	yes	yes	yes	yes	yes	yes	yes	yes	yes	yes	yes	yes	no

For more information visit Means Web Site at www.rsmeans.com

Annual Cost Guides

Means Building Construction Cost Data 2004

Available in Both Softbound and Looseleaf Editions

The "Bible" of the industry comes in the standard softcover edition or the looseleaf edition.

Many customers enjoy the convenience and flexibility of the looseleaf binder, which increases the usefulness of *Means Building Construction Cost Data 2004* by making it easy to add and remove pages. You can insert your own cost information pages, so everything is in one place. Copying pages for faxing is easier also. Whichever edition you prefer, softbound or the convenient looseleaf edition, you'll be eligible to receive *The Change Notice* FREE. Current subscribers receive *The Change Notice* via e-mail.

$108.95 per copy, Softbound
Catalog No. 60014

$136.95 per copy, Looseleaf
Catalog No. 61014

Means Building Construction Cost Data 2004

Offers you unchallenged unit price reliability in an easy-to-use arrangement. Whether used for complete, finished estimates or for periodic checks, it supplies more cost facts better and faster than any comparable source. Over 23,000 unit prices for 2004. The City Cost Indexes cover over 930 areas, for indexing to any project location in North America. Order and get *The Change Notice* FREE. You'll have year-long access to the Means Estimating **HOTLINE** FREE with your subscription. Expert assistance when using Means data is just a phone call away.

$108.95 per copy
Over 650 pages, illustrated, available Oct. 2003
Catalog No. 60014

Means Building Construction Cost Data 2004

Metric Version

The Federal Government has stated that all federal construction projects must now use metric documentation. The *Metric Version* of *Means Building Construction Cost Data 2004* is presented in metric measurements covering all construction areas. Don't miss out on these billion dollar opportunities. Make the switch to metric today.

$108.95 per copy
Over 650 pages, illus., available Dec. 2003
Catalog No. 63014

For more information visit Means Web Site at www.rsmeans.com

Annual Cost Guides

Means Mechanical Cost Data 2004

• HVAC • Controls

Total unit and systems price guidance for mechanical construction...materials, parts, fittings, and complete labor cost information. Includes prices for piping, heating, air conditioning, ventilation, and all related construction.

Plus new 2004 unit costs for:

- Over 2500 installed HVAC/controls assemblies
- "On Site" Location Factors for over 930 cities and towns in the U.S. and Canada
- Crews, labor and equipment

$108.95 per copy
Over 600 pages, illustrated, available Oct. 2003
Catalog No. 60024

Means Plumbing Cost Data 2004

Comprehensive unit prices and assemblies for plumbing, irrigation systems, commercial and residential fire protection, point-of-use water heaters, and the latest approved materials. This publication and its companion, *Means Mechanical Cost Data*, provide full-range cost estimating coverage for all the mechanical trades.

$108.95 per copy
Over 500 pages, illustrated, available Oct. 2003
Catalog No. 60214

Means Electrical Cost Data 2004

Pricing information for every part of electrical cost planning: More than 15,000 unit and systems costs with design tables; clear specifications and drawings; engineering guides and illustrated estimating procedures; complete labor-hour and materials costs for better scheduling and procurement; the latest electrical products and construction methods.

- A Variety of Special Electrical Systems including Cathodic Protection
- Costs for maintenance, demolition, HVAC/mechanical, specialties, equipment, and more

$108.95 per copy
Over 450 pages, illustrated, available Oct. 2003
Catalog No. 60034

Means Electrical Change Order Cost Data 2004

You are provided with electrical unit prices exclusively for pricing change orders based on the recent, direct experience of contractors and suppliers. Analyze and check your own change order estimates against the experience others have had doing the same work. It also covers productivity analysis and change order cost justifications. With useful information for calculating the effects of change orders and dealing with their administration.

$108.95 per copy
Over 450 pages, available Oct. 2003
Catalog No. 60234

Means Facilities Maintenance & Repair Cost Data 2004

Published in a looseleaf format, *Means Facilities Maintenance & Repair Cost Data* gives you a complete system to manage and plan your facility repair and maintenance costs and budget efficiently. Guidelines for auditing a facility and developing an annual maintenance plan. Budgeting is included, along with reference tables on cost and management and information on frequency and productivity of maintenance operations.

The only nationally recognized source of maintenance and repair costs. Developed in cooperation with the Army Corps of Engineers.

$238.95 per copy
Over 600 pages, illustrated, available Dec. 2003
Catalog No. 60304

Means Square Foot Costs 2004

It's Accurate and Easy To Use!

- **Updated 2004 price information,** based on nationwide figures from suppliers, estimators, labor experts and contractors.
- "How-to-Use" Sections, with **clear examples** of commercial, residential, industrial, and institutional structures.
- Realistic graphics, offering true-to-life illustrations of building projects.
- Extensive information on using square foot cost data, including **sample estimates** and **alternate pricing methods.**

$119.95 per copy
Over 450 pages, illustrated, available Nov. 2003
Catalog No. 60054

Annual Cost Guides

For more information visit Means Web Site at www.rsmeans.com

Means Repair & Remodeling Cost Data 2004

Commercial/Residential

You can use this valuable tool to estimate commercial and residential renovation and remodeling.

Includes: New costs for hundreds of unique methods, materials and conditions that only come up in repair and remodeling. PLUS:

- Unit costs for over 16,000 construction components
- Installed costs for over 90 assemblies
- Costs for 300+ construction crews
- Over 930 "On Site" localization factors for the U.S. and Canada.

$95.95 per copy
Over 600 pages, illustrated, available Nov. 2003
Catalog No. 60044

Means Facilities Construction Cost Data 2004

For the maintenance and construction of commercial, industrial, municipal, and institutional properties. Costs are shown for new and remodeling construction and are broken down into materials, labor, equipment, overhead, and profit. Special emphasis is given to sections on mechanical, electrical, furnishings, site work, building maintenance, finish work, and demolition. More than 45,000 unit costs, plus assemblies and reference sections are included.

$263.95 per copy
Over 1150 pages, illustrated, available Nov. 2003
Catalog No. 60204

Means Residential Cost Data 2004

Contains square foot costs for 30 basic home models with the look of today, plus hundreds of custom additions and modifications you can quote right off the page. With costs for the 100 residential systems you're most likely to use in the year ahead. Complete with blank estimating forms, sample estimates and step-by-step instructions.

$95.95 per copy
Over 550 pages, illustrated, available Oct. 2003
Catalog No. 60174

Means Light Commercial Cost Data 2004

Specifically addresses the light commercial market, which is an increasingly specialized niche in the industry. Aids you, the owner/designer/contractor, in preparing all types of estimates, from budgets to detailed bids. Includes new advances in methods and materials. Assemblies section allows you to evaluate alternatives in the early stages of design/planning.

Over 13,000 unit costs for 2004 ensure you have the prices you need...when you need them.

$96.95 per copy
Over 600 pages, illustrated, available Nov. 2003
Catalog No. 60184

Means Assemblies Cost Data 2004

Means Assemblies Cost Data 2004 takes the guesswork out of preliminary or conceptual estimates. Now you don't have to try to calculate the assembled cost by working up individual components costs. We've done all the work for you.

Presents detailed illustrations, descriptions, specifications and costs for every conceivable building assembly—240 types in all—arranged in the easy-to-use UNIFORMAT II system. Each illustrated "assembled" cost includes a complete grouping of materials and associated installation costs including the installing contractor's overhead and profit.

$179.95 per copy
Over 550 pages, illustrated, available Oct. 2003
Catalog No. 60064

Means Site Work & Landscape Cost Data 2004

Means Site Work & Landscape Cost Data 2004 is organized to assist you in all your estimating needs. Hundreds of fact-filled pages help you make accurate cost estimates efficiently.

Updated for 2004!

- Demolition features—including ceilings, doors, electrical, flooring, HVAC, millwork, plumbing, roofing, walls and windows
- State-of-the-art segmental retaining walls
- Flywheel trenching costs and details
- Updated Wells section
- Landscape materials, flowers, shrubs and trees

$108.95 per copy
Over 600 pages, illustrated, available Nov. 2003
Catalog No. 60284

Annual Cost Guides

Means Open Shop Building Construction Cost Data 2004

The latest costs for accurate budgeting and estimating of new commercial and residential construction... renovation work... change orders... cost engineering. *Means Open Shop BCCD* will assist you to...

- Develop benchmark prices for change orders
- Plug gaps in preliminary estimates, budgets
- Estimate complex projects
- Substantiate invoices on contracts
- Price ADA-related renovations

$108.95 per copy
Over 650 pages, illustrated, available Dec. 2003
Catalog No. 60154

Means Heavy Construction Cost Data 2004

A comprehensive guide to heavy construction costs. Includes costs for highly specialized projects such as tunnels, dams, highways, airports, and waterways. Information on different labor rates, equipment, and material costs is included. Has unit price costs, systems costs, and numerous reference tables for costs and design. Valuable not only to contractors and civil engineers, but also to government agencies and city/town engineers.

$108.95 per copy
Over 450 pages, illustrated, available Nov. 2003
Catalog No. 60164

Means Building Construction Cost Data 2004

Western Edition

This regional edition provides more precise cost information for western North America. Labor rates are based on union rates from 13 western states and western Canada. Included are western practices and materials not found in our national edition: tilt-up concrete walls, glu-lam structural systems, specialized timber construction, seismic restraints, landscape and irrigation systems.

$108.95 per copy
Over 600 pages, illustrated, available Dec. 2003
Catalog No. 60224

Means Heavy Construction Cost Data 2004

Metric Version

Make sure you have the Means industry standard metric costs for the federal, state, municipal and private marketplace. With thousands of up-to-date metric unit prices in tables by CSI standard divisions. Supplies you with assemblies costs using the metric standard for reliable cost projections in the design stage of your project. Helps you determine sizes, material amounts, and has tips for handling metric estimates.

$108.95 per copy
Over 450 pages, illustrated, available Dec. 2003
Catalog No. 63164

Means Construction Cost Indexes 2004

Who knows what 2004 holds? What materials and labor costs will change unexpectedly? By how much?

- Breakdowns for 316 major cities.
- National averages for 30 key cities.
- Expanded five major city indexes.
- Historical construction cost indexes.

$237.95 per year
$59.50 individual quarters
Catalog No. 60144 A,B,C,D

Means Interior Cost Data 2004

Provides you with prices and guidance needed to make accurate interior work estimates. Contains costs on materials, equipment, hardware, custom installations, furnishings, labor costs . . . every cost factor for new and remodel commercial and industrial interior construction, including updated information on office furnishings, plus more than 50 reference tables. For contractors, facility managers, owners.

$108.95 per copy
Over 550 pages, illustrated, available Oct. 2003
Catalog No. 60094

Means Concrete & Masonry Cost Data 2004

Provides you with cost facts for virtually all concrete/masonry estimating needs, from complicated formwork to various sizes and face finishes of brick and block, all in great detail. The comprehensive unit cost section contains more than 8,500 selected entries. Also contains an assemblies cost section, and a detailed reference section which supplements the cost data.

$99.95 per copy
Over 450 pages, illustrated, available Dec. 2003
Catalog No. 60114

Means Labor Rates for the Construction Industry 2004

Complete information for estimating labor costs, making comparisons and negotiating wage rates by trade for over 300 cities (United States and Canada). With 46 construction trades listed by local union number in each city, and historical wage rates included for comparison. No similar book is available through the trade.

Each city chart lists the county and is alphabetically arranged with handy visual flip tabs for quick reference.

$239.95 per copy
Over 300 pages, available Dec. 2003
Catalog No. 60124

Residential & Light Commercial Construction Standards, 2nd Edition

By RSMeans and Contributing Authors

New, updated second edition of this unique collection of industry standards that define quality construction. For contractors, subcontractors, owners, developers, architects, engineers, attorneys, and insurance personnel, this book provides authoritative requirements and recommendations compiled from the nation's leading professional associations, industry publications, and building code organizations. This one-stop reference is enhanced by helpful commentary from respected practitioners, including identification of items most frequently targeted for construction defect claims. The new second edition provides the latest building code requirements.

$59.95 per copy
600 pages, illustrated, Softcover
Catalog No. 67322A

Designing & Building with the IBC, 2nd Edition

By Rolf Jensen & Associates, Inc.

This updated comprehensive guide helps building professionals make the transition to the 2003 *International Building Code*®. Includes a side-by-side code comparison of the IBC 2003 to the IBC 2000 and the three primary model codes, a quick-find index, and professional code commentary.

- The Quick-Find Index lets you instantly identify the IBC equivalent to the model code sections you are familiar with: the *BOCA® National Building Code*, *Uniform Building Code*™, or *Standard Building Code*©.
- The Code Comparison highlights the differences between the IBC 2003, IBC 2000, and the model codes — in an easy-to-read, side-by-side format.
- The Code Commentary points out issues to watch out for in making the transition to the new international code.

With illustrations, abbreviations key, and an extensive Resource section.

$99.95 per copy
Over 875 pages
Catalog No. 67328A

Means Estimating Handbook, 2nd Edition

By RSMeans

Updated new Second Edition answers virtually any estimating technical question - all organized by CSI MasterFormat. This comprehensive reference covers the full spectrum of technical data required to estimate construction costs. The book includes information on sizing, productivity, equipment requirements, code-mandated specifications, design standards and engineering factors.

This reference—widely used in the industry for tasks ranging from routine estimates to special cost analysis projects—has been completely updated with new and expanded technical information and an enhanced format for ease-of-use.

$99.95 per copy
Over 900 pages, Hardcover
Catalog No. 67276A

For more information
visit Means Web Site
at www.rsmeans.com

Reference Books

Value Engineering: Practical Applications

. . . For Design, Construction, Maintenance & Operations

By Alphonse Dell'Isola, PE

A tool for immediate application—for engineers, architects, facility managers, owners, and contractors. Includes: Making the Case for VE—The Management Briefing, Integrating VE into Planning and Budgeting, Conducting Life Cycle Costing, Integrating VE into the Design Process, Using VE Methodology in Design Review and Consultant Selection, Case Studies, VE Workbook, and a Life Cycle Costing program on disk.

$79.95 per copy
Over 450 pages, illustrated, Hardcover
Catalog No. 67319

Facilities Operations & Engineering Reference

By the Association for Facilities Engineering and RSMeans

An all-in-one technical referance for planning and managing facility projects and solving day-to-day operations problems. Selected as the official Certified Plant Engineer reference, this handbook covers financial analysis, maintenance, HVAC and energy efficiency, and more.

$109.95 per copy
Over 700 pages, illustrated, Hardcover
Catalog No. 67318

Builder's Essentials: Advanced Framing Methods

By Scot Simpson

A highly illustrated, "framer-friendly" approach to advanced framing elements. Provides expert, but easy-to-interpret, instruction for laying out and framing complex walls, roofs, and stairs, and special requirements for earthquake and hurricane protection. Also helps bring framers up to date on the latest building code changes, and provides tips on the lead framer's role and responsibilities, how to prepare for a job, and how to get the crew started.

$24.95 per copy
250 pages, illustrated, Softcover
Catalog No. 67330

HVAC: Design Criteria, Options, Selection

Expanded 2nd Edition

By William H. Rowe III, AIA, PE

Includes Indoor Air Quality, CFC Removal, Energy Efficient Systems, and Special Systems by Building Type. Helps you solve a wide range of HVAC system design and selection problems effectively and economically. Gives you clear explanations of the latest ASHRAE standards.

$84.95 per copy
Over 600 pages, illustrated, Hardcover
Catalog No. 67306

Total Productive Facilities Management

By Richard W. Sievert, Jr.

This book provides a comprehensive program to:

- Achieve business goals by optimizing facility resources
- Implement best practices through benchmarking, evaluation & project management
- Increase your value to the organization

$29.98 per copy
Over 270 pages, illustrated, Hardcover
Catalog No. 67321

Facilities Maintenance Management

By Gregory H. Magee, PE

Now you can get successful management methods and techniques for all aspects of facilities maintenance. This comprehensive reference explains and demonstrates successful management techniques for all aspects of maintenance, repair, and improvements for buildings, machinery, equipment, and grounds. Plus, guidance for outsourcing and managing internal staffs.

$86.95 per copy
Over 280 pages with illustrations, Hardcover
Catalog No. 67249

Cost Planning & Estimating for Facilities Maintenance

In this unique book, a team of facilities management authorities shares their expertise at:

- Evaluating and budgeting maintenance operations
- Maintaining & repairing key building components
- Applying ***Means Facilities Maintenance & Repair Cost Data*** to your estimating

Covers special maintenance requirements of the 10 major building types.

$89.95 per copy
Over 475 pages, Hardcover
Catalog No. 67314

The Building Professional's Guide to Contract Documents

New 3rd Edition

By Waller S. Poage, AIA, CSI, CVS

This comprehensive treatment of Contract Documents is an important reference for owners, design professionals, contractors, and students.

- Structure your Documents for Maximum Efficiency
- Effectively communicate construction requirements to all concerned
- Understand the Roles and Responsibilities of Construction Professionals
- Improve Methods of Project Delivery

$64.95 per copy, 400 pages
Diagrams and construction forms, Hardcover
Catalog No. 67261A

Reference Books

For more information visit Means Web Site at www.rsmeans.com

Builder's Essentials: Plan Reading & Material Takeoff

By Wayne J. DelPico

For Residential and Light Commercial Construction

A valuable tool for understanding plans and specs, and accurately calculating material quantities.
Step-by-step instructions and takeoff procedures based on a full set of wórking drawings.

$35.95 per copy
Over 420 pages, Softcover
Catalog No. 67307

Planning & Managing Interior Projects

2nd Edition

By Carol E. Farren, CFM

Addresses changes in technology and business, guiding you through commercial design and construction from initial client meetings to post-project administration. Includes: evaluating space requirements, alternative work models, telecommunications and data management, and environmental issues.

$69.95 per copy
Over 400 pages, illustrated, Hardcover
Catalog No. 67245A

Builder's Essentials: Best Business Practices for Builders & Remodelers: An Easy-to-Use Checklist System

By Thomas N. Frisby

A comprehensive guide covering all aspects of running a construction business, with more than 40 user–friendly checklists. This book provides expert guidance on: increasing your revenue and keeping more of your profit; planning for long-term growth; keeping good employees and managing subcontractors.

$29.95 per copy
Over 220 pages, Softcover
Catalog No. 67329

Builder's Essentials: Framing & Rough Carpentry 2nd Edition

By Scot Simpson

A complete training manual for apprentice and experienced carpenters. Develop and improve your skills with "framer-friendly," easy-to-follow instructions, and step-by-step illustrations. Learn proven techniques for framing walls, floors, roofs, stairs, doors, and windows. Updated guidance on standards, building codes, safety requirements, and more. Also available in Spanish!

$24.95 per copy
Over 150 pages, Softcover
Catalog No. 67298A Spanish Catalog No. 67298AS

How to Estimate with Means Data & CostWorks

By RSMeans and Saleh Mubarak

New 2nd Edition!

Learn estimating techniques using Means cost data. Includes an instructional version of Means CostWorks CD–ROM with Sample Building Plans.
The step-by-step guide takes you through all the major construction items. Over 300 sample estimating problems are included.

$59.95 per copy
Over 190 pages, Softcover
Catalog No. 67324A

Unit Price Estimating Methods

New 3rd Edition

This new edition includes up-to-date cost data and estimating examples, updated to reflect changes to the CSI numbering system and new features of Means cost data. It describes the most productive, universally accepted ways to estimate, and uses checklists and forms to illustrate shortcuts and timesavers. A model estimate demonstrates procedures. A new chapter explores computer estimating alternatives.

$59.95 per copy
Over 350 pages, illustrated, Hardcover
Catalog No. 67303A

Interior Home Improvement Costs, 8th Edition

Provides 66 updated estimates for the most popular projects including estimates for home offices, in-law apartments, and remodeling for disabled residents. Includes guidance for remodeling older homes as well as attic and basement conversions, kitchen and bath remodeling, lighting and security; fireplaces; storage; stairs, new floors, and walls and ceilings.

$19.95 per copy
250 pages, illustrated, Softcover
Catalog No. 67308D

Exterior Home Improvement Costs, 8th Edition

Provides quick estimates for 64 projects including room additions and garages; windows and doors; dormers, roofs, and skylights; decks and pergolas; paving and patios; painting and siding; walls, fences, and porches; landscaping and driveways. It also contains price comparison sheets for material and equipment.

$19.95 per copy
270 pages, illustrated, Softcover
Catalog No. 67309D

Means Landscape Estimating Methods

4th Edition

By Sylvia H. Fee

This revised edition offers expert guidance for preparing accurate estimates for new landscape construction and grounds maintenance. Includes a complete project estimate featuring the latest equipment and methods, and **two chapters on Life Cycle Costing, and Landscape Maintenance Estimating.**

$62.95 per copy
Over 300 pages, illustrated, Hardcover
Catalog No. 67295B

Means Environmental Remediation Estimating Methods, 2nd Edition

By Richard R. Rast

Guidelines for estimating 50 standard remediation technologies. Use it to prepare preliminary budgets, develop estimates, compare costs and solutions, estimate liability, review quotes, negotiate settlements.

A valuable support tool for ***Means Environmental Remediation Unit Price*** and ***Assemblies*** books.

$99.95 per copy
Over 750 pages, illustrated, Hardcover
Catalog No. 64777A

Reference Books

Means Illustrated Construction Dictionary, Condensed Edition, 2nd Edition

By RSMeans

Recognized in the industry as the best resource of its kind, this has been further enhanced with updates to existing terms and the addition of hundreds of new terms and illustrations . . . in keeping with the most recent developments in the industry.
The best portable dictionary for office or field use—an essential tool for contractors, architects, insurance and real estate personnel, homeowners, and anyone who needs quick, clear definitions for construction terms.

$59.95 per copy
Over 500 pages
Catalog No. 67282A

Means Repair and Remodeling Estimating

New 4th Edition

By Edward B. Wetherill & RSMeans

Focuses on the unique problems of estimating renovations of existing structures. It helps you determine the true costs of remodeling through careful evaluation of architectural details and a site visit.
New section on disaster restoration costs.

$69.95 per copy
Over 450 pages, illustrated, Hardcover
Catalog No. 67265B

Facilities Planning & Relocation

New, lower price and user-friendly format.

By David D. Owen

A complete system for planning space needs and managing relocations. Includes step-by-step manual, over 50 forms, and extensive reference section on materials and furnishings.

$89.95 per copy
Over 450 pages, Softcover
Catalog No. 67301

Means Square Foot & Assemblies Estimating Methods

New 3rd Edition!

Develop realistic Square Foot and Assemblies Costs for budgeting and construction funding. The new edition features updated guidance on square foot and assemblies estimating using UNIFORMAT II. An essential reference for anyone who performs conceptual estimates.

$69.95 per copy
Over 300 pages, illustrated, Hardcover
Catalog No. 67145B

Means Electrical Estimating Methods

3rd Edition

Expanded new edition includes sample estimates and cost information in keeping with the latest version of the CSI MasterFormat and UNIFORMAT II. Complete coverage of Fiber Optic and Uninterruptible Power Supply electrical systems, broken down by components and explained in detail. Includes a new chapter on computerized estimating methods. A practical companion to *Means Electrical Cost Data.*

$64.95 per copy
Over 325 pages, Hardcover
Catalog No. 67230A

Means Mechanical Estimating Methods

New 3rd Edition

This guide assists you in making a review of plans, specs, and bid packages with suggestions for takeoff procedures, listings, substitutions and pre-bid scheduling. Includes suggestions for budgeting labor and equipment usage. Compares materials and construction methods to allow you to select the best option.

$64.95 per copy
Over 350 pages, illustrated, Hardcover
Catalog No. 67294A

Successful Estimating Methods: From Concept to Bid

By John D. Bledsoe, PhD, PE

A highly practical, all-in-one guide to the tips and practices of today's successful estimator. Presents techniques for all types of estimates, and advanced topics such as life cycle cost analysis, value engineering, and automated estimating. Estimate spreadsheets available at Means Web site.

$32.48 per copy
Over 300 pages, illustrated, Hardcover
Catalog No. 67287

Means Spanish/English Construction Dictionary

By RSMeans, The International Conference of Building Officials (ICBO), and Rolf Jensen & Associates (RJA)

Designed to facilitate communication among Spanish- and English-speaking construction personnel—improving performance and job-site safety. Features the most common words and phrases used in the construction industry, with easy-to-follow pronunciations. Includes extensive building systems and tools illustrations.

$22.95 per copy
250 pages, illustrated, Softcover
Catalog No. 67327

Means Productivity Standards for Construction

Expanded Edition *(Formerly Man-Hour Standards)*

Here is the working encyclopedia of labor productivity information for construction professionals, with labor requirements for thousands of construction functions in CSI MasterFormat.
Completely updated, with over 3,000 new work items.

$49.98 per copy
Over 800 pages, Hardcover
Catalog No. 67236A

Project Scheduling & Management for Construction

New 2nd Edition

By David R. Pierce, Jr.

A comprehensive yet easy-to-follow guide to construction project scheduling and control—from vital project management principles through the latest scheduling, tracking, and controlling techniques. The author is a leading authority on scheduling with years of field and teaching experience at leading academic institutions. Spend a few hours with this book and come away with a solid understanding of this essential management topic.

$64.95 per copy
Over 250 pages, illustrated, Hardcover
Catalog No. 67247A

Reference Books

For more information visit Means Web Site at www.rsmeans.com

Cyberplaces: The Internet Guide for AECs & Facility Managers
By Paul Doherty, AIA, CSI
2nd Edition
$59.95 per copy
Catalog No. 67317A

Concrete Repair and Maintenance Illustrated
By Peter H. Emmons
$69.95 per copy
Catalog No. 67146

Superintending for Contractors:
How to Bring Jobs in On-Time, On-Budget
By Paul J. Cook
$35.95 per copy
Catalog No. 67233

HVAC Systems Evaluation
By Harold R. Colen, PE
$84.95 per copy
Catalog No. 67281

Basics for Builders: How to Survive and Prosper in Construction
By Thomas N. Frisby
$34.95 per copy
Catalog No. 67273

Successful Interior Projects Through Effective Contract Documents
By Joel Downey & Patricia K. Gilbert
Now $24.98 per copy, limited quantity
Catalog No. 67313

Building Spec Homes Profitably
By Kenneth V. Johnson
$29.95 per copy
Catalog No. 67312

Estimating for Contractors
How to Make Estimates that Win Jobs
By Paul J. Cook
$35.95 per copy
Catalog No. 67160

Means Forms for Building Construction Professionals
$47.48 per copy
Catalog No. 67231

Understanding Building Automation Systems
By Reinhold A. Carlson, PE & Robert Di Giandomenico
$29.98 per copy, limited quantity
Catalog No. 67284

Managing Construction Purchasing
By John G. McConville, CCC, CPE
Now $19.98 per copy, limited quantity
Catalog No. 67302

Fundamentals of the Construction Process
By Kweku K. Bentil, AIC
Now $34.98 per copy, limited quantity
Catalog No. 67260

Means ADA Compliance Pricing Guide
$29.98 per copy
Catalog No. 67310

Means Facilities Maintenance Standards
By Roger W. Liska, PE, AIC
Now $79.95 per copy
Catalog No. 67246

How to Estimate with Metric Units
Now $9.98 per copy, limited quantity
Catalog No. 67304

Means Plumbing Estimating Methods, 2nd Edition
Now $59.95 per copy, limited quantity
Catalog No. 67283A

Preventive Maintenance for Higher Education Facilities

By Applied Management Engineering, Inc.

An easy-to-use system to help facilities professionals establish the value of PM, and to develop and budget for an appropriate PM program for their college or university. Features interactive campus building models typical of those found in different-sized higher education facilities, such as dormitories, classroom buildings, laboratories, athletic facilities, and more. Includes PM checklists linked to each piece of equipment or system in hard copy and electronic format, along with required labor hours to complete the PM tasks. Helps users select and develop the best possible PM plan within budget.

$149.95 per copy
150 pages, Hardcover
Catalog No. 67337

Preventive Maintenance Guidelines for School Facilities

By John C. Maciha

This new publication is a complete PM program for K-12 schools that ensures sustained security, safety, property integrity, user satisfaction, and reasonable ongoing expenditures.

Includes schedules for weekly, monthly, semiannual, and annual maintenance. Comes as a 3-ring binder with hard copy and electronic forms available at Means website, and a laminated wall chart.

$149.95 per copy
Over 225 pages, Hardcover
Catalog No. 67326

Historic Preservation: Project Planning & Estimating

By Swanke Hayden Connell Architects

- *Managing Historic Restoration, Rehabilitation, and Preservation Building Projects and*
- *Determining and Controlling Their Costs*

The authors explain:

- How to determine whether a structure qualifies as historic
- Where to obtain funding and other assistance
- How to evaluate and repair more than 75 historic building materials
- How to properly research, document, and manage the project to meet code, agency, and other special requirements

$99.95 per copy
Over 675 pages, Hardcover
Catalog No. 67323

Means Illustrated Construction Dictionary, 3rd Edition

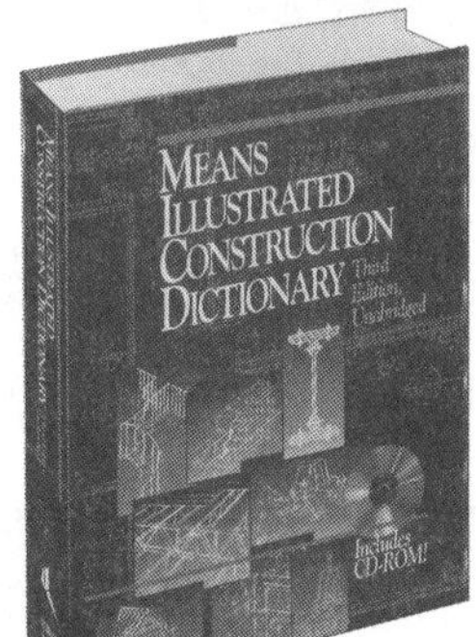

Long regarded as the Industry's finest, the Means *Illustrated Construction Dictionary* is now even better. With the addition of over 1,000 new terms and hundreds of new illustrations, it is the clear choice for the most comprehensive and current information.

The companion CD-ROM that comes with this new edition adds many extra features: larger graphics, expanded definitions, and links to both CSI MasterFormat numbers and product information.

- 19,000 construction words, terms, phrases, symbols, weights, measures, and equivalents
- 1,000 new entries
- 1,200 helpful illustrations
- Easy-to-use format, with thumbtabs

$99.95 per copy
Over 790 pages, illustrated Hardcover
Catalog No. 67292A

Seminars

Means CostWorks Training

This one-day seminar course has been designed with the intention of assisting both new and existing users to become more familiar with CostWorks program. The class is broken into two unique sections: (1) A one-half day presentation on the function of each icon; and each student will be shown how to use the software to develop a cost estimate. (2) Hands-on estimating exercises that will ensure that each student thoroughly understands how to use CostWorks. You must bring your own laptop computer to this course.

CostWorks Benefits/Features:
- Estimate in your own spreadsheet format
- Power of Means National Database
- Database automatically regionalized
- Save time with keyword searches
- Save time by establishing common estimate items in "Bookmark" files
- Customize your spreadsheet template
- Hot key to Product Manufacturers' listings and specs
- Merge capability for networking environments
- View crews and assembly components
- AutoSave capability
- Enhanced sorting capability

Unit Price Estimating

This interactive two-day seminar teaches attendees how to interpret project information and process it into final, detailed estimates with the greatest accuracy level.

The single most important credential an estimator can take to the job is the ability to visualize construction in the mind's eye, and thereby estimate accurately.

Some Of What You'll Learn:
- Interpreting the design in terms of cost
- The most detailed, time tested methodology for accurate "pricing"
- Key cost drivers—material, labor, equipment, staging and subcontracts
- Understanding direct and indirect costs for accurate job cost accounting and change order management

Who Should Attend: Corporate and government estimators and purchasers, architects, engineers...and others needing to produce accurate project estimates.

Square Foot and Assemblies Cost Estimating

This two-day course teaches attendees how to quickly deliver accurate square foot estimates using limited budget and design information.

Some Of What You'll Learn:
- How square foot costing gets the estimate done faster
- Taking advantage of a "systems" or "assemblies" format
- The Means "building assemblies/square foot cost approach"
- How to create a very reliable preliminary and systems estimate using bare-bones design information

Who Should Attend: Facilities managers, facilities engineers, estimators, planners, developers, construction finance professionals...and others needing to make quick, accurate construction cost estimates at commercial, government, educational and medical facilities.

Repair and Remodeling Estimating

This two-day seminar emphasizes all the underlying considerations unique to repair/remodeling estimating and presents the correct methods for generating accurate, reliable R&R project costs using the unit price and assemblies methods.

Some Of What You'll Learn:
- Estimating considerations—like labor-hours, building code compliance, working within existing structures, purchasing materials in smaller quantities, unforeseen deficiencies
- Identifies problems and provides solutions to estimating building alterations
- Rules for factoring in minimum labor costs, accurate productivity estimates and allowances for project contingencies
- R&R estimating examples are calculated using unite prices and assemblies data

Who Should Attend: Facilities managers, plant engineers, architects, contractors, estimators, builders...and others who are concerned with the proper preparation and/or evaluation of repair and remodeling estimates.

Mechanical and Electrical Estimating

This two-day course teaches attendees how to prepare more accurate and complete mechanical/electrical estimates, avoiding the pitfalls of omission and double-counting, while understanding the composition and rationale within the Means Mechanical/Electrical database.

Some Of What You'll Learn:
- The unique way mechanical and electrical systems are interrelated
- M&E estimates, conceptual, planning, budgeting and bidding stages
- Order of magnitude, square foot, assemblies and unit price estimating
- Comparative cost analysis of equipment and design alternatives

Who Should Attend: Architects, engineers, facilities managers, mechanical and electrical contractors...and others needing a highly reliable method for developing, understanding and evaluating mechanical and electrical contracts.

Plan Reading and Material Takeoff

This two-day program teaches attendees to read and understand construction documents and to use them in the preparation of material takeoffs.

Some of What You'll Learn:
- Skills necessary to read and understand typical contract documents—blueprints and specifications
- Details and symbols used by architects and engineers
- Construction specifications' importance in conjunction with blueprints
- Accurate takeoff of construction materials and industry-accepted takeoff methods

Who Should Attend: Facilities managers, construction supervisors, office managers...and other responsible for the execution and administration of a construction project including government, medical, commercial, educational or retail facilities.

Facilities Maintenance and Repair Estimating

This two-day course teaches attendees how to plan, budget, and estimate the cost of ongoing and preventive maintenance and repair for existing buildings and grounds.

Some Of What You'll Learn:
- The most financially favorable maintenance, repair and replacement scheduling and estimating
- Auditing and value engineering facilities
- Preventive planning and facilities upgrading
- Determining both in-house and contract-out service costs; annual, asset-protecting M&R plan

Who Should Attend: Facility managers, maintenance supervisors, buildings and grounds superintendents, plant managers, planners, estimators...and others involved in facilities planning and budgeting.

Scheduling and Project Management

This two-day course teaches attendees the most current and proven scheduling and management techniques needed to bring projects in on time and on budget.

Some Of What You'll Learn:
- Crucial phases of planning and scheduling
- How to establish project priorities, develop realistic schedules and management techniques
- Critical Path and Precedence Methods
- Special emphasis on cost control

Who Should Attend: Construction project managers, supervisors, engineers, estimators, contractors...and others who want to improve their project planning, scheduling and management skills.

Advanced Project Management

This two-day seminar will teach you how to effectively manage and control the entire design-build process and allow you to take home tangible skills that will be immediately applicable on existing projects.

Some Of What You'll Learn:
- Value engineering, bonding, fast-tracking and bid package creation
- How estimates and schedules can be integrated to provide advanced project management tools
- Cost engineering, quality control, productivity measurement and improvement
- Front loading a project and predicting its cash flow

Who Should Attend: Owners, project managers, architectural and engineering managers, construction managers, contractors...and anyone else who is responsible for the timely design and completion of construction projects.

Seminars

For more information visit Means Web Site at www.rsmeans.com

2004 Means Seminar Schedule

Location	Dates
Las Vegas, NV	March TBD
Washington, DC	April TBD
Denver, CO	May TBD
San Francisco, CA	June 7-10
Cape Cod, MA	September TBD
Washington, DC	September TBD
Las Vegas, NV	October TBD
Orlando, FL	November TBD
Atlantic City, NJ	November 15-18
San Diego, CA	December 13-16

Note: Call for exact dates and details.

Registration Information

Register Early... Save up to $125! Register 30 days before the start date of a seminar and save $125 off your total fee. *Note: This discount can be applied only once per order. It cannot be applied to team discount registrations or any other special offer.*

How to Register Register by phone today! Means toll-free number for making reservations is: **1-800-334-3509, ext. 5115.**

Individual Seminar Registration Fee $895. Individual CostWorks Training Registration Fee $349. To register by mail, complete the registration form and return with your full fee to: Seminar Division, Reed Construction Data, RSMeans Seminars, 63 Smiths Lane, Kingston, MA 02364.

Federal Government Pricing All Federal Government employees save 25% off regular seminar price. Other promotional discounts cannot be combined with Federal Government discount.

Team Discount Program Two to four seminar registrations: $760 per person–Five or more seminar registrations: $710 per person–Ten or more seminar registrations: Call for pricing.

Consecutive Seminar Offer One individual signing up for two separate courses at the same location during the designated time period pays only $1,435. You get the second course for only $540 **(a 40% discount)**. Payment must be received at least ten days prior to seminar dates to confirm attendance.

Refund Policy Cancellations will be accepted up to ten days prior to the seminar start. There are no refunds for cancellations received later than ten working days prior to the first day of the seminar. A $150 processing fee will be applied for all cancellations. Written notice of cancellation is required . Substitutions can be made at anytime before the session starts. **No-shows are subject to the full seminar fee.**

AACE Approved Courses The RSMeans Construction Estimating and Management Seminars described and offered to you here have each been approved for 14 hours (1.4 recertification credits) of credit by the AACE International Certification Board toward meeting the continuing education requirements for re-certification as a Certified Cost Engineer/Certified Cost Consultant.

AIA Continuing Education We are registered with the AIA Continuing Education System (AIA/CES) and is committed to developing quality learning activities in accordance with the CES criteria. RSMeans seminars meet the AIA/CES criteria for Quality Level 2. AIA members will receive (28) learning units (LUs) for each two-day RSMeans Course.

Daily Course Schedule The first day of each seminar session begins at 8:30 A.M. and ends at 4:30 P.M. The second day is 8:00 A.M.–4:00 P.M. Participants are urged to bring a hand-held calculator since many actual problems will be worked out in each session.

Continental Breakfast Your registration includes the cost of a continental breakfast, a morning coffee break, and an afternoon break. These informal segments will allow you to discuss topics of mutual interest with other members of the seminar. (You are free to make your own lunch and dinner arrangements.)

Hotel/Transportation Arrangements RSMeans has arranged to hold a block of rooms at each hotel hosting a seminar. To take advantage of special group rates when making your reservation, be sure to mention that you are attending the Means Seminar. You are, of course, free to stay at the lodging place of your choice. **(Hotel reservations and transportation arrangements should be made directly by seminar attendees.)**

Important Class sizes are limited, so please register as soon as possible.

Note: *Pricing subject to change.*

Registration Form

Call 1-800-334-3509 x5115 to register or FAX 1-800-632-6732. Visit our Web site www.rsmeans.com

Please register the following people for the Means Construction Seminars as shown here. Full payment or deposit is enclosed, and we understand that we must make our own hotel reservations if overnight stays are necessary.

☐ Full payment of $ ______________________ enclosed.

☐ Bill me

Name of Registrant(s)
(To appear on certificate of completion)

____

____

____

P.O. #: ______________________
GOVERNMENT AGENCIES MUST SUPPLY PURCHASE ORDER NUMBER

Firm Name ______________________

Address ______________________

City/State/Zip ______________________

Telephone No. fax No. ______________________

E-mail Address ______________________

Charge our registration(s) to: ☐ MasterCard ☐ VISA ☐ American Express ☐ Discover

Account No. ______________________ Exp. Date ______________________

Cardholder's Signature ______________________

Seminar Name City Dates

____

____

Please mail check to: Seminar Division, Reed Construction Data, RSMeans Seminars, 63 Smiths Lane, P.O.Box 800, Kingston, MA 02364 USA

MeansData™

CONSTRUCTION COSTS FOR SOFTWARE APPLICATIONS

Your construction estimating software is only as good as your cost data.

A proven construction cost database is a mandatory part of any estimating package. The following list of software providers can offer you MeansData™ as an added feature for their estimating systems. See the table below for what types of products and services they offer (match their numbers). Visit online at **www.rsmeans.com/demosource/** for more information and free demos. Or call their numbers listed below.

1. **3D International**
713-871-7000
venegas@3di.com
2. **4Clicks-Solutions, LLC**
719-574-7721
mbrown@4clicks-solutions.com
3. **Aepco, Inc.**
301-670-4642
blueworks@aepco.com
4. **Applied Flow Technology**
800-589-4943
info@aft.com
5. **ArenaSoft Estimating**
888-370-8806
info@arenasoft.com
6. **ARES Corporation**
925-299-6700
sales@arescorporation.com
7. **BSD - Building Systems Design, Inc.**
888-273-7638
bsd@bsdsoftlink.com
8. **CMS - Computerized Micro Solutions**
800-255-7407
cms@proest.com
9. **Corecon Technologies, Inc.**
714-895-7222
sales@corecon.com
10. **Estimating Systems, Inc.**
800-967-8527
esipulsar@adelphia.net
11. **G/C Emuni**
514-633-8339
rpa@gcei.com
12. **LUQS International**
888-682-5573
info@luqs.com
13. **MAESTRO Estimator**
Schwaab Technology Solutions, Inc.
281-578-3039
Stefan@schwaabtech.com
14. **Magellan K-12**
936-441-1744
sam.wilson@magellan-K12.com
15. **MC² - Management Computer**
800-225-5622
vkeys@mc2-ice.com
16. **Maximus Asset Solutions**
800-659-9001
assetsolutions@maximus.com
17. **Prism Computer Corp.**
800-774-7622
famis@prismcc.com
18. **Quest Solutions, Inc.**
800-452-2342
info@questsolutions.com
19. **Shaw Beneco Enterprises, Inc.**
877-719-4748
inquire@beneco.com
20. **Timberline Software Corp.**
800-628-6583
product.info@timberline.com
21. **TMA Systems, LLC**
800-862-1130
sales@tmasys.com
22. **US Cost, Inc.**
800-372-4003
sales@uscost.com
23. **Vanderweil Facility Advisors**
671-451-5100
info@VFA.com
24. **Vertigraph, Inc.**
800-989-4243
info-request@vertigraph.com
25. **WinEstimator, Inc.**
800-950-2374
sales@winest.com

TYPE	1	2	3	4	5	6	7	8	9	10	11	12	13	14	15	16	17	18	19	20	21	22	23	24	25
BID					●		●				●				●			●		●				●	●
Estimating		●			●	●	●	●	●	●	●	●	●		●			●	●	●		●		●	●
DOC/JOC/SABER		●			●		●			●						●			●	●					●
IDIQ		●								●						●			●						●
Asset Mgmt.														●		●	●						●		●
Facility Mgmt.	●		●										●	●		●	●				●		●		
Project Mgmt.	●	●				●			●				●	●					●	●					
TAKE-OFF					●				●			●			●			●		●		●		●	●
EARTHWORK									●			●			●			●						●	
Pipe Flow				●																					
HVAC/Plumbing					●							●													
Roofing					●																				
Design	●				●														●	●					●
Other Offers/Links:																									
Accounting/HR		●			●														●	●					
Scheduling					●								●							●		●		●	●
CAD														●						●					●
PDA																				●		●			●
Lt. Versions		●																		●		●			●
Consulting	●	●			●						●		●			●	●			●		●	●		●
Training		●			●		●		●	●			●			●	●	●	●	●	●	●	●	●	●

Reseller applications now being accepted. Call Carol Polio Ext. 5107.

FOR MORE INFORMATION

CALL 1-800-448-8182, EXT. 5107 OR FAX 1-800-632-6732

New Titles

Building Security: Strategies & Costs

By David Owen

Seeking ways to identify your facility's risk, balance security needs with safe access and a building that functions efficiently? Theft, assault, corporate espionage, unauthorized systems access, terrorism, hurricanes and tornadoes, sabotage, vandalism, fire, explosions, and other threats... All are considerations as building owners, facility managers, and design and construction professionals seek ways to meet security needs.

This comprehensive resource will help you evaluate your facility's security needs — and design and budget for the materials and devices needed to fulfill them. The text and cost data will help you to:

- Identify threats, probability of occurrence, and the potential losses— to determine and address your real vulnerabilities.
- Perform a detailed risk assessment of an existing facility, and prioritize and budget for security enhancement.
- Evaluate and price security systems and construction solutions, so you can make cost-effective choices.

Includes over 130 pages of Means Cost Data for installation of security systems and materials, plus a review of more than 50 security devices and construction solutions—how they work, and how they compare.

$89.95 per copy
Over 400 pages, Hardcover
Catalog No. 37339

Green Building: Project Planning & Cost Estimating

By RSMeans and Contributing Authors

Written by a team of leading experts in sustainable design, this new book is a complete guide to planning and estimating green building projects, a growing trend in building design and construction – commercial, industrial, institutional and residential. It explains:

- All the different criteria for "green-ness"
- What criteria your building needs to meet to get a LEED, Energy Star, or other recognized rating for green buildings
- How the project team works differently on a green versus a traditional building project
- How to select and specify green products
- How to evaluate the cost and value of green products versus conventional ones — not only for their first (installation) cost, but their cost over time (in maintenance and operation).

Features an extensive Green Building Cost Data section, which details the available products, how they are specified, and how much they cost.

$89.95 per copy
350 pages, illustrated, Hardcover
Catalog No. 67338

Life Cycle Costing for Facilities

By Alphonse Dell'Isola and Dr. Steven Kirk

Guidance for achieving higher quality design and construction projects at lower costs!

Facility designers and owners are frustrated with cost-cutting efforts that yield the cheapest product, but sacrifice quality. Life Cycle Costing, properly done, enables them to achieve both—high quality, incorporating innovative design, and costs that meet their budgets.

The authors show how LCC can work for a broad variety of projects – from several types of buildings, to roads and bridges, to HVAC and electrical upgrades, to materials and equipment procurement. Case studies include:

- Health care and nursing facilities
- College campus and high schools
- Office buildings, courthouses, and banks
- Exterior walls, elevators, lighting, HVAC, and more

The book's extensive cost section provides maintenance and replacement costs for facility elements.

$99.95 per copy
450 pages, Hardcover
Catalog No. 67341

ORDER TOLL FREE 1-800-334-3509
OR FAX 1-800-632-6732.

2004 Order Form

Qty.	Book No.	COST ESTIMATING BOOKS	Unit Price	Total
	60064	Assemblies Cost Data 2004	$179.95	
	60014	Building Construction Cost Data 2004	108.95	
	61014	Building Const. Cost Data–Looseleaf Ed. 2004	136.95	
	63014	Building Const. Cost Data–Metric Version 2004	108.95	
	60224	Building Const. Cost Data–Western Ed. 2004	108.95	
	60114	Concrete & Masonry Cost Data 2004	99.95	
	60144	Construction Cost Indexes 2004	237.95	
	60144A	Construction Cost Index–January 2004	59.50	
	60144B	Construction Cost Index–April 2004	59.50	
	60144C	Construction Cost Index–July 2004	59.50	
	60144D	Construction Cost Index–October 2004	59.50	
	60344	Contr. Pricing Guide: Resid. R & R Costs 2004	39.95	
	60334	Contr. Pricing Guide: Resid. Detailed 2004	39.95	
	60324	Contr. Pricing Guide: Resid. Sq. Ft. 2004	39.95	
	64024	ECHOS Assemblies Cost Book 2004	179.95	
	64014	ECHOS Unit Cost Book 2004	119.95	
	54004	ECHOS (Combo set of both books)	251.95	
	60234	Electrical Change Order Cost Data 2004	108.95	
	60034	Electrical Cost Data 2004	108.95	
	60204	Facilities Construction Cost Data 2004	263.95	
	60304	Facilities Maintenance & Repair Cost Data 2004	238.95	
	60164	Heavy Construction Cost Data 2004	108.95	
	63164	Heavy Const. Cost Data–Metric Version 2004	108.95	
	60094	Interior Cost Data 2004	108.95	
	60124	Labor Rates for the Const. Industry 2004	239.95	
	60184	Light Commercial Cost Data 2004	96.95	
	60024	Mechanical Cost Data 2004	108.95	
	60154	Open Shop Building Const. Cost Data 2004	108.95	
	60214	Plumbing Cost Data 2004	108.95	
	60044	Repair and Remodeling Cost Data 2004	95.95	
	60174	Residential Cost Data 2004	95.95	
	60284	Site Work & Landscape Cost Data 2004	108.95	
	60054	Square Foot Costs 2004	119.95	
		REFERENCE BOOKS		
	67147A	ADA in Practice	59.98	
	67310	ADA Pricing Guide	29.98	
	67273	Basics for Builders: How to Survive and Prosper	34.95	
	67330	Bldrs Essentials: Adv. Framing Methods	24.95	
	67329	Bldrs Essentials: Best Bus. Practices for Bldrs	29.95	
	67298A	Bldrs Essentials: Framing/Carpentry 2nd Ed.	24.95	
	67298AS	Bldrs Essentials: Framing/Carpentry Spanish	24.95	
	67307	Bldrs Essentials: Plan Reading & Takeoff	35.95	
	67261A	Bldg. Prof. Guide to Contract Documents 3rd Ed.	64.95	
	67339	Building Security: Strategies & Costs	89.95	
	67312	Building Spec Homes Profitably	29.95	
	67146	Concrete Repair & Maintenance Illustrated	69.95	
	67314	Cost Planning & Est. for Facil. Maint.	89.95	
	67317A	Cyberplaces: The Internet Guide 2nd Ed.	59.95	
	67328A	Designing & Building with the IBC, 2nd Ed.	99.95	
	67230B	Electrical Estimating Methods 3rd Ed.	64.95	

Qty.	Book No.	REFERENCE BOOKS (Cont.)	Unit Price	Total
	64777A	Environmental Remediation Est. Methods 2nd Ed.	$ 99.95	
	67160	Estimating for Contractors	35.95	
	67276A	Estimating Handbook 2nd Ed.	99.95	
	67249	Facilities Maintenance Management	86.95	
	67246	Facilities Maintenance Standards	79.95	
	67318	Facilities Operations & Engineering Reference	109.95	
	67301	Facilities Planning & Relocation	89.95	
	67231	Forms for Building Const. Professional	47.48	
	67260	Fundamentals of the Construction Process	34.98	
	67323	Historic Preservation: Proj. Planning & Est.	99.95	
	67308D	Home Improvement Costs–Int. Projects 8th Ed.	19.95	
	67309D	Home Improvement Costs–Ext. Projects 8th Ed.	19.95	
	67324A	How to Est. w/Means Data & CostWorks 2nd Ed.	59.95	
	67304	How to Estimate with Metric Units	9.98	
	67306	HVAC: Design Criteria, Options, Select. 2nd Ed.	84.95	
	67281	HVAC Systems Evaluation	84.95	
	67282A	Illustrated Construction Dictionary, Condensed	59.95	
	67292A	Illustrated Construction Dictionary, w/CD-ROM	99.95	
	67295B	Landscape Estimating 4th Ed.	62.95	
	67341	Life Cycle Costing	99.95	
	67302	Managing Construction Purchasing	19.98	
	67294A	Mechanical Estimating 3rd Ed.	64.95	
	67245A	Planning and Managing Interior Projects 2nd Ed.	69.95	
	67283A	Plumbing Estimating Methods 2nd Ed.	59.95	
	67326	Preventive Maint. Guidelines for School Facil.	149.95	
	67337	Preventive Maint. for Higher Education Facilities	149.95	
	67236A	Productivity Standards for Constr.–3rd Ed.	49.98	
	67247A	Project Scheduling & Management for Constr.	64.95	
	67265B	Repair & Remodeling Estimating 4th Ed.	69.95	
	67322A	Resi. & Light Commercial Const. Stds. 2nd Ed.	59.95	
	67327	Spanish/English Construction Dictionary	22.95	
	67145A	Sq. Ft. & Assem. Estimating Methods 3rd Ed.	69.95	
	67287	Successful Estimating Methods	32.48	
	67313	Successful Interior Projects	24.98	
	67233	Superintending for Contractors	35.95	
	67321	Total Productive Facilities Management	29.98	
	67284	Understanding Building Automation Systems	29.98	
	67303A	Unit Price Estimating Methods 3rd Ed.	59.95	
	67319	Value Engineering: Practical Applications	79.95	
		MA residents add 5% state sales tax		
		Shipping & Handling**		
		Total (U.S. Funds)*		

Prices are subject to change and are for U.S. delivery only. *Canadian customers may call for current prices. **Shipping & handling charges: Add 7% of total order for check and credit card payments. Add 9% of total order for invoiced orders.

Send Order To: **ADDV-1001**

Name (Please Print) ______________________

Company ______________________

☐ **Company**

☐ **Home** Address ______________________

____

City/State/Zip ______________________

Phone # ____________ P.O. # ____________

(Must accompany all orders being billed)

Mail To: **RSMeans Company, Inc.,** P.O. Box 800, Kingston, MA 02364-0800